CURRENT PERSPECTIVES IN

MICROBIAL ECOLOGY

CURRENT PERSPECTIVES IN
MICROBIAL
ECOLOGY

Proceedings of the Third International Symposium
on Microbial Ecology
Michigan State University
7–12 August 1983

Edited by
M. J. KLUG & C. A. REDDY

American Society for Microbiology
Washington, D.C.
1984

Copyright © 1984, American Society for Microbiology
1913 I Street, N.W.
Washington, DC 20006

Library of Congress Cataloging in Publication Data

International Symposium on Microbial Ecology (3rd :
  1983 : Michigan State University)
  Current perspectives in microbial ecology.

  Includes bibliographies and indexes.
  1. Microbial ecology—Congresses. I. Klug, M. J.
II. Reddy, C. A. III. American Society for Microbiology.
IV. Title.
QR100.I57   1983   576'.15   84-489
ISBN 0-914826-60-3

All Rights Reserved

Printed in the United States of America

CURRENT PERSPECTIVES IN MICROBIAL ECOLOGY

Contents

Continued on following page

Continued from preceding page

Continued on following page

Continued on following page

Continued from preceding page

MICROBIAL RESPONSES TO ECOSYSTEM PERTURBATIONS

METABOLISM OF NATURAL POLYMERS

BIOCONVERSION OF INORGANIC MATERIALS

ECOLOGICAL STRATEGIES FOR THE FERMENTATION INDUSTRY

Continued on following page

Continued from preceding page

BIODEGRADATION OF XENOBIOTICS

Preface

Microbial ecology is a young discipline which has recently gained prominence because of recognition of the important role that microbial processes play in all ecosystems and ultimately in maintaining the environment of our planet. In the past decade, microbial ecology has emerged into a discipline with a body of knowledge and an identity of its own. Yet microbial ecology has not and should never lose its cross-disciplinary connections with other fields of microbiology as well as those of oceanography, limnology, soil science, atmospheric chemistry, and plant and animal sciences. Since a major goal of microbial ecology is to understand microbial activity in natural habitats, knowledge of both the microorganisms and the habitat is essential. During the past decade this integration has been successfully achieved by the practitioners of microbial ecology. It is now time that there be a further integration among microbial ecologists studying different habitats, since common principles transcend the habitat being studied. To illustrate the latter, one rarely sees an interaction between those who study microbial processes in oceans and those studying microbial activities in soil, and neither of these groups frequently interact with those studying the microbiology of the gastrointestinal tract. However, all these groups have a common interest in the strategies and processes that the inhabiting microbes utilize irrespective of the habitat. Such multi-disciplinary exchange can also lead to new insights and to the transfer of methods to new fields, and hence can aid our progress in understanding microbial processes in natural habitats. It is this multi-disciplinary objective that served as a guide for planning the invited paper sessions at the Third International Symposium on Microbial Ecology.

A second objective of this meeting was to emphasize certain directions in which we felt microbial ecology would be moving in the future and the ways in which microbial ecology affects practical problems. For example, the benefits from a relationship between microbial ecology and modern genetics have just recently been recognized. The role of and mechanisms associated with genetic exchange in the environment, and the impact it has on evolution, survival of microorganisms, and ecosystem stability, are all basic questions for which little experimental evidence has been obtained. On the practical side, it is the microbial diversity in nature that yields the genes of concern to commercial interests. This diversity has not been explored in a comprehensive manner, nor has it been studied with an understanding of the ecological principles responsible.

Microbial ecology has been and will continue to be useful in biotechnology and industrial microbiology. Understanding the behavior of microbial populations requires knowledge of the growth, competition, and physiology of these populations within natural microbial communities. These principles are central to an understanding of microbial function in natural habitats as well as useful to the management of these populations within an industrial fermentation vat. Further examples of the important roles microbes play in our environment include the production and consumption of gases in our atmosphere, which can have a long-term impact upon our climate and quality of atmosphere; the degradation of numerous xenobiotic pollutants; and the bioconversion of natural and synthetic polymers. Microbes in nature also transform, corrode, and enhance the recovery of metals and are becoming increasingly important in the biological control of plant and animal pests.

The future importance of microbial ecology, at least in the commercial world, was recently emphasized by Bernard Dixon in an editorial in *Biotechnology*, in which he states that "most of the winners, I predict, would be those biotechnology teams which have had a microbial ecologist on board" (1). The knowledge of the diverse capacities of microorganisms in nature, understanding of how to recover those organisms, and experience with the physiology and growth of the more unusual organisms are all of use to a team that includes microbial geneticists and process engineers.

Invited papers for the Third Symposium on Microbial Ecology were grouped into sections representing four themes of microbial ecology: microbial adaptations, microbial interactions, microorganisms in ecosystems, and microbial bioconversions. This volume contains papers which represent highlights of oral presentations by the invited speakers on a wide range of subjects within the scope of the four themes of the symposium. These papers present the current status and projected future research directions of selected topics within these areas and are not meant to be comprehensive reviews. Subjects were selected on the basis of their current and predicted future importance and intentionally include different scales of investigations ranging from molecular to ecosystem. We hope this demonstrates how both reduction-istic and holistic approaches are utilized to unravel the complexities of cellular, population, and community interactions involved in processes within natural habitats. For example, information at the subcellular level (e.g., genetics, biochemistry) is necessary to understand many mechanisms and processes in nature. Information at the cellular and population level (e.g., growth, competition, and interaction with other organisms or with the environmental matrix) is important to understanding the success of a process or the composition of a community. Finally, information at the ecosystem level is important to understanding the impact of the process or of environmental stresses on our environment. Furthermore, it is this level which is the ultimate test of whether knowledge from the reductionist approach can reliably predict behavior at the ecosystem level.

Because of these various approaches and subject matter the format and style of the papers also vary, from those reporting original research results to those of a review nature. It is hoped that this combination provides a balance which will increase the usefulness of this volume and enhance multi-disciplinary communication, and thus lead to a better understanding of microbial ecology. The volume should be useful as a textbook for an advanced or topics course in Microbial Ecology, as well as a reference source to current literature in the multitude of presented topics.

The local organizing committee thanks the National Science Foundation for a major financial contribution to the meeting and recognizes the additional contributions from the Agricultural Experiment Station, Departments of Crop and Soil Science, Microbiology, and Public Health, and The W. K. Kellogg Biological Station, all of Michigan State University. We also thank Unesco and UNEP for support of some participants from developing nations.

We are indebted to Clayton Wells of The Kellogg Conference Center Staff for his experienced assistance in organizing the meeting. We gratefully acknowledge the contributions of Alice Gillespie, Jean Marek, Carolyn Hammarskjold, Charlotte Adams, and Ellie Tupper. Their diligent handling of correspondence, information retrieval, copy editing, and organization have allowed this volume to meet an early deadline.

Local Organizing Committee

John A. Breznak
Frank B. Dazzo
Michael J. Klug
John L. Lockwood
C. A. Reddy
James M. Tiedje

East Lansing, December 1983

LITERATURE CITED

1. **Dixon, B.** 1983. Commentary. The need for microbial ecologists. Biotechnology 1:45.

Microbes and Ecological Theory

Relevance of *r*- and *K*-Theory to the Ecology of Plant Pathogens

JOHN H. ANDREWS

Department of Plant Pathology, University of Wisconsin, Madison, Wisconsin 53706

Identifying the selection pressures responsible for evolution of a group of interrelated characteristics such as an organism's size, fecundity, and generation time presents an important and fascinating ecological challenge. One pressure that might explain many life history differences is the density of a species with respect to its resource. This is the basis for the theory of *r*- and *K*-selection.

My intent in this chapter is to discuss the origin of the theory of *r*- and *K*-selection, its current status, and its relevance to microbial ecology, particularly to the ecology of plant pathogens. The terminology and concepts are drawn largely from macroecology, but this seems appropriate since each organism, whether a bird or a bacterium, has evolved coadapted traits which promote its survival.

THE *r*- AND *K*-THEORY

It is intuitively logical that a species living near the carrying capacity (given as *K*-selected individuals) of its environment is subjected to a different form of selection pressure than one essentially in a rarefied habitat, exposed to flushes of abundant resources which provide for population expansion. The premium is on competitiveness in the former case, whereas genotypes with a high intrinsic rate of increase (*r*) are favored in the latter. These hypothetical situations are grounded in Dobzhansky's (5) observations that adaptation in the tropics is to a harsh biological environment (e.g., competition, parasitism), whereas in temperate or cold climates, characterized by fewer species, adaptation consists primarily in coping with the physical environment and securing food. He interpreted differences between tropical and temperate organisms as the outcome of different evolutionary patterns. Intricate interactions and density-dependent (DD) controls were viewed as the major factors influencing populations in the tropics, whereas in severe environments density-independent (DI) regulation selected for traits such as cold resistance or accelerated development. (DD controls cause proportionate changes in mortality or fecundity as population density changes; normally, mortality increases and percent fecundity decreases with increasing population size. DI controls operate essentially independently of population density; i.e., the same proportion of organisms is affected at any density.)

MacArthur and Wilson (15, 16) focused and formalized Dobzhansky's abstract idea in their theoretical analysis of island biogeography. They emphasized that in uncrowded environments (*r*-selection), represented by initially uncolonized islands, individuals harvesting the most food, even if wastefully, will be the fittest. Evolution here favors productivity, i.e., high *r* values. However, in time as islands "saturate" with species and food becomes limiting (*K*-selection), genotypes which can at least replace themselves even with smaller families will be superior. Evolution here favors efficiency of converting food into offspring. Hence, colonizing or pioneering organisms will be subject to *r*-selection, but as they establish, the pressure will swing to *K*-selection. This trend will be influenced by changes in the local environment (16).

Given this elegant framework, one might pose several questions, among them: (i) Which environmental conditions favor *r*- or *K*-selection? (ii) How can an organism alter its life history or other factors to maximize *r* or *K*? (iii) Can the attributes of an organism be correlated with life in the anticipated environment (9)? Not surprisingly, the apparent predictive and explanatory power of the *r*-*K* concept has been used in many contexts and has provoked much controversy. For instance, Pianka (19) proposed an *r*-*K* continuum along which species could be placed on the basis of their suites of characteristics (Table 1), and Southwood (23) related position of insect pests along such a spectrum to implications for control tactics.

Current use of the theory has been criticized as confusing and misleading. Wilbur et al. (26) pointed out that although aspects of life history strategy can be correlated with certain environmental parameters, there may not be a cause-and-effect relationship. For example, although characters such as high fecundity and short

TABLE 1. Some life history traits predicted to be associated with r- or K-selection[a]

Criterion	r-selected species	K-selected species	Plant pathogens
Mortality	Oten catastrophic; density-independent	More directed; density-dependent	Predominantly density-independent
Reproductive method	Asexual mitosis to biparental sexual	Predominantly biparental sexual	Predominantly asexual; sexual hermaphroditism common; biparental sexual occasionally
Phenotypic plasticity	Polymorphic to monomorphic	Monomorphic	Usually polymorphic
Structure of species	Monotypic to polytypic	Usually monotypic	Usually polytypic
Length of life	Short; usually less than 1 yr	Longer; usually more than 1 yr	Hours to years
Migratory tendency	High	Low	High
Utilization of resources	For productivity	For efficiency	Usually for productivity; may or may not destroy host
Population size	Variable; nonequilibrium; usually below carrying capacity; frequent recolonization necessary	Constant; equilibrium; at or near carrying capacity; no recolonization necessary	Generally variable; nonequilibrium; recolonization often necessary
Tolerance to niche overlap	Larger	Smaller	Larger

[a] Modified after Pianka (19) and Andrews and Rouse (2).

generation time in an unlimited environment will result in a high r, it is incorrect to reverse the logic and conclude that relatively nonlimiting environments are primarily responsible for evolution of organisms with such features (26). Furthermore, they argued that life history strategies cannot be fully explained as resulting from single selection pressures such as r- and K-selection.

Use of the theory has also been criticized as inexact and inappropriate. Parry (17) commented that use of the terms is now so broad that any life history dichotomy is attributed to r- or K-selection. His criticism needs to be reviewed briefly so that this chapter can be placed clearly in context.

Of the four meanings Parry discussed (Table 2), meanings III and IV are really predictions rather than explanations and can be dispensed with initially. Meaning III predicts where r- or K-selected organisms would live, i.e., in "ephemeral" or "permanent" habitats, respectively. Meaning IV suggests what they would be doing, i.e., allocating large (r) or small (K)

percentages of their resources to reproduction. Whatever else may detract from these latter two definitions as accurate descriptors (see 17), they imply merely a correlative and not necessarily a causative relationship (cf. 26). Hence, although r-selected species may occupy temporary habitats, life history features of these species may have arisen for reasons other than r-selection pressure (e.g., seasonality).

Meanings I and II are the more substantive and deal with the nature of r- and K-selection. Meaning I, stemming from MacArthur and Wilson (16), focuses on crowding. To the extent that this implies competitive pressure only, it is more restricted than meaning II, dealing with all DD and DI effects. Meaning I is acceptable, *if* it is understood that in nature K is set by the interplay of many factors and not by resources alone. Resources (primarily food) remain the center of attention in debates over r- and K-selection because the pioneering work which led to the formulation and early testing of the logistic model involved closed-system laboratory cultures of yeasts and protozoa. This emphasis

TABLE 2. Various meanings of r- and K-selection[a]

Meaning	Central aspect	Conditions under	
		r-selection	K-selection
I	Crowding	Uncrowded; productivity	Crowded; competitiveness
II	Density effects	Density-independent component	Density-dependent effects
III	Habitat	Relatively ephemeral	Relatively permanent
IV	Resource allocation	Large percent to reproduction	Small percent to reproduction

[a] Summarized from Parry (17).

continued in island biogeographic theory which implicitly or explicitly emphasized competition for food or space (e.g., 16, 18), although subsequent papers by Wilson and his colleagues (e.g., 21) discussed the role of predation and other factors in species extinction. However, MacArthur and Wilson's model did not address the factors that determine K, or fluctuations in K over time. Crowding implies the density at which adverse effects of individuals on one another become limiting, and not all such effects arise from competition (25). Other impacts include waste product accumulations, cannibalism, emigration, and increased parasitism and predation. In fact, to account for such metabolic alterations and social interactions, as well as for competition, modification of the logistic model has been suggested (22).

Provided that the broader implications of crowding are recognized, then in practice there is little difference between meaning I and meaning II, because DD controls will be more influential in crowded than uncrowded environments. Although the various sources of DD mortality (e.g., predation, metabolite accumulation) may well affect life history strategies differently, the common stimulus for this evolutionary response is crowding. Such an enlarged perspective also aligns current thinking on r- and K-theory more closely with Dobzhansky's (5) original observations. Clearly, he viewed competition as only one of many factors playing a role in complex tropical communities. The challenge in experimental design is, first, to assess the relative influence of DD versus DI mortality and, second, to identify the specific DD components involved.

In overview, r- and K-selection should be used with respect to a particular selection regime, rather than basing meanings on the individual traits predicted by the selection conditions. A consequence of arguing backwards from the traits to a definition is circular reasoning, and the theory reduces to a tautology.

TESTS OF THE THEORY

In general, there are two ways to test life history theories: the comparative approach and the predictive approach (24). The former involves comparing representative species or genera under present conditions, which are assumed to be identical to those under which they evolved, and testing hypotheses against field or laboratory observations. This provides an interesting intellectual exercise, but as Stearns (24) pointed out, the logic is weaker than that used predictively. The method is also subject to many pitfalls (12, 24). For example, differences in the trait studied may be easier to interpret between closely related species than between those that are distantly related. Despite their shortcomings, comparisons are often used in macroecology because many of the adaptations of interest lie beyond the realm of experimentation. In the predictive approach, selective differences between habitats are measured, and predictions are made about how life history differences of the species of interest might alter if the populations were interchanged or otherwise manipulated (24). If the traits do not change as anticipated, then either the experiment or the theory is flawed. Regardless of the approach taken, there are numerous ambiguities which complicate tests of r- and K-theory, and Stearns has outlined (24) a set of "reliability criteria" which could be used in the design or evaluation of experiments.

Andrews and Rouse (2) proposed that r- and K-theory is relevant to the analysis of plant pathogen life histories and used the comparative approach to rank representative pathogens along a continuum relative to each other. Two criteria were emphasized as determinants of strategy: first, the proportionate allocation of resources to maintenance, growth, and reproduction; and second, the nature of the parasitic association, particularly with respect to derivation of nutrients and impact on the host. A conceptual model was devised to describe the theoretical relationships between the population parameters biomass, or numbers, and life history strategies. A central idea was that r-selected organisms channel more resources to reproduction than K-strategists (see 7; criticized in 17). With respect to mode of nutrition, biotrophic parasites (which obtain nutrients from living host cells) were considered to be more K-selected than necrotrophic parasites and were viewed as inducing stress to host plants; by contrast, necrotrophic parasites (derivation of nutrients from killed cells), which cause "disturbance" by actively destroying plant biomass (2; cf. 10), were viewed as r-strategists. The reasoning for this distinction was that K-strategists clearly have an investment in maintaining the vigor of their hosts, with which they would tend toward coexistence. On the other hand, r-strategists can aggressively invade and destroy their hosts (the so-called "sweepstakes" strategy) and then adopt an expanding saprophytic phase (cf. insects, 23). Within the biotrophic category one might visualize a continuum ranging from parasites such as the apple scab pathogen (r-selected) to the mistletoes (e.g., *Phoradendron flavescens*), which are generally consistent with Pianka's (19) correlates for a K-strategist. In addition to being small and having a short generation time, the former channels much of its resources to asexual reproduction during the summer and can cause severe disturbance to the host. However, the mistletoes appear to empha-size efficiency over productivity. As perennial, semiparasitic evergreens, they produce carbo-hydrate photosynthetically and rely upon their host for minerals and water. They rarely destroy biomass, but act as low-grade stressors. DD controls (e.g., competition for sunlight) likely play an important role in population regulation of mistletoes, whereas populations of apple scab are markedly affected by DI factors such as temperature and rainfall.

There is an aspect of "story-telling" to com-parisons or analogies such as the above, and appealing stories can easily be concocted to fit the facts. Comparisons fall short as explanations because of the criticisms raised earlier—for one reason, they do not separate correlation from causation. (It is significant that Pianka [19] re-ferred to the features of r- and K-selection as *correlates*.) Convincing tests of r- and K-theory can only be made by identifying key components of its broad nature and then posing specific, falsifiable hypotheses which have some predic-tive value. Plant pathogens are used as examples in the following experiments, but the same con-cepts apply to free-living microbes.

A direct test of r- and K-theory is to crowd populations and see what happens (Table 3). According to the central idea, K-selection in-volves DD regulation in environments where populations approach the carrying capacity; conversely, r-selection results when DI factors maintain populations below equilibrium. Thus, one would predict that various life history traits such as fecundity and competitiveness should be influenced differently in the two environments. Microbial systems generally are much more amenable to such experimental manipulations than are plants or animals, and most of Stearns' (24) other "reliability criteria" can also be satis-fied with less difficulty. Thus, crowding or abiot-ic factors can be varied while holding other conditions constant, thereby avoiding the correl-ative-type approach and inferential evidence. The underlying premise of these tests is that there are two opposing types of selection which result in a trade-off; i.e., a K-strategist will be less fit under conditions of r-selection and vice versa. (Note that the terms K-strategist and r-strategist are relative descriptors and are appro-priately used to *compare* one organism with another.)

Direct tests are feasible with plant pathogens (especially procaryotes, fungi, and nematodes) by adapting populations in vitro or in vivo to DD and DI controls and then testing how these adapted strains perform against each other and the parental types in the two environments. DD conditions could be achieved in vitro by allow-ing populations to reach and remain at the

TABLE 3. Some possible tests of the theory of r- and K-selection[a]

Trait	Prediction under	
	DI controls	DD controls
Biomass allocation	Reproduction	Growth and maintenance
Ontogeny	Higher rate	Lower rate
Fecundity	Higher	Lower
Reserves	Lower	Higher
Nutrient respon-siveness	Faster	Slower
"Competitiveness"	Lower	Higher

[a] Hypothesis: certain life history traits result from crowding. Experiment: subject populations to crowd-ing; compare populations from, or introduce popula-tions to, crowded and uncrowded environments.

carrying capacity over many generations in fed batch or continuous culture; for DI regulation, populations could be serially transferred or sub-jected to abiotic controls (e.g., heat or cold shock). This approach was used by Luckinbill (14), who grew pairwise combinations of r- and K-adapted strains and parental types of *Esche-richia coli* in mixed culture. The proportion of competitively neutral auxotrophic and proto-trophic markers for histidine provided an indica-tion of whether adaptation to culture conditions had occurred. Experiments with plant pathogens that produce unambiguous results in vivo would be considerably more complicated to design, partly because the carrying capacity of plants for the organism of interest would need to be de-fined operationally and could be expected to be highly variable. Nevertheless, such tests are needed for a realistic appraisal of the theory, and some possibilities are outlined below.

Some of the life history traits that could be expected to vary under DD or DI controls are listed in Table 3. In theory these features are measurable; in practice they could be quantified more or less accurately depending on the trait, the species characteristics, and the assay sys-tem. For example, a measure of ontogeny could be time to sporulation, to symptom develop-ment, or to completion of life cycle. Reserves could be quantified by monitoring storage prod-ucts (e.g., lipids, poly-β-hydroxybutyrate) or calorific values (11). "Competitiveness" is a more subjective criterion and involves several possible features, among them tolerance of anti-biotics, the production of secondary substances such as inhibitory materials, siderophores, or slime, the ability to store a wide variety of materials, the ability to utilize many carbon and energy sources concurrently, and the ability to efficiently convert limiting nutrients into bio-mass. A common problem, as elaborated on elsewhere (1), seems to be the confusion of colonizing ability with competitiveness per se.

There are two corollaries to the direct test. The first is that populations isolated in nature from crowded and uncrowded environments should show traits consistent with predictions (Table 3). This has been the indirect, observational approach traditionally used in macroecology. The second corollary is that populations adapted to r- and K-selection in vitro, when reintroduced to nature, should become established best in the environments that present the corresponding selective pressure. An expectation might also be a shift in adaptive features over time; i.e., a newly colonizing species will be subjected to r-selection, but after establishment it should shift under K-selection pressure (see following comments on phenotypic and genotypic changes). This latter corollary is consistent with the biogeographic implications foreseen by MacArthur and Wilson (16). Both corollaries offer less rigorous tests of the theory than does the original proposition. The problem of correlation versus causation was noted earlier. Additionally, the corollaries are weakened by several assumptions, e.g., that r- and K-selection will be the factor determining survival of adapted populations reintroduced to nature. There is also some circular logic in that if experiments proceed according to expectations all is well, but if evidence is produced against the hypothesis one can always argue that the environments did not really present r- or K-selection pressure or, in any case, that there were too many uncontrolled variables.

If one elects to test the corollaries, a possible approach to the first would be to make inter- or intraspecific comparisons, including competitive tests, of plant pathogens based on the criteria described above, from a host species subjected to lax versus intense parasitism (e.g., early versus late in the growing season). Such comparisons should also be made on organisms from natural communities, where the artifacts of agroecosystems do not pose problems to interpretation of results. The expectations would be that parasites colonizing early have characteristics of r-strategists, whereas later, when DD controls such as competition for infection sites and resources intensify, the predominant species or biotypes are relatively K-selected. In addition to exhibiting life history traits typical of r- or K-strategists (Table 3), the early colonizers should outclass the later colonizers when tested in vitro under controlled DI conditions and vice versa (see 14). A complementary approach would be to compare characteristics of parasites, at any given time, from newly emerging roots or shoots and those attacking older portions.

The second corollary could be tested by adapting populations in vitro to DI or DD controls as described above (14) and releasing them alone and in combination to relatively "unsaturated" versus "saturated" environments. Similar studies with antagonists to the apple scab pathogen are in progress (J. H. Andrews and D. Cullen, unpublished data), but we have not extended this work yet to specifically test r- and K-concepts. If introduced populations are not sufficiently morphologically distinct (e.g., color mutants), they need to be marked (e.g., auxotrophy, antibiotic resistance) for recovery on test media. An example is to adapt in culture a physiologically marked color mutant of the scab organism to DI and DD controls. Severely infected ("saturated" environment) and slightly infected ("unsaturated" environment) apple trees in growth rooms and in an orchard could then be inoculated with these adapted strains alone and in combination. The introduced populations and their progeny could be monitored over time, distinguished from the wild type by color and from each other by nutritional requirements. The hypothesis would be that relative establishment of the K-selected strain is greater than the r-selected strain on severely infected leaves; the reverse would be expected on slightly infected leaves. Of course, the usual assumptions pertaining to marked populations, such as stability and competitive neutrality of the marker, apply and would have to be verified in advance.

IMPLICATIONS OF THE THEORY

The extent to which r- and K-selection is eventually found relevant to microbiology is secondary for the time being to the fact that microbial systems seem to have excellent potential for critical appraisals of the theory. These tests in themselves can provide for new perspectives and interpretations. If the evidence is supportive, there are numerous implications and predictions stemming from the theory. If the evidence overall is negative, then the obvious question would be: If crowding does not play a major role in shaping microbial strategies, what does? Either way, the investigator will be forced to identify the factors which control microbe populations.

If r-K theory is upheld for microbial communities, a major benefit would be the strengthening of bridges between microbial ecology and macroecology. An elegant body of ecological theory on life history tactics would be opened to microbiologists and plant pathologists. Conversely, ecologists would become better informed and appreciative of how microbes can be used to test ecological theory. Microbial systems should provide the means in large part to

quell a typical criticism, namely, that speculation flourishes in the absence of fact.

For populations shown to be r- or K-selected,
the next step might be to identify the specific
conditions that increase r or K. Is the ability to
produce several spore crops from a given conidiophore going to increase r more than increasing
conidiophore numbers by a given increment?
What factors would promote either option?
Would K be increased more by a tolerant than
by a susceptible host? What influence has the
presence of a diverse epiphytic microbial community on carrying capacity for a given pathogen and is interspecific competition a more
significant DD control than intraspecific competition?

Luckinbill (14) has posed the question of the
specific genetic mechanisms involved in r- or K-
adaptations induced in culture. To move back a
step further, one might start by asking whether
the changes induced in vitro and in vivo are
phenotypic or genotypic adaptations. The expectation would be for the former to occur
initially, followed by quick deadaptation if selection were relaxed. If the r- or K-selection pressure were severe or imposed for a long time
(typically hundreds of generations), genetic
change should occur. How will the persistence
of strains reintroduced to nature vary in the
presence or absence of the corresponding selection pressure? The issue of the kind of change is
very important because of the largely genetic
origins of r- and K-theory in island biogeography
and because the concept deals with the *evolution*
of life history traits. Gadgil and Solbrig (7)
proposed that the crucial evidence for r- and K-
selection is whether an organism maintained a
particular strategy (e.g., resource allocation) under any and all DD and DI mortality conditions.

The idea that microorganisms with distinct
phases (biphasic or heteromorphic) may be
evolving simultaneously as r-strategists under
one set of environmental conditions and as K-
strategists under another has been developed
elsewhere (1), and there are analogies with other
systems, especially seaweeds (13). Presumably,
this phenomenon is one example of Dobzhansky's (5) adaptive polymorphism whereby a
species consists of two or more types, each
having high fitness in a particular environment.
Bimodal r- and K-selection is one attractive
explanation for the life history features of many
parasites which use abundant host resources
during the summer to reproduce exponentially,
and then channel resources to vegetative growth
and maintenance in a competitive environment
during the winter. However, there are other
explanations, such as the constraints imposed
by seasonality. Definitive explanations cannot
be made until experiments are designed to test

the various hypotheses.

From the perspective of a plant pathologist,
one of the more important implications relates to
prospective strategies for biological control. Recently, there have been much attention and
debate about biological control of soil-borne and
foliar plant pathogens by microbial antagonists
(3, 4). The role of competition as a significant
organizing factor in epiphytic communities has
been questioned (4). This, in turn, throws into
question whether sustained biocontrol is realistic and, if so, what the best tactics should be.
Can r- and K-theory provide guidelines for addressing these problems? Is an r-selected antagonist which can rapidly colonize emerging plant
tissue the preferred candidate, or should an
organism with a lower r, but which is highly
competitive, be chosen? Perhaps a mixed species community with one or more members of
both groups is the preferred approach. It is too
early to resolve these questions, but r- and K-
theory does provide a stimulating framework as
a point of departure for experiments.

Finally, r- and K-selection provides a possible
interpretation for the various growth dynamics
of nonparasitic as well as parasitic microbes.
For instance, are life history traits such as
antibiotic production or slime secretion triggered by DD controls related to crowding or by
other factors? Analogies between r- and K-
strategists and allochthonous (zymogenous) or
autochthonous soil microbes, respectively, have
been drawn (8, 20). The former group comprises
transients or invaders, microorganisms which do
not participate significantly in community activity. Microbes in the latter class are residents and
at some point in their life cycle are involved in
community metabolism. Some similarities with
the r-K descriptors are readily apparent, but as
with other analogies (2, 6), these are merely
correlations. The causative mechanism for each
set of traits remains to be established. The major
challenge for the immediate future is the design
of appropriate experiments to test whether specific parameters have actually been responsible
for evolution of life history traits attributed to r-
and K-selection.

This is a contribution from the College of Agricultural and
Life Sciences, University of Wisconsin-Madison. Partial re-
search support from the National Science Foundation, grant
DEB-8110199, and from the United States Department of
Agriculture, grant 81-CRCR-1-0707, during the period in
which these ideas were formulated is gratefully acknowl-
edged.
I thank numerous colleagues, in particular F. Berbee, D.
Cullen, and two anonymous reviewers, for their comments on
the manuscript.

LITERATURE CITED

1. **Andrews, J. H.** 1984. Life history strategies of plant
   parasites. Adv. Plant Pathol. **2**:in press.

2. **Andrews, J. H., and D. I. Rouse.** 1982. Plant pathogens and the theory of r- and K-selection. Am. Nat. **120:**283–296.

3. **Blakeman, J. P., and N. J. Fokkema.** 1982. Potential for biological control of plant diseases on the phylloplane. Annu. Rev. Phytopathol. **20:**167–192.

4. **Cullen, D., and J. H. Andrews.** 1984. Epiphytic microbes as biological control agents. *In* T. Kosuge and E. W. Nester (ed.), Plant-microbe interactions, vol. 1, Molecular and genetic perspectives. Macmillan Publishing Co., Inc., New York.

5. **Dobzhansky, T.** 1950. Evolution in the tropics. Am. Sci. **38:**209–221.

6. **Esch, G. W., T. C. Hazen, and J. M. Aho.** 1977. Parasitism and r- and K-selection, p. 9–62. *In* G. W. Esch (ed.), Regulation of parasite populations. Academic Press, Inc., New York.

7. **Gadgil, M., and O. T. Solbrig.** 1972. The concept of r- and K-selection: evidence from wild flowers and some theoretical considerations. Am. Nat. **106:**14–31.

8. **Gerson, U., and I. Chet.** 1981. Are allochthonous and autochthonous soil microorganisms r- and K-selected? Rev. Ecol. Biol. Sol. **18:**285–289.

9. **Gould, S. J.** 1977. Ontogeny and phylogeny. Harvard University Press, Cambridge, Mass.

10. **Grime, J. P.** 1979. Plant strategies and vegetation processes. John Wiley & Sons, Inc., New York.

11. **Jennings, J. B., and P. Calow.** 1975. The relationship between high fecundity and the evolution of entoparasitism. Oecologia (Berlin) **21:**109–115.

12. **Lack, D.** 1968. Ecological adaptations for breeding in birds. Methuen, London.

13. **Lubchenco, J., and J. Cubit.** 1980. Heteromorphic life histories of certain marine algae as adaptations to variations in herbivory. Ecology **61:**676–687.

14. **Luckinbill, L. S.** 1978. r and K selection in experimental populations of *Escherichia coli.* Science **202:**1201–1203.

15. **MacArthur, R. H., and E. O. Wilson.** 1963. An equilibrium theory of island zoogeography. Evolution **17:**373–387.

16. **MacArthur, R. H., and E. O. Wilson.** 1967. The theory of island biogeography. Monographs in Population Biology, no. 1. Princeton University Press, Princeton, N.J.

17. **Parry, G. D.** 1981. The meanings of r- and K-selection. Oecologia (Berlin) **48:**260–264.

18. **Patrick, R.** 1967. The effect of invasion rate, species pool, and size of area on the structure of the diatom community. Proc. Natl. Acad. Sci. U.S.A. **58:**1335–1342.

19. **Pianka, E. R.** 1970. On r- and K-selection. Am. Nat. **104:**592–597.

20. **Pugh, G. J. F.** 1980. Strategies in fungal ecology. Trans. Br. Mycol. Soc. **75:**1–14.

21. **Simberloff, D. S., and E. O. Wilson.** 1969. Experimental zoogeography of islands: the colonization of empty islands. Ecology **50:**278–290.

22. **Slobodkin, L. B.** 1961. Growth and regulation of animal populations. Holt, Rinehart and Winston, New York.

23. **Southwood, T. R. E.** 1977. The relevance of population dynamic theory to pest status, p. 35–54. *In* J. M. Cherrett and G. R. Sagar (ed.), Origins of pest, parasite, disease and weed problems. Blackwell Scientific Publications, Oxford.

24. **Stearns, S. C.** 1977. The evolution of life history traits: a critique of the theory and a review of the data. Annu. Rev. Ecol. Syst. **8:**145–171.

25. **Whittaker, R. H.** 1975. Communities and ecosystems, 2nd ed. Macmillan Publishing Co., Inc., New York.

26. **Wilbur, H. M., D. W. Tinkle, and J. P. Collins.** 1974. Environmental certainty, trophic level, and resource availability in life history evolution. Am. Nat. **108:**805–817.

Microbial Diversity and Decomposer Niches

M. J. SWIFT

Department of Biological Sciences, University of Zimbabwe, Harare, Zimbabwe

Within the past two decades studies of the role of microorganisms in decomposition processes have taken two almost completely divergent paths. The first, and relatively modern approach, has been oriented toward the description of processes and has paid little attention to the roles of individual species or to the variations in behavior among them. This approach has met with a degree of success in increasing understanding of the patterns of organic decay and of mineral nutrient dynamics in the soil. A number of models have been produced which have predictive power in relation to these features (e.g., 6–8, 10, 22, 33, 42, 47), although in most cases their confidence limits are wide. All these models share two features: they are non-explicit about the composition of the decomposer community, and they are largely deterministic. Thus, the decomposer organisms are broadly pictured as "driving variables" promoting the processes of catabolism and comminution, at most being subdivided into a small number of trophic groups. The processes that are promoted are largely described by deterministic equations, regulated by variables (such as climate or nutrient concentrations) which can be smoothly integrated over definable ranges of effect.

This contrasts sharply with the approach and findings of the second school of microbial ecologists (particularly mycologists), who have been concerned above all else to demonstrate the diversity of decomposer species, of their activities, and of their patterns of distribution and abundance (e.g., 16, 21, 24, 36, 51). A general feature of many of these latter observations has been their largely indeterminate nature, which has led to a considerable reluctance to draw general conclusions concerning the ecological roles of these organisms (31).

It is possible that the limitations and wide confidence limits of the functional models of decomposition are a result of the omission of diversity-related phenomena. The bringing together of the two approaches depends, however, on the development of a rigorous conceptual basis of understanding of the structure and interactive functioning of the microbial communities of decomposer habitats. One approach to this is to develop a theoretical base by utilizing, where valuable, concepts developed in the "mainstream" of ecology.

A number of authors have cautioned against the application of theories from other areas of ecology to the ecology of fungal communities. The main reason for this caution is concern about the nature of the data available to fungal ecologists. It is argued that the very nature of the fungal growth form renders it impossible to draw valid conclusions about the distribution or abundance in space and time of individuals or populations. Without in any way challenging the validity of these contentions, it is equally possible to argue that it is this very uncertainty which necessitates the development of a theoretical base with a strong deductive content. Theories in fungal community ecology must not only be consistent with reality as it is currently understood but must have clear predictive implications. The value of theorizing lies in the questions which are thrown up and the experiments that are thus designed to test the theory. If these can be designed within the technical limits of available methodology, then progress is possible. In any event, as Parkinson (31) has said, "at worst they expand the argument on soil fungal ecology."

THE UNIT COMMUNITY

Swift (44) pictured fungal communities as having a hierarchical structure. The fundamental components of the community he termed unit communities. A unit community is a species assemblage which inhabits a volume of resource that is delimited in some clear and unequivocal way such that, whereas the species within the unit community may be expected to interact in some way, interactions between organisms in different or even neighboring unit communities will be minimized. Examples of these resource units are twigs or branches on the forest floor, dead tree boles, fecal pellets, or isolated feathers or leaves on the surface of the soil or buried within it (or in similar circumstances in aquatic environments). Where leaves form a blanket on the forest floor, the definition of boundaries may be difficult, and the distribution patterns of the fungi, like those of the general soil flora, may be continuous rather than mosaic-like. A good example of the lack of discontinuity between resource units is given by Thompson and Boddy (48), who showed how a single individual or a basidiomycete might invade a large number of resource units (tree boles) over an area of more than 100 m^2.

The hierarchy is developed from the unit community level by grouping together the communities inhabiting the same species or type of resource (e.g., to give the fungal community of beech or the fungal community of leaves) or habitat (e.g., litter layer) and finally ecosystem. Similar suggestions for hierarchical community structure have been made in other areas of ecology (e.g., for rocky shore animals [30] and for grazing-chain arthropods [39]). Lussenhop (27) discussed the concept under the terms used by Root (39) of "component" and "compound" communities.

An example of the hierarchy may be taken from the data of Hawksworth (20), who reported one of the most comprehensive surveys of the fungi of a given area yet undertaken. In the period 1968–1975, 949 species of fungi were recorded at Slapton Ley Nature Reserve, United Kingdom, although consideration of the patterns of collecting led Hawksworth to the conclusion that the true total might in fact be closer to 1,800 species. If this "compound community" is partitioned in the manner suggested above, a much more realistic view of diversity, and of the dimensions of the problem faced by the decomposition ecologist, is obtained. Of the above 949 species, 139 are reported as present within the small woodland, France Wood; of these 44 are recorded as being found in leaf litter. These species can be further subdivided in terms of the species of leaf litter so that, for example, 5 are noted as characteristic of rotting holly (*Ilex aquifolium*) leaves in France Wood (12 for the Nature Reserve as a whole). This leaf community thus represents the potential maximum size for a unit community on holly leaves. Whether any unit ever bears all 12 species at the same time is unknown from this study.

The significance of the unit community is that it defines at the same time both a floristic and a functional entity. The community of organisms inhabiting a single twig or fecal pellet is ultimately responsible for the decomposition of that twig or fecal pellet. Thus, no matter how much variation there may be in the floristic composition between unit communities, it is theoretically only necessary to understand the functioning of a single unit community to establish the principles for understanding decomposition as a whole. An experimental program which confines its attention to a single, or preferably replicated, unit community immediately dispenses with a large extent of informational "noise" compared with a less structured sampling program.

Unit community composition. Swift (46) discussed some of the aspects of experimental design involved in utilizing the unit community as a basis for community analysis, but very few studies have yet been published which have utilized this type of sampling unit. A number of recent examples have, however, contributed to understanding of the composition of unit communities and the way they are established. Communities of aquatic hyphomycetes have been studied (12, 40). C. Shearer and her colleagues (personal communication) have developed a very useful technique of leaf mapping to describe unit communities in this habitat. The unit communities of fecal pellets have also been described (28, 55), but perhaps the most exciting contributions have come from the studies of Rayner and his colleagues (5, 37, 38) on the complex mosaic of basidiomycete populations and communities inhabiting decaying wood. These complexes are maintained by a variety of genetic barriers which may exist between species, between subpopulations within species, or between individuals within populations. Unfortunately, little is known of the variations in genotype of ecologically significant features which may be correlated with these isolating factors. Presumably, it is upon these other factors that natural selection will largely operate, determining the potential niche differentiation between the species. Rayner also pointed out the existence within these communities of basidiomycetes of assemblages of microfungi.

A most important contribution to our understanding of the dynamics of unit communities was made by the experiments of Sanders and Anderson (40) on the Ingoldian Hyphomycetes colonizing blocks of wood in freshwater streams. They described the variation in species number and composition of the communities developing on wood blocks of five different sizes. The number of species increased with the increasing surface area of the blocks (Fig. 1). These data show good agreement with the predictions of island biogeography theory (29) concerning species–area relationships. H. Wildman (personal communication) has demonstrated a similar effect for the communities of fungi developing on different-sized squares of cellulose film inserted into the soil.

Resource units clearly do have the character of islands and present the same problems for colonization to fungi and other decomposers that conventional oceanic islands present to other organisms. The precepts of island biogeographic theory are known to apply to other types of microscopic "islands" inhabited by motile autotrophic (34) or heterotrophic (9) microorganisms. This, then, offers a potentially rich source of theory for framing experimental approaches to the analysis of unit communities, as shown by Sanders and Anderson and by Wildman in their pioneering studies. There is, how-

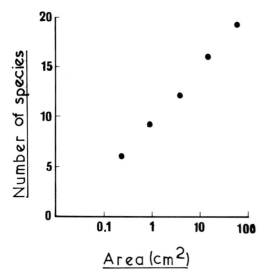

FIG. 1. Relationship between number of aquatic hyphomycete species and the area of "resource islands" (wood blocks). (Drawn from data of Sanders and Anderson [40].)

ever, one important respect in which these "resource" islands differ from "conventional" islands: they are being continually, progressively, and inevitably degraded by their inhabitants. The saprotrophic fungi, once in place on a resource island, produce extracellular enzymes that break down the organic substance of the island (resource unit) to provide the food for the fungus. Other groups of decomposers have the same effect. Thus, the chemical composition, the physical environment, and indeed the mass of the unit are progressively changed as decomposition proceeds. This is analogous in island terms to a progressive change in the size of the land area.

Classic island biogeographic theory predicts that a freshly exposed island will be progressively colonized by species until an equilibrium number is reached which is characteristic of the size of the island (Fig. 2a; 29, 41). This number will then remain more or less constant, maintained by an equilibrium between the rate of immigration to the island and extinction from it. Wildman (personal communication) showed just such an effect in his experiments with cellophane squares in soil. The number of species of fungi remained constant and characteristic of the size of the "island" over a substantial period of time, but the species composition changed, implying that some species of fungi were replaced by others. In other circumstances, however, the fungal diversity of resource islands may decline in the later stages of decomposition (44; Fig. 2b).

This is consistent with the proposition advanced above, that the effective area (= niche

volume) of the island is decreasing, a situation which imposes particular pressures on the occupying microflora. Extinction from the diminishing island demands a strategy whereby new islands of larger size can then be colonized. These islands, i.e., new leaves, twigs, etc., will be separated in both time and space involving a pattern of cyclic immigration (colonization) and extinction (dispersal) (44). Study of the pattern of change in species diversity of resource islands may therefore be predicted to include a change from a situation of net immigration to a period of net extinction, with or without an intervening period of equilibrium. Wildman (personal communication) detected such a period, but Chamier and Dixon (12) demonstrated the rapid onset of decline in species numbers in communities of Ingoldian Hyphomycetes.

The processes of colonization are affected by a variety of factors, including the size of the pool of potential species and the distance of the island from it. The models derived by Cancela da

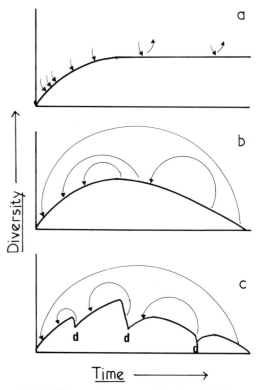

FIG. 2. Changes in species diversity with time on island communities. (a) Conventional islands (after MacArthur and Wilson [29]); (b) resource islands subject to stress (after Swift [44]); (c) resource islands subject to stress and disturbance. Arrows impinging on the curve represent immigration; arrows leaving the curve represent extinction; d = disturbance, e.g., litter fall, initiation of detritivore activity, etc.

Fonseca (11) for colonization of temporary soil habitats by arthropods might prove useful in approaching this problem.

Unit community structure. If a characteristic number of species are assembled within a unit community, then the question arises as to what roles the various organisms within the community play in the processes of decomposition. This is another way of asking what determines the differing niches they must occupy to coexist. Swift (44, 45) has suggested three dimensions to such niche structure: separation on the basis of chemical factors, mainly differences in the pattern of utilization of substrates or in response to modifier (e.g., allelopathic) chemicals in the resource; separation into different microenvironments within the unit; and separation in decomposer activities with time resulting from different reproductive strategies. Little evidence has emerged to clarify the picture of niche structure in terms of physicochemical factors, but this question has been of particular interest to a number of authors in recent years and will be taken up in a later section.

This view of community structure is, however, based on an assumption of a relatively stable or equilibrium situation. It presumes that, even within the progressively changing state of the resource, unit equilibria are reached between the occupying species in terms of their interactions so that they can coexist without niche overlap. As decay proceeds, competition for diminishing resources may increase and niche dimensions will decrease. Extinction of species will occur when this reaches an extreme position. R. Cooke and A. Rayner (manuscript in preparation) have postulated a somewhat different situation. They present the view that even in the early stages of decay when resources are abundant, and perhaps before maximum diversity is reached, there is a considerable extent of what they term "combative" competition between the occupying species, which may result in extinctions and species replacements. Nonetheless, their picture and that of most analysts of the so-called decomposer "succession" has been of a progressive set of interactions between the occupying species, leading inevitably to the extinction of species on the basis of competitive exclusion.

An alternative view of decomposer communities that may be equally valid is that, far from being an orderly and structured system, it is a relatively disordered and stochastic one. This view is consistent with the description of the constantly changing nature of the resource condition as described earlier and with the suggestion that it might resemble, in some circumstances, the situation described for phytoplankton by Hutchinson (26). Hutchinson posed the paradox of how an assemblage of species apparently requiring the same resources (seeking to occupy the same niche) could coexist. The resolution of this problem, he suggested, was that the system was not in equilibrium and that constant shifts in balance (e.g., between grazing and autotrophic growth) meant that there simply was no time for the competitive exclusion principle to work to conclusion. Huston (25) presented the same paradox in a more general way by proposing that conditions of disequilibrium are more common in nature than those of equilibrium. From this it is logical to challenge the analyses of species diversity which largely derive from assumptions of equilibrium between competing populations. This approach would appear to be a fruitful one for the analysis of decomposer communities. Beaver (1, 2) presented some interesting examples referring to decomposer animals, but in relating this approach to the fungi it is of benefit to introduce concepts from yet another area of ecology.

Grime (17) described two categories of factors which might limit the biomass of autotrophic plants in a given habitat: stress, which he defines as the external constraints which limit the rate of dry matter production of all or part of the vegetation; and disturbances, which are the mechanisms that limit the plant biomass by causing its partial or total destruction. Such categorization can usefully be applied to what we may term the condition of a resource in relation to its ability to support microbial biomass. Thus, examples of stress would be low nutrient conditions, high matric potential, or low temperature. Disturbance would be features like consumption by detritivorous animals, grazing by microtrophs, fire, sudden frost, or, as pointed out by Pugh (36), litter fall.

Stress effects are characteristic of two situations: that of the initial stages of colonization where they represent specific features of the resource type in question (e.g., the high C/N ratio and high lignification of wood in comparison with leaves), or the general situation that develops as decomposition proceeds (e.g., stress due to limitation of the concentration of available C and energy sources, i.e., generally low C/N ratios). As such, in this general situation the onset of stress is a predictable circumstance. That is to say, the effect is progressive; the clues for the decomposers of the developing situation are there in the environment and responses may be accordingly ordered (see discussion on strategies below). The same feature of predictability is largely applicable also to environmental stress.

The influence of disturbance may have a quite different pattern. That disturbances of the kinds mentioned above can drastically alter microbial communities has been amply demonstrated by a number of authors, e.g., effects on the soil flora of forests by clear-felling or fire (52, 53, 56). More specific effects on unit communities have been shown for coprophilous fungi by Yocom and Wicklow (55; microclimatic effects) and Lussenhop et al. (28; invasion by arthropods). The importance of animal grazing as a destabilizing effect on the fungal community of leaves has been shown by a number of authors (e.g., 18, 32).

A significant feature of these effects is that they intervene in a quite unpredictable fashion. This can be illustrated by branch litter fall; L. Boddy and M. J. Swift (Holarctic Ecol., in press) have shown that this may occur at states of decay varying from 10 to 80% weight loss, i.e., at extremely different times during the decomposition process. The implication of this is that, even if there is a high probability of attaining an equilibrium in the species diversity in an undisturbed resource, this will rarely be attained because of the intervention of disturbance. Even if it is attained, it will eventually be overturned. Disturbance will have the effect of completely altering the terms of whatever interactive equilibrium has been attained between the occupying species, promoting the accelerated extinction of some species and facilitating the immigration of new species. The varying patterns of microbial diversity that would be predicted on this model are illustrated in Fig. 2c. The implications are quite profound, for it postulates the likelihood of coexistence of equivalent species, i.e., that the probability of the assemblage of any particular unit community may be difficult to predict even when the potential species pool is known, and that opportunism may be a character of high selective value among the decomposer microflora.

LIFE-STYLES AND STRATEGIES OF DECOMPOSER FUNGI

The ecological success of an organism depends on the totality of the interaction between the external environment, including the presence of other organisms, and the organism's genome. The particular character of the organism's response to variations in this external environment may be described as its strategy or life-style. A productive approach in both animal and plant ecology has been to attempt to recognize patterns of strategy adapted to particular types of ecological situation. From these strategy groupings certain generalized predictions can be derived which facilitate the description of the functioning of complex communities (17, 35). In recent years a number of mycologists have attempted to borrow from these propositions of "mainstream" ecology and to determine patterns of strategy among the fungi in relation to the selective pressures of decomposer habitats (23, 36, 44, 45, 49).

The value of such an approach lies in its predictive quality. From the characteristics of an organism, determined in pure culture, from its genotype, and from its phenotypic expression under controlled and variable conditions, we should be able to predict its ecological behavior as a component of complex natural communities. Moreover, the predictive characters should be few in number and relatively easy to determine. The discussion of the structure of unit communities in the previous section highlights this approach. If it is concluded that decomposer communities are unstable, rapidly changing, and stochastic in nature, then strategy is likely to be predominant among the three niche determinants described above.

The recognition of broad differences in strategy is deeply embedded in the history of soil microbiology. For instance, Winogradzky's (54) distinction between zymogeneous and autochthonous components of the soil microflora and Garrett's (15) distinction of "sugar" fungi from other soil inhabitants clearly predate the suggestions (23, 44) that fungi can be grouped as r- or K-strategists in the sense of Pianka (35). All these categorizations recognize the distinct adaptive values of rapid growth and sporulation in the primary stages of decomposition of readily catabolized substrates. The classic study of Harper and Webster (19) on coprophilous fungi clearly established the validity of this approach. More recently, Pugh (36), Hedger (23), and in particular Cooke and Rayner (personal communication) have extended this analysis by considering fungal strategies within the framework established for autotrophic plants by Grime (17).

The strategies defined by Grime are seen as adaptations to the presence of the two selective pressures of stress and disturbance defined above. Competitors are plants adapted to stress and disturbance-free (= productive) habitats; stress tolerators have strategies particularly adapted to stressed habitats; ruderals are adapted to disturbance. Grime concluded that there was no viable strategy for habitats that are both stressed and disturbed. Grime's categories can be placed within the spectrum of strategies expressed in the terminology of animal ecology, i.e., r-selected coinciding with ruderal and K-selected with stress tolerant; the competitors (sensu Grime) occupy the middle range of this spectrum. Grime also emphasized the existence of secondary strategies, that is, the combination of competitive and ruderal strategies, for instance. The authors mentioned above (23, 36; Cooke and Rayner, personal communication) have clearly demonstrated the existence in the decomposer fungi of strategies analogous to those described by Grime for higher plants; Cooke and Rayner in particular have emphasized the role of secondary strategies.

One interesting feature of fungal behavior is that the life cycle of many species contains stages which are physiologically and morphologically distinct to the extent that they may confer adaptations to different types of pressure. This phenotypic flexibility is clearly of general advantage in the type of variable habitat described earlier. It may mean, however, that it is easier to recognize types of strategy than to ascribe specific fungi to strategy groups. For instance, the difficulty of assigning an r or K reproductive strategy to a fungus which has both asexual (conidial) and sexual (carpophoric) stages has previously been noted (45). A possibly productive approach is to consider the pattern of strategy within the framework of changes in niche dimension as described in the previous section.

Stress and disturbance in relation to niche volume. If we consider that in most cases the response of an organism to a particular environmental determinant will take the form of a Gaussian curve, then we can allocate the stressed condition of that factor to some portion of the extremes of that curve. The combination of two sets of factors to determine a two-dimensional niche will produce a set of conditions approximating those illustrated in Fig. 3, i.e., zones of productive (P) and stressed (S) conditions, among the latter being permutations of single- and double-factor stress. (This representation can of course be extended to any number of dimensions to describe the fundamental niche of the organism.)

Consider, then, the occupation of such a niche by a decomposer organism through a period of time. Initial colonization may be at a time when, in terms of the niche determinants, the resource is in a productive condition (zone A in Fig. 3). With the passage of time, the resource will change in condition and the microbe will eventually become subjected to stress. Put another way, the organism will now occupy the periphery of its fundamental niche rather than the center.

We may take as an example of this type of progression a niche determined by the two axes used in the previous example. Progressive decomposition can result in the dissipation of C and the conservation of N so that the C/N ratio drops, eventually reaching the stress zone where C is limiting (zone B in Fig. 3). At the same time, metabolically generated water or a developing rainy season, or both, may raise the moisture content of the resource to the level where periodic stress from waterlogging is experienced (zone C in Fig. 3).

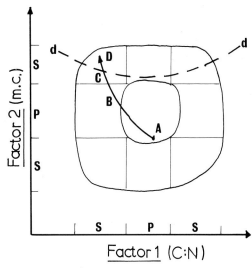

FIG. 3. Influence of stress, disturbance, and resource degradation on niche occupation. The dimensions of a two-dimensional fundamental niche (moisture content and C/N ratio of a resource) are indicated by the outermost ring; the dimensions of the productive zone of this niche are indicated by the inner ring; the zones of stress(es) for each of the factors lie in the annulus between the two. The boundary of a zone of disturbance is shown by the dashed line d–d. The shift of niche occupation by a decomposer organism as the resource changes with time is shown by the line moving from a productive zone (A) to a stressed zone (B) to a zone of double stress (C) to a zone with both stress and disturbance (D). (Further explanation in text.)

Let us further hypothesize that grazing by microtrophic Collembola may be initiated once a certain level of moisture content is reached. Thus, in zone D of Fig. 3 the fungus would be occupying a resource which exhibited a stressed *and* disturbed condition.

Two features of this scenario are particularly important to emphasize: first, that some parts of it are inevitable and predictable (e.g., the development of the stress); second, that other parts, those imposed by disturbance, may be highly probable but are not predictable. This permutation of possible selective situations provides an interesting framework from which to speculate on the possible range of fungal strategies that may have evolved in response to it.

In the broadest terms two distinct strategies may be hypothesized. These can be termed the occupation strategy and the window strategy. The basis of the former is to occupy the resource during the early stages of decomposition and to respond to the changes that occur in such a way as to maintain occupancy. The changes through which such a species would have to persist are illustrated by the pathway in Fig. 3, but it must also be reemphasized that there are many potential permutations of these changes.

The basis of the window strategy is to respond to some specific change in the condition of the resource in such a way as to secure occupancy during the process of change, i.e., when niche dimensions are changing. This may be followed either by rapid exploitation and dispersal or by a switch to an occupation strategy. The various stages of the fungal life cycle may contribute to these strategies in a variety of ways. The key to success in either case, however, is the ability to respond to change in the condition of the resource. In the case of the occupation strategy the response will be predominantly adapted to persistence. In the case of the window strategy the emphasis is on attack and retreat (i.e., on dispersal). Viewed in this way, the three strategies of Grime can be seen to operate at a secondary level for the fungi. The competitive, stress-tolerant, and ruderal are alternative versions of the window strategy, adapted to particular condition phases of the resource; within the occupation strategy they represent different phases in the overall life-style.

It is thus the contention here that the recognition of useful categories of strategy in the decomposer fungi is dependent upon the direct correlation of specific physio-morphological changes and their quantitative attributes with specific trigger phenomena emanating from the changes in condition of the resource.

CONCLUSION

The model of the decomposer community presented here is highly stochastic in nature. It predicts variation in composition and structure between unit communities of the same resource and in the development of identical unit communities with time.

The proposition advanced is that the development of communities in equilibrium with their physical and biotic environments may be relatively rare as a result of the intervention of disturbance effects. There may be important distinctions to be made here, however, between the communities of short-term resources such as leaves and fecal pellets and those of slowly decomposing resources such as the trunks or stumps (see Heal and Ineson, this volume). It has been argued that the definition of fungal strategies is a potentially rewarding approach to understanding community structure but that "forcing" decomposer fungi into the categories proposed by animal or plant ecologists (e.g., r and k, or ruderal, stress-tolerant, and competitive) may be premature. It is suggested that it is more important first to distinguish between patterns of behavior in the fungi and then to attempt to determine what categories of physiological behavior may be characteristic of these patterns.

In this respect two "supra-strategies" have been distinguished, termed the occupation and the window strategies. It is stressed that these are not suggested as necessarily correlating with any other pattern of strategy but as a starting point for the type of analysis suggested above. Such analysis depends on two areas of investigation: the analysis of resource condition and the investigation of decomposer life-styles in relation to response to stress and disturbance.

Boddy presents an example of an approach to this in which she has on the one hand described the progressive development of stress due to moisture tension in drying branches (3) and on the other described the relationship of basidiomycete activity to the same factor in pure culture (4).

Disturbance is more difficult to describe, and thus it is more difficult to structure experiments to determine its effects. Investigation of the effects of grazing or of changing environment due to litter fall are, however, entirely possible and offer rewarding opportunities for experimentation, as has already been demonstrated (18, 32, 55).

The essential information on life-styles of fungi must be investigated in a comprehensive way, placing the physiological traits of the fungi within the context of their life histories, as Cole (13) and Stearns (43) have done for animals and plants. A good deal of the necessary information is readily available, as can be seen from perusing the massive compilation of Domsch et al. (14). Wicklow (50) has provided a stimulating review of the characteristics of the conidial fungi which determine their ecological roles.

The combination of the development of theory with the use of controlled experimentation surely provides a powerful alternative to the conventional analysis of field populations as an approach to the study of decomposer communities.

I am grateful to Carole Shearer and Howard Wildman for allowing me access to their data prior to publication and to Roderic Cooke and Alan Rayner for sight of their stimulating new book in manuscript form.

LITERATURE CITED

1. **Beaver, R. A.** 1977. Non-equilibrium 'island' communities: diptera breeding in dead snails. J. Anim. Ecol. **46**:783–798.
2. **Beaver, R. A.** 1979. Non-equilibrium 'island' communities: a guild of tropical bark beetles. J. Anim. Ecol. **48**:987–1002.
3. **Boddy, L.** 1983. Microclimate and moisture dynamics of wood decomposing in terrestrial ecosystems. Soil Biol. Biochem. **15**:149–157.
4. **Boddy, L.** 1983. Effect of temperature and water potential on growth rate of wood rotting basidiomycetes. Trans. Br. Mycol. Soc. **80**:141–149.
5. **Boddy, L., and A. D. M. Rayner.** 1983. Ecological roles of

basidiomycetes forming decay communities in attached oak branches. New Phytol. **93**:77–88.

6. **Bunnell, F. L., and K. A. Scoullar.** 1975. Abisko II. A computer simulation model of carbon flux in tundra ecosystems, p. 425–448. *In* T. Rosswall and O. W. Heal (ed.), Structure and function of tundra ecosystems. Ecology Bulletin 20, Swedish Natural Science Research Council, Stockholm.

7. **Bunnell, F. L., D. E. N. Tait, and P. W. Flanagan.** 1977. Microbial respiration and substrate weight loss. II. A model of the influences of chemical composition. Soil Biol. Biochem. **9**:41–47.

8. **Bunnell, F. L., D. E. N. Tait, P. W. Flanagan, and K. van Cleve.** 1977. Microbial respiration and substrate weight loss. I. A general model of the influences of abiotic variables. Soil Biol. Biochem. **9**:33–40.

9. **Cairns, J., and J. A. Ruthven.** 1970. Artificial microhabitat size and the number of colonising protozoan species. Trans. Am. Microsc. Soc. **89**:100–109.

10. **Cale, W. G., and J. B. Waide.** 1980. A simulation model of decomposition in a shortgrass prairie. Ecol. Mod. **8**:1–14.

11. **Cancela da Fonseca, J. P.** 1979. Species colonisation models of temporary ecosystem habitats, p. 125–195. *In* G. S. Innis and R. V. O'Neill (ed.), Systems analysis of ecosystems. International Co-operative Publishing House, Fairland, Md.

12. **Chamier, A. C., and P. A. Dixon.** 1982. Pectinases in leaf degradation by aquatic hyphomycetes. I. The field study. The colonisation-pattern of aquatic hyphomycetes on leaf packs in a Surrey stream. Oecologia (Berlin) **52**:109–115.

13. **Cole, L. C.** 1954. The population consequences of life history phenomena. Q. Rev. Biol. **29**:103–137.

14. **Domsch, K. H., W. Gams, and T.-H. Anderson.** 1980. Compendium of soil fungi. Academic Press, London.

15. **Garrett, S. D.** 1951. Ecological groups of soil fungi; a survey of substrate relationships. New Phytol. **50**:149–166.

16. **Griffin, D. M.** 1972. Ecology of soil fungi. Chapman and Hall, London.

17. **Grime, J. P.** 1979. Plant strategies and vegetation processes. John Wiley & Sons, Chichester.

18. **Hanlon, R. D. G., and J. M. Anderson.** 1979. The effects of Collembola grazing on microbial activity in decomposing leaf litter. Oecologia (Berlin) **38**:93–99.

19. **Harper, J. E., and J. Webster.** 1964. An experimental analysis of the coprophilous fungal succession. Trans. Br. Mycol. Soc. **47**:511–530.

20. **Hawksworth, D. L.** 1976. The natural history of Slapton Ley Nature Reserve. X. Fungi. Field Stud. **4**:391–439.

21. **Hayes, A. J.** 1979. The microbiology of plant litter decomposition. Sci. Prog. (London) **66**:25–42.

22. **Heal, O. W., and S. F. MacLean, Jr.** 1975. Comparative productivity in ecosystems—secondary productivity, p. 89–108. *In* W. H. van Dubben and R. H. Lowe-McConnell (ed.), Unifying concepts in ecology. Dr. W. Junk br., The Hague.

23. **Hedger, J. N.** 1982. The role of basidiomycetes in composts: a model system for decomposition studies, p. 263–306. *In* J. C. Frankland, J. N. Hedger, and M. J. Swift (ed.), Decomposer basidiomycetes: their biology and ecology. Cambridge University Press, Cambridge.

24. **Hudson, H. J.** 1968. The ecology of fungi on plant remains above the soil. New Phytol. **67**:837–874.

25. **Huston, M.** 1979. A general hypothesis of species diversity. Am. Nat. **113**:81–101.

26. **Hutchinson, G. E.** 1961. The paradox of the plankton. Am. Nat. **95**:137–145.

27. **Lussenhop, J.** 1981. Analysis of microfungal component communities, p. 37–46. *In* D. T. Wicklow and G. C. Carroll (ed.), The fungal community: its organisation and role in the ecosystem. Marcel Dekker, New York.

28. **Lussenhop, J., R. Kumar, D. T. Wicklow, and J. E. Lloyd.** 1980. Insect effects on bacteria and fungi in cattle dung. Oikos **34**:54–58.

29. **MacArthur, R. H., and E. O. Wilson.** 1967. The theory of island biogeography. Princeton University Press, Princeton, N.J.

30. **Paine, R. T.** 1966. Food web complexity and species diversity. Am. Nat. **100**:65–76.

31. **Parkinson, D.** 1981. Ecology of soil fungi, p. 277–295. In G. T. Cole and B. Kendrick (ed.), Biology of conidial fungi, vol. 1. Academic Press, Inc., New York.

32. **Parkinson, D., S. Visser, and J. B. Whittaker.** 1979. Effects of collembolan grazing on fungal colonisation of leaf litter. Soil Biol. Biochem. **11**:529–535.

33. **Parnas, H.** 1975. Model for decomposition of organic material by microorganisms. Soil Biol. Biochem. **7**:161–169.

34. **Patrick, R.** 1967. The effect of invasion rates, species pool, and the size of area on the structure of the diatom community. Proc. Natl. Acad. Sci. U.S.A. **58**:1335–1342.

35. **Pianka, E. R.** 1970. On r- and K-selection. Am. Nat. **104**:592–597.

36. **Pugh, G. J. F.** 1980. Strategies in fungal ecology. Trans. Br. Mycol. Soc. **75**:1–14.

37. **Rayner, A. D. M., and N. K. Todd.** 1979. Population and community structure and dynamics of fungi in decaying wood. Adv. Bot. Res. **7**:334–421.

38. **Rayner, A. D. M., and N. K. Todd.** 1982. Ecological genetics of basidiomycete populations in decaying wood, p. 129–142. *In* J. C. Frankland, J. N. Hedger, and M. J. Swift (ed.), Decomposer basidiomycetes: their biology and ecology. Cambridge University Press, Cambridge.

39. **Root, R. B.** 1973. Organisation of a plant-arthropod association in simple and diverse habitats: the fauna of collards (*Brassica oleracea*). Ecol. Monogr. **43**:95–124.

40. **Sanders, P. F., and J. M. Anderson.** 1979. Colonisation of wood blocks by aquatic Hyphomycetes. Trans. Br. Mycol. Soc. **73**:103–107.

41. **Simberloff, D. S., and E. O. Wilson.** 1969. Experimental zoogeography of islands: the colonisation of empty islands. Ecology **50**:278–289.

42. **Smith, O. L.** 1979. An analytical model of the decomposition of soil organic matter. Soil Biol. Biochem. **11**:585–606.

43. **Stearns, S. C.** 1976. Life-history tactics: a review of the ideas. Q. Rev. Biol. **51**:3–47.

44. **Swift, M. J.** 1976. Species diversity and the structure of microbial communities, p. 185–222. In J. M. Anderson and A. MacFadyen (ed.), The role of aquatic and terrestrial organisms in decomposition processes. Blackwell Scientific Publications, Oxford.

45. **Swift, M. J.** 1982. The basidiomycete role in forest ecosystems, p. 307–338. *In* J. C. Frankland, J. N. Hedger, and M. J. Swift (ed.), Decomposer Basidiomycetes: their biology and ecology. Cambridge University Press, Cambridge.

46. **Swift, M. J.** 1983. Microbial succession during the decomposition of organic matter, p. 164–177. *In* R. G. Burns and J. H. Slater (ed.), Experimental microbial ecology. Blackwell Scientific Publications, Oxford.

47. **Swift, M. J., O. W. Heal, and J. M. Anderson.** 1979. Decomposition in terrestrial ecosystems. Blackwell Scientific Publications, Oxford.

48. **Thompson, W., and L. Boddy.** 1983. Decomposition of suppressed oak trees in even-aged plantations. II. Colonisation of tree roots by cord- and rhizomorph-producing basidiomycetes. New Phytol. **93**:277–291.

49. **Wicklow, D. T.** 1981. The coprophilous fungal community: a mycological community for examining ecological ideas, p. 47–78. *In* D. T. Wicklow and G. C. Carroll (ed.), The fungal community: its organisation and role in the ecosystem. Marcel Dekker, New York.

50. **Wicklow, D. T.** 1981. Biogeography and conidial fungi, p. 417–447. *In* G. T. Cole and B. Kendrick (ed.), Biology of conidial fungi, vol. 1. Academic Press, Inc., New York.

51. **Wicklow, D. T., and G. C. Carroll.** 1981. The fungal community: its organisation and role in the ecosystem. Marcel Dekker, New York.

52. **Wicklow, D. T., and W. F. Whittingham.** 1974. Soil microfungal changes among the profiles of disturbed conifer-hardwood forests. Ecology **55**:3–16.
53. **Widden, P., and D. Parkinson.** 1975. The effects of a forest fire on soil microfungi. Soil Biol. Biochem. **7**:125–138.
54. **Winogradzky, S.** 1924. Sur la microflore autochthone de la terre arable. C. R. Acad. Sci. **178**:1236–1239.
55. **Yocom, D. H., and D. T. Wicklow.** 1980. Community differentiation along a dune succession: an experimental approach with Coprophilous fungi. Ecology **61**:868–880.
56. **Zak, J. C., and I. T. Wicklow.** 1978. Factors influencing patterns of Ascomycete sporulation following simulated "burning" of prairie soils. Soil Biol. Biochem. **10**:533–535.

Some Potentials for the Use of Microorganisms in Ecological Theory

WILLIAM J. WIEBE

Department of Microbiology and Institute of Ecology, University of Georgia, Athens, Georgia 30602

Ecological theory at the present time is in a state of transition. The earlier hopes, for example, that stability was in some direct way related to diversity, that each ecosystem reached a unique steady-state mature assemblage, that niches could be operationally defined, that competition or predation, or both, held the key(s) to understanding the distribution and evolution of organisms, and that trophic relationships could be defined in terms of biophages and saprophages, have remained largely unfulfilled. These ideas and many others have played and continue to play an important role in the development of ecological concepts, but it is becoming clear that each offers an incomplete explanation for how ecosystems function and develop. One reason for this inadequacy may well be that the key role of microorganisms in ecological processes has not been fully recognized by the ecological theorists. Such an oversight is now less likely because microbiologists are entering the mainstream of ecological research and are not only dealing with pure cultures and gnotobiotic systems but are also examining many phenomena in situ. They are also, as exemplified in this symposium, beginning to test the relevance of ecological theories to the microbial world, as well as the relevance of microbes to ecological theory. It is an exciting period for microbiologists involved in ecological research. In this paper, I examine recent developments of three ecological concepts and discuss the roles that microbiologists can play in their evolution and validation.

SOME CONSIDERATIONS OF ECOLOGICAL THEORY

Rigler (18; Can. J. Fish Aquat. Sci., in press), in two thoughtful essays, discussed two types of scientific theories, *empirical*, which predict the future states of systems but do not offer explanations for the behavior, and *explanatory*, which propose mechanisms to account for the predictions. The first establishes order; the second defines the underlying causes for the behavior. Ecological theory, found in textbooks and in much of the literature today, deals with explanatory, but not empirical, theory. Rigler (18) believes that "... empiricism is both a challenging and essential aspect of environmental science."

Empirical theory should stimulate four types of responses: (i) "... research intended to test the validity of existing data"; (ii) "... try to find new variables or new combinations of variables that reduce the confidence limits of our predictions"; (iii) "... stimulate us to gather new data that will fill in and extend the original data base ..."; and (iv) "... stimulate some of us to move onto the next phase of science—that of replacing the empirical theory with an explanatory theory" (18). I believe that microbiologists have not made full use of this approach; some examples will be discussed in a subsequent section.

The second perspective concerning ecological theory that I wish to discuss is that advocated by E. P. Odum (*in* G. Knox and N. Polunin, ed., *Ecosystem Theory and Application*, in press; E. Odum and L. Biever, unpublished manuscript). He is concerned with the strategies that have been developed to study ecosystems. His thesis is that "... an ecosystem can neither be understood by taking it apart and studying the part piecemeal, nor can it be dealt with by conceptually enclosing it" (Odum, in press). This idea strikes at the heart of the way in which many of us do research, and I believe it needs some discussion. The basis for his thesis is that "... as components are combined to produce larger and more complex systems, new properties, often called emergent properties, appear" (Odum, in press). In other words, the whole is not equal to the sum of the parts.

How does one approach ecosystem studies operationally in this context? Odum (in press) proposes a hierarchical strategy. The first step is a "black box" approach. The system of interest is defined, and a study is made of the inputs to it and the outputs from it. In addition, major processes and components, e.g., primary production, respiration, biota, are examined. The second step involves the examination of "... operationally significant components or groups of components (populations and physical factors) as determined by observing, by modelling or by perturbing the system (as a means of identifying operationally important components). In this approach, one goes into great detail in the study of components only as far as necessary to understand or manage the system as a whole'' (Odum, in press). Odum and Biever (unpublished manuscript) point out that important components may not be only those that have high energetic flows, since the quality of the resource is also important. For example, they found that ectomycorrhiza-infected loblolly pines grew at a much greater rate than uninfected controls, far out of proportion to the energy requirements of the fungus. This also serves as an example of an emergent property; one could not predict the result of the interaction by only studying each component in isolation. In fact, on poor soil neither the tree nor the fungus can survive alone. This example also serves to remind us that mutualism has not received the attention that competition and predation have been given as control mechanisms. Energy flows involving high-quality dissolved organic matter mediated by microorganisms may prove to be more important than the more conspicuous herbivory and predation.

These two perspectives represent very different approaches to the study of ecosystems, but both have use for investigations of microorganisms and their activities in nature. And both offer challenges to us concerning how we conduct that research. Some examples of how these ideas can be incorporated into microbial ecology studies will be examined in a subsequent section. But before that, I would like to discuss two recent theories that are relevant to the topic of this paper.

CYBERNETIC CONTROL OF ECOSYSTEMS

Ecosystems, for all of the variation we see when studying them, have levels of organization and control that permit them to persist and to develop in an orderly manner through time. Control is achieved through feedback loops, and this has led ecologists to examine the possible application of cybernetic theory to ecosystems. Although there is some controversy over this approach (e.g., 6), the theory has elements worth examination.

Cybernetic systems, as defined by Weiner (26), are systems in which the input is determined in part by the output. The part of the output that affects the input is called the feedback. A corollary is that very small feedbacks can exert large effects on the input (16). At the ecosystem level the problem is to identify the key controlling factors, particularly those that have low flow but exert great effects (Odum, in press). As microbiologists, we can all think of examples of microbial processes, e.g., disease, fermentation, that fall into this category. On a much broader scale the recently proposed Gaia Hypothesis (see 14) puts forward the proposition that the biosphere is controlled not solely by abiotic properties but to a large extent by very

diffuse interactions of the biota with the physical factors. Recently, Lovelock and Whitfield (15) discussed possible limits between biological and geological control mechanisms with regard to atmospheric CO_2 (see also 28). Thus, a major issue in the study of ecosystems is the identification of the control mechanisms. With these considerations in mind, let us now examine how microorganisms can be used to study these ecological topics.

USE OF MICROORGANISMS TO DEVELOP AND TEST ECOLOGICAL THEORY

Other papers in this symposium give some excellent examples of how microbiologists are using microorganisms and microbial techniques to examine ecological theories. The large number of recent papers applying extant ecological theories to microbial systems documents the growing interest in this field. These studies represent an important extension of microbiologists' interest. In this section I will not review these studies because they have been discussed in the previous papers. Rather, I will examine how we might develop, and in some cases have begun to develop, studies that are concerned with the theories just discussed: empirical theory, ecosystem theory, and cybernetic theory.

Empirical theory. As used by Rigler (18), empirical theory only makes a prediction; it does not explain the relationship. But it does provide ''. . . a general statement that makes potentially falsifiable predictions'' (18). Whether or not one wishes to consider these general statements theories, they have provided ecologists with some of the most useful information. For example, in lakes there is a direct relationship between the abundance of phosphate and primary production (see 18). This relationship has been used for a number of years to manage lakes. Brylinsky and Mann (3) similarly have found that much of the variability in the primary production in lakes is accounted for by two factors, altitude and latitude, regardless of the type of lake. Johannes (9) reported that the body size of an organism is directly related to its rate of phosphorus excretion. Christian et al. (4) found that the microbial biomass, as measured by ATP, was constant over a wide range of habitats that had sufficient moisture and light. These include coral reef and intertidal sediments and grassland and tundra soils. Smith and Kinsey (20) found that the rates of calcium carbonate production on the windward sides of almost all undisturbed coral reefs are constant. Sheldon et al. (19) postulated that there is the same biomass of all sizes of organisms in the sea; the biomass of whales equals the biomass of bacteria.

These relationships and many others not mentioned provide important predictions for ecolo-

marine populations. Phytoplankton, with generation times on the order of 1 day, form patches 1 to 5 km in diameter; zooplankton reproduce over 50- to 100-day periods and form patches (or "ambits") ranging over 10 to 100 km; and fish reproduce even more slowly and have populations extending over thousands of kilometers (31). Theories of ecosystem variance must account for variability over several scales and for the interdependence of time and space scales.

The linkage of temporal and spatial variability has been formulated for phytoplankton populations since Kierstead and Slobodkin (19) considered the interactions between growth rates and diffusivity to explain the formation of phytoplankton patches in the sea. But useful conceptual models of the scales of variability in biological populations derive mostly from more recent work. Stommel first illustrated that sources of physical variability in the oceans range over 12 orders of magnitude from seconds to millenia and centimeters to global dimensions (32). He stated that, given such large ranges of variability and the expense of oceanographic expeditions, no single expedition or experiment could simultaneously observe phenomena over all the relevant scales. Thus, it is most important to plan expeditions with a view of defining the scales of interest concerning specific features of the oceans such as gravity waves, storm responses, large-scale currents, or even ice age variations.

The same is true for biology. Figure 1 shows a version of the Stommel Diagram, as modified to illustrate variability in zooplankton populations (15). In this diagram, the temporal and spatial scales of different sources of zooplankton variations are plotted on log scales on the horizontal axes, and the relative intensity of the variability is plotted on the vertical axes. The rough interdependence of time and space is depicted by the ridge running diagonally from the minute/10 cm range up through the peak generated by ice age variation at 10^4 km and 10^4 years. However, other ridges running parallel to the time and space axes suggest that some structures may exist over small spatial scales but persist for centuries (the variability associated with oceanic fronts of water mass boundaries at K on the diagram). Other structures extend over large distances but occur with relatively high frequency, like diel vertical migrations. The complexity of the variance structure requires that biologists also consider very rigorously those scales over which different processes can be most effectively studied.

Phytoplankton variability over the ocean surface can be intensively sampled by the electronic determination of in vivo chlorophyll fluorescence as research vessels move through the water. Tens of thousands of data points can be collected over 10 to 100 km in a few days or weeks. Since the sampling resolution is on the

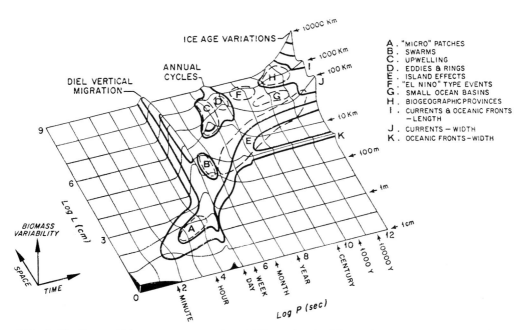

FIG. 1. The Stommel Diagram, a conceptual model illustrating the different time and space scales over which variability is distributed in the ocean. In this example zooplankton biomass variations are shown, but other parameters could be depicted. (Copyright 1978, Plenum Publishing Corp.)

order of 100 m and one per minute, variability can be observed through an "observational window" at scales from minutes to weeks and meters to kilometers (29). In addition, temperature and salinity can be sampled simultaneously, so that the relationship between biological and physical variations can be determined over these scales. Most importantly, analytical techniques exist to digest all these data. Spectral analysis, a numerical computing approach developed in communications theory to decompose the spectrum, or range of scales of periodic variations in a system, has been applied to various ecological studies over the past decade (21, 29).

Spectral analysis of data on chlorophyll and temperature variability collected together reveals regions of variability spectra in which the biological and physical properties covary, and others in which the biological signal is independent of physical variations. In general, three regions of the space spectrum characterize the horizontal variability of phytoplankton in the ocean (21). Below a critical length scale the variability in phytoplankton distribution due to biological processes of growth and grazing is smoothed by diffusive forces which tend to dissipate patches (19, 28). The critical length scale is inversely related to net growth rate (growth minus grazing). Close coupling between phytoplankton growth and herbivore grazing tends to increase the critical patch size at which biological variability surpasses that related to turbulent diffusive forces. Above the critical length scale there exists a "biological window" in the variance spectrum in which biological forces contribute more significantly to spatial organization than physical processes. At still larger scales, physical transport processes become more energetic than biological processes. Above the upper boundary of the biological window, ocean currents, meteorological forces, and variability in water mass properties control the distribution of organisms. For phytoplankton these critical length dimensions are roughly 50 to 100 m and 5 to 25 km. Marra et al. (26) have shown how these scales interact to produce chlorophyll distribution patterns at a frontal system separating the northeastern continental shelf of the United States from the deep ocean. They described a biological window (i.e., phytoplankton patches not coherent with temperature structure) of 1 to 10 km, above which variability in vertical mixing and warm core rings (see below) controlled the chlorophyll distributions.

To summarize this section briefly, we may learn more about the spatial and temporal organization of plankton populations by analyzing variability than by restricting our attention to mean values of system variables. Variation exists simultaneously at many space and time scales. With the appropriate sensors and analytical techniques, it has been shown that at different space scales various physical and biological processes assume different relative significance in controlling the spatial organization of plankton populations. At an intermediate range of the variance spectrum, the biological window, biological factors such as growth rate and predation control the spatial variability of planktonic organisms. Above this scale, mesoscale variations in water mass characteristics control the distributions of organisms. In the next section I consider the extent to which these ideas are of use in the study of bacterial distributions and dynamics in the sea.

VARIABILITY IN BACTERIAL ABUNDANCE AT THE MESOSCALE

An objective examination of bacterial distributions in the sea has become possible only within the past 10 years, with the perfection of direct enumeration techniques (16). However, even with sensitive and specific epifluorescence microscopy, sampling resolution is well below the levels described above. There is no automatic detection system for measuring bacterial abundance. Time series of data must be constructed from many dozens of discrete samples, which require labor-intensive processing. Samples can be processed at rates on the order of three data points per man-hour, but this figure does not include the original sampling operation. I am not aware of a data set from which the variability spectrum of bacterial abundance could be rigorously analyzed, even through a narrow observational window.

With current technology, it seems feasible to collect and analyze about 500 samples per man-month of microscopy labor. A data set of 500 samples collected over the course of 1 week from a ship moving at an average speed of 8 knots (i.e., 15 km h^{-1}) would provide an observational window with minimum dimensions (i.e., resolution) of 15 min and 4 km, and would admit analysis of the spectrum from 4 to perhaps 100 km. But would such a data set be useful and justify the tedium of collecting water samples flowing from a pipe every 15 min for 5 to 7 days? I suggest that it would be, and I review data on which that conclusion is based in the remainder of this paper.

My argument is based on two primary observations. First, significant variability in bacterial abundance appears to exist over the mesoscale. Second, this variability must depend upon coupling among phytoplankton, herbivores, and bacteria; however, we are largely ignorant of the scales over which these couplings are generated and maintained. I assert that a rigorous examination of the scales of bacterial variability in the sea would help to confirm the existence of these features, and more importantly, it would tell us about the basic ecological relationships among bacteria and other plankton.

The bacteriological window. The evidence for strong bacterial abundance signals in the ocean is not extensive. Bacterial abundance ranges from about 10^8 cells per liter in the upper 100 m of the open ocean to over 2×10^9 cells per liter in some productive coastal areas (17). In comparison, the range for chlorophyll is from 0.05 to over 30 µg/liter and for zooplankton perhaps another order of magnitude greater (15). Bacterial biomass levels in different environments are less well defined however, so the ranges of phytoplankton and bacterial biomass could be more similar than suggested here. But the existing data suggest that variability of bacterial stocks is more effectively damped out over large space scales than phytoplankton and zooplankton variability.

This conclusion is likely to be an artifact of infrequent sampling of bacterial abundance with poor spatial resolution. Critical length scales for bacterial patch formation could be on the order of 50 to 100 m since bacterial growth rates and diffusivity values are on the order of those for small phytoplankton. Sampling at that level of resolution has not been attempted, but evidence for elevated bacterial abundance has been revealed when sampling has been carried out with reference to hydrographic features. There is increasing evidence for large variability of bacterial abundance in frontal regions, including tidal fronts off the English coast (9), the Antarctic Polar Front southwest of Tierra del Fuego (14), and the shelf break front in the New York Bight (6, 8). Some frontal systems are only a few kilometers wide and contain concentrations of bacterial carbon exceeding 500 µg/liter (9).

For the past 2 years I have been studying bacterial distributions and production in the component water masses of warm core Gulf Stream rings (WCR). These are mesoscale eddies formed when northward meanders of the Gulf Stream pinch off, entraining parcels of warm, salty, oligotrophic Sargasso Sea water within the cooler, fresher, and more productive Slope Water between Cape Hatteras and Nova Scotia (18). They can persist in the Slope Water for periods up to 1 year or longer. Because the ring core is isolated from its source water but maintains its integrity over time and can be followed by ships and satellites, a WCR is a useful system in which to observe the temporal and spatial organization and development of entrained plankton populations. My studies of bacterial abundance in WCR 82-B between April and August 1982 suggest that bacterial levels are significantly enhanced and maintained in the

frontal regions separating the ring core (remnant Sargasso Sea water) from the surrounding Slope Water.

Figure 2 shows a summary map of WCR studied by the National Science Foundation-sponsored WCR program during 1981–1982. In particular we revisited ring 82-B in April–May, June, and August 1982, as shown by the successive positions of 82-B on days 123–125, 162–164, and 230, respectively. The ring drifted southwest at a varying rate, changed size, shape, and orientation, and eventually was eroded and reabsorbed by the Gulf Stream after 1 September (day 263). Table 1 summarizes bacterial abundance data collected in the upper 100 m of the ring, Slope Water, and Northern Sargasso Sea. The two most important features of these data are that: (i) the abundance of bacteria in the Sargasso Sea remnant ring center tripled between April and June, while numbers stayed more constant elsewhere, and (ii) abundances persisted at high levels in the high velocity region, or Gulf Stream remnant, a circular frontal zone surrounding the ring. Since these patterns were generated over ca. 40 days while bacterial growth rates were on the order of 1 to 2 per day (estimated from observing changes in cell numbers in incubations; H. W. Ducklow, unpublished data), these data constitute strong presumptive evidence of significant spatial and temporal signals in bacterial abundance in the open ocean. However, the data are derived from widely spaced samples. The hypothesis that these distributions result merely from passive accumulations of cells by physical processes could be tested by more intensive sampling and cross-spectral analysis of the biological and physical properties in each region of the ring (29).

Mechanisms regulating bacterial abundance. Let us assume that this variability in the spatial and temporal structure of bacterial abundance in the open ocean is real, not an artifact of primitive sampling techniques, and also assume that it has a biological cause. What mechanisms generate these patterns? It is generally believed that in the open ocean, heterotrophic bacterial production is sustained by dissolved organic matter flux from phytoplankton and zooplankton (1, 7). Over what scales are bacteria and phytoplankton activity coupled? Since primary production fuels bacterial activity, there should be scales over which other biological properties such as phytoplankton activity show coherence with bacteriological properties. Since there is much knowledge concerning the spatial distribution of chlorophyll a, I examine some statistical relationships between chlorophyll and bacteria below. The purpose of this discussion is to demonstrate that there are different scales at which

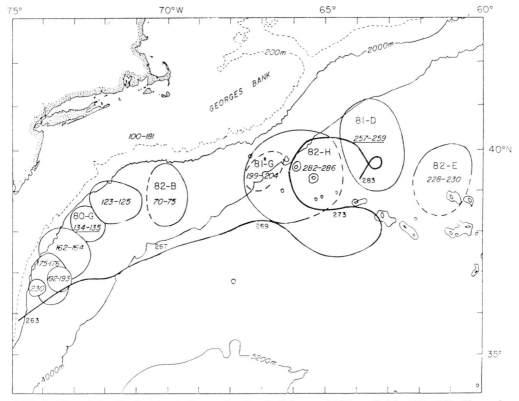

FIG. 2. Map of the North Atlantic Ocean showing Gulf Stream WCR studied in 1981–1982. The region between the 2,000-m isobath and the north wall of the Gulf Stream (the heavy line roughly paralleling the 4,000-m isobath) is the Slope Water, through which the rings drift southwest. See text for details.

these parameters are related, and that a knowledge of these scales can be used to discern the processes which generate the distribution and correlations we observe.

The abundance, biomass, and activity of heterotrophic bacteria are highest in the upper 100 m of the open ocean (the euphotic zone), suggesting the dependence of bacterial processes upon photosynthesis (17, 30). At this wide scale, extending throughout the world oceans, long-term correlations between chlorophyll stocks or photosynthetic carbon fixation rates and bacterial abundance or activity reflect the ultimate dependence of bacteria on primary production.

Statistically significant relationships between bacterial and phytoplankton parameters at this scale are found in pooled data sets covering different cruises, seasons, water mass types, and geographical locations. For example, bacterial abundance and production were highly correlated with chlorophyll and phaeopigment concentrations in samples collected from a series of transects which extended from coastal water and a sewage outfall across the Southern California Bight into blue, oligotrophic waters of the Pacific Ocean (10). Similar positive correlations

were observed for a transect of the New York Bight passing from the mouth of New York Harbor out across the shelf break front into the

TABLE 1. Bacterial abundance in water masses associated with Gulf Stream WCR 82-B, April–August 1982[a]

Region	10^8 cells per liter		
	April	June	August
Slope Water	8.1 ± 2.8 (6.2–10) $n = 11^b$	13.2 ± 6.2 (10.8–15.6) $n = 28$	ND
Ring center	3.4 ± 0.8 (3.2–3.6) $n = 71$	9.7 ± 3.3 (9.0–10.4) $n = 94$	6.9 ± 2.7 (6.0–7.8) $n = 35$
High-velocity region	17.2 ± 3.4 (15.3–19.1) $n = 15$	19.4 ± 6.3 (17.4–21.6) $n = 35$	ND
Northern Sargasso Sea	5.4 ± 0.9 (4.6–6.2) $n = 8$	ND	3.4 ± 1.0 (0.5–6.3) $n = 7$

[a] The mean value ± SD is listed; 95% confidence limits are included in parentheses. ND, No data.
[b] Number of samples.

Slope Water. This data set spanned temperature and chlorophyll ranges of 3 to 14°C and <1 to 12 µg/liter, respectively (6). Linley et al. (22) demonstrated that bacterial numbers and chlorophyll concentrations were also positively correlated in pooled data from the English Channel, the Celtic Sea off France, and the Benguela Current west of South Africa. At this scale (months to years, hundreds to thousands of kilometers) bacterial abundance appears to be set roughly by the level of phytoplankton biomass or the photosynthetic rate, or both.

Correlations between parameters or, more broadly, concordance of variability spectra at this scale reveals only a general functional relationship between bacteria and phytoplankton. But these correlations do not reveal the mechanisms coupling bacteria and phytoplankton. To learn about the specific biological and physical processes which influence bacterial ecology in the oceans, it is necessary to examine the relationships at smaller time and space scales.

There are two main hypotheses concerning mechanisms which link phytoplankton and bacteria. The first is that healthy intact phytoplankton release low-molecular-weight substances such as amino acids and simple carbohydrates, which serve as the principal carbon, nitrogen, and energy sources for bacteria (1, 20). The other mechanism concerns losses of dissolved organic matter from phytoplankton during ingestion by herbivores; this will be discussed later. In the absence of grazing by zooplankton, extracellular release of dissolved organics appears to be closely related to photosynthesis (24). If this mechanism is important for sustaining bacterial activity, then bacterial abundance and activity should be coupled to phytoplankton over the time and space scales at which photosynthesis and phytoplankton growth vary. These scales include the 1- to 100-m scale of the depth of the euphotic zone, the diurnal scale of photosynthesis, and the 10- to 20-day time scale of phytoplankton blooms. One way to test whether bacteria are closely coupled to extracellular release from phytoplankton is to examine correlations between phytoplankton and bacterial parameters at these scales.

Although there have been only a few synoptic investigations of bacterial and phytoplankton variability, the data do indicate some degree of correlation. For instance, when chlorophyll *a* concentrations decline as a linear function of depth in the euphotic zone, chlorophyll is positively correlated with bacterial abundance over individual vertical profiles 20 to 50 m deep (22). Bacterial abundance and also the frequency of dividing cells, a function of growth rate, show periodic oscillations over the diurnal cycle, with maximum values in the afternoon and minima at night (12, 20). Several investigators have shown that dissolved carbohydrate and amino acids in seawater accumulate during the day and are consumed (presumably by bacteria) at night (3, 27). Finally, bacterial abundance in the Baltic Sea responded to phytoplankton blooms, rising and falling roughly in parallel with oscillations in chlorophyll stocks, at intervals of weeks to months (20). Taken together, these observations show that the variability of phytoplankton and bacteria are roughly correlated over some of the time and space scales of relevance to photosynthetic behavior. This in turn suggests that bacteria may be coupled to phytoplankton by the release and subsequent uptake of organic compounds.

Under certain conditions, statistically significant correlations between bacterial and phytoplankton parameters are not observed (22). Some illustrations of noncorrelation of chlorophyll and bacterial numbers are shown in Fig. 3 and 4. Figure 3 shows plots of chlorophyll

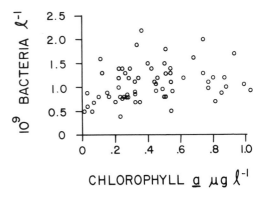

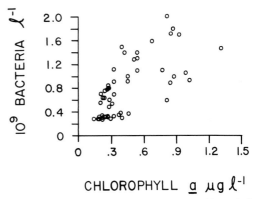

FIG. 3. Scatter plots of euphotic zone chlorophyll *a* concentration versus bacterial abundance for: (upper plot) the center of ring 82-B in June (days 162–164 in Fig. 2); (lower plot) the Sargasso Sea, ring center, and high-velocity region in April, and the high-velocity region in June.

versus bacterial abundance for different sets of data from WCR 82-B. At longer scales encompassing samples collected during April through June 1982 from three distinct water masses, the two parameters were correlated significantly (Fig. 3, lower plot). However, over a 2-week period in June, they were not correlated in samples collected within the center of the ring (Fig. 3, upper plot). The reason that correlations were not observed is apparent from the vertical profiles of bacteria and chlorophyll in the ring center.

Figure 4 shows a typical vertical profile of chlorophyll from ring center in June, with a deep chlorophyll maximum prominent at 20 to 25 m. Deep chlorophyll maxima are common features of oceanic chlorophyll profiles in summer. Bacterial abundance is more uniformly distributed, with a hint of a maximum above the chlorophyll maximum. Bacteria and chlorophyll are not correlated in these profiles ($r^2 = 0.013$), even though the general concentration of bacterial abundance within the euphotic zone is apparent. In contrast, the incorporation rate of tritiated thymidine (a function of bacterial production [6, 10]) is positively correlated with photosynthetic carbon fixation rates by the total phytoplankton assemblage ($r^2 = 0.83$) as well as by the <20 μm

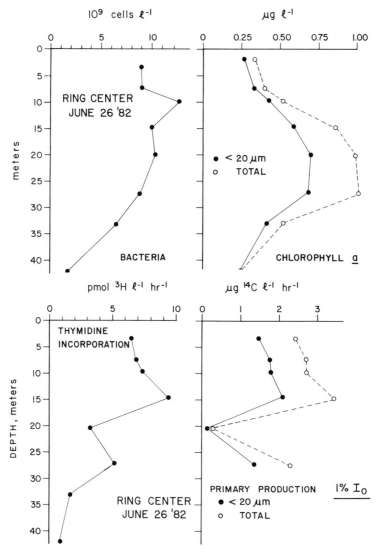

FIG. 4. Vertical profiles of bacterial abundance, [*methyl*-^{3}H]thymidine incorporation by bacteria, chlorophyll *a*, and [^{14}C]bicarbonate incorporation by phytoplankton versus depth at ring center, 0530 hours, 26 June 1982, (36°44′N latitude, 73°59′W longitude).

fraction ($r^2 = 0.90$), indicating a close coupling of bacterial and phytoplankton production in time and space (hours, meters). This in turn suggests that the bacteria could be sustained by the flux of dissolved compounds from algae, but that bacterial production is poorly coupled to increases in bacterial numbers at the same scale. Higher primary production, leading to an increased flux of substrates from the phytoplankton, stimulated bacterial production, but did not coincide with a larger standing stock of bacteria at the same depths within the water column. Thus, bacterial abundance was not related to any phytoplankton parameters inside the ring in June. However, since abundance is the result of previous growth, present abundance values could be correlated with phytoplankton properties from some previous time.

It is also apparent that chlorophyll and photosynthesis were not correlated over the profiles shown in Fig. 4. The weak coupling of production with abundance or biomass parameters may be due to removal of newly produced cells by predators (2, 5, 22). Herbivore stocks and grazing rates were higher in ring 82-B in June than in April. This may have been true for bacteriovores as well, but I have only indirect evidence for predation on bacteria in April and June. The characteristic time and space scales for predators are somewhat longer than for their prey (15, 31), suggesting that predation on phytoplankton or bacteria would disrupt correlations between those parameters at shorter scales. In Figure 3, bacterial abundance and chlorophyll are correlated over a large space scale (lower plot) which includes samples from waters in and around ring 82-B in April and June. However, at the smaller scale of the interior of 82-B (upper plot) these properties were not correlated in June. Nor were they correlated within any one water mass of those shown in the lower plot. In this instance, the identification of a scale at which bacteria and phytoplankton were *not* correlated may be as instructive as knowing the scales over which they are significantly related.

Herbivores may also influence the coupling of primary producers to bacterial processes through a process termed "sloppy feeding" (7). This mechanism operates during individual grazing incidents when herbivores break open phytoplankton cells, spilling their contents. The dissolved contents of ingested cells may become available to bacteria in the vicinity of the grazer, and composition of the released material may be quite different from that released by healthy, undamaged cells. Since releases from healthy and from partly ingested, broken cells probably occur simultaneously, distinguishing between the two mechanisms is difficult. However, each process has characteristic scales, and a consideration of them can be an aid to understanding variability in bacterial parameters and the mechanisms contributing to it.

The scales over which herbivores influence distributions of phytoplankton and bacteria are set by behavioral and developmental processes common to metazoans but largely absent from most unicellular organisms. For example, many zooplankton migrate vertically through the water column (4, 15). A swarm of copepods migrating into or out of the surface layer could conceivably modify the flux of dissolved substances to bacteria significantly. Vertical migration proceeds with a characteristic time scale of 12 to 24 h (Fig. 1) and may greatly enhance or damp out other diel effects related to photosynthesis. The ability of zooplankton to move through the water over hundreds of meters spreads out their influence over larger areas than small-scale phytoplankton distributions. Zones of locally high zooplankton biomass and grazing activity are found as alongshore bands seaward of high chlorophyll zones in areas of coastal upwelling off California (4). A band of high bacterial activity in this area could extend from the inshore high chlorophyll region through the offshore band of high grazing activity. This possibility is speculative, and testing it requires more closely spaced samples for bacterial parameters than we possess at present.

Longer generation times for herbivores may also influence bacterial variability over longer time scales. I mentioned earlier that bacterial activity seemed to oscillate in phase with phytoplankton blooms, but this need not always be true. Peaks of herbivore abundance and activity occur later than phytoplankton blooms because of the slower growth rates of the zooplankton. If herbivore grazing on phytoplankton is an important source of dissolved organic matter flux to bacteria, then peak periods of bacterial production could occur out of phase with phytoplankton blooms. Variability in bacterial populations is likely to be coupled to zooplankton populations and their grazing activities, but because of the longer time and space scales characteristic of metazoans, these couplings should generate variability at lower frequencies and spread it over larger areas. Sampling the spectrum of bacterial abundance at frequencies intermediate between those generated by phytoplankton and zooplankton processes may show little variability, whereas sampling over the time scales I have described might yield a far more dynamic picture. The same scenario is plausible in the spatial realm. The infrequent, usually widely spaced samples of bacterial abundance collected so far are adequate only to resolve variability at large scales, or that associated with water mass characteristics. I have tried to show that there are biological processes influencing bacterial growth and abundance changes which have characteristic scales of variability in time and space. Thus, we might expect to identify greater variability in bacterial parameters at these scales in a well-planned, objectively designed sampling program.

CONCLUSION: TOWARD A THEORY OF BACTERIAL DISTRIBUTIONS IN THE SEA

A good theory should explain the processes generating patterns we observe in nature and also suggest the best ways to make more profitable observations in the future. Most of the technology for measuring bacterial distributions and dynamics has only recently been developed, and it still has not been extensively tested. It is not at all surprising that there is not a body of theory for guiding the application of these new measurement techniques to questions of bacterial ecology in the sea. Bacteriologists might start by borrowing elements for such a theory from their colleagues in phytoplankton and zooplankton ecology.

A useful theory of bacterial distribution needs to explain the mechanisms coupling sources of bacterial substrates (photosynthesis and grazing) to bacterial production and the mechanisms coupling bacterial production to consumption of bacterial biomass. Furthermore, it is necessary to recognize that there are characteristic time and space scales over which these processes are coupled, and that sampling should be carried out at the proper frequency and spacing to examine these scales in a mathematically rigorous fashion. Finally, sampling should be guided by the philosophy that characterizing and decomposing variability into different frequencies will be more profitable than merely describing mean values for parameters over various arbitrarily chosen scales.

I thank Tom Malone (Horn Point Environmental Laboratory) for discussions of some of the ideas in this article. I also thank Ted Smayda and Gary Hitchcock (University of Rhode Island) for permission to use and publish their phytoplankton data, Loren Haury (Scripps Institution of Oceanography) for permission to use Fig. 1, and Terry Joyce (Woods Hole Oceanographic Institution) for permission to use Fig. 2. I am grateful to Captain Emerson Hiller and the officers and crew of RV *Knorr* for support at sea.

This work was supported by the Warm Core Rings Program and National Science Foundation grant OCE 81-17713.

LITERATURE CITED

1. Andrews, P., and P. J. LeB. Williams. 1971. Heterotrophic utilization of dissolved organic compounds in the sea. III. Measurement of the oxidation rates and concentrations of glucose and amino acids in sea water. J. Mar. Biol. Assoc. U.K. 51:111–125.
2. Azam, F., T. Fenchel, J. G. Field, J. S. Gray, L. A. Meyer-Reil, and F. Thingstad. 1983. The ecological role of water-column microbes in the sea. Mar. Ecol. Prog. Ser. 10:257–263.
3. Burney, C. M., P G. Davis, K. M. Johnson, and J. McN. Sieburth. 1982. Diel relationships of microbial trophic groups and in situ dissolved carbohydrate dynamics in the Caribbean Sea. Mar. Biol. 67:311–322.
4. Cox, J. L., L. R. Haury, and J. J. Simpson. 1982. Spatial patterns of grazing-related parameters in California coastal surface waters, 1979. J. Mar. Res. 40:1127–1153.
5. Ducklow, H. W. 1983. The production and fate of bacteria in the oceans. BioScience 33:494–501.
6. Ducklow, H. W., and D. L. Kirchman. 1983. Bacterial dynamics and distribution during a spring diatom bloom in the Hudson River plume, USA. J. Plankton Res. 5:333–355.
7. Eppley, R. W., S. G. Horrigan, J. A. Fuhrman, E. R. Brooks, C. C. Price, and K. Sellner. 1981. Origins of dissolved organic matter in southern California coastal waters: experiments on the role of zooplankton. Mar. Ecol. Prog. Ser. 6:149–159.
8. Ferguson, R. L., and A. V. Palumbo. 1979. Distribution of suspended bacteria in neritic waters south of Long Island during stratified conditions. Limnol. Oceanogr. 24:697–705.
9. Floodgate, G. D., G. E. Fogg, D. A. Jones, K. Lochte, and C. M. Turley. 1981. Microbiological and zooplankton activity at a front in Liverpool Bay. Nature (London) 290:133–135.
10. Fuhrman, J. A., J. W. Ammerman, and F. Azam. 1980. Bacterioplankton in the coastal euphotic zone: distribution, activity and possible relationships with phytoplankton. Mar. Biol. 60:201–207.
11. Gause, G. F. 1934. The struggle for existence. The Williams & Wilkins Co., Baltimore (reprinted 1971, Dover, New York).
12. Hagstrom, A., and U. Larsson. 1983. Diel and seasonal variation in growth rates of pelagic bacteria. In J. Hobbie and P. J. leB. Williams (ed.), Heterotrophic activity in the sea. Plenum Press, New York.
13. Hairston, N. G., J. D. Allan, R. K. Colwell, D. J. Futuyma, J. Howell, M. D. Lubin, J. Mathias, and J. H. Vandermeer. 1968. The relationship between species diversity and stability: an experimental approach with protozoa and bacteria. Ecology 49:1091–1101.
14. Hanson, R. B., D. Shafer, T. Ryan, D. H. Pope, and H. K. Lowrey. 1983. Bacterioplankton in Antarctic Ocean waters during late Austral winter: abundance, frequency of dividing cells, and estimates of production. Appl. Environ. Microbiol. 45:1622–1632.
15. Haury, L. R., J. A. McGowan, and P. H. Wiebe. 1978. Patterns and processes in time-space scales of plankton distributions, p. 277–328. In J. Steele (ed.), Spatial patterns in plankton communities. Plenum Press, New York.
16. Hobbie, J. H., R. J. Daley, and S. Jasper. 1977. Use of Nuclepore filters for counting bacteria by epifluorescence microscopy. Appl. Environ. Microbiol. 33:1225–1228.
17. Joint, I. R., and R. J. Morris. 1982. The role of bacteria in the turnover of organic matter in the sea. Oceanogr. Mar. Biol. Annu. Rev. 20:65–118.
18. Joyce, T., and P. Wiebe. 1983. Warm core rings of the Gulf Stream. Oceanus 26:34–44.
19. Kierstead, H., and L. B. Slobodkin. 1953. The size of water masses containing phytoplankton blooms. J. Mar. Res. 12:141–147.
20. Larsson, U., and A. Hagstrom. 1982. Fractionated phytoplankton primary production, exudate release and bacterial production in a Baltic eutrophication gradient. Mar. Biol. 67:57–70.
21. Lewis, M. R., and T. Platt. 1982. Scales of variability in estuarine ecosystems, p. 3–20. In V. Kennedy (ed.), Estuarine comparisons. Academic Press, Inc., New York.
22. Linley, E. A. S., R. C. Newell, and M. I. Lucas. 1983. Quantitative relationships between phytoplankton, bacteria, and heterotrophic microflagellates in shelf waters.

Mar. Ecol. Prog. Ser. **12**:77–89.

23. **Lotka, A. J.** 1920. Analytical note on certain rhythmic relations in organic systems. Proc. Natl. Acad. Sci. U.S.A. **6**:410.

24. **Mague, T. H., E. Friberg, D. J. Hughes, and I. Morris.** 1980. Extracellular release of carbon by marine phytoplankton; a physiological approach. Limnol. Oceanogr. **25**:262–279.

25. **Margalef, R.** 1958. Information theory in ecology. Gen. Syst. **3**:36–71.

26. **Marra, J., R. W. Houghton, D. C. Boardman, and P. J. Neale.** 1982. Variability in surface chlorophyll *a* at a shelf-break front. J. Mar. Res. **40**:575–591.

27. **Mopper, K., and P. Lindroth.** 1982. Diel and depth variations in dissolved free amino acids and ammonium in the Baltic Sea determined by shipboard HPLC analysis. Limnol. Oceanogr. **27**:336–347.

28. **Okubo, A.** 1978. Horizontal dispersion and critical scales for phytoplankton patches, p. 21–42. *In* J. Steele (ed.), Spatial patterns in plankton communities. Plenum Press, New York.

29. **Platt, T., and K. L. Denman.** 1975. Spectral analysis in ecology. Annu. Rev. Ecol. Syst. **6**:189–210.

30. **Sorokin, Y. I.** 1973. Data on biological productivity of the Western Tropical Pacific Ocean. Mar. Biol. **20**:177–196.

31. **Steele, J.** 1978. Some comments on plankton patches, p. 1–20. *In* J. Steele (ed.), Spatial patterns in plankton communities. Plenum Press, New York.

32. **Stommel, H.** 1963. Varieties of oceanographic experience. Science **139**:572–576.

Physiological and Morphological Adaptations

Role of Prostheca Development in Oligotrophic Aquatic Bacteria

JEANNE S. POINDEXTER

The Public Health Research Institute of the City of New York, Inc., New York, New York 10016

Oligotrophic freshwater habitats are generally described in terms of the flux of organic carbon, which is low. A reasonable estimate is not more than 0.1 mg of total organic carbon per liter per day (reviewed in 17). However, in illuminated surface waters, there is invariably a producer community, i.e., a source of organic carbon within the ecosystem. With adequate light, the most common limiting factor for production in freshwater ecosystems is phosphorus flux (reviewed in 13). It follows that any amount of activity by the producers enriches the habitat with organic carbon and that such enrichment further increases the relative phosphate limitation on the growth and multiplication of heterotrophs. At least, it must be acknowledged that eutrophication does not occur without phosphate enrichment relative to oligotrophic levels.

Accordingly, freshwater oligotrophs live in a phosphate-limited habitat. They must, therefore, be able to exploit not only a restricted supply of organic carbon but also phosphate limitation. In this discussion, heterotrophic prosthecate bacteria of the genera *Caulobacter* and *Hyphomicrobium*, organisms well known to be distributed in nature principally as oligotrophs, will be considered as phosphate oligotrophs.

Each individual *Caulobacter* or *Hyphomicrobium* cell occurs in three forms, in succession. The first is a motile, nonprosthecate cell that does not synthesize DNA, does not reproduce, does not grow appreciably, and synthesizes RNA and protein at relatively very low rates, as determined by incorporation of exogenous precursors (reviewed in 2). It eventually develops into the second form: the developed, prosthecate cell. With the onset of development, DNA synthesis begins, biosynthesis accelerates, and growth occurs. Eventually, this cell becomes a reproducing cell. In *Caulobacter* sp. reproduction occurs by constrictive fission; in *Hyphomicrobium* sp., reproduction occurs by budding from the distal end of the prostheca. In each organism, the products of reproduction are dissimilar in morphology, (usually) in motility, and in biosynthetic activity, since the prosthecate

offspring proceeds directly to resume biosynthesis and growth, whereas the swarmer becomes relatively inactive. The swarmer has been referred to as a "growth precursor" cell (2), and it has been suggested that it is peculiar in its ability *not* to grow (1), although it is not apparently specialized for dormancy.

There are several differences between these two organisms. Significantly, for this discussion, they are different in three ways. First, there is no evidence of DNA homology between them (10, 12), and no more rRNA cistron homology between them than between *Hyphomicrobium* sp. and photosynthetic bacteria of the genera *Rhodopseudomonas*, *Chromatium*, and *Rhodomicrobium* (the last also a prosthecate genus) (8). That is, there is no molecular evidence of phylogenetic relationship. Second, although their developmental sequences are superficially similar, the details of their respective modes of development also indicate a lack of homology. In *Caulobacter* sp., the flagellated site regularly becomes the developmental site, whereas in *Hyphomicrobium* sp., location of the developmental site varies among isolates (9). In *Hyphomicrobium* sp. T37 (21), the strain used in the studies described here, development and bud formation can occur while the flagellum is still active, and the developmental site is the previous budding site. In *Hyphomicrobium* sp. B522, on the other hand, the developmental site is the flagellated pole (11). Third, the *Caulobacter* prostheca comprises cell envelope and membranes only, whereas in *Hyphomicrobium* sp., cytoplasm is present within the appendage.

For understanding the causes and effects of development in these organisms, three approaches should, a priori, be fruitful. The first and most obvious at this point in biology would be to compare wild-type characteristics with properties of mutants unable to develop prosthecae. The major impediment to such studies has been the absence of reports of such mutants as viable clones. A second approach would be to study swarmers as they mature into growing, prosthecate cells. Generally, this approach has been pursued with the aim of identifying the (presumably genetic) signal that turns on development and biosynthesis. This approach has been extensively used and has been the source of the evidence that swarmers are "shut-down" cells. The relative inactivity of the swarmers, as well as basic biological reasoning, suggests the third approach, viz., to assume a different order of causation, to assume that development turns on biosynthesis. This would be the case if the prostheca provided the cell with some physiological capacity required for growth and biosynthesis. The question then becomes, what does the prostheca contribute to the physiology of the cell?

A clue to the possible physiological function of the prostheca has been available for more than 30 years, since the first *Caulobacter* isolate was obtained. It has been known since then that the more dilute the nutrient medium, the longer the prosthecae in a population (reviewed in 16). In 1966, Schmidt and Stanier (20) reported that it was necessary to dilute only the phosphate to stimulate prostheca elongation in *Caulobacter* and in *Asticcacaulis*, a related genus. In 1974, Haars and Schmidt (3) further showed, with chemostat-cultivated populations, that increased prostheca length in *Caulobacter* sp. would occur at lowered phosphate flux without a change in growth rate. Hirsch (5) reported that phosphate limitation stimulated prostheca elongation in *Hyphomicrobium* sp., and most recently this observation has been reported for *Rhodomicrobium* sp. (24). That is, in all four of these genera characterized by the development of elongate, cylindrical prosthecae and the production of motile, nonprosthecate cells, outgrowth of the appendage is inhibited by phosphate. Is this coincidence, or is it adaptive developmental behavior? Does the prostheca have a role in phosphate uptake and so become less important under conditions of phosphate excess?

On occasion, those studying prosthecae have been willing to suggest that the marked increase of surface-to-volume ratio that results from prostheca development is functional and that the extended surface serves to enhance the ability of the cell to capture nutrients (see, e.g., 1, 7, 16, 19), although only one research group has reported studies of the ability of (*Asticcacaulis*) prosthecae to accumulate nutrients (7, 19). Recently, I have proposed that the *Caulobacter* prostheca serves as a specialized uptake surface, that it is the principal site of entry of phosphate into the cell (J. S. Poindexter, Arch. Microbiol., in press). This proposal arose as an inference from studies that were not initiated with the aim of elucidating the physiological role of the prostheca, but whose results nevertheless implied the function. In an effort to test the hypothesis that development of an elongate,

cylindrical prostheca enhances the phosphate uptake capacity of the cell, some of those experiments have also been carried out with the clearly nonhomologous organism *Hyphomicrobium* sp. This discussion will concentrate on the results of the comparative studies as far as they have been completed, even though the evidence obtained with *Caulobacter* sp. is much more extensive than that yet available from the studies with *Hyphomicrobium* sp. Direct examination of the hypothesis, by characterization of the kinetics and sensitivities of [^{32}P]phosphate uptake by cells with and without prosthecae and by cell-free prosthecae, is only in a preliminary stage.

PHOSPHATE INHIBITION OF DEVELOPMENT AND ITS MECHANISM

Phosphate inhibition of the initiation of development in a population of prosthecate bacteria is observable as accumulation of swarmers by the end of growth in media containing various initial concentrations of phosphate. In media containing 0.2% carbon sources (equimolar mixture of glucose and glutamic acid) and an excess of nitrogen (as NH_4Cl, 0.05%), the accumulation of swarmers of a derivative of *C. crescentus* CB2 (18) increased with increasing initial phosphate concentration. (Growth yield, as dry weight, is determined by the initial concentration of carbon sources in these media [18] and so is constant among cultures at various phosphate levels.) When growth was phosphate limited (0.05 mM), only 2% of the population was accounted for by swarmer cells at the end of exponential growth; when phosphate and carbon were exhausted simultaneously (0.2 mM phosphate), 32% comprised swarmers, reflecting development of one-third of the swarmers of the last generation; and in carbon-limited populations (0.4 mM phosphate), 45% were swarmers, i.e., practically all of the last generation were undeveloped. At higher phosphate concentrations, accumulation has been observed to rise to more than 80%, at least transiently (6), reflecting inhibition of development of the swarmers of three successive generations. Similarly, in *Hyphomicrobium* sp. T37 in low-phosphate medium (0.05 mM phosphate) with 0.2% methanol, 31% of the late exponential phase population comprised swarmers, reflecting development of approximately 60% of the last generation, whereas 72% of the population (the last three generations) remained undeveloped in phosphate-excess medium (0.2 mM phosphate). (The number of undeveloped generations is calculated on the basis of swarmers accounting for 50% of the reproductive products in each generation.) Thus, although development in both organisms is a growth process and requires growth substrates, it did not occur in media still adequate for an eightfold increase in cell density when phosphate was an excess nutrient, but did occur in at least a majority of the cells in the same media when phosphate was not excessive. From such behavior, it appears that phosphate inhibits initiation of development in both organisms, and that *Hyphomicrobium* sp. T37 is more sensitive to phosphate inhibition of this stage than is *C. crescentus* CB2. (This strain difference is discussed further below.)

Phosphate inhibition of prostheca elongation is also readily observable and is the phenomenon reported for the four genera mentioned above. As illustrated in Fig. 1, the inhibitory effect of excess phosphate can be mimicked in phosphate-limited populations of *C. crescentus* CB2 by omission of calcium from the medium (Poindexter, in press). With 0.5 mM phosphate and 0.5 mM $CaCl_2$, and with 0.05 mM phosphate without $CaCl_2$, stalk length averaged approximately 2 μm; with 0.05 mM phosphate and 0.5 mM $CaCl_2$, mean stalk length was nearly 6 μm. Excess phosphate and omission of $CaCl_2$ affected hypha length similarly in *Hyphomicrobium* sp. T37 (Fig. 2). With 0.5 mM phosphate, hypha length from "mother" cell to bud averaged approximately 2 μm; with 0.05 mM phosphate without $CaCl_2$, hypha length averaged slightly less than 5 μm; and with the lower phosphate concentration and the added $CaCl_2$, hypha length generally exceeded 10 μm from "mother" cell to bud.

In both organisms, then, it appeared that phosphate inhibited development by complexing with calcium, which seemed to play a role in the process of development. In fact, *Hyphomicrobium* sp. T37 did not dependably grow in medium without added calcium and did not grow at all in medium containing 0.1 mM EGTA [ethyleneglycol-bis-(β-aminoethyl ether)N,N'-tetraacetic acid, a calcium-specific chelator]. *C. crescentus* grew in the presence of 0.5 mM EGTA without added calcium, although development was greatly reduced (Poindexter, in press). As is evident from the appearance of the populations illustrated in Fig. 1 and 2, the calcium-supported response to phosphate limitation is not a subtle phenomenon, and it is difficult to observe these populations and ignore the possibility that such an elaborate morphogenetic response could be related to the organisms' increased need for phosphate relative to other nutrients.

CALCIUM STIMULATION OF GROWTH RATE

Throughout these cultural studies, it has regularly been observed that calcium increased growth rate. The graphs in Fig. 3 illustrate the stimulatory effect of calcium on growth rates of each of the test organisms at various phosphate concentrations. The growth of T37 was more phosphate sensitive than that of CB2. However, not all isolates of *Hyphomicrobium* sp. are as sensitive as T37 (4), and the majority of *Caulobacter* species in my possession are more sensitive than *C. crescentus* strains (unpublished data). Similarly, other isolates of *Hyphomicrobium* sp. grow faster than T37 (see, e.g., 22), and other species of *Caulobacter* generally grow slower than *C. crescentus* (14). Accordingly, the differences are noted here for the two test strains only and are not intended as the basis for a general comparison between the two genera with respect to phosphate sensitivity or maximum growth rate. However, it should be stressed that, of the two test organisms employed, the one whose growth and development appeared more phosphate sensitive likewise appeared more dependent on the availability of calcium, and that growth of *C. crescentus* CB2 appeared sensitive to 0.5 and 1.0 mM phosphate only in the absence of added calcium.

There are at least two possible explanations for calcium stimulation of growth, especially at extremes of phosphate concentrations. One is that, because calcium stimulates prostheca development, it stimulates development of the surface that assists the cell in obtaining phosphate at low ambient concentrations. Similarly, high phosphate levels inhibit development of that surface by reducing the cells' access to calcium and thereby reduce their capacity for uptake of phosphate, even at higher concentrations. An alternative is that the availability of phosphate, which is demonstrably somewhat toxic for these cells (by an as yet unexplored mechanism), is in turn reduced by the addition of calcium.

PHOSPHATE UPTAKE RATES

Studies of phosphate uptake are at present being initiated. Only one feature has yet been determined and that is the influence of calcium on phosphate uptake per se. For these preliminary studies, cells were harvested, washed in basal medium without phosphate, $CaCl_2$, and carbon sources, and then resuspended in growth medium without $CaCl_2$ containing carbon source(s) and 0.25 mM phosphate. Samples were provided with carrier-free [^{32}P]phosphate after supplementation with either 0.1 mM EGTA or 0.5 mM $CaCl_2$. The linear rates of uptake observed were 260, 62, and 16 cpm/μg of protein per min for 0.05 mM phosphate-grown *C. crescentus* CB2, 0.5 mM phosphate-grown CB2, and 0.5 mM phosphate-grown *Hyphomicrobium* sp. T37, respectively. The significantly higher rate of uptake by CB2 cells grown in low-phosphate medium than by those grown in high-

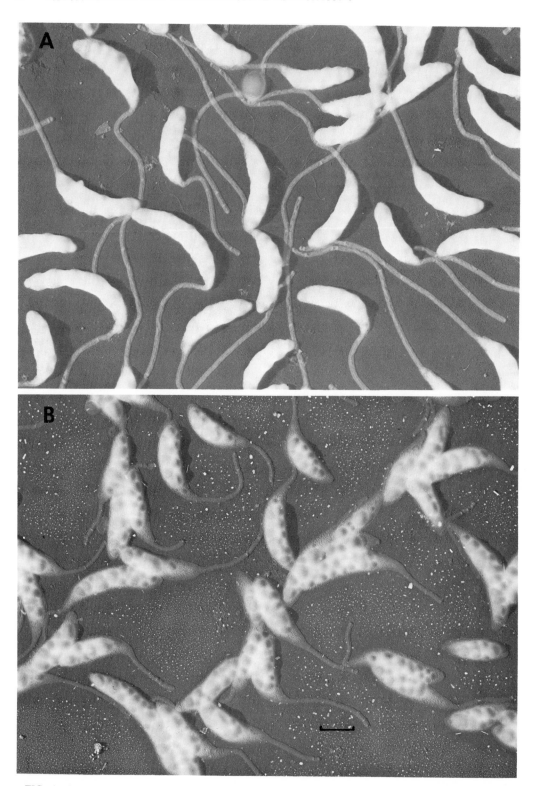

FIG. 1. *C. crescentus* CB2 at the end of exponential growth in Higg medium (15) with 0.05 mM phosphate and (A) with 0.5 mM CaCl$_2$ and (B) without added CaCl$_2$. Shadowed specimens. Marker is 1 μm.

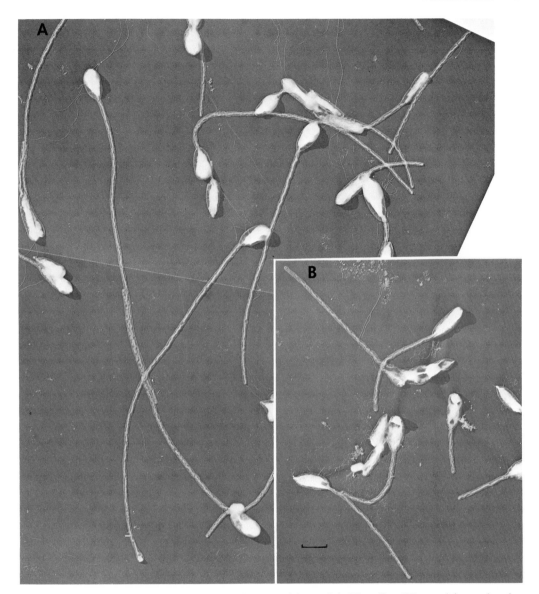

FIG. 2. *Hyphomicrobium* sp. T37 at the end of exponential growth in Hi medium (20) containing methanol as carbon source with 0.05 mM phosphate and (A) with 0.5 mM CaCl$_2$ and (B) without added CaCl$_2$. Shadowed specimens fixed with 2% formaldehyde. Marker is 1 μm.

phosphate medium could reflect a phosphate starvation-inducible system (see, e.g., 23) present in the low phosphate-grown cells, with a lower velocity system operating in high phosphate-grown cells, or it could reflect the same system present in both types of cells, but with more of it present per cell grown in low phosphate. This difference has yet to be characterized.

More importantly for the present context, for each of the three populations, the rate of uptake was identical (correlation coefficients were +0.999 to +1.000) in the presence of EGTA and of CaCl$_2$. On the basis of these initial observations, which must be considered preliminary, it is at least possible to conclude that calcium neither promotes nor retards phosphate uptake per se.

Returning, then, to the question of why calcium enhanced growth rate, it should be noted that swarmers are essentially nongrowing cells. It follows that when their proportion in the population increases, the overall rate of growth of the population will appear slower. Develop-

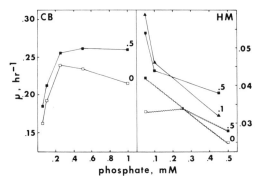

FIG 3. Influence of phosphate and calcium concentrations on growth rates of CB2 and T37 grown in Higg and Hi-methanol, respectively. Symbols: ■, media containing 0.5 mM CaCl₂; ▲, media containing 0.1 mM CaCl₂; □, media without added CaCl₂. Wavy lines: methanol provided as vapors. Straight lines: methanol fed twice daily as one drop per 20 ml of culture.

ment in *Hyphomicrobium* sp. T37 is more sensitive to phosphate than is development in *C. crescentus* CB2, and it is consistent with this that growth of T37 appeared more sensitive to phosphate than did growth of CB2. This implies that development is necessary for growth—of the cell, and of the population at its maximum possible rate.

CALCIUM AND THE UTILIZATION OF PHOSPHATE

We have also asked whether calcium stimulation of prostheca development affects the utilization of phosphate. Elsewhere, evidence has been presented that "luxury consumption" of phosphate and its storage as polyphosphate is a function of the stalked cell of *Caulobacter* (Poindexter, in press). Since similar studies have not yet been completed with *Hyphomicrobium*, only the results of two types of studies that have been carried out with both organisms will be described here. Both indicate that calcium, which promotes prostheca development, promotes the utilization of phosphate, even though calcium does not affect the rate of phosphate uptake per se.

PHB storage. As is evident in the micrographs in Fig. 1 and 2, it is characteristic of these organisms to store excess carbon as poly-β-hydroxybutyric acid (PHB) when grown to phosphate exhaustion. This occurs whether or not the cells develop excessively long prosthecae. PHB storage is indicated quantitatively in Fig. 4. The decrease in storage by CB2 as a result of the presence of calcium in the medium was approximately 50 μg/mg (dry weight) at each phosphate concentration. In T37, PHB was calculated relative to protein rather than dry

weight, but again, it is clear that phosphate greatly reduced storage, and in two of the three populations, increased calcium, from 0.1 to 0.5 mM, further reduced storage of PHB. (Lack of 100% correlation with calcium in the case of T37 was probably because this organism did not grow dependably without added calcium, and it seemed advisable to use two calcium concentrations, a procedure that reduced the contrast among the *Hyphomicrobium* populations grown with different levels of phosphate.)

In both organisms, PHB storage was reduced by phosphate, and in general, storage was further reduced at each phosphate concentration when development of longer prosthecae was allowed by the amount of calcium provided.

Arsenate inhibition. It was reasoned that, although growth of these organisms, particularly *Hyphomicrobium* sp. T37, is phosphate sensitive, phosphate should nevertheless protect against arsenate inhibition of growth. In testing this, it was found that both organisms were able to grow with arsenate up to 40 mM in appropriate media. However, growth was approximately linear, not exponential, in all cultures containing 40 mM arsenate, and to evaluate growth and growth inhibition, it was found necessary to compare growth yields rather than rates. The graphs in Fig. 5 illustrate the growth yields achieved by the two organisms in media containing 40 mM arsenate, as the proportion of yield achieved in the absence of arsenate.

C. crescentus CB2 grew with arsenate in proportion to phosphate concentration, as expected, but only with 10 times the usual level of calcium (5 mM), and not at any tested phosphate concentration with only 0.5 mM calcium. *Hyphomicrobium* sp. T37 growth was similarly dependent on calcium as well as phosphate, but quantitatively less dependent on both. This may have occurred because phosphate is both antidote and poison for this strain at higher concen-

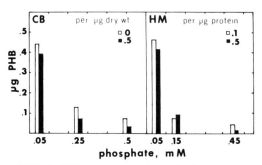

FIG. 4. PHB storage at the end of exponential growth by CB2 grown in Higg medium and by T37 grown with methanol as carbon source, added twice daily as one drop per 20 ml of culture. PHB was assayed as previously described (18).

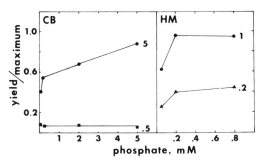

FIG. 5. Growth yield in media containing 40 mM Na_2HAsO_4 as a proportion of the maximum achieved in the absence of arsenate. Cells grown as described in the legend to Fig. 4. Proportional yield was calculated from turbidity readings in a Klett-Summerson colorimeter with a red (no. 66) filter. The figure at the end of each line indicates the millimolar $CaCl_2$ concentration.

trations, or it may indicate that T37 has a higher affinity for phosphate than does CB2.

Calcium arsenate is several times more soluble than calcium phosphate; accordingly, as the calcium and phosphate concentrations increased, phosphate should have become less available to oppose arsenate. Instead, calcium potentiated the protective effect of increased phosphate. Alternatively, if phosphate acquisition improved with calcium availability during growth, then calcium should have assisted in resistance to arsenate inhibition, as it did. Since calcium does not improve phosphate uptake per se, but did promote prostheca development, the arsenate results constitute yet another type of evidence that development improves the capacity of a population to acquire phosphate.

PROSTHECA AS SITE OF PHOSPHATE ENTRY

On the basis of these and previous observations, it seems reasonable to propose that prostheca development provides the principal site of phosphate entry into the cells of both organisms. As any decent hypothesis should, this hypothesis suggests obvious experiments, in this case, experiments to determine the relative rates and affinities of phosphate uptake by prosthecate cells, cell-free prosthecae, and non-prosthecate cells. It also carries significant implications regarding previously mysterious characteristics of the swarmer cells of *Caulobacter* and *Hyphomicrobium*, and regarding the ecology of these organisms.

With respect to swarmer behavior, if the prostheca is the principal site of entry of phosphate, and swarmers cannot acquire phosphate from low ambient levels, then in oligotrophic waters: (i) swarmer growth is negligible because growth requires net phosphorus acquisition; (ii)

swarmers do not synthesize DNA because DNA synthesis depends on a net phosphorus influx into the cell; and (iii) RNA and protein syntheses are slow in swarmers because they must proceed at the expense of inherited phosphorus. This hypothesis would explain how development turns on biosynthesis, viz., by admitting phosphate into the cell at a rate suitable for growth and biosynthesis. However, it does not yet address the question of how initiation of development is signaled. The answer may be within the syllogism above, as interpreted in the following hypothesis.

Hypothesis: The "inherited phosphorus" serves as a clock. The higher the content of inherited phosphorus, accumulated during the previous generation, the longer the swarmer will remain undeveloped, allowing dispersal of the population over a distance that increases with ambient phosphate levels. Lower levels of inherited phosphorus will precipitate earlier development, confining the population within a smaller area, but under conditions to which these organisms seem otherwise well suited and in which they appear competitive, viz., oligotrophic conditions. When activity of the swarmer reduces the inherited phosphorus to a minimal level, the swarmer begins to recycle internal phosphorus, deriving phosphate by dephosphorylation of internal membranes, specifically, membranes accumulated within a pole: the flagellated pole in *Caulobacter* sp., either pole in *Hyphomicrobium*, depending on the strain. Incidental to this attack on the membranes is a weakening of that pole, which is then pushed out, containing dephosphorylated lipoproteins that then serve to bind ambient phosphate. Any additional internal membranes that pass this point on the cell surface during subsequent development are similarly modified with respect to phosphorus content, so that the capacity for phosphate acquisition increases with prostheca length.

Some evidence in support of this hypothesis has been obtained for *Caulobacter* sp. (Poindexter, unpublished data), but the hypothesis is mainly speculative. Again, however, certain experiments become important as tests of the hypothesis, in this case, the identification and quantitation of "inherited phosphorus."

Ecologically, these prosthecate bacteria can be viewed as phosphate oligotrophs with three characteristics in common: (i) they have a specialized surface that enhances phosphate uptake; (ii) development of this surface and therefore phosphate uptake and therefore growth are inhibited by phosphate, so that these organisms are not able to multiply competitively in eutrophic, phosphate-rich waters; and (iii) they have evolved a means for dispersal in the form of a motile, vegetative, but nongrowing cell by the simple mechanism of releasing it into the environment literally cut off from its major access to a nutrient essential for growth.

Finally, it is to be expected that these organisms will otherwise be adapted for oligotrophic conditions, e.g., by an ability to exploit low concentrations of organic carbon, but those properties may, for the present, be viewed as necessary consequences of their adaptation to oligotrophic levels of phosphate. The evidence available in support of this view may at present seem slender, but the idea that has been generated by that evidence is highly explanatory and worthy of experimental investigation. The ecology of prosthecate oligotrophs is now in the desirable state of having a unitary hypothesis to guide investigations that should elucidate the relationship between their unique developmental behavior and their physiological adaptations as oligotrophs.

The experimental studies reported here were supported in part by Public Health Service grant AI-15467 from the National Institutes of Health and by National Science Foundation grant PCM 8200256.

The studies were ably assisted by James G. Hagenzieker.

LITERATURE CITED

1. **Dow, C. S., and R. Whittenbury.** 1980. Prokaryotic form and function, p. 391–417. *In* D. C. Ellwood, J. N. Hedger, M. J. Latham, J. M. Lynch, and J. H. Slater (ed.), Contemporary microbial ecology. Academic Press, Inc., New York.

2. **Dow, C. S., R. Whittenbury, and N. G. Carr.** 1983. The 'shut down' or 'growth precursor' cell—an adaptation for survival in a potentially hostile environment. Symp. Soc. Gen. Microbiol. **34:**187–247.

3. **Haars, E. G., and J. M. Schmidt.** 1974. Stalk formation and its inhibition in *Caulobacter crescentus*. J. Bacteriol. **120:**1409–1416.

4. **Harder, W., and M. M. Attwood.** 1977. Biology, physiology and biochemistry of hyphomicrobia. Adv. Microb. Physiol. **17:**303–356.

5. **Hirsch, P.** 1974. Budding bacteria. Annu. Rev. Microbiol. **28:**391–444.

6. **Iba, H., A. Fukuda, and Y. Okada.** 1975. Synchronous cell differentiation in *Caulobacter crescentus*. Jpn. J. Microbiol. **19:**441–446.

7. **Larson, R. J., and J. L. Pate.** 1976. Glucose transport in isolated prosthecae. J. Bacteriol. **126:**282–293.

8. **Moore, R. L.** 1977. Ribosomal ribonucleic acid cistron homologies among *Hyphomicrobium* and various other bacteria. Can. J. Microbiol. **23:**478–481.

9. **Moore, R. L.** 1981. The biology of *Hyphomicrobium* and other prosthecate, budding bacteria. Annu. Rev. Microbiol. **35:**567–594.

10. **Moore, R. L., and P. Hirsch.** 1972. Deoxyribonucleic acid base sequence homologies of some budding and prosthecate bacteria. J. Bacteriol. **110:**256–261.

11. **Moore, R. L., and P. Hirsch.** 1973. First generation synchrony of *Hyphomicrobium* swarmer populations. J. Bacteriol. **116:**418–423.

12. **Moore, R. L., J. Schmidt, J. Poindexter, and J. T. Staley.** 1978. Deoxyribonucleic acid homology among the caulobacters. Int. J. Syst. Bacteriol. **28:**349–353.

13. **Paerl, H. W.** 1982. Factors limiting productivity of freshwater ecosystems, p. 75–110. *In* K. C. Marshall (ed.), Advances in microbial ecology, vol. 6. Plenum Press, New York.

14. **Poindexter, J. S.** 1964. Biological properties and classification of the *Caulobacter* group. Bacteriol. Rev. **28:**231–295.

15. **Poindexter, J. S.** 1978. Selection for nonbuoyant mutants of *Caulobacter crescentus*. J. Bacteriol. **135:**1141–1145.

16. **Poindexter, J. S.** 1981. The caulobacters: ubiquitous unusual bacteria. Microbiol. Rev. **45:**123–179.

17. **Poindexter, J. S.** 1981. Oligotrophy: feast and famine existence, p. 63–89. *In* M. Alexander (ed.), Advances in microbial ecology, vol. 5. Plenum Press, New York.

18. **Poindexter, J. S., and L. F. Eley.** 1983. Combined procedure for assays of poly-β-hydroxybutyric acid and inorganic polyphosphate. J. Microbiol. Methods **1:**1–16.

19. **Porter, J. S., and J. L. Pate.** 1975. Prosthecae of *Asticcacaulis biprosthecum*: system for the study of membrane transport. J. Bacteriol. **122:**976–986.

20. **Schmidt, J. M., and R. Y. Stanier.** 1966. The development of cellular stalks in bacteria. J. Cell Biol. **28:**423–436.

21. **Tyler, P. A., and K. C. Marshall.** 1967. Pleomorphy in stalked, budding bacteria. J. Bacteriol. **93:**1132–1136.

22. **Wali, T. M., G. F. Hudson, D. A. Danald, and R. M. Weiner.** 1980. Timing of swarmer cell cycle morphogenesis and macromolecular synthesis by *Hyphomicrobium neptunium* in synchronous culture. J. Bacteriol. **144:**406–412.

23. **Wanner, B. L., and R. McSharry.** 1982. Phosphate-controlled gene expression in *Escherichia coli* using Mud*l*-directed *lacZ* fusions. J. Mol. Biol. **158:**347–363.

24. **Whittenbury, R., and C. S. Dow.** 1977. Morphogenesis and differentiation in *Rhodomicrobium vannielii* and other budding and prosthecate bacteria. Bacteriol. Rev. **41:**754–808.

Effect of Light–Nutrient Interactions on Buoyancy Regulation by Planktonic Cyanobacteria

ALLAN KONOPKA

Purdue University, Department of Biological Sciences, West Lafayette, Indiana 47907

The microbial ecologist interested in the physiology of natural populations faces a different problem from that of microbial physiologists who work on the regulation of metabolic activities in laboratory cultures. The latter can study their systems by perturbing the organisms either physiologically or genetically to produce an exaggerated response which is relatively easy to measure. However, microorganisms normally regulate metabolic activities by integrating inputs from several different environmental factors. To understand the physiological states of natural microbial populations, microbial ecologists must determine not only what factors affect microbial activities but also how these factors interact when more than one is at a subsaturating intensity.

This problem is of general importance because gradients of environmental factors occur in a variety of habitats and affect a wide range of microorganisms. These gradients have been most extensively discussed with regard to the metalimnia and hypolimnia of lakes, where gradients of temperature, light, pH, oxygen, and reduced substances such as hydrogen sulfide, ammonia, and methane occur. Sharply defined layers of specific metabolic types are found in these environments, and their vertical distribution is usually explained by overlapping gradients of environmental factors which produce suitable conditions for growth only within narrow strata.

In planktonic cyanobacteria, the interaction between environmental factors can be determined by studying buoyancy regulation. Buoyancy is provided by gas vesicles, gas-filled cytoplasmic inclusions found in a variety of procaryotic organisms. They consist of a rigid shell of protein subunits 2 nm thick which is impermeable to water. The hollow cylindrical structures found in cyanobacteria have a diameter of approximately 60 to 70 nm and a length of up to 1,000 nm. Their density is approximately 0.12 g cm^{-3} (3). The protein shell can be ruptured by pressure to irreversibly form a nonbuoyant membranous envelope. Cells are buoyant if a sufficient proportion of their volume is occupied by gas vesicles to reduce the overall density of the cell to less than that of water; the relative gas vacuole volume necessary for neutral buoyancy has been estimated to be 0.7 to 2.3% for different cyanobacterial species (20). In some cyanobacteria, buoyancy is regulated by controlling the number of intact gas vesicles in the cell, and the control mechanism is an integrative function which responds to light intensity and nutrient concentration.

REGULATION OF BUOYANCY

A model to explain how the number of gas vesicles is controlled in *Anabaena flos-aquae* was first proposed in 1972 and has been refined by Walsby and his colleagues in recent years (1, 5, 10). In this scheme (Fig. 1), cells are hypothesized to possess neutral buoyancy when the rate of energy generation balances the capacity for growth, that is, when energy can be used productively as quickly as it is generated. If the rate of photosynthetic energy generation (determined by light intensity in the short term) is less than the capacity for growth (determined by environmental factors such as temperature and nutrient concentration), the organisms will be positively buoyant. If the water column of a lake is not mixed by turbulence, these light-limited organisms will rise toward the surface. They encounter higher light intensities, and their rate of energy generation increases as long as the irradiance value is subsaturating for photosynthesis. The way in which this energy is used determines whether the organisms will lose buoyancy at a particular irradiance.

Energy is necessary to drive a number of processes, but this discussion will focus on its use in CO_2 fixation and solute uptake. As the rate of energy generation increases, more CO_2 can be fixed into low-molecular-weight organic compounds via the Calvin cycle. If there are sufficient inorganic nutrients available (primarily N and P) so that the energy can be used to produce building blocks such as amino acids and nucleotides which can be polymerized into proteins and nucleic acids, the organism remains buoyant and grows faster at the higher irradiance.

Buoyancy is lost if an environmental factor (such as the concentration of nitrogen or phosphorus) limits the growth rate of the cells to a value below the maximum possible at the incident light intensity. In this instance, energy is

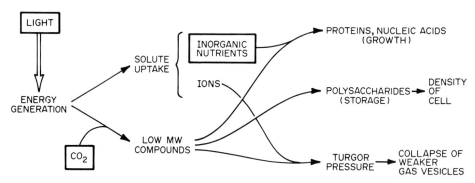

FIG. 1. Relationships among factors important to buoyancy regulation in planktonic cyanobacteria.

generated faster than it can be used for growth. Two mechanisms have been postulated to explain the loss of buoyancy which can occur after several hours of exposure to high irradiance: (i) collapse of existing gas vesicles by turgor pressure and (ii) accumulation of dense molecules in the cell. A third potential factor in buoyancy regulation is control of the rate of gas vesicle formation.

Collapse of weaker gas vesicles. The gas vesicles of several cyanobacteria can be collapsed by applying 400 to 700 kPa of pressure. These are average values, and about 10% of the vesicles are weak enough to be collapsed by pressures 100 kPa below the average. Turgor pressure can rise from about 300 to greater than 400 kPa in cells of *Anabaena, Aphanizomenon*, and *Oscillatoria* transferred from low to high light intensities. The turgor pressure at high light intensity is sufficient to collapse the weaker gas vesicles in these organisms (5, 14, 16, 28). In *Anabaena flos-aquae*, the rise in turgor pressure is due to the accumulation of low-molecular-weight organic compounds produced by the Calvin cycle (10) and K^+ ions transported by a light-dependent ion pump (1). The contributions by organic and inorganic molecules to the turgor rise are approximately equal. The intracellular accumulation of each is an energy-utilizing process which can be viewed as a mechanism to dissipate energy when energy production exceeds its utilization for growth. This would occur when nitrogen- or phosphorus-limited cells are exposed to excess irradiance.

Increasing the density of the cell. Not all cyanobacteria can collapse gas vesicles by increasing turgor pressure. In some species, the critical pressure (P_c) of their weakest gas vesicles is greater than the 300 to 600 kPa of turgor pressure which cells can generate. An extreme example is *Oscillatoria thiebautii*, which is found at depths of 200 m in the Sargasso Sea (25). Its vesicles have a P_c of 3,700 kPa, sufficient to withstand the hydrostatic pressures at those depths but impossible to collapse by turgor pressure. A red *Oscillatoria aghardii* found in Lake Gjersjoen, Norway, had a turgor pressure of 400 kPa, but its weakest gas vesicles withstood pressures of 800 kPa (29).

Oliver et al. (Abstracts, IV International Symposium on Photosynthetic Prokaryotes, Bombannes, France, 1982) have proposed that these gas vacuolate organisms might regulate their buoyancy by controlling the cellular amount of "heavy" molecules such as polysaccharides and proteins. This proposal is consistent with the relationships illustrated in Fig. 1. If nutrient-limited organisms are exposed to high light intensity, the rate of energy generation will exceed its utilization for growth. Energy can be stored by fixing CO_2 into low-molecular-weight compounds and polymerizing these into polysaccharides. The polysaccharides increase the weight of the cell but do not act as catalytically active biomass. Thus, an organism could regulate its buoyancy by controlling the amount of ballast if it cannot actively collapse gas vesicles.

Control of gas vesicle formation. Very little is known about the regulation of gas vesicle formation in cyanobacteria, although control at this level is potentially as important as, and energetically more efficient than, collapsing existing gas vesicles. Gas vesicle formation has been assumed to be constitutive and probably is so in some species, such as *Microcystis aeruginosa* (A. Konopka, unpublished data). However, there are very few direct experimental data to support this hypothesis. Strains of *Calothrix* and *Nostoc* do regulate the formation of gas vesicles. Gas vesicle production occurs only in cells that differentiate into hormogonia (22, 24). The induction of gas vesicle formation in *Nostoc mus-*

corum which occurs after the ionic strength of the medium is decreased (2) is at present the best system for studying this phenomenon. We should not exclude the idea that control of gas vesicle formation is also important in controlling buoyancy of *Anabaena* and *Aphanizomenon* grown under different conditions.

In summary, a key control element in the model of Fig. 1 is the relationship between energy generation and utilization. Light intensity has a central role in determining the buoyancy of planktonic cyanobacteria, but it is impossible to specify precisely what irradiance will result in a loss of buoyancy without knowing the physiological state of the cells. In natural populations, nitrogen or phosphorus concentration is often as important as irradiance in determining physiological state, and the concentration of limiting nutrient can modulate the effect of light intensity upon buoyancy. When growth rate is nutrient limited, the light intensity at which neutral buoyancy occurs is hypothesized to be directly related to the concentration of limiting nutrient. This relationship should hold whether buoyancy regulation occurs via collapse of gas vesicles or production of ballast molecules. It does require that cells do not produce so many gas vesicles that the regulatory mechanism is incapable of overcoming excess buoyancy.

With this model of buoyancy regulation in mind, we can examine the distribution of planktonic gas-vacuolate cyanobacteria in nature with regard to light and nutrient availability.

HABITATS OF PLANKTONIC GAS-VACUOLATE CYANOBACTERIA

There are 11 genera of cyanobacteria which contain gas-vacuolate strains (26). They include both unicellular and filamentous forms and colonial types as well as solitary ones. Some species are usually found in the epilimnia of thermally stratified mesotrophic and eutrophic lakes, whereas others generally form layers in the metalimnia or hypolimnia of such lakes (Fig. 2). Reynolds and Walsby (20) provided an excellent analysis of the vertical distribution of planktonic cyanobacteria in lakes, in which they focused attention on the relationship between the depth of the epilimnion or mixed zone (z_m) and of the euphotic zone (z_{eu}). As discussed in two cases below, this relationship determines the vertical distribution of organisms and may have an indirect effect on the species and colony size of the cyanobacteria found in the lake.

$z_{eu} < z_m$. The concentration of phototrophic biomass in eutrophic lakes can become high enough to attenuate incident radiation to less than 1% of the surface value within 2 to 4 m. If the size and exposure of the lake are large enough to create a mixed zone deeper than the

photic zone, then, under turbulent conditions, the "average" light intensity to which cells are exposed is low, because even during daylight organisms are mixed into the dark waters at the bottom of the epilimnion during part of the day. In these instances, light may become the limiting factor. Gas-vacuolate cyanobacteria respond to the low light conditions by increasing their buoyancy. If wind-induced mixing ceases, the buoyant organisms can float to the surface to form a surface bloom.

An *Aphanizomenon flos-aquae* population found in Lake Mendota, Wisconsin, during summer 1976 provided a good example of this phenomenon (16). In this lake, the euphotic zone was 4 m deep, whereas the mixed zone was 10 to 12 m. The *Aphanizomenon* colonies were always positively buoyant, and if calm conditions occurred, the population could migrate to the surface within several hours (Fig. 3).

The rapid movement of the organisms to the surface, estimated to be 275 μm/s, was due to the size of the *Aphanizomenon* colonies. The velocity of movement is described by Stokes' equation, $v = 2 \cdot gr^2 \cdot (\rho - \rho')/9\eta$, where g is the gravitational acceleration, r and ρ are the radius and the density of the particle, and ρ' and η are the density and the viscosity of the medium. Rapid rates of flotation (or sinking) can be achieved by increasing r, the particle radius. In cyanobacteria, this involves forming colonies. Large colony size is most advantageous in unstable water masses subject to intermittent periods of turbulence and calm, because the organisms can then quickly migrate both upwards and downwards over a large distance. In Lake Mendota, where light is severely attenuated in the water column, organisms which can quickly migrate to the photic zone after the onset of calm conditions would have a competitive advantage.

Although flotation of cyanobacteria to the photic zone is necessary for exposure to light under calm conditions in Lake Mendota, the light intensities found at the surface are excessively high and potentially damaging. *Aphanizomenon* in Lake Mendota regulated buoyancy under these conditions, and the population maximum descended from 0 to 4 m during the day. *Aphanizomenon* regulated buoyancy by increasing turgor pressure and collapsing weaker gas vesicles (16).

Of special interest are the conditions under which surface blooms persist; that is, the organisms at the surface are unable to regulate their buoyancy. The model for buoyancy regulation requires that cells generate energy in excess of their growth requirements in order to lose buoyancy. Energy generation would be reduced if (i) the cyanobacterial population that floated to the surface were "senescent," that is, metabolically

TURBULENCE, LIGHT, AND BLOOM FORMATION

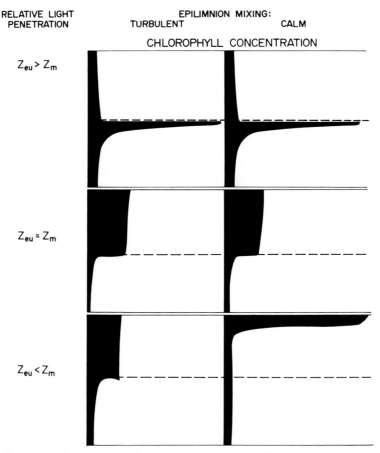

FIG. 2. Vertical distribution of gas-vacuolate cyanobacteria, as a function of wind turbulence and the relative depths of the epilimnion (z_m) and the euphotic zone (z_{eu}). After Reynolds and Walsby (20).

incompetent as a result of environmental stress, or (ii) exposure to radiation at the surface photooxidized the pigments (17). There are few experimental data to show whether senescence is the cause or a result of the inability to regulate buoyancy. Even in physiologically competent cells, photosynthesis is inhibited at the light intensities found at the lake surface on sunny days, but the photosynthetic rate necessary to drive the buoyancy-regulating mechanism is low enough (19) that photoinhibition is unlikely to explain the persistence of surface blooms.

Although the importance of photooxidation and photoinhibition in maintaining surface blooms is unclear, several investigators have shown that CO_2 limitation prevents the removal of excess buoyancy and therefore results in a persistent surface bloom. (i) Booker and Walsby (4) studied the stratification of *Anabaena flos-aquae* in a 2.2-m-deep column in the laboratory. The organisms aggregated at the surface, al-though the light intensity was 8 klx, much higher than irradiances which caused organisms grown in batch culture to sink (27). However, organisms from the surface scum lost buoyancy when exposed to 8 klx if inorganic carbon was added but remained buoyant if no carbon was added. (ii) Paerl and Ustach (18) found that the buoyancy of *Alphanizomenon flos-aquae* exposed to 2 klx irradiance was highest under CO_2-limiting conditions (that is, at pH values above 9). The capacity for energy generation may have been diminished at high pH, because net photosynthetic oxygen evolution was lower under those conditions than when tested at pH 6 to 7. (iii) The gas vacuolation of *Oscillatoria rubescens* increased by 60% within 3 days when cells were shifted from nitrogen limitation to CO_2 limitation (12). Thus, CO_2 limitation can short-circuit buoyancy regulation, but the precise mechanism is unknown. Although CO_2 limitation would reduce the rate at which osmotically

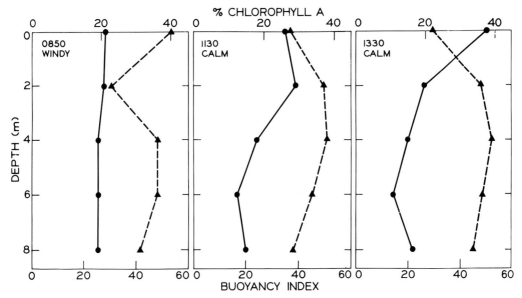

FIG. 3. Formation of a surface bloom by *Aphanizomenon flos-aquae* in Lake Mendota, Wisconsin (16). Distribution of chlorophyll *a* (○) and buoyancy (△) as a function of depth on 5 August 1976 at (a) 0850, (b) 1130, and (c) 1330 h. The prevailing wind conditions are noted in each graph.

active, low-molecular-weight organic compounds are formed, the transport of more K^+ ions into the cell could compensate for this and result in normal increases in turgor pressure. Paerl and Ustach's observation that net oxygen evolution decreased at high pH suggests that the driving force for buoyancy regulation, photosynthetic energy generation, is diminished. However, it remains to be shown that net oxygen evolution is proportional to the rate of photophosphorylation under CO_2-limited conditions.

$z_{eu} > z_m$. Sufficient light for photosynthesis often penetrates beyond the epilimnion into non-turbulent strata in oligotrophic or mesotrophic lakes. In these instances, stratified populations of diatoms, flagellated chrysophytes, or cyanobacteria with approximately neutral buoyancy are found in the metalimnion (6, 7). *Oscillatoria* species are the most common gas-vacuolate cyanobacteria found in metalimnetic layers. Their occurrence as single filaments rather than colonies is partially responsible for their success in these habitats. As mentioned above, the velocity of floating or sinking is a function of the square of the particular radius. Therefore, small particles move more slowly and are less likely to float or sink past the desired depth before the regulating mechanism can adjust the buoyancy of the organism. In the metalimnion, where gradients of light intensity and nutrient concentration are steep, only small, finely tuned, vertical migrations are necessary. Large colonies would quickly float into the epilimnion, where turbulent mixing would keep them in a high

irradiance–low nutrient environment that is less suitable for growth than the metalimnion of mesotrophic lakes.

Although *Oscillatoria* species are most commonly found in metalimnetic layers, I have found metalimnetic layers of *Aphanizomenon* and *Anabaena* in some northeastern Indiana lakes. However, the organisms in these layers did not form the large colonies typical of epilimnetic blooms of these organisms but consisted of single, relatively short filaments. Thus, they were able to accurately regulate their buoyancy, just as *Oscillatoria* sp. can. The cellular morphology of the *Aphanizomenon* and *Anabaena* filaments was similar to that of the colonial forms. The factors that determine the colonial form of cyanobacteria are unknown, although that of *Aphanizomenon* was altered by the addition of chelators or *Daphnia* (21).

As an example of a metalimnetic layer, *Oscillatoria rubescens* stratified at depths of 6 to 9 m in Crooked Lake, Indiana, where the incident light intensity averaged 2% of the surface intensity (14). Filaments regulated buoyancy by increasing their turgor pressure and collapsing gas vesicles when exposed to high irradiances. The turgor pressure decreased and buoyancy increased in samples collected from increasing depths within the metalimnion (Table 1). Note that neutral buoyancy (buoyancy index = 0) occurred between 6 and 7 m, whereas the population maximum was at 8 m. By using traps to collect floating and sinking filaments, I found that the *Oscillatoria* population was not static

TABLE 1. Buoyancy characteristics of *O. rubescens* in the water column of Crooked Lake on 10 July 1980

Depth (m)	Light intensity ($\mu E \cdot m^{-2} \cdot s^{-1}$)	Oscillatoria biomass ($\mu m^3 \cdot ml^{-1}$)	Buoyancy index	Turgor (kPa)
6	56	0.2	−8	380
7	25	0.8	11	370
8	2	10.8	27	360
9	0.4	2.8	17	350
10	0	1.3	24	330

but migrated 1 to 2 m above the depth of the population maximum. Fogg and Walsby (8) suggested that one advantage of buoyancy regulation is that organisms can migrate between depths at which different factors necessary for growth are in relatively abundant supply. In stratified lakes, the vertical distributions of light energy and inorganic nutrient concentration are inversely related, and there are especially strong gradients in the metalimnion. Although *O. rubescens* in Crooked Lake migrated only 1 to 2 m, it might assimilate inorganic nutrients more efficiently deeper in the water column and could then float to shallower depths, where higher light intensities would generate more energy to drive growth using those nutrients.

ROLE OF NUTRIENT CONCENTRATION IN BUOYANCY REGULATION

I have speculated that the concentration of nutrients such as nitrogen or phosphorus is important in buoyancy regulation, both in describing the model and in explaining vertical migration of cyanobacteria in lakes. Data from both field and laboratory studies support this speculation, but there is less work on this point than on the effect of irradiance upon buoyancy, and further studies would be useful to determine precisely how the two effects are interrelated.

Field studies of a metalimnetic population of *Oscillatoria aghardii* in Deming Lake, Minnesota, showed that NH_4^+ was the primary nutrient affecting buoyancy regulation in that system (11, 28). The irradiance at which natural samples maintained neutral buoyancy was greater than 70 μM NH_4Cl was added than in untreated controls. Furthermore, in experiments using columns of lake water isolated with Plexiglas cylinders of 25 cm diameter, the addition of NH_4^+ produced increases in *Oscillatoria* abundance at 1 m, where the light intensity was 10-fold higher than at 4 m, the depth at which *Oscillatoria* was stratified in the lake.

There have been two laboratory studies of the effect of nutrient concentration on buoyancy regulation. Walsby and Booker (27) and Booker and Walsby (4) used a thermally stratified water column to mimic in the laboratory some of the conditions found in lakes. If they used "complete" medium in the column, *Anabaena flos-aquae* was extremely buoyant and aggregated at the surface, where the irradiance was 10 klx. Excess buoyancy was probably due to CO_2 limitation, under which the organisms were unable to regulate buoyancy. When the medium composition was altered so that either K_2HPO_4 or $MgSO_4$ was the limiting nutrient, the organisms stratified at a depth of approximately 1 m in the column at an irradiance of 1.2 klx.

Although Booker and Walsby (4) were able to reproduce vertical stratification and surface bloom formation of cyanobacteria in their column, they pointed out its limitations as a model of a lake. The gradients of environmental factors are much steeper in the experimental column, and CO_2 limitation is more likely to occur at the high population densities found in the column than in a lake. Klemer et al. (12) have used a system in which the pertinent variables can be more easily manipulated. They grew *Oscillatoria rubescens* in NH_4^+-limited cyclostats. They found that if the stringency of nutrient limitation was increased by decreasing the dilution rate, the gas vacuolation of the culture decreased, as the model in Fig. 1 would predict. By systematically varying not only nutrient-limited growth rate but also the incident light intensity at which cells are grown in chemostats, the relationships between photosynthetic energy generation, its utilization for growth, and regulation of buoyancy in gas-vacuolate cyanobacteria could be determined.

Buoyancy conferred by gas vesicles is ecologically important to cyanobacteria in lakes. It prevents epilimnetic populations from sinking out of the productive zone of a lake during calm periods. Cyanobacteria can exploit the resources found at different strata in the water column by regulating buoyancy to carry out vertical migrations. Although I have described some mechanisms by which buoyant density is regulated, our knowledge is insufficient to understand the ecological distribution of gas-vacuolate cyanobacterial species. The following represent some important future topics for study.

(i) A quantitative analysis of the interaction between light intensity and nutrient concentration is needed and could be done in chemostat experiments similar to those of Klemer et al. (12). As mentioned earlier, neutral buoyancy is expected to occur at a series of combinations of light intensity and limiting nutrient concentration. Variations in physiological parameters caused by differences in growth conditions of bacteria and phytoplankton appear due to changes in relative growth rate, an expression of growth rate as a fraction of the maximum potential rate for the given environmental conditions of temperature and light (9, 15, 23). Perhaps the combinations of irradiance and nutrient concentration that result in neutral buoyancy all result in a similar relative growth rate. A prediction of this hypothesis is that metalimnetic cyanobacteria with neutral buoyancy should have moderately high relative growth rates and should not have the physiological characteristics of cells severely limited for energy or nutrients. Pigment, protein, and RNA contents, as well as alkaline phosphatase activities, of metalimnetic *Oscillatoria rubescens* in Crooked Lake are consistent with this prediction (13).

(ii) Differences in the mechanism of buoyancy regulation among cyanobacterial species could be related to their ecological distribution. Regulatory patterns for biosynthetic and catabolic pathways are known to vary in bacterial species. The collapse of gas vesicles by turgor, production of "ballast" molecules, and control of gas vesicle formation could have different degrees of importance in regulating the buoyant density of various cyanobacteria. The mechanisms used by individual species may limit the habitats in which they can be successful.

(iii) Buoyancy regulation may involve both short-term and long-term responses. The loss of buoyancy by cyanobacterial filaments exposed to high light intensity for several hours is clearly a short-term response, very often mediated by increases in turgor pressure. However, if an organism is grown for several generations under new environmental conditions, its buoyancy regulation may be accomplished directly by changes in the rate of gas vesicle formation or indirectly by changes in the macromolecular composition of the cell.

The preceding discussion emphasized the effect that physiological processes in cyanobacteria have upon vertical movements mediated by gas vesicles. However, the effect that physical mixing processes have upon the physiological ecology of not only cyanobacteria but also other phytoplankton is also important. Wind-driven circulation of isothermal water masses will generally overcome the buoyancy of gas-vacuolate organisms. Because light irradiance decreases exponentially with depth, vertical mixing imposes cycles of light irradiance that are determined by the power of the wind and the depth and light extinction coefficient of the mixed zone. In addition, periods of high wind velocity (for example, during storms) may be powerful enough to mix some nutrient-rich metalimnetic water with that of the epilimnion. This causes a pulse of nutrients to be introduced into the epilimnion. Relatively little is known about either the velocity at which physical movements circulate phytoplankton through the mixed zone

or the effect that oscillations in light intensity imposed by these movements have upon the physiological state of these organisms. Also, the effect that pulses of nutrients have upon competitive interactions between organisms appears ecologically relevant and has not been fully studied. Thus, an analysis of the interactive effects of light intensity and nutrient concentration upon the physiological state of cyanobacteria is important for two reasons. First, these factors regulate vertical movements due to cell buoyancy. Second, the vertical movements imposed by physical mixing processes generate temporal variations in light intensity and nutrient concentration.

CONCLUSIONS

Buoyancy in planktonic cyanobacteria is regulated in response to the levels of light energy and inorganic nutrients (such as nitrogen and phosphorus). These factors may alter buoyancy as a result of their effects upon the rates of energy generation and utilization. When growth rate is nutrient limited, the light intensity at which neutral buoyancy occurs is hypothesized to be directly related to the concentration of limiting nutrient. Species can mediate this response by either raising cell turgor pressure to collapse some existing gas vesicles or increasing cellular amounts of ballast molecules such as polysaccharide.

Buoyancy regulation is ecologically important to planktonic cyanobacteria which inhabit thermally stratified lakes. In these lakes, the depths at which light energy and inorganic nutrients are most abundant are very different. Gas-vacuolate organisms which can migrate between these depths may have a competitive advantage over phytoplankton which are incapable of vertical migration. Precise control of the migration rate is especially important for metalimnetic cyanobacteria, because the gradients of irradiance levels and nutrient concentration are especially steep in the metalimnion. For epilimnetic populations, the capacity to migrate quickly to the photic zone when wind-driven circulation has stopped may be very important.

Future research on buoyancy regulation in cyanobacteria should be directed at a quantitative analysis of the interactive effects of light intensity and nutrient concentration, an analysis of the regulatory mechanisms found in different species, and the physiological responses to not only transient changes in environmental factors but also long-term changes. In addition, the modulations in light intensity and nutrient concentration caused by physical mixing processes in lakes may affect the physiological state of phytoplankton, and the effect of these modulations should be determined.

LITERATURE CITED

1. **Allison, E. M., and A. E. Walsby.** 1981. The role of potassium in the control of turgor pressure in a gas-vacuolate blue-green alga. J. Exp. Bot. **32**:241–249.

2. **Armstrong, R. E., P. K. Hayes, and A. E. Walsby.** 1983. Gas vacuole formation in hormogonia of *Nostoc muscorum*. J. Gen. Microbiol. **128**:263–270.

3. **Armstrong, R. E., and A. E. Walsby.** 1981. The gas vesicle: a rigid membrane enclosing a hollow space, p. 95–129. *In* B. K. Ghosh (ed.), Organization of prokaryotic cell membranes, vol. 2. CRC Press, Boca Raton, Fla.

4. **Booker, M. J., and A. E. Walsby.** 1981. Bloom formation and stratification by a planktonic blue-green alga in an experimental water column. Br. Phycol. J. **16**:411–421.

5. **Dinsdale, M. T., and A. E. Walsby.** 1972. The interrelations of cell turgor pressure, gas-vacuolation, and buoyancy in a blue-green alga. J. Exp. Bot. **23**:561–570.

6. **Eberly, W. R.** 1964. Further studies on the metalimnetic oxygen maximum, with special reference to its occurrence throughout the world. Invest. Indiana Lakes Streams **6**:103–139.

7. **Fee, E. J.** 1976. The vertical and seasonal distribution of chlorophyll in lakes of the Experimental Lakes Area, northwestern Ontario: implications for primary production estimates. Limnol. Oceanogr. **21**:767–783.

8. **Fogg, G. E., and A. E. Walsby.** 1971. Buoyancy regulation and the growth of planktonic blue-green algae. Mitt. Int. Ver. Limnol. **19**:182–188.

9. **Goldman, J. C.** 1980. Physiological processes, nutrient availability, and the concepts of relative growth rate in marine phytoplankton ecology. *In* P. G. Falkowski (ed.), Primary productivity in the sea. Plenum Press, New York.

10. **Grant, N. G., and A. E. Walsby.** 1977. The contribution of photosynthate to turgor pressure rise in the planktonic blue-green alga *Anabaena flos-aquae*. J. Exp. Bot. **28**:409–415.

11. **Klemer, A. R.** 1976. The vertical distribution of *Oscillatoria agardhii* var. *isothrix*. Arch. Hydrobiol. **78**:343–362.

12. **Klemer, A. R., J. Feuillade, and M. Feuillade.** 1982. Cyanobacterial blooms: carbon and nitrogen limitation have opposite effects on the buoyancy of *Oscillatoria*. Science **215**:1629–1631.

13. **Konopka, A.** 1982. Physiological ecology of a metalimnetic *Oscillatoria rubescens* population. Limnol. Oceanogr. **27**:1154–1161.

14. **Konopka, A.** 1982. Buoyancy regulation and vertical migration by *Oscillatoria rubescens* in Crooked Lake, Indiana. Br. Phycol. J. **17**:427–442.

15. **Konopka, A.** 1983. The effect of nutrient limitation and its interaction with light upon the products of photosynthesis in *Merismopedia tenuissima* (Cyanophyceae). J. Phycol. **19**:in press.

16. **Konopka, A., A. E. Walsby, and T. D. Brock.** 1978. Buoyancy regulation by planktonic blue-green algae in Lake Mendota, Wisconsin. Arch. Hydrobiol. **83**:524–537.

17. **Krinsky, N. I.** 1979. Carotenoid pigments: multiple mechanisms for coping with the stress of photosensitized oxidations, p. 163–177. *In* M. Shilo (ed.), Strategies of microbial life in extreme environments. Dahlem Konferenzen, Berlin.

18. **Paerl, H. W., and J. F. Ustach.** 1982. Blue-green algal scums: an explanation for their occurrence during freshwater blooms. Limnol. Oceanogr. **27**:212–217.

19. **Reynolds, C. S.** 1975. Interrelations of photosynthetic behaviour and buoyancy regulation in a natural population of a blue-green alga. Freshwater Biol. **5**:323–338.

20. **Reynolds, C. S., and A. E. Walsby.** 1975. Water blooms. Biol. Rev. **50**:437–481.

21. **Shapiro, J.** 1979. The importance of trophic-level interactions to the abundance and species composition of algae in lakes, p. 105–116. *In* J. Barica and L. R. Mur (ed.), Hypertrophic ecosystems. Dr. W. Junk b.v., The Hague, Netherlands.

22. **Singh, R. N., and D. N. Tiwari.** 1970. Frequent heterocyst germination in the blue-green alga *Gloeotrichia ghosei* Singh. J. Phycol. **6**:172–176.

23. **Tempest, D. W.** 1976. The concept of "relative" growth rate: its theoretical basis and practical application. *In* A. C. R. Dean, D. C. Ellwood, C. G. T. Evans, and J. Melling (ed.), Continuous culture, vol. 6, Applications and new fields. Ellis Horwood, Chichester.

24. **Waaland, J. R., and D. Branton.** 1969. Gas vacuole development in a blue-green alga. Science **163**:1339–1341.

25. **Walsby, A. E.** 1978. The properties and buoyancy-providing role of gas vacuoles in *Trichodesmium* Enrenberg. Br. Phycol. J. **13**:103–116.

26. **Walsby, A. E.** 1981. Cyanobacteria: planktonic gas-vacuolate forms. *In* M. P. Starr, H. Stolp, H. G. Trüper, A. Balows, and H. G. Schlegel (ed.), The prokaryotes. Springer-Verlag, Berlin.

27. **Walsby, A. E., and M. J. Booker.** 1980. Changes in buoyancy of a planktonic blue-green alga in response to light intensity. Br. Phycol. J. **15**:311–319.

28. **Walsby, A. E., and A. R. Klemer.** 1974. The role of gas vacuoles in the microstratification of a population of *Oscillatoria agardhii* var. *isothrix* in Deming Lake, Minnesota. Arch. Hydrobiol. **74**:375–392.

29. **Walsby, A. E., H. C. Utkilen, and I. J. Johnsen.** 1983. Buoyancy changes of a red coloured *Oscillatoria agardhii* in Lake Gjersjøen, Norway. Arch. Hydrobiol., in press.

Adaptation of Bacteria to External pH

ETANA PADAN

Division of Microbial and Molecular Ecology, The Institute of Life Sciences, The Hebrew University of Jerusalem, Jerusalem, Israel

From both the intracellular and extracellular milieu the procaryotic cell, more than the eucaryotic one, is exposed to fluctuations in proton concentrations.

Inside the cell there are many metabolic reactions that must operate at constant high rates and that involve production or consumption of protons (41). A good example for H^+ addition is the production of organic acids by the anaerobic heterotrophs. These protons are liberated into the cytoplasm since there is no compartmentalization in the procaryotic cell.

The proton concentration of the extracellular environment is in direct contact with the cytoplasmic membrane of the bacterial cell. The cell wall cannot exclude molecules below 1,000 molecular weight (22). Remarkably, the proton concentration in the aquatic habitats of bacteria can vary by nine orders of magnitude. In the acidic range there are the sulfur springs with pH values of 1 to 2 (4, 5, 16, 19, 26, 30, 37, 55). At the other extreme are the soda lakes with pH values of 10 to 11 (4, 13, 18, 26, 30, 37, 50, 51). Furthermore, as a result of biological activity such as photosynthesis in eutrophic water systems, including fish ponds (43) and sewage oxidation ponds (1), the pH can change within a diurnal cycle by several units.

Fluctuations in proton concentrations represent a challenge to the bacterial cell. The proton, more than any other ion, is involved in many physicochemical as well as biochemical reactions, solvolysis, ionization, and oxidoreduction. For example, the high proton concentration of acidic environments leads to high concentrations of dissolved metals that can reach toxic levels (16). The low proton concentration of the basic environment precipitates many essential metals (30). Thus, it is not surprising that as early as 1943 Gale concluded that bacteria must be equipped with an efficient mechanism that allows them to cope with fluctuating extreme pH values (12).

pH RESISTANCE STRUCTURES IN THE PROCARYOTIC CELL

It is mainly with the acidophiles that resistance to pH was sought in structural and compositional peculiarities of the cell wall or cytoplasmic membrane, or both (for review see 26, 30).

Since most of the organisms used were *Thermoplasma, Sulfolobus*, and *Bacillus acidocaldarius*, which are both thermophiles and acidophiles, it is not clear to what extent the peculiarities found are related to acidophilicity. In addition, the first two are included in the archaebacteria (56), whereas the last is a eubacterium. Interesting molecules in point are the hydrophobic glycoprotein of *Thermoplasma acidophilum* (26, 57), the lipids containing acid-resistant ether linkages of *Sulfolobus acidocaldarius* (8, 9, 26) and *T. acidophilum* (9, 26), and the sulphonolipids containing acid-resistant C–S bonds of *B. acidocaldarius* (30). A common denominator of the acidophiles appears to be the occurrence of cyclization within the hydrocarbon portion of the membrane lipids (30). *Thermoplasma* and *Sulfolobus* lipids possess cyclohexyl rings in the alkyl chains, and *B. acidocaldarius* contains cyclohexyl fatty acids and many pentacyclic triterpene derivatives. The thiobacilli contain large amounts of cyclopropane fatty acids. It has been suggested that these constituents within the cell membrane function in a way similar to cholesterol by increasing membrane rigidity and decreasing membrane permeability (38). Obviously, additional experimental data are needed in this area of research with both acidophiles and alkalophiles. Lipids and proteins localized to the external side of the cytoplasmic membrane in particular should be studied, as these cell elements are exposed to the medium pH.

pH HOMEOSTASIS IN THE PROCARYOTIC CELL

As pertains to the cytoplasm, it has become apparent during the past years that proteins isolated from many sources, including the acidophiles and alkalophiles, exhibit a very narrow pH range, around neutrality, of optimal activity and stability (26, 30). These findings led me to suggest that, whatever the pH homeostatic mechanism, it must maintain the cytoplasmic pH constant. A straightforward experiment to test this idea was to measure the intracellular pH of the microorganisms as a function of extracellular pH. Thus, a method allowing for the determination of intracellular pH of bacteria became very crucial.

Since most bacteria, like cell organelles, are too small for insertion of electrodes, indirect methods had to be applied. Such methods have been developed for the measurement of the internal pH of chloroplast mitochondria and other organelles (36, 44). They are based on the findings that the undissociated form of a weak acid or base is freely permeable through membranes, whereas the ionized form is impermeable. If the cell is basic inside, as compared to the environment, the undissociated form of the acid will penetrate the cell, but will immediately dissociate inside, according to the pK of the acid and the cytoplasmic pH. More acid will then be allowed to penetrate the cell, and this process will continue until equilibrium is reached, at which point the acid will distribute across the membrane according to the ΔpH. At equilibrium, therefore, the ΔpH and thus the intracellular pH can be calculated from the concentration ratio of the acid. If the cell is acid inside, as compared to the environment, a base should be used.

For the application of this method to bacteria we chose *Escherichia coli*, which is known to endure a wide range of pH and to grow optimally between pH 6.5 and 8.8 (36, 37, 58). Remarkably, the intracellular pH of *E. coli* at the entire spectrum of growth-permissive pH was found constant at pH 7.6 to 7.8 (36, 37, 48).

Using this approach, many groups investigated various neutralophilic bacteria (37), including *Micrococcus lysodeikticus* (11), *Bacillus subtilis* (23), *Streptococcus faecalis* (17), and *Halobacterium halobium* (2). All were found to behave exactly like *E. coli*. Most importantly, later another independent technique to measure the intracellular pH of bacteria and organelles was developed (35, 49). It is based on the pH sensitivity of the nuclear magnetic resonance spectrum of ^{31}P in phosphate or methyl phosphonate, and it yields ΔpH values identical to those obtained with the acid and base distribution (37). Thus, it can be concluded that neutralophilic bacteria like *E. coli* do indeed possess a pH homeostatic mechanism which maintains their cytoplasmic pH constant at around pH 7.8.

PRIMARY PROTON PUMPS AND pH HOMEOSTASIS

In respiring bacteria, the pH homeostatic mechanism is found to be totally dependent on the respiratory system (36). When KCN is added or O_2 is deprived, homeostasis collapses, the protons equilibrate across the membrane, and pH_{in} equates with pH_{out} (36, 48). When respiration is restored at each pH_{out}, an appropriate ΔpH is rebuilt so that pH homeostasis is restored at pH_{in} of 7.8. The loss and regain of pH

homeostasis occurs within a time period of less than 1 min, implying that the buffer capacity of the cytoplasm may have only a transitory role in controlling intracellular pH of bacteria.

The respiratory system in the procaryotic cell, like other principal energy-converting reactions such as photosynthesis and active transport, occurs in the cytoplasmic membrane or in its invaginations. Kaback (21, 39, 40) isolated membrane vesicles which are devoid of the cytoplasm, possess the right-side-out orientation of the intact cytoplasmic membrane, and exhibit all energy transduction phenomena of the intact cell. Furthermore, Kaback and co-workers showed that this membrane preparation has the propensity to control intravesicular pH (42). The plot of ΔpH versus external pH obtained with membrane vesicles is very similar to our finding with whole cells (36). pH homeostasis is thus a membranous phenomenon which is maintained by membrane energy transduction.

As is well known today, energy transduction in membranes is the main energy-converting mechanism of the cell and occurs by the chemiosmotic mechanism advanced by Mitchell (34). Mitchell postulated that the proteins in these membranes are organized in the membrane in a very asymmetric fashion so that during catalysis they function as primary proton pumps, moving protons from one side of the membrane to the other and thereby forming a ΔpH across the membrane. Since the protons are charged, a Δψ is concomitantly formed. Thus, light or chemical energy is transduced to electrical and chemiosmotic energy and stored in the proton gradient in the form of ΔpH and Δψ, the sum of which is expressed by the following equation:

$$\Delta\bar{\mu}_{H^+}/F = \Delta\psi - Z\Delta pH$$

(*T* is the temperature in K, *R* is the gas constant, and *F* is the Faraday constant; $Z = 2.3$ RT/F). By the very same principle of membrane proteins translocating protons, the proton gradient is utilized as an energy donor for most energy-consuming and -conserving reactions of the cell, including ATP synthesis, transport, and, in bacteria, even locomotion.

The chemiosmotic energy-coupling mechanism is shared by all energy-converting membranes, including the eucaryotic ones of the chloroplast thylakoid and of the mitochondrial crysta. However, there is one characteristic which is unique to the procaryotic system. In the procaryotes the primary proton pumps are not compartmentalized in organelles, but instead are in the cytoplasmic membrane. Therefore, they are in direct contact both with the environment and with the cytoplasm, and are able to "sense" and react with the proton concentration on both sides of the cell membrane. It is this very fact that led us to suggest that the primary proton pumps of the procaryotic cells have dual functions. They are involved both in energy transduction and in pH homeostasis of the bacterial cell (36). Namely, they actively create the different ΔpH values needed at each pH_{out} to keep the cytoplasmic pH at pH 7.8.

That this ΔpH is maintained by active proton pumping in appropriate magnitude and direction was most beautifully demonstrated by an experiment in which the bacteria were shifted from acidic pH to alkaline pH (58). Cells were grown in a pH-stat at pH 7.2. As shown before (36) the ΔpH was 0.6 (basic inside), yielding a pH_{in} of 7.8. At zero time, within 30 s, pH_{out} was rapidly titrated to 8.3. At this time, however, pH_{in} was still 7.8. Hence, a ΔpH of 0.5 units—acid inside—was artificially imposed on the membrane by the shift. Subsequently, within 3 min this ΔpH decayed to 0 and the pH_{in} equated to pH_{out}. It is clear that in the first stage subsequent to the alkaline shift in pH_{out} there is a failure of pH homeostasis during which time pH_{in} is disturbed and becomes equal to pH_{out}. Following the initial failure to control pH_{in}, after 3 to 5 min, adaptation started. A ΔpH—acid inside—was actively built up, until a pH_{in} of 7.8 was reattained. It was thereafter maintained constant.

When the cells were shifted back to the acidic pH, the same sequence of events occurred, but in the opposite direction. First the ΔpH collapsed and then there was outward pumping of protons for about 3 min, after which time pH_{in} of 7.8 was restored.

Already in the early 1970s Scholes and Mitchell, by using an electrode in the extracellular fluid, showed that upon respiration at neutral extracellular pH, protons are expelled from the cytoplasm of the bacteria outward (46). The question was thus raised as to the directionality of the pumps over the entire growth-permissive pH. A straightforward way to determine this parameter of the pump is to determine the electrical potential difference it maintains across the membrane. Since protons are charged, extrusion of protons will create a potential difference which is negative inside, and if protons are taken up the reverse potential will be produced.

Since most bacteria are too small for insertion of electrodes, an indirect method had to be applied for measurement of the $\Delta\psi$. As with the determination of ΔpH, we applied to bacteria (60) the method developed in organelles (44), based on the fact that a freely permeable ion will electrophorese across a membrane according to the existing membrane potential ($\Delta\psi$) until an equilibrium is reached. If the cells are negative inside, a permeable cation will concentrate in-

side, and if the cells are positive inside, a permeable anion will be taken up. At equilibrium the concentration ratio of the ion is related to the membrane potential according to the Nernst equation and therefore allows calculation of the $\Delta\psi$. We have tried several ions. Anions were not taken up, whereas cations concentrated in the cytoplasm. The results with tetraphenyl phosphonium (TPP$^+$) and Rb$^+$ in the presence of valinomycin, which specifically renders the membrane permeable to K$^+$ and Rb$^+$, were straightforward (60). The potential is ($-$) inside over the entire pH range between 6.5 and 8.7. Furthermore, it progressively increases with pH_{out} from 80 to 90 mV at pH 6.5 to 150 mV at pH 7.5. At the more alkaline pH it remains high and almost constant. Recently, Slayman and his group (10) grew $E.$ $coli$ in the presence of 6-amidinopenicillanic acid and created giant $E.$ $coli$ cells. They succeeded in introducing microelectrodes into these cells and in measuring directly the $\Delta\psi$. Their results verified ours over the entire pH range of growth.

MODULATION OF THE PRIMARY PROTON PUMPS FOR pH HOMEOSTASIS

The outward directionality of the proton pumps implies that protons are expelled from the cytoplasm to the environment, and therefore the cytoplasm should be more alkaline than the extracellular fluid. Indeed, this is the case at the lower pH range up to 7.8, where the observed ΔpH is alkaline inside (36, 37). Furthermore, the progressive increases in the $\Delta\psi$ can also explain the progressive decrease in the ΔpH. As in any ion pump, the amount of proton pumped is determined by the potential created. The larger the potential, the larger are the electrostatic forces pulling back the proton and the fewer protons are pumped. The lesser the potential, the more protons are free to translocate across the membrane. The reason for the change of $\Delta\psi$ with pH_{out} is not clear yet, but may be related to a change of ion permeability with pH. A change of K$^+$ permeability with pH has been suggested (24, 52). In any event, it is apparent that at the acidic pH range of $E.$ $coli$ life the primary proton pumps are modulated so as to maintain the appropriate basic inside ΔpH needed for keeping the cytoplasmic pH constant at 7.8.

The situation above pH of 7.8 was at first sight completely puzzling. Here the $\Delta\psi$ hardly changes and therefore cannot account for the changes in the ΔpH. Furthermore, as mentioned above, also at the alkaline pH, the orientation of the pumps is outward and the $\Delta\psi$ is negative inside, yet the ΔpH reverses orientation and becomes acidic inside. An additional modulating device operating specifically at the basic pH has therefore been suggested (36). This device, we have postulated, is the Na^+/H^+ antiporter, which exchanges Na^+ for H^+. The protons expelled by the proton pumps are recycled by the antiporter to the cytoplasm in exchange for Na^+, and the intracellular pH is maintained at pH 7.8.

Na^+/H^+ antiporter activity was already documented by West and Mitchell in 1974 (54). Later it was demonstrated in isolated membrane vesicles (3, 47). There was, therefore, a need to prove directly the involvement of this system in pH homeostasis. A direct way to assess the participation of the Na^+/H^+ antiporter in the pH homeostasis at alkaline pH is to eliminate it by mutation and study the capacity of pH homeostasis of the mutant.

A very powerful selection procedure was afforded for such a mutant (59) since the Na^+/H^+ antiporter has an important established role in the *E. coli* cell physiology. It uses the H^+ gradient to expel Na^+ and thereby maintains a low intracellular Na^+ concentration and an Na^+ gradient directed inward. Eventually, this Na^+ gradient is needed as an energy source for two substrates that are cotransported with Na^+ into the cells: melibiose (53) and glutamate (15, 32). Hence, a mutant that is simultaneously impaired in growth capacity on both glutamate and melibiose is most probably impaired in the common denominator: the Na^+/H^+ antiporter.

Thus, after UV mutagenesis, penicillin enrichment was done on both glutamate and melibiose, and mutants that simultaneously lost the growth capacity on both substrates were selected (59). They grew normally on glycerol and glucose and very poorly on the Na^+ cotransported substrates. One of them, DZ3, was characterized genetically by both conjugation and transduction. The mutation was mapped in a new locus designated *phs*, which is 98.5 min on the *E. coli* chromosome, far apart from the operons of both glutamate and melibiose (58, 59).

Furthermore, the mutant was found impaired in its Na^+ extrusion capacity (59). Energy-depleted cells equilibrated with ^{22}Na actively extruded Na^+ upon addition of glycerol. The mutant expelled Na^+ much less efficiently than the wild type, indicating that it is indeed impaired in the Na^+/H^+ antiporter.

The mutant was next compared to the wild type with respect to growth and pH homeostasis at different external pH values (58, 59). Whereas the wild type grew optimally and exhibited pH homeostasis up to an external pH of 8.8, the mutant growth was drastically and progressively decreased beyond pH 8.2. Strikingly, the mutant accordingly lost its pH homeostasis capacity. Hence, it was concluded that the pH homeostatic capacity of the mutant is impaired at alkaline pH, strongly implying that the Na^+/H^+ antiporter is part of the pH homeostasis mechanism.

UNIFIED PRINCIPLES IN pH HOMEOSTASIS IN BACTERIA

The capacity of maintaining constant cytoplasmic pH over a wide range of external pH values is not unique to *E. coli*. As mentioned above (37), many neutralophilic bacteria behave very similarly to *E. coli*, implying that the $\Delta\bar{\mu}_{H^+}$ generated by the proton pumps of these bacteria reacts to external pH in a pattern identical to that observed in *E. coli*; $\Delta\psi$ increases with pH while ΔpH drastically decreases, and the resulting internal pH is maintained constant, in all, in the range of 7.6 to 7.8.

If this is the general mechanism chosen by nature for the control of internal pH, it is tempting to suggest that acidophiles and alkalophiles will possess essentially the same mechanism, working at their utmost capacity at the acidic range or at the alkaline range, respectively. This appears to be the case with the acidophiles. They behave in a manner similar to *E. coli* at the acidic range. The membrane potential decreases with lowering of the external pH (6, 25; for review see 26). Here it even reverses to being positive inside. The ΔpH is increased to 4.5 pH units, the maximal value observed in biological systems. The resulting internal pH of the acidophiles is maintained constant over the extreme acidic pH range, but at the lowest level of 6.5 pH_{in}. As in *E. coli* at the alkaline range, in the alkalophiles the membrane potential increases with pH to a maximum of about 130 mV or even more (26), and the ΔpH reverses, reaching a maximal reversed ΔpH of about 2.5 pH units at pH_{out} of 12. The internal pH is again kept constant over the alkaline pH range, but at the highest level of 9.5 pH_{in}. Furthermore, a mutant has been isolated by Krulwich and her colleagues (28) which lacks the Na^+/H^+ antiporter activity and, like the *E. coli* mutant, loses the capacity to grow at the extreme alkaline pH range of the respective wild-type parent.

The data show that in all bacteria pH homeostasis is brought about by modulation of both $\Delta\psi$ and ΔpH at each external pH. Therefore, the sum of these parameters, the $\Delta\bar{\mu}_{H^+}$, will also change with pH_{out}. It is the highest, but composed only of ΔpH, in the acidophiles (6, 25; but see also 33). In the neutralophiles it is somewhat lower and consists of both ΔpH and $\Delta\psi$ (37). In the alkalophiles it is very small because of the reversion of the ΔpH and is composed only of about 30 mV $\Delta\psi$ at pH_{out} of 11 (26). Since the $\Delta\bar{\mu}_{H^+}$ is the cells' main driving force, interesting bioenergetic problems are raised. The most interesting question, challenging the chemiosmotic dogma, is how the alkalophiles function with such a small $\Delta\bar{\mu}_{H^+}$.

pH HOMEOSTASIS AND THE PROCARYOTIC CELL PHYSIOLOGY

Although the principles of pH homeostasis appear identical among all bacteria, the data presented also show that the pH controlling machinery is poised at different intracellular pH_{in} in the various pH categories of the procaryotes (36). In the acidophiles it is set to pH 6.5, in the neutralophiles to pH 7.8, and in the alkalophiles to pH 9.5. An interesting question is whether the range of intracellular pH between 6.5 and 9.5 represents the range to which the cytoplasm can adapt.

Nevertheless, whatever the cytoplasmic pH setting point, the efficiency of the pH controlling devices is remarkable, allowing variation of only 0.2 pH unit. This strict constancy in pH_{in} tempts speculation that pH_{in} may have a regulatory role in the cell in addition to its importance in maintaining all proteins in their optimal pH range of activity or stability, or both. In the wild type as in the mutant, the lethal pH was 8.8 to 9. At pH_{out} of 9 viability decreased to 50% within 1 h in both strains (58). Hence, it can be suggested that at this alkaline pH there is a general deleterious effect on the cell proteins and the cells die.

Nevertheless, the described pH shift experiments within the growth-permitting pH conducted with the wild type showed that there might be tight coupling between growth and pH_{in} (58). Changes of external pH induced transient disturbance in pH homeostasis machinery. Remarkably, as long as pH_{in} was disturbed, growth was arrested. Immediately when pH homeostasis was reestablished, growth resumed at the optimal rate. The effect of pH_{in} on growth was most beautifully demonstrated with the mutant impaired in its pH homeostasis machinery. It grows optimally below pH_{out} of 7.8, where it can control its pH_{in}. Beyond this extracellular pH where it lacks pH homeostasis, it stops growing, albeit it is completely viable for at least 10 h up to pH 9. Hence, it must be concluded that disturbance of pH_{in} in the mutant unraveled a specific effect of pH_{in} on a process that is required for growth but not for viability.

Results obtained recently (D. Zilberstein, V. Agmon, S. Schuldiner, and E. Padan, unpublished data) show that this process is not energetic since respiration is normal in the mutant up to pH 8.8. The $\Delta\psi$ and therefore $\Delta\bar{\mu}_{H^+}$ is higher in the mutant than in the wild type. Even protein synthesis is normal for about 2 h, as is the induction of β-galactosidase. The synthesis of DNA is affected, but again only to 50%. Cell division is the only function that is totally arrested by the high pH. Hence, we encounter very

specific regulatory coupling between the cell cycle and pH_{in} and, therefore, via pH_{in} with the primary proton pumps and the bioenergetic state of the cell. Interestingly, very important modulatory functions of pH_{in} have been implicated recently in many eucaryotic cells (37).

Moreover, recently it became apparent that the $\Delta\psi$ may have modulatory effects on different cell functions of the procaryotic cell. These include injection of phage T4 DNA into E. coli (29), genetic transformation in E. coli (45), regulation of autolysis in Bacillus subtilis (20), solute transport reactions (31), insertion of proteins into membranes (61), and finally, a very recent finding of my laboratory, showing that $\Delta\psi$ is required as a modulator for cellulose synthesis in Acetobacter xylinum (7). Interestingly, the dependence of this reaction on $\Delta\psi$ is very similar to that of T4 DNA injection.

This entire cascade of intricate adaptations accompanying the primary modulation of the proton pumps to attain constant pH_{in} explains why bacteria are in most cases obligatory to a certain external pH range which is smaller than the nondenaturing or destructing pH range. A change in pH_{out} to a value which upsets the pH_{in} disturbs many aspects of physiology, and the cell stops growing. Thus, B. acidocaldarius when transferred from pH 2 to 6 does not die, but does not grow (14). Similar behavior is observed when B. firmus RAB is shifted from pH 10.5 to pH 9 (27).

The allowed wide range of variation in the bioenergetic parameters of the proton pumps, as opposed to the remarkable constancy in pH_{in} that they maintain, raises the speculation as to which of the functions of the primary proton pumps evolved first. Raven and Smith (41) suggested that the proton pumps primarily evolved for regulation of intracellular pH and the bioenergetic roles were acquired later in evolution.

LITERATURE CITED

1. **Abeliovich, A., and Y. Azov.** 1976. Toxicity of ammonia to algae in sewage oxidation ponds. Appl. Environ. Microbiol. 31:801–806.
2. **Bakker, E. P., H. Rottenberg, and S. R. Caplan.** 1976. An estimation of the light-induced electrochemical potential difference of protons across the membrane of Halobacterium halobium. Biochim. Biophys. Acta 440:557–572.
3. **Beck, J. C., and B. P. Rosen.** 1979. Cation/proton antiport systems in Escherichia coli: properties of the sodium/proton antiporter. Arch. Biochem. Biophys. 194:208–214.
4. **Brock, T. D.** 1969. Microbial growth under extreme conditions. Symp. Soc. Gen. Microbiol. 19:15–41.
5. **Brown, M. H., and T. Mayes.** 1980. The growth of microbes at low pH values, p. 71–98. In G. W. Gould and J. E. L. Corry (ed.), Microbial growth and survival in extremes of environment. Academic Press, London.
6. **Cox, J. C., D. G. Nicholls, and W. J. Ingledew.** 1979. Transmembrane electrical potential and transmembrane pH gradient in the acidophile Thiobacillus ferro-oxidans. Biochem. J. 178:195–200.
7. **Delmer, D. P., M. Benziman, and E. Padan.** 1982. Re-

quirement for a membrane potential for cellulose synthesis in intact cells of *Acetobacter xylinum*. Proc. Natl. Acad. Sci. U.S.A. **79**:5282–5286.

8. **DeRosa, M., S. DeRosa, A. Gambacorta, and J. D. Bu-'Lock.** 1980. Structure of calditol, a new branched chain nonitol, and of the derived tetraether lipids in thermoacidophile archaebacteria of the *Caldariella* group. Phytochemistry **19**:249–254.

9. **DeRosa, M., A. Gambacorta, B. Nicolaus, S. Sodano, and J. D. Bu'Lock.** 1980. Structural regularities in tetraether lipids of *Caldariella* and their biosynthetic and phyletic implications. Phytochemistry **19**:833–836.

10. **Felle, H., J. S. Porter, C. L. Slayman, and H. R. Kaback.** 1980. Quantitative measurements of membrane potential in *Escherichia coli*. Biochemistry **19**:3585–3590.

11. **Friedberg, I., and H. R. Kaback.** 1980. Electrochemical proton gradient in *Micrococcus lysodeikticus* cells and membrane vesicles. J. Bacteriol. **142**:651–658.

12. **Gale, E. F.** 1943. Factors influencing the enzymatic activities of bacteria. Bacteriol. Rev. **7**:139–173.

13. **Grant, W. D., and B. J. Tindall.** 1980. The isolation of alkalophilic bacteria, p. 27–38. *In* G. W. Gould and J. E. L. Corry (ed.), Microbial growth and survival in extremes of environment. Academic Press, London.

14. **Guffanti, A. A., L. F. Davidson, T. M. Mann, and T. A. Krulwich.** 1979. Nigericin-induced death of an acidophilic bacterium. J. Gen. Microbiol. **114**:201–206.

15. **Halpern, Y. S., H. Barash, S. Dover, and K. Druck.** 1973. Sodium and potassium requirements for active transport of glutamate by *Escherichia coli* K-12. J. Bacteriol. **114**:53–58.

16. **Hargreaves, J. W., E. J. H. Lloyd, and B. A. Whitton.** 1975. Chemistry and vegetation of highly acidic streams. Freshwater Biol. **5**:563–576.

17. **Harold, F. M., E. Pavlasova, and J. R. Baarda.** 1970. A transmembrane pH gradient in *Streptococcus faecalis*: origin, and dissipation by proton conductors and N,N′-dicyclohexylcarbodiimide. Biochim. Biophys. Acta **196**:235–244.

18. **Imhoff, J. F., H. G. Sahl, G. S. H. Soliman, and H. G. Truper.** 1979. The Wadi Natrun: chemical composition and microbial mass developments in alkaline brines of eutrophic desert lakes. Creomicrobiol. J. **1**:219–234.

19. **Ingledew, W. J.** 1982. *Thiobacillus ferrooxidans*. The bioenergetics of an acidophilic chemolithotroph. Biochim. Biophys. Acta **683**:89–117.

20. **Jolliffe, L. K., R. J. Doyle, and U. N. Streips.** 1981. The energized membrane and cellular autolysis in *Bacillus subtilis*. Cell **25**:753–763.

21. **Kaback, H. R.** 1971. Bacterial membranes. Methods Enzymol. **22**:99–120.

22. **Kadner, R. J., and P. J. Bassford, Jr.** 1978. The role of the outer membrane in active transport, p. 414–462. *In* B. P. Rosen (ed.), Bacterial transport. Marcel Dekker, New York.

23. **Khan, S., and R. M. Macnab.** 1980. Proton chemical potential, proton electrical potential and bacterial motility. J. Mol. Biol. **138**:599–614.

24. **Kroll, R. G., and I. R. Booth.** 1981. The role of potassium transport in generation of a pH gradient in *Escherichia coli*. Biochem. J. **198**:691–698.

25. **Krulwich, T. A., L. F. Davidson, S. J. Filip, Jr., R. S. Zuckerman, and A. A. Guffanti.** 1978. The proton motive force and β-galactoside transport in *Bacillus acidocaldarius*. J. Biol. Chem. **253**:4599–4603.

26. **Krulwich, T. A., and A. A. Guffanti.** 1983. Physiology of acidophilic and alkalophilic bacteria. Adv. Microb. Physiol. **24**:173–214.

27. **Krulwich, T. A., A. A. Guffanti, R. F. Bornstein, and J. Hoffstein.** 1982. A sodium requirement for growth, solute transport, and pH homeostasis in *Bacillus firmus* RAB. J. Biol. Chem. **257**:1885–1889.

28. **Krulwich, T. A., K. G. Mandel, R. F. Bornstein, and A. A. Guffanti.** 1979. A non-alkalophilic mutant of *Bacillus alcalophiles* lacks the Na^+/H^+ antiporter. Biochem. Biophys. Res. Commun. **91**:58–62.

29. **Labedan, B., K. B. Heller, A. A. Jasaitis, T. H. Wilson, and E. B. Goldberg.** 1980. A membrane potential threshold for phage T_4 DNA injection. Biochem. Biophys. Res. Commun. **93**:625–630.

30. **Langworthy, T. A.** 1978. Microbial life in extreme pH values, p. 279–315. *In* D. J. Kushner (ed.), Microbial life in extreme environments. Academic Press, London.

31. **Lanyi, J. K., and M. P. Silverman.** 1979. Gating effects in *Halobacterium halobium* membrane transport. J. Biol. Chem. **254**:4750–4755.

32. **MacDonald, R. E., J. K. Lanyi, and R. V. Green.** 1977. Sodium-stimulated glutamate uptake in membrane vesicles of *Escherichia coli*: the role of ion gradients. Proc. Natl. Acad. Sci. U.S.A. **74**:3167–3170.

33. **Matin, A., B. Wilson, E. Zychlinsky, and M. Matin.** 1982. The proton motive force and the physiological basis of delta pH maintenance in *Thiobacillus acidophilus*. J. Bacteriol. **150**:582–591.

34. **Mitchell, P.** 1979. Compartmentation and communication in living systems. Ligand conduction: a general catalytic principle in chemical, osmotic and chemiosmotic reaction systems. Eur. J. Biochem. **95**:1–20.

35. **Navon, G., S. Ogawa, R. G. Shulman, and T. Yamane.** 1977. High resolution ^{31}P nuclear magnetic resonance studies of metabolism in aerobic *Escherichia coli* cells. Proc. Natl. Acad. Sci. U.S.A. **74**:888–891.

36. **Padan, E., D. Zilberstein, and H. Rottenberg.** 1976. The proton electrochemical gradient in *Escherichia coli* cells. Eur. J. Biochem. **63**:533–541.

37. **Padan, E., D. Zilberstein, and S. Schuldiner.** 1981. pH homeostasis in bacteria. Biochim. Biophys. Acta **650**:151–166.

38. **Poralla, K., E. Kannenberg, and A. Blume.** 1980. A glycolipid containing hopane isolated from the acidophilic thermophilic *Bacillus acidocaldarius*, has a cholesterol-like function in membranes. FEBS Lett. **113**:107–110.

39. **Ramos, S., and H. R. Kaback.** 1977. The electrochemical proton gradient in *Escherichia coli* membrane vesicles. Biochemistry **16**:848–854.

40. **Ramos, S., and H. R. Kaback.** 1977. The relationship between the electrochemical proton gradient and active transport in *Escherichia coli* membrane vesicles. Biochemistry **16**:854–859.

41. **Raven, J. A., and F. A. Smith.** 1976. The evolution of chemiosmotic energy coupling. J. Theor. Biol. **57**:301–312.

42. **Reenstra, W. W., L. Patel, H. Rottenberg, and H. R. Kaback.** 1980. Electrochemical proton gradient in inverted membrane vesicles from *Escherichia coli*. Biochemistry **19**:1–9.

43. **Rimon, A., and M. Shilo.** 1982. Factors which affect the intensification of fish breeding in Israel. Bamidgeh **34**:87–100.

44. **Rottenberg, H.** 1979. The measurement of membrane potential and ΔpH in cells, organelles and vesicles. Methods Enzymol. **55**:547–589.

45. **Santos, E., and H. R. Kaback.** 1982. Involvement of the proton electrochemical gradient in genetic transformation in *Escherichia coli*. Biochem. Biophys. Res. Commun. **99**:1153–1160.

46. **Scholes, P., P. Mitchell, and J. Moyle.** 1969. The polarity of proton translocation in some photosynthetic microorganisms. Eur. J. Biochem. **8**:450–454.

47. **Schuldiner, S., and H. Fishkes.** 1978. Sodium-proton antiport in isolated membrane vesicles of *Escherichia coli*. Biochemistry **17**:706–711.

48. **Schuldiner, S., and E. Padan.** 1982. How does *Escherichia coli* regulate internal pH?, p. 65–74. *In* A. Martonosi (ed.), Membrane and transport. Plenum Publishing Co., New York.

49. **Slonczewski, J. L., B. P. Rosen, J. R. Alger, and R. M. Macnab.** 1981. pH homeostasis in *Escherichia coli*: measurement by ^{31}P nuclear magnetic resonance of methyl-

phosphonate and phosphate. Proc. Natl. Acad. Sci. U.S.A. **78**:6271–6275.

50. **Talling, J. F., R. B. Wood, M. V. Prosser, and R. M. Baxter.** 1973. The upper limit of photosynthetic productivity by phytoplankton: evidence from Ethiopian soda lakes. Freshwater Biol. **3**:53–76.

51. **Tindall, B. J., A. A. Mills, and W. D. Grant.** 1980. An alkalophilic halophilic bacterium with a low magnesium requirement from a Kenya soda lake. J. Gen. Microbiol. **116**:257–260.

52. **Tokuda, H., T. Nakamura, and T. Unemoto.** 1981. K^+ required for the generation of pH-dependent membrane potential and ΔpH by marine bacterium *Vibrio alginolyticus*. Biochemistry **20**:4198–4203.

53. **Tsuchiya, T., J. Raven, and T. H. Wilson.** 1977. Cotransport of Na^+ and methyl-β-thiogalactopyranoside mediated by the melibiose transport system of Escherichia coli. Biochem. Biophys. Res. Commun. **76**:26–31.

54. **West, I. C., and P. Mitchell.** 1974. Proton/sodium ion antiport in *Escherichia coli*. Biochem. J. **144**:87–90.

55. **Whitton, B. A. and B. M. Diaz.** 1981. Influence of environmental factors on photosynthetic species composition in highly acidic waters. Verh. Int. Ver. Limnol. **21**:1459–1465.

56. **Woese, C. R., L. J. Magrum, and G. E. Fox.** 1978. Archaebacteria. J. Mol. Evol. **11**:245–252.

57. **Yang, L., and A. Haug.** 1979. Purification and partial characterization of a prokaryotic glycoprotein from the plasma membrane of *Thermoplasma acidophilus*. Biochim. Biophys. Acta **556**:265–277.

58. **Zilberstein, D., V. Agmon, S. Schuldiner, and E. Padan.** 1982. The sodium/proton antiporter is part of the pH homeostasis mechanism in *Escherichia coli*. J. Biol. Chem. **257**:3687–3691.

59. **Zilberstein, D., E. Padan, and S. Schuldiner.** 1980. A single locus in *Escherichia coli* governs growth in alkaline pH and on carbon sources whose transport is sodium dependent. FEBS Lett. **116**:177–180.

60. **Zilberstein, D., S. Schuldiner, and E. Padan.** 1979. Proton electrochemical gradient in *Escherichia coli* cells and its relation to active transport of lactose. Biochemistry **18**:669–673.

61. **Zimmermann, R., C. Watts, and W. Wickner.** 1982. The biosynthesis of membrane-bound M13 coat protein energetics and assembly intermediates. J. Biol. Chem. **257**:6529–6536.

Effects of Iron on Bacterial Growth and Bioluminescence: Ecological Implications

MARGO G. HAYGOOD AND KENNETH H. NEALSON

Marine Biology Research Division, Scripps Institution of Oceanography, La Jolla, California 92093

The marine luminous bacteria include several species in three genera (1, 13, 35). The best-characterized habitats in which they occur are the planktonic, enteric, and light organ symbiotic ones, but they also occur as saprophytes and parasites (10, 19). In general, the planktonic luminous bacteria are found in low numbers (1 to a few per milliliter), whereas the habitats involving attached or associated forms are much more densely populated (19). Although planktonic luminous bacteria are ubiquitous in the world's oceans, their distribution has been intensively studied in only a few places. Those areas that have been studied, however, clearly show that these bacteria are not abundant and that the distribution of the various species is not uniform. In addition to geographical differences in occurrence, seasonal cycles in abundance have been shown in several environments (2, 23, 24, 26, 36).

The uniting feature of these bacteria is, of course, their ability to emit light, and one can reasonably ask under what circumstances this capacity is important to the ecology of the group. Our studies have thus focused on those factors that control growth and luminescence of the various luminous bacteria and on measurements of light in situ. So far, there is no evidence that planktonic luminous bacteria emit light; efforts to measure such light have shown that planktonic forms are very dim, at best <0.1% of their capacity (3; K. H. Nealson, Microb. Ecol., in press). Attached forms (saprophytes and symbionts), on the other hand, are well known to emit readily measured levels of light. Thus, the role of light emission, if any, in the ecology of these bacteria would appear to center around their symbiotic interactions with other organisms.

The following are some tentative generalizations about the planktonic habitats of the marine luminous bacteria and physiological factors postulated to influence their distribution. *Alteromonas hanedai* occurs in polar regions and grows at temperatures from 5 to 20°C (13). *Photobacterium phosphoreum* occurs in deep and cold waters, is somewhat resistant to iron starvation, as measured by the ability to grow in the presence of ethylenediamine-di-*o*-hydroxyphenyl acetic acid (EDDA), and grows from 4 to 25°C (23, 24, 29). *P. leiognathi* occurs in the Gulf of Elat, the Indo-West Pacific, and the Caribbean Sea, is resistant to iron starvation and sensitive to light, is not halotolerant, and grows from 25 to 30°C (28, 36; P. Dunlap, Abstr. West. Soc. Nat., 1981; D. H. Cohn, personal communication). The species of the *Vibrio harveyi* group (*V. harveyi, V. splendida, V. orientalis*) are widely distributed in the upper 100 m in warm coastal and warm open ocean waters, and in hypersaline lagoons and estuaries. *V. harveyi* is resistant to iron starvation, is light tolerant and halotolerant, and grows from 15 to 40°C (2, 21, 23, 26, 28, 35, 36). The species of the *V. fischeri* group (*V. fischeri* and *V. logei*) occur in cool coastal waters. *V. fischeri* is less resistant to iron starvation than *V. harveyi* and *P. leiognathi*, is sensitive to light, is not halotolerant, and grows from 4 to 30°C (24, 26, 28, 29, 36). The taxonomic methods used in these studies have not always been adequate to distinguish all of the species, and there may be as yet undefined differences between species included in "groups." Also, strains within species may vary in physiology (e.g., salt or temperature tolerance) and thus in geographical or seasonal occurrence (28). In addition, not enough environments have been carefully studied to fully support these generalizations.

Most of the distributional studies cited above have involved the isolation and characterization in the laboratory of planktonic luminous bacteria. These efforts have led to many insights and speculations as to what factors control the distribution of luminous bacteria, but the ecological interactions of these bacteria within their various habitats are undoubtedly more complex than suggested by our current knowledge. For instance, probably all luminous species engage in symbiotic associations of one or more kinds. There are two major classes: the enteric symbioses, which are relatively non-species specific, and the light organ symbioses, which are highly specific; e.g., the light organs of a given species of fish always contain the same species of bacteria. All of the symbioses occur in anatomical structures that allow release of bacteria into seawater. Therefore, the symbioses may maintain species in plankton or be in turn populated by bacteria which proliferate or are subject to selection in the planktonic habitat. The *V. harveyi* group, the *V. fischeri* group, and both *Photobacterium* groups have been found in gut tracts of fishes and crustaceans, sometimes as the dominant organism (21, 24; D. H. Cohn, personal communication). Generally, enteric populations reflect planktonic populations, although Ruby and Morin (24) found *P. phosphoreum* dominating in gut of *Chromis punctipinnus* when *V. fischeri* dominated in water off California (21, 24; D. H. Cohn, personal communication). The dynamics of the ecology of luminous bacteria are thus very complex, and symbioses may play an important role.

Three species of luminous bacteria are known to occur in light organ symbioses (19). There is a broad correlation between the temperature optima of the bacteria and the temperature of the habitat of the fish hosts. *P. phosphoreum* occurs in rectal–anal light organs in fishes living in deep and cold environments, a pattern consistent with its planktonic distribution. In fact, it has been proposed that planktonic *P. phosphoreum* cells are escaped symbionts (23, 29). The warm-water species *P. leiognathi* occurs in light organs located in a more anterior portion of the gut. The hosts, the Leiognathidae, are tropical fishes common in the Indo-West Pacific. The Monocentridae, hosts to the cool-water species *V. fischeri*, occur in temperate regions and are most abundant in Japan and Australia.

Because the various symbioses are quite diverse physiologically, we will present here a discussion of only one, keeping in mind that, although the principles may apply to other symbioses, the details will surely be quite different. The system includes the bacterium *V. fischeri* and the luminous fish *Monocentris japonicus*, which possesses two external light organs consisting of tubules containing bacteria and connected to the exterior via pores (25, 30).

. The means by which the specificity of these associations is established and maintained are unknown. In most cases very little is known about the development of light organs in larval forms of symbiotically luminous fishes. Possible sources for bacteria include direct transmission of bacteria from parent to egg, behaviorally mediated acquisition of bacteria from adult fish, or colonization of the light organ by planktonic luminous bacteria. In *M. japonicus* the first is unlikely, as eggs stripped from a gravid female are sterile. In addition, *M. japonicus* larvae do not emit light and have no light organs for at least the first 3 weeks of life (34). There is no evidence regarding the second possibility. The third seems plausible, as the symbiotic bacterial species are likely to occur in the environment of the host. This mechanism implies a selective interaction between host and symbiont, the basis for which is unknown. Although temperature selection of planktonic luminous bacteria may eliminate some possible symbionts, temperature selection alone is clearly insufficient as a mechanism.

In the symbioses involving external light organs, which can release bacteria directly into surrounding seawater, and perhaps also in the case of internal light organs which empty into the gut, we hypothesized that growth rates of bacteria in the light organ should be slow to minimize loss of excess symbionts from the light organ. Our experiments have shown that this appears to be the case in *M. japonicus* and two anomalopids (M. G. Haygood, B. M. Tebo, and K. H. Nealson, Mar. Biol., in press).

In these experiments, fish were starved to reduce contamination by enteric symbionts, washed, and placed in a container of sterile seawater. We measured release of light organ symbionts into the seawater by one or more of the following means: viable counts, particle (Coulter) counts, and epifluorescence. We measured the population size in the light organs and estimated a doubling time based on the time in which a number of bacteria equal to the light organ population were released (cells per fish divided by release rate). *M. japonicus* has approximately 1.5×10^8 *V. fischeri* cells in its light organs. We measured release rates ranging from 10^6 to 20×10^6 cells per h, which yielded estimated doubling times of 7.5 to 135 h. The doubling times estimated for *V. fischeri* released from *M. japonicus* were 10 to 100 times longer than those typically seen in batch culture in complete medium in the laboratory (45 min).

The mechanism or mechanisms responsible for limiting growth of the bacteria in the light organ must also be compatible with maximizing luminescence. In autoinduction, the bacteria produce a small molecule which accumulates in the medium and induces the luminescence system (16). Autoinduction occurs in *V. harveyi*, *V. fischeri*, *P. phosphoreum*, and *P. leiognathi* (16, 22). Autoinducer has been purified and synthesized in *V. fischeri* (8). Catabolite repression occurs in *V. harveyi* (K. H. Nealson, Ph.D. thesis, University of Chicago, Chicago, Ill., 1969), and in *V. fischeri* it occurs in a phosphate-limited chemostat (9); *P. phosphoreum* and *P. leiognathi* have not been examined for this trait. Oxygen promotes luminescence in *V. harveyi* and *P. leiognathi*, but low levels of oxygen result in maximal induction of luciferase, the light-producing enzyme, in *P. phosphoreum*, and reduced oxygen has been proposed as a symbiotic controlling mechanism (17, 18). Low osmolarity promotes luminescence in *V. harveyi* and *P. leiognathi* but not *V. fischeri* or *P. phosphoreum* (14; Dunlap, Abstr. West. Soc.

Nat., 1981). Iron repression of luminescence occurs, in various degrees, in all four species (M. G. Haygood, Abstr. Annu. Meet. Am. Soc. Microbiol. 1983, K182, p. 207; unpublished data). A physiological mechanism regulating the *M. japonicus–V. fischeri* symbiosis should maximize luminescence while slowing growth to within the range of values we measured. Of the factors described above, three appear to have these properties: oxygen limitation, low osmolarity, and iron limitation. All three can alter the growth rate of *V. fischeri* with either positive or neutral effects on luminescence. As a required nutrient, iron can actually limit growth, which should be directly related to the rate of iron supply, and the results presented below suggest that iron limitation is a good candidate for regulating the *M. japonicus–V. fischeri* symbiosis.

We decided to examine iron limitation as a candidate for control of growth and luminescence in *M. japonicus* because limitation of iron availability is a well-known mechanism of control of bacterial growth in animals (32, 33). Thus, fishes may be preadapted to use limitation of iron availability to control bacterial symbionts. Iron is required for growth of all bacteria with the possible exception of lactobacilli (20). Animals limit iron availability in blood and tissues by means of iron-binding proteins such as transferrins (which are found in fishes [27]), lactoferrin, ovo-transferrin (conalbumin), haptoglobin, etc. (7, 32, 33).

The following observations provide general evidence of the importance of this mechanism in other systems: (i) iron reverses the bacteriostatic effect of serum (4); (ii) iron overload promotes proliferation of bacteria in animals and may render less virulent strains more virulent (27); (iii) bacterial virulence is sometimes correlated with ability to grow at low iron or in the presence of chelating agents (this is due in some cases to ability to produce a siderophore) (5). This last has been shown for *V. anguillarum*, a fish pathogen (6, 31).

We thus performed experiments to determine the effect of iron on growth and luminescence of *V. fischeri*. When *V. fischeri* was grown in a series of minimal media in each of which N, P, C, or Fe was reduced, iron starvation stopped growth with markedly little effect on luminescence, especially when compared with the effects of N, P, or C starvation (Fig. 1). In a minimal medium, addition of iron caused a delay in induction (Fig. 2). In a complete seawater medium, iron can be sequestered by using EDDA, a specific iron chelator; such an experiment (Fig. 3) resulted in premature induction of luciferase. As shown in Fig. 4, this effect is at the level of luciferase synthesis. Thus, low levels of iron result in maximal luminescence per

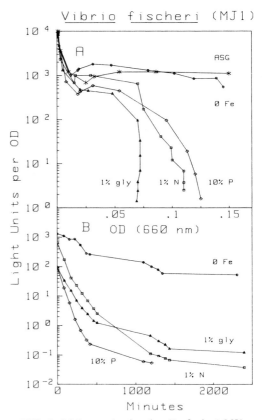

FIG. 1. Light production by *V. fischeri* MJ1, a strain isolated from *M. japonicus* (25). Light production was determined in a minimal medium (ASG: per liter, 15.5 g of NaCl, 0.75 g of KCl, 12.35 g of $MgSO_4 \cdot 7H_2O$, 2.9 g of $CaCl_2 \cdot 2H_2O$, 1 g of NH_4Cl, 0.1 g of glycerophosphate, 3 ml of glycerol, 10 mM N-2-hydroxyethylpiperazine-N'-2-ethanesulfonic acid [HEPES], pH 7.4, 2 mM $NaHCO_3$, and 16 mg of ferric ammonium citrate unless otherwise noted; HEPES, $NaHCO_3$, and ferric ammonium citrate were added as $100\times$ stock solutions after autoclaving) and in ASG with: no added iron, 10% of normal P, 1% of normal N, and 1% of normal glycerol, as indicated. ASG with 16 µg of ferric ammonium citrate per ml was used for maintenance (plates) and growth of inocula (liquid overnight cultures). Seven-milliliter cultures were inoculated with 35 µl of an overnight culture and grown in acid-washed glass tubes (18 by 150 mm) capped with plastic Morton closures at 20°C on a shaker at 80 to 100 rpm. Growth was measured as optical density (OD) at 660 nm on a Spectronic 20 spectrophotometer. (A) Light units per OD as a function of OD. (B) Light units per OD as a function of time after cessation of growth. Light was measured with an EMI type 9781A phototube and Pacific Photometrics model 110 amplifier. All light measurements are normalized.

cell. It seems clear that, if *M. japonicus* can limit iron availability in the light organ, it can use this as a mechanism to reduce the bacterial growth

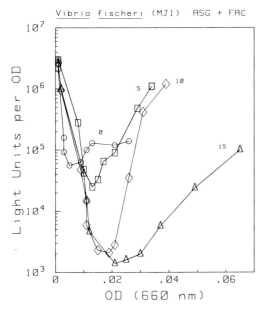

Vibrio fischeri (MJ1) ASG + FAC

FIG. 2. Light production by *V. fischeri* MJ1 grown in new, HCl-washed test tubes (18 by 150 mm) in ASG with 0, 5, 10, and 15 μg of ferric ammonium citrate per ml as indicated.

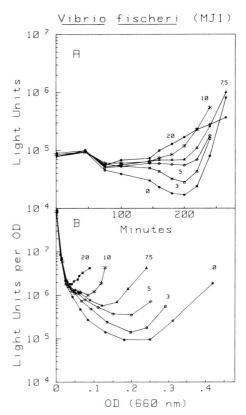

Vibrio fischeri (MJ1)

FIG. 3. Growth and light production by *V. fischeri* MJ1 grown in a complete medium (SWC) with: 0, 3, 5, 7.5, 10, and 20 μM EDDA as indicated (SWC: 75% seawater containing, per liter, 5 g of peptone, 3 g of yeast extract, and 3 ml of glycerol). (A) Light as a function of time. (B) Light per OD as a function of OD.

rate while maximizing luminescence. Thus, the physiological interactions that maintain this symbiosis may be different or more complex than previously imagined (17).

Reduced iron availability affects the various luminous bacteria differently and may, in turn, influence their ability to colonize specific habitats successfully. As described by Makemson and Hastings (15), iron represses synthesis of the luminescence system in *V. harveyi* much as it does in *V. fischeri*. Both *V. harveyi* and *V. fischeri* induce luminescence relatively late, in mid-exponential phase, perhaps as a result of their sensitivity to iron repression. By contrast, some strains of *P. phosphoreum* and *P. leiognathi* induce luminescence relatively early (optical density 0.1 or less), making it difficult to judge whether iron repression is occurring. *P. phosphoreum* strain NZ1 seems to show some iron repression effect, whereas two strains of *P. leiognathi* (PL741 and PL721A) induce very early and show little iron repression, and two others, PL840 and PL721E, induce later and show a marked iron repression. These different responses of growth and luminescence to iron may be of importance in the ecology of these bacteria, affecting their occurrence or activity in the various environments they inhabit. It would be desirable to correlate relative iron availability in seawater with the occurrence of the different species of luminous bacteria.

CONCLUSION

Several factors, including nutrient types and amounts, autoinducer, oxygen concentration, osmolarity of the medium, and iron concentration can affect the synthesis and activity of bioluminescence in marine bacteria. Understanding these factors and their interactions is central to the eventual elucidation of the role of light emission for the various bacterial species in their diverse habitats. Here we have focused on the effects of iron limitation and excess on bioluminescence of *V. fischeri*. Iron limitation provides a mechanism whereby symbiotic associations might limit growth and maximize bioluminescence. Since iron limitation results in enhanced luminescence and growth limitation to some degree in all luminous bacteria tested, this factor must be considered in any treatment of the ecology of this group. For the future, it would be desirable to correlate the occurrence and luminous activities of the different bacterial

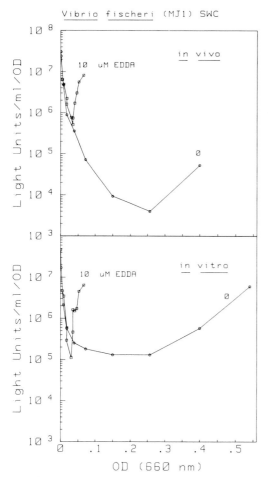

FIG. 4. In vivo light emission and in vitro luciferase in *V. fischeri* MJ1 grown in 100 ml of SWC in 300-ml nephelo flasks with 0 or 10 μM EDDA. One-milliliter samples were removed for light and luciferase measurements (11).

species with the iron availability in the different habitats. Also, since the light organs of the symbiotically luminous fishes may be continuous cultures, it will be necessary to study the effects of iron limitation in the chemostat, comparing this to limitation of other nutrients.

A number of other areas of research seem particularly important in understanding the ecology of the luminous bacteria. The means by which host fishes acquire and maintain a specific bacterial symbiont has received very little attention. The physiology of *Alteromonas hanedai* and its relative similarity to other luminous bacteria remains unknown. The enteric luminous symbionts of marine animals may play a key role in cycling luminous bacteria in and out of the plankton and light organ symbioses. Only a few species of fishes have been examined, and

the contribution of enteric symbioses to planktonic luminous bacteria remains a subject for speculation. Finally, although several excellent studies have been performed on distribution and seasonal occurrence of luminous bacteria, we can confidently discuss planktonic luminous bacteria in only a few environments; in this respect, most of the world's oceans remain unexplored.

This work was supported by Office of Naval Research contract N00014-80C-0066 to K. H. Nealson.

LITERATURE CITED

1. **Baumann, P., L. Baumann, S. S. Bang, and M. J. Woolkalis.** 1980. Reevaluation of the taxonomy of *Vibrio*, *Beneckea*, and *Photobacterium*: abolition of the genus *Beneckea*. Curr. Microbiol. **4**:127–132.
2. **Beijerinck, M. W.** 1916. Die Leuchtbacterien der Nordsee im August und September. Folia Microbiol. Delft **4**:15–40.
3. **Booth, C., and K. H. Nealson.** 1975. Bacterial bioluminescence in the open ocean. Biophys. J. **15**:53.
4. **Bullen, J. J., N. J. Rogers, and E. Griffiths.** 1974. Bacterial iron metabolism in infection and immunity, p. 517–551. *In* J. B. Neilands (ed.), Microbial iron metabolism: a comprehensive treatise. Academic Press, Inc., New York.
5. **Cox, C. D.** 1982. Effect of pyochelin on the virulence of *Pseudomonas aeruginosa*. Infect. Immun. **36**:17–23.
6. **Crosa, J. H.** 1980. A plasmid associated with virulence in the marine fish pathogen *Vibrio anguillarum* specifies an iron-sequestering system. Nature (London) **284**:566–568.
7. **Eaton, J. W., P. Brandt, J. R. Mahoney, and J. T. Lee, Jr.** 1982. Haptoglobin: a natural bacteriostat. Science **215**:691–692.
8. **Eberhard, A., A. L. Burlingame, C. Eberhard, G. L. Kenyon, K. H. Nealson, and N. J. Opperheimer.** 1981. Structural identification of autoinducer of *Photobacterium fischeri* luciferase. Biochemistry **20**:2444–2449.
9. **Friedrich, W. F., and E. P. Greenberg.** 1983. Glucose repression of luminescence and luciferase in *Vibrio fischeri*. Arch. Microbiol. **134**:87–91.
10. **Harvey, E. N.** 1952. Bioluminescence. Academic Press, Inc., New York.
11. **Hastings, J. W., T. O. Baldwin, and M. Z. Nicoli.** 1978. Bacterial luciferase: assay, purification and properties. Methods Enzymol. **57**:135–152.
12. **Hershberger, W. K., and G. A. Pratschner.** 1981. Iron-binding capacities of six transferrin phenotypes of coho salmon and potential fish health applications. Prog. Fish Cult. **43**:27–31.
13. **Jensen, M. J., B. M. Tebo, P. Baumann, M. Mandel, and K. H. Nealson.** 1980. Characterization of *Alteromonas hanedai* (sp. nov.), a nonfermentative luminous species of marine origin. Curr. Microbiol. **3**:311–315.
14. **Kossler, F.** 1970. Untersuchungen zur Salztoleranz der Atmung und Lichtemission von *Vibrio luminosus*. Z. Allg. Mikrobiol. **10**:615–625.
15. **Makemson, J. C., and J. W. Hastings.** 1982. Iron represses bioluminescence and affects catabolite repression in *Vibrio harveyi*. Curr. Microbiol. **7**:181–186.
16. **Nealson, K. H.** 1977. Autoinduction of bacterial luciferase: occurrence, mechanism and significance. Arch. Microbiol. **112**:73–79.
17. **Nealson, K. H.** 1979. Alternative strategies of symbiosis of marine luminous fishes harboring light emitting bacteria. Trends Biochem. Sci. **4**:105–110.
18. **Nealson, K. H., and J. W. Hastings.** 1977. Low oxygen is optimal for luciferase synthesis in some bacteria: ecological implications. Arch. Microbiol. **112**:9–16.
19. **Nealson, K. H., and J. W. Hastings.** 1979. Bacterial bioluminescence. Microbiol. Rev. **43**:496–518.
20. **Neilands, J. B.** 1974. Iron and its role in microbial physiol-

ogy, p. 3–34. *In* J. B. Neilands (ed.), Microbial iron metabolism: a comprehensive treatise. Academic Press, Inc., New York.

21. **O'Brien, C. H., and R. K. Sizemore.** 1979. Distribution of the luminous bacterium *Beneckea harveyi* in a semitropical estuarine environment. Appl. Environ. Microbiol. **38:**928–933.

22. **Rosson, R. A., and K. H. Nealson.** 1981. Autoinduction of bacterial bioluminescence in a carbon-limited chemostat. Arch. Microbiol. **129:**299–304.

23. **Ruby, E. G., E. P. Greenberg, and J. W. Hastings.** 1980. Planktonic marine luminous bacteria: species distribution in the water column. Appl. Environ. Microbiol. **39:**302–306.

24. **Ruby, E. G., and J. G. Morin.** 1979. Luminous enteric bacteria of marine fishes: a study of their distribution, densities, and dispersion. Appl. Environ. Microbiol. **38:**406–411.

25. **Ruby, E. G., and K. H. Nealson.** 1976. Symbiotic association of *Photobacterium fischeri* with the marine luminous fish *Monocentris japonica*: a model of symbiosis based on bacterial studies. Biol. Bull. **151:**574–586.

26. **Ruby, E. G., and K. H. Nealson.** 1978. Seasonal changes in the species composition of luminous bacteria in nearshore seawater. Limnol. Oceanogr. **23:**530–533.

27. **Sawatzki, G., F. A. Hoffman, and B. Kubanek.** 1983. Acute iron overload in mice: pathogenesis of *Salmonella typhimurium* infection. Infect. Immun. **39:**659–665.

28. **Shilo, M., and T. Yetinson.** 1979. Physiological characteristics underlying the distribution patterns of luminous bacteria in the Mediterranean Sea and the Gulf of Elat. Appl. Environ. Microbiol. **38:**577–584.

29. **Singleton, R. J., and T. M. Skerman.** 1973. A taxonomic study by computer analysis of marine bacteria from New Zealand waters. J. R. Soc. N.Z. **3:**129–140.

30. **Tebo, B. M., D. S. Linthicum, and K. H. Nealson.** 1979. Luminous bacteria and light-emitting fish: ultrastructure of the symbiosis. Biosystems **11:**269–280.

31. **Toranzo, A. E., J. L. Barja, S. A. Potter, R. R. Colwell, F. M. Hetrick, and J. H. Crosa.** 1983. Molecular factors associated with virulence of marine vibrios isolated from striped bass in Chesapeake Bay. Infect. Immun. **39:**1220–1227.

32. **Weinberg, E. D.** 1974. Iron and susceptibility to infectious disease. Science **184:**952–956.

33. **Weinberg, E. D.** 1978. Iron and infection. Microbiol. Rev. **42:**45–66.

34. **Yamada, K., M. G. Haygood, and H. Kabasawa.** 1979. On fertilization and early development in the pinecone fish, *Monocentris japonicus*. Annu. Rep. Keikyu Aburatsubo Marine Park Aquarium **10:**31–38.

35. **Yang, Y., L. P. Yeh, Y. Cao, L. Baumann, P. Baumann, J. S. Tang, and B. Beaman.** 1983. Characterization of marine luminous bacteria isolated off the coast of China and description of *Vibrio orientalis* sp. nov. Curr. Microbiol. **8:**95–100.

36. **Yetinson, T., and M. Shilo.** 1979. Seasonal and geographic distribution of luminous bacteria in the eastern Mediterranean Sea and the Gulf of Elat. Appl. Environ. Microbiol. **37:**1230–1238.

Genetic Adaptation to the Environment

The Role of Plasmids in Microbial Ecology

JOHN E. BERINGER AND PENNY R. HIRSCH

Soil Microbiology Department, Rothamsted Experimental Station, Harpenden, Hertfordshire AL5 2JQ, United Kingdom

Plasmids are circular DNA molecules which replicate as separate entities to the host chromosome(s). They are almost ubiquitous in bacteria and, although they may sometimes carry genes coding for extremely important functions, are recognized as being dispensable because there is very little evidence to suggest that in nature any plasmids carry genes whose function is essential under all conditions of growth. In bacteria plasmids range in size from about 0.05% of the chromosome (plasmid 15 of molecular weight 1.5×10^6) (10) to about 20% or possibly more (12, 29, 60, 63). A plasmid such as 15 has the capacity to code for only about three genes, whereas the very large ones found in many soil bacteria can code for 200 or more. Microorganisms may contain a number of different plasmids; for example, in *Rhizobium* seven plasmids have been observed which among them could carry about half the genetic information present on the chromosome of *Escherichia coli* (7). The intention of this review is to discuss the role of plasmids in providing genetic diversity to enable microorganisms to adapt and take advantage of a wide range of ecological niches.

RELATIONSHIP BETWEEN PLASMIDS AND OTHER EXTRACHROMOSOMAL GENETIC ELEMENTS

Brinton (9) recently compared plasmids and bacteriophages and came to the opinion that they were so similar that they should be treated as one group of genetic elements, whose role is self-replication. Phages do this through infecting bacteria and utilizing the infected host to replicate their DNA. Virulent phages kill the host. Temperate phages usually do not kill the host, and a single copy of their DNA is replicated as part of the genome of the host, usually integrated within the host chromosome, though with some phages, such as P1 (37), as a self-replicating extrachromosomal circular DNA molecule. Occasionally, temperate phages enter a virulent cycle and produce phage particles, killing the host. However, in general they are beneficial, as the host becomes immune to related phages and may also benefit from the functioning of other genes carried on the phage genome. Perhaps the best example of the latter is the acquisition of drug resistance when bacteria become infected with P1 phages that carry kanamycin or chloramphenicol resistance (24).

Plasmids are usually identified after lysing microorganisms under conditions that are designed to conserve covalently closed circular DNA molecules (plasmids) and looking for their presence after various methods of physical partitioning to separate linear from closed circular DNA. Almost by definition all covalently closed circular DNA molecules are called plasmids, though the chromosome is also included in this definition. Indeed, a major problem in genetic studies is finding techniques for isolating DNA that will separate plasmids, which may be about half the size of the chromosome itself. For small plasmids this is simple because the chromosome is broken up during the isolation procedures and thus can be separated readily. In some cases a given property, or properties, of a strain is known to be correlated with the presence or absence of a plasmid; when no characteristics are known, the plasmid is "cryptic." Even for those plasmids which code for a number of known functions, it is unusual to know the function of most of the DNA.

A further problem in clarifying the role of plasmids is that, contrary to widely held belief, it is often extremely difficult or impossible to eliminate plasmids or to transfer them to other hosts. Thus, an identification of genes carried on them can be achieved only by mutagenesis and mapping of mutations to plasmids. Furthermore, if they cannot be eliminated, it is uncertain whether they carry genes whose function is essential to the viability of the strain carrying them.

Little is known about the presence of extrachromosomal DNA in eucaryotic microorganisms, such as fungi. Strictly speaking, DNA in mitochondria, chloroplasts, and organelles is extrachromosomal, though it should not be considered as being the same as plasmid DNA in procaryotes (10, 54). There is only one well-characterized plasmid in a eucaryote, the 2μ plasmid of yeasts (26), though circular DNA molecules, which are possibly plasmids, have been reported in *Fusarium oxysporum* (25) and *Podospora anserina* (68). It is not clear whether the 2μ plasmid is beneficial to yeasts, though a role in conferring drug resistance has been reported (26).

Extrachromosomal DNA and RNA have been observed in a number of fungi and protozoa to be associated with viruses and virus-like particles (52, 53). Bacteria-like organelles have also been observed in a number of eucaryotic microorganisms, and some of these have been shown to contain DNA (55, 67). Although we know a fair amount about the killer particles in fungi and *Paramecium*, little is known about many of the other organelles which contain DNA. Because so much more is known about plasmids in procaryotes, their role in providing the extraordinary genetic variability that is found in these microorganisms will be the subject of the remainder of this review.

PLASMIDS AND CHROMOSOME REARRANGEMENTS

The reassortment of chromosomal genes that follows from gene exchange between organisms has long been recognized as an extremely important factor in the development of new types of organisms which are better adapted to particular environmental conditions. Bacteria can exchange DNA by four main processes: (i) conjugation, in which bacteria form mating pairs; (ii) transduction, in which DNA is carried from one bacterium to another by bacteriophages; (iii) transformation, in which DNA present in the medium is taken up into the cell; and (iv) protoplast fusion, in which protoplasted bacteria fuse. Protoplast fusion is the only procedure that routinely leads to true diploid formation before segregation occurs, resulting in the production of haploid recombinants. In nature it is unlikely that transformation or protoplast fusion is sufficiently common to provide an important means of genetic exchange for the great majority of microorganisms of which we are aware.

If one accepts that plasmids and the DNA of bacteriophage are analogous, both transduction and conjugation can be considered to be dependent upon "plasmids": transduction because the microbial DNA is transferred between strains as a result of its being incorporated in phage heads in place of phage DNA, and conjugation because pair formation and gene exchange in all the well-studied conjugation systems are mediated by plasmids. Bacteriophage heads are relatively small compared to the size of a bacterial chromosome, and thus it is unusual for as much as 1% of the genome to be transferred by transduction. Just how significant gene transfer by transduction is in the generation of genetic diversity in nature is unknown.

Plasmids have an extremely important role in promoting conjugation because some types (conjugative plasmids) code for transfer genes which make the host cells into donors which pair with other bacteria (usually of the same species) and transfer the plasmid. Chromosome transfer almost always occurs much less frequently than plasmid transfer (about 10^{-4} to 10^{-8} per plasmid transferred) (32), which means that plasmids which transfer at frequencies of 10^{-2} per donor cell or greater have enormous potential to bring about chromosome reassortments in the species involved. Many different plasmids isolated in nature have been shown to mobilize chromosomal DNA. Some have specific regions which make them much more efficient at mobilizing chromosomal DNA than the majority of plasmids that have been described (28, 31), R68.45 being an example of such a plasmid (28). The potential of plasmids for reassorting the gene pools of large populations of microorganisms is shown in Table 1, which demonstrates the potential of plasmids to generate genetic diversity. Large populations are needed because mating requires fairly dense cultures (ideally about 10^8 bacteria per ml) to optimize the chance of bacteria being sufficiently close to each other to form mating pairs. A good example of such an environment where transfer has been observed is the intestine (66).

Table 1 includes estimates of chromosome mobilization frequencies because this important role of bacterial plasmids is often ignored. The estimates are based on observed frequencies at which chromosomal genes are transferred and integrated in the recipient chromosome in crosses between different strains within a single species. However, it is clear that certain plasmids are freely transmissible among gram-negative bacteria and can mobilize chromosomal DNA between different species and even genera (17, 31, 58). This DNA is not able to integrate into the chromosome of the new host, but can be inherited if it is stably integrated within the plasmid which carries it. Such plasmids are called plasmid "primes" and have been made and used to study gene expression in different host backgrounds (44). Because plasmid primes are generated during "normal" matings between bacteria in vitro, it is reasonable to assume that they occur in nature. Just how common they are, and how important they have been in the development of genetic diversity in microorganisms, remains to be determined. A plasmid prime carrying a leucine metabolic gene has been isolated from a *Pseudomonas aeruginosa* strain which must have obtained it in nature before being isolated in the laboratory (33).

TABLE 1. Potential for gene transfer mediated by plasmids in rhizobia[a]

Determination	Expected frequencies of plasmid transfer		
	10^{-2}	10^{-6}	10^{-10}
Total no. of matings	10^{13}	10^{9}	10^{5}
No. of plasmid genes transferred	10^{15}	10^{11}	10^{7}
No. of matings in which chromosomal DNA is transferred			
Assuming frequency is 10^{-4} per plasmid transferred	10^{9}	10^{5}	10
Assuming frequency is 10^{-6} per plasmid transferred	10^{7}	10^{3}	10^{-1}
Assuming frequency is 10^{-8} per plasmid transferred	10^{5}	10	10^{-3}
No. of chromosomal genes transferred			
Assuming frequency is 10^{-4} per plasmid transferred	3×10^{11}	3×10^{7}	3×10^{4}
Assuming frequency is 10^{-6} per plasmid transferred	3×10^{9}	3×10^{5}	3×10^{2}
Assuming frequency is 10^{-8} per plasmid transferred	3×10^{7}	3×10^{3}	3×10^{-1}

[a] This table lists frequencies of plasmid transfer that occur between rhizobia (10^{-2}, 10^{-6}, 10^{-10}) and shows how many matings between rhizobia could be expected to occur per square kilometer of a pure crop of a legume. It also indicates the amount of chromosome transfer that could occur and gives an indication of the number of genes that can be transferred. Estimates are based on the following assumptions: (i) that there are 10^{6} legumes per km^{2} and 10^{9} rhizobia per legume (10^{15} rhizobia per km^{2}); (ii) that the *Rhizobium* chromosome carries about 3,000 genes; (iii) that the "average" sized plasmid is 1/30 the size of the chromosome; and (iv) that the "average" sized fragment of chromosomal DNA transferred is 1/10 of the total chromosome. No assumption is made about the frequency at which bacteria can mate with each other in the soil. In the laboratory pair formation and plasmid transfer seldom take as long as 30 min.

PLASMIDS AND TAXONOMY

The concept of a species is based on the idea that this is a fairly clearly defined group of organisms which is genetically separable from other groups of similar organisms. From an evolutionary point of view a species is a group of organisms that has evolved to exploit particular environments, and thus genes involved in determining the speciation are those that are particularly important for the ecology of the organisms. Classification to the species and variety level is extremely important in medicine as it allows the microbiologist to predict whether a particular microorganism is likely to be a serious pathogen. Likewise in agriculture when species of *Rhizobium* are being used to inoculate legume plants, it is important to know which species to choose for a particular host.

It is now clear that for some groups of microorganisms the taxonomic criteria used for determining species (and even for separation into genera) are determined by plasmid-borne genes and that these plasmids can be self-transmissible. "Well-known" examples are the hemolytic streptococci and members of the family *Rhizobiaceae*. Jacob et al. (38) have shown that the hemolytic and bacteriocin production properties of *Streptococcus faecalis* var. *zymogenes* are plasmid borne and that the self-transmissible plasmid carrying them is transferable to other *S. faecalis* strains. The number of other species and varieties of pathogens that are classified on similar transmissible functions is unknown.

The role of plasmids in determining the type of interaction with a host plant, and hence the classification of members of the *Rhizobiaceae*, has been studied in some detail and was reviewed recently by Beringer (6). This group of gram-negative bacteria is important because species of *Agrobacterium* are pathogens forming tumorous growths on dicotyledonous plants, and *Rhizobium* strains are beneficial because they form nitrogen-fixing nodules on leguminous plants. Agrobacteria are classified on the basis of their ability to form tumors (*A. tumefaciens*) or to induce prolific root growth (*A. rhizogenes*), or by their failure to induce any response when inoculated on plants (*A. radiobacter*). For *Rhizobium*, species identification is based on the range of legume hosts nodulated. For example, *R. trifolii* is the species which contains rhizobia that form nitrogen-fixing root nodules on clovers (*Trifolium* spp.), and *R. leguminosarum* strains form nitrogen-fixing nodules on peas *(Pisum)*, *Vicia*, *Lens*, and *Lathyrus*. Usually strains of one species do not nodulate host plants nodulated by another; they never fix nitrogen on an "incorrect" host. The ultimate criterion of the taxonomic status of members of the *Rhizobiaceae* is the response that they elicit on appropriate host plant species. However, it has been convincingly demonstrated that the ability of *R. trifolii*, *R. leguminosarum*, *A. tumefaciens*, and *A. rhizogenes* to interact with plants is determined by plasmid-borne genes. Transfer of the plasmids among these bacteria is directly correlated with transfer of host-range determination (6–8, 16, 34, 36, 43, 45). However, it should be stated that although transfer of the *Agrobacterium* tumor-forming plasmids to *Rhizobium* converts these bacteria into tumor-forming organisms, transfer of plasmids to *Agrobacterium* produces bacteria that nodulate but do not fix nitrogen. Thus, to be a nitrogen-fixing strain of *Rhizobium* it is apparently necessary to have other genes, either in the chromosome or on other plasmids. However, it may be the ability to induce and regulate these genes which is important.

Rhizobium was chosen to illustrate the potential of plasmid transfer for genetic rearrangements in nature in Table 1 because the first host-range plasmid in *Rhizobium* to be well characterized came from a field isolate and had a transfer frequency of about 10^{-2} (8). Not all plasmids conferring host-range properties in the *Rhizobiaceae* are known to be self-transmissible or to have such high natural frequencies of transfer (8, 35, 45). Indeed, in *A. tumefaciens*, plasmids which carry genes involved in tumor formation have barely detectable transfer frequencies in vitro. However, if opines, which are organic molecules produced by the plant in response to infection by *Agrobacter* strains, are provided, transfer is greatly stimulated (35).

The induction of opine synthesis in infected plants is due to the functioning of plasmid-borne genes which are transferred to the plant and become stably associated with nuclear DNA (51, 56). Thus, in *A. tumefaciens* plasmid DNA is critical to the development of the tumor which is responsible for the production of the opines. Opines have a further, and perhaps all important, role in the interaction because they act as carbon and nitrogen sources for agrobacteria carrying the appropriate tumor-inducing plasmid, but not for most other microorganisms that have been studied (59). The opines stimulate plasmid transfer to ensure that agrobacteria and presumably rhizobia (and even conceivably certain other gram-negative bacteria present) carry the appropriate genes to benefit from their synthesis. Thus, the tumor-inducing plasmids of *Agrobacterium* can be said to epitomize the concept of the ''selfish gene'' theory as expounded by Dawkins (19). They code for their own transfer and optimize their chances of survival by causing the bacteria to infect and transform plants so that opines can be produced to provide a niche which is specifically adapted for the survival of bacteria carrying the appropriate plasmid. It will be extremely interesting to see whether the few other bacterial species which can utilize opines carry the same catabolic genes as *A. tumefaciens* and indeed whether they also carry other genes which may have come from the Ti or Ri plasmids. This could readily be tested by looking for hybridization between the DNA isolated from these bacteria and fragments of Ti or Ri DNA. A positive result would provide very good evidence for the widespread dissemination of bacterial genes.

OTHER FUNCTIONS CODED FOR BY PLASMIDS

It is usual in reviews of this type to discuss the role of plasmids in conferring drug resistance, novel metabolic pathways, etc., in some detail and to stress their importance in ecology. Because we believe that for the majority of cases the value of such functions is self-evident, we have listed a range of functions known to be plasmid borne in Table 2 and will not discuss them in detail.

Perhaps the best known example of metabolic genes which are plasmid coded are those for the utilization of complex and novel substrates in *Pseudomonas* species. Classic examples are the plasmids which code for camphor and octanol utilization (CAM and OCT plasmids). Pseudomonads are interesting not only because of the extremely wide range of substrates that they can utilize as carbon and nitrogen sources, but also because they have been sufficiently well studied to enable us to learn something about the genetics of the pathways involved. Many of the catabolic genes have been shown to be plasmid borne, though some are known to be chromosomal (42, 71). However, the pseudomonads are sufficiently flexible genetically to have evolved systems of gene transfer which enable them to recombine plasmids carrying catabolic genes and to exchange specific sequences of DNA from the chromosome with plasmids (42). Thus, plasmids are used as carriers of genes and as flexible genetic elements, which enables them to reassort catabolic genes without the risk of damaging or losing essential metabolic genes resident on the chromosome.

It is clear from the list, which is far from complete, that an extremely wide range of gene functions has been shown to be plasmid borne. It is important to remember that the list is biased toward medically important and genetically well-studied bacterial species, which in general carry plasmids of molecular weight less than 100 $\times$ 10^6. As methods for isolating and purifying DNA become better, it is becoming clear that ''typical'' soil bacteria such as *Rhizobium* and *Pseudomonas* contain very large plasmids, whose presence we have only recently detected.

The observations that plasmids often carry insertion sequences and transposons (and indeed that many drug resistance genes found on plasmids are in transposons) indicate that interactions among plasmids are likely to be common and to lead to the dissemination of important genes among a wide range of genera and species (11, 13, 14, 47, 48). Indeed, it has recently been shown that genetic elements exist in *E. coli* which enhance the frequency with which plasmids recombine with each other (40).

TABLE 2. Plasmid-coded functions in
microorganisms

Function (references)	Specific examples
Resistances (18, 22, 27, 31, 49, 50, 69)	
Antibiotics	Penicillins, streptomycin, chloramphenicol
Heavy metals	Mercury, lead, cadmium
Bacteriophage	
Bacteriocins	
UV light	
Catabolism (5, 13, 20, 21, 60, 72, 73)	
Sugar utilization	Lactose, sucrose, raffinose
Opine utilization	Octopine, nopaline
Utilization of xenobiotics	Xylene, chlorobenzoates, toluene
Antagonism (1, 27, 38, 41, 46)	
Antibiotic production	
Bacteriocins	
Bacteriophage	
Toxins	Enterotoxins, plant toxins
Degradative enzymes	Hemolysins, pectinases
Gene transfer (10, 28, 31)	
Sex pili	
Transfer genes	
DNA mobilization functions	
Surface exclusion properties	
Interactions between organisms (6–8, 15, 16, 23, 34–36, 43, 45, 51, 56, 57, 65, 70)	
Host-range genes	*Rhizobium*
Tumor formation genes	*Agrobacterium*
Opine synthesis and utilization	*Agrobacterium*
Nitrogen fixation	*Rhizobium*
Auxin production	*Pseudomonas savastanoi*
Adhesion properties	K88 antigen
Susceptibility to bacteriocins and bacteriophage	
Miscellaneous (7, 39, 64)	
Restriction endonucleases	*Eco*RI
Modification enzymes	Methylases
DNA polymerase	
Pigments	Melanin in *Rhizobium*

Transposons, which are sequences of DNA containing genes for drug resistance or metabolic functions, are able to move from one DNA replicon to another as a result of the functioning of genes carried on them. They are well known because of the widespread occurrence of transposons in drug resistance plasmids, which of-ten have one or more of their resistance genes coded within transposons. The ecological significance of transposons in spreading genes around different genera is not well understood, though a major study of the occurrence of a transposon coding for ampicillin resistance (Tn*1*) in nature has been published (30). This work has shown that Tn*1* occurs all over the world in a range of different bacterial genera. Transposition from one DNA replicon to another occurs only when the two DNA replicons are present in the same cell. Thus, the widespread occurrence of Tn*1* indicates that gene transfer between bacteria is widespread (as expected) and that genes can travel around the world if there is a reasonable selection for their maintenance, as presumably occurs with Tn*1*. This survey of the distribution of Tn*1* in nature points to the type of work that could be done by ecologists who are interested in how different groups of microorganisms have evolved to be able to handle the same nutrient sources or resist particular antagonistic substances. The limited evidence we have to date suggests that the amount of gene transfer that can occur in nature is more than enough to provide a mechanism for rapid adaptation to change.

CONCLUDING REMARKS

In this review less emphasis than usual has been placed on describing plasmid-coded phenotypes, and an attempt has been made to stress the important role of plasmids as agents of gene transfer and as catalysts for genetic recombination. A number of reviews of the role of plasmids in evolution and ecology have been published (2, 3, 6, 18, 50, 61, 62, 69, 71). The survival of microorganisms in nature depends upon their ability to find suitable niches in which they can grow and survive. The definition of a suitable niche and hence relevant gene functions is not simple. At one extreme it is clear that *Pseudomonas* and *Penicillium* are very successful microorganisms; they are widespread in nature and thus are extremely adaptive. However, it is tempting to say that the bacteria which have recently been shown to be able to grow at 250°C (4) epitomize successful adaptation because they have found a niche in which competition with other organisms is effectively eliminated. Whether plasmids will be shown to be involved in any of the metabolic adaptations required in the latter case remains to be determined.

It is worth stressing the point that when the only genes carried on plasmids are those whose function is not essential to the viability of the host, plasmids will provide a reservoir of genes which can be reassorted, mutated, and experimented with. However, it should be stressed that inessential genes are carried on chromosomes and that these are just as susceptible to mutagenesis as plasmid-borne ones. Indeed, other authors in this volume review research which shows that evolutionary change can result from mutation in such genes. An obvious advantage of genes being plasmid borne is that there is greater potential for plasmids to transfer to other microorganisms and for plasmids to recombine with each other to produce new arrangements of genes. Another possible advantage is that the loss by an organism of an entire plasmid carrying unnecessary genes, when their function is no longer advantageous, is accomplished more easily than the accurate deletion of an equivalent amount of DNA from the chromosome. It is clear that for many genera of microorganisms plasmids are of little importance, as they are seldom found.

Perhaps the only way to decide whether it is advantageous to have plasmid-borne genes is to release bacteria carrying conjugative plasmids and follow the fate of whole plasmids or parts of plasmids. This can be done only when the plasmids in question are well known and fragments of known regions can be purified to use as hybridization probes against DNA prepared from isolates of other soil bacteria. If the plasmids involved carry drug resistance genes, it should be possible to follow the transfer of drug resistance in populations of microorganisms. However, as resistance can occur spontaneously and some bacteria are intrinsically resistant to particular antibodies, it would be essential to confirm that the actual DNA sequence confirming the resistance had been inherited. Experiments of this nature are fairly easy to set up, but require a large amount of work to isolate and screen representative populations of organisms.

Our knowledge of what functions are plasmid borne is colored by our interests as microbiologists and by our ignorance of the factors which are really important for survival in different habitats. For example, much is known about drug resistance plasmids in the *Enterobacteriaceae* because of the importance of antibiotics in medicine and because, in general, the plasmids involved are relatively small and have been amenable to various forms of genetic and physical manipulation for many years. On the other hand, almost nothing was known about either the presence or the functioning of the very large plasmids which are found in many soil bacteria until the recent development of suitable techniques for characterizing them. It is tempting to assume that these microorganisms, which may have between 10 and 50% of their DNA as plasmids, have transferred regions of essential "chromosomal" DNA onto plasmids and could thus be considered as procaryotes with more

than one chromosome. However, there is as yet no evidence that this has happened, which might imply that such microorganisms have the potential to be extremely adaptive through the functioning of a wide range of plasmid-borne genes. Once techniques become available to measure the "ecological competence" of microorganisms, it should become possible to look for genes which may be involved. Until that time we will be restricted to characterizing genes for metabolic pathways, drug resistance, and other functions of which we are aware and, occasionally, finding genes whose functions are unusual and may be important to the ability of that organism to exploit specific ecological niches. Perhaps the plasmid-borne genes in *Rhizobium* and *Agrobacterium*, which are involved in host–plant interactions and are the best known examples that we have as yet of the role of plasmids in carrying genes for the exploitation of a specific niche, may be found to be part of a large series of plasmid-borne functions which are of tremendous ecological advantage in a wide range of different genera.

LITERATURE CITED

1. **Akagawa, H., M. Okanishi, and A. H. Umezawa.** 1975. A plasmid involved in chloramphenicol production in *Streptomyces venezuelae*: evidence from genetic mapping. J. Gen. Microbiol. **90**:336–346.

2. **Anderson, E. S.** 1966. Possible importance of transfer factors in bacterial evolution. Nature (London) **209**:637.

3. **Anderson, E. S.** 1968. The ecology of transferable drug resistance in the enterobacteria. Annu. Rev. Microbiol. **22**:131–180.

4. **Baross, J. A., and J. W. Deming.** 1983. Growth of 'black smoker' bacteria at temperatures of at least 250°C. Nature (London) **303**:423–426.

5. **Benson, S., and J. Shapiro.** 1978. TOL is a broad-host-range plasmid. J. Bacteriol. **135**:278–280.

6. **Beringer, J. E.** 1982. The genetic determination of host-range in the Rhizobiaceae. Isr. J. Bot. **31**:89–93.

7. **Beynon, J. L., J. E. Beringer, and A. W. B. Johnston.** 1980. Plasmids and host range in *Rhizobium leguminosarum* and *Rhizobium phaseoli*. J. Gen. Microbiol. **120**:421–429.

8. **Brewin, N. J., J. E. Beringer, A. V. Buchanan-Wollaston, A. W. B. Johnston, and P. R. Hirsch.** 1980. Transfer of symbiotic genes with bacteriocinogenic plasmids in *Rhizobium leguminosarum*. J. Gen. Microbiol. **116**:261–270.

9. **Brinton, C. C.** 1971. The properties of sex pili, the viral nature of conjugal genetic transfer systems, and some possible approaches to the control of bacterial drug resistance. Crit. Rev. Microbiol. **1**:105–160.

10. **Broda, P.** 1979. Plasmids. W. H. Freeman, San Francisco.

11. **Calos, M. P., and J. H. Miller.** 1980. Transposable elements. Cell **20**:579–595.

12. **Casse, F., C. Boucher, J. S. Julliot, M. Michel, and J. Denarie.** 1979. Identification and characterization of large plasmids in *Rhizobium meliloti* using agarose gel electrophoresis. J. Gen. Microbiol. **113**:229–242.

13. **Chakrabarty, A. M., D. A. Friello, and L. H. Bopp.** 1978. Transposition of plasmid DNA segments specifying hydrocarbon degradation and their expression in various microorganisms. Proc. Natl. Acad. Sci. U.S.A. **75**:3109–3112.

14. **Cohen, S. N.** 1976. Transposable genetic elements and

plasmid evolution. Nature (London) **263:**731–738.

15. **Comai, L., and T. Kosuge.** 1980. Involvement of plasmid deoxyribonucleic acid in indoleacetic acid synthesis in *Pseudomonas savastanoi.* J. Bacteriol. **143:**950–957.

16. **Costantino, P., P. J. J. Hooykaas, H. den Dulk-Ras, and R. A. Schilperoort.** 1980. Tumor formation and rhizogenicity of *Agrobacterium rhizogenes* carrying Ti plasmids. Gene **11:**79–87.

17. **Datta, N., R. W. Hedges, E. J. Shaw, R. B. Sykes, and M. H. Richmond.** 1971. Properties of an R factor from *Pseudomonas aeruginosa.* J. Bacteriol. **108:**1244–1249.

18. **Davies, J., and D. I. Smith.** 1978. Plasmid-determined resistance to antimicrobial agents. Annu. Rev. Microbiol. **32:**469–518.

19. **Dawkins, R.** 1976. The selfish gene. Oxford University Press, Oxford.

20. **Fisher, P. R., J. Appleton, and J. M. Pemberton.** 1978. Isolation and characterization of the pesticide-degrading plasmid pJP1 from *Alcaligenes paradoxus.* J. Bacteriol. **135:**798–804.

21. **Friedrich, C. G., and B. Friedrich.** 1983. Regulation of hydrogenase formation is temperature sensitive and plasmid coded in *Alcaligenes eutrophus.* J. Bacteriol. **153:**176–181.

22. **Gaffney, D. F., T. J. Foster, and W. V. Shaw.** 1978. Chloramphenicol acetyltransferases determined by R plasmids from Gram-negative bacteria. J. Gen. Microbiol. **109:**351–358.

23. **Gantotti, B. V., S. S. Patil, and M. Mandel.** 1979. Apparent involvement of a plasmid in phaseotoxin production by *Pseudomonas phaseolicola.* Appl. Environ. Microbiol. **37:**511–516.

24. **Goldberg, R. B., R. A. Bender, and S. L. Streicher.** 1974. Direct selection for P1-sensitive mutants of enteric bacteria. J. Bacteriol. **118:**810–814.

25. **Guardiola, J., G. Grimaldi, P. Costantino, G. Micheli, and F. Cervone.** 1982. Loss of nitrofuran resistance in *Fasarium oxysporum* is correlated with loss of a 46.7 kb circular DNA molecule. J. Gen. Microbiol. **128:**2235–2242.

26. **Guerineau, M.** 1979. Plasmid DNA in yeast, p. 539–593. *In* P. A. Lemke (ed.), Viruses and plasmids in fungi. Marcel Dekker Inc., New York.

27. **Gyles, C. L., S. Palchaudhuri, and W. K. Maas.** 1977. Naturally occurring plasmid carrying genes for enterotoxin production and drug resistance. Science **198:**198–200.

28. **Haas, D., and B. W. Holloway.** 1978. Chromosome mobilization by the R plasmid R68.45: a tool in *Pseudomonas* genetics. Mol. Gen. Genet. **158:**229–237.

29. **Hansen, J. B., and R. H. Olsen.** 1978. Inc P2 group of *Pseudomonas,* a class of uniquely large plasmids. Nature (London) **274:**715–717.

30. **Heffron, F., R. Sublett, R. W. Hedges, A. Jacob, and S. Falkow.** 1975. Origin of the TEM beta-lactamase gene found on plasmids. J. Bacteriol. **122:**250–256.

31. **Hershfield, V.** 1979. Plasmids mediating multiple drug resistance in group B *Streptococcus:* transferability and molecular properties. Plasmid **2:**137–149.

32. **Holloway, B. W.** 1979. Plasmids that mobilize bacterial chromosome. Plasmid **2:**1–19.

33. **Holloway, B. W., V. Krishnapillai, and A. F. Morgan.** 1979. Chromosomal genetics of *Pseudomonas.* Microbiol. Rev. **43:**73–102.

34. **Hooykaas, P. J. J., P. M. Klapwijk, M. P. Nuti, R. A. Schilperoort, and A. Rorsch.** 1977. Transfer of the *Agrobacterium tumefaciens* Ti-plasmid to avirulent agrobacteria and *Rhizobium ex planta.* J. Gen. Microbiol. **98:**477–484.

35. **Hooykaas, P. J. J., C. Roobol, and R. A. Schilperoort.** 1979. Regulation of the transfer of Ti plasmids of *Agrobacterium tumefaciens.* J. Gen. Microbiol. **110:**99–109.

36. **Hooykaas, P. J. J., A. A. N. van Brussel, H. den Dulk-Ras, G. M. S. van Slogteren, and R. A. Schilperoort.** 1981. Sym plasmid of *Rhizobium trifolii* expressed in different rhizobial species and *Agrobacterium tumefaciens.* Nature (London) **291:**351–353.

37. **Ikeda, H., and J.-I. Tomizawa.** 1968. Prophage P1, an extrachromosomal replication unit. Cold Spring Harbor Symp. Quant. Biol. **33:**791–798.

38. **Jacob, A. E., G. J. Douglas, and S. J. Hobbs.** 1975. Self-transferable plasmids determining the hemolysin and bacteriocin of *Streptococcus faecalis* var. *zymogenes.* J. Bacteriol. **121:**863–872.

39. **Jacoby, G. A., and L. Sutton.** 1977. Restriction and modification determined by a *Pseudomonas* R plasmid. Plasmid **1:**115–116.

40. **James, A. A., P. T. Morrison, and R. Kolodner.** 1983. Isolation of genetic elements that increase frequencies of plasmid recombinants. Nature (London) **303:**256–259.

41. **Jamieson, A. F., R. L. Bieleski, and R. E. Mitchell.** 1981. Plasmids and phaseolotoxin production in *Pseudomonas syringae* pv. *phaseolicola.* J. Gen. Microbiol. **122:**161–165.

42. **Jeenes, D. J., and R. A. Williams.** 1982. Excision and integration of degradation pathway genes from TOL plasmid pWWO. J. Bacteriol. **150:**188–194.

43. **Johnston, A. W. B., J. L. Beynon, A. J. Buchanan-Wollaston, S. M. Setchell, P. R. Hirsch, and J. E. Beringer.** 1978. High frequency transfer of nodulating ability between strains and species of *Rhizobium.* Nature (London) **276:**634–636.

44. **Johnston, A. W. B., M. J. Bibb, and J. E. Beringer.** 1978. Tryptophan genes in *Rhizobium*—their organisation and their transfer to other bacterial genera. Mol. Gen. Genet. **165:**323–330.

45. **Kerr, A., P. Manigault, and J. Tempe.** 1977. Transfer of virulence *in vivo* and *in vitro* in *Agrobacterium.* Nature (London) **265:**560–561.

46. **Kirby, R., L. F. Wright, and D. A. Hopwood.** 1975. Plasmid-determined antibiotic synthesis and resistance in *Streptomyces coelicolor.* Nature (London) **254:**265–267.

47. **Kleckner, N., J. Roth, and D. Botstein.** 1977. Genetic engineering *in vivo* using translocatable drug resistance elements. New methods in bacterial genetics. J. Mol. Biol. **116:**125–159.

48. **Kopecko, D. J., J. Brevet, and S. N. Cohen.** 1976. Involvement of multiple translocating DNA segments and recombinational hotspots in the structural evolution of bacterial plasmids. J. Mol. Biol. **108:**333–360.

49. **Krishnapillai, V.** 1975. Resistance to ultraviolet light and enhanced mutagenesis conferred by *Pseudomonas aeruginosa* plasmids. Mutat. Res. **29:**363–372.

50. **Lacey, R. W.** 1975. Antibiotic resistance plasmids of *Staphylococcus aureus* and their clinical importance. Bacteriol. Rev. **39:**1–32.

51. **Larebeke, N. van, G. Engler, M. Holster, S. van den Elsacker, J. Zaenen, R. A. Schilperoort, and J. Schell.** 1974. Large plasmid in *Agrobacterium tumefaciens* essential for crown gall inducing activity. Nature (London) **252:**169–170.

52. **Leibowitz, M. J., and R. B. Wickner.** 1976. A chromosomal gene required for killer plasmid expression, mating and spore maturation in *Saccharomyces cerevisiae.* Proc. Natl. Acad. Sci. U.S.A. **73:**2061–2065.

53. **Lemke, P. A. (ed.).** 1979. Viruses and plasmids in fungi. Marcel Dekker Inc., New York.

54. **Linnane, A. W., and P. Nagley.** 1978. Mitochondrial genetics in perspective: the derivation of a genetic and physical map of the yeast mitochondrial genome. Plasmid **1:**324–345.

55. **Macdonald, R. M., M. R. Chandler, and B. Mosse.** 1982. The occurrence of bacterium-like organelles in vesicular-arbuscular mycorrhizal fungi. New Phytol. **90:**659–663.

56. **Montoya, A. L., M.-D. Chilton, M. F. Gordon, D. Sciaky, and E. W. Nester.** 1977. Octopine and nopaline metabolism in *Agrobacterium tumefaciens* and crown gall tumor cells: role of plasmid genes. J. Bacteriol. **129:**101–107.

57. **Murai, N., F. Skoog, M. E. Doyle, and R. S. Hanson.** 1980. Relationship between cytokinin production, presence of plasmids, and fasciation caused by strains of *Corynebac-*

terium fasciens. Proc. Natl. Acad. Sci. U.S.A. **77**:619–623.

58. **Olsen, R. H., and P. Shipley.** 1973. Host range and properties of the *Pseudomonas aeruginosa* R factor R1822. J. Bacteriol. **113**:772–780.

59. **Petit, A., and J. Tempe.** 1978. Isolation of *Agrobacterium* Ti-plasmid regulatory mutants. Mol. Gen. Genet. **167**:147–155.

60. **Pickup, R. W., R. J. Lewis, and P. A. Williams.** 1983. *Pseudomonas* sp. MT14, a soil isolate which contains two large catabolic plasmids, one a TOL plasmid and one coding for phenylacetate catabolism and mercury resistance. J. Gen. Microbiol. **129**:153–158.

61. **Reanney, D.** 1976. Extrachromosomal elements as possible agents of adaptation and development. Bacteriol. Rev. **40**:552–590.

62. **Richmond, M. H.** 1969. Extrachromosomal elements and the spread of antibiotic resistance in bacteria. Biochem. J. **113**:225–234.

63. **Rosenberg, C., F. Casse-Delbart, I. Dusha, M. David, and C. Boucher.** 1982. Megaplasmids in the plant-associated bacteria *Rhizobium meliloti* and *Pseudomonas solanacearum*. J. Bacteriol. **150**:402–406.

64. **Smith, H. R., G. O. Humphreys, N. D. F. Grindley, G. A. Willshaw, and E. S. Anderson.** 1976. Characterisation of plasmids coding for the restriction endonuclease *Eco*R1. Mol. Gen. Genet. **143**:319–325.

65. **Smith, H. W., and M. Linggood.** 1971. Observations of the pathogenic properties of the K88, Hly and Ent plasmids of *Escherichia coli* with particular reference to porcine diarrhoea. J. Med. Microbiol. **4**:467–485.

66. **Smith, M. G.** 1977. *In vivo* transfer of an R factor within the lower gastrointestinal tract of sheep. J. Hyg. **79**:259–268.

67. **Soldo, A. T., S. A. Brickson, and F. Larin.** 1983. The size and structure of the DNA genome of symbiont xenosome particles in the ciliate *Parauronema acutum*. J. Gen. Microbiol. **129**:1317–1325.

68. **Stahl, U., U. Kuck, P. Tudzynski, and K. Esser.** 1980. Characterisation and cloning of plasmid like DNA of the Ascomycete *Podospora anserina*. Mol. Gen. Genet. **178**:639–646.

69. **Watanabe, T.** 1963. Infectious heredity of multiple drug resistance in bacteria. Bacteriol. Rev. **27**:87–115.

70. **White, F. F., and E. W. Nester.** 1980. Hairy root: plasmid encodes virulence traits in *Agrobacterium rhizogenes*. J. Bacteriol. **141**:1134–1141.

71. **Williams, P. A.** 1982. Genetic interactions between mixed microbial populations. Philos. Trans. R. Soc. London Ser. B **297**:631–639.

72. **Williams, P. A., and D. J. Jeenes.** 1981. Origin of catabolic plasmids, p. 144–147. *In* D. Schlessinger (ed.), Microbiology—1981. American Society for Microbiology, Washington, D.C.

73. **Williams, P. A., and M. J. Worsey.** 1976. Ubiquity of plasmids coding for toluene and xylene metabolism in soil bacteria: evidence for the existence of new TOL plasmids. J. Bacteriol. **125**:818–828.

Evolution of New Phenotypes

PATRICIA H. CLARKE

Department of Biochemistry, University College London, London WC1E 6BT, United Kingdom

DIVERGENT EVOLUTION IN BIOLOGY

One of the fundamental questions of biology is the origin of new phenotypes. The classical models of evolution look for the routes whereby new species arise more or less directly from ancestral forms. The line of descent is predicted to give a linear progression in which a new species with novel attributes might eventually replace the parental stock. If this linear progression were to take place at more than one location, because of spatial separation of members of the parental species, then these events could lead to different species in each location related by having a common ancestor. From Darwin onwards, this type of "island evolution" has greatly interested naturalists who have investigated factors determining speciation in plants and animals. The island evolution provides an example of divergence from a common ancestor of descendants who live in different localities and are physically separated from one another so that genetic exchange cannot take place. Another type of divergent evolution occurs if an ancestral form gives rise to variants that can coexist with one another at the same location. These might diverge sufficiently from each other, and the ancestral type, to be identifiable as different species that are no longer capable of interbreeding. Over the long years of biological evolution events of this sort are considered to have contributed to the variety of species recognizable today, with many species of time past lost by gradual or catastrophic calamities.

MOLECULAR EVENTS

Single-site mutations. It is assumed that the changes in phenotype, resulting in the evolution of new species, are based on changes within the organism itself that may give the new variant some selective advantage. In molecular terms this has been understood as mutation within the genome followed by selection for or against the resulting phenotype. Comparative studies of the morphology and physiology of the higher plants and animals have been succeeded by comparative studies of molecular structures. The amino acid sequences of proteins carrying out similar functions in different species were found to diverge in a consistent manner. For individual proteins the homologies were greatest between species that were thought from classical evi-dence to be closely related and least between species that were thought to be distantly related. The divergence in protein sequences could be related to phylogeny on the basis of the observed amino acid differences. For individual proteins there were sequences that appeared to be highly conserved, and these were presumed to be functionally important; there were also sequences that exhibited much greater variability which could be ascribed to mutational divergence over long periods of time (11). Single amino acid differences at corresponding positions in the polypeptide could have arisen by single-site mutations within the DNA sequence of the gene for the protein in question. With increased understanding of the nature of mutation and the structure of genes, it has become evident that single-site mutations in the parental genome may be insufficient to account for the emergence of new phenotypes and the rate of evolution of new proteins (19, 20).

Gene duplication and gene clusters. Comparisons of amino acid sequences of related proteins also indicated that families of related proteins could have arisen by gene duplications. The detailed studies on the globin genes suggested that there had been duplications of ancestral globin genes followed by divergence to give a family of proteins with related functions but with significant differences in properties. The duplication could be traced back in time, and estimates could be obtained of when the duplication first appeared (19, 30). The duplication of genes would appear to have an obvious advantage in that the duplicates could evolve independently, and new proteins with new functions could appear without the loss of the original activity. In higher organisms it now appears that the genes for protein families such as the globins, the immunoglobulins, and the major incompatibility proteins are arranged in clusters and that, as well as the functional genes, there may be pseudogenes that exhibit considerable homology with the functional genes but may not be expressed or, if expressed, may have inactive products (3).

In eucaryotic organisms the coding regions are interrupted by the noncoding introns. One hypothesis suggests that the introns have important functions in separating (or bringing together) DNA sequences coding for specific protein domains. Other hypotheses suggest that the in-

71

trons are transposable genetic elements that have invaded the original completely continuous gene (19).

Concerted evolution. More detailed comparison of protein sequences led to the unexpected finding that the duplicated genes might not evolve independently. On the contrary, it is now suggested that duplicate genes, such as the globin genes, have undergone frequent genetic exchange. Zimmer et al. (30) compared the α-globin genes, and the hemoglobin α-chains, of species of apes that had a common ancestor about 10 million years ago. They found that the α-polypeptides from duplicated genes within each species were more alike than would have been predicted from 10 million years of divergent evolution and resembled those within the same species more than those of the other species. They termed this process "concerted evolution" (30). It is suggested that the related genes undergo multiple unequal crossing over, or gene conversions, and that in addition to accounting for homologies between isofunctional proteins this process can account for homologies between less closely related genes (19, 30).

GENE DUPLICATION IN BACTERIA

Bacterial systems offer a particular advantage for evolutionists in that it is possible to carry out experiments in the evolution of new enzymes as well as to compare the sequences of existing proteins. Horiuchi et al. (16) observed that duplication of the *lac* genes occurred in cultures of *Escherichia coli* maintained in chemostat culture on low concentrations of lactose. This response of duplication of genes to allow more enzyme to be produced will allow a more rapid growth rate and thus give a selected advantage to the bacteria in which the duplication has occurred. Similar duplications for other genes have been observed during growth in chemostat culture on substrates for which a single enzyme was rate limiting for growth. Growth of *Klebsiella aerogenes* in chemostat culture on the poor substrate xylitol gave rise to mutants in which the ribitol dehydrogenase gene had been duplicated (18, 26). If bacteria have the capacity to produce catabolic enzymes which have some activity for compounds related to the normal substrate, then they have the potential to utilize the poor substrate by duplication of the relevant genes. Ribitol and D-arabitol are both good growth substrates for *K. aerogenes*, and Charnetzky and Mortlock (6) found that the genes of the ribitol and D-arabitol operons were closely linked. It seemed reasonable to suppose that the genes for the catabolism of these two pentitols could have arisen by duplication of an ancestral pentitol genes. The duplication of the ribitol dehydrog-

enase gene that enabled growth on xylitol could be regarded as a first step that might eventually lead to the evolution of a new pentitol operon by subsequent divergent evolution.

CONCERTED EVOLUTION IN BACTERIA?

Catabolic pathways in bacteria are fascinating in their complexity. In a single bacterial species there may be several different pathways by means of which similar organic compounds are broken down. Reports also appear of enzymes and complex pathways for the degradation of many synthetic compounds that are never found in the natural environment, and these offer excellent systems for studying the evolution of catabolic enzymes (8). Stanier and Ornston have carried out detailed comparisons of the enzymes for the β-ketoadipate, or *ortho* cleavage, pathway by means of which many aromatic compounds are metabolized by species of *Pseudomonas* and *Acinetobacter* (27). The key metabolites prior to ring cleavage are catechol and, from some of the hydroxycompounds, protocatechuate (Fig. 1). The reactions carried out by the enzymes muconolactone isomerase and carboxymuconolactone decarboxylase both result in the formation of β-ketoadipate enol-lactone so that they have similar functions but act on different substrates. β-Ketoadipate enol-lactone is converted to β-ketoadipate by enol-lactone hydrolase (ELH), and although *Pseudomonas* species have a single enzyme for this step, there are isoenzymes in *Acinetobacter*. In *Acinetobacter* ELH I is coordinately induced with carboxymuconolactone decarboxylase by protocatechuate, and ELH II is induced coordinately with muconolactone isomerase. With this pathway it is possible to compare enzymes that carry out chemically analogous reactions in parallel pathways and also to compare enzymes that carry out successive steps in the same pathways. Further, since the pathways in *Pseudomonas* and *Acinetobacter* follow essentially the same routes, it is possible to compare the enzymes from different species.

The N-terminal sequences of the decarboxylases and isomerases from *Pseudomonas* and *Acinetobacter* have extensive regions of homology that are sufficient for them to have been derived from a common ancestor (23, 24). This interesting observation suggests that there might have been a single enzyme with both decarboxylase and isomerase activities that by duplication and mutation evolved into the decarboxylases and isomerases of present-day *Pseudomonas* and *Acinetobacter* (Fig. 2). A comparison of the enol-lactone hydrolases of *Acinetobacter* suggests that ELH I and ELH II evolved by duplication and divergence of a single gene. The enol-lactone hydrolase of *Pseudomonas putida*

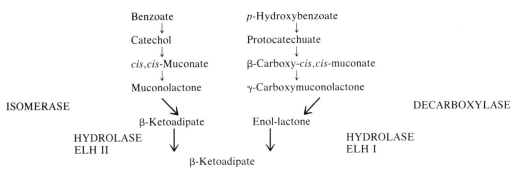

Benzoate   p-Hydroxybenzoate
↓   ↓
Catechol   Protocatechuate
↓   ↓
cis,cis-Muconate   β-Carboxy-cis,cis-muconate
↓   ↓
Muconolactone   γ-Carboxymuconolactone

ISOMERASE   DECARBOXYLASE

β-Ketoadipate   Enol-lactone

HYDROLASE   HYDROLASE
ELH II   ELH I

β-Ketoadipate

FIG. 1. Reactions of the β-ketoadipate pathways. The two convergent pathways are chemically analogous. Muconolactone isomerase and γ-carboxymuconolactone decarboxylase produce β-ketoadipate enol-lactone from their respective substrate. *P. putida* has a single β-ketoadipate enol-lactone hydrolase and *A. calcoaceticus* produces isoenzymes for this reaction. ELH I is induced coordinately with γ-carboxymuconate decarboxylase, and ELH II is induced coordinately with muconolactone isomerase (23, 24).

exhibits some homology with both ELH I and ELH II, suggesting an evolutionary relationship for all three enzymes. Such observations are compatible with gene duplication and independent evolutionary divergence, but more detailed comparisons of the amino acid sequences of these and other enzymes of the *ortho* cleavage pathway have led Ornston and Yeh to a different conclusion (23, 24, 29). The enol-lactone hydrolases of *Acinetobacter* each have sequences that resemble the muconolactone isomerase more closely than they resemble each other. Further, the sequences shared with the isomerase are not

sequences that are common to the other enol-lactone hydrolase, and this suggested that the shared isomerase hydrolase sequences could have been introduced at a later time. This interpretation of the sequence homologies envisages that duplication of genes allowed evolutionary divergence to give different enzyme functions but that in addition short DNA sequences were exchanged between the individual genes. This envisages coevolution of genes coding for the enzymes for successive steps of a catabolic pathway (Fig. 2).

The successful operation of a catabolic path-

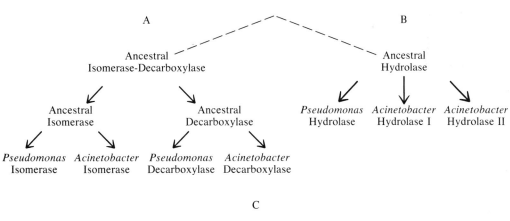

A   B

Ancestral Isomerase-Decarboxylase   Ancestral Hydrolase

Ancestral Isomerase   Ancestral Decarboxylase   *Pseudomonas* Hydrolase   *Acinetobacter* Hydrolase I   *Acinetobacter* Hydrolase II

Pseudomonas Isomerase   *Acinetobacter* Isomerase   *Pseudomonas* Decarboxylase   *Acinetobacter* Decarboxylase

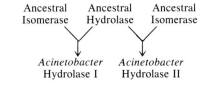

C

Ancestral Isomerase   Ancestral Hydrolase   Ancestral Isomerase

Acinetobacter Hydrolase I   *Acinetobacter* Hydrolase II

FIG. 2. Proposed evolution of enzymes. The lines of descent in A show independent evolution of the decarboxylase and isomerase genes from a common ancestor, and those in B show the evolution of the hydrolases from a common ancestor. The dotted lines indicate the improbability that the duplication of a single ancestral gene gave rise to the present day isomerases, decarboxylases, and hydrolases. The extensive divergence of the hydrolases of *Acinetobacter* and the partial homologies with isomerase suggest that these genes may have evolved in concert as shown in C (23).

way requires a complete set of enzymes, and coevolution of the genes for the enzymes concerned would appear to be an advantage to the organism. The isomerase is of no value without the hydrolase, and once the pathway had evolved it would be important to maintain genetic stability. Similar sequence homologies were observed in comparisons with other enzymes of the *ortho* cleavage pathway. Yeh and Ornston (29) concluded that in examining sequence homologies one should not assume that the homologous sequences are invariably those that have been retained from the ancestral gene but that many of the homologous sequences have been acquired subsequent to their divergence from a common ancestor. They suggested that copies of short DNA sequences were exchanged between the different structural genes as they coevolved. This hypothesis suggests that superimposed on the evolution of metabolic pathways by mutations within individual genes are the deletion, insertion, and exchange of short DNA sequences between the genes for the different enzymes.

NEW SUBSTRATE SPECIFICITIES

The enormous variety of organic compounds that could be utilized for growth by a single strain of *P. putida* led Kluyver (21) to believe it highly unlikely that for each substrate there was a specific enzyme. Thus, the only possible conclusion was that "one and the same catalyst is capable of acting on different substrates and very probably even on a large number of these." We now know that bacteria, and especially *Pseudomonas* species, are capable of producing a very large number of enzymes which are highly specific (8). However, the specificity of corresponding enzymes from different strains is not always the same, and the specificity is seldom absolute. The fact that ribitol dehydrogenase of *K. aerogenes* has weak activity for xylitol was the reason that duplication of the ribitol dehydrogenase gene allowed a reasonable rate of growth on xylitol. An alternative solution to the problem of growth on a poor substrate is to produce a more efficient enzyme by mutation in the structural gene.

In considering the coevolution of enzymes for a catabolic pathway, Yeh and Ornston (29) pointed out the constraints imposed by the requirement that all the enzymes be functional. In addition, for most catabolic pathways it is essential that an effective regulatory system be maintained. For catabolic enzymes it is clearly advantageous that enzymes should be synthesized only when the potential growth substrate is present in the environment. The regulatory systems place additional and severe constraints on the organism and make for genetic stability.

AMIDASE OF *PSEUDOMONAS AERUGINOSA*

P. aeruginosa grows on acetamide and propionamide as carbon and nitrogen sources, and these two amides are both substrates and inducers of amidase (7). Butyramide, the next amide in the aliphatic series, is a very poor enzyme substrate, does not induce amidase synthesis, and even blocks induction by substrate and nonsubstrate amide inducers. These properties lock the system, but in laboratory conditions the system can be unlocked by disconnecting the enzyme from its normal regulation. Constitutive mutants produce amidase in the absence of inducing amides, and some of these are able to grow on butyramide because, although they produce high amidase activities when grown in the absence of amides, they retain some of the regulatory characteristics of the parent strain and butyramide still blocks amidase synthesis. By a variety of selection methods it was possible to obtain a series of mutants that produced amidases with novel substrate specificities. Mutant strains were obtained that could grow on butyramide, valeramide, octanoamide, phenylacetamide, and acetanilide (Fig. 3). The novel enzymes differed from the wild-type enzyme, and from each other, in physicochemical properties and in K_m and V_{max} for their amide substrates (2). The novel enzymes were predicted to carry one, two, or three amino acid substitutions, and some of these were confirmed by sequence studies on the proteins. For example, the mutant B amidase that allows growth on

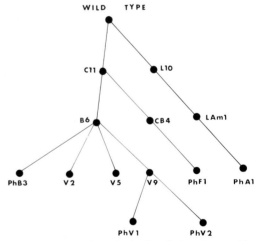

FIG. 3. Evolution of a family of phenylacetamide-utilizing mutants of *P. aeruginosa*. Mutants C11, CB4, and L10 are constitutive mutants producing wild-type amidase. Mutants B6, V2, V5, V9, PhB3, PhV1, PhV2, PhF1, and PhA1 produce amidases with altered substrate specificities that allow growth on butyramide, valeramide, and phenylacetamide, respectively (1, 4, 7, 25).

butyramide has a single amino acid substitution at position 7 from the N terminus. A serine residue of the wild-type enzyme is replaced by phenylalanine in the mutant B amidase (25). The amidase studies provide a model for the way in which changes in substrate specificity might arise during evolution by a succession of single-site mutations.

With catabolic enzymes it is important that the specificity of induction should bear some relationship to the specificity of the enzyme being induced. It is of no advantage to an organism to produce a novel enzyme if it cannot be induced by the novel substrate. In the experiments on amidase evolution, the regulatory problems were avoided by selecting the novel enzymes in constitutive strains. It was theoretically possible that a butyramide-inducible mutant could be selected directly from the wild-type strain, but none has been obtained. A butyramide-inducible mutant might be expected to arise from a constitutive B amidase strain, but this would be difficult to detect. It was thought that another way to obtain such mutants would be to start with a strain having wild-type regulation but carrying the gene for the mutant B amidase. An amidase-negative mutant with wild-type regulatory gene, $amiR^+$, and with a mutation in the $amiE$ gene was crossed with the constitutive strain PhB3, producing a mutant amidase that had activity for phenylacetamide but not for acetamide. Among the recombinants were some that carried $amiR^+$ and the $amiE16$ mutation conferring the B amidase phenotype. Strain IB10 grew on acetamide but could not grow on butyramide since, although it was capable of producing an amidase hydrolyzing butyramide, the enzyme could not be induced by butyramide (1). Butyramide-utilizing mutants were obtained from strain IB10 by selection for growth on butyramide as carbon source, and although most were constitutive in regulatory phenotype, strain BB1 was altered in substrate specificity and produced an altered amidase induced by its novel substrate (28).

The experimental evolution of amidases in the laboratory has shown that within a very short period of time it is possible for new enzyme specificities to appear as the result of single-site mutations. Further, the introduction of successive single-site mutations resulted in progressive alterations in the substrate specificity of the enzyme. This might suggest that amidase was an enzyme that would show considerable diversity among different strains of *P. aeruginosa*. When the amidases of a number of different *P. aeruginosa* strains were compared, no significant differences in the amide growth phenotypes were found. The extracts from all the strains tested showed completed immunological cross-reac-

tions with purified amidase from PAC1 used as the parent strain for the studies in experimental evolution, and there were no detectable differences in the K_m or V_{max} of the amidases examined. The inducer specificities were also similar. These results suggest that, in spite of the potential for evolving amidases with novel substrate and inducer specificities, in the absence of strong selection pressure the amidase genes remain very constant (7). The interlocking specificities for amides as substrates and inducers promote genetic stability.

GENES FROM ELSEWHERE?

Within a single bacterial species we can envisage the acquisition of new growth phenotypes by several different mechanisms. Gene duplication gives the possibility of increasing the number of available enzymes. In its most simple form a gene duplication can allow growth on a poor substrate since the increased gene dosage results in an increased amount of the original enzyme. As a permanent solution this leaves much to be desired, and such duplications are unstable and readily lost. Comparative studies have indicated homologies among proteins carrying out similar functions, and this has been taken to indicate an earlier gene duplication followed by sequence divergence to give two or more different functional proteins. Studies with eucaryotic systems have shown that in many cases families of related genes occur in clusters and may have undergone concerted evolution with partial recombination so that the homologies between them may be greater than predicted (19, 30). Ornston and colleagues have examined the ways in which the enzymes for a complex metabolic pathway are related and have concluded that gene duplication and divergent evolution cannot account for all the homologies between enzymes carrying out different functions. Earlier attempts to understand the evolution of metabolic pathways suggested that there had been tandem duplication followed by independent mutational divergence (17). Ornston and Yeh (23, 24, 29) pointed out the necessity for all the genes of a catabolic pathway to evolve in a concerted manner and accounted for homologies in genes for different but related steps of a pathway by recombination or exchange of short sequences between the genes at later stages of evolution. These events can be regarded as consisting of linear evolution with divergence of genes from common ancestors together with coevolution or horizontal evolution of genes with frequent sequence exchanges between them.

A more overt example of horizontal evolution comes from a consideration of the transfer of genetic information between bacteria belonging to different species and different genera. With the discovery that genes for many catabolic pathways were carried on transmissible plasmids, it was possible to envisage that a considerable contribution to the evolution of new catabolic pathways could be made by the transfer of blocks of genes between bacterial populations (8). These studies also raised the possibility that genes could be transferred from chromosome to plasmid, and hence mutations in chromosomal genes might be picked up on plasmids and rapidly spread through a bacterial population. This would give a much more rapid spread of new characters in a population than would be expected from single-site mutations alone. We know that the β-lactamase gene, conferring resistance to penicillin, may be within a mobile genetic element called a transposon that can move from plasmid to chromosome and plasmid to plasmid (5). Many mobile genetic elements are now known, and these entities offer a new range of evolutionary possibilities (9a). Catabolic genes may also be contained within transposable elements. Cornelis et al. (10) reported that transposon Tn951, originally identified on a plasmid from *Yersinia enterolytica*, carried an active lactose operon. The transposon contained *lacIZY* genes that were homologous with those of the *lac* operon of *E. coli*. Further, the Tn951 *lac* genes could be transferred in vivo to the drug resistance plasmid RP1, which has a very wide host range.

Plasmids of the Inc-P group are characterized by having a very broad host range. Some can act as sex factors and also form R' plasmids (12, 15). I had wondered if it would be possible to test whether catabolic genes could be transferred

from the chromosome to Inc-P plasmids and then transferred to different strains or to different genera. The amidase genes of *P. aeruginosa* are located on the chromosome in the 50-min region near to *argF*, the biosynthetic ornithine carbamoyltransferase (9, 13). By selecting for R' plasmids carrying *argF*, it has been possible to obtain plasmids carrying also the amidase genes. The *argF* gene of *P. aeruginosa* PAC1 is expressed at a similar level in *P. aeruginosa* PAO and also in *P. putida* PNN. However, the expression of *argF* from *P. aeruginosa* in *E. coli* strains is only at about 2% of that in the original *P. aeruginosa* strain (9). (Most of the catabolic plasmids found in *Pseudomonas* species are effective in both *P. aeruginosa* and *P. putida*.)

R' plasmids carrying chromosomal genes of *P. aeruginosa* have been selected in crosses in which *P. aeruginosa* strains carrying wild-type genes were the donors and the recipients were auxotrophic *P. aeruginosa* Rec⁻ strains (15), *P. putida* strains (22), or *E. coli* strains (14). Table 1 compares the ornithine carbamoyltransferase activities of *argF* auxotrophs of *P. aeruginosa*, *P. putida*, and *E. coli* carrying R' *argF* plasmids. Plasmid pAR1 was derived from a *P. aeruginosa* PAC strain and pMO778 was derived from a PAO strain.

Some of the R' plasmids selected for biosynthetic genes located in this region of the chromosome had also acquired wild-type amidase genes (22). Plasmid pMO780 *argF ami* was selected in *P. putida* and transferred to *E. coli* W4100 *argF*. In spite of the poor expression of *argF*, the ornithine carbamoyltransferase activity was sufficient to confer prototrophy. *E. coli* strains do not produce an aliphatic amidase, but the wild-

TABLE 1. Ornithine carbamoyltransferase activities[a]

Strain	Relevant genotype	Ornithine carbamoyltransferase activity in plasmid[b]:		
		None	pAR1	pMO778
P. aeruginosa				
PAO1	Wild type	23	NA	NA
PAO29	*argF10 leu-10*	NG	70	103
PAC1	Wild type	23	NA	NA
PAC51	*argF504 amiE8*	NG	26	77
P. putida				
PPN1	Wild type	16	NA	NA
PPN1022	*argF402 cys-400 ilv-400 trpB402*	NG	16	17
E. coli				
4100	*argI thi*	28	NA	NA
4100 Arg	*argF argI thi*	NG	0.7	0.7

[a] Expression of *P. aeruginosa* R' *argF* plasmids in *Pseudomonas* and *E. coli* strains.
[b] Cultures were grown in minimal salt medium for 17 to 18 h, by which time they had reached an absorbance at 670 nm of about 1.0. *P. putida* was grown at 30°C, and other strains were grown at 37°C. The specific activities are expressed in micromoles of citrulline produced per hour per milligram of protein. NG, No growth; NA, not applicable. (From Clarke and Laverack [9].)

type *ami* genes on the plasmid allowed acetamide to be used as a nitrogen source with glucose as carbon source. Growth with this medium, either on plates or in liquid medium, was very slow, but mutants could be selected that grew at a faster rate. It was found that two of the faster-growing strains were now constitutive for amidase synthesis. Other R' plasmids carrying mutant amidase genes have been constructed.

In the natural environment bacteria do not live in isolation, and there must always be the possibility of transfer of genetic material from another species. My data on the transfer of genetic information from *P. aeruginosa* to *E. coli* suggest that it is not sufficient to transfer wild-type genes for biosynthetic or catabolic pathways. Adaptation by further mutational steps, and possibly coevolution with resident genes, may be required before these *Pseudomonas* genes can function efficiently in their alien host.

CONCLUSIONS

New phenotypes evolved in the natural environment in response to selection pressures. During past millenia the biosynthetic powers of plants and animals have produced increasing numbers of organic molecules, and bacteria have evolved enzymes for degrading them. In this century the activities of the chemical industry have produced compounds that have not been found previously in nature, and this represents even more of an evolutionary challenge.

I have discussed some of the mechanisms whereby new phenotypes have emerged in laboratory experiments in evolution and some of the conclusions that have been reached from comparative studies. Many, if not all, of these mechanisms also occur in the natural environment. Single-site mutations can account for changes in enzyme specificity and in regulatory controls. Recombination within and between genes can also contribute to this. Genetic rearrangements can bring together DNA sequences corresponding to structural genes or to shorter gene regions. Genetic exchange between chromosomes and plasmids, together with the ability of the promiscuous plasmids to transfer across species boundaries, enlarges still more the possibilities of rapid evolution of new bacterial phenotypes.

LITERATURE CITED

1. **Betz, J. L., J. E. Brown, P. H. Clarke, and M. Day.** 1974. Genetic analysis of amidase mutants of *Pseudomonas aeruginosa*. Genet. Res. **23:**335–359.
2. **Betz, J. L., and P. H. Clarke.** 1972. Selective evolution of phenylacetamide-utilizing strains of *Pseudomonas aeruginosa*. J. Gen. Microbiol. **73:**161–174.
3. **Bodmer, W.** 1983. Gene clusters and genome evolution, p. 197–208. *In* D. S. Bendall (ed.), Evolution from molecules to man. Cambridge University Press, Cambridge.
4. **Brown, J. E., and P. H. Clarke.** 1970. Mutations in a

regulator gene allowing *Pseudomonas aeruginosa* 8602 to grow on butyramide. J. Gen. Microbiol. **64:**329–342.
5. **Bukhari, A. I., J. A. Shapiro, and S. L. Adhya (ed.).** 1977. DNA insertion elements, plasmids, and episomes. Cold Spring Harbor Laboratory, Cold Spring Harbor, N.Y.
6. **Charnetzky, W. T., and R. P. Mortlock.** 1974. Close genetic linkage of the determinants of the ribitol and D-arabitol pathways in *Klebsiella aerogenes*. J. Bacteriol. **119:**176–182.
7. **Clarke, P. H.** 1978. Experiments in microbial evolution, p. 137–218. *In* J. R. Sokatch and L. N. Ornston (ed.), The bacteria, vol. 6. Academic Press, Inc., New York.
8. **Clarke, P. H.** 1982. The metabolic versatility of pseudomonads. Antonie van Leeuwenhoek J. Microbiol. Serol. **48:**105–130.
9. **Clarke, P. H., and P. D. Laverack.** 1983. Expression of the *argF* gene of *Pseudomonas aeruginosa* in *Pseudomonas aeruginosa*, *Pseudomonas putida*, and *Escherichia coli*. J. Bacteriol. **154:**508–512.
9a. **Cold Spring Harbor Laboratory.** 1981. Movable genetic elements. Cold Spring Harbor Symp. Quant. Biol., vol. 45. Cold Spring Harbor Laboratory, Cold Spring Harbor, N.Y.
10. **Cornelis, G., D. Ghosal, and H. Saedler.** 1978. Tn951: a new transposon carrying a lactose operon. Mol. Gen. Genet. **160:**215–224.
11. **Dayhoff, M. O. (ed.).** 1972. Atlas of protein sequence and structure, vol. 5, National Biomedical Research Foundation, Silver Spring, Md.
12. **Haas, D., and B. W. Holloway.** 1978. Chromosome mobilization by the R plasmid R68.45: a tool in *Pseudomonas genetics*. Mol. Gen. Genet. **158:**229–237.
13. **Haas, D. B., B. W. Holloway, A. Schambock, and T. Leisinger.** 1977. The genetic organization of arginine biosynthesis in *Pseudomonas aeruginosa*. Mol. Gen. Genet. **154:**7–22.
14. **Hedges, R. W., and A. E. Jacob.** 1977. *In vitro* translocation of genes of *Pseudomonas aeruginosa* onto a promiscuously transmissible plasmid. FEMS Microbiol. Lett. **2:**15–19.
15. **Holloway, B. W.** 1978. Isolation and characterization of an R' plasmid in *Pseudomonas aeruginosa*. J. Bacteriol. **133:**1078–1082.
16. **Horiuchi, T., J. T. Tomizawa, and A. Novick.** 1962. Isolation and properties of bacteria capable of high rates of β-galactoside synthesis. Biochim. Biophys. Acta **55:**152–163.
17. **Horowitz, N. H.** 1965. The evolution of biochemical synthesis—retrospect and prospect, p. 15–23. *In* V. Bryson and H. J. Vogel (ed.), Evolving genes and proteins. Academic Press, Inc., New York.
18. **Inderlied, C. B., and R. P. Mortlock.** 1977. Growth of *Klebsiella aerogenes* on xylitol: implications for bacterial evolution. J. Mol. Evol. **9:**181–190.
19. **Jeffreys, A. J.** 1981. Recent studies of gene evolution using recombinant DNA, p. 1–48. *In* R. Williamson (ed.), Genetic engineering, vol. 2. Academic Press, Inc., London.
20. **Jeffreys, A. J., S. Harris, P. A. Barrie, D. Wood, A. Blanchetot, and S. M. Adams.** 1983. Evolution of enzyme families: the globin genes. *In* D. S. Bendall (ed.), Evolution from molecules to men. Cambridge University Press, Cambridge.
21. **Kluyver, A. J.** 1931. The chemical activities of microorganisms. University of London Press, London.
22. **Morgan, A. F.** 1982. Isolation and characterization of *Pseudomonas aeruginosa* R plasmids constructed by interspecific mating. J. Bacteriol. **149:**654–661.
23. **Ornston, L. N., and W. K. Yeh.** 1979. Origins of metabolic diversity: evolutionary divergence by sequence repetition. Proc. Natl. Acad. Sci. U.S.A. **74:**3996–4000.
24. **Ornston, L. N., and W. K. Yeh.** 1981. Towards molecular natural history, p. 140–143. *In* D. Schlessinger (ed.), Microbiology—1981. American Society for Microbiology, Washington, D.C.

25. **Paterson, A., and P. H. Clarke.** 1979. Molecular basis of altered enzyme specificities in a family of mutant amidases from *Pseudomonas aeruginosa*. J. Gen. Microbiol. **114:**75–85.

26. **Rigby, P. W. J., B. D. Burleigh, and B. S. Hartley.** 1974. Gene duplication in experimental enzyme evolution. Nature (London) **251:**200–204.

27. **Stanier, R. Y., and L. N. Ornston.** 1973. The β-ketoadipate pathway. Adv. Microb. Physiol. **9:**89–151.

28. **Turberville, C., and P. H. Clarke.** 1981. A mutant of *Pseudomonas aeruginosa* PAC with an altered amidase inducible by the novel substrate. FEMS Microbiol. Lett. **10:**87–90.

29. **Yeh, W. K., and L. N. Ornston.** 1980. Origins of metabolic diversity: substitution of homologous sequences into genes for enzymes with different catalytic activities. Proc. Natl. Acad. Sci. U.S.A. **77:**5365–5369.

30. **Zimmer, E. A., S. L. Martin, S. M. Beverley, Y. W. Kan, and A. C. Wilson.** 1980. Rapid duplication and loss of genes coding for the α chains of hemoglobin. Proc. Natl. Acad. Sci. U.S.A. **77:**2158–2162.

Adaptation by Acquisition of Novel Enzyme Activities in the Laboratory

BARRY G. HALL

Biological Sciences Group, U-44, University of Connecticut, Storrs, Connecticut 06268

Physiological adaptation involves rapid responses to environmental changes by all members of a population. Genetic adaptation, on the other hand, involves changes in only a tiny minority of the population, changes that make those few cells better adapted to their present environment or which permit them to take advantage of new environmental conditions. I am using *Escherichia coli* as a model system to study the process of genetic adaptation to altered environments. In particular, I am concerned with the process by which microorganisms acquire new enzymatic functions that permit them to utilize novel resources in their environments. By novel resources I mean those that are not utilizable by the wild-type organism.

There are four primary means by which microorganisms can acquire novel catabolic functions. They may acquire new genes from other organisms by direct transfer of DNA. This usually means acquisition of a plasmid carrying catabolic genes. Cells may, as a consequence of regulatory mutations, utilize preexisting functioning catabolic pathways to metabolize novel resources. This usually means that the novel resource resembles a normal resource sufficiently that it is a substrate for the first enzymes in the pathway, but that the novel resource does not induce synthesis of the pathway enzymes. This strategy has been reviewed recently (20), and I shall not consider it further. The third means of adaptation involves mutations that alter the catalytic specificities of existing enzymes so that they are much more active toward the novel resource. Such mutations are often accompanied by regulatory mutations that permit the altered enzyme to be synthesized in response to the presence of the novel resource in the environment. The fourth means involves "decryptifying" sets of genes for catabolic pathways that are not normally expressed during the life cycle of the cell. The majority of the work in my laboratory has explored the third means of acquiring new catabolic functions, and I summarize those studies here. Recent work has focused on the fourth strategy, decryptifying sets of preexisting unexpressed genes, and I shall describe some preliminary findings with that new system.

THE EVOLVED β-GALACTOSIDASE SYSTEM OF *E. COLI*

E. coli metabolizes lactose by the products of the genes of the *lac* operon. The *lacY* gene specifies a permease that actively transports lactose into the cell, where it is hydrolyzed to glucose and galactose by the *lacZ*-specified β-galactosidase. Strains of *E. coli* carrying deletions within the *lacZ* gene cannot hydrolyze lactose and hence cannot utilize lactose as a carbon or energy source. For such *lacZ* deletion strains lactose is indeed a novel resource. I have used a strain carrying a deletion that removes the middle one-third of the *lacZ* gene as a model system to ask how *E. coli* can re-evolve the ability to utilize lactose as a sole carbon and energy source. Because the deletion is internal to *lacZ*, that strain synthesizes a fully functional lactose permease and can thus transport lactose into the cell.

The credibility of any model system depends to some extent upon the degree to which it mimics nature. In a deliberate attempt to mimic nature these studies have relied entirely upon spontaneous mutations, and the selective conditions have been chosen to create a plausible natural situation. Strain DS4680A is streaked onto the surface of Lac-TTC plates containing broth, lactose, IPTG (isopropyl-β-D-thiogalactopyranoside, a synthetic inducer of the *lac* operon, required to turn on synthesis of the lactose permease), and the fermentation indicator triphenyl tetrazolium chloride (1, 12). Colonies grow initially at the expense of the broth and form dark-red colonies, indicative of their failure to ferment lactose. As the broth is exhausted, rare spontaneous lactose-utilizing mutants have a strong selective advantage and grow at the expense of the lactose to form white (lactose-fermenting) papillae on the surface of the colonies. Cells are restreaked from these papillae to isolate the mutants that have evolved lactose utilization.

The mutations that permit *lacZ* deletion strains of *E. coli* to utilize lactose all occur in the genes of the *ebg* (evolved β-galactosidase) operon. The *ebg* operon is located at 66 min, on the opposite side of the *E. coli* map from the *lac* operon (13) (Fig. 1). The operon consists of two structural genes and one regulatory gene. The *ebgA* gene specifies a 120-kilodalton polypeptide that is the subunit of the ebg β-galactosidase (3). The wild-type ebg β-galactosidase is virtually inactive toward lactose, but it is easily assayed by its activity toward the synthetic substrate *O*-nitrophenyl-β-D-galactoside (3, 7, 17). We designate the wild-type allele *ebgA*β and its product ebgβ enzyme. The natural function of this enzyme is unknown. The *ebgB* gene specifies a 79-kilodalton polypeptide which has been detected only by gel electrophoresis and whose function is unknown (16).

The expression of the *ebg* operon is tightly regulated (6, 8, 11). In the absence of inducer the basal level of expression amounts to only 3 to 5 molecules of ebg enzyme per cell, i.e., about one transcription of the gene per generation (11). The *ebg* operon is under negative control by a repressor that is the product of the tightly linked *ebgR* gene (13). *ebgR* mutants express the *ebg* operon constitutively (11, 13) at a level that is 2,000-fold above the basal level. When synthesized constitutively, ebg enzyme constitutes about 5% of the soluble cell protein (11). Lactose, the only effective inducer recognized by the ebg repressor, induces ebg enzyme synthesis to 100-fold above the basal level. The lactose analogs lactulose, galactosyl-β-D-arabinose, and β-methyl-galactoside are extremely weak inducers (3- to 10-fold above basal level), whereas other galactosides, including the thio-galactosides that are powerful inducers of the *lac* operon, are not recognized as inducers (11).

The *ebg* operon as described above is present in all *E. coli* K-12 strains (17). My concern is with the nature of the mutations in that operon that permit a *lacZ* deletion strain of *E. coli* to grow on lactose. A single mutation in the structural gene *ebgA* is sufficient to increase the lactase activity of ebg enzyme enough to permit lactose utilization. Those mutations improve the quality of ebg enzyme with respect to lactose hydrolysis. In addition, a mutation in *ebgR* is required to permit synthesis of a sufficient quantity of ebg enzyme for growth (11). It is useful to consider the consequences of mutations in these two genes separately.

The power of this approach to studying adaptation is that it permits direct comparisons between the initial (or ancestral) strain and the mutant (or evolved) derivatives that have acquired the ability to use the novel resource. We shall first consider those mutations that increase the catalytic efficiency of ebg enzyme with respect to β-galactoside sugars. I have purified the wild-type enzyme, ebgβ enzyme, to homogeneity and characterized it in both physical (3) and kinetic (3, 7) terms. Similarly, I have purified and characterized ebg enzyme from several evolved strains. It is useful to compare the growth rates of strains on several β-galactoside sugars with the properties of their ebg enzymes acting on those same substrates. Because several of those substrates are not effective inducers of ebg enzyme synthesis (11), it is necessary to make those comparisons in strains that synthesize ebg enzyme constitutively (*ebgR* strains).

Even when it constitutes 5% of the soluble protein, the wild-type enzyme, ebgβ, does not permit growth on any of the β-galactoside sugars shown in Table 1. The failure of the constitutive *ebgβ* strain to grow on lactose is explained by the extremely low lactase activity of that enzyme. The K_m of ebgβ enzyme for lactose is fivefold higher than the physiological internal lactose concentration generated by the Lac permease. At substrate concentrations much less than the K_m value, the velocity of an enzymatic reaction is determined almost entirely by V_{max}/K_m. That ratio is often referred to as the "specificity" of an enzyme for its substrate. The specificity of ebgβ enzyme for lactose is extremely low. By contrast, the specificity of *lacZ* β-galactosidase for lactose is 25,000 (18). Although the specificity of ebgβ enzyme for lactose is low, its specificity for the other β-galactoside sugars is even worse. None of those specificities is high enough to permit growth, but

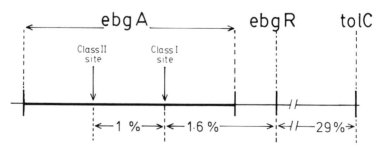

FIG. 1. Map of *ebg* operon. Distances are given in percent recombination. On the basis of recombination frequencies of other markers in this region, the class I and class II sites are estimated to be about 1,000 base pairs apart (15).

TABLE 1. Properties of ebg enzymes and growth of strains synthesizing these enzymes constitutively

Enzyme class	Substrate	Property			
		V_{max} (U/mg of pure enzyme)	K_m (mM substrate)	Specificity	Growth rate (h^{-1})
Ebg^0 ($n = 1$)[a]	Lactose	620	150	4.0	0
	Lactulose	270	180	1.5	0
	Galactosyl-arabinose	52	64	0.81	0
	Lactobionate	No detectable activity			0
Class I ($n = 3$)	Lactose	3,566	22	160	0.45
	Lactulose	69	34	2.1	0
	Galactosyl-arabinose	185	14	12.7	0.03
	Lactobionate	No detectable activity			0
Class II ($n = 3$)	Lactose	2,353	59	40	0.19
	Lactulose	1,887	26	73	0.26
	Galactosyl-arabinose	356	25	14.4	0.02
	Lactobionate	No detectable activity			0
Class IV ($n = 9$)	Lactose	1,461	0.82	1,800	0.37
	Lactulose	430	7.9	55	∙0.18
	Galactosyl-arabinose	737	3.0	244	0.13
	Lactobionate	105	15.4	6.7	0
Class V ($n = 1$)	Lactose	590	0.69	850	0.18
	Lactulose	215	6.5	33	0.10
	Galactosyl-arabinose	349	4.9	70	0.07
	Lactobionate	370	3.0	123	0.20

[a] n, Number of purified enzymes from independent mutants.

the sugars can be ordered in terms of their effectiveness as substrates as lactose > lactulose > galactosyl-arabinose >>> lactobionic acid.

Mutants selected to grow on lactose fall into two discrete classes on the basis of both growth rates and properties of their ebg enzymes. Class I $ebgA^+$ mutants represent about 90% of the isolates (3–5) and grow well on lactose. They fail to utilize lactulose or lactobionate, and the growth rate on galactosyl-arabinose is so slow (doubling time of about 24 h) that growth on galactosyl-arabinose must be considered marginal at best. Enzyme from class I strains is dramatically improved in its activity toward lactose, with both an increased V_{max} and decreased K_m contributing toward a 40-fold improvement in specificity. The failure to grow on lactulose is consistent with the absence of a significant improvement in specificity for that sugar.

Class II strains constitute the remaining 10% of strains selected to grow on lactose. These strains grow considerably slower on lactose than do class I strains, but they have the advantage of growing well on lactulose. Again, the properties of the class II ebg enzymes are consistent with the growth characteristics of class II strains. As before, improved specificities result from increases in V_{max} and decreases in K_m.

It must be emphasized that these are the only two classes of ebg^+ mutants that have been detected among several hundred isolates tested. Although class I mutants are isolated 90% of the time, there is no reason to suspect that these

mutations occur more often than class II mutations. It seems more likely that they are isolated more frequently simply because they grow faster.

When, instead of selecting for lactose utilization, we select directly for growth on lactulose, class II mutants are isolated 100% of the time (5). It has not been possible to select one-step mutants capable of growth on galactosyl-arabinose or lactobionate. Both class I and class II mutations occur spontaneously at frequencies consistent with single-point mutations (4).

The two classes were so clearly distinct that it was of interest to ask whether they represented two absolutely divergent evolutionary paths. Having acquired the property of rapid growth on lactose, were class I strains excluded from adapting to growth on lactulose? If not, would the lactulose-utilizing derivatives of class I strains be different from class II strains? As it turned out, lactulose-utilizing mutants were isolated with ease from several class I strains, and they were distinctly different from class II strains (5). Designated class IV, both their growth properties and the properties of their enzymes were different from class II. In terms of growth, the most striking (and surprising) difference was that class IV strains grew well on galactosyl-arabinose. In terms of the enzymatic characteristics, the K_m for all substrates was dramatically lowered (Table 1).

In terms of gross growth characteristics, the major difference between classes II and IV was that class II strains could not utilize galactosyl-arabinose. When galactosyl-arabinose–utilizing mutants were isolated from class II strains, they proved to have growth rates and ebg enzymes that were indistinguishable from class IV (7, 15). This led to the hypothesis that class IV strains were simply those that contained a class I and a class II mutation in the same *ebgA* gene. That hypothesis was strongly supported by showing that class IV recombinants could be recovered from matings between class I and class II strains, and that the enzymes isolated from those recombinants exhibited kinetic constants characteristic of class IV enzymes when assayed on 10 different substrates (7, 15).

Although class IV strains do not grow on lactobionate, class IV ebg enzymes do hydrolyze lactobionate at a detectable rate. In contrast to the class I and class II mutations that simply enhance activities already detectable in the unevolved enzyme, the simultaneous presence of the two mutations creates a new activity, lactobionate hydrolysis (7). The detection of this activity in vitro suggested the possibility that class IV strains might give rise to lactobionate-utilizing derivatives. Indeed, lactobionate-utiliz-

ing mutants were obtained from class IV strains at a frequency of 10^{-9}, exactly that expected for single-point mutations (5). It was not possible to select lactobionate-utilizing mutants from the wild-type, class I, or class II strains (5). The properties of one of these class V strains are shown in Table 1. Figure 2 shows the evolutionary pathway that leads from the unevolved enzyme to an enzyme that permits growth on four β-galactoside sugars.

EVOLUTION OF A FULLY REGULATED SYSTEM

The pathway shown in Fig. 1 is for the evolution of new catalytic specificities for an enzyme. Evolution of new metabolic systems involves more than just the evolution of new substrate specificities for an enzyme; it involves integrating those new enzymatic activities into the cell's regulatory network. Most metabolic systems involve a set of genes that specify all of the proteins required for a pathway to function and which are expressed in direct response to an environmental signal indicating a requirement

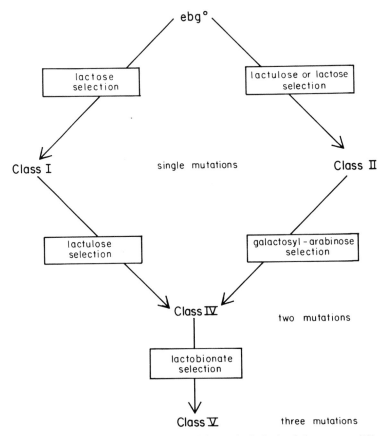

FIG. 2. Pathway for evolution of lactobionate hydrolysis of ebg enzyme (10).

for the pathway to function. The unevolved strain has a number of barriers it must overcome to use the ebg system as a regulated system for lactose metabolism. First, the ebg^0 enzyme is virtually inactive toward lactose. Second, the wild-type repressor only permits synthesis of ebg enzyme at 100-fold above the basal level (11). That level of expression is insufficient for growth even with the ebg enzymes most active toward lactose (11). Third, the lactose permease is required for the lactose catabolic pathway to function; however, its synthesis is not induced by lactose (19). Normally lactose is converted to allolactose by the *lacZ* β-galactosidase, and allolactose is the natural inducer of the *lac* operon (19). Because ebg enzyme does not carry out this transgalactosylation of lactose to generate allolactose (1, 8, 12, 23), the studies described above all included the synthetic inducer IPTG in the medium to turn on synthesis of the permease.

As a first step, a mutant was isolated that synthesized a class II enzyme under the control of the wild-type repressor. Although the activity of its ebg enzyme was increased 10-fold, the concentration of that enzyme permitted by the wild-type repressor was insufficient for growth in either the presence or the absence of IPTG (6, 11).

The wild-type repressor is insensitive to lactulose as an inducer (11), but in a second step a lactulose-sensitive repressor mutant was isolated (6). That mutation did not increase the basal level of ebg enzyme synthesis, but it made the strain sensitive to both lactulose and galactosyl-arabinose as inducers. At the same time, it increased the sensitivity of the repressor to lactose so that the lactose-induced level of ebg enzyme synthesis was now 1,000-fold above the basal level. That strain was capable of growth on lactose at the rate of $0.124 \ h^{-1}$ in the presence of IPTG, but was unable to grow in the absence of IPTG.

It is now necessary to point out another new function of class IV ebg enzymes: allolactose synthesis. This new function was discovered fortuitously (23), and it was shown that class IV enzymes convert lactose to allolactose at about 10% of the rate at which they hydrolyze lactose (8).

From these two properties of class IV enzymes, efficient galactosyl-arabinose hydrolysis and allolactose synthesis, it was predictable that a class IV mutant would be entirely self-regulated. A class IV mutant was isolated as the third step in the sequence by selecting a galactosyl-arabinose–utilizing derivative of strain 5A103 (9). The resulting strain, 5A1032, was able to grow on lactose without IPTG at the rate of $0.168 \ h^{-1}$. In the absence of lactose it expressed neither its *lacY* (permease) nor its *ebgA* (ebg β-galactosidase) gene. In the presence of lactose both genes were expressed at a rate sufficient to permit good growth on lactose. This series of mutations, then, constitutes a pathway for the evolution of an integrated metabolic system (9).

EVOLUTION BY DECRYPTIFYING GENES

There is a growing realization that many microorganisms possess genes that are not normally expressed during the life cycle of the cell, but that may be activated by mutations, by insertion elements, or by recombination (14). The existence of these genes is only revealed when microorganisms acquire new functions under deliberate selective pressures applied in the laboratory, and the only operational way to distinguish gene decryptification from simple mutations (such as those in the ebg system) is by the simultaneous expression of more than one new gene. Because simple repressor mutations could lead to the simultaneous expression of multiple genes in an operon, there is the additional empirical constraint that the new genes must be subject to environmental regulation. Even when new functions arise that require the expression of multiple regulated genes which are not expressed in the ancestral strain, there is a tendency to assume that those genes are normally expressed in nature under some condition not found in the laboratory. Because there is legitimate skepticism about the existence of cryptic genes, it is only recently that any studies have focused on the molecular basis of gene cryptification and decryptification.

It is not surprising that the most thoroughly studied case of a cryptic operon is in *E. coli*. Members of the family *Enterobacteriaceae* vary widely with respect to their abilities to utilize β-glucoside sugars. *Klebsiella* sp. can use both the aryl glucosides arbutin and salicin, and the disaccharide cellobiose (glucosyl-β-1,4-D-glucose). Most *Citrobacter* species utilize cellobiose, but not the aryl glucosides, whereas *Proteus vulgaris* uses the aryl glucosides but not cellobiose. *Salmonella* and *E. coli* are unable to grow on any β-glucosides, but *Salmonella* mutates easily to growth on cellobiose, and *E. coli* mutates to growth on arbutin and salicin at spontaneous frequencies as high as 10^{-5} (22, 24–26).

E. coli mutants that grow on these aryl glucosides are inducible for the expression of the *bgl* operon (21). The operon includes the genes *bglC*, which specifies a phosphoenolpyruvate-dependent transport protein that phosphorylates the β-glucoside sugar, *bglB*, which specifies a phospho-β-glucosidase enzyme, and *bglS*, which specifies a positive regulatory protein necessary for the expression of the operon (21). Mutations that activate the operon are in the *cis*-acting site *bglR* (21), and these *bglR+* mutations are dominant to the wild type *bglRβ* (β-glucoside–negative phenotype). It has recently been shown that these *cis*-dominant mutations are the result of insertion of the mobile DNA elements IS*1* and IS*5* into the *bglR* site (22). Schaefler and his colleagues (21) had hypothesized that mutations at *bglR* created a promoter for the operon, but neither IS*1* nor IS*5* is known to contain a promoter site or to provide promoter activity when inserted into other regions (22). The recent finding that mutations in *gyrA* and *gyrB* which affect DNA supercoiling also can activate the *bgl* operon (2) suggests that the insertions may also be acting by disrupting a local supercoiling environment that normally prevents expression of the operon.

Recent work in my laboratory has shown that *E. coli* K-12 contains a second cryptic operon for utilization of β-glucoside sugars (M. Kricker and B. G. Hall, Genetics **104**:s44, 1983; full paper to be published elsewhere). The objective of that study was to isolate a cellobiose-utilizing mutant of *E. coli*. We began by isolating a *bglR+* mutant by selection for growth on salicin. The resulting mutant, strain MK1, was unable to utilize cellobiose and was confirmed by genetic analysis to be *bglR+*. A spontaneous cellobiose-positive mutant, strain MK2, was derived from MK1 by selecting for growth on cellobiose. It was expected that cellobiose utilization would arise from mutations in the *bgl* operon that altered the specificity of either the transport protein or the phospho-glucosidase. Were this the case, the cellobiose-positive phenotype would show 40% cotransduction with *ilvD*, as does the *bglR+* mutation in MK1. Contrary to that expectation, there is no detectable cotransduction between cellobiose utilization (Cel+) and *ilvD*. When a deletion covering the entire *bgl* operon was introduced into a Cel+ strain, the growth rate on cellobiose was unaffected. Much to our surprise, the *bgl* deletion strain remained capable of growth on both arbutin and salicin. This provides strong evidence for a second cryptic operon for β-glucoside catabolism.

At this stage the use of the term "operon" is largely inferential. To determine whether the three functions (arbutin, salicin, and cellobiose utilization) were specified by separate genetic elements, a set of 15 independent cellobiose-negative mutants was isolated. All of these mutants had become negative for use of all three sugars, implying that a single gene specified all three functions. Analysis of the patterns of revertants showed that this was not the case. The majority of Cel+ revertants remained arbutin and salicin negative (Arb− Sal−). From several of these Cel+ revertants spontaneous Arb+ revertants were isolated, and a number of these remained Sal−. In a third step, Sal+ mutants were isolated from the Cel+ Arb+ strains, and these proved to be indistinguishable from the original Cel+ Arb+ Sal+ strain. This study is presently at a preliminary stage, and the nature of those mutations is still unknown. However, the stepwise reversion of the three functions indicates that those functions are specified by separate genetic elements.

CONCLUSIONS

There is now an abundance of evidence that microorganisms can evolve new metabolic capabilities as the consequence of a series of point mutations that alter either the catalytic properties of enzymes or the properties of regulatory proteins. Patricia Clarke's elegant studies, discussed by her elsewhere in this volume, show that the substrate range of the aliphatic amidase of *Pseudomonas aeruginosa* can be enormously extended by a combination of point mutations in regulatory and enzyme-coding genes. It is important to recognize that such studies tell us virtually nothing about the ways primitive enzymes might have evolved the sophisticated and specialized functions that we observe today. Both in nature and in the laboratory, evolution must work with the existing genome as its reservoir of information for generating new functions. The existing genes which are available for modification are themselves the end results of eons of selection, adaptation, and specialization. In the process of evolving to provide their current functions, much evolutionary flexibility must have been lost. It is difficult to conceive of a systematic way to probe the limits of new functions that can be acquired by simple modifications of existing enzymes. We would like to be able to predict, from the current properties of an organism, the likelihood of that organism evolving the ability to metabolize a novel resource. Although we are still far from that goal, we can say that when we can identify an existing system that metabolizes substrates that are closely related to the novel resource it is quite likely that the organism will, under sufficient selective pressure, evolve the ability to metabolize the novel resource. We can also say that a series of steps may be necessary, and that each of those steps will probably require a different selective pressure. The evolution of the ebgB system to permit lactobionate utilization is an excellent example of that situation. It was impossible to select directly for lactobionate utilization. Indeed, four steps were necessary. It was necessary to select a mutant that synthesized ebg enzyme constitutively and then to select in two successive steps a class IV *ebgA* mutant before applying any selection for lactobionate utilization per se. Using hindsight it is, of course, easy to apply exactly the correct series of selection pressures in precisely the right order. Predicting that series correctly is another matter entirely. Given sufficient knowledge about a particular enzyme and about its regulatory proteins, one might be able to make some reasonable predictions. As a practical matter, however, there is usually no detailed knowledge of either the enzyme or its regulatory system in those cases where it might be important to evolve a specific new function (for instance, selection of a mutant organism to degrade a noxious or toxic waste product). The best advice that can be given at this point would be to use as many analogs of the novel resource as possible to select a series of new mutants. As each mutant arises, one should determine the range of analogs it can utilize and subject it to selection for growth on all nonutilizable analogs. In nature, this means that the environment most likely to give rise to an organism that metabolizes a novel resource is that environment which posesses the richest variety of structurally related compounds.

Interest is now turning from adaptation via point mutations and toward adaptation via decryptification of preexisting but silent genes. Studies from Howard Slater's laboratory (discussed elsewhere in this volume) strongly suggest that dehalogenase genes may have been cryptic. Several major questions confront us concerning cryptic genes: What are the molecular mechanisms by which genes are silenced (cryptified) and later decryptified? What are the conditions that favor silencing of a gene? Why aren't cryptic genes lost due to irreversibly inactivating mutations? (For a discussion of this point, see reference 14.) How rich is this reservoir of cryptic genes? Are only certain classes of genes made cryptic? What relationship, if any, exists between environmental stress and decryptification? It now seems likely that repeated cycles of gene cryptification and decryptification may be as important a mechanism of genetic adaptation as is classical mutation by nucleotide substitutions. The next several years promise to be exciting as this new aspect of adaptive evolution is integrated into our understanding of the evolutionary process.

I am grateful to H. Slater, P. Clarke, M. Riley, H. Stokes, and M. Kricker for discussions and encouragement in developing some of the ideas presented in this article.

I am supported by Research Career Development Award 1 K04 AI 00366 from the National Institutes of Health. My research was supported by grants from the National Institutes of Health, the National Science Foundation, and the University of Connecticut Research Foundation.

LITERATURE CITED

1. **Campbell, J. H., J. Lengyel, and J. Langridge.** 1973. Evolution of a second gene for β-galactosidase in *Escherichia coli.* Proc. Natl. Acad. Sci. U.S.A. **70:**1841–1845.

2. **Dinardo, S., K. A. Voelkel, R. Sternglanz, A. E. Reynolds, and A. Wright.** 1982. *Escherichia coli* DNA topoisomerase I mutants have compensatory mutations in DNA gyrase genes. Cell **31:**43–51.

3. **Hall, B. G.** 1976. Experimental evolution of a new enzymatic function. Kinetic analysis of the ancestral (ebg⁰) and evolved (ebg⁺) enzymes. J. Mol. Biol. **107:**71–84.

4. **Hall, B. G.** 1977. Number of mutations required to evolve a new lactase function in *Escherichia coli.* J. Bacteriol. **129:**540–543.

5. **Hall, B. G.** 1978. Experimental evolution of a new enzymatic function. II. Evolution of multiple functions for EBG enzyme in *E. coli.* Genetics **89:**453–465.

6. **Hall, B. G.** 1978. Regulation of newly evolved enzymes. IV. Directed evolution of the *ebg* repressor. Genetics **90:**673–691.

7. **Hall, B. G.** 1981. Changes in the substrate specificities of an enzyme during directed evolution of new functions. Biochemistry **20:**4042–4049.

8. **Hall, B. G.** 1982. Transgalactosylation activity of ebg β-galactosidase synthesizes allolactose from lactose. J. Bacteriol. **150:**132–140.

9. **Hall, B. G.** 1982. Evolution of a regulated operon in the laboratory. Genetics **101:**335–344.

10. **Hall, B. G.** 1982. Evolution on a petri dish. The evolved β-galactosidase system as a model for studying acquisitive evolution in the laboratory. Evol. Biol. **15:**85–150.

11. **Hall, B. G., and N. D. Clarke.** 1977. Regulation of newly evolved enzymes. III. Evolution of the *ebg* repressor during selection for enhanced lactase activity. Genetics **85:**193–201.

12. **Hall, B. G., and D. L. Hartl.** 1974. Regulation of newly evolved enzymes. I. Selection of a novel lactase regulated by lactose in *Escherichia coli.* Genetics **76:**391–400.

13. **Hall, B. G., and D. L. Hartl.** 1975. Regulation of newly evolved enzymes. II. The ebg repressor. Genetics **81:**427–435.

14. **Hall, B. G., S. Yokoyama, and D. Calhoun.** 1983. Role of cryptic genes in microbial evolution. Mol. Biol. Evol. **1:**in press.

15. **Hall, B. G., and T. Zuzel.** 1980. Evolution of a new enzymatic function by recombination within a gene. Proc. Natl. Acad. Sci. U.S.A. **77:**3529–3533.

16. **Hall, B. G., and T. Zuzel.** 1980. The *ebg* operon consists of at least two genes. J. Bacteriol. **144:**1208–1211.

17. **Hartl, D. L., and B. G. Hall.** 1974. Second naturally occurring β-galactosidase in *E. coli.* Nature (London) **248:**152–153.

18. **Huber, R. E., G. Kurz, and K. Wallenfels.** 1976. A quantitation of the factors which affect the hydrolase and transgalactosylase activities of β-galactosidase (*E. coli*) on lactose. Biochemistry **15:**1994–2001.

19. **Jobe, A., and S. Bourgeois.** 1972. *lac* repressor-operator interaction. VI. The natural inducer of the *lac* operon. J. Mol. Biol. **69:**397–408.

20. **Mortlock, R. P.** 1981. Regulatory mutations and the development of new metabolic pathways in bacteria. Evol. Biol. **14:**205–267.

21. **Prasad, I., and S. Schaefler.** 1974. Regulation of the β-glucoside system in *Escherichia coli* K-12. J. Bacteriol. **120:**638–650.

22. **Reynolds, A., J. Felton, and A. Wright.** 1981. Insertion of DNA activates the cryptic *bgl* operon in *E. coli* K-12. Nature (London) **293:**625–629.

23. **Rolseth, S. J., V. Fried, and B. G. Hall.** 1980. A mutant ebg enzyme that converts lactose into an inducer of the *lac* operon. J. Bacteriol. **142:**1036–1039.

24. **Schaefler, S., and A. J. Malamy.** 1969. Taxonomic investigations on expressed and cryptic phospho-β-glucosidases in *Enterobacteriaceae.* J. Bacteriol. **99:**422–433.

25. **Schaefler, S., and I. Schenkein.** 1968. β-Glucoside perme-

ases and phospho-β-glucosidases in *Aerobacter aero-genes*: relationship with cryptic phospho-β-glucosidase in *Enterobacteriaceae*. Proc. Natl. Acad. Sci. U.S.A. **59**:285–292.

26. **Schäfler, S., and L. Mintzer.** 1959. Acquisition of lactose fermenting properties by Salmonellae. I. Interrelationship between the fermentation of cellobiose and lactose. J. Bacteriol. **78**:159–163.

Genetic Interactions in Microbial Communities

J. HOWARD SLATER

Department of Applied Biology, University of Wales Institute of Science and Technology, Cardiff CF1 3NU, Wales

How frequently do genes move between different microorganisms to produce novel genotypes? There is an attitude among microbiologists, produced by studies in microbial genetics, that many procaryotic genes are continually on complex migratory pathways linking different microbial populations. It is possible to conceive of a single genetic pool in a given habitat partitioned, perhaps weakly, by interconnecting units described at the genus and species level. From such a pool, in response to specific selection conditions, a novel organism may emerge through the realignment of different genotypes exploiting known genetic recombination processes such as conjugation, transformation, and transduction (2, 7, 8, 16). In some ways the balance of understanding has now moved too far in the direction of "unfettered movement of genes" (8) without adequately examining the question posed at the beginning of this paper. There is an attitude much like that of the three witches in Macbeth when engaged in the production of their own specialized brand of charms: all that is needed is an appropriate set of starting materials (genes in a diverse microflora), a caldron (a suitable habitat), and incantations and mixing (selection pressures).

> Double, double toil and trouble,
> Fire burn and cauldron bubble . . .
> Scale of dragon; tooth of wolf;
> Witches' mummy; maw and gulf
> Of the ravin'd salt-sea shark;
> Root of hemlock digg'd i' the dark;
> Liver of blaspheming Jew,
> Gall of goat, and slips of yew
> Sliver'd in the moon's eclipse;
> Nose of Turk, and Tartar's lips;
> Finger of birth-strangl'd babe
> Ditch-deliver'd by a drab,—
> Make the gruel thick and slab;
> And thereto a tiger's chaudron,
> For th' ingredients of our caldron.
> (W. Shakespeare, *Macbeth*, Act IV, Scene 1)

This view has come to dominate our thinking largely because we are now fully aware of the wide range of mechanisms which allow DNA sequences to move (8). In addition, powerful circumstantial evidence, derived largely from the widespread dissemination of either particular replicons (for example, plasmid RP1 has now been identified as appearing unchanged in 17 different microbial genera) or particular genes (such as the gene for β-lactamase production), shows that specific sequences of DNA can transfer widely. None of these studies has sought to investigate the frequency with which these events occur. Furthermore, it is quite apparent, especially to microbial ecologists, that genus and species boundaries are strenuously maintained and conserved: a *Flavobacterium* sp. isolated from one environment is much like that isolated somewhere else, even with respect to those properties that are not used as criteria for identification.

It is my view that under certain strongly selective conditions the potential for genetic transfer may be realized to yield novel genetic configurations. This may be especially true for microorganisms evolving novel catabolic pathways, especially in response to environmentally unusual or unique compounds (12, 16). We have suggested that mixtures of microorganisms widen the available genetic pool and may overall, but in a dispersed fashion, provide the requisite information to establish a new pathway (16). Thus, the evolution of an interacting microbial community may be a much more important prerequisite to the evolution of a novel pathway than a series of evolutionary events occurring within one organism (13, 14).

The evidence for these events is scant, and indeed little is known about the effect of environmental factors in promoting or preventing gene exchange. What information is available comes from laboratory-based studies which to a limited extent simulate some of the properties encountered by organisms in their natural habitats. But at the present time no firm conclusions can be made about the significance of gene exchange as an evolutionary event. A few recent studies have suggested that gene transfer events occur readily in chemostat mixed cultures. Knackmuss and his colleagues (4, 6) have demonstrated that the transfer of the TOL plasmid between two different pseudomonads alters their individual growth phenotypes, enabling the transconjugant to grow on additional halogenated benzoic acids. Kellog et al. (5) have suggested that the 2,4,5-T-degrading organism (that is, an organism capable of growth on 2,4,5,-T-trichlorophenoxy acetic acid) which they isolated from chemostat culture arose as a result of plasmid-mediated gene transfers. However, in both cases the detailed knowledge of the starting populations was limited, and it cannot be categorically asserted that other events, not associated with gene transfer, had not occurred.

We have sought to examine gene transfer events from a starting point which enables a detailed characterization of the initial microbial community and so all subsequent events.

GROWTH OF PSEUDOMONADS ON HALOGENATED ALKANOIC ACIDS

From a naturally isolated, complex microbial community growing on the herbicide Dalapon (2,2-dichloropropionic acid, 22DCPA), we isolated a strain of *Pseudomonas putida* designated PP3 (10). This pseudomonad grew well on 2-monochloropropionic acid (2MCPA) and was subsequently shown to synthesize two dehalogenases with different substrate specificities (17) and different mechanisms of action (18). The two enzymes principally dehalogenate mono- or disubstituted propionic or acetic acids, although they have a low level of activity against a range of different halogenated alkanoic acids (15). An even wider range of halogenated alkanoic acids act as enzyme inducers but are not substrates; that is, there is a large number of gratuitous inducers compared with acids that act as inducer and substrate (15; A. J. Weightman, A. L. Weightman, and J. H. Slater, submitted for publication). In the present context, however, the observation that some of the inducer-substrates do not support growth is important. In particular, monochloroacetic acid, dichloroacetic acid, and 2-monochlorobutanoic acid (2MCBA) are all readily dehalogenated, but none of the products, glycolate, glyoxylate, and 2-hydroxybutanoic acid (2HBA), respectively, supports growth. Thus, in cultures supplied with these substrates, the two dehalogenases are induced and the halogenated compounds are stoichiometrically converted to their degradation products: that is, *P. putida* PP3 co-metabolizes these compounds. (We have recently managed to evolve mutants of *P. putida* PP3 which can grow on glycolate and thus monochloroacetic acid and glyoxylate and consequently dichloroacetic acid. However, these are not readily obtained and require prolonged and complex selection sequences in chemostat culture [Weightman et al., submitted for publication].) The reason for these co-metabolic conversions is straightforward: both dehalogenases have poor substrate specificities and so these conversions might be fortuitous. However, there may be a selective advantage to this mechanism since halogenated alkanoic acids are toxic to *P. putida* PP3 to varying degrees, and the co-metabolic events may be considered to have an advantage by eliminating the effects of growth-inhibiting compounds.

In the laboratory these co-metabolic steps are normally viewed as unimportant curiosities. In nature they could be of much more significance, providing the basis for the establishment of interacting microbial communities (11, 12, 16). It is highly unlikely that the co-metabolic products will not be used by other populations with the capacity to utilize glycolate, glyoxylate, or 2HBA as carbon and energy sources. Accordingly, we have used this basic system to produce in the laboratory two-membered microbial communities which can be analyzed in detail.

TWO-MEMBERED MICROBIAL COMMUNITIES SHOWING A CO-METABOLIC-DEPENDENT COMMENSAL RELATIONSHIP

We have examined two systems based on glycolate production from monochloroacetic acid and 2HBA production from 2MCBA (Fig. 1). In both these communities interest centered around the frequency of dehalogenase gene transfer from the primary organism to the secondary organism, thereby evolving a novel strain with the capacity to use directly the original co-metabolite. The two communities have to be carefully established in laboratory continuous-flow culture systems and certainly do not exhibit the same high degree of stability associated with naturally isolated consortia. The most satisfactory method is to establish a steady-state culture of the primary, dehalogenating organism (for instance, *P. putida* PP3) and introduce the secondary, dependent organism (*Flavobacterium* sp.) at high population densities into the culture. The secondary organism must be taken from an exponentially growing

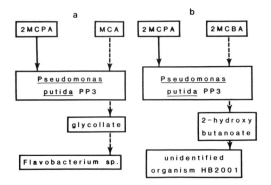

FIG. 1. Two-membered microbial communities growing on (a) 2MCPA with monochloroacetic acid or (b) 2MCPA with 2MCBA. The solid arrows indicate that the compound is used as a carbon and energy source; the dotted arrow indicates that the compound is co-metabolized.

culture, and it is advisable to allow a period of batch culture after inoculation before renewing medium flow. For unknown reasons the mono-chloroacetic acid-dependent community was less successful since, although all the monochloroacetic acid was dehalogenated, the glycolate-utilizing population did not use all the available glycolate. Thus, the glycolate-utilizing population was substantially smaller than was theoretically possible. However, this community was extensively examined by growth in an aerobic column fermentor packed with an inert support material to encourage surface growth and promote contact between the two populations. The two community members did not contain any plasmids, and under these conditions there was no evidence of dehalogenase gene transfer from *P. putida* PP3 to *Flavobacterium* sp. after over 3,500 h of continuous growth. In one experiment a monochloroacetic acid-utilizing strain of *P. putida* PP3 was isolated, but this probably was the result of selecting a mutant of PP3 which could utilize glycolate independently of a gene transfer event (Weightman et al., submitted for publication).

Similarly, we have demonstrated with the 2MCBA community that gene transfer does not occur, at least at high frequencies and within the time scale of our experiments, in the absence of a transferable plasmid in the primary population. For example, for a community growing on 2MCPA at 0.3 g of carbon per liter (to support the growth of *P. putida* PP3) and 2MCBA at 0.5 g of carbon per liter (to indirectly provide the 2HBA for the growth of *Pseudomonas* sp. HB2001), no recombinants with a 2Mcba$^+$ phenotype were selected after 700 h of continuous growth in a standard chemostat at a dilution rate of 0.05 h^{-1}. This corresponded to over 55 complete generations (12).

This group of experiments suggested that dehalogenase gene transfer, if it were to occur by transformation, did not occur. However, the information needs to be treated cautiously since a restricted range of experimental conditions were used, and these do not reflect the wide diversity of natural conditions which might promote transformation events in nature. It is also possible that 2MCBA-utilizing recombinants do arise, perhaps frequently, but that these are simply selected against by a more competitive two-membered community. This is unlikely, as will be demonstrated subsequently.

The dehalogenase genes are located on transposons (J. H. Slater, A. J. Weightman, and B. G. Hall, unpublished data). Thus, the introduction of a suitable transferable vector such as plasmid R68 might promote the transfer of dehalogenase genes at measurable frequencies. Plasmid R68.45, a derivative of R68 which has a chromosome-mobilizing potential (9, 19) was introduced into *P. putida* PP3 and, together with *Pseudomonas* sp. HB2001, a two-membered community was established as previously described. From such a system, 2MCBA-utilizing derivatives of *Pseudomonas* sp. HB2001 were readily isolated (Table 1). Indeed, the mutants were much more competitive than the original community and succeeded in displacing *Pseudomonas* sp. HB2001 entirely from the chemostat. The original primary population *P. putida* PP3 (R68.45) was not, however, replaced since this population grew on 2MCPA more efficiently than did the 2MCBA-utilizing mutant of *Pseudomonas* sp. HB2001. The reason was that only the fraction I dehalogenase was transferred to *Pseudomonas* sp. HB2001 and its μ$_{max}$ on 2MCPA was about half that of *P. putida* PP3 (R68.45), which produced both dehalogenases (see later). This experiment was not completely satisfactory since no detailed analyses were made during the 30 generations when a 2MCBA-utilizing derivative of *Pseudomonas* sp. HB2001 appeared and dominated the mixed culture. Furthermore, the putative plasmid prime carrying the dehalogenase(s) was not identified. In retrospect this experiment was also unnatural since the 2Mcba$^+$ 2Hba$^+$ phenotype appeared quickly, suggesting rapid and frequent transfer events under these selective conditions. This has not been observed subsequently.

A similar experiment is shown in Fig. 2 (D. Godwin-Thomas, A. J. Weightman, and J. H. Slater, submitted for publication). This two-membered community was established as before with the exception that the primary 2MCPA-utilizing organism, a derivative of *P. putida* PaW 340 (2Mcpa$^-$ 22Dcpa$^-$ 2Mcba$^-$ 2Hba$^-$) known as *P. putida* PPW2 contained an R prime, pUU2, which encoded for the fraction I dehalogenase (1). Plasmid pUU2 was derived from R68.44, a chromosome-mobilizing version of R68 (and *P. putida* PP3), and differed in three ways from R68.44 (Fig. 3), as follows. (i) It had lost a 2.0-kilobase (kb) fragment (segment I) which conferred its chromosome-mobilizing potential (19). (ii) It had acquired a cryptic fragment (segment II) of size 3.4 to 5.7 kb, depending on the way in which its size was calculated. This represented *P. putida* PP3 DNA but did not carry the dehalogenase genes. Together the changes in i and ii constituted plasmid pUU1 (Fig. 3). (iii) It had acquired a third section of DNA, 9.2 to 11.0 kb in size (not shown in Fig. 3), located at a site or sites between site *d* and *t* in plasmid pUU1 (Fig. 3). This encoded fraction I dehalogenase activity.

Plasmid pUU2 was stable, and it readily transferred and expressed dehalogenase I activity, especially in other *Pseudomonas* species.

TABLE 1. Growth of *P. putida* PP3 (R68.45) (2Mcpa[+] 22Dcpa[+] 2Mcba[+] 2HBa[−] Tc[r] Ap[r] Kn[r]) on 2MCPA (0.3 g of carbon per liter) and 2MCBA (0.5 g of carbon per liter) with *Pseudomonas* sp. HB2001 (2Mcpa[−] 22Dcpa[−] 2Mcba[−] 2Hba[+] Tc[s] Ap[s] Kn[s]) added to the culture after 1,080 h of growth of the primary organism[a]

Growth time (h)	Genera-tions	Culture absorbance (E_{600})	Chloride release (mM)	Viable organisms per ml			
				Total (nutrient)	PP3 (R68.45) (Tc Ap Kn)	HB2001 (2Hba[+])	Recombinants (2Mcba[+])
840	66.7	0.46	17.2	4.7×10^8	1.3×10^8	—	—
1,080	85.7	0.45	17.4	ND	ND	ND	ND
1,440	114.3	0.76	18.4	3.9×10^8	4.7×10^7	3.8×10^8	3.6×10^8
2,256	179.0	0.81	15.5	4.0×10^8	ND	6.2×10^8	6.7×10^8
2,448	194.0	1.06	21.0	3.1×10^8	9.9×10^6	4.1×10^8	4.4×10^8

[a] The recombinants had the phenotype 2Mcpa[+] 22Dcpa[+] 2Mcba[+] 3Hba[+] Tc[r] Ap[r] Kn[r]. The various populations were determined on medium as indicated in parentheses. ND, Not determined.

In the chemostat experiment shown in Fig. 2, *Pseudomonas* sp. HB2001 was carefully introduced after 300 h of growth of *P. putida* PPW2 (arrow a, Fig. 2). Recombinants were detected after 400 h (30 generations) which proved to be antibiotic-resistant derivatives of *Pseudomonas* sp. HB2001, and they were designated *Pseudomonas* sp. HB2002. This population became the dominant 2HBA-utilizing population, replacing *Pseudomonas* sp. HB2001. This evolved strain carried resistances for ampicillin, tetracycline, and kanamycin, but could not grow on 2MCBA

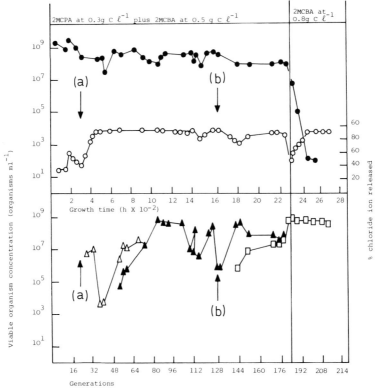

FIG. 2. The chemostat growth of a two-membered microbial community consisting of *P. putida* PPW2 containing plasmid pUU1 and *Pseudomonas* sp. HB2001, with 2MCPA (0.3 g of C per liter) and 2MCBA (0.5 g of C per liter) as the combined growth-limiting substrate. Free chloride ion as percentage of total combined (○); chloride ion supplied. Viable counts (organisms per milliliter) for *P. putida* PPW2 (●); *Pseudomonas* sp. HB2001 (△); *Pseudomonas* sp. HB2002 (▲); and *Pseudomonas* sp. HB2003 (□). Arrow a indicates the point of addition of *Pseudomonas* sp. HB2001 and arrow b indicates the second addition.

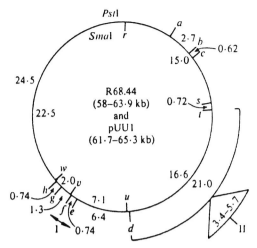

FIG. 3. Map of plasmids R68.44 and pUU1 showing the position of *Pst*I and *Sma*I restriction endonuclease digestion sites. Segment I constitutes the region of IS*21* (19) which is present in R68.44 but not pUU1. Segment II is present in pUU1 but not in R68.44; the position of the insertion is not precisely known but is located between restriction sites *d* and *t*. Plasmid pUU2 has segment III inserted at an unknown position between sites *d* and *t*. (From reference 1.)

and 2MCPA. The plasmid DNA was isolated from *Pseudomonas* sp. HB2002 and was found to be basically similar to plasmid pUU2, but with major variations which showed that the fraction I dehalogenase DNA had been lost (Fig. 4). Plasmid pUU3 was smaller than pUU2 by 8.0 kb and had lost three *Pst*I restriction endonuclease fragments of 15.0, 12.0, and 6.4 kb. These fragments correspond to the DNA between sites *c* and *e* in plasmid pUU2, the region where the dehalogenase I gene is known to be encoded (1). This DNA was not entirely lost since four new *Pst*I fragments were generated of sizes 8.0, 7.0, 5.4, and 5.0 kb (Fig. 4). Thus, the rearrangement of and loss of DNA in this region of pUU2 (leading to the formation of pUU3) produced the observed phenotype of *Pseudomonas* sp. HB2002. Moreover, we have subsequently shown that plasmid pUU3 is nontransferable, probably because of the complete or partial loss of a *tra*+ gene located within the 16.6-kb *Sma*I fragment between sites *t* and *u* (Fig. 3).

As noted above, *Pseudomonas* sp. HB2002 was more competitive than its parent strain HB2001. This is an interesting observation since a plasmid-containing population had succeeded in ousting an isogenic plasmid-minus population. This was unexpected since much of our earlier work on competition between plasmid-containing and plasmid-free populations suggested that

the plasmid-free populations were always more competitive in continuous culture (3). The results show clearly that there must be a competitive advantage for retaining pUU3, although at present that function is obscure.

The mixed culture of *P. putida* PPW2 containing plasmid pUU2 and *Pseudomonas* sp. HB2002 containing plasmid pUU3 was stably maintained for a further 1,000 h. There was no evidence of any transfer of pUU2 into strain HB2002, presumably because of plasmid incompatibility problems. Since it seemed unlikely that a *Pseudomonas* sp. with the phenotype 2Mcba+ 2Hba+ would evolve in this community, a second inoculation of *Pseudomonas* sp. HB2001 was made after 1,300 h of mixed culture growth (arrow b, Fig. 2). Within 200 h, recombinants with the expected phenotype were detected, and a novel strain designated HB2003 was isolated and shown to contain plasmid pUU2 (Fig. 4). *Pseudomonas* sp. HB2003 quickly dominated the culture, replacing strain HB2002 but not *P. putida* PPW2 because this strain was more competitive when growing on 2MCPA. When the medium was changed to contain 2MCBA alone, *P. putida* PPW2 was eliminated from the chemostat (Fig. 2). This experiment showed that it was possible to observe gene transfer and the formation of a novel organism. The novel organism was more competitive than its parental microbial community, but the frequency of its appearance was low.

CONCLUSIONS

Microbial populations and communities in nature appear to be capable of substantial movement of genes between both the same and different genera and species. It is quite clear that, for experimental evolution studies in the laboratory, under appropriate selection new phenotypes arise as a result of substantial gene rearrangement across population and community boundaries. In this respect evolution of new properties in the microbial world is considerably different from evolutionary processes in higher organisms. There is a much higher involvement of processes leading to the horizontal transfer of genetic information between existing individuals of the same generation. This contrasts with the emphasis on vertical evolution occurring between succeeding generations in most higher organism populations. On this analysis, therefore, it may be concluded that the witches brew is an acceptable picture. But what is the success rate of the potion? How frequently do genetic rearrangements occur in natural microbial populations? We do not have the quantitative answers.

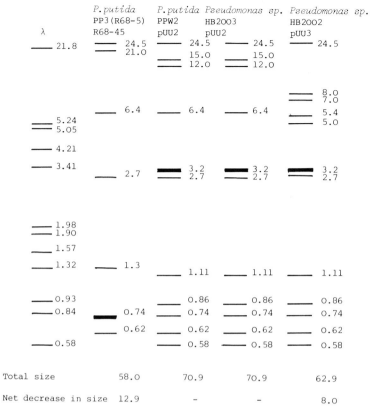

FIG. 4. *Pst*I restriction endonuclease digest fragment patterns for plasmids R68.44, two sources of pUU2 and pUU3 isolated from *P. putida* PP3 (R68.45), *P. putida* PPW2, and *Pseudomonas* sp. HB2003 and *Pseudomonas* sp. HB2002, respectively. The fragment sites are given in kilobases by calculation from a double digest of λ DNA with restriction endonucleases *Eco*RI and *Hind*III and with the use of standard fragment sizes (1). The thicker fragment lines indicate the position of two fragments of the same size.

LITERATURE CITED

1. **Beeching, J. R., A. J. Weightman, and J. H. Slater.** 1983. The formation of an R-prime carrying the fraction I dehalogenase gene for *Pseudomonas putida* PP3 using the IncP plasmid R68-44. J. Gen. Microbiol. **129:**2071–2078.
2. **Cohen, S. N.** 1976. Transposable genetic elements and plasmid evolution. Nature (London) **263:**731–738.
3. **Godwin, D., and J. H. Slater.** 1979. The influence of growth environment on the stability of a drug resistance plasmid in *Escherichia coli* K12. J. Gen. Microbiol. **111:**201–210.
4. **Hartmann, J., W. Reinekel, and H. J. Knackmuss.** 1979. Metabolism of 3-chloro-4-chloro and 3,5-dichlorobenzoate by a pseudomonad. Appl. Environ. Microbiol. **37:**421–428.
5. **Kellog, S. T., D. K. Chatterjee, and A. M. Chakrabarty.** 1981. Plasmid-assisted molecular breeding: new technique for enhanced biodegradation of persistent toxic chemicals. Science **214:**1133–1135.
6. **Knackmuss, H. J.** 1981. Degradation of halogenated and sulfonated hydrocarbons, p. 189–212. *In* T. Leisinger, A. M. Cook, R. Hutter, and J. Nuesch (ed.), Microbial degradation of xenobiotics and recalcitrant compounds. Academic Press, London.
7. **Reanney, D. C.** 1976. Extrachromosomal elements as possible agents of adaptation and development. Bacteriol. Rev. **40:**552–590.
8. **Reanney, D. C., P. C. Gowland, and J. H. Slater.** 1983. Genetic interactions among microbial communities.

Symp. Soc. Gen. Microbiol. **34:**379–422.
9. **Reiss, G., B. W. Holloway, and A. Puhler.** 1980. R68.45 a plasmid with chromosome mobilizing ability (cma) carries a tandem duplication. Genet. Res. **36:**99–109.
10. **Senior, E., A. T. Bull, and J. H. Slater.** 1976. Enzyme evolution in a microbial community growing on the herbicide Dalapon. Nature (London) **263:**476–479.
11. **Slater, J. H.** 1978. The role of microbial communities in the natural environment, p. 137–154. *In* K. W. A. Chater and H. J. Somerville (ed.), The oil industry and microbial ecosystems. Heyden and Sons, London.
12. **Slater, J. H., and A. T. Bull.** 1982. Environmental microbiology: biodegradation. Phil. Trans. R. Soc. London Ser. B **297:**575–597.
13. **Slater, J. H., and D. Godwin.** 1979. The influence of growth environment on the stability of a drug resistance plasmid in *Escherichia coli* K12. J. Gen. Microbiol. **111:**201–210.
14. **Slater, J. H., and D. Lovatt.** 1983. Biodegradation and the significance of microbial communities. *In* D. T. Gibson (ed.), Biochemistry of microbial degradation. Marcel Dekker and Sons, New York.
15. **Slater, J. H., D. Lovatt, E. Senior, A. J. Weightman, and A. T. Bull.** 1979. The growth of *Pseudomonas putida* on chlorinated aliphatic acids and its dehalogenase activity. J. Gen. Microbiol. **114:**125–136.
16. **Slater, J. H., and H. J. Somerville.** 1979. Microbial aspects of waste treatment with particular attention to the degradation of organic carbon compounds, p. 221–261. *In*

A. T. Bull, C. R. Ratledge, and D. C. Ellwood (ed.), Microbial technology: current status and future prospects. Cambridge University Press, London.

17. **Weightman, A. J., J. H. Slater, and A. T. Bull.** 1979. The partial purification of the dehalogenases from *Pseudomonas putida* PP3. FEMS Microbiol. Lett. **6:**231–234.

18. **Weightman, A. J., A. L. Weightman, and J. H. Slater.** 1982. Stereospecificity of 2-monochloropropionate dechlorination by dehalogenase enzymes from *Pseudomonas putida*. J. Gen. Microbiol. **128:**1755–1762.

19. **Willetts, N. S., C. Crowther, and B. W. Holloway.** 1981. The infection sequence I512 of R68.45 and the molecular basis for mobilization of the bacterial chromosome. Plasmid **6:**30–52.

Genetic Adaptations Involving Heavy Metals

ANNE O. SUMMERS

Department of Microbiology, University of Georgia, Athens, Georgia 30602

Living organisms have shared the world with metals since their beginnings. Some metals are essential for life, others are toxic, and some apparently have minimal effect upon either higher or lower organisms. The study of the interactions of living things with metals probably began with very early medicinal and alchemical use of such elements as mercury. With the discovery of microorganisms, the examination of their interactions not only with organic matter but also with inorganic compounds has grown steadily. It has been shown that some metals, like iron, are essential for procaryotes and lower eucaryotes even as they are for higher forms of life. The subset of metals toxic for microorganisms also overlaps those toxic for higher eucaryotes. However, there is for certain microorganisms a third category of interaction with metals not found among higher organisms, and that is the category of metabolite. The chemolithotrophic iron- and manganese-oxidizing bacteria may use these elements as a source of energy.

The ability of microorganisms to change the chemical state of metals gives them a potentially important, but currently little understood role in the geological cycling and the bioavailability of many elements. Because of concerns about the environment, in recent years attention has been focused particularly on the role of microorganisms in transforming metals either to or from more toxic or less toxic states. One of the most intriguing aspects of these studies has been the discovery that in many procaryotic microorganisms genes determining resistance to (and in at least one case, transformation of) toxic metal and metalloid compounds are carried by transmissible bacterial plasmids. The metal cation and oxyanion resistances which are known to be plasmid determined include: arsenate, arsenite, antimonate, chromate, inorganic mercury, and organic mercury in both gram-positive and gram-negative bacteria; tellurite and meta-borate in the gram-negatives; and cadmium, lead, and bismuth in *Staphylococcus* (108). Of these, the resistances to mercury compounds, to arsenate, and to cadmium are presently the best understood and will be described in detail below.

Knowledge of the occurrence of plasmid-determined metal ion resistances comes primarily from studies undertaken expressly to assess the incidence of the metal resistance phenotypes either in previously existing culture collections or in clinical or environmental material or both (5, 23, 39, 71, 72, 75, 98, 102, 104, 113, 114, 122, 123). All of these studies show that metal ion-resistant strains are sufficiently common to be cultivated without in vitro enrichment, from human and domestic animal normal flora as well as from urban and industrial sewage, soil, and both fresh and salt water. Although single and multiple metal resistances can be found in strains of bacteria which are also antibiotic resistant, only in the cases of the metals listed above has plasmid carriage actually been demonstrated by the essential criteria of transferability and curability of the phenotype and the related replicon.

If these levels of occurrence are an "adaptation," what is the selective pressure which is causing this change? For the toxic agents this question is difficult to answer, and we presently have only arguments by correlation, rather than arguments by causation, for what selects for these metal resistance phenotypes in humans, animals, and the environment. Many of them provide resistance to levels of the metal ions well above what would normally be encountered in a human or animal host. It is reasonable to argue that these resistance phenotypes are maintained in the gene pool by exposure of the bacteria to the metal ions in the environment. It is the case that high levels of multiply metal-resistant bacteria have been isolated from environments contaminated with toxic metals (83, 104, 122). In many but not all cases, the metal resistances have been shown to be plasmid determined, and frequently the plasmids carry common antibiotic resistance phenotypes as well. It is very likely, however, that the metal resistance genes themselves are not of particularly recent origin since, as noted above, living organisms have had to deal with the deleterious effects of metals from their very beginnings. This point will be noted below with particular reference to the mechanism of mercury resistance.

PLASMID-DETERMINED RESISTANCE TO MERCURY COMPOUNDS

Background. Mercury compounds have several biochemically distinct modes of toxicity. Organic and inorganic mercury compounds interact with sulfhydryl groups on both macromo-

94

lecular and low-molecular-mass compounds to form stable derivatives which are thus inactivated in their normal biological function (101, 124). Mercury compounds can also interact with carboxyl groups and imino groups; however, the interactions are less stable than those with sulfhydryl groups. Mercury markedly reduces the melting temperature of DNA; it has a high affinity for uracil and thymine (12).

Mercury has been used from ancient times to treat organic and infectious diseases (124), and medical use of mercury in disinfectants and diuretics has persisted even into this century. Modern industrial practices also contribute to the available mercury in the biosphere. There have been two occurrences of catastrophic mercury pollution in Japan when industrial mercury and mercurial emissions so heavily polluted an estuarine area that consumption of fish caught in the area led to severe neurological disorders. The use of methylmercury in the treatment of seed grain had a deleterious effect on bird populations in Sweden. Methyl mercury poisoning of humans and animals occurred in the United States and in Iraq as a result of unwitting consumption of inadequately labeled, methylmercury-treated seed grain (47).

Mercury resistance in penicillinase-producing clinical isolates of *Staphylococcus aureus* was first reported in England (24, 25, 85). Plasmid-determined mercury resistance was subsequently described in *Escherichia coli* by Smith (97) and in *Pseudomonas aeruginosa* by Loutit (53). Later work (18, 92, 126) demonstrated that there are two types of plasmid-determined resistance to mercury compounds. "Narrow-spectrum resistance" affords resistance to inorganic mercury (Hg^{2+}) and to a few organomercury compounds including merbromin (Mercurochrome) and fluorescein mercuric acetate. "Broad-spectrum resistance" affords resistance not only to the compounds listed above but also to other organomercurials including phenylmercuric acetate and ethylmercury thiosalicylate (Merthiolate or thimerosal).

Biochemical mechanism. Early studies (36–38, 44, 107, 115) on plasmid-carrying, mercury-resistant strains indicated that they converted inorganic mercury to metallic mercury vapor, which is relatively nontoxic for bacteria and diffuses rapidly from the cells. This detoxification is effected by the mercuric ion [Hg(II)] reductase (HR), a soluble, flavin adenine dinucleotide-containing, NADPH-dependent disulfide oxidoreductase (34, 93). The enzymatic activity has been found only in the cytoplasm (109), and the reaction requires a large excess of sulfhydryl compound in vitro. The active site sequence of the HR protein is nearly identical to that of the well-characterized disulfide-oxidore-

ductases glutathione reductase and lipoamide dehydrogenase (34, 35). The central 51,000 daltons of the HR structural gene of the mercury resistance transposon, Tn*501*, bears considerable resemblance to the functionally similar region of glutathione reductase (15).

The HR is synthesized as a polypeptide whose molecular mass has been reported as being from 59,000 to 69,000 daltons (15, 35, 45, 46). This polypeptide is processed slowly in vivo by the removal of 3,000 daltons from the amino terminus (46; Jackson et al., submitted for publication). The enzyme is apparently active as a dimer. Deletions (33, 105) or nonsense mutations affecting the carboxy-terminal 3,000 daltons (Jackson et al., submitted for publication) of the HR eliminate the ability to reduce inorganic mercury. The similarity between the HR and glutathione reductase does not extend to the amino or carboxy termini of the HR (15). Fox and Walsh (34, 35) have also noted no correspondence between the specific activity and the degree of processing in purified preparations of the HR and have shown that the 51,000-dalton carboxy terminus of the HR (whose amino terminus has been removed by a single chymotrypsin cleavage) still has Hg(II)-stimulated NADPH-oxidase activity. These observations suggest that the amino terminus of the HR is not essential for the enzymatic reduction of mercuric ion to elemental mercury.

Kinetic studies of the HR encoded by the plasmid R100 show that the enzyme demonstrates hysteretic behavior (86) and that it is similar but not identical to the enzyme encoded by the transposon Tn*501* in *Pseudomonas aeruginosa* (J. Williams, personal communication). The Hg(II) reductase from gram-negative bacteria appears to be quite heat stable, but that from gram-positive bacteria is not (108). There are no published data on the antigenic relationships among HR proteins from various sources; however, there may be several subgroups among gram-negative bacteria, and antisera to the gram-negative HRs do not cross-react with the enzyme from *Staphylococcus aureus* or *Bacillus* sp. (S. Silver, personal communication).

The Hg(II) uptake system. Since the HR enzyme activity is intracellular, there must be a transport system to bring Hg(II) into the cell where it can be reduced. The physiological evidence for an Hg(II) uptake system comes from a subclass of mutants which have lost mercury resistance (Hg^r) and acquired a phenotype described as "supersensitive" to mercury (Hg^{ss}). Hg^{ss} strains are more sensitive to mercury than a strain which has no plasmid at all (Hg^s; note that there are also mutants in the mercury resistance [*mer*] operon which confer only the Hg^s phenotype). The Hg^{ss} phenotype arises from a mutation which eliminates the HR enzymatic activity without altering the Hg(II) uptake activity (70). When induced by prior growth on subtoxic levels of mercuric ion, Hg^{ss} strains bind three- to fourfold more ^{203}Hg than when not so induced (32, 33, 70, 105). This Hg(II) uptake activity cannot be measured in the parental Hg^r strain because as rapidly as Hg(II) comes into the cell it is volatilized by the action of the HR enzyme and no transient accumulation of ^{203}Hg by the cells can be detected. The *mer* operon encodes three proteins of 15,100, 14,000, and 13,000 daltons (45, 46). The two larger proteins are located in the inner membrane and are not processed detectably in minicells. The 13,000-dalton peptide is soluble and is processed to a 12,000-dalton form in vivo (46). This smallest protein is found in the periplasm. We suspect that it is a binding protein for mercuric ion (Jackson and Summers, unpublished data).

However, measurable Hg(II) uptake activity is observed only in those Hg^{ss} strains in which the mutation is located in the carboxy terminus of the HR gene. Mutations in the amino terminus of the HR gene, while still Hg^{ss} in phenotype and lacking detectable HR enzymatic activty, also lack the Hg(II) uptake activity. We have found that mutants of this type do not synthesize the *mer* operon membrane-related peptides upon induction with mercury. Careful genetic and physical mapping of a large number of *mer* operon mutants and recent sequence data indicate that there is an a protein of approximately 20,000 daltons whose reading frame overlaps (but is out of phase with) that of the amino terminus of the HR. We have also recently observed a previously undescribed, faintly labeled, mercury-inducible polypeptide of approximately that size in minicell protein preparations of a strain carrying a small multicopy derivative of the operon (7). We believe this protein is a second positive regulatory function of the operon (dependent upon the previously described positive regulator, *merR*) which acts to turn on the transport function.

With the exception of the mercury resistance afforded by plasmid-determined H_2S production in certain clostridia (79), plasmid-determined mercury resistance involves enzymatic reduction of Hg(II) in all gram-positive and gram-negative genera examined (42, 43, 76, 78, 80, 106, 108). Enzymatic activity corresponding to HR has recently been described in bacteria as diverse as *Thiobacillus ferro-oxidans* (76) and *Mycobacterium scrofulaceum*, though only in the latter case is the activity known to be plasmid encoded (59). Despite the fact that it uses one NADPH for each mercury atom reduced, the HR provides a relatively inexpensive detoxification mechanism compared to antibiot-

ic resistance mechanisms which phosphorylate, acetylate, or adenylylate the antibiotic. We have not found either spontaneous or chemically induced mutants to mercury resistance in R^- strains. Apparently no single mutation in any of the related chromosomally determined disulfide oxidoreductases, nor in any transport system, can produce a detoxification system as effective as the plasmid-determined locus.

The fact that the mercury resistance mechanism causes the biotransformation of mercuric ion to relatively nontoxic, diffusible elemental mercury results in the detoxification of the medium for all of the bacteria present. Since mercury is bacteriostatic rather than bactericidal, if mercury-resistant bacteria are present they can detoxify the medium, and sensitive cells will eventually grow. In addition, because medium components can sequester the ion, rendering it considerably less toxic, cells can tolerate higher concentrations of $HgCl_2$ in richer media than in minimal media. Nonetheless, for any given medium, there is a 50- to 100-fold difference in the minimal inhibitory concentrations for sensitive and resistant strains. The minimal inhibitory concentration does not increase when the locus is carried on a multicopy plasmid (69). This is apparently because the rate of mercury transformation by the HR is limited by the ability of the Hg(II) uptake system to bring the ion into the cell. Recent observations suggest that carriage of the *mer* locus in the multicopy state is actually deleterious to the cell and results in a decrease in the efficiency of plating on mercuric chloride (73; Jackson et al., submitted for publication).

Organomercury detoxification. Plasmids conferring broad-spectrum resistance to mercury compounds encode an enzyme called the organomercurial lyase (OL) (36–38, 93) in addition to the Hg(II) reductase. The OL splits the covalent carbon-mercury bond found in such compounds as phenylmercuric acetate or methylmercury. The soluble enzyme is apparently a heterodimer composed of subunits of 22,000 and 24,000 daltons (45). The enzyme has not been extensively purified, but it appears to require only a sulfhydryl compound for its activity in vitro. The location of the enzyme within the cell is not known; attempts to demonstrate its location in either the cytosol or the periplasmic space have been inconclusive (unpublished data). The OL may be loosely associated with the membrane and thereby effect the initial degradative step before the mercurial compound enters the cell. Such an arrangement would eliminate the need for an additional organomercurial transport system. There is presently no information on this subject, other than the observation that a broad-spectrum resistance plasmid encodes only the OL peptides in addition to the ones known to be encoded by the *mer* operon alone (46).

The OL and the HR convert organomercurials to a less toxic form. Bacteria can also convert inorganic mercury to methyl mercury (11, 26, 40, 47, 51, 121, 131, 133, 134). However, the ability of bacteria to methylate mercury is not plasmid encoded but is a result of normal cellular processes which are chromosomally encoded. Formation of methyl mercury appears to be a side effect of cobalamin-mediated methionine biosynthesis (131). The rate of methylation is of the order of nanograms of Hg methylated per day (10^{-18} mol min^{-1} ml^{-1} of log-phase culture) compared to 20 nmol min^{-1} ml^{-1} for Hg(II) reduction carried out by a typical mercury-resistant *E. coli* culture. Thus, under laboratory conditions, reduction of mercuric ion takes place approximately 10^{10}-fold faster than methylation does.

Regulation and structure of the *mer* operon. Induction with mercuric chloride effects an increase of approximately 100-fold in the activity of the HR enzyme (126). The synthesis of the HR is also stimulated by the addition of cyclic AMP, and this stimulation is dependent upon the product of the *crp* locus, the catabolite activator protein. The activity of the Hg(II) uptake system is not detectably stimulated by the addition of cAMP (105). Early genetic studies showed that the *mer* operon is under the control of a positive, *trans*-acting regulatory protein (33). These initial studies, using transposon Tn*802*-generated mutants, indicated that there were only two complementation groups: group A, which encoded the *trans*-acting regulatory protein, and group B, which encoded both the transport functions and the HR structural gene. However, data on the response to catabolite activation, noted above, and more recent observations made with hydroxylamine-generated mutants show that complementation group B can be divided into at least two subgroups, one of which may have a coregulatory role as well as a role in Hg(II) uptake (Jackson et al., submitted for publication). These observations correlate well with the above-noted recent observation of a second regulatory protein responsible for turning on the transport function proteins. Studies with bacteriophage Mu *d1*–induced *mer-lac* fusions indicate that induction is responsive to as little as 10^{-8} M HgCl$_2$ in tryptone broth, and those fusions which are complementation group B mutants fall into high-expression and low-expression classes in terms of the rate of beta-galactosidase synthesis (C. Slater-Jones, unpublished data). Recent work by Ni'Bhriain et al. (73) describes a locus which they call also *merC*, but to which they were unable to attribute any function, and a second new locus mapping beyond the HR structural gene which they call *merD*. *merD* is likely an artifact of *mer*-controlled overexpression of Tn*5*-encoded genes, since the MerD phenotype is only expressed in multicopy derivatives, and the in vitro deletion mutagenesis of this region performed by Ni'-Bhriain et al. does not support the existence of any *mer*-related functions beyond the HR structural gene. Nonetheless, these observations suggest that there may be more complex interactions between the elements of the operon than was envisioned in the original model.

Early studies also assigned genotypic names to certain physiological or biochemical functions of the operon (33). Thus the regulatory function was called *merR*, the transport function was called *merT*, and the HR was called *merA*. Since we now know that there are at least three polypeptides involved in the Hg(II) uptake function, the genotypic designation *merT* was premature. We currently use this term in quotes, "*merT*," to indicate that it does not designate a single gene. The apparent order of the functions within the operon is *merR*–(promoter)–"*merT*"–*merA*. This order is still generally agreed upon, but the questions of whether there are separate promoters for the transport functions and the HR, and what is the relationship of *merR* and the newly discovered transport regulator, have not been resolved.

In the *mer* operon of plasmid NR1, the *merR* function is located immediately adjacent to the IS1b element at one end of the R-determinant region (21, 32, 33, 61, 111). Sequence available for this region (M. Chandler, personal communication) is homologous with the "left" end of the mercury resistance transposon, Tn*501* (8, 100), and includes several regions of dyad symmetry, a promoter-like region, a Shine-Dalgarno sequence, and the beginning of an open reading frame for a protein whose amino terminus is rich in arginine. The sequence is also available for a group A mutant bearing a Tn*802* insertion in this region of Tn*501*. This mutant, pUB986, has a null phenotype, and the insertion separates the promoter from the Shine-Dalgarno sequence (J. Grinsted, personal communication). The simplest, though not necessarily the only, interpretation of these data is that the promoter for the *merR* protein is at the "leftmost" side of the operon in both NR1 and Tn*501*.

In the plasmid NR1 there is a *mer* promoter located approximately 550 base pairs from IS1b. Because of its responsiveness to mercuric ion, its dependence upon the *merR* gene, and its location, we believe that this is the promoter for the Hg(II) uptake function structural genes (13). Mapping of a series of insertion mutations also demonstrated that the border between the A and B complementation groups is approximately 600 base pairs to the "right" of IS1b (7, 73).

The mercury resistance transposon, Tn501, does hybridize to the *mer* operon region of NR1 (R. Rownd and J. Miller, unpublished data); however, recent comparative work has shown that the *mer* operons of several IncFII plasmids are more closely related (by restriction analysis) to each other than any of them is to Tn501 (110, 135). Our fine-structure restriction analyses of Tn501 and the *mer* operon of NR1 (our unpublished data) and sequence data on Tn501 (15) and the *mer* operon of NR1 (our unpublished data) are also consistent with approximately 70% homology between these examples of the *mer* operon. Nonetheless, mutants in Tn501 complement those in R100 and fall into the same complementation groups as R100/NR1 derivatives (7). Mutants in R100 also complement mutations in the *mer* operon of the IncP plasmid R702 (7, 105), though it is not known how the R702 *mer* operon corresponds to either R100 or Tn501 structurally. Because of their placement with respect to transcription coming from the *mer* operon, the genes for transposase and resolvase in Tn501 are induced whenever the *mer* operon is induced (91). The transposition genes of Tn501 are related to those of both Tn1721 and Tn3. However, mutants in the transposition functions of Tn501 will complement those of Tn1721 but not those of Tn3. In this case, gross similarities are not borne out in functional complementation. Transposons conferring broad-spectrum resistance have recently been reported; in restriction analyses, they appear to differ from Tn501 (84).

PLASMID-DETERMINED RESISTANCE TO ARSENIC AND ANTIMONY COMPOUNDS

Background. Arsenate is an analog of phosphate and is taken into the cell via the phosphate transport system (2, 3, 9, 87, 88, 128, 129). It substitutes for phosphate in many enzymatic reactions; however, the arsenate derivatives formed are unstable and result in a loss of energy. Arsenite is a sulfhydryl poison, and antimonate is also thought to function in this manner (28). Arsenite also appears to have a very specific effect upon the SOS repair system in *E. coli* (89). Both organic and inorganic arsenic compounds occur in the environment. Cacodylic acid is used for pest control (28, 30, 132), and arsenite has been used in livestock dips (116).

Resistances to arsenate, arsenite, and antimonate are found on plasmids of gram-negative bacteria (41, 48, 104; T. Trezona, H. Taylor, and M. Taylor, Abstr. Annu. Meet. Am. Soc. Microbiol. 1981, H125, p. 134) and gram-positive bacteria (74). Resistance to arsenate is one of a few markers which have been associated with

strains of *S. aureus* isolated from toxic shock syndrome victims (6). Arsenate and arsenite resistances have also been found on the "lactose plasmid" of *Streptococcus lactis* (27). Large-scale epidemiological assessment of this phenotype indicates that it is not uncommon (56, 104). Resistance to arsenite mediated by an arsenite oxidase has been described in *Pseudomonas* sp. (116, 117) and *Alcaligenes* spp. (1, 77, 82), and in the latter genus this phenotype is apparently plasmid coded (M. Taylor, personal communication). Bacteria and other microorganisms are able to methylate and demethylate arsenic compounds (19, 20, 30, 57, 90, 94, 130). Many shellfish, particularly shrimp, contain arsenicals, possibly of marine microbiological origin (4, 10). It may be that many of these arsenic biotransformations, which are significant aspects of a natural "arsenic cycle" (30), are encoded by plasmids, but that remains to be determined.

Mechanism of arsenate resistance. Bacteria can become resistant to arsenate by alterations in the chromosomal genes encoding phosphate transport (2, 3, 128, 129). The plasmid-determined arsenate resistance locus also involves a transport system, rather than the transformation of arsenate to a less toxic or volatile form. The plasmid-determined resistance is additive to that provided by chromosomally determined mutations (95). Early studies with plasmids of both *E. coli* and *S. aureus* showed that resistant cells bound less radioactive arsenate than sensitive cells and that this uptake of arsenate was prevented by high levels of phosphate. The resistance phenotype and the decreased arsenate uptake phenomenon were both inducible by prior exposure of the culture to subtoxic levels of arsenate, arsenite, or antimonate (95). In fact, exposure to subtoxic concentrations of any one of these oxyanions induced the expression of the resistance phenotype for all three (as measured by ability of the strain to grow when subsequently challenged by growth in medium containing selective levels of any one oxyanion). Therefore, although their biochemical targets and mechanisms are likely quite different, these resistances are coordinately regulated.

Subsequently it has been shown that both *S. aureus* (96) and *E. coli* (66, 96) strains carrying the arsenate resistance locus on their respective plasmids synthesize an energy-dependent arsenate efflux system. In energy-starved *E. coli* cells lacking the proton-translocating ATPase, this efflux system is dependent upon added glucose but will not function with added succinate. In the presence of glucose, cyanide, an inhibitor of electron transport, has little effect on arsenate efflux, but potassium fluoride inhibits it markedly. This argues strongly that the efflux system is energized by ATP rather than by the proton gradient (66). The energetic basis for the efflux in *S. aureus* has not been established (96).

More recent work has demonstrated that in the IncFI plasmid of *E. coli*, R773, the locus for arsenate, arsenite, and antimonate resistance is carried on a 29-kilobase EcoRI fragment (65). Minicell analyses of R773 itself and of cloned derivatives carrying this fragment show that a single peptide of 64,000 daltons is synthesized in response to arsenite induction. When the size of this fragment is reduced with *Hin*dIII to 4.3 kilobases, the cloned derivative expresses this protein and resistance to arsenate and arsenite constitutively at high levels. An additional polypeptide of 16,000 daltons is also expressed constitutively by this smaller cloned fragment. This suggests that the 65,000-dalton protein is normally under negative regulation and that its repressor has been removed in the smaller recombinant derivative. Alternatively, in the smaller derivative the expression of the protein may be under control of a vehicle promoter. This is unlikely, in this particular case, because the promoter in question would be the relatively weak tetracycline resistance promoter of pBR322. The 64,000-dalton protein is found both in the membrane and in the cytoplasm, but predominantly in the latter. This observation may be an artifact of its overproduction from a multicopy plasmid. No increase in ATPase activity is associated with the appearance of this protein (65). The phenotypes and polypeptide patterns obtained with Tn5-induced mutants of the smaller derivative suggest that the arsenite resistance cistron is promoter proximal relative to arsenate resistance in the operon and that the 64,000-dalton peptide is associated with arsenite resistance. The function of the 16,000-dalton peptide is presently unknown (B. Rosen, personal communication).

Mobley et al. have recently shown that the arsenate resistance locus of the IncFI plasmid, R773, is homologous to that of the IncN plasmid, R46, under stringent hybridization conditions (67). These loci also encode quite similar arsenic-inducible polypeptides. This suggests that arsenic resistance may be carried by a transposable element on these two otherwise unrelated plasmids. DNA from 7 of 18 clinical strains originally isolated for their arsenate resistance also hybridized to DNA encoding the arsenate resistance locus of both R773 and R46. However, the remainder did not hybridize, though their arsenate resistance phenotype was stable. This suggests that, as in plasmid-encoded tetracycline resistance, there is more than one locus which confers resistance to arsenic compounds (67). Nothing is presently known of the mechanisms of resistance to arsenite or to anti-

monate which are also determined by the operon that determines arsentate resistance. It is known that strains with this locus do not have an arsenite-oxidizing activity, nor do they detoxify arsenite by sequestration (95).

PLASMID-DETERMINED RESISTANCE TO CADMIUM

Background. Unlike mercury compounds, cadmium does not appear to have a strong affinity for a specific nucleotide base, though it does bind to the bases rather than to the phosphate backbone of DNA (12). Cadmium is thought to be mutagenic, carcinogenic, and teratogenic (31, 49, 54, 55, 125, 136) and has been shown to induce single-strand breaks in *E. coli* DNA (64). It has an affinity for sulfhydryl groups which is slightly lower than that of mercury. Cadmium contamination of soil occurs in the vicinity of smelters (31) and in forest litter (52). Because of its use in agriculture, it is also found in cigarette smoke (29).

Cadmium resistance was described as a plasmid-determining marker in the late 1960s by Miller and Harmon (62) and Novick and Roth (74), and it is very common among the plasmids of *Staphylococcus aureus* (50). Since the relative minimal inhibitory concentrations for sensitive and resistant strains differ by 100-fold, cadmium resistance is a very effective selective marker in this genus. In addition, since resistance does not result in detoxifying transformation of cadmium, sensitive bacteria are not protected by the presence of resistant bacteria (unlike the case of mercury resistance).

There may be a cadmium resistance locus on the plasmids of facultative gram-negative bacteria. One report describes multiply metal- and antibiotic-resistant strains of *Pseudomonas* sp. from a hospital in Japan which were also cadmium resistant (68). However, since no curing or conjugation experiments were attempted, no linkage of this marker to other plasmid markers was demonstrated, and the association may have been fortuitous. There is also a report of transfer of a cadmium resistance locus along with loci for cobalt and nickel resistance in a strain of *Pseudomonas* (60). Since it is possible to adapt *E. coli* to growth on cadmium (63, 64), it may be that some of the observations of cadmium-resistant gram-negative bacteria (22) are the result of such a physiological adaptation, rather than of a genetic locus conferring cadmium resistance. There is a recent report of plasmid-determined cadmium resistance in the gram-negative anaerobe *Bacteroides* spp. (123), but there is presently no information on the mechanism of this resistance. In the presence of mercury, certain pseudomonads can form trace levels of methyl cadmium compounds under laboratory conditions. Methyl cadmium is volatile but also rapidly hydrolyzes in aqueous media. This phenomenon, which requires mercury, is apparently a side reaction of bacterial methylation of mercury and therefore is probably not a common mechanism for cadmium detoxification by these bacteria (F. Brinckman, personal communication). As noted above, bacterial methylation of mercury is effected by chromosomally encoded genes.

Mechanism of cadmium resistance. There are two loci conferring cadmium resistance in *Staphylococcus* plasmids. *cadA* confers high-level resistance to cadmium, and when it is mutated, the cells retain a low-level resistance to cadmium mediated by *cadB* (99). Early studies of cadmium-resistant *Staphylococcus* strains showed that they bound less radioactive cadmium than the corresponding sensitive, non-plasmid strains (16, 17, 120, 127). The details of this process (mediated by the *cadA* locus) have been worked out by Tynecka et al. (118, 119). In both sensitive and resistant strains, cadmium enters the cell via the manganese transport system, for which it has a higher affinity than does manganese itself. Cadmium uptake is irreversible in sensitive cells and ultimately leads to inhibition of respiration. The resistance mechanism apparently does not alter the manganese transport system, but rather provides a separate, energy-dependent system which expels cadmium, but not manganese, from the cell. This efflux system prevents the internal accumulation of cadmium by the cell. The system is apparently a cadmium-proton antiporter, driven by the chemiosmotic gradient. Cadmium enters the resistant cell in the same manner as in the sensitive cell, i.e., by the manganese transport system; however, the plasmid-encoded efflux system is capable of effectively excluding cadmium up to external concentrations of 100 mM. Neither growth nor respiration is affected even at this level, so the redox chain can provide protons for cadmium-proton exchange. It appears that the pH component is the driving force for the cadmium efflux system. The cadmium resistance phenotype and the efflux system are expressed constitutively.

Attempts to demonstrate *cadA*-determined cadmium efflux in membrane vesicles of *S. aureus* have not been successful, though they have demonstrated that the *cadB* locus may code an inducible cadmium-binding factor which also has some affinity for zinc (81). It is not presently known, either genetically or biochemically, how many peptides are involved in cadmium resistance.

OTHER PLASMID-DETERMINED METAL RESISTANCES

Presently nothing is known about the mechanisms of plasmid-determined resistance to the oxyanions arsenite and antimonate (95), metaborate (103), and tellurite (104). A chromate-reducing activity has been associated with strains of *Pseudomonas* capable of transferring chromate resistance (14). Tellurite resistance is known to be associated exclusively with plasmids of the IncH2 group of the *Enterobacteriaceae* and plasmids of the IncP2 group of *Pseudomonas*. There is one recent report of plasmid-carried copper resistance in *E. coli* (112) and a single report of plasmid-carried silver resistance arising in a burn patient (58). This may not be a general mechanism for silver resistance, since in more recent screening of hundreds of silver-resistant burn isolates, none have been found with transferable silver resistance, though many of these transferred antibiotic resistances at several temperatures (A. T. McManus, personal communication). The resistances to lead and to bismuth found on plasmids of *Staphylococcus* sp. have also not been examined. There are data strongly suggesting that plasmids of the facultative gram-negative bacteria confer resistances to cobalt and to nickel (60, 97) and to cadmium (22, 60, 68), but more work is needed in these areas.

CONCLUSION

In general, it appears that the most widely found instances of adaptation to metal stress in bacteria involve plasmid-determined genes. Moreover, where the mechanisms of resistance are well understood, it appears that they are biochemically and physiologically similar even in different genera of bacteria. It remains to be seen how extensive this apparent homology is at the level of regulation of gene expression and at the level of the DNA sequences themselves.

As noted above, since corresponding ecological studies to demonstrate a "cause" for metal resistances have not yet been done, we can only hypothesize that the occurrence of such plasmid-carried loci in nature is a result of direct selection for the phenotypes by toxic metals present in at least some of the many niches which these procaryote populations occupy.

Obviously there is much work yet to be done even on those systems which we currently understand fairly well, not to mention those in which at present we know nothing of the physiology or biochemistry of the resistance mechanism. The right combination of molecular and ecological studies of any one of these loci will tell us much about how the procaryote gene pool evolves in response to environmental stress.

LITERATURE CITED

1. Abdrashitova, S. A., B. N. Mynbaeva, and A. N. Ilyaletdinov. 1981. Oxidation of arsenic by the heterotrophic bacteria. Mikrobiologiya **50**:41–45.
2. Alfasi, H., D. Friedberg, and I. Friedberg. 1979. Phos-

phate transport in arsenate-resistant mutants of *Micrococcus lysodeikticus*. J. Bacteriol. **137**:69–72.

3. **Alfasi, H., and I. Friedberg.** 1977. Arsenate resistant mutants of *Micrococcus lysodeikticus* altered in phosphate transport. Isr. J. Med. Sci. **13**:630.

4. **Andreae, M. O.** 1978. Distribution and speciation of arsenic in natural waters and some marine algae. Deep-Sea Res. **25**:391–402.

5. **Austin, B., D. A. Allen, A. L. Mills, and R. R. Colwell.** 1977. Numerical taxonomy of heavy metal-tolerant bacteria isolated from an estuary. Can. J. Microbiol. **23**:1433–1447.

6. **Barbour, A. G.** 1981. Vaginal isolates of *Staphylococcus aureus* associated with toxic shock syndrome. Infect. Immun. **33**:442–449.

7. **Barrineau, P., and A. O. Summers.** 1983. A second positive regulatory function in the *mer* (mercury resistance) operon. Gene **25**:209–221.

8. **Bennett, P. M., J. Grinsted, C. L. Choi, and M. H. Richmond.** 1978. Characterisation of Tn501, a transposon determining resistance to mercuric ions. Mol. Gen. Genet. **159**:101–106.

9. **Bennett, R. L., and M. H. Malamy.** 1970. Arsenate resistant mutants of *Escherichia coli* and phosphate transport. Biochem. Biophys. Res. Commun. **40**:496–503.

10. **Benson, A. A., and R. E. Simmons.** 1981. Arsenic accumulation in Great Barrier Reef invertebrates. Science **211**:482–483.

11. **Berdicevsky, I., H. Shoyerman, and S. Yannai.** 1979. Formation of methylmercury in marine sediment under *in vitro* conditions. Environ. Res. **20**:325–334.

12. **Bloomfield, V. A., D. M. Crothers, and I. Tinoco, Jr.** 1974. Physical chemistry of nucleic acids. Harper and Row, New York.

13. **Bohlander, F. A., A. O. Summers, and R. B. Meagher.** 1981. Cloning a promoter that puts the expression of tetracycline resistance under control of the regulatory elements of the mer operon. Gene **15**:395–404.

14. **Bopp, L. H., A. M. Chakrabarty, and H. L. Ehrlich.** 1983. Chromate resistance plasmid in *Pseudomonas fluorescens*. J. Bacteriol. **155**:1105–1109.

15. **Brown, N. L., S. J. Ford, R. D. Pridmore, and D. C. Fritzinger.** 1983. Nucleotide sequence of a gene from the *Pseudomonas* transposon Tn501 encoding mercuric reductase. Biochemistry **22**:4089–4095.

16. **Chopra, I.** 1970. Decreased uptake of cadmium by a resistant strain of *Staphylococcus aureus*. J. Gen. Microbiol. **63**:265–267.

17. **Chopra, I.** 1975. Mechanisms of plasmid-mediated resistance to cadmium in *Staphylococcus aureus*. Antimicrob. Agents Chemother. **7**:8–14.

18. **Clark, D. L., A. A. Weiss, and S. Silver.** 1977. Mercury and organomercurial resistances determined by plasmids in *Pseudomonas*. J. Bacteriol. **132**:186–196.

19. **Cox, D. P., and M. Alexander.** 1974. Factors affecting trimethylarsine and dimethylselenide formation by *Candida humicola*. J. Microbiol. **1**:136–144.

20. **Cullen, W., C. L. Froese, A. Lui, B. C. McBride, D. J. Patmore, and M. Reimer.** 1977. The aerobic methylation of arsenic by microorganisms in the presence of L-methionine-methyl-d3. J. Organometal. Chem. **139**:61–69.

21. **De La Cruz, F., and J. Grinsted.** 1982. Genetic and molecular characterization of Tn21, a multiple resistance transposon from R100.1. J. Bacteriol. **151**:222–228.

22. **Devanas, M. A., C. D. Litchfield, C. McClean, and J. Gianni.** 1980. Coincidence of cadmium and antibiotic resistance in New York Bight apex benthic microorganisms. Mar. Pollut. Bull. **11**:264–269.

23. **Dos Reis, M. H. L., M. J. C. Ramos, and L. R. Trabulsi.** 1978. Resistance to mercury in enteric organisms: frequency and genetic nature. Rev. Microbiol. **9**:24–30.

24. **Dyke, K. G. H., M. T. Parker, and M. H. Richmond.** 1970. Penicillinase production and metal ion-resistance in *Staphylococcus aureus* cultures isolated from hospital patients. J. Med. Microbiol. **3**:125–136.

25. **Dyke, K. G. H., and M. H. Richmond.** 1967. Occurrence of various types of penicillinase plasmid among "hospital" staphylococci. J. Clin. Pathol. **220**:75–79.

26. **Edwards, T., and B. C. McBride.** 1975. Biosynthesis and degradation of methylmercuric in human feces. Nature (London) **253**:462–464.

27. **Efstathiou, J. D., and L. L. McKay.** 1977. Inorganic salts resistance associated with a lactose-fermenting plasmid in *Streptococcus lactis*. J. Bacteriol. **130**:257–265.

28. **Ehrlich, H. L.** 1979. How microbes cope with heavy metals, arsenic, and antimony in their environment, p. 381–408. *In* D. J. Kushner (ed.), Microbial life in extreme environments. Academic Press, Inc., New York.

29. **Ellis, K. J., D. Vartsky, I. Zanzi, and S. H. Cohn.** 1979. Cadmium: *in vivo* measurement in smokers and nonsmokers. Science **205**:323–325.

30. **Ferguson, J. F., and J. Gavis.** 1972. A review of the arsenic cycle in natural waters. Water Res. **6**:1259–1274.

31. **Fleischer, M., A. F. Sarofim, D. W. Fassett, P. Hammond, H. T. Shacklette, I. C. T. Nisbet, and S. Epstein.** 1974. Environmental impact of cadmium: a review by the panel on hazardous trace substances. Environ. Health Perspect. **7**:253–323.

32. **Foster, T. J., and H. Nakahara.** 1979. Deletions in the r-determinant *mer* region of plasmid R100-1 selected for loss of mercury hypersensitivity. J. Bacteriol. **140**:301–305.

33. **Foster, T. J., H. Nakahara, A. A. Weiss, and S. Silver.** 1979. Transposon A-generated mutations in the mercuric resistance genes of plasmid R100-1. J. Bacteriol. **140**:167–181.

34. **Fox, B., and C. T. Walsh.** 1982. Mercuric reductase—purification and characterization of a transposon-encoded flavoprotein containing an oxidation-reduction-active disulfide. J. Biol. Chem. **257**:2498–2503.

35. **Fox, B., and C. T. Walsh.** 1983. Mercuric reductase. Homology to glutathione reductase. Iodoacetamide alkylation and sequence of active site peptide. Biochemistry **22**:4082–4088.

36. **Furukawa, K., T. Suzuki, and K. Tonomura.** 1969. Decomposition of organic mercurial compounds by mercury-resistant bacteria. Agr. Biol. Chem. **33**:128–130.

37. **Furukawa, K., and K. Tonomura.** 1972. Metallic mercury-releasing enzyme in mercury-resistant *Pseudomonas*. Agr. Biol. Chem. **36**:217–226.

38. **Furukawa, K., and K. Tonomura.** 1972. Induction of metallic mercury-releasing enzyme in mercury-resistant *Pseudomonas*. Agr. Biol. Chem. **36**:2441–2448.

39. **Groves, D. J., L. Maroglio, C. W. Merriam, and F. E. Young.** 1973. Epidemiology of antibiotic and heavy-metal resistance in bacteria: a computer-based data system. Comput. Programs Biomed. **3**:123–134.

40. **Hamdy, M. K., and O. R. Noyes.** 1975. Formation of methyl mercury by bacteria. Appl. Microbiol. **30**:424–432.

41. **Hedges, R. W., and S. Baumberg.** 1973. Resistance to arsenic compounds conferred by a plasmid transmissible between strains of *Escherichia coli*. J. Bacteriol. **115**:459–460.

42. **Izaki, K.** 1981. Enzymatic reduction of mercurous and mercuric ions in *Bacillus cereus*. Can. J. Microbiol. **27**:192–197.

43. **Izaki, K.** 1978. Plasmid induced heavy metal resistance in bacteria. Jpn. J. Bacteriol. **33**:729–742.

44. **Izaki, K., Y. Tashiro, and T. Funaba.** 1974. Mechanism of mercuric chloride resistance in microorganisms. III. Purification and properties of a mercuric ion reducing enzyme of *Escherichia coli* R factor. J. Biochem. (Tokyo) **75**:591–599.

45. **Jackson, W. J., and A. O. Summers.** 1982. Polypeptides encoded by the *mer* operon. J. Bacteriol. **149**:479–487.

46. **Jackson, W. J., and A. O. Summers.** 1982. Biochemical characterization of HgCl$_2$-inducible polypeptides en-

coded by the *mer* operon of plasmid R100. J. Bacteriol. **151:**962–970.

47. **Jernelov, A., and A. Martin.** 1976. Ecological implications of metal metabolism by microorganisms. Annu. Rev. Microbiol. **29:**61–77.

48. **Joly, B. H., R. A. Cluzel, S. M. Petit, and M. E. Chanal.** 1979. Transferable sodium arsenate resistance in Enterobacteriaceae. Curr. Microbiol. **2:**151–155.

49. **Keino, H., E. Aoki-Goto, H. Yamamura, and U. Murakami.** 1977. Embryonic mortality and malformation rate in mice treated with intra amniotic injection of cadmium sulfate and zinc sulfate. Teratology **16:**111.

50. **Kondo, I., T. Ishikawa, and H. Nakahara.** 1974. Mercury and cadmium resistance mediated by penicillinase plasmid in *Staphylococcus aureus.* J. Bacteriol. **117:**1–7.

51. **Kozak, S., and C. W. Forsberg.** 1979. Transformation of mercuric chloride and methylmercury by the rumen microflora. Appl. Environ. Microbiol. **38:**626–636.

52. **Lighthart, B.** 1980. Effects of certain cadmium species on pure and litter populations of microorganisms. Antonie van Leeuwenhoek J. Microbiol. Serol. **46:**161–167.

53. **Loutit, J. S.** 1970. Investigation of the mating system of *Pseudomonas aeruginosa* strain I. IV. Mercury resistance associated with the sex factor (FP). Genet. Res. **16:**179–181.

54. **Lukasheva, L. I.** 1977. Mutagenic effect of cadmium salts on *Salmonella typhimurium.* Teor. Prakt. Vopr. Obshch. Mol. Genet., p. 60–67. Moscow, USSR.

55. **Malcolm, D.** 1979. Cadmium as a carcinogen. Mod. Pharmacol. Toxicol. **15:**173–180.

56. **Marshall, B., S. Schleuderberg, D. Rowse-Eagle, A. O. Summers, and S. B. Levy.** 1981. Ecology of antibiotic and heavy metal resistance in nature, p. 630. *In* S. B. Levy, R. C. Clowes, and E. L. Koenig (ed.), Molecular biology, pathogenicity, and ecology of bacterial plasmids. Plenum Press, New York.

57. **McBride, B. C., and T. L. Edwards.** 1977. Role of the methanogenic bacteria in the alkylation of arsenic and mercury. ERDA Symp. Ser. **42:**1–19.

58. **McHugh, G. L., R. C. Moellering, C. C. Hopkins, and M. N. Swartz.** 1975. *Salmonella typhimurium* resistant to silver nitrate, chloramphenicol and ampicillin. Lancet **i:**235–249.

59. **Meissner, P. S., and J. O. Falkinham III.** 1984. Plasmid-encoded mercuric reductase in *Mycobacterium scrofulaceum.* J. Bacteriol. **157:**669–672.

60. **Mergeay, M., C. Houba, and J. Gertis.** 1978. Extrachromosomal inheritance controlling resistance to cadmium, colbalt, and zinc ions: evidence from curing in a *Pseudomonas.* Arch. Int. Physiol. Biochim. **86:**440–441.

61. **Miki, R., A. M. Easton, and R. H. Rownd.** 1978. Mapping of the resistance genes of the R plasmid NR1. Mol. Gen. Genet. **158:**217–224.

62. **Miller, M. A., and S. A. Harmon.** 1967. Genetic association of determinants controlling resistance to mercuric chloride, production of penicillinase and synthesis of methionine in *Staphylococcus aureus.* Nature (London) **215:**531–532.

63. **Mitra, R. S., and I. A. Bernstein.** 1977. Nature of the repair process associated with the recovery of *Escherichia coli* after exposure to Cd^{2+}. Biochem. Biophys. Res. Commun. **74:**1450–1455.

64. **Mitra, R. S., and I. A. Bernstein.** 1978. Single-strand breakage in DNA of *Escherichia coli* exposed to Cd^{2+}. J. Bacteriol. **133:**75–80.

65. **Mobley, H. L. T., C.-M. Chen, S. Silver, and B. P. Rosen.** 1983. Cloning and expression of R-factor mediated arsenate resistance in *Escherichia coli.* Mol. Gen. Genet. **191:**421–426.

66. **Mobley, H. L. T., and B. P. Rosen.** 1982. Energetics of plasmid-mediated arsenate resistance in *Escherichia coli.* Proc. Natl. Acad. Sci. U.S.A. **79:**6119–6122.

67. **Mobley, H. L. T., S. Silver, F. D. Porter, and B. P. Rosen.** 1983. Homology between arsenate resistance determinants of R factors in *Escherichia coli.* Antimicrob. Agents Chemother. **25:**157–161.

68. **Nakahara, H., T. Ishikawa, Y. Sarai, I. Kondo, and S. Mitsuhashi.** 1977. Frequency of heavy-metal resistance in bacteria from inpatients in Japan. Nature (London) **266:**165–168.

69. **Nakahara, H., T. G. Kinscherf, S. Silver, T. Miki, A. Easton, and R. H. Rownd.** 1979. Gene copy number effects in the *mer* operon of plasmid NR1. J. Bacteriol. **138:**284–287.

70. **Nakahara, H., S. Silver, T. Miki, and R. H. Rownd.** 1979. Hypersensitivity to Hg^{2+} and hyperbinding activity with cloned fragments of the mercurial resistance of plasmid NR1. J. Bacteriol. **140:**161–166.

71. **Nelson, J. D., W. Blair, F. E. Brinkman, R. R. Colwell, and W. P. Iverson.** Biodegradation of phenylmercuric acetate by mercury-resistant bacteria. Appl. Microbiol. **26:**321–326.

72. **Nelson, J. D., and R. R. Colwell.** 1975. The ecology of mercury-resistant bacteria in Chesapeake Bay. Microb. Ecol. **1:**191–218.

73. **Ni'Bhriain, N. N., S. Silver, and T. J. Foster.** 1983. Tn5 insertion mutations in the mercuric resistance genes derived from plasmid R100. J. Bacteriol. **155:**690–703.

74. **Novick, R. P., and C. Roth.** 1968. Plasmid-linked resistance to inorganic salts in *Staphylococcus aureus.* J. Bacteriol. **95:**1335–1342.

75. **Olson, B. H., T. Barkay, and R. R. Colwell.** 1979. Role of plasmids in mercury transformation by bacteria isolated from the aquatic environment. Appl. Environ. Microbiol. **38:**478–485.

76. **Olson, G. J., F. D. Porter, J. Rubinstein, and S. Silver.** 1982. Mercuric reductase enzyme from a mercury-volitilizing strain of *Thiobacillus ferrooxidans.* J. Bacteriol. **151:**1230–1236.

77. **Osborne, F. H., and H. L. Ehrlich.** 1976. Oxidation of arsenite by a soil isolate of *Alcaligenes.* J. Appl. Bacteriol. **41:**295–305.

78. **Pan-Hou, H. S. K., M. Hosono, and N. Imura.** 1980. Plasmid-controlled mercury biotransformation by *Clostridium cochlearium* T-2. Appl. Environ. Microbiol. **40:**1007–1011.

79. **Pan-Hou, H. S. K., and N. Imura.** 1981. Role of hydrogen sulfide in mercury resistance determined by plasmid of *Clostridium cochlearium* T-2. Arch. Microbiol. **129:**49–52.

80. **Pan-Hou, H. S. K., Y. Kajikawa, and I. Nobumasa.** 1982. Characterization of organomercury-decomposing activity in cell extract of mercury-resistant *Clostridium cochlearium* T-2P. Ecotoxicol. Environ. Safety **6:**82–88.

81. **Perry, R. D., and S. Silver.** 1982. Cadmium and manganese transport of *Staphylococcus aureus* membrane vesicles. J. Bacteriol. **150:**973–976.

82. **Phillips, S. E., and M. L. Taylor.** 1976. Oxidation of arsenite to arsenate by *Alcaligenes faecalis.* Appl. Environ. Microbiol. **32:**392–399.

83. **Porter, F. D., S. Silver, C. Ong, and H. Nakahara.** 1982. Selection for mercurial resistance in hospital settings. Antimicrob. Agents Chemother. **22:**852–858.

84. **Radford, A. J., J. Oliver, W. J. Kelly, and D. C. Reanney.** 1981. Translocatable resistance to mercuric and phenyl mecuric ions in soil bacteria. J. Bacteriol. **147:**1110–1112.

85. **Richmond, M. H., and M. John.** 1964. Cotransduction by a staphylococcal phage of the genes responsible for penicillinase synthesis and resistance to mercury salts. Nature (London) **202:**1360–1361.

86. **Rinderle, S. J., J. E. Booth, and J. W. Williams.** 1983. Mercuric reductase from R-plasmid NR1: characterization and mechanistic study. Biochemistry **22:**869–876.

87. **Rosenberg, H. R., R. G. Gerdes, and K. Chegwidden.** 1977. Two systems for the uptake of phosphate in *Escherichia coli.* J. Bacteriol. **131:**505–511.

88. **Rosenberg, M., and I. Friedberg.** 1976. Effect of phosphate and arsenate on respiration of *Micrococcus lysodeikticus.* Isr. J. Med. Sci. **12:**1364–1365.

89. **Rossman, T. G.** 1981. Enhancement of UV-mutagenesis by low concentrations of arsenite in *E. coli*. Mutat. Res. **91**:207–211.

90. **Sanders, J. G.** 1979. Microbial role in the demethylation and oxidation of methylated arsenicals in seawater. Chemosphere **3**:135–137.

91. **Schmitt, R., J. Altenbuchner, and J. Grinsted.** 1981. Complementation of transposition functions encoded by transposons Tn*501* (HgR) and Tn*1721* (TetR), p. 359–370. *In* S. B. Levy, R. C. Clowes, and E. L. Koenig (ed.), Molecular biology, pathogenicity, and ecology of bacterial plasmids. Plenum Press, New York.

92. **Schottel, J., A. Mandal, D. Clark, and S. Silver.** 1974. Volatilisation of mercury and orgnomercurials determined by inducible R-factor systems in enteric bacteria. Nature (London) **251**:335–336.

93. **Schottel, J. L.** 1978. The mercuric and organomercurial detoxifying enzymes from a plasmid-bearing strain of *Escherichia coli*. J. Biol. Chem. **253**:4341–4349.

94. **Shariatpandhi, M., A. C. Anderson, A. A. Abdelghani, A. J. Englande, J. Hughes, and R. F. Wilkinson.** Biotransformation of the pesticide sodium arsenate. J. Environ. Sci. Health **16**:35–47.

95. **Silver, S., K. Budd, K. M. Leahy, W. V. Shaw, D. Hammond, R. P. Novick, G. R. Willsky, M. H. Malamy, and H. Rosenberg.** 1981. Inducible plasmid-determined resistance to arsenate, arsenite, and antimony (III) in *Escherichia coli* and *Staphylococcus aureus*. J. Bacteriol. **146**:983–996.

96. **Silver, S., and D. Keach.** 1982. Energy-dependent arsenate efflux: the mechanism of plasmid-mediated resistance. Proc. Natl. Acad. Sci. U.S.A. **79**:6114–6118.

97. **Smith, D. H.** 1967. R factors mediate resistance to mercury, nickel, and cobalt. Science **156**:1114–1116.

98. **Smith, H. W.** 1978. Arsenic resistance in Enterobacteria: its transmission by conjugation and by phage. J. Gen. Microbiol. **109**:49–56.

99. **Smith, K., and R. P. Novick.** 1972. Genetic studies on plasmid-linked cadmium resistance in *Staphylococcus aureus*. J. Bacteriol. **112**:761–762.

100. **Stanisich, V. A., P. M. Bennett, and M. H. Richmond.** 1977. Characterization of a translocation unit encoding resistance to mercuric ions that occurs on a nonconjugative plasmid in *Pseudomonas aeruginosa*. J. Bacteriol. **129**:1227–1233.

101. **Steel, K. J.** 1960. A note on the reaction of mercuric chloride with bacterial SH groups. J. Pharm. Pharmacol. **12**:59–61.

102. **Stickler, D. J., and B. Thomas.** 1980. Antiseptic and antibiotic resistance in Gram-negative bacteria causing urinary infection. J. Clin. Pathol. **33**:288–296.

103. **Summers, A. O., and G. A. Jacoby.** 1978. Plasmid-determined resistance to boron and chromium compounds in *Pseudomonas aeruginosa*. Antimicrob. Agents Chemother. **13**:637–640.

104. **Summers, A. O., G. A. Jacoby, M. N. Swartz, G. Mc-Hugh, and L. Sutton.** 1978. Metal cation and oxyanion resistances in plasmids of gram-negative bacteria, p. 128–131. *In* D. Schlessinger (ed.), Microbiology—1978. American Society for Microbiology, Washington, D.C.

105. **Summers, A. O., L. Kight-Olliff, and C. Slater.** 1982. Effect of catabolite repression on the *mer* operon. J. Bacteriol. **149**:191–197.

106. **Summers, A. O., and E. Lewis.** 1973. Volatilization of mercuric chloride by mercury-resistant plasmid-bearing strains of *Escherichia coli, Staphylococcus aureus*, and *Pseudomonas aeruginosa*. J. Bacteriol. **113**:1070–1072.

107. **Summers, A. O., and S. Silver.** 1972. Mercury resistance in a plasmid-bearing strain of *Escherichia coli*. J. Bacteriol. **112**:1228–1236.

108. **Summers, A. O., and S. Silver.** 1978. Microbial transformation of metals. Annu. Rev. Microbiol. **32**:637–672.

109. **Summers, A. O., and L. I. Sugarman.** 1974. Cell-free mercury(II)-reducing activity in a plasmid-bearing strain of *Escherichia coli*. J. Bacteriol. **119**:242–249.

110. **Tanaka, M., T. Yamamoto, and T. Sawai.** 1983. Evolution of complex resistance transposons from an ancestral mercury transposon. J. Bacteriol. **153**:1432–1438.

111. **Tanaka, N., J. H. Cramer, and R. H. Rownd.** 1976. EcoRI restriction endonuclease map of the composite R plasmid NR1. J. Bacteriol. **127**:619–636.

112. **Tetaz, T. J., and R. K. J. Luke.** 1983. Plasmid-controlled resistance to copper in *Escherichia coli*. J. Bacteriol. **154**:1263–1268.

113. **Thewaini, A. J., and A. A. Haroun.** 1977. Antibiotic and heavy metal resistance in intestinal and water *Escherichia coli* and non-agglutinable vibrios. J. Egypt. Public Health Assoc. **52**:111–122.

114. **Timoney, J. F., J. Port, J. Giles, and J. Spanier.** 1978. Heavy-metal and antibiotic resistance in the bacterial flora of sediments of New York Bight. Appl. Environ. Microbiol. **36**:465–472.

115. **Tonomura, K., T. Nakagami, F. Futtai, and K. Maeda.** 1968. Studies on the action of mercury-resistance microorganisms on mercurials. I. The isolation of mercury-resistance bacterium and the binding of mercurials to the cells. J. Ferment. Technol. **46**:505–512.

116. **Turner, A. W.** 1954. Bacterial oxidation of arsenite. I. Description of bacteria isolated from arsenical cattle-dipping fluids. Austral. J. Biol. Sci. **7**:452–478.

117. **Turner, A. W., and J. W. Legge.** 1954. Bacterial oxidation of arsenite. II. The activity of washed suspensions. Austral. J. Biol. Sci. **7**:479–495.

118. **Tynecka, Z., Z. Gos, and J. Zajac.** 1981. Reduced cadmium transport determined by a resistance plasmid in *Staphylococcus aureus*. J. Bacteriol. **147**:305–312.

119. **Tynecka, Z., A. Gos, and J. Zajac.** 1981. Energy-dependent efflux of cadmium coded by a plasmid resistance determinant in *Staphylococcus aureus*. J. Bacteriol. **147**:313–319.

120. **Tynecka, Z., J. Zajac, and Z. Gos.** 1975. Plasmid dependent impermeability barrier to cadmium in *Staphylococcus aureus*. Acta Microbiol. Pol. **7**:11–20.

121. **Vonk, J. W., and A. K. Sijpesteijn.** 1978. Studies on the methylation of mercuric chloride by pure cultures of bacteria and fungi. Antonie van Leeuwenhoek J. Microbiol. Serol. **39**:505–512.

122. **Walker, J. D., and R. R. Colwell.** 1974. Mercury-resistant bacteria and petroleum degradation. Appl. Microbiol. **27**:285–287.

123. **Wallace, B. L., J. E. Bradley, and M. Rogolsky.** 1981. Plasmid analyses in clinical isolates of *Bacteroides fragilis* and *Bacteroides* species. J. Clin. Microbiol. **14**:383–388.

124. **Webb, J. L.** 1966. Enzyme and metabolic inhibitors. Academic Press, Inc., New York.

125. **Webb, M. (ed.).** 1979. Topics in environmental health, part 2: The chemistry, biochemistry and biology of cadmium. Elsevier North-Holland, Amsterdam.

126. **Weiss, A. A., S. Murphy, and S. Silver.** 1977. Mercury and organomercurial resistance determined by plasmids in *Staphylococcus aureus*. J. Bacteriol. **132**:197–208.

127. **Weiss, A. A., S. Silver, and T. G. Kinscherf.** 1978. Cation transport alteration associated with plasmid-determined resistance to cadmium in *Staphylococcus aureus*. Antimicrob. Agents Chemother. **14**:856–865.

128. **Willsky, G. R., and M. H. Malamy.** 1980. Characterization of two genetically separable inorganic phosphate transport systems in *Escherichia coli*. J. Bacteriol. **144**:356–365.

129. **Willsky, G. R., and M. H. Malamy.** 1980. Effect of arsenate on inorganic phosphate transport in *Escherichia coli*. J. Bacteriol. **144**:366–374.

130. **Wong, P. T., Y. K. Chau, L. Luxon, and G. A. Bengert.** 1977. Methylation of arsenic in the aquatic environment. Trace Subst. Environ. Health **11**:100–106.

131. **Wood, J. M., F. S. Kennedy, and C. G. Rosen.** 1968. Synthesis of methylmercury compounds by extracts of a

methanogenic bacterium. Nature (London) **220:**173–174.

132. **Woolson, E. A., and P. C. Kearney.** 1973. Persistence and reactions of ^{14}C-cacodylic acid in soils. Environ. Sci. Technol. **7:**47–50.

133. **Yamada, M., and K. Tonomura.** 1972. Formation of methylmercury compounds from inorganic mercury by *Clostridium cochlearium*. J. Ferment. Technol. **50:**159–166.

134. **Yamada, M., and K. Tonomura.** 1972. Further study of formation of methylmercury from inorganic mercury by *Clostridium cochlearium* T-28. J. Ferment. Technol. **50:**893–900.

135. **Yamamoto, T., M. Tanaka, R. Baba, and S. Yamagishi.** 1981. Physical and functional mapping of Tn2603, a transposon encoding ampicillin, streptomycin, sulfonamide, and mercury resistance. Mol. Gen. Genet. **181:**464–469.

136. **Zasukhina, G. D., T. A. Sinelschikova, G. N. Lvova, and Z. S. Kirkova.** 1977. Molecular-genetic effects of cadmium chloride. Mutat. Res. **45:**169–174.

Factors Affecting Conjugal Plasmid Transfer in Natural Bacterial Communities

ROLF FRETER

Department of Microbiology and Immunology, The University of Michigan Medical School, Ann Arbor, Michigan 48109

The recent controversy generated by fears that widespread use of recombinant DNA techniques may cause the spread of dangerous genomes throughout the environment is probably the best illustration of our current ignorance concerning gene transfers in natural microbial communities. Conjugative transfers in such environments as soil, water, sewage, and the intestinal tract are difficult to demonstrate, and if they do occur, the transfer rates are generally slower than those observed in vitro. The reasons for these differences are largely unknown; we know very little about the mechanisms which promote or impede naturally occurring plasmid transfers, and there is little understanding of those factors which control the maintenance of plasmids in natural ecosystems.

There are two distinct but related aspects to the ecology of bacterial plasmids in natural environments: (i) the fate of plasmids that are newly introduced into a given ecosystem, and (ii) the evolution, maintenance, and transfer of bacterial plasmids that are "indigenous" to a given ecosystem.

Very little quantitative work has been done on the latter aspect. Indeed, one cannot be certain whether an indigenous population of plasmids can even be defined for any given natural habitat. There is, however, an increasing amount of literature describing gene transfers in model bacterial communities that are newly exposed to unusual or xenobiotic substances. The often complex interactions among gene products (enzymes) in such communities, and the resulting gene transfers, may well occur in natural ecosystems and are likely to involve genes that may be regarded as indigenous. There is also some evidence that genes indigenous to natural ecosystems, especially those of soil microorganisms, are carriers of the ancestral genes that are now commonly found in drug resistance plasmids. Various aspects of these interesting interactions have been reviewed recently by Reanney et al. (57) and in the papers in this volume by H. J. Slater, B. G. Hall, and P. A. Clarke; for this reason, they will not be discussed in the present paper. This review will deal only with interactions of the former kind. An attempt will be made to define the major known factors and

mechanisms which affect plasmid transfers and maintenance in natural environments. At the present state of the art it is not possible to single out individual parameters that would be peculiar to any single ecosystem, and the major examples illustrating the points to be made in the present paper will therefore be drawn from studies of plasmid transfer and maintenance among *Escherichia coli* strains in the mammalian gut. The literature concerning this subject is most extensive, and recent quantitative data from various laboratories concerning *E. coli* and the gut ecosystem deal with concepts which are likely to apply in principle to other ecosystems as well. As will become apparent, most of the published work on plasmid transfer in such ecosystems as soil, water, sewage, and the gut deals with *E. coli* as the host bacterium, with plasmids that specify antibiotic resistance, and with the possible transmission of these plasmids to other host bacteria. This preoccupation is no doubt a consequence of the serious impact which the transmission of resistance factors has on the incidence of human infections caused by antibiotic-resistant bacteria. Whereas hospitals and urban human populations have long been recognized as major sources for the possible dissemination of resistance factors (6, 43), the possibility that the common practice of feeding antibiotics to livestock is another such source has recently been the subject of intense and controversial discussion (49). The extensive literature dealing with the effect of selective conditions (e.g., excessive use of antibiotics) on the prevalence of plasmid-carrying bacteria in various environments will not be reviewed here. This paper will deal exclusively with the fate of plasmids and their host bacteria in various ecosystems that do not select for known plasmid markers.

DEMONSTRATIONS OF PLASMID TRANSFER IN NATURAL ECOSYSTEMS

Gene transfer has been demonstrated between *E. coli* strains (77) and between *Bacillus* species (29) in the soil, as well as between *Agrobacterium* species and among *Pseudomonas* species on plant surfaces (35, 36, 76). The process is not restricted to gene transfer within a species or between related species of the same genus, but occurs between organisms of different genera. For instance, *E. coli* plasmids are known to be transferred in vitro to *P. glycinea* (35) and to *Caulobacter crescentus* (17), both of which are soil inhabitants. *Staphylococcus aureus* plasmids have been transferred in vitro to *Bacillus subtilis* (11). Gene transfer has been demonstrated in vitro between the human pathogen *P. aeruginosa* and the plant pathogen *P. glycinea* (37). Moreover, some genetic elements are capable of being transferred successively in vitro to several species in a manner which suggests that, given the appropriate conditions, some plasmids could become established in the soil microflora (35). Conjugal gene transfer between *E. coli* strains occurred equally well in three sterile soils examined and hence appeared to be independent of the nature of most soils (77). Plasmid transfer in natural systems usually occurs at a slower rate than that observed in vitro. It is therefore noteworthy that, in one study, plasmid transfer in and on plants occurred at a higher frequency than in vitro (35).

There are numerous reports consistently documenting the presence in sewage and open waters of *E. coli* strains carrying transferable multiple antibiotic resistance markers. The source of most of these microorganisms is undoubtedly human feces. Linton et al. (43) found that sewage discharged from hospitals contained the highest percentage of coliforms with transferable drug resistance but that, quantitatively, the general population furnished most of the drug-resistant coliforms found in urban sewage. It is noteworthy that small numbers of *E. coli* strains with transmissible R-factors can also be found in wilderness rivers (6). There is considerably less evidence concerning the question whether, and to what extent, these resistance plasmids are actually being transferred among bacteria in these habitats. Suggestive evidence comes from several studies which show that the proportion of total coliform bacteria that carry multiple antibiotic resistance markers increased in sewage retention ponds (e.g., 6, 26, 28) or in the effluent of a river dam (27). Although this type of evidence is suggestive of plasmid transfer in stagnant sewage or water, i.e., under conditions of minimal agitation that otherwise could disrupt mating pairs, alternative explanations cannot be ruled out. Some workers have suspected, for example, that some antibiotic resistance plasmids may carry additional markers that specify resistance to UV light (6). Indeed, UV irradiation of sewage has been shown to favor drug-resistant *E. coli* populations (46). UV resistance may therefore favor drug-resistant coliforms in stagnant sewage or water, where UV irradiation is a likely bactericidal mechanism. As discussed by Grabow et al. (27), antibiotic resistance plasmids may also specify resistance to other antibacterial mechanisms that operate in sewage or water. Such plasmids may thereby promote increased survival of multiply resistant coliforms.

More direct evidence for actual plasmid transfer comes from experimental studies. For example, Fontaine and Hoadley (21) suspended cultures of *Salmonella choleraesuis* in raw sewage and noted transfer of tetracycline resistance markers from *E. coli* contained in the sewage to the salmonellae. Mach and Grimes (44) suspended membrane filter diffusion chambers into primary and secondary sewage clarifiers. The chambers were filled with sterile sewage, allowed to equilibrate, and then inoculated with donor and recipient bacteria (various species of the *Enterobacteriaceae*). R-plasmid transfer was noted in these experiments. Stewart and Koditshek (72) noted transfer of drug resistance among *E. coli* strains in sterile New York Bight sediment under sterile seawater. Grabow et al. (27) suspended dialysis bags containing *E. coli* donor and recipient strains in river water and noted transfer of all resistance markers tested.

Plasmid transfer has been demonstrated experimentally in various areas of the animal body, such as the kidneys (50) and the urinary bladder (73). The overwhelming majority of studies of in vivo plasmid transfers have been concerned with the gut, however. The evidence concerning plasmid transfer in the mammalian gut seems, at first, to be contradictory. Many workers have shown that plasmid transfer could occur readily in the gut of gnotobiotic animals which harbored only the donor and recipient bacteria involved in the reaction, but no indigenous microflora (e.g., 12, 33, 57, 59, 60, 78), or in newborn or very young animals where an indigenous microflora was either absent (75) or not well developed (65). In contrast, plasmid transfer in the gut of weaned animals or of adult humans could be demonstrated, if at all, only at extremely low rates. Smith (65) could easily demonstrate plasmid transfer in the intestines of 3-day-old chicks and 2-day-old calves, but he found such transfers much more difficult to detect in 10- to 14-week-old pigs. Jarolmen and Kemp (32) found R-factor transmission to be a rare occurrence in weanling pigs. In human feeding experiments, Wiedemann et al. (79), Wiedemann (78), Anderson et al. (4), Smith (66, 67), and Anderson (2) could not demonstrate any plasmid transfer at all in the gut, whereas Smith (64), Anderson (1), and Williams (80) found only a low rate of transfer in humans. This apparent inhibition of plasmid transfer in the adult intestine could easily be abolished by the administration of antibiotics (e.g., 4, 14, 19, 30, 34, 38, 61, 65). Starvation of the host also promoted plasmid transfer in the gut (68), possibly because of the larger enterobacterial populations present in starving animals.

The nature of the plasmid and that of the donor and recipient strains are important determinants of in vitro as well as in vivo plasmid transfer (60, 65). In addition, interpretation of the above-described data permits the following generalization to be made: Very little or no detectable plasmid transfer occurs in the normal gut populated by an undisturbed microflora. In contrast, when the microflora is absent, as in germfree or newborn animals, or when it is incomplete or disturbed, as in the very young or in antibiotic-treated individuals, then plasmid transfer can be observed as readily as during in vitro matings. Antibiotics, of course, can have the additional effect of providing a selective ecological advantage to transconjugants, thereby facilitating the detection of even a few conjugative events that may have occurred in vivo.

Additional examples of genetic transfers in natural microbial communities have been reviewed by Stotzky and Krasovsky (73), Reanney et al. (55, 56), and the Working Group on Revision of the Guidelines (81).

One must conclude, therefore, that plasmid transfer has been demonstrated experimentally to occur in all habitats discussed here, namely, soil, plants, water, sewage, and the gut. The significance of these findings is not always clear, however. One cannot be sure, for example, whether and to what extent plant tissue injected with suspensions of bacteria, gnotobiotic animals, or water and soil samples taken into the laboratory actually reproduce the respective natural environments. Such factors as the injection of large numbers of bacteria into tissue, the transmission of components of bacteriological media with the inoculum, and the physiological state of donor and recipient bacteria may differ in the experimental situation from the circumstances that occur naturally. It is certainly impossible at the present time to predict on the basis of the known properties of a given plasmid–host combination and on the basis of the known characteristics of a given ecosystem whether this plasmid will be able to persist and spread to other microorganisms in that habitat. These uncertainties underscore the disparity between the large amount of information presently available concerning the cellular and molecular mechanisms (i.e., the "how") of plasmid transfer and our considerable ignorance of the determinants which decide whether transfers that can be made to proceed in vitro will also take place in natural environments (i.e., our ignorance of the "why" of plasmid transfer). The following discussion is therefore concerned with the various factors that can affect plasmid transfers in natural ecosystems.

FACTORS AFFECTING PLASMID TRANSFERS IN NATURAL ECOSYSTEMS

Presence of competent recipient bacteria in the ecosystem. Many of the studies cited above were carried out by introducing plasmid donor strains into an ecosystem and observing transfer to known competent recipients which had been added simultaneously or separately from the donors. Such studies do not answer the present question, of course. It is known, however, that some E. coli and P. aeruginosa plasmids can be transferred in vitro to bacterial species indigenous to soil, water, and plants (17, 35, 37) or to bacteria indigenous to the gut (54, 74). It is not known in most cases whether such transfers will actually occur under the conditions prevailing in the natural environment, especially at natural population densities (which are usually lower than those commonly employed in laboratory experiments). The data quoted nevertheless indicate that plasmids which enter a given natural habitat in allochthonous bacterial hosts will often find competent potential recipients among the autochthonous flora. Conjugal transfer of such plasmids is therefore a distinct possibility.

Effects of the plasmid on the host bacterium. Many authors have noted an effect of plasmids on the growth rate and survival of the host bacterium in environments that do not select for known plasmid markers (e.g., 13, 15, 25, 27, 31, 58, 63, 69). Such effects are usually unfavorable to the host bacterium, or the plasmid may have no demonstrable effect. There is suggestive evidence that the presence of a plasmid may sometimes have the opposite effect as well, i.e., to increase the in vivo fitness of a bacterium (42, 80), even if such a plasmid does not specify obvious virulence determinants (16). The mechanisms underlying these phenomena have not been elucidated with certainty. In the case of antibiotic resistance plasmids the likely effects of the plasmid genes include alterations in bacterial metabolism or surface structure that develop concomitantly with antibiotic resistance. For example, Onderdonk et al. (51) found altered surface structures and a reduced ability to colonize gnotobiotic mice in E. coli strains that were chromosomally resistant to various antibiotics. Determinants in addition to drug resistance coded by the same plasmid have also been suspected of contributing to the fitness of the host bacterium. Duval-Iflah et al. (13) found that some plasmids may indeed affect the fitness of their host bacteria in disassociated mice, but noted that other, difficult to define characteristics ("adaptation") of the bacteria were even more important determinants of successful colonization of the germfree gut.

A plasmid may be lost entirely or it may fragment when the host bacterium grows in an environment which is nonselective for the plasmid markers. Such a phenomenon has been observed in the gut of human volunteers, for example (79). The resulting plasmid fragment may lose some, but not necessarily all of the antibiotic resistance markers, and it may at the same time have a less deleterious effect on the growth rate of the host bacterium (25). Theoretical mathematical considerations of the effect of plasmid carriage on host fitness in natural environments will be presented later in this review.

Physical characteristics of the habitat. The dynamic state of the habitat is of obvious importance to the ecology of bacteria and their plasmids. The rate of flow through a system has a profound effect on the kinds of mechanisms which control bacterial populations (22). The ability of bacteria to adhere to surfaces is a major mechanism which allows bacteria to reduce the rate of mechanical removal and to increase their fitness for colonizing habitats with high flow rates. It is important to note that aggregation of bacteria on surfaces also has an additional function, namely, to stimulate plasmid transfer. It is well known to plasmid geneticists that "filter matings," i.e., the aggregation of donor and recipient bacteria on a filter surface, significantly promote plasmid transfer rates over those which take effect in bacterial suspensions. Weinberg and Stotzky (77) reported that the addition of clays, particularly montmorillonite clays, increased the frequency of conjugal plasmid transfers among *E. coli* strains in soil. This suggests that the well-known adsorptive capacity of clay colloids increased the rate of gene transfer in this system. J. D. Goguen (Ph.D. thesis, University of Massachusetts, Amherst, 1980) developed a mathematical model for plasmid transfer under such conditions (among bacteria on the surface of agar plates) and showed that the limiting step in the process of conjugative plasmid transfer was the interval between the first acquisition of a plasmid by a recipient bacterium and the time at which the new host had become competent to transfer the plasmid to new recipients. This time interval was relatively short (approximately 55 min with the R1 and R1drd-19 plasmids in *E. coli*), such that it could be neglected altogether in mathematical models of plasmid transfer among suspended (rather than aggregated) bacteria (25, 41). Cullum et al. (10) estimated by means of a mathematical model the effect of a 90-min time interval of this type on plasmid transfer rates in exponentially growing broth cultures and also found it to be minor in these circumstances.

Agitation of a bacterial suspension may disrupt mating pairs and result in lower rates of plasmid transfers. Even mild stirring of a culture may have a noticeable effect (25). As reviewed by Mach and Grimes (44), pH and temperature have a profound effect on conjugal plasmid transfer rates. It is obvious that physical (and other) factors which adversely affect the viability of host bacteria (such as UV radiation, extremes of temperature or pH) will also reduce the probability for the associated plasmids to persist in a given habitat.

Effect of the physiological state of host and recipient bacteria on fertility. Levin et al. (41) made the first successful attempt to define in quantitative terms the fertility of host and recipient bacteria involved in conjugal plasmid transfer in liquid cultures. They devised a mathematical model based on mass-action kinetics with a unique and constant rate parameter. The latter "transfer rate constant" was used as a measure of fertility. This concept proved to be valuable in constructing mathematical models of the more complex plasmid transfer reactions that occur in the large intestine (25).

There is published evidence that fertility not only is dependent on the nature of the plasmid and of the interacting bacteria, but that it also depends strongly on the physiological state of the bacteria. The effect of temperature (which is one determinant of the physiological state) has already been discussed above. Antimicrobial substances are present in most natural habitats, and some of these may affect fertility, as was the case for some (but by no means all) plasmids tested in continuous-flow cultures of large intestine flora (25).

As discussed earlier in this review, the normal intestine does not readily permit plasmid transfer. A number of authors have speculated that this shortcoming may be a consequence of unfavorable conditions prevailing in the gut. Anderson (1), for example, reported studies implicating the *Bacteroides* populations of the indigenous microflora as inhibitors of in vivo plasmid transfer. Duval-Iflah et al. (14) found no such effect of *Bacteroides* on plasmid transfer in gnotobiotic mice. Wiedemann (78) discussed and tested the following mechanisms as potential inhibitors of plasmid transfer in the gut: (i) the anaerobic conditions prevalent in the normal large intestine, (ii) the effect of bile salts, (iii) metabolic products of the indigenous flora (especially acids), (iv) the growth phase of the donor and recipient bacteria, and (v) the efficiency with which potential recipients can accept a plasmid, specifically as related to the degree of roughness. This author suggested that the actual inhibition observed in vivo may be a consequence of the combined effects of all these mechanisms. Falkow (18) concluded from a review of the literature that the anaerobic conditions, low pH, and fatty acids produced by the indigenous microflora are probably responsible for the diminished transfer of R-factors that has been observed in vivo. Burman (7) determined that anaerobiosis inhibited the conjugative transfer of some plasmids, but not that of others. These findings may explain contradictory statements in the literature that anaerobiosis does inhibit conjugation (20, 48) or that it does not (3, 70). It is interesting to note in this connection that Stotzky and Krasovsky (73) described interference by unrelated bacteria with plasmid transfers in experimental infections of the urinary bladder. In contrast to the above-described hypotheses, Freter et al. (25) deducted from mathematical evaluations of in vivo and in vitro data that the low rates of plasmid transfer observed in the normal gut are to be expected on quantitative grounds alone and that such data can therefore be explained without postulating the operation of hypothetical inhibitory mechanisms.

It is well known that allochthonous bacteria that enter a given ecosystem may become "injured" such that they no longer grow on ordinary bacteriological media (e.g., *E. coli* in water; 45). Chai (9) reported that *E. coli* strains grown in Chesapeake Bay water lacked outer membrane proteins that were present after growth in rich broth. Concomitant with the loss of outer membrane proteins was a reduced sensitivity of the bacteria to colicins and bacteriophages. It is conceivable that similar changes in cell surface structure brought about by growth in natural habitats will also affect the fertility for conjugal plasmid transfer and that such changes may be in the direction of increased fertility as well as toward decreased fertility.

The physiological state of bacteria is of course the major determinant of their rate of multiplication. As will be discussed below, this rate not only determines the fitness of bacteria to colonize a given habitat but also correlates with fertility. One of the major differences between plasmid-transfer experiments in the laboratory and plasmid transfers in natural ecosystems is the fact that most ecosystems are stable. Stability, of course, implies that microorganisms that are newly introduced into the system ("invaders") are prevented from colonizing it by the antagonistic action of the existing indigenous flora ("residents"). Remarkably, this holds true even if the invader is a strain of an autochthonous species which has no less potential ability to colonize the system than the resident strain that is already established in the indigenous flora (24). A number of authors have published data which document another puzzling feature of such systems: When the experiment is changed such that the former "invader" strain is introduced first and members of the formerly indigenous flora are introduced later, the former "invader" is often able to colonize the system. In other words, whether a strain which is potentially able to compete with other flora and to colonize a given habitat will actually do so often depends on the order in which the different bacteria are introduced. This has been shown, for example, for diflora continuous-flow cultures by Ozawa and Freter (53), for growth of bacteria on burn wounds by Anthony and Wannamaker (5), for staphylococci in gnotobiotic animals by Orcutt and Schaedler (52), for oral flora in rats by Mikx et al. (47), and for *E. coli* in gnotobiotic animals by Duval-Iflah et al. (13).

An explanation for these phenomena has recently been advanced (22) which applies to systems such as the large intestine, where adhesion of the indigenous flora is of importance but where, on the other hand, the rate of flow out of the system is lower than the maximal potential growth rate of the bacteria (such that even nonadherent bacteria could potentially colonize the system). In addition to the large gut, other systems with similar characteristics might include the flora of the mouth and throat, sewage, and some open (polluted) waters (especially their sediments). The explanation was arrived at by means of a mathematical model which is based on the theory that the major mechanism which controls the complex bacterial populations in such systems is competition for the numerous metabolic substrates present (22). In this view each indigenous bacterial species can colonize the system because it is more efficient than the other bacteria in utilizing one or a few of these substrates, such that the population size of each species is controlled by this one substrate (which is thereby the limiting substrate for this species). Evidence supporting the theory that competition for limiting substrates controls the flora of the large intestine has been presented (23). The mathematical model employs Monod's classical equation which relates the concentration of a limiting nutrient to the bacterial growth rate constant. The following assumptions are made in the model: (i) A resident strain colonizes the large intestine. An invader strain is ingested once in relatively large numbers. (ii) Resident and invader strains have exactly the same properties. (iii) Both of these strains compete for the same adhesion sites on the gut wall. (iv) Offspring of adherent strains occupy additional sites or, when most sites are filled, are shed into the lumen. (v) All adherent bacteria are slowly shed into the lumen by mucus flow. (vi) Both resident and invader *E. coli* strains compete for the same limiting nutrient. Numerical values for the parameters and mathematical details of the model have been published (22). The results of a computer simulation of this model are shown in Fig. 1 (upper section). As can be seen, when an invader strain is introduced at 50 h, it is eliminated rapidly, in spite of the fact that its properties are exactly the same as those of the resident. The large number of invaders causes a temporary decrease in the concentration of the limiting nutrient (Fig. 1, bottom), thereby causing a temporary decrease in the number of residents as well. The explanation for this phenomenon is given in Fig. 2, which relates the growth rate constant of bacterial multiplication to the concentration of the limiting nutrient by means of Monod's equation. The dashed lines mark the rates of multiplication which are required for a bacterium to colonize (i.e., to maintain constant populations) in the same system that is pictured in Fig. 1. As is apparent, the adherent (resident) strain requires a much lower rate of multiplication for colonization than a strain which is suspended, because the washout rate for adherent bacteria is lower. Consequently, an adherent resident strain can expand its population size until the concentration of the limiting substrate reaches a relatively low value (dashed line in Fig. 2). A suspended strain, in contrast, is washed out of the system at a higher rate than an adherent strain and, for this reason, would require a higher concentration of the limiting nutrient to maintain a constant population (dashed line in Fig. 2). Consequently,

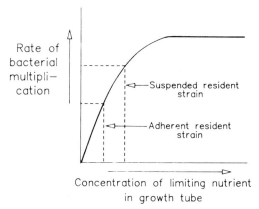

FIG. 2. Relationship between multiplication rate and limiting nutrient concentration at equilibrium in a continuous-flow culture system for adherent and suspended resident bacterial strains (based on the Monod equation).

when an invader strain (which, of course, is suspended initially) enters a system which already harbors a similar adherent resident strain, the concentration of the limiting nutrient in that system is too low to permit sufficiently rapid multiplication of the invader to prevent washout. In contrast, if the invader strain is implanted first, it can find a sufficient number of adhesion sites to be able to colonize. This, then, explains why the order of implantation can be important in colonization of a given ecosystem, and why an invader strain may not be able to colonize a natural habitat, even though it has all the attributes that are necessary for colonization.

Another phenomenon which reduces the chance for an invader bacterium to colonize a given system is the fact that the physiological adaptation that is often necessary before multiplication can begin in a new environment results in very long lag phases. As described for the large intestine (23), such lag phases can be sufficiently protracted that most of the invaders are eliminated before they can begin multiplication.

For the reasons outlined above, the introduction of new potential plasmid donors into natural ecosystems differs from the conditions that obtain in the usual laboratory experiments in that the donor bacteria remain in the lag phase for prolonged periods of time. Even if multiplication commences, its rate is often insufficient to compensate for elimination either by physical means or by a variety of antibacterial mechanisms.

Levin et al. (41) have shown that the multiplication rates of donor and recipient bacteria strongly correlate with fertility, the latter being much reduced at slow multiplication or under

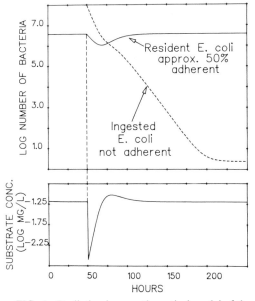

FIG. 1. Prediction by a mathematical model of the fate of an E. coli strain which invades the large gut of an animal that already harbors an adherent E. coli resident strain (upper panel). The lower panel shows the concentration of the limiting substrate in the system (based on the Monod equation).

static conditions. The same was noted in more complex models of plasmid transfers in the gut (25): The rate constant of plasmid transfer to resident recipient *E. coli* strains from donor *E. coli* strains entering the gut (which did not multiply initially and consequently were rapidly washed out) was lower by several orders of magnitude than the rate constant of transfer from newly formed transconjugants (which, being of resident *E. coli* origin, did multiply in the system). It is reasonable to conclude, therefore, that the low rates of bacterial multiplication that prevail in most natural ecosystems (and especially the still lower rates of multiplication of newly invading potential donors) may be regarded as major impediments to efficient plasmid transfer in nature.

QUANTITATIVE STUDIES OF PLASMID TRANSFER

The mathematical models of plasmid transfer developed by Levin et al. (41), which are based on a simple mass action model, have already been discussed earlier in this paper. These authors recommended the use of the transfer rate constant as the measure of fertility of conjugative plasmids. To my knowledge, this constant is the only measure of fertility that is independent of the donor and recipient concentrations which happen to be employed in a given experiment. In contrast, more conventional methods for determining fertility (e.g., those based on the proportion of transconjugants appearing in a mixed culture—the so-called "transfer frequency") are entirely dependent on the population densities of donors and recipients, and are therefore reproducible only under similar experimental conditions. Consequently, such methods cannot be regarded as quantitative in any rigorous sense of that term, even though the results are expressed in numerical form. For this reason, the transfer rate constant defined by Levin et al. (41) is also the only parameter which can predict (at least in theory) the rates of plasmid transfers in ecosystems where the bacterial densities differ from those employed in the laboratory tests that had been conducted to determine fertility. Because of this feature, the transfer rate constant can also be used to estimate rates of plasmid transfer in ecosystems where donor and recipient population densities are so low as to make direct experimental studies impractical (25). Stewart and Levin (71), Levin and Rice (39), and Levin and Stewart (40) extended these studies by developing more detailed mathematical models for conjugative and for mobilizable nonconjugative plasmids. They demonstrated that there are conditions in which such plasmids can indeed become established in a continuous-flow system even when the bacteria carrying them have no

advantage or have a reduced reproductive fitness as compared to plasmid-free bacteria. In their models, plasmids become established when the rate of plasmid transfer by conjugation or mobilization, or both, is sufficiently high to compensate for the loss of plasmid-carrying bacteria that occurs as a result of selection against them. The authors considered the conditions required for maintenance of such plasmids to be sufficiently stringent to make it unlikely that these could be realized in many natural ecosystems. They concluded, therefore, that only plasmids which confer a selective advantage upon the bacterial host are likely to be maintained in natural bacterial populations. Freter et al. (25) devised models of plasmid ecology in the intestine, using parameters of fertility, plasmid segregation, etc., derived from actual in vivo experiments with wild-type *E. coli*. These models showed that long-term maintenance of plasmids in the human and animal gut is indeed a likely possibility even in the absence of positive selection.

In several important studies Selander and Levin (62) and Caugant et al. (8) determined the electrophoretic variations and plasmid profiles of *E. coli* strains isolated from normal human populations. They concluded from statistical evaluations of their data that the genetic diversity of *E. coli* strains which can be observed even in a single human host is not likely to be the result of genetic recombination, but appears to be a consequence of successive colonization by different *E. coli* genotypes. These authors were surprised by the large number of plasmids carried by individual *E. coli* genotypes, even though the human hosts had no history of antibiotic treatment. They interpreted their data to be compatible with the assumption that plasmid transfer does indeed occur among natural *E. coli* populations.

In a recent study from my laboratory a series of mathematical models of increasing complexity were devised to assess the importance of a variety of parameters in the transfer and maintenance of R-plasmids in the mouse gut and in continuous-flow culture models of the mouse large-intestinal ecosystem (25). Among other contributions the mathematical analysis was able to distinguish between two suspected mechanisms (segregation versus depression of the bacterial growth rate by the plasmid) as an explanation for the experimental finding that plasmid-bearing and plasmid-free *E. coli* strains coexisted in the gut for long periods of time in spite of undiminished fertility. In general, the predictions furnished by the mathematical models were consistent with the conclusion that the long-term interactions observed experimentally were often the consequences of minor quantitative differences in parameters such as growth rates, fertility, rates of segregation, etc., which were too small to be detected except by precise mathematical analysis of long-term experiments, but which were nevertheless decisive determinants of the ultimate fate of the plasmids and their hosts. The major imperfection of the current models is the failure to consider reactions among adherent bacteria. Estimates of the error likely to be introduced by this shortcoming show that it may be of minor importance. Nevertheless, the modeling of interactions among adherent bacterial populations remains a major challenge which must be met in the future.

CONCLUSIONS

A considerable amount of published evidence supports the assumption that conjugal plasmid transfers occur in many natural environments. The factors which determine the likelihood of such transfers, and the mechanisms which affect the subsequent fate of the plasmids and their new host bacteria, are numerous and are interrelated in many complex ways. For this reason it is not possible at the present time to predict whether any given plasmid–host combination will be able to enter any specific natural environment and whether the plasmid will be able to persist there. A number of examples are described of mechanistic mathematical models which have been able to evaluate the nature of some such interactions and to make choices between alternative explanations for experimentally observed phenomena. Although mathematical models cannot furnish definitive proof for the reality of any given mechanism of interaction, they can give important leads for promising experimental exploration which often would be difficult to obtain in any other way. It seems likely, therefore, that progress in this area will be optimal if experimental studies are based on ideas derived from the predictions of mathematical models, and if the experimental results, in turn, inspire the construction of improved mathematical models to suggest further avenues of experimentation, and so on in an iterative process which eventually converges to a complete understanding of the problem. The magnitude of the problem is such, however, that this convergence is unlikely to occur at a time in history when it will threaten today's generation of ecologists with obsolescence.

Experimental studies from my laboratory described in this review were supported by U.S. Public Health Service contract NO1-AI-62518 and by grant 1 RO1 AI 17154 from the National Institutes of Health.

LITERATURE CITED

1. **Anderson, E. S.** 1975. Viability of, and transfer of a plasmid from, *E. coli* K12 in the human intestine. Nature (London) **255:**502–504.

2. **Anderson, E. S.** 1978. Plasmid transfer in *Escherichia coli.* J. Infect. Dis. **137:**686–687.

3. **Anderson, J. D.** 1975. Factors that may prevent transfer of antibiotic resistance between gram-negative bacteria in the gut. J. Med. Microbiol. **8:**83–88.

4. **Anderson, J. D., W. A. Gillespie, and M. H. Richmond.** 1973. Chemotherapy and antibiotic-resistance transfer between enterobacteria in the human gastro-intestinal tract. J. Med. Microbiol. **6:**461–473.

5. **Anthony, B. F., and L. W. Wannamaker.** 1967. Bacterial interference in experimental burns. J. Exp. Med. **125:**319–336.

6. **Bell, J. B., G. E. Elliott, and D. W. Smith.** 1983. Influence of sewage treatment and urbanization on selection of multiple resistance in fecal coliform populations. Appl. Environ. Microbiol. **46:**227–232.

7. **Burman, L. G.** 1977. Expression of R-plasmid functions during anaerobic growth of an *Escherichia coli* K-12 host. J. Bacteriol. **131:**69–75.

8. **Caugant, D. A., B. A. Levin, and R. K. Selander.** 1981. Genetic diversity and temporal variation in the *E. coli* population of a human host. Genetics **98:**467–490.

9. **Chai, T. J.** 1983. Characteristics of *Escherichia coli* grown in bay water as compared with rich medium. Appl. Environ. Microbiol. **45:**1316–1323.

10. **Collum, J., J. F. Collins, and P. Broda.** 1978. The spread of plasmids in model populations of *Escherichia coli* K12. Plasmid **1:**545–556.

11. **Dancer, B. N.** 1980. Transfer of plasmids among bacilli. J. Gen. Microbiol. **121:**263–266.

12. **Ducluzeau, R., and A. Galinha.** 1967. Recombinaison *in vivo* entre une souche Hfr et une souche F⁻ d'Escherichia coli K12, ensemencees dans le tube digestif de souris axeniques. C.R. Acad. Sci. Paris **264:**177–180.

13. **Duval-Iflah, Y., P. Raibaud, and M. Rousseau.** 1981. Antagonisms among isogenic strains of *Escherichia coli* in the digestive tracts of gnotobiotic mice. Infect. Immun. **34:**957–969.

14. **Duval-Iflah, Y., P. Raibaud, C. Tancrede, and M. Rousseau.** 1980. R-plasmid transfer from *Serratia liquefaciens* to *Escherichia coli* in vitro and in vivo in the digestive tract of gnotobiotic mice associated with human fecal flora. Infect. Immun. **28:**981–990.

15. **Dykhuizen, D. E., and D. L. Hartl.** 1983. Selection in chemostats. Microbiol. Rev. **47:**150–168.

16. **Elwell, L. P., and P. L. Shipley.** 1980. Plasmid-mediated factors associated with virulence of bacteria to animals. Annu. Rev. Microbiol. **34:**465–496.

17. **Ely, B.** 1979. Transfer of drug resistance factors to the dimorphic bacterium *Caulobacter crescentus.* Genetics **91:**371–380.

18. **Falkow, S.** 1975. Infectious multiple drug resistance, p. 234. Pion, Ltd., London.

19. **Farrar, W. E., M. Eidson, P. Guerry, S. Falkow, L. M. Drusin, and R. B. Roberts.** 1972. Interbacterial transfer of an R-factor in the human intestine: *in vivo* acquisition of R-factor-mediated kanamycin resistance by a multi-resistant strain of *Shigella sonnei.* J. Infect. Dis. **126:**27–33.

20. **Fisher, K. W.** 1957. The role of the Krebs cycle in conjugation in Escherichia coli. J. Gen. Microbiol. **16:**120–135.

21. **Fontaine, T. D., III, and A. W. Hoadley.** 1976. Transferable drug resistance associated with coliforms isolated from hospital and domestic sewage. Health Lab. Sci. **13:**238–245.

22. **Freter, R.** 1984. Mechanisms that control the microflora in the large intestine. *In* D. J. Hentges (ed.), Human intestinal flora in health and disease. Academic Press, Inc., New York.

23. **Freter, R., H. Brickner, M. Botney, D. Cleven, and A. Aranki.** 1983. Mechanisms which control bacterial populations in continuous flow culture models of mouse large intestinal flora. Infect. Immun. **39:**676–685.

24. **Freter, R., H. Brickner, J. Fekete, M. M. Vickerman, and K. E. Carey.** 1983. Survival and implantation of *Escherichia coli* in the intestinal tract. Infect. Immun. **39**:686–703.

25. **Freter, R., R. R. Freter, and H. Brickner.** 1983. Experimental and mathematical models of *Escherichia coli* plasmid transfer in vitro and in vivo. Infect. Immun. **39**:60–84.

26. **Grabow, W. O. K., and I. G. Middendorff.** 1973. Survival in maturation ponds of coliform bacteria with transferable drug resistance. Water Res. **7**:1589–1597.

27. **Grabow, W. O. K., O. W. Prozesky, and J. S. Burger.** 1975. Behaviour in river and dam of coliform bacteria with transferable or non-transferable drug resistance. Water Res. **9**:777–782.

28. **Grabow, W. O. K., and M. van Zyl.** 1976. Behaviour in conventional sewage purification processes of coliform bacteria with transferable or nontransferable drug-resistance. Water Res. **10**:717–723.

29. **Graham, J. B., and C. A. Istock.** 1979. Gene exchange and natural selection cause *Bacillus subtilis* to evolve in soil culture. Science **204**:637–639.

30. **Guinee, P. A. M.** 1965. Transfer of multiple drug resistance from *Escherichia coli* to *Salmonella typhimurium* in the mouse intestine. Antonie van Leeuwenhoek J. Microbiol. Serol. **31**:314–322.

31. **Helling, R. B., T. Kinney, and J. Adams.** 1981. The maintenance of plasmid-containing organisms in populations of *Escherichia coli*. J. Gen. Microbiol. **123**:129–141.

32. **Jarolmen, H., and G. Kemp.** 1969. R factor transmission in vivo. J. Bacteriol. **99**:487–490.

33. **Jones, R. T., and R. Curtiss III.** 1970. Genetic exchange between *Escherichia coli* strains in the mouse intestine. J. Bacteriol. **103**:71–80.

34. **Kasuya, M.** 1964. Transfer of drug resistance between enteric bacteria introduced in the mouse intestine. J. Bacteriol. **88**:322–328.

35. **Lacy, G. H., and J. V. Leary.** 1975. Transfer of antibiotic resistance plasmid RP1 into *Pseudomonas glycinea* and *Pseudomonas phaseolicola in vitro* and *in planta*. J. Gen. Microbiol. **88**:49–57.

36. **Larebeke, N. van, C. Genetello, J. Shell, R. A. Schilperoort, A. K. Hermans, J. P. Hernalsteens, and M. van Montagu.** 1975. Acquisition of tumour-inducing ability by non-oncogenic agrobacteria as a result of plasmid transfer. Nature (London) **255**:742–743.

37. **Leary, J. V.** 1979. Transfer and integration of chromosomal genes from *Pseudomonas glycinea* into *Pseudomonas aeruginosa*. Can. J. Microbiol. **25**:637–640.

38. **Lebek, G.** 1963. Ueber die Entstehung mehrfachresistenter Salmonellen. Ein experimenteller Beitrag. Zentralbl. Bakteriol. Parasitenkd. Infektionskr. Hyg. Abt. 1 Orig. **188**:494–505.

39. **Levin, B. R., and V. A. Rice.** 1980. The kinetics of transfer of nonconjugative plasmids by mobilizing conjugative factors. Genet. Res. **35**:241–259.

40. **Levin, B. R., and F. M. Stewart.** 1980. The population biology of bacterial plasmids: a priori conditions for the existence of mobilizable nonconjugative factors. Genetics **94**:425–443.

41. **Levin, B. R., F. M. Stewart, and V. A. Rice.** 1979. The kinetics of conjugative plasmid transmission: fit of a simple mass action model. Plasmid **2**:247–260.

42. **Levy, S. B., B. Marshall, and D. Rowse-Eagle.** 1980. Survival of *Escherichia coli* host-vector systems in the mammalian tissue. Science **209**:391–394.

43. **Linton, K. B., M. H. Richmond, R. Bevan, and W. A. Gillespie.** 1974. Antibiotic resistance and R factors in coliform bacilli isolated from hospital and domestic sewage. J. Med. Microbiol. **1**:91–103.

44. **Mach, P. A., and D. J. Grimes.** 1982. R-plasmid transfer in a wastewater treatment plant. Appl. Environ. Microbiol. **44**:1395–1403.

45. **McFeters, G. A., S. C. Cameron, and M. W. LeChevallier.** 1982. Influence of diluents, media and membrane filters on detection of injured waterborne coliform bacteria. Appl. Environ. Microbiol. **43**:97–103.

46. **Meckes, M. C.** 1982. Effects of UV light disinfection on antibiotic-resistant coliforms in wastewater effluents. Appl. Environ. Microbiol. **43**:371–377.

47. **Mikx, F. H. M., J. S. van der Hoeven, A. J. M. Plasschaert, and K. G. Konig.** 1975. Effect of *Actinomyces viscosus* on the establishment and symbiosis of *Streptococcus mutans*? and *Streptococcus sanguis* in SPF rats on different sucrose diets. Caries Res. **9**:1–20.

48. **Moodie, H. S., and D. R. Woods.** 1973. Anaerobic R factor transfer in *Escherichia coli*. J. Gen. Microbiol. **76**:437–440.

49. **Novick, R. P.** 1981. The development and spread of antibiotic-resistant bacteria as a consequence of feeding antibiotics to livestock. Ann. N.Y. Acad. Sci. **368**:23–59.

50. **Novick, R. P., and S. I. Morse.** 1967. In vivo transmission of drug resistance factors between strains of *Staphylococcus aureus*. J. Exp. Med. **125**:45–59.

51. **Onderdonk, A., B. Marshall, R. Cisneros, and S. B. Levy.** 1981. Competition between congenic *Escherichia coli* K-12 strains in vivo. Infect. Immun. **32**:74–79.

52. **Orcutt, R., and R. W. Schaedler.** 1973. Control of staphylococci in the gut of mice, p. 435–440. *In* J. B. Heneghan (ed.), Germfree research. Biological effect of gnotobiotic environments. Academic Press, Inc., New York.

53. **Ozawa, A., and R. Freter.** 1964. Ecological mechanism controlling growth of *Escherichia coli* in continuous flow cultures and in the mouse intestine. J. Infect. Dis. **114**:235–242.

54. **Privitera, G., A. Dublanchet, and M. Sebald.** 1979. Transfer of multiple antibiotic resistance between subspecies of *Bacteroides fragilis*. J. Infect. Dis. **139**:97–101.

55. **Reanney, D. C., W. P. Roberts, and W. J. Kelly.** 1982. Genetic interactions among microbial communities, p. 287–322. *In* A. T. Bull and J. H. Slater (ed.), Microbial interactions and communities, vol. 1. Academic Press, London.

56. **Reanney, D. C., P. C. Gowland, and H. Slater.** 1983. Genetic interactions among microbial communities, p. 379–421. *In* J. H. Slater, R. Whittenbury, and J. W. T. Wimpenny (ed.), Microbes in their natural environment. Cambridge University Press, Cambridge.

57. **Reed, N. D., D. G. Sieckmann, and C. E. Georgi.** 1969. Transfer of infectious drug resistance in microbially defined mice. J. Bacteriol. **100**:22–26.

58. **Richmond, M. H.** 1977. The survival of R-plasmids in the absence of antibiotic selection pressure, p. 61–70. *In* J. Drews and G. Hogenauer (ed.), R-factors: their properties and possible control. Springer Verlag, Vienna.

59. **Salzman, C. T., and L. Klemm.** 1968. Transfer of antibiotic resistance (R factor) in the mouse intestine. Proc. Soc. Exp. Biol. Med. **128**:392–398.

60. **Sansonetti, P., J. P. Lafont, K. Jaffe-Brachet, J. F. Guillot, and E. Chaslus-Dancla.** 1980. Parameters controlling interbacterial plasmid spreading in a gnotoxenic chicken gut system: influence of plasmid and bacterial mutations. Antimicrob. Agents Chemother. **17**:327–333.

61. **Schneider, H., S. B. Formal, and L. S. Baron.** 1961. Experimental genetic recombination *in vivo* between *Escherichia coli* and *Salmonella typhimurium*. J. Exp. Med. **114**:141–148.

62. **Selander, R. K., and B. R. Levin.** 1980. Genetic diversity and structure in *Escherichia coli* populations. Science **210**:545–547.

63. **Slater, J. H., and D. Godwin.** 1980. Microbial adaptation and selection, p. 137–160. *In* D. C. Ellwood, J. N. Hedger, M. J. Latham, J. M. Lynch, and J. H. Slater (ed.), Contemporary microbial ecology. Academic Press, Inc., New York.

64. **Smith, H. W.** 1969. Transfer of antibiotic resistance from animal and human strains of *Escherichia coli* to resident *E. coli* in the alimentary tract of man. Lancet **1**:1174–1176.

65. **Smith, H. W.** 1970. The transfer of antibiotic resistance

between strains of enterobacteria in chicken, calves and pigs. J. Med. Microbiol. **3:**165–180.

66. **Smith, H. W.** 1978. Is it safe to use *Escherichia coli* K12 in recombinant DNA experiments? J. Infect. Dis. **137:**655–660.

67. **Smith, M. G.** 1976. R factor transfer *in vivo* in sheep with *E. coli* K12. Nature (London) **261:**348.

68. **Smith, M. G.** 1977. Transfer of R factors from *Escherichia coli* to salmonellas in the rumen of sheep. J. Med. Microbiol. **10:**29–35.

69. **Smith, P. R., E. Ferrell, and K. Dunican.** 1974. Survival of R⁺ *Escherichia coli* in sea water. Appl. Microbiol. **27:**983–984.

70. **Stallions, D. R., and R. Curtiss III.** 1972. Bacterial conjugation under anaerobic conditions. J. Bacteriol. **111:**294–295.

71. **Stewart, F. M., and B. R. Levin.** 1977. The population biology of bacterial plasmids: a priori conditions for the existence of conjugationally transmitted factors. Genetics **87:**209–228.

72. **Stewart, K. R., and L. Koditshek.** 1980. Drug-resistance transfer in *Escherichia coli* in New York Bight sediment. Mar. Pollut. Bull. **11:**130–133.

73. **Stotzky, G., and V. N. Krasovsky.** 1981. Ecological factors that affect the survival, establishment, growth and genetic recombination of microbes in natural habitats, p. 31–42. *In* S. B. Levy, R. C. Clowes, and E. L. Koenig (ed.), Molecular biology, pathogenicity and ecology of bacterial plasmids. Plenum Press, New York.

74. **Tally, F. P., D. R. Snydman, S. L. Gorbach, and M. H. Malany.** 1979. Plasmid-mediated, transferable resistance to clindamycin and erythromycin in *Bacteroides fragilis*. J. Infect. Dis. **139:**83–88.

75. **Walton, J. R.** 1966. *In vivo* transfer of infectious drug resistance. Nature (London) **211:**312–313.

76. **Watson, B., T. C. Currier, M. P. Gordon, M. Chilton, and E. W. Nester.** 1975. Plasmid required for virulence of *Agrobacterium tumefaciens*. J. Bacteriol. **123:**255–264.

77. **Weinberg, S. R., and G. Stotzky.** 1972. Conjugation and genetic recombination of *Escherichia coli* in soil. Soil Biol. Biochem. **4:**171–180.

78. **Wiedemann, B.** 1972. Die Uebertragung extrachromosomaler Resistenzfaktoren in der Darmflora und ihre Hemmung. Zentralbl. Bakteriol. Parasitenkd. Infektionskr. Hyg. Abt. 1 Orig. **220:**106–123.

79. **Wiedemann, B., H. Knothe, and E. Doll.** 1970. Uebertragung von R-Faktoren in der Darmflora des Menschen. Zentralbl. Bakteriol. Parasitenkd. Infektionskr. Hyg. Abt. 1 Orig. **213:**183–193.

80. **Williams, P. H.** 1977. Plasmid transfer in the human alimentary tract. FEMS Microbiol. Lett. **2:**91–95.

81. **Working Group on Revision of the Guidelines.** 1981. Evaluation of the risks associated with recombinant DNA research. Recomb. DNA Tech. Bull. **4:**166–179.

Mechanisms of Microbial Adhesion to Surfaces

Direct Ultrastructural Examination of Adherent Bacterial Populations in Natural and Pathogenic Ecosystems

J. WILLIAM COSTERTON

Department of Biology, University of Calgary, Calgary, Alberta T2N 1N4, Canada

MECHANISMS OF BACTERIAL ADHERENCE

Whereas workers investigating bacterial adhesion have evidenced the reductionist tendency to regard the particular single mechanism that they study as of paramount importance in the adherence of bacteria to surfaces, the past 4 years have taught us that bacteria recognize no such constraints and often use many different mechanisms to achieve and maintain their attachment to surfaces. A case in point is the attachment of cells of enterotoxigenic strains of *Escherichia coli* to the ileal mucosa of the newborn calf. We know that the K99 pilus of these organisms is the primary agent of pathogenic adherence (29), and we (3) have visualized this very specific adhesion directly on the infected tissue (Fig. 1). Nevertheless, these bacteria can use several other pili for this pathogenic adhesion (35), and pilus-minus strains have been shown to cause fatal infections ·of colostrum-deprived calves (4). In situ examination of K99[+] and pilus-minus strains on the infected tissue, with the use of specific antibodies against the K30 glycocalyx component of these cells, has shown that cells of both are surrounded by a very extensive capsular structure (Fig. 2) that may act as a low-affinity adhesion factor and as an essential element for persistence in the surfactant-rich milieu of the gut. Similarly, there is no doubt that the P pilus is an avid adhesion ligand in uropathogenic strains of *E. coli* (46), but our recent cooperative examination (with C. Svanborg-Eden) of P pilus-bearing strains directly on tissue surfaces of the infected kidney (Fig. 3) showed that adherent cells are surrounded by very large amounts of an amorphous matrix whose reaction with ruthenium red indicates that it is an acid polysaccharide (26). This extracellular polysaccharide glycocalyx has been shown to protect bacteria from surfactants (19), antibodies (2), and phagocytes (43), and it may be a pivotal factor in bacterial persistence on the infected tissue.

Recent very extensive examinations of bacteria growing in a wide variety of natural (1, 5, 15, 17) and pathogenic (8, 22, 32) situations have shown that virtually all of these cells are surrounded by extracellular glycocalyx structures (11, 12) of considerable thickness (>0.1 μm in most cases). Thus, the effective surface of "wild" bacteria capable of survival in natural and pathogenic situations consists largely of their highly hydrated exopolysaccharide matrix (45) and of protruding proteinaceous pili. This essentially fibrous bacterial cell boundary becomes increasingly regular as both the glycocalyx and the pili may be lost upon subculture in vitro, and successive layers of inner envelope structures are exposed at the modified cell surface. Of course, each of these newly exposed cell envelope layers will show a complex pattern of adhesion to various substrata, but this observation cannot be extrapolated to yield information on pathogenesis or bacterial survival in natural environments because these cell envelope structures may not actually be present at the bacterial cell surface in "wild" strains of the organism that are capable of survival and competition in actual ecosystems. These in vitro-adapted strains of bacteria may yield very consistent adhesion data, because a multiplicity of adhesion factors has been reduced to one or two, but it seems axiomatic that studies of pathogenic and natural adhesion, and of the biophysics of cell–surface interactions, must be conducted using ecologically competent wild bacterial cells. Because of this insidious "drift" in the surface structures of bacteria that are removed from their natural environments, I favor the analysis of the surface composition of bacteria by direct examination within their peculiar ecological niche, using powerful modern immunological tools such as monoclonal antibodies (Fig. 1 and 2). These direct observations of bacteria in situ are especially attractive in view of observations that pilus formation may be conditioned upon environmental factors (35), and in view of the developing perception that the surface architecture of bacterial cells may undergo a progressive series of changes during the etiological development of a disease (34) or during the sequential colonization of an ecological niche (6).

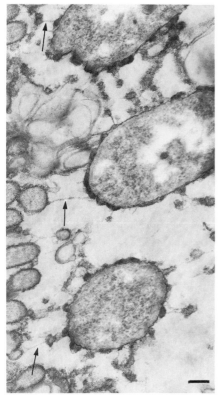

FIG. 1. Cells of an enterotoxigenic strain of *E. coli* adherent to the ileal mucosa of a newborn calf by means of K99 pili. These pili have been thickened by reaction with monoclonal antibodies to make them visible (arrows) in sectioned material. Bar, 0.1 μm.

COMPLEX PATTERNS OF BACTERIAL ADHESION

Although reproducible experimentation requires that variable parameters be minimized, we must be aware that short-term studies of the adhesion of bacteria from phenotypically stable cultures to the surfaces of derived or cultured tissue cells in vitro provide an inadequate basis for extrapolation to the colonization of tissues in natural situations. As a sine qua non, bacteria must survive in an environment, before and after adhesion to surfaces, and they must therefore possess a full complement of protective surface structures, just as cells within a tissue are covered by their own glycoprotein glycocalyces (37). Direct examination of well-developed autochthonous (40, 42) bacterial populations on animal tissue surfaces (6) reveals that these populations are often highly structured in that certain bacterial morphotypes are directly adherent to the tissue surface whereas others grow in successive layers with an adherent biofilm as much as 50 μm and 20 bacterial cells in width.

Thus, whereas the initial colonizers adhere directly to the tissue glycocalyx, most bacteria in these tissue-adherent populations actually attach themselves to other bacteria in a highly structured series of microbial consortia. Simple consortia have been defined on inert surfaces in that the initial colonizer of a nutritive substratum (e.g., cellulose) adheres to its specific substrate (Fig. 4) and then attracts a secondary colonizer to its surface by the provision of

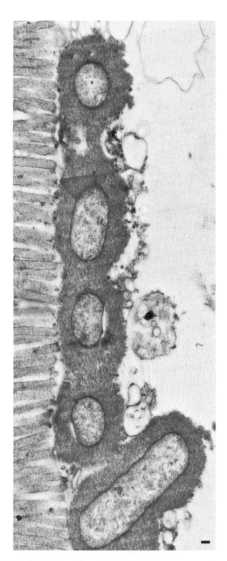

FIG. 2. Cells of a K99⁻K30⁺ enterotoxigenic strain of *E. coli* adherent to the ileal mucosa of a newborn calf. The K30 exopolysaccharide antigen of these cells has been stabilized against dehydration collapse, like that seen in the K99⁺K30⁺ cell in Fig. 1, by reaction with anti-K30 antibodies. This K30 material is seen to form a very extensive capsular glycocalyx surrounding these bacterial cells. Bar, 0.1 μm.

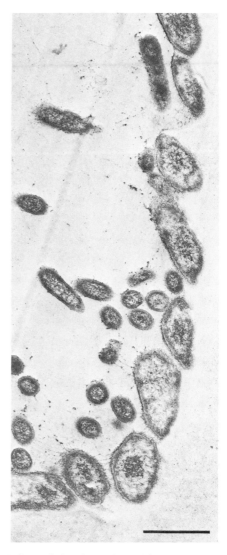

mogenization and careful selection of media (33). The availability of gnotobiotic animals allows us to reconstitute tissue surface consortia in vivo (7), and these studies have already revealed that these organisms cooperate physiologically both with other bacterial members of the consortium and with the tissues themselves

FIG. 3. Cells of a P pilus-bearing strain (1682) of uropathogenic *E. coli* growing as an adherent microcolony on an epithelial surface in the kidney of an infected mouse (collaboration with Svanborg-Eden). These bacteria, whose primary adherence to this tissue surface depends on the P pilus, have surrounded themselves with such large amounts of exopolysaccharide glycocalyx that they are embedded (arrow) to various degrees within this matrix. Bar, 1.0 μm.

FIG. 4. Cells of two bacterial morphotypes that have been selectively attracted from a mixed microbial population to form a simple consortium on the surface of a cellulosic fragment of plant cell wall in the bovine rumen. The primary colonizer is *Bacteroides succinogenes*, whose cellulolytic enzymes have digested shallow "pits" in the cellulose surface; a very thin (ca. 0.2-μm) rod-shaped organism has been attracted to colonize the surface of the *Bacteroides* biofilm and thus become the second member of a simple bacterial consortium. Bar, 1.0 μm.

ligands and specific nutrients (6). Thus, perhaps, the optimum strategy to select and study an adhesion system that actually operates in nature involves an initial morphological examination of the complexity and approximate bacterial composition of a natural tissue-adherent population. The isolation of the members of this community, for individual taxonomic identification and physiological examination, often requires tissue ho-

(6). In one specific case, the bovine rumen epithelium specifically attracts several bacterial species that colonize this tissue and provide, throughout the life of the animal, an enzyme (urease) that is required for the normal function of this organ (7). A developing understanding of the composition and physiological coordination of autochthonous tissue-adherent bacterial populations has already allowed the successful manipulation of the bacterial populations of newborn ruminants to promote weight gain and to prevent neonatal intestinal disease.

In addition to the complexities of the myriad ecological microniches that are sequentially colonized by bacteria, and in addition to the complex physiologically integrated consortia within which most autochthonous bacteria live, bacterial growth on tissue surfaces is made more refractory to analysis by characteristics of the tissue surface itself. With the exception of the highly specialized bacteria that actually insert themselves into the intestinal epithelium (41), most bacteria in the intestine lie in the mucus blanket (16) that moves slowly over this tissue, and they have very little apparent contact (Fig. 5) with the tissue itself (38). The administration of lectins can radically alter these spatial relationships and bind certain bacteria from the mucus phase to the tissue surface (Fig. 6)—with severe pathogenic consequences. We should, perhaps, bear in mind that the phenomenal genotypic and phenotypic plasticity evidenced by bacteria is actually a strategic device to enable these organisms to find, colonize, and compete within an infinite series of very highly specialized ecological microniches.

FIG. 5. Bacterial and protozoan cells within the antibody-stabilized mucus blanket covering the surface of the ileal mucosa of the mouse. The microvillar surface of the tissue itself is uncolonized by bacteria except for occasional trichome-forming organisms that are actually inserted into the intestinal tissue (not seen here). Bar, 5 μm.

IMPORTANCE OF ADHESION IN MICROBIAL ECOLOGY

The numerical and physiological predominance of sessile (attached) bacteria was initially established in flowing aquatic systems (10) where more than 99% of these organisms were found to grow in confluent adherent biofilms within which each cell has almost complete access to organic substrates that "load" the glycocalyx like an ion-exchange resin. Many important chemical transformations in aquatic environments have been located entirely in the epilithic surface and sediment zones (47), and in the neuston layer (44), where complex bacterial biofilms develop and proliferate (25). This "biofilm reactor" model of aquatic microbial ecology has now been extended to account for the depletion of organic compounds in groundwater (21) and the proliferation of heterotrophic bacteria in the proximal zones of rock formations in water-injection oil-recovery operations (13). In industrial aquatic systems, sessile biofilm populations have also been found to predominate and to be instrumental in both corrosion and flow reduction (36). Bacteria within these biofilms have been shown to be much more resistant to commercial chemical biocides (39) than their floating planktonic counterparts, and biofilm bacteria in the hospital environment have been shown to resist the action of antiseptics (27) at concentrations that easily kill dispersed cells of the same species. Because sessile bacteria predominate in all of the natural and industrial aquatic systems examined to date, and because these biofilm bacteria differ from planktonic forms in many important parameters, it is now clear that we cannot continue to extrapolate data obtained by using either planktonic or in vitro-grown cells in our attempts to understand microbial processes in these systems (12).

The consequences of bacterial growth in adherent biofilms are equally important in the microbial ecology of the bacterial colonization of tissues and inert surfaces in the human body. Biofilm development on epithelial and tooth surfaces of the mouth has long been recognized

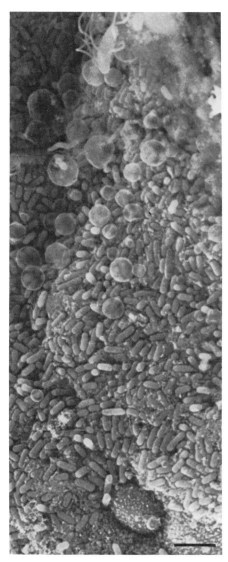

FIG. 6. Bacteria growing directly on the microvillar surface of the ileum of the mouse, and causing severe pathological changes in this tissue (viz., the large vesicles seen in this micrograph), when the animals are treated orally with a lectin that appears to bind them very firmly to the tissue surface. Bar, 5 μm.

and is very well studied (18), and we have recently determined that many body tissues (31) and metal and plastic (8, 24, 29) surfaces of modern prostheses also develop extensive adherent biofilms. The biofilm mode of growth protects adherent bacteria from host defense factors such as surfactants (19), antibodies (2), and phagocytes (43), and this is especially true on plastic and metal surfaces which, themselves, inhibit the phagocytic activity of macrophages (23). Skin bacteria, such as *Staphylococcus epi-*

dermidis, rapidly colonize the plastic surfaces of sutures that traverse the human skin and produce extensive biofilms on these inert surfaces (Fig. 7), but inflammation and infection rarely develop around these simple prostheses. Similarly, the placement of plastic intrauterine contraceptive devices in the uterus leads to the development of very thick confluent bacterial biofilms on both their plastic and metal (copper) surfaces (Fig. 8) and to a slightly elevated incidence of infections of both the fallopian tubes and the peritoneum (30). The identification of fallopian tube pathogens (actinomycetes) within the biofilm on these devices suggests that these adherent bacteria constitute a harmless reservoir population, as long as they remain within the biofilm, but that their spread onto the sur-

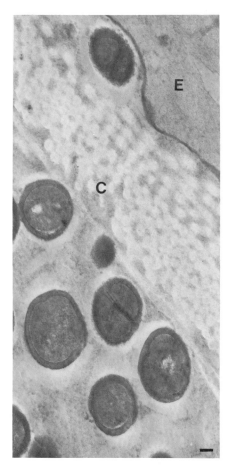

FIG. 7. Cells of the autochthonous skin organism *S. epidermidis* growing as extensive adherent biofilms on the surfaces of plastic sutures that traverse the human skin. These gram-positive cocci grow in very large glycocalyx-enclosed masses adjacent to the collagen layers (C) and epithelial cells (E) of the skin, but they very rarely cause inflammation or infection. Bar, 0.1 μm.

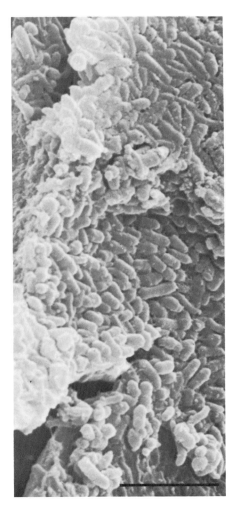

FIG. 8. Bacterial cells shown to have produced very thick biofilms on the surfaces of intrauterine contraceptive devices recovered from the uteri of entirely asymptomatic reproductive-age females. These mixed populations produce such phenomenal amounts of glycocalyx material that it "buries" bacterial cells at the surface of the biofilm (top) and produces an adherent layer that often exceeds 0.5 mm in thickness. The wearing of these devices does not usually cause infection, but it is associated with a slightly elevated incidence of infection of the fallopian tubes and of the peritoneum. Bar, 5 μm.

faces of adjacent, normally sterile tissues initiates a pathogenic process (30). The more invasive use of plastic prostheses, such as the installation of indwelling intracardial (Hickman) and intraperitoneal (Tenchoff) catheters through a cuff in the skin and into deep tissues, has often been complicated by enigmatic bacterial infections. Although these infections often involve saprophytic skin bacteria such as *S. epidermidis*, the involvement of the prosthesis as an

invasion tract seemed to be negated by consistently negative laboratory cultures of these devices upon their removal from the body. Direct examination of both of these types of catheters (Fig. 9) has shown that gram-positive cocci develop very thick confluent biofilms on their plastic surfaces. The cohesion of these adherent

FIG. 9. Cells of *S. epidermidis* and *Candida albicans* growing within a very thick mixed biofilm in the lumen of the cardiac tip of a Hickman catheter used for 14 months to gain access directly to the heart of a patient with acute nonlymphocytic leukemia. This patient had episodes of both staphylococcal and *Candida* bacteremia and later developed a mycotic aneurism of the aorta that yielded cultures of *C. albicans*. Because of the biofilm mode of growth of these organisms, routine cultures of blood removed via the catheter, and of the catheter itself after its removal, failed consistently to detect these adherent organisms that obviously constituted a dangerous reservoir of potential pathogens. Bar, 1 μm.

biofilms precludes the recovery of their component cells by routine microbiological analysis and also limits their pathogenic impact on the host in that only a small proportion of patients whose protheses are colonized actually develop infections. We have adapted methods for the recovery of sessile bacteria from inert surfaces in aquatic environments (17), and from animal tissues (6), to accomplish the detection and quantitative recovery of these potentially pathogenic biomaterial-colonizing bacteria. When a prosthesis is implanted entirely within normally sterile areas of the human body, it can become colonized by bacteria as a result of lapses in surgical asepsis or of hematogenous spread of bacteria in a transient bacteremia. When the plastic and metal surfaces of deep prostheses, such as joints and cardiac pacemakers, become colonized by bacteria, these organisms form very thick biofilms (Fig. 10) within which their cells are enveloped in a very extensive protective matrix of the fibrous anionic exopolysaccharide of their glycocalyces (3, 28). These "foreign body" infections often necessitate the removal of the prosthesis because the bacteria within these adherent biofilms are highly resistant to antibiotics (17) at concentrations that easily kill tha same organisms growing as planktonic cells in derived laboratory cultures. Antibiotic chemotherapy of these deep "foreign body" infections is further complicated by sampling problems in that samples of body fluids (e.g., synovial fluid) often yield cultures of secondary planktonic pathogens whereas the primary adherent pathogen often escapes detection, and therapy is therefore targeted to the wrong organism. Thus, it is clear that prosthetic biomaterials are readily colonized by bacteria whose development of adherent glycocalyx-enclosed biofilms protects them from both host clearance factors and antibiotics.

CONCLUSIONS

In addition to retaining the bacterial cell in a suitable environmental microniche, the general process of adhesion positions the cell within the surface zone which is rich in nutrients because organic molecules are attracted to these interfaces (48). Once attached to a surface, the bacterium surrounds itself with a fibrous anionic exopolysaccharide matrix that acts as an ion-exchange "resin" to attract and concentrate charged nutrients, and this matrix burgeons, with cell growth, until a thick coherent biofilm is formed. This sessile biofilm mode of growth has been shown to protect adherent cells from surfactants, bacteriophage, phagocytic amoebae, chemical biocides, antiseptics, antibiotics, antibodies, and phagocytic leukocytes; adherent growth has also been shown to protect bacteria

FIG. 10. Cells of *S. aureus* that had colonized the metal portion of an endocardial pacemaker, following bacteremic spread from a traumatic olecranon bursitis. These organisms have formed a biofilm of macroscopic proportions within which the bacterial cells are embedded in large masses of amorphous exopolysaccharide glycocalyx material (bottom). This "foreign body" infection was treated for 6 weeks with 12 g of cloxacillin per day, which yielded serum concentrations far in excess of the in vitro minimal inhibitory concentration for the organisms, but the biofilm continued to serve as a reservoir for acute recurring bacteremic infections, and the removal of this cardiac prosthesis was eventually necessary. Bar, 1.0 μm.

from removal by host defense mechanisms such as mucocilliary bronchial clearance (9). Because both the biofilm-forming sessile phenotype and the mobile planktonic phenotype of the same bacterial strains are seen in natural systems, we must presume that these organisms are capable of producing a phenotypic spectrum of growth types from adherent biofilms to flagellated mobile "swarmer" cells. This spectrum is forced toward its sessile extreme by concerted antibacterial factors, like those present in the uncompromised animal body, while these same organisms drift gradually (14) toward the planktonic extreme (9) when growing in compromised tissues (e.g., burned skin) or in vitro cultures. The observed predominance of sessile biofilm-forming bacterial populations in a wide variety of natural, industrial, and tissue-surface environments argues for a generally protective role of their mode of growth and of their enveloping glycocalyces. The observed drift away from complete and intact cell surface structures upon in vitro cultivation in the absence of antibacterial factors indicates that the sessile mode of growth is not inherently energetically favorable and that it is the general requirement for protection that induces most bacteria in natural, industrial, autochthonous, and pathogenic populations to grow within adherent biofilms.

From the direct observational data reviewed herein, it is patent that bacteria growing in natural and pathogenic environments differ radically, especially in their cell surface characteristics, from the same genotype growing in vitro in batch culture. Thus, extrapolation from in vitro studies to environmental conclusions should be tempered with direct observations to ascertain, as a minimum requirement, that factors invoked as adhesion mechanisms are actually present in the natural population being studied.

We acknowledge the joint support of the Natural Sciences and Engineering Research Council of Canada and of the Medical Research Council of Canada in the form of a Special Collaborative Grant.

LITERATURE CITED

1. **Bae, H. C., E. H. Cota-Robles, and L. E. Casida, Jr.** 1972. Microflora of soil as viewed by transmission electron microscopy. Appl. Microbiol. **23:**637–648.
2. **Baltimore, R. S., and M. Mitchell.** 1980. Immunologic investigations of mucoid strains of *Pseudomonas aeruginosa*: comparison of susceptibility by opsonic antibody in mucoid and nonmucoid strains. J. Infect. Dis. **141:**238–247.
3. **Chan, R., S. D. Acres, and J. W. Costerton.** 1982. Use of specific antibody to demonstrate glycocalyx, K99 pili, and the spatial relationship of K99+ enterotoxigenic *Escherichia coli* in the ileum of colostrum-fed calves. Infect. Immun. **37:**1170–1180.
4. **Chan, R., C. J. Lian, J. W. Costerton, and S. D. Acres.** 1983. The use of specific antibodies to demonstrate the glycocalyx and spatial relationships of a K99−, F41− enterotoxigenic strain of *Escherichia coli* colonizing the ileum of colostrum-deprived calves. Can. J. Comp. Med. **47:**150–156.
5. **Cheng, K.-J., and J. W. Costerton.** 1975. Ultrastructure of cell envelopes of bacteria of the bovine rumen. Appl. Microbiol. **29:**841–849.
6. **Cheng, K.-J., R. T. Irvin, and J. W. Costerton.** 1981. Autochthonous and pathogenic colonization of animal tissues by bacteria. Can. J. Microbiol. **27:**461–490.
7. **Cheng, K.-J., and R. J. Wallace.** 1979. The mechanism of passage of endogenous urea through the rumen wall and the role of ureolytic epithelial bacteria in the urea flux. Br. J. Nutr. **42:**553–557.
8. **Christensen, G. D., W. A. Simpson, A. I. Bisno, and E. H. Beachey.** 1982. Adherence of slime-producing strains of *Staphylococcus epidermidis* to smooth surfaces. Infect. Immun. **47:**318–326.
9. **Costerton, J. W., M. R. W. Brown, and J. M. Sturgess.** 1979. The cell envelope: its role in infection, p. 41–62. *In* R. G. Doggett (ed.), *Pseudomonas aeruginosa*: clinical manifestations and current therapy. Academic Press, Inc., New York.
10. **Costerton, J. W., and R. R. Colwell.** 1979. Native aquatic bacteria: enumeration, activity and ecology. ASTM Press, Philadelphia.
11. **Costerton, J. W., R. T. Irvin, and K.-J. Cheng.** 1981. The role of bacterial surface structures in pathogenesis. Crit. Rev. Microbiol. **8:**303–338.
12. **Costerton, J. W., R. T. Irvin, and K.-J. Cheng.** 1981. The bacterial glycocalyx in nature and disease. Annu. Rev. Microbiol. **35:**299–324.
13. **Costerton, J. W., and E. S. Lashen.** 1983. The inherent biocide resistance of corrosion-causing biofilm bacteria. Corrosion 83, paper no. 246, p. 1–11. Nace, Anaheim, Calif.
14. **Doggett, R. W., G. M. Harrison, and E. S. Wallis.** 1964. Comparison of some properties of *Pseudomonas aeruginosa* isolated from infections of persons with and without cystic fibrosis. J. Bacteriol. **87:**427–431.
15. **Fletcher, M., and G. D. Floodgate.** 1973. Electron microscopic demonstration of an acidic polysaccharide involved in the adhesion of a marine bacterium to solid surfaces. J. Gen. Microbiol. **72:**325–334.
16. **Freter, R., P. C. M. O'Brien, and M. S. Macsai.** 1981. Role of chemotaxis in the association of motile bacteria with intestinal mucosa: in vivo studies. Infect. Immun. **34:**234–240.
17. **Geesey, G. G., R. Mutch, J. W. Costerton, and R. B. Green.** 1978. Sessile bacteria: an important component of the microbial population in small mountain streams. Limnol. Oceanogr. **23:**1214–1223.
18. **Gibbons, R. J., and J. Van Houte.** 1975. Bacterial adherence in oral microbial ecology. Annu. Rev. Microbiol. **29:**19–44.
19. **Govan, J. R. W.** 1975. Mucoid strains of *Pseudomonas aeruginosa*: the influence of culture medium on the stability of mucus production. J. Med. Microbiol. **8:**513–522.
20. **Isaacson, R. E., P. C. Fusco, C. C. Brinton, and H. W. Moon.** 1978. In vitro adhesion of *Escherichia coli* to porcine small intestinal epithelial cells: pili as adhesive factors. Infect. Immun. **21:**392–397.
21. **Ladd, T. I., R. M. Ventullo, P. M. Wallis, and J. W. Costerton.** 1982. Heterotrophic activity and biodegradation of labile and refractory compounds by groundwater and stream microbial populations. Appl. Environ. Microbiol. **44:**321–329.
22. **Lam, J., R. Chan, K. Lam, and J. W. Costerton.** 1980. Production of mucoid microcolonies by *Pseudomonas aeruginosa* within infected lungs in cystic fibrosis. Infect. Immun. **28:**546–556.
23. **Leake, E. S., and M. J. Wright.** 1979. Variations in the form of attachment of rabbit alveolar macrophages to various substrata as observed by scanning electron microscopy. RES J. Reticuloendothel. Soc. **25:**417–441.
24. **Locci, R., G. Peters, and G. Pulverer.** 1981. Microbial colonization of prosthetic devices. III. Adhesion of staphylococci to lumina of intravenous catheters perfused with bacterial suspensions. Zentralbl. Bakteriol. Parasitenkd. Infektionskr. Hyg. Abt. 1 Orig. Reihe B **173:**300–307.
25. **Lock, M. A., R. R. Wallace, J. W. Costerton, R. M. Ventullo, and S. E. Charlton.** 1983. River epilithon: towards a structural-functional model. Oikos **41:**in press.
26. **Luft, J. H.** 1971. Ruthenium red and ruthenium violet. I. Chemistry, purification, methods of use for electron microscopy, and mechanism of action. Anat. Rec. **171:**347–368.

27. **Marrie, T. J., and J. W. Costerton.** 1981. An electron microscopic study of the prolonged survival of *Serratia marcescens* in chlohexidine. Appl. Environ. Microbiol. **42:**1093–1102.

28. **Marrie, T. J., and J. W. Costerton.** 1982. A scanning and transmission electron microscopic study of an infected endocardial pacemaker lead. Circulation **66:**1339–1343.

29. **Marrie, T. J., and J. W. Costerton.** 1983. A scanning electron microscopic study of the adherence of uropathogens to a plastic surface. Appl. Environ. Microbiol. **45:**1018–1024.

30. **Marrie, T. J., and J. W. Costerton.** 1983. A scanning and transmission electron microscopic study of the surfaces of intrauterine contraceptive devices. Am. J. Obstet. Gynecol. **146:**384–394.

31. **Marrie, T. J., G. K. M. Harding, and A. R. Ronald.** 1978. Anaerobic and aerobic urethral flora in healthy females. J. Clin. Microbiol. **8:**67–72.

32. **Marrie, T. J., G. K. M. Harding, A. R. Ronald, J. Dikkema, J. Lam, J. Hoban, and J. W. Costerton.** 1979. Influence of mucoidy on antibody coating of *Pseudomonas aeruginosa.* J. Infect. Dis. **139:**357–361.

33. **McCowan, R. P., K.-J. Cheng, and J. W. Costerton.** 1979. Colonization of a portion of the bovine tongue by unusual filamentous bacteria. Appl. Environ. Microbiol. **37:**1224–1229.

34. **McGee, Z. A., A. P. Johnson, and D. Taylor-Robinson.** 1981. Pathogenic mechanisms of *Neisseria gonorrhoeae*: observations on damage to human fallopian tubes in organ culture by gonococci of colony type 1 or type 4. J. Infect. Dis. **143:**413–422.

35. **Morris, J. A., C. Thorns, A. C. Scott, W. J. Sojka, and G. A. Wells.** 1982. Adhesion in vitro and in vivo associated with an adhesive antigen (F41) produced by a K99 mutant of the reference strain *Escherichia coli* B41. Infect. Immun. **36:**1146–1153.

36. **Picoloqlou, B. F., N. Zelver, and W. G. Characklis.** 1980. Biofilm growth and hydraulic performance. J. Hydraul. Div. Am. Soc. Civ. Eng. **106:**733–746.

37. **Roseman, S.** 1974. Complex carbohydrates and intercellular adhesion, p. 317–337. *In* E. Y. C. Lee and E. E. Smith (ed.), Biology and chemistry of eukaryotic cell surfaces. Academic Press, London.

38. **Rozee, K. R., D. Cooper, K. Lam, and J. W. Costerton.** 1982. Microbial flora of the mouse ileum mucous layer and epithelial surfaces. Appl. Environ. Microbiol. **43:**1451–1463.

39. **Ruseska, I., J. Robbins, E. S. Lashen, and J. W. Costerton.** 1982. Biocide testing against corrosion causing oilfield bacteria helps control plugging. Oil Gas J. **80:**253–264.

40. **Savage, D. C.** 1977. Microbial ecology of the gastrointestinal tract. Annu. Rev. Microbiol. **31:**107–133.

41. **Savage, D. C., and R. V. H. Blumershine.** 1974. Surface-surface association in microbial communities populating epithelial habitats in the murine gastrointestinal ecosystem: scanning electron microscopy. Infect. Immun. **10:**240–250.

42. **Savage, D. C., R. Dubos, and R. W. Schaedler.** 1968. The gastrointestinal epithelium and its autochthonous bacterial flora. J. Exp. Med. **127:**67–76.

43. **Schwarzmann, S., and J. R. Boring, III.** 1971. Antiphagocytic effect of slime from a mucoid stain of *Pseudomonas aeruginosa.* Infect. Immun. **3:**762–767.

44. **Sieburth, J. M., P. J. Willis, K. M. Johnson, C. M. Burney, D. M. Lavoie, K. R. Hinga, D. A. Caron, F. W. French, III, P. W. Johnson, and P. G. Davis.** 1976. Dissolved organic matter and heterotrophic micronueston in the surface microlayers of the North Atlantic. Science **194:**1415–1418.

45. **Sutherland, I. W. (ed.).** 1977. Bacterial exopolysaccharides—their nature and production in surface carbohydrates of the prokaryotic cell. Academic Press, London.

46. **Svanborg-Eden, C., and H. A. Hanson.** 1978. *Escherichia coli* pili as mediators of attachment to human urinary tract epithelial cells. Infect. Immun. **21:**229–237.

47. **Wyndham, R. C., and J. W. Costerton.** 1981. Heterotrophic potentials and hydrocarbon biodegradation potentials of sediment microorganisms within the Athabasca oil sands deposit. Appl. Environ. Microbiol. **41:**783–790.

48. **Zobell, C. E.** 1943. The effect of solid surfaces upon bacterial activity. J. Bacteriol. **46:**39–56.

Microbial Attachment to Nonbiological Surfaces

MADILYN FLETCHER AND S. McELDOWNEY

Department of Environmental Sciences, University of Warwick, Coventry CV4 7AL, United Kingdom

When a bacterium encounters a surface, initial adhesion is dependent upon the adsorption of bacterial surface polymers or structures. This involves a complex interaction of (i) the bacterial surface components, (ii) substratum surface properties (e.g., charge, presence of polar or nonpolar groups; see 12, 24), (iii) adsorbed substances on the substratum (e.g., hydrated macromolecules) (see 11), (iv) dissolved medium components (see 9), and (v) water. Once adhesion has occurred, physiological activity can then lead to changes in cell surface components, resulting in the accumulation of extracellular polymers, in firm attachment with no accumulation of polymers, or in detachment from the surface.

Although initial adhesion to solid surfaces is strongly dependent upon the surface properties of the bacterium, little is known about the surface chemical characteristics which promote or inhibit adhesion or about the chemical basis for nonspecific adhesive interactions. It is feasible that, in each adhesive event, one or more of a number of different types of interaction (e.g., hydrophobic or hydrogen-bonding, electrostatic interactions) may be involved, depending upon the chemical "compatibility" of the bacterial and substratum surfaces, and that different types of interaction will be dominant with various species or strains. Clearly, the potential for adhesive interactions varies with different organisms, and accordingly attachment ability differs enormously (22).

However, not only do various bacterial strains differ in their ability to attach to solid surfaces, but also the attachment of a given bacterium may vary with culture conditions. In this laboratory, studies with strains of *Pseudomonas fluorescens, Enterobacter cloacae,* a *Flexibacter* sp., and a *Chromobacterium* sp. demonstrated that the attachment ability of each organism was modified by growing the cells in continuous or batch culture under either carbon or nitrogen limitation or by varying the carbon source (glucose, glucose and glycerol, lactose, galactose, mannose, or sucrose, all at 0.1%, wt/vol, concentration) (S. McEldowney, unpublished data). Attachment to polystyrene was allowed for 1 h and was evaluated by crystal violet staining as described in reference 22. Although growth conditions affected attachment ability, each organism illustrated a different attachment response to the various culture conditions. For example, carbon-limited growth tended to promote attachment of the *Chromobacterium* sp., whereas it had little effect on the adhesion of *P. fluorescens* and reduced the attachment of the *Flexibacter* sp. It seems likely that these variations in attachment ability induced by the different growth regimens were due, at least in part, to associated differences in cell surface composition. However, it has not been possible to establish a relationship between adhesiveness and gross evaluations of bacterial surface properties, i.e., measurement of liquid contact angles on lawns of cells (2, 26) and adsorption of the bacteria to cation- and anion-exchange resins and to hydrophobic interaction chromatography gels (7).

Clear evidence of the significance of bacterial surface components in adhesive interactions was provided by a study of the attachment ability of a *P. fluorescens* strain and two mutants with altered cell surface properties (J. H. Pringle, M. Fletcher, and D. C. Ellwood, J. Gen. Microbiol., in press). The wild type attached to solid surfaces in high numbers, as compared to a mucoid mutant which produced an alginate exopolysaccharide and whose attachment ability was considerably less (Table 1). The second type of mutant, a "rough" form, had proportionally less polysaccharide in the lipopolysaccharide (LPS) and an increased attachment ability. Thus, attachment was inhibited by the production of an acidic, thus highly hydrated, exopolymer and was facilitated by a reduction in the polysaccharide component of the LPS. Moreover, these attachment properties were also illustrated by the distributon of the three types of cell within a continuous culture system; the wild type and rough strains tended to be concentrated on the fermentor walls, whereas the mucoid organism was concentrated in the bulk phase (Pringle et al., in press). This study illustrates two very important points. First, the attachment ability of bacteria depends not only upon the species and on phenotypic changes in the cell surface composition under different growth regimens, but may also be modified by mutations and genotypic modifications of surface composition. Second, although bacterial surface polymers, such as LPS, membrane proteins, or exopolysaccharides, may act as adhesives in adsorption interactions, an increase in

TABLE 1. Relationship between surface compositions of *P. fluorescens* strains and their attachment abilities

Strain	Yield of crude LPS (% cell dry wt)	Yield of lipid-free polysaccharide (% wt of LPS)	Yield of crude exo-polysaccharide (% cell dry wt)	Attachment to polysaccharide[a]
Wild type	2.9	17.3	1.3	0.190
Mucoid mutant	2.3	20.2	173.0	0.051
Rough mutant	2.0	15.4	0.9	0.216

[a] Absorbance at 590 nm of crystal violet-stained attached film.

quantity of polymer does not necessarily promote adhesion and may indeed prevent initial attachment.

However, in contrast to the reduction in adhesion by increased production of polymer, there is also abundant evidence that exopolymers, particularly polysaccharides, are involved in adhesion (5, 14). How can these two apparently inconsistent types of data be reconciled? It is likely that many observations of exopolymers associated with attached cells are accumulations of polymer which have occurred over some time. This was illustrated by the appearance of a hydrated, polymeric gel which occurred in association with microcolonies of a marine pseudomonad, whereas recently attached cells, which had not had time to grow, were apparently attached to the surface by a compact cell surface polymer (possibly LPS) (10). Certainly the polymeric gels in biofilm slime layers take time for their development, and time-dependent attachment involving polymer production at the surface has been discussed by ZoBell (28) and Marshall et al. (18). Thus, the initial adsorption interaction between the bacterium and solid surface may not require hydrated cell surface polysaccharides, and may indeed be favored by their absence, whereas biofilm formation, which involves cell–cell adhesion, requires exopolymer production.

It is clear that, if we are to understand the role of bacterial surface polymers in the promotion or prevention of adhesion, it is necessary to examine the adhesion process over a period of time spanning the moment of initial interaction between the two surfaces and subsequent modifications of bacterial surface properties or production of exopolymers. Thus, initial adhesion may be strengthened, or, on the other hand, weakened, resulting in detachment of the bacterium.

One way in which the time dependence of adhesion can be studied is by determining whether numbers of attached cells increase with the time which has been allowed for attachment to the substratum. A number of workers (4, 16, 19) have found that numbers of attached cells increase with time up to a maximum, which may reflect the carrying capacity of the substratum (8). Although this increase in attached numbers with time may be a function of the increased number of bacterium–surface collisions, it could also be due to a time-dependent increase in strength of adhesive binding associated with physiological processes of the bacteria at the surface. If the bacteria make a physiological contribution to time-dependent attachment, then the extent of increase in attached numbers with attachment time might be expected to depend somewhat on the physiological status of the cells. This was examined with four strains (*Flexibacter* sp., *Chromobacterium* sp. *P. fluorescens*, *E. cloacae*) grown in continuous culture at dilution rates of 0.05, 0.1, 0.15, or 0.2 h^{-1}. Samples of the bacteria were then allowed to attach to two types of polystyrene, i.e., petri dish polystyrene (PD), which is hydrophobic, and a more hydrophilic form, purchased commercially as tissue culture polystyrene (TCD). (For method, see 22.) Exposure of the cell suspensions to the substrata was for 5 or 60 min. Although there was usually an increase in numbers of attached cells with the longer attachment time (60 min), the extent of this increase varied considerably with both the organism and the dilution rate (Fig. 1). With both the *Flexibacter* sp. and the *Chromobacterium* sp., the difference between numbers attached at 5 and 60 min of incubation increased with dilution, and thus growth, rate. Possibly, the faster-growing cells were also better able to strengthen the adhesive bond, e.g., by producing suitable adhesives, over the 60-min exposure time. However, with the *Chromobacterium* sp., flocculation was induced at the higher growth rates, and this contributed to the corresponding larger number of attached cells. By contrast, *P. fluorescens* showed little difference in numbers attached after 5 and 60 min and no change in numbers attached with dilution rate. Similarly, there was no appreciable difference between numbers of attached cells of *E. cloacae* after 5 and 60 min, but there was some reduction with dilution rate at 60 min of attachment time. Such results emphasize the different properties of individual strains and illustrate the impossibility of generalizing about particular factors and their influence on bacterial attachment.

Although changes in bacterial surface properties may result in a strengthening of adhesion with time, they may also lead to a weakening of attachment and to desorption (29). Bacteria which at one time were apparently firmly at-

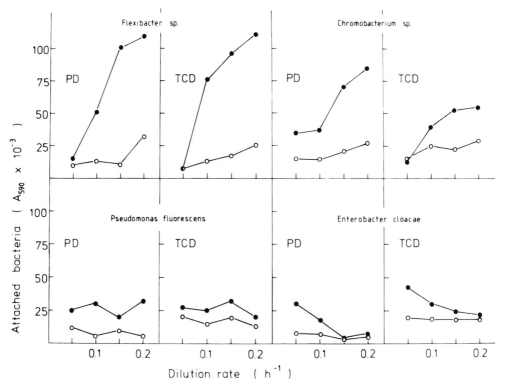

FIG. 1. Relationship between the attachment of bacteria grown in continuous culture and the dilution rate. Attachment was measured as the absorbance of the crystal violet-stained attached film at 590 nm. The substrata were hydrophobic (PD) and more hydrophilic, tissue culture (TCD) polystyrene. Incubation with the substrata was for 5 (○) or 60 (●) min.

tached and resisted washing off were found to detach from test substrata over incubation times up to 2 h (3). Moreover, the extent of detachment was dependent upon the composition of the substratum, so that detachment increased with increase in substratum hydrophobicity. The strength of adhesion, as reflected by amount of detachment, may also be related to the physiological status of the cells. For example, a relationship between the culture age of *Bacillus cereus* and its resistance to removal by shear stress has been found (20). Also, there was an inverse relationship between the proportion of cells assimilating substrate (measured by microautoradiography) and the proportion of cells which became detached over 2 h (3). These may be illustrations of time-dependent attachment in which physiological processes in some way lead to a strengthening of attachment and a reduction in detachment.

However, a serious, but unavoidable, deficiency in most experimental and ultrastructural studies of attachment processes is that the attached bacteria are usually fixed before counting or observation. To study the progression of attachment, replicate samples are usually pre-

pared and removed at intervals, but rarely has the attachment of specific cells been monitored over an extended time period. Some light microscope studies have provided continuous observations of apparent attachment and detachment in the laboratory and have provided valuable information on the sequence of adhesion events and on the orientation of attaching cells (17, 18, 25). However, ordinary light microscopy of living material can give no information on the nature of the interaction between the bacterial and substratum surfaces or on the presence or production of bacterial exopolymers.

Recently, a light microscopic method, interference reflection microscopy (IRM), has been developed, which allows detection of points of contact between attached cells and the underlying surface, as well as measurement of the distance between the two surfaces, under the right conditions and if the relevant parameters are known (6, 15). In this method, the cell is allowed to attach to a cover glass which is inverted and viewed by epi-illumination. Some of the incident collimated light is reflected at the cover slip–liquid boundary, while light is also reflected by the cell surface–liquid boundary.

The degree of interference between the two reflected components is dependent upon the distance between the cover glass and cell surfaces and on the refractive index (n) of each phase through which the collimated light travels (6). The general result is that the image is darkest where the cell is at its closest approach to the cover glass and increases in brightness to a maximum at which the distance between the two surfaces is $\lambda/4n$. For monochromatic light of 546 nm and assuming a refractive index of the separating medium to be ~1.3, this distance of maximum image brightness is ~105 nm. If the distance between the cell and cover slip is further increased, the image approaches a second dark minimum at a distance of $\lambda/2n$ (~210 nm), and the cycle continues to repeat with increasing distance. With each repeat of the cycle, however, the dark minimum is considerably reduced in intensity, because the incident light is not precisely monochromatic or coherent and its angle of incidence varies through several degrees (1).

IRM has been used to identify the adhesive sites on animal tissue cells (1) and amoebae (21, 27), but the technique is more difficult to apply to bacteria because of their small size. In this laboratory, IRM was used to study the adhesion to glass cover slips of several bacterial strains (*Pseudomonas* sp., *Acinetobacter* sp., *Flexibacter* sp.) in liquid culture and of an *Acinetobacter* sp. grown in slide culture (13). A Zeiss Standard 18 microscope, fitted with an epifluorescent condenser (IV Fl; Zeiss, Oberkochen, West Germany) with an interference green filter (BP 546/9; Zeiss) in the exciter position and a barrier filter (41; Zeiss), was used. The objective was either a Neofluar 100/1.3 Ph (phase) or Neofluar 100/1.3 (fluorite achromat) (Zeiss), and the iris field stop was closed right down to collimate the light as much as possible. Use of the phase objective allowed alternate viewing of the same cells illuminated by either IRM or phase. However, the Neofluar objective without phase produced a superior image and was used for photography (film: Technical Pan, Kodak, Rochester, N.Y.).

Continuous observation of suspended cells provided some fascinating information on the nature of bacterial approach to a surface and possible subsequent attachment. Some cells which appeared to be attached, that is, they no longer exhibited Brownian motion, produced a bright image with IRM, indicating that their distance from the surface was on the order of 100 nm. With time, there sometimes was the appearance of a dark area (the dark minimum) on the cell, indicating closest approach to the surface. The size of the dark minimum area could increase until it filled a symmetrical area

corresponding to the shape of the cell, but never completely filling the cell outline because of the cell curvature. Rod-shaped cells sometimes had a dark area down the length of the cell. These areas of dark minimum, or closest approach, were interpreted as areas of firm adhesion. However, such "firm" adhesion was not necessarily permanent, and occasionally cells, for no apparent reason or, more often, because of disturbance by adjacent motile cells, gradually (over several minutes) detached from the glass, the dark area becoming smaller and smaller as desorption occurred. The final break was sometimes quite sudden.

Such observations were impossible to photograph in this system because of the continuous movement of attaching cells and the exposure time (2 s) needed for maximum resolution of the photographic image. Thus, slide cultures (13) of an *Acinetobacter* sp. were prepared, so that the bacteria which attached to a glass cover slip were sandwiched between the cover slip and a glass slide precoated with agar containing nutrients. The bacteria were first cultured for 22 h in nutrient broth, in nutrient broth with 0.2% glucose, or in a minimal medium containing glucose, glycerol, glucose/glycerol (1:1), sucrose, or sodium lactate (all at 2.2%, wt/vol, concentration) as the carbon source. Suspended cells were centrifuged and suspended in N-2-hydroxyethyl-piperazine-N-2-ethanesulfonic acid (HEPES) buffer (pH 7.4) to ~2×10^7 to 13×10^7 ml^{-1}, and 5-μl portions were placed on glass slides precoated (by dipping at ~60°C) with the corresponding agar (appropriate medium plus 1.0% agar; Difco Laboratories, Detroit, Mich.). A sterile cover slip was placed on the drop, allowing it to spread. These slide cultures were incubated at 15 or 4°C and examined periodically by IRM to observe attachment and colonization of the cover slips and any changes or differences in the interference patterns obtained.

With this technique, bacteria cultured with nutrient broth or nutrient broth with glucose at 15°C appeared to attach readily and firmly to the glass cover slip, as was indicated by the considerable dark area in the center of the cell outline (Fig. 2). With growth, new cells were also firmly attached. However, when cells were cultured at 4°C, as expected they grew more slowly, and during the first 24 h or so of culture, the dark interference image was frequently patchier than that which occurred at 15°C, indicating less close association between the two surfaces.

With IRM, the bacterial "surface" is the boundary of the refractive indices of the bacterium and the adjacent medium. This boundary may be altered or become more diffuse through the production of exopolymers, which may also modify the value of n for the medium separating

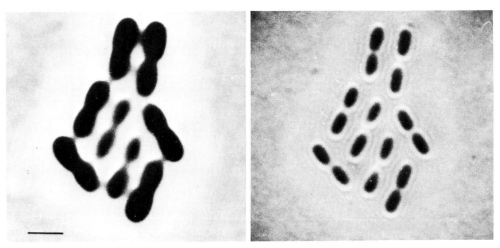

FIG. 2. Slide culture of *Acinetobacter* sp. photographed with phase contrast (left) and IRM (right). The medium was nutrient broth, and incubation was at 15°C for 6 h. Bar represents 2 μm.

the two surfaces. Production of exopolymers may also prevent close approach of the cell to the surface. As slow-growing bacteria (24, 26) and those cultured at low temperatures (23) have been found to produce larger quantities of exopolymer than faster-growing cells, such increased polymer production could explain the patchier interference pattern observed with cells grown at 4°C (as compared with those at 15°C).

With the cells grown in the defined media, all demonstrated moderately dark interference patterns, except for those grown in sodium lactate, which had a much brighter interference pattern, indicating less close association with the substratum (Fig. 3). This was supported by parallel attachment experiments, in which bacteria from the various media were allowed to attach for 2 h to PD and TCD polystyrene (for method, see

22). Numbers attached to the surfaces were counted microscopically, after rinsing off loosely attached cells, fixing with Bouin's fixative, and staining with crystal violet. Attachment of bacteria grown in lactate to either PD or TCD was negligible, whereas bacteria grown in the other defined media attached to PD or to both surfaces in low, but measurable, numbers. The reason for poor attachment of lactate-grown cells is not known, but the relationship between negligible attachment to polystyrene and lack of dark interference patterns with bacteria attached to glass suggests that IRM does demonstrate differences in degrees of adhesive interaction.

Although much progress has been made over the past 10 years in elucidating the mechanism(s) of bacterial adhesion to surfaces, it now seems likely that a number of complex adhesive inter-

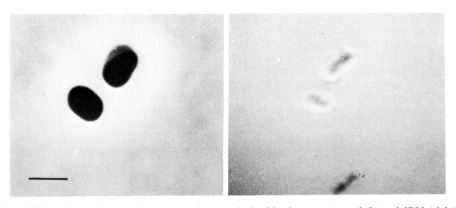

FIG. 3. Slide culture of *Acinetobacter* sp. photographed with phase contrast (left) and IRM (right). The medium contained sodium lactate as the carbon source, and incubation was at 15°C for 3.5 h. Bar represents 2 μm.

actions may be involved and that bacterial strains probably differ considerably in their attachment processes. Moreover, cell surface properties, and thus adhesion ability, of specific strains may be considerably affected by growth conditions or mutations. Bacterial cell surface polymers play an important role in adhesion by (i) determining the chemistry, and thus potential adhesive interactions, of the cell surface, (ii) inhibiting initial adhesion, or (iii) promoting cell aggregation, and thus biofilm formation. It is now important that we determine the specific chemical characteristics of these bacterial surface components which affect adhesive interactions, as well as the time-scale and conditions under which they are produced. Thus, we shall gain an understanding of the chemical bases for initial adhesion, the strengthening of the attachment bond, and detachment.

Some of the work described was supported by a Science and Engineering Research Council CASE Studentship.

LITERATURE CITED

1. **Abercrombie, M., and G. A. Dunn.** 1975. Adhesions of fibroblasts to substratum during contact inhibition observed by interference reflection microscopy. Exp. Cell Res. **92:**57–62.
2. **Absolom, D. R., C. J. Van Oss, R. J. Genco, D. W. Francis, and A. W. Neumann.** 1980. Surface thermodynamics of normal and pathological human granulocytes. Cell Biophys. **2:**113–126.
3. **Bright, J. J., and M. Fletcher.** 1983. Amino acid assimilation and electron transport system activity in attached and free-living marine bacteria. Appl. Environ. Microbiol. **45:**818–825.
4. **Clark, W. B., and R. J. Gibbons.** 1977. Influence of salivary gland components and extracellular polysaccharide synthesis from sucrose on the attachment of *Streptococcus mutans* 6715 to hydroxyapatite surfaces. Infect. Immun. **18:**514–523.
5. **Costerton, J. W., G. G. Geesey, and K.-J. Cheng.** 1978. How bacteria stick. Sci. Am. **238:**86–95.
6. **Curtis, A. S. G.** 1964. The mechanism of adhesion of cells to glass. J. Cell. Biol. **20:**199–215.
7. **Dahlbeck, B., M. Hermansson, S. Kjelleberg, and B. Norkrans.** 1981. The hydrophobicity of bacteria—an important factor in their initial adhesion at the air-water interface. Arch. Microbiol. **128:**267–270.
8. **Fletcher, M.** 1977. The effects of culture concentration and age, time and temperature on bacterial attachment to polystyrene. Can. J. Microbiol. **23:**1–6.
9. **Fletcher, M.** 1983. The effects of methanol, ethanol, propanol and butanol on bacterial attachment to surfaces. J. Gen. Microbiol. **129:**633–641.
10. **Fletcher, M., and G. D. Floodgate.** 1973. An electron-microscope demonstration of an acidic polysaccharide involved in the adhesion of a marine bacterium to solid surfaces. J. Gen. Microbiol. **74:**325–334.
11. **Fletcher, M., and K. C. Marshall.** 1982. A bubble contact angle method for evaluating substratum interfacial characteristics and its relevance to bacterial attachment. Appl. Environ. Microbiol. **44:**184–192.
12. **Fletcher, M., and K. C. Marshall.** 1982. Are solid surfaces of ecological significance to aquatic bacteria?, p. 199–236. In K. C. Marshall (ed.), Advances in microbial ecology, vol. 6. Plenum Press, New York.
13. **Fry, J. C., and T. Zia.** 1982. A method for estimating viability of aquatic bacteria by slide culture. J. Appl. Bacteriol. **53:**189–198.
14. **Geesey, G. G.** 1982. Microbial exopolymers: ecological and economic considerations. ASM News **48:**9–14.
15. **Gingell, D., and I. Todd.** 1979. Interference reflection microscopy. A quantitative theory for image interpretation and its application to cell-substratum separation measurement. Biophys. J. **26:**507–526.
16. **Harber, M. J., R. Mackenzie, and A. W. Asscher.** 1983. A rapid bioluminescence method for quantifying bacterial adhesion to polystyrene. J. Gen. Microbiol. **129:**621–632.
17. **Marshall, K. C., and R. H. Cruickshank.** 1973. Cell surface hydrophobicity and the orientation of certain bacteria at interfaces. Arch. Mikrobiol. **91:**29–40.
18. **Marshall, K. C., R. Stout, and R. Mitchell.** 1971. Mechanism of the initial events in the sorption of marine bacteria to surfaces. J. Gen. Microbiol. **68:**337–348.
19. **Ørstavik, D.** 1977. Sorption of *Streptococcus faecium* to glass. Acta Pathol. Microbiol. Scand. Sect. B **85:**38–46.
20. **Powell, M. S., and N. K. H. Slater.** 1982. Removal rates of bacterial cells from glass surfaces by fluid shear. Biotechnol. Bioeng. **24:**2527–2537.
21. **Preston, T. M., and C. A. King.** 1978. Cell-substrate associations during the amoeboid locomotion of *Naegleria*. J. Gen. Microbiol. **104:**347–351.
22. **Pringle, J. H., and M. Fletcher.** 1983. Influence of substratum wettability on attachment of freshwater bacteria to solid surfaces. Appl. Environ. Microbiol. **45:**811–817.
23. **Rose, A. H.** 1967. Thermobiology. Academic Press, London.
24. **Sutherland, I. W.** 1977. Microbial exopolysaccharide synthesis, p. 40–57. In P. A. Sanford and A. Laskin (ed.), Extracellular microbial polysaccharides. American Chemical Society, Washington, D.C.
25. **Tadros, T. F.** 1980. Particle-surface adhesion, p. 93–116. In R. C. W. Berkeley, J. M. Lynch, J. Melling, P. R. Rutter, and B. Vincent (ed.), Microbial adhesion to surfaces. Ellis Horwood, Chichester, U.K.
26. **Uhlinger, D. J., and D. C. White.** 1983. Relationship between physiological status and formation of extracellular polysaccharide glycocalyx in *Pseudomonas atlantica*. Appl. Environ. Microbiol. **45:**64–70.
27. **Van Oss, C. J., and C. F. Gillman.** 1972. Phagocytosis as a surface phenomenon. I. Contact angles and phagocytosis of non-opsonized bacteria. RES J. Reticuloendothel. Soc. **12:**283–292.
28. **ZoBell, C. E.** 1943. The effect of solid surfaces upon bacterial activity. J. Bacteriol. **46:**39–56.
29. **Zvyagintsev, D. G., V. S. Guzev, and I. S. Guzeva.** 1977. Relationship between adsorption of microorganisms and the stage of their development. Microbiology (USSR) **46:**245–249.

Attachment of Nitrogen-Fixing Bacteria to Plant Roots

FRANK B. DAZZO

Department of Microbiology and Public Health, Michigan State University, East Lansing, Michigan 48824

The process of attachment between microorganisms and higher plants is receiving considerable attention in light of its effect on plant morphogenesis, nutrition, symbiosis, and infectious disease. These positive cellular recognitions are believed to arise from a specific union, reversible or irreversible, between chemical receptors on the surface of interacting cells (7). This hypothesis implies that communication occurs when cells that recognize one another come into contact, and therefore the complementary components of the cell surfaces have naturally been the focus for most biochemical studies. Such is the case for studies on the infection of legume root hairs by the nitrogen-fixing symbiont *Rhizobium*. According to the lectin-recognition hypothesis (6, 15), specific, complementary lectin–polysaccharide interactions serve as a basis of host specificity in this nitrogen-fixing symbiosis. During early stages of the infection process, these bacteria attach via hapten-reversible interactions and then later become irreversibly anchored to the host cell. The ability of the bacteria to attach to root hairs is controlled by Roa (root hair attachment) genes, which occur on large, transmissible plasmids (49). The Roa$^-$ phenotype is illustrated by noninfective mutant strains which are defective in hapten-specific attachment steps (15, 32, 40, 45, 46, 49). However, genetic studies have clearly shown that successful infection of root hairs by rhizobia requires additional events of cellular recognition (11, 27, 31, 40, 46). Thus, root hair attachment is viewed as only one of the several events of cellular recognition leading to successful infection.

Light and electron microscopy of the *Rhizobium*–clover symbiosis has revealed multiple mechanisms of bacterial attachment to the root hairs (11, 15, 16, 27, 32, 34, 37, 39, 48; F. B. Dazzo et al., in preparation). A nonspecific mechanism allows all species of rhizobia to attach in low numbers (16). In addition, a specific mechanism allows selective attachment in significantly larger numbers under identical conditions (16). Host-specific attachment has also been demonstrated in *R. japonicum*–soybean (45), *R. leguminosarum*–pea (32), and *R. meliloti*–alfalfa (C. Lafreniere and L. M. Bordeleau, in preparation; C. Caniole-Arias and G. Favelukas, in preparation) root systems. However, specificity was not found in quantitative root attachment studies in which very high densities of radiolabeled rhizobia (10^9 to 10^{10} cells per seedling) were used (8), but many unattached bacteria could have contributed to this result. In a recent study employing marble chips to dislodge bacteria attached to the root system and quantitative plating assays, "firm" attachment was found to be host specific in *R. trifolii*–clover and *R. meliloti*–alfalfa systems, and "loose" attachment was nonspecific (48). It is therefore apparent that bacterial attachment to host roots is an early expression of cellular recognition in the *Rhizobium*–legume symbiosis.

Transmission electron microscopy (15) disclosed that the initial bacterial attachment step consisted of contact between the fibrillar capsule of *R. trifolii* and electron-dense globular aggregates lying on the outer periphery of the clover root hair cell wall. This "docking" stage is the first point of physical contact between the microbe and the host (phase I attachment), and occurs within minutes after inoculation of encapsulated cells of *R. trifolii* on the plant.

To identify the molecules involved in phase I attachment, we examined the surface components of the bacterium and the host that interact with the same order of specificity as is observed with the adhesion of the bacterial cells. Immunochemical and genetic studies have demonstrated that the surfaces of *R. trifolii* and clover epidermal cells contain a unique carbohydrate antigen that is immunochemically cross-reactive (11, 15), suggesting its structural relatedness on both symbionts. This antigen contains receptors that bind hapten reversibly to a multivalent clover lectin called trifoliin A (originally trifoliin) which has been isolated from seeds, seedling roots, and root exudate of white clover plants (12, 21). This *Rhizobium*-binding lectin is distributed on root hair surfaces in a gradient which is greatest at the growing tip and least at the base of the root hair (21). Recently, we demonstrated the de novo synthesis of trifoliin A in white clover seedling roots (J. E. Sherwood et al., in preparation) and excretion of large electron-dense aggregates of trifoliin A into root exudate (G. L. Truchet et al., in preparation). A specific hapten inhibitor of the interaction between trifoliin A and surface polysaccharides from *R. trifolii* is 2-deoxy-D-glucose (9, 11, 15), and this hapten specifically inhibits 90+% of attachment of *R. trifolii* to clover root hairs (16, 49). The remaining background level of attachment is characteristic of heterologous rhizobia and is insensitive to 2-deoxy-D-glucose (16). This hapten sugar specifically facilitates the elution of trifoliin A from the intact clover root (9, 21), suggesting that the lectin is anchored to glycosylated receptors on the root surface. Similarly, lectins on pea, alfalfa, and soybean roots accessible for binding to the appropriate rhizobia have been demonstrated (25, 26, 32, 33, 35, 40, 45). Hapten sugars elute *Rhizobium*-binding lectins from these intact roots (25, 35) and inhibit attachment of the bacterial symbiont to root hairs of the plant host (32, 33, 45). Receptor sites on clover, peas, and wild soybean which specifically bind capsular polysaccharides from the corresponding rhizobial symbiont are located on root hairs (9, 30, 32), exactly matching the distribution of the *Rhizobium*-binding lectin on the root surface (21, 26, 45). Again, binding of these capsular polysaccharides to the root is inhibited specifically by the hapten sugars of the host lectin (9, 30, 33). Considered collectively, these results strongly suggest that certain legume lectins are involved in the specific attachment of the rhizobia to the host root hairs.

Several immunochemical and genetic studies suggest that trifoliin A binds to the surface polysaccharides on *R. trifolii* which are antigenically cross-reactive with a surface antigen on clover roots. The relevant findings are as follows: (i) anti-clover root antibody and trifoliin A bind specifically to the same isolated polysaccharides (11, 29); (ii) this interaction is specifically inhibited by the hapten 2-deoxy-D-glucose; (iii) the genetic markers of *R. trifolii* that bind trifoliin A and the antibody cotransform into *Azotobacter vinelandii* with 100% frequency (5); and (iv) monovalent Fab fragments of immunoglobulin G from anti-clover root antibody strongly block the binding of trifoliin A to *R. trifolii* (11). These studies suggest that *R. trifolii* and clover roots have similar saccharide receptors for trifoliin A. However, the definitive test of their identity as antigenically related structures will require knowledge of the minimal saccharide sequence that binds the clover lectin.

Other studies have presented the following evidence that these unique surface receptors for trifoliin A on *R. trifolii* are responsible for binding these bacteria to clover root hairs (11): (i) Fab fragments of anti-clover root immunoglobulin G block phase I attachment of *R. trifolii* to clover root hairs; (ii) only the *A. vinelandii* hybrid transformants that carry the trifoliin A receptor bind to clover root hairs in phase I attachment assays; (iii) in competition assays the *R. trifolii* polysaccharides that bind trifoliin A show the highest affinity for clover root hairs. On the basis of these studies, we proposed a

model to explain this early recognition event on the clover root hair surface (15). According to this hypothesis, the multivalent trifoliin A lectin cross-bridges antigenically similar saccharide residues on *R. trifolii* and clover to form the correct interfacial structure which favors specific adsorption of the bacteria to the root hair surface. The lectin may also function as a "cell recognition molecule" since it could feasibly influence which cells associate in sufficient proximity to the root hairs to allow subsequent specific recognition steps to occur. Figure 1 (17) is a revision of the original cross-bridging model (15), which takes into account the more recent findings that *R. trifolii* has multiple receptors for trifoliin A (11, 29) and that this lectin is involved in tip adhesion of adjacent root hairs, suggesting an excess of trifoliin A receptors (18).

After this initial attachment, there is a firm anchoring of the bacterial cell to the root hair surface (phase II adherence) (Dazzo et al., submitted for publication). This may be important in maintaining the firm contact between the bacterium and the host root hair necessary for triggering the tight root hair curling (shepherd's crook formation) and successful penetration of the root hair cell wall during infection (39).

During phase II adherence, fibrillar materials recognized by scanning electron microscopy are characteristically found associated with the adherent bacteria (Dazzo et al., submitted for publication). The nature of these microfibrils is unknown. One possibility is that they are bundles of cellulose microfibrils, known to be produced by many rhizobia (22, 39). Another possibility is that they are collections of pili, which have recently been demonstrated in *Rhizobium* (34; Am. Phytol. Soc. Abstr. 328, 1981).

The presence of trifoliin A in clover root exudate which can bind to receptors on *R.*

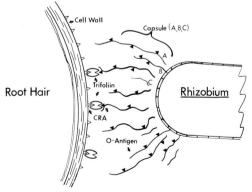

FIG. 1. Proposed cross-bridging of *R. trifolii* receptors to clover root hairs by the host lectin trifoliin A. From Dazzo and Truchet (17), courtesy of the editors of *Journal of Membrane Biology*.

trifolii provides supporting evidence for a lectin recognition model proposed by Solheim (44). According to this model, a glycoprotein lectin excreted from the legume root binds to the rhizobia. This active complex then combines with a receptor site on the root. Thus, both partners in the symbiosis could benefit from the discriminatory reaction of a cross-bridging lectin which could be either bound to a glycosylated receptor on the root hair cell wall or released from the root to bind to the rhizobial cell (44). This event would help to ensure that only the symbiotic bacterium could establish the proper intimate contact with the host cell required to trigger other recognition events that lead to successful infection.

Several observations indicate that attachment alone is not the key event required to initiate root hair infection. Very few root hairs to which infective rhizobia attach eventually become infected. This may be due to a transient susceptibility of the root hairs to infection by the rhizobial symbiont (2). The hybrids of *A. vinelandii* which bind trifoliin A (5) attach specifically to clover root hairs (11) but do not infect them. Finally, although mutant strains of *R. trifolii* and *R. meliloti* which fail to bind the host lectin neither attach well to the host root hairs nor infect them (16, 40), another class of noninfective mutant strains has been shown to bind the host lectin and attach to the host root hairs (31, 40). Each of these cases serves to illustrate the importance of lectin-mediated root hair attachment to the infection process, but makes it clear that other post-attachment events of cell recognition must occur to advance the infection process to the stage of root hair penetration.

The attachment process is regulated by at least three different mechanisms. One mechanism involves a plant component, a second mechanism involves a bacterial component, and a third mechanism involves the interaction of both the bacterium and the plant in situ in the rhizosphere.

A mechanism regulated by the plant host seems to be a growth response of the root to an exogenous supply of nitrate (10, 14). The specific attachment of *R. trifolii* 0403 to clover root hairs and the levels of trifoliin A on these epidermal cells decline in parallel as the nitrate concentration is increased from 1 to 15 mM in the rooting medium (10). This is not caused by direct interaction of nitrate with trifoliin A or its saccharide receptors on the bacterium and the plant (13). When grown at 15 mM nitrate, root cell walls bind less trifoliin A (14) (suggesting inaccessibility of lectin receptors) and roots release less trifoliin A into the root exudate (12). In addition, the chemistry of clover and pea root cell walls changes when the plants are grown with adequate nitrate supply (14, 23). Whether these chemical changes in root cell walls are directly responsible for the lack of lectin accumulation on the root hairs and of attachment of the bacterial symbiont is not yet known.

A mechanism regulated by the bacterium involves its growth phase-dependent accumulation of the saccharide receptor for the lectin. These receptors are transient (4, 20) and are explained by chemical modifications of the surface polysaccharides which accompany culture aging (29, 38; Sherwood et al. in preparation). For instance, methylation of galactose residues in the capsular polysaccharide of *R. japonicum* occurs as the culture enters stationary phase, reducing its affinity for the galactose-specific soybean lectin (38). For *R. trifolii* 0403, growth phase-dependent increases in the quinovosamine content of its lipopolysaccharide (29) and the uronic acid-pyruvate-acetate content of its capsular polysaccharides (Sherwood et al., in preparation) occur at the same time in the growth curve that these surface polysaccharides on the cells become optimally reactive with trifoliin A. Potential importance of these results to the infection process is suggested by the finding that inocula harvested at an age when most cells bind the host lectin lead to more root infections than inocula harvested at other culture ages (3, 29).

The presence of multiple lectin receptors on *R. trifolii* and the related species *R. leguminosarum* raises the question of whether each one has a different role in root hair infection. Infection studies performed with mutant strains of *R. leguminosarum* suggest that its capsular polysaccharides are responsible for attachment to pea host root hairs via cross-bridging lectins as an early recognition event, followed by secondary recognition events requiring the host range-specific binding of lectin on localized sites of the root hair to lipopolysaccharide, which triggers subsequent invasive steps (31). This hypothesis is currently under investigation.

The ultimate level of regulation of attachment involves the interaction of the bacterial and plant symbionts in the rhizosphere. In the *R. japonicum*–soybean symbiosis, certain *R. japonicum* strains express lectin receptors in the rhizosphere but not in laboratory culture media (1). In the *R. trifolii*–clover system, the plant root releases enzymes which alter the lectin-binding capsule of the bacterial inoculum in a way which favors a preferred polar attachment (19). Based on kinetics of attachment (Dazzo et al., submitted for publication) and this enzymatic alteration of the bacterial capsule in the rhizosphere (19), we have subdivided phase I attachment of encapsulated *R. trifolii* to clover root hairs into the following sequential events (17; Dazzo et al., submitted for publication) (Fig. 2). (i) Most encapsulated cells which have trifoliin A receptors around the entire cell surface bind within minutes in a random orientation to clover root hair tips where trifoliin A accumulates. (ii) Cells which do not immediately contact the root hairs encounter enzymes in root exudate which modify their surface polysaccharides so that they become progressively less reactive with trifoliin A. This alteration proceeds less rapidly at one cell pole. Newly synthesized lectin-binding polysaccharide may also be deposited at one pole of nonencapsulated cells at a rate which may keep pace with the exudate enzyme-mediated modifications. (iii) Some cells with trifoliin A or its saccharide receptors, or both, bound to one pole eventually contact the cell wall along the sides of the root hair, where they then attach end-on.

Genes in *R. trifolii* which are important to trifoliin A receptor synthesis and attachment to clover root hairs are encoded on the same large symbiotic plasmid. For instance, the genes required for the 2-deoxy-D-glucose inhibitable attachment are on the nodulation plasmid designated pWZ2 (49), and incorporation of the trifoliin A binding sugar quinovosamine into the lipopolysaccharide of *R. trifolii* is controlled by the clover nodulation plasmid (42). Conjugal transfer of the nodulation plasmid from *R. trifolii*

into Ti plasmid-cured *Agrobacterium tumefaciens* results in a hybrid which nodulates clover roots (28). In addition, this hybrid specifically binds trifoliin A and attaches to, markedly curls, and infects clover root hairs (Dazzo, et al., submitted for publication). These studies show that the "symbiotic" plasmids of *R. trifolii* carry genetic information required to express all of the root hair recognition events known to exist in the *R. trifolii*–clover symbiosis, and these events can be expressed, albeit at lower levels, in an *A. tumefaciens* background.

The most important evidence countering the lectin-recognition hypothesis is that which identifies five lines of soybean that lack the *N*-acetylgalactosamine-specific, 120,000-dalton soybean lectin in their seeds (41). *R. japonicum* can still infect the root hairs and form nitrogen-fixing nodules on roots of these soybean lines (41). Thus, the presence of this lectin in soybean seeds is not necessary for root nodulation by *R. japonicum*. Recently, however, lectins which bind *R. japonicum* have been detected in seeds of these same lines (24, 43), and soybean lectin mRNA has been detected in their roots (M. Goldberg et al., in press). Thus, soybean has multiple lectins which interact with *R. japonicum*, and the seed and root lectins seem to be under separate genetic control, despite their chemical similarity.

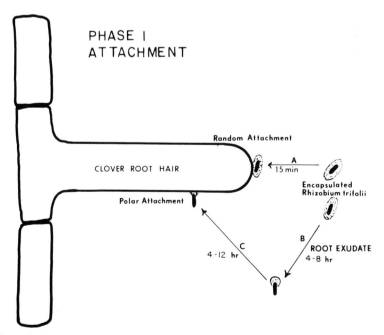

FIG. 2. Schematic diagram of the sequential events of phase I attachment of *R. trifolii* to clover root hairs, with an inoculum of fully encapsulated cells. From Dazzo and Truchet (17), courtesy of the editors of *Journal of Membrane Biology*.

ATTACHMENT OF *AZOSPIRILLUM* TO GRASS ROOTS

The attachment of *Azospirillum brasilense* Sp7 to root hairs of pearl millet in hydroponic culture is similar in some respects to attachment of *R. trifolii* 0403 to clover root hairs (47). Such attachment can be viewed as favoring the colonization of the rhizoplane by azospirilla to occupy niches which take advantage of root exudation. Adherent azospirilla were associated with granular material on root hairs and fibrillar material on undifferentiated epidermal cells. Significantly fewer cells of azospirilla attached to millet root hairs when the roots were grown in culture medium containing 5 mM nitrate, whereas attachment to undifferentiated epidermal cells was unaffected. Root exudate from millet contained nondialyzable and protease-sensitive (proteins) which bound to azospirilla in the root environment and promoted their attachment to millet root hairs, but not to undifferentiated epidermal cells. Millet root hairs absorbed more cells of azospirilla than of *R. trifolii, Pseudomonas* sp., *Azotobacter vinelandii, Klebsiella pneumoniae*, and *Escherichia coli*. It is clear from these studies that attachment of azospirilla to root hairs of millet is accomplished by mechanisms which differ from those that attach azospirilla to undifferentiated epidermal cell surfaces.

It has recently been shown that the strains of *Azospirillum* which successfully occupy niches on the grass rhizoplane attach better to the grass roots than do other strains which occupy a more invasive niche and colonize the middle lamella of the cortex within the root (D. J. Jain and D. G. Patriquin, in preparation). Thus, the ability of *Azospirillum* to attach to the root surface may be very important to establishing which of these two habitats and niches can be occupied by the bacteria.

Korhonen et al. (36) have recently examined the in vitro adhesion of associative nitrogen-fixing *Klebsiella* spp. to grass roots. *Klebsiella* cells were labeled with ^{3}H-amino acids and incubated with the grass roots. Their attachment was dependent on inoculum density, incubation time and temperature, pH and ionic strength of the incubation buffer, and the growth phase of the bacterial cells. Type 1 (mannose-sensitive) and type 3 fimbriae (pili) were isolated from the labeled bacteria and examined for their possible role in root attachment. The binding of type 1 pili to roots was inhibited by the hapten α-methylmannoside and by Fab monovalent fragments directed against the purified pili. However, nonpiliated mutant strains attached very well to the whole grass roots. More work on the role of pili in root attachment is needed and will require detailed microscopy to sort out the mechanisms of attachment of these nitrogen-fixing organisms to grass root hairs and other epidermal cells.

CONCLUDING REMARKS

The above discussion provides substantial evidence in favor of the lectin-recognition hypothesis as an important determinant of bacterial attachment to plant roots and host specificity in the nitrogen-fixing *Rhizobium*–legume symbiosis. However, it seems unlikely that acceptance or rejection of the lectin-recognition hypothesis will ever become universal until the recognition code is deciphered and mutant strains of both symbionts—bacterium and plant—with altered lectin/lectin receptors and no pleiotropic effects are available for analysis. In addition, *Azospirillum*–grass root associations are beginning to emerge as useful systems to further advance our knowledge of mechanisms of microbial attachment to plant roots.

Portions of the work described here were supported by USDA-CRGO competitive grant 82-CRCR-1-1040, National Science Foundation grant 80-21906, and Project 1314H from the Michigan Agricultural Experiment Station. This is Journal Article no. 10907 from the Michigan Agricultural Experiment Station.

LITERATURE CITED

1. **Bhuvaneswari, T. V., and W. D. Bauer.** 1978. The role of lectins in plant-microorganism interactions. III. Influence of rhizosphere/rhizoplane culture conditions on the soybean lectin-binding properties of rhizobia. Plant Physiol. **62:**71–74.
2. **Bhuvaneswari, T. V., A. A. Bhagwat, and W. D. Bauer.** 1981. Transient susceptibility of root cells in four common legumes to nodulation by rhizobia. Plant Physiol. **68:**1144–1149.
3. **Bhuvaneswari, T. V., K. K. Mills, D. K. Crist, W. R. Evans, and W. D. Bauer.** 1982. Effect of culture age on infection of soybeans by *Rhizobium japonicum*. J. Bacteriol. **153:**443–451.
4. **Bhuvaneswari, T. V., S. G. Pueppke, and W. D. Bauer.** 1977. Role of lectins in plant-microorganism interactions. I. Binding of soybean lectin to rhizobia. Plant Physiol. **60:**486–491.
5. **Bishop, P. E., F. B. Dazzo, E. R. Applebaum, R. J. Maier, and W. J. Brill.** 1977. Intergeneric transformation of genes involved in the *Rhizobium*-legume symbiosis. Science **198:**938–939.
6. **Bohlool, B. B., and E. L. Schmidt.** 1974. Lectins: a possible basis for specificity in the *Rhizobium*-legume root nodule symbiosis. Science **185:**269–271.
7. **Burnet, F.** 1971. Self-recognition in colonial marine forms and flowering plants in relation to evaluation of immunity. Nature (London) **232:**230–235.
8. **Chen, A. P., and D. A. Phillips.** 1976. Attachment of *Rhizobium* to legume roots as the basis for specific associations. Physiol. Plant. **38:**83–88.
9. **Dazzo, F. B., and W. J. Brill.** 1977. Receptor site on clover and alfalfa roots for *Rhizobium*. Appl. Environ. Microbiol. **33:**132–136.
10. **Dazzo, F. B., and W. J. Brill.** 1978. Regulation by fixed nitrogen of host-symbiont recognition in the *Rhizobium*-clover symbiosis. Plant Physiol. **62:**18–21.
11. **Dazzo, F. B., and W. J. Brill.** 1979. Bacterial polysaccha-

ride which binds *Rhizobium trifolii* to clover root hairs. J. Bacteriol. **137**:1362–1373.

12. **Dazzo, F. B., and E. M. Hrabak.** 1981. Presence of trifoliin A, a *Rhizobium*-binding lectin, in clover root exudate. J. Supramol. Struct. Cell. Biochem. **16**:133–138.

13. **Dazzo, F. B., and E. M. Hrabak.** 1982. Lack of a direct nitrate-trifoliin A interaction as related to the *Rhizobium*-clover symbiosis. Plant Soil **69**:259–264.

14. **Dazzo, F. B., E. M. Hrabak, M. R. Urbano, J. E. Sherwood, and G. L. Truchet.** 1981. Regulation of recognition in the *Rhizobium*-clover symbiosis, p. 292–295. *In* A. H. Gibson and W. E. Newton (ed.), Current perspectives in nitrogen fixation. Australian Academy of Science, Canberra.

15. **Dazzo, F. B., and D. H. Hubbell.** 1975. Cross-reactive antigens and lectin as determinants of symbiotic specificity in the *Rhizobium*-clover association. Appl. Microbiol. **30**:1017–1033.

16. **Dazzo, F. B., C. A. Napoli, and D. H. Hubbell.** 1976. Adsorption of bacteria to roots as related to host specificity in the *Rhizobium*-clover association. Appl. Environ. Microbiol. **32**:168–171.

17. **Dazzo, F. B., and G. L. Truchet.** 1983. Interactions of lectins and their saccharide receptors in the *Rhizobium*-legume symbiosis. J. Membr. Biol. **73**:1–16.

18. **Dazzo, F. B., G. L. Truchet, and J. W. Kijne.** 1982. Lectin involvement in root hair tip adhesions as related to the *Rhizobium*-clover symbiosis. Physiol. Plant. **56**:143–147.

19. **Dazzo, F. B., G. L. Truchet, J. E. Sherwood, E. M. Hrabak, and A. E. Gardiol.** 1982. Alteration of the trifoliin A-binding capsule of *Rhizobium trifolii* 0403 by enzymes released from clover roots. Appl. Environ. Microbiol. **44**:478–490.

20. **Dazzo, F. B., M. R. Urbano, and W. J. Brill.** 1979. Transient appearance of lectin receptors on *Rhizobium trifolii*. Curr. Microbiol. **2**:15–20.

21. **Dazzo, F. B., W. E. Yanke, and W. J. Brill.** 1978. Trifoliin: a *Rhizobium* recognition protein from white clover. Biochim. Biophys. Acta **536**:276–286.

22. **Deinema, M., and L. P. T. Zevenhuizen.** 1971. Formation of cellulose fibrils by gram negative bacteria and their role in bacterial flocculation. Arch. Mikrobiol. **78**:42–57.

23. **Diaz, C., J. W. Kijne, and A. Quispel.** 1981. Influence of nitrate on pea root cell wall composition, p. 426. *In* A. H. Gibson and W. E. Newton (ed.), Current perspectives in nitrogen fixation. Australian Academy of Science, Canberra.

24. **Dombrink-Kurtzman, M. A., W. E. Dick, K. A. Burton, M. C. Cadmus, and M. E. Slodki.** 1983. A soybean lectin having 4-O-methylglucuronic acid specificity. Biochem. Biophys. Res. Commun. **111**:798–803.

25. **Gade, W., M. A. Jack, J. B. Dahl, E. L. Schmidt, and F. Wolf.** 1981. The isolation and characterization of a root lectin from soybean (*Glycine max* L.) cultivar Chippewa. J. Biol. Chem. **256**:12905–12910.

26. **Gatehouse, J. A., and D. Boulter.** 1980. Isolation and properties of a lectin from the roots of *Pisum sativum*. Physiol. Plant. **49**:437–442.

27. **Higashi, S., and M. Abe.** 1980. Scanning electron microscopy of clover root hair infection by *Rhizobium trifolii*. Appl. Environ. Microbiol. **40**:1094–1099.

28. **Hooykaas, P. J., A. A. N. van Brussell, H. den Hulk-Ras, G. M. van Slogteren, and R. A. Schilperoort.** 1981. Sym plasmid of *Rhizobium trifolii* expressed in different rhizobia species and *Agrobacterium tumefaciens*. Nature (London) **291**:351–353.

29. **Hrabak, E. M., M. R. Urbano, and F. B. Dazzo.** 1981. Growth-phase dependent immunodeterminants of *Rhizobium trifolii* lipopolysaccharide which bind trifoliin A, a white clover lectin. J. Bacteriol. **148**:697–711.

30. **Hughes, T. A., and G. H. Elkan.** 1981. Study of the *Rhizobium japonicum*-soybean symbiosis. Plant Soil **61**:87–91.

31. **Kamberger, W.** 1979. Role of cell surface polysaccharides in the *Rhizobium*-pea symbiosis. FEMS Microbiol. Lett. **6**:361–365.

32. **Kato, G., Y. Maruyama, and M. Nakamura.** 1980. Role of bacterial polysaccharides in the adsorption process of the *Rhizobium*-pea symbiosis. Agric. Biol. Chem. **44**:2843–2855.

33. **Kato, G., Y. Maruyama, and M. Nakamura.** 1981. Involvement of lectins in *Rhizobium*-pea recognition. Plant Cell Physiol. **22**:759–771.

34. **Kijne, J. W., I. A. M. van der Schaal, C. L. Diaz, and F. van Iren.** 1982. Mannose-specific lectins and the recognition of pea roots by *Rhizobium leguminosarum*, p. 521–529. *In* T. C. Bog-Hansen and G. A. Spengler (ed.), Lectins, vol. 3. W. De Gruyter, Berlin.

35. **Kijne, J. W., I. A. M. van der Schaal, and G. E. de Vries.** 1980. Pea lectins and the recognition of *Rhizobium leguminosarum*. Plant Sci. Lett. **18**:65–74.

36. **Korhonen, T. K., K. Haahtela, A. M. Ahonen, M. Rhen, A. Pere, and E. Tarkka.** 1982. In vitro adhesion of associative nitrogen-fixing klebsiellas to plant roots, p. 143–150. *In* P. Uomala (ed.), Second National Symposium on Nitrogen Fixation. Finnish National Fund for Research and Development, Helsinki, Finland.

37. **Kumarasinghe, M. K., and P. S. Nutman.** 1977. *Rhizobium*-stimulated callose formation in clover root hairs and its relation to infection. J. Exp. Bot. **28**:961–976.

38. **Mort, A. J., and W. D. Bauer.** 1980. Composition of the capsular and extracellular polysaccharides of *Rhizobium japonicum*: changes with culture age and correlations with binding of soybean seed lectin to the bacteria. Plant Physiol. **66**:158–163.

39. **Napoli, C. A., F. B. Dazzo, and D. H. Hubbell.** 1975. Production of cellulose microfibrils by *Rhizobium*. Appl. Microbiol. **30**:123–131.

40. **Paau, A. S., W. T. Leps, and W. J. Brill.** 1981. Agglutinin from alfalfa necessary for binding and nodulation by *Rhizobium meliloti*. Science **213**:1513–1515.

41. **Pull, S. P., S. G. Pueppke, T. Hymowitz, and J. H. Orf.** 1978. Soybean lines lacking the 120,000 dalton seed lectin. Science **200**:1277–1279.

42. **Russa, R., T. Urbanik, E. Kowalczuk, and Z. Lorkiewicz.** 1982. Correlation between occurrence of plasmid pUCS202 and lipopolysaccharide alterations in *Rhizobium*. FEMS Microbiol. Lett. **13**:161–165.

43. **Schmidt, E. L.** 1979. Initiation of plant root-microbe interactions. Annu. Rev. Microbiol. **33**:335–376.

44. **Solheim, B.** 1975. Possible role of lectin in the infection of legumes by *Rhizobium trifolii* and a model of the recognition reaction between *Rhizobium trifolii* and *Trifolium repens*. NATO Advanced Study Institute Symposium on Specificity in Plant Diseases, Sardinia.

45. **Stacey, G., A. S. Paau, and W. J. Brill.** 1980. Host recognition in the *Rhizobium*-soybean symbiosis. Plant Physiol. **66**:609–614.

46. **Stacey, G., A. S. Paau, D. Noel, R. J. Maier, L. E. Silver, and W. J. Brill.** 1982. Mutants of *Rhizobium japonicum* defective in nodulation. Arch. Microbiol. **132**:219–224.

47. **Umali-Garcia, M., D. Hubbell, M. Gaskins, and F. Dazzo.** 1980. Associations of *Azospirillum* with grass roots. Appl. Environ. Microbiol. **39**:219–226.

48. **van Rensberg, H. J., and B. Strijdom.** 1982. Root surface association in relation to nodulation of *Medicago sativa*. Appl. Environ. Microbiol. **44**:93–97.

49. **Zurkowski, W.** 1980. Specific adsorption of bacteria to clover root hairs, related to the presence of the plasmid pWZ2 in cells of *Rhizobium trifolii*. Microbios **27**:27–32.

Mechanisms of the Attachment of Bacteria to Animal Cells

G. W. JONES

Department of Microbiology and Immunology, The University of Michigan Medical School, Ann Arbor, Michigan 48109

Colonization of the animal body is a hazardous event for most bacteria and is particularly so for pathogens. They are confronted not only by the host's normal constitutive protective mechanisms but, on most body surfaces, also by severe competition from the normal, indigenous flora. Microbes attempting to establish populations on the living mucosal surfaces of the body must contend with host mechanisms designed to cleanse the surfaces physically of detritus and microbes. Mechanisms such as peristalsis in the small intestine, micturition in the urinary tract, and the mucociliary escalator of the respiratory tract are particularly effective in this regard. Colonization in the face of physical removal is further exacerbated by the slow growth of most organisms in vivo and, in the case of pathogens, by their induction of the host's immunological protective responses.

Although the attachment of bacteria to surfaces was alluded to some 70 years ago and Theobald Smith presented pictorial evidence of the phenomenon in 1925 (43), it was not until 30 years ago that the phenomenon was studied seriously and the proposal made that this property contributed to bacterial infectivity or colonizing ability. Gibbons and van Houte (13) observed that the specific attachment of bacteria to various surfaces of the oral cavity in vitro reflected their propensity to colonize these sites in vivo and proposed attachment as a mechanism by which these bacteria overcome removal by the continuous flow of saliva. Earlier, Duguid and Old (5) had established that many members of the *Enterobacteriaceae* were demonstrably adhesive when tested in vitro, and they speculated on the role these properties played in the establishment of infections.

ADHESION AND COLONIZATION

The essential contribution of adhesive activity to colonization was first demonstrated in *Escherichia coli* strains which cause infections of the intestine of the neonatal pig and are characterized by the production of an antigen designated K88. These *E. coli* strains differ in behavior from the *E. coli* strains of the normal flora of the intestine in that they colonize the small intestine and are present in the gut in many-fold greater numbers. For bacteria to colonize such an environment so successfully, one must suppose that they overcome removal by peristalsis, perhaps by adhering to the mucosa. The observation that the K88 protein was encoded on a plasmid which was maintained by bacterial strains in vivo, but not always by cultures grown in vitro, and the demonstration that the K88 antigen of all strains possessed adhesive properties demonstrable as hemagglutinating activity (27) made the proposition that K88 was a specific adhesive protein essential for colonization of the small intestine a tenable one (26).

It was readily demonstrated that K88-producing bacteria adhered to the epithelial cell surface of the small intestine, but the same strain cured of the plasmid, and hence unable to produce K88, did not do so (26). Of critical importance was the demonstration that K88 was produced in copious amounts in vivo and bound to the epithelial cells, but not the mucus gel (26). These results were reproduced in vitro with slices of intestinal tissue, in which test system it was shown that phenotypically K88-negative cultures grown at 18°C did not adhere and that K88 antibody alone inhibited attachment.

The ecological significance of K88 to these *E. coli* cells was apparent, however, only in conventional newborn pigs. Enterotoxin synthesis is required for disease in such animals (42), but only in those animals infected with the toxigenic K88-producing strain did this occur (26, 42); no disease or colonization resulted from challenges with toxigenic K88-negative mutant strains (26). In the germfree piglet, in contrast, both K88-producing *E. coli* strains and K88-negative strains derived from them colonized the intestine and caused disease. The growth of K88-negative strains, however, was confined entirely to the lumen of the gut, whereas K88-producing bacteria predominated on the mucosa (26). This significant difference in the behavior of K88-negative strains in conventional and germfree pigs is probably due to a much reduced peristalsis (and hence lower clearance) of the intestine of the latter animals. It is well established that the normal flora of the large bowel has a marked suppressive influence on the populations of *E. coli* cells in the gut, exerting its influence possibly through direct competition for substrate (9) and by the production of substances such as fatty acids (32) and H_2S (9). Few bacteria are resident in the small intestine, however, and microbial competition, therefore, must be negligible. Colonization of the small intestine may well circumvent the problem of competition from the normal flora, but at the same time confronts the microbial cell with possible removal by fluid shear and flow generated by peristalsis, and hence imposes on the bacterium the need to adhere to the mucosa. In these organisms it is apparent that K88 serves this function.

It has become well established that adhesiveness is a common attribute of both gram-positive and gram-negative pathogens which specifically, and naturally, colonize various surface sites of several animal species, and many adhesive activities can now be assigned to particular surface components (1, 18, 22). In only a few instances, however, have direct connections between adhesiveness, particular surface components, and colonization been established. The better examples of these remain those studies on E. coli enteropathogenic for the pig which illustrate, moreover, that adhesion is achieved by different types of E. coli through the agencies of quite distinct adhesive appendages (17, 33) which recognize and bind to different surface receptors (16). In many other cases, however, the role of an adhesin in the natural habitat can only be assumed from in vitro observations, and such assumptions are often confounded by the existence of more than a single adhesive activity in a culture (see below).

Mucosal and other surfaces are the sites most commonly colonized by bacteria. Some species of bacteria, such as Salmonella typhimurium, however, possess the ability to penetrate into the living cells of the host and to survive as intracellular parasites. The bacteria first enter the intestinal epithelial cells and then pass from these cells into other cells of the body, including macrophages (46). The bacteria play no active role in the internalization process, which seems to be a function entirely of the animal cell's endocytic mechanism (29), but do need to become intimately associated with the animal surface prior to internalization. The latter can be a function of the animal cell, but one may speculate that intracellular parasites have evolved their own means of attachment which promote efficient internalization and perhaps subsequent intracellular survival.

S. typhimurium does possess an adhesive property (25), encoded on a plasmid (23), which fulfills these requirements. The chromosomally encoded type 1 fimbrial adhesin is much less effective in this regard and promotes little or no adhesion or internalization (24). Curing of the adhesin plasmid not only results in the loss of adhesive and invasive properties, but also in the reduced ability to colonize the intestine and in the loss of virulence (23).

Salmonellae are ill suited to the intestinal environment, surviving poorly within the gut lumen (3). The salmonella adhesin, however, allows these bacteria to escape the strictures imposed on growth in the intestine by promoting colonization of the animal cell where competition from other microorganisms is minimized. (The bacteria in this habitat are confronted by other bactericidal and growth-inhibiting mechanisms originating in the host cells and tissue fluids; these the bacteria counteract with other microbial systems also apparently encoded on the plasmid.) Although the consequences of S. typhimurium and E. coli adhesion are quite different, the final results are alike in the sense that both promote colonization of sites where the populations of indigenous bacteria are small or nonexistent and microbial competition, in consequence, is reduced.

MULTIPLE ADHESIN PRODUCTION

About three decades ago Duguid and others (see 5) showed that some E. coli cultures produced more than one adhesin. This is now a well-established phenomenon found in cultures of several diverse bacterial species. The conditions of culture, the type of in vitro test employed, the animal cell used as the substratum, or the addition to the test of specific inhibitors of certain adhesins allows one adhesive activity of a culture to be detected in the presence of another (11). The expression of multiple adhesive activities by a culture can be due to the simultaneous production of at least two adhesins by single cells (36), although by and large this phenomenon is unexplored.

The ability of many bacteria to produce more than one adhesin in vitro has created some controversy about the precise role each plays in colonization of a given habitat. This type of controversy has surrounded the role in infection of type 1 fimbriae of the Enterobacteriaceae which may be present in cultures that are also expressing other adhesive activities. Although readily produced in culture, type 1 fimbriae play no role in the colonization of the pig intestine by K88-producing E. coli cells. There is some statistical evidence that strains of S. typhimurium with fimbriae remain in the intestine longer than strains without fimbriae (4), but there is no evidence that the fimbriae are involved in the adhesive event that leads to invasion of the animal cell (24). We have found that nonfimbriate wild-type and mutant strains are as adhesive and invasive as wild-type fimbriate strains and that the introduction of the structural gene (pil) for type 1 fimbriae reduces the adhesive activity of a normally nonfimbriate strain; the latter is reversed by D-mannose (24), a specific inhibitor of type 1 fimbrial adhesion (see below). Type 1 fimbriae probably impede the activity of the plasmid-encoded adhesin of *S. typhimurium* by causing the bacterium to be displaced too far from the surface of the animal cell. This adhesin is particularly sensitive to such displacements, becoming inactive when the distance of separation between bacterium and animal cell reaches approximately 15 nm (i.e., approximately twice the diameter of type 1 fimbriae [25]).

The reasons for the production of more than one distinctly different adhesin are again unclear and largely unexplored. One may speculate that some adhesins provide auxiliary or alternative systems, or indeed are involved at a stage of the interaction which precedes the final attachment via the principal adhesin. Hydrophobic surface properties would promote most adhesive interactions in this way, and such properties have been found to be greater in adhesive bacteria than in nonadhesive mutant strains (7, 44). *Streptococcus sanguis*, an organism that colonizes the buccal and tooth surfaces, exhibits the extremes of hydrophobicity (35, 48); more than 98% of the bacteria partition in the alkane phase in a two-phase hexadecane–phosphate buffer system (unpublished data). The question which naturally arises is whether these hydrophobic properties account for the total binding activity of the bacterial cell or whether they reflect the presence of a specific adhesin. Recent studies with mutant strains have shown (Table 1) that in stationary-phase culture specific attachment to the saliva pellicle of a hydroxyapatite surface and hydrophobicity are associated properties in only some strains and that both of these are independent of attachment to hydroxyapatite. Attachment to buccal cells is probably due to a different adhesive system also; the adhesion of stationary-phase cultures of mutant strains to

buccal cells is about 10 to 20% of the parental value, but this relative value increases to 75% in log-phase cultures. Attachment to hydroxyapatite is not apparent in either mutant or parent cultures during log phase. These organisms, therefore, appear to possess several adhesive/adsorptive activities, one or more of which may function as the main adhesive activity on a specific surface.

An alternative explanation for multiple adhesin production by bacteria such as *Vibrio cholerae* (19) and salmonellae (14), which exist in extracorporal as well as corporal habitats, is that one adhesin functions in the former and another in the latter location. M. A. Savageau (Am. Nat., in press) has presented an argument for the colonization of more than one habitat by *E. coli* based on demand theory analysis of biochemical pathways (38). One may argue on this basis that the positive regulation of K88 adhesin synthesis not only shows a high demand for this adhesin in vivo but also implies low demand elsewhere. Conceivably, some adhesins may be redundant vestiges of some bacteria which serve a real function only in related species that colonize other, perhaps extracorporal habitats.

IS ADHESION NECESSARY FOR COLONIZATION?

Since the first demonstration of the essential roles some bacterial adhesins play in the colonization of mucosal and other surfaces, there has been a less than critical evaluation of the roles of other adhesins in this initial phase of microbial colonization. The contributions of such bacterial activities as motility, chemotaxis, and adsorption (i.e., interactions not involving specific adhesin–receptor interactions, such as the hydrophobic binding of *S. sanguis*) have been largely ignored, yet it is now apparent that such activities may facilitate or even be essential for the binding of adhesins; in one instance at least, such activities can promote colonization in the absence of any detectable adhesive activity.

The roles these activities may play in colonization of mucosal surfaces became apparent during studies on the *V. cholerae* adhesin. It had been established that a classical strain of *V. cholerae* associated exclusively with the mucosal surface of intestinal tissue. *V. cholerae* also attached to L-fucose receptors of the glycocalyx of the free brush border membranes of epithelial cells, but needed the presence of divalent cations to do so; moreover, only motile vibrios and those grown in liquid culture were adhesive. Vibrios did not attach to the mucus gel which normally overlays the epithelial cells, but could migrate rapidly through it (20).

When the parameters which defined the adhesion of *V. cholerae* cells to brush border mem-

TABLE 1. Relative hydrophobic and adhesive properties of *S. sanguis* wild-type and mutant strains

Strain	Origin	Hydropho-bicity[a]	Adhesion to		
			S-HA[b]	HA[b]	BC[c]
FC1	Parent	1.0	1.0	1.0	1.0
FC1/2	FC1	0.40	0.05	0.37	0.3
C5	Parent	1.0	1.0	1.0	1.0
C5/2	C5	0.19	0.29	0.27	0.4

[a] Measured in phase partition test against hexadecane; >98% of parental bacteria enter alkane phase.

[b] Saliva-coated (S-HA) or uncoated (HA) hydroxyapatite; 30 to 60% of parental strains attach in both tests.

[c] Between 100 and 200 parental bacteria adhered per buccal cell (BC).

branes were compared with those which govern their association with the mucosal surface, marked discrepancies became apparent (10). None of the parameters cited above applied to mucosal association, suggesting that the L-fucose receptor was not involved. The only similarities between adhesion to brush borders and association with the mucosa were that both were inhibited by peptic digests of the mucosa, and in both systems nonmotile vibrios failed to interact. Further studies (8) demonstrated that those fractions of the peptic digest of mucosa which inhibited mucosal association were most active as chemoattractants of vibrios. Following this, it became apparent that chemotaxis was the bacterial attribute which promoted mucosal association and that the vibrios accumulated in the mucus gel and not on the glycocalyx. This raises the question of whether the attractant-sensing systems of bacteria allow them to distinguish between surfaces, perhaps by responding to specific attractants. One piece of evidence suggests that some specificity in the response may be involved (47). The collision frequencies of chemotactic and nonchemotactic *S. typhimurium* cells with HeLa cells are entirely random. However, chemotactic salmonellae are attracted to debilitated cells which increase their collision frequencies threefold. This response can be inhibited competitively by the addition of glycine, asparate, asparagine, alanine, and serine but not by the addition of other amino acids, carbohydrates, or citric acid cycle intermediates. From blocking experiments conducted with a standard chemotaxis test system, it was found that only the bacterial response to glycine mimicked precisely their response to HeLa cells in the presence of potential inhibitors of chemotaxis (Table 2). From this it was deduced that glycine was the attractant released by the HeLa cell.

Chemotaxis toward mucosae and animal cells, therefore, appears to be an ecologically relevant phenomenon, being a more important determinant of the ability of *V. cholerae* to colonize the mouse intestine than the vibrio adhesins (8). Although one may question the relevance of such observations on the grounds that humans, not mice, are the natural host of *V. cholerae*, the accidental laboratory infection of a human subject by a chemotactic, nonadhesive vibrio strain goes some way to substantiate the role of chemotaxis. The chemotactic targeting of bacteria may be one of the earliest of the events which decide the final site of colonization. For example, some types of adhesive *E. coli* cells have the capacity to adhere to epithelial cells from all sites of the small intestine, yet they adhere only to the terminal small intestinal cells (33). Significantly, these strains are nonmotile and hence presumably depend on chance contact with the

TABLE 2. Comparative inhibition of the chemotaxis of *S. typhimurium* to HeLa cells and to amino acids which block this response

Com-petitor	Inhibition of chemotaxis to:					
	HeLa cells	Gly-cine	Aspar-tate	Aspara-gine	Ala-nine	Ser-ine
Glycine	+[a]	+	+	+	+	−
Alanine	+	+	−	−	+	−
Citrate	−	−	+	+	+	+
Maltose	−	−	−	+	−	−

[a] Symbols: +, inhibition of chemotaxis; −, no inhibition of chemotaxis.

mucosa, which increases as the organism is swept along the intestinal tract. Motile chemotactic strains, in contrast, adhere to sites distributed along the entire length of the small intestine (26).

ADHESIN RECEPTORS OF THE ANIMAL CELL SURFACE

The best evidence suggests that the sites of attachment of many bacterial adhesins are particular carbohydrate residues of the animal cell surface (18, 22); other adhesins, however, probably bind by intercalating into the lipid bilayer, by interacting with surface proteins such as fibronectin, and yet others bind to insoluble carbohydrate polymers deposited on the surface by the bacterial cell (1). The specific receptor concept provides the reason for the selective attachment of bacteria to different animal cells and, in part, for the correlation between this and the colonization of these cells in vivo.

The studies of Duguid and Old (5) on the inhibition by D-mannose of the attachment of bacteria by means of type 1 fimbriae first implicated carbohydrates as possible adhesin receptors of erythrocytes. Later, detailed investigations by Old (34) revealed the degree of specificity of such inhibitors, defined more precisely the molecular configuration necessary for inhibitory activity, and suggested that the mannose receptor residue would be a terminal sugar. Work on the K88 adhesin and its receptor, however, was the first attempt at defining a receptor isolated from the organism's natural habitat (12). On the basis of the initial observation that glycoproteins from the pig intestine inhibited the adhesive activity of the K88 adhesin, serum glycoproteins and glycoproteins isolated from epithelial surfaces were analyzed for adhesion inhibition. Existing terminal receptors were destroyed and new receptors were generated from subterminal sites of the oligosaccharide chain by chemical treatment. Although the typical microheterogeneity of such oligosaccharides must result in a mixture of terminal groups both before and after each step of degradation, it was nevertheless possible to discern the involvement of D-galactose or a derivative, possibly linked to a hexosamine, in the K88 receptor function; subsequent support for this conclusion came from studies with mono-, di-, tri-, and tetrasaccharide inhibitors (40).

From simple inhibition assays it is only possible to speculate on the probability that the inhibitor competes with the surface receptor for the binding site of the adhesin and hence resembles the receptor in its molecular configuration. Clearly, a more direct approach is to show the binding of the inhibitor to the adhesin or, better still, to demonstrate that the binding of the bacterium to an inhibitor immobilized on a surface provides sufficient binding energy to hold the bacterium on that surface. Such has been demonstrated and lends credence to the argument that the inhibitor constitutes at least an important part of the natural receptor. Jones and Freter (21) observed that the attachment of *V. cholerae* to brush border membranes was inhibited by L-fucose, but not by D-fucose, D-galactose, or L-galactose, and that L-fucose was present on the brush border membranes and hence was very possibly the *V. cholerae* receptor. The latter conclusion was supported by the observation that *V. cholerae* adhered to Sepharose beads to which L-fucose had been covalently linked, but not when D-fucose was employed. This confirmed that L-fucose not only bound the vibrios but did so with sufficient avidity to anchor the bacteria onto the surface. Essentially similar studies were done later with a receptor/inhibitor of an *E. coli* adhesin, isolated from a reactive animal cell. The inhibitor, after reinsertion into the membrane of a normally inert animal cell, was shown to cause bacterial adhesion to the latter (30).

Tissue cells which carry adhesin receptor may not be components of the surface colonized by the bacteria (e.g., erythrocytes). Adhesion to such cells, therefore, is relevant only insofar as they provide a suitable in vitro model. More significant to microbial ecology are the converse observations that tissue cells of a surface normally colonized by a bacterium can in some individuals lack the receptor (41). This is not surprising if one reflects on the fact that colonization by some bacteria is detrimental to the host, and one may thus expect the selection of individuals lacking the receptor to occur. Such individuals should be resistant to colonization and to disease. This was found originally to be true of the K88 adhesin receptor when no adhesion was observed in two usually reproducible adhesion tests which employed slices of intestinal tissue and brush border membranes of the epithelial cells. From the examination of the ancestry of animals from which the tissues have been obtained and from subsequent breeding experiments, it was deduced that the inheritable receptorless character was recessive. The profound influence of this genetic trait on the intestinal ecology of K88-producing *E. coli* cells was clearly apparent in infection experiments when receptorless animals were found to be resistant to colonization by these bacteria (37). The resistance of some humans to colonization of their urinary tract by specific types of *E. coli* also has been shown to be due to an absence of suitable receptors (31).

ADHESINS

The majority of adhesins which have been characterized in detail are proteins (1, 18, 22), although some organisms such as *Streptococcus pyogenes* have evolved other mechanisms (2). Superficially, the proteinaceous adhesins take one of two basic forms: fimbriae such as type 1 fimbriae are distinct rigid or flexible rods up to 1 μm or more in length and of uniform diameter; the K88 and similar adhesins, in contrast, appear less regular in form since they consist of thin filaments which form tangled masses. Jones and Isaacson (22) deduced recently that fimbriae are formed by the tight helical winding of a polymer composed of the adhesin subunits and that adhesins such as K88 and K99 are basically the same except that the polymers are wound in open helices.

The long filamentous form of the adhesin is probably mandated by two requirements. First, the adhesive appendage needs to bridge the gap between the bacterial cell and the animal cell which is maintained by the mutual repulsion of the two bodies. Second, the small radius of curvature of the adhesin results in its experiencing less repulsion than the bacterial cell (18, 22). Likewise, the general hydrophobic properties of adhesins, noted above, would promote adhesin–receptor interactions by exclusion of water as a separation factor. Although not proven in all cases, it appears that each subunit of the adhesin has a binding site and that a single adhesin filament aligned lengthwise on a surface attaches at multiple points (45).

Only N-terminal amino acid sequences are known in any detail and then only for a few adhesins (18, 22). Nevertheless, remarkable similarities between some of these adhesins can be discerned. Those of *Neisseria gonorrhoeae*, *Moraxella liquefaciens*, and *Pseudomonas aeruginosa* differ by only two amino acids in the first 20 N-terminal amino acids (22). Type 1 fimbriae of *E. coli* and *Klebsiella pneumoniae* and the F7 adhesin of *E. coli* are morphologically almost indistinguishable and are almost identical in N-terminal amino acid sequences (22). The F7 adhesin, however, differs from the others in its binding properties, the most obvious of which is its insensitivity to D-mannose. The significance of compositional similarities but functional differences is not known.

Although striking similarities often exist in morphology, amino acid sequences, and function of the adhesins of both related and unrelated bacteria, subtle differences occur in the antigenicity of the same adhesin in different strains and, indeed, in the periodic production of the same adhesin in different antigenic forms by the same strain. The reasons are basically ecological ones. N. gonorrhoeae, for example, depends entirely on its human host for survival. At the same time N. gonorrhoeae is a pathogen which induces potential protective host defense mechanisms which are partially directed against the adhesins. The continual antigenic drift of the adhesin, therefore, allows some bacteria to avoid these anti-adhesive antibodies and to maintain themselves on the mucosal surface. The antigenic modification of an adhesin subunit probably occurs at a single, variable, immunodominant site, whereas subunit sites involved in the formation of the adhesin filament and in binding to the surface are probably conserved (39).

Environmental control is a characteristic feature of adhesins produced by most bacteria. Qualitative phase variation is the classical phenomenon of cultural conditions favoring either production or nonproduction of the adhesin. This is somewhat different from quantitative phase variation which is characterized by the production of different amounts of adhesin (22). Not only do bacteria differ in their response to cultural conditions in this respect, but also the adhesin a strain produces is synthesized or assembled preferentially under different conditions (see below).

One assumes that the regulated production of an adhesin is ecologically significant for organisms in corporal and extracorporal habitats. The 987P adhesin of E. coli, which is necessary for mucosal colonization, is readily produced in vivo, but production in vitro is difficult to maintain (17, 33). K88 is positively regulated (28), substantiating the imperative of K88 production in vivo. Significantly perhaps, at the extracorporal temperature of 18°C, K88, K99, and the S. typhimurium (24) adhesins are not produced. Recently, Isaacson (15) showed that in the case of the K99 adhesin this is due to repression of the synthesis of K99 subunits, possibly on an inner membrane site. In contrast to these adhesins, type 1 fimbriae are produced at 18°C; in vitro at 37°C marked phase variation, controlled

at the transcriptional level (6), is highly characteristic of the adhesin.

It is obvious that the cultural conditions of bacteria can be manipulated so that they produce a demonstrable adhesin or even adhesins. As stressed previously, this is in itself significant, but in no way does it mean that this adhesin has any function in the colonization of the bacterium's habitat. Clearly, not only must the adhesin be necessary for the colonization of the latter, but the bacteria must be capable of producing the adhesin under in vivo conditions. Frankly, I do not understand the significance of the regulation of adhesin production by the environment. It is satisfying to believe that each adhesin has some specific function and that its production is precisely controlled in such a way that it is synthesized when required by the bacteria in a corporal or extracorporal habitat.

CONCLUSIONS

Although successful colonization of cellular and inert surfaces of the body can be initiated by several different forms of adhesive interaction (1), the more commonly recognized type involves adhesive proteins. Whereas much is known about the composition of these adhesive proteins (22), little is understood about such things as the tertiary structure responsible for receptor recognition and binding, the modulation of adhesin antigenic forms and expression, the reasons for the apparent relatedness of the adhesins of unrelated species, or the precise configuration of their carbohydrate receptors. All of these present major challenges to be faced in the arenas of molecular biology and molecular genetics. Questions concerned with the function(s) of an adhesin in a given habitat and how these functions may be integrated into the entire colonization process, or the sequence and relevance of events and interactions leading to final adhesin–receptor binding, will be solved in a different manner, but are equally thought provoking. Perhaps central to the ecology of adhesive microbes is the phenomenon of multiple adhesin production and the influence of environmental factors on switching from the production of one adhesin to another.

LITERATURE CITED

1. Beachey, E. H. (ed.). 1980. Bacterial adherence. Receptors and recognition, series B, vol. 6. Chapman and Hall, London.
2. Beachey, E. H., A. W. Sampson, and I. Ofek. 1980. Interactions of surface polymers of Streptococcus pyogenes with animal cells, p. 389–405. In R. C. W. Berkeley, J. M. Lynch, J. Melling, P. R. Rutter, and B. Vincent (ed.), Microbial adhesion to surfaces. Ellis Harwood, Chichester.
3. Carter, P. B., and F. M. Collins. 1974. The route of enteric infection in normal mice. J. Exp. Med. 193:1189–1203.

4. **Duguid, J. P., M. R. Dureckar, and D. W. F. Wheater.** 1976. Fimbriae and infectivity in *Salmonella typhimurium*. J. Med. Microbiol. **9:**459–473.

5. **Duguid, J. P., and D. C. Old.** 1980. Adhesive properties of *Enterobacteriaceae*, p. 185–218. *In* E. H. Beachey (ed.), Bacterial adherence. Receptors and recognition, series B, vol. 6. Chapman and Hall, London.

6. **Eisenstein, B. I.** 1981. Phase variation of type 1 fimbriae in *Escherichia coli* is under transcriptional control. Science **216:**337–339.

7. **Faris, A., T. Wadstrom, and J. M. Freer.** 1981. Hydrophobic adsorptive and hemagglutinating properties of *Escherichia coli* possessing colonizing factor antigens (CFA/I or CFA/II), type 1 pili and other pili. Curr. Microbiol. **5:**67–72.

8. **Freter, R.** 1982. Bacterial association with the mucus gel system of the gut, p. 278–281. *In* D. Schlessinger (ed.), Microbiology—1982. American Society for Microbiology, Washington, D.C.

9. **Freter, R., H. Brickner, M. Botney, D. Cleven, and A. Aranki.** 1983. Mechanisms which control bacterial populations in continuous flow culture models of mouse large intestinal flora. Infect. Immun. **39:**676–685.

10. **Freter, R., and G. W. Jones.** 1976. Adhesive properties of *Vibrio cholerae*: nature of the interaction with intact mucosal surfaces. Infect. Immun. **14:**246–256.

11. **Freter, R., and G. W. Jones.** 1983. Models for studying the role of bacterial attachment in virulence and pathogenesis. Rev. Infect. Dis. **5:**S647–S658.

12. **Gibbons, R. A., G. W. Jones, and R. Sellwood.** 1975. An attempt to identify the intestinal receptor for the K88 adhesin by means of a haemagglutination inhibition test using glycoproteins and fractions from some colostrum. J. Gen. Microbiol. **86:**228–240.

13. **Gibbons, R. J., and J. van Houte.** 1975. Bacterial adherence in oral microbial ecology. Annu. Rev. Microbiol. **29:**19–44.

14. **Hendricks, C. W.** 1971. Increased recovery rate of salmonellae from stream bottom sediments versus surface waters. Appl. Microbiol. **21:**379–380.

15. **Isaacson, R. E.** 1983. Regulation of expression of *Escherichia coli* pilus K99. Infect. Immun. **40:**633–639.

16. **Isaacson, R. E., P. C. Fusco, C. C. Brinton, and H. W. Moon.** 1978. In vitro adhesion of *Escherichia coli* to porcine small intestinal epithelial cells: pili as adhesive factors. Infect. Immun. **19:**392–397.

17. **Isaacson, R. E., B. Nagy, and H. W. Moon.** 1977. Colonization of porcine small intestine by *Escherichia coli*: colonization and adhesion factors of pig enteropathogens that lack K88. J. Infect. Dis. **135:**531–539.

18. **Jones, G. W.** 1977. The attachment of bacteria to the surfaces of animal cells, p. 139–176. *In* R. L. Reissig (ed.), Microbial interactions. Receptors and recognition, series B, vol. 3. Chapman and Hall, London.

19. **Jones, G. W.** 1980. The adhesive properties of *Vibrio cholerae* and other *Vibrio* species, p. 219–249. *In* E. H. Beachey (ed.), Microbial interactions. Receptors and recognition, series B, vol. 6. Chapman and Hall, London.

20. **Jones, G. W., G. D. Abrams, and R. Freter.** 1976. Adhesive properties of *Vibrio cholerae*: adhesion to isolated rabbit brush border membranes and hemagglutinating activity. Infect. Immun. **14:**232–239.

21. **Jones, G. W., and R. Freter.** 1976. Adhesive properties of *Vibrio cholerae*: nature of the interaction with isolated rabbit brush border membranes and human erythrocytes. Infect. Immun. **14:**240–245.

22. **Jones, G. W., and R. E. Isaacson.** 1983. Proteinaceous bacterial adhesins and their receptors. Crit. Rev. Microbiol. **10:**229–260.

23. **Jones, G. W., D. K. Rabert, D. M. Svinarich, and H. J. Whitfield.** 1982. Association of adhesive, invasive, and virulent phenotypes of *Salmonella typhimurium* with autonomous 60-megadalton plasmids. Infect. Immun. **38:**476–486.

24. **Jones, G. W., and L. A. Richardson.** 1981. The attachment to, and invasion of HeLa cells by *Salmonella typhimurium*: the contribution of mannose-sensitive and mannose-resistant haemagglutinating activities. J. Gen. Microbiol. **127:**361–370.

25. **Jones, G. W., L. A. Richardson, and D. Uhlman.** 1981. The invasion of HeLa cells by *Salmonella typhimurium*: reversible and irreversible bacterial attachment and the role of bacterial motility. J. Gen. Microbiol. **127:**351–360.

26. **Jones, G. W., and J. M. Rutter.** 1972. Role of the K88 antigen in the pathogenesis of neonatal diarrhea caused by *Escherichia coli* in piglets. Infect. Immun. **6:**918–927.

27. **Jones, G. W., and J. M. Rutter.** 1974. The association of K88 antigen with haemagglutinating activity in porcine strains of *Escherichia coli*. J. Gen. Microbiol. **84:**135–144.

28. **Kehoe, M., R. Sellwood, P. L. Shipley, and G. Dougan.** 1981. Genetic analysis of K88-mediated adhesion of enterotoxigenic *Escherichia coli*. Nature (London) **291:**122–126.

29. **Kihlstrom, E., and L. Nilsson.** 1977. Endocytosis of *Salmonella typhimurium* 395M and MR10 by HeLa cells. Acta Pathol. Microbiol. Scand. Sect. B **85:**322–328.

30. **Leffler, H., and C. Svanborg-Eden.** 1980. Chemical identification of a glycosphingolipid receptor for *Escherichia coli* attaching to human urinary tract epithelial cells and agglutinating human erythrocytes. FEMS Microbiol. Lett. **8:**127–134.

31. **Lomberg, H., U. Jordal, C. Svanborg-Eden, H. Leffler, and B. Samuelsson.** 1981. P1 blood group and urinary tract infection. Lancet **i:**551–552.

32. **Meynell, G. G., and T. V. Subbarah.** 1963. Antibacterial mechanisms of the mouse gut. Br. J. Exp. Pathol. **44:**197–208.

33. **Moon, H. W., R. E. Isaacson, and J. Pohlenz.** 1979. Mechanisms of association of enteropathogenic *Escherichia coli* with intestinal epithelium. Am. J. Clin. Nutr. **32:**119–127.

34. **Old, D. C.** 1972. Inhibition of the interaction between fimbrial haemagglutinins and erythrocytes by D-mannose and other carbohydrates. J. Gen. Microbiol. **71:**149–157.

35. **Olsson, J., and G. Westergren.** 1982. Hydrophobic surface properties of oral streptococci. FEMS Microbiol. Lett. **15:**319–323.

36. **Orskov, I., F. Orskov, and A. Birch-Andersen.** 1980. Comparison of *Escherichia coli* fimbrial antigen F7 with type 1 fimbriae. Infect. Immun. **27:**657–666.

37. **Rutter, J. M., M. R. Burrows, R. Sellwood, and R. A. Gibbons.** 1975. A genetic basis for resistance to enteric disease caused by *E. coli*. Nature (London) **257:**135–136.

38. **Savageau, M. A.** 1977. Design of molecular control mechanisms and the demand for gene expression. Proc. Natl. Acad. Sci. U.S.A. **74:**5647–5651.

39. **Schoolnik, G. K., J. Y. Tai, and E. C. Gotschlich.** 1982. Receptor binding and antigenic domains of gonococcal pili, p. 312–316. *In* D. Schlessinger (ed.), Microbiology—1982. American Society for Microbiology, Washington, D.C.

40. **Sellwood, R.** 1980. The interaction of the K88 antigen with porcine intestinal epithelial cell brush-borders. Biochim. Biophys. Acta **632:**326–335.

41. **Sellwood, R., R. A. Gibbons, G. W. Jones, and J. M. Rutter.** 1975. Adhesion of enteropathogenic *Escherichia coli* to pig intestinal brush-borders: the existence of two pig phenotypes. J. Med. Microbiol. **8:**405–411.

42. **Smith, H. W., and M. A. Lingood.** 1971. Observations on the pathogenic properties of the K88, H1Y and Ent plasmids of *Escherichia coli* with particular reference to porcine diarrhoea. J. Med. Microbiol. **4:**467–486.

43. **Smith, T., and M. L. Orcutt.** 1925. The bacteriology of the intestinal tract of young calves with special reference to the early diarrhea ("scours"). J. Exp. Med. **41:**89–106.

44. **Smyth, C., P. Jonsson, E. Olsson, O. Sonderlund, J. Rosengren, S. Hjesten, and T. Wadstrom.** 1978. Differences in hydrophobic surface characteristics of porcine enteropathogenic *Escherichia coli* with and without K88 antigen as revealed by hydrophobic interaction chroma-

tography. Infect. Immun. **22:**462–472.

45. **Sweeney, G., and J. H. Freer.** 1979. Location of binding sites on common type 1 fimbriae from *Escherichia coli.* J. Gen. Microbiol. **112:**321–328.

46. **Takeuchi, A.** 1967. Electron microscope studies of experimental salmonella infection. I. Penetration into the intestinal epithelium by *Salmonella typhimurium.* Am. J. Pathol. **50:**109–136.

47. **Uhlman, D. L., and G. W. Jones.** 1982. Chemotaxis as a factor in interactions between HeLa cells and *Salmonella typhimurium.* J. Gen. Microbiol. **128:**415–418.

48. **Westergren, A., and J. Olsson.** 1983. Hydrophobicity and adherence of oral streptococci after repeated subculture in vitro. Infect. Immun. **40:**432–435.

Mechanisms of Specific Bacterial Adhesion to Cyanobacterial Heterocysts

F. S. LUPTON AND K. C. MARSHALL

Department of Bacteriology, University of Wisconsin, Madison, Wisconsin 53706, and School of Microbiology, University of New South Wales, Kensington, New South Wales, Australia

The adhesion of microorganisms to surfaces is of enormous significance to the functions of microorganisms in all microbial ecosystems (2–4, 22, 26). The specific adhesion of microorganisms describes the highly selective and sometimes exclusive adhesion of microorganisms to particular surfaces and has only been observed in organism–organism interactions (10, 22, 38, 39).

The nonspecific adsorption of microorganisms to surfaces can be accounted for in terms of long-range interactive forces such as the DLVO theory of double-layer interactions (27, 35). Specific adhesion, however, may arise from detailed short-range stereochemical interactions involving hydrophobic, dipole, electrostatic, and chemical bonding forces (35). To initiate these short-range interactions, complementary molecular configurations or groups, adhesins, must occur on both the bacterial and adsorbent surfaces (4, 22).

There appear to be many examples of specific adhesion among microorganisms and between micro- and macroorganisms in natural environments (2, 4), and it is likely that many of these examples of specific microbial adhesion must share similar adhesive mechanisms. It is very probable, therefore, that environmental factors as well as physicochemical forces have some influence on the observed specificity of adhesion in some systems. Examples of specific microbial adhesion include: the adhesion of lactobacilli to gastrointestinal epithelial cells (16, 36–39, 42); the adhesion of oral bacteria to teeth (20, 29, 30); the adhesion of enteropathogenic *Escherichia coli* to the intestinal epithelial cells of several animal species (14, 37); the adhesion of *Rhizobium* spp. to specific legumes (10, 23); and the adhesion of certain *Xanthomonas* plant pathogens to their hosts (11, 12, 28).

Bacterial attachment to the terminal heterocysts of *Cylindrospermum* was reported as early as 1932 (18). Schwabe and Mollenhauer (40) commented that bacteria are especially abundant on the heterocysts of many *Cylindrospermum* species. Paerl (31) described apparently specific associations between bacteria on the heterocysts of *Anabaena* and *Aphanizomenon* species during cyanobacterial blooms. The bac-

terial adhesion appears to be very selective for heterocysts, as the attached bacteria are almost exclusively located on *Anabaena* heterocysts, even though heterocysts comprise only 5 to 10% of cells in *Anabaena* filaments (32). Paerl and Kellar (33, 34) have proposed that the specific association of bacteria with heterocysts of planktonic *Anabaena* species directly promotes nitrogen fixation in the cyanobacteria by creating a reduced microzone around the heterocysts, thus protecting the oxygen-sensitive nitrogenase from the high oxygen tensions which occur during cyanobacterial blooms. Thus, the adhesion of bacteria to heterocysts indicates a close symbiotic relationship between these two procaryotic organisms which may have a marked effect on nitrogen fixation in some aquatic ecosystems. The aim of this project was to investigate the phenomenon of specific bacterial adhesion with regard to the basis of specificity.

MATERIALS AND METHODS

Organisms. Two bacteria, *Pseudomonas* sp. SL10 and *Zoogloea* sp. SL20, were isolated from *Anabaena* cultures and maintained on a minimal basal medium (25). *Anabaena* cultures were maintained on a minimal salts medium in the absence of inorganic nitrogen (21).

Measurement of adhesion. (i) Isotope labeling. The adhesion of *Pseudomonas* sp. SL10 and *Zoogloea* sp. SL20 to heterocysts was assayed by use of ^{14}C-labeled bacteria. The adhesion of *Zoogloea* sp. SL20 was assayed with [^{14}C]thymidine-labeled bacteria. *Pseudomonas* sp. SL10 demonstrated a poor uptake of [^{14}C]thymidine label, and adhesion was assayed by using bacteria labeled with a mixture of L-U-^{14}C-amino acids. The specificity of adhesion was confirmed by phase-contrast microscopy. For *Pseudomonas* sp. SL10, 30 µl of a 50-µCi/0.5-ml solution (ethanol-water, 2:9.8) of an L-U-^{14}C-amino acid mixture (specific activity, 55 mCi/matom of carbon, New England Nuclear Corp., Boston, Mass.) was added to 100 ml of minimal basal medium. Cultures were grown in medium containing labeled amino acids for 48 h at 30°C. For *Zoogloea* sp. SL20, 50 µl of a 50-µCi/ml aqueous solution of [2-^{14}C]thymidine (specific activity, 50 to 60 mCi/mmol, Radiochemical Centre, Amersham, England) was added to 100 ml of minimal basal medium and cultures were grown in medium with labeled thymidine for 48 h at 30°C.

(ii) Adhesion assay. Cyanobacterial cultures were centrifuged at $1,100 \times g$ for 1 min in 50-ml plastic centrifuge tubes and washed three times in 30 ml of fresh minimal medium; the final pellet was resuspended in 20 ml of minimal medium. A 100-ml amount of bacterial culture was centrifuged at $10,000 \times g$ for 5 min and washed three times in 30 ml of fresh medium; the final pellet was resuspended in 20 ml of minimal medium.

A 1-ml amount of cyanobacterial inoculum was mixed in a tube with 1 ml of labeled bacterial inoculum and 3 ml of minimal medium without phosphate, and the mixture was incubated in the light at 25°C for 24 h. Since the assay medium was both P and N limited, little or no growth of bacteria occurred after attachment to heterocyst surfaces. After incubation, the cultures were washed three times in 5 ml of minimal medium, by centrifugation at $220 \times g$ for 1 min. This treatment was effective in removing unattached bacteria, as confirmed by direct microscopic examination of the washed cyanobacteria. After the last washing, the cyanobacterial pellet was resuspended in 5 ml of filter-count scintillation solution (Packard Instrument Co., Inc., Downers Grove, Ill.) and placed in 24-ml scintillation vials. Disintegrations per minute were determined on a Packard Tri-Carb C2425 scintillation counter.

The heterocyst concentration of the cyanobacterial inoculum was determined from chlorophyll extraction and absorbance at 663 nm by means of standard curves of dry weight of heterocysts versus absorbances at 663 nm. Various concentrations of cyanobacteria in 1-ml amounts were centrifuged ($1,100 \times g$, 1 min), the supernatant was discarded, and 5 ml of methanol was added. The pellet was resuspended by use of a Vortex mixer, and the suspension was incubated at 4°C for 10 min. After centrifugation, the supernatant was collected and the absorbance at 663 nm was determined in a Rye Unicam SP6-400 spectrophotometer. Heterocysts were isolated and purified from equal volumes of cyanobacterial suspensions by use of the ultrasonication technique of Fay and Lang (15), and standard curves were constructed to determine milligrams (dry weight) of heterocysts from the absorbance read at 663 nm. Growth of the cyanobacteria in an N-free medium ensured that a relatively constant heterocyst/total cell ratio was obtained. Isolated heterocysts were resuspended in 1 ml of distilled water, and the number of heterocysts was determined by use of a Petroff-Hausser counting chamber. One milli-

gram (dry weight) of heterocysts was found to contain $52 \times 10^6 \pm 9 \times 10^6$ individual heterocysts.

The activity of the bacterial inoculum was determined by adding 50 µl of each ^{14}C-labeled cell suspension in 5 ml of filter-count, and the disintegrations per minute were counted. The concentration of viable cells was determined by using a standard pour plate technique, whereas the total cell concentration was determined by using a Petroff-Hausser counting chamber. The activity of bacterial inocula ranged from 9,500 to 61,000 dpm/10^9 bacterial cells. Adhesion was expressed as bacteria attached per milligram (dry weight) of heterocysts.

(iii) Quantification of adhesion. The adhesion of some bacteria to solid surfaces can be expressed quantitatively in terms of adsorption isotherms (1, 7). The initial adhesion of bacteria to surfaces is considered to be a diffusion-controlled process requiring a certain period of time to reach equilibrium (9). Adsorption isotherms define the relationship between the adsorption capacity of a solid surface and the concentration of the microbial cells at equilibrium (26). Type I (Langmuin-type) isotherms exhibit a rapid rise in the numbers of attached cells with increasing cell concentration, up to a limiting value which indicates that adsorption is restricted to a monolayer. This isotherm can be expressed mathematically as: $q = CN/(C + K_d)$, where C is cell concentration at equilibrium (cells per milliliter), q is the number of cells adsorbed per unit weight of adsorbent material, N is the adsorption capacity of adsorbent material (binding sites per milligram of adsorbent), and K_d is the equilibrium constant (cells per milliliter).

One linearization of the type I adsorption isotherm is: $C/q = (C/N) + (K_d/N)$, whereby a plot of C/q versus C results in a straight line with X and Y intercepts of $-K_d$ and K_d/N, respectively.

The reciprocal of K_d is the affinity constant, K_a. Applebaum et al. (1) used the product of K_a and N to give an index of the adhesion of streptococci to saliva-coated hydroxyapatite beads, as it can be shown that $K_aN = q([1/C] + K_a)$, and thus, K_aN is a linear function of q and is independent of N, regardless of the range of cell concentrations used. The values of K_d and N can be determined by regression analysis using the method of least squares. The sigmoid type II isotherm represents multilayer adsorption to nonporous solids (26) with the infection point representing the formation of a complete monolayer.

Adsorption isotherms for the adhesion of *Pseudomonas* sp. SL10 and *Zoogloea* sp. SL20 were determined by placing a volume of cyanobacterial culture equal to 1 mg (dry weight) of heterocysts into 26-ml plastic centrifuge tubes and washing the cyanobacteria once with 5 ml of minimal medium without phosphate, before adding 1 ml of labeled bacterial suspension. The cultures were then transferred to 30-ml Pyrex test tubes and incubated for 24 h at 25°C. Fifty microliters of supernatant fluid was removed and counted to determine the number of free cells per milliliter at equilibrium (C). The cyanobacteria were washed three times in 5 ml of minimal medium and the adhesion to heterocysts was determined to give the number of cells bound at equilibrium (q).

pH and electrolyte concentration. The influence of pH on adhesion was determined by suspending bacteria and cyanobacteria in minimal medium with 0.02 M phosphate buffer, pH 6 to 8, and determining the number of bacteria attached. Bacteria and cyanobacteria were suspended in minimal medium with sodium chloride concentrations ranging from 0 to 0.8% (wt/vol).

Inhibition studies. The adhesion of bacteria to heterocysts was determined in the presence of 0.01 M EDTA and 1 mg of Tween 80 per liter.

To determine the effects of methylation of surface carboxyl groups on adhesion, bacterial and heterocyst surfaces were methylated by the procedure of Zvyaginstev and Guzev (44).

Sodium borate forms a complex with adjacent sugar hydroxyls which are in a *cis* configuration, such as galactose and mannose, but not with those in a *trans* configuration, such as glucose or xylose (13, 24). Bacteria and cyanobacteria were incubated in minimal medium containing 0.015 M sodium borate, and the adhesion to heterocysts was determined.

Lectins are plant proteins which demonstrate specific binding activity with certain sugar residues (41). The adhesion of bacteria to heterocysts was determined in the presence of 100-µg/ml concentrations of the following lectins: concanavalin A, α-D-mannose and α-D-glucose; garden pea agglutinin, α-D-mannose; peanut agglutinin, β-D-galactose and N-acetyl-α-D-galactosamine; and soybean agglutinin, N-acetyl-α,β-D-galactosamine.

TABLE 1. Measurement of specific bacterial adhesion to heterocysts of *Anabaena* species

Bacterium	*Anabaena* species	Available binding sites/mg (dry wt) of heterocysts (N)	Adhesion index (K_aN)
Pseudomonas sp. SL10	*A. cylindrica*	2.10×10^7	0.42
	A. azollae	7.87×10^7	0.24
Zoogloea sp. SL20	*A. cylindrica*	3.42×10^7	0.10
	A. azollae	3.51×10^7	0.09

TABLE 2. Effect of pH and sodium chloride concentration on the adhesion of *Pseudomonas* sp. SL10 and *Zoogloea* sp. SL20 to heterocysts of *A. cylindrica*

Variable	Bacteria attached ($\times 10^7$)/mg (dry wt) of heterocysts	
	Pseudomonas sp. SL20	*Zoogloea* sp. SL20
pH		
6.0	0.92 ± 0.16	5.38 ± 4.20
6.5	0.88 ± 0.06	11.05 ± 0.73
7.0	1.02 ± 0.16	14.60 ± 6.82
7.5	1.07 ± 0.22	13.18 ± 4.88
8.0	1.68 ± 0.43	10.05 ± 5.26
NaCl concentration (%, wt/vol)		
0.00	1.41 ± 0.19	4.46 ± 1.35
0.10	1.53 ± 0.46	7.28 ± 2.55
0.40	1.04 ± 0.34	6.30 ± 0.75
0.76	0.43 ± 0.23	3.66 ± 1.01

Periodate cleaves sugar residues between adjacent hydroxyls to produce formic acid and aldehyde (8, 43). Bacteria and cyanobacteria were treated with 0.01 M sodium periodate at 4°C in the dark for 20 min, after which time the cells were washed and adhesion was determined.

Bacteria and cyanobacteria were treated with trypsin (1 mg/ml, 37°C, 20 min) or pronase (1 mg/ml, 20°C, 20 min), and adhesion was measured.

RESULTS

The adhesion parameters for the attachment of *Pseudomonas* sp. SL10 and *Zoogloea* sp. SL20 to *Anabaena* heterocysts, calculated from adsorption isotherms, are significantly different (Table 1). There are more potential binding sites for *Pseudomonas* sp. SL10 on *A. azollae* heterocysts, but the affinity for these sites is lower than that for sites on *A. cylindrica*, as shown by the adhesion indices. There appears to be no difference between the number of binding sites or affinities for the adhesion of *Zoogloea* sp. SL20 cells to *A. cylindrica* and *A. azollae* heterocysts. *Pseudomonas* sp. SL10 has a greater affinity for its binding sites than does *Zoogloea* sp. SL20.

The adhesion of *Pseudomonas* sp. SL10 to *Anabaena* heterocysts appears to be enhanced at high pH (Table 2) but is inhibited by high sodium chloride concentrations (Table 2). Adhesion was not inhibited by the surfactant Tween 80 but was inhibited in the presence of EDTA (Table 3). Periodate treatment of heterocysts but not *Pseudomonas* cells significantly inhibited adhesion, whereas pronase treatment of *Pseudomonas* cells but not heterocysts decreased adhesion (Table 3). Trypsin treatment of both sur-

TABLE 3. Effect of treatments and inhibitors on the adhesion of *Pseudomonas* sp. SL10 to heterocysts of *A. cylindrica*

Treatment or inhibitor	Treated surface	Percentage of control adhesion
Treatment		
Trypsin	Bacterium	79 ± 13
	Heterocyst	142 ± 17
Pronase	Bacterium	35 ± 13
	Heterocyst	170 ± 86
Periodate	Bacterium	119 ± 13
	Heterocyst	28 ± 20
Methylation	Bacterium	23 ± 9
	Heterocyst	9 ± 7
Inhibitor		
EDTA		63 ± 20
Tween 80		123 ± 19
Borate		73 ± 45
Concanavalin A		50 ± 26
Peanut agglutinin		97 ± 25
Soybean agglutinin		78 ± 18
Garden pea agglutinin		76 ± 14

faces appeared to have no effect on adhesion nor did treatment with sodium borate.

The adhesion of *Zoogloea* sp. SL20 to *Anabaena* heterocysts was inhibited by periodate treatment of both the bacterial and heterocyst surfaces (Table 4). Both trypsin and pronase treatments of the bacterial cells inhibited adhesion (Table 4). Periodate and trypsin treatment of *Zoogloea* sp. SL20 removed the matrix of extracellular fibrils surrounding the cells (25). Adhesion of *Zoogloea* sp. SL20 was inhibited in the presence of sodium borate (Table 4).

DISCUSSION

There appears to be no significant difference in the adhesion of *Zoogloea* sp. SL20 to heterocysts of different *Anabaena* species between the number of potential binding sites and affinities as calculated from the reciprocal langmuir plots of adsorption isotherms. This suggests that the heterocyst adhesins involved in the adhesion of *Zoogloea* sp. SL20 are common to many *Anabaena* species and are evenly distributed on the surface of the heterocysts. *Pseudomonas* sp. SL10, however, does show significant differences in its adhesion to the heterocysts of different *Anabaena* species. This may reflect different adhesins between *Anabaena* species, as adhesion appears to reflect antigenic differences between cyanobacterial species. *A. azollae* is morphologically and antigenically distinct from *A. cylindrica* (17). Furthermore, adsorption isotherms suggest that not only does the number of binding sites vary between heterocysts of different *Anabaena* species but also the affinity of *Pseudomonas* sp. SL10 for these sites varies.

The difference in adhesion of *Pseudomonas* sp. SL10 to different heterocysts could also result from variations in the distribution and abundance of a common adhesin between heterocysts of different species, as the adhesion of *Pseudomonas* sp. SL10 to heterocysts always involves polar attachment, and treatments which inhibit adhesion to *A. cylindrica* heterocysts also inhibit adhesion to heterocysts of *A. azollae* (F. S. Lupton, Ph.D. thesis, University of New South Wales, Kensington, New South Wales, Australia, 1982).

It appears that the initial adsorption of *Pseudomonas* sp. SL10 to heterocysts involves ion triplet formation between carboxyl groups on both surfaces and divalent cations as: (i) adhesion is promoted at high pH when ionization of carboxyl groups is favored, (ii) adhesion is inhibited by the presence of monovalent cations, (iii) EDTA, a chelator of divalent cations, decreases adhesion, and (iv) methylation of both bacterial and heterocyst surfaces inhibits adhesion.

Although ion triplet formation appears to be involved in the initial adhesion process, it is readily reversible (9, 26). Nonviable cells could be readily desorbed from heterocysts by washing with distilled water and EDTA, whereas viable cells could not be desorbed after the initial adsorption as a result of ion triplet formation (Lupton, Ph.D. thesis, University of New South Wales).

The adhesin present on the bacterial surface appears to contain protein, as pronase treatment but not periodate treatment inhibits adhesion. The failure of trypsin to inhibit adhesion could be due to a lack of arginine or lysine residues at the active site of the adhesin or steric hindrance by other cell wall constituents. The heterocyst adhesin appears to contain carbohydrate, as periodate oxidation but neither trypsin nor pro-

TABLE 4. Effect of treatments and inhibitors on the adhesion of *Zoogloea* sp. SL20 to heterocysts of *A. cylindrica*

Treatment or inhibitor	Treated surface	Percentage of control adhesion
Treatment		
Trypsin	Bacterium	32 ± 12
	Heterocyst	80 ± 15
Pronase	Bacterium	15 ± 4
	Heterocyst	70 ± 23
Periodate	Bacterium	58 ± 15
	Heterocyst	8 ± 4
Methylation	Bacterium	—
	Heterocyst	76 ± 6
Inhibitor		
EDTA		57 ± 9
Tween 80		30 ± 22
Borate		41 ± 19
Concanavalin A		122 ± 27

nase treatments inhibited adhesion. The inhibition of adhesion by concanavalin A suggests that either terminal glucosyl or mannosyl residues are involved, but since neither garden pea agglutinin nor sodium borate treatments inhibited adhesion, glucose is likely to be the active residue of heterocyst polysaccharide. It may be significant that glucose has been demonstrated to be more common as a terminal sugar residue of heterocyst polysaccharides compared to polysaccharides of vegetative cell sheaths (5, 6).

The specific adhesion of *Zoogloea* sp. SL20 involves an adhesion associated with extracellular bacterial fibrils. These fibrils can be removed by both periodate oxidation and trypsin treatments, suggesting that they are composed of both protein and carbohydrate. The heterocyst adhesin appears to contain carbohydrate but not protein.

The different processes by which *Pseudomonas* sp. SL10 and *Zoogloea* sp. SL20 attach specifically to heterocysts is reflected in the modes of adhesion observed for the two bacteria. *Pseudomonas* sp. SL10 always attaches to the envelope in a perpendicular manner with the polar end of the bacterium in contact with the fibrous envelope of the heterocyst. A bulge is often observed in the polar end of the bacterial cell wall at the point of contact with the heterocyst envelope. The perpendicular mode of adhesion shown by *Pseudomonas* sp. SL10 appears similar to that of other bacteria which attach specifically to certain surfaces, such as *Rhizobium trifolii* adsorbing to white clover root hairs and other examples of specific bacterial adhesion to plant surfaces (10). A low radius of curvature reduces the potential energy barrier between the two surfaces and promotes short-range stereochemical interactions due to the relatively closer contact of the two surfaces (9). Furthermore, the adhesion of *Pseudomonas* sp. SL10 to heterocysts is similar to that of *R. trifolii* to root hairs in that it involves two phases (10). However, whereas the adhesion of *R. trifolii* to root hairs involves a reversible but specific phase mediated by lectin-specific sugar interactions followed by a nonspecific but irreversible adhesion mediated by cellulose-like extracellular fibrils, the adhesion of *Pseudomonas* sp. SL10 to heterocysts involves a reversible and nonspecific interaction between carboxyl groups on both surfaces and divalent cations. This precedes a specific and irreversible adhesion between proteinaceous adhesins associated with the polar end of the bacterial cell wall and polysaccharide-containing adhesins associated with the outer layer of the heterocyst envelope. This mechanism of adhesion appears similar to that of *P. tolaasii* to barley roots, which appears to be dependent on the presence of divalent

cations but not monovalent cations and involves an initial reversible nonspecific phase of adhesion followed by an irreversible specific phase (10). Similar mechanisms of adhesion may be found between bacteria and surfaces in different environments. A microorganism may be able to attach specifically to a particular surface in a given habitat in the absence of certain other bacteria, which may also possess the physicochemical properties to adsorb and attach to the surface but which are not naturally found in that habitat as a result of environmental influences. Such may be the case between *Pseudomonas* sp. SL10 attaching to heterocysts and *P. tolaasii* attaching to barley root hairs. It is very probable, therefore, that ecological factors as well as physicochemical factors influence the specificity of adhesion.

The process of specific adhesion between *Zoogloea* sp. SL20 cells and heterocysts is quite unlike that of *Pseudomonas* sp. SL10 and involves interactions between protein associated with bacterial fibrils and heterocyst envelope polysaccharide or between bacterial and heterocyst polysaccharides. This interaction appears not to involve electrostatic interactions between charged groups or hydrophobic forces. Although similar mechanisms of adhesion may occur between bacteria and surfaces in different environments, if two bacteria are to attach to the same surface with a minimum of competition, then different mechanisms of adhesion must be employed, as appears to be the case for the adhesion of *Pseudomonas* sp. SL10 and *Zoogloea* sp. SL20 to heterocysts.

There are many areas in which this study could be extended in future investigations. The proteinaceous adhesin from the cell wall of *Pseudomonas* sp. SL10 could be isolated with the use of nonionic surfactants or sonication to disrupt the bacterial cell wall and release the adhesin in an active form. The adhesin may be purified by using column chromatography with bound heterocyst envelope polysaccharides. Once isolated, the binding of the adhesin to terminal glucosyl residues could be confirmed and the localization of the bacterial adhesin at the polar end of the bacterial cell wall could then be investigated by using ferritin or peroxidase labeled antibodies, prepared against the bacterial adhesin. It may be possible to determine the nature of macromolecules associated with the heterocyst surface which apparently contain free carboxyl groups. The heterocyst envelope could be disintegrated by use of physical techniques such as sonication, and ion-exchange chromatography could be used to separate the anionic constituents. Specific interactions between polysaccharides from *Zoogloea* sp. SL20 and heterocyst envelope polysaccharides could be investigated by using the rheological, chiroptical, and nuclear magnetic resonance techniques of Dea et al. (11, 12). The degree of bacterial adhesion obtained under laboratory conditions appears much reduced compared to that reported by Paerl (31, 32) for natural blooms in freshwater lakes. There is a need to investigate the environmental conditions occurring during cyanobacterial blooms to determine how they influence bacterial adhesion. Finally, the adhesion of *Pseudomonas* sp. SL10 and *Zoogloea* sp. SL20 to other heterocystous cyanobacteria could be investigated to determine the range of intraspecies specificity. There is a need to determine how widespread the phenomenon of specific bacterial adhesion to heterocysts is in natural environments, including marine and hypersaline environments where heterocystous cyanobacteria occur in significant numbers, and to show that adhesion has ecological significance in these environments.

This study was supported by a grant from the Australian Research Grants Committee.
We thank B. Kefford for helpful discussions.

LITERATURE CITED

1. **Applebaum, B., E. Bolub, S. C. Holt, and B. Rosan.** 1979. In vitro studies of dental plaque formation: adsorption of oral streptococci to hydroxyapatite. Infect. Immun. 25:717–728.
2. **Beachey, E. H.** 1980. Bacterial adherence, vol. 6, series B, Receptors and recognition, Chapman and Hall, London.
3. **Berkeley, R. C. W., J. M. Lynch, J. Melling, P. R. Rutler, and B. Vincent.** 1980. Microbial adhesion to surfaces. Ellis Horwood, Chichester, England.
4. **Bitton, G., and K. C. Marshall.** 1980. Adsorption of microorganisms to surfaces. Wiley-Interscience, New York.
5. **Cardemil, L., and C. P. Wolk.** 1976. The polysaccharides from heterocyst and spore envelopes of a blue-green alga: methylation analysis and structure of the backbones. J. Biol. Chem. 251:2967–2975.
6. **Cardemil, L., and C. P. Wolk.** 1979. The polysaccharides from heterocyst and spore envelopes of a blue-green alga: structure of basic repeating unit. J. Biol. Chem. 254:736–741.
7. **Clark, W. B., L. L. Bammann, and R. J. Gibbons.** 1977. Comparative estimates of bacterial affinities and adsorption sites of hydroxyapatite surfaces. Infect. Immun. 19:846–853.
8. **Crandall, M. A., and T. D. Brock.** 1968. Molecular basis of mating in the yeast, *Hansenula wingei*. Bacteriol. Rev. 32:139–163.
9. **Daniels, S. L.** 1980. Mechanisms involved in sorption of microorganisms to solid surfaces, p. 7–58. *In* G. Bitton and K. C. Marshall (ed.), Adsorption of microorganisms to surfaces. John Wiley & Sons, Inc., New York.
10. **Dazzo, F.** 1980. Adsorption of microorganisms to roots and other plant surfaces, p. 253–316. *In* G. Bitton and K. C. Marshall (ed.), Adsorption of microorganisms to surfaces. John Wiley & Sons, Inc., New York.
11. **Dea, I. C. M., and E. R. Morris.** 1977. Synergistic xanthan gels, p. 174–182. *In* P. A. Sandford and A. Laskin (ed.), Extracellular microbial polysaccharides. American Chemical Society, Washington, D.C.
12. **Dea, I. C. M., E. R. Morris, D. A. Rees, E. J. Welsh, H. A. Barnes, and J. Price.** 1977. Associations of like and unlike polysaccharides: mechanism and specificity in ga-

lactomannans interacting bacterial polysaccharides and related systems. Carbohydr. Res. 57:249–272.
13. **DeVries, A. I.** 1971. Glycoproteins as biological antifreeze agents in Antarctic fishes. Science 172:1152–1155.
14. **Duguid, J. P., and D. C. Old.** 1980. Adhesive properties of the *Enterobacteriaceae*, p. 185–217. *In* E. H. Beachey (ed.), Bacterial adherence. Chapman and Hall, London.
15. **Fay, P., and N. J. Lang.** 1971. The heterocysts of blue-green algae. I. Ultrastructural integrity after isolation. Proc. R. Soc. London Ser. B 178:185–192.
16. **Fuller, R., and B. E. Brooker.** 1974. Lactobacilli which attach to the crop epithelium of the fowl. Am. J. Clin. Nutr. 27:1305–1312.
17. **Gates, J. E., R. W. Fisher, T. W. Goggin, and N. I. Azrolan.** 1980. Antigenic differences between *Anabaena azollae* fresh from the *Azolla* fern cavity and free living cyanobacteria. Arch. Microbiol. 128:126–129.
18. **Geitler, L.** 1932. Cyanophyceae, p. 1196. *In* L. Rabenhorst (ed.), Rabenhorsts Kryptogamenflora, vol. 14. Akademische Verlagsgesellschaft, Leipzig.
19. **Gibbons, R. J., E. C. Moreno, and D. M. Spinell.** 1976. Model delineating the effects of a salivary pellicle on the adsorption of *Streptococcus miteor* onto hydroxyapatite. Infect. Immun. 14:1109–1112.
20. **Gibbons, R. J., and J. Van Houte.** 1980. Bacterial adherence and the formation of dental plaques, p. 61–104. *In* E. H. Beachey (ed.), Bacterial adherence. Chapman and Hall, London.
21. **Gorham, P. R., J. McLachlan, J. Hammer, and W. K. Kim.** 1964. Isolation and culture of toxic strains of *Anabaena flos-aquae* (lyngb) de Beib. Verh. Int. Ver. Limnol. 15:796–804.
22. **Jones, G. W.** 1977. The attachment of bacteria to the surfaces of animal cells, p. 139–176. *In* J. L. Ressing (ed.), Microbial interactions, vol. 3, Receptors and recognition, series B. Chapman and Hall, London.
23. **Lippincott, J. A., and B. Lippincott.** 1980. Microbial adherence in plants, p. 375–398. *In* E. H. Beachey (ed.), Bacterial adherence. Chapman and Hall, London.
24. **Little, C. A.** 1951. Borate and substances of biological interest. Adv. Enzymol. 12:493–527.
25. **Lupton, F. S., and K. C. Marshall.** 1981. Specific adhesion of bacteria to heterocysts of *Anabaena* spp. and its ecological significance. Appl. Environ. Microbiol. 42:1085–1092.
26. **Marshall, K. C.** 1976. Interfaces in microbial ecology. Harvard University Press, Cambridge, Mass.
27. **Marshall, K. C., R. Stout, and R. Mitchell.** 1971. Mechanisms of the initial events in the sorption of marine bacteria to surfaces. J. Gen. Microbiol. 68:337–348.
28. **Morris, E. R., D. A. Rees, G. Young, M. D. Walkinshaw, and A. Drake.** 1977. Order-disorder transition for a bacterial polysaccharide in solution. A role for polysaccharide conformation in recognition between *Xanthomonas* pathogen and its plant host. J. Mol. Biol. 110:1–16.
29. **Newman, H. N.** 1980. Retention of bacteria on oral surfaces, p. 207–251. *In* G. Bitton and K. C. Marshall (ed.), Adsorption of microorganisms to surfaces. Wiley-Interscience, New York.
30. **Ofek, I., E. H. Beachey, and N. Sharon.** 1978. Surface sugars of animal cells as determinants of recognition in bacterial adherence. Trends Biochem. Sci. 3:159–160.
31. **Paerl, H. W.** 1976. Specific associations of the blue-green algae *Anabaena* and *Aphanizomenon* with bacteria in freshwater blooms. J. Phycol. 12:431–435.
32. **Paerl, H. W.** 1978. Role of heterotrophic bacteria promoting N_2 fixation by *Anabaena* in aquatic habitats. Microb. Ecol. 4:215–231.
33. **Paerl, H. W., and P. E. Kellar.** 1978. Optimization of N_2-fixation in O_2-rich waters, p. 68–75. *In* M. W. Loutit and J. A. R. Miles (ed.), Microbial ecology. Springer-Verlag, Berlin.
34. **Paerl, H. W., and P. E. Kellar.** 1978. Significance of bacterial *Anabaena* (Cyanophyceae) associations with

respect to N$_2$-fixation in freshwater. J. Phycol. **14**:254–260.

35. **Pethica, B. A.** 1980. Microbial and cell adhesion, p. 19–45. *In* R. C. W. Berkeley, M. J. Lynch, J. Melling, P. R. Rutler, and B. Vincent (ed.), Microbial adhesions to surfaces. Ellis Horwood, Chichester, England.

36. **Savage, D. C.** 1977. Microbial ecology of the gastrointestinal tract. Annu. Rev. Microbiol. **31**:107–133.

37. **Savage, D. C.** 1979. Introduction to mechanisms of association of indigenous microbes. Am. J. Clin. Nutr. **32**:113–118.

38. **Savage, D. C.** 1980. Colonization by and survival of pathogenic bacteria on intestinal mucosal surfaces, p. 175–206. *In* G. Bitton and K. C. Marshall (ed.), Adsorption of microorganisms to surfaces. Wiley-Interscience, New York.

39. **Savage, D. C.** 1980. Adherence of normal flora to mucosal surfaces, p. 31–59. *In* E. H. Beachey (ed.), Bacterial

adherence, vol. 6, series B, Receptors and recognition. Chapman and Hall, London.

40. **Schwabe, G. H., and R. Mollenhauer.** 1967. Uber den Einfluss der Begleitbakterien auf das Lagerbild von *Nostoc sphaericum*. Nova Hedwigia **13**:77–80.

41. **Sharon, N., and H. Lis.** 1972. Lectins: cell-agglutinating and sugar-specific proteins. Science **177**:949–959.

42. **Suegara, N., M. Morotomi, T. Watanabe, Y. Kawai, and M. Mutai.** 1975. Behavior of microflora in the rat stomach: adhesion of lactobacillii to the keratinized epithelial cells of the rat stomach in vitro. Infect. Immun. **12**:173–179.

43. **White, A., P. Handler, and E. L. Smith.** 1973. Principles of biochemistry, 5th ed. McGraw-Hill, Kogakusha, Tokyo.

44. **Zvyaginstev, D. G., and U. S. Guzev.** 1971. Concentration and separation of bacteria on dowex anionite. Microbiology (USSR) **40**:123–126.

Effects of Interfaces on Survival Mechanisms of Copiotrophic Bacteria in Low-Nutrient Habitats

STAFFAN KJELLEBERG

Department of Marine Microbiology, University of Göteborg, S-413 19 Göteborg, Sweden

Starvation survival is important in microbial ecology because it is a mechanism that perpetuates the species. The genome is able to express itself once the environmental conditions become favorable (2, 54). The existence of different nutritional types of bacteria in overall oligotrophic systems has long been assumed from results based on physiological studies on model bacteria (e.g., 17, 26, 28, 32, 47, 48, 62, 64). These data should be compared with results from studies of natural ecosystems. I will first introduce some reported data which strongly indicate the existence of bacteria with widely different substrate affinity and uptake systems. These findings, in combination with concepts of different concentration regimes, the chemical activity of a substrate, and the biological activity (liability to biological degradation of the substrate), form the logical justification for both studies on oligo- and copiotrophic bacteria in natural systems and studies on the significance of the effects of surfaces.

This paper provides an introduction to present studies on the concept of starvation survival in ecology and the diversity in the microbial responses to nutrient limitation.

NUTRITIONAL GROUPS OF BACTERIA

Isolations and definitions. The levels of organic material in seawater are low, ranging from an average of 1.0 mg of carbon per liter in surface waters to 0.5 mg of carbon per liter of deep seawater (53). In fact many natural environments appear to be oligotrophic in character and subject to rapid changes in nutritional conditions. These are thus best described in terms of flux of nutrients (63). Complex natural situations do not select for a steady state (40). In 1968, Jannasch (38) obtained data on growth characteristics of heterotrophic marine bacteria which suggest that part of the population is adapted to growth, and able to compete successfully with species growing less efficiently, under low substrate concentrations. Organisms of the latter group may occur either as dormant stages (67) or associated with particles supplying nutrient concentrations higher than those in the surrounding seawater (4). Subsequent reports on successful isolations of bacteria with widely divergent nutritional requirements have appeared frequently.

Carlucci and Shimp (11, 18, 19) isolated a marine bacterium capable of growing, but not showing turbidity in the growth medium, at extremely low concentrations of dissolved organic matter. Several obligate, as well as facultative, oligotrophs have now been characterized (e.g., 35–37, 71). Two recent surveys on a number of marine isolates have also demonstrated survival strategies typical of copiotrophic bacteria (2, 34). It seems valid to distinguish between those bacteria, the oligotrophs, that are adapted to grow at low substrate concentrations (32) without responding to high nutrient zones and the copiotrophic bacteria, which require relatively high nutrient levels for growth (63).

Very briefly, an ideal oligotroph would be a small sphere or rod, possibly equipped with an appendage, that displays a high affinity for uptake of many different substrates. Its transport system has a low specificity and a high substrate affinity, transport being directed toward the formation of reserve material. This implies a low growth rate. A detailed description of these bacteria is presented by Poindexter (62). Copiotrophic bacteria, as described in more detail below, upon starvation turn into small spheroids which exhibit properties that enable them to survive in periods measured in weeks and months rather than days (15, 50).

Distribution. The division into these groups is naturally not as clear as this presentation might imply. Carlucci and Shimp (11), by sampling at various near- and offshore locations, showed the existence of at least obligate, facultative, and copiotrophic bacteria in all samples. Recently, Azam and Hodson (4) presented results on multiphasic kinetics for glucose uptake by marine bacteria that were largely caused by the presence of bacteria with very low (nanomolar) to very high (millimolar) constants expressing affinity.

Furthermore, growth of free-living marine bacteria (10, 24, 39), as well as division and the formation of smaller cells rather than growth (57–59), has been demonstrated in transfer of isolates or seawater mixed cultures into unsupplemented sterile seawater. The growth of seawater cultures could reflect the bacterial consumption of dissolved organic matter in a manner analogous to that of natural populations of marine bacteria in seawater. In water taken from near shore, rod-shaped cells and not cocci were responsible for growth after transfer into unsupplemented seawater (J. W. Ammerman, J. A. Fuhrman, A. Hagstöm, and F. Azam, submitted for publication), even though very small free-living bacteria are most commonly found in marine waters (14, 20, 33, 49, 52, 70, 78). Nutritional factors are, however, well known to convert rods to spheres and vice versa (23, 45, 57).

The small free-living cells were suggested by Wright (75, 76) to be dormant and were shown to change in size with both enrichment and dilution of nutrients. The fact that the sea is not dominated by large rod-shaped bacteria could depend on predation by bacteriovores, predominantly heterotrophic flagellates that feed more efficiently on particles >2 μm than on smaller ones (16, 17, 73, 74, 77). Bacterial populations in coastal and offshore water, being in a state of dormancy or of an oligotrophic nature, would not sustain a grazing microzooplankton fauna, since a threshold concentration of bacteria for grazing is required. Predation might thus control the size distribution of bacteria in the sea, whereas nutrient availability limits the population density (1). It was suggested by Ishida et al. (34) that obligate oligotrophs have planktonic properties and that facultative oligotrophs tend to attach on particles in water. Wilson and Stevenson (74) presented evidence that epibacteria are easily cultured, saprophytic bacteria being markedly influenced by short-term phenomena, whereas planktobacteria are more uniform in their distribution and are not markedly influenced by short-term events such as nutrient variability. These results support the concept of more nutritionally demanding bacteria at microzones and interfaces.

SUBSTRATE CONCENTRATION REGIMES

On the basis of quantitative results, three different concentration regimes may be considered (4): low concentrations in the bulk phase, episodic high-concentration microzones, and sustained high-concentration microzones. The response of bacteria to sites of nutrient enrichment and thus the importance of surfaces exhibiting conditioning films in microbial ecology have always been emphasized. Hard data are, however, rarely presented in this context.

There are two approaches that convincingly argue for the need to carefully examine effects of surfaces in studies of starvation survival. Several investigators (referred to in reference 4) have found discrepancies between the natural concentrations (Sn) of amino acids and glucose obtained by chemical methods and the sum of the transport constant K_t of a compound and Sn

determined by microbiological analysis. The chemical determination of Sn gives a higher value than the parameter $K_t + Sn$ given by uptake kinetic measurements of natural marine bacteria. Part of the substrate could therefore be either adsorbed and not equally available for bacterial uptake or utilized within enriched microzones by bacteria with low substrate affinities (copiotrophic in character). Bacteria with high K_t and high maximum uptake velocity might accumulate in such microzones. A large part of dissolved organic matter would thus be utilized in microzones and does not diffuse into the bulk. Tracer techniques would therefore also underestimate substrate uptake rates should the tracer specific activity be the same in the microzone or at the surface as in the bulk phase.

This leads to a second point of importance. Dissolved organic matter consists of a wide range of molecules exhibiting a broad range of surface-active characteristics. Lipids, for example, are normal constituents in the environment, found in high concentrations in surface microlayers (56), phytoplankton production zones (8), and sediments (61). Long-chain fatty acids, which are a very suitable substrate when immobilized at surfaces, appear toxic when free in the bulk phase. Configuration rearrangements of molecules allow for different enzyme activities, some of which are significant only should the substrate be surface localized (22, 72). It is likely that microbial degradation of dissolved organic matter occurs at any point in a system, the organisms obtaining carbon and energy from different types of molecules. One should therefore distinguish between chemical and biological activity. A leucine molecule will have different chemical activities in the bulk phase than in an interface (7; M. Hermansson and B. Dahlbäck, submitted for publication) as a consequence of its hydrophobic character. The number of collisions between the substrate and the organism is reduced at surfaces, and this lowers the chemical activity of the substance. Because of substrate rearrangements at the interface, however, the biological activity might be increased.

EFFECTS OF STARVATION ON COPIOTROPHIC BACTERIA

For the reasons given above, the behavior of copiotrophic bacteria, a normal constituent in natural habitats, has been studied with emphasis given to interfacial phenomena. The general behavior of copiotrophic bacteria exposed to starvation conditions in the environment must first be presented. The research performed in this area is concerned with three aspects: demonstration of the effects of starvation, determination of the physiological mechanisms for starvation-induced responses, and consideration of the relevance of starvation in population dynamics and ecology. I apologize to many of my colleagues for not covering the rather extensive literature on physiologically related experiments and results in nonmarine bacteria.

The rapid changes in morphology, cell volume, and numbers reported here seem to be general phenomena in a wide range of microbial habitats, essentially all those with an intermittent supply of nutrients. They are frequently observed not only in aquatic systems but also in soil (12), the gut (P. Conway, personal communication), and the oral cavity (J. Olsson, personal communication), for example.

Short-term starvation. Oligotrophic and copiotrophic bacteria exhibit different survival mechanisms, although some responses to nutrient limitation appear to be similar. Most of the information available concerning these responses deals with the behavior of copiotrophic bacteria. Upon starvation these bacteria undergo a sequence of processes. It is feasible to distinguish between changes that occur under a short-term exposure and effects from long-term starvation. A rapid active reorganization within a few hours leads to the production of high numbers of small, metabolically competent cells ("dwarfs") (34, 45). These small cells are the result of fragmentation, which can be defined as division without growth (58), and continuous size reduction of fragmented cells (S. Kjelleberg and M. Hermansson, submitted for publication). With time, a drop in cell volume corresponded to a relatively constant number of cells, which reveals a continuous size reduction (Fig. 1). Continuous size reduction was recently proved to be responsible for a significant proportion of the total drop in cell volume as seen by determining the biovolume (the average cell volume multiplied by the number of bacteria) of a population at various times of starvation (46). There is normally a change from rods to coccobacilli under this process. The endogenous respiration drops, possibly to maintain intracellular material. The initial large increase in the number of viable cells, due to fragmentation, is followed by a decline, usually until a constant level results (2, 59). A psychrophilic marine bacterium (54) showed behavior similar to that of two mesophilic bacteria (45, 46), all studied in some detail, although the initial short-term starvation phase lasted for a longer time period. In addition, Dawson et al. (15) reported increased adhesive properties and morphology changes involving an apparent reduction in the amount of mucopeptide in the wall without a corresponding reduction in the outer membrane.

These drastic changes that occur in a few hours after transfer of a log-phase culture into starvation medium indicate that a merely passive response would not fully account for the dwarfing process, e.g., the rapid loss of the cell wall. Subsequent studies (34, 46) have shown that the initial starvation response is an active process because: the rate of dwarfing was reduced with increased age of the bacterial cell; dwarfing was reversibly inhibited by low temperature, low pH, and inhibitors which disrupt the electron-transport chain and the proton flow at the cytoplasmic membrane; and inhibitors of protein synthesis had no effect, which implies that the enzymes for the dwarfing process are constitutive and that the cells are able to respond to the rapid changes that occur in the environment.

Metabolic capability. Physiological responses upon starvation seem to relate to the utilization or production of ATP. The ATP level per viable cell initially decreased but thereafter in various patterns showed a steady increase (3, 50) or remained at constant levels (55; S. Kjelleberg and B. Dahlbäck, in preparation). Increased levels of ATP per viable and respiring cell most probably depended on loss in viability of the population and therefore was a consequence of cryptic growthlike effects (64). Protein is known to be metabolized during starvation (3, 27, 42), and it shows a pattern of rapid decrease in levels with starvation time similar to that of RNA and DNA. Whereas the protein and DNA values stabilized per viable cell, the RNA content subsequently increased in *Vibrio* sp. Ant-300, a psychrophilic marine organism (3). Koch (47, 48) stated that cells in a carbon-limited state of slow and balanced growth have elevated levels

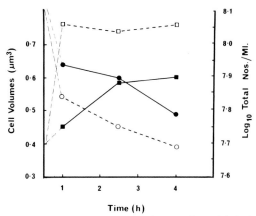

FIG. 1. Short-term changes in cell volume (circles) and cell numbers (squares) upon starvation of *Vibrio* sp. DW1 (closed symbols) and *Pseudomonas* sp. S9 (open symbols). These changes reveal the processes of fragmentation and continuous size reduction. Time is measured from commencement of starvation. The results of a typical experiment are given. After Kjelleberg et al. (46).

of mRNA, tRNA, amino acids, and RNA polymerase. Starving or slow-growing cells divert their energy toward synthesis of the macromolecules that are necessary for protein synthesis.

The high levels of ATP in starving cells reinforce the concept that these cells are physiologically capable of metabolism if a substrate becomes available (54). The ATP is necessary for initial active transport. Moreover, chemotaxis is observed in the early stages of starvation survival of *Vibrio* sp. Ant-300. This organism furthermore binds four amino acids tested to a significantly greater extent after 30 days of starvation than in a nonstarved state (54).

The ability of starved cells to respond to changes in the environment is well documented. Although these cells exhibit low endogenous respiration, they show an instantaneous response to exogenous energy substrates (2, 3, 45, 54). This rapid response, which is more pronounced for starved than nonstarved cells, indicates that the concept of flux of nutrients and intermittent inputs is valid for the bulk phase in natural systems. High levels of nutrient inputs occur from time to time in oligotrophic habitats.

EFFECTS OF INTERFACES ON STARVATION SURVIVAL

It was previously known that the metabolic activity of attached bacteria is different from that of bacteria in the liquid phase (21, 25, 27, 29–31, 44). Carbon limitation appears to promote adhesion to surfaces (9) and better growth of attached bacteria than of those in the liquid phase (41). The presence of interfaces appears also to have a strong influence on starvation survival responses.

A hydrophilic marine vibrio produced cells that decreased in cell volume faster with time at an interface than in the bulk liquid during starvation (45). The difference in size between bacteria at the surface and in the bulk lasted during the entire starvation period of the experiment (22 h). The interface thus triggered the starvation survival process to go faster. Measurements of the endogenous metabolism of these cells revealed the active process involved in producing small cells during starvation survival. The oxygen uptake measured during the initial starvation phase is twice as high per hydrophilic cell at the interface compared to a hydrophobic bacterium (46). Furthermore, K. C. Marshall and B. A. Humphrey (in preparation), using microcalorimetric techniques, found that the hydrophilic marine vibrio also exhibited a significantly higher heat output per attached cell than per free-living cell when low levels of glutamate were used as substrate. Hydrophilic bacteria starve at a faster rate at nutrient-limited interfaces,

whereas relatively hydrophobic bacteria diminish in size more rapidly in the bulk than at the interface (34).

The question arose whether more efficient scavenging and better survival occur for hydrophobic bacteria because they get closer to the surface and capture whatever contaminating nutrients might be available. *Vibrio* sp. DW1, a marine strain, showed very efficient uptake and regrowth of starved cells at interfaces (45). Cells starved for 18 h, producing a smaller cell volume at the air–water interface compared to the bulk phase, were supplied with nutrients at the interface. A tryptone-yeast extract mixture that was held up at the interface with the help of half a monolayer of a lipid film was supplied. The concentration of nutrients was per unit volume, too low to permit growth in the bulk liquid. The surface-associated bacteria showed a large increase in size, and as nutrients were used up the starvation process started again (Fig. 2). Such efficient scavenging of nutrients at interfaces was seen also by following the behavior of bacteria at the solid–water interface of a dialysis microchamber (45).

In this technique, bacteria are trapped between a cover slip and a dialysis membrane below which the diluent flows. The bacteria on the membrane have the advantage of nutrient accumulation at a surface. Any contaminating

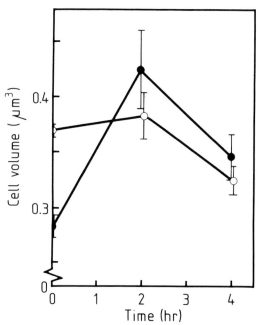

FIG. 2. Changes with time in cell volume of 18-h small starved cells of a *Vibrio* sp. DW1 at the air–water interface conditioned with nutrients (●) and in the aqueous subphase (○). After Kjelleberg et al. (45).

nutrient is likely to be adsorbed to the membrane and used by the bacteria. Many differences were observed between the behavior of bacteria on the dialysis membrane and those in bulk liquid in the absence of added nutrient. Motility of *Vibrio* sp. DW1 persisted for 2 days, and division and size reduction were still occurring at 3 days, some 70 h after such activity had ceased in the bulk phase.

Furthermore, as previously noted in the bulk liquid conditions, the small starved cells rapidly responded to nutrients, produced flagella, and moved away from the interface. A conventional marine growth medium at 1/1,000 strength was added to the interface, which per unit volume did not give a growth response in the bulk liquid. Higher levels of ATP per biovolume have been recorded for three marine strains at interfaces than in the bulk phase during both short- and long-term starvation (Kjelleberg and Dahlbäck, in preparation). This corresponds to the immediate response of starved cells to added nutrients at interfaces.

Cell surface morphology and degree of adhesion. Efficient scavenging of surface-localized nutrients was thus established and appeared to give a pronounced advantage to more hydrophobic bacteria. There is now substantial evidence for the relation between low nutrient conditions and the degree of adhesion. Already in 1971,

Marshall et al. (52) found small dwarflike bacteria at interfaces which were suggested to be primary colonizers. Adhesion was also suggested by Dawson et al. (15) to represent a tactic in the survival strategy.

How does starvation survival influence the bacterial surface; i.e., what are the prerequisites for increased adhesion? Let us first consider apparent changes in the outer layer. Figure 3 shows an example of the drastic morphological changes upon exposure to 5 h of a starvation regime for a marine vibrio. The drastic reduction in cell volume induces folding and filament production by surface-active components in the outer layer (15). This was reflected as an increase in surface roughness in scanning electron microscopy preparations for a number of marine strains tested (Kjelleberg and Hermansson, submitted for publication). *Vibrio* sp. DW1 seemingly sticks to the substratum by virtue of the fibrillar formations (Fig. 3).

Secondly, upon starvation, some strains can display an increased liability to irreversible binding (15; Kjelleberg and Hermansson, submitted for publication). The number of irreversibly adhered bacteria increased with length of exposure to a starvation regime for five of seven strains of marine bacteria tested. Measurements of bacterial surface characteristics under such conditions revealed a clear correlation between de-

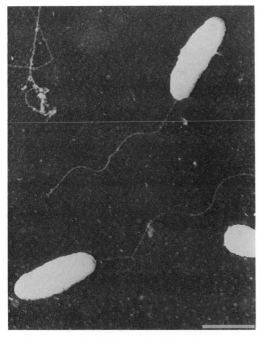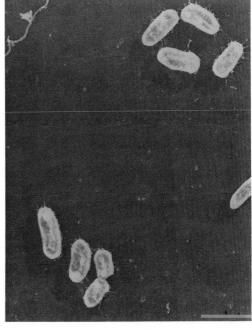

FIG. 3. Starvation-induced effects on *Vibrio* sp. DW1, as seen by transmission electron micrographs of critical-point-dried specimens. (a) Log-phase bacteria washed in artificial seawater. (b) Bacteria starved for 5 h in artificial seawater. Bar = 1.0 μm. After Dawson et al. (15).

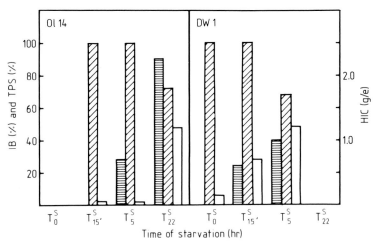

FIG. 4. Degree of irreversible binding (IB; ▤), two-phase separation in a hexadecane–water system (TPS; ▨) expressed as the percentage of cells that remain in the aqueous phase (100% denotes hydrophilic bacteria with this method), and hydrophobic interaction chromatography (HIC) results expressed as the ratio gel/eluate (the higher the value of gel/eluate, the higher is the tendency to hydrophobic interaction) for cells of *Spirillum* sp. O114 and *Vibrio* sp. DW1 examined at various times during exposure to a starvation regime. T_0^S designates start of the starvation process. After Kjelleberg and Hermansson (submitted for publication).

gree of hydrophobicity, as measured in hydrophobic interaction chromatography and two-phase separation in a hexadecane-water system, and the degree of irreversible binding. Figure 4 shows these effects for marine strains of *Spirillum* and *Vibrio*. Charge effects as measured by electrostatic interaction chromatography were not obviously related to the increase in irreversible binding (Kjelleberg and Hermansson, submitted for publication).

Cell surface characteristics and scavenging of surface-localized nutrients. Would a high degree of hydrophobic character lead to a survival advantage at surfaces?

Bacteria bind at surfaces as a result of both long- and short-range forces. The strength and balance of these forces result in a reversible or an irreversible binding. The short-range forces consist mainly of dipole–dipole interactions, chemical bonding, and what has been designated as hydrophobic interaction (68). The last is, however, not a force that depends on the interaction between the bodies (5, 6). The concept of hydrophobicity reflects properties of the diluent, the aqueous phase in which bacteria are suspended. All available experiments indicate that hydrophobic interaction, defined as the decrease in free energy associated with bringing two bodies into contact, depends to a large extent on the aqueous phase itself, possibly as a result of the strong cohesive forces between water molecules or the unique structure of water (5, 69). A thermodynamic model presented by Absolom et al. (1) relates bacterial adhesion to the interfacial tensions between all three interacting compo-

nents in a system: the cell surface, the solid substratum, and the liquid phase. The model predicts, for example, a minimum attachment of a surface tension value of the aqueous phase that is similar to the surface free energy of the bacteria. Furthermore, a higher surface tension of the aqueous phase than of the bacterial surface, normally found in natural waters, leads to a higher degree of adhesion to hydrophobic than to hydrophilic surfaces. With respect to the bacterial surface, a low surface free energy or hydrophobic character simply indicates a low degree of polarity.

The question at the beginning of this section could be answered as follows. The past few years have resulted in the following pieces of information:

(i) Hydrophobicity has often been shown to be a major physicochemical parameter for adhesion of bacteria in a wide range of systems (13, 51, 60, 65, 73). These include air–water, solid–water, and oil–water interfaces in marine systems as well as various interfaces in higher organisms. A recent paper by Pringle and Fletcher (65), showing the preference of freshwater isolates for hydrophobic surfaces, accentuates how important changes in bacterial surface characteristics are before and during attachment. Such alterations strongly influence the degree and strength of adhesion. The methods used to measure hydrophobicity detect the availability of quantitatively and qualitatively different residues on the bacterial surface. It is not clear whether these influence the bacteria–substrata interaction in identical ways. It is possible that receptor-mediated bindings occur in conjunction with adjacent residues for hydrophobic interaction. The latter stabilize an otherwise weak binding complex (16). The occurrence of hydrophobic interactions for an individual organism at very different substrata (66), as well as the strong correlation between adhesion and degree of hydrophobicity in large surveys of natural isolates (13, 66), point toward the importance of the thermodynamic model.

(ii) The degree of hydrophobic interaction is related to the degree of irreversible binding (43; Kjelleberg and Hermansson, submitted for publication).

(iii) Surface-active materials, localized at solid–liquid interfaces, are scavenged by irreversibly bound bacteria, rod-shaped marine bacteria, acting as "interface specialists" among the surface-associated population (43, 46). Uptake experiments to follow either the removal of labeled stearic acid localized at a solid surface or the production of labeled CO_2 were performed with hydrophilic and hydrophobic bacteria under oligotrophic bulk-phase conditions. A rapid and immediate removal of surface-bound [^{14}C]stearic acid by the hydrophobic wild type of *Serratia marcescens* was seen. The nonpigmented hydrophilic mutant of the strain did not scavenge the surface-localized fatty acids before 14 h of incubation in the experimental system, but thereafter showed the same rate of removal as did the wild-type form (41). A second pair of strains with similar fatty acid uptake rates, the hydrophilic marine strain *Vibrio* sp. DW1 and the hydrophobic marine strain *Pseudomonas* sp. S9, displayed a pattern in scavenging of surface-localized stearic acid corresponding to that of the strains of *Serratia*. Considerably fewer viable hydrophobic cells at the surface, compared to the numbers of viable hydrophilic bacteria, carried out a more efficient scavenging of surface-localized nutrients (Kjelleberg and Hermansson, submitted for publication). In both studies the degree of irreversible binding was significantly higher for the hydrophobic cells compared to the hydrophilic ones, and scavenging of surface-localized nutrients was therefore in proportion to the number of irreversibly adhered bacteria. Although conventional rod-shaped bacteria behave in this way, it should be noted that bacteria with different modes of mobility, such as reversibly adhering "crawling" cells of *Leptospira*, also display an efficient scavenging of surface-localized fatty acids (43).

CONCLUSIONS

In conclusion, starvation of copiotrophic bacteria imposes rapid changes in cell volumes and numbers. This active process leads, via fragmentation and continuous size reduction, to the formation of small cells that show an immediate response to added nutrients both in bulk liquid and at an interface. Elevated levels of ATP and RNA, as well as an increased number of substrate-binding sites of starved bacteria, agree with this metabolic capacity. In some strains small starved cells increase their degree of hydrophobic character. This decrease in polarity of the cells occurs concomitantly with an increase in irreversible binding to solid surfaces. This is a clear survival strategy of copiotrophic bacteria in low-nutrient oligotrophic habitats. It is possible that a portion of dissolved organic matter in natural systems is degraded in microzones and at interfaces. The degree of irreversible binding has been shown to relate to bacterial scavenging efficiency of surface-localized nutrients. Electron micrographs show large changes in texture of the outer surface during the starvation process. We can also conclude that interfaces lacking a sufficient amount of nutrients to support bacterial growth trigger cells to become even smaller than in the bulk phase during starvation conditions, thus further increasing the surface-to-volume ratio and the packing density.

A basis for the understanding of "interfaces in microbial ecology" is the strong influence of environmental conditions on bacterial behavior. Future research may pay particular attention to the effects of comparative carbon-, nitrogen-, and phosphorus-limiting situations. Such alterations will also lead to studies of bacterial cell surface characteristics and detailed analysis of structures that mediate adhesion.

I gratefully acknowledge helpful discussions with L. Adler, P. Conway, B. Dahlbäck, and M. Hermansson.

This work was supported by a grant from the Swedish Natural Science Research Council.

LITERATURE CITED

1. **Absolom, D. R., F. V. Lamberti, Z. Policova, W. Zingg, C. J. van Oss, and A. W. Neumann.** 1983. Surface thermodynamics of bacterial adhesion. Appl. Environ. Microbiol. **46:**90–97.
2. **Amy, P. S., and R. Y. Morita.** 1983. Starvation-survival patterns of sixteen freshly isolated open-ocean bacteria. Appl. Environ. Microbiol. **45:**1109–1115.
3. **Amy, P. S., C. Pauling, and R. Y. Morita.** 1983. Starvation-survival processes of a marine vibrio. Appl. Environ. Microbiol. **45:**1041–1048.
4. **Azam, F., and R. E. Hodson.** 1981. Multiphasic kinetics for D-glucose uptake by assemblages of natural marine bacteria. Mar. Ecol. Prog. Ser. **6:**213–222.
5. **Ben-Naim, A.** 1976. Hydrophobic interaction. Colloq. Int. C.N.R.S. **246:**215–221.
6. **Ben-Naim, A.** 1980. Hydrophobic interactions. Plenum Press, New York.
7. **Bright, J. J., and M. Fletcher.** 1983. Amino acid assimilation and electron transport system activity in attached and free-living marine bacteria. Appl. Environ. Microbiol. **45:**818–825.
8. **Brockmann, U. H., G. Kattner, G. Hentzschel, K. Wandschneider, H. D. Junge, and H. Hühnerfuss.** 1976. Natürliche Oberflächenfilme im Seegebiet vor Sylt. Mar. Biol. **36:**135–146.

9. **Brown, C. M., D. C. Ellwood, and J. R. Hunter.** 1977. Growth of bacteria at surfaces. Influence of nutrient limitation. FEMS Microbiol. Lett. **1**:163–166.

10. **Carlucci, A. F., and S. L. Shimp.** 1974. Isolation and growth of a marine bacterium in low concentrations of substrate, p. 363–367. *In* R. R. Colwell and R. Y. Morita (ed.), Effect of the ocean environment on microbial activities. University Park Press, Baltimore.

11. **Carlucci, A. F., and S. L. Shimp.** 1976. Distribution, isolation, and cultural characteristics of low-nutrient marine bacteria, p. 827–875. FCRG Annual Report, University of California, San Diego.

12. **Casida, L. E., Jr.** 1977. Small cells in pure cultures of *Agromyces ramosus* and in natural soil. Can. J. Microbiol. **23**:214–216.

13. **Dahlbäck, B., M. Hermansson, S. Kjelleberg, and B. Norkrans.** 1981. The hydrophobicity of bacteria: an important factor in their initial adhesion at the air-water interface. Arch. Microbiol. **128**:267–270.

14. **Daley, R. J., and J. E. Hobbie.** 1975. Direct count of aquatic bacteria by a modified epifluorescent technique. Limnol. Oceanogr. **20**:875–881.

15. **Dawson, M. P., B. Humphrey, and K. C. Marshall.** 1981. Adhesion: a tactic in the survival strategy of a marine vibrio during starvation. Curr. Microbiol. **6**:195–198.

16. **Doyle, R. J., W. E. Nesbitt, and K. G. Taylor.** 1982. On the mechanism of adherence of *Streptococcus sanguis* to hydroxylapatite. FEMS Microbiol. Lett. **15**:1–5.

17. **Druilhet, R. E., and J. M. Sobek.** 1976. Starvation survival of *Salmonella enteritidis*. J. Bacteriol. **125**:119–124.

18. **Fenchel, T.** 1980. Suspension feeding in ciliated protozoa: feeding rates and their ecological significance. Microb. Ecol. **6**:13–25.

19. **Fenchel, T.** 1982. Ecology of heterotrophic microflagellates. IV. Quantitative occurrence and importance as bacterial consumers. Mar. Ecol. Prog. Ser. **9**:35–42.

20. **Ferguson, R. L., and P. Rublee.** 1976. Contribution of bacteria to standing crop of coastal plankton. Limnol. Oceanogr. **21**:141–145.

21. **Fletcher, M.** 1979. A microautoradiographic study of the activity of attached and free-living bacteria. Arch. Microbiol. **122**:271–274.

22. **Fletcher, M., and K. C. Marshall.** 1982. Are solid surfaces of ecological significance to aquatic bacteria?, p. 199–236. *In* K. C. Marshall (ed.), Advances in microbial ecology, vol. 6. Plenum Press, New York.

23. **Fontana, R., P. Canepari, and G. Satta.** 1979. Alterations in peptidoglycan chemical composition associated with rod-to-sphere transition in a conditional mutant of *Klebsiella pneumoniae*. J. Bacteriol. **139**:1028–1038.

24. **Fuhrman, J. A., and F. Azam.** 1980. Bacterioplankton secondary production estimates for coastal waters of British Columbia, Antarctica, and California. Appl. Environ. Microbiol. **39**:1085–1095.

25. **Goulder, R.** 1977. Attached and free bacteria in an estuary with abundant suspended solids. J. Appl. Bacteriol. **43**:399–405.

26. **Gray, T. R. G., and J. R. Postgate.** 1976. The survival of vegetative microbes. Symp. Soc. Gen. Microbiol. **26**:1–432.

27. **Gronlund, A. F., and J. J. R. Campbell.** 1963. Nitrogenous substrates of endogenous respiration in *Pseudomonas aeruginosa*. J. Bacteriol. **86**:58–66.

28. **Harrison, A. P., and F. R. Lawrence.** 1963. Phenotypic, genotypic, and chemical changes in starving populations of *Aerobacter aerogenes*. J. Bacteriol. **85**:742–750.

29. **Harvey, R. W., and L. Y. Young.** 1980. Enrichment and association of bacteria and particulates in salt marsh surface water. Appl. Environ. Microbiol. **39**:894–899.

30. **Harvey, R. W., and L. Y. Young.** 1980. Enumeration of particle-bound and unattached respiring bacteria in the salt marsh environment. Appl. Environ. Microbiol. **40**:156–160.

31. **Hendricks, C. W.** 1974. Sorption of heterotrophic and enteric bacteria to glass surfaces in the continuous culture of river water. Appl. Microbiol. **28**:572–578.

32. **Hirsch, P.** 1979. Life under conditions of low nutrient concentrations, p. 357–372. *In* M. Shilo (ed.), Strategies of microbial life in extreme environments. Dahlem Konfr. Life Sci. Res. Report 13. Verlag Chemie, Weinheim.

33. **Hobbie, J. E., R. J. Daley, and S. Jasper.** 1977. Use of Nuclepore filters for counting bacteria by fluorescence microscopy. Appl. Environ. Microbiol. **33**:1225–1228.

34. **Humphrey, B. A., S. Kjelleberg, and K. C. Marshall.** 1983. Responses of marine bacteria under starvation conditions at a solid-water interface. Appl. Environ. Microbiol. **45**:43–47.

35. **Ishida, Y., I. Imai, T. Miyagaki, and H. Kadota.** 1982. Growth and uptake kinetics of a facultatively oligotrophic bacterium at low nutrient concentrations. Microb. Ecol. **8**:23–32.

36. **Ishida, Y., and H. Kadota.** 1981. Growth patterns and substrate requirements of naturally occurring obligate oligotrophs. Microb. Ecol. **7**:123–130.

37. **Ishida, Y., K. Shibahara, H. Uchida, and H. Kadota.** 1980. Distribution of obligately oligotrophic bacteria in Lake Biwa. Bull. Jpn. Soc. Sci. Fish. **46**:1151–1158.

38. **Jannasch, H. W.** 1968. Growth characteristics of heterotrophic bacteria in seawater. J. Bacteriol. **95**:722–723.

39. **Jannasch, H. W.** 1969. Estimations of bacterial growth rates in natural waters. J. Bacteriol. **99**:156–160.

40. **Jannasch, H. W.** 1974. Steady state and the chemostat in ecology. Limnol. Oceanogr. **19**:716–720.

41. **Jannasch, H. W., and P. H. Pritchard.** 1972. The role of inert particulate matter in the activity of aquatic microorganisms. Mem. Ist. Ital. Idrobiol. Suppl. **29**:289–308.

42. **Jones, K. L., and M. E. Rhodes-Roberts.** 1981. The survival of marine bacteria under starvation conditions. J. Appl. Bacteriol. **50**:247–258.

43. **Kefford, B., S. Kjelleberg, and K. C. Marshall.** 1982. Bacterial scavenging: utilization of fatty acids localized at a solid/liquid interface. Arch. Microbiol. **133**:257–260.

44. **Kirchman, D., and R. Mitchell.** 1982. Contribution of particle-bound bacteria to total microheterotrophic activity in five ponds and two marshes. Appl. Environ. Microbiol. **43**:200–209.

45. **Kjelleberg, S., B. A. Humphrey, and K. C. Marshall.** 1982. The effects of interfaces on small starved marine bacteria. Appl. Environ. Microbiol. **43**:1166–1172.

46. **Kjelleberg, S., B. A. Humphrey, and K. C. Marshall.** 1983. Initial phases of starvation and activity of bacteria at surfaces. Appl. Environ. Microbiol. **46**:978–984.

47. **Koch, A. L.** 1971. The adaptive responses of *Escherichia coli* to a feast and famine existence. Adv. Microb. Physiol. **6**:147–217.

48. **Koch, A. L.** 1979. Microbial growth in low concentrations of nutrients, p. 261–279. *In* M. Shilo (ed.), Strategies of microbial life in extreme environments. Dahlem Konfr. Life Sci. Res. Report 13. Verlag Chemie, Weinheim.

49. **Kogure, K., U. Simidu, and N. Tage.** 1979. A tentative direct microscopic method for counting living marine bacteria. Can. J. Microbiol. **25**:415–420.

50. **Kurath, G., and R. Y. Morita.** 1983. Starvation-survival physiological studies of a marine *Pseudomonas* sp. Appl. Environ. Microbiol. **45**:1206–1211.

51. **Marshall, K. C., and R. H. Cruickshank.** 1973. Cell surface hydrophobicity and the orientation of certain bacteria at interfaces. Arch. Microbiol. **91**:29–40.

52. **Marshall, K. C., R. Stout, and R. Mitchell.** 1971. Selective sorption of bacteria from seawater. Can. J. Microbiol. **17**:1413–1416.

53. **Menzel, D. W., and J. H. Ryther.** 1970. Distribution and cycling of organic matter in the oceans, p. 31–54. *In* D. W. Hood (ed.), Organic matter in natural water. Institute of Marine Science, College, Alaska.

54. **Morita, R. Y.** 1982. Starvation-survival of heterotrophs in the marine environment, p. 171–198. *In* K. C. Marshall (ed.), Advances in microbial ecology, vol. 6. Plenum Press, New York.

55. **Nelson, L. M., and D. Parkinson.** 1978. Effect of starva-

tion on survival of three bacterial isolates from an Arctic soil. Can. J. Microbiol. 24:1460–1467.

56. **Norkrans, B.** 1980. Surface microlayers in aquatic environments, p. 51–58. *In* M. Alexander (ed.), Advances in microbial ecology, vol. 4. Plenum Publishing Corp., New York.

57. **Novitsky, J. A., and R. Y. Morita.** 1976. Morphological characterization of small cells resulting from nutrient starvation of a psychrophilic marine vibrio. Appl. Environ. Microbiol. 32:617–662.

58. **Novitsky, J. A., and R. Y. Morita.** 1977. Survival of a psychrophilic marine vibrio under long-term nutrient starvation. Appl. Environ. Microbiol. 33:635–641.

59. **Novitsky, J. A., and R. Y. Morita.** 1978. Possible strategy for the survival of marine bacteria under starvation conditions. Mar. Biol. 48:289–295.

60. **Öhman, L., K.-E. Magnusson, and O. Stendahl.** 1982. The mannose-specific lectin activity of *Escherichia coli* type 1 fimbriae assayed by agglutination of glycolipid-containing liposomes, erythrocytes, and yeast cells and hydrophobic interaction chromatography. FEMS Microbiol. Lett. 14:149–153.

61. **Parker, R. I., and J. Taylor.** 1983. The relationship between fatty acid distributions and bacterial respiratory types in contemporary marine sediments. Estuarine Coastal Shelf Sci. 16:173–189.

62. **Poindexter, J. S.** 1981. Oligotrophy: fast and famine existence, p. 63–89. *In* M. Alexander (ed.), Advances in microbial ecology, vol. 5. Plenum Press, New York.

63. **Poindexter, J. S.** 1981. The caulobacters: ubiquitous unusual bacteria. Microbiol. Rev. 45:123–179.

64. **Postgate, J. R., and J. R. Hunter.** 1962. The survival of starved bacteria. J. Gen. Microbiol. 29:233–263.

65. **Pringle, J. H., and M. Fletcher.** 1983. Influence of substratum wettability on attachment of freshwater bacteria to solid surfaces. Appl. Environ. Microbiol. 45:811–817.

66. **Rosenberg, M., E. Rosenberg, H. Judes, and E. Weiss.** 1984. Hypothesis. Bacterial adherence to hydrocarbons and to surfaces in the oral cavity. FEMS Microbiol. Lett., in press.

67. **Stevenson, L. H.** 1978. A case for dormancy in aquatic systems. Microb. Ecol. 4:127–133.

68. **Tadros, T. F.** 1980. Particle-surface adhesion, p. 93–116. *In* R. C. W. Berkeley, J. M. Lynch, J. Melling, P. R. Rutter, and B. Vincent (ed.), Microbial adhesion to surfaces. Ellis Horwood Publisher, Chichester.

69. **Tanford, C.** 1979. Interfacial free energy and the hydrophobic effect. Proc. Natl. Acad. Sci. U.S.A. 76:4175–4176.

70. **Torella, F., and R. Y. Morita.** 1981. Microcultural study of bacterial size changes and microcolony and ultramicrocolony formation by heterotrophic bacteria in seawater. Appl. Environ. Microbiol. 41:518–527.

71. **Van der Kooij, D., and W. A. M. Hijnen.** 1983. Nutritional versatility of a starch-utilizing flavobacterium at low substrate concentrations. Appl. Environ. Microbiol. 45:804–810.

72. **Verger, R., M. C. E. Mieras, and G. H. de Haas.** 1973. Action of phospholipase A at interfaces. J. Biol. Chem. 248:4023–4034.

73. **Weiss, E., M. Rosenberg, H. Judes, and E. Rosenberg.** 1982. Cell-surface hydrophobicity of adherent oral bacteria. Curr. Microbiol. 7:125–128.

74. **Wilson, C. A., and L. H. Stevenson.** 1980. The dynamics of the bacterial population associated with a salt marsh. J. Exp. Mar. Biol. Ecol. 48:123–138.

75. **Wright, R. T.** 1978. Measurements and significance of specific activity in the heterotrophic bacteria of natural waters. Appl. Environ. Microbiol. 36:297–305.

76. **Wright, R. T., and R. B. Coffin.** 1983. Planktonic bacteria in estuaries and coastal waters of northern Massachusetts: spatial and temporal distribution. Mar. Ecol. Prog. Ser. 11:205–216.

77. **Wright, R. T., R. B. Coffin, C. P. Ersing, and D. Pearson.** 1982. Field and laboratory measurements of bivalve-filtration of natural marine bacterioplankton. Limnol. Oceanogr. 27:91–98.

78. **Zimmermann, R., and L.-A. Meyer-Reil.** 1974. A new method for fluorescence staining of bacterial populations on membrane filter. Kiel. Meeresforsch. 30:24–27.

New and Unusual Microorganisms and Niches

Stem-Nodulating Rhizobia

B. L. DREYFUS, D. ALAZARD, AND Y. R. DOMMERGUES

Laboratoire de Microbiologie des Sols, ORSTOM B.P 1386, Dakar, Senegal

In symbiosis with *Rhizobium* sp., most leguminous plants form nitrogen-fixing nodules on their roots. Only a few legume species also bear nodules on their stems. Stem nodulation was first reported in *Aeschynomene aspera* (17), *A. paniculata* (28), and *Neptunia oleracea* (25) and later in *A. indica* (2, 30, 29), *A. elaphroxylon* (19), *A. evenia*, and *A. filosa* (4).

Recently, Dreyfus and Dommergues (10, 11) reported the discovery of profuse stem nodulation in *Sesbania rostrata*, a fast-growing annual tropical legume which can reach a height of 3 to 5 m within 3 to 4 months. The stem nodules of *S. rostrata* were found to actively fix N_2 and to be induced by a specific strain of *Rhizobium* sp. able to form nodules on both stem and roots. Stem-nodulating rhizobia have also been isolated from stem nodules of *A. aspera* (26), *A. indica* (14; D. Alazard, unpublished data), *A. afraspera*, *A. crassicaulis*, *A. elaphroxylon*, *A. schimperi*, *A. sensitiva* (D. Alazard and B. L. Dreyfus, 5th Int. Symp. Nitrogen Fixation, Noordwijkerhout, The Netherlands, 1983), and *Neptunia oleracea* (B. Dreyfus, unpublished data). Stem nodulation has been reported in other *Aeschynomene* species, namely, *A. denticulata*, *A. pratensis*, *A. rudis*, and *A. scabra* (14).

All known stem-nodulated legumes belong to the three genera *Sesbania*, *Aeschynomene*, and *Neptunia*, and have in common the ability to grow in waterlogged soils. In nature, the stem nodules of these plants are often restricted to the submerged stem and just above the flood water level.

The aim of this paper is to review the available information on stem nodulation, with special emphasis on the unusual properties of the infecting rhizobia.

STEM NODULATION SITE

The most distinctive characteristic of all stem-nodulating legumes is the presence of predetermined nodulation sites on the stems. The formation of these sites is independent of the *Rhizobium* infection. Until recently, the nature of the nodulation site was not clearly defined. Some authors have proposed that it is a lenticel in *S. rostrata* (11) and *A. paniculata* (28). Others have suggested that infection occurs on emerging adventitious roots appearing on submerged stems (2, 17, 26). Recent studies show that the infection is always initiated at the base of preformed incipient primordia on the stem. These primordia are in a dormant state, most likely as a result of a phytohormonal effect. Depending on the host plant, either they remain under the cortex or they are hidden by an external structure (lenticel or epidermal dome) or slightly pierce this structure, showing a protruding dormant apex. Anatomical study shows that these dormant, protruding or hidden primordia exhibit a typical root structure, a fact confirmed by the ability of these primordia to develop into adventitious roots when stems are immersed in water (13, 14). For that reason we call these structures dormant root primordia or root primordia. We shall see later that the distinction between protruding and hidden root primordia is important because this character governs the host sensitivity to rhizobial infection.

In *S. rostrata* (Fig. 1A) nodulation sites are distributed evenly in three or four vertical lines all along the stem (11, 13). The root primordium always pierces the stem epidermis, emerging 0.1 to 0.3 cm from an epidermal dome and forming a circular fissure (Fig. 2). Since the root primordia are protruding, the aerial infection by *Rhizobium* sp. occurs readily, producing the nitrogen-fixing nodules. The nodulation sites of *S. rostrata* are formed continuously throughout the growth of the stem and remain sensitive to *Rhizobium* infection during the whole life of the plant. In the absence of *Rhizobium* infection, the dormancy of the root primordia can be broken by immersing the stem in water, which induces adventitious roots, as indicated above, or by placing an internodal stem cutting into a nutrient medium (M. Barreto, personal communication). In the latter case, the upper primordium of the cutting develops into a shoot bud, indicating the triple potential of the nodulation site.

The genus *Aeschynomene* includes species with nodulation sites ranging from the *Sesbania* type with protruding root primordia to the hidden root primordia type. Stem-nodulated *Aeschynomene* can be divided into three subgroups.

FIG. 1. Different types of stem-nodulated legumes. (A) *S. rostrata*, the most evolved type, with a profusion of nodules distributed along vertical lines. (B) *A. afraspera*, also an evolved type of stem-nodulated legume, with profuse nodulation. (C) *A. elaphroxylon*, the least evolved *Aeschynomene* sp., with a few nodules at the base of the stem. (D) *N. oleracea*, with nodules occurring only on developed adventitious roots formed at the level of the stem node. In all cases the bar represents 0.5 cm.

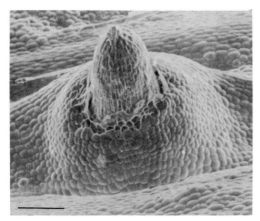

FIG. 2. Uninoculated infection site of *S. rostrata*. The root primordium always protrudes through the epidermis, forming a fissure that can be subsequently invaded by rhizobia. The bar represents 25 μm. From Duhoux and Dreyfus (13).

Subgroup 1 comprises species with the most evolved root primordia, as susceptible to infection as those of *S. rostrata*. *A. afraspera* is of this type (Fig. 1B) (D. Alazard and E. Duhoux, unpublished data).

Subgroup 2, which can be considered as intermediate between subgroups 1 and 3, comprises species with root primordia less developed than those of subgroup 1. The root primordia scarcely penetrate the epidermis and are often located in the center of lenticels. Two typical species of this subgroup are *A. scabra* (14) and *A. indica* (14; Alazard and Duhoux, unpublished data). Their root primordia are still accessible to *Rhizobium* infection, and these species nodulate, though less readily than *S. rostrata* or *A. afraspera*. *A. paniculata* and *A. sensitiva* (14, 28) probably also belong to the same subgroup. Like *A. scabra* and *A. indica*, *A. paniculata* was reported to have lenticels (28).

Subgroup 3 comprises species with the least evolved root primordia. Two typical species are *A. crassicaulis* and *A. elaphroxylon* (Fig. 1C), the latter species being an aquatic tree (26; Alazard and Duhoux, unpublished data). In these species, stem nodules are restricted to the lower and submerged section of the stem, where the conditions break the dormancy of the root primordia and induce their development into adventitious roots susceptible to rhizobial infection.

In contrast to *S. rostrata* and all the *Aeschynomene* species mentioned above, the nodulation sites of *N. oleracea* are located only in the vicinity of the stem nodes. When the plant grows on drained soils, the root primordia remain embedded in the stem cortex and no nodules are formed. When *N. oleracea* grows in waterlogged soils, the stems float at the surface of the water, and the root primordia pierce the stem epidermis and develop into typical adventitious roots, which are then susceptible to infection by *Rhizobium* sp. Nodules are usually found at the base of these roots (Fig. 1D).

Thus, differences in the structure of the nodulation site appear to explain the variations observed in the stem-nodulating ability of different species. Stem-nodulating legumes form a continuum between those with the most developed root primordia (e.g., those of *S. rostrata* and *A. afraspera*) and those with the least evolved ones (those of *A. elaphroxylon* and *N. oleracea*).

STEM INFECTION BY *RHIZOBIUM* SP.

Root nodulation has been very well documented (e.g., in Dart's excellent review [8]), but it is only recently that studies on stem nodulation have been initiated. Since these investigations concerned mainly *S. rostrata* (27) and little is known about the infection process in *Aeschynomene* and *Neptunia* species, the present discussion will be devoted mainly to the development of stem nodules in *S. rostrata*.

Before dealing with this specific topic, it is necessary to review briefly our current knowledge on the initiation of root nodules. Two different types of root infection by *Rhizobium* sp. have been reported for root nodules. In most legumes, especially temperate legumes, the infection process starts in root hairs, with the formation of an infection thread. The infection thread containing the *Rhizobium* cells passes through the root hair into the root cortex. After the infection threads have entered the meristematic cells of the cortex, the rhizobia are released into the cytoplasm of the host cells (8, 22). The second type of root infection observed in a few tropical legumes, such as *Arachis hypogaea* (6), starts by direct intercellular infection. The rhizobia penetrate at the lateral root junctions between the basal cells of root hairs and proliferate between these cells, forming intercellular zones of infection. This intercellular phase is followed by an intracellular invasion which starts with the invagination of the plant cell wall. Infection threads are not formed. The intracellular infection then develops by successive divisions of host cells, with each daughter cell receiving the rhizobia (1, 6). Similar to the *Arachis* type of infection is that of *Stylosanthes* (7), where direct cell invasion occurs but root hairs are not involved and rhizobia do not form intercellular zones of infection as in *A. hypogaea* (7). The *Arachis* and *Stylosanthes* (both in the same tribe as *Aeschynomene*) types of infection thus differ from the typical temperate legume infection by two characters: absence of root hair penetration and absence of infection threads.

In *S. rostrata* (27) the rhizobia reach the base of root primordium through the fissure encircling it. This fissure provides an optimal micro-environment for the appropriate specific *Rhizobium* sp. which colonizes it. Nodule genesis then consists of three distinct stages.

(i) **Intercellular infection.** The rhizobia penetrate the intercellular space of the root primordium basal cells where they multiply in large numbers, forming pockets of infection (Fig. 3). At the same time, differentiation of some cortical cells of the root primordium occurs through an activating mechanism which is still not known. The resulting meristematic cells are then infected by the rhizobia. No root hairs have been observed either in the fissure or on the root primordium.

(ii) **Development of infection threads.** The large intercellular pockets (up to 50 *Rhizobium* cells wide) progressively get narrower and narrower in a funnel-like manner, ending in an infection thread (one or two cells wide) which divides and penetrates the meristematic cells.

(iii) **Intracellular infection.** Rhizobia are released from the infection threads into the cell cytoplasm and are promptly individually surrounded by the membrane envelope (Fig. 4). At this stage (4 to 5 days after inoculation), the rhizobia (then at the bacteroid stage) begin to fix N_2 symbiotically, and the nodule exhibits the red color characteristic of leghemoglobin. Bacteroids which are originally alone in the membrane envelope later divide so that there can be up to 20 *Rhizobium* cells per membrane envelope (11). The mode of infection of *S. rostrata* stems is thus unique among the known legumes, as it involves both an intercellular invasion by *Rhizobium* sp. similar to that of *Arachis hypogaea* (6) and the development of infection threads as in temperate legumes.

In several *Aeschynomene* species the mode of infection seems to be related more to that of *Stylosanthes* sp. than to that of *Arachis* sp. since neither root hairs nor infection threads are found, and the rhizobia are spread by cell division after direct intercellular infection (2, 21).

In *N. oleracea*, according to Schaede (25), nodules are formed from infection threads initiated at the base of adventitious roots. No root hairs are found. The mode of infection in *N. oleracea* could thus be similar to that of *S. rostrata*.

The question has been raised as to whether rhizobia could reach the nodulation sites from the roots via the stem vascular bundles (17). Such an infection has never been found.

In nature, nodulation sites of *S. rostrata* are irregularly infected so that nodules appear to be distributed at random along the stem and branches. Dust seems to play a significant role in the nodulation of stems, as plants that grow along dirt roads are generally more nodulated than other plants. Other vectors that may be significant include insects and rain. Since only some of the nodulation sites are spontaneously infected, the number, and consequently the total weight, of stem nodules varies from one plant to another. Inoculating the stems greatly increases the number and weight of stem nodules of *S. rostrata*, so that nearly all the nodulating sites are nodulated (Fig. 1A).

Inoculation of *Aeschynomene* sp. of subgroup 1 always results in significant improvement of nodulation. Repeated and heavy inoculation is required for maximum nodulation of *Aeschynomene* sp. of subgroups 2 and 3.

MATURE STEM NODULE AND ITS LEGHEMOGLOBIN CONTENT

Two days after stem inoculation, there are structural changes in the root primordia of *S. rostrata* (27). Within 1 week, one can recognize the typical morphology of nodules and detect acetylene reduction activity (11). Stem nodules of *S. rostrata* are generally spherical, sometimes irregular. Except for their green chloroplast-containing cortex, they resemble typical round or oval root nodules found on *Glycine max* or *Vigna unguiculata* (11). Their diameter ranges from 0.3 to 0.8 cm, and they are easily detachable from the stem. In contrast to stem nodules, root nodules appear elongated, with an apical meristem. Plants of *S. rostrata* 3 to 4 m high can bear stem nodules of up to 40 g of fresh weight (8 g of dry weight), whereas the fresh weight of root nodules reaches only 2 to 4 g per plant.

In *Aeschynomene* spp., nodules were reported to appear within 8 days (*A. scabra*) or 12 days (*A. denticulata, A. indica, A. pratensis, A. rudis,* and *A. evenia*) after stem inoculation (14). Stem nodules of *Aeschynomene* are usually ellipsoidal (ca. 4 by 5 mm) or spherical (3 to 7 mm in diameter). They form more or less prominent swellings under the stem epidermis and are not easily detached from the stems. The dry weight of stem nodules of *A. scabra* can reach 0.5 g per plant (14).

Stem nodules of *N. oleracea* are elongated as a result of the presence of an apical meristem and measure up to 12 mm long. Unlike other stem nodules, those of *N. oleracea* do not harbor chloroplasts in their cortex.

In *S. rostrata* and *Aeschynomene* spp., the green cortex with chloroplasts surrounds the red-pigmented central zone. As expected, the red pigment of stem nodules was shown to be leghemoglobin (R. P. Legocki, A. R. J. Eaglesham, and A. A. Szalay, *in* A. Puhler, ed.,

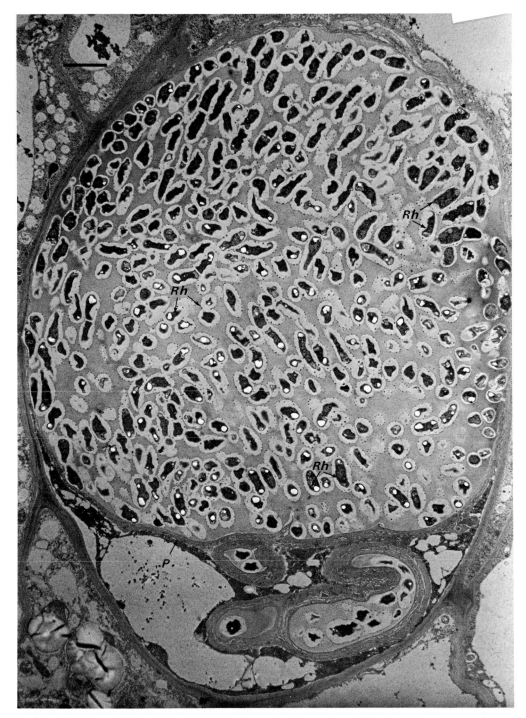

FIG. 3. Intercellular infection in *S. rostrata*. Multiplication of rhizobia (Rh) in intercellular spaces between the basal cells of the root primordium. The bar represents 5 μm. (Micrograph kindly supplied by H. C. Tsien.)

Proceedings of the First International Sympo- sium on Molecular Genetics of the Bacteria- Plant Interaction, in press; D. Bogusz, personal communication). In *A. scabra* (Legocki et al., in press), there are two distinct molecular species of leghemoglobin, designated Lbα and Lbβ. Stem and root nodules of *A. scabra* contained the same amount of Lbα, but Lbβ was substan-

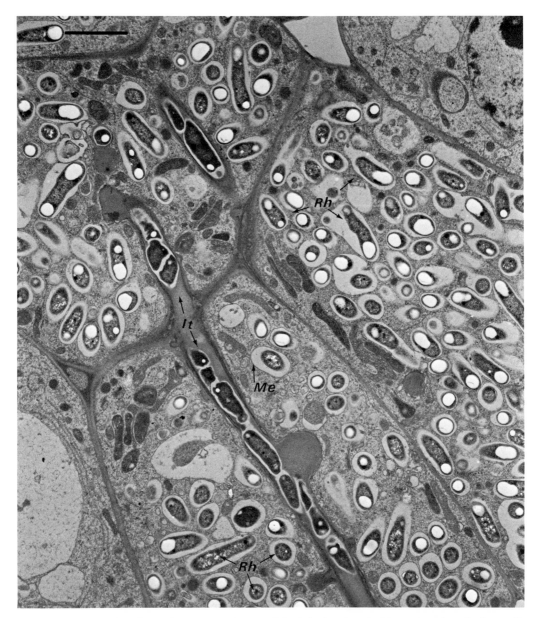

FIG. 4. Intracellular infection in *S. rostrata*. Intercellular infection threads (It) penetrate into the host cells where the rhizobia are released and immediately enclosed in the membrane envelope (Me). The bar represents 5 μm. From Tsien et al. (27).

tially more abundant in stem nodules. Recently, Bogusz (personal communication) found that the leghemoglobin from stem nodules of *S. rostrata* was composed of four different molecular species, whereas that of root nodules was composed of only three. Three root molecular species were common to stem and root nodules, and the supplementary fourth molecular species (called IIIS) was specific to the stem nodules. The differences in leghemoglobin composition

between stem and root nodules of both *S. rostrata* and *A. scabra* might be in relation to the increased oxygen level resulting from the photosynthetic activity in the cortical tissues of stem nodules.

NITROGEN FIXATION

Nitrogen fixation by stem nodules of *A. scabra* and *S. rostrata* was measured by the acetylene reduction method (18). The specific activity of stem nodules was very similar in the two species, 259 μmol of C_2H_4 g^{-1} (dry weight) h^{-1} for *S. rostrata* (11) and 270 μmol of C_2H_4 g^{-1} (dry weight) h^{-1} for *A. scabra* (Legocki et al., in press). These activities are comparable to those of root nodules of other legumes such as soybean (20) or cowpea (3). When expressed on the whole plant basis, acetylene reduction activity was 600 and 165 μmol of C_2H_4 per plant per h in *S. rostrata* and *A. scabra*, respectively. The higher activity found in *S. rostrata* is easily explained by its greater size at the time of evaluation of acetylene reduction activity. Nitrogen fixation by field-grown *S. rostrata* was shown to be about 200 kg ha^{-1} for a period of 7 weeks (24), indicating that this is one of the most powerful nitrogen-fixing legumes. Because of this potential, *S. rostrata* has been successfully used as green manure in paddy fields in Senegal, increasing the rice grain yield two- to threefold.

Combined nitrogen is well known to affect nodulation and nitrogen fixation of root-nodulated legumes, through processes that are not yet fully understood except perhaps for infection, which is known to be locally inhibited by NO_3^- (23). When grown in the presence of combined nitrogen (6 mM) in hydroponic solution or 200 kg of N ha^{-1} in soil, stem nodulation and related nitrogen fixation were not affected, whereas root nodulation was completely inhibited (9). Higher concentrations of combined nitrogen (10 to 15 mM) did not reduce the number of nodules but did affect their weight and nitrogenase activity (G. Rinaudo, personal communication). Similarly, Eaglesham and Szalay (14) found that, in *A. scabra*, N concentrations of 17 mM did not alter the number of nodules but strongly inhibited their growth and nitrogenase activity. Other *Aeschynomene* species appeared to be less tolerant to combined nitrogen than *A. scabra*, since no nodules were found when the same concentration (17 mM) was used (14).

The threshold of tolerance to combined nitrogen thus appears to be much higher in *S. rostrata* and some other stem-nodulated legumes than in legumes that have only root nodules. This characteristic confers to these stem-nodulated legumes a specific agronomic advantage.

STEM-NODULATING RHIZOBIA

Host specificity. Since we never found any cross-inoculation between rhizobia nodulating stems of *S. rostrata*, *Aeschynomene* spp., and *N. oleracea*, we propose to classify the stem-nodulating strains according to their host relatedness.

S. rostrata group. Rhizobia from *S. rostrata* stem nodules (type strain ORS571) are fast growers (generation time, 3 h) but exhibit many physiological characteristics of the cowpea group (Garcia et al., in preparation).

Aeschynomene group. From a study of 15 tropical strains of rhizobia isolated from stem nodules of different *Aeschynomene* species, Alazard and Dreyfus (5th Int. Symp. Nitrogen Fixation, 1983) suggested that the *Aeschynomene* rhizobia be classified into three subgroups.

Subgroup 1 contains rhizobia isolated from *A. afraspera*, which seem to have a narrow host range since they were found to effectively nodulate only *A. afraspera*.

Subgroup 2 contains rhizobia from *A. indica*, *A. schimperi*, and *A. sensitiva*, which form a relatively homogeneous cross-inoculation subgroup. Strain BTAi1, isolated from stem nodules of *A. indica* (Legocki et al., in press), probably belongs to this subgroup. Strain BTAi1, a fast-growing *Rhizobium* strain, was shown to be an intermediate type of *Rhizobium* sp. with characteristics of both fast and slow growers (M. D. Stowers and A. R. J. Eaglesham, J. Gen. Microbiol., in press).

Subgroup 3 contains rhizobia from *A. elaphroxylon* and *A. crassicaulis*, probably related to the cowpea group, since they are typically slow growers (generation time, 10 h) and effectively nodulate the roots of *Macroptilium atropurpureum*, the test host for the cowpea rhizobia.

This classification of *Aeschynomene* rhizobia fits reasonably well with the classification of their hosts of origin based on the structure of the infection site.

Neptunia group. The *Neptunia* rhizobia are fast growers and probably are closely related to *R. meliloti* since they nodulate alfalfa, though ineffectively, and they contain a megaplasmid of the same size as that of *R. meliloti* (C. Rosenberg, personal communication).

Root-stem specificity. In *S. rostrata*, two types of strains have been isolated: stem-nodulating strains (called stem strains), capable of nodulating both stems and roots, and root-nodulating strains (called root strains), which nodulate only roots (Dreyfus, unpublished data). Both types are fast growers, but according to a taxonomic study (Garcia et al., in preparation), stem strains are closer to the cowpea rhizobia than to the root strains. These results suggest the involvement of specific stem-nodulating genes in the stem strains, but to date there is no information on what these genes might encode for. Since stem nodulation is dependent on the presence of roots (as dormant root primordia), this root-stem specificity is surprising. Furthermore, this specificity seems to be restricted to *S. rostrata*, since it has not been observed in other stem-nodulated legumes.

Nitrogen-fixing growth of the stem-nodulating rhizobia. For many years, it was thought that rhizobia fixed atmospheric N_2 only as bacteroids in the nodules. In 1975, several laboratories (see Gibson's review [16]) simultaneously discovered that the cowpea *Rhizobium* sp. strain 32H1 could produce nitrogenase in culture when grown under very low oxygen tension (1 µM dissolved O_2). Later, this faculty was found in a few other strains, all of the slow-growing type. In spite of their nitrogenase activity, none of these strains was able to grow solely on N_2; addition of combined nitrogen was required. Recently, we reported that strain ORS571 from stem nodules of *S. rostrata* not only exhibited high rates of nitrogenase activity in culture but also presented the unique property among rhizobia of growing in a nitrogen-free liquid or solid medium, at reduced pO_2 (12). Since then, C. Gebhardt, G. L. Turner, B. Dreyfus, and F. J. Bergerson (J. Gen. Microbiol., in press) were able to grow strain ORS571 as an N_2-fixing continuous culture. According to these authors, under optimal conditions of O_2 supply (9 µM dissolved O_2), nitrogenase activity reached 2,000 nmol of C_2H_4 mg^{-1} (dry weight) h^{-1}, and strain ORS571 showed a greater tolerance for dissolved O_2 (9 µM) than strain CB756 (1 µM) (5). This high tolerance to O_2 could result from the adaptation of the *Rhizobium* sp. to the relatively high pO_2 in the stem nodules in relation to the photosynthetic activity of chloroplasts in the cortex. N_2 fully supported the growth of ORS571, but nicotinic acid was required as a growth factor at a rate about 10 times higher than that required in the presence of combined nitrogen (C. Elmerich, B. Dreyfus, and J. P. Aubert, FEMS Microbiol. Lett., in press).

Recently, we have shown (Alazard and Dreyfus, 5th Int. Symp. Nitrogen Fixation, 1983) that tropical *Rhizobium* strains isolated from stem nodules of *Aeschynomene* subgroups 1 and 2 also exhibited significant nitrogenase activity in culture in the absence of combined nitrogen at an O_2 tension lower than for *Sesbania* sp. strain ORS571 (2 µM dissolved O_2). Thus, the ability to grow on N_2 in culture could be a widespread feature among stem-nodulating rhizobia.

Genetic analysis of nitrogen fixation in the stem-nodulating rhizobia. Following studies which have elucidated physiological traits of stem-nodulating rhizobia, molecular genetics investigations have recently been initiated in several laboratories.

Taking advantage of the property of *Sesbania* sp. strain ORS571 to grow on N_2 as the sole nitrogen source, Dreyfus et al. (12) obtained Nif$^-$ mutants of strain ORS571 by using standard genetic methods. These mutants, unlike the parent strain, were unable to grow on agar plates under conditions of nitrogen fixation and did not

fix nitrogen in the nodules of *S. rostrata*. Concomitantly, by hybridizing the total DNA of strain ORS571 with the *nif* KDH probe of *Klebsiella pneumoniae*, Elmerich et al. (15) could detect the presence of nitrogen fixation (*nif*) genes on a 13-kilobase *Bam*HI fragment. This fragment was cloned in vector pRK 290 to yield plasmid pRS1. The pRS1 plasmid was introduced into the *nif* mutants by conjugation, and genetic complementation was observed in one of the Nif$^-$ mutants in both free-living and symbiotic states.

By hybridization with *nif* KDH of *K. pneumoniae*, *nif* genes were located by Legocki et al. (in press) on 28- and 6-kilobase *Eco*RI fragments of *A. indica* strain BTAi1. A 6.7-kilobase KDH part of the 28-kilobase fragment was cloned and used for isolation of nitrogenase structural genes and their promoters. To the best of our knowledge, Nif$^-$ mutants of *Aeschynomene* strains have not yet been obtained.

Up to now, attempts to detect megaplasmids have been unsuccessful either in strain ORS571 (C. Rosenberg, personal communication) or in strain BTAi1 (A. Szalay, personal communication), which might mean that *nif* and possibly *nod* genes are located on the chromosome of these stem-nodulating strains, despite their being fast growers.

CONCLUSIONS

Rhizobia specific to stem-nodulated legumes differ in two respects from other known rhizobia: (i) their nitrogenase appears to be well protected against O_2, which could be an adaptation to their location in a photosynthetic nodule, and (ii) they are able to grow in the free-living state at the expense of N_2, which may confer upon these bacteria a selective saprophytic advantage.

In stem-nodulated legumes such as *S. rostrata* and *Aeschynomene* sp. of group 1, nodulation sites are well above the soil and water level, which circumvents the problems of competition from indigenous rhizobia when the stems are inoculated.

The unique characteristic of stem-nodulated plants is the presence of predetermined nodulation sites. Each site is composed of a dormant root primordium whose base is infected by the rhizobia. The structure of the nodulation site varies between two extreme types, forming a continuum between the most evolved type (*Sesbania* and *A. afraspera*) and the least evolved (*A. elaphroxylon* and *N. oleracea*). Interestingly, some Asian soybean cultivars produce dormant primordia on the lower stem, and when submerged, these primordia form adventitious roots susceptible to rhizobial infection at the stem–root junction (A. R. J. Eaglesham and A.

Ayanaba, *in* N. S. Subba Rao, ed., *Selected Topics in Biological Nitrogen Fixation*, in press). This observation suggests that root primordia are not restricted to typical stem-nodulated legumes and that, by breeding or manipulating the plant genome, one could possibly induce other legumes to produce root primordia on their stems, making them susceptible to infection by rhizobia.

Such an achievement would have important implications in agriculture because stem nodulation confers upon the host several advantages. In waterlogged soils low oxygen levels are well known to reduce or inhibit root nodulation of legumes. Stem-nodulated legumes therefore have the advantage in having functional nodules in the air or at the surface of the water where the oxygen tension is still satisfactory. Since the stem nodule is a photosynthetic organ, presumably less energy is drained from the leaves to the N_2 fixation sites, so that stem-nodulated legumes are probably more efficient than other legumes. Besides these characteristics, stem-nodulated legumes have the unusual capability of actively fixing N_2 even in the presence of high rates of combined nitrogen in the soil, which allows more nitrogen to be added to and less removed from the cropping system.

The published and unpublished studies carried out in the ORSTOM laboratory at Dakar were supported by the French Ministry of Industry and Research, grant 83.V.0400.

LITERATURE CITED

1. **Allen, O. N., and E. K. Allen.** 1940. Response of the peanut plant to inoculation with rhizobia with special reference to morphological development of the nodules. Bot. Gaz. (Chicago) **102:**121–142.
2. **Arora, N.** 1954. Morphological development of the root and stem nodules of *Aeschynomene indica* L. Phytomorphology **4:**211–216.
3. **Ayanaba, A., and T. L. Lawson.** 1977. Diurnal changes in acetylene reduction in field-grown cowpeas and soybeans. Soil Biol. Biochem. **9:**125–129.
4. **Barrios, S., and V. Gonzalez.** 1971. Rhizobial symbiosis on Venezuelan savanas. Plant Soil **34:**707–719.
5. **Bergersen, F. J., and G. L. Turner.** 1976. The role of O_2-limitation in control of nitrogenase in continuous cultures of *Rhizobium* sp. Biochem. Biophys. Res. Commun. **73:**524–531.
6. **Chandler, M. R.** 1978. Some observations on infection of *Arachis hypogaea* L. by *Rhizobium*. J. Exp. Bot. **29:**749–755.
7. **Chandler, M. R., R. A. Date, and R. J. Roughley.** 1982. Infection and root nodule development in *Stylosanthes* species by *Rhizobium*. J. Exp. Bot. **33:**47–57.
8. **Dart, P. J.** 1977. Infection and development of leguminous nodules, p. 367–472. *In* R. W. F. Hardy and W. S. Silver (ed.), A treatise on dinitrogen fixation, section III. John Wiley & Sons, Inc., New York.
9. **Dreyfus, B., and Y. R. Dommergues.** 1980. Non-inhibition de la fixation d'azote atmospherique par l'azote combiné chez une legumineuse à nodules caulinaires, *Sesbania rostrata*. C. R. Acad. Sci. Ser. D **291:**767–770.
10. **Dreyfus, B. L., and Y. R. Dommergues.** 1981. Stem nodules on the tropical legume, *Sesbania rostrata*, p. 471. *In*

11. **Dreyfus, B., and Y. R. Dommergues.** 1981. Nitrogen-fixing nodules induced by *Rhizobium* on the stem of the tropical legume *Sesbania rostrata*. FEMS Microbiol. Lett. **10:**313–317.
12. **Dreyfus, B., C. Elmerich, and Y. R. Dommergues.** 1983. Free-living *Rhizobium* strain able to grow under N_2 as the sole nitrogen source. Appl. Environ. Microbiol. **45:**711–713.
13. **Duhoux, E., and B. Dreyfus.** 1982. Nature des sites d'infection par le *Rhizobium* de la tige de la legumineuse *Sesbania rostrata* Brem. C.R. Acad. Sci. Ser. III **294:**407–411.
14. **Eaglesham, A. R. J., and A. A. Szalay.** 1982. Aerial stem nodules on *Aeschynomene spp*. Plant Sci. Lett. **29:**265–272.
15. **Elmerich, C., B. Dreyfus, G. Reysset, and J. P. Aubert.** 1982. Genetic analysis of nitrogen fixation in a fast-growing *Rhizobium*. EMBO J. **1:**499–503.
16. **Gibson, A. H., W. R. Scowcroft, and J. D. Pagan.** 1977. Nitrogen fixation in plants: an expending horizon?, p. 387–417. *In* W. Newton, J. R. Postgate, and C. Rodriguez-Barrueco (ed.), Recent developments in nitrogen fixation. Academic Press, London.
17. **Hagerup, O.** 1928. En hygrofil baelgplante (*Aeschynomene aspera* L.) med bakterieknolde paa staengelen. Dan. Bot. Ark. **15:**1–9.
18. **Hardy, R. W. F., R. D. Holsten, E. K. Jackson, and R. C. Burns.** 1968. The acetylene-ethylene assay for N_2-fixation: laboratory and field evaluation. Plant Physiol. **43:**1185–1207.
19. **Jenik, J., and J. Kubikova.** 1969. Root system of tropical trees. 3. The heterorhizis of *Aeschynomene claphroxylon* (Guill. et Perr.) Taub. Preslia **41:**220–226.
20. **Lawn, R. J., and W. A. Brun.** 1974. Symbiotic nitrogen fixation in soybeans. I. Effect of photosynthetic source-sink manipulations. Crop Sci. **14:**11–16.
21. **Napoli, C. A., F. Dazzo, and D. H. Hubbell.** 1975. Ultrastructure of infection and "common antigen" relationships in *Aeschynomene*, p. 35–37. *In* J. Vincent (ed.), Proceedings of the Fifth Australian Legume Nodulation Conference, Brisbane. Commonwealth Scientific and Industrial Research Organization, Melbourne.
22. **Newcomb, W.** 1981. Nodule morphogenesis and differentiation. Int. Rev. Cytol. Suppl. **13:**247–297.
23. **Pate, J. S.** 1977. Functional biology of dinitrogen fixation by legumes, p. 473–517. *In* R. W. F. Hardy and W. F. Silver (ed.), A treatise on dinitrogen fixation, section III. John Wiley & Sons, Inc., New York.
24. **Rinaudo, G., B. Dreyfus, and Y. R. Dommergues.** 1983. *Sesbania rostrata* green manure and the nitrogen content of rice crop and soil. Soil Biol. Biochem. **15:**111–113.
25. **Schaede, R.** 1940. Die knollchen der adventiven wasserwurzeln von *Neptunia oleracea* und ihre bakteriensymbiose. Planta **31:**1–21.
26. **Subba Rao, N. S., K. V. Tilak, and C. S. Singh.** 1980. Root nodulation studies in *Aeschynomene aspera*. Plant Soil **56:**491–494.
27. **Tsien, H. C., B. L. Dreyfus, and E. L. Schmidt.** 1983. Initial stages in the morphogenesis of the nitrogen-fixing stem nodules of *Sesbania rostrata*. J. Bacteriol. **156:**888–897.
28. **Von Suessenguth, K., and R. Beyerle.** 1936. Uber bakterien-knollchen am spross von *Aeschynomene paniculata* Willd. Hedwigia **75:**234–237.
29. **Yatazawa, M., and H. Susilo.** 1980. Development of upper stem nodules in *Aeschynomene indica* under experimental conditions. Soil Sci. Plant Nutr. (Tokyo) **26:**317–319.
30. **Yatazawa, M., and S. Yoshida.** 1979. Stem nodules in *Aeschynomene indica* and their capacity of nitrogen fixation. Physiol. Plant. **45:**293–295.

Recent Progress in the Microbiology of Hydrothermal Vents†

H. W. JANNASCH AND D. C. NELSON

Woods Hole Oceanographic Institution, Woods Hole, Massachusetts 02543

Hot springs provide geothermic energy in the form of reduced inorganic compounds for microbial growth. The presence of chemosynthetic microorganisms, especially sulfur-oxidizing bacteria, in and around fumaroles, geysers, and other volcanic emanations is long known (30). However, their role as primary producers of organic carbon received little attention because of its relative insignificance in the presence of photosynthetic organisms.

Deep-sea hydrothermal springs or vents differ in two respects. They support extensive populations of certain invertebrates in the virtual absence of photosynthetic input, and, according to recent data, they appear to be quantitatively more important than terrestrial hot springs. Deep-sea vents occur at tectonic ocean floor spreading zones extending for about 25,000 miles within the Atlantic, Pacific, and Indian Oceans. From heat-flow measurements the quantity of vent water emitted has been estimated to amount to about 0.5% of the rate at which the oceans receive water from the earth's rivers (8).

At the ocean floor spreading centers freshly extruded lava contracts on cooling, allowing seawater to permeate several kilometers downward into the newly formed crust. Reacting with basaltic rock at temperatures exceeding 350°C, the seawater transforms into highly reduced "hydrothermal fluid" enriched in metals, hydrogen sulfide, and hydrogen. On rising back to the sea floor, it meets with dissolved free oxygen, either when it mixes with seawater in the porous subsurface layer of pillow lava and emerges as "warm" vents (8 to 23°C) at rates of 1 to 2 cm/s, or when it is discharged directly into bottom seawater as "hot" vents (ca. 350°C) at rates of up to 2 m/s. Hot vents form characteristic chimney-like pipes from precipitates of anhydrite and metal sulfides, as well as black or grey suspensions of polymetal sulfides in the emitted waters. All vent sites studied so far are geochemically different. This is reflected in an equally different occurrence and abundance of micro- and macroorganisms. Comprehensive accounts of the biologically important geochemical features of the deep-sea vents are presented by Edmond et al. (8, 9).

Direct and circumstantial evidence for a number of microbial processes has been reviewed (17). Diving expeditions to the deep-sea vents have been few, especially for biological research. Therefore, much of the research is slowed by the lack of fresh material and, with the exception of work on pure culture isolates, the impossibility of following up new and unexpected observations without long delays. This situation is the major reason for the preliminary and scattered nature of some of the information available. Many colleagues have received samples from diving expeditions. Their published and some unpublished (with permission) results are included in this progress report.

CHEMOSYNTHESIS

The microbial process of greatest importance at the vents, the principal life support for the rich invertebrate populations in the absence of light, is chemosynthesis. Since the inorganic sources of energy for microbial growth at the vents are of geothermic, i.e., abiotic origin, the term "chemosynthetic primary production" has been used (18, 21). So far it has been demonstrated for sulfur-oxidizing, for methanogenic, and possibly for manganese-oxidizing bacteria by pure culture isolations. The bacterial production of sulfide by using geothermal hydrogen has not yet been found but is likely to occur. No ammonia- or nitrite-oxidizing bacteria could be found by a most-probable-number technique (34) at the 21°N (East Pacific Rise) vent site (J. B. Waterbury, personal communication). There is a possibility, as preliminary chemical studies have shown, that nitrification plays a major role at a newly discovered vent site (Guaymas Basin, Gulf of California).

If part of the methane found in the hydrothermic fluid is of truly geothermic origin, its bacterial oxidation could, by definition, be included into primary production. On the other hand, biological methanogenesis as an anaerobic chemosynthetic process has been suggested by experimental observations (3) and by the isolation of a thermophilic methane-producing bacterium from the 21°N vent site (W. J. Jones, J. A. Leigh, F. Mayer, C. R. Woese, and R. S. Wolfe, Arch. Microbiol., in press). Acetogenic bacteria have not yet been isolated. The first

† Contribution no. 5463 of the Woods Hole Oceanographic Institution and Contribution no. 17 of the 21°N Biology Expedition, 1982.

aerobic hydrogen-oxidizing bacterium was recently obtained from tube worm tissue (see below).

For vent water samples the in situ CO_2-fixation data of Tuttle et al. (33) indicate that the spatial distribution of chemosynthesis is extremely variable in and near warm vents. This underscores the fact that incomplete mixing between vent and ambient waters and the uneven distribution of particles within the vent plumes hinder a quantitative assessment of chemosynthesis.

SULFUR BACTERIA

Most of the obligately autotrophic sulfur bacteria (22) isolated from the vicinity of the vents were physiologically and morphologically indistinguishable from *Thiomicrospira pelophila* (28), and strains were isolated repeatedly from the two sites visited to date. In a study at the 21°N vent site the distribution of this highly sulfide-tolerant species was found to be patchy. Whenever found (i.e., two of six dilution series from samples at that site), its abundance was relatively high (Nelson and Jannasch, unpublished data). *Thiomicrospira pelophila* L-12, isolated from a Galapagos Rift vent, has been characterized by its 5S rRNA sequence as different, perhaps significantly, from the type strain of that species (D. Stahl, D. Lane, G. Olsen, and N. Pace, Nature [London], in press). Vent isolates tested were found to be mesophilic in their temperature optima (25 to 33°C); they have doubling times as short as 1 h at 1 atm of pressure, and in situ pressure reduced their growth rates by about one-fourth (27; Nelson and Jannasch, unpublished data). An equivalent inhibition by in situ pressure of total CO_2 incorporation was observed in vent water samples amended by the addition of thiosulfate (33).

Obligately autotrophic vent isolates have been tested to determine whether excretion of organic carbon compounds accompanies balanced growth. In batch cultures approximately 8 to 9% of total [^{14}C]bicarbonate incorporated into cells was subsequently excreted into the medium as acid nonvolatile compounds (27; C. O. Wirsen and H. W. Jannasch, unpublished data). Of the excreted organic matter approximately 80% had a molecular weight of less than 1,000, and virtually all of it could be metabolized by a facultatively autotrophic bacterium also isolated from vent waters. These observations indicate another possible trophic link in vent communities.

Filamentous sulfur bacteria are a conspicuous component of vent communities. At the 21°N vent site, *Thiothrix* sp. appears to predominate, attached to the rock surfaces which are abundant in close proximity to flowing, sulfide-containing waters. By contrast, in the Guaymas Basin, *Beggiatoa* sp. can be found in large aggregates at the surface of sediments percolated with warm sulfide-containing water. In preserved material from that site, filaments of *Beggiatoa* sp. of several width classes ranging up to 150 μm in diameter have been observed (17). The abundance of *Beggiatoa* sp., coupled with recent demonstrations of nitrogen fixation and chemoautotrophy in this organism (24; D. C. Nelson and H. W. Jannasch, Arch. Microbiol., in press), suggests that it may be an important source of fixed carbon and nitrogen in these vent communities. N_2-fixing ability has not been established within the genera *Thiomicrospira* and *Thiobacillus*.

Several mixotrophic oxidizers of reduced-sulfur compounds have been isolated. These strains, although acid producers, are not capable of strictly autotrophic growth but are significantly enhanced by the addition of thiosulfate to media of low organic content (28). They also fix some cell carbon from CO_2. Two of the trophosome isolates (see below) which are stimulated by sulfur compounds may be of this metabolic type.

Enrichments for chemoautotrophic sulfur bacteria, when inoculated with a wide variety of vent materials, most frequently yielded base-producing heterotrophs. These bacteria oxidize thiosulfate to polythionates without an accompanying fixation of CO_2 or increment to growth (28). In water samples from the Galapagos Rift vents, most-probable-number estimates indicated that such base producers comprise from 0.1% to more than 10% of total direct cell counts (33).

OTHER MICROORGANISMS NEWLY OBSERVED OR ISOLATED

Methane was found early in the emitted hydrothermal fluid of most vent types investigated (29, 35). Other dissolved gases present included H_2, CO, and N_2O (7, 23). The microbial generation of some of these gases was suggested by experiments incubating water samples with the addition of some inorganic and organic compounds at high temperatures (3).

Recently the first isolate of a new methanogenic bacterium from samples collected at the base of a "hot smoker" vent at the 21°N vent site was reported (Jones et al., in press). This obligately anaerobic chemosynthetic organism uses H_2 and CO_2 as substrates, requires no organic compounds for growth, and is not stimulated by acetate, formate, yeast extract, or Trypticase. It grows optimally at pH 6 and 85°C. Under these conditions, its generation time is 26 min, a rate unusually high for methanogens so far reported. Its guanine-plus-cytosine (G+C) ratio is 31 mol%. Under UV illumination the colonies of this organism show the bluish-green fluorescence characteristic of methanogens.

This new organism has been classified as a species of the genus *Methanococcus*. It differs from its closest taxonomic neighbor, *Methanococcus thermolithotrophicus* (16), isolated from a shallow (0.5 m) geothermal marine sediment, by its optimum growth conditions (20°C higher and a full pH unit lower) and by a different 16S rRNA oligonucleotide catalog. Whereas all known methanogens contain tetraether or noncyclic diether lipids, the new isolate forms a cyclic diether with a molecular weight two mass units less than the noncyclic diether (P. B. Comita and R. B. Gagosian, Science, in press). According to these authors it is possible that this compound plays a role in the growth of this organism at high temperatures and pressures.

Up to 20% of the cells observed by transmission electron microscopy of microbial mats collected from the immediate surroundings of the vent openings showed extensive cytoplasmic membrane systems. From their specific morphological structure it appears that all of these cells represent type I methylotrophic bacteria (R. Whittenbury, personal communication). This was confirmed by a number of isolations from mat material, water samples, and symbiotic animal tissue (R. S. Hanson, personal communication). Growth was observed on methane and methylamine, but not on methanol. Most strains resemble *Methylococcus capsulatus*, but some of the isolates are represented by long rods and motile, vibrio-shaped cells. Although isolations from *Riftia* sp. trophosome material were positive, no C_1-enzyme activity could be detected in samples taken from frozen ($-10°C$) tissue (M. Lidstrom, personal communication).

Over the last 3 years, a number of new groups of thermophilic microorganisms have been isolated and described from land-based and shallow marine hydrothermal emissions (36, 37). When incubated at normal pressure, these bacteria showed growth optima at temperatures from 65 to near 100°C. Under pressure, growth was observed at up to 110°C with an optimum at 105°C (13, 31). Some of these new organisms are chemolithoautotrophic; others are heterotrophic. Most of the former are sulfur and sulfate reducing or, as in the case of the above-mentioned *Methanococcus thermolithotrophicus* (16), CO_2 reducing.

An indication that such organisms might indeed occur at the vents was suggested by the above-mentioned experiments (3) demonstrating gas production in high-temperature incubations of vent water samples. Recently the incorporation of [³H]adenine into ATP, RNA, and DNA was measured in samples of freshly collected hot smoker minerals at various temperatures and normal pressure. From the data for 90°C the specific growth rate was calculated to be 0.55 h^{-1}, one of the highest values reported to date for a natural population of microorganisms. No isolation and description of highly thermophilic bacteria from warm or hot vents has been published at this time, with the exception of the new *Methanococcus* species mentioned above (Jones et al., in press). Recently microbial growth at at least 250°C was reported (2), a feature to be dealt with elsewhere in this symposium.

Another conspicuously large proportion of cells in microbial mats collected from the immediate vicinity of warm vents represent stalk-forming bacteria. Strains of the genera *Hyphomonas* and *Hyphomicrobium* were subsequently isolated by J. Poindexter (personal communication). Presently these isolates are being characterized and compared with the hitherto known marine strains of these genera (R. M. Weiner, personal communication).

Spirochetes were also readily observed by phase and scanning electron microscopy in freshly collected microbial mats from the Galapagos Rift vent site. After persevering attempts, a slowly growing spirochete was isolated in pure culture (14). It represents the first anaerobic, free-living spirochete from a deep-sea habitat.

Manganese-oxidizing bacteria were isolated from warm and hot water vents as well as from water samples (10). In contrast to strains obtained from ferromanganese nodules, the vent isolates were able to oxidize free Mn^{2+} and did not depend on a prior binding of Mn^{2+} to certain specific absorbents. Furthermore, the Mn^{2+} oxidase was found to be inducible in vent isolates and not constitutive as found with other manganese-oxidizing bacteria.

Still unexplained are the cyanobacteria-like trichomes, resembling the genus *Calothrix*, that were abundant in the microbial mat material collected at the Galapagos Rift vent site (19). It is most likely that their metabolism is chemosynthetic, but in addition to sulfur oxidation, iron oxidation at a low pH maintained within the sheaths has also been proposed, as suggested by the type of heavy iron oxide deposits in the immediate vicinity of the trichomes.

CHEMOSYNTHETIC SYMBIONTS

The major, most immediate, and presumably most efficient transfer of chemosynthetically produced organic carbon to those invertebrates predominantly clustering around the vents is believed to take place by symbiosis. There is no evidence that the copious populations of vestimentiferan tube worms (*Riftia pachyptila*; 20) and the white giant clams (*Calyptogena magnifica*; 4) are receiving a substantial or even minor supply of a photosynthetic food source. At depths of 2,500 to 2,600 m, the animal biomass is commonly orders of magnitude lower than observed at the vents. The major support for the hypothesis of a symbiotic mode of nutrition of the predominant vent invertebrates is based so far on histological (5, 6) and enzymological (11, 12) evidence.

The tube worms (up to 2.5 m long and 5 cm wide) are mouthless and gutless. Their body cavity is filled with a brown spongy tissue which extends over about 75% of the worm's length and may represent more than 50% of an individual specimen's weight. Transmission electron microscopy showed the trophosome to consist largely of procaryotic cells; i.e., no internal membrane-bound organelles, a non-membrane-bound nuclear region, and a gram-negative-type cell wall were found. Lipopolysaccharide tests were strongly positive (6) and indicated the presence of this typical outer cell wall constituent of gram-negative bacteria.

In *Riftia* sp. trophosomes the Calvin cycle enzymes ribulose biphosphate carboxylase and ribulose 5-phosphate kinase were found (11). Both of these enzymes occur only in autotrophic organisms. Also detected were high activities of the following enzymes involved in the transformation of sulfur compounds: rhodanase, cleaving thiosulfate into sulfite and sulfur and catalyzing the transfer of sulfur from thiosulfate to glutathione (32); ATP sulfurylase, activating sulfate by formation of adenosine 5′-phosphosulfate (APS) (25); APS reductase, catalyzing the reversible formation of APS from AMP and sulfite (26). Of these three enzymes, only APS reductase is strictly limited to occurrence in bacteria and, thus, has value for indicating the presence of procaryotic symbionts in animal tissue (C. Cavanaugh, manuscript in preparation). In combination with the finding of the Calvin cycle enzymes in trophosome tissue, enzymatic evidence for the proposed symbiosis

between *Riftia* sp. and chemoautotrophic sulfur bacteria is strong.

These findings do not exclude the possibility that other bacteria may occur and actively metabolize within the trophosome. Scanning electron microscopy showed a predominance of large spherical cells with a diameter of 3 to 5 μm, but also a large variety of smaller long or short rods (6, 17). This variability depends on the location of the sample within the length of the trophosome. It cannot be decided at this point whether the observed variety of shapes and sizes represents polymorphism of a single unique symbiont or an actual morphological and metabolic diversity of several bacterial types present.

Growth of obligately or facultatively chemoautotrophic, sulfur-oxidizing symbionts in the trophosome depends on peculiar properties of the animals' blood system, namely, mechanisms preventing autooxidation and poisoning. Indeed, studies on the extracellular annelid-type hemoglobin of *Riftia* sp. resulted in the finding of a high oxygen affinity and capacity (1) and the presence of a sulfide-binding protein.

Of the isolates thus far obtained (Table 1) from high dilutions of fresh trophosome material, one strain (designated Hsp) resembles the above-mentioned large spherical cells when grown under certain pure culture conditions in the presence of organic compounds. This isolate, however, appears to be a hydrogen-oxidizing bacterium which is apparently not stimulated by reduced sulfur compounds. Two other isolates (E7 and A7) are stimulated by reduced sulfur compounds. The 5S rRNA sequences of these three isolates do not match the predominant procaryotic sequence obtained from frozen (−10°C) trophosome tissue (D. Stahl, personal communication). To date, however, this sequence analysis has been performed on tissue

TABLE 1. Isolates obtained from dilution series of homogenized fresh trophosome material and their preliminary characterization[a]

Strain designation	Isolation dilution[b] (ml of trophosome per tube)	Isolation conditions[c]	DNA (mol% G+C)	Remarks
E7	10^{-5}	Sulfide gradient; 1 mM thiosulfate	63	Sulfide enhances yield
E5	10^{-5}	Sulfide gradient; 1 mM thiosulfate	27	*Cytophaga* sp. likely
G3	10^{-5}	1 mM thiosulfate	39	
Hsp	10^{-6}	1 mM thiosulfate	36	Hydrogen enhances yield
A7	10^{-6}	1 mM thiosulfate (full air)	36	Thiosulfate enhances yield

[a] Nelson, Waterbury, and Jannasch, unpublished data.

[b] Dilution series of trophosome homogenate (in sterile seawater) were done in 1:10 steps. Dilutions presented in this column were the greatest ones showing bacterial growth (except for G3) and yielded the strains indicated.

[c] All isolation media were based on natural seawater and contained 0.2% agar and dilute complex organics and vitamins. Except as noted, the incubation gas mix was 50% methane, 2% oxygen, 0.1% carbon dioxide, and the balance nitrogen.

from a single worm which was not the source of the bacterial isolates.

The fact that the strains of Table 1 have not been shown to be simple chemoautotrophs (prototrophs) does not rule them out as potential trophosome symbionts. It can be argued that an auxotroph might be more readily regulated by the host. Several of these strains (E7, A7, and Hsp) are currently being tested for a significant CO_2 fixation potential and the presence of ribulose bisphosphate carboxylase, both of which appear to be necessary attributes of a worm symbiont (see above).

The *Riftia* sp. trophosome contains in excess of 10^9 procaryotic cells per ml (6). If the majority of these cells are culturable, dilutions greater by about 3 orders of magnitude than those providing successful isolations (Table 1) should yield cultures in an appropriate medium. On the other hand, if the great majority of the procaryotic cells are not culturable, either as a result of damage during trophosome homogenization or as an intrinsic property of the symbionts, contaminants become a major problem. Isolate E5,

for example, based on its pigmentation, cell morphology, and DNA base ratio, is probably a *Cytophaga* sp. and a contaminant. Under such circumstances, the nucleotide characterizations of both host tissues and isolates, as discussed above and below, take on primary significance in verifying the isolation of the actual symbiont(s). Furthermore, symbiont nucleotide analyses performed on appropriate tissues of several worms and clams from different vent locations might provide valuable information on the uniformity of symbionts within a host species.

DNA has been extracted from trophosome tissue ($-80°C$ frozen) and analyzed for base composition by thermal denaturation. The simplest interpretation of the resultant biphasic melting curve (Fig. 1) is that there is, in addition to DNA of a typical invertebrate ratio, a major procaryotic component containing approximately 60 mol% G+C. This base ratio is characteristic of some species of *Thiobacillus* (22). If the genome size of the procaryotic component does not prove to be typically bacterial, the taxonomic inferences based on the G+C ratio would be

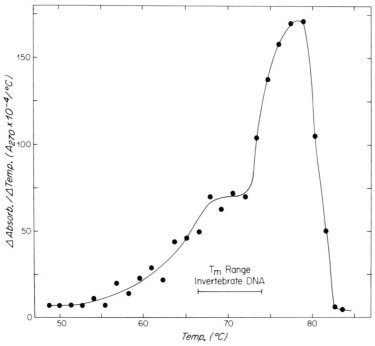

FIG. 1. Thermal transition of DNA extracted from *Riftia pachyptilla* trophosome tissue. From raw data (Abs$_{270}$ versus Temp) the derivative was estimated as $\Delta Abs_{270}/\Delta Temp$ at 1.4°C intervals. This indicates the biphasic nature of the thermal denaturation. For the buffer system employed (0.1× SSC [1× SSC = 0.15 M NaCl plus 0.015 M sodium citrate]), the equation: mol% G+C = $2.44(T_m - 53.9)$ was derived from ~6 DNA standards. Accordingly, the derivative maximum of this figure (estimated $T_m = 78.4°C$) corresponds to a base ratio of 59.8% G+C. The shoulder at approximately 70°C is within the T_m range, which corresponds to the base ratio range (32 to 50 mol% G+C) of invertebrates (34). DNA was extracted from 2.5 g of frozen ($-80°C$) tissue and thermally denatured (heating at 0.25°C min^{-1}) by published methods (15). (Data from Nelson, Waterbury, and Jannasch, manuscript in preparation.)

minimized, and the chances of isolating a culturable symbiont may actually be less than so far assumed.

Analysis of the 5S rRNA of *Calyptogena magnifica* gill tissue indicates a single procaryotic sequence which is significantly different from that of the *Riftia* sp. Likewise, an analysis of the symbiont-containing tissue of a nonvent clam, *Solemya velum*, indicates that its symbiotic component has a 5S sequence as dissimilar from the symbiont of *Riftia* sp. and the symbiont of *Calyptogena* sp. as the latter two are from each other. For reference, the dissimilarity of the 5S sequence of *Escherichia coli* from that of *Photobacterium phosphoreum* is of the same magnitude (Stahl et al., manuscript in preparation).

Presently, emphasis of continuing research in deep-sea vent microbiology is directed toward bacterial thermophilism, nucleotide sequence analysis, and chemosynthetic symbiosis.

Our research is supported by the National Science Foundation (grants OCE80-24253 and OCE81-17560).

LITERATURE CITED

1. **Arp, A. G., and J. J. Childress.** 1983. Sulfide binding by the blood of the hydrothermal vent tube worm *Riftia pachyptila.* Science **219:**295–297.
2. **Baross, J. A., and J. W. Deming.** 1983. Growth of 'black smoker' bacteria at temperatures of at least 250°C. Nature (London) **303:**423–426.
3. **Baross, J. A., M. D. Lilley, and L. I. Gordon.** 1982. Is the CH₄, H₂ and CO venting from submarine hydrothermal systems produced by thermophilic bacteria? Nature (London) **298:**366–368.
4. **Boss, K. J., and R. D. Turner.** 1980. The giant white clam from the Galapagos Rift, *Calyptogena magnifica* species novum. Malacologia **20:**161–194.
5. **Cavanaugh, C. M.** 1983. Symbiotic chemotrophic bacteria in marine invertebrates from sulfide-rich habitats. Nature (London) **302:**58–61.
6. **Cavanaugh, C. M., S. L. Gardiner, M. L. Jones, H. W. Jannasch, and J. B. Waterbury.** 1981. Procaryotic cells in the hydrothermal vent tube worm *Riftia pachyptila* Jones: possible chemoautotrophic symbionts. Science **213:**340–341.
7. **Corliss, J. B., J. Dymond, L. I. Gordon, J. M. Edmond, R. P. von Herzen, R. D. Ballard, K. Green, D. Williams, A. Bainbridge, K. Crane, and T. H. van Andel.** 1979. Submarine thermal springs on the Galapagos Rift. Science **203:**1073–1083.
8. **Edmond, J. M., and K. Von Damm.** 1983. Hot springs on the ocean floor. Sci. Am. **248:**78–93.
9. **Edmond, J. M., K. L. Von Damm, R. E. McDuff, and C. I. Measures.** 1982. Chemistry of hot springs on the East Pacific Rise and their effluent dispersal. Nature (London) **297:**187–191.
10. **Ehrlich, H.** 1983. Manganese oxidizing bacteria from a hydrothermally active area on the Galapagos Rift. Ecol. Bull. **35:**357–366.
11. **Felbeck, H.** 1981. Chemoautotrophic potentials of the hydrothermal vent tube worm, *Riftia pachyptila* (Vestimentifera). Science **213:**336–338.
12. **Felbeck, H., J. J. Childress, and G. N. Somero.** 1981. Calvin-Benson cycle and sulfide oxidation enzymes in animals from sulfide-rich habitats. Nature (London) **293:**291–293.
13. **Fischer, F., W. Zillig, K. O. Stetter, and G. Schreiber.** 1983. Chemolithoautotrophic metabolism of anaerobic extremely thermophilic archaebacteria. Nature (London) **301:**511–513.
14. **Harwood, C. S., H. W. Jannasch, and E. Canale-Parole.** 1982. An anaerobic spirochaete from deep sea hydrothermal vents. Appl. Environ. Microbiol. **44:**234–237.
15. **Herdman, M., M. Janvier, J. B. Waterbury, R. Rippka, R. Y. Stanier, and M. Mandel.** 1979. Deoxyribonucleic acid base composition of cyanobacteria. J. Gen. Microbiol. **111:**63–71.
16. **Huber, H., M. Thomm, H. König, G. Thies, and K. O. Stetter.** 1982. *Methanococcus thermolithotrophicus,* a novel thermophilic lithotrophic methanogen. Arch. Microbiol. **132:**47–50.
17. **Jannasch, H. W.** 1984. Microbial processes at deep sea hydrothermal vents, p. 677–710. *In* R. A. Rona (ed.), Hydrothermal processes at sea floor spreading centers. Plenum Publishing Corp., New York.
18. **Jannasch, H. W., and C. O. Wirsen.** 1979. Chemosynthetic primary production at East Pacific sea floor spreading centers. BioScience **29:**492–498.
19. **Jannasch, H. W., and C. O. Wirsen.** 1981. Morphological survey of microbial mats near deep-sea thermal vents. Appl. Environ. Microbiol. **41:**528–538.
20. **Jones, J. L.** 1980. *Riftia pachyptila,* n. gen., n. sp., the vestimentiferan worm from the Galapagos Rift geothermal vents (Pogonophora). Proc. Biol. Soc. Wash. **93:**1295–1313.
21. **Karl, D. M., C. O. Wirsen, and H. W. Jannasch.** 1980. Deep sea primary production at the Galapagos hydrothermal vents. Science **207:**1345–1347.
22. **Kuenen, J. G., and O. H. Tuovinen.** 1981. The genera *Thiobacillus* and *Thiomicrospira,* p. 1023–1036. *In* M. P. Starr, H. Stolp, H. G. Trüper, A. Balows, and H. G. Schlegel (ed.), The prokaryotes. Springer, Berlin.
23. **Lilley, M. D., M. A. deAngelis, and L. I. Gordon.** 1982. CH₄, H₂, CO and N₂O in submarine hydrothermal vent waters. Nature (London) **300:**48–50.
24. **Nelson, D. C., J. B. Waterbury, and H. W. Jannasch.** 1982. Nitrogen fixation and nitrate utilization by marine and freshwater *Beggiatoa.* Arch. Microbiol. **133:**172–177.
25. **Peck, H. D.** 1962. Comparative metabolism of inorganic sulfur compounds in microorganisms. Bacteriol. Rev. **26:**67–94.
26. **Peck, H. D., T. E. Deacon, and J. T. Davidson.** 1965. Studies on adenosine-5'-phosphosulfate reductase from *Desulfovibrio desulfuricans* and *Thiobacillus thioparus.* Biochim. Biophys. Acta **96:**429–446.
27. **Ruby, E. G., and H. W. Jannasch.** 1982. Physiological characteristics of *Thiomicrospira* sp. isolated from deep sea hydrothermal vents. J. Bacteriol. **149:**161–165.
28. **Ruby, E. G., C. O. Wirsen, and H. W. Jannasch.** 1981. Chemolithotrophic sulfur-oxidizing bacteria from the Galapagos Rift hydrothermal vents. Appl. Environ. Microbiol. **42:**317–324.
29. **Spiess, F. N., K. C. MacDonald, T. Atwater, R. Ballard, A. Carranza, D. Cordoba, C. Cox, V. M. Diaz Garcia, J. Francheteau, J. Guerrero, J. Hawkins, R. Haymon, R. Hessler, T. Juteau, M. Kastner, R. Larson, B. Luyendyk, J. D. Macdougall, S. Miller, W. Normark, J. Orcutt, and C. Rangin.** 1980. East Pacific Rise: hot springs and geophysical experiments. Science **207:**1421–1433.
30. **Stanier, R. Y., E. A. Adelberg, and J. Ingraham.** 1976. The microbial world. Prentice-Hall, Englewood Cliffs, N.J.
31. **Stetter, K. O.** 1982. Ultrathin mycelia-forming organisms from submarine volcanic areas having an optimum growth temperature of 105°C. Nature (London) **300:**258–260.
32. **Suzuki, I.** 1974. Mechanisms of inorganic oxidation and energy coupling. Annu. Rev. Microbiol. **28:**85–101.
33. **Tuttle, J. H., C. O. Wirsen, and H. W. Jannasch.** 1983. Microbial activities in the emitted hydrothermal waters of the Galapagos Rift vents. Mar. Biol. **73:**293–299.

34. **Watson, S. W., F. W. Valois, and J. B. Waterbury.** 1981. The family Nitrobacteraceae, p. 1005–1022. *In* M. P. Starr, H. Stolp, H. G. Trüper, A. Balows, and H. G. Schlegel (ed.), The prokaryotes. Springer, Berlin.

35. **Welhan, J. A., and H. Craig.** 1979. Methane and hydrogen in East Pacific Rise hydrothermal fluid. Geophys. Res. Lett. **6:**829.

36. **Zillig, W., A. Gierl, G. Schreiber, S. Wunderl, D. Janeko-** vic, **K. O. Stetter, and H. P. Klenk.** 1983. The archaebacterium *Thermofilum pendens* represents a novel genus of the thermophilic, anaerobic, sulfur respiring Thermopto-teales. Syst. Appl. Microbiol. **4:**79–87.

37. **Zillig, W., R. Schnabel, and J. Tu.** 1982. The phylogeny of archaebacteria, including novel anaerobic thermoacido-philes in the light of RNA polymerase structure. Natur-wissenschaften **69:**197–204.

Endolithic Microorganisms in Extreme Dry Environments: Analysis of a Lithobiontic Microbial Habitat

E. IMRE FRIEDMANN AND ROSELI OCAMPO-FRIEDMANN

Department of Biological Science, Florida State University, Tallahassee, Florida 32306

LITHOBIONTIC MICROBIAL HABITATS: TERMINOLOGY, DISTRIBUTION

Rock-inhabiting (lithobiontic) microorganisms are widespread in nature, yet information on this group, despite their obvious ecological and geological interest, is comparatively sparse and widely scattered in the literature. Microorganisms colonize rock surfaces, or the rock fabric adjacent to the surface, in a wide variety of climates and environments. These range from hot deserts to Antarctica, from Alpine to submerged marine and freshwater rocks, and from caves to the surfaces of buildings and monuments. Perhaps the only general statement that can be made of these habitats is that lithobiontic microorganisms colonize rock surfaces that are exposed, i.e., not covered by soil, or occupied by more aggressive higher organisms such as large lichens, mosses, or sessile animals. Endolithic microorganisms appeared very early in the history of Earth: they are known since the pre-Cambrian period (2).

As we will discuss later, the key for understanding the ecology of lithobiontic microorganisms often lies in the physical parameters of their immediate environment. This environment is determined by the climate in the millimeter range, for which the term "nannoclimate" is proposed. The nannoclimate of lithic microbial habitats can be very different not only from the large-scale macroclimate but also from the microclimate, traditionally defined as the climate in the size range of plants and animals. Yet, for most lithic microbial environments, information on the nannoclimate is scarce or nonexistent.

In the past, terms describing various lithobiontic habitats have been used by different authors with divergent meanings. In Table 1, we attempt to summarize the recent terminology (19, 25), keeping well in mind that by defining types we do not wish to imply that all lithobiontic microbial habitats can be neatly classified in these categories, as, in fact, intermediary forms exist. In such cases, the categories are still useful as descriptive terms to characterize intermediary types. The listings of representative occurrences and of types of microorganisms are not exhaustive, nor are the references, which are provided only as a help to the reader in finding further information in the literature.

ENDOLITHIC MICROORGANISMS IN DESERTS

The following survey is limited to endolithic microorganisms in extreme arid environments, including both hot deserts and frigid polar deserts.

Among the endolithic types listed in Table 1, euendoliths seem to be absent in extreme dry environments, apparently because rock-boring activity requires water. Euendolithic lichens ("endolithic" lichens of the lichenological literature) occur mostly in arid to semiarid climates, but cannot exist in extreme arid conditions.

Such extreme arid rocks may be colonized by chasmo- and cryptoendolithic microorganisms. The following short list of localities is an indication of their widespread occurrence: chasmoendoliths—Negev (19), southwestern United States and Mexico (7, 15), central Asia (23, 34), Arctic (28, 37), Antarctica (1, 8, 12), Namibia, Sinai, central Australia, Atacama (unpublished data); cryptoendoliths—Negev (6, 19), Antarctica (8, 12, 20), Arctic (5), Sinai, southwestern United States (unpublished data).

The standing biomass is relatively large: biomass estimates based on Kjeldahl nitrogen, ATP, and chlorophyll determinations ranged from 32.25 to 176.75 g m^{-2} of rock surface in Antarctica (18) and from 37.5 to 185.0 g m^{-2} of rock surface in the Negev and Sonoran deserts (11).

ROCKS AS SUBSTRATES FOR ENDOLITHIC GROWTH

Not all types of rocks can be colonized by endolithic microorganisms; colonization is dependent upon certain physical properties of the rock (such as color, opacity, and porosity or presence of fissures) rather than its chemical composition. As endolithic microbial communities always seem to include photosynthetic primary producers (cyanobacteria or algae), only light-colored rocks and rocks translucent to some extent are suitable for colonization. The availability of space for microbial colonization inside the rock is another fundamental necessity: thus, narrow fissures and cracks of weathered rocks (granite, calcite, quartzite, dolomite) may serve as a substrate for chasmoendoliths,

TABLE 1. Lithobiontic habitats

Habitat type and description		Representative occurrences and references	Representative microorganisms[a]
Hypolithic: under pebbles and small stones lying on soil		Hot deserts (desert pavements) (15, 19)	Cyanobacteria, green algae
Endolithic: inside rocks	*Euendolithic*: in spaces partly or wholly excavated by organisms actively penetrating into the rock fabric	Aquatic (marine and freshwater) (26, 27)	Cyanobacteria, green algae, red algae, brown algae, filamentous fungi
		Semiarid to arid, alpine, tropical (3, 24)	Lichens, cyanobacteria
		Surfaces of buildings and monuments (4)	Cyanobacteria, filamentous fungi, yeasts
	Chasmoendolithic: in fissures and cracks	Semiarid to extreme arid, hot deserts (7, 15, 19)	Cyanobacteria
		Semiarid to extreme arid, polar deserts (1, 6, 8, 12)	Cyanobacteria, green algae, *Xanthophyceae*
		Alpine	Cyanobacteria, green algae
	Cryptoendolithic: in structural cavities of porous rocks, typically under a rock crust	Extreme arid, hot deserts (6, 7, 15, 19)	Cyanobacteria
		Extreme arid, polar deserts (5, 8, 12, 20)	Cyanobacteria, lichens, green algae, filamentous fungi, yeasts
Epilithic: on surface of rocks		Aquatic (marine and freshwater)	Cyanobacteria, various algae
		Caves (10, 22, 24)	Cyanobacteria, red algae, green algae
		Extreme arid, hot deserts (desert varnish) (31, 38)	Fungi
		Alpine (24)	Cyanobacteria, green algae

[a] Unidentified heterotrophic bacteria are present in most or all habitat types.

and porous rocks (such as well-sorted sandstones with ample air spaces) can be inhabited by cryptoendoliths. A further condition of microbial colonization seems to be a certain minimal density of the substrate, as rocks with low specific gravity, such as rhyolite, do not support endolithic growth.

Because of these rather stringent requirements, the occurrence of endolithic microorganisms and the distribution of the chasmoendolithic and cryptoendolithic types of colonization are ultimately dependent on (besides climatic factors) the presence of suitable rock types as surface outcrops.

The morphology of endolithic colonization in cold and hot deserts is remarkably similar (Fig. 1–3). The uppermost 1 to 3 mm of the rock are free from microorganisms. Colonies are present in the next few millimeters, where the microorganisms adhere to or grow between the rock crystals (Fig. 4 and 5). The depth of the occupied zone depends on physical properties of the rock and the complexity of the microbial community. Chasmoendolithic and cryptoendolithic habitats

FIG. 1. Chasmoendolithic microbial colonization (cyanobacteria and eucaryotic algae) in vertically fractured friable marble. Coastal desert, Southern Victoria Land, Antarctica. After Friedmann (9).

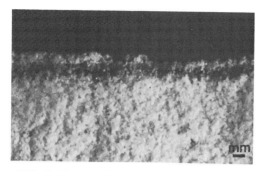

FIG. 2. Cryptoendolithic microbial colonization (cyanobacteria) in vertically fractured Nubian sandstone. Sinai desert. After Friedmann (11).

are different in that the fissures of the former are (to varying degrees) open to the outer surface of the rock, whereas in the latter case the airspace system of the porous rock is sealed from the outside by a continuous surface crust. This crust is to some extent permeable to gases and liquids, but not to cellular organisms (6, 20). The impact of this insulating effect on the internal airspace system will be discussed later.

ENDOLITHIC MICROORGANISMS IN HOT DESERTS

In spite of the morphological similarities between hot and cold desert endolithic habitats, the composition of the communities shows considerable differences. In hot deserts, all endolithic samples we examined contained only procaryotes, which form a narrow green zone a few millimeters below and parallel to the rock crust (Fig. 2). The photosynthetic primary producers are cyanobacteria, nearly always members of the genus *Chroococcidiopsis* (or related genera from the order *Pleurocapsales*; the taxonomy of this group needs critical revision). A coccoid cyanobacterium, morphologically corresponding to *Chroococcus turgidus* (Kütz.) Näg. (6, 36), was found in a single cryptoendolithic sample in the Negev desert (Fig. 4). The cyanobacteria form mostly monospecific populations in the rock, accompanied by some small colorless or orange-pigmented nonphotosynthetic bacteria (unidentified rods and cocci).

ENDOLITHIC MICROORGANISMS IN THE ANTARCTIC COLD DESERT

The endolithic microbial communities in polar deserts mostly show a higher level of complexity than those in hot deserts. Although the first cryptoendolithic sample we described from the Antarctic cold desert (20) was similar to those found in hot deserts and contained a single cyanobacterium (*Chroococcidiopsis* sp.; unpublished data), later extensive studies (9, 13)

showed that these simple cyanobacterial endolithic communities are rare in Antarctica. In most cases, endolithic communities contain a number of organisms, including eucaryotes, which are dominant. This is true for both chasmoendolithic and cryptoendolithic communities (8, 12).

In this paper, we will discuss mainly the cryptoendolithic lichen communities that have been the subject of our detailed studies. These communities live in the frigid Ross Desert (= "dry valleys") in Southern Victoria Land, Antarctica, colonizing the porous Beacon sandstones. They can be compared with hot desert cryptoendolithic communities in the Negev and Sinai, which colonize the morphologically similar Nubian sandstones.

The cryptoendolithic lichen communities in sandstone rocks in Southern Victoria Land form

FIG. 3. Cryptoendolithic lichen community in Beacon sandstone, McMurdo upland, Southern Victoria Land, Antarctica. (a) Surface of rock showing biogenous exfoliative weathering. Original. (b) Vertically fractured rock showing colonization by cryptoendolithic lichen (upper, thin black zone followed by white zone) and by *Hemichloris antarctica* and cyanobacteria (lower, wide green zone). After Friedmann (8).

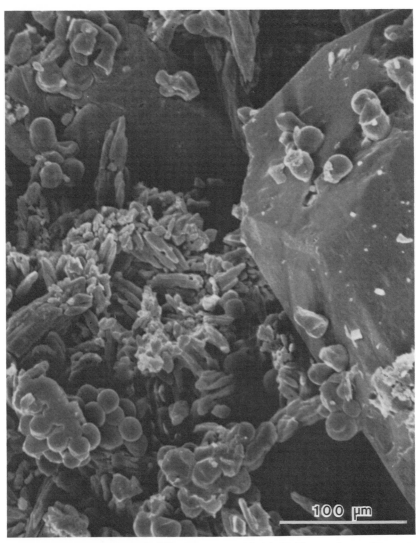

FIG. 4. Cryptoendolithic cyanobacteria (*Chroococcus turgidus*) in Nubian sandstone, Negev desert. Fractured rock surface with large hexagonal quartz crystals, small platelet-like matrix crystals, and spherical bacterial cells. Scanning electron micrograph. After Friedmann (6).

multicolored parallel bands under the surface (Fig. 3b). A typical growth shows an upper black zone about 1 mm thick, then a white zone about 2 to 4 mm thick, and below this an often indistinct green zone of varying thickness. The black and white zones are formed by filamentous fungi (heavily pigmented in the black zone) and algae from the genera *Trebouxia* and *Pseudotrebouxia* (P. Archibald, unpublished data). The fungi and algae together form a symbiotic lichen association that is very different from the usual thalloid lichen organization. Here the loose filaments and cell clusters grow between and around the crystals of the rock substrate (Fig. 5), as if the lichen were embedded in the rock matrix. The

green zone was interpreted earlier (12) as part of the lichen. Recently, it has been recognized (E. Tschermak-Woess and E. I. Friedmann, submitted for publication) that this zone is formed by a free-living green alga, *Hemichloris antarctica* (proposed name). Despite the presence of fungal filaments, this alga does not form a lichen association. In culture, *H. antarctica* is damaged by temperatures above 20°C and can therefore be considered a psychrophile. It is often accompanied by similarly unlichenized coccoid cyanobacteria from the genera *Chroococcidiopsis* or *Gloeocapsa*. In the specimen illustrated in Fig. 3b, two nonlichenized photosynthetic microorganisms occupy different positions in the green

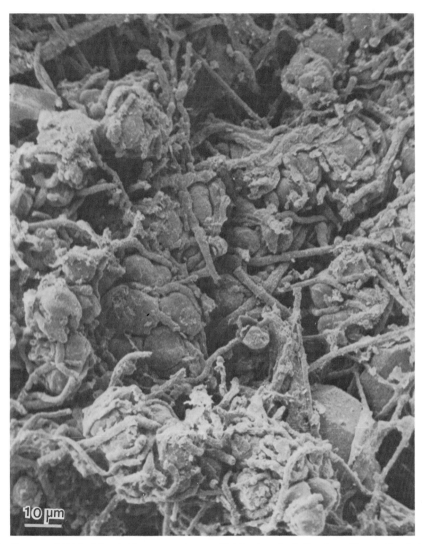

FIG. 5. Cryptoendolithic lichen in Beacon sandstone, McMurdo upland, Southern Victoria Land, Antarctica. Fractured rock surface with spherical algal cells and surrounding fungal hyphae growing around and between quartz crystals. Scanning electron micrograph. (After Friedmann [12]; copyright 1982 by the American Association for the Advancement of Science.)

(lowermost) zone. In this zone, the upper level is colonized by *Hemichloris antarctica* and the lower, by *Chroococcidiopsis* sp. These organisms may have been forced by the more aggressive lichens to the lower, highly shaded zone, an unadvantageous position. The cryptoendolithic lichen community also contains colorless and pigmented nonphotosynthetic bacteria (12).

MICROORGANISM–SUBSTRATE INTERACTION

The activity of the cryptoendolithic lichen community results in the mobilization of iron compounds in the rock as well as in a character-

istic exfoliative weathering pattern. Beacon sandstones often contain iron compounds. These coat the surface of the quartz crystals, coloring the rock from straw color to deep reddish brown. In the lichen zone, iron compounds are, as a rule, absent, and the rock crystals here appear white (12). It is not known at this time which of the microorganisms is active in the solubilization of iron compounds.

The exfoliative weathering on the rock surface (Fig. 3a) is apparently also due to the activity of the organisms. The cementing substance between the sandstone grains is solubilized at the level of the lichen, and the covering rock crust peels off. Portions of the lichen thus exposed are dispersed by wind and may serve as reproductive structures, but the bulk probably falls on adjacent soils where it becomes the source of organic matter for decomposers. Evidence in support of this assumption is the repeated appearance of lichen constituent fungi and algae in sprinkle-plate cultures of soils from the vicinity of rocks colonized by cryptoendolithic lichens (H. S. Vishniac, personal communication).

After peeling of the rock crust, the fungus hyphae penetrate deeper into the substrate, and a new crust is formed on the surface. Subsequent layers are "sliced off" in this way, resulting in the characteristic exfoliating weathering pattern (Fig. 3a). There are no reliable estimates for the time scale of this process at this time, but rock crust formation, being a geological process, is certainly very slow compared with the biological processes. It is therefore likely that the organisms of the lichen community living *under* such a crust are of considerable age.

ENDOLITHIC ENVIRONMENT IN THE ANTARCTIC DESERT: NANNOCLIMATE

To understand the adaptive strategies of the Antarctic cryptoendolithic lichen community, it is necessary to consider the physical and chemical parameters of their environment.

The climate of the ice-free mountainous areas of the McMurdo upland in Southern Victoria Land is one of the most inhospitable on our globe. Summer air temperatures are mostly between -15 and $0°C$, and relative humidity ranges between about 15 and 75%. The aridity of the region is due to the dry katabatic (downslope) winds descending from the Antarctic ice plateau, which contribute greatly to the harshness of the climate. Except for the very infrequent fruiting lichens in rock fissures or in small protected patches on the rocks (16), there are no living organisms on the surface. Animal life seems to be completely absent.

The nannoclimate of the rock surface and the upper few millimeters below is a controlling factor in the pattern of microbial colonization. Thus, the abiotic conditions on the rock surface can be attributed mainly to the rapid temperature oscillations at around $0°C$, which result in a rapid sequence of freezing and thawing. These fluctuations are caused by frequent gusts of katabatic winds. The surface, warmed by solar radiation to well above $0°C$, rapidly freezes at each wind gust (14). At the same time, however, the endolithic zone under the surface, protected by the rock crust, is little affected. There, solar radiation raises the temperature to well above the ambient, up to about $10°C$. It should be pointed out that such favorable conditions exist only at rock faces exposed to the sun (mainly facing north); surfaces facing south remain at subzero air temperatures and are not colonized by cryptoendoliths. During a typical sunny summer day, temperatures in the endolithic lichen zone may remain above $0°C$ for about 9 h and above $-6°C$ (lower limit of photosynthetic activity) for about 13 h (29, 30).

Solar radiation (quantum flux density) in the photosynthetically active range at the surface may amount to about 2,000 $\mu E\ m^{-2}\ s^{-1}$, comparable to values measured in hot deserts. Inside the rock substrate, there is a steep light gradient. According to preliminary results of ongoing research (E. I. Friedmann and J. Nienow, unpublished data), approximately 0.1% of the ambient light energy reaches the upper level of the lichen zone (immediately under the rock crust), and approximately 0.01% reaches the green (*Hemichloris*) zone under the lichen. These values refer to dry rocks, a condition under which photosynthesis may be minimal. Wetting the rock increases the rate of transmission by about 10-fold (to ca. 1% under the rock crust and ca. 0.1% in the lower green zone).

Availability of water is, of course, of critical significance. Snowfall is irregular, and much of the snow sublimates or is blown away by wind without melting. Under certain weather conditions, however, snow does melt on the rock surface (9), and the water imbibed by the rocks is retained for extended periods of time, even during very low ambient relative humidities (30).

Inorganic nutrients seem to be available from the rock substrate (12). Specifically, nitrates and ammonium compounds are generally abundant in the upper few centimeters of the rocks, a fact that accounts for the absence of N fixers in the endolithic microbial communities (17).

ADAPTATIONS IN THE ANTARCTIC DESERT

This short survey of nannoclimatic parameters explains one of the remarkable features of the Antarctic cryptoendolithic microhabitat: in the midst of a truly inhospitable macroenvironment, there exists a microscopic environmental pocket where conditions are comparatively mild. These favorable conditions inside the rock are in sharp contrast to the abiotic rock surface. In fact, CO_2 exchange measurements in the laboratory (29) confirmed that the temperature range of photosynthesis in cryptoendolithic lichens (-9 to $-6°C$ minimum, about 10 to $9°C$ optimum) is similar to that of the crustose epilithic Antarctic lichens living in less extreme climates (32). These data indicate that cryptoendolithic lichens do not show any unusual physiological adaptation to their environment.

The problem organisms face in this environment is how to gain access to the microscopic niches with a favorable microclimate. The narrow cryptoendolithic airspace is accessible only to unicellular organisms, not to thallose lichens. The singular adaptive achievement of the Antarctic cryptoendolithic lichens is the change of their growth form from thallose to filamentous, which enables them to penetrate the porous rock substrate. This process requires a rather unique morphogenetic (rather than physiological) adaptation—a change of growth form that resulted in an organization unique among lichens.

ADAPTATIONS IN HOT DESERTS

Information on the endolithic nannoclimate in hot deserts is sparse. Rock temperature in the endolithic zone reaches 47°C on a typical summer day in the Sonoran desert (7, 11), and this measurement may not even represent the maximum. Water from the generally low and infrequent rainfall is probably not effectively utilized by the endolithic microbial community. The main source of water seems to be dew condensation, which, in many desert areas, occurs with regular frequency and is more or less evenly distributed over the whole year (15). Water condensation, or at least a significant rise in relative humidity to the point of availability for microorganisms, is probably very frequent during the cool hours of the night, followed again by a sharp drop in humidity as rock temperature rises at daylight.

The effect of water stress on five *Chroococcidiopsis* strains and *Chroococcus* sp. N41, all isolated from cryptoendolithic or hypolithic (growing under the surface of stones) habitats in the Negev desert, was investigated in the laboratory (35). These organisms appear to have different adaptive responses. Strains of *Chroococcidiopsis* (the type generally present in hot desert endolithic habitats) show no particular adaptation to grow or photosynthesize at lower water potentials, whereas the less widespread *Chroococcus* sp. N41 does. Both types share the ability to tolerate extreme desiccation and to resume photosynthesis within a very short time after rewetting.

COMPARATIVE ASPECTS: HOT AND COLD DESERTS

Endolithic microorganisms in hot deserts and in Antarctica share a common survival strategy: they are capable of "switching" their metabolic activities on and off in response to rapid changes in environmental conditions. In the Antarctic environment in winter, or in the temporary absence of direct insolation, the organisms are inactive in a frozen state, which they can apparently tolerate without damage. When temperatures rise as a result of insolation, metabolic activity is resumed. Under these conditions it is not surprising that eucaryotic organisms such as green algae and fungi find accommodation in the endolithic environment.

In hot deserts, conditions are different. A favorable combination of available water and moderate temperature range occurs in early morning after dew fall. Later in the day, the temperature rises and the relative humidity drops drastically. The combination of temperature increase in a hydrated state and the concomitant loss of water imposes a severe environmental stress that eucaryotic organisms do not seem to tolerate. This is probably the reason that, in hot deserts, endolithic microbial communities are composed of procaryotes only, and the number of species is considerably lower than it is in the Antarctic desert.

It seems, therefore, that the endolithic environment in hot deserts is much harsher and more extreme than its counterpart in the Antarctic polar desert. This is true in spite of the fact that the outside environment in the Antarctic desert is more "hostile" to life than is that in hot desert areas (11).

ENDOLITHIC HABITAT AS A CLOSED SYSTEM

A peculiar feature of the cryptoendolithic habitat in both hot and cold deserts is the rigid limitation imposed by the fact that the organisms exist inside a finite volume of space. This factor has a profound controlling effect on the entire endolithic ecosystem.

The sharp upper and lower boundaries of the cryptoendolithic zone in the rock substrate (Fig. 2 and 3) are apparently due to physical gradients such as light and temperature. Within these confines, the volume of the airspace available for the organisms is limited. As pointed out earlier, the presence of a rock crust covering suggests that these endolithic communities are of considerable age, and it is therefore easy to see that, even at slow growth rates, available space sooner or later becomes saturated. It seems to be safe to assume that a steady state ensues in which photosynthetic activity in the ecosystem is balanced by respiration. These conditions create a "closed ecosystem," a concept discussed in some detail by Maguire et al. (33). In such a system, the degree of closure may range from biological closure (closed to passage of organisms) to adiabatic isolation (closed to the flow of matter and energy).

On this scale, cryptoendolithic communities may belong to the most highly closed ecosystems that exist in nature. They are closed biologically (at least on the biological time scale), and at the same time the flow of matter is significantly restricted. CO_2 gas-exchange measurements in intact and broken Antarctic rocks (29) as well as measurement of $^{14}CO_2$ uptake both in nature and in the laboratory (J. R. Vestal, unpublished data) show that CO_2 exchange through the rock crust is very low, indicating that an internal CO_2 pool exists in the endolithic airspace. This may well serve as the control mechanism in regulating photosynthesis, a necessity in the physically limited space of the cryptoendolithic microenvironment (13).

It is evident from this short survey that the presence of environmental extremes and some peculiar adaptations is not the only salient feature of the endolithic habitat in deserts. Equally significant is the fact that this habitat is an insulated and protected niche within an extreme and inhospitable macroenvironment, a singularly closed system, and an integral microscopic world upon which outside conditions have only limited influence.

The interest of these unique habitats for general ecological theory is manifold. Because of their insulated nature, the endolithic habitats in deserts may, in many respects, resemble naturally occurring "microcosms." Desert cryptoendolithic communities can also serve as models for extremely simple ecosystems, as they lack secondary consumers and predators. The pattern of nutrient flow in the Antarctic cryptoendolithic habitat reveals an unusually simple ecosystem (12), and hot desert cryptoendolithic ecosystems, which consist only of primary producers (cyanobacteria) and decomposers (heterotrophic bacteria), are even more simple.

As these insulated microbial habitats may represent the last footholds of life in a gradually deteriorating environment (12), they can serve as models for certain exobiological scenarios, as discussed elsewhere (21). Changes in microbial colonization patterns in desert rocks can be used as indicators for paleoclimate (3). The geomicrobiological significance of biogenous rock weathering due to the activity of cryptoendolithic microorganisms is evident (12).

It is obvious from this short survey that many aspects of the biology of lithobiontic microorganisms await further study. It is nearly certain that future search will reveal still unknown and unusual habitats, organisms, and adaptations. Similarly, progress in nannometeorological instrumentation and theory is likely to open new perspectives in understanding patterns of distribution and ecology.

Thanks are due to Anne B. Thistle for critically editing the manuscript.

This research was supported by National Aeronautics and Space Administration grant NSG 7337 and by National Science Foundation grant DPP 80-17581.

LITERATURE CITED

1. **Broady, P. A.** 1981. The ecology of chasmolithic algae at coastal locations in Antarctica. Phycologia **20:**259–272.
2. **Campbell, S. E.** 1982. Precambrian endoliths discovered. Nature (London) **299:**429–431.
3. **Danin, A., R. Gerson, K. Marton, and J. Garty.** 1982. Patterns of limestone and dolomite weathering by lichens and blue-green algae and their palaeoclimatic significance. Palaeogeogr. Palaeoclimatol. Palaeoecol. **37:**221–233.
4. **Eckhardt, F. E. W.** 1978. Microorganisms and weathering of a sandstone monument, p. 675–686. *In* Wolfgang Krumbein (ed.), Proceedings of the 3rd International Symposium on Environmental Biogeochemistry and Geomicrobiology, 1977. Ann Arbor Science Publishers, Inc., Ann Arbor, Mich.
5. **Eichler, H.** 1981. Kleinformen der hocharktischen Verwitterung im Bereich der Oobloyah Bay, N-Ellesmere Island, N.W.T., Kanada-Formengenese und Prozesse. Heidelb. Geographische Arbetien **69:**465–486.
6. **Friedmann, E. I.** 1971. Light and scanning electron microscopy of the endolithic desert algal habitat. Phycologia **10:**411–428.
7. **Friedmann, E. I.** 1972. Ecology of lithophytic algal habitats in Middle Eastern and North American deserts, p. 182–185. *In* L. E. Rodin (ed.), Ecophysiological foundation of ecosystems productivity in arid zones. Nauka, U.S.S.R. Acad. Sci., Leningrad.
8. **Friedmann, E. I.** 1977. Microorganisms in Antarctic desert rocks from dry valleys and Dufek Massif. Antarct. J. U.S. **12:**26–30.
9. **Friedmann, E. I.** 1978. Melting snow in the dry valleys is a source of water for endolithic microorganisms. Antarct. J. U.S. **13:**162–163.
10. **Friedmann, E. I.** 1979. The genus *Geitleria* (Cyanophyceae or Cyanobacteria): distribution of *G. calcarea* and *G. floridana* n.sp. Plant Syst. Evol. **131:**169–178.
11. **Friedmann, E. I.** 1980. Endolithic microbial life in hot and cold deserts. Origins Life **10:**223–235.
12. **Friedmann, E. I.** 1982. Endolithic microorganisms in the Antarctic cold desert. Science **215:**1045–1053.
13. **Friedmann, E. I., R. O. Friedmann, L. Kappen, C. P. McKay, and J. R. Vestal.** 1983. The cryptoendolithic microbial community: a "closed" ecosystem in nature, p. 101. *In* D. L. DeVincenzi and L. G. Pleasant (ed.), First Symposium on Chemical Evolution and the Origin and Evolution of Life. NASA Conf. Publ. 2276. NASA, Moffett Field, Calif.
14. **Friedmann, E. I., R. O. Friedmann, and C. P. McKay.** 1981. Adaptations of cryptoendolithic lichens in the Antarctic desert, p. 65–60. *In* Colloque sur les Écosystèmes Subantarctiques. Paimpont, C.N.F.R.A., No. 51. Comité National Français des Recherches Antarctiques, Paris.
15. **Friedmann, E. I., and M. Galun.** 1974. Desert algae, lichens, and fungi, p. 165–212. *In* G. W. Brown (ed.), Desert biology, vol. 2. Academic Press, Inc., New York.
16. **Friedmann, E. I., J. Garty, and L. Kappen.** 1980. Fertile stages of cryptoendolithic lichens in the dry valleys of southern Victoria Land. Antarct. J. U.S. **15:**166–167.
17. **Friedmann, E. I., and A. P. Kibler.** 1980. Nitrogen economy of endolithic microbial communities in hot and cold deserts. Microb. Ecol. **6:**95–108.
18. **Friedmann, E. I., P. A. LaRock, and J. O. Brunson.** 1980. Adenosine triphosphate (ATP), chlorophyll, and organic nitrogen in endolithic microbial communities and adjacent soils in the dry valleys of southern Victoria Land. Antarct. J. U.S. **15:**164–166.
19. **Friedmann, E. I., Y. Lipkin, and R. O. Paus.** 1967. Desert algae of the Negev (Israel). Phycologia **6:**185–200.
20. **Friedmann, E. I., and R. Ocampo.** 1976. Endolithic blue-green algae in the dry valleys: primary producers in the Antarctic desert ecosystem. Science **193:**1247–1249.
21. **Friedmann, E. I., and R. Ocampo-Friedmann.** 1984. The Antarctic cryptoendolithic ecosystem: relevance to exobiology. Origins of Life (in press).

22. **Friedmann, I.** 1961. Ecology of the atmophytic nitrate-alga *Chroococcidiopsis kashaii* Friedmann. Studies in cave algae from Israel IV. Arch. Microbiol. **42:**42–45.

23. **Glazovskaya, M. A.** 1950. Vyvetrivanie gornykh porod v nival'nom poyase tsentral'nogo Tyan-Shanya (Rock weathering in the arable belt of Central Tyan-Shan). Tr. Pochv. Inst., Akad. Nauk SSSR **34:**28–48.

24. **Golubić, S.** 1967. Algenvegetation der Felsen. Schweizerbart, Stuttgart.

25. **Golubić, S., I. Friedmann, and J. Schneider.** 1981. The lithobiontic ecological niche, with special reference to microorganisms. J. Sediment. Petrol. **51:**475–478.

26. **Golubić, S., R. D. Perkins, and K. J. Lukas.** 1975. Boring microorganisms and microborings in carbonate substrates, p. 229–259. *In* R. W. Frey (ed.), The study of trace fossils. Springer-Verlag, New York.

27. **Golubić, S., and J. Schneider.** 1979. Carbonate dissolution, p. 107–129. *In* P. A. Trudinger and D. J. Swaine (ed.), Biogeochemical cycling of mineral-forming elements. Elsevier, Amsterdam.

28. **Gromov, B. V.** 1957. Mikroflora skal'nykh porod i primitivnykh pochv nekotorykh severnykh raionov SSSR (The microflora of rock layers and primitive soils of some northern districts of the U.S.S.R.). Mikrobiologya **26:**52–59.

29. **Kappen, L., and E. I. Friedmann.** 1983. Ecophysiology of lichens in the dry valleys of Southern Victoria Land, Antarctica. II. CO_2 gas exchange in cryptoendolithic lichens. Polar Biol. **1:**227–232.

30. **Kappen, L., E. I. Friedmann, and J. Garty.** 1981. Ecophysiology of lichens in the dry valleys of Southern Victoria Land, Antarctica. I. Microclimate of the cryptoendolithic lichen habitat. Flora (Jena) **171:**216–235.

31. **Krumbein, W. E., and K. Jens.** 1981. Biogenic rock varnishes of the Negev desert (Israel), an ecological study of iron and manganese transformation by cyanobacteria and fungi. Oecologia (Berlin) **50:**25–38.

32. **Lange, O. L., and L. Kappen.** 1972. Photosynthesis of lichens from Antarctica, p. 83–95. *In* G. A. Llano (ed.), Antarctic terrestrial biology, Antarct. Res. Ser. vol. 20. American Geophysical Union, Washington, D.C.

33. **Maguire, B., Jr., L. B. Slobodkin, H. J. Morowitz, B. Moore III, and D. B. Botkin.** 1980. A new paradigm for the examination of closed ecosystems, p. 30–68. *In* J. P. Giesy, Jr. (ed.), Microcosms in ecological research. D.O.E. Symp. Ser. 52. U.S. Department of Energy, Washington, D.C.

34. **Odintsova, S. V.** 1941. Obrazovanie selitry v pustyne (Niter formation in deserts). Dokl. Akad. Nauk SSSR **32:**578–580.

35. **Potts, M., and E. I. Friedmann.** 1981. Effects of water stress on cryptoendolithic cyanobacteria from hot desert rocks. Arch. Microbiol. **130:**267–271.

36. **Potts, M., R. O. Friedmann, M. A. Bowman, and B. Tozün.** 1983. *Chroococcus* S24 and *Chroococcus* N41 (cyanobacteria): morphological, biochemical and genetic characterization and effects of water stress on ultrastructure. Arch. Microbiol. **135:**81–90.

37. **Royzin, M. B.** 1960. Mikroflora skal i primitivnykh pochv vysokogornoi arkticheskoi pustyni (The microflora of rocks and primitive soils of the high altitude arctic desert). Bot. Zh. (Leningrad) **45:**997–1008.

38. **Staley, J. T., F. Palmer, and J. B. Adams.** 1982. Microcolonial fungi: common inhabitants on desert rocks? Science **215:**1093–1095.

Evidence for Microbial Growth in High-Pressure, High-Temperature Environments

JOHN A. BAROSS, JODY W. DEMING, AND ROBERT R. BECKER

School of Oceanography and Department of Biochemistry, Oregon State University, Corvallis, Oregon 97331, and Chesapeake Bay Institute, Johns Hopkins University, Shady Side, Maryland 20764

It was Brock's contention that conditions providing for the occurrence of liquid water, and not some absolute temperature maximum, define the environmental limits of life (6). This hypothesis was based on observations at Yellowstone National Park that bacteria existed in most thermal environments, including boiling water. Until recently, however, the upper temperature limit of environments harboring living organisms had rarely exceeded 100°C (2, 6, 42). Bacteria capable of growing at 105°C were isolated from marine solfatara sites, and other extreme thermophiles have been cultured from geothermal wells (42, 43).

The discovery of 350°C deep ocean springs, where 350°C hydrothermal fluids are kept liquid as a result of hydrostatic pressure, presented the first opportunity to test Brock's hypothesis at temperatures well above 100°C. These superheated vents, or "smokers," have been found along the East Pacific Rise, in the Guaymas Basin, and at the Gorda Ridge (14, 17, 37). At the depths of the vents (>2,600 m), seawater will remain liquid at temperatures of 460°C or greater, depending on the exact depth (and, therefore, hydrostatic pressure) and the salinity of the hot water (4, 9). Field experiments to test for microbial activity in undiluted hydrothermal fluid at in situ temperature (350°C) and pressure (265 atm) are as yet impractical, given the limitations of deep-sea submersible work and high flow rates (meters per second) of the fluids emerging from the smokers. However, it has been possible to collect samples of the superheated fluids, although unavoidably mixed with various amounts of ambient seawater, for shipboard studies (1). Thermophilic bacterial communities that grow in a temperature range of 75 to 100°C at 1 atm were cultured from a set of seven such water samples that were collected from smokers at 21°N, inoculated into inorganic media, and incubated at 86°C and 1 atm (1). With the availability of cultures (albeit mixed) of thermophilic bacteria from these extreme smoker environments, we were able to test Brock's hypothesis in the laboratory. The fate of the bacterial community derived from the purest (304°C) sample of hydrothermal fluid was monitored with time of incubation at temperatures of

150, 200, 250, and 300°C under the in situ vent pressure of 265 atm.

The resulting growth curves, based on direct counts made by use of epifluorescence microscopy and the DNA-specific stain 4',6-diamidino-2-phenyl indole (DAPI), and on colorimetric assays of trichloroacetic acid-precipitable protein, have been published (2). Presented along with these growth curves were (i) a representative amino acid analysis of the particulate protein collected after 4.7 h at 250°C, (ii) representative transmission electron micrographs of the bacterial community after 5.9 h at 250°C, (iii) concentrations of soluble DNA measured in the medium after incubation at 300°C, (iv) results of an inoculated sample treated with Hg and incubated at 200°C in which numbers of DAPI-stainable particles decreased rapidly with time, and (v) results of an uninoculated medium control run at 200°C in which no DAPI-stainable particles or trichloroacetic acid-precipitable protein were detected. On the basis of all of these data, as well as an additional amino acid analysis that was not published, we concluded that bacterial growth had occurred at temperatures of at least 250°C.

Trent et al. (J. Trent, R. Chastain, and A. Yayanos, submitted for publication), from the laboratory at the Scripps Institution of Oceanography where we conducted our experiments, do not accept the possibility that bacteria grew, or even survived, at these temperatures and pressure. They submit, instead, that all of our evidence is based on the abiotic generation of chemical artifacts from the medium or on contamination introduced during sampling processing. Noting that we had run control experiments only at 200°C, they ran additional control experiments at 250°C. Using the same culturing apparatus described earlier (2), they incubated the uninoculated medium we used (2), as well as the same medium inoculated with a culture of thermosensitive bacteria (*Vibrio harveyi*), at 250°C and 265 atm of pressure for 6 h. They reported the generation of significant, though erratic, numbers of abiotic particles that autofluoresced in the same wavelength used to measure DAPI epifluorescence, as well as the formation of substances that gave a false-positive reaction to colorimetric protein assays. In no case did their particle counts or false-positive protein reactions duplicate our data; i.e., they did not observe increased concentration as a function of incubation time. The abiotic nature of the particles and substances detected in their experiments was confirmed by the absence of any significant, or even detectable, levels of amino acids. In fact, their results prove that the concentrations and complex spectrum of amino acids we measured in a sample removed from the culture incubation at 250°C could not have been generated from components of the medium (e.g., N-2-hydroxyethylpiperazine-N'-2-ethane-sulfonic acid [HEPES] buffer, yeast extract, or the remaining inorganic compounds). Trent et al. attribute our electron micrographs of intact bacteria, and presumably our amino acid and soluble DNA measurements, to contamination introduced after the samples were cooled, decompressed, and fixed in glutaraldehyde.

In this paper we present the complete amino acid analyses of particulate protein in the inoculated samples from our original experiments, as a function of incubation time, at 200, 250, and 300°C under 265 atm of pressure. These results substantiate the conclusion that "black smoker" bacteria had grown at temperatures of at least 250°C and render moot other interpretations that are presented by Trent et al.

We also review sampling procedures used in the original experiments (2), describe unpublished results of preliminary experiments, provide the available measurements of soluble DNA from the 300°C samples, and present current micrographs of the bacterial community which is being maintained at temperatures of 95 to 107°C. We conclude with a discussion of some of the possible ecological and evolutionary implications of the existence of these unique microorganisms.

SAMPLING AND METHODS

Hot water samples from 21°N were collected during dives 978 to 982 of the DSRV *Alvin* in November 1979 by use of specially constructed gold-plated stainless-steel water samplers (31). The extent of dilution of the hydrothermal fluid with ambient seawater is reflected in the temperature and concentration of metals and gases in the water sample, as discussed below. Procedures for the chemical and initial microbiological analyses of the samples have been published (1). Composition of the growth medium and procedures for culturing the bacterial community, derived from the sample of almost pure hydrothermal fluid, in a titanium syringe at in situ vent pressure and temperatures up to 300°C, have also been published (2). All samples removed from the superheated culturing vessel during

time-course experiments were rapidly cooled during withdrawal, decompressed, and fixed in 2% electron microscope-grade glutaraldehyde (Tousimis). Prior to each sampling event, the culture was mixed and the bleed-off line was flushed with at least five times its volume. Fixed samples were examined within minutes of collection by phase microscopy with methylene blue stain, and within hours by epifluorescence microscopy with acridine orange. Fixation in glutaraldehyde permitted sample storage and

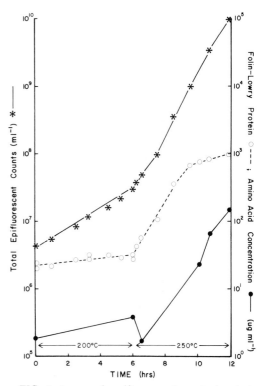

FIG. 1. Increase in epifluorescent counts, in colorimetric assays performed by the Folin-Lowry method, and in concentration of the total amino acids from particulate protein determined with a Durram amino acid analyzer, as a function of incubation time at 200 and 250°C in a medium inoculated with black smoker bacteria (see reference 2 for experimental details). Not all of the trichloroacetic acid-extracted precipitate was soluble in 1 N NaOH. This precipitate was removed by centrifugation and hydrolyzed separately from the NaOH-soluble fraction. In most cases approximately 50% of the total amino acids were found in the NaOH-insoluble fraction. The total amino acid protein reported here is the sum of both fractions, except in the 12-h sample where only the NaOH fraction was analyzed. The protein concentration of the final sample at 250°C (12 h), then, is an underestimate. The unknown amino acid peaks reported in our original paper (2) were present in all of the samples and did not disappear after hydrolysis for 72 h.

transport to Oregon State University, where the microscopic counts were repeated with DAPI stain.

Protein concentrations were measured by the colorimetric Folin-Lowry methods as previously described (2). The amino acid content of particulate protein in samples incubated for 4.7 and 5.9 h at 250°C was also determined, although only the results from the 4.7-h sample were published (see reference 2, legend to Fig. 1, and footnote to Table 2). Ultimately, all samples from the 200, 250, and 300°C experiments in the titanium syringe, for which sufficient volumes remained, were analyzed for amino acid content of the particulate protein. Portions of the same samples were concentrated by centrifugation for DNA analyses. The supernatant fractions were examined fluorometrically for soluble DNA after staining with Hoechst 33258 dye (Calbiochem), by the procedures of Paul and Myers (38), with purified *Escherichia coli* B DNA as a standard. A variety of extraction methods were applied to the pelleted fraction, even though the improbable success of extracting DNA from glutaraldehyde-fixed materials was predictable.

The experiments we were able to accomplish at Scripps Institution of Oceanography, using in situ vent pressure and temperatures between 150 and 300°C, were completed in two stages. The first set of experiments involved a glass syringe as the culturing chamber, temperatures of 150, 200, and 250°C, sampling intervals of 2 to 12 h, and a medium containing 0.05% $NaHCO_3$ as an inorganic carbon source. The number of epifluorescent particles in samples from these experiments increased with time at all three temperatures, but the increase at 200 and 250°C appeared as a short (4 to 6 h) spurt with limited yield. Therefore, in the second set of experiments, which involved the more thermotolerant titanium syringe and temperatures of 200, 250, and 300°C, we sampled more frequently (at least at hourly intervals) and increased the concentration of $NaHCO_3$ in the medium 10-fold to 0.5%, the percentage reported from the medium composition (2). Only those experiments for which a significant number of samples were taken during the period of increase in microscopic counts were published (2).

The same mixed culture used as the inoculum in the pressurized, high-temperature experiments is currently maintained in shaking water baths at 100 to 107°C under positive gas pressure in serum bottles. The culturing medium, as previously described (1), has been modified as follows. A minimal salt solution (1) is supplemented with yeast extract (0.001 g liter^{-1}), trace element mix (1), K_2HPO_4 (5 ml of a 5% solution, sterilized separately, per liter), Fe^{2+}-EDTA (3 ml of a solution containing 77 mg of

$FeSO_4 \cdot 7H_2O$ + 103 mg of sodium EDTA in 50 ml of distilled water and sterilized separately), $Na_2MoO_4 \cdot 2H_2O$ (1 ml of a 10% solution sterilized separately), H_3BO_3 (0.003 g liter^{-1}), $(NH_4)_2SO_4$ (0.1 g liter^{-1}), and HEPES buffer (Sigma, 2 g liter^{-1}), and the pH is adjusted to 6.5 to 6.8. This basic medium can be supplemented with various carbon substrates and energy sources. A sulfur medium is prepared by adding 5 g of $Na_2S_2O_3 \cdot H_2O$, 0.03 g of $MnSO_4 \cdot H_2O$, and 7 ml of a 10% solution of $NaHCO_3$ (sterilized separately) per liter. Alternative sources of carbon and energy are added to a final concentration of 0.5% except when the substrates or energy sources are gases. In the latter case, a 50-ml culture in a 100-ml capped serum bottle is flushed with Ar for 2 min and then warmed to 90°C in a water bath. Gas pressure in the bottle is released by use of a syringe, and an additional 25 ml of gas is removed. Sixty milliliters of the gas or gas mixtures used for culturing is added, creating a positive pressure in the serum bottles. The bacterial communities have been cultured successfully in CH_4, CO, and a mixture of H_2 and CO_2. All cultures were incubated anaerobically in 100-ml serum bottles closed with butyl rubber caps and aluminum seals.

Bacterial cultures at 100 to 107°C were prepared for electron microscopy by sampling in late log phase and fixing with electron microscope-grade glutaraldehyde (Tousimis) to a final concentration of 2%. The fixed samples were resuspended in 0.2 M cacodylate buffer, postfixed with 1% uranyl acetate, and stained with 1% OsO_4. Thin sections were examined with a Philips electron microscope model 300.

RESULTS AND DISCUSSION

Characteristics of the smoker environment. Submarine hot springs occur along ridge crests where cold seawater circulates through crustal cracks, reacting at depth with basalt at temperatures of at least 350°C (12, 14, 17) and ascending back to the sea floor. The temperature of the hydrothermal waters when they reach the sea floor can vary from a few degrees above ambient seawater to greater than 350°C (14), depending in part on degree of subsurface mixing and rate of ascent. Since the depth of hot springs at 21°N is greater than 2,500 m, seawater remains liquid at temperatures of at least 460°C (4, 9). A characteristic of the rapidly rising, 350°C fluids is their formation of polymetallic sulfide chimneys or "smokers." The inorganic chemical composition of the hot waters and associated chimneys has been described (14, 17, 31, 32, 34, 47).

Table 1 summarizes some of the preliminary microbiological data and the biologically impor-

TABLE 1. Microbiological and related chemical properties of hot water venting from sulfide chimneys at 21°N along the East Pacific Rise (November 1979)

Dive/ sample	Vent field[a]	Magnesium temp (°C)[b]	pH	Si (μM)	Concn of biologically important constituents (μM)[c]			Dissolved gases[d]				Microbial characteristics[e]		
					H₂S	Mn	Fe	CH₄ (μM)	H₂ (μM)	CO (nM)	N₂O (nM)	Total no. of bacteria (ml⁻¹) (direct counts)	Thermophiles present	Microbial production of CH₄, H₂, and CO at 100°C
978 (bag)	NGS	2	—[f]	—	—	—	—	0.5	0.02	3	18	8.6×10^5	—	Not tested
979 11/12	OBS	2	7.5	200	1.2	17	56	0.2	5.3	26	22	1.8×10^5	+	Not tested
979 1/2	OBS	2	7.0	390	3.4	46	14	0.1	13.5	8	27	3.6×10^5	+	Not tested
978 7/8	NGS	40	6.0	1,300	600	120	20	8.8	—	56	ND[g]	7.4×10^5	+	Not tested
980 7/8	SW	44	6.3	2,839	475	68	—	—	—	—	—	3.5×10^5	+	Not tested
979 3/4	OBS	152	4.9	9,600	1,300	286	206	30.2	639	134	ND	2.5×10^5	+	+
978 5/6	NGS	189	5.1	11,700	7.5	1,055	153	—	—	—	—	1.4×10^4	+	+
982 7/8	NGS	304	4.2	18,320	4,300	800	—	53.4	562	95	ND	4.7×10^5	+	+
Ambient[h]		1.7	7.8	160	0	—	—	0.0003	0.0004	0.3	20	Not tested	—	Not tested

[a] NGS, National Geographic Society site; OBS, Ocean Bottom Seismometer site; SW, Southwest site.

[b] The temperatures were calculated on the basis of the concentrations of Mg since the hot hydrothermal fluids are depleted in Mg. The Mg data as well as the equation relating temperature to Mg were taken from McDuff and Edmond (35).

[c] Chemical data from John Edmond (personal communication and reference 14).

[d] Dissolved gas data from Lilley and co-workers (31, 32).

[e] Microbiological data from Baross et al. (1, 31).

[f] Sample not analyzed chemically.

[g] ND, Not determined.

[h] Bottom water collected away from vent field.

tant chemical and physical properties of the hot waters sampled in November 1979 from three vent sites located at 21°N along the East Pacific Rise. The differences in the temperature and concentrations of metals and gases in these samples are due to the extent of dilution of the hydrothermal water with ambient seawater. This is reflected in the concentrations of Mg. Since Mg is depleted in hydrothermal water at high temperature, as a result of water-basalt reactions, the extent to which the hot water contains entrained seawater Mg reflects the extent of dilution and is thus reflected in the calculated temperature. This relationship between Mg and temperature was discussed by McDuff and Edmond (35). One of the important ramifications of this information is that, since the 350°C water is depleted in Mg and other seawater components, the smoker vents must be conduits extending from the point at which seawater is heated (to at least 350°C) to the sea floor. Thus, any bacteria present in undiluted smoker fluids must have originated from, or at least circulated through, the source depth of the hot water.

All of the hot water samples were greatly enriched in H_2S, Mn, Fe, Si, and other minerals and gases, primarily as a result of water/rock reactions at high temperatures. It has been demonstrated, for example, that the principal source of H_2S-S in the hot "chimney" waters at 21°N is basaltic rock and not seawater SO_4 (41, 44). In addition to the transition metals and H_2S, concentrations of the dissolved gases CH_4, CO, and H_2, potential energy sources for bacteria (31, 32, 47), are significantly higher than saturation. At the present time there is preliminary evidence for CH_4 oxidation in these environments (31; H. Craig, Eos Trans. Am. Geophys. Union, 62:893, 1981).

The source of these gases in hydrothermal waters is not yet clear. Both H_2 and CO, for example, are common gases associated with volcanic activity (7, 8, 15, 16, 22, 23, 26, 36). It is generally believed that the major source of H_2 in volcanic systems is from the oxidation of magnetite to hematite, whereas CO could originate within the magma from C–O–H–S equilibria (24, 31). On the other hand, the source of CH_4 in volcanic systems is not well understood, and there is some controversy as to the source and levels of CH_4 entrapped in the mantle (25). The conditions necessary to produce CH_4 abiogenically are harsh: in the absence of organic precursors, temperatures in excess of 500°C for extended periods of time (weeks or months) are necessary (28). In hydrothermal systems, CH_4 generally follows ^3He levels (^3He is a conserved gas of magmatic origin) to an extrapolated end member of 350°C (34). It has generally been believed that living organisms could not exist at

this temperature, thus signifying an abiogenic source for CH_4.

It now appears possible that a portion of these hydrothermal gases, at least in lower to intermediate temperature vents, may be produced by thermophilic bacteria. The thermophilic communities isolated from black smoker water samples produced CH_4, H_2, and CO when cultured at 100°C in either formate or sulfur media (Table 1; reference 1). Very high concentrations of these gases were also detected in a growth experiment at 300°C and 265 atm (2). Furthermore, some of the bacteria in this community can utilize various C_1 compounds including CH_4, methanol, and formate. Therefore, it is possible that both CH_4 production and consumption are occurring in the same environments. Any conclusions about the source of CH_4 based on $^{13}CH_4$ levels could be misleading, since methanogens and methylotrophs fractionate the stable carbon isotopes in an opposite manner (10, 18, 21, 39, 40). Obviously, further work is necessary before we can assess the extent to which thermophiles are influencing the chemistry of hydrothermal waters, and at what temperatures.

Thermophilic bacteria and growth at high temperatures and pressures. The thermophilic communities cultured from the two hottest water samples (978 5/6 and 982 7/8; see Table 1), and from a sample of chimney rock, were found to grow in a sulfur and metal medium at 100°C with generation times of 37 to 66 min (1). No growth was observed below 70 to 75°C.

The source of the thermophilic bacteria from the water samples must have been either the 350°C hydrothermal fluid itself or the 2°C seawater entrained in the hot water plume as it emerged from the chimney. There are several lines of evidence that argue against ambient seawater as the natural habitat for these extremely thermophilic bacteria: (i) thermophilic bacteria were cultured successfully only from those water samples containing some portion of hydrothermal fluid (1; Table 1); (ii) an ambient seawater sample collected in a Niskin bag at a vent field yielded no thermophiles (1; Table 1); and (iii) the available cultures of the smoker thermophiles fail to grow at temperatures below 70°C (1). If the hydrothermal fluid was the source of these bacteria, then the question arises as to whether they grow at 350°C at depth in the sea floor crust or originate from some lower-temperature environment, such as a seawater intake site, and thus just survive exposure to 350°C.

In an effort to learn whether some of these thermophilic bacteria could survive or grow at 350°C in medium kept liquid as a result of elevated hydrostatic pressure, the thermophilic bacterial community from vent sample 982 7/8

TABLE 2. Total amino acid content (nanograms per milliliter of sample) of particulate protein from samples of black smoker bacteria incubated at 200, 250, and 300°C at 265 atm of pressure with increasing time

Amino acid	200°C		250°C				300°C		
	Inoculum	6 h	0.5 h	4.2 h	4.7 h	5.9 h	0 h	1.25 h	3.3 h
Aspartic acid	120.8	241.2	177.8	1,197.1	4,078.7	4,761.4	198.6	56.0	356.6
Threonine	47.0	102.4	57.2	473.2	1,721.5	2,380.0	63.6	20.2	155.6
Serine	139.6	234.7	85.8	1,199.8	4,361.0	7,565.3	227.6	61.4	363.6
Glutamic acid	176.5	433.1	188.4	2,530.8	6,698.3	6,747.3	324.6	71.8	451.0
Proline[a]					2,100.7				
Glycine	187.4	419.7	158.2	3,037.0	9,800.0	47,587.5	339.4	105.4	391.0
Alanine	57.4	119.7	64.4	700.0	2,411.9	4,094.0	129.6	33.8	177.6
Valine	85.4	163.9	102.0	722.6	1,189.5	1,901.3	137.2	98.6	209.4
Methionine	Trace	34.7	Trace	199.8	571.2	2,395.5	Trace	Trace	41.4
Isoleucine	92.8	214.4	102.8	679.0	1,235.8	1,395.2	127.8	99.0	119.8
Leucine	87.7	190.8	81.0	917.8	2,362.4	2,403.9	125.0	34.6	180.8
Tyrosine	20.9	79.8	39.2	380.6	331.8	1,276.1	78.6	Trace	93.8
Phenylalanine	50.4	132.7	42.6	656.0	1,309.0	1,287.0	64.6	Trace	108.8
Histidine	446.6	608.3	459.8	7,359.0	22,733.3	59,675.0	759.8	228.8	1,658.6
Lysine[b]	32.0	244.1	Trace	537.1	1,352.9	1,752.0	162.6	Trace	162.0
Arginine	196.7	231.5	44.6	2,317.7	2,668.0	Trace	57.8	Trace	242.8
Total (µg/ml)	1.76	3.47	1.60	23.00	65.19	145.22	2.77	0.81	4.71

[a] All but one of these analyses were done on a Durrum amino acid analyzer that does not have a detection system for proline. The one proline concentration that was determined was done with a modified Beckman 120B amino acid analyzer.

[b] The lysine levels are not as dependable as the other amino acid concentrations because the cells were treated with glutaraldehyde.

(Table 1) was incubated at in situ vent pressure (265 atm) and at temperatures ranging from 150 to 300°C (2). The available culture apparatus did not permit a test at 350°C. The resulting growth curves at 150, 200, 250, and 300°C have been published (2). The generation times at 265 atm ranged from 8 h at 150°C to 40 min at 250°C, with equivocal results at 300°C. Results of earlier, unpublished experiments at 200 and 250°C, when the bacteria were supplied with a lower concentration of $NaHCO_3$ as a potential carbon source (0.05% rather than 0.5%), indicated similar doubling times but lower cell yields (measured by epifluorescence microscopy) by an order of magnitude.

During the first 4 h of incubation at 250°C, concentrations of the total particulate protein as measured by the Folin-Lowry assay doubled at a rate similar to the microscope counts (2; Fig. 1). Trent et al. have argued that these two measurements of growth, Folin-Lowry assay and epifluorescence microscopy, were detecting increases in chemical artifacts from the medium and not the growth of bacteria. It is possible that some of the microscopic counts may have been artificially high, as a result of the inadvertent inclusion of abiotic, autofluorescing particles in the counts, although we have no evidence that this occurred. To determine what portion of the substances measured by the Folin-Lowry assay was complete protein, we determined the amino acid content of particulate protein (trichloroace-

tic acid precipitable) in the inoculated sample as a function of incubation time at 200, 250, and 300°C at 265 atm of pressure. The results from the 200 and 250°C experiments are presented in

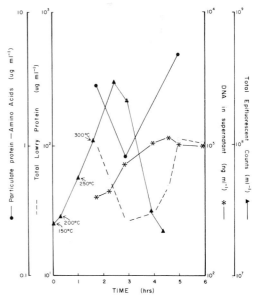

FIG. 2. Changes in concentrations of epifluorescent particles, Folin-Lowry protein, total amino acids, and DNA detected in the medium when a culture of thermophilic bacteria (982 7/8) was heated at 300°C and 265 atm of pressure.

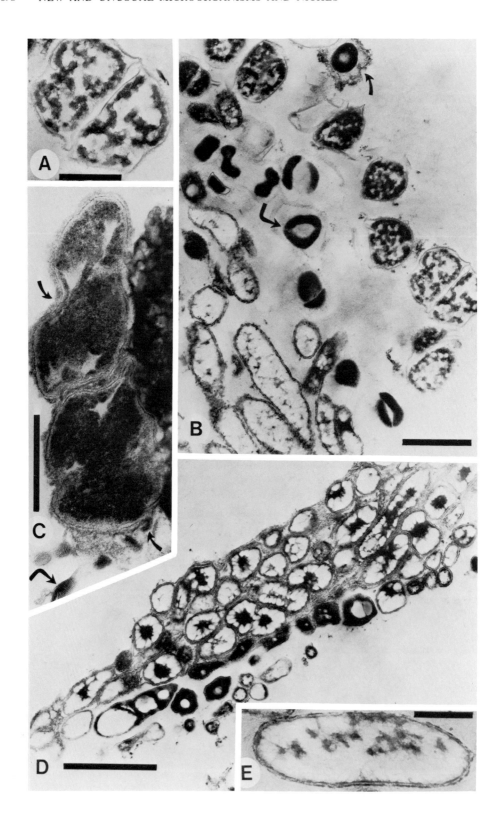

FIG. 3. Transmission electron micrographs of ultrathin sections of thermophilic bacterial communities cultured at 107°C and positive gas pressure in serum bottles in a shaking water bath containing commercial coolant (DuPont). Bacteria representative of the community from hot water sample 982 7/8 are shown in A, B, D, and E. When cultured in a CH_3OH medium (see text for details), three morphologically distinct organisms dominated the community (B). Morphological types identified by arrows (B) were similar to those seen by epifluorescence microscopy when the culture was incubated at 150 to 250°C. When cultured in CH_4, the common rod-shaped organisms were more clearly vacuolated (D). An organism derived from solid pyritic material, sampled from the outer surface of an active black smoker, and cultured at 107°C in an atmosphere of H_2 and CO_2 is shown in C. Arrows indicate the unusual outer membrane layers characteristic of this organism. Bars in A, C, and E are 0.5 μm; in B, 1 μm; and in D, 2μm.

Fig. 1. All of the amino acids detected by the standard methods we used were present in each sample and showed increases in concentration, with time, at all three temperatures (Table 2). At 250°C, the doubling time of the concentration of total, as well as individual, amino acids was approximately 1 h, as also indicated by the Folin-Lowry assay and microscopic counts (Fig. 1). The results of these amino acid analyses further substantiate the original conclusion that the black smoker bacteria grew at 250°C.

Results of the growth experiment conducted at 300°C, the temperature limit of the culturing chamber, were not as clear. However, microscopic counts and levels of particulate protein, measured by both Folin-Lowry assay and amino acid analyses (Fig. 2), suggest that there may have been brief periods of growth at 300°C at the beginning and end of the experiment. It is conceivable that the intervening drop in cell counts and protein levels represented a selective elimination of some portion of the bacterial community at this extreme temperature. After 3 h at 300°C, membranous spheres in transmission electron micrographs, and increasing concentrations of DNA in the medium, were observed. One possible explanation for these data is that the organism(s) responsible for increased protein levels after 3 h at 300°C was sensitive, and lysed upon exposure, to the extreme temperature differential or the decompression required to collect the samples (2). Another source of stress to the organisms after 3 h at 300°C may have been the development of an acidic pH in the culturing chamber. Although we did not measure the pH of the medium at this point, other experiments with uninoculated seawater in pressurized vessels have shown that increases in temperature will cause the pH to drop from neutrality at room temperature to 6 at 200 to 250°C and to 3 at 300°C (5).

Is 250 or 300°C the maximum growth temperature for these thermophilic bacteria? The in situ temperature of the apparent source of bacteria, hydrothermal fluids emerging from a black smoker, was 350°C. Chemically, the actual solutions in which the bacteria may be growing in situ (whether attached to rock surfaces or free living) are considerably different from seawater (or a seawater-based culturing medium) as a result of passage through and reaction with basaltic rock at temperatures as high as 500°C (27). Determination of the upper temperature and pressure limits for growth of these bacteria may ultimately require a culturing system that more closely resembles actual hydrothermal systems in which seawater is continuously passed through basalt.

At present we do not know whether all 350°C vents contain thermophilic communities capable of growing at 250°C under hydrostatic pressure, and if so whether they are composed of identical or different groups of bacteria. A study of the thermophilic communities from different vents within a vent field as well as at different ridge zones could have important implications for understanding hydrothermal circulation and some of the geochemistry associated with these environments.

Morphology of thermophiles. Cultures obtained from each of the waters tested (Table 1) contained three or more morphologically distinct types of bacteria. Attempts to obtain pure cultures have been unsuccessful. One of the approaches we took was to vary the nutritional and physical conditions of the medium to make it more selective. In cases in which growth occurred, the community remained intact. The three thermophilic communities that have been studied in detail (1) were found to grow on C_1 compounds including CH_4, CO, methanol, and formate, as well as CO_2, in a reduced sulfur and metal medium.

Electron micrographs of ultrathin sections of morphological types present in one of the thermophilic communities, when cultured at 107°C, are shown in Fig. 3. This culture served as the inoculum for pressurized growth experiments described earlier (2). Although rod-shaped bacteria containing multiple membranes, and organisms with thick cell walls and pinched ends, were observed in the culture when incubated at 250°C (2), the general morphological characteristics of the same community when cultured at 107°C were somewhat different. Also shown in Fig. 3C is a multimembraned organism isolated from a portion of rock from an active black smoker and cultured in CO_2 and H_2 at 107°C.

Evolutionary implications. Volcanism was the major source of inorganic elements and compounds, organic gases, and possibly more complex organic compounds to the early ocean and atmosphere, and eventually to continental ecosystems. At the time when the first ocean appeared, the earth had an oxidizing atmosphere consisting principally of CO_2 (13, 45). Recent evidence suggests that this early ocean was anoxic and had temperatures of at least 50 to 70°C, even as recently as 1 billion years ago (3, 13, 20, 29). The Archaean ocean may even have had temperatures as high as 200°C, with the ocean waters kept liquid as a result of the high partial pressure of CO_2 in the atmosphere (19). Before the formation of islands and continents, the earth's surface was probably covered with a shallow ocean (R. Hart, personal communication) characterized by extensive hydrothermal systems. The first evidence of microbial life appears in the fossil record from that period, approximately 3.5 billion years ago (30, 33, 46). In some cases these fossil-containing sediments were hydrothermal and marine (11). This set of observations implies that the microorganisms which developed on earth during the first 2 billion years were thermophilic (1, 20, 29) and probably utilized magma-derived nutrients. Since present-day hydrothermal environments may be analogous to environments in which early life forms evolved (3, 11), a study of thermophiles from these systems could increase our understanding of the kinds and activities of the first bacterial communities to develop on earth.

Support for this research was provided by grant N00014-79C-004 from the Biological Oceanography Division of the Office of Naval Research.

We thank A. A. Yayanos for use of the high-temperature/high-pressure facilities in his laboratory. We also acknowledge M. Lilley, C. Dahm, and M. de Angelis for advice and editorial comments, M. Lilley, L. Gordon, and J. Edmond for permission to use data, R. Y. Morita for use of laboratory facilities, A. Soeldner and C. Weiss for electron microscopy, T. Bailey for performing the amino acid analyses, and D. Powers, C. Anfinsen, and A. Komoriya for critical evaluation of the protein analyses.

LITERATURE CITED

1. **Baross, J. A., M. D. Lilley, and L. I. Gordon.** 1982. Is the CH_4, H_2 and CO venting from submarine hydrothermal systems produced by thermophilic bacteria? Nature (London) **298:**366–368.
2. **Baross, J. A., and J. W. Deming.** 1983. Growth of "black smoker" bacteria at temperatures of at least 250°C. Nature (London) **303:**423–426.
3. **Berry, W. B. N., and P. Wilde.** 1983. Evolutionary and geologic consequences of organic carbon fixing in the primitive anoxic ocean. Geology **11:**141–145.
4. **Bischoff, J. L.** 1980. Geothermal system at 21°N, East

Pacific Rise: physical limits on geothermal fluid and role of adiabatic expansion. Science **207:**1465–1469.
5. **Bischoff, J. L., and R. J. Rosenbauer.** 1983. A note on the chemistry of seawater in the range of 350°–500°C. Geochim. Cosmochim. Acta **47:**139–144.
6. **Brock, T. D.** 1978. Thermophilic microorganisms and life at high temperatures. Springer-Verlag, New York.
7. **Casadevall, T. J., and L. P. Greenland.** 1981. The chemistry of gases emanating from Mount St. Helens, May–September 1980, p. 221–226. In P. W. Lipman and D. R. Mullineaux (ed.), The 1980 eruptions of Mount St. Helens, Washington. Geological Survey Professional Paper 1250, Washington, D.C.
8. **Chaigneau, M., and K. Conrad.** 1970. Volcanic gas from Vulcano (Lipari Islands). C.R. Acad. Sci. **271:**165–167.
9. **Chen, C. T.** 1981. Geothermal system at 21°N. Science **211:**298.
10. **Coleman, D. D., J. B. Risatti, and M. Schoell.** 1981. Fractionation of carbon and hydrogen isotopes by methane-oxidizing bacteria. Geochim. Cosmochim. Acta **45:**1033–1035.
11. **Corliss, J. B., J. A. Baross, and S. E. Hoffman.** 1981. An hypothesis concerning the relationship between submarine hot springs and the origin of life on Earth. Oceanologica Acta, No. Sp. 59–69.
12. **Corliss, J. B., J. Dymond, L. I. Gordon, J. M. Edmond, R. P. von Herzen, R. D. Ballard, K. Green, D. Williams, A. Bainbridge, K. Crane, and T. H. van Andel.** 1979. Submarine thermal springs on the Galapagos Rift. Science **203:**1073–1083.
13. **Costa, U. R., W. S. Fyfe, R. Kerrich, and M. W. Nesbitt.** 1980. Archean hydrothermal talc evidence for high ocean temperatures. Chem. Geol. **30:**341–349.
14. **Edmond, J. M., K. L. Von Damm, R. E. McDuff, and C. I. Measures.** 1982. Chemistry of hot springs on the East Pacific Rise and their effluent dispersal. Nature (London) **297:**187–191.
15. **Evans, W. D., N. G. Banks, and L. D. White.** 1981. Analyses of gas samples from the summit crater, p. 227. In P. W. Lipman and D. R. Mullineaux (ed.), The 1980 eruption of Mount St. Helens, Washington. Geological Survey Professional paper 1250, Washington, D.C.
16. **Finlayson, J. B., I. L. Barnes, and J. J. Naughton.** 1968. Developments in volcanic gas research in Hawaii. In L. Knopoff, C. L. Drake, and P. J. Hart (ed.), The crust and mantle of the Pacific Basin. Geophys. Monogr. Am. Geophys. Union **12:**428–438.
17. **Francheteau, J., H. D. Needham, P. Choukroune, T. Juteau, M. Séguret, R. D. Ballard, P. J. Fox, W. Normark, A. Carranza, D. Cordoba, J. Guerrero, C. Rangin, M. Bougault, P. Cambon, and R. Hekinian.** 1979. Massive deep-sea sulphide ore deposits discovered on the East Pacific Rise. Nature (London) **277:**523–528.
18. **Fuchs, G., R. Tauer, H. Ziegler, and W. Stichler.** 1979. Carbon isotope fractionation by Methanobacterium thermoautotrophium. Arch. Microbiol. **120:**135–139.
19. **Fyfe, W. S.** 1980. Archaean hot oceans, p. 1–4. In Proceedings 3rd Colloquium on Planetary Water, State University of New York, Buffalo. NASA, Washington, D.C.
20. **Fyfe, W. S.** 1981. The environmental crisis: quantifying geosphere interactions. Science **213:**105–110.
21. **Games, L. M., J. M. Hayes, and R. P. Gunsalus.** 1978. Methane-producing bacteria: natural fractionations of the stable carbon isotopes. Geochim. Cosmochim. Acta **42:**1295–1297.
22. **Gerlach, T. M.** 1980. Evaluation of volcanic gas analyses from Kilauea volcano. J. Volcanol. Geotherm. Res. **1:**295–317.
23. **Gerlach, T. M.** 1980. Evaluation of volcanic gas analyses from Surtsey volcano, Iceland, 1964–1967. J. Volcanol. Geotherm. Res. **8:**191–198.
24. **Gerlach, T. M., and B. E. Nordlie.** 1975. The C-H-O-S gaseous system. II. Temperature, atomic composition, and molecular equilibria in volcanic gases. Am. J. Sci. **275:**377–394.

25. **Gold, T. J.** 1982. Earth outgassing of methane, p. 45–69. *In* P. McGeer and E. Durbin (ed.), Methane—fuel for the future. Plenum Press, New York.

26. **Graeber, E. J., P. J. Modreski, and T. M. Gerlach.** 1979. Compositions of gases collected during the 1977 east rift eruption, Kilauea, Hawaii. J. Volcanol. Geotherm. Res. **5:**337–344.

27. **Gregory, R. T., and H. P. Taylor.** 1981. An oxygen isotope profile in a section of cretaceous oceanic crust, Samail ophiolite, Oman: evidence for δ^{18} oxygen buffering of the oceans by deep (> 5 km) seawater-hydrothermal circulation in mid-ocean ridges. J. Geophys. Res. **86:**2737–2755.

28. **Gür, T. M., and R. A. Huggins.** 1983. Methane synthesis on nickel by a solid-state ionic method. Science **219:**967–969.

29. **Hoyle, F., and N. C. Wickramasinghe.** 1979. Life cloud. Sphere Books, Ltd., London.

30. **Knoll, A. H., and E. S. Barghoorn.** 1977. Archean microfossils showing cell division from the Swaziland system of South Africa. Science **198:**396–398.

31. **Lilley, M. D., J. A. Baross, and L. I. Gordon.** 1983. Reduced gases and bacteria in hydrothermal fluids: the Galapagos Spreading Center and 21°N East Pacific Rise. *In* K. Bostrom, K. Smith, L. Laubier, and P. Rona (ed.), Hydrothermal processes at seafloor spreading centers. Plenum Press, New York.

32. **Lilley, M. D., M. A. de Angelis, and L. I. Gordon.** 1982. Methane, hydrogen, carbon monoxide and nitrous oxide in submarine hydrothermal vent waters. Nature (London) **300:**48–50.

33. **Lowe, D. R.** 1980. Stromatolites 3,400 Myr old from the Archean of Western Australia. Nature (London) **284:**441–443.

34. **Lupton, J. E., R. F. Weiss, and H. Craig.** 1977. Mantle helium in hydrothermal plumes in the Galapagos Rift. Nature (London) **267:**603–604.

35. **McDuff, R. E., and J. M. Edmond.** 1982. On the fate of sulfate during hydrothermal circulation at mid-ocean ridges. Earth Planet. Sci. Lett. **57:**117–132.

36. **Muenow, D. W.** 1973. High temperature mass spectrometric gas-release studies of Hawaiian volcanic glass: Pele's tears. Geochim. Cosmochim. Acta **37:**1551–1561.

37. **Normark, W. R., J. L. Morton, R. A. Koski, D. A. Clague, and J. R. Delaney.** 1983. Active hydrothermal vents and sulfide deposits on the southern Juan de Fuca Ridge. Geology **11:**158–163.

38. **Paul, J. H., and B. Myers.** 1982. Fluorometric determination of DNA in aquatic microorganisms by use of Hoechst 33258. Appl. Environ. Microbiol. **43:**1393–1399.

39. **Rosenfeld, W. D., and R. S. Silverman.** 1959. Carbon isotope fractionation in bacterial production of methane. Science **130:**1658–1659.

40. **Schoell, M.** 1980. The hydrogen and carbon isotopic composition of methane from natural gases of various origins. Geochim. Cosmochim. Acta **44:**649–661.

41. **Shanks, W. L., J. L. Bischoff, and R. J. Rosenbauer.** 1981. Seawater sulfate reduction and sulfur isotope fractionation in basaltic systems: interaction of seawater with fayalite and magnetite at 200–350°C. Geochim. Cosmochim. Acta **45:**1977–1995.

42. **Stetter, K. O.** 1982. Ultrathin mycelia-forming organisms from submarine volcanic areas having an optimum growth temperature of 105°C. Nature (London) **300:**258–260.

43. **Stetter, K. O., M. Thomm, J. Winter, G. Wildgruber, H. Huber, W. Zillig, D. Janecovic, H. Konig, P. Palm, and S. Wunderi.** 1981. *Methanothermus fervidus*, sp. nov., a novel extremely thermophilic methanogen isolated from an Icelandic hot spring. Zentralbl. Bakteriol. Parasitenkd. Infektionskr. Hyg. Abt. 1 Orig. Reihe C **2:**166–178.

44. **Styrt, M. M., A. J. Brackmann, H. D. Holland, B. C. Clark, V. Pisutha-Arnond, S. C. Eldridge, and H. Ohmoto.** 1981. The mineralogy and the isotopic composition of sulfur in hydrothermal sulfide/sulfate deposits on the East Pacific Rise, 21°N latitude. Earth Planet. Sci. Lett. **53:**382–390.

45. **Walker, J. C. G.** 1983. Possible limits on the composition of the Archaean Ocean. Nature (London) **302:**518–520.

46. **Walter, M. R., R. Buick, and J. S. R. Dunlop.** 1980. Stromatolites 3,4000–3,5000 Myr old from the North Pole Area, Western Australia. Nature (London) **284:**443–445.

47. **Welhan, J. A., and H. Craig.** 1979. Methane and hydrogen in East Pacific Rise hydrothermal fluids. Geophys. Res. Lett. **6:**829–831.

Habitats of *Chloroflexus* and Related Organisms

RICHARD W. CASTENHOLZ

Department of Biology, University of Oregon, Eugene, Oregon 97403

This has been an exciting past decade for the discovery of new bacteria, not merely slight variations of well-known species but entirely new physiological, ecological, and genetic categories, including the recognition of the Archaebacteria. Among the phototrophs, too, many of the discoveries have been major, including the involvement of an entirely novel pigment, bacteriorhodopsin, in the generation of ATP.

Chloroflexus sp., discovered 10 years ago by Pierson and Castenholz (15), is now an organism of interest to a variety of microbiologists and biochemists, and more specifically to photobiologists and phylogenists interested in early microbial evolution (Fig. 1). *Chloroflexus* sp. and other members of the *Chloroflexaceae* are anoxygenic phototrophic eubacteria, as are the green and purple bacteria (Table 1). A more detailed overview of all of these categories was presented by Trüper and Pfennig (20).

C. aurantiacus, the first organism described for this category of phototrophs, has been isolated from alkaline pH hot springs the world over. It is known to grow to 66°C in culture and to 70 to 72°C in the springs. What appears to be the same species is found abundantly in these habitats, often, in fact, in nearly monospecific and certainly obvious populations. In other habitats commonly examined, it simply was not apparent. But this has now become another age of searching for new microbes, and *Chloroflexus*-like organisms, not surprisingly, turn out not to have restricted themselves to the freshwater thermal environment.

By *Chloroflexus*-like, I mean septate filamentous bacteria that have the ability to glide and perform anoxygenic photosynthesis using bacteriochlorophyll *a* rather than chlorophyll *a* (see 3).

Chloroflexus was thought unique enough for Trüper (19) to propose the family *Chloroflexaceae* to accommodate *C. aurantiacus* and other such organisms that presumably would be discovered later. The filamentous, gliding nature of *Chloroflexus* sp. was in itself sufficient to set it apart from the *Chlorobiaceae* (green sulfur bacteria) as this group was understood in 1974. In addition, the isolates were photoheterotrophs, but were facultatively so, allowing aerobic chemoheterotrophy as well (15, 16). A considerable range of organic substrates was usable, but it appears that acetate, glycerol, glucose, pyruvate, and glutamate were most easily metabolized (12, 15). In the *Chlorobiaceae*, sulfide-, sulfur-, thiosulfate-, or H_2-dependent photoautotrophy is still the rule, although acetate or other reduced carbon sources may be used to make up part of the cell carbon (10). *"Chloropseudomonas ethylica,"* the supposed acetate-dependent flagellated green bacterium still included in *Bergey's Manual* (1), has subsequently been proved to be a co-culture of *Prosthecochloris* (green sulfur bacterium) and the sulfur-reducing anaerobe *Desulfuromonas* (13).

The presence of chlorosomes (chlorobium vesicles) that contain mainly bacteriochlorophyll *c* and some carotenoids has been the main link between the *Chloroflexaceae* and *Chlorobiaceae* (Table 1). Both *Chlorobium* sp. and *Chloroflexus* sp. contain photosynthetic reaction centers in a nonconvoluted cytoplasmic membrane (17, 18a). The predominance or importance of monocyclic carotenoids (e.g., chlorobactene, γ-carotene), including carotenoid glycosides (mainly within the membrane) (18a), in both groups has pointed to a closer relationship between these two groups than to any others except perhaps to the nonphotosynthetic flexibacteria (e.g., *Herpetosiphon*), which possess some similar types of carotenoids (11).

However, the catalogs of 16S rRNA oligonucleotides that have been used by Woese and others to judge relatedness among procaryotes have shown a great distance between *Chloroflexus* and *Chlorobium* ($S_{ab} < 0.2$) and indicated a slightly closer relationship between *Chlorobium* sp. and some cyanobacteria (6). Was the bifurcation or divergence among chlorosome-containing eubacteria very ancient, was evolutionary divergence between these groups more rapidly paced than between many other groups, or was the ability to synthesize chlorosomes and their characteristic pigments coded for by a plasmid genome capable of "lateral" transfer between distantly related eubacteria? Present work (17, 18) has shown a photosynthetic reaction center in *Chloroflexus* sp. that has characteristics more similar to those of the *Rhodospirillaceae* than to those of the green sulfur bacteria.

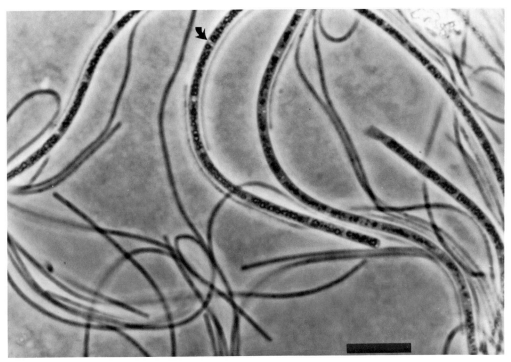

FIG. 1. Mixture of *C. aurantiacus* J-10-f1 (narrower filament) and feral *F-1*. Septa can be seen in *F-1* (arrow). Phase-contrast microscopy. Bar = 10 μm.

MESOPHILIC *CHLOROFLEXACEAE*

Most of the work on the *Chloroflexaceae* has been with two clonal strains of thermophilic *C. aurantiacus*, OK-70-f1 and J-10-f1, both of which have growth rate maxima between 52 and 60°C. The search for mesophilic *Chloroflexaceae* has not been intense, but it is now on the verge of fruition. Gorlenko (7) described a mesophilic strain *C. aurantiacus* which grew best between 20 and 25°C. However, it appears that this strain is .no longer available in axenic culture (V. M. Gorlenko, personal communication). A marine isolate has not been reported, although Gorlenko has said that marine forms exist. Recently J. Stolz and Y. Cohen (separate papers in Y. Cohen, R. W. Castenholz, and R. W. Halvorson, ed., *Microbial Mats: Stromatolites*, in press) have published electron micrographs of marine intertidal and hypersaline microbial mats, respectively, which show definite septate filaments with chlorosomes. *Chloroflexus* sp. or a similar organism appears to form an interwoven matrix which supports the mat-forming cyanobacterium *Microcoleus chthonoplastes* in Solar Lake, Sinai (Cohen, in press). R. Castenholz, B. Pierson, and J. Gibson have recently observed *Chloroflexus*-like filamentous organisms containing typical pigments from intertidal microbial mats and mat enrichments near

Woods Hole, Massachusetts (unpublished data).

A few other mesophilic organisms that are *Chloroflexus*-like have been described. *Chloronema*, from freshwater plankton, is similar to *Chloroflexus* but contains gas vesicles and a sheath (5). *Oscillochloris chrysea* from anoxic freshwater mats is a relatively wide diameter trichome (ca. 5 μm) which resembles a rapidly gliding gas-vesiculate *Oscillatoria* sp. (cyanobacteria) but with chlorosomes containing bacteriochlorophyll *c* lining not only the outer cytoplasmic membrane but also the membranes of the complete and incomplete septa (8). Both *Chloronema* sp. and *Oscillochloris* sp. appear to be photoheterotrophs, although none of the species involved has been reported as an axenic culture.

Thus, at this time rather little is known of the nonthermal habitats of members of the *Chloroflexaceae*, but it is reasonable to predict that some types may be major components of the anoxic layer of some inland waters and of some marine microbial mats.

The marine phototroph *Chloroherpeton thalassium*, recently described by J. Gibson, N. Pfennig, and J. B. Waterbury (Arch. Microbiol., in press), is a gliding, sinuous, elongate rod which is not septate. It is about 1.0 μm wide and contains gas vesicles and chlorosomes with bacteriochlorophyll *c*. Like the green sulfur bacte-

TABLE 1. Anoxygenic photosynthetic procaryotes

Category	Form	Type of motility	Chlorophyll[a]	Usual C source	Reductant	Chemo-aerobic growth	Major carotenoids
Chromatiaceae (purple sulfur bacteria)	Unicell	Flagellated	Bchl *a* (or *b*)	CO_2	S^{2-}	Some species	Spirilloxanthin, rhodopinal, okenone
Rhodospirillaceae (purple nonsulfur bacteria)	Unicell	Flagellated	Bchl *a* (or *b*)	Organic C	Organic C	Some species	Spirilloxanthin, rhodopinal, spheroidene
Chlorobiaceae (green sulfur bacteria)	Unicell	One species glides	Bchl *c*, *d*, or *e* + Bchl *a*	CO_2	S^{2-}	No	Isorenieratene, chlorobactene
Chloroflexaceae	Filament	Gliding	Bchl *c* or *d* + Bchl *a*	Organic C (CO_2)	Organic C (S^{2-})	Some species	Anaerobic: β- and γ-carotene, hydroxy-γ-carotene glucosides Aerobic: echinenone, oxo-γ-carotene, and glucosides
"*Heliothrix*" type (= *F-1*)	Filament	Gliding	Bchl *a*	Organic C	Organic C	?	Oxo-γ-carotene and glucosides

[a] Bchl, Bacteriochlorophyll.

ria, it is a sulfide-dependent unicellular photoautotroph, but unlike the others it possesses gliding motility. However, the 16S rRNA catalog indicates that *Chloroherpeton* belongs to the *Chlorobium* branch (Gibson et al., in press).

ADDITIONS TO THE THERMOPHILIC *CHLOROFLEXACEAE*

Chloroflexus sp., with the attributes of the originally described species, is the prime undermat or second-layer microbe of hot spring mats, where it obtains organic carbon in probable forms such as glucose or glycolate from the cyanobacterial photoautotrophs and forms such as acetate, propionate, butyrate, and lactate from the fermentation cycle of the mat, which involves buried cyanobacteria and *Chloroflexus* cells as substrates (D. M. Ward, E. Beck, N. P. Revsbech, K. A. Sandbeck, and M. R. Winfrey, in Y. Cohen, R. W. Castenholz, and H. Halvorson, ed., *Microbial Mats: Stromatolites*, in press). Living phototrophic *Chloroflexus* sp. may be exposed to highly aerobic conditions during daylight hours when O_2 diffuses from oxygenic cyanobacteria, but is within anaerobic conditions at night when O_2-sensitive pigment synthesis can take place. This is the situation in hot springs the world over when there is little primary sulfide in the source waters or in water below 70°C.

It is in sulfide-rich waters almost everywhere (>0.05 mM) that cyanobacterial species characteristic of temperatures over about 56 to 57°C

disappear, presumably because of sulfide intolerance; in their place are nearly pure mats of *Chloroflexus* sp., supported by the sulfide and CO_2 instead of by organic compounds from the cyanobacteria (2). Thus, sulfide relieves the dependence of *Chloroflexus* sp. on organic carbon, allowing conversion to photoautotrophy. Instead of inhibiting, sulfide allows *Chloroflexus* sp. to grow to 65 to 66°C, nearly its usual upper temperature limit, but as pure masses uncontaminated by any other phototrophs. Such mats, with anoxygenic phototrophs comprising the base of the food web, are common in Iceland, New Zealand, and a few areas of North America (2).

The upper, sulfide-dependent *Chloroflexus* mats of the Mammoth Terraces (Yellowstone) have been shown to photoincorporate CO_2, and as expected, this process was significantly stimulated by sulfide (unpublished data). Photoautotrophic growth with sulfide was slow in culture isolates from this mat, as it was in other strains of *Chloroflexus* sp. (2). Sulfide microelectrode measurements in the natural mat have shown that sulfide consumption in the upper illuminated layers occurs during light hours and that sulfide is restored rapidly from deeper in the mat during darkness (D. M. Ward and N. P. Revsbech, Abstr. Annu. Meet. Am. Soc. Microbiol. 1983, I176, p. 168). Since the entire system is continuously anaerobic, it is possible that this upward diffusion of sulfide is also continuous. Primary sulfide is the rate-limiting reductant supporting the primary production of the surface layer. Consumers in the mat below must include fermenters and sulfate reducers (these springs have a high sulfate concentration), but also possibly reducers of elemental sulfur (the product of sulfide oxidation by photoautotrophic *Chloroflexus* sp.). It also seems likely that the binding layers of *Chloroflexus* sp. below the thin (<1 mm) surface layer of photoautotrophic *Chloroflexus* sp. are photoheterotrophic "consumers" even though they may be the same organism as in the top mat layer. Thus, a whole sulfur and carbon cycle appears in the form of a relatively simple anaerobic microbial mat, a possible analog of the most primitive microbial mats or stromatolites that occurred even before oxygenic photosynthesis and an aerobic atmosphere.

In isolates of these mats from each area, photoheterotrophy still appears to be the preferred type of metabolism, at least for rapid growth. In isolates from the upper *Chloroflexus*-based mats of Mammoth Springs in Yellowstone Park, the faculty for aerobic chemoheterotrophy is absent (R. W. Castenholz and S. J. Giovannoni, Abstr. Annu. Meet. Am. Soc. Microbiol. 1981, I79, p. 100). Although these strains can tolerate normal O_2 levels and consume O_2, no growth occurs in darkness, and in light it continues for only a few hours, presumably by phototrophic means until pigment content declines substantially as a result of photodynamic damage. In contrast, "typical" isolates of *Chloroflexus* sp., J-10-fl and OK-70-fl, consume oxygen in darkness and simultaneously increase dry weight and protein.

AEROBICALLY ADAPTED
CHLOROFLEXACEAE

Although the type strain of *Chloroflexus* sp. (J-10-fl) is capable of continued aerobic chemoheterotrophic growth, it is mainly an anaerobic photoheterotroph, and although many microbial mats of hot springs are partially or regularly exposed to O_2, the aerobic abilities of these types probably play a secondary role in survival. However, this physiological versatility at least allows energy-yielding reactions in darkness. Doemel and Brock (4) have shown that in some hot spring pools, *Chloroflexus* cells migrate upward through the cyanobacterial mat layer at night, presumably reaching microaerobic conditions.

However, in a few hot springs, mainly in Oregon, another *Chloroflexus*-like organism grows in great abundance (Fig. 1). The organism has been referred to as *F-1* since first discovered (14) and as "*Heliothrix oregonensis*" more recently, although a formal description has not yet been made (B. K. Pierson, S. J. Giovannoni,

and R. W. Castenholz, Appl. Environ. Microbiol., in press).

In the daytime, at least, and perhaps continuously, *F-1* is exposed to O_2, often at saturation levels. It grows from only about 56°C to below 40°C, not as an under mat but as a top mat with a layer of oxygenic cyanobacteria under it! In the hot springs where abundant, it contains a high carotenoid content which acts as a light shield for the cyanobacteria which live below. This is confirmed by the high chlorophyll *a* and c-phycocyanin contents of the cyanobacteria, levels typical of shade-adapted cells (unpublished data).

F-1 contains bacteriochlorophyll *a* in the cytoplasmic membrane but no chlorosomes nor the accompanying accessory bacteriochlorophyll (Table 1). It is, nevertheless, likely to be related to *Chloroflexus*; it is a gliding, septate filament of indefinite length bearing a high content of carotenoid pigments similar to those of aerobically grown *Chloroflexus* sp. (e.g., 4-oxo-γ-carotene and myxobactone).

Little is known about the physiology of *F-1*. Photoheterotrophy again seems to be the principal mode, with a relatively high light intensity requirement as evidenced by light intensity versus [^{14}C]acetate incorporation experiments and by the positive correlation between abundance and unshaded situations during spring and summer (Pierson et al., in press). A high light requirement for phototrophy is understandable since the bacteriochlorophyll *a* content (1 µg/mg, dry weight) is restricted to an unconvoluted cytoplasmic membrane. However, the efficiency of the extensive carotenoid pigments as a light-harvesting system is not yet known.

F-1 has been difficult to culture by methods usable for *Chloroflexus* sp. At present only a co-culture is available. *F-1* is dependent on the obligate aerobe *Isocystis pallida*, which is a nonphotosynthetic oligotroph living on glucose (S. J. Giovannoni and E. Schabtach, Abstr. Annu. Meet. Am. Soc. Microbiol. 1983, I27, p. 144). Under conditions of co-culture, *Heliothrix* still synthesizes bacteriochlorophyll *a*. Whether *Heliothrix* requires O_2 or is merely tolerant of O_2 is not easily determinable under the current culture restrictions. It is possible, however, that aerobic chemoheterotrophy is a necessity under low light intensity or darkness or as a supplement to photoheterotrophy even under high light. An O_2 requirement, even for bacteriochlorophyll synthesis, is now not without precedent. *Erythrobacter longus*, apparently related to some of the purple bacteria (C. Woese, personal communication), requires high O_2 for bacteriochlorophyll *a* synthesis, but as yet the importance of this pigment in the energy metabolism of this organism has not been demonstrated (9).

CONCLUSIONS

It appears now that *Chloroflexus*-type organisms may range in their metabolic diversity from strict anaerobic photoheterotrophy and facultative sulfide-dependent photoautotrophy to facultative, or possibly obligate, aerobic chemoheterotrophy. It should now be determined by nucleotide sequence data whether some of the aerobic flexibacteria (sense of Soriano [18b]) are direct descendants of the *Chloroflexaceae* that have lost bacteriochlorophylls entirely.

It seems almost certain that many more examples of *Chloroflexus*-related organisms will be discovered in the near future. Although the best known examples now are derived from hot springs, this is probably because the persons most interested in this group in the past have worked on hot spring microbial ecology. With a greater interest in this group, the range of habitats will undoubtedly be greatly expanded. The metabolic diversity of some members of the *Chloroflexaceae* resembles closely that of the purple nonsulfur bacteria. Therefore, one might expect to find *Chloroflexus*-like organisms in similar habitats. These habitats are very diverse, but few support conspicuous populations of purple nonsulfur bacteria. Thus, *Chloroflexus*-like organisms may also be hidden in numerous niches and could perhaps be revealed by enrichment techniques. Methods for isolation have been described by Castenholz and Pierson (3), but additional innovative techniques will probably have to be applied.

LITERATURE CITED

1. **Buchanan, R. E., and N. E. Gibbons (ed.).** 1974. Bergey's manual of determinative bacteriology, 8th ed. The Williams & Wilkins Co., Baltimore.
2. **Castenholz, R. W.** 1973. The possible photosynthetic use of sulfide by the filamentous phototrophic bacteria of hot springs. Limnol. Oceanogr. **18:**863–876.
3. **Castenholz, R. W., and B. K. Pierson.** 1981. Isolation of members of the family Chloroflexaceae, p. 290–298. *In* M. P. Starr, H. Stolp, H. G. Trüper, A. Balows, and H. G. Schlegel (ed.), The prokaryotes, vol. 1. Springer-Verlag, Berlin.
4. **Doemel, W. N. and T. D. Brock.** 1974. Bacterial stromatolites: origin of laminations. Science **184:**1083–1085.
5. **Dubinina, G. A., and V. M. Gorlenko.** 1975. New filamentous photosynthetic green bacteria containing gas vacuoles. Microbiology (USSR) **44:**452–458.
6. **Fox, G. E., E. Strackebrandt, R. B. Hespell, J. Gibson, J. Maniloff, T. A. Dyer, R. S. Wolfe, W. E. Balch, R. S. Tanner, L. J. Magrum, L. B. Zoblen, R. Blakemore, R. Gupta, L. Bonen, G. J. Lewis, D. A. Stahl, K. R. Luehrsen, K. N. Chu, and C. R. Woese.** 1980. The phylogeny of prokaryotes. Science **209:**356–463.
7. **Gorlenko, V. M.** 1975. Characteristics of filamentous phototrophic bacteria from freshwater lakes. Microbiology (USSR) **44:**682–684.
8. **Gorlenko, V. M., and T. A. Pivovarova.** 1977. On the belonging of blue-green alga *Oscillatoria coerulescens* Gicklhorn, 1921 to a new genus of Chlorobacteria *Oscillochloris* nov. gen [In Russian, with English summary.] Isvestiya Akademii Nauk SSR, Seriya Biologicheskaya **3:**396–409.
9. **Harashima, K., M. Nakagawa, and N. Murata.** 1982. Photochemical activities of bacteriochlorophyll in aerobically grown cells of aerobic heterotrophs, *Erythrobacter* species (OCh 114) and *Erythrobacter longus* (OCh 101). Plant Cell Physiol. **23:**185–194.
10. **Kelly, D. P.** 1974. Growth and metabolism of the obligate photolithotroph *Chlorobium thiosulfatophilum* in the presence of added organic nutrients. Arch. Microbiol. **100:**163–178.
11. **Kleinig, H., and H. Reichenbach.** 1977. Carotenoid glucosides and menaquinones from the gliding bacterium *Herpetosiphon giganteus* Hp a2. Arch. Microbiol. **112:**307–310.
12. **Madigan, M. T., R. S. Petersen, and T. D. Brock.** 1974. Nutritional studies on *Chloroflexus*, a filamentous, photosynthetic, gliding bacterium. Arch. Microbiol. **100:**97–103.
13. **Olson, J. M.** 1978. Confused history of *Chloropseudomonas ethylica* 2K. Int. J. Syst. Bacteriol. **28:**128–129.
14. **Pierson, B. K., and R. W. Castenholz.** 1971. Bacteriochlorophylls in gliding filamentous prokaryotes from hot springs. Nature (London) **233:**25–27.
15. **Pierson, B. K., and R. W. Castenholz.** 1974. A phototrophic gliding filamentous bacterium of hot springs. *Chloroflexus aurantiacus*, gen. and sp. nov. Arch. Microbiol. **100:**5–24.
16. **Pierson, B. K., and R. W. Castenholz.** 1974. Studies of pigments and growth in *Chloroflexus aurantiacus*, a phototrophic filamentous bacterium. Arch. Microbiol. **100:**283–305.
17. **Pierson, B. K., and J. P. Thornber.** 1983. Isolation and spectral characterization of photochemical reaction centers from the thermophilic green bacterium *Chloroflexus aurantiacus* strain J-10-f1. Proc. Natl. Acad. Sci. U.S.A. **80:**80–84.
18. **Pierson, B. K., J. P. Thornber, and R. E. B. Seftor.** 1983. Partial purification, subunit structure and thermal stability of the photochemical reaction center of the thermophilic green bacterium *Chloroflexus aurantiacus*. Biochim. Biophys. Acta **723:**322–326.
18a.**Schmidt, K.** 1980. A comparative study on the composition of chlorosomes (Chlorobium vesicles) and cytoplasmic membranes from *Chloroflexus aurantiacus* strain Ok-70-fl and *Chlorobium limicola* f. *thiosulfatophilum* strain 6230. Arch. Microbiol. **124:**21–31.
18b.**Soriano, S.** 1973. Flexibacteria. Annu. Rev. Microbiol. **27:**155–170.
19. **Trüper, H. G.** 1976. Higher taxa of the phototrophic bacteria: *Chloroflexaceae* fam nov., a family for the gliding, filamentous, phototrophic "green" bacteria. Int. J. Sys. Bacteriol. **26:**74–75.
20. **Trüper, H. G., and N. Pfennig.** 1981. Characterization and identification of the anoxygenic phototrophic bacteria, p. 299–312. *In* M. P. Starr, H. Stolp, H. G. Trüper, A. Balows, H. G. Schlegel (ed.), The prokaryotes, vol. 1. Springer-Verlag, Berlin.

Infectious Processes in Plants

Development of Vesicular-Arbuscular Mycorrhizae

GLYNN D. BOWEN

Division of Soils, CSIRO, Glen Osmond, South Australia 5064

I regard infection processes in vesicular-arbuscular (VA) mycorrhizae to include arrival of the fungus at the root, penetration and development of the infection, and then its spread to other parts of the root. Resistance to infection is the exception rather than the rule—not only are most families of higher plants susceptible but also several pteridophytes and bryophytes are infected (19). However, variation occurs between fungus–host combinations in the development of the infection. Compared with plant pathogens and legume nodules, research on VA mycorrhizae is recent, having developed mainly over the past decade. Despite some excellent electron microscope studies, the physiological bases of infection are not well understood. In this paper I comment on some of the approaches to analyzing the biology of the infection and on some of the physiological considerations.

The three cardinal phases of infection are indicated in Fig. 1a, from Furlan and Fortin (18), in which temperature is shown to affect all three phases: a lag phase (I), a period of rapid development of the infection (II), and a "plateau" phase (III) indicating the maximum percentage of root infected for that plant species (47). Figure 1b, from Abbott and Robson (1), shows that with some fungus–plant combinations infection is so rapid that phase I hardly exists.

PREINFECTION

The preinfection phase involves germination of spores (or other propagules), growth to the root, and possibly growth in the rhizosphere. However, growth in the rhizosphere is not demonstrably necessary for infection, although it will probably be involved in secondary infection along the root.

Soil factors such as pH, temperature, and nutrient level affect the amount of root infected with VA endophytes (18, 21, 25, 35, 44), but usually the effects on the individual steps—preinfection, the entry of the fungus, and spread—have not been defined. Overall environmental effects on the first two of these, which largely constitute phase I, can be initially examined together by using an experimental approach (e.g., 44) or by the mathematical analysis of infection (45). Experimentally it is important that numbers of infection points, or "infection units" (i.e., infections arising from the one entry point), be counted rather than the percentage of the root which is mycorrhizal—the usual measurement. The number of infection units on a root does not increase proportionally beyond a certain number of propagules (11) (which must be determined for each fungus–plant combination), and thus it is important that the inoculum be standardized to be somewhat below that point so that maximum sensitivity is obtained in experiments on environmental effects. Soil temperature strongly affects phase I (Fig. 1a; 44), as does soil moisture; Reid and Bowen (39) showed that a reduction of soil moisture from -0.19 MPa (22% moisture) to -0.43 MPa (15% moisture) reduced the number of infection points in *Medicago truncatula* by 77%. In clover, 500 mg of superphosphate per 3 kg of soil reduced numbers of infection points by 81 to 94% (25). Soil pH also affects phase I; e.g., it markedly effects germination of spores of VA fungi in soil (15).

The fungus species and the type of propagule can affect phase I greatly. Considerable variation occurs in the time for germination of spores; e.g., spores of *Acaulospora laevis* may take 1 to 2 months to germinate (38) whereas inocula consisting of infected roots can infect rapidly.

Experimental approaches to examining the individual factors in the preinfection phase (spore germination, germ tube growth, rhizosphere growth) are simple to devise and have been discussed earlier (8). Infection can be blocked at any one of these stages. I therefore stress only that studies in laboratory media can be quite misleading if extrapolated to soil.

The plant's role in stimulation of VA endophytes in soil is still unclear. Spores germinate in moistened soil in the absence of plant roots, but there is no comprehensive study of whether roots can enhance this. However, with spores on agar-covered glass slides in soil, directed growth of hyphae to onion root (presumably along a chemical gradient) occurred for some 3 to 4 mm with one fungus species and for 1.6 mm from spores of two other species (38). Koske (29) found attraction of germ tubes of *Gigaspora*

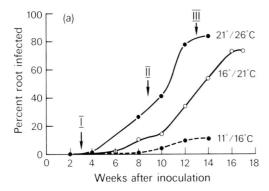

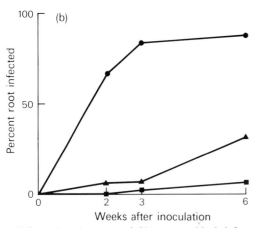

FIG. 1. Development of VA mycorrhizal infections. (a) Effect of 3 day/night temperatures (indicated) on infection of *Allium cepa* by *Gigaspora calospora*. Infection phases I, II, and III are indicated for 21°C/26°C. From Furlan and Fortin (18). (b) Development of infection by three endophytes on *Trifolium subterraneum*: ●, *Glomus fasciculatus*; ▲, fine endophyte; ■, *Acaulospora* sp. From Abbott and Robson (1).

gigantea to roots of bean and of corn by volatile agents from the roots.

INFECTION PROCESS

There are few studies of the infection process at the light microscope level, and there are almost no studies of the initial penetration of the root. Infection is often (not always) preceded by appressorium formation. Hyphae then penetrate the outer cells (sometimes through a root hair) or (commonly) pass between epidermal cells and become intracellular on reaching the second layer of cortical cells. Structural changes occur in the hyphae inside the root, where growth can be intercellular or intracellular.

Detailed reviews of the ultrastructure of VA mycorrhizae have been given by Carling and Brown (12) and by Scannerini and Bonfante-

Fasolo (42). Most, if not all, studies have been on a few species of *Glomus*, only one of four or five genera producing VA mycorrhizae. Most of the ultrastructural studies are on the penetration of cortical cells and the subsequent formation of the arbuscule, the highly branched organ involved in the exchange of nutrients and metabolites with the plant. The other major structural body, the vesicle, is a one- to several-walled multinucleate ovoid body, usually some 50 by 70 μm, in which lipid storage is prominent and in which glycogen sometimes occurs. Vesicles may develop both within and between plant cells and appear to be incipient chlamydospores.

In arbuscule development, penetration starts as a constricted hyphal peg causing the cortical cell wall to stretch and invaginate (24). After penetration, the hypha resumes its normal size and dichotomizes frequently to produce the arbuscule. This is multinucleate, is always enclosed in the fungal cell wall (sometimes very thin), and has a relatively short life span, usually from 4 to 15 days (5, 14). A stretching of the host cell wall and alteration of the middle lamella during penetration (27, 28) suggest a fungus enzyme action in penetration of the cell wall (and possibly a plant reaction also) rather than a mechanical rupture of the cell wall. During arbuscule development the fungus plasmalemma invaginates and produces multivesiculate paramural bodies or lomasomes, which are particularly abundant in the arbuscule branch. These are rich in polysaccharides and have been suggested as important in exchanges of metabolites between fungus and plant (16). The arbuscule has many organelles and has inclusions of glycogen, lipid, and polyphosphate (12, 14, 28).

What of the plant's reaction to the infection? Many, but not all, of the reactions to biotrophic pathogens occur; e.g., papilla formation (3, 17) does not occur as a host response. The trunk of the arbuscule, however, is surrounded by a "collar," continuous with the plant cell wall and containing glycoproteins which are the same as those of the cell wall (16, 42). As the arbuscule develops, this collar becomes thinner and is replaced by a matrix of plant cell origin consisting largely of disorganized polysaccharide fibrils (16) that surrounds the entire arbuscule. The amount of this interfacial matrix diminishes with development of the arbuscule. Histochemical tests show that the matrix also contains some protein and possibly glycoprotein (6). Host cell wall deposition is not rapid enough to contain the fungus, which apparently disorients the ability of the plant to organize the fibrillar network into a cell wall. When growth of the arbuscule ceases, the fibrillar layer encases it.

The plant cell's reaction to the infection is slight; although phenolic-type deposition does not occur (24), there are several detectable changes in the host cytoplasm. The highly invaginated host plasmalemma is elaborated into paramural bodies which can contain cytoplasm and which are cut off and occur near the fungus. Host cytoplasm is increased by up to 20-fold (14); the nucleus becomes highly polyploid and the nucleolus is greatly enlarged. Golgi activity increases markedly (no doubt involved in the secretion of the polysaccharide matrix), and the endoplasmic reticulum is also greatly increased. These changes reflect great metabolic activity of the host cell: VA mycorrhizae have been found to respire at a rate six times that of roots lacking the VA endophyte (F. E. Sanders, J. K. Martin, and G. D. Bowen, unpublished data). Plastid development in the cell is blocked at the protoplastid phase or plastids turn into chromoplasts, which suggests a modification of the carbohydrate metabolism of the root cell; Scannerini and Bonfante-Fasolo (42) pointed out that, although starch is usually absent from invaded cells, minute amyloplasts have been observed and the fungus may not prevent starch accumulation totally. In sharp contrast to plant pathogenic infections, no observable cytochemical or morphological difference in the host plasmalemma occurs after infection (6), a fundamental difference which may be crucial in the maintenance of a two-directional transfer of nutrients and metabolites between fungus and root.

PHYSIOLOGY OF INFECTION

In contrast to most plant pathogens, VA mycorrhiza fungi have an extremely wide range of hosts. With VA mycorrhizae the major phenomena to explain are the lack of specificity, the absence of typical infections in a few plant families, and the inhibition of infection by high soil P and by other environmental factors such as low light intensity.

Recognition and lack of specificity. The VA mycorrhiza symbiosis is very old indeed. Fossils of some of the earliest land plants indicate them to have had VA mycorrhizae (36). During evolution, this has conferred an effective selective advantage on mycorrhizal individuals in nutrient-poor sites. Infection is usually absent in wet conditions (21, 39), and if the dry land forms of families such as *Cyperaceae* and *Commelinaceae* (predominantly wet-site families) evolved separately from related plants such as the *Graminaeae*, this could explain the general absence of susceptibility to VA infection in *Cyperaceae* and *Commelinaceae*. These would then have evolved other mechanisms for coping with low-nutrient soils, e.g., production of many fine roots.

Nothing is really known about recognition phenomena in VA mycorrhizae and reasons for the lack of specificity. Physiological studies on this are difficult (but not impossible) because the fungi cannot yet be grown in vitro, although limited growth can be obtained from germinating spores (22). A first site of recognition may be at the cell wall, which could assist penetration by the stimulation of pectin-degrading (or other) enzymes in the fungus, causing the cell wall changes which occur during penetration. Enzyme activities of extra-matrical hyphae not involved in penetration, needed as a reference point, have not been performed. As with some plant pathogens, appressorium formation indicates a type of recognition, possibly caused by lectin-type reactions or polysaccharide–polysaccharide interactions; again, this has not been studied. However, appressorium formation could be only a thigmotrophic response, and in any case recognition does not necessarily incur appressorium formation; often appresoria do not occur with VA infection.

The most important site of recognition is probably between the fungus and the host plasmalemma, and this leads to secretion of the matrix and the increased host metabolic activities indicated above. As hypothesized by Sequeira (43) for *Agrobacterium*, a wide host range would be consistent with nonspecific recognition phenomena involving polymers common to many plant species. The stimulus for the host response is not known with VA mycorrhizae; e.g., it is not known whether recognition involves lectins of the host, lectins of the fungi, or polysaccharide–polysaccharide interactions. Polysaccharide and protein complexes of VA fungal cell walls can differ among closely related species (6), but obviously this does not result in differing specificity.

The fungus apparently prevents the host from assembling a cell wall from the matrix produced by the cell; more detailed study of this apparent inhibition may be rewarding. Also, VA mycorrhizal fungi should be examined for their ability to elicit phytoalexins; it is possible that they do not, for this would also necessitate either their indifference to a wide range of phytoalexins produced in higher plants or an ability to break down a wide range. The many and varied changes in the root cell with arbuscule formation are interesting mainly in their amount, rather than in their nature. This and the reversion of the host cell to "normality" after arbuscule degeneration suggest that the fungus may affect the production of growth regulators. The production of cytokinin and growth regulator compounds by VA mycorrhizal fungi is indirectly evidenced by the enlarged nucleolus of the plant cell and a possible effect on stomatal physiology (4). A hormone interaction with the plasmalemma, should it occur, would also enhance exchange of nutrients and metabolites via effects on permeability.

The apparent lack of infection of *Chenopodiaceae, Brassicaceae, Caryophyllaceae*, and a few other families could be due to any of numerous reasons, ranging from a physical barrier of the cell wall, to an absence of essential nutrients, to production of toxins by the plant. A lack of nutrients is not likely to be the major reason, for VA endophytes fungi can grow in the rhizosphere of nonhost plants (5, 15, 37). The occurrence of a wide range of S compounds in *Brassicaceae* and of betalins (related to phenols and anthocyanins) in *Chenopodiaceae*, with fungistatic activity, might be important. Infection of some species in these two families has been recorded (23, 30, 34, 37, 48), but in all but one case (48) infection was light; in no case did arbuscules form, and usually infection was in older roots and in the presence of a susceptible plant species, a most interesting observation.

The nonoccurrence of arbuscules in *Chenopodiaceae, Brassicaceae*, and some other plants, and the restriction of arbuscule formation to inner cortical cells in some plants (12, 24, 28), indicate some very specific cell recognition, fungus–host interactions within a root system. In what ways do some interior cortical cells differ physiologically and physically (e.g., oxygen level) from outer cortical cells? If single-gene mutants could be found which produce only vesicles in susceptible plant species, much might be learned about recognition and stimuli for arbuscules. If single-gene mutants of the cruciferous plant *Arabidopsis thaliana*, used so extensively in biochemical genetics research, were found which produce normal infections, they would be valuable material. Some infection of this plant has been recorded (30).

Effects of environment. Of the environmental factors which markedly affect formation of VA mycorrhizae, soil phosphate has received most study. Figure 2, from Jasper et al. (25), shows that increasing soil phosphate markedly decreases VA mycorrhizal infection and that VA mycorrhizal fungi differ in their sensitivity to increasing phosphate. The phosphate acts via the plant and not on the soil phases of the fungi (33).

Phosphate deficiency leads to an increase in the loss of sugars and amino acids from the root into the rhizosphere partly as a result of an increase in these compounds in deficient roots (7, 20) and partly as a result of increases in plasmalemma permeability (20) which sometimes, but not always, occurs (7). This greater "exudation" by low-phosphate plants has been correlated with a greater proportion of the root infected (some weeks later), and a causal relation with infection has been claimed with exu-

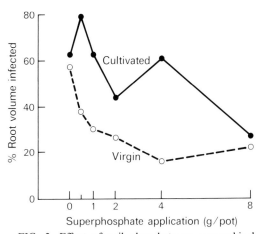

FIG. 2. Effect of soil phosphate on mycorrhizal infection of *Lolium rigidum* by the natural populations of: ●, a previously phosphate-fertilized, cultivated soil (endophyte: *Glomus monosporus*); ○, adjacent virgin soil (endophyte: *Acaulospora laevis*). From Jasper, Robson, and Abbott (25).

date composition rather than cell composition. The acceptance of this poses difficulties (20). In some instances there has been a correlation between percentage of the root which is mycorrhizal and total sugars in exudates and in other cases with amino acids (20, 26). Also, analysis of the whole root is a poor indicator of the chemical composition of the cytoplasm. "Exudates" probably reflect the composition of the cytoplasm, not the whole cell (with which correlations would be expected to be poorer). Losses of organic acids from the root often exceed those of sugars and amino acids (46), and these may warrant study. Furthermore, substances normally recorded as "exudates" come from several parts of the root (32). Although root exudation almost certainly assists spread of the fungus in the rhizosphere, there is as yet no good evidence that rhizosphere growth is necessary for infection. Neither does the plasmalemma–permeability–exudate theory comfortably accommodate a finding (40) that small amounts of phosphate increase the percentage of the root infected. If exudates are involved in the infection process, it is possible that it will not be by such common groups of compounds as those above but by specific substances (? elicitors of enzyme activity).

Jasper et al. have suggested (25) that phosphorus nutrition acts via effects on concentration of soluble carbohydrates within the cell. A good inverse relation was found between soluble carbohydrates of clover roots at various phosphate levels and the number of infection units 21 days later, but the relationship was less convincing for infection numbers at 14 days. In later studies

(40) a relationship was observed between soluble carbohydrates in roots (manipulated by temperature, light, and plant defoliation) and numbers of infections 6 days later. However, the strongest correlations were between carbohydrate in the whole root and the percentages of root infected 12 days later.

The best correlations obtained by protagonists of either of the two theories above are with the percentage of the root infected, which consists of infection and growth and spread in the root after infection. I am attracted to the "carbohydrate" theory as being important in growth in the root and spread of the infection. However, we are dealing with correlations (moreover, at a coarse level), and phosphate deficiency can affect many components of the cell. It may be interesting, for example, to examine effects of phosphate nutrition on glycerols and lipids, which may be important in the nutrition of *Endogonaceae* and related fungi (22, 31).

VA mycorrhizal fungi have defied all attempts to grow them. If one then hypothesizes that the plant provides not only energy sources but also (so far) unknown growth factors, it may well be that phosphate nutrition affects not only the level of soluble carbohydrates and other energy sources but also production of this factor(s)— whatever it might be.

SPREAD OF INFECTION

Plant response is often correlated with infection level of the root (2, 41), as this provides the energy sources for growth of the fungi into soil (and nutrient absorption). In field soil that contains VA fungus propagules, continued development of the infection occurs from spread of existing infections and from new infections. The slopes of phase II in Fig. 1a are a combination of these two phenomena. These must be separated in analyzing plant response to infection and its possible management. In inoculation programs, the fungus is added to a discrete part of the young root system (or the seed), and the spread of the infection, often in competition with indigenous VA mycorrhizal fungi, is important to a plant response. Competition between mycorrhizal fungi has been examined by Abbott and Robson (1) and by Wilson and Trinick (49) and will not be discussed further here.

Spread of the infection has received little study, although satisfactory methods for doing so experimentally are very simple. Some consider internal spread to be limited and refer to a study (13) with onion and one VA mycorrhiza. fungus, in which a spread of less than 5 mm in each direction was recorded. However, other studies (5) have indicated that infection of onion by two fungus species was quite localized but that considerable spread occurred in clover.

Two complementary approaches to investigating spread have been developed. In the first (G. D. Bowen, unpublished data) plants are grown in soil in boxes with removable Plexiglas sides so that root segments of known ages can be inoculated locally and subsequent spread internally and externally can be examined. In the second approach, spread is derived from mathematical analyses (with certain assumptions) of the increase of infection of roots with time (9, 10, 45). Although the mathematical approach has considerable power, especially in distinguishing between spread and increased infection via new infections from soil (45), it cannot be stressed too strongly that hypotheses generated by such models must be verified by experimental approaches.

Smith and Walker (45) have mathematically dissected the spread of existing infection concomitant with new infections (from soil) in clover. The basic assumptions included that (within limits) any uninfected portion of the root is equally likely to become infected regardless of age and that all infections grow at the same rate regardless of age. These assumptions need direct experimental study. The number of infections, the total length of root, and the total length of root infected with time are used to derive the numbers of new infections per centimeter per day and the spread of the infection. The model indicated different effects of nitrogen and of chloride additions to soil on these two properties. The analyses of test data for the particular fungus–plant–soil combination suggested that secondary infections were of little consequence; this may well be so when high inoculum densities of fungi are used with which new infections occur rapidly, but if these conditions are not met, secondary infections may be very important and confound the analysis.

A second type of analysis by Buwalda et al. (9, 10) examined the spread of inoculum along roots from a point source. The assumption (and biological interpretation of the analyses) is made that this is mainly by growth in the rhizosphere leading to secondary infections (an assumption which needs testing). The equation used is formally identical to that of Smith and Walker (45). Measurements of total root length and infected root length at various times are used to calculate a spread rate S and a correction term n, accommodating the observation of several authors that different species have upper limits to the infection, phase III of Fig. 1a. The effects of phosphorus nutrition on these two parameters in wheat and in leek (*Allium porrum*), independent of the effect on root growth, were examined. Rate of spread in wheat was 37% faster than in leek and was not generally affected by phosphate level within species. The factor n was affected markedly both by species and by soil phosphorus level: at low P, almost all the leek root but only half of the wheat root became infected; at higher soil P three-quarters of the leek root but less than one-tenth of the wheat root was infected. One area of concern is that the two models discussed here, developed with different assumptions of the biology, have essentially the same form. This reinforces the need to examine the biology of the system as well as the mathematics and to test generated hypotheses experimentally.

Little is known of the physiology of spread and of the determinants of maximum percentage of infected root. External spread in the rhizosphere will probably be affected both by translocation of substrates from the existing arbuscules and by the nature of the rhizosphere "exudates." The apparent lack of effect of soil P level on spread of the fungus in the studies of Buwalda et al. (10) (which was assumed to be external) is in conflict with data indicating greater "exudation" under low P conditions (20). It is highly likely, however, that spread both internally and externally will be affected by internal energy sources (e.g., soluble sugars) available in the root. The factor n may, as Buwalda et al. (10) indicated, be of major importance in the physiology of the symbiosis. Differences in n could indicate that root segments of some species are susceptible for only a short time. However, they might also reflect the cardinal importance of "balanced development," which is important in a mutualistic symbiosis. Factors such as the distribution of plant assimilate between host cell and fungus possibly determine both n and the longevity of the arbuscule; this, I believe, deserves more detailed physiological study.

CONCLUSIONS

To understand the factors involved in infection processes in VA mycorrhizae, it will be necessary to use experimental approaches to analyze the three separate phases: preinfection growth to the root, initiation of the infection, and subsequent spread (and growth of the fungus into soil). The construction of curves relating soil populations of the fungi to the number of infections is basic to defining factors affecting the first two of these. Mathematical analysis of the increase of infection with time is a powerful emerging tool in distinguishing spread of existing infections from increases due to new infections and in indicating the effects of soil and plant factors on these. However, the assumptions used to develop the analyses, and the hypotheses they generate, must be tested experimentally and not embraced uncritically. Such methods being developed for "epidemiology" analysis of VA mycorrhizal infections will also be directly applicable to the analysis of root pathogenic infections.

The great lack of specificity and the mildness (and subtlety) of the plant cell reaction to infection are outstanding physiological phenomena in VA mycorrhizal infections. Relatively little is known about the biochemistry of penetration and "recognition" phenomena. An inhibition of host cell wall assembly appears confirmed. There are indications that the fungus increases polysaccharide production by the host and also increases host cell metabolism, but there is little knowledge of the mechanism of this or of whether the host cell produces reaction substances such as phytoalexins. There is also little knowledge of possible growth factors produced by plant cells which enable their growth of the fungus, in sharp contrast to our inability to grow the fungi in the absence of a plant.

Much of the data from physiology studies relating infection to environmental factors, such as plant phosphate status, relate to development and spread of infection rather than the infection processes themselves. The spread of infection in the root and the longevity of arbuscules are basic to abundant growth of hyphae from VA mycorrhizae into soil (and nutrient uptake for the plant). Understanding the physiological control of these, probably in relation to photosynthate distribution along the root, is an important area for future research.

It is a pleasure to acknowledge helpful discussions with L. K. Abbott, R. C. Foster, J. L. Parke, J. M. Trappe, and A. D. Robson.

LITERATURE CITED

1. **Abbott, L. K., and A. D. Robson.** 1982. Infectivity of vesicular arbuscular mycorrhizal fungi in agricultural soils. Aust. J. Agric. Res. **33**:1049–1059.
2. **Abbott, L. K., and A. D. Robson.** 1982. The role of vesicular arbuscular mycorrhizal fungi in agriculture and the selection of fungi for inoculation. Aust. J. Agric. Res. **33**:389–408.
3. **Aist, J. R.** 1976. Papillae and related wound plugs of plant cells. Annu. Rev. Phytopathol. **14**:145–163.
4. **Allen, M. F., and M. G. Boosalis.** 1983. Effects of two species of v.a. mycorrhizal fungi on drought tolerance of winter wheat. New Phytol. **93**:67–76.
5. **Bevege, D. I., and G. D. Bowen.** 1975. Endogone strain and host plant differences in development of vesicular-arbuscular mycorrhizas, p. 77–86. *In* F. E. Sanders, B. Mosse, and P. B. Tinker (ed.), Endomycorrhizas. Academic Press, London.
6. **Bonfante-Fasolo, P., J. Dexheimer, S. Gianinazzi, V. Gianinazzi-Pearson, and S. Scannerini.** 1981. Cytochemical modifications in the host-fungus interface during intracellular interactions in vesicular-arbuscular mycorrhizae. Plant Sci. Lett. **22**:13–21.
7. **Bowen, G. D.** 1969. Nutrient status effects loss of amido- and amino acids from pine roots. Plant Soil **30**:139–142.
8. **Bowen, G. D.** 1980. Misconceptions, concepts and approaches in rhizosphere biology, p. 283–304. *In* D. C. Ellwood, J. N. Hedger, M. J. Latham, J. M. Lynch, and J. H. Slater (ed.), Contemporary microbial ecology. Academic Press, London.

9. **Buwalda, J. G., G. J. S. Ross, D. P. Stribley, and P. B. Tinker.** 1982. The development of endomycorrhizal root systems. III. The mathematical representation of the spread of vesicular-arbuscular mycorrhizal infection in root systems. New Phytol. **91:**669–682.

10. **Buwalda, J. G., G. J. S. Ross, D. P. Stribley, and P. B. Tinker.** 1982. The development of endomycorrhizal root systems. IV. The mathematical analysis of effects of phosphorus on the spread of vesicular-arbuscular mycorrhizal infection in root systems. New Phytol. **92:**391–399.

11. **Carling, D. E., M. F. Brown, and R. A. Brown.** 1979. Colonisation rates and growth responses of soybean plants infected by vesicular-arbuscular mycorrhizal fungi. Can. J. Bot. **57:**1769–1772.

12. **Carling, D. E., and M. F. Brown.** 1982. Anatomy and physiology of vesicular-arbuscular and nonmycorrhizal roots. Phytopathology **72:**1108–1114.

13. **Cox, G., and F. E. Sanders.** 1974. Ultrastructure of the host-fungus interface in a vesicular-arbuscular mycorrhiza. New Phytol. **73:**901–912.

14. **Cox, G., and P. B. Tinker.** 1976. Translocation and transfer of nutrients in vesicular-arbuscular mycorrhiza. I. The arbuscule and phosphorus transfer: a quantitative ultrastructural study. New Phytol. **77:**371–378.

15. **Daniels, B. A., and J. M. Trappe.** 1980. Factors affecting spore germination of the vesicular-arbuscular mycorrhizal fungus, *Glomus epigaeus.* Mycologia **72:**457–471.

16. **Dexheimer, J., S. Gianinazzi, and V. Gianinazzi-Pearson.** 1979. Ultrastructural cytochemistry of the host-fungus interfaces in the endomycorrhizal association *Glomus mosseae/Allium cepa.* Z. Pflanzenphysiol. **92:**191–206.

17. **Dickinson, C. H., and J. A. Lucas.** 1982. Plant pathology and plant pathogens, 2nd ed., p. 168. Blackwell Scientific Publications, Oxford.

18. **Furlan, V., and J. A. Fortin.** 1973. Formation of endomycorrhizae by *Endogone calospora* on *Allium cepa* under three temperature regimes. Nat. Can. **100:**467–477.

19. **Gerdemann, J. W.** 1975. Vesicular-arbuscular mycorrhizae, p. 575–591. *In* J. G. Torrey and D. F. Clarkson (ed.), The development and function of roots. Academic Press, London.

20. **Graham, J. H., R. T. Leonard, and J. A. Menge.** 1981. Membrane mediated decrease in root exudation responsible for phosphorus inhibition of vesicular arbuscular mycorrhiza formation. Plant Physiol. **68:**548–552.

21. **Hayman, D. S.** 1983. The physiology of vesicular-arbuscular mycorrhizal symbiosis. Can. J. Bot. **61:**944–965.

22. **Hepper, C. M.** 1983. Limited independent growth of a vesicular-arbuscular mycorrhizal fungus *in vitro.* New Phytol. **93:**537–542.

23. **Hirrel, M. C., H. Mehravaran, and J. W. Gerdemann.** 1978. Vesicular-arbuscular mycorrhizae in the Chenopodiaceae and Cruciferae: do they occur? Can. J. Bot. **56:**2813–2817.

24. **Holley, J. D., and R. L. Peterson.** 1979. Development of a vesicular-arbuscular mycorrhiza in bean roots. Can. J. Bot. **57:**1960–1978.

25. **Jasper, D. A., A. D. Robson, and L. K. Abbott.** 1979. Phosphorus and the formation of vesicular-arbuscular mycorrhizas. Soil Biol. Biochem. **11:**501–505.

26. **Johnson, C. R., J. A. Menge, S. Schwab, and I. P. Ting.** 1982. Interaction of photoperiod and vesicular-arbuscular mycorrhizae on growth and metabolism of sweet orange. New Phytol. **90:**665–669.

27. **Kaspari, J.** 1973. Elektronenmikroscopische Untersuchung zur Feistruktur der endotrophen Tabakmykorrhiza. Arch. Mikrobiol. **92:**201–207.

28. **Kinden, D. A., and M. F. Brown.** 1975. Electron microscopy of vesicular-arbuscular mycorrhizae of yellow poplar. III. Host-endophyte interactions during arbuscular development. Can. J. Microbiol. **21:**1930–1939.

29. **Koske, R. E.** 1982. Evidence for a volatile attractant from plant roots affecting germ tubes of v.a. mycorrhizal fungus. Trans. Br. Mycol. Soc. **79:**305–310.

30. **Kruckelmann, H. W.** 1975. Effects of fertilizers, soils, soil tillage and plant species on the frequency of *Endogone* chlamydospores and mycorrhizal infection in arable soils, p. 511–525. *In* F. E. Sanders, B. Mosse, and P. B. Tinker (ed.) Endomycorrhizas. Academic Press, London.

31. **Lewis, D. H.** 1975. Comparative aspects of the carbon nutrition of mycorrhizas, p. 119–148. *In* F. E. Sanders, B. Mosse, and P. B. Tinker (ed.) Endomycorrhizas. Academic Press, London.

32. **Martin, J. K.** 1978. The variation with plant age of root carbon available to soil microflora, p. 299–302. *In* M. W. Loutit and J. A. R. Miles (ed.), Microbial ecology. Springer-Verlag, Berlin.

33. **Menge, J. A., D. Steirle, D. J. Bagyaraj, E. L. V. Johnson, and R. T. Leonard.** 1978. Phosphorus concentration in plants responsible for inhibition of mycorrhizal infection. New Phytol. **80:**575–578.

34. **Miller, R. M.** 1979. Some occurrences of vesicular-arbuscular mycorrhiza in natural and disturbed ecosystems of the Red Desert. Can. J. Bot. **57:**619–623.

35. **Mosse, B.** 1972. The influence of soil type and *Endogone* strain on the growth of mycorrhizal plants in phosphate deficient soils. Rev. Ecol. Biol. Sol **9:**529–537.

36. **Nicolson, T. H.** 1975. Evolution of vesicular-arbuscular mycorrhizas, p. 25–34. *In* F. E. Sanders, B. Mosse, and P. B. Tinker (ed.), Endomycorrhizas. Academic Press, London.

37. **Ocampo, J. A., J. Martin, and D. S. Hayman.** 1980. Influence of plant interactions on vesicular-arbuscular mycorrhizal infections. I. Host and non-host plants grown together. New Phytol. **84:**27–35.

38. **Powell, C. L.** 1976. Development of mycorrhizal infections from *Endogone* spores and infected root segments. Trans. Br. Mycol. Soc. **66:**439–445.

39. **Reid, C. P. P., and G. D. Bowen.** 1979. Effects of soil moisture on v.a. mycorrhizal formation and root development in *Medicago*, p. 211–219. *In* J. L. Harley and R. S. Russell (ed.), The root-soil interface. Academic Press, London.

40. **Same, B. I., A. D. Robson, and L. K. Abbott.** 1983. Phosphorus, soluble carbohydrates and endomycorrhizal infection. Soil Biol. Biochem. **15:**593–597.

41. **Sanders, F. E., P. B. Tinker, R. L. B. Black, and S. M. Palmerley.** 1977. The development of endomycorrhizal root systems. I. Spread of infection and growth promoting effects with four species of vesicular arbuscular endophyte. New Phytol. **78:**257–268.

42. **Scannerini, S., and P. Bonfante-Fasolo.** 1983. Comparative ultrastructural analysis of mycorrhizal associations. Can. J. Bot. **61:**917–943.

43. **Sequeira, L.** 1980. Defenses triggered by the invader: recognition and compatibility phenomena, p. 179–200. *In* J. G. Horsfall and E. B. Cowling (ed.), Plant disease, vol. 5. Academic Press, Inc., New York.

44. **Smith, S. E., and G. D. Bowen.** 1979. Soil temperature, mycorrhizal infection and nodulation of *Medicago truncatula* and *Trifolium subterraneum.* Soil Biol. Biochem. **11:**469–473.

45. **Smith, S. E., and N. A. Walker.** 1981. A quantitative study of mycorrhizal infection in *Trifolium*: separate determination of the rates of infection and of mycelial growth. New Phytol. **89:** 225–240.

46. **Smith, W. H.** 1977. Tree root exudates and the forest soil ecosystem: exudate chemistry, biological significance and alteration by stress, p. 289–301. *In* J. K. Marshall (ed.), The below ground ecosystem; a synthesis of plant associated processes. Colorado State University, Fort Collins.

47. **Sutton, J. C.** 1973. Development of vesicular-arbuscular mycorrhizae in crop plants. Can. J. Bot. **51:**2487–2493.

48. **Williams, S. E., A. G. Woolum II, and E. F. Aldon.** 1974. Growth of *Atriplex canescens* improved by formation of vesicular-arbuscular mycorrhizal. Proc. Soil Sci. Soc. Am. **38:**962–965.

49. **Wilson, J. M., and M. J. Trinick.** 1983. Infection development and interactions between vesicular-arbuscular mycorrhizal fungi. New Phytol. **93:**543–553.

Molecular Basis for the Specificity of Plant Pathogenic Microorganisms for Their Hosts

R. N. STRANGE

University College London, London, United Kingdom

Plants are great synthesizers: from atmospheric carbon dioxide, which they fix at the rate of about 3×10^{11} tonnes/year (15), and a few minerals taken from the soil, sometimes with the aid of microorganisms, they are able to make an almost infinite array of compounds. They are also ubiquitous, few land surfaces of the earth being devoid of plant life. Because of these two attributes plants represent a potential ecological niche for microorganisms unparalleled in its nutritional diversity and abundance. Nevertheless, for the vast majority of microorganisms, plants remain forbidden territory. In this paper I attempt to examine some of the reasons for this situation, together with its converse in which a microorganism successfully infects and parasitizes plant tissue. Space dictates that I be selective, and I shall accordingly draw most of my examples from the more recent literature.

THE GENETIC PERSPECTIVE

A primary reason for the resistance of plants to parasites is their genetic variability. There are over 5×10^5 plant species (71), and within these species there is often considerable polymorphism. Thus, a microorganism which has adapted successfully to a parasitic mode of life with one plant genotype may be unable to overcome the resistance of a quite closely related plant. Conversely, if the parasite is too virulent, it may kill its host before either parasite or plant has had time to reproduce, with the consequence of extinction for both partners. The hazardous course of survival for the parasite therefore lies between the Scylla of effective host resistance and the Charybdis of overvirulence. Some parasites have managed to steer this course with skill and have co-evolved with their hosts in an uneasy equilibrium, an equilibrium which was to be rudely jolted in domestic crops by the plant breeder in his quest for resistance.

Prior to the beginning of this century, there were some conscious efforts to select resistant cultivars. For example, in 1799 Knight reported on the resistance of some of his hybrid wheats to "blight" probably caused by *Puccinia striiformis*, but it was Biffen's (8, 9) demonstration that resistance in this host–parasite combination obeyed Mendel's laws of heredity that marked the beginning of the development of scientific breeding for resistance. Although Biffen's work showed that resistance was inherited as a recessive character, much subsequent study has demonstrated that resistance is more generally dominant and is often conferred by one or a few genes. As a result of these discoveries, considerable effort has been expended on incorporating resistance genes into commercial cultivars. These cultivars not surprisingly become popular with farmers and are soon grown over wide areas. Unfortunately, variants of the parasite arise which have acquired the ability to attack the new cultivars, frequently with devastating results. For example, ten Houten (67) reported the demise of a promising wheat cultivar, Heine VII, in these terms: "When inoculated with the known strains of stripe rust (*Puccinia striiformis*) it remained resistant in our field experiments both in 1950 and 1951; when introduced commercially in 1952, 14% of the total wheat area in the Netherlands was sown with it, and no stripe rust occurred. In 1953 the area sown with Heine VII increased to 43% of the total and only one small locus of infection was found in a breeder's farm. In 1955 when 81% of the total wheat area had been sown with Heine VII it was everywhere heavily infected with stripe rust, and in 1956 70% of the winter wheat area was destroyed."

The hazard of genetic uniformity in crop plants was even more dramatically highlighted by the southern corn leaf blight epidemics of 1970–1971 (68). In the 1950s and 1960s considerable use had been made of a male sterility factor in corn breeding programs since it facilitated the production of hybrid lines. By 1970 about 85% of the hybrid seed corn in the United States contained this cytoplasmically inherited gene (68). Unfortunately, it had a pleiotropic expression which took the form of acute susceptibility to a variant of the fungus *Helminthosporium maydis*. Losses totaled at least $1 billion (66).

Both of these examples show that plant parasites have the ability to vary in pathogenicity. Studies of this variation were pioneered by Flor, who concluded that for each gene that conditions resistance in the host there is a corresponding gene that conditions pathogenicity in the parasite (23). Generally, pathogenicity and virulence are inherited as recessive characters. Correspondingly non-pathogenicity or avirulence is dominant.

The southern corn leaf blight disaster prompted a study of the genetic vulnerability of major crop plants (53), and the trend is now away from resistance which is controlled by only one or a few genes to resistance controlled by multiple genes which it is hoped will prove more durable (38).

INCONGRUITY AND INCOMPATIBILITY

Although in some crop plants (61), as a result of the efforts of the plant breeder, we now have good information about the genetics of resistance, we are still largely ignorant of the expression of effective resistance genes except in terms of the inability of a given parasite to grow well in the plant that possesses them. Moreover, information is available only for genes which confer resistance to parasites which have co-evolved with their hosts. We have no knowledge of the genetics of "nonhost" resistance since the crosses of the plant species implied by this term are not possible by conventional breeding techniques. Hogenboom (35) has introduced some useful terms to describe these two types of resistance. He has pointed out that "nonhost" resistance is likely to be caused by "incongruity" between the partners, whereas "host" resistance is caused by "incompatibility" and implies a disturbance of the infection process.

Examples of incongruity. Heath (31) has studied incongruity in rust infections. There are several stages through which rust fungi must pass to establish compatible relations with their hosts: spore germination, orientation of the germ tube on the leaf surface in a direction which maximizes its chance of encountering a stoma, the formation of infection structures (appressorium, infection peg, substomatal vesicle, infection hyphae), penetration of cell walls by the young haustorium, and establishment of a successful metabolic relation by the haustorium with the host cell. All these stages represent points at which the relationship may fail. For example, germination of uredospores of the cowpea rust fungus (*Uromyces phaseoli* subsp. *vignae*) on leaves of cabbage, pea, and tomato is low. Tomato leaves appear to possess an inhibitor, but the leaf surfaces of the other two species are difficult to wet. Uredospores are also hydrophobic, yet the presence of free water is necessary for their germination since it is required to leach out the spores' own germination inhibitor (44). Appressoria on cabbage and pea, when formed, frequently failed to coincide with stomata. Those that were located over stomata were examined further. In cabbage 14% ceased to grow during the formation of the penetration peg or substomatal vesicle and the remaining 86% ceased to grow during the formation of infection hyphae. On the other hosts, growth continued to the formation of the haustorial mother cell before being halted (31).

Examples of incompatibility. Incompatible associations of parasites with host plants always seem to involve intimate contact of the two partners, whereas this need not necessarily be true of incongruent associations. For example, in cowpeas resistant to *U. phaseoli* subsp. *vignae*, growth normally ceased only after the formation of at least one haustorium (31). Again, in other rust infections this appears to be the stage at which resistance is expressed (29). Similarly, resistance of soybeans to an incompatible race of *Phytophthora megasperma* subsp. *glycinea* (= *Phytophthora megasperma* subsp. *sojae*) was expressed after penetration (64).

DEFENSE MECHANISMS

Theoretical considerations. If they are to survive the constant onslaught of microorganisms from the soil and the atmosphere, it is axiomatic that plants must possess mechanisms of defense. These should be adequate to inhibit infections by both incongruent and incompatible parasites. The possession by plants of preformed structural barriers and antimicrobial compounds may be sufficient to defend the plant against incongruent parasites, but since they represent complex biosyntheses, they would a priori be unlikely to have a role to play in resistance to incompatible parasites determined by only one or a few genes. In incompatible resistance, defense always seems to be an active process, whereas in incongruent resistance defense may be active or passive. Active defense must be triggered, and this process may be brought about specifically by incompatible parasites, presumably as a result of the interaction of products of avirulence and resistance genes. Incongruent parasites may trigger the same defense response by nonspecific means, i.e., by causing injury. In both cases "triggering" may cause the release of a "constitutive elicitor" (4) or "second messenger" (39) which brings about the defense response.

Experimental data. There are many reports concerning preformed structural and biochemical barriers to infection, but these will be considered later in the context of a parasite's ability to overcome them. There are also a variety of active responses to infection which have been considered as possible defense mechanisms. These include the deposition of cell wall material (75), silicification (76), lignification (7), hypersensitivity (32), and the synthesis of low-molecular-weight antimicrobial compounds termed phytoalexins (literally plant defenders). Of these responses, phytoalexins have received the most attention. They are synthesized in both incongruent (46) and incompatible (47) associations and in response to many nonspecific agents. These include treatment of plant tissues with the salts of heavy metals (30), UV irradiation (43), chloroform vapor (5), and physical wounding (62). The chemical structures of phytoalexins are various and include flavonoids, chalcones, stilbenes, terpenoids, and polyacetylenic compounds (6). These diverse modes of phytoalexin elicitation and chemical structure, coupled with their failure in some instances to inhibit parasites, have led some plant pathologists to question whether they play any role in resistance. This may now be regarded as an extreme view since in many host–parasite interactions, particularly those involving members of the *Leguminosae*, the concentrations of phytoalexins attained are sufficient to explain the cessation of growth of the parasite (46). The isolation and chemical characterization of phytoalexins are therefore important endeavors, but it is even more crucial that events which regulate their accumulation are understood.

There are many examples of phytoalexin elicitation in incongruent associations. For example, Ingham has made extensive use of *Helminthosporium carbonum*, a maize parasite, to elicit phytoalexins in many members of the *Leguminosae* (37).

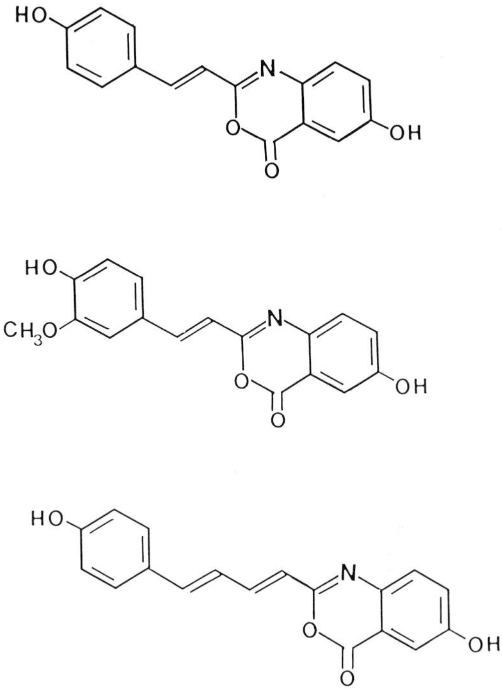

At the time of writing, the best example of mediation of resistance to an incompatible parasite by phytoalexins is found in the work of Mayama et al. (47–52). These authors isolated and characterized three novel phytoalexins in oats challenged by incompatible races of *Puccinia coronata* f. sp. *avenae* (Fig. 1). They showed that the concentration of phytoalexins increased at the same time as the fungus became inhibited and that this relationship held for oat cultivars containing 21 different genes for resistance. Other cultivars lacking resistance genes did not accumulate the phytoalexins.

Recently, we have isolated a constitutive elicitor of phytoalexin synthesis from pigeon pea (*Cajanus cajan*) (14). Aqueous ethanolic extracts of leaves or seeds of the plant were made, and the ethanol was removed by evaporation. The remaining aqueous solution was applied as droplets to superficially wounded leaves of the plant. Three phytoalexins accumulated: one was the known compound pinostrobin chalcone, and the other two were novel isomeric stilbene carboxylic acids (13) (Fig. 2). The elicitor extract was purified by thin-layer chromatography, and the elicitor was identified by ^{13}C nuclear magnetic resonance as sucrose. Maximum accumulations of phytoalexins (200 to 450 $\mu g \ g^{-1}$ of fresh weight of leaves, depending on the compound) were attained 72 h after the application of sucrose, and the concentrations of the phyto-

FIG. 1. Avenalumins, three phytoalexins from oats (52).

alexins were linearly related to the logarithm of the dose of sucrose applied in the range of 1 to 100 $\mu g \ ml^{-1}$. When actinomycin D, cycloheximide, or puromycin at concentrations of 10 $\mu g \ ml^{-1}$ or more was incorporated with sucrose in the elicitor assay, no phytoalexin accumulation occurred, strongly suggesting that RNA and protein syntheses were required for phytoalexin synthesis. We are now interested in determining how sucrose causes this gene derepression.

Other workers have shown that sucrose elicits phytoalexin accumulation in pea leaf disks (60), and it may be pertinent that Russell (6) and Cartwright and Russell (11) reported increased resistance of plants sprayed with sucrose prior to inoculation with fungal parasites.

Sucrose is normally abundant in plants and may be found in high concentrations in the phloem as well as in the apoplast and vacuoles (2, 26), but it is not clear whether it is generally excluded from the cytoplasm. If this is so, then the disruption of either the plasma membrane or the tonoplast would allow sucrose to diffuse not only into the cytoplasm of the affected cell but also through plasmodesmata to the cytoplasm of neighboring cells. How this process of sucrose dislocation then triggers phytoalexin synthesis is unknown, although binding of sucrose to a receptor may well be involved.

Lignification is another defense mechanism

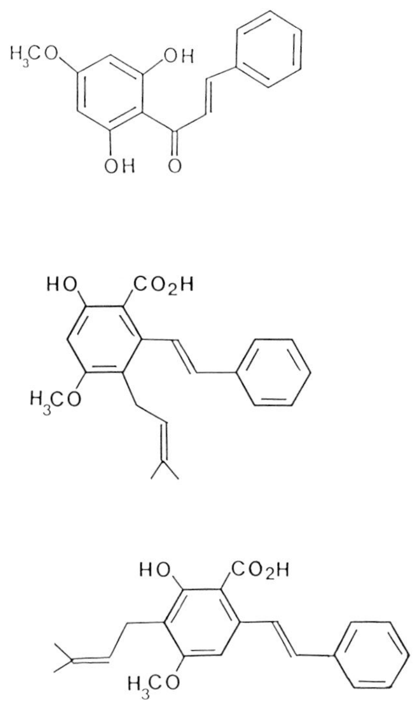

FIG. 2. Leaf phytoalexins of pigeon pea (13).

which may involve gene derepression (69) and has been particularly studied by Ride and co-workers. He showed that lignin was deposited around small local wounds in wheat leaves inoculated with two nonpathogens of the plant, *Botrytis cinerea* and *Mycosphaerella pinodes*, but not when the fungal challenge was omitted (56). In nonwounded leaves challenged with *B. cinerea*, lignified papillae (wall-like depositions of the host [1]) and haloes (59) occurred, and these were highly resistant to fungal degradation (57). Unlike phytoalexin accumulation in the *Leguminosae*, abiotic treatments, with the exception of mercuric ions, did not elicit the response (54), but chitin, an important component of most fungal cell walls, was effective (55). The induced lignin, which differed chemically from the lignin of challenged leaves, was extremely resistant to degradation by a commercial preparation of cell wall-degrading enzymes (56). Ride suggested that lignification may hinder fungal infections in several ways; these include increased resistance to enzymatic (57) and mechanical penetration, restriction of the flow of metabolites between fungus and plant, lignification of the parasite, and the accumulation of lignin precursors and free radicals which might be toxic to the challenging organism (58). In connection with the latter point, Hammerschmidt and Kuć (28) showed that the lignin precursor coniferyl alcohol was extremely toxic

to two species of *Colletotrichum*, 0.3 µg being sufficient to inhibit the growth of these fungi in one assay. This work arose out of a study of acquired resistance by Kuć and co-workers. They found that prior challenge of cucumbers with either *Colletotrichum lagenarium* or *C. cucumerinum* induced systemic resistance to disease caused by either fungus and that this resistance may be related to the ability of the protected tissue to lignify more rapidly and to a greater extent than controls which had not received the initial challenge (28).

MECHANISMS OF PATHOGENICITY AND VIRULENCE

Theoretical considerations. For an organism to be a successful plant parasite, it must be able to breach the host defenses and grow in the host. These attributes may be both nonspecific and specific. For example, there is evidence (see below) that parasites which enter their hosts through the cuticle require cutinases and cell wall-degrading enzymes to achieve infection whereas wound parasites do not need cutinases. The chemical composition of cutin and cell walls does not vary greatly from species to species, and therefore the possession of cutinases and cell wall-degrading enzymes may be regarded as nonspecific virulence attributes. In contrast, as discussed above, plants do contain both preinfection- and postinfection-formed antimicrobial compounds with diverse structures which vary according to plant species. The ability to degrade such compounds may reside in specific enzymes, and such enzymes can therefore be regarded as specific virulence factors. Additionally, many plant parasites produce toxins which may be specifically poisonous only to their host species or may be active against a wider range of plants. Such compounds have many interesting effects on sensitive tissue, but perhaps their most important role is to inhibit active defense mechanisms.

Parasites that have long co-evolved with higher plants have developed subtle relationships with their hosts which are favorable to themselves. These include alteration of translocation patterns so that nutrients are diverted to the parasite and even genetic engineering with the result that the plant produces metabolites which may be used exclusively by the parasite.

Experimental data: breaching the cuticle and the cell wall. The recent studies of Kolattukudy and co-workers have firmly established the importance of cutinases in penetration (20, 41, 45). They investigated the role of cutinolytic enzymes from *Fusarium solani* f. sp. *pisi* and *Colletotrichum gloeosporioides* in infections of garden pea (*Pisum sativum*) and papaya (*Carica papaya*), respectively. Specific inhibition of the cutinase of a virulent isolate of F. *solani* f. sp. *pisi* either by diisopropyl fluorophosphate or by rabbit antiserum prepared against the enzyme drastically decreased infection of intact pea stems (45). The antiserum, however, had no effect if the stem tissue was wounded. Another isolate of this pathogen which lacked cutinase was virulent for wounded but not intact stems. The virulence of this isolate for intact stems could be increased by the addition of cutinase to inocula, but only reached that found in wounded stems when the cell wall-degrading enzymes pectinesterase, pectinase, and cellulase were added (41). An enzyme was obtained from C. *gloeosporioides* which hydrolyzed cutin but differed immunologically and in some of its catalytic properties from the enzyme obtained from F. *solani* f. sp. *pisi*. The two enzymes resembled each other in that they were both inhibited by diisopropyl fluorophosphate and their homologous antisera (50). Dickman et al. (20) also performed an interesting virulence enhancement experiment. A species of *Mycosphaerella*, which is a wound pathogen of papaya, was able to infect and form lesions on intact fruit which had been previously treated with the cutinase from C. *gloeosporioides*.

These experiments clearly show that the secretion of cutinases is important for at least some parasites which infect plants through the cuticle. The preliminary data from the isolate of F. *solani* f. sp. *pisi* lacking cutinase also suggest that cell wall-degrading enzymes may be essential, and it is hoped that this work will be followed up with an equally clear demonstration of the role of these enzymes in the infection process.

Overcoming preinfection- and postinfection-formed antimicrobial compounds. Saponins are antimicrobial compounds that are widely distributed in plants (63). They are toxic to fungi as a result of their ability to form complexes with sterols of the plasma membrane. One of the most intensively studied saponins is tomatine, and recently Défago and Kern (17) have obtained good evidence for its role in the protection of green tomatoes against *Fusarium solani*. They mutagenized a wild-type isolate which was inhibited in vitro by 100 ppm of tomatine and obtained five mutants which were able to grow on media containing 800 ppm of tomatine. All five mutants caused a severe rot in green tomatoes which contain tomatine, whereas the wild type did not. Red tomatoes, which do not contain tomatine, were as susceptible to the mutant as they were to the wild-type strain. Genetic analysis of two of the mutants (18) showed that pathogenicity for green tomatoes, low sterol content, and insensitivity to tomatine were always inherited together. Thus, it appears that

the virulence of the mutants may be attributed to their low sterol content and consequent insensitivity to tomatine. Another mechanism for overcoming the antifungal activity of tomatine is found in the tomato pathogen *Fusarium oxysporum* f. sp. *lycopersici*. This organism detoxifies the compound by the hydrolytic activity of a specific enzyme (3, 24).

A number of mechanisms for the fungal detoxification of phytoalexins have been described, and this ability is probably an important determinant of pathogenicity in at least some instances. Recently Van Etten et al. (70) surveyed isolates from mating population VI of *Nectria haematococca* (the perfect stage of *Fusarium solani* f. sp. *pisi*) for virulence to pea, sensitivity to the pea phytoalexin, pisatin, and the ability to detoxify this compound by demethylation (Fig. 3). All isolates which could not demethylate pisatin were sensitive to it and were nonpathogenic or of low virulence.

When isolates differing both in their ability to demethylate pisatin and their virulence on pea were crossed, three types of progeny were obtained: (i) progeny which had low virulence and were unable to demethylate pisatin, (ii) progeny which had high or moderate virulence and which could demethylate pisatin and were insensitive to it, and (iii) progeny which had low virulence although they were tolerant of pisatin and able to demethylate it. The existence of this last class demonstrated that an ability to demethylate pisatin is not the only factor required for high virulence on pea and that other virulence factors are required which are inherited independently of this trait.

Since a complex series of events including gene derepression is required for phytoalexin

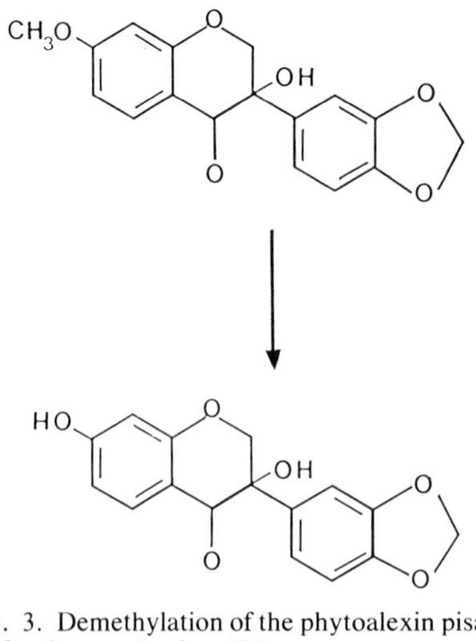

FIG. 3. Demethylation of the phytoalexin pisatin, a detoxification mechanism (70).

synthesis, there are opportunities other than detoxifying the end product open to an infectious agent for rendering the phytoalexin response ineffective. First, the plant may fail to recognize the parasite, and consequently the mechanism of phytoalexin synthesis may not be triggered. Second, the parasite may have acquired the ability to suppress phytoalexin synthesis. Evidence for water-soluble glucans which have this property has been obtained from *Phytophthora infestans* (21). Significantly, the glucans from compatible races were more effective than those from incompatible races. Precisely how such compounds act and their role as determinants of virulence must await a more complete understanding of phytoalexin elicitation. A third possibility is that constitutive elicitors in the plant may be destroyed before they have triggered phytoalexin synthesis. In this connection it is interesting to note that different races of *Phytophthora megasperma* f. sp. *glycinea* secrete different forms of invertases (77). The invertase of compatible races suppressed the phytoalexin response but those of incompatible races did not. However, the notion that compatibility may be caused by the hydrolysis of sucrose (which may function as a constitutive elicitor in this plant; J. S. Dahiya and R. N. Strange, unpublished data) is dashed by the finding that when the catalytic activity of the invertases was destroyed by heat they still suppressed phytoalexin accumulation.

Poisoning the plant. A large number of compounds which have deleterious effects on plants have been obtained from plant pathogenic microorganisms (22). Fifteen of these obtained from fungi are specific in the sense that they exert their effect only on plant genotypes susceptible to the parasites that produce them (42). One way of studying their role in pathogenicity is to compare mutants which have lost the ability to produce toxin with wild-type strains. This approach has been adopted by Scheffer and co-workers in their investigations of *Helminthosporium victoriae* and *H. carbonum* (12, 74). These fungi are pathogenic for certain genotypes of oats and maize, respectively, and secrete toxins which selectively affect their hosts. In both instances strains which did not produce the toxins achieved far fewer penetrations than toxigenic isolates. Furthermore *H. victoriae* and a Tox⁻ mutant of *H. carbonum* colonized maize leaves of a sensitive cultivar when the *H. carbonum* toxin was added, strongly suggesting that the toxin was necessary for pathogenicity.

Gnanamanickam and Patil also found evidence that a toxin from *Pseudomonas phaseolicola* allowed an incompatible strain of this bacterium to multiply in bean leaves, and it appeared that this virulence enhancement was mediated through suppression of the phytoalexin response (27).

One early effect of a number of toxins is to cause the loss of selective permeability of membranes. Recently, we devised a rapid and convenient assay to measure this phenomenon (65). Protoplasts were isolated from cotyledons of cucumber, a species which is susceptible to a devastating wilt caused by *Phytophthora drechsleri*, and were incubated for 3 h with a toxin obtained from the fungus. Protoplasts were scored for viability with fluorescein diacetate used as a vital stain (73). The relationship between probit percent protoplast death and the logarithm of toxin concentration was linear.

The evidence from many studies suggests that toxins are often important determinants of pathogenicity or virulence, but there have been few genetic investigations of this thesis relative to the number of toxins described. Analysis of the progeny from crosses of toxigenic and nontoxigenic strains for toxin production and pathogenicity or virulence should help to resolve this point.

Subverting the metabolism of the host. An important concept in parasitism is that the parasitic mode of life offers the advantage of a ready supply of nutrients in the form of host metabolites. Obligate parasites such as rusts have established an ingenious mechanism by which they attract the host's nutrients toward themselves. This appears to be mediated through elevated cytokinin levels in the vicinity of the infection which mobilize the host's metabolites (19).

An even more remarkable example of the hijacking of host metabolism by a plant parasite is the crown gall disease caused by *Agrobacterium tumefaciens*. Virulent strains of this bacterium cause tumors on many dicotyledonous plants (16). All virulent strains contain a large plasmid, a segment of which (tDNA) is transferred to the nuclear genome of the plant (36). This segment directs, among other activities, the synthesis of nonprotein amino acids, termed opines, which serve as carbon and nitrogen sources for the bacterium. The opines also promote conjugation, with the result that the plasmid may spread rapidly through a mixed population of bacteria.

CONCLUSIONS

The recent work of Kolattukudy's group emphasizes the importance of cutinases in infections which take place through the cuticle (20, 41, 45). These enzymes may be inhibited by diisopropyl fluorophosphate, which is therefore an effective antipenetrant. Possibly other compounds with similar properties as well as compounds that inhibit cell wall-degrading enzymes may be developed as crop protectants.

Normally, incompatible pathogens and many incongruent ones are inhibited after the penetration stage either by preformed compounds or by postinfection-synthesized phytoalexins or lignin. We need to know far more about these natural methods of plant resistance so that we can exploit them. Incompatibility involves two phases: recognition of the invader and activation of the defense response (40). At present, there is little understanding of the recognition process at the biochemical level, although at the genetic level it is usually conditioned by complementary dominant alleles in both partners. There is, however, some evidence that the recognition event may be manipulated. Cartwright et al. (10) showed that the rice blast-specific dichlorocyclopropane fungicide WL 28325 allowed leaves of genetically susceptible plants to react in a typically resistant manner to the rice blast pathogen *Pyricularia oryzae*, i.e., by local hypersensitive necrosis and the production of phytoalexins.

After recognition has occurred, a complex sequence of reactions is initiated, possibly involving the release of a constitutive elicitor, gene derepression, and the synthesis of phytoalexins or lignin. It is important that these processes are understood so that they may be optimized by plant breeding. For example, it is conceivable that some plants contain suboptimal concentrations of constitutive elicitor or may only respond to it weakly.

Where toxins play an important role in pathogenicity or virulence, it is possible to identify disease-resistant plants by selecting individuals which are insensitive to the toxin. This approach was first used by Wheeler and Luke (72), who drenched 4.5×10^7 oat seedlings with a crude preparation of the *H. victoriae* toxin. Those that looked normal (ca. 1 in 10^5) were inoculated with the fungus and planted out. A month later 92% of these plants were disease-free, whereas all those from a control batch which had not been through the toxin selection procedure had died. More recent applications have involved protoplasts and tissue culture. For example, Gengenbach et al. (25) cultured calli derived from immature embryos of maize susceptible to southern corn leaf blight in the presence of the toxin produced by the fungus. After five transfers, plants were regenerated which were toxin insensitive and resistant to the fungus.

It is now known that it is feasible to delete large portions of the tDNA of *A. tumefaciens*, including those sequences specifying tumorigenic functions, and substitute other base sequences without affecting integration into the plant genome (33). Thus, the Ti plasmid is a ready-made vector for the introduction of foreign DNA into plants. One problem is that these

sequences may remain silent. A solution to this is to insert the foreign sequences close to a promoter which normally directs the expression of opines (34). One possible use of this technology is to incorporate genes specifying products which would interact with an essential constituent of compatible parasites and, as a result, trigger the defense system. If this proved possible, then the resistance would probably be more durable than that obtained by conventional breeding techniques since mutations of the locus specifying the essential constituent of the parasite would be lethal.

LITERATURE CITED

1. **Aist, J. R.** 1976. Papillae and related wound plugs of plant cells. Annu. Rev. Phytopathol. **14:**145–163.
2. **Akazara, T., and K. Okamoto.** 1980. Biosynthesis and metabolism of sucrose, p. 199–220. *In* J. Preiss (ed.), The biochemistry of plants, a comprehensive treatise, vol. 3, Carbohydrates: structure and function. Academic Press, Inc., New York.
3. **Arneson, P. A., and R. D. Durbin.** 1967. Hydrolysis of tomatine by *Septoria lycopersici*: a detoxification mechanism. Phytopathology **57:**1358–1359.
4. **Bailey, J. A.** 1982. Mechanism of the phytoalexin accumulation, p. 289–318. *In* J. A. Bailey and J. W. Mansfield (ed.), Phytoalexins. Blackie, Glasgow.
5. **Bailey, J. A., and M. Berthier.** 1981. Phytoalexin accumulation in chloroform-treated cotyledons of *Phaseolus vulgaris*. Phytochemistry **20:**187–188.
6. **Bailey, J. A., and J. W. Mansfield (ed.).** 1982. Phytoalexins. Blackie, Glasgow.
7. **Beardmore, J., J. P. Ride, and J. W. Granger.** 1983. Cellular lignification as a factor in the hypersensitive resistance of wheat to stem rust. Physiol. Plant Pathol. **22:**209–220.
8. **Biffen, R. H.** 1905. Mendel's laws of inheritance and wheat breeding. J. Agric. Sci. **1:**4–48.
9. **Biffen, R. H.** 1912. Studies in inheritance of disease resistance. II. J. Agric. Sci. **4:**421–449.
10. **Cartwright, D. W., P. Langcake, and J. P. Ride.** 1980. Phytoalexin production in rice and its enhancement by a dichlorocyclopropane fungicide. Physiol. Plant Pathol. **17:**259–267.
11. **Cartwright, D., and G. E. Russell.** 1980. European and Mediterranean Cereal Rusts Conference, Bari, Italy, p. 25–28.
12. **Comstock, J. C., and R. P. Scheffer.** 1973. Role of host-selective toxin in colonization of corn leaves by *Helminthosporium carbonum*. Phytopathology **63:**24–29.
13. **Cooksey, C. J., J. S. Dahiya, P. J. Garratt, and R. N. Strange.** 1982. Two novel stilbene-2-carboxylic acid phytoalexins from *Cajanus cajan*. Phytochemistry **21:**2935–2938.
14. **Cooksey, C. J., J. S. Dahiya, P. J. Garratt, and R. N. Strange.** 1983. Sucrose: a constitutive elicitor of phytoalexin synthesis. Science **220:**1398–1400.
15. **Coombs, J., and D. O. Hall.** 1982. Techniques in bioproductivity and photosynthesis, p. 159. Pergamon Press, Oxford.
16. **DeCleene, M., and J. DeLey.** 1977. The host range of crown gall. Bot. Rev. **42:**389–466.
17. **Défago, G., and H. Kern.** 1983. Induction of *Fusarium solani* mutants insensitive to tomatine their pathogenicity and aggressiveness to tomato fruits and pea plants. Physiol. Plant Pathol. **22:**29–37.
18. **Défago, G., H. Kern, and L. Sedlar.** 1983. Genetic analysis of tomatine insensitivity, sterol content and pathogenicity for green tomato fruits in mutants of *Fusarium solani*. Physiol. Plant Pathol. **22:**39–43.

19. **Dekhuijzen, H. M.** 1976. Endogenous cytokinins in healthy and diseased plants, p. 526–559. *In* R. Heitefuss and P. H. Williams (ed.), Physiological plant pathology (Encyclopedia of plant physiology, vol. 4). Springer-Verlag, Berlin.

20. **Dickman, M. B., S. S. Patil, and P. E. Kolattukudy.** 1982. Purification, characterization and role in infection of an extracellular cutinolytic enzyme from *Colletotrichum gloeosporioides* Penz. on *Carica papaya* L. Physiol. Plant Pathol. **20:**333–347.

21. **Doke, N., N. A. Garas, and J. Kuć.** 1979. Partial characterization and aspects of the mode of action of a hypersensitive inhibiting factor (HIF) isolated from *Phytophthora infestans*. Physiol. Plant Pathol. **15:**127–140.

22. **Durbin, R. D.** 1981. Toxins in plant disease, p. 1–513. Academic Press, Inc., New York.

23. **Flor, H. H.** 1971. Current status of the gene-for-gene concept. Annu. Rev. Phytopathol. **9:**275–296.

24. **Ford, J. E., D. J. McCance, and R. B. Drysdale.** 1979. The hydrolysis of tomatine by an inducible extracellular enzyme from *Fusanium oxysporum* f. sp. *lycopersici*. Linnean Soc. London Symp. Ser. **7:**237–239.

25. **Gengenbach, B. G., C. E. Green, and L. M. Donovan.** 1977. Inheritance of selected pathotoxin resistance in maize plants regenerated from cell cultures. Proc. Natl. Acad. Sci. U.S.A. **74:**5113–5117.

26. **Giaquinta, R. T.** 1980. Translocation of sucrose and oligosaccharides, p. 271–320. *In* J. Preiss (ed.), The biochemistry of plants, a comprehensive treatise, vol. 3, Carbohydrates: structure and function. Academic Press, Inc., New York.

27. **Gnanamanickam, S. S., and S. S. Patil.** 1977. Phaseotoxin suppresses bacterially induced hypersensitive reaction and phytoalexin synthesis in bean cultivars. Physiol. Plant Pathol. **10:**169–179.

28. **Hammerschmidt, R., and J. Kuć.** 1982. Lignification as a mechanism for induced systemic resistance in cucumber. Physiol. Plant Pathol. **20:**61–71.

29. **Harder, D. E., D. J. Samborski, R. Rohringer, S. R. Rimmer, W. K. Kim, and J. Chong.** 1979. Electron microscopy of susceptible and resistant near isogenic (sr 6/Sr 6) lines of wheat infected by *Puccinia graminis tritici*. III. Ultrastructure of incompatible interactions. Can. J. Bot. **57:**2626–2634.

30. **Hargreaves, J. A.** 1979. Investigations into the mechanism of mercuric chloride stimulated phytoalexin accumulation in *Phaseolus vulgaris* and *Pisum sativum*. Physiol. Plant Pathol. **15:**279–287.

31. **Heath, M. C.** 1974. Light and electron microscope studies of the interactions of host and non-host plants with cowpea rust—*Uromyces phaseoli* var. *vignae*. Physiol. Plant Pathol. **4:**403–414.

32. **Heath, M. C.** 1976. Hypersensitivity, the cause or the consequence of rust resistance? Phytopathology **66:**935–936.

33. **Hernalsteens, J. P., F. Van Vliet, M. De Beuckeleer, A. Depicker, G. Engler, M. Lemmers, M. Holsters, M. Van Montagu, and J. Schell.** 1980. The *Agrobacterium tumefaciens* Ti plasmid as a host vector system for introducing foreign DNA in plant cells. Nature (London) **287:**654–656.

34. **Herrera-Estrella, L., A. Depicker, M. van Montagu, and J. Schell.** 1983. Expressure of chimaeric genes transferred into plant cells using a Ti-plasmid-derived vector. Nature (London) **303:**209–213.

35. **Hogenboom, N. G.** 1983. Letter to the editor. Bridging a gap between related fields of research: pistil-pollen relationships and the distinction between incompatibility and incongruity in nonfunctioning host-parasite relationships. Phytopathology **73:**381–383.

36. **Holsters, M., J. P. Hernalsteens, M. van Montagu, and J. Schell.** 1982. Ti plasmids of *Agrobacterium tumefaciens*: the nature of TIP, p. 269–298. *In* G. Kahl and J. S. Schell (ed.), Molecular biology of plant tumors. Academic Press, Inc., New York.

37. **Ingham, J. L.** 1981. Phytoalexin induction and its taxonomic significance in the Leguminosae (sub-family Papilionoideae), p. 599–626. *In* R. M. Polhill and P. H. Rowen (ed.), International Legume Conference, Kew, 1978. Advances in Legume Systematics. Royal Botanic Gardens, Kew, Surrey.

38. **Johnson, R.** 1981. Letter to the editor. Durable resistance: definition of genetic control, and attainment in plant breeding. Phytopathology **61:**567–568.

39. **Keen, N. T.** 1982. Specific recognition in gene-for-gene host-parasite systems. Adv. Plant Pathol. **1:**35–82.

40. **Keen, N. T., and B. Bruegger.** 1977. Phytoalexins and chemicals that elicit their production in plants, p. 1–26. *In* P. A. Hedin (ed.), Host plant resistance to pests. American Chemical Society, Washington, D.C.

41. **Köller, W., L. R. Allan, and P. E. Kolattukudy.** 1982. Rôle of cutinase and cell wall degrading enzymes in infection of *Pisum sativum* by *Fusarium solani* f. sp. *pisi*. Physiol. Plant Pathol. **20:**46–60.

42. **Kono, Y., H. W. Knoche, and J. M. Daly.** 1981. Structure: fungal host-specific, p. 221–257. *In* R. D. Durbin (ed.), Toxins in plant disease. Academic Press, Inc., New York.

43. **Langcake, P., and R. J. Pryce.** 1977. The production of resveratrol and the viniferins of grape vines in response to ultraviolet irradiation. Phytochemistry **16:**1193–1196.

44. **Macko, V., R. C. Staples, P. J. Allen, and J. H. A. Renwick.** 1971. Identification of germination self-inhibitor from wheat stem rust uredospores. Science **173:**835–836.

45. **Maiti, I. B., and P. E. Kolattukudy.** 1979. Prevention of fungal infection of plants by specific inhibition of cutinase. Science **205:**507–508.

46. **Mansfield, J. W.** 1982. The rôle of phytoalexins in disease resistance, p. 253–288. *In* J. A. Bailey and J. W. Mansfield (ed.), Phytoalexins. Blackie, Glasgow.

47. **Mayama, S., S. Hayashi, R. Yamamoto, T. Tani, U. Ueno, and H. Fukami.** 1982. Effects of elevated temperature and α amino oxyacetate on the accumulation of averalumins in oat leaves infected with *Puccinia coronata* f. sp. *avenae*. Physiol. Plant Pathol. **20:**305–312.

48. **Mayama, S., Y. Matsuura, H. Iida, and T. Tani.** 1982. The role of avenalumin in the resistance of oat to crown rust, *Puccinia coronata* f. sp. *avenae*. Physiol. Plant Pathol. **20:**189–199.

49. **Mayama, S., and T. Tani.** 1982. Microspectrophotometric analysis of the location of avenalumin in oat leaves in response to fungal infection. Physiol. Plant Pathol. **21:**141–149.

50. **Mayama, S., T. Tani, Y. Matsuura, U. Ueno, and H. Fukami.** 1981. The production of phytoalexins by oat in response to crown rust, *Puccinia coronata* f. sp. *avenae*. Physiol. Plant Pathol. **19:**217–226.

51. **Mayama, S., T. Tani, Y. Matsuura, T. Ueno, K. Hirabayashi, H. Fukami, Y. Mizuno, and H. Irie.** 1981. Biologically active substances related to the induced resistance in the oat crown rust. J. Agric. Chem. Soc. Jpn. **198:**697–704.

52. **Mayama, S., T. Tani, T. Ueno, K. Hirabayashi, T. Nakashima, H. Fukami, Y. Mizuno, and H. Irie.** 1981. Isolation and structure elucidation of genuine oat phytoalexin, avenalumin I. Tetrahedron Lett. **22:**2103–2106.

53. **National Academy of Science.** 1972. Genetic vulnerability of major crops. National Academy of Science, National Research Council, Washington, D.C.

54. **Pearce, R. B., and J. P. Ride.** 1980. Specificity of induction of the lignification response in wounded wheat leaves. Physiol. Plant Pathol. **16:**197–204.

55. **Pearce, R. B., and J. P. Ride.** 1982. Chitin and related compounds as elicitors of the lignification response in wounded wheat leaves. Physiol. Plant Pathol. **20:**119–123.

56. **Ride, J. P.** 1975. Lignification in wounded wheat leaves in response to fungi and its possible rôle in resistance. Physiol. Plant Pathol. **5:**125–134.

57. **Ride, J. P.** 1980. The effect of induced lignification in the resistance of wheat cell walls to fungal degradation. Physiol. Plant Pathol. **16:**187–196.

58. **Ride, J. P.** 1982. The rôle of cell wall alterations in resistance to fungi. Ann. Appl. Biol. **89:**302–306.

59. **Ride, J. P., and R. B. Pearce.** 1979. Lignification and papilla formation at sites of attempted penetration of wheat leaves by non-pathogenic fungi. Physiol. Plant Pathol. **15**:79–92.

60. **Robinson, T. J., and R. K. S. Wood:** 1976. Factors affecting accumulation of pisatin by pea leaves. Physiol. Plant Pathol. **9**:285–297.

61. **Russell, G. E.** 1967. Report of the Plant Breeding Institute, Cambridge, p. 85–86.

62. **Sakai, S., K. Tomiyama, and N. Doke.** 1979. Synthesis of a sesquiterpenoid phytoalexin rishitin in non-infected tissue from various parts of potato plants immediately after slicing. Ann. Phytopathol. Soc. Jpn. **33**:216–222.

63. **Schönbeck, F., and E. Schlöesser.** 1976. Preformed substances as potential protectants, p. 653–678. *In* R. Heitefuss and P. H. Williams (ed.), Physiological plant pathology. Springer-Verlag, Berlin.

64. **Stössl, P., G. Lazarovits, and E. W. B. Ward.** 1981. Electron microscope study of race-specific and age-related resistant and susceptible soybeans *Phytophthora megasperma* var. *sojae.* Phytopathology **71**:617–623.

65. **Strange, R. N., D. J. Pippard, and G. A. Strobel.** 1982. A protoplast assay for phytotoxic metabolites produced by *Phytophthora drechsleri* in culture. Physiol. Plant Pathol. **20**:359–364.

66. **Tatum, L. A.** 1971. The southern corn leaf blight epidemic. Science **171**:1113–1116.

67. **ten Houten, J. G.** 1974. Plant pathology: changing agricultural methods and human society. Annu. Rev. Phytopathol. **14**:1–11.

68. **Ullstrup, A. J.** 1972. The impact of the southern corn leaf blight epidemics of 1970–1971. Annu. Rev. Phytopathol. **10**:37–50.

69. **Vance, C. P., T. K. Kirk, and R. T. Sherwood.** 1980. Lignification as a mechanism of disease resistance. Annu. Rev. Phytopathol. **18**:259–288.

70. **Van Etten, H. D., D. E. Matthews, and D. A. Smith.** 1982. Metabolism of phytoalexins, p. 181–217. *In* J. A. Bailey and J. W. Mansfield (ed.), Phytoalexins. Blackie, Glasgow.

71. **Weier, T. E., C. R. Stocking, and M. G. Barbour.** 1974. Botany: an introduction to plant biology, 5th ed., p. 7. John Wiley & Sons, Inc., New York.

72. **Wheeler, H. E., and H. H. Luke.** 1955. Mass screening for disease-resistant mutants in oats. Science **122**:1229.

73. **Widholm, J. M.** 1972. The use of fluorescein diacetate and phenosafranine for determining viability of cultured plant cells. Stain Technol. **47**:189–194.

74. **Yoder, O. C., and R. P. Scheffer.** 1969. Role of toxin in early interactions of *Helminthosporium victoriae* with susceptible and resistant oat tissue. Phytopathology **59**:1954–1959.

75. **Zeyen, R. J., and R. W. Bushnell.** 1979. Papilla response of barley epidermal cells caused by *Erysiphe graminis:* rate and method of deposition determined by microcinematography and transmission electron microscopy. Can. J. Bot. **57**:898–913.

76. **Zeyen, R. J., T. L. W. Carver, and G. G. Ahlstrand.** 1983. Relating cytoplasmic detail of powdery mildew infection to presence of insoluble silicon by sequential use of light microscopy, SEM, and X-ray microanalysis. Physiol. Plant Pathol. **22**:101–108.

77. **Ziegler, E., and R. Pontzen.** 1982. Specific inhibition of glucan-elicited glyceollin accumulation in soybeans by an extracellular mannan-glycoprotein of *Phytophthora megasperma* f. sp. *glycinea.* Physiol. Plant Pathol. **20**:321–331.

Infection Process in the *Rhizobium*–Legume Symbiosis

BJØRN SOLHEIM

Institute of Biology and Geology, University of Tromsø, N-9001 Tromsø, Norway

Bacteria of the genus *Rhizobium* form nitrogen-fixing nodules on several economically important legumes and are an important source of biologically fixed nitrogen in natural ecosystems. The degree of specificity in the symbiotic relationship between legumes and rhizobia varies to a great extent. Usually, a high degree of specificity is found when the host is infected through the root hairs, whereas less specific processes are likely to be involved when entry is direct, through cracks in the root epidermis (14). In the three best-studied host/symbiont relationships, clover–*R. trifolii*, pea–*R. leguminosarum*, and soybean–*R. japonicum*, the infections take place through the root hair cells. The following discussion on the infection process in the *Rhizobium*–legume symbiosis is mainly based on results obtained with these three systems.

Several recent reviews on the subject have been published (2, 11, 48). The specificity of the infection process is probably governed by a series of events leading up to the formation of the infection thread. Recognition reactions start in the rhizosphere even before contact between the two organisms. Bauer (2) presented a general multistep recognition model that can be broken down into individual signal-and-response steps. The interaction between *Rhizobium* spp. and legumes is very complex, and we have just started to characterize some of the individual steps in the infection process.

PREINFECTION EVENTS AND SPECIFICITY IN *RHIZOBIUM*–LEGUME SYMBIOSIS

Legume roots exude into the soil a wide variety of compounds (3), and *Rhizobium* species are attracted by a wide range of low-molecular-weight components of root exudate (26). From birdsfoot trefoil (*Lotus corniculatus*), a high-molecular-weight glycoprotein has been purified that strongly attracts trefoil *Rhizobium* sp. (17). Differences in motility (34) and chemotactic responses (1) might be important for competitions between strains of *Rhizobium* sp., but it is unlikely that chemotaxis is a major factor in host–symbiont specificity.

A detailed description of the attachment of rhizobia to host root hairs is presented by Dazzo in this volume and therefore will not be repeated here.

The first visible response of the host to the rhizobia is a deformation of root hairs (27, 31, 49, 53, 54). Culture filtrates from compatible rhizobia cause deformation of root hairs. Closely related, but incompatible, rhizobia have also been found to cause these reactions (27, 54). Genes controlling root hair deformation have been found to be located on the plasmid harboring other genes for symbiosis (28, 43), and marked root hair deformation seems to be a prerequisite for infection to occur. The chemical nature of the deformation factor is not established, but there is evidence that polysaccharides from the capsule of rhizobia might be active (31). Solheim and Raa (49) found that compatible rhizobia grown in the rhizosphere of host plants produced much more potent deformation factors than when grown in pure culture. Recently, a factor causing branching of root hairs has been isolated and partially characterized from the filtrate of *R. trifolii* cultured in association with clover plants (T. V. Bhuvaneswari and B. Solheim, Abstr. Plant Physiol. **69**:22, 1982).

PENETRATION—FORMATION OF INFECTION THREAD

Penetration by rhizobia is determined by the host and can be either through infection threads or through openings between epidermal cells as in peanut and *Stylosanthes* (13, 14). We know very little about the mechanism and control of this process or about the subsequent development of the nodule (see 2, 51).

Penetration of root hairs by rhizobia has been studied by several investigators, using both light and electron microscopy (11). Until the recent work of Callaham and Torrey (12), these studies confirmed Nutman's (44) original hypothesis of invagination of the root hair wall at the infection site and formation of an infection thread. However, Callaham and Torrey (12) found breaks and changes in the hair cell wall at the site of penetration. In response to the penetration, the host deposits fibrillar cell wall-like materials around the bacteria, and a tubular structure called the infection thread is formed. It seems that the bacteria have to become entrapped in a pocket formed between host cell walls, either by a curling/deformation reaction (11) or at the junction between hair cell and adjacent epidermal cell (B. G. Turgeon, Ph.D. thesis, University of Dayton, Dayton, Ohio, 1982), before penetration takes place. Rhizobia are able to produce low levels of pectinase, hemicellulase, and cellulase (33, 42), and it is possible that when the bacteria are entrapped in pockets between cell walls the concentration of these enzymes builds up, and a local degradation of the cell wall takes place. It has, however, been difficult to verify these results (2).

The host cell nucleus is in close contact with the tip of the growing infection thread (24, 27, 45). The thread again penetrates the root hair cell wall at the base of the root hair, passing into a neighboring cell. The bacteria are released from the infection thread enclosed in a membrane of host origin into cortical cells. These cells and neighboring cells are induced to divide, and the bacteria differentiate into larger bodies called bacteroids. The result is an organized structure in the root called a nodule, in which infected cells contain bacteroids capable of dinitrogen fixation.

ROLE OF LECTINS IN THE RECOGNITION PROCESS

On the basis of the striking correlation between binding of soybean seed lectin and infectivity in *R. japonicum*, the symbiont of soybean, Bohlool and Schmidt (9) formulated their hypothesis for host specificity in the *Rhizobium*–legume symbiosis. Others (10, 15, 40) were unable to demonstrate this correlation. An explanation of these contradictory results is probably found in the demonstration that for some strains of *R. japonicum* lectin binding was transient and dependent on growth phase (7) and for others lectin binding could be demonstrated only when the bacteria were cultured in the presence of host root exudate or in the rhizosphere of the host (5). A similar correlation of lectin binding and host specificity has been found in the clover–*R. trifolii* interaction (18, 20). Lectins identical or closely related to the seed lectins have been found on roots of clover (19, 23), soybean (25), and pea (29, 38). Several models for the involvement of lectins in the infection process have been proposed (20, 32, 46). Common for these models is that lectins are mediating as some sort of binding of the bacteria to the host surface. Studies of interactions of isolated lectins with isolated polysaccharides from fast-growing rhizobia indicate that an initial binding of the lectin to the capsular polysaccharide cannot alone account for the specificity of the infection process (30, 35). *R. japonicum* mutants unable to synthesize detectable lectin-binding capsule at any culture age were not distinguishable from the parental strain by light microscopy with respect to the rate, the mode, or the location of their attachment (40, 41). About twice as

many cells of the parental strain attached to the root compared with the mutants, and the mutants were less efficient than the capsulated parent in nodulating soybean cultivar Williams (41). Infective as well as noninfective strains of *R. leguminosarum* bind host lectin and attach to host roots (47). Reaction between host lectin and symbiont lipopolysaccharide correlates with infectivity in *R. leguminosarum* (35). A transient ability of *R. trifolii* to bind clover lectin correlates with appearance of a specific lipopolysaccharide in the bacterial cell wall when grown in broth (30) and with capsular polysaccharide when grown on plants (22). There are several reports about specific inhibition of attachment of bacteria to the root surface by simple hapten sugars (21, 50, 55) and by lectin-binding capsular polysaccharide and lipopolysaccharide from infective bacteria (36). But there is no report of inhibition of infection by lectin-binding polysaccharides. On the contrary, culture filtrate containing exopolysaccharide (47) and partially purified oligosaccharide-containing fractions from the filtrate of *R. trifolii* cultured in association with clover plants (T. V. Bhuvaneswari and B. Solheim, unpublished data) stimulate infection. An enhancement of nodulation in soybean was found when the roots were pre-treated with *R. japonicum* culture filtrate or capsular/exopolysaccharide (W. D. Bauer et al., Abstr. Plant Physiol. 63:135, 1979). The hapten sugar *N*-acetylgalactosamine was also effective. It has been suggested (48) that the stimulation of the infection process by lectin-binding carbohydrates might be due to induction of the root hair and its adjacent cells to produce more lectin (still to be examined), which takes part in a later event of the infection process. Preincubation of *Rhizobium* sp. in host root exudates (4) or addition of concentrated root exudates to the growth medium of inoculated plants (47) initiates earlier infection after inoculation. We found (Solheim and Fjellheim, unpublished data) that both clover roots and pea roots contain enzymes that are able selectively to degrade polysaccharides isolated from *R. trifolii* and *R. leguminosarum*, respectively. Dazzo et al. (22) reported enzymes in clover root exudate that altered the capsule of *R. trifolii*. Some host specificity in this reaction was found. The enzymes gradually stripped off the capsule of the symbiont. The polar region of the rod-shaped bacteria was the last part affected. At the same time as the capsule was lost, the bacteria lost their ability to bind clover lectin. How these specific host enzymes participate in the infection process is under investigation. They might be important in at least two ways. They change the surface structure of the symbiont, making it able to polar bind to the root hair surface and initiate infection. Also the oligo/polysaccharides that are released to the rhizosphere might be important in initiating the next step in the infection process.

Bhuvaneswari et al. (8) found for soybean that the infectibility of given host cells was a transient property that appeared and then was lost within a few hours. This transient susceptibility was found in preemergent and developing root hair cells. The same nodulation pattern was found in cowpea and alfalfa, while in clover it appeared to be an induced susceptibility of mature root hair cells in addition to the pattern found in the other host plants (6). A cell-free bacterial exudate preparation from *R. trifolii* rendered mature root hair cells of white clover more rapidly susceptible to nodulation. When clover is inoculated with an infective *R. trifolii* strain, two main types of root hair deformations occur. One is a curling of root hair tips; the other is branching of root hairs. The infection of a root hair might either start from the tip or from the branch, and when mature root hairs are infected, the infection starts in the branch (Bhuvaneswari and Solheim, Abstr. Plant Physiol. **69**:22, 1982). Infection initiated in developing root hairs, on the other hand, originated close to the tips of unbranched and curled root hairs. It was possible to isolate from the growth medium of inoculated clover plants fractions containing oligosaccharides and small polysaccharides with molecular weight between 1,200 and 10,000 that induce branching responses in white clover root hairs (Bhuvaneswari and Solheim, Abstr. Plant. Physiol. **69**:22, 1982). The branching factor has been purified by solvent extraction, ion-exchange chromatography, gel filtration, and high-pressure liquid chromatography. The sugar composition of the fractions has been determined by gas chromatography followed by mass spectrometry. The active fractions cause root hair branching at concentrations as low as 10^{-8} to 10^{-9} M in the plant growth medium (Solheim and Bhuvaneswari, unpublished data).

These biologically active low-molecular-weight polysaccharides are a product of the interaction of infective *R. trifolii* and white clover roots. They might be synthesized by the bacteria in response to the host exudates or by enzymes from the host roots liberating them from the bacterial capsule or cell wall. The branching factors induce local cell wall growth, which might produce an area on the root hair cell at a suitable developmental stage for initiation of infection. The mechanism of root hair deformation itself is unknown, but there are probably receptors on the root hair surface that can transmit signals over the cell membrane in response to symbiont-produced molecules, resulting in pre-penetration host reactions, such as root hair deformation. These receptors might be available for receiving the signal only at a certain transient developmental stage. Lectins might function as receptors for relatively low-molecular-weight signal molecules rather than binding the bacteria specifically to the host surface. The bacteria might also produce lectins. Kijne et al. (39) found that *R. leguminosarum* produced a mannose-specific agglutinin and suggested that this agglutinin might be involved in bacterial adhesion to host roots.

In several cell–cell interactions oligo- and polysaccharides have been found to act as signal molecules (52). A symbiotic infection process is very similar to a pathogenic infection process, even if different selection pressure operates during evolution. Host enzymes can release biologically active glucans from the cell wall of fungal pathogens (16, 37). These glucans elicit phytoalexin production in the host, which is probably a defense reaction against invasion of the pathogen. In the *Rhizobium*–legume system similar reactions might induce susceptibility in the host, favoring the establishment of a symbiotic relationship.

CONCLUSIONS

The infection of legumes by rhizobia is a complicated process governed by a series of events leading up to an organized structure in the root capable of fixing atmospheric nitrogen. Recognition reactions start in the rhizosphere even before contact between the two organisms. Both bacteria and host cells have to be at a certain developmental stage for infection to take place. This developmental stage can be modified or induced by the other partner in the symbiosis. The bacteria produce polysaccharides that stimulate the infection of the host while the host exudes enzymes to the rhizosphere that change the surface composition of the bacteria. In the rhizosphere biologically active components, probably oligosaccharides, are formed as a result of the interaction, causing deformation and branching of the root hairs. A marked root hair deformation seems to be a prerequisite for infection to occur. Infective bacteria produce polysaccharides that specifically bind to host lectin, but the function of this binding is still under investigation.

It is necessary to know the exact timing of the different steps in the infection process. A powerful tool for studying the interaction between host and symbiont will be a series of well-defined mutants with single mutations at the different steps in the process. Studies with such mutants will lead to a better understanding of the biochemical mechanism of host specificity. By cloning the genes involved in the infection process, the primary gene products from the bacteria will be found. Later, when the recombinant DNA technique routinely can be applied to plants, the corresponding gene products from the plant can be determined.

I am grateful to W. D. Bauer and T. V. Bhuvaneswari for critically reading this manuscript and to F. B. Dazzo for stimulating discussions on the specificity of the infection process.

LITERATURE CITED

1. **Ames, P., S. A. Schluederberg, and K. Bergman.** 1980. Behavioral mutants of *Rhizobium meliloti*. J. Bacteriol. **141:**722–727.
2. **Bauer, W. D.** 1981. Infection of legumes by rhizobia. Annu. Rev. Plant Physiol. **32:**407–449.
3. **Beringer, J. E., N. Brewin, A. W. B. Johnston, H. M. Schulman, and D. A. Hopwood.** 1979. The *Rhizobium*-legume symbiosis. Proc. R. Soc. London Ser. B **204:**219–233.
4. **Bhagwat, A. A., and J. Thomas.** 1982. Legume-*Rhizobium* interactions: cowpea root exudate elicits faster nodulation response by *Rhizobium* species. Appl. Environ. Microbiol. **43:**800–805.
5. **Bhuvaneswari, T. V., and W. D. Bauer.** 1978. Role of lectins in plant-microorganism interactions. III. Influence of rhizosphere/rhizoplane culture conditions on the soybean lectin-binding properties of rhizobia. Plant Physiol. **62:**71–74.
6. **Bhuvaneswari, T. V., A. A. Bhagwat, and W. D. Bauer.** 1981. Transient susceptibility of root cells in four common legumes to nodulation by rhizobia. Plant Physiol. **68:**1144–1149.
7. **Bhuvaneswari, T. V., S. G. Pueppke, and W. D. Bauer.** 1977. Role of lectins in plant-microorganism interactions. I. Binding of soybean lectin to rhizobia. Plant Physiol. **60:**486–491.
8. **Bhuvaneswari, T. V., B. G. Turgeon, and W. D. Bauer.** 1980. Early events in the infection of soybean (*Glycine max.* L. Merr) by *Rhizobium japonicum*. I. Localization of infectible root cells. Plant Physiol. **66:**1027–1031.
9. **Bohlool, B. B., and E. L. Schmidt.** 1974. Lectins: a possible basis for specificity in the *Rhizobium*-legume root nodule symbiosis. Science **185:**269–271.
10. **Brethauer, T. S., and J. Paxton.** 1977. The role of lectin in soybean-*Rhizobium japonicum* interactions, p. 381–388. *In* B. Solheim and J. Raa (ed.), Cell wall biochemistry related to specificity in host-plant pathogen interactions. Universitetsforlaget, Oslo.
11. **Broughton, W. J.** 1978. Control of specificity in legume-*Rhizobium* associations. J. Appl. Bacteriol. **45:**165–194.
12. **Callaham, D. A., and J. B. Torrey.** 1981. The structural basis for infection of root hair of *Trifolium repens* by *Rhizobium*. Can. J. Bot. **59:**1647–1664.
13. **Chandler, M. R.** 1978. Some observations on infection of *Arachis hypogaea* L. by *Rhizobium*. J. Exp. Bot. **29:**749–755.
14. **Chandler, M. R., R. A. Date, and R. J. Roughley.** 1982. Infection and root-nodule development in *Stylosanthes* species by *Rhizobium*. J. Exp. Bot. **33:**47–57.
15. **Chen, A. T., and D. A. Phillips.** 1976. Attachment of *Rhizobium* to legume roots as the basis for specific interactions. Physiol. Plant. **38:**83–88.
16. **Cline, K., and P. Albersheim.** 1981. Host-pathogen interactions. XVII. Hydrolysis of biologically active fungal glucans by enzymes isolated from soybean cells. Plant Physiol. **68:**221–228.
17. **Currier, A. W., and G. A. Strobel.** 1981. Characterization and biological activity of trefoil chemotactin. Plant Sci. Lett. **21:**159–165.
18. **Dazzo, F. B.** 1980. Determinants of host specificity in the *Rhizobium*-clover symbiosis, p. 165–187. *In* W. E. Newton and W. H. Orme-Johnsen (ed.), Nitrogen fixation, vol. 2. University Park Press, Baltimore.
19. **Dazzo, F. B., and W. J. Brill.** 1977. Receptor site on clover and alfalfa roots for *Rhizobium*. Appl. Environ. Microbiol. **33:**132–136.
20. **Dazzo, F. B., and D. H. Hubbell.** 1975. Cross-reactive antigens and lectin as determinants of symbiotic specificity in the *Rhizobium*-clover association. Appl. Microbiol. **30:**1017–1033.
21. **Dazzo, F. B., C. Napoli, and D. H. Hubbell.** 1976. Adsorption of bacteria to roots as related to host-specificity in the *Rhizobium*-clover symbiosis. Appl. Environ. Microbiol. **32:**166–171.
22. **Dazzo, F. B., G. L. Truchet, J. E. Sherwood, E. M. Hrabak, and A. E. Gardiol.** 1982. Alternation of the Trifoliin A-binding capsule of *Rhizobium trifolii* 0403 by enzymes released from clover roots. Appl. Environ. Microbiol. **44:**478–490.
23. **Dazzo, F. B., W. E. Yanke, and W. J. Brill.** 1978. Trifoliin: a *Rhizobium* recognition protein from white clover. Biochim. Biophys. Acta **539:**276–286.
24. **Fåhraeus, G.** 1957. The infection of clover root hairs by nodule bacteria studied by a simple glass slide technique. J. Gen. Microbiol. **16:**374–381.
25. **Gade, W., M. A. Jack, J. B. Dahl, E. L. Schmidt, and F. Wold.** 1981. The isolation and characterization of a root lectin from soybean (*Glycine max* (L), cultivar Chippewa). J. Biol. Chem. **256:**12905–12910.
26. **Gaworzewska, E. T., and M. J. Carlile.** 1982. Positive chemotaxis of *Rhizobium leguminosarum* and other bacteria towards root exudates from legumes and other plants. J. Gen. Microbiol. **128:**1179–1188.
27. **Haak, A.** 1964. Über den einfluss der Knöllchenbakterien auf die Wurzelhaare vol Leguminosen und nicht Leguminosen. Zentralbl. Bakteriol. Parasitenkd. Infektionskr. Hyg. Abt. 2 **117:**343–361.
28. **Hooykaas, P. J. J., A. A. N. van Brussel, H. den Dulk-Ras, C. M. S. van Slogteren, and R. A. Schilperoot.** 1981. Sym plasmid of *Rhizobium trifolii* expressed in different rhizobial species and *Agrobacterium tumefaciens*. Nature (London) **291:**351–353.
29. **Hosselet, M., E. Van Driessche, M. Van Poucke, and L. Kanarek.** 1983. Purification and characterization of an endogenous root lectin from *Pisum sativum* L., p. 549–558. *In* T. C. Bøg-Hansen and G. A. Spengler (ed.), Lectins—biology, biochemistry, clinical biochemistry, vol. 3. W. de Gruyter, Berlin.
30. **Hrabak, E. M., M. R. Urbano, and F. B. Dazzo.** 1981. Growth-phase-dependent immunodeterminants of *Rhizobium trifolii* lipopolysaccharide which bind trifoliin A, a white clover lectin. J. Bacteriol. **148:**697–711.
31. **Hubbell, D. H.** 1970. Studies on the root hair "curling factor" of *Rhizobium*. Bot. Gaz. **131:**337–342.
32. **Hubbell, D. H.** 1981. Legume infection by *Rhizobium*: a conceptual approach. BioScience **31:**832–837.
33. **Hubbell, D. H., V. M. Morales, and M. Umali-Garcia.** 1978. Pectolytic enzymes in *Rhizobium*. Appl. Environ. Microbiol. **35:**210–213.
34. **Hunter, W. J., and C. J. Fahring.** 1980. Movement by *Rhizobium* and nodulation of legumes. Soil Biol. Biochem. **12:**537–542.
35. **Kamberger, W.** 1979. Role of cell surface polysaccharides in *Rhizobium*-pea symbiosis. FEMS Microbiol. Lett. **6:**361–365.
36. **Kato, G., Y. Maruyama, and M. Nakamura.** 1980. Role of bacterial polysaccharides in the adsorption process of the *Rhizobium*-pea symbiosis. Agric. Biol. Chem. **44:**2843–2855.
37. **Keen, N. T., and M. Yoshikawa.** 1983. β-1.3-Endoglucanase from soybean releases elicitor-active carbohydrates from fungus cell walls. Plant Physiol. **71:**460–465.
38. **Kijne, J. W., I. A. M. van der Schaal, and G. E. de Vries.** 1980. Pea lectins and the recognition of *Rhizobium leguminosarum*. Plant Sci. Lett. **18:**65–74.
39. **Kijne, J. W., I. A. M. van der Schaal, C. L. Diaz, and F.**

van Iren. 1983. Mannose-specific lectins and the recognition of pea roots by *Rhizobium leguminosarum*, p. 521–529. *In* T. C. Bøg-Hansen and G. A. Spengler (ed.), Lectins—biology, biochemistry, clinical biochemistry, vol. 3. W. de Gruyter & Co., Berlin.

40. **Law, I. J., and B. W. Strijdom.** 1977. Some observations on plant lectins and *Rhizobium* specificity. Soil Biol. Biochem. **9**:79–84.

41. **Law, I. J., Y. Yamamoto, A. J. Mort, and W. D. Bauer.** 1982. Nodulation of soybean by *Rhizobium japonicum* mutants with altered capsule synthesis. Planta **154**:100–109.

42. **Martinez-Molina, E., V. M. Morales, and D. H. Hubbell.** 1979. Hydrolytic enzyme production by *Rhizobium*. Appl. Environ. Microbiol. **38**:1186–1188.

43. **Morrison, N. A., C. Y. Hau, M. J. Trinick, J. Shine, and B. G. Rolfe.** 1983. Heat curing of a sym plasmid in a fast-growing *Rhizobium* sp. that is able to nodulate legumes and the nonlegume *Parasponia* sp. J. Bacteriol. **153**:527–531.

44. **Nutman, P. S.** 1956. The influence of the legume in root-nodule symbiosis. A comparative study of host determinants and functions. Biol. Rev. Cambridge Philos. Soc. **31**:109–151.

45. **Nutman, P. S.** 1959. Some observations on root-hair infection by nodule bacteria. J. Exp. Bot. **10**:250–262.

46. **Raa, J., B. Robertsen, B. Solheim, and A. Tronsmo.** 1977. Cell surface biochemistry related to specificity of pathogenesis and virulence of microorganisms, p. 11–30. *In* B. Solheim and J. Raa (ed.), Cell wall biochemistry related to specificity in host-plant pathogen interactions. Universitetsforlaget, Oslo.

47. **Solheim, B.** 1983. Possible role of lectins in binding rhizobia to host roots, p. 539–547. *In* T. C. Bøg-Hansen and G. A. Spengler (ed.), Lectins—biology, biochemistry, clinical biochemistry, vol. 3. W. de Gruyter & Co., Berlin.

48. **Solheim, B., and J. Paxton.** 1981. Recognition in *Rhizobium*-legume interaction, p. 71–83. *In* R. C. Staples and G. Toenniesson (ed.). John Wiley & Sons, Inc., New York.

49. **Solheim, B., and J. Raa.** 1973. Characterization of the substances causing deformation of root-hairs of *Trifolium repens* when inoculated with *Rhizobium trifolii*. J. Gen. Microbiol. **77**:241–247.

50. **Stacey, G., A. S. Paau, and W. J. Brill.** 1980. Host recognition in the *Rhizobium*-soybean symbiosis. Plant Physiol. **66**:609–614.

51. **Vance, C. P., and L. E. B. Johnson.** 1981. Nodulation: a plant disease perspective. Plant Dis. **65**:118–124.

52. **Woodward, J. R., P. J. Keane, and B. A. Stone.** 1980. β-Glucans and β-glucan hydrolases in plant pathogenesis with special reference to wilt-inducing toxins from *Phytophthora* species, p. 113–141. *In* P. A. Sandford and K. Matsuda (ed.), Fungal polysaccharides. Adv. Chem. Ser. American Chemical Society, Washington, D.C.

53. **Yao, P. Y., and J. M. Vincent.** 1969. Host specificity in root hair ''curling factor'' of *Rhizobium* spp. Aust. J. Biol. Sci. **22**:413–423.

54. **Yao, P. Y., and J. M. Vincent.** 1976. Factors responsible for the curling and branching of clover root hairs by *Rhizobium*. Plant Soil **45**:1–16.

55. **Zurkowski, W.** 1980. Specific adsorption of bacteria to clover root hairs, related to the presence of the plasmid pWZ2 in cells of *Rhizobium trifolii*. Microbios **27**:27–32.

The Actinorhizal Infection Process: Review of Recent Research

ALISON M. BERRY

Biology Department, Carleton University, Ottawa, Ontario, Canada

Actinorhizal root nodules are nitrogen-fixing, modified roots which develop as a result of infection of host woody plant species by *Frankia* (*Actinomycetales*). The first successful isolation of the *Frankia* organism into pure culture was reported by Callaham et al. (13), and this has proved a watershed for further isolation and structural investigation of *Frankia* strains in culture and within the nodule symbiosis. The purpose of this review is to summarize areas of this recent research which bear upon the actinorhizal infection process, particularly the structural information concerning *Frankia* and the host nodule tissue.

The infection process is a series of cell–cell interactions which leads to the establishment and continued development of the symbiotic state, the root nodule. The events of infection begin in the rhizosphere, a transitional zone between soil and root tissue. The process appears to involve progressive changes in cell structure and physiology of both host and endophyte. This implies the operation of some sort of recognition phenomenon, as well as mutual regulatory interactions.

Much of our knowledge to date concerning infection and nodulation derives from cytological studies, and this research will be emphasized in the current review. There have been several earlier reviews of the actinorhizal literature which include information on the infection process (2, 9, 26, 48). The general cellular characteristics of root hair infection, host cortical cell modification, and invasion by the endophyte, discussed below, closely parallel the events in the legume infection process, as reviewed or reported on recently by several authors (7, 16, 42, 51; Solheim, this volume).

FRANKIA: MORPHOLOGICAL CHARACTERISTICS

Frankia is a branched septate filamentous actinomycete. In culture *Frankia* (Fig. 1a) forms dense hyphal mats. Sporangia are produced via hyphal septation (30, 34), and these exhibit considerable morphological diversity. Spores are thick walled with a distinctive triple-track outer envelope (30) and may represent a resting stage in the life cycle.

Morphologically specialized structures known as vesicles develop in vitro after a short period of hyphal proliferation following transfer. Vesicles are perhaps the most prominent structural feature of *Frankia* in the nodule tissue of most actinorhizal genera as well (Fig. 1b). Ontogenetically, vesicles represent short branch hyphae which expand apically and subsequently undergo internal wall ingrowths and septations. Acetylene reduction (nitrogenase activity) in culture correlates precisely with vesicle differentiation (46), and several other lines of evidence (22, 31, 44; A. D. L. Akkermans, Doctoral thesis, University of Leiden, Leiden, The Netherlands, 1971) support the assumption that the vesicle is the usual site of nitrogenase activity in *Frankia* both in vitro and in the nodule. Endophytes within nodules of *Casuarina* spp., which are effective in nitrogen fixation but which lack a morphologically distinct vesicle stage (see 47), present a puzzling exception. The site of nitrogenase activity in *Frankia* still needs to be located definitively at the subcellular level.

Using freeze-fracture and other techniques, Torrey and Callaham (50) demonstrated the heavily laminated nature of the vesicle wall, and they suggested that this might represent a glycolipid layer which protects nitrogenase against oxygen destruction and which is analogous to the cyanobacterial heterocyst wall.

SOIL ECOLOGY OF NODULATION

Frankia has not yet been successfully isolated or identified from native soils. Research concerning the persistence and distribution of the organism in the soil has been reviewed by van Dijk (53). Individual nodules may persist for 3 to 5 years in the field (44; Akkermans, Doctoral thesis, University of Leiden, 1971), adding new lobes during each growing season. Van Dijk (53) presented indirect evidence that the proliferation of *Frankia* in such nodules is the major source of soil inoculant. Schwintzer et al. (44), on the other hand, found that an extensive seasonal turnover of *Frankia* occurs in nodules of *Myrica gale*, with only a limited carry-over into the following season of hyphae which function in internal reinfection of the new nodule lobes. Spores may be produced in quantity within senescing nodule tissue (54), and these have

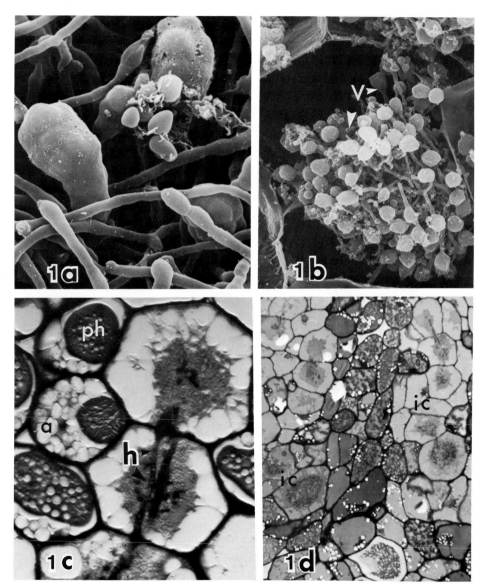

FIG. 1. (a) *Frankia* sp. ArI3 (11), in vitro culture: sporangia and spores within sporangial matrix. Scanning electron micrograph. ×5,100. (b) ArI3 within *Alnus* cortical cell: vesicles (v). Scanning electron micrograph. ×1,900. (c) Septate *Frankia* invasion hyphae (h) within file of cortical cells of *M. gale*. These cells have overwintered in this condition, and the hyphae will reinfect the next season's growth. Adjacent noninfected cortical cells are smaller, containing multiple amyloplasts (a) and phenolic-filled vacuoles (ph). Nomarski interference contrast with toluidine blue staining. ×3,000. (d) Longitudinal section of *M. gale* nodule lobe, showing files of infected cortical cells (ic), with acropetal *Frankia* infection and subsequent differentiation within enlarging host cells. Light micrograph with toluidine blue staining. ×750. Light micrographs by the method of Schwintzer et al. (44); scanning electron micrographs by the method of Berry and Torrey (11). *M. gale* material from field collection of C. Schwintzer.

been proposed as infective propagules for subsequent nodulation. Spores are of variable occurrence in *Alnus* or *Myrica* field nodule populations, however (38, 44, 52), and so cannot be the sole means of soil propagation for *Frankia*.

Research regarding the relationship between nodulation capacity of soils and edaphic factors (e.g., soil type, moisture, pH, nitrogen regime, temperature, and presence of soil phenolics) is necessarily correlative (23, 40). There is some information, again correlative, on soil nodulation capacity in relation to the presence or absence of the host species at a given site (e.g., 1, 56). In general, in the absence of the host species, soil nodulation capacity appears to decrease with time.

In addition, the soil or rhizosphere environment is influenced by the biotic community. Rose and Youngberg (43) reported increased nodulation and improved productivity when *Ceanothus* was inoculated with mycorrhizal strains and *Frankia*. Such tripartite associations were successfully initiated in sterile soil cultures.

Knowlton et al. (24) found that soil bacteria other than *Frankia* (e.g., pseudomonads) stimulated nodulation when these were co-inoculated in agar slant culture. P. Perinet and M. Lalonde (Can. J. Bot., in press) demonstrated that nodulation of plantlet cultures is not dependent upon the presence of such "helper" organisms. In a complex rhizosphere environment, however, soil bacteria can produce substrates, lytic enzymes, and hormones (12) which may affect either host plant or *Frankia*. The role played by soil microorganisms in the infection process may thus be significant, although no obligate "helper" relationship occurs.

Recently, advances have been made in methodology for rapid and repeatable isolations of *Frankia* from nodules, primarily of alder (5, 28; D. R. Benson and D. Hanna, Can. J. Bot., in press), and these are useful in screening successful symbionts. Further progress in investigating the ecology of *Frankia* in the soil will be made when a means of recovering the organisms directly from the soils becomes available. Although attempts at soil isolations by use of fractionation techniques have not yet yielded *Frankia* (5), some modification of these procedures is likely to meet with success (D. Baker, personal communication).

ROOT HAIR INFECTION

Frankia hyphae enter the host root initially by penetration of a deformed root hair, as reported in *Alnus* species, in *Comptonia* and *Myrica*, and in *Casuarina cunninghamiana* (3, 14, 15, 26, 41; A. M. Berry, Ph.D. thesis, University of Massachusetts, Amherst, 1983). The infection occurs in a highly deformed region of the root hair wall (Fig. 2d), apparently within a crypt formed by the junction of several hair walls (14; Berry, Ph.D. thesis, University of Massachusetts, 1983). The crypt may provide an important microenvironment for *Frankia* at the rhizoplane prior to wall penetration. Host wall deposition in the vicinity of the infection site is extensive, and this appears to be a continuation of the root hair wall rather than, for example, a callosic plug (14; Fig. 2e). Once within the root hair cell, *Frankia* hyphae extending into the cell base are surrounded by a pectinaceous encapsulating matrix, of host derivation according to Lalonde and Knowles (29). Electron micrographs (Fig. 2f) reveal a fibrillar component of the encapsulation within the root hair, apparently in two oriented wall-like layers.

In the rhizosphere, *Frankia* is pleomorphic in the sense that all stages of the in vitro life cycle can be observed at the host root surface (28; Berry, unpublished data). However, there is as yet insufficient direct evidence to support the hypothesis that a novel morphological stage occurs outside the hair at the infection site, surrounded by host-derived "exoencapsulation" (26).

ROOT HAIR DEFORMATION

The massive distortion of numerous growing root hair cells on actinorhizal roots is a distinctive characteristic of the host cellular response to inoculation. This striking deformation of host root hair cells precedes actual infection and may be a necessary precondition for successful infection by *Frankia*. Root hairs of axenically grown *Alnus* seedlings are not deformed; however, inoculation of such seedlings either with *Frankia* or with miscellaneous noninfective soil bacteria, including pseudomonads, resulted in an extensive deformation response (24). Root hair deformation is host specific in that neither pseudomonads nor *Frankia* induced deformational changes in *Betula* seedlings under identical growth and inoculating conditions (A. M. Berry and J. G. Torrey, Can. J. Bot., in press).

In plant cellular terms, deformation of root hairs occurs as a function of root hair differentiation; the developmental locus where deformation occurs on an inoculated root is equivalent to the locus of root hair elongation in axenic roots (Berry, Ph.D. thesis, University of Massachusetts, 1983; Berry and Torrey, in press). Ultrastructural comparison of cell walls from axenic and inoculated, deformed root hair walls (A. M. Berry, J. G. Torrey, and M. E. McCully, *in* R. B. Goldberg, ed., *Plant Molecular Biology*, in press) indicated that changes occurred in the wall matrix of inoculated root hairs, although the effects wrought by the *Pseudomonas* strain used were different from those due to *Frankia*. These results may correlate with earlier histochemical work (Berry, Ph.D. thesis, University of Massachusetts, 1983; Berry and Torrey, in press) performed with the fluorochrome acridine orange, which suggested that production of polyuronidic wall fractions was altered from axenic to inoculated conditions. These lines of evidence

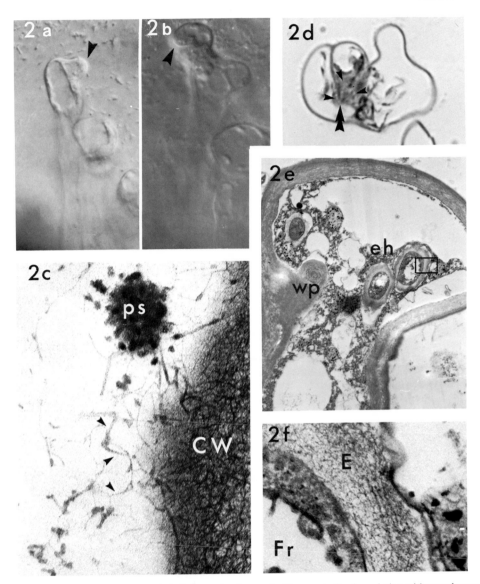

FIG. 2. (a, b) Time course of root hair deformation in axenic *Alnus* root after inoculation with pseudomonads: (a) 2 h after inoculation, branch tip (arrow); (b) 8 h after inoculation, second branch (arrow) and note expansion of first branch tip. Nomarski interference optics. Both ×500. (c) Section of pseudomonad outer wall (ps) at the surface of the root hair wall (CW), *Alnus*. Note periodic patterning of strands between bacterial section and wall (arrows). Transmission electron micrograph. ×80,000. (d) Infected root hair of *Alnus*, section through infection site. Note apparently septate hyphae (arrows) radiating from point near junction of bases of two adjacent lobes (double arrow). Light micrographs, toluidine blue O staining. ×1,600. (e) Infected root of *Alnus*, with wall proliferation (wp) and encapsulated hyphae (eh). Transmission electron micrograph. ×7,000. (f) Enlarged area from 2e, showing fibrillar layers of encapsulation (E) surrounding *Frankia* (Fr). ×75,000. Light micrographs by the method of Schwintzer et al. (44); transmission electron micrographs by the method of Berry et el. (in press). *Alnus* material grown as in Berry (Ph.D. thesis, University of Massachusetts, 1983).

indicate that the root hair deformation process is a modification of root hair tip growth, which appears to be a result of two phenomena: (i) root hair branching (Fig. 2a and b) and (ii) alteration of wall structure or composition. Further ultrastructural studies are in progress.

Exogenously applied auxins did not affect root hair differentiation in *Alnus* axenic seedlings, nor did bacterial extracts such as lysates, exudates, or pasteurized fractions stimulate root hair deformation in this system (Berry and Torrey, in press). It is difficult, however, to draw conclusions from such data, since bacterially produced exopolymers or plant growth substances are probably part of a local or continuous interaction at the region of hair cell growth, leading to root hair deformation and wall penetration.

BINDING AND RECOGNITION AT THE RHIZOPLANE

Recent ultrastructural studies (Berry et al., in press) have demonstrated the presence of broad, heterogeneous strands which run between the root hair wall surface and the bacterial surface of both *Frankia* and pseudomonads (Fig. 2c). The strands exhibit an internal periodic staining pattern, the significance of which is uncertain at this time, but which suggests the presence of glycoprotein rather than, for example, polysaccharides such as comprise the strands of root cap mucilage. These latter are finer and show a staining pattern of irregularly oriented dots (e.g., see Fig. 1 of reference 23a), probably representing ferulic acid deposits. The internally patterned strands at the root hair surface may function as lectins. They are also present at the surface of axenically grown root hairs, indicating that they are of host origin, but here they do not have any discernible linking function. Instead they appear to be a component of the surface mucilage.

At the sites of bacterial attachment to the wall surface, the strands are particularly preserved. The bacteria may contribute a stabilizing factor to form a linked network. Further characterization of these structures is currently being carried out.

Lectins have been demonstrated to bind to the *Frankia* cell surface (A. Chaboud and M. Lalonde, Can. J. Bot., in press), and some correlation was found between patterns of lectin binding and host plant specificity for several strains of *Frankia*. The sugar 2-*O*-methyl-D-mannose has been reported in whole cell hydrolysates of 12 diverse *Frankia* strains and is apparently unique to the *Frankiaceae* (32).

THE *FRANKIA* ENDOPHYTE

The growth and differentiation of *Frankia* in the nodule structurally parallels that of the organism in culture (Fig. 1b). Several actinorhizal genera have been characterized in the nodule (4, 6, 8, 37, 42, 43; K. A. VandenBosch and J. G. Torrey, Can. J. Bot., in press). Both vesicles and sporangia will occur in vitro even for strains which do not express these structures in the nodule (25), but in one case (4) the suppression of vesicles in the nodule appears to be due to an endophyte factor. Furthermore, there is some evidence that the physiological requirements of

Frankia in vitro differ from those in the nodule (19; A. D. L. Akkermans et al., Can. J. Bot., in press). These lines of evidence seem to indicate degrees of mutual physiological regulation during *Frankia* differentiation in the nodule, but the matter is still speculative.

Indeed, the entire question of interactive patterns in nodulation, for example, determinants of host range in cross-inoculation groups, is currently an open one. Certainly the current classification of *Frankia* based on such criteria as cross-inoculation tests (6) will require major revisions in the future.

NODULE DEVELOPMENT

The structural aspects of prenodule formation and subsequent nodule development have been established for a number of actinorhizal species (15, 17, 33, 35–37, 45, 47, 49; Berry, Ph.D. thesis, University of Massachusetts, 1983).

From these studies a generally consistent set of cytological characteristics for nodule formation can be assembled. An actinorhizal root nodule is a series of modified lateral roots. Each successive nodule lobe develops from the pericycle of a previous lobe and retains the major tissue regions of a primary root: epidermis, cortex, and central vascular cylinder. Within the nodule tissue *Frankia* invades and proliferates in cells of the cortex (Fig. 1c). Infected host cells exhibit distinctive cell morphology as compared with cortical cells of uninfected roots. Cells are hypertrophied, and the nuclei within the cells are characteristically large, with prominent nucleoli. In the genera *Coriaria* and *Datisca*, multiple nuclei arise within a single infected cell (35). Mitochondria are present in high numbers, indicating that the cells are metabolically very active (37).

Plant hormones, especially auxins and cytokinins, are considered to play a role in the physiological regulation of nodule development in legumes. Elevated levels of auxins, cytokinins, and gibberellins have been found in actinorhizal nodules (55). Whether these hormones are derived from host tissue or from the *Frankia* microsymbiont is not yet known.

Within the cortex, infected cells occur in axial files or irregular sectors which alternate with uninfected cells (Fig. 1d). Nearby uninfected cortical cells often contain heavy deposits of phenolic compounds, as judged by green to blue color with toluidine blue O staining and osmium binding. Amyloplasts are commonly present in high numbers in uninfected cells (Fig. 1c). Infected cells do not commonly contain amyloplasts nor such heavy phenolic deposits, although phenolics have been reported in *Comptonia* and in vacuoles of infected cells of *Coriaria* (35). There may be deposits of lignin in walls of infected cells of *M. gale* (W. Newcomb, personal communication). Phenolic compounds as determined by histochemical techniques accumulate in quantity in the vacuoles of (uninfected) endodermal cells; likewise, the periderm tissues are tannin filled.

PHENOLIC COMPOUNDS

The remarkable extent of phenolic accumulation in uninfected nodule cells may represent a partial plant cellular defense response to infection by *Frankia*. The antimicrobial effects of plant phenolic accumulation in response to perturbation have been documented in other systems (19). Functionally, phenolics are probably one factor restricting *Frankia* infection to specific locations within the nodule cortex. Additionally, the presence of phenolic compounds could preserve the nodule structure against destruction by pathogenic microorganisms, as suggested by Newcomb et al. (37).

Of relevance is recent work by Y. Perradin, M. J. Mottet, and M. Lalonde (Can. J. Bot., in press) which indicates that common plant phenolic compounds inhibit the growth of several *Frankia* strains in vitro (see also 18). Certain of these compounds also have effects upon the course of *Frankia* differentiation, in some cases increasing apparent vesicle number or suppressing sporangium formation. This suggests a more elaborate regulatory role in the nodule for phenolic compounds. The biochemical identification and quantitation of the various nodule phenolics would permit the use of specific in vitro assays for *Frankia* morphogenesis, but this remains to be done.

INEFFECTIVE NODULATION

Evidence of other types of incompatibility responses can be derived from examining host–endophyte combinations which result in nodules deficient in nitrogenase activity (ineffective) or nodules with low levels of activity (low effectivity; see 20). These provide an important comparative tool for interpreting the cellular basis of effectivity in nodules as well. The cytological features of two such combinations have been recently described (in *Elaeagnus* [6] and in *M. gale* [VandenBosch and Torrey, in press]), and the literature on the subject has been cogently reviewed (VandenBosch and Torrey, in press).

The salient structural feature of these ineffective nodules is the absence of *Frankia* vesicles. In addition, *Frankia* does not proliferate as extensively within the infected host cells and indeed occurs commonly either in the middle lamellar region or along the inner face of the infected cell walls, an unusual location for the endophyte under effective conditions. In both *M. gale* and *Elaeagnus* ineffective nodule combinations, sporangia differentiate at high frequency relative to hyphal proliferation in the infected zone, both intra- and intercellularly. Infected host cells were found to undergo senescence even though *Frankia* proliferation within these cells was not extensive. Ineffective nodules in general are small and limited in the number of lobes formed. VandenBosch and Torrey (in press) suggested that the starch grains present in infected cells, another feature of ineffective nodules, are evidence that host carbon metabolism is redirected away from endophyte vesicle formation (see also 37).

Tjepkema et al. (46) reported that succinate, malate, or fumarate plus EDTA is required for vesicle differentiation on defined media. They also found that NH_4Cl (1 mM) prevents vesicle formation in the same growth conditions. The presence of ammonium in the growth medium can eliminate nitrogenase activity in effective nodules also (22), and this effect correlates with the occurrence of damage to preexisting vesicles and vesicle-containing host cells.

There are thus several possible metabolic mechanisms for regulating *Frankia* differentiation and the expression of nitrogenase activity during the nodulation process. We need more information concerning host–endophyte interactions in terms of carbon and nitrogen metabolism in the functioning nodule.

CONCLUSIONS

The concept of compatibility or incompatibility is complex in the nodule symbiosis, encompassing recognition phenomena and perhaps host defense mechanisms which may parallel plant–pathogen interactions (25). At the same time, symbiotic compatibility includes an element of continuing and mutual host–endophyte regulation which is both physiological and cell developmental. We do not yet have information concerning the biochemical or molecular bases for such interactions.

During root hair deformation and subsequent infection, *Frankia* effects changes in host cell wall structure and deposition. Ultrastructural and histochemical evidence suggests that the matrix component of the root hair wall is altered. Binding of bacteria to the outer wall surface appears to involve an element of the host cell surface mucilage, which exhibits internal structure possibly consistent with that of glycoprotein. This surface element appears at the electron microscopic level to attach to *Frankia* and to *P. cepacia*.

The actinorhizal symbiosis is an important area for future research because of the diversity of host genera which *Frankia* nodulates successfully and because of the striking similarities between actinorhizal and *Rhizobium*–legume symbioses, in both infection and nitrogen fixation. There is still very little information of any kind on the genetics of either the plant hosts or the *Frankia* endophytes. Preliminary genetic analysis of plasmids in *Frankia* has recently been reported (39). Recent cytological research on the infection process has focused on only a few of the actinorhizal genera; additional patterns of infection may occur, for example, in *Elaeagnus* or *Casuarina*. There is thus a need for continuing research at the cytological level, with the use of carefully defined conditions of growth and inoculation.

The potential for applications of actinorhizal species in silvicultural practice worldwide, for example, for land reclamation, as interplanted crops, or in biomass plantations, appears extremely promising. We still lack much basic information on the soil ecology of *Frankia* and on the nodulation and performance of actinorhizal hosts in a range of soil habitats. The development of tripartite associations (host plant, actinorhizae, and mycorrhizae) is still in the early stages of understanding and manipulation, but offers exciting possibilities for forest production. On a practical level, such soil ecological studies might be most effective when integrated with current forest management goals and practices.

I thank W. Newcomb and M. McCully, and anonymous reviewers, for suggestions during manuscript preparation; L. McIntyre, W. Ormerod, and E. Seling for fine technical assistance; and the researchers who graciously sent comments and manuscripts.

LITERATURE CITED

1. **Akkermans, A. D. L., and A. Houwers.** 1979. Symbiotic nitrogen fixers available for use in temperate forestry, p. 23–35. *In* J. C. Gordon, C. T. Wheeler, and D. A. Perry (ed.), Symbiotic nitrogen fixation in the management of temperate forests. Forest Research Laboratory, Oregon State University, Corvallis.

2. **Akkermans, A. D. L., and C. van Dijk.** 1979. Non-leguminous root-nodule symbioses with actinomycetes and *Rhizobium*. *In* W. Broughton (ed.), The ecology of N_2 fixation, Oxford University Press, London.

3. **Angulo Carmona, A. F.** 1974. La formation des nodules fixateurs d'azote chez *Alnus glutinosa* (L.) Vill. Acta Bot. Neerl. 23:257–303.

4. **Baker, D., W. Newcomb, and J. G. Torrey.** 1980. Characterization studies of an ineffective actinorhizal microsymbiont. *Frankia* sp. EuI1 (Actinomycetales). Can. J. Microbiol. 26:1072–1089.

5. **Baker, D., and J. G. Torrey.** 1979. The isolation and cultivation of actinomycetous root nodule endophytes, p. 38–56. *In* J. C. Gordon, C. T. Wheeler, and D. A. Perry (ed.), Symbiotic nitrogen fixation in the management of temperate forests. Forest Research Laboratory, Oregon State University, Corvallis.

6. **Baker, D., and J. G. Torrey.** 1980. Characterization of an effective actinorhizal microsymbiont, *Frankia* sp. AvcI1 (Actinomycetales). Can. J. Microbiol. 26:1066–1071.

7. **Bauer, W. D.** 1981. Infection of legumes by rhizobia. Annu. Rev. Plant Physiol. 32:407–449.

8. **Becking, J. H.** 1974. Frankiaceae (Becking 1970), p. 701–706. *In* R. E. Buchanan and N. E. Gibbons (ed.), Bergey's manual of determinative bacteriology, 8th Ed. The Williams & Wilkins Co., Baltimore.

9. **Becking, J. H.** 1975. Root nodules in non-legumes, p. 507–566. *In* J. G. Torrey and D. T. Clarkson (ed.), The development and function of roots. Academic Press, London.

10. **Benson, D. R., and D. E. Eveleigh.** 1979. Ultrastructure of the nitrogen-fixing symbiont of *Myrica pensylvanica* L. (Bayberry) root nodules. Bot. Gaz. 140(Suppl.):S15–S21.

11. **Berry, A., and J. G. Torrey.** 1979. Isolation and characterization *in vivo* and *in vitro* of an actinomycetous endophyte from *Alnus rubra* Bong, p. 69–83. *In* J. C. Gordon, C. T. Wheeler, and D. A. Perry (ed.), Symbiotic nitrogen fixation in the management of temperate forests. Forest Research Laboratory, Oregon State University, Corvallis.

12. **Bowen, G. D., and A. D. Rovira.** 1961. The effects of microorganisms on plant growth. I. Development of roots and root hairs in sand and agar. Plant Soil 15:166–188.

13. **Callaham, D., P. Del Tredici, and J. G. Torrey.** 1978. Isolation and cultivation *in vitro* of the actinomycete causing root nodulation in *Comptonia*. Science 199:899–902.

14. **Callaham, D., W. Newcomb, J. G. Torrey, and R. L. Peterson.** 1979. Root hair infection in actinomycete-induced root nodule initiation in *Casuarina, Myrica* and *Comptonia*. Bot. Gaz. 140(Suppl.):S51–S59.

15. **Callaham, D., and J. G. Torrey.** 1977. Prenodule formation and primary nodule development in roots in *Comptonia* (Myricaceae). Can. J. Bot. 55:2306–2318.

16. **Callaham, D. A., and J. G. Torrey.** 1981. The structural basis for infection of root hairs of *Trifolium repens* by *Rhizobium*. Can. J. Bot. 59:1647–1664.

17. **Calvert, H. E., A. H. Chaudhary, and M. Lalonde.** 1979. Structure of an unusual root nodule symbiosis in a nonleguminous herbaceous dicotyledon, p. 474–475. *In* J. C. Gordon, C. T. Wheeler, and D. A. Perry (ed.), Symbiotic nitrogen fixation in the management of temperate forests. Forest Research Laboratory, Oregon State University, Corvallis.

18. **Dawson, J. O., S. Knowlton, and S. H. Sun.** 1981. The effect of juglone concentration on the growth of *Frankia in vitro*. Forestry Research Report No. 81-2. University of Illinois, Champaign-Urbana.

19. **Deverall, B. J.** 1977. Defence mechanisms in plants. Cambridge University Press, Cambridge.

20. **Dillon, J. T., and D. Baker.** 1982. Variations in nitrogen fixing efficiency among pure-cultured *Frankia* strains tested on actinorhizal plants as an indication of symbiotic compatibility. New Phytol. 92:215–219.

21. **Huss-Dannell, K., W. Roelofsen, A. D. L. Akkermans, and P. Meijer.** 1982. Carbon metabolism of *Frankia* spp. in root nodules of *Alnus glutinosa* and *Hippophae rhamnoides*. Physiol. Plant. 54:461–499.

22. **Huss-Dannell, K., A. Sellstedt, A. Flower-Ellis, and M. Sjostrom.** 1982. Ammonium effects on function and structure of nitrogen-fixing root nodules of *Alnus incana* (L.) Moench. Planta 156:332–340.

23. **Jobidon, R., and J. R. Thibault.** 1982. Allelopathic growth inhibition of nodulated and unnodulated *Alnus crispa* var. *mollis* Fern. seedlings by *Populus balsamifera*. Am. J. Bot. 69:1213–1223.

23a.**Juniper, B. E., and R. M. Roberts.** 1966. Polysaccharide synthesis and the fine structure of root cells. J.R. Microsc. Soc. 85:63–72.

24. **Knowlton, S., A. Berry, and J. G. Torrey.** 1980. Evidence that associated soil bacteria may influence root hair infection of actinorhizal plants by *Frankia*. Can. J. Microbiol. 26:971–977.

25. **Kosuge, T.** 1981. Carbohydrates in plant-pathogen interactions, p. 584–615. *In* W. Tanner and F. A. Loewus (ed.), Encyclopedia of plant physiology, N.S., vol. 13B, Plant carbohydrates II: extracellular carbohydrates. Springer Verlag, Berlin.

26. Lalonde, M. 1977. The infection process of *Alnus* root nodule symbiosis, p. 569–589. *In* W. Newton, J. R. Postgate, and C. Rodriguez-Barrueco (ed.), Recent developments in nitrogen fixation. Academic Press, London.

27. Lalonde, M. 1978. Confirmation of the infectivity of a free-living actinomycete isolation from *Comptonia peregrina* root nodules by immunological and ultrastructural studies. Can. J. Bot. 56:2621–2635.

28. Lalonde, M., H. E. Calvert, and S. Pine. 1981. Isolation and use of *Frankia* strains in actinorhizae formation, p. 296–299. *In* A. H. Gibson and W. E. Newton (ed.), Current perspectives in nitrogen fixation. Australian Academy of Science, Canberra.

29. Lalonde, M., and R. Knowles. 1975. Ultrastructure, composition, and biogenesis of the encapsulation material surrounding the endophyte in *Alnus crispa* var. *mollis* root nodules. Can. J. Bot. 53:1951–1971.

30. Lechevalier, M., and H. Lechevalier. 1979. The taxonomic position of the actinomycetic endophytes, p. 111–122. *In* J. C. Gordon, C. T. Wheeler, and D. A. Perry, (ed.), Symbiotic nitrogen fixation in the management of temperate forests. Forest Research Laboratory, Oregon State University, Corvallis.

31. Mian, S., and G. Bond. 1978. The onset of nitrogen fixation in young alder plants and its relation to differentiation in the nodular endophyte. New Phytol. 80:187–192.

32. Mort, A., P. Normand, and M. Lalonde. 1983. 2-O-methyl-D-mannose, a key sugar in the taxonomy of *Frankia*. Can. J. Microbiol. 29:993–1002.

33. Newcomb, W. 1981. Fine structure of the root nodules of *Dryas drummondii* Richards (Rosaceae). Can. J. Bot. 59:2500–2514.

34. Newcomb, W., D. Callaham, J. G. Torrey, and R. L. Peterson. 1979. Morphogenesis and fine structure of the actinomycetous endophyte of nitrogen-fixing root nodules of *Comptonia peregrina*. Bot. Gaz. 140(Suppl.):S22–S34.

35. Newcomb, W., and C. E. Pankhurst. 1982. Fine structure of actinorhizal root nodules of *Coriaria arborea* (Coriariaceae). N.Z. J. Bot. 20:93–103.

36. Newcomb, W., and C. E. Pankhurst. 1982. Ultrastructure of actinorhizal root nodules of *Discaria toumatou* Raoul (Rhamnaceae). N.Z. J. Bot. 20:105–113.

37. Newcomb, W., R. L. Peterson, D. Callaham, and J. G. Torrey. 1978. Structure and host-actinomycete interactions in developing root nodules of Comptonia peregrina. Can. J. Bot. 56:502–531.

38. Normand, P., and M. Lalonde. 1982. Evaluation of *Frankia* strains isolated from provenances of two *Alnus* species. Can. J. Microbiol. 28:1133–1142.

39. Normand, P., P. Simonet, J. L. Butour, C. Rosenberg, A. Moiroud, and M. Lalonde. 1983. Plasmids in *Frankia* sp. J. Bacteriol. 155:32–35.

40. Perry, D. A., C. T. Wheeler, and O. T. Helgerson. 1979. Nitrogen-fixing plants for silviculture: some genecological considerations, p. 243–252. *In* J. C. Gordon, C. T. Wheeler, and D. A. Perry (ed.), Symbiotic nitrogen fixation in the management of temperate forests. Forest Research Laboratory, Oregon State University, Corvallis.

41. Pommer, E. H. 1956. Beitrage zur Anatomie und Biologie der Wurzel knollchen von *Alnus glutinosa* Gaertn. Flora (Jena) 143:604–634.

42. Robertson, J. G., P. Lyttleton, and C. E. Pankhurst. 1981. Preinfection and infection processes in the legume-*Rhizobium* symbiosis, p. 280–291. *In* A. H. Gibson and W. E. Newton (ed.), Current perspectives in nitrogen fixation. Australian Academy of Science, Canberra.

43. Rose, S. L., and C. T. Youngberg. 1981. Tripartite association in snowbrush (*Ceanothus velutinus*): effect of vesicular-arbuscular mycorrhizae on growth, nodulation, and nitrogen fixation. Can. J. Bot. 59:34–39.

44. Schwintzer, C. R., A. M. Berry, and L. D. Disney. 1982. Seasonal patterns of root nodule growth, endophyte morphology, nitrogenase activity and shoot development in *Myrica gale*. Can. J. Bot. 60:746–757.

45. Strand, R., and W. M. Laetsch. 1977. Cell and endophyte structure of the nitrogen-fixing root nodules of *Ceanothus integerrimus* H. and A. I. Fine structure of the nodule and its endosymbiont. Protoplasma 93:165–178.

46. Tjepkema, J. D., W. Ormerod, and J. G. Torrey. 1981. Factors affecting vesicle formation and acetylene reduction (nitrogenase activity) in *Frankia* sp. CpI1. Can. J. Microbiol. 27:835–843.

47. Torrey, J. G. 1976. Initiation and development of root nodules of *Casuarina* (Casuarinaceae). Am. J. Bot. 63:335–344.

48. Torrey, J. G. 1978. Nitrogen fixation by actinomycete-nodulated angiosperms. BioScience 28:586–592.

49. Torrey, J. G., and D. Callaham. 1979. Early nodule development in *Myrica gale*. Bot. Gaz. 140(Suppl.):S10–S14.

50. Torrey, J. G., and D. Callaham. 1982. Structural features of the vesicle of *Frankia* sp. CpI1 in culture. Can. J. Microbiol. 28:749–757.

51. Turgeon, B. G., and W. D. Bauer. 1982. Early events in the infection of soybean by *Rhizobium japonicum*: time course and cytology of the initial infection process. Can. J. Bot. 60:152–161.

52. van Dijk, C. 1978. Spore formation and endophyte diversity in root nodules of *Alnus glutinosa* (L.) Vill. New Phytol. 81:601–615.

53. van Dijk, C. 1979. Endophyte distribution in the soil, p. 84–94. *In* J. C. Gordon, C. T. Wheeler, and D. A. Perry (ed.), Symbiotic nitrogen fixation in the management of temperate forests. Forest Research Laboratory, Oregon State University, Corvallis.

54. van Dijk, C., and E. Merkus. 1976. A microscopical study of the development of a spore-like stage in the life cycle of the root-nodule endophyte of *Alnus glutinosa* (L.) Gaertn. New Phytol. 77:63–91.

55. Wheeler, C. T., I. E. Henson, and M. E. McLaughlin. 1979. Hormones in plants bearing actinomycete nodules. Bot. Gaz. 140(Suppl.):S52–S57.

56. Youngberg, C. T., A. G. Wollum, and W. Scott. 1979. *Ceanothus* in Douglas-fir clearcuts: nitrogen accretion and impact on regeneration, p. 224–233. *In* J. C. Gordon, C. T. Wheeler, and D. A. Perry (ed.), Symbiotic nitrogen fixation in the management of temperate forests. Forest Research Laboratory, Oregon State University, Corvallis.

Adherence and Host Recognition in *Agrobacterium* Infection

JAMES A. LIPPINCOTT, BARBARA B. LIPPINCOTT, AND JONATHAN J. SCOTT

Department of Biochemistry, Molecular Biology and Cell Biology, Northwestern University, Evanston, Illinois 60201

Agrobacterium tumefaciens is the causal agent of the tumorous disease of higher plants called crown gall. Once initiated by introduction of these bacteria into a plant wound, the tumorous plant cells proliferate autonomously in the absence of the infecting bacteria, although with appropriate selection and treatment they may be induced to differentiate and form normal-appearing plants (12). It was shown in recent years that the large DNA plasmid (Ti plasmid) carried by the bacterium is essential for *Agrobacterium* virulence. Part of this plasmid DNA becomes incorporated in the chromatin of the transformed plant cell, where it is replicated and passed on to daughter tumor cells with each cell division (21). Much of the current excitement in this field centers on the fact that the Ti plasmid–bacterium complex provides a vector to introduce new genetic material into susceptible plant hosts.

The ecological importance of the interaction between these bacteria and their hosts has long been considered one of protection and nutrition. The discovery of the Ti plasmid and its role, however, suggests an even more important aspect of this interaction; i.e., it provides a suitable means for bacterial genetic exchange (8, 9, 24). Some of the agrobacteria introduced into a plant wound transfer their plasmid DNA to plant cells. The resulting tumorous cells produce aphrodesiacs called opines, the prototypes being lysopine and octopine (N^α-propionyl–substituted lysine and arginine), which promote genetic exchange between plasmid-containing agrobacteria in the wound and related plasmid-free bacteria. This genetic aspect of the bacterium–host niche thus provides a compelling explanation for the selection and maintenance of the Ti plasmid and of tumorigenicity.

It would not be surprising, therefore, to find that the agrobacteria have developed mechanisms to recognize and adhere to their hosts, considering the need to transmit plasmid DNA to host cells and the potential genetic value of this relationship to the bacterium. The agrobacteria are motile and show a positive chemotaxis to plant cell products such as might occur in the vicinity of a wound (1). In this response no particular specificity relative to host versus non-host has been demonstrated, however. Results obtained in our laboratory beginning in the late 1960s have demonstrated what appears to be the first interaction between the bacterium and the plant host that is selective. The bacterium must adhere to a specific wound site to initiate a tumor. The specificity of this interaction suggests that it is critical to the host–bacterium recognition process and consequently to the establishment, replication, and genetic exchange that occurs among agrobacteria in this specialized niche.

SITE ADHERENCE IN *AGROBACTERIUM* INFECTION

The requirement for bacterial site attachment in crown gall formation by *A. tumefaciens* and the basic characteristics of this response were first shown by Lippincott and Lippincott (10) and have since been confirmed by a number of investigators (reviewed by Lippincott and Lippincott [15]). The number of tumors formed on bean leaves by a given inoculum is reduced by the addition of certain avirulent agrobacteria and by heat- or UV-killed virulent bacteria that compete for attachment sites. Attachment is completed during the first 15 min after inoculation in this system, and competing cells inhibit only when they accompany or precede the tumorigenic cells (10). The response measured in this assay shows considerable specificity, as cells of *Corynebacterium fasciens, Bacillus megaterium, B. pumilis, Escherichia coli, Pseudomonas savastanoi, Rhizobium meliloti,* and certain strains of *A. radiobacter* all fail to compete for these sites. Matthysse et al. (18) and Ohyama et al. (23) have developed in vitro assays for *Agrobacterium* adherence to tissue-cultured cell suspensions which confirm many of the findings obtained from in vivo competition data. Some caution is necessary, however, in viewing the results obtained in the latter type of experiments since they measure all adhering bacteria rather than only those bacteria that adhere in a way that is effective in causing tumor initiation (5, 21).

BACTERIAL COMPONENT OF ADHERENCE

Whatley et al. (32) showed that bacterial cell envelope preparations and envelope lipopolysaccharide (LPS) obtained from site-binding agrobacteria inhibit tumor initiation in a manner

TABLE 1. Inhibition of strain B6 tumor initiation by LPS isolated from strain B6 and by the polysaccharide (PS) obtained from this LPS

Amt/ml of inoculum[a]	% Inhibition of tumor formation[b]	
	LPS	LPS-PS
5.0 mg	79	
0.5 mg	75	71
50.0 μg	69	64, 56
5.0 μg	64	59
0.5 μg	57, 41	42, 45
50.0 ng	34	38
0.5 ng	27	33
0.005 ng	10	

[a] LPS or LPS-PS mixed 1:1 with bacteria just prior to inoculation and ca. 0.1 ml of inoculum applied to each leaf.

[b] Two experiments shown for each.

similar to intact dead bacteria, whereas LPS from avirulent agrobacteria that do not compete for attachment sites is noninhibitory. Application of the inhibitory LPS shortly after the bacteria has no effect on tumor initiation, consistent with the proposed competition for adherence sites. A concentration of 1 μg/ml, or 0.1 μg per leaf as inoculations are carried out, is sufficient to give about 50% inhibition of tumor initiation (Table 1). The polysaccharide derived from LPS by hydrolysis in 1% acetic acid is nearly as effective an inhibitor as LPS, whereas the lipid A moiety (solubilized with bovine serum albumin) is noninhibitory. Cell envelope peptidoglycan is also noninhibitory.

The capacity of agrobacteria to adhere to plant sites can be determined by the Ti plasmid. When the Ti plasmid is introduced into the non-site-binding A. radiobacter strains S1005, 6467, and TR1, they become virulent, and both whole cells and LPS from these plasmid-containing derivatives now show site attachment (33). This activity is lost when these strains are cured of the plasmid. Several natural tumorigenic strains, however, retain site attachment ability even after they are cured of the Ti plasmid. The insertion of transposon Tn5 at some chromosomal locations can cause loss of Agrobacterium virulence even in the presence of an active Ti plasmid (5). This is correlated in several such strains with a reduced ability to adhere to cultured cells and a loss of the ability to inhibit tumor initiation, providing further evidence that the bacterial chromosome is involved in determining host adherence. These experiments employed a strain of Agrobacterium that has both chromosomal and Ti plasmid determinants for adherence specificity. Since none of the strains made avirulent by Tn5 insertion into the Ti plasmid was found to have lost adherence, these

results are in agreement with previous data (33) showing that plasmid-cured strains still exhibit adherence and hence that the bacterial chromosome is sufficient to determine adherence specificity. Similar Tn5 insertions into a Ti plasmid introduced into an A. radiobacter strain that is not site adhering in the absence of the plasmid will be necessary to demonstrate plasmid genes for adherence by this method.

Both chromosomal genes and Ti plasmid genes, each capable of determining site attachment, therefore appear to be present in most natural tumorigenic strains. The two genes or sets of genes may not be responsible for identical polysaccharide structures, however, as the nature of the O-antigen polysaccharide sequence has not been determined in either case and the degree of variability in polysaccharide structure consistent with adherence is also unknown.

The above results demonstrate that attachment is essential for tumor formation and that both site attachment and tumorigenicity are correlated with the presence or absence of the Ti plasmid in strains which initially could not adhere to attachment sites. These differences are also correlated with the ability of LPS from these strains to inhibit tumor initiation, but as yet efforts to demonstrate specific alterations in the LPS responsible for these changes have not been successful.

The role and specificity of LPS in adherence have also been demonstrated in a very different way. The introduction of the E. coli plasmid pSa into virulent strain 15955 of Agrobacterium, resulting in a strain carrying both the pSa and Ti plasmids, causes loss of tumorigenicity (6). We have confirmed these results on the more sensitive bean leaf assay. These pSa-containing bacteria also fail to compete with virulent bacteria for infection sites, and when LPS is isolated from strains carrying both plasmids it is noninhibitory (22). The pSa-induced loss of virulence is thus correlated with loss of LPS site-binding ability (22). When cured of the pSa plasmid, these strains regain tumorigenicity, indicating that no permanent change in the Ti plasmid or in the genes governing site attachment has occurred. It appears that the presence of the pSa plasmid results in a sufficiently altered LPS structure that it no longer supports bacterial adherence and by this means causes loss of tumorigenicity.

Secondary events that serve to anchor the bacterium to the plant cell more firmly have also been observed. Matthysse et al. (17) showed that most strains of Agrobacterium produce cellulose microfibrils after adherence, and these presumably add to the stability of the bacterium–host complex and can serve to entrap additional bacteria. Mutants that have lost the ability

TABLE 2. Evidence that the *Agrobacterium* adherence site is part of the host cell wall

1. Cell wall-containing fractions from susceptible bean leaves compete for agrobacteria in tumor initiation; membrane preparations and cell organelles do not.
2. A 50% reduction in tumor initiation is obtained with only 0.1 mg of purified plant cell walls per ml.
3. The cell walls are inhibitory when applied before or with the bacteria but do not inhibit when applied 15 min after the bacteria when effective adherence is complete.
4. Pretreating the cell walls with dead site-binding bacteria or LPS from these strains largely neutralizes the inhibitory activity of the cell walls. Neither whole cells nor LPS from non-site-binding agrobacteria have this neutralizing effect.
5. The cell walls from all host dicots tested are inhibitory; those from nonhost monocots, embryonic dicot tissues, and crown gall tumors are noninhibitory.
6. A particular class of cell wall compounds, pectins and especially PGA, inhibit tumor initiation, the latter at concentrations of 1 ng or less per ml.

to produce cellulose, however, are still virulent and can bind to host cells, indicating that this is not an essential element in this pathogen–host relationship.

PLANT SITE OF *AGROBACTERIUM* ADHERENCE

Of various fractions from homogenized bean leaves of optimal sensitivity to *A. tumefaciens*, only those containing plant cell walls are active in competing with natural wound sites for bacteria in the infectivity bioassay (11, 13). Evidence that the plant cell wall is the site of *Agrobacterium* adherence is summarized in Table 2. The comparison of isolated cell walls from hosts versus certain nonhosts or nonsusceptible portions of hosts indicates that there is specificity at this level as well as that shown by the LPS (14, 15). Since the inhibitory effect of the cell walls is decreased by LPS from site-binding strains but not by LPS from non-site-binding strains, the isolated cell walls exhibit the same selectivity as the in vivo infection site.

Mild extraction of inhibitory plant cell walls to remove pectin leaves behind a particulate fraction composed of cellulose plus hemicellulose that is noninhibitory, whereas the extracted pectin shows inhibition (23). Polygalacturonic acid (PGA), a major pectin component, shows inhibition at concentrations as low as 1 ng per inoculated leaf (Table 3); however, commercial pectin (PGA with about 30% methyl esterified carboxyl groups) is about 10^4 times less effective as an inhibitor (13). PGA is noninhibitory when added to leaves 15 min after they are inoculated with bacteria, evidence that it acts by adhering to the

bacterium before it reaches the plant site, thus blocking successful attachment. This binding appears tight since bacteria treated with PGA and subsequently washed five times by centrifugation and resuspension remain inhibited. The inhibitory effect of pectin and PGA on tumorigenesis has also been shown by use of a potato disk assay for *Agrobacterium* infection (25). This assay does not provide a 1:1 proportionality between inoculum concentration and tumor number, however, and concentrations of pectin of 0.1 to 1.0 mg ml^{-1} are required for 50% inhibition of tumor initiation as opposed to 0.1 to 1.0 μg ml^{-1} in the bean leaf assay.

Several polymeric plant carbohydrates, gums, and lectins (jack bean, kidney bean, wheat germ, gorse seed) have been tested, and all are noninhibitory at concentrations of 10 mg ml^{-1} (carbohydrates and gums) or 0.1 mg ml^{-1} (lectins). A preparation of potato lectin (courtesy of L. Sequeira) which differentiates pathogenic and nonpathogenic *Pseudomonas solanacearum* strains failed to agglutinate either site-binding or non-site-binding strains of *Agrobacterium*. Also, treatment of inhibitory cell walls with trypsin or pronase had little effect on their activity, whereas pectinase treatment quickly destroyed this activity. All evidence currently available, therefore, supports the conclusion that only the pectic compounds in the plant cell wall are involved in the initial adherence of *Agrobacterium*.

Cell walls isolated from crown gall tumor tissues and parts of developing seed embryos (pinto bean, peas, castor bean) lack *Agrobacterium* binding sites (14). These differences between normal, embryonic, and tumor tissues in vivo are maintained when each is grown in tissue culture. Thus, there is a change in cell wall

TABLE 3. Effect of a PGA series of different degrees of methylation on tumor initiation by *A. tumefaciens* strain 15955

% Methylation of PGA inoculated with bacteria[a]	Mean no. of tumors per leaf	
	Expt 1	Expt 2
46[b]	17.8	55.4
35	18.6	67.6
30	14.7	50.5
25	10.4	51.8
18	7.1	35.0
14.5	—	31.7
12	6.2	—
2.4	5.2	20.6

[a] Leaves inoculated with a 1:1 mixture of PGA and bacteria providing inocula containing 2×10^8 bacteria per ml and 1 mg of PGA uronate per ml esterified as indicated.
[b] Initial PGA from which all others were derived by saponification in 10 mM NaOH.

metabolism or structure in early bean development so that by 7 days the bean seedling produces cell walls with attachment sites. A similar but opposite change occurs in the transformation of normal cells to crown gall tumor cells, as cell walls from the latter are noninhibitory. This type of change may be of ecological importance to the bacterium since the multiplying bacteria will be surrounded and nourished by cells to which they do not adhere, and thus they can escape. Cell walls from monocots, which as a group are not susceptible to *Agrobacterium* infection (4), also lack sites to which this bacterium adheres.

Evidence indicating that the degree of methylation of plant cell wall PGA may account for the differences observed in the binding of *Agrobacterium* to various preparations has been obtained (26). Treatment of inhibitory cell walls with pectin methyl transferase results in loss of inhibitory activity, and this change is reversible by pectinesterase treatment. Similarly, pectinesterase treatment of noninhibitory cell walls obtained from either monocots, tumors, or embryonic dicot tissues makes them inhibitory, and this change can be reversed by treatment with pectin methyl transferase. Thus, the galacturonate residues in PGA which are accessible to the bacterium in cell walls from monocotyledons and from embryonic and tumorous dicotyledons may be more highly methylated than those in vegetative adult tissues of dicotyledons. This difference may account for their very different effects on *Agrobacterium* infectivity and is sufficient reason to account for the failure of *Agrobacterium* to infect monocots or embryonic dicot tissues.

We have prepared a series of PGA samples with different degrees of methyl esterification by treating a highly esterified PGA at 4°C with 10 mM NaOH for various periods of time. The lower the degree of methylation, the more inhibitory the PGA per unit weight, and at 35% methylation and above there is little or no inhibition with 1 mg of PGA per ml (Table 3). These changes in inhibitory activity occur over a range of methyl ester content which is common in plant pectins.

Direct analysis of the total alkali-labile methyl ester content of inhibitory and noninhibitory cell walls of the same species, however, shows no significant differences per unit of dry weight, possibly indicating that only a special portion of the pectic material of the cell wall is responsible for the apparent site differences. Determinations of the amount of uronic acid in several representatives of the various types of cell wall show that there is considerable PGA in all of the cell walls.

We have recently developed an extraction and fractionation procedure which gives a specific highly inhibitory material from host cell walls, whereas comparable fractions from monocot, embryonic, and tumorous dicot cell walls are noninhibitory. The inhibitory material is extracted in a portion of the pectin fraction that seems to be tightly calcium bound.

Extraction of host cell walls three times for 1 h each in 0.25 M sodium acetate solubilizes about 0.5 to 1% of the cell wall. This pectic material has a ratio of uronate to neutral sugar between 1 and 2 and is noninhibitory when tested in the infectivity assay. Subsequent extraction of the cell wall residue with 0.05 M EDTA for 4 h at 35°C removes a second pectic fraction that contains most of the inhibitory activity of the cell wall. This fraction constitutes about 2% of the initial cell wall and has a uronate/neutral sugar ratio of 3 to 5. Similar fractions are obtained from noninhibitory cell walls, but these are noninhibitory. However, when the latter are treated with pectinesterase they become inhibitory, again suggesting that methyl esterification of uronate residues may account for these differences.

The inhibitory EDTA-extracted pectin from mature host cell walls has a molecular weight of 200,000 or greater since most of the material elutes from a Sephadex G-200 column at the void volume. About 20% of this pectin elutes in the void volume from a column of Bio-Gel A50m, indicative of a molecular weight in the millions. The separation on these molecular sieve columns does not, however, yield fractions which differ either in their ability to inhibit tumor initiation or to any great extent in their uronate to neutral sugar ratio, suggesting that they are part of a continuous spectrum of molecular sizes all of fairly similar composition. The EDTA fraction from noninhibitory cell walls cannot be distinguished from the EDTA fraction from inhibitory cell walls, either on these columns or on DEAE-Sephadex ion-exchange columns.

The EDTA pectin fraction from host bean leaf tissue is even more inhibitory in the infection assay than either commercial PGA or isolated *Agrobacterium* LPS (Table 4). Again, inhibition increases linearly with the log of pectin concentration, with about 30% inhibition at 0.01 ng ml^{-1} and 90% inhibition at 1 mg ml^{-1}. If one assumes an average molecular weight of 6×10^5 for this pectin, then 1.0 ng ml^{-1} would contain about 10^9 molecules of pectin. This concentration is sufficient to provide about 50% inhibition in the presence of 10^8 bacteria per ml, the concentration of bacteria used in these inhibition-titration experiments. Since significant inhibition is observed with one-hundredth this amount of the EDTA pectin fraction, some cooperative interaction between each molecule

TABLE 4. Inhibition of tumor initiation by
commercial PGA and by EDTA-extracted pectin
from bean leaves

PGA (Citrus, Sigma Chemical Co.)		EDTA extract of bean leaf cell walls	
Concn in inoculum[a]	% Inhibition of tumors	Concn in inoculum[a]	% Inhibition of tumors
1.0 mg	97, 94	5.0 mg	94
0.1 mg	80, 74	0.5 mg	83
10 μg	73, 68	50 μg	80, 73
1.0 μg	59, 57	5.0 μg	76
0.1 μg	47, 55	0.5 μg	72, 66
10 ng	27	50 ng	60
1.0 ng	32	0.5 ng	47
0.1 ng	2	50 pg	41
		5.0 pg	27

[a] Amount per milliliter. Bacteria and pectin compounds mixed 1:1 just prior to inoculation and ca. 0.1 ml of inoculum applied per leaf.

of inhibitor and many bacteria seems necessary to account for this remarkable efficiency.

AGROBACTERIUM–MOSS INTERACTIONS

In collaboration with L. Spiess we have also looked at the effect of Agrobacterium on moss plants (28–30). Certain mosses respond to Agrobacterium by greatly increased bud formation and the development of these buds into normal gametophores, the principal macroscopic phase of the moss life cycle. Although virulent agrobacteria are more effective in this regard then avirulent agrobacteria, the response is not limited to virulent agrobacteria since both rhizobia and many bacterial isolates adhering to moss in nature show similar activity. In all but a few of the latter isolates, the effect depends upon direct association of the bacteria with the moss since dead bacteria, LPS, or cell walls from moss or higher plants, as well as PGA, inhibit this response. The response is not obtained when the moss and bacteria are separated by a membrane filter in a parabiotic chamber. These effects on the moss thus seem to depend on bacterial adherence to the moss which, to the extent examined, is comparable to that in the higher plant infection process.

In studies of bacterial isolates found adhering in nature to three different species of moss, on these same moss species grown in axenic culture most of the isolates induced bud development only on the species that was its host in nature. It appears, therefore, that these associations are not fortuitous and may well contribute to the ability of moss to colonize particular niches. Comparable promotive effects by various soil rhizobacteria have been reported by Burr et al. (2) and Schroth and Hancock (27) for higher plants although it is not known whether an

adherence requirement is essential for these effects.

CONCLUSIONS

The initial specificity-determining event in Agrobacterium–host interactions depends on bacterial adherence to a site in a host wound. The polysaccharide portion of the bacterial LPS seems to be the primary determinant of adherence on the part of the bacterium, and both chromosomal and Ti plasmid genes as well as other plasmids and transposons can affect the adherence activity of this LPS. All evidence thus far indicates that the plant site is a portion of the host plant cell wall that is exposed during the wound-inoculation procedure. Fractionation of host cell walls suggests that the host site is in the pectic portion of the wall. The activity of various compounds on Agrobacterium infection assays indicates that uronate-rich polysaccharides are involved and that the degree of methylation of the uronate residues can affect this activity. Although other components in the bacterial outer membrane and the plant cell wall may well contribute to this initial interaction, there is currently no evidence for such involvement. The necessary event between Agrobacterium and its host that initially specifies whether an infection will occur, then, is seen as a specific bacterial carbohydrate–host carbohydrate interaction.

Although many additional steps beyond adherence are clearly required before Agrobacterium tumor initiation is complete, the adherence step seems to provide an initial screen in the host recognition process. Additional plasmid determinants clearly affect host range (16, 31) and the type of host developmental response (19), some of which apparently depend on the levels of specific plant hormones produced by the transformed plant cell. In the final analysis, therefore, a series of necessary events must occur before a tumor is formed, and all of these are potential contributors to the phenomenon of host range and specificity.

The direction of future studies on Agrobacterium host adherence is relatively straightforward. They will necessarily depend on more direct measurements of the adherence of components to the bacterium, to plant cells, or to plant cell wall preparations, and suitable assays need to be developed. The chemical features of these host and bacterial polysaccharides essential for specificity, as well as the physical characterization of the interaction, remain to be determined. The major portion of the genetic information necessary for LPS production must lie in the bacterial chromosome. However, those genes that directly govern adherence specificity in both the Ti plasmid and the chromosome need to be characterized. Since surface adherence events are frequently associated with the generation of intracellular signals, it is possible that *Agrobacterium* adherence per se initiates specific metabolic activity on the part of the bacterium, the host, or both. Relatively little work in this area has been attempted, and these remain intriguing possibilities. Finally, knowledge of the exact adherence mechanism and its signal function, if any, may have application in the extension of *Agrobacterium* infection to plant cells with walls that do not support adherence so that the gene transfer activity of the bacterium could be extended to these plants.

The protein–polysaccharide type of adherence found in certain host–pathogen relations provides elegant testimony to the specificity possible with these components. The very broad range of hosts susceptible to *Agrobacterium*, however, may be considered to indicate a different adherence mechanism. Highly specific interactions between unlike polysaccharides that result in gel formation have been described (3, 20). Such interactions can provide a suitable model for the type of association proposed for the *Agrobacterium* host adherence. Other broad host range pathogens, as well as saprophytic microorganisms that colonize plants, may be found to adhere by similar mechanisms, as is suggested in the bacterium–moss interactions described here. Marine pseudomonads adhere to a variety of plastics, glass, and even metal surfaces, and in many cases bacterial polysaccharides seem to account for this adherence (7). A polysaccharide–polysaccharide type of adherence mechanism thus presents no conceptual difficulties and may prove to be a common type of adherence among microorganisms that colonize a variety of plants.

LITERATURE CITED

1. **Beiderbeck, R., and R. Hohl.** 1978. Attraction von *Agrobacterium tumefaciens* durch Pflanzenzellen. Phytopathol. Z. **92**:184–187.
2. **Burr, T. J., M. N. Schroth, and T. Suslow.** 1978. Increased potato yields by treatment of seed pieces with specific strains of *Pseudomonas fluorescens* and *P. putida*. Phytopathology **68**:1377–1383.
3. **Dea, I. C. M., E. R. Morris, D. A. Rees, E. J. Welsh, H. A. Barnes, and J. Price.** 1977. Associations of like and unlike polysaccharides: mechanism and specificity in galactomannans, interacting bacterial polysaccharides, and related systems. Carbohydr. Res. **57**:249–272.
4. **De Cleene, M., and J. De Ley.** 1976. The host range of crown gall. Bot. Rev. **42**:389–466.
5. **Douglas, C. J., W. Halperin, and E. W. Nester.** 1982. *Agrobacterium tumefaciens* mutants affected in attachment to plant cells. J. Bacteriol. **152**:1265–1275.
6. **Farrand, S. K., C. I. Kado, and C. R. Ireland.** 1981. Suppression of tumorigenicity by the IncW R plasmid pSa. Mol. Gen. Genet. **181**:44–51.
7. **Fletcher, M.** 1980. Adherence of marine micro-organisms to smooth surfaces, p. 345–374. *In* E. H. Beachey (ed.), Bacterial adherence. Chapman and Hall, London.

8. **Genetello, C., N. Van Larebeke, M. Holsters, A. De Picker, M. Van Montagu, and J. Schell.** 1977. Ti plasmids of *Agrobacterium* as conjugative plasmids. Nature (London) **265**:561–563.
9. **Kerr, A., P. Manigault, and J. Tempe.** 1977. Transfer of virulence *in vivo* and *in vitro* in Agrobacterium. Nature (London) **265**:560–561.
10. **Lippincott, B. B., and J. A. Lippincott.** 1969. Bacterial attachment to a specific wound site as an essential stage in tumor initiation by *Agrobacterium tumefaciens*. J. Bacteriol. **97**:620–628.
11. **Lippincott, B. B., M. H. Whatley, and J. A. Lippincott.** 1977. Tumor induction by *Agrobacterium* involves attachment of the bacterium to a site on the host plant cell wall. Plant Physiol. **59**:388–390.
12. **Lippincott, J. A., and B. B. Lippincott.** 1976. Morphogenic determinants as exemplified by the crown-gall disease, p. 356–388. *In* R. Heitefuss and P. H. Williams (ed.), Encyclopedia of plant physiology, new series, vol. 4. Physiological plant pathology. Springer-Verlag, Berlin.
13. **Lippincott, J. A., and B. B. Lippincott.** 1977. Nature and specificity of the bacterium-host attachment in *Agrobacterium* infection, p. 439–451. *In* B. Solheim and J. Raa (ed.), Cell wall biochemistry related to specificity in host-plant pathogen interactions. Norway Universitetsforlaget, Oslo.
14. **Lippincott, J. A., and B. B. Lippincott.** 1978. Cell walls of crown-gall tumor and embryonic plant tissues lack *Agrobacterium* adherence sites. Science **199**:1075–1078.
15. **Lippincott, J. A., and B. B. Lippincott.** 1980. Microbial adherence in plants, p. 375–398. *In* E. H. Beachey (ed.), Bacterial adherence. Chapman and Hall, London.
16. **Loper, J. E., and C. I. Kado.** 1979. Host range conferred by the virulence-specifying plasmid of *Agrobacterium tumefaciens*. J. Bacteriol. **139**:591–596.
17. **Matthysse, A. G., K. V. Holmes, and R. H. G. Gurlitz.** 1981. Elaboration of cellulose fibrils by *Agrobacterium tumefaciens* during attachment to carrot cells. J. Bacteriol. **145**:583–595.
18. **Matthysse, A. G., P. M. Wyman, and K. V. Holmes.** 1978. Plasmid-dependent attachment of *Agrobacterium tumefaciens* to plant tissue culture cells. Infect. Immun. **22**:516–522.
19. **Moore, L., G. Warren, and G. Strobel.** 1979. Involvement of a plasmid in the hairy root disease of plants caused by *Agrobacterium rhizigenes*. Plasmid **2**:617–626.
20. **Morris, E. R., D. A. Rees, G. Young, M. D. Walkinshaw, and A. Darke.** 1977. Order-disorder transition for a bacterial polysaccharide in solution. A role for polysaccharide conformation in recognition between Xanthomonas pathogen and its plant host. J. Mol. Biol. **110**:1–16.
21. **Nester, E. W., and T. Kosuge.** 1981. Plasmids specifying plant hyperplasias. Annu. Rev. Microbiol. **35**:531–565.
22. **New, P. B., J. J. Scott, C. R. Ireland, S. K. Farrand, B. B. Lippincott, and J. A. Lippincott.** Plasmid pSa causes loss of LPS-mediated adherence in *Agrobacterium*. J. Gen. Microbiol. (in press).
23. **Ohyama, K., L. E. Pelcher, A. Schaeffer, and L. C. Fowke.** 1979. *In vitro* binding of *Agrobacterium tumefaciens* to plant cells from suspension culture. Plant Physiol. **63**:382–387.
24. **Petit, A., J. Tempe, A. Kerr, M. Holsters, M. Van Montagu, and J. Schell.** 1978. Substrate induction of conjugative activity of *Agrobacterium tumefaciens* Ti plasmids. Nature (London) **271**:570–572.
25. **Pueppke, S. G., and U. K. Benny.** 1981. Induction of tumors on *Solanum tuberosum* L. by *Agrobacterium*: quantitative analysis, inhibition by carbohydrates and virulence of selected strains. Physiol. Plant Pathol. **18**:169–179.
26. **Rao, S. S., B. B. Lippincott, and J. A. Lippincott.** 1982. *Agrobacterium* adherence involves the pectic portion of the host cell wall and is sensitive to the degree of pectin methylation. Physiol. Plant. **56**:374–380.
27. **Schroth, M. N., and J. G. Hancock.** 1981. Selected topics

in biological control. Annu. Rev. Microbiol. **35:**453–476.

28. **Spiess, L. D., B. B. Lippincott, and J. A. Lippincott.** 1971. Development and gametophore induction in the moss *Pylaisiella selwynii* as influenced by *Agrobacterium tumefaciens.* Am. J. Bot. **58:**726–731.

29. **Spiess, L. D., B. B. Lippincott, and J. A. Lippincott.** 1976. The requirement of physical contact for moss gametophore induction by *Agrobacterium tumefaciens.* Am. J. Bot. **63:**324–328.

30. **Spiess, L. D., B. B. Lippincott, and J. A. Lippincott.** 1981. Bacteria isolated from moss and their effect on moss development. Bot. Gaz. (Chicago) **142:**512–518.

31. **Thomashow, M. F., C. G. Panagopoulos, M. P. Gordon, and E. W. Nester.** 1980. Host range of *Agrobacterium tumefaciens* is determined by the Ti plasmid. Nature (London) **283:**794–796.

32. **Whatley, M. H., J. S. Bodwin, B. B. Lippincott, and J. A. Lippincott.** 1976. Role of *Agrobacterium* cell envelope lipopolysaccharide in infection site attachment. Infect. Immun. **13:**1080–1083.

33. **Whatley, M. H., J. B. Margot, J. Schell, B. B. Lippincott, and J. A. Lippincott.** 1978. Plasmid or chromosomal determination of *Agrobacterium* adherence specificity. J. Gen. Microbiol. **107:**395–398.

Future Paths in Rhizosphere Studies†

The rhizosphere is the root itself and the volume of soil influenced by root. However, the nature of the root/soil interface and the loss of substrates (both soluble and insoluble) from roots influencing the adjacent soil change greatly with time and space as the root ages. The root/soil interface is clearly defined only with young parts of roots; in older roots the interface is less clearly defined, and saprophytes and minor pathogens invade the cortex. Even in young roots individual cells can be short-lived and are heavily colonized by soil microorganisms. A number of organisms can reduce root growth without causing recognizable disease. It is necessary to further identify such organisms and to define their roles. We need to be able to distinguish the different microbial substrates (and environments) at different root ages, preferably under soil conditions. These changes may be related to microbial activity. Because of the multiplicity of microbial environments, more progress will be made if specific questions are addressed with particular species.

MATHEMATICAL MODELS AND CHEMOSTATS

Despite the power of chemostats in defining growth rates and nutrient fluxes in organisms (especially at low substrate levels), they are limited in application to rhizosphere studies because of the difficulty in incorporating realistic surfaces. Growth rates on surfaces, and diffusion of nutrients to and from surfaces, probably are major determinants of microbial growth in the rhizosphere. More information on these factors is necessary to construct good models of microbial growth in the rhizosphere.

Model construction is of great value in defining the parameters which need to be investigated and measured in experimental studies. Models can be extremely useful in indicating the relative importance of different plant, microbial, and environmental factors affecting microbial growth in the rhizosphere. However, it is important to find ways to validate the predictions of models of rhizosphere growth.

† Conclusions of the Rhizosphere Round Table at the Third International Symposium on Microbial Ecology, August, 1983. *Panel:* G. D. Bowen (Co-convenor), CSIRO, Australia; R. Campbell, University of Bristol, U.K.; D. C. Coleman, Colorado State University, U.S.A.; M. P. Greaves, ARC Weed Research Organization, U.K.; J. M. Lynch (Co-convenor), Glasshouse Crops Research Institute, U.K.

RELEASE OF CARBON COMPOUNDS

The "roles" of exudates, exfoliates, and other labile materials from roots need further study, as their impact may vary from reducing nutrient supply to the plant (via increased microbial growth) to increasing nutrient availability, to detoxifying metals, such as aluminum, to assisting in defense against soil-borne pathogens. There is little good information on substrate from anything but young laboratory-grown plants. There is a need to study the impact on plants of large losses of assimilated carbon from their roots (up to 30% of the plant's dry matter production).

LABORATORY MEDIA

The test systems to be used in rhizosphere ecology depend on the nature of the study. For some basic studies on mechanisms, highly simplified, gnotobiotic liquid culture or tissue culture systems may be appropriate initially. However, in the end, the system must be referred back to plant roots growing in natural soils. Great care is needed in extrapolating from artificial systems (e.g., agar culture) to soil systems, especially when interactions between organisms and community dynamics are being examined. Even greenhouse and growth room studies do not adequately reflect the field system.

ELECTRON MICROSCOPY

Electron microscopy studies are appropriate to many rhizosphere studies but great care is needed in sample preparation and data interpretation because of many potential artifacts (especially in dehydrated material). There is adequate general descriptive work, and electron microscopes should now be used to answer specific questions. Several techniques developed in other fields will be extremely useful in rhizosphere studies, but these have yet received little attention, e.g., electron probe analysis, high energy dispersive X-ray analysis, enzyme localization. However, every effort should be made to integrate electron microscope findings with data obtained at other levels of study to give a more comprehensive overall picture of the plant–soil interaction.

SITES OF COLONIZATION

Available evidence suggests that lateral colonization from soil of the growing root dominates the composition of the rhizoplane, and this is governed largely by the populations of particular microorganisms and rooting intensity. However, there is a need to elucidate the extent to which migration of existing organisms to new roots can occur in natural soil.

There is a need for increased study of the roles of protozoans and other micro- and mesofauna (such as nematodes, microarthropods) in regulation of microbial population and composition and in mineralization or immobilization of nutrients (e.g., N). There is evidence that protozoans on roots can keep pace with growing root tips, but this is unlikely to occur with fungi and bacteria on actively growing roots. Studies of colonization will be most effective when they focus on particular organisms of interest.

OTHER PHOTOTROPHS

Photoautotrophs other than higher plants (e.g., algae and cyanobacteria) may have a significant impact on microbial activity in soil, because of their abundance in the surface 2 mm and their fixation of CO_2 and sometimes N. Considerable portions of fixed material may be released to soil as both soluble and insoluble substrates. Some photosynthetic bacteria also occur in the rhizosphere. More work is needed on phototrophs, other than higher plants, in soil.

GENERAL CONCLUSIONS

The rhizosphere is the area of microbial influence on plants which has not yet been explored or exploited fully. The full potential of manipulation of microbial populations around roots has yet to be realized in agriculture, horticulture, and forestry. There is a need to formulate specific questions of importance and to use all available techniques and experimental approaches to their solution.

G. D. Bowen and D. C. Coleman prepared the manuscript.

Gastrointestinal Microecology

Physiological Basis for Interactions Among Rumen Bacteria: *Streptococcus bovis* and *Megasphaera elsdenii* as a Model

J. B. RUSSELL AND M. S. ALLEN

U.S. Department of Agriculture and Department of Animal Science, Cornell University, Ithaca, New York 14853

In ruminant animals such as cattle and sheep, feedstuffs are subjected to fermentation in the rumen prior to gastric and intestinal digestion. During this process, microbial cells, organic acids (primarily acetic, propionic, and butyric), carbon dioxide, methane, ammonia, and heat are produced. The animal uses the organic acids and microbial protein as sources of energy and amino acids, respectively, but methane, heat, and ammonia production can cause a loss of energy and nitrogen. The efficiency of ruminant nutrient utilization is dependent on the balance of these products, and this balance is regulated by the types and activities of rumen microorganisms (see 39 and references therein).

Diversity within the rumen microbial ecosystem is very great. Over 200 species of bacteria have been isolated from the rumen, and although some of these species are undoubtedly casual passengers brought in with the feed, more than 25 types are able to achieve counts greater than 10^7/ml. Protozoa are rarely present at greater than 10^6/ml, but they can account for more than half of the microbial mass in the rumen. There are at least 13 genera of rumen protozoa (see 7–9, 15–17).

Few studies have enumerated rumen microorganisms to the species level, but it is clear that relative numbers are not static (9, 11, 46). The greatest fluctuations are observed after abrupt shifts in diet (18, 42), but the proportions can and do change during the course of a single feeding cycle (21, 22). Realistic models of rumen fermentation should seek to explain why these population shifts occur (39).

There are numerous reports in the literature that the numbers of a particular rumen bacterium increased dramatically when a specific diet was fed (16). Most of these observations have been explained by the increased amount of a specific feed component (i.e., pectin, hemicellulose, cellulose, starch, soluble sugars, lactate, etc.) that could be utilized by the bacterium. Such explanations are satisfying because bacterial growth is always dependent on the presence of a suitable energy source, but they do not indicate why other organisms capable of using the same energy source did not proliferate to as great an extent.

The rumen bacteria have evolved together over millions of years, and there are numerous interrelationships among them. Particular species may be involved in commensal, competitive, amensal, mutualistic, and even predator–prey relationships with other rumen microorganisms. Relative growth is dependent on these interactions, and these interactions are ultimately controlled by the physiological characteristics of the individual organisms (39).

The type of energy sources that bacteria can utilize are obviously important, but the rate and efficiency of energy utilization can also be crucial. Success is likewise dependent on the requirement and availability of other essential nutrients. During periods of nutrient starvation or exposure to toxic compounds, resistance becomes an effector of overall growth. If more than one energy source is available, it may be advantageous to select particular ones. Protozoal predation or physical conditions may make attachment to feed particles desirable.

LACTIC ACIDOSIS

Many ruminant animals (especially beef cattle in feedlots) are grown on rations containing greater than 50% starch. High levels of starch are added to the ration to ensure high energy intakes, promote rapid growth, and minimize the maintenance energy of the animal (45). These rations are, however, often associated with indigestion. In extreme cases, rumen ulceration, founder, liver abscess, and even death of the animal can result (41). Early work by Hungate et al. (18) demonstrated that the impairment to animal performance was invariably associated with an initial overgrowth of *Streptococcus bovis*, a low rumen pH, and the inability of lactate-utilizing bacteria to ferment all the lactic acid that was produced. Recent experiments by Counotte et al. (13) indicated that *Megasphaera elsdenii* was the primary lactic acid-fermenting bacterium in starch-fed cattle. In view of the evidence suggesting that population changes of major consequence to the animal may be regulated by relative growth rates of *S. bovis* and *M. elsdenii*, it seems appropriate to consider the physiological basis for interactions between these organisms. The following is thus a discussion of experimental evidence concerning these interactions and the development of a computer model based on known physiological properties of the organisms.

S. BOVIS AND M. ELSDENII

In forage-fed ruminants, *S. bovis* counts are rarely greater than 10^7/ml, but its numbers can increase to 10^{10}/ml if large quantities of starch are added to the diet (1, 2, 18, 42). Many of the rumen bacteria ferment starch (7, 8, 16), so it appears that the relative success of *S. bovis* on starch rations is not solely due to its amylolytic ability. *S. bovis* has a very high maximum growth rate, and it can produce cell material at a faster rate than other rumen bacteria when nutrients are in excess (34). During periods of rapid growth, lactic acid production by *S. bovis* can cause rumen pH to decline to the point where the growth rates of most species of normal rumen bacteria are inhibited and acid-tolerant lactobacilli become predominant (18, 38, 40, 42). *M. elsdenii* can ferment maltose, but it is unable to grow on dextrins or intact starch molecules (18, 26). Increased numbers of *M. elsdenii* on starch-containing rations have usually been explained by its ability to use lactic acid (6, 14, 18).

COMPETITION STUDIES

When *S. bovis* and *M. elsdenii* were cocultured in chemostats with starch as the limiting energy source (37), relative numbers were dependent on both pH and dilution rate (Fig. 1). In general, higher dilution rates were correlated with higher ratios of *S. bovis* to *M. elsdenii*. This was particularly evident at the 0.35 h^{-1} dilution rate and pH greater than 6.0. When the dilution rate was 0.35 h^{-1} and the pH was decreased from 6.0 to 5.2, there was a dramatic decrease in the *S. bovis* to *M. elsdenii* ratio. If the dilution rate was 0.11 or 0.21 h^{-1}, similar decreases in pH resulted in only a small decline in the ratio of *S. bovis* to *M. elsdenii*. At pH values lower than 5.0, a pronounced increase in relative numbers of *S. bovis* occurred at all three dilution rates.

PHYSIOLOGICAL CHARACTERISTICS

When *S. bovis* was grown in batch cultures containing starch, significant amounts of maltose could be detected in cell-free medium (Fig. 2). These data suggested that maltose was the primary extracellular product of starch hydroly-

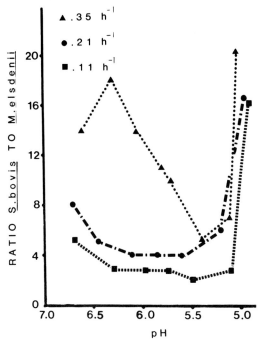

FIG. 1. Relative numbers of *S. bovis* and *M. elsdenii* cells in coculture chemostat incubations with soluble starch as the limiting substrate. Dilution rates were (▲) 0.35 h^{-1}, (●) 0.21 h^{-1}, or (■) 0.11 h^{-1}. From Russell et al. (37).

sis and that maltose would be available to either *S. bovis* or *M. elsdenii* in coculture. Maltose affinities were then determined in separate chemostat incubations, and Lineweaver-Burk plots of residual maltose versus dilution rate indicated that *S. bovis* had a higher affinity (lower affinity constant, K_s) than *M. elsdenii* (Fig. 3a–b).

To estimate the relative efficiency of cell growth on maltose, we determined cell yields in continuous culture (see Fig. 4a–b). Double reciprocal plots of 1/dilution rate versus 1/yield indicated that *M. elsdenii* had a higher maintenance energy expenditure (slope of the plot) than *S. bovis*, but cell yields were similar over the dilution rate range of 0.12 to 0.36 h^{-1} and at pH of 6.6 or 5.7. *S. bovis* had an even lower apparent maintenance energy requirement at pH 5.2, but cell yields were noticeably lower. Decreases in cell yield were due to a decline in the theoretical maximum growth yield (1/intercept to the ordinate). *M. elsdenii* cell yields were similar at pH 5.2, 5.7, and 6.6.

For many years, it was assumed that *S. bovis* carried out a homolactic fermentation (47), but more recent chemostat experiments indicated that it could produce significant quantities of acetate and ethanol if growth rate was slow (36, 37). At pH values above 6.0, *S. bovis* produced

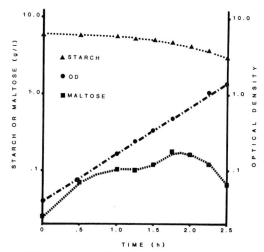

FIG. 2. Growth of *S. bovis* in batch cultures containing soluble starch. Changes in (▲) starch, (●) optical density, and (■) maltose are in $\log_{10}$ scale.

ranged from 4.6 to 4.4. *M. elsdenii*, by contrast, was unable to maintain a growth rate equal to the dilution rate only if the pH was greater than 4.80.

QUALITATIVE ASSESSMENTS

Competition between *S. bovis* and *M. elsdenii* can largely be explained by physiological characteristics of the two organisms. When *S. bovis* and *M. elsdenii* were cocultured on starch (Fig. 1), there were always more *S. bovis* cells. Examination of Michaelis constants (Fig. 3a–b) likewise indicated that *S. bovis* was always a more rapid utilizer of maltose. At pH 6.6 there was a marked increase in the *S. bovis* to *M. elsdenii* ratio as the dilution rate increased (Fig.

little lactate at dilution rates from 0.12 to 0.36 h^{-1} (Fig. 5a). Decreasing the pH below 6.0 resulted in more lactate production, but the ratio of lactate produced to maltose fermented was dependent on the dilution rate. When the dilution rate was 0.36 h^{-1}, there was an exponential increase in lactate as the pH was lowered from 6.0 to 4.95. At dilution rates of 0.12 and 0.21 h^{-1} there was little increase in the lactate to maltose ratio until the pH was less than 5.3.

Lactate fermentations yield less ATP than acetate and ethanol production, and this difference can explain the change in theoretical maximum growth yield for *S. bovis* (see Fig. 4a). When the pH was 5.7 or higher, acetate and ethanol were the predominant products at all dilution rates tested, and more ATP was available to drive cell synthesis (see Fig. 5a). At pH 5.2 lactate production (particularly at faster dilution rates) increased while ATP and cell yields decreased.

The growth of *M. elsdenii* on lactate was once again dependent on both pH and dilution rate (Fig. 5b). When pH was near neutral, higher dilution rates resulted in greater cell yields, but this difference declined as pH within the chemostat vessel was lowered. When the pH was lower than 5.1, cultures grown at the slowest dilution rate (0.12 h^{-1}) exhibited a higher yield of cells than cultures grown at the two higher dilution rates. Cell yields on lactate were always lower than yields from maltose.

Previous reports indicated that *S. bovis* was more resistant to low pH than *M. elsdenii* (36, 40, 44), and such a difference was evident in these experiments. When *S. bovis* was grown on maltose in continuous culture, the washout pH

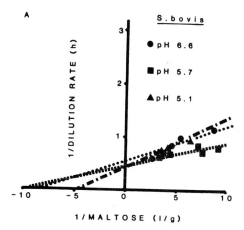

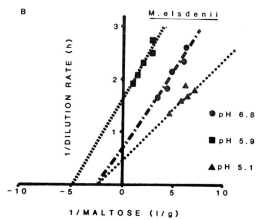

FIG. 3. Lineweaver-Burk transformation of dilution rate versus residual maltose remaining unfermented in the chemostat for: (A) *S. bovis* at pH (●) 6.6, (■) 5.7, and (▲) 5.1; (B) *M. elsdenii* at pH (●) 6.8, (■) 5.9, and (▲) 5.1. Some data points are from Russell et al. (37); other data points are unpublished results.

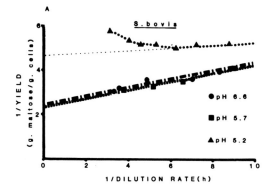

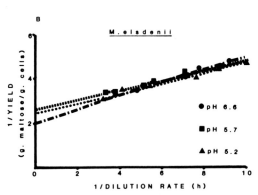

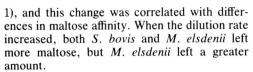

FIG. 4. Maintenance energy plots for growth of (A) *S. bovis* and (B) *M. elsdenii* in maltose-limited chemostats at pH (●) 6.6, (■) 5.7, and (▲) 5.2. Some data points are from Russell et al. (37); other data points are unpublished results.

1), and this change was correlated with differences in maltose affinity. When the dilution rate increased, both *S. bovis* and *M. elsdenii* left more maltose, but *M. elsdenii* left a greater amount.

Above pH 6.0, the cross-feeding of lactate from *S. bovis* to *M. elsdenii* would have been minimal, because *S. bovis* produced little lactate at any of the dilution rates tested (Fig. 5a). When the dilution rate was 0.36 h^{-1}, there was an exponential increase in *S. bovis* lactate production as pH decreased. This increase in lactate production was correlated with a pronounced decrease in the *S. bovis* to *M. elsdenii* ratio. At dilution rates of 0.12 and 0.21 h^{-1}, *S. bovis* produced little lactate until pH 5.2 (Fig. 5a), and from pH 6.6 to 5.2 there was little change in the *S. bovis* to *M. elsdenii* ratio (Fig. 1).

Maltose yields for *S. bovis* and *M. elsdenii* were similar at pH 5.7 and 6.6, but the *S. bovis* yield declined significantly at pH 5.2 (Fig. 4a–b). *M. elsdenii* cell yields from maltose were as high as pH 5.2, but yields from lactate declined at the

two higher dilution rates (Fig. 5b). The pronounced increase in the *S. bovis* to *M. elsdenii* ratio at pH lower than 5.2 (Fig. 1) correlates with differences in pH resistance and chemostat washout.

COMPUTER MODEL

Because the competition between *S. bovis* and *M. elsdenii* was influenced by so many physiological parameters (maltose affinity, maximum growth rate, maintenance energy, yield, lactate production, lactate utilization, and resistance to low pH), it was difficult to make intuitive judg-

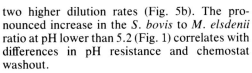

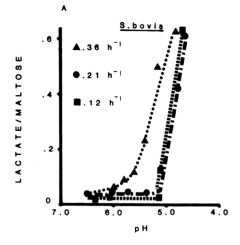

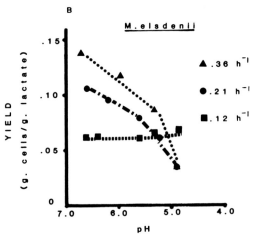

FIG. 5. (A) Ratio of lactate production to maltose utilization by *S. bovis* in chemostats at dilution rates of (▲) 0.36 h^{-1}, (●) 0.21 h^{-1}, and (■) 0.12 h^{-1}. From Russell et al. (37). (B) Yield of *M. elsdenii* from lactate at various pH values (unpublished data). Lactate-limited chemostats were run at dilution rates of (▲) 0.36 h^{-1}, (●) 0.21 h^{-1}, and (■) 0.11 h^{-1}.

TABLE 1. Computer model for the competition of *S. bovis* and *M. elsdenii* in co-continuous culture[a]

```
5 LPRINT "Competition model for S. bovis and M. elsdenii"
10 PH=6.85
20 DIM Z(4)
30 LPRINT "              S. bovis to M. elsdenii ratio"
35 LPRINT " pH          D=.11/h        D=.21/h        D=.35/h"
40 PH=PH-.05
50 DILUTIONRATE=.11:GOTO 80
60 DILUTIONRATE=.21:GOTO 80
70 DILUTIONRATE=.35
80 IF PH=>5.6 THEN STREPSLOPE=.04*PH-.174
90 IF PH<5.6 THEN STREPSLOPE=-.02*PH+.162
100 IF PH=>5.6 THEN STREPINTERCEPT=-.02*PH+.532
110 IF PH<5.6 THEN STREPINTERCEPT=-.18*PH+1.428
120 IF PH=>5.9 THEN MEGASLOPE=.29
130 IF PH<5.9 THEN MEGASLOPE=.1125*PH-.37375
140 IF PH=>5.9 THEN MEGAINTERCEPT=-.133*PH+2.576
150 IF PH<5.9 THEN MEGAINTERCEPT=1.6625*PH-8.019
160 STREPMALTOSELEFT=(1/(1/DILUTIONRATE-STREPINTERCEPT))*STREPSLOPE
170 MEGAMALTOSELEFT=(1/(1/DILUTIONRATE-MEGAINTERCEPT))*MEGASLOPE
180 MALTOSEDIFFERENCE=MEGAMALTOSELEFT-STREPMALTOSELEFT
190 IF DILUTIONRATE=.11 THEN 200 ELSE 210
200 MALTOSEDIFFERENCE11=MALTOSEDIFFERENCE
210 RELATIVEINCREASE=MALTOSEDIFFERENCE/MALTOSEDIFFERENCE11
220 MEGAMALTOSEUSED=100/(RELATIVEINCREASE+1)
230 STREPMALTOSEUSED=MEGAMALTOSEUSED*RELATIVEINCREASE
240 IF DILUTIONRATE=.11 THEN 270 ELSE 250
250 IF DILUTIONRATE=.21 THEN 350 ELSE 260
260 IF DILUTIONRATE=.35   THEN 430
270 LACTATETOMALTOSERATIO=.02+8.712*10^27*EXP(-13.5175*PH)
280 IF PH=>5.5 THEN STREPMALTOSEYIELD=-.0078*PH+.294
290 IF PH<5.5 THEN STREPMALTOSEYIELD=.23125*PH-1.0175
300 IF PH=>5.2 THEN MEGAMALTOSEYIELD=-.0065*PH+.2607
310 IF PH<5.2 THEN MEGAMALTOSEYIELD=.5775*PH-2.772
320 IF PH=>5! THEN MEGALACTATEYIELD=-.0025*PH+.0783
330 IF PH<5! THEN MEGALACTATEYIELD=.35*PH-1.68
340 GOTO 500
350 LACTATETOMALTOSERATIO=.035+3.579*10^24*EXP(-11.97*PH)
360 IF PH=>5.7 THEN STREPMALTOSEYIELD=-.0156*PH+.39
370 IF PH<5.7 THEN STREPMALTOSEYIELD=.2527*PH-1.1315
380 IF PH=>5.2 THEN MEGAMALTOSEYIELD=.0058*PH+.2462
390 IF PH<5.2 THEN MEGAMALTOSEYIELD=.705*PH-3.384
400 IF PH=>5! THEN MEGALACTATEYIELD=.0322*PH-.1032
410 IF PH<5! THEN MEGALACTATEYIELD=.35*PH-1.68
420 GOTO 500
430 LACTATETOMALTOSERATIO=7583*EXP(-1.93*PH)
440 IF PH=>5.7 THEN STREPMALTOSEYIELD=-.02*PH+.453
450 IF PH<5.7 THEN STREPMALTOSEYIELD=.3073*PH-1.4183
460 IF PH=>5.2 THEN MEGAMALTOSEYIELD=.0204*PH+.2086
470 IF PH<5.2 THEN MEGAMALTOSEYIELD=.8025*PH-3.852
480 IF PH=>5! THEN MEGALACTATEYIELD=.04*PH-.1232
490 IF PH<5! THEN MEGALACTATEYIELD=.34*PH-1.632
500 MEGALACTATEUSED=STREPMALTOSEUSED*LACTATETOMALTOSERATIO
505 IF STREPMALTOSEYIELD<=0 THEN STREPMALTOSEYIELD=.00001
510 RELATIVENUMBEROFSTREP=STREPMALTOSEUSED*STREPMALTOSEYIELD
515 IF MEGAMALTOSEYIELD<=0 THEN MEGAMALTOSEYIELD=.00001
520 RELATIVENUMBEROFMEGAMALTOSE=MEGAMALTOSEUSED*MEGAMALTOSEYIELD
525 IF MEGALACTATEYIELD<=0 THEN MEGALACTATEYIELD=.00001
530 RELATIVENUMBEROFMEGALACTATE=MEGALACTATEUSED*MEGALACTATEYIELD
540 RELATIVENUMBEROFMEGA=RELATIVENUMBEROFMEGAMALTOSE+RELATIVENUMBEROFMEGALACTATE
550 SIZEADJUST=2.5
560 STREPTOMEGARATIO=(RELATIVENUMBEROFSTREP/RELATIVENUMBEROFMEGA)*SIZEADJUST
570 IF DILUTIONRATE=.11 THEN 580 ELSE 590
580 Z(1)=PH:Z(2)=STREPTOMEGARATIO:GOTO 60
590 IF DILUTIONRATE=.21 THEN 600 ELSE 610
600 Z(3)=STREPTOMEGARATIO:GOTO 70
610 Z(4)=STREPTOMEGARATIO
620 LPRINT Z(1),Z(2),Z(3),Z(4)
630 IF PH 4.8 THEN 40 ELSE 640
640 END
```

[a] BASIC—80 REV. 5.21, Microsoft Corp., Bellevue, WA 98004.

TABLE 2. Glossary of variable names in computer model shown in Table 1

STREPSLOPE: Slope of the Lineweaver-Burk plot shown in Fig. 3a.
STREPINTERCEPT: Intercept to the ordinate of the Lineweaver-Burk plot shown in Fig. 3a.
MEGASLOPE: Slope of the Lineweaver-Burk plot shown in Fig. 3b.
MEGAINTERCEPT: Intercept to the ordinate of the Lineweaver-Burk plot shown in Fig. 3b.
STREPMALTOSELEFT: Amount of maltose remaining unfermented in *S. bovis* continuous cultures.
MEGAMALTOSELEFT: Amount of maltose remaining unfermented in *M. elsdenii* continuous cultures.
MALTOSEDIFFERENCE: Difference in amounts of maltose left unfermented by *M. elsdenii* and *S. bovis*.
RELATIVEINCREASE: Relative increase in the difference in maltose left by each species compared to dilution
    rate 0.11 h^{-1}.
MEGAMALTOSEUSED: Amount of maltose used by *M. elsdenii*.
STREPMALTOSEUSED: Amount of maltose used by *S. bovis*.
LACTATETOMALTOSERATIO: Ratio of lactate production to maltose fermentation by *S. bovis*.
STREPMALTOSEYIELD: Dry-weight cell yield of *S. bovis* on maltose.
MEGAMALTOSEYIELD: Dry-weight cell yield of *M. elsdenii* on maltose.
MEGAGALACTATEYIELD: Dry-weight cell yield of *M. elsdenii* on lactate.
RELATIVENUMBEROFSTREP: Mass of *S. bovis* cells in coculture.
RELATIVENUMBEROFMEGAMALTOSE: Mass of *M. elsdenii* cells that were produced from maltose in
    coculture.
RELATIVENUMBEROFMEGALACTATE: Mass of *M. elsdenii* cells that were produced from lactate in
    coculture.
RELATIVENUMBEROFMEGA: Total mass of *M. elsdenii* cells that were produced in coculture.
SIZEADJUST: Adjustment factor to convert ratio of cell mass to ratio of cell numbers.
STREPTOMEGARATIO: Ratio of *S. bovis* and *M. elsdenii* cells in coculture.

ments about the relative importance of each one. To assess the data more completely, we constructed a computer model (see Tables 1 and 2). Equations to run the model were derived from the data in Fig. 3 through 5.

Amounts of maltose remaining unfermented in the chemostat vessel were calculated from the Michaelis constants shown in Table 1 (lines 80 to 150). Discrete values of K_s/k_{max} and $1/k_{max}$ were taken from Fig. 3, and intermediate pH values were extrapolated linearly. As the dilution rate increased, both *S. bovis* and *M. elsdenii* left more maltose (lines 160 to 170), but the amount left by *M. elsdenii* was always greater (line 180). By comparing amounts of maltose left (lines 190 to 200), it was possible to estimate the relative increase in maltose that would be left by each species (line 210). Assuming that residual maltose was an indication of actual affinity (35), and assuming that the difference in maltose left represented the relative increase in maltose that would be available to *S. bovis*, it was then possible to derive the amount of maltose that was used by each species (lines 220 to 230).

Maltose fermentation by *S. bovis* sometimes yielded lactate that was in turn available to *M. elsdenii*. Lactate production was dependent on pH and dilution rate (Table 1, lines 270, 350, 430) as well as the amount of maltose that was fermented by *S. bovis* (line 500). Since *S. bovis* was unable to ferment lactate, and because lactate was never detected in coculture until *M. elsdenii* washed out, it was not necessary to determine lactate affinity constants.

Bacterial growth is always dependent on the presence of a suitable energy source, but utilization is not always proportional to growth. Variations in cell yields are caused by differences in ATP synthesis or maintenance energy (29, 43). Equations describing the effect of pH on maltose yields were developed from the data in Fig. 4. With *S. bovis*, yield remained constant until pH 5.7 or 5.5, but then declined linearly as pH approached chemostat washout (Table 1, lines 280, 290, 360, 370, 440, 450). *M. elsdenii* yields from maltose remained constant until pH 5.2, and we assumed that yield would once again decline linearly to zero as pH decreased from 5.2 to chemostat washout (lines 300, 310, 380, 390, 460, 470). *M. elsdenii* yields from lactate (Fig. 5b) were represented by pairs of linear equations that were associated with specific pH ranges (lines 320, 330, 340, 410, 480, 490). Relative numbers of *S. bovis* and *M. elsdenii* cells were computed by multiplying maltose or lactate utilization and the appropriate yield term (lines 510 to 540).

Relative numbers in the coculture incubations (Fig. 1) were based on direct microscopic counts (number) rather than cell weight. To test the model with the data in Fig. 1, we used an adjustment factor (Table 1, line 550). The number 2.5 was used because *M. elsdenii* is 2.5 times as large as *S. bovis* (16, 27).

Computer simulation of the competition between *S. bovis* and *M. elsdenii* (Fig. 6) was in most cases comparable to the actual data (Fig. 1). Some deviation between computed and actual data was observed above pH 6.0, but the model adequately predicted the large change in *S. bovis* to *M. elsdenii* ratio at pH lower than 6.0 and dilution rate 0.35 h^{-1}. There was only a slight overprediction of the pH at which *M. elsdenii* washed out of the co-continuous culture.

SENSITIVITY ANALYSES

The model also enabled us to test the impact of each physiological characteristic on the overall competition. When the amount of maltose left by either *S. bovis* or *M. elsdenii* was increased by 10%, there was little change (less than 0.20 U) in the predicted ratio of the two organisms. A 10% increase in lactate production from maltose fermentation by *S. bovis* had virtually no influence at the two lowest dilution rates, but the *S. bovis* to *M. elsdenii* ratio was decreased by 0.28 U at pH 5.9 and dilution rate 0.35 h^{-1}. Similar increases in either *S. bovis* or *M. elsdenii* yield from maltose resulted in a change as great as 1.4 U, with greatest effect at faster dilution rates. *M. elsdenii* yield from lactate was less sensitive to a 10% increase; the relative numbers of *M. elsdenii* increased by less than 0.25 U. The model was most affected by the minimum pH at which *M. elsdenii* could grow. Any change in the point where *M. elsdenii* cell yields declined to zero had a pronounced influence on relative numbers, provided that pH was lower than 5.0.

SIMPLIFICATIONS

Computations predicting the relative numbers of *S. bovis* and *M. elsdenii* in coculture considered a variety of factors, but some simplifications were made. All incubations were conducted in media that contained 0.5 g of yeast extract, 1.0 g of Trypticase, and 2.0 g of acetate per liter. Rather rich medium was used because Forsberg demonstrated that these nutrients may be required by *M. elsdenii* (14). *S. bovis*, by contrast, has simple nutritional requirements and is able to grow in simple salts medium that contains ammonia (10, 48). Growth factor limitation is more apt to limit *M. elsdenii*, and during such times the balance could be shifted in favor of *S. bovis*.

In coculture maltose (resulting from starch hydrolysis) was the only energy source available to *S. bovis*, but *M. elsdenii* was able to use lactate or maltose. Bacteria often use some energy sources to the exclusion of others (23, 28, 33), and preferential utilization of either lactate or maltose by *M. elsdenii* could have affected overall partitioning of energy sources between *S. bovis* and *M. elsdenii*. Such complications were unlikely because previous experiments indicated that *M. elsdenii* was capable of utilizing maltose and lactate simultaneously (34).

It is well documented that rumen protozoa ingest and digest a significant number of rumen bacteria, and some investigations have suggest-

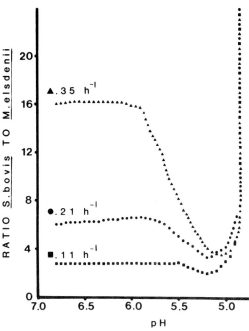

FIG. 6. Computer simulation of the growth of *S. bovis* and *M. elsdenii* in co-continuous culture with starch as the limiting substrate at dilution rates of (▲) 0.35 h^{-1}, (●) 0.21 h^{-1}, and (■) 0.11 h^{-1}.

ed that selective ingestion could alter the balance of rumen bacteria (12). Rumen protozoa are, however, very sensitive to low pH and are usually found in low numbers on starch rations (18, 19). Given this, the absence of protozoa in our in vitro experiments does not represent a serious deviation from reality.

The rumen is often used as an example of a natural chemostat, but there are kinetic differences between the two systems. In a chemostat, nutrients are present in a soluble form, and all components including bacteria pass out at the same dilution rate. In the rumen there are at least two major dilution rates. The solid fraction (large feed particles) turns over at a slower rate than the liquid and small-particle pool (45). Rumen bacteria are found in both phases, but a majority (approximately three-fourths) are associated with the solid fraction. Latham studied attachment of rumen bacteria and found *S. bovis* and *M. elsdenii* almost exclusively in the rumen liquor, where conditions would more closely resemble a laboratory chemostat (20).

It is generally assumed that there is minimal bacterial attachment in a laboratory chemostat, but "wall growth" is an insidious problem that can affect kinetic interpretations. At low dilution rates (less than 0.3 h^{-1}) little wall growth was observed, but at higher dilution rates thin films of bacteria were sometimes observed on the chemostat vessel. Wall growth may have contributed to high theoretical maximum growth rates for *M. elsdenii* (Fig. 3b). Complications from wall growth in the computer model may have been minimized because wall growth was also possible in coculture.

Even though the rumen usually contains an abundance of solid feed, many soluble energy sources are present at low or negligible levels during a portion of the feeding cycle (39). Recently, Mink and Hespell examined the effect of energy source starvation in rumen bacteria and found that *M. elsdenii* was able to maintain significant levels of viability for greater than 13 h (24, 25). The viability of *S. bovis* has not been studied as critically, but standard laboratory reinoculation procedures indicate that *S. bovis* is also able to tolerate significant periods of starvation. Considering the apparent resistance of each organism, it is doubtful that energy source starvation would play a significant role in their interaction.

OTHER MODELS

With the development of computer technology over the past two decades, it has been possible to integrate simultaneously a large number of variables, and these techniques have been readily adapted to the chemical and physical sciences. Complexity is often as great (or greater) in biology, but the use of computers has not received such wide acceptance. Baldwin and his colleagues applied simulation techniques to feed digestion in the rumen and constructed several dynamic models of the processes involved (3–5, 30–32). In one of these models, the rumen microflora was subdivided into eight metabolic groups, but during solutions, "considerable simplification . . . occurred" (31). The authors concluded that "current data and concepts . . . do not accommodate adequately competitions among the several rumen microbial species and, thus that additional data and concepts regarding microbial interactions are required."

CONCLUSIONS

As pointed out by Hungate in 1960, a complete analysis of any natural habitat requires an elaboration of: (i) the kinds and numbers of organisms that are present, (ii) the activities of the organisms, and (iii) the extent to which the activities are expressed (15). Since the first isolations of strictly anaerobic bacteria, most, if not all, of the predominant rumen bacteria have been isolated, and pure culture studies have given us a good indication of their activities. The extent to which these activities are expressed is largely determined by their physiological characteristics.

The present model (Table 1), although simplified when compared to the in vivo rumen, indicates that physiological characteristics can indeed provide realistic assessments of the interactions of rumen bacteria. The problem at this time lies in the biology, not the mathematics. Few experiments have described and quantitated the physiology of the various rumen bacteria.

This research was supported by the United States Dairy Forage Research Center.

LITERATURE CITED

1. **Allison, J. J.** 1976. Population control in the rumen and microbial adaptation to change in diet, p. 10–18. *In* M. S. Weinberg and A. L. Sheffner (ed.), Buffers in ruminant physiology and metabolism. Church and Dwight Co., Inc., New York.
2. **Allison, M. J., J. A. Bucklin, and R. W. Dougherty.** 1964. Ruminal changes after overfeeding with wheat and the effect of intraruminal inoculation on adaptation to a ration containing wheat. J. Anim. Sci. **23**:1164–1171.
3. **Baldwin, R. L., and L. J. Koong.** 1977. A dynamic model of ruminant digestion or evaluation of factors affecting nutritive value. Agric. Syst. **2**:255–274.
4. **Baldwin, R. L., L. J. Koong, and M. J. Ulyatt.** 1977. The formation and utilization of fermentation end-products: mathematical models. *In* R. T. J. Clark and T. Bauchop (ed.), Microbial ecology of the gut. Academic Press, Inc., New York.
5. **Baldwin, R. L., H. L. Lucas, and R. Cabrera.** 1969. Energetic relationships in the formation and utilization of fermentation end-products, p. 319–334. *In* A. T. Phillipson (ed.), Physiology of digestion and metabolism in the ruminant. Oriel Press, Newcastle upon Tyne, England.
6. **Baldwin, R. L., W. A. Wood, and R. S. Emery.** 1965. Lactate metabolism by *Peptostreptococcus elsdenii*: evidence for lactyl coenzyme A dehydrase. Biochim. Biophys. Acta **97**:202–213.
7. **Bryant, M. P.** 1959. Bacterial species of the rumen. Bacteriol. Rev. **23**:125–153.
8. **Bryant, M. P.** 1963. Symposium on microbial digestion in ruminants: identification of groups of anaerobic bacteria active in the rumen. J. Anim. Sci. **22**:801–813.
9. **Bryant, M. P., and L. A. Burkey.** 1953. Numbers and some predominant groups of bacteria in the rumen of cows fed different rations. J. Dairy Sci. **36**:218–224.
10. **Bryant, M. P., and I. M. Robinson.** 1961. Studies on the nitrogen requirements of some ruminal cellulolytic bacteria. Appl. Microbiol. **9**:96–103.
11. **Bryant, M. P., and I. M. Robinson.** 1968. Effects of diet, time after feeding, and position sampled on numbers of viable bacteria in the bovine rumen. J. Dairy Sci. **51**:1950–1955.
12. **Coleman, G. S.** 1980. Rumen ciliate protozoa. Adv. Parasitol. **18**:121–173.
13. **Counotte, G. H. M., R. A. Prins, R. A. M. Janssen, and M. J. A. de Bie.** 1981. Role of *Megasphaera elsdenii* in the fermentation of DL-[2-¹³C]lactate in the rumen of dairy cattle. Appl. Environ. Microbiol. **42**:649–655.
14. **Forsberg, C. W.** 1978. Nutritional characteristics of *Megasphaera elsdenii*. Can. J. Microbiol. **24**:981–985.
15. **Hungate, R. E.** 1960. Microbial ecology of the rumen. Bacteriol Rev. **24**:353–364.
16. **Hungate, R. E.** 1966. The rumen and its microbes, p. 8–90 and p. 245–280. Academic Press, Inc., New York.
17. **Hungate, R. E.** 1975. The rumen microbial ecosystem. Annu. Rev. Ecol. Syst. **6**:39–66.
18. **Hungate, R. W., R. W. Dougherty, M. P. Bryant, and R. M. Cello.** 1952. Microbial and physiological changes

associated with acute digestion in sheep. Cornell Vet. **42:**423–449.

19. **Hungate, R. E., M. P. Bryant, and R. A. Mah.** 1964. The rumen bacteria and protozoa. Annu. Rev. Microbiol. **18:**131–166.

20. **Latham, M. J.** 1980. Adhesion of bacteria to plant cell walls, p. 339–350. *In* R. Berkely, A. Lych, J. Melling, P. Rutter, and B. Vincent (ed.), Microbial adhesion to surfaces. Ellis Horwood Ltd., West Sussex, England.

21. **Mackie, R. I., and F. M. C. Gilchrist.** 1979. Changes in lactate-producing and lactate-utilizing bacteria in relation to pH in the rumen of sheep during stepwise adaptation to a high-concentrate diet. Appl. Environ. Microbiol. **38:**422–430.

22. **Mackie, R. I., and S. Heath.** 1979. Enumeration and isolation of lactate-utilizing bacteria from the rumen of sheep. Appl. Environ. Microbiol. **38:**416–421.

23. **Megasanik, B.** 1961. Catabolite repression. Cold Spring Harbor Symp. Quant. Biol. **26:**249–256.

24. **Mink, R. W., and R. B. Hespell.** 1981. Survival of *Megasphaera elsdenii* during starvation. Curr. Microbiol. **5:**51–56.

25. **Mink, R. W., and R. B. Hespell.** 1981. Long-term nutrient starvation of continuously cultured (glucose-limited) *Selenomonas ruminantium.* J. Bacteriol. **148:**541–550.

26. **Miura, H., M. Horiguchi, and T. Matsumoto.** 1980. Nutritional interdependence among rumen bacteria, *Bacteroides amylophilus, Megaspaera elsdenii,* and *Ruminococcus albus.* Appl. Environ. Microbiol. **40:**294–300.

27. **Ogimoto, K., and S. Imai.** 1981. Atlas of rumen microbiology, p. 71–124. Japan Scientific Societies Press, Tokyo.

28. **Peterkofsky, A.** 1976. Cyclic nucleotides in bacteria. Adv. Cyclic Nucleotide Res. **7:**1–48.

29. **Pirt, S. J.** 1965. The maintenance energy of bacteria in growing cultures. Proc. R. Soc. Lond. Ser. B **163:**224–231.

30. **Reichl, J. R., and R. L. Baldwin.** 1975. Rumen modeling: rumen input-output balance models. J. Dairy Sci. **58:**879–890.

31. **Reichl, J. R., and R. L. Baldwin.** 1976. A rumen linear programming model for evaluation of concepts of rumen microbial function. J. Dairy Sci. **59:**439–454.

32. **Rice, R. W., J. G. Morris, B. T. Maeda, and R. L. Baldwin.** 1974. Simulation of animal functions in models of production systems: ruminants on the range. Fed. Proc. **33:**188–195.

33. **Rickenberg, H. V.** 1974. Cyclic AMP in prokaryotes. Nucleic Acid Res. **17:**353–369.

34. **Russell, J. B., and R. L. Baldwin.** 1978. Substrate preferences in rumen bacteria: evidence of catabolite regulatory mechanisms. Appl. Environ. Microbiol. **36:**319–329.

35. **Russell, J. B., and R. L. Baldwin.** 1979. Comparison of substrate affinities among several rumen bacteria: a possible determinant of rumen bacterial competition. Appl. Environ. Microbiol. **37:**531–536.

36. **Russell, J. B., and R. L. Baldwin.** 1979. Comparison of maintenance energy expenditures and growth yields among several rumen bacteria grown in continuous culture. Appl. Environ. Microbiol. **37:**537–543.

37. **Russell, J. B., M. A. Cotta, and D. B. Dombrowski.** 1981. Rumen bacterial competition in continuous culture: *Streptococcus bovis* versus *Megasphaera elsdenii.* Appl. Environ. Microbiol. **41:**1394–1399.

38. **Russell, J. B., and D. B. Dombrowski.** 1980. Effect of pH on the efficiency of growth by pure cultures of rumen bacteria in continuous culture. Appl. Environ. Microbiol. **39:**604–610.

39. **Russell, J. B., and R. B. Hespell.** 1981. Microbial rumen fermentation. J. Dairy Sci. **64:**1153–1169.

40. **Russell, J. B., W. M. Sharp, and R. L. Baldwin.** 1979. The effect of pH on maximum bacterial growth rate and its possible role as a determinant of bacterial competition in the rumen. J. Anim. Sci. **48:**251–255.

41. **Slyter, L. L.** 1976. Influence of acidosis on rumen function. J. Anim. Sci. **43:**910–929.

42. **Slyter, L. L., M. P. Bryant, and M. J. Wolin.** 1966. Effect of pH on population and fermentation in a continuously cultured rumen ecosystem. Appl. Microbiol. **14:**573–578.

43. **Stouthamer, A. H.** 1979. The search for correlation between theoretical and experimental growth yields, p. 1–47. *In* J. R. Quayle (ed.), International review of biochemistry. Microbial biochemistry, vol. 21. University Park Press, Baltimore.

44. **Therion, J. A., A. Kistner, and J. H. Kornelius.** 1982. Effect of pH on growth rates of rumen amylolytic and lactylic bacteria. Appl. Environ. Microbiol. **44:**428–431.

45. **Van Soest, P. J.** 1982. Nutritional ecology of the ruminant, p. 212–231 and 294–309. O and B Books, Inc., Corvallis, Ore.

46. **Warner, A. C. I.** 1962. Enumeration of rumen microorganisms. J. Gen. Microbiol. **28:**119–128.

47. **Wolin, M. J.** 1964. Fructose-1,6-diphosphate requirement of streptococcal lactic dehydrogenases. Science **146:**775–777.

48. **Wolin, M. J., G. B. Manning, and W. O. Nelson.** 1959. Ammonium salts as a sole source of nitrogen for the growth of *Streptococcus bovis.* J. Bacteriol. **78:**147–149.

Adaptations of Gastrointestinal Bacteria in Response to Changes in Dietary Oxalate and Nitrate†

M. J. ALLISON AND C. ADINARAYANA REDDY

*National Animal Disease Center, Agricultural Research Service, U.S. Department of Agriculture, Ames,
Iowa 50010; and Department of Microbiology and Public Health, Michigan State University, East Lansing,
Michigan 48824*

Interactions between gastrointestinal microbes and their host animal can have profound effects on the well-being of the host. The result of a given microbe's activities may be harmful or even lethal to the host. Other microbes, through their own struggle for survival, perform beneficial functions, often to the extent that their activities are critical to the ecology and survival of the host. Numerous examples of beneficial microbial activities have been obtained in studies of the rumen, which is the best known of gastrointestinal ecosystems. Metabolic activities similar to those carried out by rumen microbes are also found in other gut environments, and the appreciation of this increases with knowledge of these habitats.

Although interactions between gut microbes and toxic substances have long been recognized, these have, in general, received little detailed study. The importance of the role of microbes in gastrointestinal detoxifications is readily appreciated in the rumen, where dietary substances are subjected to microbial attack prior to gastric and small intestinal digestion. In this regard, ruminal microbes may perhaps be considered as a first line of defense for the animal (1).

In this paper we consider the degradation of a toxic substance (oxalic acid) by a new group of anaerobic bacteria that were first isolated from the rumen and are now known to be present in other gut environments. We also discuss the activities of the ruminal population in the production of a toxic substance, nitrite (NO_2^-), from dietary nitrate (NO_3^-), and the subsequent detoxication of NO_2^- by its further reduction to NH_4^+. In both cases, microbial populations in the gut change substantially in response to dietary intake of specific substrates, and the changes are adaptations of great benefit to the host animal.

OXALIC ACID IN PLANTS AND ANIMALS

A wide variety of plants eaten by humans and animals contain low levels (<1 g/100 g) of oxalic acid. Other plants considered as high oxalic acid producers contain >5 g of oxalic acid per 100 g

(dry weight). Oxalic acid may be present in plants as the free acid, but is more frequently present as a soluble (sodium, potassium, or ammonium) or insoluble (especially calcium) salt. The oxalate content of plants may vary considerably with anatomic site, plant age, and nutritional parameters. Its role may vary with plant species and environmental conditions from that of a useless product of metabolism to a useful excretory product or a substance used as a nutrient or energy reserve. Oxalate is also a common and major metabolic product of various fungi, and calcium oxalate crystals have been observed in algae from several orders (20).

Animal tissues are exposed to oxalate arising from endogenous tissue metabolism of various substrates (e.g., glycine, serine, glyoxylate, and ascorbic acid) as well as to oxalate in the diet. Our concern here is with the interaction that occurs between gastrointestinal microbial populations and dietary oxalate.

Toxicity of oxalic acid centers upon the formation in tissues of calcium oxalate salts of low solubility, especially the monohydrate (Whewellite, Ksp. 37°C = 8.6). Pathological signs include physical damage to tissues by these crystals as well as tetany in acute toxicity due to complexing with, and removal of, circulating ionized calcium (20, 24). Extensive deposits of calcium oxalate crystals in renal tissue are a usual sign of toxicity, and crystal formation in absorptive tissues may lead to hemorrhagic gastroenteritis.

ADAPTATION OF RUMINAL MICROBES TO OXALATE

Ruminants fed gradually increasing amounts of oxalate adapt to and develop a tolerance to quantities of oxalate that would be lethal to animals that have not been adapted (24). This adaptation is due to development of an increased capacity for oxalate degradation by ruminal microbes (31, 44). There is evidence that the increased rates of oxalate degradation associated with adaptation lead to increased absorption of calcium from the calcium oxalate in tropical grasses (6).

The transition from low (<0.05 μmol/ml per h) to high (>0.2 μmol/ml per h) oxalate degradation

† Article no. 11125 from the Michigan Agricultural Experiment Station.

248

rates by microbes in rumen fluid from sheep fed gradually increasing amounts of *Halogeton glomeratus* (halogeton; 12 to 15% oxalic acid as the K salt) occurred in 3 to 4 days. A similar increase in oxalate degradation rates occurs when oxalate is fed as sodium oxalate rather than as halogeton. This transition is not accompanied by changes that can be detected by microscopic examinations of the mixed population, nor are there detectable related changes in volatile fatty acids in the rumen fluid (4).

Tests with fractions of the mixed ruminal population indicated that oxalate was degraded by bacteria rather than by rumen protozoa. Although aerobic oxalate-degrading bacteria have been isolated from the rumen (32, 35), no evidence was given that the isolates were active in the rumen. We reasoned that it was unlikely that aerobes were important as oxalate degraders in the rumen in view of the anaerobic nature of the ruminal environment and the substantial evidence that most of the functional rumen bacteria are obligate anaerobes. Furthermore, our comparisons of [^{14}C]oxalate degradation rates by mixed cultures from the rumen incubated anaerobically versus cultures exposed to oxygen (4) also supported the concept that the active agent was an anaerobe.

None of the rumen bacteria tested from an extensive culture collection (including many of the organisms described by M. P. Bryant and associates) degraded oxalate. Neither was oxalate degraded by any of 99 strains isolated from medium 98-5 (12) as the predominant organisms from the rumen of a sheep adapted to a diet containing halogeton.

When an in vitro continuous culture system (3) inoculated with rumen contents from sheep fed alfalfa hay was fed gradually increasing amounts of oxalate, rates of oxalate degradation increased 10- to 100-fold, reaching 2 to 9 μmol/ml per h. Measurements from one of the in vitro adaptation experiments are shown in Fig. 1. When high levels of oxalate were fed abruptly (rather than increasing gradually over a 3- to 7-day period), this selection of a population with increased oxalate degradation rates was not observed. Methanogenesis in the mixed continuous culture population was positively correlated with amounts of oxalate degraded, and inhibitors of methanogenesis, such as the hemiacetal of chloral and starch, also inhibited oxalate degradation (Fig. 1). In one experiment an oxalate-degrading population was maintained when the oxalate feed rate was 6.25 μmol/ml per h, but at the same dilution rate (d) (0.08 h^{-1}) when oxalate feed was changed to 8.3 μmol/ml per h, the capacity of the system was exceeded and the oxalate degradation rate rapidly decreased.

One liter of contents from an oxalate-adapted fermentor was placed into the rumen of a sheep that had not been adapted to oxalate. At 0.25 and 24 h after this inoculation, halogeton and alfalfa (400 g each) were placed in the rumen. The halogeton supplied a dosage of 1.1 g of oxalic acid per kg of body weight, which is approximately the lethal dose for a sheep that has not been conditioned to oxalate (24). Inoculation of the rumen of this sheep with fermentor contents, however, apparently substituted for the gradual adaptation to oxalate, and signs of toxicity were not noted even though the "le-

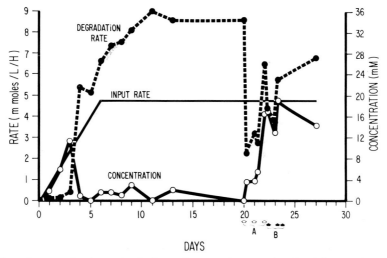

FIG. 1. Adaptation to oxalate by a mixed culture of rumen microorganisms in a 1-liter continuous culture (d = 0.062 h^{-1}). The fermentor was fed alfalfa (25 g/day), and at points labeled A, 30 mg of the hemiacetal of chloral and starch (Smith Kline and French Laboratories, Philadelphia, Pa.) was added to the fermentor. At points labeled B, twice that amount was added. From Allison et al. (3).

thal'' dose of halogeton was fed on 2 successive days. Measurements of oxalate degradation rates in rumen fluid from this sheep are shown in Fig. 2.

An enrichment culture with 45 mM oxalate as the principal energy source was established by Dawson et al. (15) from an adapted continuous culture. In this culture, 1 mol of CH_4 was produced per 3.8 mol of oxalate degraded. Hydrogen and formate inhibited oxalate degradation, but not methanogenesis by this culture, whereas inhibitors of methanogenesis (dilute chloroform and benzyl viologen) inhibited oxalate degradation. When the enrichment was cultured in a continuous culture system at $d = 0.078$ h^{-1}, a mixed population that degraded oxalate but no longer produced CH_4 was obtained. With this simplified mixed population, benzyl viologen, H_2, or formate no longer inhibited oxalate degradation. The first isolation of an obligately anaerobic, oxalate-degrading gastrointestinal bacterium (isolate OxB) was subsequently obtained from this population (14).

The relationship between CH_4 production and oxalate degradation in the mixed cultures is partially explained by the fact that the oxalate degrader OxB produces approximately 1 mol of formate per mol of oxalate degraded. This, however, does not explain the inhibition of oxalate degradation by benzyl viologen, H_2, or formate when, and only when, methanogenesis was "coupled" to oxalate degradation. One possibility is that predominant oxalate degraders in the "coupled" population differ from isolate OxB

and that the former exist in a syntrophic relationship with methanogens that is analogous to the syntrophisms described by Bryant and co-workers (29, 30).

A number of the properties of isolate OxB fit well with our observations of adaptations by ruminants to dietary oxalate and with most of our other results with mixed cultures. This gram-negative, obligate anaerobe failed to use any of a wide variety of substrates other than oxalate. Growth rates as high as $\mu = 0.3\ h^{-1}$ and cell yields of 1.1 g/mol of oxalate have been observed. The limited substrate range indicates that selection of increased numbers of oxalate degraders would indeed be closely regulated by oxalate availability. The inhibition of oxalate degradation in mixed cultures fed high levels of oxalate, as well as experiments with high levels of oxalate in isolation media, suggests, however, that most ruminal oxalate degraders are unable to grow well at the high levels of oxalate ($\geqq100$ mM) that are well tolerated by isolate OxB. The stoichiometry of CH_4 production by mixed cultures agrees with data indicating that OxB produces approximately 1 mol of formate per mol of oxalate degraded (14). These oxalate degraders do not appear to fit into any existing taxonomic group, and a proposal that a new genus and species be established is in preparation.

Studies with isolate OxB indicated that oxalate-degrading bacteria could be detected in anaerobic roll tubes on the basis of formation of cleared zones around colonies when the culture medium contained calcium oxalate. Work is in

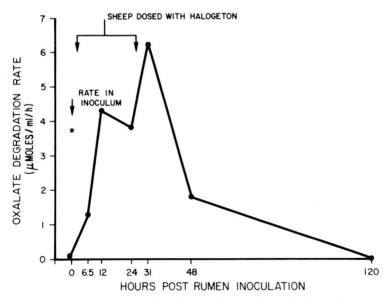

FIG. 2. Oxalate degradation rates in rumen fluid from a sheep (45 kg) that had been inoculated with 1 liter of fluid from an oxalate-adapted fermentor. Halogeton (400 g) was subsequently placed in the rumen at times as indicated by the arrows.

progress (S. L. Daniel, M. J. Allison, and H. M. Cook) to use this method to determine concentrations of oxalate-degrading bacteria in the rumen. Preliminary results indicated that oxalate degraders were present at 10^5 to 5×10^5/g in rumen contents of a sheep that was adapted to halogeton and that had an oxalate degradation rate of 0.8 to 3.3 μmol/g per h. Further work is needed to determine the optimal levels of oxalate in media used for such tests and to determine the reliability of such estimates of numbers of oxalate degraders.

OXALIC ACID DEGRADATION IN NONRUMINAL GUT HABITATS

Anaerobic, oxalate-degrading bacteria are not limited to the ruminal environment. Shirley and Schmidt-Nielsen (40) presented evidence that oxalate was extensively degraded by gut microbes in certain desert rodents whose diets contained plants with high levels of oxalate. We found that rates of oxalate degradation by microbes in large bowel contents from rabbits, guinea pigs, swine, and a horse increased markedly when the oxalate content of the diet increased (2). With laboratory rats, however, this change in response to dietary oxalate was not observed. Further work indicates that wild rats harbor gastrointestinal oxalate degraders and that these will colonize the gut of white laboratory rats (S. L. Daniel, M. J. Allison, and P. A. Hartman, Abstr. Annu. Meet. Am. Soc. Microbiol. 1983, I118, p. 159).

Oxalate degradation by bacteria in human feces was demonstrated in 1940 (5), but this activity has received little attention since then. This paucity of information is somewhat surprising in view of the importance of absorbed oxalate in urinary stone disease and of evidence that the colon is the main site of oxalate absorption.

We found a wide range of oxalate degradation rates (0.1 to 4.8 μmol/g per h) by bacterial populations in fecal samples from humans on normal but uncontrolled diets. Oxalate-degrading bacteria isolated from human feces (M. J. Allison and H. M. Cook, Abstr. Int. Congr. Microbiol., 23rd, 1982, p. 84) resemble isolate OxB in many respects. As they use few, if any, substrates other than oxalate, populations of these organisms in the human colon are probably limited by the availability of oxalate. We propose that these bacteria may influence concentrations of soluble oxalate at absorptive sites, and thus the amount of dietary oxalate that is absorbed. If this is true, it is logical to further propose that loss of the function of these organisms may be an important factor contributing to hyperoxaluria associated with enteric resections and disease (enteric hyperoxaluria). Further

studies of these organisms and their role in the human gut are clearly needed.

ADAPTATION OF DIETARY NITRATE BY RUMINANTS

Ruminant intoxication by nitrates present in vegetation has been recognized for at least 50 years (7, 8, 37), and the frequency of reports in current literature indicates that nitrate (NO_3^-) poisoning is still a common problem worldwide. Oat hay, sorghums, sugar beet tops, corn forage, and various grasses and weeds are among the plants that at times contain quantities of NO_3^- that are sufficient to cause intoxication (36, 41). Animal husbandry decisions are complicated by the fact that the NO_3^- content of plants may vary greatly with stage of growth, moisture availability, fertilization level, and treatment with herbicides.

Sapiro et al. (39) showed that NO_3^- was reduced by ruminal microbes to nitrite (NO_2^-). Lewis (26) found that NO_3^- could subsequently be reduced to ammonia, but that the accumulation and absorption of the intermediate, NO_2^-, could lead to production of methemoglobin, causing a reduction of the oxygen-transporting capacity of blood. The relative rates of reduction of NO_3^- and NO_2^- are thus critical. Factors which are probably most important in regulating these rates include the availability of substrates yielding hydrogen for the reduction, physiological conditions (e.g., pH) in the rumen, and, of course, the enzymatic capacity of the mixed microbial population.

Carbohydrates, formate, pyruvate, succinate, lactate, citrate, malate, and glycine are among the substrates that can be utilized for the reduction of NO_3^- or NO_2^-, or both (21, 25, 27). Diets that include adequate amounts of readily fermented carbohydrate thus tend to promote more rapid reductions and allow the animal to tolerate higher levels of dietary NO_3^- (33, 43). The optimal pH values for NO_3^- and NO_2^- reduction by rumen bacteria have been reported to be 6.5 to 6.6 and 5.6 to 5.8, respectively (27, 42).

Several studies (13, 18, 33, 34) indicate that increased dietary levels of NO_3^- induce qualitative or quantitative changes, or both, in rumen microbial populations that lead to increased tolerance to high NO_3^- diets. It has also been suggested that adaptation to NO_3^- is accompanied by increased rates of detoxication (reduction) of 3-nitropropanol, which is a toxic constituent of *Astragalus* species (28). The changes in microbial populations that are responsible for adaptation are not understood. Our experiments were thus conducted to provide quantitative information of the effect of dietary NO_3^- on rumen populations, and to examine these popu-

TABLE 1. Rates of reduction of nitrate and nitrite by microbes in rumen fluid from two sheep

Sheep	Nitrate fed[a]	Rate of reduction (nmol/min per ml)[b]	
		Nitrate	Nitrite
1	0	4.5 ± 3.6 (3)	25 ± 26 (3)
1	0.18	56 ± 26 (6)	69 ± 40 (6)
3	0.16	59 ± 33 (5)	54 ± 31 (3)
3	0.32	61 ± 7 (2)	72 ± 28 (2)
3	0.47	117 ± 17 (6)	62 ± 21 (5)

[a] Calculated as grams of NO_3^- per kilogram of body weight per day, as $NaNO_3$ mixed with diet and fed at 12-h intervals.

[b] Values are mean ± SD (n), measured in rumen fluid obtained 4 h postfeeding after incubation (38°C under CO_2) with 2, 4, or 10 mM $NaNO_3$.

lations to obtain information that would help to characterize the microbial changes involved in the adaptation to NO_3^-.

Adult sheep (35 to 46 kg) prepared with ruminal cannulae were fed alfalfa meal–ground corn (60:40), 1 kg/day (two feedings, 12-h interval), unless otherwise indicated. Sodium nitrate was added to the diet as required in individual experiments. In vitro rates of NO_3^- and NO_2^- reduction by microbes in rumen fluid that had been strained through cotton gauze were determined during incubation at 38°C under a CO_2 atmosphere. Nitrate and nitrite were measured by an automated colorimetric method based on a diazo coupling of sulfonilamide with N-(1-naphthyl)-ethylenediamine (Technicon AutoAnalyzer Methodology Bulletin, Tarrytown, N.Y.), and standard curves of levels in rumen fluid were prepared daily.

Adaptation of ruminal microbes to dietary NO_3^- was monitored by measuring the rate of NO_3^- or NO_2^- reduction during in vitro incubation with strained rumen fluid obtained from sheep fed sodium nitrate (0 to 0.47 g of NO_3^- per kg of body weight per day). There was a dramatic increase (15- to 25-fold) in the rates of NO_3^- reduction in the animal adapted to ≥0.18 g of NO_3^- per kg as compared to the rates observed with strained rumen fluid of animals fed no NO_3^- (Table 1). Also, note that the rate of NO_3^- reduction was substantially higher than the rate of NO_2^- reduction when the animal was fed 0.47 g of NO_3^- per day. Concomitant with the NO_3^- disappearance, there was transient accumulation of NO_2^-, which peaked in the first 60 min of incubation. NO_3^- and NO_2^- were stoichiometrically reduced to NH_4^+ (data not shown) during incubation. These results are in agreement with our findings that the specific activities of dissimilatory NO_3^- and NO_2^- reductases in cell extracts of mixed rumen bacteria from nitrate-adapted sheep increased >10-fold in comparison

to those from sheep not receiving dietary nitrate (C. A. Reddy and M. J. Allison, Abstr. Annu. Meet. Am. Soc. Microbiol. 1981, K50, p. 145). The latter studies also showed that NO_3^- reductase activity is membrane bound and is inhibited by azide and hydroxyquinoline-N-oxide, indicating that it is a respiratory-type nitrate reductase. This enzymatic activity was associated with the bacterial, but not the protozoal, fraction of the rumen population.

Additional studies showed that the rate of NO_3^- reduction by microbes in strained rumen fluid collected from a sheep increased >15-fold 4 h after the first feeding of NO_3^- in comparison to the rate observed with strained rumen fluid from the same sheep prior to NO_3^- feeding (results not shown). These results suggested a rapid induction of the NO_3^--reducing enzymes in the mixed rumen biota. Other results showed that there were no great differences in the rates of NO_3^- or NO_2^- reduction by the mixed rumen bacteria in the rumen contents of sheep collected 1, 2, 4, or 12 h after feeding NO_3^- to a NO_3^--adapted sheep.

To determine the time course of adaptation, sheep were fed 0.17 g of NO_3^- per kg of body weight per day, and rumen samples were collected 1 day prior to NO_3^- feeding and the first several days after NO_3^- feeding. On each day samples were collected 1, 2, 3, and 4 h after the morning feeding and 1 h prior to feeding. The data on in vivo NO_3^- concentrations 1 h after feeding on each of the first 6 days during adaptation and the day prior to the start of adaptation are presented in Table 2. Neither NO_3^- nor NO_2^- was detectable in the samples collected the day prior to the first NO_3^- feeding. There was an appreciable concentration of NO_3^- present in the 1-h postprandial rumen sample on the first day of NO_3^- feeding. However, on each successive day of adaptation, the NO_3^- concentration decreased, and no NO_3^- was detected on

TABLE 2. Nitrate and nitrite concentrations in the rumen of a sheep during adaptation to nitrate

Day[a]	Concn (mM) at 1 h postfeeding[b]	
	Nitrate	Nitrite
−1	<0.01	<0.01
0	1.7	0.26
1	1.6	0.40
2	0.9	0.42
6	<0.01	<0.01

[a] On days 0 to 6, $NaNO_3$ was added to the diet of a sheep (40.7 kg) at the rate of 0.18 g of NO_3^- per kg of body weight per day.

[b] Rumen samples obtained prefeeding (one-half of daily ration) and at 2, 3, and 4 h postfeeding contained less than 0.01 mM NO_3^- and NO_2^- on all days.

the sixth day of adaptation. Similarly, NO_2^- was also not detected 1 h after feeding NO_3^- on the sixth day of adaptation. These results thus support conclusions reached on the basis of our measurements of NO_3^- and NO_2^- reduction rates by strained rumen fluid in vitro, indicating that the ruminal population acquired a greater capacity for NO_3^- and NO_2^- reduction during the adaptation period. Optimal adaptation to this level of NO_3^- appears to have been attained within 6 days.

The effects of added NO_3^- and NO_2^- on rates of formation of fermentation products during in vitro incubation of rumen fluid were examined (M. J. Allison, C. A. Reddy, and H. M. Cook, J. Anim. Sci. 53[Suppl. 1]:283, 1981). Incubations were for 2 h, and NO_3^- and NO_2^- were added at concentrations of 4 and 10 mM to rumen fluid from sheep adapted to dietary NO_3^- and from sheep that had not been fed NO_3^-.

Methanogenesis was severely inhibited by the addition of either NO_3^- or NO_2^-, suggesting that these two electron acceptors effectively compete with methanogenesis as sinks for electrons generated during fermentation. These results are in agreement with the free energy change (ΔGó) in the reactions concerned:

$$
\begin{array}{ll}
& \Delta\text{Gó (kJ/reaction)} \\
NO_3^- + 2H^+ + 4H_2 \rightarrow & -598 \\
\quad NH_4^+ + 3H_2O & \\
4H_2 + CO_2 \rightarrow CH_4 + & -131 \\
\quad 2H_2O &
\end{array}
$$

The fact that inhibition of methanogenesis was more pronounced with microbial populations from the NO_3^--adapted animal is not surprising in view of the marked increase in rates of both NO_3^- and NO_2^- reduction by NO_3^--adapted bacterial populations.

Addition of NO_3^- had a pronounced effect on the rates of volatile fatty acid production by mixed rumen bacteria in vitro. In the presence of NO_3^-, both NO_3^--adapted and unadapted populations showed higher acetate production and decreased rates of butyrate and propionate production. These effects were much greater in

animals fed 30 or 60 g of NO_3^- per day as compared to those in animals fed no NO_3^-. Nitrite had a similar effect on volatile fatty acid production when added to the NO_3^--adapted population. However, when NO_2^- was added to rumen contents from sheep fed no NO_3^-, there was a marked increase in lactate and a marked decrease in the production not only of propionate and butyrate but also of acetate.

Increased acetate production associated with NO_3^- reduction is not an unexpected finding as this has been observed in vitro with rumen populations from animals not adapted to NO_3^- (9) and with pure cultures of several anaerobes (16, 19, 23, 45). A trend noted in our experiments toward increased proportions of acetate in the rumen of sheep fed NO_3^- (data not shown) is in agreement with findings of others (10, 17).

Adaptation of the ruminal population to NO_3^- was also examined by using a chemostat continuous culture system. The culture (1 liter of fluid volume in a 2-liter fermentor) was started on day 0 with 1 liter of coarsely strained rumen fluid from a sheep fed the alfalfa–corn (60:40) diet. The methods and culture medium were essentially those described by Isaacson et al. (22) except that the medium contained 5 mM $(NH_4)_2SO_4$ but no urea and that glucose (40 mM) served as the energy source. The dilution rates used were of the same order as those measured in vivo, and both inflow and outflow were controlled by a peristaltic pump. The pH in the fermentor ranged between 6.5 and 7.0. The results in Table 3 indicate that addition of sodium nitrate to the medium significantly reduced the rate of CH_4 production. This finding agrees with results obtained during the in vitro short-term incubations described above.

In contrast to the results with short-term incubations of rumen fluid, no obvious effects on acetate production in the chemostat culture were caused by feeding NO_3^-. The production of propionate, however, was less when the chemostat culture was fed NO_3^-. Butyrate production increased markedly when the dilution rate was reduced from 0.06 to 0.03 h^{-1}. Measure-

TABLE 3. Effects of nitrate on fermentation of glucose by mixed ruminal microbes in a continuous culture system[a]

Days	d (h^{-1})	NO_3^- (mM)	CH_4[b]	Acetate[b]	Propionate[b]	Butyrate[b]	Cells[c]
6–9	0.06	0	28.1	113	10	7.6	45
12–15	0.06	10	12.0	99	4.0	3.8	31
26–28	0.06	20	0.5	103	1.2	3.0	46
36–38	0.03	20	1.9	101	2.6	32	52
48–51	0.03	0	50.4	112	8.2	28	36

[a] Chemostat culture (1 liter), medium of Isaacson et al. (22).
[b] Micromoles produced per 100 μmol of glucose.
[c] Dry weight, grams per mole of glucose.

ments of NH_4^+ in fermentor effluent gave values ranging from 2.8 mM during the initial period to 16.1 mM during the 36- to 38-day period when nitrate was present at 20 mM in the medium.

One goal of this experiment was to test the hypothesis that, with NO_3^- as N source, increases in cell yield could be demonstrated which might reflect an increase in ATP yield through the function of dissimilatory NO_3^- reductase. Although the highest cell yields (Table 3) were obtained when the medium fed contained the highest level of NO_3^- (20 mM), differences between estimates of yield were not significant ($P > 0.16$). We believe, however, that this preliminary experiment was not an adequate test of the hypothesis.

Nitrite is a highly reactive substance, and maintenance of a low E_h in culture media is difficult in the presence of NO_2^-. Nitrite toxicity to anaerobes such as *Clostridium botulinum* is well documented, but little is known concerning the effects of NO_2^- on specific rumen bacteria. With rumen contents from a sheep that had not been adapted to NO_3^-, addition of 2 mM NO_2^- to a roll tube culture medium (98-5 plus xylose [12]) led to a 64% reduction in colony counts. This level of NO_2^- caused only a 22% reduction in colony counts with rumen contents from a NO_3^--adapted sheep. Culture counts from both NO_3^--adapted and nonadapted rumen populations were reduced about 80% when 10 mM NO_2^- was added to the medium (A. Marinho and M. Allison, unpublished data). We suggest that adaptation to dietary NO_3^- may involve selection of organisms that are able to tolerate low levels of NO_2^-, but further work to test this is needed.

We obtained evidence for a rapid induction of NO_3^- and NO_2^- reductase activities in rumen microbes when sheep are fed NO_3^-. Besides this induction, there apparently is a more gradual selection of microbial groups that have the ability to reduce NO_3^- and NO_2^-. To further test this, concentrations of nitrate-reducing bacteria in rumen contents of NO_3^--adapted and unadapted sheep were compared. Total cultivable bacteria were similar in the rumen contents of sheep adapted to NO_3^- (56×10^8/g) and in unadapted sheep (32×10^8 to 49×10^8/g). However, there were large differences in our estimates of the concentrations of nitrate-reducing bacteria in these sheep. The estimates (calculated as most probable numbers) were based on tests for nitrate reduction in a broth medium (with glucose, cellobiose, xylose, soluble starch, lactate, and formate as substrates) that had been inoculated with dilutions of rumen contents. In NO_3^--adapted sheep the NO_3^- reducers comprised 18.2% (10.2×10^8/g) of the total cultivable bacteria, whereas in sheep fed no NO_3^- these

numbers comprised only 1.8%.

Predominant microbes in the rumen of animals fed normal diets have been well characterized (11, 38). However, similar information concerning populations from the rumen of animals adapted to high levels of dietary NO_3^- is lacking. Characteristics of the five bacterial groups that were the predominant isolates from a NO_3^--adapted sheep (0.47 g of NO_3^- per kg of body weight per day) are given in Table 4. Group I and group II were motile vibrios which were presumptively identified as selenomonads, and they were the single most predominant group. The group I and group II strains are very similar except that the group I strains are able to metabolize formate. Group III strains were a homogeneous group: they were relatively large, nonmotile, nonsporing, gram-positive rods with rounded ends; they reduced NO_3^-, metabolized formate, and produced predominantly lactate and minor amounts of acetate and succinate when grown with glucose as the energy source. The fourth group, consisting of gram-negative motile vibrios, was presumptively identified as *Anaerovibrio*. All of the latter strains were negative for NO_3^- reduction and produced propionic and acetic acids as the major products of metabolism of glucose or lactate. Group V included gram-positive, nonmotile, nonsporing, lactate-fermenting short rods that produced major amounts of acetate and minor amounts of succinate. They were unable to utilize glucose as an energy source and were variable in their ability to reduce nitrate. On the basis of this and other available evidence, both group III and group V organisms appear to belong to new genera of anaerobes, but further detailed studies are needed before these organisms can be named. A sixth group, designated "miscellaneous," contains a number of genera including *Bacteroides, Ruminococcus, Butyrivibrio*, and others usually known to be predominant in normal sheep, i.e., sheep fed no NO_3^-.

CONCLUSIONS

Adaptation to dietary oxalate occurs in the rumen through selection of increased proportions of a recently identified group of anaerobic bacteria which use oxalate as a major source of carbon and energy. Animals with adapted ruminal populations are able to tolerate amounts of oxalate that would be lethal to nonadapted animals.

Oxalate-degrading anaerobes similar to those isolated from the rumen are normally present in the large bowel of humans and other animals. In several herbivores, a selection of these organisms that is apparently similar to that observed in the rumen occurs with addition of oxalate to the diet.

TABLE 4. Characteristics of predominant bacterial groups isolated from NO_3^--adapted sheep[a]

Group	Morphotype	Gram reaction	Motility	NO_3^- reduction[b]	Growth and products from Glucose[c]	Growth and products from Lactate[c]	Formate metabolism[d]	Presumptive identification
I	Large vibrios	−	+	+	+, Laps	−	+++	Selenomonads
II	Large vibrios	−	+	+	+, L(as)	−	+	Selenomonads
III	Large thick rods with rounded ends	+	−	+	+, Las	−	+++	Unclassified
IV	Medium-sized to large vibrios	−	+	−	+, Pa	+, Pa	−	*Anaerovibrio*
V	Short thick rods with rounded ends	+	−	8/16	−	+, As	−	Unclassified
VI	Variable characteristics							Miscellaneous

[a] None of the isolates grew aerobically or produced spores.

[b] In group V, only 8 of 16 strains tested were positive for NO_3^- reduction.

[c] Growth was represented as "no growth" (−) or "growth" (+). Metabolic products: L, lactic acid; a, acetic acid; p, propionic acid; s, succinic acid. Upper-case letters represent major products, and lower-case letters refer to minor products.

[d] Formate metabolism was measured by monitoring the disappearance of formate with time during incubation.

There was a pronounced adaptation response to dietary NO_3^-, as evidenced by: (i) greatly increased rates of NO_3^- and NO_2^- reduction; (ii) a marked increase in the most probable number of NO_3^- reducers; and (iii) selection of different predominant bacterial groups in the NO_3^--adapted animal.

NO_3^- and NO_2^- reduction effectively compete with methanogenesis for electrons generated during fermentation in the rumen and also affect the rates of volatile fatty acid production.

LITERATURE CITED

1. **Allison, M. J.** 1978. The role of ruminal microbes in the metabolism of toxic constituents from plants, p. 101–118. *In* R. F. Keeler, K. R. vanKampen, and L. F. James (ed.), Effects of poisonous plants on livestock. Academic Press, Inc., New York.
2. **Allison, M. J., and H. M. Cook.** 1981. Oxalate degradation by microbes of the large bowel of herbivores: the effect of dietary oxalate. Science **212:**675–676.
3. **Allison, M. J., H. M. Cook, and K. A. Dawson.** 1981. Selection of oxalate-degrading rumen bacteria in continuous culture. J. Anim. Sci. **53:**810–816.
4. **Allison, M. J., E. T. Littledike, and L. F. James.** 1977. Changes in ruminal oxalate degradation rates associated with adaptation to oxalate ingestion. J. Anim. Sci. **45:**1173–1179.
5. **Barber, H. H., and E. J. Gallimore.** 1940. The metabolism of oxalic acid in the animal body. Biochem. J. **34:**144–148.
6. **Blaney, B. J., R. J. W. Gartner, and T. A. Head.** 1982. The effects of oxalate in tropical grasses on calcium phosphorus and magnesium availability to cattle. J. Agric. Sci. **99:**533–540.
7. **Bradley, W. B., H. F. Eppson, and O. A. Beath.** 1939. Nitrate as the cause of oat hay poisoning. J. Am. Vet. Med. Assoc. **94:**541–542.
8. **Bradley, W. B., H. F. Eppson, and O. A. Beath.** 1940. Methylene blue as an antidote for poisoning by oat hay and other plants containing nitrates. J. Am. Vet. Med. Assoc. **96:**41–42.
9. **Bryant, A. M.** 1965. The effect of nitrate on the in vitro fermentation of glucose by rumen liquor. N.Z. J. Agric. Res. **8:**118–125.
10. **Bryant, A. M., and M. J. Ulyatt.** 1965. Effects of nitrogenous fertilizer on the chemical composition of short rotation rye grass and its subsequent digestion by sheep. N.Z. J. Agric. Res. **8:**109–117.
11. **Bryant, M. P.** 1977. Microbiology of the rumen, p. 278–302. *In* M. J. Swenson (ed.), Dukes' physiology of domestic animals, 9th ed. Cornell University Press, Ithaca, N.Y.
12. **Bryant, M. P., and I. M. Robinson.** 1961. An improved nonselective culture medium for ruminal bacteria and its use in determining diurnal variation in numbers of bacteria in the rumen. J. Dairy Sci. **44:**1446–1456.
13. **Davison, K. L., W. Hansel, M. J. Wright, and K. McEntee.** 1962. Adaptation to high nitrate intake by cattle, p. 44–51. *In* Proceedings of the Cornell Nutrition Conference for Feed Manufacturers. Cornell University Press, Ithaca, N.Y.
14. **Dawson, K. A., M. J. Allison, and P. A. Hartman.** 1980. Isolation and some characteristics of anaerobic oxalate-degrading bacteria from the rumen. Appl. Environ. Microbiol. **40:**833–839.
15. **Dawson, K. A., M. J. Allison, and P. A. Hartman.** 1980. Characteristics of anaerobic oxalate-degrading enrichment cultures from the rumen. Appl. Environ. Microbiol. **40:**840–846.
16. **DeVries, W., T. R. M. Reitveld-Struijk, and A. H. Strouthamer.** 1977. ATP formation associated with fumarate and nitrate reduction in growing cultures of *Viellonella alcalescens*. Antonie van Leeuwenhoek J. Microbiol. Serol. **43:**153–167.
17. **Farra, P. A., and L. D. Satter.** 1971. Manipulation of the ruminal fermentation. Effect of nitrate on ruminal volatile fatty acid production and milk composition. J. Dairy Sci. **54:**1018–1024.
18. **Geurink, J. H., A. Malestein, A. Kemp, and A. T. van't Klooster.** 1979. Nitrate poisoning in cattle. 3. The relationship between nitrate intake with hay of fresh forage and the speed of intake on the formation of methemoglobin. Neth. J. Agric. Sci. **27:**268–276.
19. **Hasan, S. M., and J. B. Hall.** 1975. The physiological function of nitrate reduction in *Clostridium perfringens*. J. Gen. Microbiol. **87:**120–128.
20. **Hodgkinson, A.** 1977. Oxalic acid in biology and medicine. Academic Press, Inc., New York.
21. **Holtenius, P.** 1957. Nitrite poisoning in sheep, with special reference to the detoxification of nitrite in the rumen. Acta Agric. Scand. **7:**113–163.
22. **Isaacson, H. R., F. C. Hinds, M. P. Bryant, and F. N. Owens.** 1975. Efficiency of energy utilization by mixed

rumen bacteria in continuous culture. J. Dairy Sci. **58:**1645–1659.

23. **Ishimoto, M., M. Umeyama, and S. Chiba.** 1974. Alteration of fermentation products from butyrate to acetate by nitrate reduction in *Clostridium perfringens.* Z. Allg. Mikrobiol. **14:**115–121.

24. **James, L. F., and J. E. Butcher.** 1972. Halogeton poisoning of sheep: effect of high level oxalate intake. J. Anim. Sci. **35:**1233–1238.

25. **Jones, G. A.** 1972. Dissimilatory metabolism of nitrate by the rumen microbiota. Can. J. Microbiol. **18:**1783–1787.

26. **Lewis, D.** 1951. The metabolism of nitrate and nitrite in the sheep. 1. The reduction of nitrate in the rumen of the sheep. Biochem. J. **48:**175–180.

27. **Lewis, D.** 1951. The metabolism of nitrate and nitrite in the sheep. 2. Hydrogen donators in nitrate reduction by rumen microorganisms in vitro. Biochem. J. **49:**149–153.

28. **Majak, W., K. J. Cheng, and J. W. Hall.** 1982. The effect of cattle diet on the metabolism of 3-nitropropanol by ruminal microorganisms. Can. J. Anim. Sci. **62:**855–860.

29. **McInerney, M. J., R. I. Mackie, and M. P. Bryant.** 1981. Syntrophic association of a butyrate-degrading bacterium and *Methanosarcina* enriched from bovine rumen fluid. Appl. Environ. Microbiol. **41:**826–828.

30. **Montfort, D. O., and M. P. Bryant.** 1982. Isolation and characterization of an anaerobic syntrophic benzoate-degrading bacterium from sewage sludge. Arch. Microbiol. **133:**249–256.

31. **Morris, M. P., and J. Garcia-Rivera.** 1955. The destruction of oxalate by rumen contents of cows. J. Dairy Sci. **38:**1169.

32. **Müller, H.** 1950. Oxalsaure als Kohlenstoffquelle fur Mikroorganismen. Arch. Mikrobiol. **15:**137–148.

33. **Nakamura, Y., Y. Tada, H. Shibuya, J. Yoshida, and R. Nakamura.** 1979. The influence of concentrates on the nitrate metabolism of sheep. Jpn. J. Zootech. Sci. **50:**782–789.

34. **Nakamura, Y., J. Yoshida, R. Nakamura, and H. Horie.** 1976. Nitrate metabolism of microorganisms in the rumen of sheep fed high nitrate forages. Jpn. J. Zootech. Sci. **47:**63–67.

35. **O'Halloran, M. W.** 1962. The effect of oxalate on bacteria isolated from the rumen. Proc. Aust. Soc. Anim. Prod. **4:**18–21.

36. **O'Hara, P. J., and A. J. Fraser.** 1975. Nitrate poisoning in cattle grazing crops. N.Z. Vet. J. **23:**45–53.

37. **Rimington, C., and J. I. Quin.** 1933. The presence of lethal factor in certain members of the plant genus *Tripulus.* S. Afr. J. Sci. **30:**472–482.

38. **Russell, J. B., and R. B. Hespell.** 1981. Microbial rumen fermentation. J. Dairy Sci. **64:**1153–1169.

39. **Sapiro, M. L., S. Hoflund, R. Clark, and J. I. Quin.** 1949. Studies on the alimentary tract of the Merino sheep in South Africa. XVI. The fate of nitrate in ruminal ingesta as studied in vitro. Onderstepoort J. Vet. Sci. **22:**357–372.

40. **Shirley, E. K., and K. Schmidt-Nielsen.** 1967. Oxalate metabolism in the pack rat, sand rat, hamster and white rat. J. Nutr. **91:**496–502.

41. **Singer, R. H.** 1972. The nitrate poisoning complex. Proc. 76th Annu. Meet. U.S. Anim. Health Assoc., p. 310–322.

42. **Takahashi, J., Y. Masuda, and E. Miyaga.** 1978. Effect of pH and level of nitrate on the reduction of nitrate and nitrite in vitro. Jpn. J. Zootech. Sci. **49:**1–5.

43. **Takahashi, J., T. Masuko, S. Endo, K. Dodo, and H. Fujita.** 1980. Effects of dietary protein and energy levels on the reduction of nitrate and nitrite in the rumen and methemoglobin formation in sheep. Jpn. J. Zootech. Sci. **51:**626–631.

44. **Watts, P. S.** 1957. Decomposition of oxalic acid in vitro by rumen contents. Aust. J. Agric. Res. **8:**266–270.

45. **Yamamoto, I., T. Mitsui, and M. Ishimoto.** 1982. Nitrate reduction in *Bacteroides multiacidus*: its effect on glucose metabolism and some properties of the reductase. J. Gen. Appl. Microbiol. **28:**451–463.

Role of Mycophagy and Bacteriophagy in Invertebrate Nutrition

MICHAEL M. MARTIN AND JEROME J. KUKOR

Division of Biological Sciences, University of Michigan, Ann Arbor, Michigan 48109

Microorganisms are routinely ingested by many groups of invertebrates. Organisms that consume plant detritus, decaying fruit, rotting wood, and herbivore dung ingest a complex assemblage of bacteria, protozoa, and fungi along with the plant tissue that constitutes the bulk of their food. Aquatic filter feeders and periphyton scrapers include significant quantities of bacteria and microscopic algae in their diets. Fungal tissue is a major food of many arthropods, including species that graze selectively on the vegetative hyphae growing on dead plant tissue or in the soil and also species that consume the fruiting bodies of macrofungi. There are several groups of insects that go to great lengths to culture the fungus that they consume.

Microbial tissues and microbial cells are, of course, potential sources of the various macro- and micronutrients generally needed for normal existence by virtually all animals. Thus, if ingested microorganisms are digested and assimilated, they can contribute to meeting the nutritional requirements of the organism that consumes them. However, in addition to being biomass with potential nutritive value, microorganisms are also active catalytic agents with diverse metabolic capabilities. Thus, if ingested microorganisms survive and proliferate in the digestive tract, or if they liberate enzymes that remain active in the gut milieu, they can augment or extend the digestive and metabolic capabilities of an organism that consumes them. This paper explores the extent to which the nutritive value and metabolic potential of ingested microorganisms are exploited by invertebrates.

ASSIMILATION OF INGESTED MICROBES

Microbes have high nutritive potential. Bacteria, algae, yeasts, and filamentous fungi are among the most nitrogen-rich natural foods commonly exploited by invertebrates (Table 1). In addition to being a rich source of protein, microbes contain comparatively high levels of other macronutrients, such as lipids and carbohydrates, and are also good sources of critical micronutrients, such as unsaturated fatty acids and vitamins (22, 31, 37, 65). Sterols, which are required by insects for normal growth and repro-

duction, are present in fungi, yeasts, and algae (22, 37, 65), but occur in only a few groups of bacteria, such as the mycoplasmas and methane-oxidizing forms (15).

Assimilation efficiency (AD), also known as approximate digestibility, is a standard measure of the efficiency with which a food is used by an organism. It is defined as the fraction (or percentage) of ingested material that is not egested, and is generally calculated on a dry weight basis (68):

$$AD = \frac{\text{dry wt of food ingested} - \text{dry wt of feces}}{\text{dry wt of food ingested}} \times 100$$

The average assimilation efficiency of the amphipod crustacean *Gammarus pseudolimnaeus* feeding on fungal mycelium is 70% (7) (Table 2). In quantitative studies of the assimilation of yeasts, bacteria, and diatoms, it has been convenient to use isotopically labeled cells. In such studies, AD refers to the percentage of label, rather than the percentage of biomass, assimilated.

$$AD = \frac{\text{total radioactivity ingested} - \text{total radioactivity in feces}}{\text{total radioactivity ingested}} \times 100$$

The assimilation efficiency of the isopod crustacean *Tracheoniscus rathkei* feeding on ^{14}C-labeled yeast cells is 72% (55) (Table 2). The assimilation efficiencies of the amphipod *Hyalella azteca* (23), the isopod *T. rathkei* (57), and several species of snails (16, 32) feeding on various species of labeled bacteria are even higher, averaging 80%. These three types of microbial tissue are, therefore, assimilated as efficiently as other high-quality foods, such as arthropod tissues, seeds, and grains, and considerably more efficiently than foliage (62) (Table 2). The radioactive label in diatoms is also assimilated efficiently by amphipods (23), snails (16, 32), and blackfly larvae (61). Blue-green algae appear to be the only type of microorganism that is not effectively assimilated by its predators.

Comparison of the viability of heterotrophic bacteria in the food and gut of the naid oligochaete *Nais variabilis* indicates a reduction in

TABLE 1. Nitrogen levels of various natural
materials exploited as food by invertebrates

Material	Nitrogen content (% of dry wt)	Reference
Bacteria	11.5–12.5	31
Algae	7.5–10.0	31
Yeast	7.5–8.5	31
Arthropod tissue	6.2–14.0	3, 19
Filamentous fungi	2.0–8.0	31
Pollen	2.0–7.0	63
Seeds	1.0–7.0	46
Cambium	0.9–5.0	3, 46
Live foliage	0.7–5.0	3, 46
Leaf litter	0.5–2.5	3, 46
Soil	0.1–1.1	3
Wood	0.03–0.2	3
Phloem sap	0.004–0.6	46
Xylem sap	0.0002–0.1	46

viability of 99.6% during passage through the gut
(24). Although loss of viability does not prove
that digestion has occurred, it is nonetheless
compatible with the idea that ingested bacteria
are used as food. A reduction in the abundance
or variety, or both, of bacteria during the passage of food through the gut has also been
reported in tubificid oligochaetes (69), in the
isopod *T. rathkei* (57), in the snail *Helix pomatia*
(8), and in the locust *Schistocerca gregaria* (53).
Using a fluorescent-antibody technique, Johnson and O'Keefe (29) showed that intact *Rhizobium leguminosarum* bacteroids ingested by the
pea leaf weevil, *Sitona lineatus*, are more prevalent in the anterior and middle portions of the
midgut, whereas bacteroids with disrupted cell
walls were observed more often in the posterior
portions of the midgut. No intact bacteroids
were observed in the hindgut.

The demonstration of efficient assimilation of
microorganisms by various invertebrates implies
the presence of digestive enzymes active against
some of the major constituents of microbial
cells. Although the biochemical basis for a capacity to digest microbes has received little
systematic attention, several investigations of
the digestive enzymes of invertebrates have
provided useful information. Digestive β-1,3-
and α-1,4-glucanases occur widely in invertebrates that commonly ingest fungi and algae (34,
40, 41, 44, 45, 51, 52, 64). Since β-1,3-glucans
are cell wall components in most fungi and are
storage products in diatoms and other algae, and
α-1,4-glucans are the reserve carbohydrates of
fungi as well as all chlorophyll-containing plants,
such enzymes are clearly of adaptive value.
Enzymes active against chitin, a structural polysaccharide present in both fungi diatoms, also
occur widely in litter- and soil-feeding annelids
and molluscs (27, 28, 34, 51). Chitinases appear

to play a less important role in the digestive
processes of detritus-feeding insects, where activity is rarely detectable (40, 44, 45, 51). Even
in beetles that are exclusively mycophagous,
chitinase activity has been detected in only a few
of the species examined, and then only at very
low levels (41). Chitinase activity has been reported in seven species of astigmatid mites (2).

The enzymatic basis for a capacity to digest
and assimilate bacteria is less well understood.
Activity similar to that of lysozyme, an enzyme
causing the lysis of gram-positive bacteria by
hydrolyzing β-1,4-glucosidic linkages in cell wall
peptidoglycans, has been detected in the gut
fluids of many insects (36), ticks (54), mites (17),
and molluscs (47, 58). However, since lyso-
zymes from many sources can also degrade
chitin (50), it is not clear whether lysozymal
activity in the gut fluids reflects an adaptation to
bacteriophagy, mycophagy, or both. Although
blackfly larvae are generally thought to derive
significant nutritional benefit from the digestion
of bacteria attached to the fine particulate matter
they ingest (5, 21), no lysozyme activity could be
detected in the gut fluids of *Prosimulium mix-
tum/fuscum* (M. M. Martin, unpublished data).
Other enzymes that might participate in bacteri-
olysis include proteases, which are present at
high levels in blackfly larvae (Martin, unpublished data), and β-glucuronidase, which has
been detected in a marine nematode believed to
feed on bacteria, but not in another species that
feeds on algae (26). Bacterial products might
also be liberated from bacteria exposed to osmotic or pH stress, or to the effects of surfactants in the gut fluids.

Ingested microorganisms are thought to play a
particularly significant role in the diets of detritus feeders, and many studies have demonstrated a clear preference on the part of detritivores
for litter that supports a rich culture of microorganisms, especially fungi (18). Although microbial biomass is the most nutritious and digestible

TABLE 2. Assimilation efficiencies of invertebrates
on different types of food

Type of food	AD (%)		References
	Mean (n)	Range	
Bacteria	80 (7)	60–97	16, 23, 32, 57
Arthropod tissue	80 (32)	37–98	62
Seeds/grain	72 (9)	46–96	62
Yeast	72 (1)	—	55
Fungal mycelium	70 (10)	65–77	7
Diatoms	70 (10)	61–88	16, 23, 32, 61
Forb foliage	53 (38)	16–97	62
Grass	43 (15)	19–81	62
Tree foliage	39 (99)	2–94	62
Blue-green algae	23 (5)	6–49	23, 32, 61

component of detritus, the amount of microbial tissue present is too small to make a significant quantitative contribution to the macronutrient requirements of a detritus-feeding invertebrate. Cummins and Klug (18) estimated that fungal tissue may make up as much as 10% of the weight of leaf detritus after a residence time of several weeks in a stream. However, even if this fungal tissue were assimilated with 100% efficiency, it would account for only 8.3% of the growth of larvae of the crane fly, *Tipula abdominalis*, feeding upon such leaf litter.

The quantitative contribution of bacterial biomass to the nutrition of detritus feeders is even more problematical. Perhaps selective grazing of bacteria is possible for some small invertebrates, such as rotifers, nematodes, and harpacticoid copepods, but most invertebrates ingest bacteria by consuming mineral or detrital particles to which bacteria are attached (20). Even on heavily colonized surfaces, the amount of bacterial biomass is small. Microscopic observations have shown that detrital fragments contain 2 to 15 bacteria per 100 μm^2 of surface area, corresponding to 5×10^8 to 1×10^{10} bacteria per g (dry weight) (20). Assuming that an average bacterium weighs 2×10^{-7} to 4×10^{-7} μg (dry weight), a fragment of detritus may contain 0.1 to 4.0 mg of bacteria per g. Baker and Bradnam (5) estimated that a filter-feeding blackfly larva ingests 1,500 μg of detrital particles per day, of which 10%, or 150 μg, is assimilated. They further estimated that the particulate matter ingested in a day contains 2.8×10^7 bacteria, weighing a total of 5.6 μg. Thus, even if the ingested bacteria are assimilated with 100% efficiency, they would account for less than 4% of the total biomass assimilated. Calculations of this sort suggest that bacteria make a relatively minor quantitative contribution to the nutrition of invertebrates that ingest them, even to species that specialize upon substrates that are fairly heavily colonized. Perhaps the capacity to digest bacteria is more important as a defense against pathogens than as a mechanism for obtaining nutrients.

PROLIFERATION OF INGESTED BACTERIA

Death, lysis, and assimilation are not the only fates possible for ingested microbes. Some bacteria find the gut environment favorable for growth and reproduction. Although Reyes and Tiedje (57) demonstrated a reduction in the number of cells of *Pseudomonas* and *Flavobacterium* in the feces of *T. rathkei* compared to the food, indicating digestion and assimilation of these two strains of bacteria, these same authors observed that the total number of viable bacteria of all kinds that could be isolated from the gut of this isopod was actually larger than the number

that could be isolated from the food. There was an even greater increase in total viable counts between the food and the feces, indicating net microbial proliferation during passage of food through the animal. They further suggested that the relatively constant number of bacteria platable from the gut of this animal represented a steady-state bacterial population resulting from approximately equal rates of multiplication and of digestion and elimination. Net increases in bacterial standing crops or colony-forming units between anterior and posterior segments of the alimentary tract, suggesting microbial proliferation as food passes along the gut, have also been reported in soil-feeding annelids (33), in dipteran larvae (33, 66), in the termite *Procubitermes aburiensis* (11), in the wood-feeding larvae of the beetle *Oryctes nasicornis* (9, 59), and in the litter-feeding millipede *Glomeris marginata* (4). The extent of microbial proliferation will, of course, be limited by the residence time in the segment of the gut where conditions are favorable for growth. Residence time will be influenced by both feeding rate and certain aspects of gut morphology, such as the relative sizes of the different segments of the gut, the presence of ceca, and the occurrence of spines or other structures that might serve as sites of attachment for the bacteria. Whether the multiplication of ingested microbes is nutritionally beneficial or detrimental to the animal ingesting them depends upon whether the process results in net addition to or subtraction from the pool of nutrients available for assimilation. Reyes and Tiedje (56, 57) concluded that in the gut of the isopod *T. rathkei* ingested microbes compete for the more readily digestible substrates rather than assist in the digestion of refractory substances. On the other hand, the presence of high levels of volatile fatty acids in the midgut of the beetle *O. nasicornis* suggests that fermentation processes, presumably attributable to either ingested or resident bacteria, have generated products of potential use to the beetle (10). A measure of the rate of volatile fatty acid production would be required to evaluate the quantitative contribution of fermentation products to the energy requirements of the beetle.

In addition to digestion and fermentation, it is also reasonable to suppose that ingested microbes might metabolize secondary metabolites, including potential toxins, present in the food. Support for this intriguing possibility is provided by the demonstration that the capacity of the isopod *Oniscus asellus* to cleave aromatic rings is attributable to ingested microorganisms with dioxygenase activity (30).

The possible involvement of ingested bacteria in the digestion of refractile substrates, the fermentation of digestion products, and the metabolism of potential toxins deserves much more attention than it has received to date. The microbial ecologist's view of the invertebrate gut as an ecosystem in which interacting populations of microorganisms process matter and energy provides a useful context in which to formulate future research proposals in this important area.

ACQUISITION OF DIGESTIVE ENZYMES FROM FUNGI

There is no indication that fungal cells commonly retain their viability after ingestion. However, fungal enzymes, liberated from ingested fungal tissue, have been found to be stable and active in the gut fluids of a number of invertebrates. Fungi produce extracellular enzymes active against a wide spectrum of natural substrates, such as cellulose, hemicellulose, pectin, chitin, cutin, lignin, tannins, and other polyphenols (37). The ingestion of fungal tissue is, therefore, a mechanism by which an organism might augment its digestive capabilities and expand the range of natural substrates suitable for exploitation as nutrient sources (39). The significance of acquired fungal enzymes to the digestive capabilities of an insect was first documented by Martin and Martin (42, 43), in studies of the digestive enzymes in the midgut fluids of the fungus-growing termite *Macrotermes natalensis*.

The fungus-growing termites are higher termites which lack the assemblage of xylophagous protozoa that bring about cellulose digestion in the hindguts of the lower termites. These termites have long fascinated biologists because of their symbiotic association with fungi that grow in their nests on structures referred to as "fungus combs" (60). The surface of the comb is covered with a sparse growth of mycelium and numerous small white spheres or nodules. These nodules are the conidia and conidiophores of a basidiomycetous fungus, *Termitomyces* sp., believed to be restricted to the nests of the *Macrotermitinae*. The mycelium is a mixture of *Termitomyces* and various ascomycetous species in the family *Xylariaceae*. By consuming small quantities of the *Termitomyces* nodules, the termites acquire hydrolytic enzymes active against cellulose. Cellulose digestion in insects is generally accomplished by a "cellulase complex" consisting of three classes of hydrolytic enzymes: C_x-cellulases (endoglucanases), C_1-cellulases (cellobiohydrolases), and β-glucosidases (cellobiases) (38). The entire complex is present in the midgut fluids of *M. natalensis* workers. The C_1-cellulases are acquired enzymes derived entirely from ingested *Termitomyces* nodules, whereas the C_x-cellulases originate in part from the termites and in part from ingested fungal tissue (42, 43). Abo-Khatwa (1), studying the closely related termite *M. subhyalinus*, obtained comparable results.

Kukor and Martin (35) have investigated the role of acquired fungal enzymes in the digestive processes of the siricid woodwasp *Sirex cyaneus*, another wood-feeder that is associated with a fungal symbiont. The larvae develop in timber in galleries lined with the mycelium of *Amylostereum chailletii*, the fungal symbiont introduced into the wood along with the egg at the time of oviposition. The normal larval diet is a mixture of wood and fungal mycelium. Müller (49) demonstrated significant digestion and assimilation of wood constituents, including cellulose and hemicellulose, by larvae of *S. gigas* and *S. phantoma*. The midgut of *S. cyaneus* larvae contains enzymes active against cellulose, carboxymethylcellulose, and larchwood xylan, a representative hemicellulose (Table 3). The enzymes are present in larvae which have been

TABLE 3. Enzymatic activity toward cellulose, carboxymethylcellulose, and larchwood xylan in extracts of gut contents of *S. cyaneus* larvae and cultures of *A. chailletii*[a]

Source of extract	Activity (U/mg [dry wt] of dissolved solids)		
	Cellulase complex	C_x-cellulase	Xylanase
S. cyaneus midguts			
Larvae collected from natural galleries in balsam fir	0.33 ± 0.13	8.86 ± 6.20	11.59 ± 7.21
Larvae cultured on balsam fir chips permeated by *A. chailletii*	0.48 ± 0.01	20.17 ± 0.59	33.41 ± 3.97
Larvae cultured on sterile balsam fir chips	0.00	0.09	1.39
Fungus and substrates			
Broth from liquid culture of *A. chailletii* growing on cellulose	0.33 ± 0.00	7.45 ± 0.33	9.24 ± 0.33
Extract of balsam fir chips permeated by *A. chailletii* mycelium	0.37 ± 0.00	1.58 ± 0.65	6.78 ± 2.53
Extract of sterile balsam fir chips	0.00	0.00	0.00

[a] Five replicates were performed except in the case of the larvae cultured on sterile balsam fir chips, in which case a single determination was made on midguts pooled from five survivors. Entries are $\bar{x}$ ± SEM. A unit of activity is the amount of enzyme required to liberate 1 μmol of maltose equivalents per h (37°C, pH 5, incubation volume of 1.0 ml) (35).

reared in the laboratory on balsam fir chips permeated by the mycelium of *A. chailletii*, as well as larvae collected from their natural galleries in the trunks of standing balsam fir trees. Enzymatic activity toward these same substrates is also evident in both the fluid from a liquid culture of *A. chailletii* growing on microcrystalline cellulose and in an extract of balsam fir chips permeated by the fungus. When larvae are fed a diet of symbiont-free balsam fir chips, extracts of which are enzyme-free, larval mortality is high and the level of gut enzymes decreases dramatically (Table 3). These results suggested to us that the C_x-cellulases and xylanases present in the larval gut fluids were probably fungal enzymes acquired when the larvae ingested the mycelium of their fungal symbiont. Enzymes responsible for C_x-cellulase and xylanase activity in cultures of *A. chailletii* and extracts of *S. cyaneus* guts were purified by a protocol involving concentration, dialysis, ion-exchange chromatography on DEAE-Sepharose CL6B, gel permeation chromatography on Sephadex G-75, and chromatofocusing on Polybuffer Exchanger 94. The identities of the fungal and larval enzymes were established by analytical isofocusing on ultrathin polyacrylamide gels, and the results confirmed our hypothesis that the larval enzymes were ingested fungal enzymes (35).

Acquired fungal enzymes play a very different role in the biology of the attine ants (13, 14). These fascinating insects culture fungi in their nests, using leaves, flowers, plant debris, insect frass, and insect carcasses as substrates. The larvae feed only on their fungal symbiont, whereas the adult workers consume small amounts of fungal tissue while subsisting largely on plant juices. Thus, unlike the fungus-growing termites and siricid woodwasps, the diet of the fungus-growing ants does not include highly refractory polysaccharides, and acquired fungal enzymes would seem to have no role in the digestive processes of these insects. However, Boyd and Martin (13, 14) demonstrated that fungal enzymes ingested by the workers survive passage through the gut and are still active in the fecal material, which is regularly applied to the materials used by the ants as substrates in their fungus gardens. Thus, the feeding and defecation behavior of the workers can be viewed as a mechanism that has replaced the process of enzyme secretion by the fungus. Furthermore, the ants achieve a distribution of enzymes highly conducive to rapid initial growth by the fungus, since they feed in mature portions of the fungus garden, where enzymes are produced in abundance, and defecate at the site of inoculation, where only small quantities of enzymes can be produced by the small mycelial fragments placed there by the ants. In this truly remarkable symbiosis, the ants acquire digestive enzymes from their fungal associate, only to return them in a fashion that enhances the competitive ability and fitness of the fungal donor.

In the macrotermitine termites, siricid woodwasps, and attine ants, the acquisition and utilization of fungal enzymes is an integral part of a complex, specialized, highly coevolved mutualism. However, it is also possible to envision a critical role for acquired fungal enzymes in the digestive processes of detritus feeders, where the fungi that are the potential sources of enzymes are not symbionts, but rather are normal colonizers of the detrital food. Indeed, low to moderate levels of C_x-cellulase have been detected in the gut fluids of the detritus-feeding immature forms of a number of insect species in the orders Plecoptera (45; R. L. Sinsabaugh, E. F. Benfield, and A. E. Linkins, Abstr. Annu. Meet. N. Am. Benthol. Soc., 29th, Provo, Utah, 1981), Trichoptera (12, 40, 48), and Diptera (12, 67; Sinsabaugh et al., Abstr. Annu. Meet. N. Am. Benthol. Soc., 29th, 1981). On the basis of a correlation between the level of C_x-cellulase activity in the gut and the presence or level of this enzyme in the food, it has been proposed that ingested fungal enzymes are responsible, at least in part, for the cellulolytic capacity of the terrestrial isopod *Philoscia muscorum* (25) and the aquatic amphipod *G. fossarum* (6).

We have recently undertaken a study of the contribution of ingested enzymes to the digestive capabilities of the litter-feeding isopod *T. rathkei*. Both C_x-cellulase and xylanase activity occur in the digestive tract of this isopod and in the leaf litter on which it normally feeds. Isopods reared on a microbe-free diet show a marked reduction in the levels of these two classes of hydrolytic activity, suggesting that activity may be due in part to ingested enzymes. Using a purification scheme similar to the one employed in our investigation of siricid woodwasps (35), we have shown that some, but not all, of the C_x-cellulases and xylanases in the digestive tract of *T. rathkei* have chromatographic properties and isoelectric pH values identical to those of enzymes extracted from the litter, providing further evidence consistent with the hypothesis that ingested enzymes remain active in the gut of this detritivore. Experiments are in progress to ascertain the extent to which acquired enzymes actually contribute to the digestive efficiency of these organisms when they are feeding on their normal diet of leaf litter.

FUTURE DIRECTIONS

There is a growing realization that the ingestion of microorganisms by invertebrates can have significant implications beyond the mere provision of a supplemental source of readily digested and assimilated biomass. In our research we are particularly interested in the functional significance of microbial enzymes liberated in the gut by the lysis of ingested microbial cells. We have firmly established the significance of enzymes acquired in this fashion to several groups of insects involved in highly specialized symbiotic relationships with fungi. By directing future studies toward invertebrate species with more casual associations with free-living fungi, we hope to assess the generality of this mechanism for acquiring a digestive capacity. Although our interests have been primarily in invertebrate mycophagy, investigations in invertebrate bacteriophagy are also likely to be rewarded by discoveries of considerable significance. Since bacteria are virtually everywhere, bacteriophagy must be the rule among invertebrates. Future studies that will delineate the fates of ingested bacteria in different groups of insects with different dietary habits seem likely to turn up cases in which bacteria that proliferate in the gut act as pathogens, parasites, competitors, commensals, and mutualists. It does not seem unreasonable to expect that there may be cases in which the suitability of a given substrate as food may be determined by the metabolic activities of the associated bacteria after ingestion. It is apparent to us that future investigations that clarify the significance to invertebrates of the proliferation of ingested bacteria may provide major insights, not only into the nature of animal–microbe interactions, but also into more general aspects of the nutritional ecology of invertebrates.

We acknowledge with gratitude the generous support of our research on insect–fungus associations that has been provided by the National Institutes of Health (grant AI-07386), the National Science Foundation (grants GB-31581, PCM-78-22733, PCM-82-03537, DEB-80-22634, and a Faculty Fellowship in Science), C.S.I.R., Pretoria (grant 398-77-21-REI), the Sloan Foundation, and the Horace Rackham School of Graduate Studies of the University of Michigan.

LITERATURE CITED

1. **Abo-Khatwa, N.** 1978. Cellulase of fungus-growing termites: a new hypothesis on its origin. Experientia **34:**559–560.

2. **Akimov, I. A., and V. N. Barabanova.** 1978. Vliyanie osobennostei pitaniya akaroidnykh kleshchei na aktivnost ikh nekotorykh pischchevaritel nykh fermentov. Ekologiya **2:**27–31.

3. **Allen, S. E. (ed.).** 1974. Chemical analysis of ecological materials, p. 242 and 507–511. Blackwell Scientific Publications, Oxford.

4. **Anderson, J. M., and D. E. Bignell.** 1980. Bacteria in the food, gut contents and faeces of the litter-feeding millipede *Glomeris marginata* (Villers). Soil. Biol. Biochem. **12:**251–254.

5. **Baker, J. H., and L. A. Bradnam.** 1976. The role of

6. **Bärlocher, F.** 1982. The contribution of fungal enzymes to the digestion of leaves by *Gammarus fossarum* Koch. Oecologia (Berlin) **52:**1–4.

7. **Bärlocher, F., and B. Kendrick.** 1975. Assimilation efficiency of *Gammarus pseudolimnaeus* on fungal mycelium or autumn-shed leaves. Oikos **26:**55–59.

8. **Bayne, C. J.** 1973. Molluscan internal defense mechanism: the fate of ^{14}C-labelled bacteria in the land snail *Helix pomatia* L. J. Comp. Physiol. **86:**17–25.

9. **Bayon, C.** 1981. Modifications ultrastructurales des parois végétales dans le tube digestif d'une larve xylophage *Oryctes nasicornis* (Coleoptera, Scarabaeidae): rôle des bacteries. Can. J. Zool. **59:**2020–2029.

10. **Bayon, C., and J. Mathelin.** 1980. Carbohydrate fermentation and by-product absorption studied with labelled cellulose in *Oryctes nasicornis* larvae (Coleoptera: Scarabaeidae). J. Insect Physiol. **26:**833–840.

11. **Bignell, D. E., H. Oskarsson, and J. M. Anderson.** 1980. Distribution and abundance of bacteria in the gut of a soil-feeding termite *Procubitermes aburiensis* (Termitidae, Termitinae). J. Gen. Microbiol. **117:**393–403.

12. **Bjarnov, N.** 1972. Carbohydrases in *Chironomus, Gammarus* and some Trichopteran larvae. Oikos **23:**261–263.

13. **Boyd, N. D., and M. M. Martin.** 1975. Faecal proteinases of the fungus-growing ant, *Atta texana*: properties, significance and possible origin. Insect Biochem. **5:**619–635.

14. **Boyd, N. D., and M. M. Martin.** 1975. Faecal proteinases of the fungus-growing ant, *Atta texana*: their fungal origin and ecological significance. J. Insect Physiol. **21:**1815–1820.

15. **Brock, T. D.** 1979. Biology of microorganisms, 3rd ed., p. 139. Prentice-Hall, Inc., Englewood Cliffs, N.J.

16. **Calow, P., and C. R. Fletcher.** 1972. A new radiotracer technique involving ^{14}C and ^{51}Cr, for estimating the assimilation efficiencies of aquatic, primary consumers. Oecologia (Berlin) **9:**155–170.

17. **Childs, M., and C. E. Bowman.** 1981. Lysozyme activity in six species of economically important astigmatid mites. Comp. Biochem. Physiol. B **70:**615–617.

18. **Cummins, K. W., and M. J. Klug.** 1979. Feeding ecology of stream invertebrates. Annu. Rev. Ecol. Syst. **10:**147–172.

19. **DeFoliart, G. R.** 1975. Insects as a source of protein. Bull. Entomol. Soc. Am. **21:**161–163.

20. **Fenchel, T. M., and B. B. Jørgensen.** 1977. Detritus food chains of aquatic ecosystems: the role of bacteria. Adv. Microb. Ecol. **1:**1–58.

21. **Fredeen, F. J. H.** 1964. Bacteria as food for blackfly larvae (Diptera: Simuliidae) in laboratory cultures and in natural streams. Can. J. Zool. **42:**527–548.

22. **Griffin, D. H.** 1981. Fungal physiology, p. 18–39. John Wiley & Sons, Inc., New York.

23. **Hargrave, B. T.** 1970. The utilization of benthic microflora by *Hyalella azteca* (Amphipoda). J. Anim. Ecol. **39:**427–437.

24. **Harper, R. M., J. C. Fry, and M. A. Learner.** 1981. A bacteriological investigation to elucidate the feeding biology of *Nais variabilis* (Oligochaeta: Naididae). Freshwater Biol. **11:**227–236.

25. **Hassall, M., and J. B. Jennings.** 1975. Adaptive features of gut structure and digestive physiology in the terrestrial isopod *Philoscia muscorum* (Scopoli) 1763. Biol. Bull. (Woods Hole, Mass.) **49:**348–364.

26. **Jennings, J. B., and A. Deutsch.** 1975. Occurrence and possible adaptive significance of β-glucuronidase and arylamidase ("leucine aminopeptidase") activity in two species of marine nematodes. Comp. Biochem. Physiol. A **52:**611–614.

27. **Jeuniaux, C.** 1963. Chitine et chitinolyse. Masson, Paris.

28. **Jeuniaux, C.** 1966. Chitinases. Methods Enzymol. **8:**644–650.

29. **Johnson, M. P., and L. E. O'Keefe.** 1981. Presence and possible assimilation of *Rhizobium leguminosarum* in the

bacteria in the nutrition of aquatic detritivores. Oecologia (Berlin) **24:**95–104.

gut of pea leaf weevil, *Sitona lineatus*, larvae. Entomol. Exp. Appl. **29**:103–108.

30. **Kaplan, D. L., and R. Hartenstein.** 1978. Studies on monooxygenases and dioxygenases in soil macroinvertebrates and bacterial isolates from the gut of the terrestrial isopod, *Oniscus asellus* L. Comp. Biochem. Physiol. B **60**:47–50.

31. **Kihlberg, R.** 1972. The microbe as a source of food. Annu. Rev. Microbiol. **26**:427–466.

32. **Kofoed, L. H.** 1975. The feeding biology of *Hydrobia ventrosa* (Montagu). I. The assimilation of different components of the food. J. Exp. Mar. Biol. Ecol. **19**:233–241.

33. **Kozlovskaja, L. S.** 1971. Der Einfluss der Wirbellosen auf die Tätigkeit der Mikroorganismen in Torfboden. Organismes du sol et production primaire. IV. Colloquium pedobiologiae, Dijon 14/19-IX-1970, p. 81–88.

34. **Kristensen, J. H.** 1972. Carbohydrases of some marine invertebrates with notes on their food and on the natural occurrence of the carbohydrates studied. Mar. Biol. **14**:130–142.

35. **Kukor, J. J., and M. M. Martin.** 1983. Acquisition of digestive enzymes by siricid woodwasps from their fungal symbiont. Science **220**:1161–1163.

36. **Malke, H.** 1965. Uber das Vorkommen von Lysozym in Insekten. Z. Allg. Mikrobiol. **5**:42–47.

37. **Martin, M. M.** 1979. Biochemical implications of insect mycophagy. Biol. Rev. Cambridge Philos. Soc. **54**:1–21.

38. **Martin, M. M.** 1983. Cellulose digestion in insects. Comp. Biochem. Physiol. A **75**:313–324.

39. **Martin, M. M.** 1984. The role of ingested enzymes in the digestive processes of insects, p. 155–172. *In* J. M. Anderson, A. D. M. Rayner, and D. Walton (ed.), Animalmicrobial interactions. Cambridge University Press, Cambridge.

40. **Martin, M. M., J. J. Kukor, J. S. Martin, D. L. Lawson, and R. W. Merritt.** 1981. Digestive enzymes of larvae of three species of caddisflies (Trichoptera). Insect Biochem. **11**:501–505.

41. **Martin, M. M., J. J. Kukor, J. S. Martin, T. E. O'Toole, and M. W. Johnson.** 1981. Digestive enzymes of fungus-feeding beetles. Physiol. Zool. **54**:137–145.

42. **Martin, M. M., and J. S. Martin.** 1978. Cellulose digestion in the midgut of the fungus-growing termite *Macrotermes natalensis*: the role of acquired digestive enzymes. Science **199**:1453–1455.

43. **Martin, M. M., and J. S. Martin.** 1979. The distribution and origins of the cellulolytic enzymes of the higher termite *Macrotermes natalensis*. Physiol. Zool. **52**:1–11.

44. **Martin, M. M., J. S. Martin, J. J. Kukor, and R. W. Merritt.** 1980. The digestion of protein and carbohydrate by the stream detritivore, *Tipula abdominalis* (Diptera, Tipulidae). Oecologia (Berlin) **46**:360–364.

45. **Martin, M. M., J. S. Martin, J. J. Kukor, and R. W. Merritt.** 1981. The digestive enzymes of detritus-feeding stonefly nymphs (Plecoptera; Pteronarcyidae). Can. J. Zool. **59**:1947–1951.

46. **Mattson, W. J.** 1980. Herbivory in relation to plant nitrogen content. Annu. Rev. Ecol. Syst. **11**:119–161.

47. **McHenry, J. G., T. H. Birkbeck, and J. A. Allen.** The occurrence of lysozyme in marine bivalves. Comp. Biochem. Physiol. B **63**:25–28.

48. **Monk, D. C.** 1976. The distribution of cellulase in freshwater invertebrates of different feeding habits. Freshwater Biol. **6**:471–475.

49. **Müller, W.** 1934. Untersuchungen über die Symbiose von Tieren mit Pilzen und Bakterien. Arch. Mikrobiol. **5**:84–147.

50. **Muzzarelli, R. A. A.** 1979. Chitin. Pergamon Press, Oxford.

51. **Nielsen, C. O.** 1962. Carbohydrases in soil and litter invertebrates. Oikos **13**:200–215.

52. **Nielsen, C. O.** 1963. Laminarinases in soil and litter invertebrates. Nature (London) **199**:1001.

53. **Payne, D. W., and L. M. Davidson.** 1974. Cellulose digestion in the locust, *Schistocerca gregaria*. J. Entomol. Ser. A **48**:213–215.

54. **Podnoboronov, V. M., G. I. Stephanchenok-Rudnik, and I. M. Grokhovsskaya.** 1975. Izuchenie antibakterial nogo deistviya organov i thanei kleschchei *Ornithodoros papillipes* Birula. Medskaya Parazitol. **44**:29–33.

55. **Reyes, V. G., and J. M. Tiedje.** 1973. Metabolism of ^{14}C-uniformly labelled organic material by woodlice [Isopoda: Oniscoidea] and soil microorganisms. Soil. Biol. Biochem. **5**:603–611.

56. **Reyes, V. G., and J. M. Tiedje.** 1976. Metabolism of ^{14}C-labelled plant materials by woodlice [*Tracheoniscus rathkei* Brandt] and soil microorganisms. Soil. Biol. Biochem. **8**:103–108.

57. **Reyes, V. G., and J. M. Tiedje.** 1976. Ecology of the gut microbiota of *Tracheoniscus rathkei* (Crustacea, Isopoda). Pedobiologia **16**:67–74.

58. **Rodrick, G. E., A. L. Vincent, and W. A. Sodeman, Jr.** 1980. Selected enzyme activities in *Physa gyrina* (Gastropoda) tissues. Comp. Biochem. Physiol. B **65**:177–180.

59. **Rossler, M. E.** 1961. Ernährungsphysiologische Untersuchungen an Scarabaedenlarven (*Oryctes nasicornis* L., *Melolontha melolontha* L.). J. Insect Physiol. **6**:62–80.

60. **Sands, W. S.** 1969. The association of termites and fungi, p. 495–524. *In* K. Krishna and F. M. Weesner (ed.), Biology of termites, vol. 1. Academic Press, Inc., New York.

61. **Schröder, P.** 1981. Zur Ernährungsbiologie der Larven von *Odagmia ornata* Meigen (Diptera: Simuliidae). Arch. Hydrobiol. Suppl. **59**:97–133.

62. **Slansky, F., and J. M. Scriber.** 1982. Selected bibliography and summary of quantitative food utilization by immature insects. Bull. Entomol. Soc. Am. **28**:43–55.

63. **Southwick, E. E., and D. Pimentel.** 1981. Energy efficiency of honey production by bees. BioScience **31**:730–732.

64. **Sova, V. V., L. A. Elyakova, and V. E. Vaskovsky.** 1970. The distribution of laminarinases in marine invertebrates. Comp. Biochem. Physiol. **32**:459–464.

65. **Stewart, W. D. P. (ed.).** 1974. Algal physiology and biochemistry, p. 1–85, 206–280, 757. University of California Press, Berkeley.

66. **Szabo, I., M. Marton, and I. Buti.** 1969. Intestinal microflora of the larvae of St. Mark's fly. IV. Studies on the intestinal bacterial flora of the larval population. Acta Microbiol. Acad. Sci. Hung. **16**:381–397.

67. **Terra, W. R., C. Ferreira, and A. G. DeBianchi.** 1979. Distribution of digestive enzymes among the endo- and ectoperitrophic spaces of midgut cells of *Rhynchosciara* and its physiological significance. J. Insect Physiol. **25**:487–494.

68. **Waldbauer, G. P.** 1968. The consumption and utilization of food by insects. Adv. Insect Physiol. **5**:229–288.

69. **Wavre, M., and R. O. Brinkhurst.** 1971. Interactions between some tubificid oligochaetes and bacteria found in the sediments of Toronto Harbour, Ontario. J. Fish. Res. Board. Can. **28**:335–341.

Influence of Plasmids on the Colonization of the Intestine by Strains of *Escherichia coli* in Gnotobiotic and Conventional Animals

Y. DUVAL-IFLAH AND J. P. CHAPPUIS

Laboratoire d'Ecologie Microbienne, I.N.R.A.-C.N.R.Z., 78350 Jouy-en-Josas, France

Plasmids are extrachromosomal DNA elements of bacteria that constitute a dispensable gene pool. Under most circumstances, they are not essential for the successful growth and metabolism of their host bacterium. However, they carry genes for supplementary activity that may allow their host to better survive in adverse environments. The best-known plasmids are those that specify antibiotic resistances (13, 29, 33, 56). Many plasmids have also been found to contribute to the pathogenicity of a variety of microorganisms (13). Some strains of the common intestinal bacterium *Escherichia coli* have been identified as the causative agents of a variety of diarrheal illnesses in young and adult humans and in young domestic animals (31, 38, 42, 45, 55, 57). These enteropathogenic strains are able to elaborate one or both of two types of enterotoxins, the heat-labile toxin designated LT and the heat-stable toxin designated ST. They also elaborate proteinaceous surface antigens called colonization or adherence antigens. Genetic analysis of the enterotoxigenic *E. coli* strains isolated from pigs and calves indicated that enterotoxin production (13, 19, 48, 50, 52, 53) and elaboration of surface antigens (30, 34, 43, 52, 54) were carried on plasmids. In enterotoxigenic strains isolated from humans, the same plasmid may encode for both the ST toxin and the colonization factor antigen CFA/I (5, 15, 27), or for both ST and LT toxins and colonization factor antigen CFA/II (5, 36). Other virulence plasmids identified in *E. coli* strains include ColV, which confers upon its host a greater ability to survive in the blood, peritoneal fluid, and alimentary tract of a variety of test animals (46, 49, 50), and *Vir* plasmids, which specify antigen and a toxin lethal for rabbits, mice, and chickens (46).

The colonization of the gut by strains of *E. coli* has attracted the attention of numerous investigators for many reasons. Strains of this species constitute a typical and constant member of the gut flora in adults, and yet its population of between 10^5 and 10^8 cells per g of feces is but a small proportion of a total fecal bacterial population, which is in excess of 10^{11}/g (18, 21). A second reason is that certain strains of this species may be pathogenic for both human newborns and adults and for young animals. The pathogenicity of such strains is mostly plasmid associated, as indicated before. The question which arises is to know whether the possession of extrachromosomal elements confers on the host bacterium an ecological advantage or disadvantage. It is now admitted that the plasmids responsible for the elaboration of specific adherence antigens such as K88, K99, CFA/I, and CFA/II confer on the host bacterium the ability to proliferate in sites of the small intestine which are not normally colonized by bacteria. There is considerable evidence that the K88 antigen, peculiar to *E. coli* strains enterotoxigenic for swine, is adhesive for swine intestinal epithelium in neonatal piglets (2, 23, 48, 50, 51). The K88$^+$ *E. coli* cells appear to proliferate in the anterior small intestine where normally very few bacteria are present, instead of being carried along with the normal movement of the chyme. Smith and co-workers (47, 52) emphasized the importance of K99 antigen in facilitating adhesion of bacteria on the anterior intestinal epithelium of experimentally infected calves and lambs as well as in the posterior small intestine of pigs (32, 50). Volunteer studies and experiments in rabbits have confirmed the hypothesis that CFA/I plays a significant role as a virulence factor in human diarrhea. The volunteers who ingested CFA/I-positive *E. coli* shed the organism in the stool for more than 7 days, whereas those ingesting the CFA/I-negative strain shed the organism for a maximum of 3 days (14). The colonization of the anterior small intestine is an abnormal event. Therefore, adhesion to epithelial cells of the small intestine may constitute a pathological event even though the adhering bacteria produce no toxins.

Experiments with R factor-carrying *E. coli* in the human gastrointestinal tract have shown that, in general, antibiotic-resistant populations survive less well than their R$^-$ counterparts (1, 20, 25). Thus, strains of *E. coli* that colonize the gastrointestinal tract of humans in the absence of antibiotic selection pressure are not usually R$^+$. It can be postulated, therefore, that R plasmid-carrying strains are at an ecological disadvantage in the gastrointestinal tract when faced with the antagonistic effects of a complex normal flora or that plasmid-free strains may inhibit the establishment of their corresponding plasmid-carrying strains in the gastrointestinal tract ecosystem.

The aim of our work was to demonstrate the existence of intraspecific interactions between plasmid-free and plasmid-carrying strains of *E. coli* in the digestive tracts of gnotobiotic mice and piglets and of conventional human newborns. We also investigated the role of K88 in the pathogenicity of a K88$^+$ *E. coli* strain in gnotobiotic piglets.

Axenic C3H mice (CNRS, Centre de Sélection et d'Elevage des Animaux de Laboratoire, Orléans, France) were maintained in plastic-film isolators (La Calhene, Bezons, France) and fed pelleted commercial diet sterilized by γ irradiation (4 Mrad). Piglets were obtained by spontaneous delivery as previously described (9) and were maintained in plastic-film isolators. They were colostrum deprived and fed autoclaved concentrated cow milk supplemented with glucose and vitamins.

Forty-six human newborns from Antoine Béclère Maternity (France) were selected for this study. One milliliter of an 18-h *E. coli* culture (5 × 10^8 viable cells) was administered by mouth to adult mice and to 2-day-old piglets. The human newborns received 2 ml of a 10^{-2} dilution of the 18-h bacterial culture within the first 2 h of life. Bacterial counts were performed from freshly passed feces and stools and from small intestine segment contents and tissues. Bacterial strains and media used in selective counts are described in Table 1. Further details were previously described (10–12).

ANTAGONISMS AMONG ISOGENIC STRAINS OF *E. COLI* WHICH DIFFER THROUGH R PLASMID CARRIAGE IN THE DIGESTIVE TRACT OF GNOTOBIOTIC MICE

The *E. coli* strains EM0, EM, and EM4, when monoassociated with axenic mice, established themselves at a high population level of 10^9 to 10^{10} viable bacteria per g of feces. These strains, when first implanted in the digestive tract of mice, inhibited the further establishment of each one of their respective transconjugant strains EM9, EM8, and EM5, which carry the plasmid pYD1 responsible for resistance to 14 antibiotics. At the steady state, the ratio between the pYD1$^-$ strains and the pYD1$^+$ strains ranged from 10:3 to 10:2 when expressed as the log$_{10}$. This antagonism was no more effective when mice were first associated with the pYD1-carrying strains. In this reverse order of inoculation, the latter strains established themselves at popu-

lation levels of 10^9 to 10^{10}/g of feces and the pYD1-free strains were at 10- to 100-fold lower levels. Inhibition of the establishment of pYD1-carrying strains was restored when the mice first associated with these latter strains received 1 ml of a 10^{-2} dilution of feces collected from mice which were colonized with the pYD1-free strains for more than 30 days. These results are illustrated in Fig. 1. This figure represents the antagonism between strain EM4, a pYD1-free strain resistant to rifampin, and EM8, a pYD1-carrying strain. When strain EM4 was first established in mice, it inhibited the establishment of EM8 (Fig. 1a). The latter strain spontaneously lost its pYD1 plasmid, and its pYD1$^-$ derivatives became established in the dominant population. In vivo conjugation occurred between EM4 and EM8, and the colonization of the resulting transconjugants was inhibited in the same manner as that of strain EM8. The antagonisms were different when mice were first colonized with EM8 (Fig. 1b, c, and d). A different type of antagonism was observed between pYD3-positive and -negative strains. Strain EM0, a pYD3-free strain, was shown to inhibit colonization by EM, the pYD3-carrying strain. This antagonism was not affected by the order of inoculation of the strains into mice. Contrary to plasmids pYD1 and pYD3, the defective mutant pYD4 was shown to confer on the carried strain EM6 an ecological advantage since this latter strain repressed, but did not inhibit, the establishment of its parental pYD4-free strain EM0. This type of interaction is designated as a permissive barrier effect. Strain EM6 was again at a disadvantage when faced with FEM0, which was EM0 associated for at least 30 days with mice.

Antagonism between pYD1-free strain EM and its transconjugant pYD1-carrying strain EM8 was also studied in the digestive tract of mice which were associated with a complex human fecal flora or with some of its components. In the presence of complex human fecal flora, strain EM established itself at a population level of 10^5 to 10^6/g of feces and strain EM8 was at a 100-fold lower level. In the presence of a simplified human fecal flora, composed of 41 strains (2 *Streptococcus*, 2 *Peptostreptococcus*, 2 *Bifidobacterium*, 4 *Eubacterium*, 9 *Bacteroides*, 22 *Clostridium*), strain EM became established at a high population level of 10^8/g of feces, whereas strain EM8 was not able to establish itself at a detectable level (below 10^2/g of feces). In the presence of a high population of *Bacteroides*, 10^9 to 10^{10}/g of feces, strains EM and EM8 established themselves at respective population levels of 10^8 to 10^9 and 10^3 to 10^4/g of feces. In all of the previous experiments strain EM rapidly decreased and was replaced by

TABLE 1. Strains used

Strain[a]	Characteristics[b]	Plasmid(s)	Source	Colicin[c]	Ent[d]	Medium used for selective counts[e]
EM	*lacZ* Lac$^+$ Raf$^-$ Tra$^+$ Su	pYD3	Human fecal flora	+	—	DCA-Raf or Drig-Su
EM0	*lacZ* Lac$^+$ Raf$^+$	None	EM6 cured of plasmid pYD4	–	—	DCA-Raf
EM1	*lacZ rpoB nalA* Lac$^-$ 42$^-$ Tra$^+$ Col Min Ap Cb Cp Sm Bt Nm Km Lv Gm Tm Cm Su Tc Pm	pYD1	Human urine	–	—	MH-Ap-Col
EM4	*lacZ ropB* Lac$^+$ Raf$^+$ Tra$^+$ Su	pYD3	Mutant of EM resistant to rifampin	+	—	MH-Rif
EM5	*lacZ ropB* Lac$^+$ Raf$^-$ Tra$^+$ Ap Cb Cp Sm Bt Nm Km Pm Lv Gm Tm Cm Su Tc	pYD1, pYD3	Conjugation, EM1 × EM4	+	—	MH-Rif-Tc
EM6	*lacZ* Lac$^+$ Raf$^+$ Tc	pYD4	Transconjugant carrying R64 *drd*11 (39)	–	—	Drig-Tc
EM8	*lacZ* Lac$^+$ Raf$^-$ Tra$^+$ Ap Cb Cp Sm Bt Nm Km Pm Lv Gm Tm Cm Su Tc	pYD1, pYD3	Conjugation, EM1 × EM	+	—	Drig-Ap
EM9	*lacZ* Lac$^+$ Raf$^+$ Tra$^+$ Ap Cb Cp Sm Bt Nm Km Pm Lv Gm Tm Cm Su Tc	pYD1	Conjugation, EM1 × EM0	–	—	DCA-Raf-Ap
EPEC	*lacZ* Lac$^+$ Raf$^+$ Tra$^+$ Km Lv Tc Ent$^+$ K88$^+$ 987P$^-$	Ent, K88	Diarrheal piglets[f]	ND	LT, ST	DCA-Raf-Tc
EEC	*lacZ* Lac$^+$ Raf$^-$ Tra$^+$ Km Lv Tc Ent$^+$ 987P$^-$	Ent	K88-free EPEC derivative	ND	LT, ST	DCA-Raf-Tc
JP1 (K88$^+$)	*lacZ* Lac$^+$ Raf$^+$ Tra$^+$ K88$^+$	K88	K88-carrying transconjugant	ND	—	DCA-Raf

[a] All the strains are *E. coli* except EM1, which is *Serratia liquefaciens*.

[b] Lac$^-$, Lactose not fermented; Lac$^+$, lactose fermented; Raf$^-$, raffinose not fermented; Raf$^+$, raffinose fermented; 42$^-$, no growth at 42°C; Min, resistance to minocyline; Col, resistance to colistin. Other symbols are as in Bachmann et al. (3) and in Novick et al. (33).

[c] Colicin production tested with *E. coli* K-12 as the sensitive strain; ND, not done.

[d] Enterotoxin LT, thermolabile, tested with Y1 cells (26) and with permeability of rabbit skin (16); ST, thermostable, tested with infant mice (7).

[e] The basic media were MH, Mueller-Hinton; Drig, Drigalski agar medium which contained lactose and crystal violet for the indication of lactose fermentation; DCA-Raf, modified deoxycholate agar medium supplemented with raffinose and neutral red for the indication of raffinose fermentation. These media were supplemented with one or two of the following antibiotics: ampicillin (Ap) at 100 µg/ml in MH and 10 µg/ml in Drig; colistin (Col) at 200 µg/ml; rifampin (Rif) at 30 µg/ml; tetracycline (Tc) at 16 µg/ml; sulfonamide (Su) at 100 µg/ml.

[f] Kindly supplied by L. Renault.

strain EM8 under ampicillin intake (500 µg/ml of drinking water).

An experiment was done to determine whether the antibiotic pressure selected for a specific plasmid or for special strains carrying this plasmid. Gnotobiotic mice associated with the simplified human fecal flora were inoculated with strain EM8 and with a strain of *Serratia liquefaciens* which harbored plasmid pYD1. Neither strain was steadily established. However, both strains established themselves at a high population level in mice which were given ampicillin orally. But later on, during a period of 20 days following the end of antibiotic intake, the persistence of strain EM8 was observed whereas the strain of *Serratia* was eliminated again. Thus, the antibiotic exerted its selective pressure on the strain, not on the plasmid.

PRECOCIOUS IMPLANTATION OF A PLASMID-FREE STRAIN OF *E. COLI* IN THE DIGESTIVE TRACT OF CONVENTIONAL HUMAN NEWBORNS: EFFECT ON THE KINETICS OF THE ESTABLISHMENT OF ANTIBIOTIC-RESISTANT *E. COLI*

The results with gnotobiotic mice have shown that previous establishment of a plasmid-free strain, EM0, generally inhibited the further establishment of R-carrying *E. coli*. The present work was undertaken to determine whether a precocious implantation of EM0 into newborns

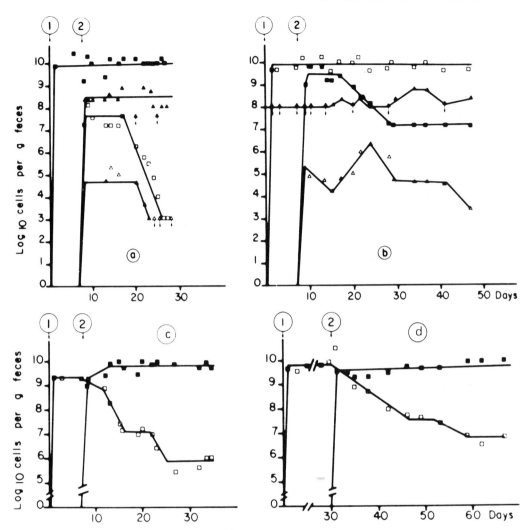

FIG. 1. In vivo antagonism between *E. coli* strains EM4 and EM8. (a) Axenic mice were successively inoculated with 1 ml of an 18-h culture of strains EM4 (1) and EM8 (2). (b, c, d) Axenic mice were first inoculated with 1 ml of an 18-h culture of strain EM8 (1). Then 7 days later (b) mice received 1 ml of an 18-h culture of EM4 (2); (c, d) 7 and 30 days later, respectively, mice received 1 ml of a 10^{-2} dilution of feces collected from mice which had been associated with strain EM4 for more than 30 days (2). Symbols: (a, b) ■, strain EM4; □, strain EM8; △, EM8 (pYD1), transconjugant resulting from in vivo mating between EM4 and EM8; ▲, EM8 (pYD1⁻); ▲ and △ with arrows, strains were not detectable; (c, d) □, strain EM8; ■, EM4 associated with axenic mice for more than 30 days.

might inhibit or reduce the spontaneous establishment of antibiotic-resistant *E. coli*.

Twenty-two healthy human newborns were inoculated within 2 h of life with *E. coli* strain EM0. This strain was previously shown to be nontoxigenic by both in vitro and in vivo tests and was sensitive to 18 usual antibiotics. Strain EM0 became established at a high population level of 10^7 to 10^{10}/g of feces in the 22 newborns who ingested it. It remained at these levels in 19 of them during the following 4 days. Other types of *E. coli* also established themselves at compa-

rable levels. However, EM0 was predominant in 9 of the 22 infants. *E. coli* strains resistant to ampicillin or tetracycline, or both, appeared early at high levels in six 2-day-old newborns, but later were not demonstrable in any of them. In 24 control noninoculated newborns, *E. coli* spontaneously became established at levels comparable to that of the inoculated infants. In seven of the control newborns antibiotic-resistant *E. coli* appeared early in the dominant population during the observation period (up to 7 days).

ANTAGONISMS BETWEEN HETEROGENIC STRAINS OF *E. COLI*, WHICH DIFFER THROUGH VIRULENCE PLASMIDS, IN THE DIGESTIVE TRACT OF GNOTOBIOTIC PIGLETS

In view of the previous results, which showed that some strains of *E. coli* may inhibit the establishment of isogenic or heterogenic R plasmid-carrying strains, it was of interest to know whether such antagonisms might occur between strains of *E. coli* which differ through virulent plasmid carriage, such as Ent and K88.

Effect of the monoassociation of axenic piglets with *E. coli* strains harboring Ent or K88, or both plasmids. Axenic colostrum-deprived piglets maintained in plastic-film isolators were used. Two piglets monoassociated with the entero-pathogenic (EPEC) strain C5148 (Ent$^+$, K88$^+$) rapidly developed diarrhea and died within 24 to 48 h. Four other piglets which were inoculated with the enterotoxigenic (EEC) strain C5148 (Ent$^+$, K88$^-$), which is the K88-free derivative of the previous EPEC strain, also had watery feces and died within 2 to 6 days. At autopsy piglets of both groups had obvious signs of enteritis: congestion and sometimes hemorrhagic lesions of the stomach and proximal small intestine. Bacterial counts were performed from the proximal (S1) and the distal (S3) small intestinal segment contents. The number of the EPEC and the EEC strains was always $\geq 10^8$/g of both S1 and S3 segments.

Nine piglets were monoassociated with strain JP1 (K88$^+$), which harbored only the K88 plasmid. Six of them survived and were healthy for more than 10 days and were sacrificed at 11, 12, 13, 14, 17, and 19 days after ingestion of the JP1 (K88$^+$). At autopsy, no signs of enteritis were observed. The remaining three piglets became ill, but never had diarrhea. One of them died on day 7, the two others were sacrificed when dying at 4 and 10 days after ingestion of JP1 (K88$^+$). At autopsy, congestion and necrosis of the distal part of the small intestine were observed. In two

of the piglets the stomach was necrotized, and the colon and the cecum were congested. Bacterial counts were performed from contents and intestinal tissues of S1 and of the middle part of the small intestine (S2). The number of JP1 (K88$^+$) in the S1 and S2 tissues was $>10^7$/g in the three ill piglets and $<10^5$/g in the six healthy piglets. Its number in S1 and S2 contents was variable in the healthy piglets, but in most of the cases lower than in the ill piglets (Table 2). Thus, EEC and EPEC strains of *E. coli* both provoked diarrhea and death of all the tested gnotobiotic piglets. By contrast, strain JP1 (K88$^+$) never provoked diarrhea. However, when the small intestinal tissues were efficiently colonized by strain JP1 (K88$^+$), the gnotobiotic piglets became ill.

Effect of diassociation of axenic piglets with EM0 and EEC strains. Five axenic colostrum-deprived piglets were inoculated with the plasmid-free strain EM0, which rapidly became established at a high population level of 10^{10}/g of feces. Then 48 h later, the EEC strain C5148 (Ent$^+$ K88$^-$) was inoculated into the piglets. All the diassociated piglets survived more than 10 days after ingestion of the EEC strain. One piglet remained healthy with normal feces and was sacrificed 15 days after the EEC ingestion. A second one also had normal feces during the first 9 days, then developed diarrhea and died 2 days later. The three remaining piglets had light diarrhea the day after EEC ingestion. One of them recovered 13 days later and was sacrificed when healthy after 5 additional days; another one died on day 11, and the last one was sacrificed on day 13. When the feces were normal, the numbers of EM0 and EEC strains were 10^{10} and $<10^9$/g of feces, respectively. When the feces were watery, the population levels of both strains were never greater than 10^9/g of feces (Table 3). At autopsy, the dead piglets had obvious signs of enteritis, whereas those which were sacrificed had normal intestines. Bacterial counts were performed from the S1 and S3 segment contents. The EEC strain

TABLE 2. Population level of JP1 (K88$^+$) and EPEC (Ent$^+$ and K88$^+$) in small intestine contents and tissues of healthy and ill gnotobiotic piglets

Strain	Mean (SD) of $\log_{10}$ *E. coli* cells							
	Contents				Tissues			
	Healthy piglets[a]		Dead or dying piglets		Healthy piglets		Dead or dying piglets	
	S1	S2	S1	S2	S1	S2	S1	S2
JP1 (6/9)[b]	6.3 (1.4)	7.3 (1.2)	8.3 (0.6)	10.1 (0.1)	4.8 (1.5)	4.5 (1.0)	7.3 (1.7)	9.2 (0.3)
JP1 +	6.5 (0.4)	6.8 (1.5)	7.0 (0.6)	7.2 (0.6)	4.7 (0.7)	4.3 (0.8)	7.3 (0.9)	6.9 (0.9)
EPEC (7/9)	5.6 (0.6)	5.8 (1.3)	7.5 (0.0)	7.7 (0.2)	4.0 (0.5)	3.9 (0.7)	7.8 (0.5)	7.3 (0.6)

[a] S1 and S2, Proximal and middle part of the small intestine, respectively.
[b] Number of healthy piglets/total number of piglets tested.

TABLE 3. Kinetics of the establishment of *E. coli* in the diassociated piglets[a]

Piglets diassociated with:	Mean (SD) of $\log_{10}$ *E. coli* cells			
	Healthy pigs		Dead or diarrheal piglets	
	1–2 days[b]	3–18 days	1–2 days	3–18 days
JP1(K88[+]) +	9.9 (±0.3)	9.3 (±0.3)	9.9 (±0.2)	6.4 (±0.6)
EPEC (7/9)[c]	7.5 (±0.5)	9.0 (±0.2)	6.7 (±1.0)	6.8 (±0.2)
EM0 +	9.9 (±0.2)	10.0 (±0.2)	8.4 (±0.6)	8.7 (±0.6)
EEC (2/5)	8.6 (±0.7)	8.6 (±0.6)	8.9 (±0.1)	8.5 (±0.5)

[a] Diassociated piglets: gnotobiotic piglets colonized with two bacterial strains.
[b] Number of days after EPEC or EEC ingestion.
[c] Healthy piglets/total tested piglets.

was present at 10^9/g of S3 contents of the dead piglets, whereas this number was relatively low (5×10^5 to 3×10^7) in the S3 contents of the sacrificed piglets as well as in S1 segments of both the healthy and the ill piglets. Thus, associating piglets first with EM0 then with the EEC strain prevented severe diarrhea and protracted the survival time.

Effect of diassociation of axenic piglets with JP1 (K88[+]) and EPEC strains. Nine piglets were first inoculated with strain JP1 (K88[+]), which rapidly became established at a high population level of 10^{10}/g of feces. Three or more days later they were inoculated with an EPEC strain. Kinetics of the establishment of both strains and appearance of diarrhea were observed. During the following 2 days the EPEC became established at a low population level of about 10^7/g of feces in all nine piglets. Then the feces became soft in seven piglets. These piglets remained healthy until they were sacrificed at 3, 4, 8, 9, and 10 days after EPEC ingestion. The population levels of both strains were at about 10^9/g of feces (Table 3). At autopsy no signs of enteritis were observed. The remaining two piglets rapidly developed watery feces and were dying on the second and on the third days after EPEC ingestion. The total population levels of both strains were $<10^7$/g of feces (Table 3). At autopsy no signs of enteritis were observed. Bacterial counts were performed from the contents and tissues of S1 segments and of the middle part (S2) of the small intestine. The number of EPEC and JP1 (K88[+]) in S1 and S2 tissues was $>10^6$/g in the dying piglets and $<10^5$ in the healthy ones. Their number in the S1 and S2 contents was variable in the healthy piglets, but generally lower than in the dying ones (Table 2). These results indicated that two of the nine piglets were efficiently colonized with both EPEC and JP1 (K88[+]) strains and developed diarrhea in the same manner as the EPEC monoassociated piglets.

Our results show that all of the *E. coli* strains used in our experiments could establish at high population levels in the digestive tracts of monoassociated gnotobiotic mice and piglets. Interactions between isogenic strains when associated with mice, however, were readily apparent. When the parental strain had become established before its plasmid-carrying derivative, the latter had a marked ecological disadvantage. Figure 1a shows that, in the case of in vivo mating during the assay, colonization by the transconjugant harboring pYD1 was also inhibited by the dominant parental flora. Similar but less marked effects were observed with strains carrying plasmid pYD3. The outcome of these interactions, however, is dependent on the order of inoculation of the strains. When plasmid-carrying strains were established first, it was observed that the antagonisms were changed and the plasmid-free strains no longer exerted a drastic barrier effect against their plasmid-harboring derivatives. These results might indicate that when the animals are colonized first with a plasmid-free strain, it becomes established in such large numbers that, just by the effect of its numbers, it prevents subsequent colonization by plasmid-containing strains. Results obtained by Ducluzeau et al. (8) and by Hudault et al. (22) have shown that two target strains, *Shigella flexneri* and *Lactobacillus casei*, were eliminated by an association of *E. coli* plus a strain of *Clostridium* and by *E. coli*, respectively, even when these target strains first colonized the gnotobiotic mice. This indicated that large initial numbers of the target organism may not affect the expression of the barrier effect. Moreover, our results show that the barrier effect exerted by plasmid-free strains was restored when the latter were "adapted" to the animal host. This adaptation was achieved when the strains were maintained for 30 days in the digestive tracts of gnotobiotic mice. This observation indicated that when the strains were in the same physiological state, the plasmid-free strains were at an ecological advantage even when they were inoculated later. Observation with an electron microscope showed large differences after adaptation to the host. The presence of blebs around the adapted cells might be explained by an increased synthesis of lipopolysaccharides and phospholipids (24). This has been previously described and discussed in detail (11).

Antibiotic intake resulted in a rapid increase of the plasmid-carrying strains. However, it was also shown that the antibiotic selection pressure depends on the characteristics of the bacterial host rather than on those of the plasmid that is carried. For example, E. coli and S. liquefaciens, which harbor the same R plasmid, are not maintained at the same level under the same regimen. Similar observations have been noted in humans (20, 37, 44).

At birth, the digestive tracts of human newborns are rapidly colonized with members of the Enterobacteriaceae (4, 35, 40), especially E. coli, which constitute the dominant, and almost the exclusive, population during the first days. Our results show that E. coli strains spontaneously establish themselves at high levels of 10^{10}/g of feces 24 or 48 h after birth, and, unlike in adults (1, 25), the dominant E. coli strains might be antibiotic resistant. It was also shown that an exogenous E. coli strain (EM0) may be steadily established in the digestive tract of human newborns when inoculated early after birth. The establishment of this strain at high population levels of 10^9 to 10^{10}/g of feces contributed to repression of the further establishment of multiply antibiotic-resistant E. coli strains which presumably harbored plasmids (10). A plasmid-free strain might therefore be able to compete efficiently with plasmid-carrying strains in gnotobiotic and even in conventional conditions. The ecological importance of this phenomenon is obvious, and it is hoped that inoculation of strains like EM0 would be generalized to contribute to an early establishment of a selected beneficial flora.

Competition against virulence plasmid-harboring strains in the digestive tract of gnotobiotic piglets was shown to be more subtle. EEC strains harboring only the Ent plasmid were found to provoke diarrhea and death of monoassociated piglets. These results corroborate those of Miniats and Gyles (28). However, such strains were described as nonpathogenic for conventional weaned piglets (51). In experiments on diassociation of piglets with E. coli EM0 and EEC, our results indicated that the pathogenesis of EEC is to some extent reduced. All the piglets diassociated with EM0 and EEC survived longer than their counterparts monoassociated with EEC. The diassociated piglets had

no diarrhea for variable periods of time. During these periods, the population level of EM0 in the feces was maximal, 10^{10}/g, and that of EEC was lower and rarely reached 10^9/g. This antagonism was aleatory since in some piglets, which concomitantly had a light diarrheal episode, the total E. coli population shifted down. It is therefore assumed that EM0 confers on the diassociated piglets a protection against EEC. But it is difficult to conclude whether this protection is due to reduction of the population level of EEC or to reduction of the available amounts of enterotoxin(s) in vivo. Association of the axenic piglets with JP1 (K88$^+$), which harbored only the K88 plasmid, showed that the piglets might be divided into two groups. The piglets of one group were shown to be efficiently colonized with JP1 (K88$^+$) since this strain was found in large number throughout the small intestinal contents and tissues. These piglets probably belong to the phenotype that permits adhesion of the K88$^+$ organisms to the small intestinal epithelium (17, 41) and have been classified as 88S (S = susceptible) in contrast with those that do not permit adhesion of the K88$^+$ organisms and have been classified as 88R (R = resistant). The piglets of the second group probably belong to the 88R phenotype. The piglets which did not permit adhesion of JP1 (K88$^+$) did not develop diarrhea, but they became ill after a time lapse of 4 to 10 days. Autopsy showed lesions that are different from those due to enteritis and especially necrosis of the stomach and of the distal part of the small intestine and of the colon. The number of 88S we found is lower than that reported by others (41). Experiments are in progress to test piglets originating from different herds. The number of JP1 (K88$^+$) cells in the small intestinal tissues of the 88R piglets was always low (Table 2). Moreover, these piglets appeared healthy up to 19 days and their digestive tracts had a normal appearance. The diassociation of piglets with JP1 (K88$^+$) and EPEC revealed a similar phenomenon. Two piglets among the nine tested might be considered as 88S since they rapidly developed diarrhea and their small intestinal tissues were efficiently colonized by both strains (Table 2). The other seven piglets did not permit adhesion of any of the strains and therefore might be considered as 88R. The original finding is that the association of JP1 (K88$^+$) reduced to a large extent the pathogenesis of the EPEC strain in these 88R piglets since they survived and were healthy up to 10 days. These results might be compared to those obtained with the diassociation of piglets with strains EM0 and EEC. In the 88R piglets, JP1 (K88$^+$) behaved toward the EPEC strain as did strain EM0 toward the EEC strain. By contrast, bacterial competition cannot be achieved in 88S piglets. These findings contradict those previously described by others (6).

Our findings show that, in general, the plasmids confer on the host bacterium an ecological disadvantage. However plasmids such as K88, K99, CFA/I, and CFA/II confer on their host bacterium the ability to proliferate in the small intestine, where normally few bacteria are present. But these strains are pathogenic and may kill their animal host. They consequently destroy their own ecosystem. Therefore, this can also be considered a type of ecological disadvantage. Our results also show the existence of intraspecific interactions which we believe are important in newborns who are colonized with few bacterial varieties. Precocious implantation of a mixture of strains selected for their ability to compete with other undesirable strains might constitute a way to prevent establishment of pathogenic strains during the first days of life. Work is in progress to look for strains of *E. coli* more competitive than those described here.

LITERATURE CITED

1. **Anderson, J. D.** 1974. The effect of R-factor carriage on the survival of *Escherichia coli* in the human intestine. J. Med. Microbiol. **7:**85–90.
2. **Arbuckle, J. B. R.** 1970. The location of *Escherichia coli* in the pig intestine. J. Med. Microbiol. **3:**333–340.
3. **Bachmann, B. J., K. B. Low, and A. L. Taylor.** 1976. Recalibrated linkage map of *Escherichia coli* K-12. Bacteriol. Rev. **40:**116–167.
4. **Borderon, J. C., E. Borderon, and Y. A. Chabbert.** 1974. Etude quantitative de la colonisation par entérobactéries multirésistantes des enfants prématurés. Ann. Microbiol. (Paris) **125B:**45–57.
5. **Craviato, A. M., M. McConnell, B. Rowe, S. M. Scotland, H. R. Smith, and G. A. Willshaw.** 1980. Studies of colonisation factors in *Escherichia coli* strains isolated from humans: identification and plasmid analysis, p. 530–532. *In* R. C. W. Berkely, J. M. Lynch, J. Melling, P. R. Rutter, and B. Vincent (ed.), Microbial adhesion to surfaces. Ellis Horwood Ltd., Chichester, England.
6. **Davidson, J. N., and D. C. Hirsh.** 1976. Bacterial competition as a means of preventing neonatal diarrhea in pigs. Infect. Immun. **13:**1773–1774.
7. **Dean, A. G., Y. C. Ching, R. G. Williams, and L. B. Harden.** 1972. Test for *Escherichia coli* enterotoxin using infant mice: application in a study of diarrhea in children in Honolulu. J. Infect. Dis. **125:**407–411.
8. **Ducluzeau, R., and P. Raibaud.** 1974. Interaction between *Escherichia coli* and *Shigella flexneri* in the digestive tract of "gnotobiotic" mice. Infect. Immun. **9:**730–733.
9. **Ducluzeau, R., P. Raibaud, B. Lauvergeon, P. Gouet, Y. Riou, C. Griscelli, and J. C. Ghnassia.** 1976. Immediate postnatal decontamination as a means of obtaining axenic animals and human infants. Can. J. Microbiol. **22:**563–566.
10. **Duval-iflah, Y., M. F. Ouriet, C. Moreau, N. Daniel, J. C. Gabilan, and P. Raibaud.** 1982. Implantation précoce d'une souche de *Escherichia coli* dans l'intestin de nouveau-nés humains: effet de barrière vis-à-vis de souches de *E. coli* antibiorésistantes. Ann. Microbiol. (Paris) **133A:**393–408.
11. **Duval-iflah, Y., P. Raibaud, and M. Rousseau.** 1981. Antagonisms among isogenic strains of *Escherichia coli* in the digestive tracts of gnotobiotic mice. Infect. Immun. **34:**957–969.
12. **Duval-iflah, Y., P. Raibaud, C. Tancrede, and M. Rousseau.** 1980. R-plasmid transfer from *Serratia liquefaciens* to *Escherichia coli* in vitro and in vivo in the digestive tract of gnotobiotic mice associated with human fecal flora. Infect. Immun. **28:**981–990.
13. **Elwell, L. P., and P. L. Shipley.** 1980. Plasmid-mediated factors associated with virulence of bacteria to animals. Annu. Rev. Microbiol. **34:**465–496.
14. **Evans, D. G., D. J. Evans, Jr., W. S. Tjoa, and H. L. Dupont.** 1978. Detection and characterization of colonization factor of enterotoxigenic *Escherichia coli* isolated from adults with diarrhea. Infect. Immun. **19:**727–736.
15. **Evans, D. G., R. P. Silver, D. J. Evans, Jr., D. G. Chase, and S. L. Gorbach.** 1975. Plasmid-controlled colonization factor associated with virulence in *Escherichia coli* enterotoxigenic for humans. Infect. Immun. **12:**656–667.
16. **Evans, D. J., D. G. Evans, and S. L. Gorbach.** 1973. Production of vascular permeability factor by enterotoxigenic *Escherichia coli* isolated from man. Infect. Immun. **8:**725–730.
17. **Gibbons, R. A., R. Selwood, M. Burrows, and P. A. Hunter.** 1977. Inheritance of resistance to neonatal diarrhea in the pig: examination of the genetic system. Theor. Appl. Genet. **51:**65–70.
18. **Gustafsson, B. E.** 1982. The physiological importance of colonic microflora. Scand. J. Gastroenterol. **77(Suppl.):**117–131.
19. **Gyles, C. L., M. So, and S. Falkow.** 1974. The enterotoxin plasmids of *Escherichia coli*. J. Infect. Dis. **130:**40–49.
20. **Hartley, C. L., and M. H. Richmond.** 1975. Antibiotic resistance and survival of *Escherichia coli* in the alimentary tract. Br. Med. J. **4:**71–74.
21. **Holdeman, L. V., I. J. Good, and W. E. C. Moore.** 1976. Human fecal flora: variation in bacterial composition within individuals and a possible effect of emotional stress. Appl. Environ. Microbiol. **31:**359–375.
22. **Hudault, S., R. Ducluzeau, F. Dubos, P. Raibaud, J. C. Ghnassia, and C. Griscelli.** 1976. Elimination du tube digestif d'un enfant "gnotoxénique" d'une souche de *Lactobacillus casei* issue d'une préparation commerciale: démonstration chez des souris gnotoxéniques du rôle antagoniste d'une souche de *Escherichia coli* d'origine humaine. Ann. Microbiol. (Paris) **127B:**75–82.
23. **Jones, G. W., and J. M. Rutter.** 1972. Role of K88 antigen in the pathogenesis of neonatal diarrhea caused by *Escherichia coli* in piglets. Infect. Immun. **6:**918–927.
24. **Knox, K. W., M. Vesk, and E. Work.** 1966. Relation between excreted lipopolysaccharide complexes and surface structures of a lysine limited culture of *Escherichia coli*. J. Bacteriol. **92:**1206–1217.
25. **Lemozy, J.** 1976. La résistance des bactéries aux antibiotiques en population urbaine. Sci. Tech. Anim. Lab. **1:**170–171.
26. **Mathieu, D., D. Boulier, and P. Tournier.** 1977. L'entérotoxine thermolabile d'*Escherichia coli* (LT). Mise en évidence *in vitro*. Med. Mal. Infect. **7:**486–490.
27. **McConnell, M. M., H. R. Smith, G. A. Willshaw, A. M. Field, and B. Rowe.** 1981. Plasmids coding for colonization factor antigen I and heat-stable enterotoxin production isolated from enterotoxigenic *Escherichia coli*: comparison of their properties. Infect. Immun. **32:**927–936.
28. **Miniats, O. P., and C. L. Gyles.** 1972. The significance of proliferation and enterotoxin production by *Escherichia coli* in the intestine of gnotobiotic pigs. Can. J. Comp. Med. **36:**150–159.
29. **Mitsuhashi, S.** 1969. The R factors. J. Infect. Dis. **119:**89–100.
30. **Mooi, F. R., and F. K. De Graaf.** 1979. Isolation and characterization of K88 antigens. FEMS Microbiol. Lett. **5:**17–20.
31. **Moon, H. W., E. M. Kohler, R. A. Schneider, and S. C. Whip.** 1980. Prevalence of pilus antigens, enterotoxin types, and enteropathogenicity among K88-negative enterotoxigenic *Escherichia coli* from neonatal pigs. Infect. Immun. **27:**222–230.

32. **Moon, H. W., B. Nagy, R. E. Isaacson, and I. Ørskov.** 1977. Occurrence of K99 antigen on *Escherichia coli* isolated from pigs and colonization of pig ileum by K99⁺ enterotoxigenic *E. coli* from calves and pigs. Infect. Immun. **15:**614–620.

33. **Novick, R. P., R. C. Clowes, S. N. Cohen, R. Curtiss III, N. Datta, and S. Falkow.** 1976. Uniform nomenclature for bacterial plasmids: a proposal. Bacteriol. Rev. **40:**168–189.

34. **Ørskov, I., and F. Ørskov.** 1966. Episome-carried surface antigen K88 of *Escherichia coli*. I. Transmission of the determinant of the K88 antigen and influence of the transfer of chromosomal markers. J. Bacteriol. **91:**69–75.

35. **Patte, C., C. Tancrede, P. Raibaud, and R. Ducluzeau.** 1979. Premières étapes de la colonisation bactérienne du tube digestif du nouveau-né. Ann. Microbiol. (Paris) **130A:**69–84.

36. **Penaranda, M. E., M. B. Mann, D. G. Evans, and D. J. Evans, Jr.** 1980. Transfer of an ST:LT: CFA/II plasmid into *Escherichia coli* K-12 strain RR1 by co-transformation with PSC 301 plasmid DNA. FEMS Microbiol. Lett. **8:**251–254.

37. **Petrocheilou, V., M. H. Richmond, and P. M. Bennett.** 1977. Spread of a single plasmid clone to an untreated individual from a person receiving prolonged tetracycline therapy. Antimicrob. Agents Chemother. **12:**219–225.

38. **Sack, R. B.** 1975. Human diarrheal disease caused by enterotoxigenic *Escherichia coli*. Annu. Rev. Microbiol. **29:**333–353.

39. **Sansonetti, P., J. P. Lafont, A. Jaffe-Brachet, J. F. Guillot, and E. Chaslusdancla.** 1980. Parameters controlling interbacterial plasmid spreading in a gnotoxenic chicken gut system: influence of plasmid and bacterial mutations. Antimicrob. Agents Chemother. **17:**327–333.

40. **Seeliger, H. P. R., and H. Werner.** 1963. Recherches qualitatives et quantitatives sur la flore intestinale de l'homme. Ann. Inst. Pasteur Paris **105:**911–936.

41. **Selwood, R., R. A. Gibbons, G. W. Jones, and J. M. Rutter.** 1975. Adhesion of enteropathogenic *Escherichia coli* to pig intestinal brush borders. The existence of two pig phenotypes. J. Med. Microbiol. **8:**405–411.

42. **Shimizu, M., and T. Terashima.** 1982. Appearance of enterotoxigenic *Escherichia coli* in piglets with diarrhea in connection with feed changes. Microbiol. Immunol. **26:**467–477.

43. **Shipley, P. L., C. L. Gyles, and S. Falkow.** 1978. Characterization of plasmids that encode for the K88 colonization antigen. Infect. Immun. **20:**559–566.

44. **Smith, H. W.** 1969. Transfer of antibiotic resistance from animal and human strains of *Escherichia coli* to resident *E. coli* in the alimentary tract of man. Lancet **2:**1174–1176.

45. **Smith, H. W.** 1976. On acute diarrhoeas in childhood. Ciba Found. Symp. **42:**45–72.

46. **Smith, H. W.** 1978. Transmissible pathogenic characteristics of invasive strains of *Escherichia coli*. J. Am. Vet. Med. Assoc. **173:**601–607.

47. **Smith, H. W., and S. Halls.** 1967. Observations by the ligated intestinal segment and oral inoculation methods in *Escherichia coli* infections in pigs, calves, lambs and rabbits. J. Pathol. Bacteriol. **93:**499–529.

48. **Smith, H. W., and S. Halls.** 1968. The transmissible nature of the genetic factor in *Escherichia coli* that controls enterotoxin production. J. Gen. Microbiol. **52:**319–334.

49. **Smith, H. W., and M. B. Huggins.** 1976. Further observations on the association of the col V plasmid of *Escherichia coli* with pathogenicity and with survival in the alimentary tract. J. Gen. Microbiol. **92:**335–350.

50. **Smith, H. W., and M. B. Huggins.** 1978. The influence of plasmid-determined and other characteristics of enteropathogenic *Escherichia coli* on their ability to proliferate in the alimentary tracts of piglets, calves and lambs. J. Med. Microbiol. **11:**471–492.

51. **Smith, H. W., and M. A. Linggood.** 1971. Observations on the pathogenic properties of the K88, Hly and Ent plasmids of *Escherichia coli* with particular reference to porcine diarrhoea. J. Med. Microbiol. **4:**467–485.

52. **Smith, H. W., and M. A. Linggood.** 1972. Further observations on *Escherichia coli* enterotoxins with particular regard to those produced by atypical piglet strains and by calf and lamb strains: the transmissible nature of these enterotoxins and of a K antigen possessed by calf and lamb strains. J. Med. Microbiol. **5:**243–250.

53. **So, M., W. S. Dallas, and S. Falkow.** 1978. Characterization of an *Escherichia coli* plasmid encoding for synthesis of heat-labile toxin: molecular cloning of the toxin determinant. Infect. Immun. **21:**405–411.

54. **Van Embden, J. D. A., F. K. De Graaf, L. M. Schouls, and J. S. Treppa.** 1980. Cloning and expression of a deoxyribonucleic acid fragment that encodes for the adhesive antigen K99. Infect. Immun. **29:**1125–1133.

55. **Wachsmuth, I. K., S. Falkow, and R. W. Ryder.** 1976. Plasmid-mediated properties of a heat-stable enterotoxin-producing *Escherichia coli* associated with infantile diarrhea. Infect. Immun. **14:**403–407.

56. **Watanabe, T.** 1963. Infective heredity of multiple drug resistance in bacteria. Bacteriol. Rev. **27:**87–115.

57. **WHO Scientific Working Group.** 1980. *Escherichia coli* diarrhea. Bull. WHO **58:**23–36.

Importance of Microbial Nitrogen Metabolism in the Ceca of Birds

ATLE MORTENSEN

Department of Arctic Biology and Institute of Medical Biology, University of Tromsø, Tromsø, Norway

The alimentary canal in birds normally includes two ceca, protruding from the ileo-colonic junction. In general, the ceca are well developed in herbivorous birds, and this is especially so among galliforms (29). According to Leopold (25) the size of the ceca in this group of birds is strongly related to the normal diet of the birds. Thus, browsing species such as the willow ptarmigan and the capercaille have significantly longer ceca than the seed-eating species such as the bobwhite quail and the ring-necked pheasant. The diet of the browsing birds is characterized by a high content of fibers and a low content of nitrogen and phosphorus.

The ceca of most birds contain about 10^{11} bacteria per g (wet weight). The cecal flora is dominated by obligate anaerobic, non-spore-forming, gram-positive and -negative rods and cocci, together with different types of clostridia (E. M. Barnes, Intersect. Congr. IAMS, 1st, Tokyo, 1974). There is a remarkable lack of protozoa in the ceca of domesticated birds. In the ceca of captive willow ptarmigan, for instance, no protozoa are present, whereas the ceca of free-ranging willow ptarmigan contain both flagellates and amoebae (17). These organisms are also found in the ceca of wild red grouse (11), and it is known that California quail ceca contain a rich protozoan fauna (27). The small intestine of herbivorous birds normally has a very low number of bacteria. This is also true in the case of wild willow ptarmigan (17).

ROUTES OF ENTRY INTO THE CECA

Nitrogenous compounds enter the ceca through three different routes and are of different origin. First, a large part of the cecal nitrogen content is derived from proteins which have escaped digestion and absorption in the small intestine. These proteins include both dietary proteins and proteins released by intestinal secretion. Second, urinary nitrogen gains access to the ceca in the cecal filling process, and third, it has recently been demonstrated that nitrogen in the form of urea enters the ceca directly through the cecal wall (A. Mortensen and A. Tindall, unpublished data).

Cecal filling has been the subject of a number of investigations, and ureographic studies have revealed that when the colon is filled to a certain degree with content from the small intestine, a retrograde peristalsis is elicited in the colon and the ceca, at the same time that antegrade peristalsis persists in the small intestine (12). The proximal part of the cecum is very narrow with long villi, and it is believed that only very small particles can pass this segment. The result is that, when filling starts, the liquid part of the colonic content, together with fine particles, is squeezed into the ceca, whereas fibers and bigger particles are left behind in the colon. The fibrous remainder is voided in the form of dry pellets, which in browsing galliforms are called woody droppings. The retrograde peristalsis starts so far distal that the contents of the cloaca are also brought into the colon and later into the ceca. The result is, therefore, that cecal microorganisms are provided with nitrogenous compounds both from the feed and from the urine (1, 12, 22). For most birds the proportion of urine entering the ceca is unknown, but in the turkey it is known that up to 35% of the urine is transported into the ceca (6).

There is reason to believe that the separation of the digesta, which is caused by the squeezing of the colonic content, means that material entering the ceca has a higher protein content than the very crude fibers which are left in the colon. It is also probable that the material entering the ceca contains carbohydrates which are more suitable for microbial degradation than those remaining in the colon.

It has recently been demonstrated that nitrogen in the form of urea enters the ceca by diffusion directly through the cecal wall. When radioactively labeled urea was given intraperitoneally to captive willow ptarmigan, radioactive carbon dioxide was rapidly released in the expired air. Since the ingress of urea through other routes to the ceca was prevented, and only microorganisms possess urease, this is taken as evidence that urea entered the ceca directly through the wall (Mortensen and Tindall, unpublished data). It is also worth noticing that only a small proportion (less than 5%) of the injected activity was recovered in the urine.

In ruminants, it has been proposed that bacteria attached to the rumen wall play a major role in the generation of the concentration gradient necessary for the passive diffusion of urea through the rumen wall (10). The thick layer of bacteria normally found adjacent to the epithelial surface in the ceca of domestic fowl (13), and the large population of spirochetes and amoebae found adjacent to and embedded in the epithelium of the ceca of willow ptarmigan (17), may function in a similar way.

FATE OF CECAL NITROGEN

Protein. Several investigators have demonstrated proteolytic activity in the ceca of birds (5, 17, 36). That the birds are able to utilize the resultant amino acids is, however, doubtful. Cecal absorption of amino acids could not be detected in willow ptarmigan (33), and according to Holdsworth and Wilson (19), active transport of glycine is lost a few days after hatching in chickens. Similar results concerning methionine and alanine have been presented (26), and it is suggested that the ceca may play a role in the absorption of yolk material in prehatched and neonatal birds. A higher nitrogen retention is demonstrated in normal than in cecectomized chickens (38), but this may be a result of the absorption of ammonia rather than amino acids. This idea is supported by experiments performed by Nesheim and Carpenter (35) and Salter and Coates (39).

Urinary nitrogen compounds. Uric acid is the dominant nitrogen compound in the urine of birds, and it is known that most avian cecal bacteria are able to utilize uric acid as a source of nitrogen. In young chickens the predominant bacteria utilizing uric acid were *Streptococcus faecalis* and *Escherichia coli* (30). Up to 1.8×10^{10} uric acid-utilizing bacteria per g (wet weight) of cecal content are found in chickens, turkeys, ducks, pheasants, and guinea fowl (4). None of these bacteria has been shown to have an absolute requirement for uric acid. Normally, the utilization of uric acid includes the decomposition of the compound into ammonia and carbon dioxide. Acetic acid may also be one of the end products of the decomposition, as shown for *Clostridium acidiurici*, which is the only bacterium with an absolute requirement for uric acid (3).

By tracer techniques, rapid decomposition of uric acid has been demonstrated in the ceca of both captive and wild willow ptarmigan. In these experiments, which were performed in vivo, one of the end products of the decomposition was carbon dioxide (32).

Urea. Urea is rapidly hydrolyzed in the ceca of captive willow ptarmigan (Mortensen and Tindall, unpublished data). Since urea is a normal intermediate compound in the degradation of uric acid, the liberation of radioactive carbon dioxide from the decomposition of uric acid (32) indicates that urease is also functional in the cecum of free-ranging willow ptarmigan. Evidence for hydrolysis of urea by intestinal microorganisms in domestic fowl has been presented by Okumura and his co-workers (37). They found that growth of conventional chicks given a basal diet containing adequate amounts of all the essential amino acids, but none of the nonessential amino acids, was improved by the supplementation of dietary urea. Germfree chicks, on the other hand, did not benefit from the urea supplement. They concluded that the intestinal bacteria were responsible for the release of ammonia from the urea and that the bird in turn used the ammonia for the synthesis of nonessential amino acids.

Ammonia assimilation. So far, it appears that all the nitrogen compounds entering the ceca may result in the formation of ammonia. The importance of cecal nitrogen metabolism therefore becomes strongly dependent upon the further fate of the released ammonia. As already mentioned there are results pointing to the possibility that ammonia is utilized as a source of nitrogen for the synthesis of nonessential amino acids in chicks (7, 23, 37, 40). Maximum utilization of ammonium compounds presupposes, however, that the diet contains all the essential amino acids in sufficient amounts and that there is a deficit in nonessential amino acids (8). This is a situation more likely to occur in the laboratory than in nature, when the birds are eating the normal diet to which they have been adapted for generations.

In the ceca of both wild and captive willow ptarmigan, ammonia is assimilated into new amino acids by the action of microbial glutamic dehydrogenase (33). After 0.5 h of in vivo incubation of radioactively labeled α-ketoglutarate in the ceca, a significant part of the radioactivity was found in glutamic acid, whereas almost all the radioactive α-ketoglutarate had disappeared. This result by no means indicates that ammonia assimilation by other enzyme systems does not take place. It is, on the contrary, very likely that some of the representatives of the complex population of microorganisms also assimilate ammonia by the use of the glutamine synthetase/glutamate synthetase pathway. It is also possible that microorganisms, which do not possess glutamic dehydrogenase, assimilate ammonia by action of alanine dehydrogenase. This enzyme catalyzes the formation of alanine from pyruvate and ammonia.

In general, assimilation of ammonia, glutamic dehydrogenase, and alanine dehydrogenase normally occurs when ammonia is in excess. The glutamine synthetase/glutamate synthetase system has a higher affinity for ammonia than the two other enzymes and therefore tends to be active when the concentration of ammonia is low (34). From Table 1 it can be seen that the amino acid composition of the pool of free amino acids in cecal content of wild willow ptarmigan is somewhat different from the composition of the pool including all the amino acids in the ceca of the same birds. Most remarkable is the high contribution from glutamic acid in the free amino acid pool. (By the present method of analysis glutamine could not be distinguished from glutamic acid, so what is called glutamic acid represents both.) Also, the contribution from proline and alanine was remarkably high in this pool. The relatively high concentrations of the mentioned amino acids indicate a high rate of de novo synthesis of amino acids from ammonia and keto acids. It is, however, not possible from these results to establish which system is the more active in the ammonia assimilation. There is still the possibility of a high initial production of glutamic acid because of the high contribution from proline. Proline is normally synthesized from glutamic acid via a very short pathway including one spontaneous reaction and one enzyme-catalyzed reaction (24). The high contribution from alanine could be due either to an assimilation of ammonia by the alanine dehydrogenase pathway or to a transamination reaction involving glutamic acid. It is worth noticing that the highest percentage contribution from glutamic acid, proline, and alanine was obtained in the afternoon. According to McBee and West (28) this is the time of the day when the production of volatile fatty acids is highest, whereas their concentration is lowest. The results are thus in accordance with the idea that there is a positive correlation between the rate of ammonia assimilation and the rate of fermentation.

The actual concentrations of free amino acids were very low in the ptarmigan ceca, being 9.6 mM in the morning and only 1.2 mM in the afternoon. Similar concentrations and also a similar composition of free amino acids have been found in the rabbit cecum (20). The free amino acids in the ptarmigan ceca, measured by the present method, also include amino acids from the intracellular compartment of the microorganisms. It is therefore uncertain that the values in Table 1 represent amino acid concentrations which are actually exposed to the cecal epithelium and are therefore potentially absorbable. If this were the case, however, the amino acids that could be of significance for the bird would be mainly the nonessential amino acids, which could easily be synthesized by the bird itself.

FUNCTIONS OF THE CECA

Protein digestion and nitrogen recycling. The data so far presented indicate that the cecal

TABLE 1. Mean percentage distribution of free and combined amino acids in the ceca of wild willow ptarmigan[a]

Amino acid	Combined amino acids (n = 9)	Free amino acids	
		Morning shot (n = 3)	Afternoon shot (n = 2)
Aspartic acid	9.0	11.4	5.1
Threonine	6.0	3.7	5.4
Serine	6.3	5.9	0.4
Glutamic acid	13.1	18.8	29.1
Glycine	9.2	6.0	0.8
Alanine	9.6	12.1	21.5
Valine	6.2	0.8	0.9
Methionine	2.4	5.3	3.5
Isoleucine	5.3	3.2	0.9
Leucine	8.1	6.0	2.2
Tyrosine	2.7	1.4	0.9
Phenylalanine	3.6	2.1	1.0
Histidine	2.9	0.6	0.3
Lysine	6.0	3.0	3.1
Arginine	4.2	1.0	—[b]
Proline	6.3	15.4	25.2
Total concentration	—	9.6 mM	1.2 mM

[a] Free amino acids were obtained by suspending cecal content with three parts of ice-cold ethanol, followed by centrifuging. The perfusate was then dried under vacuum, and the amino acids were resuspended in a phosphate buffer for amino acid analysis. The combined amino acids were obtained by hydrolyzing the cecal content with concentrated HCl under vacuum for 24 h. After neutralization with NaOH the hydrolysate was filtered and suspended in phosphate buffer. The amino acids were quantitatively measured by use of an automatic amino acid analyzer (JLC-6AH JEOL).

[b] Not measured.

microorganisms are able to utilize the major nitrogen compounds which enter the ceca as a source of nitrogen for de novo synthesis of amino acids. It is likely, but not yet verified, that the cecal microorganisms directly incorporate into their own cell material amino acids released by the hydrolysis of proteins. Further, it is indicated that ammonia produced in the ceca may be absorbed by the bird and may serve as a source of nitrogen for the synthesis of nonessential amino acids. Active absorption of amino acids from the ceca of grown birds has never been demonstrated. Although absorption of peptides has not been examined, it could be suggested that digestion of protein is not a useful function of the avian ceca. For the same reason the attractive idea that the ceca participate in recycling of excretory nitrogen must be rejected. It is, however, possible that ammonia of urinary origin is absorbed by the bird, but this would mean that ammonia enters a futile cycle which is hardly of benefit to the bird.

In general, it seems that the cecal microflora plays no role in making nitrogen more available for the host bird. But this does not mean that cecal nitrogen metabolism is irrelevant for the bird. To realize the significance of cecal nitrogen metabolism, we should keep two important points in mind. First, there is a competition for substrate between the microorganisms and the host bird, and when both the bird and the microbes are interested in the same compounds, the bird is bound to lose. Second, because the ceca are situated posterior to the small intestine, the cecal microorganisms live in an environment where the supply of nitrogen may be a serious problem. Nitrogen, in any form, is therefore valuable to the microorganisms and is not released for the benefit of the bird. There is thus no need for active transport mechanisms for amino acids in the ceca, and these mechanisms are consequently lost a few days after hatching when the ceca are invaded by microorganisms.

Supplement of fermentation products. What is made available for the bird are the waste products from the energy metabolism of the cecal microorganisms. These waste products are shown to be mainly volatile fatty acids in the birds so far examined (2, 14, 15, 28). The birds' ability to perform aerobic metabolism enables them to utilize these acids as substrates for their energy metabolism.

A useful contribution from the ceca to the bird is dependent upon a rapid conversion of energy by the microorganisms, which means a relatively high growth rate. A high growth rate requires a rapid synthesis of new cell material, which to a great extent consists of protein. Remembering the nitrogen-poor substrate entering the ceca from the small intestine, I could suggest that nitrogen is a limiting factor for the growth of the cecal microflora. This hypothesis has not yet been the subject of direct experimental investigation in birds. In ruminants, however, it is well documented that there is a close relationship between the availability of nitrogen and the rate of fermentation of polymeric carbohydrates (9, 31). In pigs it has also been shown that the activity of the cecal microorganisms could be increased by the addition of certain nitrogen compounds (21).

Detoxification. The indications of a direct passage of urea from the body fluids into the ceca of willow ptarmigan (Mortensen and Tindall, unpublished data) have encouraged the idea that the cecal microorganisms participate in the detoxification of the bird's nitrogenous waste products. In addition to urea, this detoxification process may also involve ammonia. Evidence for the absorption of ammonia from the avian ceca has already been presented (35, 39). Movement of ammonia in the opposite direction has

been found in the willow ptarmigan during perfusion experiments with perfusate with low pH (about 5) (Mortensen, unpublished data). The pH of the cecal content of the birds is normally lower than the pH of the blood, and in some birds (for instance, the turkey) a pH down to 5.5 in the cecal content has been measured (5). Free ammonia can easily penetrate intestinal walls, whereas the same intestinal walls are almost completely impermeable to the ammonium ion. Thus, cecal contents with low pH would act as an ammonia trap.

The significance of the detoxification is still unknown. What we know is that nephritis caused by the precipitation of urates is one of the main mortality factors in populations of captive willow ptarmigan. A satisfactory explanation of the cause of the disease has not been presented (18), but it is possible that the disease could be due partly to the reduced removal of urea and ammonia by the ceca, thereby forcing the equilibrium in the direction of increased uric acid synthesis. Bringing the willow ptarmigan into captivity causes the ceca to undergo remarkable changes, which certainly affect its normal function. The size of the ceca is reduced, the epithelium undergoes significant changes, and a microflora completely different from the original is established (16, 17).

From the microorganism's point of view, there is another aspect of the detoxification. By absorbing the volatile fatty acids and other fermentation products, the host bird permits the microorganisms to continue their energy metabolism at a high rate. Without the removal of the fermentation products, it is probable that the fermentation rate would be reduced because of end-product inhibition.

CONCLUDING REMARKS

The relationship between the cecal microorganisms and their host bird represents a well-developed and functional symbiosis. Central to this symbiosis is a mutual exchange of certain compounds, which for one part are important substrates and for the other are waste products. This is illustrated in Fig. 1, which shows some of the major routes of nitrogen and energy compounds in a simplified system consisting of a bird with a cecum and a cecal microorganism. The cecal microorganisms provide the host bird with energy-rich fermentation products, but to do so they need a supply of extra nitrogen, which is furnished by the bird. The mutual detoxification effect of the exchange of material between the microorganisms and the bird is likely to have promoted the development of the present system.

The existence of many processes in birds is

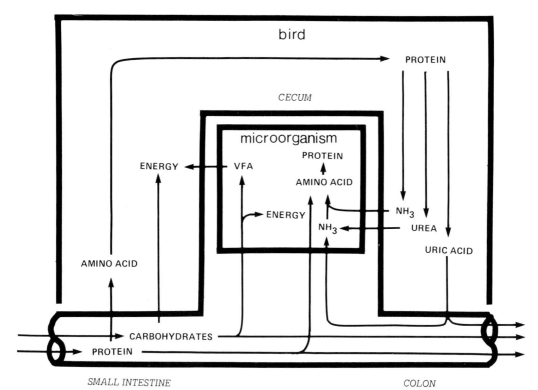

FIG. 1. Flow of nitrogen and energy compounds in a symbiotic system consisting of a bird and a cecal microorganism.

not in doubt, but more detailed knowledge of them is conspicuous by its absence. A major goal for future work is therefore to gain more insight into the quantitative aspects of the different processes. In this respect it is important to determine how far the cecal microorganisms' access to nitrogen affects the rate of fermentation in the ceca. Concerning fermentation, much remains to be learned about the biochemical pathways that result in the formation of volatile fatty acids. There is, for instance, still no direct evidence linking the formation of volatile fatty acids to the degradation of cellulose. Moreover, in future investigations on the energetics of cecal fermentation, proper care should be taken that experiments are performed under conditions in which the supply of nitrogen and the removal of fermentation products resemble the natural situation. Since this is very difficult to achieve by in vitro techniques, it would probably be more fruitful to employ in vivo experiments to solve these problems.

I am greatly indebted to A. Tindall for his advice during the preparation of the manuscript.

The work has been supported by the Norwegian Research Council for Science and the Humanities.

LITERATURE CITED

1. **Akester, A. R., R. S. Anderson, K. J. Hill, and G. W. Osbaldiston.** 1967. A radiographic study of urine flow in the domestic fowl. Br. Poult. Sci. **8:**204–215.
2. **Annison, E. F., K. J. Hill, and R. Kenworthy.** 1968. Volatile fatty acids in the digestive tract of the fowl. Br. J. Nutr. **22:**207–216.
3. **Barker, H. A.** 1961. Fermentation of nitrogenous organic compounds, p. 151–201. In I. C. Gunsalus and R. Y. Stanier (ed.), The bacteria, vol. 2. Metabolism. Academic Press, Inc., New York.
4. **Barnes, E. M., and C. S. Impey.** 1974. The occurrence and properties of uric acid decomposing anaerobic bacteria in the avian caecum. J. Appl. Bacteriol. **37:**393–409.
5. **Bedbury, H. P., and G. E. Duke.** 1983. Cecal microflora of turkeys fed low or high fiber diets: enumeration, identification, and determination of cellulolytic activity. Poultry Sci. **62:**675–682.
6. **Björnhag, G., and I. Sperber.** 1977. Transport of various food components through the digestive tract of turkeys, geese and guinea fowl. Swed. J. Agric. Res. **7:**57–66.
7. **Blair, R., D. W. F. Shannon, J. M. McNab, and D. J. W. Lee.** 1972. Effects on chick growth on adding glycine, proline, glutamic acid or diammonium citrate to diets containing crystalline essential amino acids. Br. Poult. Sci. **13:**215–228.
8. **Blair, R., and R. J. Young.** 1970. Utilization of ammonium nitrogen by growing and laying birds, p. 93–98. Proceedings from Cornell Nutrition Conference for Feed Manufacturers.
9. **Chalupa, W.** 1972. Metabolic aspects of nonprotein nitrogen utilization in ruminant animals. Fed. Proc. **31:**1152–1164.

10. Cheng, K.-J., and R. J. Wallace. 1979. The mechanism of passage of endogenous urea through the rumen wall and the role of urolytic epithelial bacteria in the urea flux. Br. J. Nutr. 42:553–557.

11. Fantham, H. B. 1911. Observations on the parasitic protozoa of the red grouse (Lagopus scoticus), with a note on the grouse fly, p. 692–708. In The grouse in health and in disease. Smith, Elder & Co., London.

12. Fenna, L., and D. A. Boag. 1974. Filling and emptying of the galliform caecum. Can. J. Zool. 52:537–540.

13. Fuller, R., and A. Turvey. 1971. Bacteria associated with the intestinal wall of the fowl (Gallus domesticus). J. Appl. Bacteriol. 34:6–7.

14. Gasaway, W. C. 1976. Seasonal variation in diet, volatile fatty acid production and size of the cecum of rock ptarmigan. Comp. Biochem. Physiol. A 53:109–114.

15. Gasaway, W. C. 1976. Volatile fatty acids and metabolizable energy derived from cecal fermentation in the willow ptarmigan. Comp. Biochem. Physiol. A 53:115–121.

16. Hanssen, I. 1979. Micromorphological studies on the small intestine and caeca in wild and captive willow grouse (Lagopus lagopus lagopus). Acta Vet. Scand. 20:351–365.

17. Hanssen, I. 1979. A comparison of the microbial conditions in the small intestine and caeca of wild and captive willow grouse (Lagopus lagopus lagopus). Acta Vet. Scand. 20:365–371.

18. Hanssen, I. 1982. Nephritis and uric acid diathesis in captive willow ptarmigan (Lagopus l. lagopus). Acta Vet. Scand. 23:446–455.

19. Holdsworth, C. D., and T. H. Wilson. 1967. Development of active sugar and amino acid transport in the yolk sac and intestine of the chicken. Am. J. Physiol. 212:233–240.

20. Hoover, W. H., and R. N. Heitmann. 1975. Cecal nitrogen metabolism and amino acid absorption in the rabbit. J. Nutr. 105:245–252, 1975.

21. Just, A., H. Jørgensen, and J. A. Fernández. 1981. The digestive capacity of the caecum-colon and the value of the nitrogen absorbed from the hind gut for protein synthesis in pigs. Br. J. Nutr. 46:209–219.

22. Koike, T. I., and L. Z. McFarland. 1966. Urography in the unanesthetized hydropenic chicken. Am. J. Vet. Res. 27:1130–1133.

23. Lee, D. J. W., and R. Blair. 1972. Effects on chick growth of adding various non-protein nitrogen sources or dried autoclaved poultry manure to diets containing crystalline essential acids. Br. Poult. Sci. 13:243–249.

24. Lehninger, A. L. 1975. Biochemistry. Worth Publishers, Inc. New York.

25. Leopold, A. S. 1953. Intestinal morphology of gallinaceous birds in relation to food habits. J. Wildl. Manage. 17:197–203.

26. Lerner, S., P. Sattelmeyer, and R. Rush. 1975. Kinetics of methionine influx into various regions of chicken intestine. Comp. Biochem. Physiol. A 50:113–120.

27. Lewin, V. 1963. Reproduction and development of young in a population of California quail. Condor 65:249–278.

28. McBee, R. H., and G. C. West. 1969. Cecal fermentation in the willow ptarmigan. Condor 71:54–58.

29. McNab, J. M. 1973. The avian caeca: a review. World's Poultry Sci. J. 29:251–263.

30. Mead, G. C., and B. W. Adams. 1975. Some observations on the caecal microflora of the chick during the first two weeks of life. Br. Poult. Sci. 16:169–176.

31. Moir, R. S., and L. E. Harris. 1962. Ruminal flora studies on the sheep. X. Influence of nitrogen intake upon ruminal function. Nutrition 77:285–298.

32. Mortensen, A., and A. Tindall. 1981. Caecal decomposition of uric acid in captive and free ranging willow ptarmigan (Lagopus lagopus lagopus). Acta Physiol. Scand. 111:129–133.

33. Mortensen, A., and A. Tindall. 1981. On caecal synthesis and absorption of amino acids and their importance for nitrogen recycling in willow ptarmigan (Lagopus lagopus lagopus). Acta Physiol. Scand. 113:465–469.

34. Murrell, J. C., and A. Dalton. 1983. Ammonia assimilation in Methylococcus capsulatus (Bath) and other obligate methanotrophs. J. Gen. Microbiol. 129:1197–1206.

35. Nesheim, M. C., and K. J. Carpenter. 1967. The digestion of heat-damaged protein. Br. J. Nutr. 21:299–411.

36. Nitsan, Z., and E. Alumot. 1963. Role of the cecum in the utilization of raw soybean in chicks. J. Nutr. 80:299–304.

37. Okumura, J., D. Hewitt, D. N. Salter, and M. E. Coates. 1976. The role of the gut microflora in the utilization of dietary urea by the chick. Br. J. Nutr. 36:265–272.

38. Payne, W. L., R. R. Kifer, D. G. Snyder, and G. F. Combs. 1971. Studies of protein digestion in the chicken. 1. Investigation of apparent amino acid digestibility of fish meal protein using cecectomized, adult male chickens. Poult. Sci. 50:143–150.

39. Salter, D. N., and M. E. Coates. 1971. The influence of the microflora in the alimentary tract on protein digestion in the chick. Br. J. Nutr. 26:55–69.

40. Sugahara, M., and S. Ariyoshi. 1967. The nutritional value of the individual nonessential amino acids as the nitrogen source in the chicken nutrition. Agric. Biol. Chem. 31:1270–1275.

Microbial Competition

Competition Among Bacteria: an Overview

H. VELDKAMP, H. van GEMERDEN, W. HARDER, and H. J. LAANBROEK

Department of Microbiology, University of Groningen, Haren, The Netherlands

Stanier et al. (65) calculated that "after 48 hours of exponential growth, a single bacterium with a doubling time of 20 minutes would produce a progeny of 2.2×10^{31} g, or roughly 4,000 times the weight of the earth." In reality, of course, this can never happen because in nature the specific growth rate of bacteria is rarely maximal and remains so only for very short periods. The main factor governing specific growth rates of microbes in nature is the concentrations of solutes essential for growth, and these are extremely low. For instance, the concentrations of available carbon and energy sources for chemoorganotrophic bacteria generally are in the nanomolar range. In aerobic environments most of these bacteria are highly versatile and can use a large variety of organic compounds as a C and energy source. One example concerns the ammonifying bacteria occurring in a eutrophic body of water in The Netherlands. Of around 200 different species or strains isolated, more than 90% were able to utilize such commonly occurring substrates as glucose, glutamate, and aspartate as the only source of carbon and energy (58). Thus, a vigorous competition for these and other nutrients should occur in such an environment.

Being successful in capturing available solute molecules is of major importance in the struggle for existence, and nutrient limitation undoubtedly has been a dominating selective pressure to which microbes have been exposed during evolution. Competitive abilities with respect to nutrient uptake form part of a complex of properties which together determine the survival of species. Another property of major importance is the ability to survive starvation periods. For most bacteria the environment is energy limited, and during periods of total energy depletion the cells must depend on mechanisms which prevent the proton motive force across the cell membrane from becoming zero since this force plays an essential role in the uptake of most solutes. Complete dissipation of the proton motive force would mean that the cells would lack competitive abilities when substrates such as carbon and energy sources reappeared in the environment (26, 27). Other components of the overall surviv-

al strategy may be mechanisms to escape predation or to escape competition for available nutrients by chemotactic movement along concentration gradients of nutrients, hydrogen ions, oxygen, or toxic compounds. In summary, being successful in competing for limiting nutrients under conditions when growth is possible is only part of the overall achievement in surviving in a potentially hostile environment.

As yet, it is very difficult and for most bacterial species even impossible to study competitive abilities appropriately in the natural environment. For this reason our present knowledge has been derived almost entirely from laboratory experiments in which mixtures of different species are generally exposed to conditions which differ from those occurring in nature. In the first half of this century, batch culture techniques were applied for this purpose. The obvious disadvantage of these is that nutrient concentrations have to be applied which are very much higher than those occurring in nature. In the early enrichment cultures selection therefore occurred on the basis of maximum specific growth rates exposed under conditions of nutrient sufficiency. Winogradski (76) recognized this artificial approach and its limitations. He therefore made silica gel plates impregnated with low concentrations of nutrients and inoculated these with soil particles. This allowed him to observe the development of what he called "autochthonous" bacteria which came to the fore at very low nutrient concentrations. However, at the time hardly anything could be done to study the properties of these bacteria any further. It was not until the chemostat was introduced (44, 47) that experiments could be carried out under well-defined, though still unnatural, conditions which allowed bacterial growth at extremely low concentrations of growth rate-limiting nutrients. Jannasch (18) was the first to confirm in this way the existence of bacteria which show a selective advantage under these conditions. Since then, the chemostat has been the most powerful tool for studying microbial competition in the laboratory. Results of competition experiments with the chemostat with aerobically grown bacteria have been reviewed by Harder et al. (12) and

Kuenen and Harder (30). Genetic aspects of selection in chemostats have been discussed by Dykhuizen and Hartl (5).

In this introduction to the symposium on microbial competition, restrictions are made with respect to both organisms and techniques. Only bacteria are dealt with, and with respect to techniques emphasis has been given to the "chemostat approach" of studying bacterial competition. In the next section the kinetics of bacterial growth are discussed, and this is followed by results obtained in laboratory studies with phototrophs and with anaerobic bacteria occurring in the sediments of natural waters. Finally, a few concluding remarks are made regarding virtues and limitations of these kinds of studies.

KINETICS OF NUTRIENT-LIMITED COMPETITION

In the past decades many attempts have been made to predict the dynamics of microbial populations in nature in terms of competition for limiting nutrients (or light in the case of phototrophs) (30, 62, 71, 72). Since both nutrient uptake rates and growth rates of microbes have often been found to approximate hyperbolic functions of external nutrient concentrations, the kinetic basis for these predictions in most cases has been the equation:

$$\mu = \frac{\mu_{max} \cdot S}{K_s + S} \qquad (1)$$

where μ is the specific growth rate, μ_{max} is the maximum specific growth rate, S is the concentration of the growth-limiting nutrient, and K_s is a saturation constant numerically equal to S at 0.5 μ_{max}. When written as in equation 1, the equation is generally referred to as the Monod equation (43), but it has equally been applied to the kinetics of nutrient uptake and then the term Michaelis-Menten equation is often used, by analogy with enzyme kinetics (42). In contrast to the kinetic description of enzymatic processes which is derived from a mechanistic analysis of enzyme-catalyzed reactions, the equations describing whole-cell processes are entirely empirical and are accepted only because they seem to fit experimental data adequately.

On the basis of equation 1, competition of different organisms for a common growth-limiting substrate can be rationalized as follows (30, 52, 54). Consider two organisms A and B possessing growth characteristics for the growth-limiting nutrients as shown in Fig. 1a. When organism A is grown in a chemostat, the concentration of the growth-limiting substrate in steady state is maintained at a level characteristic for that organism at the particular dilution rate (= growth rate) applied. If, in a steady-state culture

of A, cells of organism B are introduced, the specific growth rate of this organism will always be lower than that of organism A, and as a result organism B is washed out of the culture because it cannot grow at the rate required by the dilution rate. However, when the growth characteristics are as shown in Fig. 1b, μ_B will be higher than μ_A at higher substrate concentrations (at higher dilution rates), and then organism A will be washed out. The reverse will happen at substrate concentrations below the crossing point of the two curves. Theoretically, coexistence of organisms A and B can occur at the crossing point, but it has been shown mathematically that this is essentially an unstable condition (6).

This analysis shows that the rate of growth of an organism at the prevailing concentration of the limiting nutrient (which is determined by both the organism's K_s and μ_{max}) is decisive in the outcome of competition. Since continuous culture theory predicts that the steady-state concentration of the limiting nutrient is independent of the yield (Y = weight of organisms formed per unit amount of the limiting nutrient consumed), Y has no influence on the outcome of the competition in a chemostat. This probably also holds for single nutrient-limited competition in nature (71). It is important to point out that the above reasoning is valid only if no other interactions between organisms A and B occur.

The theory can easily be extended to competition between more than two organisms for one growth-limiting substrate. Descriptions of competition between different organisms for two or more nutrients have also been published (8, 66, 77). These mathematical analyses which attempt to provide some measure of understanding of competition as it probably occurs in many natural environments (11) are also based on the Monod equation, but now incorporate modifications to account for the influence of the metabolism of one substrate on that of another. An important conclusion from these mathematical exercises is that competition between different organisms for limiting concentrations of multiple substrates is not merely based on the organism's growth characteristics K_s and μ_{max}, but also on the yield of cells on these substrates. A further conclusion is that, in the absence of other interactions, coexistence of different species is possible up to a number equal to the number of limiting nutrients provided. In contrast to competition for a single growth-limiting nutrient, stable mixed cultures therefore are possibly the rule, a condition which is also predicted for many natural ecosystems.

The above rationalization of microbial growth and competition in a chemostat, as based on the Monod equation, often accounts for the behav-

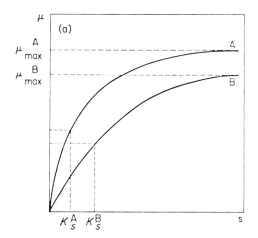
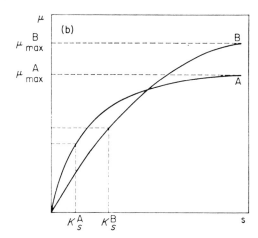

FIG. 1. Relationship of μ to s for organisms A and B. (a) $K_s^A < K_s^B$ and $\mu_{max}^A > \mu_{max}^B$; (b) $K_s^A < K_s^B$ and $\mu_{max}^A > \mu_{max}^B$. (After Veldkamp [70]).

ior of pure and mixed cultures (7, 29, 30, 72) in spite of the fact that the reasoning is based on an empirical relationship between rate of growth and external (limiting) nutrient concentration and that further simplifications have been incorporated (30, 36). In view of this, the following considerations appear to be in order, particularly since they have led to some measure of confusion in discussions on competition.

In the above analysis we have seen that the organism's characteristics μ_{max} and K_s are decisive in the outcome of single nutrient-limited competition, because these parameters determine the specific growth rate at the prevailing concentration of that nutrient. The apparent analogy of K_s for growth and K_m of enzymes has led a number of workers to infer that species with low K_s values for a particular nutrient will always be at an advantage under conditions of low nutrient concentrations compared to species with higher K_s values (13). This ignores the fact that the rate of nutrient uptake at low concentrations is determined by the slope of the Monod equation at those concentrations. This slope, as S approaches zero, becomes equal to μ_{max}/K_s. Thus, it has been suggested that this ratio should be used, instead of K_s, when comparing the competitive ability of different microbes under conditions in which they have to compete for a common limiting nutrient (13). Indeed, this appears to be a more valid approach. It only holds, however, for conditions in which the limiting nutrient concentration is much lower than K_s. A further complication relates to the fact that microorganisms have generally been treated as homogeneous catalysts to which Monod-type kinetics can be directly applied, rather than as phenotypically highly variable organisms (W. Harder and L. Dijkhuizen, Annu. Rev. Microbi-

ol., in press) that therefore may require a more involved kinetic theory. There is an abundance of information in the literature (for reviews see 36; Harder and Dijkhuizen, in press) that demonstrates that kinetic constants of growth and nutrient uptake are highly dependent upon the "physiological state" of the cells (both with respect to the chemical and physical composition of the environment during growth of organisms and pretreatment of cells prior to kinetic measurements), because microorganisms are very much a product of their environment (14). Thus, to come to a meaningful understanding of the behavior of microbial cells under specified environmental conditions, one must determine the kinetic constants of the cells, using organisms grown under those conditions. Thus, the practice of determining K_s values of organisms in batch culture experiments (56) may lead to unjustified conclusions. Moreover, the determination of growth parameters does not abolish the need to perform actual competition experiments.

In recent years attempts have been made to develop a mechanistic theory of microbial nutrient uptake and growth (2, 3). This theory takes into account that the elegant continuous culture concept as developed by Monod (44) cannot be extended to very dilute (oligotrophic) environments such as much of the open oceans without due experimental justification. The relationship proposed by Button (3) for the specific growth rate of an organism in the steady state (coupled to uptake of the limiting nutrient A) is

$$\mu_A = a_A \cdot A \cdot Y_{XA} \qquad (2)$$

where a_A is the specific affinity of the organism for the substrate A, A is the steady-state concentration of the substrate, and Y_{XA} is the yield of cell material produced from substrate utilized. The specific affinity is a function of the rate of substrate accumulation by the organism's transport systems for the limiting nutrient and cell biomass and takes into account any phenotypic adaptation of solute transport systems that organisms may express in response to nutrient limitation. The equation also includes the variable Y_{XA}, which is known to change with growth rate. This change can often be expressed in terms of a linear relationship, and incorporation of this relationship in equation 2 gives an improved expression of the kinetics of nutrient-limited growth. If the change in a_A with growth rate is also known and a formal expression of this relationship is included in equation 2, a comprehensive expression is obtained which has a sound theoretical base (2, 3). As pointed out by Button (3), the value of using the specific affinity parameter is particularly apparent in the case of multiple substrate utilization, as it occurs in many (oligotrophic) natural environments and in the estimation of numbers of organisms in a mixed population which are capable of using a particular substrate.

COMPETITION AMONG PHOTOTROPHIC BACTERIA

On the basis of their general physiology, three groups of phototrophic bacteria can be recognized. The first group comprises the purple sulfur bacteria and the green sulfur bacteria. These organisms are able to assimilate only a very limited number of organic compounds, and their abundance in nature can mainly be explained by their photolithotrophic features. Their characteristic habitats are stagnant and anaerobic bodies of water.

The second group of phototrophic bacteria is best characterized as the photo-organotrophic counterparts of the purple and green sulfur bacteria. This group comprises the *Rhodospirillaceae* and the *Chloroflexaceae*. Types of metabolism observed in the *Rhodospirillaceae* include fermentation, respiration, and phototrophic growth in the presence and absence of oxygen. In general, these organisms are flexible and far more versatile than the purple and green sulfur bacteria. Some representatives have even been shown to grow at the expense of sulfide oxidation, despite their trivial name of purple nonsulfur bacteria (T. A. Hansen, Ph.D. thesis, University of Groningen, Groningen, The Netherlands, 1974).

The cyanobacteria are the third group of phototrophic procaryotes. These organisms differ from the first two groups mentioned with respect to their photosystem. *Chromatiaceae*, *Chlorobiaceae*, *Rhodospirillaceae*, and *Chloroflexa-*

ceae have only one photosystem, whereas the cyanobacteria are equipped with both photosystems I and II. As a rule, these organisms use water as the electron donor—and as a consequence produce oxygen—but the utilization of sulfide as electron donor in photosynthesis has been reported (4). Under such conditions they may live in competition with other phototrophic bacteria, for which the presence of oxygen can be very disadvantageous.

There can be little doubt that phototrophic bacteria compete with heterotrophic bacteria for such nutrients as phosphate and ammonium. Also, purple sulfur bacteria not only will compete with cyanobacteria for, e.g., sulfide, but as well with chemolithotrophic bacteria like *Thiobacillus* spp. From an ecological viewpoint, such interactions are most important. Nevertheless, competition experiments between these groups of organisms have not yet been performed. A few experiments carried out with members of the *Chromatiaceae*, *Chlorobiaceae*, and *Rhodospirillaceae* which illustrate factors involved in the competition between phototrophic bacteria are described below. A general overview of the ecology of phototrophic bacteria has been given by van Gemerden and Beeftink (69).

Competition for sulfide between different purple sulfur bacteria. In the family *Chromatiaceae* species may differ considerably with respect to their cell dimensions. Being large implies that the surface-to-volume ratio is low, and it has been demonstrated that this ratio is a determinative factor for the outcome of competition between organisms with comparable nutritional requirements (29). By growing different species of *Chromatium*, the consistent observation was made that the small *Chromatium vinosum* grew faster than the large *Chromatium weissei* regardless of the concentration of sulfide. However, in natural blooms the large forms (*Chromatium weissei*, *Chromatium okenii*) often dominate over small species like *Chromatium vinosum*. Evidently, other factors are involved.

On the basis of the selective enrichment of the large forms, Pfennig (49) deduced that, in addition to the concentration of sulfide, important parameters were temperature, pH, and light intensity. Also, a day–night rhythm, rather than continuous illumination with incandescent light, was in favor of the large forms. In the laboratory, competition experiments have been performed which emphasized in particular the relevance of alternating light and dark periods (68). In these experiments, temperature, pH, and intensity of light were deliberately chosen to favor the small *Chromatium vinosum* rather than to enrich for the large *Chromatium weissei*.

The oxidation of sulfide by phototrophic bacteria in anaerobic habitats is a light-driven process, whereas the production of sulfide continues in the dark. Consequently, a day–night regimen results in diel fluctuations in the concentration of sulfide. Such fluctuations have been reported in deep lakes (1, 64), but can be observed as well in small stratified ponds (22). In laboratory experiments these conditions were mimicked by alternately incubating sulfide-limited continuous cultures in the light and in the dark. Sulfide, accumulated in the preceding dark period, is rapidly oxidized in the first hours of the subsequent light period. Both in *Chromatium vinosum* and in *Chromatium weissei*, this results in the intracellular deposition of glycogen. However, in *Chromatium weissei* this is predominantly accomplished by the oxidation of sulfide to sulfur, whereas in *Chromatium vinosum* a substantial amount of sulfide is oxidized to sulfate. As a consequence of the fact that the rate of glycogen synthesis is about the same in the two strains, sulfide is depleted more rapidly in pure cultures of *Chromatium weissei* than in those of *Chromatium vinosum* of identical density. In mixed cultures of the two bacteria, *Chromatium weissei* can only take advantage of its capacity to store sulfur when elevated concentrations of sulfide indeed occur. As is to be expected on the basis of the growth rate responses to sulfide concentration, the continuous illumination of cultures inoculated with both strains resulted in complete exclusion of *Chromatium weissei*. However, light–dark cycles resulted in stable coexistence of the small *Chromatium vinosum* and the large *Chromatium weissei*. One example is shown in Fig. 2. It was found that longer nights, i.e., more sulfide accumulation at the end of the dark periods, resulted in higher proportions of the large bacterium (68). These data show that some organisms are better adapted to fluctuations than others. It is of interest that the specific rate of carbon dioxide fixation exhibited by *Chromatium weissei* in the periods of nonlimiting sulfide concentrations is higher than the organism's maximum specific growth rate. This once more illustrates that the organism is well adapted to fluctuations in the environmental concentration of sulfide.

Competition for sulfide and acetate between purple nonsulfur bacteria (*Rhodospirillaceae*) and purple sulfur bacteria (*Chromatiaceae*). *Rhodospirillaceae* are ubiquitous but never bloom, in contrast to *Chromatiaceae* and *Chlorobiaceae*. It is generally accepted that the toxicity of sulfide is one of the key factors to explain this phenomenon. Sulfate is invariably present in anaerobic habitats in which the nutritional requirements of the purple nonsulfur bacteria are fulfilled. The presence of sulfate-reducing bacteria inevitably results in the production of sulfide.

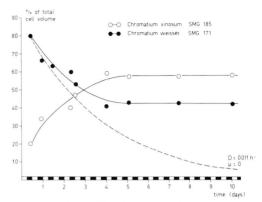

FIG. 2. Competition between *Chromatium vinosum* and *Chromatium weissei* in a sulfide-limited chemostat run at a dilution rate of 0.011 h^{-1}. The light regimen was 6 h light–6 h dark. The dashed line represents the theoretical washout of *Chromatium weissei* in the absence of growth (modified after van Gemerden [68]).

This compound on the one hand inhibits the growth of *Rhodospirillaceae* and on the other hand promotes the development of the *Chromatiaceae* (50, 51).

It has been demonstrated that several representatives of the *Rhodospirillaceae* are able to grow at the expense of sulfide oxidation, provided the concentration is kept low (10; Hansen, Ph.D. thesis, University of Groningen, 1974). However, *Rhodopseudomonas capsulata* was an exception in this respect. This organism was found to be remarkably resistant to elevated sulfide concentrations and in fact could be grown in media designed for genuine purple sulfur bacteria provided thiamine had been added. It thus appears that for this organism the toxicity concept offers no plausible explanation for its low numbers in natural habitats. By growing *R. capsulata* in continuous culture, it was found that the organism not only has a high tolerance toward sulfide, but also grows fast at low concentrations of sulfide (Dijkstra et al., in preparation). On the basis of the relation between growth rates and sulfide concentration, one would expect *R. capsulata* to outcompete *Chromatium vinosum* at all sulfide concentrations below 0.3 mmol/liter.

However, *R. capsulata* is only able to oxidize sulfide to sulfur, which is deposited outside the cells (Hansen, Ph.D. thesis, University of Groningen, 1974). In mixed cultures, *Chromatium* can thus freely oxidize all the elemental sulfur. It should be kept in mind that in the oxidation of sulfur to sulfate six electrons are released, but only two are released in the oxidation of sulfide to sulfur. The yield per millimole of reducing power is similar in the two organisms. With sulfide as the limiting factor, stable coexistence of *R. capsulata* and *Chromatium vinosum* was indeed observed. However, at a dilution rate of 0.06 h^{-1} *R. capsulata* formed no more than 5% of the total biomass. This indicates that *Chromatium* must have oxidized a substantial proportion of the sulfide in spite of its lower affinity for this compound as compared to its competitor (D. J. Wijbenga and H. van Gemerden, in preparation).

With acetate as the limiting substrate, *R. capsulata* outcompeted *Chromatium vinosum* regardless of the dilution rate. During mixotrophic growth on equimolar mixtures of sulfide and acetate, coexistence was again established. Under these conditions, no more than 15% of the total biomass could be attributed to *R. capsulata* in spite of its higher affinity for both sulfide and acetate. These data may illustrate the severe drawback of the inability to oxidize sulfur, but an even higher selective pressure is created by the inability to respond fast when substrate concentrations fluctuate or when a substrate previously absent is introduced into the culture medium. It was found in sulfide-limited cultures of *R. capsulata* that the injection of acetate directly into the culture vessel (final concentration 0.2 mmol/liter) resulted in much reduced rates of sulfide oxidation (73). The addition of acetate (final concentration, 10 mmol/liter) to batch cultures of *R. capsulata* growing on sulfide even resulted in complete cessation of sulfide oxidation; instead, sulfide was produced by reduction of sulfur (D. J. Wijbenga and H. van Gemerden, 3rd International Symposium on Photosynthetic Prokaryotes, Oxford, 1979). Finally, acetate-grown cells of *R. capsulata* show low rates of sulfide oxidation, not so much because they cannot oxidize sulfide, but rather because the Calvin cycle enzymes are adjusted to low levels under heterotrophic conditions (17, 59). Such phenomena are not observed in cultures of *Chromatium vinosum*. As a consequence, competition between *R. capsulata* and *Chromatium vinosum* under fluctuating conditions results in virtually complete exclusion of *R. capsulata*. The availability of sulfide is of crucial importance in the development of blooms of phototrophic bacteria. It can be deduced from the present data why *R. capsulata* is easily isolated from so many habitats, but is invariably found in low numbers.

Quantity and quality of light as a selective factor for anoxygenic phototrophic bacteria. In stratified lakes, light is a factor of utmost importance in the development of blooms of phototrophic bacteria. On several occasions, such blooms have been shown to be light limited (1, 64; Guerrero et al., in preparation). Light with wavelengths shorter than 400 nm or longer than 700 nm does not penetrate very deeply in water (50), and consequently carotenoid-containing species (purple sulfur bacteria, brown-colored green sulfur bacteria) are best equipped. Such organisms are able to trap the available light with their accessory pigments, and often the bloom consists of purple sulfur bacteria and brown sulfur bacteria (1, 67). In blooms of phototrophic bacteria, usually one species becomes dominant. Yet, other types of phototrophic bacteria can be enriched from the layer. Conceivably, the quantity or quality of light is the decisive factor in the growth of the different types. In the bloom observed during the summer in Lake Kinneret (Israel), the brown *Chlorobium phaeobacteroides* is the dominant species. At the same time, lower numbers of the purple sulfur bacterium *Thiocapsa roseopersicina* are observed (1). In the water layer where the bloom was observed, the concentrations of sulfide and acetate were 1 to 4 mg/liter and 0.2 mg/liter, respectively; the maximum light intensity was about 1 μE m^{-2} s^{-1}.

It is of interest to know why the *Chlorobium* species becomes dominant. Being available simultaneously, sulfide and acetate can be expected to be utilized by both *Chlorobium* and *Thiocapsa*. Under identical conditions with respect to medium composition, pH, temperature, and illumination, the affinity for sulfide of *Chlorobium phaeobacteroides* was much higher than that of *T. roseopersicina*; for acetate the opposite was found (Veldhuis et al., in preparation). As a consequence, one would expect coexistence of the two strains at saturating light intensities. At a dilution rate of 0.06 h^{-1} this was indeed observed, *Thiocapsa* being the most abundant species.

A preference for either sulfide or acetate was assessed in pure cultures by decreasing the light intensity in such a way that the sulfide/acetate limitation was replaced by a more and more severe light limitation. It was found that, at decreasing light intensities, the brown *Chlorobium* continued to oxidize all of the sulfide, but progressively less acetate was assimilated. *Thiocapsa* was found to behave in an opposite fashion (Veldhuis et al., in preparation). Thus, under light-limiting conditions, the two bacteria have different preferences and consequently can be expected to influence each other to a limited extent only. However, in co-cultures of *Chlorobium* and *Thiocapsa* growing mixotrophically on sulfide and acetate, a similar stepwise decrease in the light intensity resulted in progressively increasing ratios of *Chlorobium* numbers over *Thiocapsa* numbers. The explanation for this is that *Chlorobiaceae* are far better adapted to low light intensities than *Chromatiaceae*. Judged from the relation between the specific growth rate and the light intensity, *Chlorobium* not only grows faster at lower light intensities but also has remarkably low maintenance requirements. The maintenance rate coefficient for *Thiocapsa* was found to be −0.010/h, whereas that of *Chlorobium* was −0.001/h (Veldhuis et al., in preparation). Similarly, in mixed cultures of *Chlorobium phaeovibrioides* and *Thiocystis violacea* incubated at various depths, the latter was selectively enriched only in those bottles that were incubated close to the surface. At greater depths, *Chlorobium phaeovibrioides* became the dominant organism (41).

COMPETITION BETWEEN ANAEROBIC BACTERIA INVOLVED IN MINERALIZATION IN SEDIMENTS OF NATURAL WATERS

Mineralization processes in anaerobic sediments have been the subject of many studies by several research groups in recent years. Klug and his co-workers made extensive in situ studies of the pathways of anaerobic mineralization in sediments of eutrophic lakes. In these sediments, organic matter was largely fermented to acetate and hydrogen, with smaller amounts of propionate (38), whereas lactate became a more important fermentation product when only glucose was used as substrate (25). Degradation of these fermentation products resulted largely in the production of methane (25, 38, 40). However, although inorganic sulfate concentrations were very low in the sediments, a substantial part of the organic matter was oxidized by sulfate reduction (24, 60). Propionate and lactate, for example, were at least 50% oxidized by sulfate-reducing bacteria (61). Apparently, competition for available substrates, in particular acetate and hydrogen, between sulfate-reducing and methanogenic bacteria occurred in the sediments. The success of sulfate-reducing bacteria in scavenging for these substrates even at low sulfate concentrations could be explained by a relatively high affinity for these substrates (37, 39), which was confirmed in laboratory studies with pure cultures of sulfate-reducing and methanogenic bacteria (28, 56). Addition of sulfate to sulfate-limited freshwater sediments stimulated sulfate reduction, whereas methanogenesis was inhibited (75, 78). The success of sulfate-reducing bacteria in competition for fermentation products was also obvious from in situ experiments with carbon-limited, sulfate-rich marine sediments. In these sediments sulfate reduction was the major terminal step in the pathway of anaerobic mineralization (15, 16, 20, 21, 46, 57, 63). However, a minor production of methane could still occur as a result of degradation of compounds inaccessible for sulfate-reducing bacteria such as methanol and methylamines

(48, 57, 74). In marine sediments with a relative shortage of sulfate due to the input of high amounts of easily degradable organic compounds, methanogenesis came to the fore (9, 45).

The success of sulfate-reducing bacteria in competition for small amounts of available substrates can be explained not only by their high affinity for substrates but also by their versatility. Compared to methanogenic bacteria, sulfate-reducing bacteria are able to use a larger range of substrates for growth (35, 53).

An example of the advantage of versatility in the presence of mixed substrates was found in competition experiments between two species of glutamate-fermenting clostridia isolated from an anaerobic digestor applied for purification of industrial wastewater rich in glucose and amino acids, among which glutamate occurred in high concentration (34). One of the two species, *Clostridium cochlearium*, was a specialist which used only glutamate and two related substrates for growth, whereas the other, *Clostridium tetanomorphum*, was more versatile and could also use glucose. Competition for glutamate in a glutamate-limited chemostat between the two species always resulted in predominance of the specialist and washout of the more versatile species. Addition of glucose as well as glutamate to the reservoir of the chemostat resulted in coexistence in which *Clostridium tetanomorphum* consumed all of the glucose and also some of the glutamate.

The occurrence of lactate-consuming anaerobes in an estuarine sediment was established by making dilution series from mud in anaerobic agar shake cultures (H. J. Laanbroek, H. J. Geerligs, A. A. C. M. Peynenburg, and J. Siesling, submitted for publication). From the highest dilution two fermentative bacteria (*Veillonella alcalescens* and *Acetobacterium* sp.) and the sulfate-reducing *Desulfovibrio baculatus* were isolated. Their maximum specific growth rates on lactate were determined in washout experiments in a chemostat, and competition experiments were carried out in a lactate-limited chemostat at two dilution rates. A diagrammatic representation of the relationship of specific growth rate and lactate concentration is given in Fig. 3. In mixed cultures of *Veillonella* and *Acetobacterium* in a lactate-limited chemostat, the former organism consumed practically all of the lactate, although *Acetobacterium* was not washed out as it could maintain itself on the hydrogen and carbon dioxide produced by *Veillonella*. Addition of formate (20 mM) to the coexisting populations resulted in an increase in the population density of *Acetobacterium*; the propionate level of the culture hardly changed, however, indicating that lactate consumption by

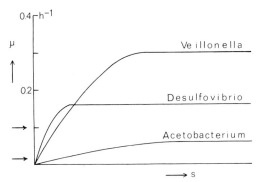

FIG. 3. Specific growth rate (μ) of *Veillonella alcalescens*, *Acetobacterium* sp., and *Desulfobulbus baculatus* as a function of lactate concentration (s) (schematic). Maximum specific growth rates were calculated from differences between dilution rates and washout rates in a chemostat as described by Hannasch (19). Arrows indicate dilution rates at which competition experiments were carried out.

Veillonella was not affected (Fig. 4). The observations summarized in Fig. 3 were confirmed by results obtained with enrichment cultures in lactate-limited chemostats with excess sulfate inoculated with estuarine mud. In these cultures *Desulfovibrio* became dominant, provided the sulfide concentration remained low enough to

prevent iron from becoming growth-limiting for the sulfate reducer. Similar observations were made in enrichments with ethanol as the growth-limiting substrate in the presence of excess sulfate (31). However, when, as a result of sulfide production, a dual limitation of ethanol and iron was introduced, *Desulfobulbus propionicus*, which can ferment ethanol in the presence of carbon dioxide, became dominant (31). *Desulfovibrio* was not completely washed out, however, and addition of iron to the mixed culture led to a rapid increase in the *Desulfovibrio* population and complete washout of *Desulfobulbus*. Coexistence of the two species was easily obtained in mixed cultures fed with growth-limiting amounts of both ethanol and propionate in the presence of excess sulfate and iron, where *Desulfobulbus* used all of the propionate and perhaps also a small amount of ethanol (H. J. Laanbroek, H. J. Geerligs, L. Sijtsma, and H. Veldkamp, submitted for publication). A third sulfate-reducing species, *Desulfobacter postgatei*, isolated from the same intertidal sediments (33), was also unsuccessful in competing with *Desulfovibrio baculatus* for ethanol in an ethanol-limited chemostat. However, as a result of its ability to utilize the acetate produced by *Desulfovibrio baculatus*, coexistence was reached.

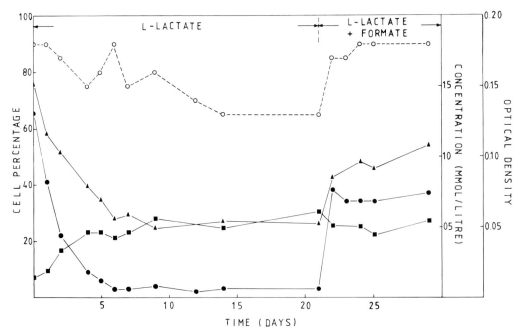

FIG. 4. Competition for L-lactate between *Veillonella alcalescens* NS.L49 and *Acetobacterium* NS.L40 in an anaerobic energy-limited chemostat at a dilution rate of 0.02 h⁻¹, pH 7.0, and 30°C. The medium fed into the chemostat contained 20 mM L-lactate as energy source, and after 21 days 20 mM formate was added as additional energy source. Symbols: ●, cell percentage of *Acetobacterium* NS.L40; ○, optical density at 660 nm; ▲, acetate; ■, propionate (Laanbroek et al., submitted for publication).

In a sulfate-limited chemostat with ethanol as C and energy source *Desulfovibrio* grew faster than the other ethanol-utilizing species, *Desulfobacter postgatei* and *Desulfobulbus propionicus* (Laanbroek et al., submitted for publication). Thus, both sulfate and ethanol limitation favored *Desulfovibrio baculatus*. The latter species occurred in approximately the same numbers as *Desulfovibrio desulfuricans* in the estuarine mud, as determined by colony counts in anaerobic agar shake cultures. Both species are similar with respect to utilization of carbon and energy sources, the main difference being the ability of *D. baculatus* to use sulfur as electron acceptor.

Competition between *Desulfobacter* and *Desulfobulbus* for sulfate in the sulfate-limited chemostat provided with acetate and propionate as C and energy sources, respectively, was won by *Desulfobulbus*, which produces acetate from propionate during sulfate reduction. The outcome of none of the competition experiments between the sulfate reducers was affected by inclusion of clay particles (illite; see 32) in the culture medium.

All competition experiments with the sulfate-reducing species were carried out at 30°C, a temperature which is easily reached in the upper layers of the blackish sediment at low tide during summer. The pH was 7.0, and this pH value has been repeatedly measured in the interstitial water of the sediments. The sulfide concentration in the medium was kept below 0.5 mM (a realistic value in this sediment) by flushing with nitrogen, unless purposely higher sulfide concentrations were allowed to be reached. In the latter case iron limitation can be easily introduced. The dilution rate applied in all these competition experiments was $0.02 \ h^{-1}$.

Care should be taken to extrapolate results of competition experiments of the type described above to field conditions. First, the bacterial species involved were isolated from dilution series of mud in anaerobic agar shake cultures. It can, of course, not be excluded that other bacteria occurred which were not encountered in this way. This holds in particular for bacteria which have syntrophic relations with others. Second, environmental parameters fluctuate continually, and although the combination of pH, pS^{2-}, temperature, and growth limitations used undoubtedly does occur, the species involved are exposed in the sediments to a variety of other combinations of these factors, and other important variables may also be involved which affect bacterial properties (such as K_s and μ_{max}) and therefore their competitive abilities.

CONCLUDING REMARKS

The aim of the competition experiments described in the previous sections has been to obtain insight into possible selective advantages of individual species in their natural environment when competing for nutrients (or light in the case of phototrophs) involved in their energy metabolism. The niches of many bacteria occurring in the same habitat often show overlap with respect to the ability to use energy substrates. One important piece of information is, of course, to what extent substrates such as ethanol and lactate in sediments are actually available to the bacteria competing for them. Also, to make competition experiments as ecologically relevant as possible, secondary environmental factors such as pH, Eh, and temperature should be realistic. In other words, the organisms exposed to competition in the laboratory should be isolated from the same habitat and should be subjected to conditions which are likely to occur there. This also means that the fluctuations of environmental parameters have to be taken into account in such chemostat experiments. These may reveal selective advantages, as was illustrated in the experiments discussed above in which a light and dark rhythm was introduced in competition for sulfide among *Chromatium* species. Another example of the successful application of fluctuation of environmental parameters (nutrient concentrations) in the laboratory to reveal selective advantages is that of competition experiments between specialized and versatile colorless sulfur bacteria (Kuenen and Robertson, this volume).

One important point is that predictions which can be made from chemostat experiments should be verified in the natural environment. As yet, this has been done too little. Not only must laboratory experiments be carried out as much as possible under ecologically realistic conditions, but once a selective advantage of a species has been discovered in the laboratory, whether the occurrence and activity of a species is actually due to the predicted factor should be checked in nature. With some groups of bacteria this is easier than with others. This is relatively easy for phototrophic bacteria which can be localized in a limited area in stratified lakes. Here, the organisms can often be recognized microscopically, and their occurrence can be determined by direct counts and pigment analysis. Light conditions can be measured, and the concentration and turnover rate of the limited number of potential energy substrates present can be determined. For other groups of bacteria these measurements in the field of parameters which determine their survival may be more difficult, although promising progress has been made, as is illustrated by the work of Jørgensen and Revsbech (23) on colorless sulfur bacteria.

Trying to discover by means of chemostat studies the strong points of individual species with respect to competition and other interactions is one aspect of laboratory studies. Another and more elusive one is the summing up of such studies to understand nature's complexity. This is an entirely different matter which has to be handled with extreme care. It would seem naive to think that isolating a few organisms seemingly more important than others because of their relative abundance, and adding up their properties as established in the laboratory, would lead to an understanding of the complex reality in nature. A more appropriate approach to this latter aim seems to be to avoid the mixing of pure cultures and start from the more complex end by introducing in the laboratory (either in a chemostat or in any other appropriate device) complex communities exposed to more complex conditions than single or dual substrate limitations. An early and very successful approach in this respect was the "Winogradski column" (55).

It has been stated above that competition experiments with the chemostat should be carried out with organisms that exist in the same habitat and by applying conditions which occur there. This does not, of course, mean that a chemostat ever mimics a natural environment exactly. In a chemostat, organisms which grow slower than others are eliminated, but in nature different conditions prevail. When looking at a photograph of a sample of soil or sediment made with a scanning electron microscope, one can see microcolonies adhering to particles which very often are far away from others. Short periods of relatively fast growth are followed by slow growth or dormancy. And the ability to utilize more than one energy source, probably often even several energy sources simultaneously, in addition to differences in optimal conditions for growth with respect to secondary factors such as pH, pO_2, and temperature, explains why the diversity of microbes encountered in nature is much larger than that which might be expected on the basis of laboratory experiments carried out thus far. Nevertheless, competition experiments with the chemostat have revealed selective principles and selective advantages of individual species which could not have been discovered otherwise. When used for this purpose, the chemostat will remain a useful tool for future experiments, especially when applied more often in connection with field work and with other laboratory techniques aimed at unraveling nature's complexity.

We thank Jan Gottschal for valuable discussions and Marry Pras for typing the manuscript.

LITERATURE CITED

1. **Bergstein, T., Y. Henis, and B. Z. Cavari.** 1979. Investigations on the photosynthetic sulfur bacterium *Chlorobium phaeobacteroides* causing seasonal blooms in Lake Kinneret. Can. J. Microbiol. **25:**999–1007.

2. **Button, D. K.** 1978. On the theory of control of microbial growth kinetics by limiting nutrient concentrations. Deep Sea Res. **25:**1163–1177.

3. **Button, D. K.** 1983. Differences between the kinetics of nutrient uptake by micro-organisms, growth and enzyme kinetics. Trends Biochem. Sci. **8:**121–124.

4. **Cohen, Y., B. B. Jørgensen, E. Padan, and M. Shilo.** 1975. Sulphide-dependent anoxygenic photosynthesis in the cyanobacterium *Oscillatoria limnetic.* Nature (London) **257:**489–492.

5. **Dykhuizen, D. E., and D. L. Hartl.** 1983. Selection in chemostats. Microbiol. Rev. **47:**150–168.

6. **Frederickson, A. G.** 1977. Behaviour of mixed cultures of microorganisms. Annu. Rev. Microbiol. **31:**63–87.

7. **Gottschal, J. C., and J. G. Kuenen.** 1980. Selective enrichment of facultatively chemolithotrophic Thiobacilli and related organisms in continuous culture. FEMS Microbiol. Lett. **7:**241–247.

8. **Gottschal, J. C., and T. F. Thingstad.** 1982. Mathematical description of competition between two and three bacterial species under dual substrate limitation in the chemostat: a comparison with experimental data. Biotechnol. Bioeng. **24:**1403–1418.

9. **Gunnarsson, L. Å. H., and P. H. Rönnow.** 1982. Interrelationships between sulfate-reducing and methane-producing bacteria in coastal sediments with intense sulfide production. Mar. Biol. **69:**121–128.

10. **Hansen, T. A., and H. van Gemerden.** 1972. Sulfide utilization by purple non-sulfur bacteria. Arch. Mikrobiol. **86:**49–56.

11. **Harder, W., and L. Dijkhuizen.** 1982. Strategies of mixed substrate utilization in microorganisms. Phil. Trans. R. Soc. London Ser. B **297:**459–480.

12. **Harder, W., J. G. Kuenen, and A. Matin.** 1977. Microbiol selection in continuous cultures. A review. J. Appl. Bacteriol. **43:**1–24.

13. **Healey, F. P.** 1980. Slope of the Monod equation as an indicator of advantage in nutrient competition. Microb. Ecol. **5:**281–286.

14. **Herbert, D.** 1961. The chemical composition of microorganisms as a function of their environment. Symp. Soc. Gen. Microbiol. **11:**391–418.

15. **Howarth, R. W., and A. Giblin.** 1983. Sulfate reduction in the salt marshes at Sapelo Island, Georgia. Limnol. Oceanogr. **28:**70–82.

16. **Howarth, R. W., and J. M. Teal.** 1979. Sulfate reduction in a New England salt marsh. Limnol. Oceanogr. **24:**999–1013.

17. **Hurlberg, R. E., and J. Lascelles.** 1963. Ribulose diphosphate carboxylase in Thiorhodaceae. J. Gen. Microbiol. **33:**445–458.

18. **Jannasch, H. W.** 1967. Enrichment of aquatic bacteria in continuous culture. Arch. Mikrobiol. **59:**165–173.

19. **Jannasch, H. W.** 1969. Estimation of bacterial growth rates in natural waters. J. Bacteriol. **99:**156–160.

20. **Jørgensen, B. B.** 1977. The sulfur cycles of a coastal marine sediment (Limfjorden, Denmark). Limnol. Oceanogr. **22:**814–832.

21. **Jørgensen, B. B.** 1980. Mineralization and the bacterial cycling of carbon, nitrogen and sulphur in marine sediments, p. 239–251. *In* D. C. Ellwood, J. N. Hedger, M. J. Latham, J. M. Lynch, and J. H. Slater (ed.), Contemporary microbial ecology. Academic Press, London.

22. **Jørgensen, B. B., J. G. Kuenen, and Y. Cohen.** 1979. Microbial transformations of sulfur compounds in a stratified lake (Solar Lake, Sinai). Limnol. Oceanogr. **24:**799–822.

23. **Jørgensen, B. B., and N. P. Revsbeck.** 1983. Colorless sulfur bacteria, *Beggiatoa* spp. and *Thiovulum* spp., in O_2 and H_2S microgradients. Appl. Environ. Microbiol. **45:**1261–1270.

24. **King, G. M., and M. J. Klug.** 1982. Comparative aspects of sulfur mineralization in sediments of a eutrophic lake

basin. Appl. Environ. Microbiol. **43**:1406–1412.

25. **King, G. M., and M. J. Klug.** 1982. Glucose metabolism in sediments of a eutrophic lake: tracer analysis of uptake and product formation. Appl. Environ. Microbiol. **44**:1308–1317.

26. **Konings, W. N., and H. Veldkamp.** 1980. Phenotypic responses to environmental change, p. 161–191. *In* D. C. Ellwood, J. N. Hedger, M. J. Latham, J. M. Lynch, and J. H. Slater (ed.), Contemporary microbial ecology. Academic Press, London.

27. **Konings, W. N., and H. Veldkamp.** 1983. Energy transduction and solute transport mechanisms in relation to environments occupied by microorganisms. Symp. Soc. Gen. Microbiol. **34**:153–187.

28. **Kristjansson, J. K., P. Schönheit, and R. K. Thauer.** 1982. Different K_s values for hydrogen of methanogenic bacteria and sulfate-reducing bacteria: an explanation for the apparent inhibition of methanogenesis by sulfate. Arch. Microbiol. **131**:278–282.

29. **Kuenen, J. G., J. Boonstra, H. G. J. Schröder, and H. Veldkamp.** 1977. Competition for inorganic substrates among chemoorganotrophic and chemolithotrophic bacteria. Microb. Ecol. **3**:119–130.

30. **Kuenen, J. G., and W. Harder.** 1982. Microbial competition in continuous culture, p. 342–367. *In* R. G. Burns and J. H. Slater (ed.), Experimental microbial ecology. Blackwell Scientific Publications, Oxford.

31. **Laanbroek, H. J., T. Abee, and I. M. Voogd.** 1982. Alcohol conversions by *Desulfovibrio propionicus* Lindhorst in the presence and absence of sulfate and hydrogen. Arch. Microbiol. **133**:178–184.

32. **Laanbroek, H. J., and H. J. Geerligs.** 1983. Influence of clay particles (Illite) on substrate utilization by sulfate-reducing bacteria. Arch. Microbiol. **134**:161–163.

33. **Laanbroek, H. J., and N. Pfennig.** 1981. Oxidation of short-chain fatty acids by sulfate-reducing bacteria in freshwater and in marine sediments. Arch. Microbiol. **128**:330–335.

34. **Laanbroek, H. J., A. J. Smit, G. Klein Nulend, and H. Veldkamp.** 1979. Competition for L-glutamate between specialised and versatile *Clostridium* species. Arch. Microbiol. **120**:61–66.

35. **Laanbroek, H. J., and H. Veldkamp.** 1982. Microbial interactions in sediment communities. Phil. Trans. R. Soc. London Ser. B **297**:533–550.

36. **Law, A. T., B. R. Robertson, S. S. Dunker, and D. K. Button.** 1976. On describing microbial growth kinetics from continuous culture data: some general considerations, observations, and concepts. Microb. Ecol. **2**:261–283.

37. **Lovley, D. R., D. F. Dwyer, and M. J. Klug.** 1982. Kinetic analysis of competition between sulfate reducers and methanogens for hydrogen in sediments. Appl. Environ. Microbiol. **43**:1373–1379.

38. **Lovley, D. R., and M. J. Klug.** 1982. Intermediary metabolism of organic matter in the sediments of a eutrophic lake. Appl. Environ. Microbiol. **43**:552–560.

39. **Lovley, D. R., and M. J. Klug.** 1983. Sulfate reducers can outcompete methanogens at freshwater sulfate concentrations. Appl. Environ. Microbiol. **45**:187–192.

40. **Lovley, D. R., and M. J. Klug.** 1983. Methanogenesis from methanol and methylamines and acetogenesis from hydrogen and carbon dioxide in the sediments of a eutrophic lake. Appl. Environ. Microbiol. **45**:1310–1315.

41. **Matheron, R., and R. Baulaigue.** 1979. Influence de l'intesité lumineuse sur la teneur en bactériochlorophylle et sur la croissance des sulfobactéries phototrophes marines. Can. J. Microbiol. **26**:464–467.

42. **Michaelis, L., and M. M. L. Menten.** 1913. Die Kinetik der Invertinwirkung. Biochem. Z. **49**:333–369.

43. **Monod, J.** 1942. Recherches sur la croissance des cultures bactériennes. Hermann, Paris.

44. **Monod, J.** 1950. La technique du culture continue; théorie et applications. Ann. Inst. Pasteur Paris **79**:390–410.

45. **Mountfort, D. O., and R. A. Asher.** 1981. Role of sulfate reduction versus methanogenesis in terminal carbon flow in polluted intertidal sediment in Waimea Inlet, Nelson, New Zealand. Appl. Environ. Microbiol. **42**:252–258.

46. **Mountfort, D. O., R. A. Asher, E. L. Mays, and J. M. Tiedje.** 1980. Carbon and electron flow in mud and salt sandflat intertidal sediments at Delaware Inlet, Nelson, New Zealand. Appl. Environ. Microbiol. **39**:686–694.

47. **Novick, A., and L. Szilard.** 1950. Experiments with the chemostat on spontaneous mutations of bacteria. Proc. Natl. Acad. Sci. U.S.A. **36**:708–715.

48. **Oremland, R. S., and S. Polcin.** 1982. Methanogenesis and sulfate reduction: competitive and noncompetitive substrates in estuarine sediments. Appl. Environ. Microbiol. **44**:1270–1276.

49. **Pfennig, N.** 1965. Anreicherungskulturen für rote und grüne Schwefelbakteriën. Zentr. Bakteriol. Parasitenkd. Infektionskr. Hyg. Abt. Suppl. 1, p. 179–189.

50. **Pfennig, N.** 1967. Photosynthetic bacteria. Annu. Rev. Microbiol. **21**:285–324.

51. **Pfennig, N.** 1978. General physiology and ecology of photosynthetic bacteria, p. 3–18. *In* R. K. Clayton and W. R. Sistrom (ed.), The photosynthetic bacteria. Plenum Press, New York.

52. **Pfennig, N., and H. W. Jannasch.** 1962. Biologische Grundfragen bei der homokontinuierlichen Kultur van Mikroorganismen. Ergeb. Biol. **25**:93–105.

53. **Pfennig, N., F. Widdel, and H. G. Trüper.** 1981. The dissimilatory sulfate-reducing bacteria, p. 928–940. *In* M. P. Starr, H. Stolp, H. G. Trüper, A. Balows, and H. G. Schlegel (ed.), The prokaryotes. A handbook of habitats, isolation and identification of bacteria. Springer-Verlag, Berlin.

54. **Powell, E. O.** 1958. Criteria for the growth of contaminants and mutants in continuous culture. J. Gen. Microbiol. **18**:259–268.

55. **Schlegel, H. G., and N. Pfennig.** 1961. Die Anreicherungskultur einiger Schwefelpurpurbakteriën. Arch. Mikrobiol. **38**:1–39.

56. **Schönheit, P., J. K. Kristjansson, and R. K. Thauer.** 1982. Kinetic mechanism for the ability of sulfate reducers to out-compete methanogens for acetate. Arch. Microbiol. **132**:285–288.

57. **Senior, E., E. B. Lindström, J. M. Banat, and D. B. Nedwell.** 1982. Sulfate reduction and methanogenesis in the sediment of a salt marsh on the east coast of the United Kingdom. Appl. Environ. Microbiol. **43**:987–996.

58. **Sepers, A. B. J.** 1981. Diversity of ammonifying bacteria. Hydrobiologia **83**:343–350.

59. **Slater, J. H., and I. Morris.** 1973. Photosynthetic carbon dioxide assimilation by *Rhodospirillum rubrum*. Arch. Mikrobiol. **88**:213–223.

60. **Smith, R. L., and M. J. King.** 1981. Reduction of sulfur compounds in the sediments of a eutrophic lake basin. Appl. Environ. Microbiol. **41**:1230–1237.

61. **Smith, R. L., and M. J. Klug.** 1981. Electron donors utilized by sulfate-reducing bacteria in eutrophic lake sediments. Appl. Environ. Microbiol. **42**:116–121.

62. **Smouse, P. E.** 1980. Mathematical models for continuous culture growth dynamics of mixed populations subsisting on a heterogeneous resource base. I. Single competition. Theor. Popul. Biol. **17**:16–36.

63. **Sørensen, J., B. B. Jørgensen, and N. R. Revsbech.** 1979. A comparison of oxygen, nitrate and sulfate respiration of coastal marine sediments. Microb. Ecol. **5**:105–115.

64. **Sorokin, Y.** 1970. Interrelations between the sulphur and carbon turnover in leromictic [sic] lakes. Arch. Hydrobiol. **66**:391–446.

65. **Stanier, R. Y., E. A. Adelberg, and J. L. Ingraham.** 1976. The microbial world. Prentice-Hall Inc., Englewood Cliffs, N.J.

66. **Taylor, P. A., and P. J. LeB. Williams.** 1975. Theoretical studies on the coexistence of competing species under controlled flow conditions. Can. J. Microbiol. **21**:90–98.

67. **Trüper, H. G., and S. Genovese.** 1968. Characterization of photosynthetic sulfur bacteria causing red water in lake

Faro (Messina, Sicily). Limnol. Oceanogr. **13:**225–232.

68. **van Gemerden, H.** 1974. Coexistence of organisms competing for the same substrate: an example among the purple sulfur bacteria. Microb. Ecol. **1:**104–119.

69. **van Gemerden, H., and H. H. Beeftink.** 1983. Ecology of phototrophic bacteria, p. 147–186. *In* J. G. Ormerod (ed.), The anoxygenic phototrophic bacteria. Blackwell Scientific Publications, Oxford.

70. **Veldkamp, H.** 1970. Enrichment cultures of prokaryotic organisms, p. 305–361. *In* J. Norris and D. W. Ribbons (ed.), Methods in microbiology, vol. 3A. Academic Press, London.

71. **Veldkamp, H.** 1977. Ecological studies with the chemostat. Adv. Microb. Ecol. **1:**59–94.

72. **Veldkamp, H., and H. W. Jannasch.** 1972. Mixed culture studies with the chemostat. J. Appl. Chem. Biotechnol. **22:**105–123.

73. **Wijbenga, D. J., and H. van Gemerden.** 1981. The influence of acetate on the oxidation of sulfide by *Rhodopseudomonas capsulata*. Arch. Microbiol. **129:**115–118.

74. **Winfrey, M. R., and D. M. Ward.** 1983. Substrates for sulfate reduction and methane production in intertidal sediments. Appl. Environ. Microbiol. **45:**193–199.

75. **Winfrey, M. R., and J. G. Zeikus.** 1977. Effect of sulfate on carbon and electron flow during microbial methanogenesis in freshwater sediments. Appl. Environ. Microbiol. **33:**275–281.

76. **Winogradski, S.** 1949. Microbiologie du sol. Oeuvres complètes. Masson, Paris.

77. **Yoon, H., G. Klinzing, and H. W. Blanch.** 1977. Competition for mixed substrates by microbial populations. Biotechnol. Bioeng. **19:**1193–1210.

78. **Zaiß, U.** 1981. Seasonal studies of methanogenesis and desulfurication in sediments in the river Saar. Zentralbl. Bakteriol. Parasitenkd. Infektionskr. Hyg. Abt. 1 Orig. Reihe C **2:**76–89.

Growth and Interactions of Microorganisms in Spatially Heterogeneous Ecosystems

J. W. T. WIMPENNY, J. P. COOMBS, AND R. W. LOVITT

Department of Microbiology, University College, Cardiff CF2 1TA, Wales, United Kingdom

Microbes in many natural habitats occupy particular positions in space in response to a variety of different physical and chemical gradients. Furthermore, neighboring spaces contain different microbial types, and this makes possible a range of interactions, including competition, between them. The result is a heterogeneous microbial ecosystem in which diffusion is a major mechanism for solute transfer. There are many examples of such ecosystems, ranging from microbial film which may be only a few micrometers thick to lakes which are stratified over a depth of several meters. Such systems have been reviewed (21, 27, 28, 32).

The proliferation of microbes at fixed points in space is a feature that is lost in laboratory culture systems which are deliberately stirred or agitated. Homogeneity may lead to a loss of information or even to the generation of misleading information when results of such experiments are extrapolated to conditions found outside the laboratory. For example, two organisms which may occupy different spaces and interact only slightly in their natural habitat will, if grown together in a homogeneous culture vessel, be forced to compete for nutrients or even to inhibit each other. Results obtained in a chemostat might lead one to predict that one organism will totally outgrow the other, but in nature this may not be the case. Again, two organisms may interact naturally through diffusion, but may occupy mutually exclusive habitat domains; for example, one may be an obligate aerobe whereas the other has a strictly anaerobic physiology.

A further aspect of spatial organization is that, in nature, an organism may require two essential nutrients which are only available from sources located in opposing directions. The ability to use these nutrients will depend on the position of the organism. This vectorial element cannot be reproduced in a single stirred vessel. An example is the growth of sulfur organisms such as *Beggiatoa* and *Thiovulum* at the oxic/anoxic interface where oxygen and sulfide diffuse from opposite directions (12).

Bazin and his colleagues (3) noted two other problems that arise when homogeneous culture systems are used to investigate heterogeneous ecosystems. First, in cases in which the analysis is based on steady-state conditions, growth and the subsequent occupation of space by microbes are not taken into account. Second, microbes in solute gradients are frequently in transient states, and in such states they may exhibit behavior quite different from that displayed when they are at or near steady state.

The chemostat has a role to play in microbial ecology, but to accept it as capable of answering all questions is to accept the reductionist argument that any system may be understood if enough is known about its fundamental units. This approach, it must be said, yields much essential information, but it tells us little about the behavior of larger assemblages of the basic unit or about the system as a whole. This happens to be true in most areas of science, not least in microbial ecology. As we have said before (32), it is not that the information derived from the reductionist approach is wrong, but merely that it may be irrelevant or inapplicable when considering the whole system.

Why use laboratory models to study natural environments? Many difficulties arise during in vivo investigations of natural ecosystems. There may be a lack of uniformity about a natural system which leads to poor reproducibility of measurements. Ecosystems are frequently very complex, with a multiplicity of processes occurring in a small volume. This means that it may be physically difficult to make accurate measurements, and it may not be easy to determine the cause of particular effects observed. The complexities present in a small volume are magnified when dealing with larger systems. Investigations into larger systems may be simpler technically, but they generate a mean or average result which may conceal large regional differences in composition. Such results tend to generate "black" or "gray" box models of the particular ecosystem and to say little about the finer detail. Parkes (20) has pointed out that the value of laboratory systems lies in the degree of control which can be attained over the important variables, but he stresses the need for selecting the variables correctly.

Another problem with making direct measurements is that there is no way of knowing to what extent the system is perturbed when the measurements are made. The assumption has to be made that the system behaves in the same way when it is measured as when it is not. There is no easy answer here because any attempt to determine parameters in an ecosystem necessarily depends on an interaction between the observer and the system.

Views on the relative merits of laboratory systems vary from those which condemn all models as artifacts to those which consider the chemostat capable of solving most microbiological problems. We consider the use of models to be justified providing that (i) the model is rigorously tested and rejected or modified if found to be unsuitable, and (ii) the results obtained from models are seen as corroborating results from in situ studies.

If the importance of spatial heterogeneity is assumed and the value of model systems is accepted, then it follows that heterogeneous laboratory models are required. Model systems of this sort developed by us and by others are briefly reviewed below.

LINKED HOMOGENEOUS MODEL SYSTEMS

Feeding the output of one stirred fermentor to the input of another is to start to incorporate spatial heterogeneity into individually homogeneous systems. Multistage chemostats have been used in microbial ecology by, among others, John Parkes and David Nedwell (personal communication) in the United Kingdom, both of whom have established laboratory models of estuarine sediment ecosystems.

Whereas multistage systems allow flow in a single direction only, flow in two directions is more characteristic of natural solute transfer. A bidirectional system was described by Cooper and Copeland (9) to generate a salinity gradient for the investigation of an estuarine ecosystem. Lovitt and Wimpenny (13, 14), approaching the problem via fermentation technology, described a system of interconnected glass fermentor vessels which they called the "gradostat" (Fig. 1).

Gradostat. The gradostat consists normally of a series of five linked fermentor vessels located on a stepped platform. The system is fed at each end from reservoirs containing different sources, and medium is pumped up the array in one direction and flows back down the system over a series of weirs. Flow in both directions is therefore equal. If a red dye enters the system from one end and a blue dye from the other, opposing gradients of "redness" and "blueness" will become established. Strictly speaking, these are not continuous but rather are stepped gradients. However, it is clear that unique conditions appear in each of the vessels. If the dyes are replaced by essential nutrients and the system is inoculated, the organism will grow best in those vessels where nutrient conditions are optimal. One of the earliest experiments carried out was the growth of *Paracoccus denitrificans* in coun-

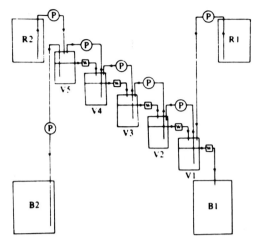

FIG. 1. Gradostat. Five culture vessels (V1 to V5) are fed with medium from reservoir R1 in an upward direction, with transfer between the vessels by a series of pumps (P). Medium from reservoir R2 is fed down the sequence of vessels over a series of weirs (w). B1 and B2 are the effluent-collecting vessels.

tergradients of succinate and nitrate (14). At suitable concentrations of nutrients, the organism grew best in the middle vessel of the array, and numbers decreased toward either end.

In another experiment (15), *Escherichia coli* was grown in opposing gradients of oxidant (oxygen) and reductant (glucose). The organisms showed a range of biochemical adaptations to gradients of electron donors and acceptors, notably in their ability to oxidize substrates and in the levels both of oxidative enzyme activities and of cytochrome pigments.

Competition studies are possible with the gradostat (32). A system containing *E. coli* was established that was similar to the one just described. An obligate aerobe, *Pseudomonas aeruginosa*, was introduced, and this displaced *E. coli* from the aerobic vessels. The system was then inoculated with an obligate anaerobe, *Clostridium acetobutyricum*, and this displaced *E. coli* from the anaerobic end of the array. This experiment supports the view that a "specialist" in its own niche will always outcompete a "generalist."

The main advantage of the gradostat is that it is an open system capable of reaching a steady state and hence allowing a reliable analysis of its behavior. The volume of the system (about 600 ml in each vessel) means that adequate material is available for analysis. Bidirectional flow in the gradostat also means that it is possible for organisms occupying different parts of the system to interact. The gradostat is therefore a powerful tool since it permits the linkage of metabolic processes which normally occur in mutually

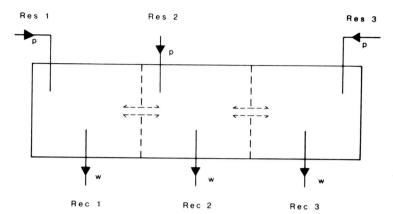

FIG. 2. Herbert model. Membranes separate the three vessels, allowing the passage of solutes but not organisms. Each vessel is fed from a separate reservoir (Res) by pumps (p), and a constant volume of medium is maintained by weirs (w). Spent medium is collected in receivers (Rec).

exclusive habitats. Examples include the cyclical interactions of the nitrogen and sulfur cycles. Alteration in particular vessel volumes and in the ratio of flow rates in each direction can increase the flexibility of the gradostat by altering the concentration profiles and suggests applications in genetics and in the metabolism of inhibitory compounds.

Herbert model. One feature of the gradostat which is not typical of many natural ecosystems is that both solutes and organisms are transferred simultaneously. This problem is overcome in a model conceptually similar to the "Ecologen," devised by R. A. Herbert (personal communication). In this system, three linked vessels are separated from one another by membrane filters (Fig. 2). Here, solutes but not organisms can diffuse between vessels. In a typical experiment, *Clostridium butyricum* was grown on a glucose–salts medium. The fermentation products diffused across the membrane into the middle vessel, where a *Desulfovibrio* sp. used them as carbon and energy sources for the reduction of sulfate. The sulfide generated diffused across the membrane into the third vessel, which was illuminated. Here, a *Chromatium* sp. oxidized the sulfide to sulfate, and the latter diffused back into the middle vessel. In this way a functional sulfur cycle was established in the second and third vessels, "driven" by the fermentation products of the *Clostridium* in the first vessel and by light energy in the third. This system is potentially a most useful one and has recently been used to grow nitrifying and denitrifying organisms together. The gradostat and Herbert's model represent two extremes in the disposition of organisms and substrates. The "natural" situation will often fall somewhere between the two.

GEL-STABILIZED SYSTEMS

Linked homogeneous systems have a major virtue: they are "open." However, they have poor resolving ability. If greater resolution is needed, the addition of a gelling agent to a liquid system allows the establishment of continuous solute gradients generated solely by molecular diffusion between source and sink regions. This principle was first used by Beijerinck in 1889 in the technique called "auxonographie" (4). Since then, agar-stabilized gradient systems have largely been confined to antibiotic susceptibility testing and to immunological investigations.

A simple laboratory model system can be established in test tubes or beakers containing a source layer which contains full-strength agar, basal nutrients, and a diffusing solute (Fig. 3a). A semisolid agar layer containing basal nutrients and an inoculum of organisms, but no diffusing solute, is poured above this. Gradients of solute and oxygen from the atmosphere are thus set up in the semisolid layer. Gels can be analyzed with needle electrodes to measure pH, oxygen, and redox potential, for example. In addition, cores may be removed and either scanned to assess growth or sliced for chemical or biological analysis.

When such a system was set up with a basal medium containing amino acids, yeast extract, and mineral salts, and with glucose as the diffusing solute in the source layer, a number of organisms grown in the system generated discrete multiple bands of growth (30). Of these, *Bacillus cereus* consistently produced several bands and was chosen for detailed study (8). Oxygen and glucose gradients were needed for band formation. Subsurface bands appeared in order, lowest first, over a period of 2 to 14 days.

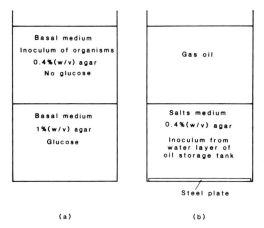

FIG. 3. (a) Standard gel-stabilized nutrient gradient system. (b) Gel-stabilized system used to model the growth of organisms from the water layer of an oil storage tank.

Increasing glucose or decreasing the basal medium strength moved the band position higher up the gel layer. Steep pH gradients developed in the growth layer. These ranged from about 7.6 at the surface to about 5.3 at 5 to 10 mm below the surface. Two experiments confirmed the importance of pH in band formation. First, the pH profile and band position could be altered by adding suitable concentrations of phosphate buffer. Second, no pH gradient was detected and no bands were observed in a gel incubated anaerobically. However, when a layer of agar containing alkali was poured above the gel surface, even if the system was incubated anaerobically, a pH gradient was established and band formation was restored.

Despite these experiments, the exact mechanism for discrete band formation is not clear. Computer simulations predict that bands will be formed if an asymmetric threshold model is incorporated. A physiological interpretation is that the organism stops growing at a particular low pH, but does not start to grow again until a much higher pH is reached. We have yet to test this hypothesis experimentally. Hoppensteadt and Jager (11) analyzed a similar example of band formation numerically. Consecutive growth rings were observed when a histidine-requiring *Salmonella* strain was grown as a lawn on a glucose–salts agar plate and a drop of concentrated histidine solution was placed at the center of the dish. Their conclusion that an asymmetric activation threshold must operate agrees in principle, though not in detail, with our own.

Other examples of band formation have from time to time appeared in the literature (10, 19,

22–26) and have been noted but not formally reported by other workers including H. Veldkamp, W. D. Grant, C. M. Brown, and H. Abdolahi. These observations were reviewed in more detail elsewhere (32).

Gel systems have other uses which may be of more value to microbial ecology: for example, modeling interactions in vertically stratified ecosystems. We have used a model of this sort to investigate the ecology of the water base found in oil storage tanks (30). This habitat is of economic importance because sulfate-reducing bacteria may be responsible for corroding the steel base of the tank. The water layer is normally sampled either by lowering a tube and withdrawing a sample or through a drain cock at the base of the tank. In either case, the spatial heterogeneity of the water layer is lost, and mixing of aerobic and anaerobic layers will almost certainly lead to errors in assessing the numbers of oxygen-sensitive bacteria.

In the gel-stabilized model system, an inoculum from the water base was mixed with a semisolid agar layer above a steel plate and beneath a layer of gas oil (Fig. 3b). The system became spatially organized, showing aerobic and anaerobic regions and gradients of oxygen, redox potential, sulfide, and iron. This model shows how such an ecosystem could develop in the absence of mechanical disturbance, with diffusion as the sole means of solute transfer.

Two-dimensional gel-stabilized model systems. The gel systems described so far have incorporated diffusion gradients in one dimension only. They have also been systems in which the concentration of solute at any one point was varied with time. An attempt to overcome this problem was made by Caldwell and co-workers (5, 6), who developed a two-dimensional plate system in which steady-state gradients were established (Fig. 4).

The system was used to study the growth of *E. coli* in radioactively labeled acetate and to study the growth of *Rhodomicrobium*, *Thiopedia*, and *Hyphomicrobium* in gradients of methylamine and sodium sulfide. Segregation of these organisms was shown.

The system was also used to investigate the growth of plankton from a forest pond in gradients of acetate and mud "pore water." A major advantage of the Caldwell system is that it is capable of segregating "stenobiotic" organisms, i.e., those with narrow, well-defined habitat requirements, from the more versatile "eurybiotic" organisms with less exacting requirements. The disadvantages of the Caldwell system, however, are that it takes about 2 days for a steady state to be reached and the working area of the plate is relatively small (about 2.5 cm square).

We have established two-dimensional gradi-

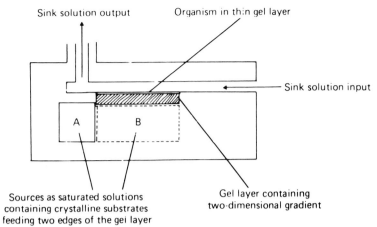

FIG. 4. Steady-state two-dimensional gradient plate described by Caldwell et al. (6). Solutes A and B, present as crystalline suspensions in reservoirs, diffuse into neighboring edges of an agar gel layer. The system is open since solutes diffuse into an aqueous solution which passes over the surface of the agar. Organisms are grown in a thin surface agar layer after the whole system reaches steady state.

ent systems in 10- and 12-cm square plastic petri dishes (Fig. 5). Plates were constructed by use of the wedge plate technique first described by W. Szybalski (Bacteriol. Proc., p. 36, 1952). A traditional wedge plate containing acid and alkaline wedges was established in the bottom of the

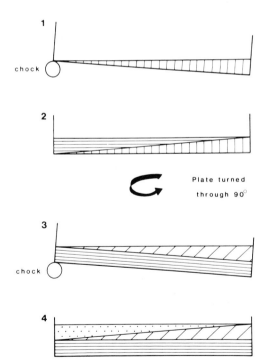

FIG. 5. Stages in establishing two-dimensional wedge plates. Each wedge of agar differs with respect to one component, e.g., wedge 1, acidic; 2, alkaline; 3, saline; 4, nonsaline.

square plate and allowed to set. The plate was then rotated about 90°, and a second gradient was developed at right angles to the first by use of wedges of growth medium with and without sodium chloride. Such salt–pH gradient systems were shown to be effective for comparing the habitat ranges of different bacterial species. In such experiments it was clear that species of *Staphylococcus* were extremely salt tolerant, whereas other organisms, for example, *Chromobacterium violaceum*, were very sensitive to low salt concentrations (Fig. 6a). In many experiments salt tolerance was associated with rather narrow environmental pH ranges. Such two-dimensional gradient plates offer a versatile and simple technique for examining some of the environmental constraints which determine the ecology of different species.

In a few of these two-dimensional experiments, mixed cultures were investigated. In one clear example a culture of the red-pigmented organism *Serratia marcescens* was mixed with the yellow organism *Micrococcus luteus* before plating on a salt–pH gradient plate. Separately, the two species showed considerable overlap in habitat range. However, competitive exclusion was apparent when the two were grown together. Thus, there was a clear boundary between the two growth zones, each species occupying less space than either would if grown individually.

There appears to be considerable promise from taxonomic, diagnostic, and ecological points of view in using simple methods like two-dimensional plates. Taxonomists should be able to distinguish between related strains and species by their behavior in appropriate gradient systems, and there are many possibilities for the

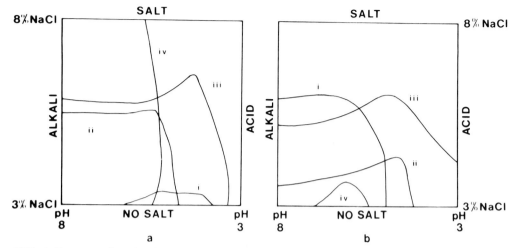

FIG. 6. Summary of results of growing a number of bacteria on replicate two-dimensional wedge plates. (a) Limits of growth for (i) *Chromobacterium violaceum*, (ii) *Flavobacterium* sp., (iii) *Serratia marcescens*, (iv) *Staphylococcus albus*. (b) Limits of growth for (i) *Bacillus cereus*, (ii) *B. marcerans*, (iii) *B. megaterium*, (iv) *B. cereus* var. *mycoides*.

examination of the habitat attributes of organisms isolated from particular ecosystems. Plates need not be restricted to salt–pH gradients; any solute of interest can be incorporated into a gradient. It is quite possible to include a third solute or physical condition such as temperature in a series of two-dimensional gradient plates. For example, if 10 identical salt–pH gradient plates are poured and incubated at 10 different temperatures, the resultant patterns of growth can be represented as a three-dimensional object in space. Similarly, a third solute can be incorporated into replicate plates at different concentrations. Such three-dimensional structures will give an important indication of the potential habitat domain of a particular organism.

MICROBIAL FILM FERMENTORS

Given an appropriate environment containing sufficient water and nutrients, solid surfaces can soon be covered by a sheet or film of microbial growth. Such films are very common and may have beneficial or harmful economic consequences, depending on where they appear. Useful films include those found in sewage-treatment plants, where they grow on the substrata of trickling filter beds or on the disks of rotating disk aerators. Less welcome films form on the surface of human teeth, where they can cause caries, or on the surfaces of marine equipment and installations, where they initiate fouling or even anaerobic corrosion cells.

It is not clear from the literature that microbial film has ever been systematically investigated as an ecosystem in its own right. It is our view that microbial film represents an important family of heterogeneous ecosystems whose end product is the result of the ordered development in space and in time of a community of microbial species. Much remains to be done before the biology and physiology of such communities can be understood.

We (7, 28, 32) have described a laboratory film fermentor designed to generate film of constant thickness under controlled conditions (Fig. 7). The fermentor is set up inside a sterile enclosure to facilitate experiments with pure cultures. Film forms in plastic film pans which are themselves located in a rotating porous Teflon disk, irrigated with sterile medium. Each pan can be removed without affecting its neighbors and replaced by a plastic blank. In operation the film pans pass beneath a porous Teflon scraper bar to ensure that films develop to no more than a maximum thickness. At present the fermentor has been used to establish films of pure or mixed cultures of dental organisms. Growth rates of such films have been monitored, and films have been examined by transmission electron microscopy. This is only the first stage, however. Such a system should allow a much more detailed examination of the properties of films. The deployment of microelectrodes for oxygen, pH, Eh, and sulfide is possible. Film can be embedded and sectioned either for electron microscopy or to determine the profile of elements across it by use of electron microprobe analysis, or it can be freeze-sectioned in either dimension for examination of the distribution of microbial species, enzymatic activities, or solutes across the

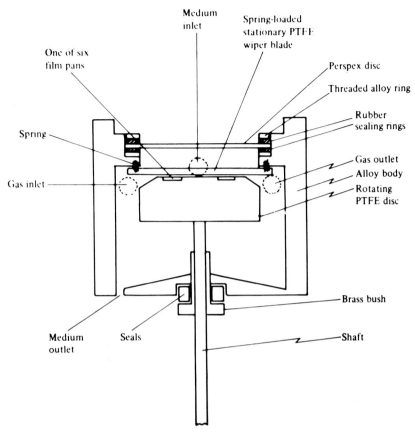

FIG. 7. Thin film fermentor developed by Coombe et al. (7). The rotating porous Teflon (PTFE) disk contains six pans of 300-μm depth in which microbial film is allowed to develop. A constant film depth is maintained by the spring-loaded PTFE wiper blade. The film is supplied with a suitable nutrient medium and gas mixture.

array. Much remains to be done to determine the vital role of the polymer-producing bacteria which are responsible for maintaining the structural integrity of microbial films. The ability to investigate pure cultures of film-producing organisms is fundamentally important to any model system from this point of view.

OTHER MODEL SYSTEMS

There are other laboratory models which have important parts to play in the study of structured microbial communities. One of the more important is the percolating column. Percolating columns are good analogs of one of the most important natural habitats, the soil. They share in common the downward flow of solutions from the surface to deeper horizons. Simple soil columns have been used by different workers (1, 18), and systems have been simplified considerably by using glass beads instead of soil particles (2, 16, 17). The subject has been discussed in more detail by Bazin et al. (3).

The bacterial colony is a good example of a structured microbial community with special characteristics. It is usually a pure culture and often is a clone derived from a single cell. Large colonies seen under optimal laboratory conditions are rare in nature, suggesting that the laboratory version should be regarded as a "model" or paradigm for natural structured ecosystems. Thus, microbial film has much in common with the flatter bacterial colony. In addition, colonies demonstrate a primitive form of morphogenesis since they become recognizable morphological entities whose form is controlled by a range of internal and external factors. The most important of the latter are physicochemical gradients which are a dominant feature of colonial growth since the cell packing densities and hence rates of metabolism are very high. For example, colony respiration rates increase rapidly during growth (31), leading to steep oxygen gradients within the structure (29). The influence of these solute gradients on colony development merits further investigation.

NUMERICAL MODELS

The addition of spatial heterogeneity in natural or artificial microcosms increases the complexity of the dynamics of each system. Reaction-diffusion equations are essential for describing such ecosystems quantitatively. It is not easy to solve these equations analytically, and numerical methods using computer techniques are needed.

Two numerical models were developed for us by S. Jaffe. The first was used to predict the behavior of the gradostat. The program determined changes in up to 10 cells plus solutes as a function of time. Systems modeled included the single- or multistage chemostat as well as the gradostat. This model allowed us to predict some of the simpler dynamic properties of the gradostat without the need to set it up and carry out experiments, and hence saved considerable time. Experiments in which *Paracoccus denitrificans* was grown in the gradostat in countergradients of succinate and nitrate confirmed the computer predictions even when the relative concentrations of the two substrates were altered.

A second simulation model was used to predict changes in up to 10 cells plus solutes in gel-stabilized diffusion systems. We used this model (32) to try to explain the periodic growth of *B. cereus* in the gel-stabilized system. The program has also been used to identify the factors which determine where, in a spatially organized system, a cell actually grows. It appears that solute concentration, diffusion coefficient, distance, and yield coefficients are important spatial determinants, whereas growth rate and substrate affinity values are not.

CONCLUSIONS

We have tried in this paper to indicate the vital importance of spatial heterogeneity in microbial ecosystems. We have also sought to convince readers of the need for laboratory and numerical models which incorporate elements of heterogeneity. Such vectorial models, though much more complex than homogeneous systems such as stirred laboratory fermentors, give information on the spatial organization of ecosystems which is otherwise very hard to determine. The types of systems that have been used are summarized below.

The gradostat represents a linked homogeneous system in which steady-state countergradients can be established in a linear array of stirred vessels. Such a system allows the existence of a number of different habitats linked together by flow between vessels. The virtues of the gradostat are that it operates under steady-state conditions and that it generates sufficient material for biochemical analyses to be performed. Its main disadvantages are that it lacks spatial resolution and that it is complex to set up and operate.

Where separation of species is thought valuable, the system described by R. A. Herbert (personal communication), in which adjacent stirred vessels are separated by membranes, promises to be most useful in microbial ecosystem research.

In gel-stabilized ecosystems, solute transfer can be restricted to molecular diffusion by the addition of gelling agents to nutrient solutions. One-dimensional systems with single or bidirectional solute source gradients are easy to establish. Such systems are closed for solutes and cells, and they are capable of very high spatial resolution, as is clear from our studies of growth-band formation in cultures of bacteria like *B. cereus*. Such models have already played a part in the analysis of growth in the water phase of oil storage tanks, and they promise to be a useful tool for investigating a wide range of natural ecosystems.

Two-dimensional gel systems have another important part to play in assessing the habitat requirements of different species and also in assessing the habitat range of a group of organisms isolated from the same natural ecosystem. A systematic examination of different ecosystems by use of such a technique is badly needed.

Microbial film appears widely under appropriate conditions, but little is known about the complex interactions which take place in such natural ecosystems. Model film fermentors will play an important part in such research.

Certain ecosystems, in particular soil, develop under conditions where nutrients are transported largely in one direction by water moving downward under the influence of gravity. Percolating columns can simplify such systems, at the same time retaining many of their natural characteristics.

The bacterial colony is an example of a structure whose morphology and physiology are dominated by solute availability and distribution. Such multicellular arrays are natural paradigms of simple morphogenetic systems and can at the same time give useful information on interactions between organisms of the same species where spatial heterogeneity is present.

Future developments in the area of heterogeneous microbiology are likely to be exciting as more scientists appreciate the importance of spatial organization in natural microbial ecosystems. The collection of model systems described here marks the start of an era in which such models will proliferate and become far more refined and sophisticated. We believe that only then will many of the secrets of microbial ecology yield to experimentation.

We gratefully acknowledge grants from NERC and SERC which have supported much of the work described in this report.

In addition, we thank our colleagues H. Abdollahi, R. A. Coombe, S. Jaffe, A. C. Peters, A. Tatevossian, P. Waters, and S. G. Whittaker for their valuable contributions.

LITERATURE CITED

1. **Ardakani, M. S., J. T. Rehbock, and A. D. McLaren.** 1973. Oxidation of nitrite to nitrate in a soil column. Soil Sci. Soc. Am. Proc. **37:**53–56.
2. **Bazin, M. J., and P. T. Saunders.** 1973. Dynamics of nitrification in a continuous flow system. Soil Biol. Biochem. **5:**531–543.
3. **Bazin, M. J., P. T. Saunders, and J. I. Prosser.** 1976. Models of microbial interaction in the soil. Crit. Rev. Microbiol. **4:**463–498.
4. **Beijerinck, M.** 1889. Auxanography, a method useful in microbiological research, involving diffusion in gelatin. Arch. Neerl. Sci. Exactes Nat. Haarlem **23:**367–372.
5. **Caldwell, D. E., and P. Hirsch.** 1973. Growth of microorganisms in two-dimensional steady-state diffusion gradients. Can. J. Microbiol. **19:**53–58.
6. **Caldwell, D. E., S. H. Lai, and J. M. Tiedje.** 1973. A two-dimensional steady-state diffusion gradient for ecological studies. Bull. Ecol. Res. Commun. (Stockholm) **17:**151–158.
7. **Coombe, R. A., A. Tatevossian, and J. W. T. Wimpenny.** 1982. Bacterial thin films as *in vitro* models for dental plaque, p. 239–249. *In* R. M. Frank and S. A. Leach (ed.), Surface and colloidal phenomena in the oral cavity: methodological aspects. IRL Press, London.
8. **Coombs, J. P., and J. W. T. Wimpenny.** 1982. Growth of *Bacillus cereus* in a gel-stabilized nutrient gradient system. J. Gen. Microbiol. **128:**3093–3101.
9. **Cooper, D. C., and B. J. Copeland.** 1973. Responses of continuous-series estuarine microecosystems to point-source input variations. Ecol. Monogr. **43:**213–236.
10. **Evans, J. B., and W. E. Kloos.** 1972. Use of shake cultures in a semi-solid thioglycolate medium for differentiating staphylococci from micrococci. Appl. Microbiol. **23:**326–331.
11. **Hoppensteadt, F. C., and W. Jager.** 1979. Pattern formation by bacteria, p. 68–81. *In* W. Jager, H. Rost, and P. Taufu (ed.), Biological growth and spread, Lecture notes in biomathematics, vol. 38. Springer-Verlag, Berlin.
12. **Jorgensen, B. B.** 1982. Ecology of the bacteria of the sulphur cycle with special reference to anoxic-oxic interface environments. Philos. Trans. R. Soc. London Ser. B **298:**543–561.
13. **Lovitt, R. W., and J. W. T. Wimpenny.** 1979. The gradostat: a tool for investigating microbial growth and interactions in solute gradients. Soc. Gen. Microbiol. Q. **6:**80.
14. **Lovitt, R. W., and J. W. T. Wimpenny.** 1981. The gradostat: a bidirectional compound chemostat, and its application in microbiological research. J. Gen. Microbiol. **127:**261–268.
15. **Lovitt, R. W., and J. W. T. Wimpenny.** 1981. Physiological behaviour of *Escherichia coli* grown in opposing gradients of glucose and oxygen plus nitrate in the gradostat. J. Gen. Microbiol. **127:**269–276.
16. **Macura, J., and F. Kunck.** 1965. Continuous flow method in soil microbiology. IV. Decomposition of glycine. Folia Microbiol. **10:**115–124.
17. **Macura, J., and F. Kunck.** 1965. Continuous flow method in soil microbiology. V. Nitrification. Folia Microbiol. **10:**125–135.
18. **Macura, J., and I. Malik.** 1958. Continuous flow method for the study of microbiological processes in soil samples. Nature (London) **182:**1796–1797.
19. **Nitsch, B., and H. J. Kutzner.** 1973. Wachstum von Streptomycetin in Schuttelagarkultur: eine neue methode zur festellung des c-quellen-spektrums, p. 481–486. *In* Symposium on Technische Mikrobiologie. Berlin.
20. **Parkes, R. J.** 1982. Methods for enriching, isolating, and analysing microbial communities in laboratory systems, p. 45–102. *In* A. T. Bull and J. H. Slater (ed.), Microbial interactions and communities, vol. 1. Academic Press, London.
21. **Perfil'ev, B. V., and D. R. Gabe.** 1969. Capillary methods of investigating microorganisms (English translation). Oliver & Boyd, Edinburgh.
22. **Tschapek, M., and N. Giambiagi.** 1954. The formation of Liesegang rings by *Azotobacter* under oxygen inhibition / Die bildung von Liesegang'schen ringen durch Azotobakter bei O2-hemmung. Kolloid Z. **135:**47–48.
23. **Williams, J. W.** 1938. Bacterial growth "spectrum" analysis. I. Methods and application. Am. J. Med. Technol. **4:**58–61.
24. **Williams, J. W.** 1938. Bacterial growth "spectrums." II. Their significance in pathology and bacteriology. Am. J. Pathol. **14:**642–645.
25. **Williams, J. W.** 1939. Growth of microorganisms in shake cultures under increased oxygen and carbon dioxide tensions. Growth **3:**21–33.
26. **Williams, J. W.** 1939. The nature of gel mediums as determined by various gas tensions and its importance in growth of microorganisms and cellular metabolism. Growth **3:**181–196.
27. **Wimpenny, J. W. T.** 1981. Spatial order in microbial ecosystems. Biol. Rev. **56:**295–342.
28. **Wimpenny, J. W. T.** 1982. Responses of microorganisms to physical and chemical gradients. Philos. Trans. R. Soc. London Ser. **297:**497–515.
29. **Wimpenny, J. W. T., and J. P. Coombs.** 1983. Penetration of oxygen into bacterial colonies. J. Gen. Microbiol. **129:**1239–1242.
30. **Wimpenny, J. W. T., J. P. Coombs, R. W. Lovitt, and S. G. Whittaker.** 1981. A gel-stabilized model ecosystem for the investigation of microbial growth in spatially ordered solute gradients. J. Gen. Microbiol. **127:**277–287.
31. **Wimpenny, J. W. T., and M. W. A. Lewis.** 1977. The growth and respiration of bacterial colonies. J. Gen. Microbiol. **103:**9–18.
32. **Wimpenny, J. W. T., R. W. Lovitt, and J. P. Coombs.** 1983. Laboratory models for the investigation of spatially and temporally organized ecosystems. Symp. Soc. Gen. Microbiol. **34:**66–117.

Evaluation of Competition in *Rhizobium* spp.

NOËLLE AMARGER

Laboratoire de Microbiologie des Sols, Institut National de la Recherche Agronomique, BV 1540, 21034 Dijon-Cedex, France

Bacteria of the genus *Rhizobium* are present in the majority of soils. Their numbers as well as the type of species encountered are dependent both on the biotic and abiotic soil environment and on the legume species, whether cultivated or wild, which develop on these soils. The symbiotic properties of indigenous rhizobia specific for one legume species may be more or less satisfactory. In effect, the different strains capable of forming nodules on one legume variety may not fix nitrogen or may fix nitrogen in more or less important quantities, resulting in more or less important yields.

At nodule formation, the legume may favor one of a number of strains of *Rhizobium* to form the nodules (7, 14, 22, 33). *Rhizobium* spp. may therefore differ in their capacity to be selected by the plant host, i.e., in their nodulating competitiveness.

The choice made by the plant is not dependent on the nitrogen-fixing ability of the strain (2, 23, 26). A great number of nodules may be formed by a strain fixing little or no nitrogen, even in the presence of effective strains (32).

Under natural conditions, different situations may be encountered when legumes are planted. If the specific *Rhizobium* spp. are absent, the seeds can be successfully inoculated with an effective strain of *Rhizobium*. If *Rhizobium* spp. are present in the soil, nitrogen fixation by indigenous strains of *Rhizobium* may be enough to support legume growth if the strains are effective. However, if indigenous strains are ineffective, nitrogen may limit legume growth. In such soils, seed inoculation at planting time with an effective strain may not have the desired effect as a result of interstrain competition with indigenous *Rhizobium* spp. (16, 18, 22, 24). Although it is not yet understood why certain strains should have a competitive advantage, competitiveness is a characteristic that should be evaluated so that nodulation by inoculating strains will be ultimately favored.

WHAT IS MEANT BY COMPETITION

The term "competition" when applied to *Rhizobium* spp. usually means competition for nodule formation between strains of *Rhizobium* from the moment they are together in the same environment until they are inside the nodule. There is, therefore, competition at several different levels. Strains compete to survive and multiply in the soil prior to seed germination and in the rhizosphere after germination (9, 10, 13, 25). The term saprophytic competence has been proposed by Chatel (8) for this initial level of competition. Surviving bacteria will then compete for available root infection sites (22, 33) and during root penetration, phenomena which themselves probably include several stages. Finally, since not all primary infections lead to nodule formation, there is most likely competition during nodule development.

It is difficult to differentiate between the levels of competition, particularly in field studies where it is difficult to enumerate the *Rhizobium* population. Competition studies have generally characterized only the number of nodules formed from a known inoculum population, as these are two quantifiable parameters.

METHODS

The number of nodules formed by a particular strain in the presence of one or more strains can be evaluated by direct or indirect methods.

The direct method consists of identifying the particular strain which is at the origin of a nodule, assuming that only one bacterium was at the initial site of infection. This method relies on some characteristic(s) of interest; serological techniques and their derivatives are the most widely used techniques for strain identification (11, 27, 29), but antibiotic-resistant mutants and phage typing have also been successfully used (6, 28, 31).

In the indirect method, strains capable of inducing both nodulation and a second, easily measurable reaction from the plant host are used. For example, Johnson and Means (21) described strains capable of inducing chlorosis in soybean plants, the degree of chlorosis being correlated with the number of nodules formed. Thus, chlorosis can be used as a marker for evaluating competition between this strain and one or more other strains added in an inoculating mixture.

A second indirect method consists of measuring the yield of plants grown in nitrogen-free media inoculated with a mixture of one effective strain and one ineffective strain (3). Plant growth is correlated with the number of effective nodules formed as the plant is nitrogen limited. The competitiveness of different effective strains can be compared by the differences in growth of plants inoculated with a mixture composed of a standard ineffective strain and a standard population size of the effective strain of interest. There are some advantages to this method, as ineffective strains are readily available (30) whereas chlorosis-inducing strains are not.

Strains of *Rhizobium* species which induce the formation of dark nodules on cowpea have been described by Eaglesham (12). The identification of the nodules formed by one of these strains in the presence of other strains is therefore greatly simplified, and competitiveness should be easily measurable.

QUANTITATIVE RELATIONSHIP

The competition between strains of *Rhizobium* has been studied both in the laboratory and in the field. Laboratory studies have employed axenic seedlings which are inoculated with a mixture of usually equal proportions of two strains, one of which is an identifiable reference strain. If the two strains are different in competitiveness, significantly more nodules will be formed by the more competitive strain (17, 20, 22).

In field experiments, seeds inoculated with a known number of an identifiable strain are planted in soils containing indigenous *Rhizobium* spp. The competitiveness of different strains used to inoculate legume seeds can be compared by determining the proportion of nodules formed by the indigenous rhizobia and by the inoculant strain (5, 11, 15, 16, 22, 26).

In field experiments, the number of nodules formed by the inoculant strain is not directly proportional to the inoculum size; Kapusta and Rouwenhorst (24) recovered 6, 10, and 37% of *R. japonicum* 138 inocula in nodules of soybeans inoculated at doses of 10^3, 10^6, and 10^9 cells per cm of row, respectively. In a similar experiment with soybean, the percentage of nodules formed by inoculum rhizobia increased from two to six times when the inoculum size was increased from 10^4 to 10^8 cells, the magnitude of the increase being dependent on the soil used (34). In some cases, increasing the inoculum size had no effect on the recovery of the inoculum strain in the nodules (5, 22).

Results from the different experiments reported in the literature are generally qualitative and cannot easily be compared. Only a few strains, primarily of *R. japonicum* and *R. trifolii*, have been used for most of the work done to date in this area. Furthermore, although in many cases the strains found to be the more competitive in greenhouse tests are also the more competitive in the field (3, 21), this is not always the case (20).

It would be desirable to understand the quantitative relationship between the number of bacteria in the inoculum and the proportion of nodules formed by these bacteria, to predict what dose would be required for successful inoculation in soils containing indigenous *Rhizobium* spp. It would also be useful to be able to estimate the competitiveness of *Rhizobium* strains to choose the best competitor and to study the influence of environmental factors on competitiveness.

The percentage of nodules formed by the inoculum strain has been found to be correlated with the logarithm of the number of bacteria in the inoculum for both soybean (34, 35) and fava bean (1). The slopes of curves derived from these two parameters were higher for the more competitive strain than for the less competitive strain. However the relationship between the slope and competitiveness could not be generalized for all soil conditions (4).

A more widely applicable quantitative relationship of the form $y = ax^n$ was found between the ratio of the number of bacteria of the competing strains and the ratio of the nodules made by each strain (4):

$$N_A/N_B = C_{AB} (I_A/I_B)^k \qquad (1)$$

the log transformation of which gives a linear relationship

$$\log (N_A/N_B) = \log (C_{AB}) + k \log (I_A/I_B) \quad (2)$$

where A is the competitor strain, B is indigenous soil strains in field experiments or the reference strain in an inoculum mixture used in greenhouse or laboratory experiments, N_A and N_B are the numbers of nodules formed by A and B, respectively, I_A and I_B are the numbers of bacteria A and B, respectively, at the moment of inoculation, C_{AB} is equal to N_A/N_B when I_A is equal to I_B, and k is the slope.

When determining the competitiveness of strain A with respect to strain B in an inoculum mixture containing the two, strain A is defined as the "competitor strain" and strain B, as the "reference strain." Likewise, when determining the competitiveness of an inoculum strain in soil, the inoculum strain is defined as the "competitor strain" and the indigenous rhizobial population specific for the host legume, as the "reference strain."

The value C_{AB} in equation 1 defines the competitiveness of equal populations of the competitor strain A versus the reference strain(s) B. If $C_{AB} > 1$, strain A is more competitive, and if $C_{AB} < 1$ strain A is less competitive than strain(s) B.

The slope of the regression line, k, was found

to vary according to the experiment; the extreme values were 0.15 and 1.22, but most of the values fell between 0.3 and 0.5.

DIRECT MEASUREMENT OF COMPETITIVENESS OF *R. LEGUMINOSARUM* STRAINS

To verify the validity of the relationship expressed in equation 2, the competitiveness of five effective strains of *R. leguminosarum* resistant to one antibiotic was estimated on fava bean in greenhouse studies. Seven different combinations of two strains each were inoculated onto seedlings grown in sand on a nitrogen-free medium (3). Eight different proportions of the two strains were used for each combination, with three replicates each. The nodules were identified after 2 months of growth (3).

For the seven pairs studied, the linear regressions of $\log (N_A/N_B)$ as a function of $\log (I_A/I_B)$ were significant ($P < 0.01$). The slopes of the lines were not significantly different, and the mean value was 0.65. Values of C_{AB} were estimated as the antilog of the y intercept of the line and are tabulated in Table 1.

The number of nodules formed by strain 1 varied according to which reference strain was used in the mixture. For example, strain 1 was 2.4 times more competitive than strain 5 but less competitive than the other three strains. Strains 2 and 3 formed 14 and 7 times more nodules than strain 1, respectively. Therefore, if we consider strain 1 as the reference strain, strain 2 is twice as competitive as strain 3. When strains 2 and 3 were mixed, a $C_{2,3}$ of 2.3 was found. There is thus good agreement between direct comparisons of two strains and comparisons of each with a third reference strain as methods for determining competitiveness. A comparison of the competitiveness of two strains, 1 and 2, can therefore be made by comparing values obtained for each with a third reference strain (i.e., $C_{1,2} \simeq C_{1,3}:C_{2,3}$). This is advantageous since only the reference strain and not the two wild competitor strains need be marked.

For the five strains studied, the biggest differ-

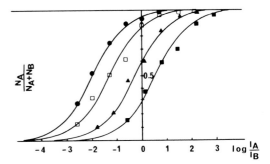

FIG. 1. Proportion of nodules formed on fava bean by *R. leguminosarum* strain 2 (●), 3 (□), 4 (▲), or 5 (■). N_A, mixed with strain 1; N_B, as a function of the logarithm of the number of bacteria of the two strains in the inoculum, I_A/I_B.

ence in competitiveness was observed between strains 2 and 5, strain 2 being 30 times more competitive than strain 5. When mixed, 250 times more cells of strain 5 than of strain 2 would be needed to form an equal number of nodules. Figure 1 illustrates these differences in competitiveness.

INDIRECT EVALUATION OF COMPETITIVENESS

As previously described, the competitiveness of strains of *Rhizobium* can be compared by using ineffective (2) or chlorosis-inducing strains (21); the quantification of the nodules formed is then replaced by the measurement of dry matter or degree of chlorosis of the plant. If the quantitative relationship between the amount of dry matter and the number of effective nodules formed is known, one can measure the competitiveness by this indirect method.

The dry matter produced by alfalfa inoculated with a mixture of effective and ineffective strains was found to be directly proportional to the log of the number of effective nodules formed on the roots (2). According to equation 2, the dry matter produced must therefore be directly proportional to $\log (I_A/I_B)$, and the y intercept of this line must be a function of the competitiveness of the effective strain in comparison to the ineffective strain. If one has two or more competitor strains of identical effectiveness, each mixed with the same ineffective reference strain and inoculated onto plants, the differences in growth (i.e., the y intercept) will therefore be a function of the difference in competitiveness between the strains.

Preliminary results with fava bean showed that the amount of dry matter produced by plants inoculated with mixtures of effective (*A*) and ineffective (*B*) strains was directly proportional to $\log (I_A/I_B)$. It would be desirable to

TABLE 1. Competitiveness (C_{AB}) of five strains of
R. leguminosarum inoculated in different
combinations on fava bean

Strains ($A{:}B$)	C_{AB}
1:2	0.07
1:3	0.14
1:4	0.70
1:5	2.4
2:3	2.3
3:5	12.3
4:5	2.8

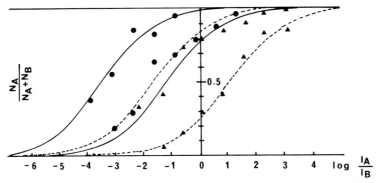

FIG. 2. Proportion of nodules formed by *R. leguminosarum* strain 2 (●) and strain 4 (▲) $[N_A/(N_A + N_B)]$ inoculated at increasing ratios (I_A) onto seeds of fava bean planted in soil M (———) and soil N (- - - -) containing indigenous *Rhizobium* spp.

demonstrate that the y intercept is a function of the competitiveness and that differences in its values are indicative of differences in competitiveness.

INFLUENCE OF SOIL USED ON COMPETITIVENESS

Equation 2 applies equally when a competitor strain (A) is inoculated onto seeds planted in soils containing indigenous rhizobia (B). The population of *Rhizobium* specific for one legume behaves as a single strain. Similar results were found for field-grown fava beans and fava beans greenhouse-grown in pots filled with the field soil (1). We should therefore be able to conduct competition experiments on different soils in the greenhouse and extrapolate the results to the field.

Increasing amounts of strains 2 and 4 (the same strains as in Table 1) were inoculated onto seeds of fava beans grown in two different soils, each containing 7.2×10^3 *R. leguminosarum* cells per g. Equation 2 was verified in the four cases (Fig. 2). The slopes of the regression lines (k) were not significantly different ($P < 0.01$), and the mean value was 0.49, which was significantly different from the value obtained when the two strains were mixed. This value is very close to the values found in similar experiments (4).

Complementary studies are necessary to check whether the value of k is constant in this type of competition experiment which uses the method of *Rhizobium* inoculation commonly employed in agronomic practice. If the value of k is constant, only one inoculum dose is needed to determine strain competitiveness. On the other hand, if the value of k varies with strains or with soils used, k will have to be determined in each case by using three or four inoculum doses. This single- or multiple-level determination of

the inoculant strain's competitiveness will permit us to predict what number of bacteria is necessary to produce a given proportion of nodules in the soil tested. It should therefore be possible to determine whether commercial inoculant can be made concentrated enough to achieve the desired effect.

The results in Fig. 2 show that strain 2, which was found to be 10 times more competitive than strain 4 by direct evaluation, also formed more nodules in the two soils. When the soil strains were considered as the reference strain (B) to be compared with the two inoculant competitor strains, strain 2 was found to be 13 and 22 times more competitive than strain 4 in soils M and N, respectively. Soils can therefore have a significant effect on the expression of competitiveness of strains of *Rhizobium*. This effect may be due to the biotic or abiotic properties of the soils or to the nature of the indigenous *Rhizobium* population specific to the legume.

Moreover, although the number of indigenous rhizobia was the same in both soils, the two inoculant strains formed more nodules in soil M than in soil N. Two hundred times more bacteria must be added to soil N than to soil M for 50% of the nodules recovered to contain the inoculum strain. Here also, the effect of the soil used on competitiveness could be due either to differences in the competitiveness of indigenous strains or differences in the abiotic properties of the soils.

CONCLUSIONS

Results obtained so far indicate that the relationship between the ratio of the number of nodules formed by two strains of *Rhizobium* (N_A/N_B) and the ratio of the number of bacteria in the inoculum mixture (I_A/I_B) is a function of type $N_A/N_B = C_{AB} (I_A/I_B)^k$, which also corresponds to the ratio of two Freundlich adsorption isotherms. The number of nodules formed by a strain varies as a function of the number of bacteria in the medium in the same way that the amount of a substance that is adsorbed varies as a function of its concentration in solution. If we take the analogy further, we can imagine that there is a selective adsorption of the different strains of the specific *Rhizobium* spp. on the roots of the legume. This is in agreement with the findings of Jansen van Rensburg and Strijdom (19), who observed that the more competitive strains of *R. meliloti* were also the ones that adhered most firmly to the roots.

The determination of the two parameters C_{AB} and k (this latter perhaps being a constant) allows one to measure the competitiveness of one strain (the competitor) as compared to another inoculum strain or to soil strains (the reference strain[s]). It is thus possible to predict the chances of success with a particular inoculum. It should also be possible to study the different factors which directly or indirectly alter the competitiveness of rhizobia and thus to allow a more precise approach to studying a little-understood phenomenon.

I thank E. Topp, Department of Microbiology, McGill University, for his help in the redaction of the manuscript.

LITERATURE CITED

1. **Amarger, N.** 1974. Compétition pour la formation des nodosités sur la féverole entre souches de *Rhizobium leguminosarum* et souches du sol. C.R. Acad. Sci. **279**:527–530.
2. **Amarger, N.** 1981. Competition for nodule formation between effective and ineffective strains of *Rhizobium meliloti*. Soil Biol. Biochem. **13**:475–480.
3. **Amarger, N.** 1981. Selection of Rhizobium strains on their competitive ability for nodulation. Soil Biol. Biochem. **13**:481–486.
4. **Amarger, N., and J. P. Lobreau.** 1982. Quantitative study of nodulation competitiveness in *Rhizobium* strains. Appl. Environ. Microbiol. **44**:583–588.
5. **Boonkerd, N., D. F. Weber, and D. F. Bezdicek.** 1978. Influence of *Rhizobium japonicum* strains and inoculation methods on soybeans grown in Rhizobia-populated soil. Agron. J. **70**:547–549.
6. **Brockwell, J., E. A. Schwinghamer, and R. R. Gault.** 1977. Ecological studies of root nodule bacteria introduced into field environments. V. A critical examination of the stability of antigenic and streptomycin resistance markers for identification of strains of *Rhizobium trifolii*. Soil Biol. Biochem. **9**:19–24.
7. **Caldwell, B. E., and G. Vest.** 1968. Nodulation interactions between soybean genotypes and serogroups of *Rhizobium japonicum*. Crop Sci. **8**:680–682.
8. **Chatel, D. L., R. M. Greenwood, and C. A. Parker.** 1968. Saprophytic competence as an important character in the selection of *Rhizobium* for inoculation, p. 65–73. *In* Transactions of the Ninth International Congress of Soil Science 4V Adelaide, Australia.
9. **Chatel, D. L., and C. A. Parker.** 1973. The colonization of host-root and soil by Rhizobia. I. Species and strain differences in the field. Soil Biol. Biochem. **5**:425–432.
10. **Chatel, D. L., and C. A. Parker.** 1973. The colonization of host-root and soil by Rhizobia. II. Strain differences in the

11. species *Rhizobium trifolii*. Soil Biol. Biochem. **5**:433–440.
11. **Dudman, W. F., and J. Brockwell.** 1968. Ecological studies of root nodule bacteria introduced into field environments. I. A survey of field performance of clover inoculants by gel immune diffusion serology. Aust. J. Agric. Res. **19**:739–47.
12. **Eaglesham, A. R. J., M. H. Ahmad, S. Hassouna, and B. J. Goldman.** 1982. Cowpea rhizobia producing dark nodules: use in competition studies. Appl. Environ. Microbiol. **44**:611–618.
13. **Franco, A. A., and J. M. Vincent.** 1976. Competition amongst rhizobial strains for the colonization and nodulation of two tropical legumes. Plant Soil **45**:27–48.
14. **Gareth Jones, D., and G. Hardarson.** 1979. Variation between white clover varieties in their preference for strains of *Rhizobium trifolii*. Ann. Appl. Biol. **92**:221–228.
15. **Gibson, A. H., R. A. Date, J. A. Ireland, and J. Brockwell.** 1976. A comparison of competitiveness and persistence amongst five strains of *Rhizobium trifolii*. Soil Biol. Biochem. **8**:395–401.
16. **Ham, G. E., V. B. Cardwell, and H. W. Johnson.** 1971. Evaluation of *Rhizobium japonicum* inoculants in soils containing naturalized populations of rhizobia. Agron. J. **63**:301–303.
17. **Hardarson, G., and D. Gareth Jones.** 1979. Effect of temperature on competition amongst strains of *Rhizobium trifolii* for nodulation of two white clover varieties. Ann. Appl. Biol. **92**:229–236.
18. **Holland, A. A.** 1970. Competition between soil and seed borne *Rhizobium trifolii* in nodulation of introduced of *Trifolium subterraneum*. Plant Soil **32**:293–302.
19. **Jansen van Rensburg, H., and B. W. Strijdom.** 1982. Root surface association in relation to nodulation of *Medicago sativa*. Appl. Environ. Microbiol. **44**:93–97.
20. **Jansen van Rensburg, H., and B. W. Strijdom.** 1982. Competitive abilities of *Rhizobium meliloti* strains considered to have potential as inoculants. Appl. Environ. Microbiol. **44**:98–106.
21. **Johnson, H. W., and V. M. Means.** 1964. Selection of competitive strains of soybean nodulating bacteria. Agron. J. **56**:60–62.
22. **Johnson, H. W., U. M. Means, and C. R. Weber.** 1965. Competition for nodule sites between strains of *Rhizobium japonicum* applied as inoculum and strains in the soil. Agron. J. **57**:178–185.
23. **Johnston, A. W. B., and J. E. Beringer.** 1976. Mixed inoculations with effective and ineffective strains of *Rhizobium leguminosarum*. J. Appl. Bacteriol. **40**:375–380.
24. **Kapusta, G., and D. L. Rouwenhorst.** 1973. Influence of inoculum size on *Rhizobium japonicum* serogroup distribution frequency in soybean nodules. Agron. J. **65**:916–919.
25. **Marques Pinto, C. P., P. Y. Yao, and J. M. Vincent.** 1974. Nodulating competitiveness amongst strains of *Rhizobium meliloti* and *R. trifolii*. Aust. J. Agric. Res. **25**:317–329.
26. **Materon, L. A., and C. Hagedorn.** 1982. Competitiveness of *Rhizobium trifolii* strains associated with red clover (*Trifolium pratense* L.) in Mississippi soils. Appl. Environ. Microbiol. **44**:1096–1101.
27. **Means, U. M., H. W. Johnson, and R. A. Date.** 1974. Quick serological method of classifying strains of *Rhizobium japonicum* in nodules. J. Bacteriol. **87**:547–553.
28. **Obaton, M.** 1971. Utilisation de mutants spontanés résistants aux antibiotiques pour l'étude écologique des *Rhizobium*. C.R. Acad. Sci. **272**:2630–2633.
29. **Schmidt, E. L., R. O. Bankole, and B. B. Bohlool.** 1968. Fluorescent antibody approach to study of rhizobia in soil. J. Bacteriol. **95**:1987–1992.
30. **Schwinghamer, E. A.** 1967. Effectiveness of *Rhizobium* as modified by mutation for resistance to antibiotics. Antonie van Leeuwenhoek J. Microbiol. Serol. **33**:121–136.
31. **Schwinghamer, E. A., and W. F. Dudman.** 1973. Evaluation of spectinomycin as a marker for ecological studies with Rhizobium spp. J. Appl. Bacteriol. **36**:263–272.
32. **Vincent, J. M.** 1954. The root nodule bacteria as factors in

clover establishment in the red basaltic soils of the Lismore district, N.S.W. I. A survey of native strains. Aust. J. Agric. Res. **5:**55–60.

33. **Vincent, J. M., and L. M. Waters.** 1953. The influence of the host on competition amongst clover root nodule bacteria. J. Gen. Microbiol. **9:**357–370.

34. **Weaver, R. W., and L. R. Frederick.** 1974. Effect of inoculum rate on competitive nodulation of *Glycine max.* L. Merrill. I. Green house studies. Agron. J. **66:**229–232.

35. **Weaver, R. W., and L. R. Frederick.** 1974. Effect of inoculum rate on competitive nodulation of *Glycine max.* L. Merrill. II. Field studies. Agron. J. **66:**233–236.

Competition Among Chemolithotrophic Bacteria Under Aerobic and Anaerobic Conditions

J. G. KUENEN AND L. A. ROBERTSON

Laboratorium voor Microbiologie, Delft University of Technology, Delft, The Netherlands

When studying any microbiological process or the role of a particular group of organisms in nature, one immediately realizes that most microbial communities are very complex and cannot simply be regarded as the sum of a set of pure cultures. Each microbial cell is involved in a network of interactions with its biotic and abiotic environment. However, the study of the ecophysiology of pure cultures of microorganisms is one of the first steps on the long road to the understanding of the contributions made by the different species to the complex microbial community. Once the physiological basis of the different types of metabolism is known, the studies can be extended to the investigation of microbial interactions by using simple, defined mixtures of known organisms. Ideally, the results of these studies can clarify the most important interactions of species with their natural environment and with each other, making possible the understanding of how such organisms survive in nature. One of the most important of the environmental factors which determine the success or survival of an organism is the limitation of growth by vital nutrients. For the examination of the growth and competition of species under such limitation, the chemostat has proved to be extremely useful, and it has yielded much of the important information needed to understand the response of organisms to nutrient limitation. Also, in the examination of competition between bacteria for growth-limiting substrates, the chemostat has been an invaluable tool, allowing the simulation of prolonged intervals of substrate limitation and thereby artificially amplifying selective pressures as they may occur in nature (13). The beauty of this approach is that it not only provides the possibility of testing predictions based on physiological studies, but also may reveal complicating factors which confound the theoretical predictions and thus may deepen our understanding of that particular microbial interaction. Results so obtained can then, as we will show here, be extrapolated to predict the outcome of enrichment cultures and the behavior of microbial communities in the wild.

Various factors control the outcome of competition for a single growth-limiting substrate. At very low substrate concentrations the dominant feature is the affinity (μ_{max}/K_s, that is, the initial slope of the μ-S curve) of the organisms competing for the substrate in question, but other factors such as pH or oxygen availability will modify this effect. In theory it should be possible to determine the affinity or the μ-S curves for each substrate and each organism separately, but this is often very difficult, if not impossible, as the concentrations at which limitation occurs are usually below the detection levels. The most direct method of comparing the affinities of the different organisms for their substrates is, therefore, to use the organisms in competition experiments in chemostats with the appropriate nutrient limitation. By using two or three different dilution rates, the relative slopes of the μ-S curves for the different organisms can be drawn and compared.

Our interest in the general principles of selection was aroused during a study of the physiology and ecology of the colorless sulfur bacteria (for review see 11). This group of organisms is very suitable for the study of a number of principles regarding competition and other interactions for several reasons. As their major environmental impact—the oxidation of reduced sulfur compounds—is well defined, their activities are easily recognized. At the same time, the colorless sulfur bacteria make up a large group of organisms which, except in their sulfur-oxidizing capacities, are physiologically very different (8, 14). This makes the group particularly interesting for the investigation of how the range of physiological differences governs the competitiveness and survival of these bacteria in nature. The best known genus, *Thiobacillus*, includes obligate chemolithotrophs which are able to grow only with inorganic energy sources and carbon dioxide for carbon, versatile species able to use the autotrophic and heterotrophic modes of life, and chemolithoheterotrophs which can derive energy from reduced sulfur compounds but which require an organic carbon source. Together, the colorless sulfur bacteria play an important role in the global sulfur cycle, providing a counterbalance for the activities of the sulfate-reducing bacteria (7, 17). Although the separation of the niches occupied by some groups is obvious (for example, *T. ferrooxidans* requires a much more acidic environment than most of the other chemolithotrophs), other species are apparently very similar and their occurrence, often in the same ecosystem, requires explanation.

This paper will therefore focus on some of the principles behind the competitiveness and survival of colorless sulfur bacteria and show how simple laboratory continuous-flow systems can be used to investigate the ecological niches of these species. We will also show that the principles derived in this work can be applied to more complex systems and can shed some light on the roles of different types of physiology in the global sulfur cycle.

A typical example of the use of the chemostat to determine relative μ-S curves can be found in a study of the ecological niches of two obligate chemolithotrophs isolated from mud taken from the Waddenzee (15). These species, *Thiobacillus thioparus* and *Thiomicrospira pelophila*, appeared on examination to have remarkably similar physiologies. It was suggested that their coexistence might be due to different K_s values and thus different responses to the limitation of essential nutrients in their environment. As sulfide, or another reduced sulfur compound, and oxygen are simultaneously needed by these bacteria for growth, their activity is often restricted to the narrow interface between aerobic and anaerobic conditions. Thus, sulfide and oxygen would be obvious examples of possible limiting factors, but iron is another possibility. The thiobacilli have a relatively high iron requirement and, at the approximate pH of their natural environment (7, 8) and in the presence of sulfide, the solubility of iron is very low. In a series of

experiments *Thiobacillus thioparus* and *Thiomicrospira pelophila* were studied in a chemostat while competing for thiosulfate or iron (12). The outcome of the experiment was determined by the pH and the concentration of iron in the medium. Above pH 7.5, or at very low iron concentrations, *T. pelophila* was dominant; below pH 6.5, or at higher iron concentrations, the situation was reversed (Fig. 1). *T. pelophila* also had a higher tolerance for sulfide. These observations led to the speculation that the niche occupied by *T. pelophila* at the sulfide-oxygen interface in the mud might be the point in the gradient where the sulfide concentration was highest. At this position, the iron concentration would be reduced because of the low solubility of ferrous sulfide, and as the pH was around 8.0, conditions would clearly favor the latter species. If sulfuric acid was produced, the pH and sulfide concentration would fall, and *T. thioparus* would then come into its own. Subsequent field studies showed that *T. pelophila* was more abundant in muds rich in sulfide than in sediments with lower sulfide levels. This would support the hypothesis given here.

For many years the coexistence of the obligate and facultatively chemolithotrophic thiobacilli in the same environment has posed a complex and intriguing question. From batch culture experiments using mixtures of known bacteria and enrichment cultures, it appeared that the specialized obligate chemolithotrophs always grew much faster than the facultative, or versatile, species. To throw some light on this ques-

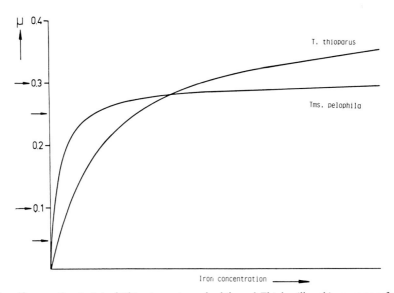

FIG. 1. Specific growth rate (μ) of *Thiomicrospira pelophila* and *Thiobacillus thioparus* as a function of the growth-limiting iron concentration. The bacteria were grown in a thiosulfate–minerals medium in a chemostat at 25°C. The arrows indicate the dilution rates at which competition experiments were carried out (12).

tion (3), competition for growth-limiting thiosulfate or acetate among an obligate chemolithotroph (*Thiobacillus neapolitanus*), a versatile species able to grow autotrophically, heterotrophically, or mixotrophically (*Thiobacillus versutus*; formerly known as *Thiobacillus A2*) (6, 22), and an obligate heterotroph (spirillum G7) was studied. It was found that when a single substrate was supplied continuously, or if the substrates were not growth limiting, the specialist species were able to outcompete the mixotroph because of their higher specific growth rate on their specialist substrate (Fig. 2A). However, if the substrates were mixed and supplied as growth-limiting compounds in the chemostat, the outcome of the experiment was dependent on the ratio between the two (Fig. 2B). When the supply of either substrate greatly exceeded that of the other, the specialists were successful, but when the quantities of thiosulfate and acetate provided were of the same order of magnitude *T. versutus* dominated. From the results of a series of chemostat experiments with pure cultures of *T. versutus*, it was found that the physiological explanation of this phenomenon lies in the ability of *T. versutus* to grow mixotrophically rather than sequentially on the thiosulfate and acetate as long as the energy or carbon source, or both, was growth limiting (4). Under these conditions *T. versutus* is able to reduce the concentrations of both acetate and thiosulfate to a level at which the two specialists can no longer maintain the required growth rate. Essentially, this means that in *T. versutus* the apparent K_s values for either acetate or thiosulfate are reduced by the second substrate. In two-membered cultures the coexistence of two species could be observed, but at a certain threshold concentration *T. versutus* was able to outcompete either specialist entirely. This was not the case in three-membered cultures, where the versatile species dominated the mixture over a wide range of acetate/thiosulfate ratios, but failed to displace the specialists completely. Similar results were obtained by Smith and Kelly (20), using *T. neapolitanus* and *T. versutus* in competition for thiosulfate or a thiosulfate and glucose mixture. However, they introduced pH as an additional variable and established that this has as profound an effect on the outcome of the experiments as it does for *Thiobacillus thioparus* and *Thiomicrospira pelophila*.

These results clearly pointed to a niche for facultatively chemolithotrophic thiobacilli in the natural environment, where mixtures of organic compounds and inorganic reduced sulfur compounds are usually available.

Although a stable supply of a mixture of nutrients in the environment may occasionally occur, it is obvious that circumstances are often such that the nutrient supply fluctuates. The effect of this in a nutrient-limited continuous culture of the same three organisms used in the continuous-flow experiments described above was therefore also studied (5). When the alternate feeds each contained a single substrate (i.e., thiosulfate or acetate), the versatile *T. versutus* was able to maintain itself in two-

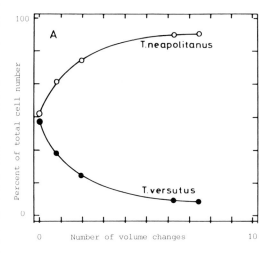

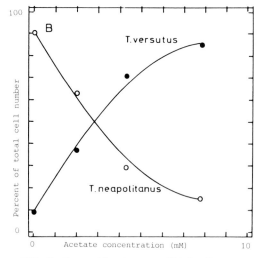

FIG. 2. Competition between *Thiobacillus versutus* and *T. neapolitanus* for growth-limiting substrates in a chemostat at 28°C. *T. versutus* (●), *T. neapolitanus* (○). (A) Medium contained 40 mM thiosulfate, dilution rate = 0.05 h^{-1}. Graph shows percentage of total cell number against number of volume changes. (B) Medium contained 40 mM thiosulfate and different amounts of acetate, dilution rate = 0.07 h^{-1}. Graph shows steady-state percentage of total population against increasing acetate concentration. (Adapted from reference 5.)

membered cultures by growing on the substrate not used by the specialist (e.g., acetate when cultured with *T. neapolitanus*), but when both specialists were present in a three-membered culture, this was not possible, and the mixotroph washed out. When different mixtures of thiosulfate and acetate were alternately supplied to the three-membered cultures, the outcome was determined by the concentrations involved (Fig. 3). If the thiosulfate or acetate concentration in the alternate mixture was low, *T. versutus* was able to maintain a small population. However, as the concentrations of the two types of substrate in the alternate supplies approached the same order of magnitude, the versatile species rapidly outnumbered both specialists since it was able to grow mixotrophically throughout the experiment, rather than having to switch from

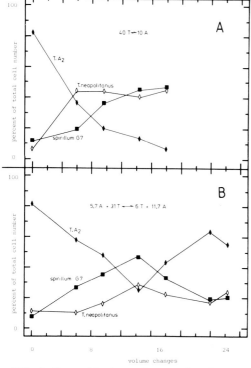

FIG. 3. Competition in continuous culture between *Thiobacillus versutus, T. neapolitanus*, and spirillum G7 for thiosulfate and acetate as growth-limiting substrates. The dilution rate was 0.05 h^{-1}, and two different substrate supplies were supplied in 4-h alternating periods. (A) One medium contained 10 mM acetate and the other contained 40 mM thiosulfate. (B) One medium contained 5.7 mM acetate plus 31.0 mM thiosulfate and the other contained 6 mM thiosulfate plus 11.7 mM acetate. *T. versutus* (◆), spirillum G7 (■), and *T. neapolitanus* (◇) as the percentage of the total cell number present in the culture. (Adapted from reference 4.)

autotrophy to heterotrophy or undergo periods of starvation, as did the specialists.

This interpretation was backed by a detailed study of the physiologies of both *T. neapolitanus* and *T. versutus* in pure culture during alternating growth and starvation or alternating growth on thiosulfate and acetate (1). It appeared that during an acetate period *T. versutus* repressed too much of its autotrophic potential to be able to respond rapidly at the onset of an autotrophic period. On the other hand, the specialist *T. neapolitanus* retained its full potential during starvation periods so that it could immediately grow at the required rate as soon as thiosulfate supplies recommenced. This rapid response has been called the greater "reactivity" of *T. neapolitanus*.

From the results of the experiments described above, it is possible to summarize some of the environmental factors influencing the outcome of competition for one or more substrates:

(i) When the substrates are not growth limiting, or when a continuous supply of one substrate is provided, the appropriate specialist organism will be favored by reason of its superior growth rate.

(ii) When growth-limiting concentrations of mixed substrates are available, and the supplies of organic and inorganic components are in the same order of magnitude, the ability of the versatile species to grow mixotrophically enables it to outgrow or dominate specialists, but when either substrate greatly exceeds the other, the specialists will be favored.

(iii) During mixed substrate supply, the coexistence of all participating species is possible and depends on the ratios between the organic and inorganic energy sources.

(iv) The outcome of competition for fluctuating substrates is dependent on the composition of the alternate feeds. As pointed out by Beudeker et al. (1), the physiological basis for the success of the specialists lies in their greater reactivity under conditions of substrate fluctuation. This implies that only when the conditions are suitable for mixotrophic growth can a versatile species outcompete a specialist for substrate.

(v) The results of any competition experiment can be modified by other factors such as pH if any of the participating species has a vastly different requirement.

These principles can be checked for their applicability to more complex microbial communities by means of enrichments in continuous culture.

Table 1 shows the results of various enrichments done with thiosulfate, sulfide, or a mixture of thiosulfate and acetate. It can be seen that the type of metabolism shown by the major-

TABLE 1. Results of enrichment cultures done in a chemostat with growth-limiting substrates under various regimes

Inoculum source	Culture conditions	Substrates (mM)		Dominant population	Predominant secondary population	Reference
		S^{2-} or $S_2O_3^{2-}$	Acetate			
Freshwater ditch	Aerobic, constant feed, mixed substrate	20	5	Mixotroph	Mixotroph	4
		20	10	Mixotroph	Heterotroph	4
		10	15	Chemolitho-heterotroph	Heterotroph	4
	Aerobic, alternate feeds, single substrates	40 (4 h)	10 (4 h)	Mixotroph	Chemolitho-heterotroph	5
Marine mud (Waddenzee)	Aerobic, constant feed, single substrate	32	0	Obligate autotroph	—	15
	Anaerobic (denitrifying), constant feed, single substrate	40	0	Obligate autotroph	Heterotroph	24
Effluent treatment system	Anaerobic (denitrifying), constant feed, mixed substrate	10	10	Mixotroph	Heterotroph	18

ity organism in each case could have been predicted from the principles given above. Where only the inorganic substrate was supplied, chemolithotrophs were obtained under both aerobic and anaerobic conditions. If these concentrations were in the same order of magnitude, a more versatile species appeared. Another enrichment in which 10 mM thiosulfate and 15 mM acetate were used yielded a chemolithoheterotroph which was capable of using thiosulfate for energy, but not of fixing carbon dioxide. At this ratio between the substrates, although the thiosulfate served as a valuable energy source, the relative supply of organic carbon was high enough to render carbon dioxide fixation energetically undesirable. One interesting deviation from the predicted results was found in an enrichment culture provided with alternating 4-h feeds of thiosulfate and acetate. In this culture, a facultatively chemolithotrophic *Thiobacillus* became dominant. This organism stored large quantities of poly-β-hydroxybutyrate during the periods of acetate feed, thereby obviating the need for immediate full operation of the Calvin cycle. This reserve material was then used to provide energy for carbon dioxide fixation at the start of the autotrophic period. Enrichments from marine sources gave unpredictable results (4, 21).

It should be mentioned here that continuous-flow enrichments have not so far led to pure cultures. In all cases, mixed secondary populations were able to maintain themselves at levels ranging from a few percent to more than half of the population.

Our studies have now been extended to the investigation of part of an anaerobic industrial wastewater treatment system (U.S. patent SN 317585, 1983; European patent EP 0051888A1, 1982). The factory effluent contains large amounts of organic material, sulfate, and ammonium. Sulfide is produced by sulfate-reducing bacteria in the preliminary reactor, where the organic material is broken down into the simple molecules utilized by the methanogens, and in the methane fermentor. The sulfide and the ammonium both constitute environmental hazards, and the remainder of the treatment plant is therefore devoted to dealing with them. Effluent from the methane reactor is passed into an anaerobic fluidized bed column where the sulfide is oxidized by using nitrate as the electron acceptor. The nitrate is produced in the final column in the series by microbiological oxidation of the effluent ammonium. This is then recirculated to the sulfide-oxidizing, denitrifying column. The diversity and behavior of the microbial community within this column are currently under study in our laboratory. In addition to sulfide and nitrate, the inflow to the denitrifying column contains a certain amount of organic material. This is mostly acetate, with some butyrate and propionate. The amounts of inorganic reduced sulfur compounds and organic materials are in the same order of magnitude, although the situation is complicated by biomass washing through from the preceding columns. If the precepts discussed above can be applied to an anaerobic system of this complexity, versatile, mixotrophic bacteria must form a significant proportion of the population. This has proved to be the case (L. A. Robertson and J. G. Kuenen, unpublished data). We have been unable to isolate obligate chemolithotrophs from either the column or enrichments based on material from the column. About 30% of the population could be termed facultatively chemolithotrophic. Most of the heterotrophs present were found to require complex additives such as yeast extract for growth, and none of them can oxidize reduced sulfur compounds, suggesting that cell lysis products from the inflowing biomass may contribute a significant proportion of their nutritional requirement. Enrichment cultures made in an anaerobic chemostat with an inoculum from the column, together with a feed containing 10 mM thiosulfate and 10 mM acetate, resulted in an isolate which could grow aerobically or anaerobically, using oxygen or nitrate as the electron acceptor. This organism was able to grow autotrophically on reduced sulfur compounds or hydrogen and heterotrophically on a wide variety of substrates. Not unexpectedly, it could also grow mixotrophically on mixtures of organic and inorganic substrates. The isolate, a gram-negative coccus which usually grows in chains, has been called *Thiosphaera pantotropha* (18). The physiology of this organism exactly fits the predictions made for a typical species dominant in the column. Furthermore, its distinctive morphology has made it possible to identify this organism in microscope preparations made from samples direct from the column, indicating that it forms an important part of the population.

During preliminary comparisons of the denitrifying ability of the new isolate with that of the better-known species *T. versutus* (*Thiobacillus* A2), it was noticed that, although aerobically grown cultures of *T. versutus* required the expected 4- to 5-h lag phase before nitrogen production began, this was not the case with *Thiosphaera pantotropha*, which began producing gas immediately. Further investigations showed that in batch culture with the dissolved oxygen maintained at 80 to 90% of air saturation, *T. pantotropha* cultures supplied with nitrate grew more rapidly ($\mu = 0.347$ with, 0.257 without) and gave a yield 75% of that obtained if the nitrate was omitted. The finding of dissimilatory nitrate and nitrite reductases in the aerobically grown cells, together with the disappearance of nitrate from the medium, confirmed the initial impression that this organism is able to denitrify in the presence of substantial concentrations of dissolved oxygen. Similar results were obtained by other workers (9, 10) with isolates from activated sludge, one of which still actively denitrified at oxygen concentrations twice that of air saturation. By considering the conditions existing in the column and the principles of selection discussed above, it can be seen that this ability to denitrify in the presence of appreciable amounts of oxygen can be explained. In practice, the column receives some inflows, especially that from the nitrifying column, which have not been made anaerobic. Any organism which could co-

respire nitrate and oxygen, rather than having to reinduce its denitrifying system after contact with oxygen, would have an obvious selective advantage. Additionally, the selective pressure in the column would favor cells which could grow rapidly while denitrifying, and the speed of growth in the absence of nitrate would be largely irrelevant. Preliminary results indicate that the majority of other isolates obtained from the column also have this ability.

Thus far, our work was concentrated on the effect of mixtures of inorganic sulfur compounds and organic substrates on the coexistence of different species. The question arose as to whether the metabolism of organic sulfides such as the methyl sulfides [CH_3SH, $(CH_3)_2S$, $(CH_3)_2S_2$] would follow a similar pattern. Such compounds seemed to be ideal enrichment substrates for the versatile, mixotrophic, sulfur-oxidizing species. The microbial breakdown of methyl sulfides is not only of scientific interest, but is also of importance in the paper-making industry. Effluent from the manufacture of paper often contains dimethyl sulfide, which is malodorous and toxic, and which cannot be discharged into the environment. Therefore, an investigation into the possibilities of the microbiological removal of dimethyl sulfide is proceeding in this laboratory. An early report in this field (19) claimed that an obligately chemolithotrophic *Thiobacillus* species, strain MS1, was able to utilize the dimethyl sulfide, deriving carbon and energy, if supplied with thiosulfate as an additional energy source. Subsequent screening of the known thiobacilli (16; G. M. H. Scheuderman-Suylen and J. G. Kuenen, unpublished data) for this ability has proved negative, and *Thiobacillus* strain MS1 has so far proved elusive despite efforts to reisolate it from samples supplied by Sivelä. Another obligate chemolithotroph isolated from these samples was unable to utilize dimethyl sulfide or any of its related compounds. In our experience, the toxicity of dimethyl sulfide is too great to permit successful enrichment in batch culture, and attempts at continuous enrichment with growth-limiting dimethyl sulfide have thus far been unsuccessful because of its volatility. From the work of de Bont et al. (2), it has become known that dimethyl sulfide is the initial product of dimethyl sulfoxide metabolism, and therefore an enrichment was attempted under conditions of growth-limiting dimethyl sulfoxide in an aerobic chemostat. The mixed culture so obtained was able to oxidize dimethyl sulfide at the same high rate as it could dimethyl sulfoxide (1,004 and 1,068 nmol of oxygen/min per mg of protein, respectively). Sulfide and formaldehyde were also oxidized fairly rapidly (605 and 225, respectively), but the oxidation rate for formate was much lower (47). Sulfate was produced. However, rather than a *Thiobacillus*-like organism, a species of *Hyphomicrobium* constituted over 90% of the community. A pink methylotroph and a yellow heterotroph made up most of the remainder (Scheuderman-Suylen and Kuenen, unpublished data). As dimethyl sulfide could be considered two parts organic (methyl) to one part sulfide, it might be predicted that an enrichment using dimethyl sulfide or dimethyl sulfoxide as the sole source of carbon and energy would produce a chemolithoheterotroph, a description that probably fits this *Hyphomicrobium* sp. as well as it does the *Hyphomicrobium* sp. isolated by another team from batch enrichments on dimethyl sulfoxide (2). This isolate, *Hyphomicrobium* strain S, was able to oxidize reduced sulfur compounds, deriving energy therefrom, but was not able to fix carbon dioxide.

CONCLUSION

The examples discussed above demonstrate some of the important mechanisms of competition between the different types of colorless sulfur bacteria while growing under substrate limitation. The results obtained from these continuous-flow experiments demonstrate that the ability to grow mixotrophically is a very important property in the outcome of competition in situations where growth-limiting mixtures of substrates are concerned, the outcome being determined by the ratio between the inorganic and organic types of substrate. It is clear that these results can be used to predict the outcome of enrichments and, in fact, to permit the design of enrichment cultures for organisms which cannot be enriched for in batch culture, especially the mixotrophs. This was shown by the isolation of a number of facultatively chemolithotrophic, aerobic and denitrifying sulfur oxidizers from chemostat enrichments designed for the isolation of such organisms by using the appropriate mixtures of substrates. The reisolation of the denitrifying species, *Thiosphaera pantotropha*, from the denitrifying, desulfurizing effluent treatment system which provided the original inoculum for the enrichment, together with the fact that more than 30% of the community found in the column were mixotrophs, give a further indication that these principles can apply in more complex systems. The enrichment of the chemolithoheterotrophic, dimethyl sulfide-utilizing *Hyphomicrobium* from a culture based on organic sulfides indicates that the "mixotrophic" metabolism is more advantageous than the cooperative breakdown of the methyl sulfides by a methylotroph and a *Thiobacillus*.

Whether the principles of competition and selection established with the colorless sulfur bacteria can be extended to predict the behavior of other types of bacteria now remains to be established. From the few examples available, this seems to be the case (Veldkamp et al., this volume).

The availability in nature of suitable mixtures of inorganic reduced sulfur compounds and organic material indicates that not only the commonly known obligate chemolithotrophs but also the mixotrophs and chemolithoheterotrophs may play an important role in the natural recycling of reduced sulfur compounds. Bacterial counts made in the field by some investigators tend to confirm this, and it is now obvious that continuous-flow competition studies are essential in the search for an understanding of ecological balances and the niches occupied by the obviously different and apparently similar species found in any ecosystem. A new awareness that ecological understanding is a prerequisite for the solution of applied problems such as that of wastewater management has resulted in an upsurge of interest in ecology. The increasing sophistication of waste treatment, especially that of industrial effluent, requires a detailed knowledge of the efficiency, stability, and resistance to fluctuation of the bacterial community. Our successful extrapolation of results obtained in laboratory chemostats to predict the composition of the community in the denitrifying, desulfurizing treatment system under study demonstrates that continuous-flow competition studies can make a valuable contribution to the solution of complex ecological problems.

Some of the work reported here was done by G. M. H. Scheulderman-Suylen, R. F. Beudeker, and J. C. Gottschal, to whom we are grateful.

LITERATURE CITED

1. **Beudeker, R. F., J. C. Gottschal, and J. G. Kuenen.** 1982. Reactivity versus flexibility in *Thiobacilli*. Antonie van Leeuwenhoek J. Microbiol. Serol. **48:**39–51.
2. **de Bont, J. A. M., J. P. van Dijken, and W. Harder.** 1981. Dimethyl sulphoxide and dimethyl sulphide as a carbon, sulphur and energy source for growth of *Hyphomicrobium* S. J. Gen. Microbiol. **127:**315–323.
3. **Gottschal, J. C., S. de Vries, and J. G. Kuenen.** 1979. Competition between the facultatively chemolithotrophic *Thiobacillus* A2, an obligately chemolithotrophic *Thiobacillus* and a heterotrophic spirillum for inorganic and organic substrates. Arch. Microbiol. **121:**241–249.
4. **Gottschal, J. C., and J. G. Kuenen.** 1980. Selective enrichment of facultatively chemolithotrophic *Thiobacilli* and related organisms in continuous culture. FEMS Microbiol. Lett. **7:**241–247.
5. **Gottschal, J. C., H. J. Nanninga, and J. G. Kuenen.** 1981. Growth of *Thiobacillus* A2 under alternating growth conditions in the chemostat. J. Gen. Microbiol. **126:**85–96.
6. **Harrison, A. P., Jr.** 1983. Genomic and physiological comparisons between heterotrophic thiobacilli and *Acidiphilium cryptum, Thiobacillus versutus* sp. nov., and *Thiobacillus acidophilus* nom. rev. Int. J. Syst. Bacteriol. **33:**211–217.
7. **Jørgensen, B. B.** 1982. Ecology of the bacteria of the

sulphur cycle with special reference to anoxic-oxic interface environments. Philos. Trans. R. Soc. London Ser. B **298:**543–561.

8. **Kelly, D. P.** 1982. Biochemistry of the chemolithotrophic oxidation of inorganic sulphur. Philos. Trans. R. Soc. London Ser. B **298:**499–528.

9. **Krul, J. M.** 1976. Dissimilatory nitrate and nitrite reduction under aerobic conditions by an aerobically and anaerobically grown *Alcaligenes* sp. and by activated sludge. J. Appl. Bacteriol. **40:**245–260.

10. **Krul, J. M., and R. Veeningen.** 1977. The synthesis of the dissimilatory nitrate reductase under aerobic conditions in a number of denitrifying bacteria, isolated from activated sludge and drinking water. Water Res. **11:**39–43.

11. **Kuenen, J. G., and R. F. Beudeker.** 1982. Microbiology of *Thiobacilli* and other sulphur-oxidizing autotrophs, mixotrophs and heterotrophs. Philos. Trans. R. Soc. London Ser. B **298:**473–497.

12. **Kuenen, J. G., J. Boonstra, H. G. J. Schroder, and H. Veldkamp.** 1977. Competition for inorganic substrates among chemoorganotrophic and chemolithotrophic bacteria. Microb. Ecol. **3:**119–130.

13. **Kuenen, J. G., and J. C. Gottschal.** 1982. Competition among chemolithotrophs and methylotrophs and their interactions with heterotrophic bacteria, p. 153. *In* A. T. Bull and J. H. Slater (ed.), Microbial interactions and communities, vol. 1. Academic Press, Inc., New York.

14. **Kuenen, J. G., and O. H. Tuovinen.** 1981. The genera *Thiobacillus* and *Thiomicrospira*, p. 1024. *In* M. P. Starr, H. Stolp, H. G. Truper, A. Balows, and H. G. Schlegel (ed.), The prokaryotes. A handbook on habitats, isolation, and identification of bacteria. Springer-Verlag, Berlin.

15. **Kuenen, J. G., and H. Veldkamp.** 1972. *Thiomicrospira pelophila*, gen. nov., sp. n., a new obligately chemolithotrophic colourless sulfur bacterium. Antonie van Leeuwenhoek J. Microbiol. Serol. **38:**241–256.

16. **Milde, K., W. Sand, W. Wolff, and E. Bock.** 1983. *Thiobacilli* of the corroded concrete walls of the Hamburg sewer system. J. Gen. Microbiol. **129:**1327–1333.

17. **Pfennig, N., and F. Widdel.** 1982. The bacteria of the sulphur cycle. Philos. Trans. R. Soc. London Ser. B **298:**433–441.

18. **Robertson, L. A., and J. G. Kuenen.** 1983. *Thiosphaera pantotropha* gen. nov. sp. nov., a facultatively anaerobic, facultatively autotrophic sulphur bacterium. J. Gen. Microbiol. **129:**2847–2855.

19. **Sivelä, S., and V. Sundman.** 1977. Demonstration of *Thiobacillus*-type bacteria which utilize methyl sulphides. Arch. Microbiol. **103:**303–304.

20. **Smith, A. L., and D. P. Kelly.** 1979. Competition in the chemostat between an obligately and a facultatively chemolithotrophic *Thiobacillus*. J. Gen. Microbiol. **115:**377–384.

21. **Smith, D. W., and S. F. Finazzo.** 1981. Salinity requirements of a marine *Thiobacillus intermedius*. Arch. Microbiol. **129:**199–203.

22. **Taylor, B. F., and D. S. Hoare.** 1969. New facultative *Thiobacillus* and a reevaluation of the heterotrophic potential of *Thiobacillus novellus*. J. Bacteriol. **100:**487–497.

23. **Timmer-ten Hoor, A.** 1975. A new type of thiosulphate oxidizing, nitrate reducing microorganism: *Thiomicrospira denitrificans* sp. nov. Neth. J. Sea Res. **9:**344–350.

Freshwater Algal Ecology: Taxonomic Trade-Offs in the Temperature Dependence of Nutrient Competitive Abilities

DAVID TILMAN AND RICHARD L. KIESLING

Department of Ecology and Behavioral Biology, University of Minnesota, Minneapolis, Minnesota 55455

The communities of photosynthetic microorganisms which coexist in lakes display a wide array of apparently repeatable patterns. Some of these patterns are seasonal. For instance, many north-temperate lakes are dominated by diatoms in early spring as solar radiation increases and as overturn leads to high levels of silicate (3). During the time of diatom dominance, there is a successional sequence of diatom species apparently controlled by silicon (4, 7) and phosphorus (13). As the water column warms and nutrients, especially silicon (6), are depleted, diatoms decline and are replaced by chrysophytes and green algae (3). With further warming and possibly further depletion of nutrients, especially nitrogen, green algae and chrysophytes tend to be replaced by blue-green algae. Late in the season, often at the time of autumnal isothermal mixing with its associated cooler water temperatures and increased silicon concentrations, diatoms may reach a second peak (3). Phytoplanktonic communities also have structure along spatial gradients. Within large lakes, there are frequently marked differences in the nearshore and offshore phytoplanktonic communities. Even more striking differences exist along the nutrient and temperature gradients which are established as rivers flow into and are mixed with large lakes, such as the Great Lakes (E. E. Kopczynska, Ph.D. thesis, University of Michigan, Ann Arbor, 1973). There are also vertical, small-scale spatial differences in the dominance of algal species in lakes. Lakes and different areas within lakes may also differ consistently in the number of co-occurring algal species. Understanding the causes of such patterns should be a major goal of aquatic microbial ecology.

Numerous hypotheses have been proposed as explanations for such patterns. The hypotheses have taken both "equilibrium" and "nonequilibrium" perspectives, and have tended to stress (i) the importance of competition for nutrients, (ii) the importance of mortality processes, especially herbivory and sinking, or (iii) the importance of various physical factors such as temperature, pH, and incident light intensity. With few exceptions there has been little effort to simultaneously determine the relative importance of several processes or to study the interactive effects of several processes. However, the work done to date (see review in 13) suggests that each of these factors is, at times, an important determinant of the structure of phytoplankton communities. These factors are not mutually exclusive and there is no reason to treat them as such, except for the hope, expressed in Ockham's Razor, that the natural world may be explained by simple models. Indeed, all three approaches, be they based on "equilibrium" or "nonequilibrium" assumptions, share a common element. To explain the long-term coexistence of numerous species, each approach must assume "trade-offs" in the traits of the phytoplankton species. This means that a species which gains an advantage relative to another species because of one of its traits must be at a disadvantage compared to the other species because of some other trait (12). Without such trade-offs, one species would be a "superspecies" (sensu Tilman [12]) and displace all others. If such trade-offs exist, recent theory has shown that species should be separated along gradients of the relevant variables in an order determined by the characteristics of the individual species (10, 12).

In microbial ecology, these trade-offs are most often physiological. If, for instance, silicon and phosphorus are limiting resources in a lake, the coexisting species should differ in their requirements for Si and P such that a species which is a superior competitor for P is an inferior competitor for Si. Physiological data supporting the existence of such a trade-off have been reported (13). Similarly, when N and P are limiting, an inferior competitor for N must be a superior competitor for P if it is to coexist in a habitat in which N and P are limiting. If there is a single limiting resource, but a physical factor such as temperature or pH changes spatially or seasonally, a species which is a superior competitor for the resource under one environmental condition must be an inferior competitor under other conditions if competition for that resource is the explanation for the long-term coexistence of the species. Similarly, if herbivory and nutrient competition are to explain algal coexistence, a species which is a superior nutrient competitor must also be more susceptible to herbivory (5). Clearly, such trade-offs can occur in more than two dimensions. As a first approximation to studying such trade-offs and to determining the possible relative importance of nutrient competition and a physical factor, we report in this paper results of experiments in which N/P and Si/P gradients were established at various temperatures. The data collected in these experiments are analyzed both as a complete set by use of multiple regression, to determine the relative importance of these factors, and as selected subsets, to analyze supply ratio or temperature effects with other variables experimentally controlled.

METHODS

All of the experiments used planktonic algal assemblages collected at 10 stations along a transect in Lake Superior from the mouth of the St. Louis River at Duluth, Minnesota, to a point 40 km east by northeast in the western arm of the lake. Such collections were made in the spring and summer of 1980, 1981, and 1982. Water samples from depths of 2 and 15 m at each of these stations were mixed, and zooplankton grazers were removed with nylon screening and micropipettes. The remaining algal assemblage was then used as the inoculum for a series of chemostat (cyclostat) experiments. None of the rotifers, protozoa, or other zooplankton remaining in the inoculum survived in our chemostats for more than 10 days. The algal assemblages from Lake Superior always contained a wide array of algal species, including 8 to 12 genera of diatoms, several chrysophytes, 8 to 10 genera of green algae, 4 genera of blue-green algae, and 1 or 2 genera of dinoflagellates.

The chemostat reaction vessels were 500-ml polycarbonate flasks sealed with silicone stoppers, through each of which were pushed two stainless-steel tubes which served as inflow and outflow/sampling ports. The reaction vessels contained 300 ml of medium. A flow rate of 0.3 day^{-1} was maintained with a Manostat peristaltic pump which fed sterile medium into the flasks through silicone tubing. The contents of the reaction vessels were mixed at 180 orbits min^{-1} for 10 s every min and kept at a predetermined temperature in a Percival incubator. Light was provided by Cool White bulbs at 300 μE m^{-2} s^{-1} on a 14 h/10 h light/dark cycle. The influent medium was WC (2) modified as described by Tilman (11), with NO_3, PO_4, and SiO_2 adjusted to give desired ratios.

Two types of experiments were performed. One type consisted of a series of chemostats with influent concentrations of Si and P adjusted to give 11 points along an Si/P gradient. Absolute concentrations of Si and P ranged from 184 μmol of SiO_2 and 0.7 μmol of PO_4 per liter to 27 μmol of SiO_2 and 29 μmol of PO_4 per liter, with supply Si/P ratios ranging from 267:1 to 1:1. The concentrations of Si, P, and other nutrients in the medium were such that only Si and P should have been limiting at any point along the gradi-

ent. NO_3 was kept at 1,000 μmol/liter. These experiments used an algal assemblage collected in July 1980, were performed simultaneously at 9 and 15°C, and lasted 35 days. The second type of experiment was an N/P gradient. For the N/P gradient, SiO_2 was kept at 150 μmol/liter, but N and P were adjusted so that one or both would be limiting. Absolute concentrations of N and P ranged from 2 μmol of N and 10 μmol of P per liter to 100 μmol of N and 0.2 μmol of P per liter. N/P ratios ranged from 0.2:1 to 500:1. The N/P experiments were performed with algal assemblages collected in October 1980 (4 N/P ratios at 15°C, experiments lasting 55 days), July 1981 (3 N/P ratios at 15°C, experiments lasting 50 days), and July 1982 (10 N/P ratios replicated at 10, 17, and 24°C, experiments lasting 58 days).

Algal counts were performed microscopically by use of a Sedgwick-Rafter chamber. All algae were identified from preserved material, at least to genus and in most cases to species. Algal counts were performed throughout the experiments, but only those from the final day of each experiment were used in the analyses presented here. Critical linear dimensions of each taxon were determined and used to calculate its biovolume. All algal counts were analyzed as biovolume, not as density (cells per milliliter). For the analyses presented here, algal counts were lumped by taxon (green, blue-green, and diatom). Although chrysophytes and dinoflagellates were present in our initial Lake Superior samples, they did not survive in our chemostats. All regression analyses used version 4 of MULTREG (15). For regression analyses, proportional abundances were arc-sine-square root transformed to reduce heteroscedasticity. The Si/P experiments and the N/P experiments were analyzed both separately and together. Si/P and N/P ratios were analyzed as log_{10} of the supply ratios. Regression analysis was also performed at individual temperatures to determine the effects of the gradients independent of temperature.

RESULTS

Multiple regression. Multiple regression analysis was performed on the entire data set as a whole. Algae were grouped by taxon. Temperature, influent Si/P ratios, and influent N/P ratios were the independent variables. Diatoms were significantly negatively related to temperature and significantly positively related to Si/P ratios. Each of these factors explained about 30% of the variance in the relative abundance of diatoms, as judged by the cumulative r^2 (Table 1). These results indicate that diatoms were increasingly dominant at lower temperatures and for lower availabilities of phosphate relative to silicate. Green algae were significantly negatively related

TABLE 1. Multiple regression of transformed proportional abundances of major algal taxa against temperature (T), $\log_{10}$ of influent silicate-to-phosphate ratios (Si/P), and $\log_{10}$ of influent nitrate-to-phosphate ratios (N/P)[a]

Taxon	Factor	Regression slope	Cumulative r^2
Diatoms	T	−0.08***	0.30
	Si/P	0.42***	0.59
	N/P	0.005 NS	0.59
Green algae	Si/P	−0.39***	0.35
	T	0.01 NS	0.36
	N/P	0.04 NS	0.37
Blue-green algae	T	0.06***	0.53
	N/P	−0.05 NS	0.55
	Si/P	−0.02 NS	0.55

[a] Stepwise regression was used, with the variables entered in order of significance of their simple correlations. The total data set was analyzed. Significance levels given for regression slopes are based on t tests with 53 degrees of freedom, with NS meaning $P > 0.05$ and *** meaning $P \leq 0.001$.

to Si/P ratio, but not significantly related to N/P ratio or temperature. Thus, green algae were most abundant when silicon was in low supply relative to phosphorus. Blue-green algae were significantly positively related only to temperature, being increasingly abundant at higher temperatures (Table 1).

Si/P gradients. Because the total data set is a mixture of two types of experiments, these two types were also analyzed separately. For the 22 chemostats in which NO_3 was 1,000 µmol/liter and Si and P were adjusted so that Si or P would be limiting, blue-green algae were very rare at the termination of the experiments and showed no significant response to either Si/P ratio or temperature (Table 2, A). Diatoms and green algae were much more abundant. Both had highly significant responses to Si/P ratio. Diatoms dominated the higher Si/P ratios for which all algal taxa present should have been phosphorus limited. Green algae dominated the low Si/P ratios for which the diatoms should have been silicon limited. Diatoms tended to be more abundant at the lower temperature and green algae tended to be more abundant at the higher temperature, but these effects were not significant (Table 2, A). When the results at a single temperature were analyzed (15°C), the effects of Si/P ratio on the relative dominance by diatoms or green algae was even more dramatic. The correlation between Si/P ratio and diatoms was +0.94 and that between Si/P ratio and green algae was −0.92 at 15°C.

N/P gradients. The five sets of experiments spanned temperatures from 10 to 24°C. Over this range, temperature was the most significant

variable. Diatoms were highly significantly negatively correlated with temperature, and blue-green algae were highly significantly positively correlated with temperature (Table 2, B). Green algae were not significantly correlated with temperature. Graphs of the relative abundance of these three taxa against temperature (Fig. 1) show that the dominance of blue-green algae increased almost exponentially with temperature over the range used in these experiments. Diatoms were much more abundant at lower temperatures. Green algae tended to dominate intermediate temperatures, their peak abundance occurring at about 15°C. Inspection of Fig. 1 suggests that much of the additional variation observed can be explained by N/P ratio, though not using the assumption of linear, additive effects inherent in multiple regression (compare with Table 1). Although diatoms were positively and blue-green algae were negatively correlated with N/P supply ratios (Table 2, B), neither relationship was significant. The abundance of green algae was statistically independent of N/P ratios, but seemed to reach its peak at intermediate N/P ratios. At a single temperature (24°C), blue-green algae and diatoms showed much stronger effects of N/P supply ratios on their relative abundance, with a correlation coefficient of +0.59 for diatoms and of −0.59 for blue-green algae.

DISCUSSION

The results of all of our nutrient gradient experiments on algal assemblages from Lake Superior strongly suggest that composition patterns in the algal community cannot be explained solely in terms of a single factor or

TABLE 2. Simple regression analysis of end results of Lake Superior chemostat experiments[a]

Taxon	Factor	Correlation coefficient
A. Si/P experiments (9 and 15°C)		
Diatoms	Si/P	+0.73**
	T	−0.31 NS
Green algae	Si/P	−0.73**
	T	+0.26 NS
Blue-green algae	Si/P	−0.08 NS
	T	+0.22 NS
B. N/P experiments (10, 15, 17, 24°C)		
Diatoms	N/P	+0.30 NS
	T	−0.66**
Green algae	N/P	−0.16 NS
	T	−0.01 NS
Blue-green algae	N/P	−0.21 NS
	T	+0.76**

[a] All regressions were performed on transformed proportional abundances with log transformed Si/P or N/P ratios. T is temperature; ** means $P \leq 0.01$, NS means $P > 0.05$.

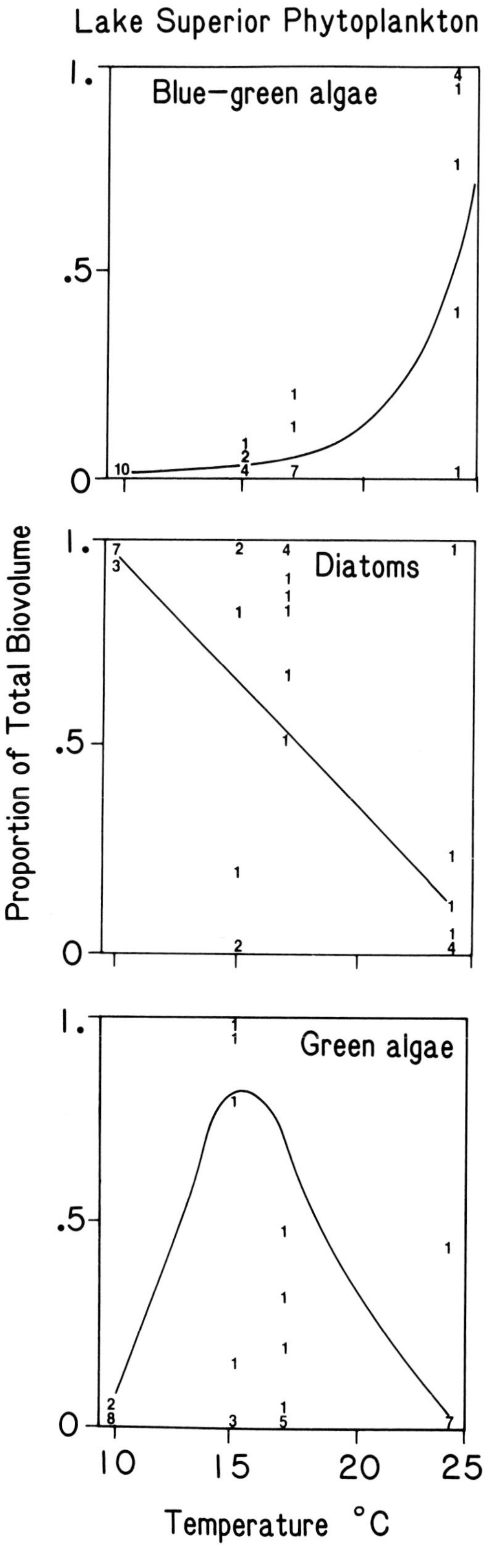

FIG. 1. Temperature dependence of the proportional abundance of blue-green algae, diatoms, and green algae in a series of chemostats in which various nitrogen-to-phosphorus ratios were used. Part B of Table 2 gives linear regression coefficients for these data. Curves shown were hand drawn to data. The numbers in each figure indicate how many data points fell at a given spot. See also Table 1.

process. Si/P ratios, N/P ratios, and temperature all affected the relative abundance of the major algal taxa. Temperature had an unexpectedly great effect on the relative abundance of diatoms and blue-green algae. Although some studies of temperature physiology have been performed on freshwater algae (8, 9, 14), there has been little suggestion in the literature that major taxa may be so apparently distinct in their temperature optima. The striking effect of temperature on the dominance patterns of the three major taxa of these experiments (Fig. 1) suggests that the species of each taxon which are present in Lake Superior and which can reproduce in our laboratory chemostats tend to be relatively similar in their temperature optima. These results suggest that the Lake Superior diatom species in our experiments are increasingly superior phosphorus (and possibly nitrogen) competitors at lower temperatures, especially at temperatures less than about 12°C. The species of blue-green algae present in our chemostats seem to be superior nitrogen competitors at temperatures generally above about 20°C. The species of green algae seem to be superior competitors at intermediate temperatures if there are low Si/P ratios or intermediate N/P ratios.

These results, however, should not be interpreted as suggesting that all diatom species are superior competitors only at lower temperatures or that all blue-green algal species are superior competitors only at warmer temperatures in all lakes. There are too many exceptions for such a generalization to be valid (p. 427–443 of reference 3). The results do, though, raise some interesting questions.

Is the apparent temperature separation of these taxa an artifact of our collecting or culture methods? Chrysophytes and microflagellates never dominated our laboratory chemostats, nor did "cold-water, low-light" blue-green algae such as *Oscillatoria rubescens* (1, as cited in 3). We do not know the reasons for this. It may be that these taxa are incapable of living under our culture conditions. Alternatively, might the apparent temperature separation of diatoms, green algae, and blue-green algae be caused by regular, repeatable correlations between nutrient limitation patterns and water temperature in Lake Superior? Changes in relative limitation by silicon, phosphorus, and nitrogen are seasonally correlated with changes in temperature in Lake Superior and probably other north-temperate lakes. If phosphorus is most limiting relative to nitrogen and silicon in the spring when water temperatures are cool, of all the species which are superior competitors for phosphorus, those which had their maximum nutrient-saturated growth rate near that temperature would probably be favored. Similarly, if nitrogen tends to be most limiting relative to phosphorus and silicon when water temperatures are near their maximum, of all the species which are superior N competitors, those which dominated at that time would likely be the algae which reached their maximal nutrient-saturated growth rate at higher temperatures. If silicon limitation of diatoms tends to reduce phosphorus consumption to the point where there is sufficient phosphorus for the reproduction of other species, and if this occurs at intermediate temperatures, of all the species which were the next best competitors for phosphorus, those with an intermediate temperature optimum would probably be favored. If such patterns recurred year after year, species which did not have the favored sets of traits would be likely to become locally extinct. Conversely, it could also be argued that the species which have the superior ability to grow at low temperatures should be the superior competitors for phosphorus because phosphorus tends to be limiting at that time. Similarly, the reverse of our earlier argument could be constructed for species with superior ability to grow at high temperatures. Whichever way the argument is stated, the conclusions are the same: when several important environmental variables change in concert in a regular, predictable manner, the species favored are those which respond best to particular combinations of the factors. The end result of such selection would be species with a series of correlated traits, as observed in our experiments.

The observed responses of natural assemblages of diatoms, green algae, and blue-green algae to Si, P, N, and temperature suggest that there are three major ways that these taxa exploit the north-temperate lake habitat; (i) by species which are superior competitors for phosphorus and have low temperature optima; (ii) by species which have low or no requirements for silicon, are superior competitors for N, and have high temperature optima; and (iii) by species which have low or no requirements for silicon, are the second-best competitors for phosphorus, and have intermediate temperature optima.

As already discussed, diatoms tend to be dominant in lakes during early spring when water temperatures are low and in autumn when water temperatures are again low. Green algae

tend to dominate later in the season when lakes are warmer. Blue-green algae are often dominant in mid to late summer when water temperatures reach their seasonal high. It would be tempting to ascribe the seasonal successional sequence solely to temperature on the basis of the similarity of the major patterns observed in lakes to the patterns observed in these experiments. However, these experiments also have shown significant effects of nutrients on dominance by these taxa. The decline in diatoms could just as easily be ascribed to declining silicon levels in lakes, and the increase in abundance of green algae in lakes could be ascribed to changing Si/P ratios (see Tables 1 and 2). Similarly, the increase in abundance of blue-green algae observed in late summer could be ascribed to changing N/P ratios because N/P ratios are often at their seasonal low late in the summer. However, the regular seasonal correlations between nutrients and temperature in lakes mean that such simple approaches may miss the mechanisms controlling the observed seasonal dynamics of algal populations. Temperature-dependent growth, nutrient competition, and their interactions are all likely to be important. The apparent ability of any one factor, alone, to explain events may well be an artifact of repeatable correlations among these variables in lakes.

For two species to coexist stably, it is necessary that they have "trade-offs" in their requirements for, or responses to, various limiting factors or resources (12). If there are such trade-offs, species should be separated along gradients (12). Species separation was observed along both the Si/P and the N/P gradients. For instance, for the N/P experiments at 24°C, the blue-green alga *Anaebaena flos-aquae* dominated N/P ratios from 0.2:1 to 10:1, often comprising 95 to 99% of the total community biovolume. *Dactylococcopsis* cf. *acicularis* was absent at those ratios, but was the most dominant blue-green alga at an N/P ratio of 20:1. *Oscillatoria* cf. *limnetica* was greater than 99% of the biovolume at an N/P ratio of 40. Blue-green algae were less than 1% at an N/P ratio of 500:1. The flask with an N/P ratio of 500 was dominated by two diatoms, *Synedra radians*, which averaged 110 μm in length, and *Synedra* cf. *tenera*, which averaged 60 μm in length. These and other *Synedra* species dominated all the other flasks with high N/P ratios at both 17 and 10°C, and were abundant at low N/P ratios at these temperatures.

CONCLUSIONS

Natural algal communities collected in Lake Superior were used in chemostat competition experiments performed at various temperatures and with different influent N/P and Si/P ratios.

Multiple regression analyses of the data from these 56 chemostats revealed that temperature was generally the most important variable controlling dominance by major taxonomic groups of algae. Temperature was the most important and N/P supply ratios the next most important variable controlling the dominance of the cyanobacteria (blue-green algae), which were dominant at high temperatures and low N/P ratios. Temperature and Si/P supply ratios were equally important in determining the dominance of *Bacillariophyceae* (diatoms), which were increasingly dominant at lower temperatures and higher Si/P ratios. *Chlorophycophyta* (green algae) were dominant only under conditions of both intermediate temperature and low Si/P or intermediate N/P ratios.

These results suggest that there may be "trade-offs" in the physiology of these taxa, with each taxon being a superior competitor for a particular temperature range and pattern of nutrient limitation. If such trade-offs are found to occur in other habitats, in general, some questions in algal ecology might be more easily answered. For instance, it might be possible to predict which taxon would dominate a lake at a particular time without the need to perform detailed studies of the temperature and nutrient physiology of all the dominant species. If several aspects of the physiology of algal species are found to be consistently correlated, less physiological information might need to be collected to answer specific questions.

However, such generalizations will not be possible until studies like those reported in this paper are performed on algal assemblages from a wide variety of lakes. Ideally, such future work should include zooplankton grazing as well as such "nonequilibrium" effects as nutrient fluctuations. Once such work has been performed on an array of species from a variety of lakes, the validity of taxonomically based generalizations concerning nutrient competitive abilities and temperature optima can be determined.

We thank Bruce Monson and Robert Sterner for help with these experiments and Kiesy Strauchon for help with the preparation of this manuscript.

This work is the result of research sponsored by the Minnesota Sea Grant Program, supported by the National Oceanic and Atmospheric Administration Office of Sea Grant, Department of Commerce, under grant NA82AA-D-00039, and is Research Contribution No. 141.

LITERATURE CITED

1. **Findenegg, I.** 1943. Untersuchungen uber die Okolgie und die Produktionsverhaltniss des Planktons in Karntner Seengebiete. Int. Rev. Gesamten Hydrobiol. Hydrogr. 43:368–429.

2. **Guillard, R. R. L.** 1975. Culture of phytoplankton for feeding marine invertebrates, p. 29–60. *In* W. L. Smith and M. H. Chanley (ed.), Culture of marine invertebrate animals. Plenum Press, New York.

3. **Hutchinson, G. E.** 1967. A treatise on limnology, vol. 2. John Wiley & Sons, Inc., New York.

4. **Kilham, P.** 1971. A hypothesis concerning silica and the freshwater planktonic diatoms. Limnol. Oceanogr. 16:10–18.

5. **Levin, B. R., F. M. Stewart, and L. Chao.** 1977. Resource-limited growth, competition, and predation: a model and experimental studies with bacteria and bacteriophage. Am. Nat. 111:3–24.

6. **Lund, J. W. G., F. J. H. Mackereth, and C. H. Mortimer.** 1963. Changes in depth and time of certain chemical and physical conditions and of the standing crop of *Asterionella formosa* Hass. in the North Basin of Windermere in 1947. Philos. Trans. R. Soc. London Ser. B 246:255–290.

7. **Pearsall, W. H.** 1930. Phytoplankton in the English Lakes. I. The proportions in the water of some dissolved substances of biological importance. J. Ecol. 18:306–320.

8. **Rhee, G.-Y., and I. J. Gotham.** 1981. The effect of environmental factors on phytoplankton growth: temperature and the interactions of temperature with nutrient limitation. Limnol. Oceanogr. 26:635–648.

9. **Rodhe, W.** 1948. Environmental requirements of freshwater plankton algae. Experimental studies in the ecology of phytoplankton. Symb. Bot. Ups. 10(1):1–149.

10. **Tilman, D.** 1980. Resources: a graphical-mechanistic approach to competition and predation. Am. Nat. 116:362–393.

11. **Tilman, D.** 1981. Experimental tests of resource competition theory using four species of Lake Michigan algae. Ecology 62:802–815.

12. **Tilman, D.** 1982. Resource competition and community structure. Princeton University Press, Princeton, N.J.

13. **Tilman, D., S. S. Kilham, and P. Kilham.** 1982. Phytoplankton community ecology: the role of limiting nutrients. Annu. Rev. Ecol. Syst. 13:349–372.

14. **Tilman, D., M. Mattson, and S. Langer.** 1981. Competition and nutrient kinetics along a temperature gradient: an experimental test of a mechanistic approach to niche theory. Limnol. Oceanogr. 26:1020–1033.

15. **Weisberg, S.** 1981. MULTREG users manual version 4.0. University of Minnesota School of Statistics Technological Report no. 298. University of Minnesota, Minneapolis.

Microbes as Predators or Prey

Heterotrophic, Free-Living Protozoa: Neglected Microorganisms with an Important Task in Regulating Bacterial Populations

MARIANNE CLARHOLM

Department of Microbiology, Swedish University of Agricultural Sciences, S-750 07 Uppsala, Sweden

"A microorganism is an organism of microscopic or ultramicroscopic size—the expression is used especially of bacteria and protozoa." (Webster's *Third International Dictionary of the English Language*, 1976.)

Despite this definition, protozoa are not particularly well known to most ecological microbiologists, and traditionally they have been handled by the zoologists. Today, microbiology is establishing itself as an ecological discipline with a key role in ecosystem studies. This means that in the future it will be necessary for many more microbiologists to deal also with heterotrophic, free-living protozoa. This important component of the ecosystem can no longer be ignored. In an arable field soil, the standing crop biomass of protozoa is reported to be around 5 g (dry weight) m^{-2} (46) or approximately the same as that of earthworms (36). Protozoa, however, have much shorter turnover times, a fact which indicates their major influence at the ecosystem level (51).

Protozoa are the major consumers of bacteria (12, 14). Calculations made by Stout and Heal (52) for an arable field soil estimated that protozoa consumed 150 to 900 g of bacteria m^{-2} year^{-1}, which is equal to a production of 15 to 85 times the standing crop. Both bacteria and protozoa thus have small standing crop values relative to a high, but in time unevenly distributed, production. Their ecological significance is enhanced by their ability to survive adverse conditions (15) and their use as a food source for animals. In spite of their predicted importance, conspicuously little information is available concerning the heterotrophic free-living protozoa. One reason lies in the technical difficulties encountered in the quantitative estimations, but also in that protozoa have largely been ignored by ecological microbiologists, who are the ones trained in the techniques needed to investigate smaller forms of protozoa.

PROTOZOA

Protozoa are unicellular, eucaryotic, water-dependent organisms ranging in size from 2 to 1,000 μm. Their major food is bacteria (12), but many species also consume algae (12, 55), detritus (47), and spores (40) or hyphae (1, 12) of fungi. Some heterotrophic protozoa can utilize dissolved organic matter (53); others are predatory, feeding on smaller protozoa, as well as on nematodes or other metazoans (45). Protozoa multiply through division, with recorded minimum generation times of around 2 or 3 to >12 h, depending on species. Under prolonged periods of unfavorable conditions, many protozoa have the ability to survive as resting cysts (11). Many species, not the least those occurring in soil, are opportunistic and can quickly adjust to an increased or decreased bacterial production.

Free-living protozoa belong to either the ciliates, the flagellates, or the rhizopods; these groups are characterized by their locomotory organelles among other features.

Ciliates. The ciliates comprise mainly larger forms, mostly longer than 25 μm. Most forms use cilia to swim or creep along surfaces, but temporarily or permanently sessile forms are known. Representatives of several groups are ciliary suspension feeders. Different species select for bacteria of different sizes, and they thus avoid competition for the same food source (21). This may explain the coexistence of several species in the same habitat. Ciliates need relatively large amounts of water to be able to swim (16) and fairly high concentrations of bacteria (e.g., >10^7 bacteria ml^{-1} [4, 22]) to feed and reproduce. High ciliate activity is therefore generally excluded in normal, aerated soils with high enough bacterial numbers because the water content is insufficient. Bactivorous forms are also rare in the open waters where the bacterial levels are too low to promote growth, but here alga-consuming and other predatory forms are common instead. Important habitats for ciliates feeding on bacteria include sediments, eutrophicated waters, sewage plants, and persistently wet soils.

Flagellates. The flagellates consist of many small, generally free-swimming species ranging from 3 to 10 μm in size. Most of them have one or two flagella used for locomotion, for temporary attachment, and for trapping bacteria, which are engulfed at the base of the flagellum. Some small forms are able to utilize dissolved organic carbon directly from the water (53). Phototrophic nutrition is common, either as an obligatory function or as an alternative to phagotrophy. Minimum generation times are reported to be within the range of 2 to 5 h (23). Flagellates are frequent in both terrestrial and aquatic habitats; they are the dominant protozoa in free waters containing low levels of bacteria (23).

Rhizopods. The rhizopods can be divided into naked amoebae and thecate amoebae, which are shelled forms. The latter have minimum generation times of 2 to 10 days (48), the time it takes to make a theca; the smallest naked amoebae can multiply within hours. Naked amoebae are the dominant bacterial consumers in soil (7, 19). They are 10 to 300 μm in diameter, but only about 2 μm thick. This shape is particularly suitable for a life in the water films of solid surfaces in the soil. Here, naked amoebae slowly move within the water layer, ingesting surface-living bacteria through phagocytosis. The majority of the naked amoebae are located in the rhizosphere (5, 17), where most of the bacterial population also occurs (36). Sediments and the surfaces of seaweeds are also reported to harbor naked amoebae, but no quantitative estimates seem to be available.

METHODS FOR QUANTITATIVE ESTIMATIONS

In water samples, small protozoa can be counted directly in an epifluorescence microscope after collection on a filter following fixation in formaldehyde and staining with a fluorescent dye (6, 20, 30). Samples from soils and sediments contain too much debris for accurate direct observation. Indirect most-probable-number (MPN) methods are therefore applied when protozoa are enumerated from these habitats and also when their numbers are too low for direct counting. Today, most researchers make the necessary dilutions for MPN estimations in microtiter trays with some suitable bacteria as the food source (18). After dilution of the sample, the trays are incubated, and later the wells (8 or 12 for each dilution) are examined for the occurrence of the different groups of protozoa. The number of wells containing protozoa is estimated in the three most diluted rows where protozoa are found, and the MPN is obtained from a suitable table based on Poisson distributions (e.g., 43). The MPN is thereafter corrected for the dilutions made before the sample was introduced into the tray. The MPN method gives only the total number of protozoa present and does not indicate whether the protozoa were encysted or in an active state when sampled. MPN estimations of protozoa are time-consuming, and the values obtained may not be valid for more than 1 day if their growth conditions are altered (1).

PROTOZOA IN SEWAGE

Sewage contains both readily available energy and inorganic nutrients, and thus bacteria are able to grow well in the absence of protozoa. A healthy proliferation of protozoa is, however, necessary for the proper function of sewage treatment plants and infiltration units. In the absence of protozoa, the bacterial numbers will increase dramatically and the effluent will be turbid and of inferior quality (13). Protozoa also cause flocculation of solids, which is necessary for effective cleaning of the water. By consuming the bacteria, the protozoa oxidize part of the bacterial carbon to CO_2; at the same time they release nitrogen as ammonia for nitrifiers and subsequently for the denitrifiers to transform into gaseous forms. Here protozoa thus contribute to the purification of water by acting as a link between the heterotrophic bacteria on the one hand, which decompose the organic nitrogen-containing materials, and the metabolically specialized denitrifying bacteria on the other hand, which eventually remove the nitrogen from the water.

PROTOZOA IN SOIL

In natural habitats the most limiting factor for bacterial growth is the lack of a utilizable energy source (8, 50). In soil the largest input over the growing season comes from the plant roots in the form of both soluble and particulate carbon (37, 44). Water availability is in turn the most important and uncontrollable factor influencing plant growth in the field. During dry periods, the growth activities of both plants and microorganisms are reduced. When the soil is moistened, there will be a burst in growth activities starting with a bacterial population and followed by a subsequent increase in protozoa, mainly naked amoebae (7, 19). If the soil is influenced by a root, the bacterial production will be much larger than in root-free soil (7, 17, 36).

The wet-dry cycles which normally occur in nature make the activities of the microorganisms irregular and therefore difficult to observe and understand. Laboratory experiments can be used to study the interactions of protozoa, bacteria, and plants without the disturbing influence of changing abiotic factors. Darbyshire and Greaves (17) followed bacterial and protozoan populations in the rhizosphere and the nonrhizosphere of potted white mustard plants with weekly samplings over the full growth period

(Fig. 1). For bacteria, they recorded only one high value, which occurred at the time of budding, whereas naked amoebae scored high at four successive samplings. The last peak value for amoebae was found 1 week after the bacterial peak. Since protozoa in turn are consumed by many larger animals (9, 38, 39), this means that for more than 3 weeks there had been a bacterial production large enough to sustain the high level of naked amoebae observed. Most of these bacteria were, however, consumed very rapidly, and thus the bacterial production could only be registered indirectly, via the protozoa.

The energy to support the bacterial growth in the rhizosphere came from the roots of the mustard (44). Judging from the development of the microorganisms, the carbon input appears to have increased throughout to flowering. Thereafter, the decrease appears to have been drastic, resulting in the steep decreases in numbers observed for both bacteria and naked amoebae. The growth period could thus be divided into two phases. The first one, lasting up to flowering, is characterized by high levels of fine-root activity, high levels of microbial growth around the roots, and by a large plant uptake of nitrogen. At the time of flowering, annual plants have

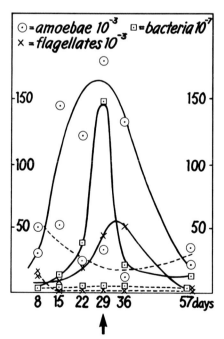

FIG. 1. Development of bacteria, naked amoebae, and flagellates in the rhizosphere (solid lines) and in the nonrhizosphere (broken lines) of potted white mustard plants. The arrow indicates the start of budding. The figure is based on Table 1 of Darbyshire and Greaves (17). Values are given per gram of dry soil.

taken up 80% or more of their total nitrogen requirement (33). During the second phase, they direct most of their synthesized carbon into seed production relying mostly on translocations within the plant to supply the seeds with nitrogen (33). The switch to energy-rich seed production leaves less of the available carbon to be transported to the roots. This leads to a diminished production and activity of the fine feeding roots and in turn also to decreased microbial activities as shown in Fig. 1.

PROTOZOA AS MINERALIZING AGENTS

In most places where there are bacterial populations of any size, there will also occur a consumption of these bacteria by protozoa (39). The consumption will result in a production of bacterial grazers concomitant with a release of inorganic nutrients, e.g., ammonium (26, 29, 31, 49). It has been proposed for some time now that bacterial grazers serve as mineralizing agents (29, 31, 32, 35). The idea has, however, not gained wide acceptance, because of the lack of convincing experimental evidence. Early papers were also published by researchers with other main interests (29) or by non-microbiologists (31). All the reports of increased ammonia release in connection with protozoan grazing came from aquatic environments, where the technical problems are less than in soil. Habte and Alexander (27, 28) demonstrated the capacity of protozoa to decrease the number of rhizobia in soil down to a certain level but did not comment on nutrient release associated with the grazing.

The mineralizing function of bacterial grazers in the soil was examined by Coleman et al. (10). Their studies were conducted in sterilized soil systems, so-called microcosms, in which certain parameters could be excluded or included and the effects could be studied under controlled but artificial conditions. The group found increased levels of ammonia in soils with naked amoebae present, but since the root-derived carbon was replaced by addition of glucose at the start, the relevance of the interactions for the plant was demonstrated only indirectly.

POSSIBLE IMPLICATIONS OF PLANT–MICROORGANISM INTERACTIONS

Plant breeders have suggested that the carbon transported below ground is wasted and that efforts should be made to minimize these losses. This will not work, since even in fertilized crops, around half of the nitrogen found in the plant originates from the soil organic matter. To make this nitrogen available for plant uptake, the root-derived carbon is needed to activate the bacteria. Early investigations with ^{15}N (3) showed that the mineralization of nitrogen from soil organic matter is much higher in the presence of plants. This indicates that the plant has a way to increase the mineralization compared with that in fallow soil. Estimation of the size of the organic nitrogen pool that is mineralized under these conditions and thus available to the crop for the next growing season is at present a topic of growing scientific interest.

Microcosms have been used to estimate the role protozoans play in releasing nitrogen from the soil organic matter. Wheat plants were grown in large test tubes with sterilized soil and then inoculated with a small inoculum of native soil bacteria or with bacteria plus protozoa (Clarholm, submitted for publication). After 6 weeks, the plants grown with protozoa present weighed 80% more and contained 20% more nitrogen (Fig. 2). The amount of nitrogen added with the organisms was negligible in comparison with the amounts taken up by the plants. If the nitrogen in the seed was subtracted, the plants grown with protozoa had an 80% higher nitrogen content. Part of the nitrogen taken up by the plants grown in the ungrazed soil with bacteria

only had most probably been released by ammonifying bacteria from the microorganisms killed at autoclaving. This was a one-time source, whereas the bacteria–protozoan interactions formed a working unit leading to a continuous release of nitrogen from the organic matter as long as it was fueled by the root-derived carbon. In soil without plants in the same experiment, no increase in inorganic nitrogen was observed, a result that stressed the importance of a suitable energy source for the bacteria to be able to mineralize organic nitrogen in soil.

The organic nitrogen will be mineralized through activities of heterotrophic microorganisms and the root-derived carbon will contribute much of the necessary energy for bacterial growth (36). Using carbon from the roots as a carbon and energy source, the otherwise energy-limited bacteria around the roots are now able to release nitrogen and also other essential nutrients from the organic matter, but just to make it part of their own increasing bacterial biomass. The bacterial production attracts bacterial feeders, of which protozoa are the most prevalent and effective (19, 52). When the bacteria are consumed, part of the bacterial nitrogen is excreted as ammonia (29, 49), and as such it could be taken up by the roots or be transformed to nitrite and nitrate by chemotrophic bacteria.

The plant thus seems to be able to induce a two-step reaction involving bacteria and bacterial grazers, which as a result will make organic nitrogen available for the plant. Another reported effect of the presence of protozoa in soil is an increased rate of uptake of inorganic nitrogen by plants (20).

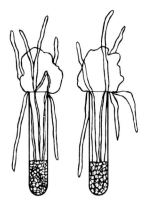

3 wheat plants grown with	bacteria	bacteria +protozoa
dry weight shoots	116	213
roots	65	113
shoots + roots	181	326
N content in shoots	3.92	4.39
roots	0.93	1.41
N in shoots+roots	4.85	5.80
−N in seeds	3.24	3.24
N mineralized from org. matter in soil	1.61	2.56

FIG. 2. Dry weights and nitrogen contents of three wheat plants grown for 6 weeks in sterilized soil reinoculated with bacteria only or with bacteria + protozoa. Figures are given in milligrams per microcosm.

PROTOZOA IN WATER

In water, the grazer food chain has traditionally been the most emphasized pathway for transferral of energy and nutrients (54). Although recognized, the detrital pathway has been considered to play a less important role (54). Starting with studies conducted in the 1970s (42, 44), a general picture is now emerging with evidence for major flows through the microheterotrophic organisms, mainly the bacteria, accumulating (2, 54).

The interactions described above, leading to nitrogen mineralization, are found also in aquatic habitats (Fig. 3). In water, phytoplankton perform most of the photosynthesis. Characteristically, the losses of carbon from the phytoplankton to the dissolved organic matter pool is estimated to be 30 to 40% of the primary production (34, 54). In experiments investigating nitrogen shortage conditions, phytoplankton grown under increased nitrogen stress have increased the excreted dissolved organic matter to nearly 80% of the photosynthetically fixed carbon (2).

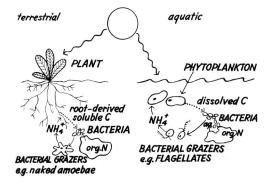

FIG. 3. Illustration of the interactions between photosynthesizing organisms, bacteria, and bacterial grazers (protozoa) leading to mineralization of nitrogen from the organic matter in a terrestrial and an aquatic system.

These results suggest a connection between available energy and the mineralization of organic nitrogen. It is also an indication of the fact that the primary producers seem to have ways to regulate the mineralization by increasing the bacterial activities, which are mainly energy limited (6, 50). Using the dissolved organic matter as an energy source, bacteria can utilize nitrogen-containing organic material in the water for their biosynthetic needs. Protozoa, e.g., flagellates, then feed on the bacteria and excrete ammonia which can then be taken up by the phytoplankton.

OBSERVATIONS OF MICROORGANISM INTERACTIONS IN THE FIELD

Since it is very difficult to make accurate in situ production measurements of the microorganisms involved, it has so far not been possible to evaluate the absolute importance of the bacteria–protozoan interactions. Some field observations are now available with short enough time between observations (1 to 3 days) to match the short generation times of the organisms investigated. In the humus layer of a podzolized forest soil, where most of the feeding fine roots are located, an increase in bacterial populations was recorded 2 days after a rainfall. It was followed by a peak of naked amoebae on day 5 (7). The observed increase in protozoa was large enough to explain at least 60% of the bacterial decrease to be the result of protozoan consumption of bacteria.

From the marine environment there are observations of a cyclic behavior involving bacteria and heterotrophic flagellates (24). The growth cycle found consisted of a maximum and a minimum in standing crop biomass within 15 days, with the flagellates lagging 4 days behind the bacteria. Estimations of the filtration rates of

the most common species of flagellates indicated that 12 to 67% of the whole water volume was filtered through the protozoa per day (24). The few measurements obtained from the field so far thus indicate a strong interaction between the two groups, pointing at an important regulating role for the protozoa in the environment.

It has been argued that "The rate of organic nitrogen mineralization has not been extensively studied probably because there is little that one can do to influence the rate even if one wanted to do this" (25). This might be true for natural, balanced ecosystems with a tight circulation and small losses of nitrogen. For the agricultural situation, extensively manipulated by humans, one would like to understand better how organic nitrogen is mineralized and what factors govern the release. This knowledge could lead to better agricultural management, thereby minimizing the losses of expensive nitrogen fertilizers. When the "Story of Nitrogen Mineralization" is written, ammonifying bacteria and decomposing fungi will likely play a major role, but the roles for bacteria and protozoa will not be minor, particularly when related to crop growth.

Thanks are due to Tom Fenchel for his constructive criticism and encouragement while this paper was being written and to Walter Dobrogosz for helpful suggestions to improve the language.

This paper was written while I was working within the project "Ecology of Arable Land. The Role of Organisms in Nitrogen Cycling" financed by Swedish Council for Planning and Coordination of Research, Swedish Council for Forestry and Agricultural Research, Swedish Environmental Protection Board, and Swedish Natural Science Council.

LITERATURE CITED

1. **Alabouvette, C., I. Lemaitre, and M. Pussard.** 1981. Densité de population de l'amibe mycophage *Thecamoeba granifeia* s. sp. *minor* (*Amoebida, Protozoa*). Mesure et variations expérimentales dans le sol. Rev. Ecol. Biol. Sol **18:**179–192.
2. **Azam, F., T. Fenchel, J. G. Field, L. A. Meyer-Reil, and F. Thingstad.** 1983. The ecological role of water-column microbes in the sea. Mar. Ecol. Prog. Ser. **10:**257–263.
3. **Bartholomew, W. V., and F. E. Clark.** 1950. Nitrogen transformations in soil in relation to the rhizosphere microflora, p. 112–113. *In* Trans. Fourth Int. Congr. Soil Sci., vol. 2. Hoitsema Bros., Groningen, The Netherlands.
4. **Berk, S. G., R. R. Colwell, and E. B. Small.** 1976. A study of feeding responses to bacterial prey by estuarine ciliates. Trans. Am. Microsc. Soc. **95:**514–520.
5. **Biczok, F.** 1965. Protozoa in the rhizosphere, p. 120. *In* Progress in protozoology. London International Congress Ser. No. 91. Excerpta Medica Foundation, Amsterdam.
6. **Caron, D. A.** 1983. Technique for enumeration of heterotrophic and phototrophic nanoplankton, using epifluorescence microscopy, and comparison with other procedures. Appl. Environ. Microbiol. **46:**491–498.
7. **Clarholm, M.** 1981. Protozoan grazing of bacteria in soil—impact and importance. Microb. Ecol. **7:**343–350.
8. **Clark, F. E.** 1967. Bacteria, p. 15–49. *In* A. Burges and F. Raw (ed.), Soil biology. Academic Press, Inc., New York.
9. **Coleman, D. C., R. V. Anderson, C. V. Cole, E. T. Elliott, L. Woods, and M. Campion.** 1978. Trophic interactions in soil as they affect energy and nutrient dynamics. IV. Flow

of metabolic and biomass carbon. Microb. Ecol. **4**:373–380.

10. **Coleman, D. C., C. V. Cole, R. V. Anderson, M. Blaha, M. K. Campion, M. Clarholm, E. T. Elliott, H. W. Hunt, B. Shaefer, and J. Sinclair.** 1977. An analysis of rhizosphere-saprophage interactions in terrestrial ecosystems. Ecol. Bull. (Stockholm) **25**:299–309.

11. **Corliss, J. O., and S. C. Esser.** 1974. Comments on the role of the cyst in the life cycle and survival of free-living protozoa. Trans. Am. Microsc. Soc. **93**:578–593.

12. **Coûteau, M.-M., and M. Pussard.** 1982. Nature du régime alimentaire des protozaires du sol, p. 179–194. *In* P. H. Lebrun, M. André, A. de Medts, C. Grégoire-Wibo, and G. Wauthy (ed.), New trends in soil biology. Dieu-Brichart, Ottignies-Lovain-la-Neuve.

13. **Curds, C. R.** 1982. The ecology and role of protozoa in aerobic sewage treatment processes. Annu. Rev. Microbiol. **36**:27–46.

14. **Curds, C. R., and M. J. Bazin.** 1977. Protozoan predation in batch and continuous culture, p. 115–176. *In* M. R. Droop and H. W. Jannasch (ed.), Advances in aquatic microbiology. Academic Press, London.

15. **Cutler, C. W.** 1927. Soil protozoa and bacteria in relation to their environment. J. Quekett Microsc. Club Ser. 2 **15**:309–330.

16. **Darbyshire, J. F.** 1976. Effect of water suctions on the growth in soil of the ciliate *Colpoda steini* and the bacterium *Azotobacter chroococcum.* J. Soil Sci. **27**:369–376.

17. **Darbyshire, J. F., and M. P. Greaves.** 1967. Protozoa and bacteria in the rhizosphere of *Sinapis alba* L., *Trifolium repens* L. and *Lolium perenne* L. Can. J. Microbiol. **13**:1057–1068.

18. **Darbyshire, J. F., R. F. Wheatley, M. P. Greaves, and R. H. E. Inkson.** 1974. A rapid micromethod for estimating bacterial and protozoan populations in soil. Rev. Ecol. Biol. Sol **11**:465–475.

19. **Elliott, E. T., and D. C. Coleman.** 1977. Soil protozoan dynamics in a shortgrass prairie. Soil Biol. Biochem. **9**:113–118.

20. **Elliott, E. T., D. C. Coleman, and C. V. Cole.** 1979. The influence of amoebae on the uptake of nitrogen by plants in gnotobiotic soil, p. 221–229. *In* J. D. Harley and R. S. Russell (ed.), The soil root interface. Academic Press, Inc., New York.

21. **Fenchel, T.** 1980. Suspension feeding in ciliated protozoa: functional response and particle size selection. Microb. Ecol. **6**:1–11.

22. **Fenchel, T.** 1980. Suspension feeding in ciliated protozoa: feeding rates and their ecological significance. Microb. Ecol. **6**:13–25.

23. **Fenchel, T.** 1982. Ecology of heterotrophic microflagellates. II. Bioenergetics and growth. Mar. Ecol. Prog. Ser. **8**:225–231.

24. **Fenchel, T.** 1982. Ecology of heterotrophic microflagellates. IV. Quantitative occurrence and importance as consumers of bacteria. Mar. Ecol. Prog. Ser. **9**:35–42.

25. **Fenchel, T., and T. H. Blackburn.** 1979. Bacteria and mineral cycling. Academic Press, London.

26. **Fenchel, T., and P. Harrison.** 1976. The significance of bacterial grazing and mineral cycling for the decomposition of particulate detritus, p. 285–300. *In* J. M. Anderson and A. Macfadyen (ed.), The role of terrestrial and aquatic organisms in decomposition processes. Blackwell Scientific Publications, Oxford.

27. **Habte, M., and M. Alexander.** 1975. Protozoa as agents responsible for the decline of *Xanthomonas campestris* in soil. Appl. Microbiol. **29**:159–164.

28. **Habte, M., and M. Alexander.** 1977. Further evidence for the regulation of bacterial populations in soil by protozoa. Arch. Microbiol. **113**:181–183.

29. **Hardin, G.** 1944. Physiological observations and their ecological significance: a study of the protozoan *Oikomonas termo.* Ecology **25**:192–201.

30. **Hobbie, J. E., R. J. Daley, and S. Jasper.** 1977. Use of Nuclepore filters for counting bacteria by fluorescence microscopy. Appl. Environ. Microbiol. **33**:1225–1228.

31. **Johannes, R. E.** 1965. The influence of marine protozoa on nutrient regeneration. Limnol. Oceanogr. **10**:434–442.

32. **Johannes, R. E.** 1968. Nutrient regeneration in lakes and oceans, p. 203–214. *In* M. R. Droop and E. J. Ferguson Wood (ed.), Advances in microbiology of the sea, vol. 1. Academic Press, Inc., New York.

33. **Knowles, F., and J. E. Watkin.** 1931. The assimilation and translocation of plant nutrients in wheat during growth. J. Agric. Sci. **2**:612–632.

34. **Larsson, U., and Å. Hagström.** 1979. Phytoplankton exudate release as an energy source for the growth of pelagic bacteria. Mar. Biol. **52**:199–206.

35. **Lofs-Holmin, A., and U. P. Boström.** 1982. Role of earthworms in carbon and nitrogen cycling, p. 160–176. *In* T. Rosswall (ed.), Ecology of arable land: the role of organisms in nitrogen cycling. Progress Report 1981. Swedish University of Agriculture, Uppsala.

36. **Lynch, J. M.** 1976. Products of soil microorganisms in relation to plant growth. Crit. Rev. Microbiol. **5**:67–107.

37. **Martin, J. K., and R. J. Kemp.** 1980. Carbon loss from roots of wheat cultivars. Soil Biol. Biochem. **12**:551–554.

38. **Miles, H. B.** 1963. Soil protozoa and earthworm nutrition. Soil Sci. **95**:407–409.

39. **Noland, L. F.** 1924. Factors influencing the distribution of fresh water ciliates. Ecology **6**:437–452.

40. **Old, K. M., and J. F. Darbyshire.** 1978. Soil fungi as food for giant amoebae. Soil Biol. Biochem. **10**:93–100.

41. **Pomeroy, L. R.** 1970. The strategy of mineral cycling. Annu. Rev. Ecol. Syst. **1**:171–190.

42. **Pomeroy, L. R.** 1974. The ocean's food web, a changing paradigm. BioScience **24**:499–504.

43. **Rowe, R., R. Todd, and J. Waide.** 1977. Microtechnique for most-probable-number analysis. Appl. Environ. Microbiol. **33**:675–680.

44. **Sauerbeck, D. R., and B. G. Johnen.** 1977. Root formation and decomposition during plant growth, p. 141–148. *In* Proceedings of the International Symposium on Soil Organic Matter Studies, Vienna, IAEA-SM-211/16, vol. I. International Atomic Energy Agency [Unipub], Vienna.

45. **Sayre, R. M.** 1973. *Theratromyxa weberi,* an amoeba predatory on plant-parasitic nematodes J. Nematol. **5**:258–264.

46. **Schnürer, J., M. Clarholm, and T. Rosswall.** 1982. Microorganisms, p. 77–88. *In* T. Rosswall (ed.), Ecology of arable land. The role of organisms in nitrogen cycling. Progress Report 1981. Swedish University of Agriculture, Uppsala.

47. **Schönborn, W.** 1965. Untersuchungen uber die Ernährung bodenbewohnender Thestacean. Pedobiologia **5**:205–210.

48. **Schönborn, W.** 1975. Ermittlung der Jahresproduktion von Boden-Protozoen. I. *Euglyphidae (Rhizopoda, Testacea).* Pedobiologia **15**:415–424.

49. **Sherr, B. F., E. B. Sherr, and T. Berman.** 1983. Grazing, growth and ammonium excretion rates of a heterotrophic microflagellate fed with four species of bacteria. Appl. Environ. Microbiol. **45**:1196–1201.

50. **Stotzky, G., and A. G. Norman.** 1963. Factors limiting microbial growth activities in soil. III. Supplementary substrate additions. Can. J. Microbiol. **10**:143–147.

51. **Stout, J. D.** 1980. The role of protozoa in nutrient cycling and energy flow. Adv. Microb. Ecol. **4**:1–50.

52. **Stout, J. D., and O. W. Heal.** 1967. Protozoa, p. 149–195. *In* A. Burges and F. Raw (ed.), Soil biology. Academic Press, Inc., New York.

53. **Umorin, P. P.** 1976. Relationships between bacteria and flagellates in destruction of organic matter. Zh. Obshch. Biol. **37**:831–835 (in Russian, English summary).

54. **Williams, P. J. leB.** 1981. Incorporation of microheterotrophic processes into the classical paradigm of the planktonic food web. Kiel. Meeresforsch. Sonderh. **5**:1–28.

55. **Wright, S. J. L., K. Redhead, and H. Maudsley.** 1981. *Acanthamoeba castellanii,* a predator of cyanobacteria. J. Gen. Microbiol. **125**:293–300.

Fungal Development, Predacity, and Recognition of Prey in Nematode-Destroying Fungi

BIRGIT NORDBRING-HERTZ AND HANS-BÖRJE JANSSON

Department of Microbial Ecology, University of Lund, S-223 62 Lund, Sweden

The nematode-destroying fungi are natural enemies of nematodes, capable of capturing, killing, and digesting them. The nematodes have a dual function in this relationship: they not only serve as prey but they also trigger fungal development in such a way that infective structures (traps or conidia) are formed.

Most of the investigations of these interrelationships have been carried out in the laboratory; there are few conclusive results from studies of the nematode-destroying fungi as predators and parasites of nematodes in soil. Several reviews of our present knowledge of the biology and ecology of these fungi have appeared during recent years (3–5, 18), and the potential of biological control of nematode pests by natural enemies has been excellently discussed (19).

In this paper we review our results on those fungi which attack free-living nematodes. The formation of infective structures is a prerequisite for these fungi to capture living nematodes. Consequently, predacity, defined as the ability to reduce nematode numbers in nature as well as in laboratory experiments, also depends on fungal growth and development (Fig. 1). Many environmental factors affect the development of the fungi and, indirectly, capture of nematodes and predacity. Furthermore, evidence of the ability of the fungi to recognize their prey has come from studies of the attraction of nematodes toward the fungi and, more directly, from the adhesion of nematodes to the infective structures and its molecular background.

FUNGAL GROWTH AND DEVELOPMENT

The nematode-destroying fungi are a heterogeneous group with respect to saprophytic/predacious ability. On the basis of their ability to form infective structures and their saprophytic/predacious ability, they were divided into three ecological groups. The first two groups contain the nematode-trapping (predatory) fungi, and the third group contains the endoparasitic fungi (13). Table 1 gives a brief description of the three groups.

The ability to attack living nematodes is restricted to a certain phase of the fungal development among the nematode-destroying fungi. In the endoparasitic species, conidia infect nematodes either by ingestion or by adhesion to the nematode cuticle. *Harposporium anguillulae* and *Meria coniospora* infect the nematodes by nonadhesive and adhesive conidia, respectively (Fig. 2C and F). In these fungi the nematode harbors the entire mycelium, conidiophores appearing outside the host after the nematode has been consumed (Fig. 2E). The more saprophytic species show a variety of adhesive or nonadhesive structures. *Arthrobotrys oligospora* forms adhesive networks (Fig. 2A) in which nematodes are trapped by adhesion (Fig. 2B). Adhesive knobs are used by *Dactylaria candida* (Fig. 2D).

The tendency to form trapping structures varies greatly among species of predatory fungi. Some form traps spontaneously in all environments, whereas others need special substances for induction of trap formation. Nematodes trigger development in these fungi which, in turn, use them as prey. The induction of trap formation by nematodes is a complex event, and the actual triggering mechanism is still obscure. Nematodes provide proteinaceous material (22, 30) and volatile compounds (6, 22, 27) which all play a part in the induction process. In the more saprophytic species like *A. oligospora*, small peptides with a high proportion of nonpolar and aromatic amino acid residues in combination with a dilute nutrient status cause heavy trap formation (21, 24). This has provided a useful laboratory system for comparing the characteristics of traps and hyphae on different levels of complexity (for review see reference 23a).

During these studies it became apparent that especially the initial phases of trap formation were strongly influenced by environmental factors such as nutrient level, colony age, moisture content, pH, etc. (e.g., 22). The induction of trap formation by living nematodes under suitable conditions was a rapid process (1 to 2 h) compared with peptide-induced trap formation (6 to 12 h) (22). Furthermore, the nematode-induced trap formation in some cases occurred rhythmically (17), implying the involvement of differing membrane characteristics of the fungal colony (18).

PREDACITY

In the laboratory most of the nematophagous fungi can be grown on ordinary substrates.

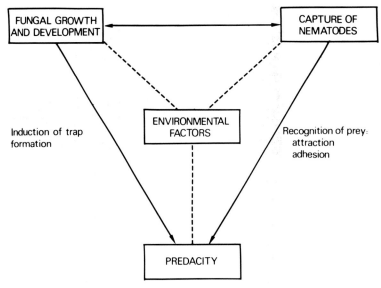

FIG. 1. Relationship of predacity to other factors.

When nematodes are added to an agar culture, all of them are rapidly captured and ingested by the fungi, provided that infective structures are present. In soil the fungi do not appear to completely destroy a nematode population (9, 12), as is the case with other successful microbial predators and parasites which do not exterminate their hosts (1, 2).

Different species of nematophagous fungi had different capacities to destroy nematodes (predacity) in a sterilized soil microcosm (12). The most predacious species killed about 90% of added nematodes in less than 2 days; other fungi destroyed very few nematodes in that system. The lowest predacity was shown by the network-forming fungi, followed by fungi with oth-

er types of traps, and the highest predacity was shown by the endoparasitic fungi (12; Table 1). Similar results were presented earlier by Cooke (9), who showed that nematode-trapping fungi with constricting ring traps were poor saprophytes in soil, but were able to decrease nematode numbers, in contrast to fungi with the network type of traps.

ATTRACTION

Predacity. The ability of the nematode-destroying fungi to attract different species of nematodes has been studied by use of an agar plate assay (13, 14). On the whole, 75% of the nematode-destroying fungi attracted the nema-

TABLE 1. Ecological groups of nematophagous fungi[a]

Characteristic	Nematode-trapping fungi		Endoparasitic fungi, group 3
	Group 1	Group 2	
Trapping organs	Adhesive networks; inducible by nematodes or with chemicals	Adhesive knobs, adhesive branches, (non)-constricting rings; spontaneously produced	Conidia, either adhering to nematode cuticles or ingested by nematodes
Growth pattern	Fast growing and relatively good saprophytes; weak predacious ability	Relatively slow-growing; weak saprophytes; great predacious ability	Mostly obligate parasites; very slow growing
Type species	*Arthrobotrys oligospora* *A. conoides*	*Dactylaria candida* *Monacrosporium cionopagum* *D. gracilis*	*Meria coniospora* *Harposporium anguillulae*

[a] From H. B. Jansson, Ph.D. dissertation, University of Lund, Lund, Sweden, 1982. Both predacity and attraction increase from group 1 to group 3.

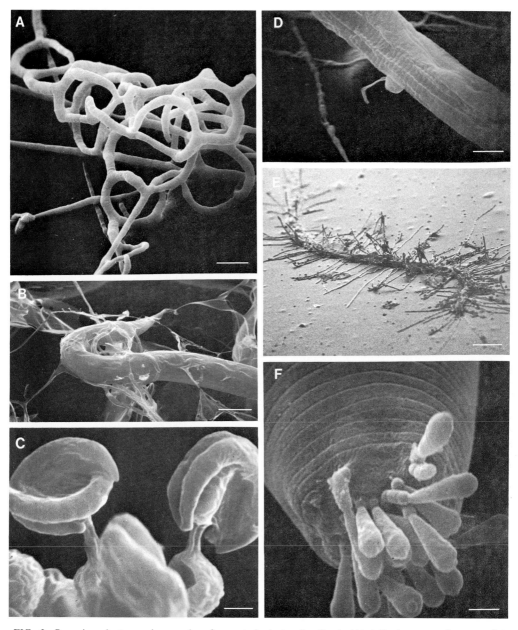

FIG. 2. Scanning electron micrographs of nematode-destroying fungi. (A) Adhesive network traps of *A. oligospora*. (Bar: 10 μm.) (B) Nematode captured in adhesive trap of *A. oligospora*. (Bar: 20 μm.) (C) Conidia of the endoparasitic fungus *H. anguillulae*. (Bar: 1 μm.) (D) Adhesive knob trap of *D. candida* adhered to nematode cuticle. (Bar: 5 μm.) (E) Nematode consumed by *M. coniospora*. (Bar: 50 μm.) (F) Nematode infected in the mouth region by conidia of *M. coniospora*. (Bar: 2 μm.) Figures 2A, 2D, and 2F are from references 18, 25, and 15, respectively. By courtesy of GIT Verlag Ernst Giebeler, W. de Gruyter & Co., and Society for General Microbiology.

todes. There were different attraction patterns for different nematode species: fungus-feeding nematodes were attracted to all fungi tested, whereas other plant-parasitic species were attracted to very few fungi (14). The various attraction patterns in different types of nematodes may represent a possible specificity in choice of prey by the nematode-destroying fungi.

When different types of nematophagous fungi

were compared with respect to their ability to attract nematodes, the most saprophytic nematophagous fungi (group 1) and also the non-nematophagous fungi tested showed the weakest attraction, and the endoparasitic fungi (group 3) showed the strongest attraction; the fungi of group 2 were in an intermediary position (13, 14; Table 1).

In the predacity experiments reported above, the more predacious group 2 and 3 fungi very rapidly destroyed the majority of the nematodes in the sterile soil. These fungi spontaneously produce traps and infective conidia, respectively. In group 1 fungi the traps are not produced spontaneously. Consequently, capture of nematodes cannot occur instantly. After a lag phase that allows the nematodes to induce trap formation, these fungi also start to destroy nematodes (12). Thus, there was a very good correlation between the ability to attract nematodes and predacity (12), which is indicated in Table 1. It therefore appears that the attraction reflects the dependence of the fungi on nematodes for nutrients (13).

Infective structures. The role of the adhesive infective structures in attracting nematodes became evident from the following studies.

The network-forming fungus *A. oligospora* can easily be manipulated to produce traps (21). A nematode-trapping fungus thus enters the predacious phase from a saprophytic one when traps are formed. *A. oligospora* with traps on the hyphae increased its ability to attract nematodes by a factor of two (12). This supports the hypothesis that a higher state of predacity increases the ability to attract nematodes.

Since capture organs of nematode-trapping fungi increase attraction intensity, it was argued that the corresponding infective structures in the endoparasites could be responsible for the strong attraction in these fungi. Only adhesive conidia which adhere to nematode cuticles attracted the nematodes, whereas nonadhesive conidia which are ingested by the worms and conidia of nematode-trapping fungi were not attractive (11). These results indicate two different infection strategies for the endoparasitic nematophagous fungi. In the fungi with nonadhesive conidia (e.g., *H. anguillulae*, Fig. 2C) which are swallowed by the bacterium-feeding nematodes, the conidia are ingested randomly together with food, e.g., bacteria and yeasts. On the other hand, the fungi with adhesive conidia (e.g., *M. coniospora*, Fig. 2F) actively attract their nematode hosts, and the conidia easily adhere to the nematode cuticle.

RECOGNITION OF PREY

As pointed out above, the presence of traps or infective conidia is a prerequisite for predacity, and the presence of these structures also increases attraction of nematodes to the fungi. The dependence on a certain developmental phase of the fungal mycelium for capture of nematodes is thus obvious, and this is also true for the recognition of prey on the molecular level (Nordbring-Hertz, in press). We believe that the capture process is initiated by a lectin-carbohydrate interaction between the fungal infective structure and the nematode surface. This initial recognition step is the signal leading to penetration and digestion of the nematode. Evidence for this view has come from biological–biochemical studies mainly on three different fungi: *A. oligospora*, *D. candida*, and *M. coniospora*, representing different saprophytic/predacious ability (Table 1) and different methods of attacking nematodes. The results are summarized in Table 2.

TABLE 2. Evidence of lectin–carbohydrate interaction in nematode-destroying fungi[a]

Characteristic	Lectin–carbohydrage recognition in:		
	A. oligospora	*D. candida*	*M. coniospora*
Capture of nematodes	No specificity	No specificity	Probably some specificity
Hapten inhibition of capture	GalNAc	2-Deoxyglucose	Sialic acid[b]
Demonstration of carbohydrate on nematode surface	GalNAc	ND	Sialic acid[b]
Binding of cells	Erythrocytes	*R. trifolii* 0403	ND
Inhibition of capture by treatment of fungus with trypsin and glutaraldehyde	+	+	+
Inhibition of capture by treatment of nematode with lectins	ND	ND	+ (limulin)
Isolation and characterization of carbohydrate-binding protein	+ (GalNAc specific; Ca^{2+} dependent; subunit molecular weight 20,000)	ND	ND

[a] ND, Not done.
[b] N-Acetyl-neuraminic acid.

The background to these studies was that nematodes, added to a trap-containing culture of *A. oligospora*, are immediately captured. The experience of this trapping was further substantiated by studies of the interaction on the ultra-structural level (28). We suggested that, if a lectin–carbohydrate interaction is involved in adhesion, then preexposure of the lectin-carrying surfaces to specific carbohydrates would prevent attachment of the nematode to the traps (26). Using a dialysis membrane technique (23), we treated a trap-containing colony with different carbohydrate solutions before additions of nematodes. Only *N*-acetyl-D-galactosamine (GalNAc) at 20 mM inhibited capture substantially, suggesting the involvement of a fungal carbohydrate-binding protein. GalNAc residues were demonstrated on the surface of the nematode by the aid of a peroxidase–lectin conjugate. A model prey, consisting of erythrocytes, accumulated around the traps, blood group A with GalNAc terminally showing a tendency to attach more easily than B or O (26). The protein nature of the trap binding structure was further demonstrated by the treatment of traps with trypsin and glutaraldehyde, again leading to inhibition of capture in these in vivo experiments (25; Table 2).

In vitro experiments were performed in which intact fungal material was surface labeled with [^{125}I]iodosulfanilic acid and then homogenized. The labeled homogenate bound to both GalNAc-Sepharose and to nematodes, giving further evidence for the presence of a carbohydrate-binding protein (20). The conclusive evidence for a trap lectin, however, was demonstrated recently. Liquid culture-grown fungal material with a high traps/hyphae ratio was surface labeled as before and homogenized. The labeled fungal homogenate was first purified by affinity chromatography on GalNAc-Sepharose. The GalNAc-specific protein, eluted from this column, bound to Ca^{2+} on a metal chelate affinity column (8). The carbohydrate-binding protein thus obtained was GalNAc specific and Ca^{2+} dependent, and it had a subunit molecular weight of ~20,000, as demonstrated by sodium dodecyl sulfate electrophoresis and confirmed by autoradiography (C. A. K. Borrebaeck, B. Mattiasson, and B. Nordbring-Hertz, unpublished data).

The results show clearly that *A. oligospora*, a fungus which is a fairly good saprophyte with modest ability to attract nematodes and to reduce their numbers, exhibits a specificity on the molecular level. Subsequent hapten inhibition experiments with other species and strains have shown different carbohydrate specificities (25; D. Ackroyd and W. D. Rosenzweig, personal communication) or have failed to show a speci-ficity in binding simple sugars. The reason for these varying specificities is not known, but they might reflect a poor host specificity of predatory fungi.

In *D. candida*, a very efficient nematode trapper under laboratory conditions, capture of nematodes was inhibited by treatment with trypsin and glutaraldehyde as in *A. oligospora*, indicating the protein nature of the binding structure. In hapten inhibition experiments, 2-deoxyglucose was the only simple carbohydrate that inhibited capture. *Rhizobium trifolii* 0403 binds to clover roots through 2-deoxyglucose residues (for review see 10). There was a tendency in *R. trifolii* to attach to traps of *D. candida* (25, F. B. Dazzo, personal communication). However, further evidence for a carbohydrate-binding protein in *D. candida* is still lacking (Table 2).

As shown in Table 1, there is an increasing attraction and predacity along with an increased need for nematodes as nutrient. There are also indications of increased specificity in the endoparasites. Evidence for this has come from studies on adhesion of conidia of *M. coniospora* to specific sites on the nematode cuticle. The infection sites are different in bacterium-feeding and plant-parasitic nematodes.

In *M. coniospora* the adhesion of conidia takes place preferentially in the mouth region of bacterium-feeding nematodes (11) and more specifically at the sensory organs (15). Plant-parasitic nematodes were infected also in other parts of the cuticle. The sensory organs are situated around the mouth of the nematodes and at the tail of male specimens and contain the receptors for chemoattraction (e.g., 31). When nematode sensory organs were blocked with conidia, the nematodes lost their ability to be attracted. As in *A. oligospora* and *D. candida*, treatment of the conidia of *M. coniospora* with trypsin and glutaraldehyde reduced adhesion, indicating the protein nature of the structure involved in adhesion of conidia to the nematode cuticle (Table 2).

As in the predatory fungi, adhesion was inhibited by treatment of the infective structures with sugar haptens. In the case of *M. coniospora* only the sialic acid *N*-acetyl-neuraminic acid inhibited conidial adhesion, indicating that sialic acids might be located on nematode sense organs and be responsible for the adhesion. When the nematodes were treated with the enzyme sialidase, which splits off sialic acids, there was a reduced adhesion (15), pointing to the importance of sialic acids in this fungus–nematode interaction. Furthermore, treatment of the nematodes with the lectin limulin, which binds specifically to sialic acids, also reduced adhesion (Table 2). Other lectins tested did not do so (H. B. Jansson and B. Nordbring-Hertz, J. Gen. Microbiol., in press). These results indicate that a carbohydrate-binding protein specific for sialic acid is present on the adhesive conidia of *M. coniospora*.

CONCLUDING REMARKS

The nematode-destroying fungi attack and consume a wide range of nematodes (14). This lack of host specificity within a group of fungi with such different strategies to capture prey increases the chances of survival and reduces the risk of self-elimination (1). Furthermore, in the more saprophytic species the possibility of switching from a saprophytic mode of life to a predacious one gives these fungi a competitive advantage.

In spite of the poor host specificity, there is a recognition on the molecular level in some species of the fungi in that carbohydrate-binding proteins present on the fungal structures bind to carbohydrates on the nematode surface. In *A. oligospora*, the presence of a lectin recognition is considered as just one step in a series of signal and response reactions (7). Information on the further events in the interaction process comes from studies on the biochemical, cytochemical, and ultrastructural levels (29).

Also, in the endoparasite *M. coniospora* there seems to be a lectin-mediated interaction between the fungus and the nematode. Conidia, adhering to the chemotactic apparatus, interfered with the nematode's ability to be attracted. This connection between attraction and adhesion is linked by the presence of sialic acid residues on the nematode sensory organs (15; Jansson and Nordbring-Hertz, in press). As in *A. oligospora* the interaction between a carbohydrate on the nematode surface and a fungal protein appears to be the signal to the further events in the capture process. Taken together, our results contribute one possible explanation of the function of lectins in nature.

The possible increased host specificity in the endoparasitic fungi might be due to the differing distribution of carbohydrates on nematode cuticles. This opens up possibilities for studies of the more specific associations represented by the fungi that parasitize cyst nematodes (16, 19).

This work was supported by grants from the Swedish Natural Science Research Council.

LITERATURE CITED

1. **Alexander, M.** 1981. Why microbial predators and parasites do not eliminate their prey and hosts. Annu. Rev. Microbiol. **35**:113–133.
2. **Baker, K. F., and R. J. Cook.** 1974. Biological control of plant pathogens. W. H. Freeman, San Francisco.
3. **Barron, G. L.** 1977. The nematode-destroying fungi. Topics in mycobiology, no 1. Canadian Biological Publications Ltd., Guelph, Ontario, Canada.
4. **Barron, G. L.** 1981. Predators and parasites of microscopic animals, p. 167–200. *In* G. T. Cole, and B. Kendrick

(ed.). Biology of conidial fungi, vol. 2. Academic Press, Inc., New York.
5. **Barron, G. L.** 1982. Nematode-destroying fungi, p. 533–552. *In* R. G. Burns and J. H. Slater (ed.), Experimental microbial ecology. Blackwell Scientific Publications, London.
6. **Bartnicki-Garcia, S., J. Eren, and D. Pramer.** 1964. Carbon dioxide-dependent morphogenesis in *Arthrobotrys conoides*. Nature (London) **204**:804.
7. **Bauer, W. D.** 1981. Infection of legumes by Rhizobia. Annu. Rev. Plant Physiol. **32**:407–449.
8. **Borrebaeck, C. A. K., B. Lönnerdahl, and M. E. Etzler.** 1981. Metal chelate affinity chromatography of the *Dolichos biflorus* seed lectin and its subunits. FEBS Lett. **130**:194–196.
9. **Cooke, R. C.** 1963. Ecological characteristics of nematode-trapping hyphomycetes. I. Preliminary studies. Ann. Appl. Biol. **52**:431–437.
10. **Dazzo, F. B.** 1980. Adsorption of microorganisms to roots and other plant surfaces, p. 253–316. *In* G. Bitton and K. C. Marshall (ed.), Adsorption of microorganisms to surfaces. John Wiley & Sons, Inc., New York.
11. **Jansson, H. B.** 1982. Attraction of nematodes to endoparasitic nematophagous fungi. Trans. Br. Mycol. Soc. **79**:25–29.
12. **Jansson, H. B.** 1982. Predacy by nematophagous fungi and its relation to the attraction of nematodes. Microb. Ecol. **8**:233–240.
13. **Jansson, H. B., and B. Nordbring-Hertz.** 1979. Attraction of nematodes to living mycelium of nematophagous fungi. J. Gen. Microbiol. **112**:89–93.
14. **Jansson, H. B., and B. Nordbring-Hertz.** 1980. Interactions between nematophagous fungi and plant-parasitic nematodes: adhesion, induction of trap formation, and capture. Nematologica **26**:383–389.
15. **Jansson, H. B., and B. Nordbring-Hertz.** 1983. The endoparasitic nematophagous fungus *Meria coniospora* infects nematodes specifically at the chemosensory organs. J. Gen. Microbiol. **129**:1121–1126.
16. **Kerry, B.** 1980. Biocontrol: fungal parasites of female cyst nematodes. J. Nematol. **12**:253–259.
17. **Lysek, G., and B. Nordbring-Hertz.** 1980. An endogenous rhythm of trap formation in the nematophagous fungus *Arthrobotrys oligospora*. Planta **152**:50–53.
18. **Lysek, G., and B. Nordbring-Hertz.** 1983. Die Biologie nematodenfangender Pilze. Forum Mikrobiol. **6**:201–208.
19. **Mankau, R.** 1980. Biological control of nematode pests by natural enemies. Annu. Rev. Phytopathol. **18**:415–440.
20. **Mattiasson, B., P. A. Johansson, and B. Nordbring-Hertz.** 1980. Host-microorganism interaction. Studies on the molecular mechanisms behind the capture of nematodes by nematophagous fungi. Acta Chem. Scand. Ser. B **34**:539–540.
21. **Nordbring-Hertz, B.** 1973. Peptide-induced morphogenesis in the nematode-trapping fungus *Arthrobotrys oligospora*. Physiol. Plant. **29**:223–233.
22. **Nordbring-Hertz, B.** 1977. Nematode-induced morphogenesis in the predacious fungus *Arthrobotrys oligospora*. Nematologica **23**:443–451.
23. **Nordbring-Hertz, B.** 1983. Dialysis membrane technique for studying microbial interactions. Appl. Environ. Microbiol. **45**:290–293.
23a.**Nordbring-Hertz, B.** 1983. Mycelial development and lectin-carbohydrate interactions in nematode-trapping fungi. *In* D. H. Jennings and A. D. M. Rayner (ed.), Ecology and physiology of the fungal mycelium. Cambridge University Press, New York.
24. **Nordbring-Hertz, B., and C. Brinck.** 1974. Qualitative characterization of some peptides inducing morphogenesis in the nematode-trapping fungus *Arthrobotrys oligospora*. Physiol. Plant. **31**:59–63.
25. **Nordbring-Hertz, B., E. Friman, and B. Mattiasson.** 1982. A recognition mechanism in the adhesion of nematodes to nematode-trapping fungi, p. 83–90. *In* T. C. Bøg-Hansen

(ed.), Lectins—biology, biochemistry and clinical biochemistry, vol. 2. W. de Gruyter, Berlin.

26. **Nordbring-Hertz, B., and B. Mattiasson.** 1979. Action of a nematode-trapping fungus shows lectin-mediated host-microorganism interaction. Nature (London) **281:**477–479.

27. **Nordbring-Hertz, B., and G. Odham.** 1980. Determination of volatile nematode exudates and their effects on a nematode-trapping fungus. Microb. Ecol. **6:**241–251.

28. **Nordbring-Hertz, B., and M. Stålhammar-Carlemalm.** 1978. Capture of nematodes by *Arthrobotrys oligospora*, an electron microscope study. Can. J. Bot. **56:**1297–1307.

29. **Nordbring-Hertz, B., M. Veenhuis, and W. Harder.** 1984. Dialysis membrane technique for ultrastructural studies of microbial interactions. Appl. Environ. Microbiol. **47:**195–197.

30. **Pramer, D., and S. Kuyama.** 1963. Symposium on biochemical bases of morphogenesis in fungi. II. Nemin and the nematode-trapping fungi. Bacteriol. Rev. **27:**282–292.

31. **Ward, S.** 1978. Nematode chemotaxis and chemoreceptors, p. 141–168. *In* G. L. Hazelbauer (ed.), Taxis and behavior. Receptors and recognition, series B, vol. 5. Chapman and Hall, London.

Bdellovibrio as a Predator

M. SHILO

Division of Microbial and Molecular Ecology, The Institute of Life Sciences, The Hebrew University of Jerusalem, Jerusalem, Israel

Members of the genus *Bdellovibrio* are obligate predators, requiring for growth and multiplication gram-negative bacteria, which serve as their prey. The predation process involves attachment to the prey cell, penetration into its periplasmic space, elongation and multiple fission, and swarming out after lysis of the ghost of the prey cell (reviewed in 29, 35, 36, 39, 45, 57).

Bdellovibrio species are widespread in aquatic and terrestrial ecosystems, yet little is known about their ecology and role in nature and their importance in affecting the dynamics of microbial populations. Despite the great interest in their unique life cycle, the few research groups involved in their study have investigated in detail only a few bdellovibrio strains among the numerous isolates obtained from different ecosystems.

Taxonomically, all bdellovibrios are presently considered to belong to a single genus containing several species (4). Their classification into one group was based on morphological characteristics, as well as their predatory capacity and unique life cycle. However, these are all properties inextricably connected with the predatory mode of life.

A more detailed study of several strains isolated from different ecosystems shows that, in spite of the general similarities, profound differences exist among different bdellovibrio types. The guanine-plus-cytosine ratios of their DNA differ widely (50% for *B. bacteriovorus*, 42 to 43.5% for *B. starrii* and *B. tolpii*, and 33% for the marine strains). Moreover, the marine strains have a unique and absolute requirement for the four cations prevalent in the marine environment (Na^+, K^+, Mg^{2+}, and Ca^{2+}) (4, 21, 22). Thus, what in fact we seem to be dealing with is a heterogeneous group which will have to be reclassified in the future.

For such an understanding, the bdellovibrio group might be considered a conglomerate resulting from the convergent evolutionary development of propensities toward a predatory mode of life in different organisms which come to share in common characteristics necessary for predation of bacterial prey cells (57).

The bdellovibrio group is characterized by a biphasic developmental cycle in which freely motile cells of the attack phase and intraperiplasmic, multiplication-phase cells alternate. The attack-phase cell is incapable of DNA synthesis, whereas the multiplication-phase cell shows DNA replication, elongation, and fission, and produces within the periplasmic space of the prey cell motile, attack-phase progeny which swarm out and are capable themselves of attacking new prey cells. Attack of the prey cell and multiplication within its periplasmic space are both necessary and sufficient for completion of the developmental cycle of *Bdellovibrio* cells (1). The process can take place without addition of any external nutrients: even heat- or UV-killed prey cells suffice to allow completion of the cycle (55).

The bacterial prey of *Bdellovibrio* spp. is usually only slightly larger in mass and size than the predator cell. Yet it comprises an extremely minute nutrient package sufficient for completion of the growth cycle of the bdellovibrio organism and progeny production. This explains the extreme efficiency in energy and building-block conservation in the biosynthesis of the bdellovibrio cell, as shown by the requirement for tight regulation in the timing of prey cell breakdown and the high Y_{ATP} (15, 30). Several unique mechanisms which partly explain this efficient recycling of prey components by bdellovibrios have been unraveled by the work of Rittenberg and his co-workers. These involve uptake and utilization of nucleotides in DNA and RNA synthesis (26, 30), which preserves the energy-rich phosphate bond of the nucleotide, incorporation of unchanged fatty acids and lipids in the membranes of the bdellovibrio cells (20), and the use of lipid A components of the membrane lipopolysaccharide of the prey in the synthesis of the bdellovibrio cell (24, 25). Moreover, there is the incorporation, recently shown, of unchanged, prey-derived, major outer membrane protein (OmpF) into the outer membrane matrix of the *Bdellovibrio* cells growing in *Escherichia coli*, *Salmonella typhimurium*, or *Klebsiella pneumoniae* prey cells (7, 13).

A strict regulation controls the attachment and recognition of the prey cell (56, 57) and the penetration of the peptidoglycan (42), lipopolysaccharide (24, 25), and protein layers of the cell wall of the prey cell. An additional unique feature of the intraperiplasmic phase of *Bdellovibrio* cells, namely, their capacity to direct the biosynthetic modification of the outer envelope of the invaded prey cell, has been described by Thomashow and Rittenberg (42). The penetrating bdellovibrio cell affects the chemical composition of the cell wall of the invaded prey cell by deacetylation of the entire prey cell wall (43). The covalent acylation of the peptidoglycan layer (44) strengthens and stabilizes the bdelloplast. The purpose of these changes is yet unclear, but very likely securing of the bdelloplast as a growth chamber, through stabilization of permeability and osmotic properties, as well as protection of the cell from multiple attack (45), is involved.

Other special enzyme systems of *Bdellovibrio* cells seem to involve the efficient breakdown of prey cell components (14, 23, 31) and their recycling for bdellovibrio synthesis and growth. Perisemic signals control and trigger the replication of DNA (12, 17, 28) and the elongation and multiple fission of the multiplication-phase cell. The nature of the prey cell, its size, and the state of breakdown of its protoplast were found to be important in controlling the bdellovibrio development and its progeny number (18).

In some bdellovibrio strains, capable of bdellocyst formation (2, 46, 47), the state of the penetrated prey cell seems to regulate the bdellovibrio developmental events. A starved prey induces the sequence leading to cyst formation, whereas production of multiplication-phase cells seems to occur only in well-fed prey cells (48).

Our understanding of the interaction of *Bdellovibrio* cells within prey cells is based on light microscope and electron microscope studies of the complex sequence of penetration into the prey cell, intraperiplasmic growth and fission, and swarming out by the bdellovibrio progeny (1, 3, 33). The main phases of the life cycle and the different development alternatives are depicted schematically in Fig. 1. The kinetics of attachment and penetration stages in the bdellovibrio life cycle can be measured by use of differential centrifugation and filtration combined with radioactive labeling, which separate the free, attack-phase cells from bdellovibrio cells attached to prey or penetrated into the periplasmic space of prey cells (54, 55). By use of different mutant lipopolysaccharide sequences of prey bacteria, the specificity of bdellovibrio attachment could be analyzed (56).

Before the regulatory mechanisms controlling bdellovibrio growth and its developmental cycle could be studied, separation of the two alternating phases was required. Furthermore, experimental means to obtain transformation under controlled conditions and in the absence of prey cells had to be developed. Two different approaches have been used experimentally to transform one phase to the other. Both approaches lead essentially to similar results.

The first approach, which involves synchronous transformation of the free, attack-phase cells into the multiplication-phase cells in rich medium in the absence of prey, is achieved by addition of a macromolecular prey component (12, 17, 28). The active factor obtained from *E. coli* cells had a molecular weight of 50,000 (28) and was rapidly inactivated by both RNase and pronase (17). When added to an axenic culture of prey-dependent *Bdellovibrio* cells, this factor transformed motile attack-phase cells into multiplication-phase cells and fully simulated the transformation observed in the intraperiplasmic growth up to the fission stage. Thus, the factor triggered loss of the flagellum uptake of [^{14}C]thymidine, DNA synthesis, and cell elongation, but inhibited fission. However, fission was achieved when the active factor was removed at the elongation stage by dilution, by centrifugation, or by use of enzymes such as RNase (17). The newly formed, motile bdellovibrio progeny cells were capable of immediately attacking prey cells. Addition of the *E. coli* factor allowed bdellophage replication, which takes place only after bdellovibrio penetration into prey cells in the axenic, bdellophage-infected bdellovibrio (52).

The second approach was based on the premature release of intraperiplasmic multiplication-phase cells by lysis of the bdelloplast wall from without, achieved by use of the lytic enzyme produced by the *Bdellovibrio* cell in the presence of EDTA at the termination of its intraperiplasmic cycle (32).

With both these methods, it has been clearly demonstrated that only the multiplication-phase cell is capable of thymidine uptake from the milieu and of DNA replication.

The high frequency of mutations of prey-independent predatory *Bdellovibrio* species (6, 34, 37, 53) has afforded a powerful tool in the study of bdellovibrio physiology and in understanding the absolute prey dependence of *Bdellovibrio* sp. in nature. Many of these mutants can grow in rich media in the absence of prey, but have retained the capability of attacking and preying on other bacteria. Various phases in the growth cycle, such as elongation, septation, and multiple fission, could be studied in synchronized axenic cultures of such mutants. The action of factors which trigger division (8, 51) and of inhibitors specifically affecting septation and multiple fission of *Bdellovibrio* cells, such as the antibiotic virginiamycin S (51), could be shown with these mutants.

These facultative mutants, though easily obtained at high mutation frequencies in many of the bdellovibrio strains, seem to be limited to laboratory conditions; they have never been found in nature. In competition experiments in

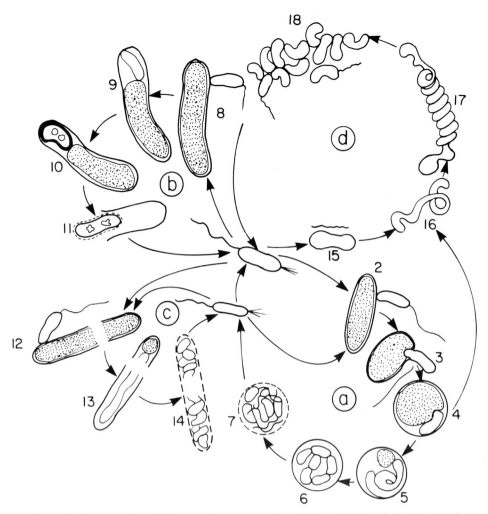

FIG. 1. Life cycle of *Bdellovibrio* spp. (a) Growth of *Bdellovibrio* spp. in a normal-sized prey bacterium such as *E. coli*. The free attack-phase bdellovibrio cell is a small, curved rod bearing a single sheathed flagellum at one pole of the cell and a number of fibers at the opposite pole (stage 1). It may elongate to a limited extent in its free form before attaching to the host (stage 2). After attachment, the *Bdellovibrio* cell penetrates through a pore formed in the prey cell wall, losing its flagellum in the process (stage 3). The pore is smaller than the diameter of the *Bdellovibrio* cell, and the latter becomes constricted as penetration proceeds (Varon et al., submitted for publication). The infected host cell is converted into a swollen spherical body, the "bdelloplast" (stage 4) (33, 40). The envelope of the prey cell at this stage differs from that of the uninfected prey in its morphology (38) and composition (43). According to Abram and Davis (1), the penetrating *Bdellovibrio* cell is attached to the protoplast of the infected prey. This assumption may be supported by scarlike structures on the protoplast observed by scanning electron microscopy (38). Intraperiplasmic growth and elongation of the bdellovibrio multiplication-phase cells into a C- or helical-shaped filament (33, 40) follow stage 5. During the growth phase, there is progressive shrinkage of the prey protoplast. Finally, the long bdellovibrio filament divides by multiple fission (stage 6); each resulting progeny cell acquires a flagellum during or just after division (stage 7). Shortly before lysis of the ghost has been completed, the bdellovibrio progeny can microscopically be seen swimming around inside the now-empty bdelloplast. The newly released progeny (stage 1) can immediately attach to new prey cells. (b) Cyst formation and germination of *Bdellovibrio* sp. W in *Rhodospirillum rubrum*. Under special conditions (starvation), *Bdellovibrio* sp. W enters a cycle of cyst development (47). After attachment (stage 8) and penetration (stage 9), the encysting bdellovibrio organism increases in size, accumulates inclusion bodies, and becomes kidney shaped and surrounded by a heavy outer layer (stage 10). During germination (stage 11), the outer wall is broken down, the inclusion material changes shape, and the germinant elongates, acquires a flagellum, and is transformed into the attack-phase, free bdellovibrio cell (stage 1). (c) Development of *Bdellovibrio* cells in filamentous, multinucleate *E. coli*. When a bdellovibrio cell infects a filamentous, multinucleate bacterium (stage 12), the prey does not become spherical (18), though a local "ballooning" may be seen (40). The predator elongates into a very long filament (stage 18) and divides upon exhaustion of the host protoplast. The number of progeny is in this case much higher (stage 14) than that obtained in a normal-sized host. (d) Development in rich medium of axenic cultures of prey-independent bdellovibrio mutants (3) and prey-dependent bdellovibrio strains after addition of prey cell factor (17, 28). When bdellovibrios are grown in the absence of prey cells under special conditions (rich medium, addition of prey cell factor), the bdellovibrio cell loses its flagellum (stage 15), elongates to a filament (stages 16 and 17), and undergoes multiple fission to form attack-phase progeny (stage 18). Transfer of released multiplication-phase cells (32) into the rich medium alone allows for development to attack-phase cells (arrow from stage 4 to stage 16).

continuous culture with the prey-dependent, wild-type *Bdellovibrio* in the presence of prey (49, 50), mutants were always inferior and were overgrown by the wild type. It was, therefore, suggested (6, 36) that prey dependence involves control of initiation of growth only when triggered by attachment of the predator cell to a suitable prey cell. Thus, only the wild-type *Bdellovibrio* is restricted to conditions in which completion of the cycle is secured, and there is no possibility of incomplete development triggered by transient nutrient fluxes. The facultative mutant, on the other hand, will be triggered to lose irreversibly its flagellum and to start growth even under conditions insufficient for completion of the cycle.

Studies of the interaction of *Bdellovibrio* with its prey in nature have been difficult as a result of the lack of methods for following the predator's activity in situ. An easy and extremely sensitive method which overcomes these obstacles involves the attack of marine *Bdellovibrio* cells on luminous marine bacterial prey (21, 22, 59). It consists of measurement of light emission, which is temporarily enhanced after attachment of the bdellovibrio cell on the *Photobacterium leignathii* prey cell. This period is followed by a rapid decay of light emission after penetration into the periplasmic space. It was possible to monitor directly the bdellovibrio–prey cell interaction and to develop a mathematical model fitting the Lotka-Voltera model. The model seems to indicate that chance collision with no chemotaxis occurs and that only 3% of the collisions between bdellovibrio cells and prey cells lead to penetration. Calculations based on these experiments showed that 7×10^5 prey cells per ml are required to obtain stable predator and prey populations. Using the *Bdellovibrio–Photobacterium* system, various parameters affecting the interaction were established, including the effect of external pollutants and inhibitors (58), predator and prey densities (M. Varon, M. Fine, and A. Stein, submitted for publication), and the relative affinity of different prey organisms for bdellovibrio attack (49).

The recent introduction of the genes coding for the subunits of luciferase into *E. coli* and the amplification of light emission by introducing lambda phage promoter now make it possible to extend the marine *Bdellovibrio–Photobacterium* system to *Bdellovibrio* species from soil and freshwater ecosystems with luminous *E. coli* used as prey cells (9).

The widespread distribution of *Bdellovibrio* spp. not only in the nutrient-rich environments, such as sewage treatment plants, but also in oligotrophic environments in oceans, lakes, rivers, and soil led to the suggestion that the predatory way of life could be an escape mechanism from a nutrient-poor environment into the "cozy" intraperiplasmic space of the prey cells (29, 57). Penetration into the prey cell affords a sufficient nutrient supply for bdellovibrio multiplication and also protects bdellovibrio cells from the lethal effect of photooxidation (11), from attack of bdellophages (53), and, probably, from inhibitory pollutants in the environment (58).

Studies carried out in batch culture (19), as well as those in continuous culture (10), showed that successful continuous maintenance of the bdellovibrio population required a minimal prey density of more than 10^5 cells per ml.

Such density allows the number of collisions between predator and prey sufficient for successful penetration. This requirement is not met in most oligotrophic environments. Furthermore, when a suspension of prey cells already penetrated by predator cells was diluted to a low density, the intraperiplasmic bdellovibrio multiplication did not take place. This is possibly a result of slow leakage of a critical component and its dissipation below threshold level, as indicated by the protective effect of spermine (M. Varon, M. Fine, and A. Stein, Arch. Microbiol., in press).

The discrepancy between the high prey density requirement and the ubiquitous presence of bdellovibrio cells in oligotrophic conditions can possibly be explained by the suggestion that *Bdellovibrio* sp. multiplies at interfaces, such as the air–water or the benthic sediment–water interfaces, or on the surface of suspended particles (57) where there are high prey concentrations. All of these interfaces are characterized by relatively rich bacterial communities, with numbers many orders of magnitude higher than in their surroundings. Indeed, in Chesapeake Bay, *Bdellovibrio* numbers were highest in shoreline samples and in winter were high only at the soil–water interface (60).

Still other considerations rule out the possibility for bdellovibrio development in the oligotrophic water column as a whole. These are the absence of an effective search mechanism for prey (57), such as chemotaxis, the requirement for numerous collisions for penetration to take place (59), and the fact that the search period must be brief since the energy reserves of attack-phase bdellovibrio cells are rapidly exhausted (16).

An analogous situation indicating that predators (flagellates and ciliates) in oligotrophic environments may be confined to microniches at interfaces having relatively high prey populations has recently been described (5). Marine snow, as well as detrital aggregates, was shown to concentrate 100-fold or even a 1,000-fold the prey cells adhering to particles. Under these conditions, predators were concentrated in numbers far above that in the surrounding water column (5).

The significance of interface surfaces for growth in predator–prey systems in continuous culture has been confirmed experimentally with *Tetrahymena* sp. and its bacterial prey. Here, silicon treatment of the growth vessel prevented the wall growth of the bacteria, which in turn affected predator growth (27). In other predator–prey systems studied in controlled chemostat experiments, *Tetrahymena* sp. fed on bacteria was shown to reflect a strategy of exploitation of patchy transient environments (41), and bacteriovorous ciliates adapted to niches in which they could prey on temporally or spatially patchy resources (10).

Experiments in continuous cultures of luminous bacteria at low density (10^4 cells per ml) with marine *Bdellovibrio* spp. often showed that low, steady-state bdellovibrio populations could be sustained for long periods (Varon et al., submitted for publication) in contrast to what was predicted from experimental models in batch conditions (19, 59). This would be expected if wall growth and irregularities in the chemostat vessel's surface allowed locally dense growth of prey cell populations.

The role of *Bdellovibrio* spp. in controlling or affecting the dynamics of communities of gram-negative bacteria in nature is still controversial (10). And yet, the fact that *Bdellovibrio* spp. are ubiquitous in aquatic and terrestrial ecosystems, together with their obligate dependence on prey cells, must mean that large numbers of prey cells are decomposed by bdellovibrio cells under many and various environmental conditions. One reason for the difficulties encountered in understanding the role of bdellovibrios in nature may be that the wrong environment has been sampled and measured. Normally, investigators have focused on what happens in the water column of oceans, lakes, or rivers, whereas the really relevant niches for bdellovibrio activity may well be located at the various interfaces.

LITERATURE CITED

1. **Abram, D., and B. K. Davis.** 1970. Structural properties and features of parasitic *Bdellovibrio bacteriovorus*. J. Bacteriol. **104:**948–965.
2. **Burger, A., G. Drews, and R. Ladwig.** 1968. Wirtskreise und Infektionscyclus eines neu isolierten *Bdellovibrio bacteriovorus* Stammes. Arch. Mikrobiol. **61:**261–279.
3. **Burnham, J. C., T. Hashimoto, and S. F. Conti.** 1968. Electron microscopic observations on the penetration of *Bdellovibrio bacteriovorus* into gram-negative bacterial hosts. J. Bacteriol. **96:**1366–1381.
4. **Burnham, J. C., and J. Robinson.** 1974. Families and genera of uncertain affiliation: genus *Bdellovibrio*, p. 212–214. *In* R. E. Buchanan and N. E. Gibbons (ed.), Bergey's manual of determinative bacteriology, 8th ed. The Williams & Wilkins Co., Baltimore.
5. **Caron, D. A., P. G. Davis, L. P. Madin, and J. McN.**

Sieburth. 1982. Heterotrophic bacteria and bacteriovorous protozoa in oceanic macroaggregates. Science **218:**795–797.
6. **Diedrich, D. L., C. F. Denny, T. Hashimoto, and S. F. Conti.** 1970. Facultative parasitic strain of bdellovibrio bacteria. J. Bacteriol. **101:**989–996.
7. **Diedrich, D. L., C. A. Portnoy, and S. F. Conti.** 1983. *Bdellovibrio* possesses a prey-derived OmpF protein in its outer membrane. Curr. Microbiol. **8:**51–56.
8. **Eksztejn, M., and M. Varon.** 1977. Elongation and cell division in bdellovibrio bacteria. Arch. Microbiol. **114:**175–181.
9. **Engelbrecht, J., K. Nealson, and M. Silverman.** 1983. Bacterial bioluminescence: isolation and genetic analysis of functions from *Vibrio fischeri*. Cell **32:**773–781.
10. **Fenchel, T. M., and B. B. Jorgensen.** 1977. Detritus food chains of aquatic ecosystems: the role of bacteria, p. 1–58. *In* M. Alexander (ed.), Advances microbial ecology, vol. 1. Plenum Press, New York.
11. **Friedberg, D.** 1977. Effect of light on *Bdellovibrio bacteriovorus*. J. Bacteriol. **131:**399–404.
12. **Friedberg, D.** 1978. Growth of host dependent *Bdellovibrio* in host cell free system. Arch. Microbiol. **116:**185–190.
13. **Guerrini, F., V. Romano, A. Bazzocchi, and M. Valenzi.** 1980. Parasitic modification of the membrane proteins of *E. coli* and other gram-negative hosts to infection by *Bdellovibrio bacteriovorus*. Atti Assoc. Genet. Ital. **26:**175–177.
14. **Hespell, R. B., G. F. Miozzari, and S. C. Rittenberg.** 1975. Ribonucleic acid destruction and synthesis during intraperiplasmic growth of *Bdellovibrio bacteriovorus*. J. Bacteriol. **123:**481–491.
15. **Hespell, R. B., R. A. Rosson, M. F. Thomashow, and S. C. Rittenberg.** 1973. Respiration of *Bdellovibrio bacteriovorus* strain 109J and its energy substrates for intraperiplasmic growth. J. Bacteriol. **113:**1280–1288.
16. **Hespell, R. B., M. F. Thomashow, and S. C. Rittenberg.** 1974. Changes in cell composition and viability of *Bdellovibrio bacteriovorus* during starvation. Arch. Microbiol. **97:**313–327.
17. **Horowitz, A. T., M. Kessel, and M. Shilo.** 1974. Growth cycle of predaceous bdellovibrios in a host-free extract system and some properties of the host extract. J. Bacteriol. **117:**477–480.
18. **Kessel, M., and M. Shilo.** 1976. Relationship of bdellovibrio elongation and fission to host cell size. J. Bacteriol. **128:**477–480.
19. **Keya, S. O., and M. Alexander.** 1975. Regulation of parasitism by host-density: the *Bdellovibrio-Rhizobium* interrelationship. Soil Biol. Biochem. **7:**231–237.
20. **Kuenen, J. G., and S. C. Rittenberg.** 1975. Incorporation of long-chain fatty acids of the substrate organism by *Bdellovibrio bacteriovorus* during intraperiplasmic growth. J. Bacteriol. **121:**1145–1157.
21. **Marbach, A., and M. Shilo.** 1978. Dependence of marine bdellovibrios on potassium, calcium, and magnesium ions. Appl. Environ. Microbiol. **36:**169–177.
22. **Marbach, A., M. Varon, and M. Shilo.** 1976. Properties of marine bdellovibrios. Microb. Ecol. **2:**284–295.
23. **Matin, A., and S. C. Rittenberg.** 1972. Kinetics of deoxyribonucleic acid destruction and synthesis during growth of *Bdellovibrio bacteriovorus* strain 109D on *Pseudomonas putida* and *Escherichia coli*. J. Bacteriol. **111:**667–673.
24. **Nelson, D. R., and S. C. Rittenberg.** 1981. Partial characterization of lipid A of intraperiplasmically grown *Bdellovibrio bacteriovorus*. J. Bacteriol. **147:**809–874.
25. **Nelson, D. R., and S. C. Rittenberg.** 1981. Incorporation of substrate cell lipid A components into the lipopolysaccharide of intraperiplasmically grown *Bdellovibrio bacteriovorus*. J. Bacteriol. **147:**860–868.
26. **Pritchard, M. A., D. Langley, and S. C. Rittenberg.** 1975. Effects of methotrexate on intraperiplasmic and axenic growth of *Bdellovibrio bacteriovorus*. J. Bacteriol. **121:**1131–1136.

27. **Ratnam, D. A., S. Pavlou, and A. G. Fredrickson.** 1982. Effects of attachment of bacteria to chemostat walls in a microbial predator-prey relationship. Biotechnol. Bioeng. **24:**2675–2694.

28. **Reiner, A. M., and M. Shilo.** 1969. Host-independent growth of *Bdellovibrio bacteriovorus* in microbial extracts. J. Gen. Microbiol. **59:**401–410.

29. **Rittenberg, S. C.** 1979. *Bdellovibrio*: a model of biological interactions in nutrient-impoverished environments?, p. 305–322. *In* M. Shilo (ed.), Strategies of microbial life in extreme environments. Dahlem Konferenzen, Berlin. Verlag Chemie, New York.

30. **Rittenberg, S. C., and R. B. Hespell.** 1975. Energy efficiency of intraperiplasmic growth of *Bdellovibrio bacteriovorus*. J. Bacteriol. **121:**1137–1144.

31. **Rosson, R. A., and S. C. Rittenberg.** Regulated breakdown of *Escherichia coli* deoxyribonucleic acid during intraperiplasmic growth of *Bdellovibrio bacteriovorus* 109J. J. Bacteriol. **140:**620–633.

32. **Ruby, E. G., and S. C. Rittenberg.** 1983. Differentiation following premature release of intraperiplasmically growing *Bdellovibrio bacteriovorus*. J. Bacteriol. **154:**32–40.

33. **Scherff, R. H., J. E. DeVay, and T. W. Carroll.** 1966. Ultrastructure of host-parasite relationships involving reproduction of *Bdellovibrio bacteriovorus* in host bacteria. Phytopathology **56:**627–632.

34. **Seidler, R. J., and M. P. Starr.** 1969. Isolation and characterization of host-independent bdellovibrios. J. Bacteriol. **100:**769–785.

35. **Shilo, M.** 1969. Morphological and physiological aspects of the interaction of *Bdellovibrio* with host bacteria. Curr. Top. Microbiol. Immunol. **50:**174–209.

36. **Shilo, M.** 1973. Rapports entre *Bdellovibrio* et se hôtes. Nature de la dependance. Bull. Inst. Pasteur Paris **71:**21–31.

37. **Shilo, M., and B. Bruff.** 1965. Lysis of gram-negative bacteria by host-independent ectoparasitic *Bdellovibrio bacteriovorus* isolates. J. Gen. Microbiol. **40:**317–328.

38. **Snellen, J. E., and M. P. Starr.** 1974. Ultrastructural aspects of localized membrane damage in *Spirillum serpens* VHL early in its association with *Bdellovibrio bacteriovorus* 109D. Arch. Mikrobiol. **100:**179–195.

39. **Starr, M. P.** 1975. *Bdellovibrio* as symbiont: the associations of bdellovibrios with other bacteria interpreted in terms of a generalized scheme for classifying organismic associations. Symp. Soc. Exp. Biol. **29:**93–124.

40. **Starr, M. P., and N. L. Baigent.** 1966. Parasitic interaction of *Bdellovibrio bacteriovorus* with other bacteria. J. Bacteriol. **91:**2006–2017.

41. **Swift, S. T., I. Y. Najita, K. Ohtaguchi, and A. G. Fredrickson.** 1982. Some physiological aspects of the autecology of the suspension-feeding protozoan *Tetrahymena pyriformis*. Microb. Ecol. **8:**201–216.

42. **Thomashow, M. F., and S. C. Rittenberg.** 1978. Intraperiplasmic growth of *Bdellovibrio bacteriovorus*: solubilization of *Escherichia coli* peptidoglycan. J. Bacteriol. **135:**998–1007.

43. **Thomashow, M. F., and S. C. Rittenberg.** 1978. Intraperiplasmic growth of *Bdellovibrio bacteriovorus* 109J: N-deacetylation of *Escherichia coli* peptidoglycan amino sugars. J. Bacteriol. **135:**1003–1014.

44. **Thomashow, M. F., and S. C. Rittenberg.** 1978. Intraperiplasmic growth of *Bdellovibrio bacteriovorus* 109J: attachment of long-chain fatty acids to *Escherichia coli* peptidoglycan. J. Bacteriol. **135:**1015–1023.

45. **Thomashow, M. F., and S. C. Rittenberg.** 1979. The intraperiplasmic growth cycle—the life style of the bdellovibrios, p. 115–138. *In* J. H. Parish (ed.), Developmental biology of prokaryotes. Blackwell Scientific Publications, Oxford.

46. **Tudor, J. J., and S. F. Conti.** 1977. Characterization of bdellocysts of *Bdellovibrio* sp. J. Bacteriol. **131:**314–322.

47. **Tudor, J. J., and S. F. Conti.** 1977. Ultrastructural changes during encystment and germination of *Bdellovibrio*. J. Bacteriol. **131:**323–330.

48. **Tudor, J. J., and S. F. Conti.** 1978. Characterization of germination and activation of *Bdellovibrio* bdellocysts. J. Bacteriol. **133:**130–138.

49. **Varon, M.** 1981. Interaction of *Bdellovibrio* with its prey in mixed microbial populations. Microb. Ecol. **7:**97–105.

50. **Varon, M.** 1979. Selection of predation-resistant bacteria in continuous culture. Nature (London) **277:**386–388.

51. **Varon, M., C. Cocito, and J. Seijffers.** 1976. Effect of virginiamycin on the growth cycle of *Bdellovibrio*. Antimicrob. Agents Chemother. **9:**179–188.

52. **Varon, M., and R. Lewisohn.** 1972. Three-membered parasitic system: a bacteriophage *Bdellovibrio bacteriovorus* and *Escherichia coli*. J. Virol. **9:**519–525.

53. **Varon, M., and J. Seijffers.** 1975. Symbiosis-independent and symbiosis-incompetent mutants of *Bdellovibrio bacteriovorus* 109J. J. Bacteriol. **124:**1191–1197.

54. **Varon, M., and M. Shilo.** 1968. Interaction of *Bdellovibrio bacteriovorus* and host bacteria. I. Kinetic studies of attachment and invasion of *Escherichia coli* B by *Bdellovibrio bacteriovorus*. J. Bacteriol. **95:**744–753.

55. **Varon, M., and M. Shilo.** 1969. Interaction of *Bdellovibrio bacteriovorus* and host bacteria. II. Intracellular growth and development of *Bdellovibrio bacteriovorus* in liquid cultures. J. Bacteriol. **99:**136–141.

56. **Varon, M., and M. Shilo.** 1969. Attachment of *Bdellovibrio bacteriovorus* to cell wall mutants of *Salmonella* spp. and *Escherichia coli*. J. Bacteriol. **97:**977–979.

57. **Varon, M., and M. Shilo.** 1979. Ecology of aquatic bdellovibrios, p. 1–48. *In* M. R. Droop and H. W. Jannasch (ed.), Advances in aquatic microbiology, vol. 2. Academic Press, London.

58. **Varon, M., and M. Shilo.** 1981. Inhibition of the predatory activity of *Bdellovibrio* by various environmental pollutants. Microb. Ecol. **7:**107–111.

59. **Varon, M., and B. P. Zeigler.** 1978. Bacterial predator-prey interaction at low prey density. Appl. Environ. Microbiol. **36:**11–17.

60. **Williams, H. N., W. A. Fakler, Jr., and D. E. Shay.** 1982. Seasonal distribution of bdellovibrios at the mouth of the Patuxent River in the Chesapeake Bay. Can. J. Microbiol. **28:**111–115.

Natural Bacteria as Food Resources for Zooplankton

KAREN G. PORTER

Department of Zoology, University of Georgia, Athens, Georgia 30602

The classic description of trophic interactions of plankton has been a food chain consisting of algae fed upon by crustacean zooplankton which, in turn, are fed upon by fish. Recently, other resources such as bacteria and detritus and additional consumers such as microflagellates, protozoans, rotifers, and mucus net feeders have been shown to be ecologically significant (26, 31; Pomeroy, this volume). The possible importance of bacterial grazing was suggested when algal production was found to be insufficient to support observed zooplankton production (21). Recently, planktonic grazers have been proposed as the major factor producing the discrepancy between bacterial biomass and heterotrophic activity that increases with increasing eutrophy in both marine and fresh waters (13; numerous papers in this volume). This paper reviews the rates and the mechanisms by which filter-feeding zooplankton collect bacteria and compares them with those for other food items. The nutritional quality of bacteria, once collected, for zooplankton growth and reproduction is considered. Bacterial and zooplankton abundances often show close correlations (22), suggesting that zooplankton may use bacteria as a major resource in nature. The bias in this review is toward recent studies of freshwater crustacean zooplankton and natural bacterioplankton. The older literature, including that on marine and benthic feeding primarily on cultured bacteria, has been reviewed elsewhere (13, 32).

FEEDING ON BACTERIA

Our understanding of the mechanisms by which "filter feeders" capture food particles is currently under revision. Suspension feeding by fine-particle capturers such as grazing copepods and *Daphnia* sp. was once analogized to sieving: particles with dimensions that exceeded the distance between meshes on the filtering appendages were intercepted by adjacent mesh structures and captured; those that were smaller than the smallest meshes passed through and were lost. Under these conditions, natural free-living bacteria, which are generally smaller than 1 μm, were expected to pass through the meshes of most filter feeders. It is now recognized that in many cases particle capture occurs under conditions of low Reynolds number viscous flow with little or no water flowing through the filtering appendages and with a relatively thick boundary layer surrounding the appendages (7, 16, 27). Beating of the filtering appendages entrains water containing food particles into the filtering chamber. Copepods have been observed to seize large algal particles and prey that are entrained (16). Capture of smaller particles in the size range of bacteria cannot be observed by currently available techniques. However, the efficiency with which they are captured is lower than that for the larger algae and can be affected by changing their surface properties such as charge and wettability (7). This suggests that ionic attraction and hydrophobic-hydrophilic interactions between fine particles and the surfaces of the filtering appendages may be working in the process of small-particle capture. This may also be the case for microflagellates, mollusks, and mucus net spinners that capture fine particles on a mucus sheet or a finely meshed net. For the microbiologist familiar with the processes of bacterial attachment to surfaces, the interface between suspension feeding and cell surface properties is an exciting area for potential cooperative research.

The use of the literature as a source for rates of bacterial collection by zooplankton that can be applied as cropping rates for natural populations is suspect. Most studies of feeding on bacteria have been conducted with cultured laboratory strains of bacteria, which are generally larger than 1 μm and may have occurred as flocs or films (2, 3, 5, 8, 10, 11, 14, 17–20, 32–36). This makes them larger than most natural bacteria, which occur as 0.1- to 1.0-μm unattached cells (12), a size range in which capture efficiency by most zooplankton is reduced. Cultured bacteria may also differ from natural bacteria in their surface properties and nutritional quality.

Even when grazing rates are determined with natural plankton associations, the methods used to measure and calculate bacterial feeding rates can influence the absolute value of the rates (13, 25, 27). Radiotracer studies which label bacteria focus mainly on macrozooplankton that can be hand picked for the determination of activity accumulated by individuals during the feeding period. Label may also be lost from the bacteria or exchanged with other ingestible particles during the process of labeling and feeding. The alternative cell count method requires long incubations to get differences between containers with and without grazers. They may then suffer from "bottle" effects. Also, the control without grazers may not reflect true cell growth during the feeding period because those cells are cut off from the nutrients regenerated by the grazers. Methods that remove grazers by dilution may enhance bacterial growth by enrichment. Computational difficulties also arise when compensating for cell growth (27). These factors make the rates reported in the literature highly variable within species and within and between investigations.

However, when natural bacterioplankton that has been separated from larger particles by filtration is used in feeding rate experiments (13, 15, 24, 25, 27, 37; Table 1), bacterial feeding rates are clearly lower than rates measured on larger particles such as algae and yeast. In a controlled comparative study of daphnids (27), bacterial filtering rates were found to range from 0.1 to 3.0 cm^{-3} per animal per h. They increased with increasing body size of the feeder according to the relationship FR = 0.133W$^{0.654}$, where filtering rate (FR) is expressed as cubic centimeters swept clear of bacteria per animal per hour, and body weight (W) is in micrograms (dry weight). Within a species, however, the efficiency of bacterial capture was 20 to 30% that of algal capture and decreased with increasing body size (27). The general increase in filtering rate with increasing body size reflects the great-

er amount of water that is processed by larger crustacean zooplankton. Differences in filtering efficiency among species could not be predicted by differences in filtering mesh sizes, further suggesting that sieving is not the mechanism of fine-particle capture. The decrease in capture efficiency with increasing body size probably reflects the reduced probability of encounter of bacteria with the filter surface as the volume filtered increases by the cube while filter surface area only increases by the square of the body length.

Bacteria have long been recognized as a major food resource for ciliated protozoans (3); however, little is known about actual protozoan grazing rates in nature. Fenchel (6) used bacterium-sized latex beads to estimate clearance rates which ranged from 10^{-6} to 10^{-4} cm^{-3} per animal per h, but these rates may be biased by the surface properties of the inert particles (7). Clearance rates of approximately 10^{-3} cm^{-3} per animal per h, or 10^4 bacteria per animal per h, were found for *Paramecium caudatum* (36), and rates of 23 to 244 bacteria per animal per h were found for the marine ciliate *Uronema nigricans* (4). However, these and other measurements of bacterial grazing rates used large, cultured bacteria at high concentrations of 10^7 to 10^8 bacteria per cm^3. Using bacterioplankton at natural concentrations of 10^6 bacteria per cm^3, in experiments similar in design to those of Porter et al. (27), R. Sanders (unpublished data) found clearance rates ranging from 10^{-6} to 10^{-5} cm^{-3} per animal per h for *Tetrahymena pyriformis*. This corresponds to 3 to 37 bacteria per ciliate per h. Ingestion rates were directly related to bacterial density, another factor to consider when comparing laboratory feeding experiments. Also, feeding thresholds, food concentrations below which feeding ceases, may occur as they do in copepods. Algal feeding by protozoans is common, larger species being more effective at collecting larger particles (6). Therefore, bacteria probably constitute the major diet of only the smaller (<150 µm) protozoans.

Another factor that affects the process of bacterial collection appears to be the presence of other particles in the feeding mixture. Collection rates of what appeared to be free-living bacteria in epifluorescent counts are elevated in the presence of algae in a process called "piggybacking" (27). Piggybacking occurs either by the loose attachment or association of bacteria with the other particles in suspension or by the enhancement of bacterial collection by the filtering appendages once other particles have been intercepted by them. It should not be confused with the well-documented process of feeding on bacteria firmly attached to detrital particles. It also did not reflect a feeding threshold, as it was

TABLE 1. Filtering rates of some freshwater and marine invertebrates fed natural bacterioplankton

Animal	Size (mm)	Filtering rate[a]	Source of data
Crustacea			
Daphnia middendorfiana	1.3–2.8	0.3–1.3	24
D. pulex	1.3–2.9	0.9–1.1	24
D. longiremis	1.5–1.8	0.6–1.3	24
D. magna	2.4	2.8	27
D. parvula	0.6–1.0	0.2–0.3	27
Ceriopdaphnia lacustris	0.4–0.7	0.1–0.3	27
Bosmina longirostris	0.4	0.21	27
Polyartemiella hazeni	9.0	2.8	24
Larvacea			
Oikopleura dioica	0.2–1.5	0.02–8.4	15
Protozoa			
Helicostomella sublata	—	10^{-4}	13
Tetrahymena pyriformis	0.05–0.06	10^{-6}–10^{-5}	—[b]
Mollusks			
Donax gouldii	6.0–10.5	4.6	13
Mytilus edulis	6.5–26	6.5	13

[a] Cubic centimeters per animal per hour.
[b] R. Sanders, unpublished data.

exhibited by cladocerans that do not reduce feeding activities at low food levels (28). The association of bacteria with other particles makes it essential to measure rates of bacterial collection by animals in whole, natural water if the rates are to be used in determinations of trophic dynamics in the field.

Despite these problems, recent studies (5, 27, 33, 37), including several reported in this volume, allow us to make the generalization that bacteria are filtered at lower rates than larger algal particles by daphnids, copepods, mollusks, and rotifers. Planktonic copepods seem least able to filter bacteria, whereas ciliate protozoans, microflagellates, and perhaps larvaceans are most efficient.

NUTRITIONAL QUALITY OF BACTERIA

The assimilation efficiency of bacteria is about equivalent to that of algae and is reviewed elsewhere (32). A more important criterion for food quality is whether, once it is ingested and assimilated, it provides nutrients that promote optimal growth at the individual and population level. High concentrations of cultured bacteria (10^7 to 10^8 cells per cm^3) have been used to maintain planktonic protozoans, rotifers, and some crustaceans in the laboratory (3, 33, 34, 36); however, few studies with natural bacteria have been done. Natural bacterioplankton have a lower abundance (10^5 to 10^7 cells per cm^3), are possibly in a starved nutritional state, and are filtered with low efficiency. Most authors calculate that bacterioplankton biomass can supply only a fraction of the daily carbon, energy, or biomass requirements for even the most efficient of bacterial feeders (15, 22, 24).

To test the nutritional adequacy of natural bacterioplankton, we fed two co-occurring freshwater crustacean zooplankton continuously renewed supplies of bacteria from the <1-μm filtrate of lake water that contained 10^6 bacteria per cm^3 (24). The zooplankton grew more than on bacteria-free water (0.22-μm filtrate). However, they always did best in whole lake water with its natural phytoplankton or with *Chlamydomonas* sp. added (Table 2). Reproduction occurred in both zooplankton species on the bacterial diet, but only one species, *Ceriodaphnia lacustris*, was able to maintain a growing population as evidenced by the positive growth exponent, r, in Table 2. Differences in filtering rates between the two species could not explain the differences in growth and reproduction. This study was repeated throughout the year and showed that neither bacterioplankton nor whole lake water was sufficient to support maximal growth rates. The two cladocerans were food limited at all times of the year.

The quality of the culture medium, especially its vitamin and trace metal content, can affect the nutritional quality of algal foods for sustained crustacean growth (9, 30). Maternal effects, e.g., the nutritional status mothers impart to offspring through reserves in the egg, also can give a false picture of the nutritional quality of a test food source (9). A. Tessier (unpublished data; Table 3), using offspring of female *Daphnia magna* acclimated to bacteria or high-quality algal food for three generations, conducted a controlled comparative study by the methods of Goulden et al. (9). Bacteria were cultured in a yeast extract and nutrient broth with vitamins, presumably to provide them in the best possible condition as food. Algal and bacterial concentrations were optimal and equivalent on a dry-weight basis. Groups of 20 to 30 offspring were isolated, fed, and examined daily. Body growth and fecundity (Table 3) of the bacteria-fed isolates were found to be reduced compared to the isolates fed algae. In another study, Tessier found that the addition of algae to the bacterial diet enhanced growth and reproduction, but that these parameters reflected the algal abundance alone. Similar findings were reported by Lee and Taga at this Symposium. Since *D. magna* can filter bacteria (Table 2), the cultured bacteria were concluded to be nutritionally inadequate. In a review of the literature, G. Goulden and L. L. Henry (*in* D. Meyers and J. R. Strickler, ed., *Trophic Interactions Within Aquatic Ecosystems*, in press) conclude that bacteria and procaryotes in general are deficient in certain fatty acids, such as linoleic and linolenic acids, that are essential for reproduction in many crustaceans. For most zooplankton, it appears that bacteria can only serve as a supplemental food source and that algal intake is essential to supply both their total carbon and energy needs and their requirements for certain essential nutrients.

As yet we have no evidence of a resident microflora in the guts of zooplankton such as *Daphnia* sp., copepods, or rotifers. The passage of intact viable bacterial cells through the guts of these zooplankton is common (24; unpublished data). Other groups that maintain a gut flora were reviewed by Martin and Kukor at this Symposium.

Ciliate protozoans are commonly cultured on bacteria and require a critical density ranging from 10^6 to 10^8 cells per cm^3 for positive population growth (2, 34). The exact density is dependent on the ciliate species, the bacterial strain, and growth conditions, with the food value of a given bacterium varying among ciliate species (34). In aquatic systems, bacterial densities as high as this are found only in sediments, eutrophic lakes, and polluted waters. This suggests that bacteria alone may not be able to support

TABLE 2. Different abilities of two co-occurring species of cladocerans to grow and reproduce on natural bacterioplankton in the 1-μm filtrate of lake water[a]

Species and conditions	Body size at death (mm)	Life span (days)	No. of young per female	r (day^{-1})
Ceriodaphnia lacustri				
Bacteria-free water	0.30	2.2	0	—
1-μm filtrate	0.60	13.9	2.2	0.07
Lake water	0.74	28.7	19.8	0.24
Chlamydomonas + lake water	0.76	16.5	16.6	0.31
Daphnia parvula				
Bacteria-free water	0.49	3.6	0	—
1-μm filtrate	0.67	12.1	0.4	−0.05
Lake water	1.14	26.2	23.5	0.23
Chlamydomonas + lake water	1.48	25.7	66.3	0.37

[a] Bacteria-free 0.22-μm filtrate, whole lake water, and lake water enriched with *Chlamydomonas* sp. are included for comparison (summarized from 23).

the population growth of ciliated microzooplankton in most planktonic environments. Fenchel (6) hypothesized that bacterial densities are maintained at low levels by heterotrophic microflagellates and that these and small autotrophic cells are the main diets of ciliates, with bacteria serving as a supplement.

COMMUNITY STRUCTURE

In fresh waters, the abundances of certain small crustaceans, rotifers, and protozoans often increase with increasing eutrophy (4, 22). This is especially true of ciliated protozoans (Table 4). A shift from crustacean macrozooplankton to these microzooplankton also occurs seasonally within a lake and often coincides with increased bacterial numbers (22, 23). This correlation suggests that microzooplankton may utilize bacteria either as a sole source or a major

component of the food required to sustain positive population growth. These protozoans are, in turn, fed upon and reduced in numbers by crustacean macrozooplankton (1, 27, 29). The impact of macrozooplankton grazing on bacterial biomass will be less than that on algal standing crops as a consequence of their reduced filtering efficiency on ultrafine particles. Macrozooplankton can filter as much as 400%, and generally do filter 25%, of the water in a eutrophic lake in 1 day (11). With a 30% bacterial filtering efficiency, they may clear 8 to 130% of the water for bacteria. This may have a major impact on bacterial biomass and, if feeding is selective, can influence bacterial community structure. In marine plankton the impact on bacterial biomass of even the most efficient bacterial feeding macrozooplankter is minor (15; numerous studies reported in this volume). Instead, microflagellates are believed to be the major consumers of bacteria, exerting a significant control on the standing stock of marine bacteria (6). As in most grazer systems, modest cropping stimulates prey activity through the recycling of nutrients. This is believed to be occurring in the bacterioplankton-grazer system as well (12).

CONCLUSION

It appears that natural, free-living bacterioplankton provide only a fraction of the nutritional requirements of macrozooplankton and many microzooplankton in marine and fresh waters. Supplements of algae or other zooplankton are required for prolonged fecundity. This is due, in part, to the low abundance, nutritional inadequacy, and reduced filtering efficiency of bacteria. The ability of some suspension feeders to collect bacteria seems to be influenced by the surface properties of the bacteria. This suggests that a high degree of selective cropping may occur. Differential digestion of ingested bacteria may impose further selective mortality, and nutrient regeneration may enhance surviving bacterial strains. Current literature suggests that the impact of grazing is greatest on the bacterioplankton of eutrophic fresh waters. However,

TABLE 3. Growth and early reproduction of *Daphnia magna* fed equivalent dry weights of algae and *Aerobacter aerogenes*[a]

Feed stock	Body wt (μg, dry wt)		Clutch size (eggs per clutch)		
	Day 4	Day 6	First	Second	Third
Bacteria (10^6 cells per cm^3)	15.5 ± 0.9 (20)	17.9 ± 1.5 (30)	3.82 ± 0.336 (11)	4.82 ± 0.46 (11)	6.83 ± 0.86 (6)
Algae (10^4 cells per cm^3)	64.8 ± 3.7 (20)	162.8 ± 3.3 (30)	8.28 ± 0.25 (96)	21.25 ± 0.49 (89)	20.8 ± 0.35 (80)

[a] Values are means ± SE for the number of individuals shown in parentheses. From A. Tessier and C. Goulden (unpublished data).

TABLE 4. Densities of planktonic ciliates found in lakes of various trophic states (summarized from 22)

Trophic state	Annual mean density[a] ($\times 10^3$ per liter)
Ultraoligotrophic	2.4
Oligotrophic	2.3–10.8
Mesotrophic	25–91
Meso-eutrophic	18
Eutrophic	56–145
Hypereutrophic	156

the most recent studies reported in this volume indicate that zooplankton, especially microzooplankton, can limit the numbers of both free-living and attached bacteria while enhancing their productivity through nutrient recycling in all of the planktonic systems studied to date.

I am sincerely grateful for the sharing of information and unpublished data by Allen Tessier and Clyde Goulden of the Philadelphia Academy of Sciences and Robert Sanders of the University of Georgia. Discussions with them, Michael Pace, and Lawrence Pomeroy were especially helpful.

The study was supported by National Science Foundation grant DEB 8203254 and is Contribution Number 18 of the Lake Oglethorpe Limnological Association.

LITERATURE CITED

1. **Berk, S. G., D. C. Brownlee, R. H. Heinle, H. J. King, and R. R. Colwell.** 1977. Ciliates as a food source for marine planktonic copepods. Microb. Ecol. 4:27–40.
2. **Berk, S. G., R. R. Colwell, and E. B. Small.** 1976. A study of feeding response to bacterial prey by estuarine ciliates. Trans. Am. Microsc. Soc. 95:514–520.
3. **Burbank, W. D.** 1942. Physiology of the ciliate *Colpidium colpoda*. I. The effect of various bacteria as food on the division rate of *C. colpoda*. Physiol. Zool. 15:342–362.
4. **Crisman, T. L., J. R. Beaver, and J. S. Bays.** 1982. Examination of the relative impact of microzooplankton and macrozooplankton on bacteria in Florida lakes. Verh. Int. Ver. Limnol. 21:359–368.
5. **DeMott, W. R.** 1982. Feeding selectivities and relative ingestion rates of *Daphnia* and *Bosmina*. Limnol. Oceanogr. 27:518–527.
6. **Fenchel, T.** 1980. Relation between particle size selection and clearance in suspension feeding ciliates. Limnol. Oceanogr. 25:733–738.
7. **Gerritsen, J., and K. G. Porter.** 1982. Fluid mechanics, surface chemistry, and filter feeding by *Daphnia*. Science 216:1225–1227.
8. **Gophen, M., B. Z. Cavari, and T. Berman.** 1974. Zooplankton feeding on differentially labeled algae and bacteria. Nature (London) 247:393–394.
9. **Goulden, C. E., R. M. Comotto, J. A. Hendrickson, Jr., L. L. Hornig, and K. L. Johnson.** 1982. Procedures and recommendations for the culture and use of *Daphnia* in bioassay studies, p. 139–160. *In* J. G. Pearson, R. B. Foster, and W. E. Bishop (ed.), Aquatic toxicology and hazard assessment. Fifth Conference, American Society for Testing and Materials STP 766, Washington, D.C.
10. **Hadas, O., B. Z. Cavari, Y. Kott, and V. Bachrach.** 1982. Preferential feeding behavior of *Daphnia magna*. Hydrobiology 89:49–52.
11. **Haney, J. F.** 1973. An *in situ* examination of the grazing activities of natural zooplankton communities. Arch. Hydrobiol. 72:87–132.
12. **Hobbie, J. E.** 1979. Activity and bacterial biomass. Arch. Hydrobiol. Beih. Ergebn. Limnol. 12:59–63.
13. **Hollibaugh, J. T., J. A. Fuhrman, and F. Azam.** 1980. A technique to radioactively label natural assemblages of bacterioplankton for use in trophic studies. Limnol. Oceanogr. 25:172–181.
14. **Huq, A., E. B. Small, P. A. West, M. Huq, R. Rahman, and R. R. Colwell.** 1983. Ecological relationships between *Vibrio cholerae* and planktonic crustacean copepods. Appl. Environ. Microbiol. 45:275–283.
15. **King, K. R., J. T. Hollibaugh, and F. Azam.** 1980. Predator-prey interactions between the larvacean *Oikopleura dioica* and bacterioplankton in enclosed water columns. Mar. Biol. 56:49–57.
16. **Koehl, M. A. R., and J. R. Strickler.** 1981. Copepod feeding currents: food capture at low Reynolds number. Limnol. Oceanogr. 26:1062–1073.
17. **Lampert, W.** 1974. A method for determining food selection by zooplankton. Limnol. Oceanogr. 19:995–998.
18. **Malovitskaya, L. M., and Y. I. Sorokin.** 1961. An experimental study of the feeding of *Diaptomus* (Crustacea, Copepoda). Tr. Inst. Biol. Vodokhran 4:262–272.
19. **McMahon, J. W., and F. H. Rigler.** 1965. Feeding rate of *Daphnia magna* Staus in different foods labeled with radioactive phosphorus. Limnol. Oceanogr. 10:105–113.
20. **Monakov, A. V., and Y. I. Sorokin.** 1961. Quantitative data on the feeding of *Daphnia*. Tr. Inst. Biol. Vodokhran. 4:251–261.
21. **Nauwerck, A.** 1963. Die Beziehungen zwischen Zooplankton und Phytoplankton in See Erken. Symb. Bot. Ups. 17:1–163.
22. **Pace, M. L.** 1982. Planktonic ciliates: their distribution, abundance, and relationship to microbial resources in a monomictic lake. Can. J. Fish. Aquat. Sci. 39:1106–1116.
23. **Pace, M. L., and J. D. Orcutt, Jr.** 1981. The relative importance of protozoans, rotifers, and crustaceans in a freshwater zooplankton community. Limnol. Oceanogr. 26:822–830.
24. **Pace, M. L., K. G. Porter, and Y. S. Feig.** 1983. Differential utilization of bacterial resources by two co-occurring cladocerans, *Daphnia parvula* and *Ceriodaphnia lacustris*. Ecology 64:1145–1156.
25. **Peterson, B. J., J. E. Hobbie, and J. F. Haney.** 1978. *Daphnia* grazing on natural bacteria. Limnol. Oceanogr. 23:1039–1044.
26. **Pomeroy, L. R.** 1974. The ocean's food web: a changing paradigm. BioScience 24:499–504.
27. **Porter, K. G., Y. S. Feig, and E. F. Vetter.** 1983. Morphology, flow regimes, and filtering rates of *Daphnia*, *Ceriodaphnia*, and *Bosmina* fed natural bacteria. Oecologia (Berlin) 58:156–163.
28. **Porter, K. G., J. D. Orcutt, and J. Gerritsen.** 1983. Functional response and fitness in a generalist filter feeder, *Daphnia magna* (Cladocera: Crustacea). Ecology 64:735–742.
29. **Porter, K. G., M. L. Pace, and J. F. Battey.** 1979. Ciliate protozoans as links in freshwater planktonic food chains. Nature (London) 277:563–565.
30. **Provasoli, L., D. E. Conklin, and A. S. D'Agostino.** 1970. Factors inducing fertility in aseptic crustacea. Helgol. Wiss. Meersunters. 20:443–454.
31. **Saunders, G. W.** 1972. The transformation of artificial detritus in lake water. Mem. Ist. Ital. Idrobiol. 29(Suppl.):261–288.
32. **Sorokin, Yu. I.** 1978. Decomposition of organic matter and nutrient regeneration, p. 501–616. *In* O. Kinne (ed.), Marine ecology, vol. 4. Wiley, Chichester.
33. **Starkweather, P. L., J. J. Gilbert, and T. M. Frost.** 1979. Bacterial feeding by *Brachionus calyciflorus*: clearance and ingestion rates, behavior and population dynamics. Oecologia (Berlin) 44:26–30.
34. **Taylor, W. D.** 1979. Overlap among cohabiting ciliates in their growth response to various prey bacteria. Can. J. Zool. 57:949–951.
35. **Tezuka, Y.** 1971. Feeding of *Daphnia* on planktonic

bacteria. Jpn. J. Ecol. **21**:127–134.

36. **Tezuka, Y.** 1974. An experimental study on the food chain among bacteria, *Paramecium* and *Daphnia*. Int. Rev. Gesamten Hydrobiol. **59**:31–37.

37. **Wright, R. T., R. B. Coffin, C. P. Ersing, and D. Pearson.** 1982. Field and laboratory measurements of bivalve filtration of natural marine bacterioplankton. Limnol. Oceanogr. **27**:91–98.

Simulation Model of a Food Web with Bacteria, Amoebae, and Nematodes in Soil

H. W. HUNT, D. C. COLEMAN, C. V. COLE, R. E. INGHAM, E. T. ELLIOTT, AND L. E. WOODS

Natural Resource Ecology Laboratory, Colorado State University, Fort Collins, Colorado 80523

The role of microbivorous fauna in nutrient mineralization in soil is receiving considerable attention (D. C. Coleman, C. P. P. Reid, and C. V. Cole, Adv. Ecol. Res., in press; E. T. Elliott, K. Horton, J. C. Moore, D. C. Coleman, and C. V. Cole, Proceedings of the Conference on Biological Processes and Soil Fertility, in press; M. J. Mitchell and J. P. Nakas, ed., *Microfloral and Faunal Interactions in Natural and Agroecosystems*, in press), although its importance is not universally accepted (4). The obvious direct effect of predation is that the prey population loses biomass and the predator population obtains food. Indirect effects of predation include tying up nutrients in predator biomass and products (feces and dead animals) and the excretion of nutrients available to the prey. The effect of release of wastes by predators will depend on the factor limiting growth of the prey (19). For example, if bacteria are carbon limited, release of soluble organic nitrogenous wastes will have a feedback effect on bacterial growth, whereas release of ammonia will not.

The net outcome of the various direct and indirect effects of predation is not simple to predict. Predation on microbes may increase total CO_2 evolution and O_2 consumption (8, 17, 27). Amphipods feeding on eelgrass increased decomposition (18), but microbial biomass was not measured, and it was not known whether the effect was direct or indirect. Grazing by collembolans decreased fungal biomass and increased bacterial biomass, but the effect on total microbial biomass was variable (17). Millipedes reduced microbial biomass but not microbial production and mineral turnover (30). Anderson et al. (1) noted that predation by either amoebae or nematodes reduced bacterial biomass. In the same experiment, it was shown that predation increased mineralization of C (8), P (7), and N (31). In some cases, predation has increased bacterial populations (2).

The objective of the present research was to formulate a simulation model to help interpret a series of laboratory experiments on the role of amoebae and nematodes in nitrogen mineralization. We attempted to construct the simplest possible model that satisfactorily mimics observations on CO_2 evolution, NH_4^+ levels, and biomass without contradicting what is known

about the ecology and physiology of the organisms involved.

We know of no previous models for nutrient cycling in soils representing a food web. Newman and Watson (3) modeled bacterial growth in the rhizosphere, but did not include predators. Witkamp and Frank (30) included millipedes in a model for ^{137}Cs transfers during leaf decomposition, but did not distinguish microbial biomass from leaf material. McBrayer et al. (23) constructed a model of forest litter decomposition that distinguished fungi, fungivores, detritivores, and predators. Nutrients (N, P, etc.) were assumed to passively follow biomass, and none of the organic and inorganic fractions of nutrients in soil was included.

Most models simulating food chains or simple food webs in continuous culture (5, 11–13, 21, 22, 29) disregard the release by the predator of nutrients available to the prey. Taub (28) presented a model for ciliates grazing on algae in a two-stage continuous culture in which the ciliates mineralized N, which became available for algal growth.

Hunt et al. (19) described a physiologically based model of bacteria that predicted the effects of limiting nutrient and grazing on bacterial composition, and the kinds of waste products produced by bacteria growing in different media. The present model is derived from our previous work in that the C/N ratio varies with conditions, but it treats microbial growth more simply by using the overall C/N ratio to regulate uptake instead of representing the levels of various constituents within cells.

MODEL STRUCTURE

Figure 1 shows the relationships among modeled variables. The bacterial growth portion of the model is similar to the variable-ratio model of H. W. Hunt, C. W. Cole, and E. T. Elliott (Soil Sci., in press), except that the maintenance energy requirement is assumed to decline in starving populations and slightly different numerical values are used for most model parameters. Amoebae and nematodes are described in a fashion similar to those in the model of Cole (7) and Coleman (10). State variables for glucose and glycol are included because the model has been adapted for laboratory experiments in which glucose was added to soil sterilized with propylene oxide, which leaves a residue of propylene glycol available to the bacteria (1). Soluble and particulate native organic matter include microbial biomass killed by soil sterilization and wastes released by predators.

Uptake by bacteria. Use of various substrates is assumed to be regulated to maintain the bacterial C/N ratio between 3.5 and 10.0. Between these limits, uptake proceeds at the maximum rate, determined by a Michaelis-Menten equation:

$$U_i = \frac{V_i \cdot S_i \cdot B}{K_i + S_i} \qquad (1)$$

where U_i is rate of uptake of substrate i at concentration S_i (micrograms of C or N per gram of soil), B is bacterial biomass (micrograms of C per gram of soil), and V_i and K_i are the maximum uptake rate and half-saturation constants for uptake of substrate i. Values of V_i for the four substrates (NH_4^+, glucose, glycol, and soluble native organic matter) are 0.07, 0.3, 0.035, and 2.0 μg of substrate C or N per μg of bacterial C per h, respectively, and the corresponding values for K_i are 100 μg of N per g of soil and 580, 370, and 90 μg of C per g of soil. Hunt et al. (in press) argued that the low half-saturation constants determined from liquid culture are inapplicable for bacterial growth in soil, where uptake is more affected by diffusion.

Bacteria obtain C and N from four different pools. With three exceptions, uptake from each pool proceeds independently at the rate determined by equation 1. The first exception is that the presence of glucose represses the use of a poorer energy source (25), glycol in this case. Thus, glycol uptake in the model is set to zero until glucose falls below 2 μg of C per g of soil. The second exception occurs if uptake at the maximum rates results in a C/N ratio greater than 10. In this case, uptake of glucose or glycol is reduced to the level necessary to hold the C/N ratio below 10. The third exception occurs if uptake at the maximum rates results in a bacterial C/N ratio lower than 3.5. In this case, NH_4^+ uptake is reduced enough to keep the C/N ratio above 3.5. If reducing NH_4^+ uptake does not maintain the C/N ratio above 3.5, as when bacteria grow on N-rich organic material, excess organic N is released as NH_4^+ to prevent the C/N ratio from falling below 3.5. Since native organic matter is both a C and an N source, its use is assumed to proceed at the maximum rate under all conditions. The mathematical equations necessary to accomplish regulation of uptake as outlined above are straightforward (Hunt et al., in press).

Bacterial respiration. Respiration by bacteria is assumed to consist of growth and maintenance components. The rate of CO_2 evolution R_B (micrograms of CO_2-C per gram of soil per hour) is calculated as

$$R_B = (1 - Y_B) \cdot Uc + B \cdot M_B \qquad (2)$$

where Uc is total C uptake from glucose, glycol, and native organic matter; Y_B, the maximal possible yield, is 0.7 (6); B is bacterial biomass; and M_B (0.037 h^{-1}) is the maintenance rate. Ensign (15) found that the respiration rate of starving *Arthrobacter* cells fell almost exponentially by 0.14 h^{-1} for 1 day. In the model, the value of M_B decreases by 0.14 h^{-1} whenever either the C or N supply is insufficient to support growth of at least 0.015 h^{-1}. The decrease in M_B continues for 26 h to a minimum of 0.001 h^{-1}, a value chosen to allow the model to correctly predict CO_2 evolution in the later stages of our soil incubations.

Growth of amoebae. The model is formulated for situations in which amoebae feed only on bacteria. A type III sigmoid functional response (26) was used for the effect of bacterial density on feeding rate, P_{ab} (micrograms of C per gram of soil per hour):

$$P_{ab} = V_{ab} \cdot \frac{1}{1 + (K_{ab}/B)^P} \cdot A \qquad (3)$$

where A is amoebal biomass (micrograms of C per gram of soil), V_{ab} (0.03 h^{-1}) is the maximum feeding rate, K_{ab} (80 μg of C per g of soil) is the bacterial density at which feeding rate is half its maximum, and P (10.0) is a nondimensional shape parameter. According to equation 3, feeding rate is less than 1% of maximum at a bacterial density of 50 μg of C per g of soil. Although consumption is not entirely eliminated at low levels of bacteria, equation 3 allows bacterial populations preyed on by amoebae to persist at about 40 to 50 μg of C per g of soil.

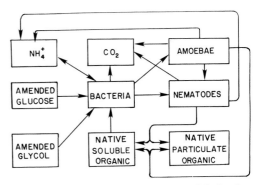

FIG. 1. Compartment diagram for model of carbon and nitrogen transfers in an axenic soil incubation.

Amoebae are assumed to digest 95% of bacteria eaten. The undigested fraction is returned to the native organic matter pools, 35% soluble and 65% particulate. The C/N ratios of the digested, soluble undigested, and particulate undigested fractions are all assumed equal to that of the bacteria eaten.

Respiration of amoebae was modeled by use of the same equation as for bacteria (equation 2) except that the maximal yield of amoebal C from digested C was taken as 0.6, and the maintenance rate was 0.001 h^{-1}.

The amoebal C/N ratio was assumed constant at 7. Because bacteria are relatively rich in N, amoebae are usually limited by available C and excrete excess digested N as ammonia (27).

Growth of nematodes. Nematodes and amoebae were modeled similarly, except that different values were used for the constants in the equations for growth, excretion, etc. Because nematodes are relatively large, a fraction of soil pores is inaccessible to them (20). Since nematodes are larger than amoebae, we assumed that bacteria are less available to nematodes than to amoebae (14). Consumption of bacteria by nematodes is the same as that predicted by equation 3, except that the maximum feeding rate (3.6 h^{-1}) and half-saturation constant (300 µg of C per g of soil) are higher and the shape parameter P (7.0) is lower. The rate of feeding is 1% of maximum at a bacterial density of 156 µg of C per g of soil, which results in minimal bacterial populations of around 150 µg of C per g of soil.

The nematodes modeled feed on both bacteria and amoebae. Since amoebae are mobile, moving in and out of water-filled pores large enough to admit nematodes, we assumed that there is no threshold density of amoebae and have used a type II functional response (26) for the rate of consumption, P_{na} (micrograms of C per gram of soil per hour), of amoebae by nematodes:

$$P_{na} = \frac{V_{na} \cdot A \cdot T}{K_{na} + A} \qquad (4)$$

where A is amoebal biomass, T is nematode biomass (micrograms of C per gram of soil), V_{na} (0.025 h^{-1}) is the maximum rate, and K_{na} (10 µg of C per g of soil) is the amoebal density at which feeding rate is half its maximum value.

We assumed that amoebae are more digestible than bacteria (95% versus 41%) because amoebae lack cell walls. As in the case of amoebae feeding on bacteria, the C/N ratio of digested and undigested fractions is equal to that of the prey eaten, and the undigested fraction is divided between soluble (35%) and particulate (65%) native organic matter. The nematode C/N ratio is held constant at 10, in the range reported by Myers and Krusberg (24), and excess digested N

is excreted as ammonium. Nematode respiration is predicted by using equation 2, as for amoebae and bacteria, except that the maximum yield is only 0.055 and the maintenance rate is 0.0007 h^{-1}.

Data employed. Numerical values for parameters were chosen by trial and error to achieve a fit to published data on growth of single species of bacteria (*Pseudomonas cepacia*), amoebae (*Acanthamoeba polyphaga*), and nematodes (*Mesodiplogaster lheritieri*) in sterilized soil (1, 7, 8, 10, 31; E. T. Elliott, M.S. thesis, Colorado State University, Fort Collins, 1978; E. T. Elliott, C. V. Cole, B. C. Fairbanks, L. E. Woods, R. Bryant, and D. C. Coleman, Soil Biol. Biochem., in press). The experiment simulated included a treatment in which glucose and NH$_4^+$ were added to the microcosms at 0, 72, and 144 h, as well as an unamended control. Therefore, the glucose and NH$_4^+$ state variables were augmented by 200 µg of C per g of soil and 2 µg of N per g of soil at h 72 and 144 to simulate the amended treatment.

Initial biomass of bacteria (2.5 µg of C per g of soil), amoebae (2 µg of C per g of soil), and nematodes (1 µg of C per g of soil) were estimated from their known individual sizes and by assuming some mortality on inoculation. The initial value for NH$_4^+$ was set at its measured value of 47 µg of N per g of soil. Although glycol was estimated at 1,600 µg of C per g of soil (1), it was initialized at half this value to prevent excessive CO$_2$ evolution in some treatments. The initial value for soluble native organic matter (100 µg of C per g of soil), chosen to allow rapid early growth, is compatible with the hypothesis that this material originates from microbes killed during sterilization (Hunt et al., in press). Its initial C/N ratio was set at 7.

MODEL PERFORMANCE

Figure 2 gives an example of how well the model mimicked the temporal dynamics. Carbon dioxide evolution and NH$_4^+$ levels, both simulated and observed, showed similar dynamics during the first 100 h for all four combinations of organisms, but treatment differences existed on the final sample date (Table 1). The model was within all the 95% confidence intervals on day 24 for CO$_2$, NH$_4^+$, and amoebal biomass. The confidence intervals for nematodes were quite small, and the model prediction was outside the interval in all four instances. Nevertheless, the model accounted for 82% ($R^2 = 0.82$) of the observed variation in nematode biomass over all 20 observations (five times × four treatments).

Test of predictive power. Since the data in Table 1 and Fig. 2 were used to help determine parameter values in the model, the ability of the model to fit those data is not a test of the

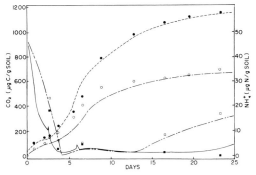

FIG. 2. Comparison of model (lines) with data (points) of Elliott (M.S. thesis, Colorado State University, 1978) in treatment with bacteria, amoebae, and nematodes all present. (– – – –, ●) Cumulative CO_2, C amended. (— · — · —, ○) Cumulative CO_2, unamended. (———, ■) NH_4^+, C amended. (— · · —, □) NH_4^+, unamended.

predictive power of the model, for which independent data are needed. Therefore, the model was used to simulate a similar experiment done with the same species of bacteria and nematodes (2). The soil was sterilized by autoclaving, so that glycol was initialized to zero. The glucose amendment was 500 µg of C per g of soil, and ammonium was 80 µg of N per g of soil. The inoculum size was 1.0 µg of C per g of soil for nematodes and 25 µg of C per g of soil for bacteria. The model underestimated nematode biomass but correctly predicted rapid NH_4^+ immobilization from an initial value of 80 to an observed minimum of 7 µg of N per g of soil (18 predicted) on the second day in either the presence or the absence of nematodes. Immobilization was followed immediately by mineralization, and NH_4^+ increased on day 7 to 39 µg of N per g of soil in the presence of nematodes (39 predicted) and to 15 µg of N per g of soil in their absence (23 predicted). By day 35, observed

NH_4^+ was 59 and 42 µg of N per g of soil in the presence or absence of nematodes (58 and 52 predicted). The model predicted complete utilization of the original soluble organic N (100 µg of C per g of soil) before 21 h, with a subsequent increase to a peak of about 3 µg of N per g of soil on days 2 and 3, but only in the presence of nematodes. This agrees qualitatively with the observed amino acid peaks (2) of 20 and 3 µg of N per g of soil around day 5 in the presence or absence of nematodes. Some of the discrepancy between predicted and observed nematode biomass may result from use of a fixed body size in converting numbers to biomass, whereas juvenile body size ranges over an order of magnitude. Furthermore, the model treats nematodes as if they were all active, whereas inactive stages may predominate (2).

Importance of amoebae to nematodes. Nematode populations grew larger in the presence of amoebae than when limited to a diet of bacteria (Table 1). The explanation for this trend is not obvious, since amoebae compete with nematodes for food as well as provide another food source. Elliott et al. (14) proposed that amoebae consume bacteria in water-filled pores too small to admit nematodes, are eaten by nematodes on exiting the small pores, and thereby increase total food availability to nematodes. The model incorporates this hypothesis, since the feeding threshold is about 50 µg of bacterial C for amoebae and 150 µg for nematodes; but the model still underestimates the increase in nematode biomass when amoebae are present (Table 1). Lack of response of nematodes to amoebae in the model has several explanations. First, nematodes eat fewer bacteria when amoebae are present because amoebae eat bacteria and reduce the bacterial population. Second, the difference between feeding thresholds assumed for nematodes and amoebae (~100 µg of bacterial C per g of soil) is small compared with the predict-

TABLE 1. Comparison of data (Elliott, M.S. thesis, Colorado State University, 1978) with model predictions after 24 days of incubation[a]

Treatment		Variable (µg/g of dry soil)			
Amendment[b]	Biota[c]	CO_2-C	NH_4^+-N	Nematode C	Amoeba C
Amended	B	640/700	0/0	—	—
	AB	1,043/930	2/7	—	340/240
	NB	1,010/1,040	0/0	17/18	—
	NAB	1,138/1,160	3/0	17/26	11/41
Unamended	B	450/530	0/0	—	—
	AB	630/610	17/32	—	220/170
	NB	590/680	1/0	9/10	—
	NAB	660/700	15/16	11/18	19/43

[a] Predicted values are to the left of the slash and observed values to the right.
[b] Amended on days 0, 3, and 6 with glucose (200 µg of C per g of soil) and ammonium (2 µg of N per g of soil).
[c] The letters N, A, and B designate the presence of nematodes, amoebae, and bacteria, respectively.

ed amount of bacteria eaten by nematodes in the absence of amoebae (944 μg of C per g of soil in the amended treatment), so that the increase in available food is relatively small.

Elliott et al. (14) found that nematode populations grew larger in the presence of amoebae on agar plates, where the pore size hypothesis would not apply. Apparently, amoebae stimulate nematode growth by some mechanism not incorporated in the model, such as better nutritional quality or more efficient use of amoebae (14).

Reality of feeding thresholds. For the simulation results described above under "Test of predictive power," the model included a threshold density for bacteria fed on by nematodes. The effect of eliminating the threshold was determined by assigning a value of 1.0 to the shape parameter P in equation 3. The model then failed to predict the experimental result (2) of an abrupt halt (at about 90 μg of bacterial C per g of soil) in the decrease of bacteria preyed on by nematodes. According to the model, most of the readily available substrate was consumed by bacteria before the nematode population was big enough to reduce bacterial biomass, so that by the time bacteria were declining, there was too little substrate left to replace biomass consumed by nematodes, and bacteria became virtually extinct. Thus, a feeding threshold is necessary to account for population dynamics of bacteria preyed on by nematodes.

A similar argument for the reality of feeding thresholds can be developed by using data (9) on bacteriophagic amoebae. Habte and Alexander (16) concluded that physical barriers in soil could not account for bacterial persistence, because predation practically eliminated bacteria when bacterial growth was inhibited by an anti-

biotic. However, in their experiments, the bacteria were grown in solution culture and added to soil, whereas in our experiments the bacteria were grown in situ, which might account for the apparent difference in protection from predators.

Factors controlling mineralization. A series of model runs was made to determine the effects of nematodes and substrate C/N ratios on N mineralization. Glycol was set to zero, glucose was initialized at 2,000 μg of C per g of soil, and no other variables were changed. Table 2 shows the effect of varying the initial NH_4^+ supply on ammonification and NH_4^+ level after 300 h. In the absence of nematodes, ammonification by 300 h occurred only if the initial C/N ratio of substrates was less than 10 and bacteria became C limited, as indicated by the bacterial C/N ratio reaching its minimum of 3.5.

Predation increased production at all N levels (Table 2). Regardless of whether bacteria were limited by N or by C, nematodes made limiting nutrients available to bacteria in the form of NH_4^+ and soluble organic N, thus promoting recycling of both C and N. Predation increased bacterial ammonification by stimulating bacterial growth and hastening exhaustion of the C supply. Ammonification by nematodes and by bacteria occurred at different times. For example, with an initial NH_4^+ supply of 75 μg of N, 92% of the N ammonified by nematodes was released before 100 h, and virtually all of this was taken up by bacteria. Bacteria did not begin to release NH_4^+ until after 100 h, but were the source of more than half of the final NH_4^+ pool of 9.2 μg of N. Thus, an increase in NH_4^+ level in the presence of nematodes may not be entirely attributable to ammonification by nematodes. Low NH_4^+ levels observed while nematode

TABLE 2. Simulated effects of N supply and predation on N mineralization and bacterial production after 300 h

Determination	Initial NH_4^+ (μg of N/g of soil)				
	25	50	75	200	250
Initial C/N ratio of substrate[a]	53	33	24	10	8
Bacteria without nematodes					
Bacterial production (μg of C per g of soil)	398	648	898	815	762
Bacterial C/N ratio	10	10	10	3.8	3.5
Ammonification by bacteria (μg of N per g of soil)	0	0	0	0	47
Final NH_4^+ (μg of N per g of soil)	0	0	0	0	47
Bacteria with nematodes					
Bacterial production (μg of C per g of soil)	526	961	1,051	1,086	1,029
Bacterial C/N ratio	10	4.1	3.5	3.5	3.5
Ammonification by bacteria (μg of N per g of soil)	0	0	5.1	16	39
Ammonification by nematodes (μg of N per g of soil)	25	38	51	106	106
Final NH_4^+ (μg of N per g of soil)	<0.2	<0.2	9.2	83	132

[a] Including glucose (2,000 μg of C per g of soil) and soluble organic N (100 μg of C per g of soil) with a C/N ratio of 7.

populations are expanding have been interpreted as evidence against the release of NH_4^+ by nematodes (3), but such release may go undetected if bacteria have a readily available C source.

CONCLUSIONS

The model successfully accounted for CO_2 ($R^2 = 0.86$) and NH_4^+ ($R^2 = 0.64$) dynamics and treatment response. Nematode biomass was predicted fairly well ($R^2 = 0.82$), except that the model underestimated nematode response to the presence of amoebae. Amoebal dynamics were not predicted well ($R^2 = 0.10$), probably in part because amoebal biomass estimates were so variable (coeffecient of variation = 69%) and a constant cell size was assumed in converting from numbers to biomass, even though cell size is quite variable (E. Ingham, personal communication). Some confidence in the model's validity is engendered by its successful prediction of N mineralization in an experiment done with different C supply and inoculation size for bacteria.

The model by no means incorporates everything known about the ecology and physiology of the organisms. It would be beneficial to distinguish amoebal trophs from cysts, which are not eaten by *M. lheritieri*, and to distinguish juvenile from adult nematodes, which differ in their ability to ingest amoebae.

Model behavior reaffirmed the credibility of the ideas that predators increase productivity of their prey by recycling limiting nutrients (19) and that ammonification by the predator contributes to increased NH_4^+ levels observed when predators are added to bacterial cultures. On the other hand, ammonification by bacteria may be greater in the presence of predators, and NH_4^+ released by predators will not build up until bacteria become C limited.

Results of model exercises suggested new interpretations of several observations. The model behaved unrealistically when the feeding threshold was deleted, which supports the view that thresholds are real, at least for bacteria grown in situ. For interactions between amoebae and nematodes, the model assumed: (i) that both amoebae and nematodes feed on bacteria, but with a lower threshold for amoebae, and (ii) that nematodes feed on amoebae with zero threshold. With these simple assumptions, the model failed to predict the observed stimulation of nematode growth in the presence of amoebae. Therefore, interactions between amoebae and nematodes appear to be more complicated, and some other factor may be important, such as better nutritive value of amoebae than of bacteria.

This research was supported by National Science Foundation grant DEB-8004193-01.

LITERATURE CITED

1. **Anderson, R. V., E. T. Elliott, J. F. McClellan, D. C. Coleman, C. V. Cole, and H. W. Hunt.** 1978. Trophic interactions in soils as they affect energy and nutrient dynamics. Microb. Ecol. 4:361–371.
2. **Anderson, R. V., W. D. Gould, L. E. Woods, C. Cambardella, R. E. Ingham, and D. C. Coleman.** 1983. Organic and inorganic nitrogenous losses by microbivorous nematodes in soil. Oikos 40:75–80.
3. **Anderson, R. V., and T. B. Kirchner.** 1982. A simulation model for life-history strategies of bacteriophagic nematodes, p. 157–177. In D. W. Freckman (ed.), Nematodes in soil ecosystems. University of Texas Press, Austin.
4. **Bååth, E., U. Lohm, B. Lundgren, T. Rosswall, B. Soderstrom, and B. Sohlenius.** 1981. Impact of microbial-feeding animals on total soil activity and nitrogen dynamics: a soil microcosm experiment. Oikos 37:257–264.
5. **Bader, F. G., A. G. Fredrickson, and H. M. Tsuchiya.** 1976. Dynamics of an algal-protozoan grazing interaction, p. 257–279. In R. P. Canale (ed.), Modeling biochemical processes in aquatic ecosystems. Ann Arbor Science, Ann Arbor, Mich.
6. **Camp, T. R., and R. L. Meserve.** 1974. Water and its impurities. Dowden, Hutchinson and Ross, Stroudsburg, Pa.
7. **Cole, C. V., E. T. Elliott, H. W. Hunt, and D. C. Coleman.** 1978. Trophic interactions in soils as they affect energy and nutrient dynamics. V. Phosphorus transformations. Microb. Ecol. 4:381–387.
8. **Coleman, D. C., R. V. Anderson, C. V. Cole, E. T. Elliott, L. Woods, and M. K. Campion.** 1978. Trophic interactions in soils as they affect energy and nutrient dynamics. IV. Flows of metabolic and biomass carbon. Microb. Ecol. 4:373–380.
9. **Coleman, D. C., C. V. Cole, R. V. Anderson, M. Blaha, M. Campion, M. Clarholm, E. T. Elliott, H. W. Hunt, B. Shaefer, and J. Sinclair.** 1977. An analysis of rhizosphere-saprophage interactions in terrestrial ecosystems. Ecol. Bull. (Stockholm) 25:299–309.
10. **Coleman, D. C., C. V. Cole, H. W. Hunt, and D. A. Klein.** 1978. Trophic interactions in soils as they affect energy and nutrient dynamics. I. Introduction. Microb. Ecol. 4:345–349.
11. **Curds, C. R.** 1971. A computer-simulation study of predator-prey relationships in a single-stage continuous-culture system. Water Res. 5:793–812.
12. **Curds, C. R.** 1971. Computer simulations of microbial population dynamics in the activated-sludge process. Water Res. 5:1049–1066.
13. **Drake, J. F., J. L. Jost, A. G. Fredrickson, and H. M. Tsuchiya.** 1968. The food chain, p. 87–95. In Conference Bioregenerative Systems. NASA SP-165. National Aeronautics and Space Administration, Washington, D.C.
14. **Elliott, E. T., R. V. Anderson, D. C. Coleman, and C. V. Cole.** 1980. Habitable pore space and microbial trophic interactions. Oikos 35:327–335.
15. **Ensign, J. C.** 1970. Long-term starvation survival of rod and spherical cells of *Arthrobacter crystallopoietes*. J. Bacteriol. 103:569–577.
16. **Habte, M., and M. Alexander.** 1978. Mechanisms of persistence of low numbers of bacteria preyed upon by protozoa. Soil Biol. Biochem. 10:1–6.
17. **Hanlon, R. D. G., and J. M. Anderson.** 1979. The effects of collembola grazing on microbial activity in decomposing leaf litter. Oecologia (Berlin) 38:93–99.
18. **Harrison, P. G.** 1977. Decomposition of macrophyte detritus in seawater: effects of grazing by amphipods. Oikos 28:165–169.
19. **Hunt, H. W., C. V. Cole, D. A. Klein, and D. C. Coleman.** 1977. A simulation model for the effect of predation on

bacteria in continuous culture. Microb. Ecol. **3**:259–278.

20. **Jones, F. G. W., and A. J. Thomasson.** 1976. Bulk density as an indication of pore space in soils usable by nematodes. Nematologica **22**:133–137.

21. **Jost, J. L., J. F. Drake, H. M. Tsuchiya, and A. G. Fredrickson.** 1973. Microbial food chains and food webs. J. Theor. Biol. **41**:461–484.

22. **Jost, J. L., J. F. Drake, A. G. Fredrickson, and H. M. Tsuchiya.** 1973. Interactions of *Tetrahymena pyriformis, Escherichia coli, Azotobacter vinelandii,* and glucose in a minimal medium. J. Bacteriol. **113**:834–840.

23. **McBrayer, J. F., D. E. Reichle, and M. Witkamp.** 1974. Energy flow and nutrient cycling in a cryptozoan food-web. Publ. No. 575. Environmental Science Division, Oak Ridge National Laboratory, Oak Ridge, Tenn.

24. **Myers, R. F., and L. R. Krusberg.** 1965. Organic substances discharged by plant parasitic nematodes. Phytopathology **55**:429–437.

25. **Pastan, I., and R. Perlman.** 1970. Cyclic adenosine monophosphate in bacteria. Science **169**:339–344.

26. **Smith, R. L.** 1980. Ecology and field biology, 3rd ed. Harper & Row, New York.

27. **Stout, J. D.** 1973. The relationship between protozoan populations and biological activity in soils. Am. Zool. **13**:193–201.

28. **Taub, F. B.** 1973. Biological models of freshwater communities. Ecological Research Series no. EPA-660/3-73-008. Environmental Protection Agency, Washington, D.C.

29. **Tsuchiya, H. M., J. F. Drake, J. L. Jost, and A. G. Fredrickson.** 1972. Predator-prey interactions of *Dictyostelium discoideum* and *Escherichia coli* in continuous culture. J. Bacteriol. **110**:1147–1153.

30. **Witkamp, M., and M. L. Frank.** 1970. Effects of temperature, rainfall, and fauna on transfer of ^{137}Cs, K, Mg and mass in consumer-decomposer microcosms. Ecology **51**:465–474.

31. **Woods, L. E., C. V. Cole, E. T. Elliott, R. V. Anderson, and D. C. Coleman.** 1982. Nitrogen transformations in soil as affected by bacterial-microfaunal interactions. Soil Biol. Biochem. **14**:93–98.

Biological Control

Ecological Principles of Biocontrol of Soilborne Plant Pathogens: *Trichoderma* Model

YIGAL HENIS

Faculty of Agriculture, Rehovot 76-100, Israel

The use of fungal antagonists to control soilborne plant pathogens has received much attention in recent years (6, 37, 43, 54, 58). Although antagonism to various fungi is easily demonstrable on growth media and in sterile soil, attempts to control disease by direct inoculation of natural soil either have failed or have given inconsistent results. This situation was elegantly summarized by Garrett (33), who related this failure to the "little chance that an alien organism has in establishing itself in a foreign environment which is already colonized to its full ecological capacity." A similar opinion was expressed by Alexander (1), who concluded, in the light of many reported unsuccessful attempts to improve plant growth by inoculation, that "such attempts were doomed to fail on the ground of the inoculant containing an alien organism, which by its nature was inferior to the native soil microflora." However, in spite of these rather pessimistic views, one cannot ignore the increasing numbers of reports on successful control of diseases caused by soilborne pathogens through direct application of antagonists (3, 4, 6, 26–32, 44, 51, 63; Y. Henis, J. A. Lewis, and G. C. Papavizas, Soil Biol. Biochem., in press). These frequent "exceptions to the rule" must be carefully examined, because they may supply us with the key to the understanding of the conditions which may lead to a consistently successful biological control, in spite of the "alien" theory.

Among the few antagonists which have proved successful under both laboratory and field conditions, isolates of *Trichoderma* spp. have been studied in detail (21–24, 42) and have shown promising results, especially in controlling *Rhizoctonia solani* and *Sclerotium rolfsii* (3, 26–28, 30, 63; J. Gressel and W. Rau, *in Encyclopedia of Plant Physiology NS*, in press). Similarly, soil suppressiveness against plant pathogens has been associated with the presence of *Trichoderma* spp. under both laboratory (12, 40, 41) and field (13) conditions. Isolates of this genus produce antibiotics (21, 22, 24, 53) and lytic enzymes (23, 27, 28, 53) and compete with other fungi for nutrients (48).

In spite of the widespread investigation of *Trichoderma* isolates as biocontrol agents (7, 24–28, 30, 35, 44, 46, 48, 50, 53, 54, 56, 63, 65), the information available to date regarding their possible mode of action and the actual mode of action of *Trichoderma* antagonists in the soil is surprisingly meager. We have been studying the interaction between *Trichoderma* isolates and sclerotia of *S. rolfsii* in culture and in natural soil, in an attempt to develop a screening procedure for the evaluation of *Trichoderma* isolates as effective biocontrol agents under field conditions.

THE PATHOGEN

S. rolfsii Sacc. is a polyvorous, widespread pathogen, which may rarely develop a perfect stage, *Athelia rolfsii* (Curzi) Tu and Kimbrough. Normally, it appears either as a vegetative mycelium or as sclerotia, which are multicellular vegetative structures highly resistant to adverse conditions (17).

Sclerotia. When placed on a glucose-mineral medium, fresh sclerotia of *S. rolfsii* germinate. However, only a small proportion of the potentially germinable sclerotial cells are usually involved in germination on growth media, leaving the sclerotial tissues (rind, cortex, and medulla) relatively intact and the sclerotium capable of regermination (17). In contrast to rich media, poor ones such as water agar do not normally support vigorous germination of *S. rolfsii* fresh sclerotia, unless they are dried first, to trigger germination (18, 60–62). Yet, the inner sclerotial tissues (cortex, medulla) consist of cells rich in reserve materials (15), and the sclerotia constantly excrete amino acids and sugars to their surroundings (16, 18, 61). The emerging mycelium, whose walls may be lysed by glucanase and chitinase (14, 26–28), spreads into the agar and on its surface, either in the form of single hyphae or as mycelial strands. Formation of those strands may be induced by certain isolates of *Trichoderma* (36). When the agar medium is covered with mycelium, within few days young, white sclerotial initials ("pinheads") appear, which soon turn dark brown and mature.

When placed in natural soil, fresh sclerotia of *S. rolfsii* will neither germinate nor be attacked by the soil microflora (60; Y. Henis and G. C. Papavisas, Phytopathology **72**:1010, 1982, and in press). Their constantly excreted metabolites are probably used up by their surrounding microbial population. Fresh sclerotia, however, can be induced to germinate vigorously (and at the same time to become susceptible to degradation by the surrounding microflora) by drying (17, 18, 60–62; Henis and Papavisas, Phytopathology **72**:1010, 1982, and in press). The dried sclerotia excrete large amounts of sugars and amino acids (17, 18, 60, 61) and germinate eruptively (55, 56). The eruptive germination is characterized by direct emergence of parallel-growing hyphae in the form of mycelial strands followed by an immediate production of new sclerotia. At the final stage of this process, the germinating sclerotium becomes empty, leaving behind the melanin-rich external rind as an empty shell.

It was shown by Smith (61) that overnight immersion of sclerotia of *S. rolfsii* in water reduced the extent of degradation of sclerotia by the soil microflora, but only slightly affected their germination. According to Smith, sclerotial excretions supported the attack of the sclerotium by the surrounding microflora. It was recently found (Y. Henis, J. A. Lewis, P. B. Adams, and G. C. Papavizas, Phytopathology **72**:107, 1982; Henis and Papavizas, in press) that fresh sclerotia of *S. rolfsii* could be induced to germinate and be made susceptible to degradation by *T. harzianum* in natural soil not only by drying (Fig. 1) but also by short exposure to Vapam (sodium *N*-metho-di-thiocarbonate) (Fig. 2) and by localized heat treatments (Fig. 3 and Table 1). These treatments also increased sclerotial germinability in the absence of the antagonist. These results could be explained by assuming that in the fresh sclerotial tissues nutrients are kept in a form unavailable both to the sclerotium and to its surrounding antagonists. These nutrients are released after a physical or chemical shock and are then subjected to competition between the sclerotium and its antagonists. This hypothesis corroborates the observation that neither the emerging mycelium nor the germinating sclerotium is attacked by *Trichoderma* in natural soil (Henis and Papavizas, Phytopathology **72**:1010, 1982, and in press). The apparent discrepancy between this hypothesis and the little effect of overnight dipping in water on sclerotial germination (61) may be explained by assuming that the newly available nutrients are present in two fractions: an extractable, easily diffusible one, and a tissue-bound one, the latter being available only to the sclerotial germinating cells. Furthermore, the elimination of the activity of the sclerosphere microflora by continued washings enables the sclerotium to germinate without being interrupted by external antagonism. *Trichoderma* may

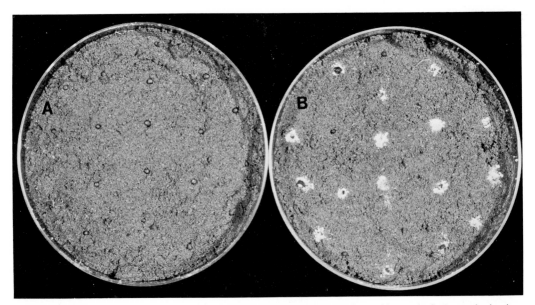

FIG. 1. Effect of drying on susceptibility of PDA-produced sclerotia of *S. rolfsii* strain Sr-3 to colonization and degradation by *T. harzianum* strain WT-6 after 7 days of incubation in natural soil. (A) Fresh sclerotia; (B) sclerotia incubated overnight under anhydrous $CaSO_4$ at room temperature. Both treatments were dipped in a spore suspension of *T. harzianum* containing 1.5×10^7/ml and pushed into the soil by means of a glass rod. (Henis and Papavizas, unpublished data.)

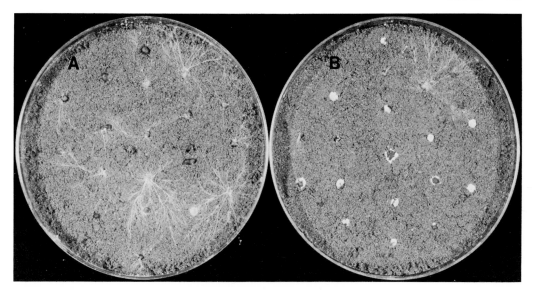

FIG. 2. Effect of exposure of *S. rolfsii* strain Sr-3 sclerotia to metham-sodium (1,000 μg/ml) in soil for 1 h on their subsequent germinability and susceptibility to colonization and degradation by *T. harzianum* strain WT-6 in natural soil. (A) No WT-6. (B) Treated sclerotia were dipped in a spore suspension (1.5 × 10⁷/ml). Sclerotia were pushed into soil and incubated for 7 days. (Henis and Papavizas, unpublished data.)

require the external nutrients for germination and penetration, and the internal ones for sclerotial degradation.

Mycelium. When placed on an agar medium, disks of *S. rolfsii* culture behave in a way similar to that of fresh sclerotia, linear spread of mycelium being the same for both and mycelial density depending on nutrient concentration, mainly carbon source. Because of the relatively uniform growth obtained with this type of inoculum, it is usually preferred in antagonism tests on agar plates. It should be noted, however, that differences between sclerotia and agar disks as inocula may exist. Thus, it was recently found (38; Henis et al., Phytopathology **72:**707, 1982) that sclerotial germination but not mycelial growth from agar disks could still take place in a medium containing 8-hydroxyquinoline at a concentration of 10 μg/ml. Another form of mycelial inoculum of *S. rolfsii* is mycelial mat obtained from shallow, standing liquid cultures of the pathogen. Washed mycelial mat placed on natural soil will give rise to a fast linear spread of the pathogen on the soil surface, without any external nutrient source. Growth vigor and resistance to microbial attack of mycelial mat culture of *S. rolfsii* in natural soil depend on mycelial aging and can be increased by localized heat treatment and cold storage (Henis and Papavizas, in press). A fourth form of *S. rolfsii* inoculum is inoculated oats (Fig. 4), which behave in a similar way to dried sclerotia. When placed on soil, they germinate vigorously, mainly in the form of mycelial strands, and form sclerotia on their surface and at the meeting points of the strands. It appears that there is some kind of thigmotropism which causes the strands to grow toward each other. The oat culture responds to *Trichoderma* inoculation in a way similar to the sclerotia: young cultures are not attacked, but partially heated ones (Henis and Papavizas, in press) and old ones, which have been incubated in the cold (5°C) for 2 weeks or more, turn susceptible to attack and degradation by *T. harzianum* (Y. Henis, R. D. Lumsden, and G. C. Papavizas, unpublished data).

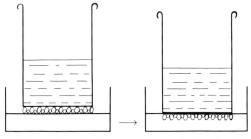

FIG. 3. Localized heat treatment of sclerotia of *S. rolfsii*. Fresh PDA-produced sclerotia were placed on the surface of agar medium and pressed for 15 s into the medium with the bottom of a beaker containing boiling water. Except for the point of contact with the beaker, the sclerotia were protected from the heat by the agar. When incubated in natural soil, they germinated eruptively and became susceptible to attack by *Trichoderma* sp. (Henis and Papavizas, in press.)

TABLE 1. Effect of partial heat treatment and puncturing on *S. rolfsii* sclerotia vulnerability to attack by *T. harzianum* WT-6

Treatment	Germination (%)		Attack (%) by		New sclerotia in soil
	On PDA	In soil	WT-6	Indigenous microflora	
Control	100	0	0	0	0
Punctured	100	0	0	0	0
Heated	100	60	85	7	11

THE ANTAGONIST

Like *S. rolfsii*, linear growth rate and mycelium density of the spreading colony of *Trichoderma* spp. depend on nutrient concentration and medium composition. Sporulation is observed later on and is favored by light (Gressel and Rau, in press). Usually the spores are of all shades of green, but some may be white as well (Henis and Papavizas, Phytopathology **72**:1010, 1982, and in press). When grown on a poor medium (i.e., water agar), linear spread is much less affected by nutrient concentration than mycelial density. Sporulation often takes place at the colony margin rather than in the colony center, away from the inoculation point. This behavior has been observed both on agar media and in soil. Phialospores of *Trichoderma* introduced into natural soil are subjected to soil fungistasis (47) and will normally germinate poorly unless exposed to an external nutrient source (20) under proper conditions (19). On the other hand, partially or fully sterilized soil appears to be a suitable growth medium for the antagonist, which will also sporulate on it. It appears that *Trichoderma* sp. can utilize nutrients at relatively low level. However, this speculation and its possible relationship to biological control have never been examined in detail.

INTERACTIONS BETWEEN *S. ROLFSII* AND *TRICHODERMA* SPP.

Growth media. A commonly used method of observing interactions between two organisms is by oppositely inoculating them on an agar medium. Inhibition of one organism by another can occur as a result of exolysis (27, 28, 52, 57), antibiosis (21, 22, 53), competition (43, 48), change of the environment, e.g., pH (12, 33), or direct parasitism (2, 3, 6, 11, 20, 24, 31, 37, 38, 39, 42, 51; Henis and Papavizas, in press). These interactions have been reviewed by several authors (1, 6, 8, 11, 33, 34, 39, 44, 47, 49, 54, 58). Except for competition, all those mechanisms have been reported for *Trichoderma* sp., although not necessarily in the same organism. In addition *S. rolfsii* may respond to *T. harzianum* by forming mycelial strands (36). Among the nine recognized species and many yet unexplored strains of *Trichoderma*, there is a great variability in antibiotic production, parasitism, and antagonistic spectrum toward other organisms (21–23, 53; Y. Henis, J. A. Lewis, and G. C. Papavizas, Soil Biol. Biochem., in press). Examples of direct parasitism of *R. solani* and *S.*

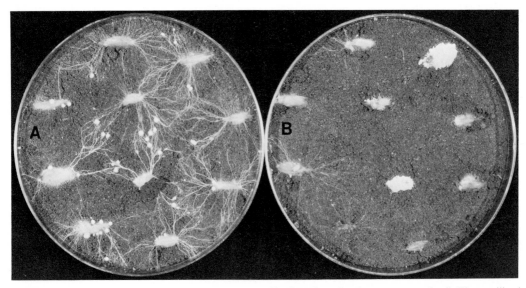

FIG. 4. Effect of *T. harzianum* on outgrowth of *S. rolfsii* from inoculated oats on natural soil. The sterilized oats were inoculated with *S. rolfsii* and incubated for 7 days at 30°C. After being kept at 5°C for 14 days, they were incubated for 7 days on natural soil, either with no further treatment (A) or after being dipped in a spore suspension (1.5×10^7/ml) of *T. harzianum* WT-6 (B). (Henis, Loumsden, and Papavizas, unpublished data.)

rolfsii by *T. harzianum* and *T. hamatum* on agar media as viewed under scanning and transmission electron microscopes are shown in Fig. 5 and 6. However, demonstration of the antagonistic capacity (e.g., lysis, antibiosis, etc.) on an agar plate should not be taken as a proof that it is also functioning under field conditions (3). Another limitation of the agar test is the inability to observe parasitism in the presence of antibiosis, which may keep the mycelia of the antagonist and its host away from each other or kill the host's mycelium before it is reached by the antagonist. In a recent experiment (Henis et al., in press), 10 *Trichoderma* isolates of three different species were compared for their antagonistic activity towards *S. rolfsii* on potato dextrose agar (PDA) and in natural soil with their biocontrol capacity against root rot caused by *S. rolfsii* to bean seedlings. A correlation coefficient of $r = -0.844$ between inhibition of sclerotial germination by *Trichoderma* in natural soil and plant stand in *Trichoderma*-inoculated soil was obtained, but no correlation could be calculated with antagonism as expressed by antibiosis and overgrowth on PDA plates. When meeting with each other on the plates, the pathogen and its antagonist either remained at a small distance from each other, implying mutual antibiosis, or grew into each other's colony. In this area, direct parasitism could have taken place. None of the 10 *Trichoderma* isolates tested overgrew the pathogen's colony (10). Instead, four of the *Trichoderma* colonies were overgrown by *S. rolfsii*. Growth of the pathogen's mycelium into the antagonist's colony was in the form of strands, carried over densely branched mycelial

"fingers." Selection of *Trichoderma* isolates as biocontrol agents is often based on antagonistic capacity in agar culture (10, 30, 56). This may explain the often contradicting results reported by authors who selected their isolates on the basis of agar plate tests.

When fresh sclerotia of *S. rolfsii* are inoculated with *Trichoderma* sp. and placed on water agar, the sclerotia are penetrated by the antagonist within a few days (38, 41). However, they remain intact and viable, except for a small portion (ca. 10%) which loses its firmness, darkens, and finally degrades. In the degraded sclerotia, the medulla disappears, as a result of a lytic activity of the antagonist, which also undergoes autolysis, leaving behind chlamydospores inside the sclerotium and phialospores outside it (38; Henis et al., Phytopathology **72**:707, 1982). As in the case of biocontrol efficiency, *Trichoderma* isolates differ greatly in their penetration capacity into *S. rolfsii* sclerotia. Apparently, penetration is a necessary, but not the only, property which determines *Trichoderma* biocontrol capacity.

Interactions in soil. Predried sclerotia of *S. rolfsii* readily germinate in soil and, when inoculated with *Trichoderma* sp., are degraded (Henis and Papavizas, Phytopathology **72**:1010, 1982, and in press). According to Smith (60) germination of predried sclerotia is independent of their increased degradation in natural soil. However, the results obtained with predried, *Trichoderma*-inoculated sclerotia suggest that both germination and degradation processes may depend on nutrient sources which become available as a result of the drying process. These nutrients

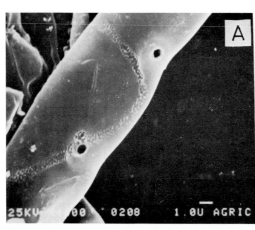 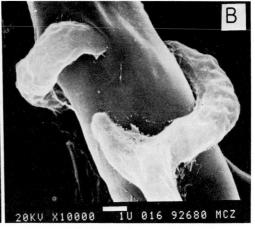

FIG. 5. Scanning electron micrographs of *Trichoderma* spp. hyphae interacting with *S. rolfsii* (A) and of *T. hamatum* interacting with *R. solani* (B). (A) Hyphae of *S. rolfsii* from which a coiling hypha of *T. harzianum* was removed, showing digested zone with penetration sites caused by the antagonist (×4,700). (B) Hyphae of *T. hamatum* coiling around and penetrating *R. solani* hypha. ×7,100. (Taken from reference 27, by permission of *Phytopathology*.)

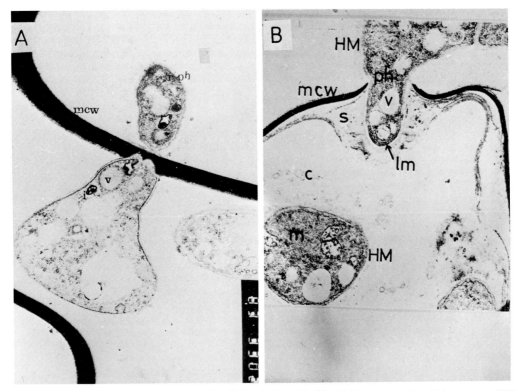

FIG. 6. Transmission electron micrographs of *R. solani* attacked by *T. hamatum* and by *T. harzianum*. HM, *T. hamatum*; HR, *T. harzianum*; mcw, melanin-containing cell wall; ph, penetrating hyphae; v, vacuole; m, mitochondria; c, cytoplasm. (A) *T. hamatum* penetrating through melanin-rich cell walls of *R. solani*. ×13,350. (B) *T. hamatum* penetrating *R. solani*. The host responds by formation of "sheath matrix." ×10,200. (Taken from reference 26, by permission of *Phytopathologische Zeitschrift*.)

may be located either inside or outside the dried sclerotia. In the first case, the penetrating *Trichoderma* hyphae interact with the pathogen's germinating cells inside the sclerotium, whereas in the latter case, both the antagonist and the pathogen interact with the soil microflora as well. No information is available as to the minimum nutrient levels required by *Trichoderma* spores to overcome soil fungistasis and to suppress or degrade the pathogen's propagules. In this struggle for limited nutrients the inoculum potential of the antagonist, as well as the level and activity of the soil microflora, may play an important role. Inoculum potential has long been recognized as an important factor in pathogenesis and biocontrol (5, 7, 33, 63). Few attempts, however, have been made so far to elucidate the mechanism underlying its effect.

Can *Trichoderma* parasitize living mycelium of *S. rolfsii* in soil? In a recent experiment (Henis and Papavizas, unpublished data) mycelial mat of the pathogen was produced by growing *S. rolfsii* in a shallow layer culture of potato dextrose broth for 7 days at 30°C, washed, and cut into 2- by 2-cm pieces. When placed on the

surface of smoothened fresh soil plates, these pieces germinated readily, giving rise to new young mycelium within 24 h. Isolates of *T. harzianum* which inhibited sclerotial germination and reduced disease caused by *S. rolfsii* to bean seedlings in the greenhouse were not capable of attacking this mycelium, when applied as a suspension containing 1.5×10^7 spores per ml. However, they did attack and sporulate on heat-killed mycelium, and they attacked living mycelium which had been brought in contact with another piece of killed mycelium. It appears, therefore, that *Trichoderma* prefers "weakened" mycelium of *S. rolfsii* to a living, active one. This observation is supported further by the effect of partial (or localized) heat treatment of *S. rolfsii*-inoculated oats, which is similar to the effect of localized heat treatment of sclerotia of *S. rolfsii* (Henis and Papavizas, in press; Henis et al., in press). The capacity of *Trichoderma* to lyse *S. rolfsii* mycelium may depend on the pathogen's age and growth conditions. Thus, the increase in lytic enzymes observed with *Trichoderma* grown in autoclaved sand on *S. rolfsii* (29) mycelium was probably due to aging of the pathogen's mycelium during the incubation period. This does not mean that living and active mycelia of all fungi are immune to all *Trichoderma* isolates, but it strongly supports the commonly accepted view that *Trichoderma*, as well as many other necrotrophic fungi, requires a food base for its antagonistic activity in soil. Barnett and Binder (8) distinguish between necrotrophic, destructive parasites, which kill their host, and biotrophic, balanced ones, which merely obtain nutrients from their hosts. They regard "obligate" mycoparasitism as an arbitrary, temporary condition which merely reflects our lack of knowledge of the growing conditions in the absence of the host. Moreover, these authors consider most of the necrotrophic mycoparasites as facultative or opportunistic, being capable of utilizing a wide variety of substrates. We suggest that one should distinguish between (i) obligate saprophytes which are incapable of attacking living mycelium under any circumstances, (ii) facultative or opportunistic mycoparasites which attack living mycelium when furnished with an external food source, such as soluble or particulate organic matter and "weakened" or dead mycelium, and (iii) obligatory mycoparasites which are capable of attacking living structures of other fungi in soil without being assisted by an external food base. This last group includes fungi such as *Conyothyrium minitans* (64), *Sporidesmium sclerotiovorum* (3), and possibly *Gliocladium roseum* (9, 42). According to this definition, it is the independence of the mycoparasite on soil organic matter rather than its inability to grow on growth media (8) which determines its definition as an obligatory mycoparasite. The fact that *Trichoderma* belongs to group ii explains the necessity to add a food base to preparations used for biocontrol of soilborne plant pathogens (4, 5, 6, 29, 30, 35, 65). Unfortunately, in spite of its apparent simplicity, this suggested classification has its limits and complexities: a certain fungus may be obligatory toward one host and facultative toward another. Thus, one may imagine a situation in soil in which a *Trichoderma* isolate is colonizing living mycelium of a saprophytic fungus for which it is an obligate parasite and using it as a food base to attack *S. rolfsii* mycelium for which it is a facultative parasite.

Interactions with soil fauna. Jager et al. (42) observed an increased eating activity of insects on hyphae of *Glycladium roseum* in natural soil. We have observed an increased activity of nematodes and mites on and around dried sclerotia of *S. rolfsii* either before or after it was attacked by *Trichoderma* sp. in soil (Y. Henis, G. C. Papavizas, J. A. Lewis, and R. D. Lumsden, unpublished data). No information is available as to the mechanisms (6) of attraction of the soil fauna to the dried sclerotia; are they attracted to the leading nutrients, to the multiplying microflora, to *Trichoderma* mycelium and spores, or to all these components? The involvement of the soil fauna in spreading the antagonist or in any other aspect of its action in soil still awaits clarification. It may be concluded that, in spite of these difficulties, the soil environment remains the only reliable medium for the evaluation of the antagonistic capacity of a mycoparasite in situ.

Role of inoculum potential. Much has been said on the importance of inoculum potential as a factor which determines success of infection of a plant host by its pathogen. It is agreed that the same priniciple applies to the host-mycoparasite relationship, as is the case with *Armillaria mellea* and *Trichoderma viride* (6, 33, 63).

Antagonism of *S. rolfsii* on natural soil plates by *Trichoderma* sp., as revealed by inhibition of germination of sclerotia, and on inoculated oats did not take place at spore concentrations lower than 10^5/ml. It is likely that soil fungistasis (47) is involved in the requirement for high inoculum potential. This would explain the different results obtained in sterile and nonsterile soils (59) and after partial sterilization (28, 31, 45, 50, 51). What difference does it make to the mycoparasite whether it is present at a low or a high spore concentration? What is the physiological basis of the infection–inoculum potential relationship? Unfortunately, very little has been done so far to answer these interesting questions. Mutual synergism between the mycoparasite's propagules during germination is one possible mechanism. The presence of chlamydospores (46) in spore suspensions of *Trichoderma* sp. and their possible role in the successful attack and in the final degradation of the host should also be considered. Understanding of the factors involved may help us in improving our preparations by increasing their biocontrol potential at lower inoculum levels.

CONCLUSIONS

The interaction between *Trichoderma* spp. and *S. rolfsii* in soil is a rather complicated process whose outcome depends on both the antagonist and its host. *S. rolfsii* response to *Trichoderma* sp. depends on its developmental stage, its age, and its physiological state. Young hyphae and fresh sclerotia of *S. rolfsii* are not attacked by *Trichoderma* sp. in natural soil. However, physiological shocks such as drying, heat damage, or contact with a fungicide turn the relatively resistant sclerotium into a susceptible one and at the same time induce its germination. Penetration capacity of the antagonist into the sclerotium is an important, but not the only, factor involved. Thus, sclerotia may carry *Trichoderma* spp. and yet appear intact and capable of germination. It is not clear whether inhibition of sclerotial germination by *T. harzianum* in soil results from hyphal degradation, from competition for nutrients, or from both. However, no antibiotic production seems to be involved in the system studied. The experimental evidence indicates that *Trichoderma* is an opportunistic, rather than an obligate parasite, which requires a food base to attack living hyphae in soil. Inoculum potential and capacity to overcome the competing soil microflora and soil fungistasis, as well as external nutrients, production of lytic enzymes, and recognition of specific sites on the host's mycelium may all play a role in the struggle between the antagonist and its host.

LITERATURE CITED

1. **Alexander, M.** 1971. Microbial ecology. John Wiley & Sons, Inc., New York.
2. **Alexander, M.** 1981. Why microbial predators and parasites do not eliminate their prey and hosts. Annu. Rev. Microbiol. **38**:113–134.
3. **Ayers, W. A., and P. B. Adams.** 1981. Mycoparasitism and its application to biological control of plant diseases, p. 91–106. *In* G. C. Papavizas (ed.), Biological control in crop production. BARC Symposium No. 5. Allanheld, Osmun & Co., Totowa, N.J.
4. **Backman, P. A., and R. Rodziguez-kabana.** 1975. A system for the growth and delivery of biological control agents to the soil. Phytopathology **65**:819–821.
5. **Baker, K. F., and R. J. Cook.** 1974. Biological control of plant pathologens. W. H. Freeman Co., San Francisco.
6. **Baker, R.** 1968. Mechanisms of biological control of soilborne plant pathogens. Annu. Rev. Phytopathol. **6**:263–294.
7. **Baker, R.** 1978. Inoculum potential, p. 137–157. *In* J. G. Horsfall and E. B. Couling (ed.), Plant pathology: an advanced treatise, vol. 2. Academic Press, Inc., New York.
8. **Barnett, H. L., and I. L. Binder.** 1973. The fungal host-parasite relationship. Annu. Rev. Phytopathol. **11**:273–291.
9. **Barnett, H. L., and V. G. Lilly.** 1962. A destructive mycoparasite Gliocladium roseum. Mycologia **54**:72–77.
10. **Bell, O. K., H. D. Wells, and C. R. Markham.** 1982. In vitro antagonism of *Trichoderma* species against six fungal plant pathogens. Phytopathology **72**:379–382.
11. **Boosalis, M. G., and R. Mankau.** 1965. Parasitism and predation of soil microorganisms, p. 374–391. *In* K. F. Baker and W. C. Snyder (ed.), Ecology of soil-borne plant pathogens. University of California Press, Berkeley.
12. **Chet, I., and R. Baker.** 1980. Induction of suppressiveness to Rhizoctonia solani in soil. Phytopathology **70**:994–998.
13. **Chet, I., and R. Baker.** 1981. Isolation and biocontrol potential of Trichoderma hamatum from soil naturally suppressive to Rhizoctonia solani. Phytopathology **71**:286–290.
14. **Chet, I., and Y. Henis.** 1969. Effect of catechol and disodium EDTA on melanin content of hyphal and sclerotial walls of Sclerotium rolfsii Sacc. and the role of melanin in the susceptibility of these walls to β-1,3-glucanase and chitinase. Soil Biol. Biochem. **1**:131–138.
15. **Chet, I., Y. Henis, and N. Kislev.** 1969. Ultrastructure of sclerotia and hyphae of Sclerotium rolfsii Sacc. J. Gen. Microbiol. **57**:143–147.
16. **Chet, I., D. Timar, and Y. Henis.** 1977. Physiological and ultrastructural changes occurring during germination of sclerotia of Sclerotium rolfsii Sacc. Can. J. Bot. **55**:1137–1142.
17. **Coley Smith, J. R., and R. C. Cooke.** 1971. Survival and germination of fungal sclerotia. Annu. Rev. Phytopathol. **9**:65–92.
18. **Coley Smith, J. R., A. Ghaffar, and Z. V. R. Javed.** 1974. The effect of dry conditions and subsequent leakage and rotting of fungal sclerotia. Soil Biol. Biochem. **6**:307–312.
19. **Danielson, R. M., and C. B. Davey.** 1973. Non nutritional factors affecting the growth of *Trichoderma* in culture. Soil Biol. Biochem. **5**:495–504.
20. **Danielson, R. M., and C. B. Davey.** 1973. Carbon and nitrogen nutrition of *Trichoderma*. Soil Biol. Biochem. **5**:505–515.
21. **Dennis, C., and J. Webster.** 1971. Antagonistic properties of species-groups of *Trichoderma*. I. Production of non-volatile antibiotics. Trans. Br. Mycol. Soc. **57**:25–39.
22. **Dennis, C., and J. Webster.** 1971. Antagonistic properties of species-groups of *Trichoderma*. II. Production of volatile antibiotics. Trans. Br. Mycol. Soc. **57**:41–48.
23. **Dennis, C., and J. Webster.** 1971. Antagonistic properties of species groups of *Trichoderma*. III. Hyphal interaction. Trans. Br. Mycol. Soc. **57**:363–369.
24. **Dubos, B., J. J. Guillaumin, and M. Schubert.** 1978. Action du *Trichoderma viride* Pers., apporté avec divers substrats organiques sur l'initiation et la croissance des rhizomorphes d'*Armillaria mellea* (Vahl.) Karst. dans deux types des sols. Ann. Phytopathol. **10**:187–196.
25. **Durrell, L. W.** 1968. Hyphal interaction by *Trichoderma viride*. Mycopathol. Mycol. Appl. **35**:138–144.
26. **Elad, Y., R. Barak, I. Chet, and Y. Henis.** 1983. Ultrastructural studies of interaction between *Trichoderma* spp. and plant pathogenic fungi. Phytopathol. Z. **107**:168–175.
27. **Elad, Y., I. Chet, P. Boyle, and Y. Henis.** 1983. The parasitism of *Trichoderma* spp. on plant pathogens. Ultrastructural studies and detection by FITC lectins. Phytopathology **73**:85–88.
28. **Elad, Y., I. Chet, and Y. Henis.** 1982. Degradation of plant pathogenic fungi by *Trichoderma harzianum*. Can J. Microbiol. **28**:719–725.
29. **Elad, Y., I. Chet, and Y. Henis.** 1981. Biological control of *Rhizoctonia solani* in strawberry fields by *Trichoderma harzianum*. Plant Soil **60**:245–254.
30. **Elad, Y., I. Chet, and J. Katan.** 1980. *Trichoderma harzianum*: a biocontrol agent effective against *Sclerotium rolfsii* and *Hizoctonia solani*. Phytopathology **70**:119–121.
31. **Elad, Y., Y. Hadar, I. Chet, and Y. Henis.** 1981. Prevention with *Trichoderma harzianum* Rifai aggr. of reinfestation by *Sclerotium rolfsii* Sacc. and *Rhizoctonia solani* Juhn of soil fumigated with methyl bromide, and improvement of disease control in tomatoes and peanuts. Crop Protect. **1**:199–211.
32. **Elad, Y., Y. Hadar, E. Hadar, I. Chet, and Y. Henis.** 1981. Biological control of *Rhizoctonia solani* by *Trichoderma harzianum* in carnation. Plant Dis. **65**:675–677.
33. **Garrett, S. D.** 1965. Toward biological control of soilborne plant pathogens, p. 4–17. *In* K. F. Baker and W. C. Snyder (ed.), Ecology of soil-borne plant pathogens. University of California Press, Berkeley.
34. **Griffin, G. J., T. S. Hova, and R. Baker.** 1975. Soil fungistasis: elevation of exogenous carbon and nitrogen requirements for spore germination by fungastatic volatiles in soil. Can. J. Microbiol. **21**:1468–1475.
35. **Hadar, Y., I. Chet, and Y. Henis.** 1979. Biological control of *Rhizoctonia solani* damping off with wheat bran culture of *Trichoderma harzianum*. Phytopathology **69**:64–68.
36. **Hadar, Y., Y. Elad, I. Chet, and Y. Henis.** 1981. Induction of macroscopic strand formation in *Sclerotium rolfsii* by *Trichoderma harzianum*. Isr. J. Bot. **30**:156–164.
37. **Hashioka, Y.** 1973. Mycoparasitism in relation to phytopathogens. Shokubutsu Byogai Kenkuyu **8**:179–190.
38. **Henis, Y., P. B. Adams, J. A. Lewis, and G. C. Papavizas.** 1983. Penetration of sclerotia of *Sclerotium rolfsii* by *Trichoderma* spp. Phytopathology **73**:1043–1046.

39. **Henis, Y., and I. Chet.** 1975. Microbial control of plant pathogens. Adv. Appl. Microbiol. **19:**85–111.

40. **Henis, Y., A. Ghaffar, and R. Baker.** 1978. Integrated control of *Rhizoctonia solani* damping-off of radish: effect of successive plantings, PCNB and *Trichoderma harzianum* on pathogen and disease. Phytopathology **68:**900–907.

41. **Henis, Y., A. Ghaffar, and R. Baker.** 1979. Factors affecting suppressiveness to *Rhizoctonia solani* in soil. Phytopathology **69:**1164–1169.

42. **Jager, G., A. Ten Hooper, and H. Velvis.** 1979. Hyperparasites of *Rhizoctonia solani* in Dutch potato fields. Neth. J. Plant Pathol. **85:**253–268.

43. **Kelly, W. D., and R. Rodrigues-Kabana.** 1976. Competition between *Phytophthora cinnamomi* and *Trichoderma* spp. in autoclaved soil. Can. J. Microbiol. **22:**1120.

44. **Kommedahl, T., and C. Windels.** 1981. Introduction of microbial antagonists to specific courts of infection: seeds, seedlings and wounds, p. 227–248. *In* G. C. Papavizas (ed.), Biological control in crop production. BARC Symposium No. 5. Allanheld, Osmun & Co., Totowa, N.J.

45. **Kreutzer, W. A.** 1965. The reinfestation of treated soil, p. 495–507. *In* K. F. Baker and W. C. Snyder (ed.), Ecology of soil-borne plant pathogens. University of California Press, Berkeley.

46. **Lewis, J. A., and G. C. Papavizas.** 1983. Production of chlamydospores and conidia by *Trichoderma* spp. in liquid and solid growth media. Soil Biol. Biochem. **15:**351–357.

47. **Lockwood, J. L.** 1977. Fungistasis in soils. Biol. Rev. **52:**1–43.

48. **Marques, R., and A. Touvet.** 1972. Comparison between biological control and chemical treatments against root parasites of trees in relation to the climate. C. R. Acad. Sci. Ser. D **274:**234–239.

49. **Mitchell, J. E.** 1973. The mechanism of biological control of plant diseases. Soil Biol. Biochem. **5:**721–728.

50. **Munnecke, D. E., M. J. Kolbezen, and W. D. Wilbur.** 1973. Effect of methylbromide or carbon disulfide on *Armillaria* and *Trichoderma* growing on agar medium and relation to survival of *Armillaria* in soil following fumigation. Phytopathology **63:**1352–1357.

51. **Ohr, H. D., D. E. Munnecke, and J. Bricker.** 1973. The interaction of *Armillaria mellea* and *Trichoderma* spp. as modified by methyl bromide. Phytopathology **63:**965–973.

52. **Papavizas, G. C., and J. A. Lewis.** 1981. Introduction and augmentation of microbial antagonists for the control of soilborne plant pathogens, p. 305–322. *In* G. C. Papavizas (ed.), Biological control in crop production. BARC Symposium No. 5. Allanheld, Osmun & Co., Totowa, N.J.

53. **Papavizas, G. C., and J. A. Lewis.** 1983. Physiological and biocontrol characteristics of stable mutants of *Trichoderma viride* resistant to MBC fungicides. Phytopathology **73:**407–411.

54. **Papavizas, G. C., and R. D. Lumsden.** 1980. Biological control of soilborne fungal propagules. Annu. Rev. Phytopathol. **18:**389–413.

55. **Punja, Z. K., and R. G. Grogan.** 1981. Eruptive germination of sclerotia of *Sclerotium rolfsii*. Phytopathology **71:**1092–1099.

56. **Punja, Z. K., R. G. Grogan, and T. Unruh.** 1982. Comparative control of *Sclerotium rolfsii* on golf green in northern California with fungicides, inorganic salts, and *Trichoderma* spp. Plant Dis. **66:**1125–1128.

57. **Rodrigues-Kabana, R., W. D. Kelley, and E. A. Carl.** 1978. Proteolytic activity of *Trichoderma viride* in mixed culture with *Sclerotium rolfsii* in soil. Can. J. Microbiol. **24:**487–490.

58. **Schroth, M. N., and J. G. Hancock.** 1981. Selected topics in biological control. Annu. Rev. Microbiol. **35:**453–476.

59. **Singh, R. S.** 1964. Effect of *Trichoderma viride* on the growth of Phythium sp. in sterilized and unsterilized soil. Naturwissenschaften **51:**173.

60. **Smith, A. M.** 1972. Drying and wetting of sclerotia promotes biological control of *Sclerotium rolfsii* Sacc. Soil Biol. Biochem. **4:**119–123.

61. **Smith, A. M.** 1972. Nutrient leakage promotes biological control of dried sclerotia of *Sclerotium rolfsii* Sacc. Soil Biol. Biochem. **4:**125–130.

62. **Smith, A. M.** 1972. Biological control of fungal sclerotia in soil. Soil Biol. Biochem. **4:**131–134.

63. **Smith, K. T., R. O. Blanchard, and W. C. Shortle.** 1979. Effect of spore load of *Trichoderma harzianum* on wood-invading fungi and volume of discolored wood associated with wounds in Acer rubrum. Plant Dis. Rep. **63:**1070–1071.

64. **Tribe, H. T.** 1957. On the parasitism of Sclerotinia trifoliorum by Coniothyrium minitans. Trans. Br. Mycol. Soc. **40:**489–499.

65. **Wells, H. D., D. K. Bell, and C. A. Jaworski.** 1972. Efficacy of *Trichoderma harzianum* as biological control for *Sclerotium rolfsii*. Phytopathology **62:**44.

Bacteria as Biocontrol Agents of Plant Disease

M. N. SCHROTH, J. E. LOPER, AND D. C. HILDEBRAND

College of Natural Resources, Agricultural Experiment Station, Department of Plant Pathology, Berkeley, California 94720

Plant root health in nature primarily depends upon the inherent resistance of a plant to microbial invasion and the biological equilibrium between competing beneficial and deleterious microorganisms in the rhizosphere as mediated by the environment. Rhizosphere bacteria play a major role in this equilibrium by their interactions with various pathogenic agents. However, much of the evidence for bacteria serving as deterrents of plant disease is relatively circumstantial and indirect, and often is derived from experimentation under controlled environmental conditions in greenhouse or laboratory settings. Nevertheless, the positive nature of this evidence, coupled with the remarkable, widespread commercial success in controlling the ubiquitous crown gall disease by an avirulent strain of *Agrobacterium tumefaciens* (also referred to as *A. radiobacter*) (34, 35), has focused attention on the use of bacteria as biocontrol agents. Interest also has been generated by reports showing that the application of root-colonizing bacteria to seeds can increase plant growth and yields (8, 15, 40, 47, 48, 70), apparently because of substantial changes in rhizosphere populations of microorganisms (39, 69; T. V. Suslow and M. N. Schroth, Phytopathology, in press). The hope and expectation of biological control proponents are that the efficacy of bacteria as biocontrol agents will be increased by genetic engineering when the key factors determining bacterial colonization of plant surfaces and antagonism against pathogens are identified.

Because the subject of biological control is of such current interest, there have been a number of recent reviews (2, 3, 21, 56, 64, 68; T. J. Burr and A. J. Calsar, Crit Rev. Microbiol., in press). Two of these (68; Burr and Calsar, in press) have focused on beneficial rhizobacteria and have comprehensively covered much of the recent information. To avoid redundancy, this paper will include some review, present some recent research on rhizobacteria, and attempt to develop a synthesis of useful precepts for using bacteria as biocontrol agents.

STATUS OF BACTERIA AS BIOCONTROL AGENTS

Biocontrol of crown gall. The use of an avirulent strain of *A. tumefaciens*, strain K84, to control crown gall represents the only widespread commercial application of a biocontrol agent for plant disease control to date (35). As has been noted (64), crown gall is a unique disease in that the pathogen and antagonist interact in a relatively exclusive, isolated ecological niche. The wound, the site of infection, must be protected by K84 only until it heals. Strain K84 has the ability to colonize the wound site in competition with pathogenic strains and to produce an agrocin that is inhibitory to many pathogenic strains (25).

The resistance of some strains of *A. tumefaciens* to strain K84 reduces the effectiveness of the biocontrol in some areas (1, 54, 55, 67). In California, the use of K84 has had mixed success. For example, in one experiment (M. N. Schroth and A. H. McCain, unpublished data) half of 10,000 Mazzard cherry root stock were inoculated with a heavy suspension of strain K84 immediately before planting. Six months later, 40 and 42% of treated and untreated root stock, respectively, exhibited the crown gall disease. The pathogenic *A. tumefaciens* strain isolated from crown gall tissue was resistant to the K84 agrocin. It also produced a substance that was inhibitory to K84. Similar results were obtained in the same location with pear and prune root stock, and in Texas with willow root stock (George Philly, personal communication).

Strain K84 is ineffective in controlling the principal strains of *A. tumefaciens* that infect grapevine, *Vitis vinifera* L. (55). This is unfortunate, since crown gall has become a major problem throughout the world consonant with the rapid increase in grapevine plantings. *Agrobacterium* also is systemic in the vines (14), thus complicating development of a successful biological control program. One suggestion for controlling crown gall of grapevine is to alter a grapevine-virulent strain by deleting oncogenicity genes on the Ti plasmid and inserting DNA fragments with genes determining the production of an agrocin or another antibiotic substance. The resulting avirulent strain, exhibiting inhibitory action against virulent strains, represents an example of the potential use of genetic engineering technology in the development of improved biocontrol agents.

Other biocontrol agents. Most reports concerning biocontrol agents describe laboratory and greenhouse experimentation. It has been difficult to obtain consistent, positive results in the field with procedures that are both economically sound and compatible with commercial farming practices, regardless of the nature of the biocontrol agent. Table 1 lists examples of bacteria that have shown promise in replicated field experiments, either in disease control or in plant growth promotion.

MUTATION AND POPULATION CHANGES

A review of biological control literature reveals a recurrent pattern of events progressing from an initial report claiming the successful use of a biocontrol agent to a period of inconsistent results and, later, abandonment of the program. A major problem in the study and development of biocontrol agents may be caused by the general instability of these bacteria in culture, aggravated by repeated transfer and culturing practices. The instability associated with culture and storage of bacterial biocontrol strains is

TABLE 1. Examples of bacteria used to control diseases in the field or to elicit a plant growth response

Plant	Pathogen[a]	Inoculated bacterium
Apple	Erwinia amylovora	Erwinia herbicola (59)
Apple	Nectria galligena	Bacillus subtilis (72)
Cabbage	UK	B. subtilis (8)
Barley	UK	B. subtilis (47), Streptomyces griseus (48)
Carrot	UK	B. subtilis, S. griseus (53)
Corn	UK	B. subtilis (17, 41)
Flax	Fusarium oxysporum f. sp. lini	Bacillus sp. (74)
Mushroom	Pseudomonas tolaasii	Pseudomonas sp. (51)
Oats	UK	B. subtilis, S. griseus (48)
Onion	Sclerotium cepivorum	B. subtilis (75)
Potato	UK	Pseudomonas spp. (15, 30, 40)
Radish	UK	Pseudomonas spp. (38)
Sugar beet	UK	Pseudomonas spp. (70)
Tobacco	Pseudomonas solanacearum	Avirulent strain of P. solanacearum (18)
Wheat	Gaeumannomyces graminis	Pseudomonas sp. (76)
Wheat	UK	B. mycoides (16), B. subtilis (47, 48), S. griseus (48)

[a] UK, Unknown.

probably analogous to that observed with phytopathogenic bacteria. It is well known that many bacterial plant pathogens lose virulence if not frozen or lyophilized. For example, loss of virulence occurs rapidly with *Pseudomonas solanacearum*, occurs less rapidly with organisms such as *P. syringae* pv. *savastanoi*, *Xanthomonas campestris* pv. *vesicatoria*, and *X. campestris* pv. *campestris*, and rarely occurs with *Erwinia* soft rotting bacteria and *A. tumefaciens*. However, the stability of characters determining the ecological behavior of an organism rarely has been examined. The virulence of a pathogen is relatively simple to test by directly inoculating a plant, but testing for "ecological competence," or the ability of an organism to survive and compete in nature, presents a different problem. Presumably, there are a large number of determinants that enable microorganisms, including phytopathogens and biocontrol agents, to compete successfully in the rhizosphere. The loss of any one determinant could substantially alter their ability to incite or control a disease.

Assay procedures, analogous to the pathogenicity tests of bacterial phytopathogens, are needed for the routine testing of the efficacy of biocontrol agents. These procedures, however, are relatively cumbersome in comparison to pathogenicity tests since they must address "ecological competence" and a quantitative reduction in disease incidence or severity.

Bacterial strains, when removed from a natural competitive environment in which selective pressures allow survival of only "fit" mutants, may change when repeatedly subcultured in vitro. Mutants frequently arise in culture which may be less "ecologically competent" in nature. For example, bacteria no longer need to produce a glycocalyx in culture, as they do in nature, and therefore mutants that do not produce the extracellular polysaccharide frequently grow more rapidly and predominate (23). The properties of these mutants, as reviewed by Costerton et al. (23, 24), may differ from those of the wild-type strains in phage sensitivity, antigenic and immunological properties (58), colicin sensitivity, antibiotic susceptibility, mating ability, and ability to adhere to surfaces. Costerton et al. (23) speculated that the loss of the bacterial glycocalyx in culture, as they do in nature, is standard in culture and is responsible for much of the inconsistency in data on adhesion of bacteria to a variety of animal tissue cells.

Mutants deficient in extracellular polysaccharide production and other mutants which occur in vitro may play a similar role in the inconsistency of results obtained by the use of biocontrol agents in plant pathology. Impaired ability to resist such factors as antibiotics, phage, bacteriocins, and moisture stress, and the loss of lipopolysaccharides or proteins that determine attachment and motility, may profoundly alter the ecological capability of rhizosphere bacteria. Another common mutation in vitro which should affect "ecological competence" is a change from smooth to rough colony types. The change most likely would be accompanied by a loss of important characteristics related to colonization ability, since, generally, the greatest biochemical activity of bacteria is suggested to be associated with the smooth phases (42).

Other molecular events which could result in mutant populations displacing the wild-type strain in culture have been identified. Plasmid loss commonly occurs in culture when selection for a plasmid-determined phenotype is lifted (65). Transposon insertions induce mutations resulting in loss of function, as demonstrated by the loss of indoleacetic acid production and attenuation of virulence with *P. syringae* pv. *savastanoi* (20). The probability of selecting a mutant during subculture is increased when the mutant has a selective advantage in culture. Examples include the selection of auxotrophic mutants under conditions in which loss of synthesis enables the auxotrophs to grow rapidly in comparison to the wild type because of conservation of energy (46, 77). In the case of old cultures, selection may favor mutants with increased tolerance to accumulated metabolic products that are toxic to the wild type (26).

In our experience with fluorescent pseudomonads as biocontrol agents, there has been a general loss in efficacy with repeated transfers, frequently accompanied by changes in colony morphology, the development of bacterial chains without septations, loss or reduction in motility, reduction in intensity of fluorescent pigment, loss of or reduction in antibiosis, and changes in osmoregulation. It is essential for the development of bacteria as biological control agents that caution be exercised in culturing these organisms and that every effort be made to immediately freeze or lyophilize strains of possible interest.

BACTERIZATION

Bacterization is the process of improving plant growth by applying specific bacterial strains to seeds or roots of agricultural crops (9). It has been tested sporadically with variable success for several decades (22, 49, 60, 68; Burr and Calsar, in press). Replicated field trials carried out in California and Idaho (15, 40, 70) demonstrated that the application of specific bacterial strains to seed or seed pieces resulted in significantly increased yields of sugar beet and potato.

The success of bacterization has been tempered by unexplained failures to achieve consistent yield increases under field or greenhouse conditions. The lack of understanding of mechanisms by which beneficial rhizobacteria promote plant growth limits bacterization technology and development, including the screening, selection, and maintenance of efficacious strains. Little information is available concerning how soil type, host plant genotype, and environmental conditions affect the ability of rhizobacteria to colonize and persist on roots. Other factors such as bacterial production of physiologically active metabolites also may be important. Failures of a beneficial bacterium to promote plant growth, therefore, could be the result of any one of numerous factors that affect its physiological activity and population dynamics in the rhizosphere.

Both indirect and direct mechanisms have been suggested to explain the positive influence that beneficial bacteria have on plant growth. One indirect effect is the population reduction of various minor fungal and bacterial pathogens in the root system (39, 69; Suslow and Schroth, in press). The inoculation of seed with beneficial rhizobacteria resulted in changes in the composition of root microflora including both bacteria and minor fungal pathogens (39, 69; Suslow and Schroth, in press) sensu Salt (61). Possible direct effects include enhanced nutrient uptake or solubilization (4, 11, 53) and enhanced levels of phytohormones in the root zone (10, 12, 13, 31–33).

Phytohormones. Many of the bacteria isolated from roots caused a reduction in root elongation and increased lateral branching, reminiscent of symptoms induced with exogenous sources of the phytohormone indole-3-acetic acid (IAA). In field soil, these deleterious bacteria caused a more subtle general stunting of plant growth. The literature describing the influence of bacterial elaboration of IAA on plant growth is contradictory (13, 31, 53). Since the elaboration of IAA by rhizobacteria may play a role in plant growth promotion or in rhizosphere colonization, a study was initiated to examine and quantify IAA production by several rhizobacterial strains known to influence plant growth.

Seventeen rhizobacterial strains beneficial to potato, sugar beet, or radish, and 13 strains deleterious on sugar beet, were screened to identify those which produce indole compounds in vitro. Indole production was detected by the observation of a red color upon addition of Salkowski reagent to culture filtrates, as described by Gordon and Weber (27). Seven of the 17 beneficial strains (41%) and all of the 13 deleterious strains produced detectable levels of indole compounds in vitro when grown in standard King's B medium. None of the four beneficial strains (E6, SH5, RV3, B4) which promoted yields of sugar beet under field conditions produced detectable levels of indole compounds. In all cases, the indole-positive compound produced was identified as IAA on the basis of high-pressure liquid chromatography with the use of two separate and selective detectors connected in series as described previously (71; M. Hein, Ph.D. thesis, University of Minnesota, Minneapolis, 1983). In most cases, IAA was the only Ehrlich reagent-reactive compound produced in culture filtrates, as determined by thin-layer chromatography in two separate solvent systems (S.-T. Liu, Ph.D. thesis, University of California, Berkeley, 1977).

Quantification of IAA by high-pressure liquid chromatography techniques indicated that beneficial strains produced relatively low levels of IAA in vitro (≤ 3 µg/ml in 5-day cultures grown in King's B medium), whereas deleterious strains produced higher concentrations (up to 20 µg/ml). Definitive evidence as to the contribution of elaborated IAA to the observed symptoms induced by deleterious bacteria will be provided only by further genetic and biochemical analysis. Nevertheless, we suggest that the relatively large concentrations of IAA produced by several deleterious bacterial strains have the potential to markedly influence plant growth.

We examined the influence that bacterial IAA may have on the growth of sugar beet in experiments with the IAA-producing phytopathogenic bacterium P. syringae pv. savastanoi (45). Seed inoculation with pv. savastanoi strain 2009 resulted in decreased root elongation of sugar beet in comparison with uninoculated controls. However, seed inoculation with mutant strains 2009-3 and 2009-561, which were deficient in IAA production (19), did not decrease root elongation, indicating that IAA production by strain 2009 was responsible for the observed inhibition of root elongation (45).

The influence of exogenous sources of IAA on plant growth and development is known to be host, tissue, and concentration dependent (6, 7, 73). The differential sensitivities of host plants, the varied concentrations of IAA produced by soil bacteria, and the variation in experimental conditions may explain the apparent inconsistencies in the literature describing the influence of bacterial sources of IAA on plant growth. Low concentrations of IAA may stimulate root elongation, whereas higher concentrations may be inhibitory.

We speculate, without direct evidence, that bacterial production of IAA may increase root cell permeability, with a subsequent increase in root exudation. If so, IAA production by a rhizosphere bacterium might play an ecological role, elevating levels of root exudates and thereby increasing the nutritive base. Although no evidence exists directly linking IAA production

by rhizosphere bacteria to increased root exudation, the influence of IAA on cell wall plasticity, membrane permeability, and subsequent proton excretion is well documented (50). Also, Libbert and Silhengst (44) demonstrated that IAA elaborated by epiphytic bacteria was transferred to plant tissue. Three studies showed that the presence of rhizosphere bacteria increased root exudation in contrast to gnotobiotic control plants (5, 6, 57). Furthermore, seed inoculation with several of the deleterious IAA-producing rhizobacterial strains described above resulted in increased fungal colonization of sugar beet roots (68, 69), possibly because of increased nutrient availability in the rhizosphere. Production of IAA by rhizosphere bacteria may have an important role in the susceptibility of plant roots to bacterial and fungal colonization.

Inhibitory substances. Antibiotic compounds represent a second type of bacterial metabolites of potential importance in plant growth promotion and biocontrol (36, 37, 62, 63). In a recent study, we investigated the possible relationships among in vitro antibiosis, effect on plant growth, and taxonomic groupings of bacteria. No attempt was made to differentiate between classical antibiotic activity and inhibitory effects due to the elaboration of siderophores. The production of siderophores by fluorescent pseudomonads and their role in biocontrol have been treated in recent reviews (64, 68; Burr and Calsar, in press).

In vitro antibiosis by fluorescent pseudomonads associated with plant roots was examined with 26 strains isolated from the rhizospheres of bean plants grown at two locations. These strains were tested for inhibitory activity against members of six bacterial strains (A. tumefaciens, Erwinia carotovora, Escherichia coli, Pseudomonas marginalis, P. syringae pv. syringae, and P. syringae pv. phaseolicola) cultured on two different media (potato dextrose agar and King's medium B) and against two fungal species (Pythium ultimum and Rhizoctonia solani) cultured on potato dextrose agar. Antibiosis was rated on a scale of 0 to 3 based on diameters of inhibitory zones, and the results were averaged for each strain. The strains were briefly characterized by use of standard physiological and nutritional tests (66) to determine whether there was a relationship between antibiotic activity and species. The effect of each strain on plant growth was also tested by previously detailed methods (15, 40, 70).

The pseudomonads were classified as either P. fluorescens or P. putida on the basis of physiological and nutritional test results (66). The P. fluorescens strains in general were gelatin, trehalose, levan, and denitrification positive, and benzoate negative (Table 2). Interest-

TABLE 2. Characterization of antagonistic fluorescent pseudomonads and their effect on plant growth

Strain	Antibiosis rating[a]	Physiological and nutritional properties					Effect on plant growth	
		Levan production	Gelatin hydrolysis	Denitrification	Trehalose utilization	Benzoate utilization	Stimulation	Inhibition
A124	0.6	−	−	+	−	+	+	−
B123	0.9	+	+	+	+	−	−	−
A126	1.11	+	+	+	−	−	−	−
A214	1.1	+	+	+	−	−	−	+
A211	1.2	+	+	+	+	−	−	−
B226	1.2	+	+	+	−	−	+	−
A117	1.3	+	+	+	+	−	−	+
B222	1.3	+	+	+	+	−	+	−
A213	1.5	+	+	+	−	−	−	−
A225	1.5	+	+	−	−	−	−	+
B122	1.7	+	+	+	−	−	+	−
A216	1.8	−	−	+	+	−	+	−
A217	1.8	−	−	+	+	−	−	−
A215	1.9	+	+	+	+	−	+	−
B112	2.0	+	+	+	+	−	−	+
B121	2.1	−	−	−	−	+	+	−
A122	2.1	−	−	−	−	+	+	−
B216	2.1	−	−	−	−	+	−	−
B113	2.2	−	−	+	−	+	−	−
A112	2.4	−	−	+	−	+	−	+
B111	2.4	−	−	−	−	+	−	+
A212	2.4	−	−	−	−	+	−	+
B117	2.5	−	−	−	−	+	−	+
B211	2.5	−	−	−	−	+	−	+
B212	2.6	−	−	−	−	+	−	+
B214	2.6	−	−	−	−	+	−	+

[a] Average antibiosis rating against six bacterial and two fungal species based on a scale of 0 (none) to 3 (severe).

ingly, the antibiosis rating (AR) appeared to be an additional factor which could distinguish *P. fluorescens* and *P. putida* strains. With one exception, the AR of the *P. putida* strains was at the high end of the scale (2.1 to 2.6 range). In contrast, the AR of the *P. fluorescens* strains was low, ranging from 0.6 to 2.0. Also, a possible relationship was noted between the AR of a strain and its effect on plant growth. Five of seven strains with an AR of 2.4 or higher exhibited an inhibitory effect on plant growth, and none of the seven promoted growth, whereas the seven growth-promoting strains all had ARs of 2.1 or lower. Only four strains in this category inhibited plant growth, and four had no effect.

Fluorescent pseudomonads produce a variety of secondary metabolites (43), some of which have broad inhibitory activity to bacteria or fungi (28, 29) or toxicity to plants (43). The production of large concentrations of these substances in the rhizosphere might lead to reduction in plant growth. It is only when antibiotic compounds are produced in amounts toxic to pathogens or deleterious minor pathogens but not to the plant that plant growth increases could occur.

It has become apparent that no single bacterial metabolite or single class of bacterial metabolites, such as phytohormones or antibiotics, will provide a simple explanation of the mechanism of plant growth promotion.

BACTERIA AS BIOCONTROL AGENTS AND ROOT COLONIZERS

It would be helpful to the development and use of bacteria as biocontrol agents if it were possible to synthesize, unify, and compose a framework of useful precepts and postulations. It is apparent, however, that the present state of knowledge makes this a risky task. Nevertheless, the following are some considerations and speculations.

(i) Biocontrol agents and root-colonizing bacteria belong to many different genera.

(ii) Rhizosphere bacteria populations are highly variable among plants in the field and are log-normally distributed (J. E. Loper, T. V. Suslow, and M. N. Schroth, Symp. 4th Int. Congr. Plant Pathol., Melbourne, Australia, 1983). Therefore, beneficial bacteria when utilized as seed inoculants will provide unequal protection among plants.

(iii) Root colonization should be visualized as

an interaction between the colonizer and the root, and may result in a "cost" to the plant manifested by increased root exudation.

(iv) Effective use of biocontrol agents may require some changes in cultural practices.

(v) Transplant systems offer excellent opportunities for use of bacterial inoculants for biocontrol or plant growth promotion.

(vi) The most successful use of bacteria will probably be in moist soils, although actinomycetes and other bacteria which can withstand low soil moistures may prove effective against fungi such as the fusaria, which tolerate low soil moistures.

(vii) Phytohormones produced by bacteria may influence plant growth directly or indirectly. However, this influence will differ among plant species, and with concentration.

(viii) Specific plant growth-promoting bacterial strains will be more effective on crops of short growing season such as radishes. The pressures to return to a biological equilibrium come into effect with long-season crops.

(ix) Early-season growth promotion will be manifested by an increased yield with some crops, but not with others.

(x) Root-colonizing bacteria may be categorized as beneficial, neutral, and deleterious. Deleterious strains are more commonly isolated from roots than are proven beneficial ones.

(xi) A bacterium beneficial to one plant species may be deleterious to another.

(xii) Rhizobacteria may be relatively plant specific, cultivar specific, or nonspecific in root-colonizing ability.

(xiii) Populations of rhizosphere microorganisms vary among plant cultivars as do the quantity and composition of root exudates.

(xiv) Large populations of bacteria on seeds or seedlings may be deleterious because of competition for nutrients, O_2, or supraoptimal concentrations of bacterial metabolites.

(xv) Beneficial rhizobacteria probably occur on roots of all plants in all soil types. However, they may occur in higher proportion in suppressive soils.

(xvi) It is unlikely that one organism alone is responsible for a suppressive soil.

(xvii) The most easily controlled diseases are likely to be those in which host susceptibility is of short duration.

(xviii) Root pathogens of major importance will be more difficult to control than those of minor importance because of such factors as superior ability to survive, compete, resist antibiosis, or quickly infect susceptible tissues.

(xix) Biocontrol agents are more effective against pathogens where the inoculum occupies relatively restricted niches and is not widely distributed.

(xx) The most effective biocontrol agent may be an avirulent, antagonistic strain derived from or related to the pathogen.

(xxi) In vitro antibiosis is difficult to use as a criterion for selecting biocontrol agents for several reasons: its absence may merely be a reflection of the substrate; some antibiotics are not readily soluble in water; an antibiotic produced in vitro may be adsorbed or inactivated in a soil environment, or in sufficient quantities to affect a target organism; an antibiotic may affect only the pathogenic capability of the target organism, but not its ability to grow; and an antibiotic may be toxic to both the plant and the rhizosphere microflora.

(xxii) It is unlikely that the efficacy of a biocontrol agent can be attributed to any one property, such as the production of an inhibitory substance. As with pathogenesis, a number of biological processes may determine the observed events.

(xxiii) An inoculum containing a mixture of biocontrol agents with varied properties may be more effective than a single strain.

Although books and many papers have been written on the subject of biocontrol in plant pathology, much of the material is of an empirical nature. With most causal agents of plant disease, the associated microflora that affect their pathogenic and saprophytic behavior are not known. Even when identified, there is scant information available concerning the ecological, physiological, or biochemical characteristics that influence their potential to effect a biocontrol. This large group of organisms has been neglected principally because they are not of major interest to either plant pathologists or microbiologists. To develop and employ biocontrol agents effectively, there must be a much better understanding of the factors that favor their colonization of plant surfaces, survival, and dissemination, and an identification of mechanisms of antagonism. The role of biocontrol agents in the management of plant disease will assume greater significance commensurate with the effort that is placed on the study of their biology.

We thank Greg Davis, Mich Hein, and Michael Carnes of Monsanto Agricultural Products Company for useful discussions and phytohormone analysis.

LITERATURE CITED

1. **Alconero, R.** 1980. Crown gall of peaches from Maryland, South Carolina, and Tennessee and problems with biological control. Plant Dis. **64**:835–838.
2. **Baker, K. F.** 1980. Microbial antagonism—the potential for biological control, p. 327–347. *In* D. C. Ellwood, M. J. Latham, J. N. Hedger, J. M. Lynch, and J. H. Slater (ed.), Contemporary microbial ecology. Academic Press, London.
3. **Baker, K. F., and R. J. Cook.** 1974. Biological control of plant pathogens. W. H. Freeman and Co., San Francisco.

4. **Barber, D. A.** 1978. Nutrient uptake, p. 131–162. *In* Y. R. Dommergues and S. V. Krupa (ed.), Interactions between non-pathogenic soil microorganisms and plants. Elsevier Scientific Publishing Co., New York.

5. **Barber, D. A., and J. M. Lynch.** 1977. Microbial growth in the rhizosphere. Soil Biol. Biochem. **9:**305–308.

6. **Batra, M. W., K. L. Edwards, and T. K. Scott.** 1975. Auxin transport in roots: its characteristics and relationship to growth, p. 299–325. *In* J. G. Torrey and D. T. Clarkson (ed.), The development and function of roots. Academic Press, London.

7. **Biale, J. B., and F. F. Halma.** 1937. The use of heteroauxin in rooting of subtropicals. Proc. Am. Soc. Hortic. Sci. **35:**433–447.

8. **Broadbent, B., K. F. Baker, N. Franks, and J. Holland.** 1977. Effect of *Bacillus* spp. on increased growth of seedlings in steamed and in nontreated soil. Phytopathology **67:**1027–1034.

9. **Brown, M. E.** 1974. Seed and root bacterization. Annu. Rev. Phytopathol. **12:**181–197.

10. **Brown, M. E., and S. K. Burlingham.** 1968. Production of plant growth substances by *Azotobacter chroococcum.* J. Gen. Microbiol. **53:**135–144.

11. **Brown, M. E., S. K. Burlingham, and R. M. Jackson.** 1964. Studies on *Azotobacter* species in soil. III. Effects of artificial inoculation on crop yields. Plant Soil **20:**194–214.

12. **Brown, M. E., R. M. Jackson, and S. K. Burlingham.** 1968. Effects produced on tomato plants, *Lycopersicon esculentum,* by seed or root treatment with gibberellic acids and indolyl-3-acetic acid. J. Exp. Bot. **19:**554.

13. **Brown, M. E., and N. Walker.** 1970. Indoyl-3-acetic acid formation by *Azotobacter chroococcum.* Plant Soil **32:**250–253.

14. **Burr, T. J., and B. H. Katz.** 1983. Isolation of *Agrobacterium tumefaciens* biovar 3 from grapevine galls and sap, and from vineyard soil. Phytopathology **73:**163–165.

15. **Burr, T. J., M. N. Schroth, and T. V. Suslow.** 1978. Increased potato yields by treatment of seedpieces with specific strains of *Pseudomonas fluorescens* and *P. putida.* Phytopathology **68:**1377–1383.

16. **Campbell, R., and J. L. Faull.** 1979. Biological control of *Gaeumannomyces graminis:* field trials and the ultrastructure of the interaction between the fungus and a successful antagonistic bacterium, p. 603–609. *In* B. Schippers and G. Gams (ed.), Soil-borne plant pathogens. Academic Press, Inc., New York.

17. **Chang, I.-P., and T. Kommedahl.** 1968. Biological control of seedling blight of corn by coating kernels with antagonistic microorganisms. Phytopathology **58:**1395–1401.

18. **Chen, W., E. Echandi, and H. W. Spurr, Jr.** 1982. Protection of tobacco plants from bacterial wilt with avirulent bacteriocin-producing strains of *Pseudomonas solanacearum,* p. 482–492. *In* J. C. Lozano (ed.), Proceedings of the 5th International Conference on Plant Pathogenic Bacteria, 16–21 August 1981, Cali, Colombia. University of Missouri Press, Kansas City.

19. **Comai, L., and T. Kosuge.** 1980. Involvement of plasmid deoxyribonucleic acid in indoleacetic acid synthesis in *Pseudomonas savastanoi.* J. Bacteriol. **142:**950–957.

20. **Comai, L., and T. Kosuge.** 1983. Transposable element that causes mutations in a plant pathogenic *Pseudomonas* sp. J. Bacteriol. **154:**1162–1167.

21. **Cook, R. J.** 1977. Management of the associated microbiota, p. 145–166. *In* J. G. Horsfall and E. B. Cowling (ed.), Plant disease—an advanced treatise, vol. 1. Academic Press, Inc., New York.

22. **Cooper, R.** 1959. Bacterial fertilizers in the Soviet Union. Soils Fert. **22:**327–333.

23. **Costerton, J. W., and R. T. Irvin.** 1981. The bacterial glycocalyx in nature and disease. Annu. Rev. Microbiol. **35:**299–324.

24. **Costerton, J. W., R. T. Irvin, and K. J. Cheng.** 1981. The role of bacterial surface structure in pathogenesis. Crit. Rev. Microbiol. **8:**303–338.

25. **Ellis, J. G., A. Kerr, M. van Montagu, and J. Schell.** 1979. *Agrobacterium:* genetic studies on agrocin 84 production and the biological control of crown gall. Physiol. Plant Pathol. **15:**311–319.

26. **Frobisher, M.** 1968. Fundamentals of microbiology, 8th ed. W. B. Saunders Co., Philadelphia.

27. **Gordon, S. A., and R. P. Weber.** 1951. Colorimetric estimation of indoleacetic acid. Plant Physiol. **26:**192–195.

28. **Howell, C. R., and R. D. Stipanovic.** 1979. Control of *Rhizoctonia solani* on cotton seedlings with *Pseudomonas fluorescens* and with an antibiotic produced by the bacterium. Phytopathology **69:**480–482.

29. **Howell, C. R., and R. D. Stipanovic.** 1980. Suppression of *Phythium ultimum*-induced damping-off of cotton seedlings by *Pseudomonas fluorescens* and its antibiotic, pyoluteorin. Phytopathology **70:**712–715.

30. **Howie, W. J., and E. Echandi.** 1983. Rhizobacteria: influence of cultivar and soil type on plant growth and yield of potato. Soil Biol. Biochem. **15:**127–132.

31. **Hussain, A., and V. Vancura.** 1970. Formation of biologically active substances by rhizosphere bacteria and their effect on plant growth. Folia Microbiol. (Prague) **11:**468.

32. **Jackson, R. M., M. E. Brown, and S. K. Burlingham.** 1964. Similar effects on tomato plants of *Azotobacter* inoculation and application of gibberellins. Nature (London) **203:**851–852.

33. **Katznelson, H., and S. E. Cole.** 1965. Production of gibberellin-like substances by bacteria and actinomycetes. Can. J. Microbiol. **11:**733–741.

34. **Kerr, A.** 1972. Biological control of crown gall: seed inoculation. J. Appl. Bacteriol. **35:**493–497.

35. **Kerr, A.** 1980. Biological control of crown gall through production of agrocin 84. Plant Dis. **64:**24–30.

36. **Kloepper, J. W., J. Leong, M. Teintze, and M. N. Schroth.** 1980. *Pseudomonas* siderophores: a mechanism explaining disease suppressive soils. Curr. Microbiol. **4:**317–320.

37. **Kloepper, J. W., J. Leong, M. Teintze, and M. N. Schroth.** 1980. Enhanced plant growth by siderophores produced by plant growth-promoting rhizobacteria. Nature (London) **286:**885–886.

38. **Kloepper, J. W., and M. N. Schroth.** 1979. Plant growth promoting rhizobacteria on radishes, p. 879–882. *In* Station de Pathologie Végétale et Phytobactériologie (ed.), Proceedings of the IV International Conference on Plant Pathogenic Bacteria, vol. 2. I.N.R.A. Route de Saint-Clément Beaucouzé, 49000 Angers, France.

39. **Kloepper, J. W., and M. N. Schroth.** 1981. Relationship of *in vitro* antibiosis of plant growth-promoting rhizobacteria to plant growth and the displacement of root microflora. Phytopathology **71:**1020–1024.

40. **Kloepper, J. W., M. N. Schroth, and T. D. Miller.** 1980. Effects of rhizosphere colonization by plant growth-promoting rhizobacteria in potato plant development and yield. Phytopathology **70:**1078–1082.

41. **Kommedahl, T., and I. C. Mew.** 1965. Biocontrol of corn root infections in the field by seed treatment with antagonists. Phytopathology **65:**296–300.

42. **Lamanna, C., M. F. Mallette, and L. N. Zimmerman.** 1973. Basic bacteriology; its biological and chemical background, 4th ed., p. 624. The Williams & Wilkins Co., Baltimore.

43. **Leisinger, T., and R. Margraff.** 1979. Secondary metabolites of the fluorescent pseudomonads. Microbiol. Rev. **43:**422–442.

44. **Libbert, E., and P. Silhengst.** 1970. Interactions between plants and epiphytic bacteria regarding their auxin metabolism. VIII. Transfer of ^{14}C-indoleacetic acid from epiphytic bacteria to corn coleoptiles. Physiol. Plant. **23:**480–487.

45. **Loper, J. E., M. N. Schroth, and N. J. Panopoulos.** 1982. Influence of bacterial sources of indole-3-acetic acid (IAA) on root elongation of sugar beet. Phytopathology **72:**997.

46. **Mason, T. G., and J. H. Slater.** 1979. Competition between an *Escherichia coli* tyrosine auxotroph and a proto-

trophic revertant in glucose- and tyrosine-limited chemostats. Antonie van Leewenhoek, J. Microbiol. Serol. **45**:253–263.

47. **Merriman, P. R., R. D. Price, and K. F. Baker.** 1974. The effect of inoculation of seed with antagonists of *Rhizoctonia solani* on the growth of wheat. Aust. J. Agric. Res. **25**:213–218.

48. **Merriman, P. R., R. D. Price, J. F. Kollmorgen, T. Piggott, and E. H. Ridge.** 1974. Effect of seed inoculation with *Bacillus subtilis* and *Streptomyces griseus* on the growth of cereals and carrots. Aust. J. Agric. Res. **25**:219–226.

49. **Mishustin, E. N., and A. N. Naumova.** 1962. Bacterial fertilizers, their effectiveness and mode of action. Transl. Mikrobiol. **31**:543–555.

50. **Moore, T. C.** 1979. Biochemistry and physiology of plant hormones. Springer-Verlag, Inc., New York.

51. **Nair, N. G., and P. C. Fahy.** 1976. Commercial application of biological control of mushroom bacterial blotch. Aust. J. Agric. Res. **27**:415–422.

52. **Nair, S. K., and P. Tauro.** 1979. Effect of inoculation with *Pseudomonas azotogensis* on the yield and straw weight of wheat. Plant Soil **52**:453–455.

53. **Nakas, J. P., and D. A. Klein.** 1980. Mineralization capacity of bacteria and fungi from the rhizosphere-rhizoplane of a semiarid grassland. Appl. Environ. Microbiol. **39**:113–117.

54. **Panagopoulos, C. G., P. G. Psallidas, and A. S. Alivizatos.** 1978. Studies on biotype 3 of *Agrobacterium radiobacter* var. *tumefaciens*, p. 221–228. *In* Station de Pathologie Végétale et Phytobactériologie (ed.), Proceedings of the IV International Conference on Plant Pathogenic Bacteria, vol. 1. I.N.R.A. Route de Saint-Clément Beaucouzé, 49000 Angers, France.

55. **Panagopoulos, C. G., P. G. Psallidas, and A. S. Alivizatos.** 1979. Evidence of a breakdown in the effectiveness of biological control of crown gall, p. 569–578. *In* B. Schippers and G. Gams (ed.), Soil-borne plant pathogens. Academic Press, Inc., New York.

56. **Papavizas, G. C., and R. D. Lumsden.** 1980. Biological control of soilborne fungal propagules. Annu. Rev. Phytopathol. **18**:589–413.

57. **Prikryl, Z., and V. Vančura.** 1980. Root exudates of plants. VI. Wheat root exudation as dependent on growth, concentration gradient of exudates, and the presence of bacteria. Plant Soil **57**:69–83.

58. **Reynolds, B. L., and H. Pruul.** 1971. Protective role of smooth lipopolysaccharide in the serum bactericidal reaction. Infect. Immun. **4**:764–771.

59. **Riggle, J. H., and E. J. Klos.** 1972. Relationship of *Erwinia herbicola* to *Erwinia amylovora*. Can. J. Bot. **50**:1077–1083.

60. **Rovira, A. D.** 1963. Microbial inoculation of plants. I. Establishment of free-living nitrogen-fixing bacteria in the rhizosphere and their effects on maize, tomato, and wheat. Plant Soil **19**:304–314.

61. **Salt, G. A.** 1979. The increasing interest in "minor pathogens," p. 289–312. *In* B. Schippers and G. Gams (ed.), Soil-borne plant pathogens. Academic Press, Inc., New York.

62. **Scher, F. M., and R. Baker.** 1980. Mechanism of biological control in a *Fusarium*-suppressive soil. Phytopathology **70**:412–417.

63. **Scher, F. M., and R. Baker.** 1982. Effect of *Pseudomonas putida* and a synthetic iron chelator on induction of soil suppressiveness to *Fusarium* wilt pathogens. Phytopathology **72**:1567–1573.

64. **Schroth, M. N., and J. W. Hancock.** 1981. Selected topics in biological control. Annu. Rev. Microbiol. **35**:453–476.

65. **Slater, J. H., and D. Godwin.** 1980. Microbial adaptation and selection, p. 137–160. *In* D. C. Ellwood, J. N. Hedger, M. J. Latham, J. M. Lynch, and J. H. Slater (ed.), Contemporary microbial ecology. Academic Press, London.

66. **Stanier, R. Y., N. J. Palleroni, and M. Doudoroff.** 1966. The aerobic pseudomonads: a taxonomic study. J. Gen. Microbiol. **43**:159–271.

67. **Sule, S.** 1978. Pathogenicity and agrocin 84 resistance of *Agrobacterium tumefaciens*, p. 255–257. *In* B. Schippers and G. Gams (ed.), Soil-borne plant pathogens. Academic Press, Inc., New York.

68. **Suslow, T. V.** 1982. Role of root-colonizing bacteria in plant growth, p. 187–223. *In* M. S. Mount and G. H. Lacy (ed.), Phytopathogenic prokaryotes, vol. 1. Academic Press, Inc., New York.

69. **Suslow, T. V., and M. N. Schroth.** 1982. Role of deleterious rhizobacteria vs. minor pathogens in reducing crop growth. Phytopathology **72**:111–115.

70. **Suslow, T. V., and M. N. Schroth.** 1982. Rhizobacteria of sugar beets: effects of seed application and root colonization on yield. Phytopathology **72**:199–206.

71. **Sweetser, P. B., and D. G. Swartzfager.** 1978. Indole-3-acetic acid levels of plant tissue as determined by a new high performance liquid chromatographic method. Plant Physiol. **61**:254–258.

72. **Swinburne, T. R., and A. E. Brown.** 1976. A comparison of the use of *Bacillus subtilis* with conventional fungicides for the control of apple canker (*Nectria galligena*). Ann. Appl. Biol. **82**:365–368.

73. **Thimann, K. V.** 1937. On the nature of inhibition caused by auxin. Am. J. Bot. **24**:407–412.

74. **Tu, C. C., Y. H. Cheng, and Y. C. Chang.** 1978. Antagonistic effect of some bacteria from *Fusarium* wilt-suppressive soil and their effect on the control of flax wilt in the field. Taiwan J. Agric. Res. China **27**:245–258.

75. **Utkhede, R. S., and J. E. Rahe.** 1980. Biological control of onion white rot. Soil Biol. Biochem. **12**:101–104.

76. **Weller, D. M., and J. R. Cook.** 1983. Suppression of take-all of wheat by seed treatments with fluorescent pseudomonads. Phytopathology **73**:463–469.

77. **Zamenhof, S., and H. H. Eichhorn.** 1967. Study of microbial evolution through loss of biosynthetic functions: establishment of defective mutants. Nature (London) **216**:456–458.

Biocontrol of *Botrytis cinerea* on Grapevines by an Antagonistic Strain of *Trichoderma harzianum*

BERNADETTE DUBOS

I.N.R.A., Station de Pathologie Végétale, Domaine de la Grande Ferrade, 33140 Pont de la Maye, France

The integration of biological treatments in protection of vineyards is attractive in view of the number of chemical operations used during the vegetative period of the vine as well as the drawbacks this type of protection can create (pollution, biological balance, disruption, etc.).

Research carried out in Bordeaux since 1974 has shown (6) that the fungus *Trichoderma harzianum* is antagonistic on the aerial parts of the vine against *Botrytis cinerea*, an important parasite which causes gray mold. In trials in the vineyards near Bordeaux, treatments composed of a diluted mixture of *T. harzianum* spores grown on an oat agar nutritive medium, adjusted to a concentration of 10^8 conidia per ml and administered four times (the "standard method" [4]), achieved a control efficiency of 60 to 70% over several years (see Table 1). Though excellent, these results are inferior to those obtained by a chemical treatment using specific, particularly active fungicides such as procymidone, vinchlozoline, or iprodione, with which a control efficiency of 80 to 95% can be achieved. Unlike these fungicides, however, *T. harzianum* does not lead to the development of any resistant strains of *B. cinerea*.

Combining biological and chemical forms of control may reduce the drawbacks of each as well as providing improved protection for the grapes and reducing the possibility of development of chemically resistant strains of the parasite. To realize this new strategy the action of *T. harzianum* against *B. cinerea* in vitro as well as in vivo must be examined.

IN VITRO ACTION OF *T. HARZIANUM* AGAINST *B. CINEREA*

In vitro interactions of pathogenic fungi and antagonists including *T. harzianum* are frequently studied and will be mentioned here only briefly. The latest and best works in this field are those of Dennis and Webster (4, 5) and Henis and Chet (7).

The mechanisms usually considered are: (i) antibiosis, the production of substrates with a fungistatic or fungicidal effect; (ii) competition, the capacity to exploit environmental factors to the detriment of the parasite; and (iii) mycoparasitism, the destruction or deterioration of the parasite by predation or enzymatic lysis.

Although it has not been demonstrated that the antagonist's behaviors in the laboratory and in nature are identical, in vitro study of the above mechanisms does aid in determining an antagonist's potential. These general forms of action provided the basis for selection of the most appropriate strain of *Trichoderma* spp.

We have determined that the interactions of the pair *Trichoderma* sp./*Botrytis* sp. are strictly dependent on temperature. Although practically every strain of *Trichoderma* sp. possesses antagonistic properties between 15 and 25°C, above and below these limits only a few exhibit these properties. Although in vitro studies are often elaborate and technically very clever, a factor as simple yet significant as temperature is rarely mentioned, except by Tronsmo and Dennis (8).

MODES OF ACTION OF *T. HARZIANUM* AGAINST *B. CINEREA* IN THE VINEYARD

The antagonistic effect of *T. harzianum* interferes with successive phases of the development of *B. cinerea*.

Effect on the establishment of the parasite. Research of the most suitable periods of intervention in the vineyard shows that treatment at the end of flowering is obviously important (Table 2). There is no precise period for the administration of the antagonist; it can last from the beginning of flowering until the floral parts fall.

One of the characteristics of *B. cinerea* is its saprophytic phase, which is as important as the parasitic phase. The fungus colonizes various plant fragments, where it develops abundantly and from this nutritive base infects sound green organs. It had been suspected that *T. harzianum* could first colonize the senescent floral parts adhering to the berries, thus preventing the saprophytic establishment of *B. cinerea*. This hypothesis has now been proven by many experiments.

Isolations made from floral parts collected in plots treated or not treated with *T. harzianum*. Two lots (treated and control), each composed of 600 floral parts, were picked at random (100 from each elementary plot in a trial including six repetitions) and were cultivated on a nutritive salt agar medium. The percentages of floral parts

TABLE 1. Biological control of *B. cinerea* by *T. harzianum* over 6 years of experimentation[a]

Years	Overall% of rot (treated)[b]	Overall % of rot (control)[b]	% Efficiency[c]
1976	9.4	31.4	70
1977	9.5	33.6	71.7
1978	7.6	24.6	69.3
1979	4.2	22.5	81.1
1980	7	15.9	55.9
1981	9.7	24	59
1982	4	27	84

[a] Concentration of the spore preparation was 10^8 conidia per ml. Treatment was applied at four points: the end of flowering, bunching of the grapes, the onset of ripening, and 3 weeks before harvesting.

[b] For 100 bunches of grapes examined per plot of land.

[c] Percentage of efficiency obtained according to Abbott's formula (1).

colonized by various fungi are compared in Table 3.

The results show that 60.5% of the floral parts in the control lots carried *B. cinerea*, compared with 11.5% of the treated floral parts. Scanning electron microscopy of the floral parts showed that the conidia of the antagonist germinate and colonize the tissues; moreover, the fruiting structures of *T. harzianum* can be observed in the vineyard in favorable climatic conditions (when the weather is wet and mild).

Protection of the grapes at knotting (end of flowering). Bulit and Lafon (2) have shown that direct infection of the young berries by a suspension of conidia of *B. cinerea* does not occur. On the other hand, infection via contaminated floral parts is a total success, demonstrating the importance of the saprophytic phase.

When the floral parts are rendered non-colonizable by the parasite, the possibility of disease developing on the grapes is reduced.

TABLE 2. Effect of treatment with *T. harzianum* at flowering[a]

Years	Efficiency of treatment at the end of flowering (%)	Efficiency of four applications made according to standard method (%)
1976	39.8	70
1977	39.2	71.7
1978	40.1	69.3
1979	81.1	81.1
1980	60.4	57
1981	ND	59.6

[a] Data expressed as percentage of efficiency obtained, according to Abbott's formula (1) for 100 bunches of grapes examined per plot of land. Treatment was applied at four stages: the end of flowering, bunching of the grapes, the onset of ripening, and 3 weeks before harvesting. ND, Not done.

TABLE 3. Percentage of grape floral parts colonized by various fungi

Colonizer	% of flower parts colonized	
	Treated	Control
T. harzianum	86.2	5
T. harzianum + *B. cinerea*	8	0.5
B. cinerea	3.5	60
Miscellaneous fungi	2.3	34

For this experiment, 300 grapes visibly unaffected by gray mold and still bearing their corollas (the senescent parts of the flower) were picked when their diameter reached 3 to 5 mm. They were then placed pedicel down in petri dishes of agar water. The berries were divided into three parts: the corollas of the first lot were extracted and dipped for 15 min into a spore suspension of *T. harzianum*; the corollas of the second lot were treated like the first but were also dipped into a culture filtrate of *T. harzianum* grown on a malt liquid medium; and the third lot was untreated. In the treated plots (1 and 2) the floral parts were inoculated with a drop of suspension of *B. cinerea* conidia (500,000 conidia per ml). The petri dishes were incubated at 20°C. The results, recorded within 2 weeks, demonstrated that colonization of floral parts by *B. cinerea* is impeded when the parts have been either previously colonized by the antagonist or treated with a culture filtrate of *T. harzianum* (Table 4).

Action on the development of the parasite. The disease generally starts spreading at the onset of the ripening. As ripening ends and maturation begins, the grapes can be contaminated either by the conidia of the parasite present in the vineyard atmosphere (3) or by mycelium which spreads from the first rotting grapes to the sound grapes. Thus, receptivity of the grapes in conjunction with the quantity of inoculum can lead to an explosion of gray mold in a very short time.

The action of *T. harzianum* on the development of the parasite is illustrated by comparing the percentage of gray mold in the treated and

TABLE 4. Colonization of young berries through their corollas

Lot	Treatment[a]	*B. cinerea* colonization (%)	
		Floral parts	Berries
1.	*T. harzianum* + *B. cinerea*	17	12
2.	*T. harzianum* + *B. cinerea* + *T. harzianum* filtrate	22	19
3.	*B. cinerea* only	95	89

[a] For details, see the text.

control plots used in the vineyard trials (Table 5). From this table it is evident that although the percentages of affected bunches did not differ statistically, the percentages of gray mold did.

At the end of flowering, when the inoculum is abundant, although the antagonist does not limit the number of infection foci it does limit their development. This hypothesis was checked in the vineyard, where the antagonist was observed to fructify on the grapes on both the affected and sound parts, barring development of mycelium of the parasite. *T. harzianum* apparently acts as a hyperparasite; the classical forms of parasitism can be observed near the necrosis by a microscope.

Action on parasite conservation. When the weather turns cold *B. cinerea* no longer parasitizes the fruit, and it begins to form sclerotia. The sclerotia are sometimes abundant on the vine shoots, especially at the ends where maturation is defective. It is possible to get rid of some but not all sclerotia by pruning the vine.

The action of *T. harzianum* on the formation of sclerotia on the vine was evaluated as shown in Table 6. The results, which were statistically confirmed, show clearly that treatments applied during vegetative growth (at the period of the standard method), whether chemical or biological, did not deter sclerotia formation, nor did treatment with *T. harzianum* after harvesting.

Action on parasite dissemination. Late in winter or in early spring, when the weather is wet and mild, the sclerotia carried by the bunches are covered with conidia. Generally during mid-April and May the sclerotia develop more conidia (2), at least in the southwest of France. This production of conidia is responsible for the first attack of the parasite on the leaves.

The action of *T. harzianum* on the conidia which developed from the sclerotia was studied in natural conditions. Vine shoots bearing sclerotia were cut into pieces of 20 cm. These were laid on the ground in a circle around a spore trap. Two lots were observed. One plot was treated during the second half of March with a *T. harzianum* spore suspension. The second was left untreated to serve as control. The develop-

TABLE 6. Action of *T. harzianum* on formation of sclerotia of *B. cinerea*

Variable (no. of treatments)	Avg. no. of sclerotia
4 (standard method).....................	15.4
3 (early maturation until fall of leaves)....	8.9
2 (after grape harvest; during leaf fall)	17.1
Dichlofluanide per standard method	14.9
Control...............................	16.3

ment of the conidia population produced by the sclerotia was observed from April to June, using methods described by Bulit and Verdu (3).

Germination of the sclerotia begins on wet days, with maximum production during the second half of May. *T. harzianum* did not delay fructification but did considerably reduce the number of conidia produced (about 40%). Microscopic observations and cultivation in petri dishes proved that many sclerotia were destroyed by *T. harzianum*, as were conidiophores before they were able to produce conidia.

DISCUSSION

Experiments in the vineyard since 1974 with a biopreparation with a *T. harzianum* base have led to a definition of the conditions for use of this fungus in the fight against gray mold (*B. cinerea*). Studies of this antagonist's actions on the parasite indicate that *T. harzianum* can repress *B. cinerea* at every stage of its development on the vine, especially the following.

(i) Installation of the parasite. Application of the antagonist at flowering permits colonization of the senescent floral parts adhering to the berries. This colonization impedes the saprophytic phase of the development and *B. cinerea* slows the appearance of its first centers in the vineyard.

(ii) Development of the parasite. With its hyperparasitic ability, *T. harzianum* can limit the development of *B. cinerea* on the grapes.

(iii) Preservation of the parasite. Application of *T. harzianum* at the end of vegetation significantly diminishes the frequency of sclerotia on the wood.

(iv) Dissemination of the parasite. Study of conidia indicates that although the antagonist does not slow the appearance of fructification on the sclerotia, it does significantly reduce the number of conidia produced.

These last two properties of the antagonist are capable of diminishing the inoculum responsible for early parasite attacks in the vineyard as spring begins.

These results naturally lead to two questions: how can *T. harzianum* best be put to use, and more particularly, how can it be used in vine-

TABLE 5. Comparison between the percentage of affected bunches and the percentage of berries with gray mold

Treatment[a]	Infected bunches (%)	Berries per bunch with gray mold (%)
Treated with *T. harzianum*	78.2	9.4
Control	93	31.4

[a] Four treatments: at the end of flowering, at bunching of grapes, at the onset of ripening, and 3 weeks before harvesting.

yards throughout France? These questions require a look backwards.

The modern period of the fight against *B. cinerea* in the vineyard has consisted of two phases. In the first the phenological method of treatments, the standard method which consists of the systematic application of at least four chemical treatments, was employed. In the second phase, which is now being developed, a justification is sought for each of the treatments in the standard method. This phase was made possible because of the shape of the *Botrytis* epidemics. The mathematical model takes into account the interrelations among climate, parasite, and plant. This step may anticipate a new phase in the evolution of the fight against *Botrytis*, a phase which combines biological intervention with a reduced number of chemical treatments. This phase will call for the introduction of another biotic factor in the climate-parasite-plant system, so that the system will become fourfold: climate, parasite, antagonist, and plant.

Conceptualization of the model must provide for integration of the new factor, and the mathematical model must take into account the action of the climate on the antagonist as well as on the parasite.

The integration of *T. harzianum* in a well-thought-out fight against gray mold in the vineyard will contribute to the elaboration of treatment methods which are not only profitable but also less damaging to the environment than purely chemical methods. Its significance is now recognized; at present French industries (Société Orsan) are able to produce functioning formula of the antagonist on a large scale.

LITERATURE CITED

1. **Abbott, W. S.** 1925. A method of computing the effectiveness of an insecticide. J. Econ. Entomol. **18**:265–267.
2. **Bulit, J., and R. LaFon.** 1972. Biologie due *Botrytis cinerea* Pers. et le développement de la pourriture grise de la vigne. Rev. Zool. Agric. Pathol. Veg. **71**:1–10.
3. **Bulit, J., and D. Verdu.** 1973. Annual variations in the aerial sporing of *Botrytis cinerea* Pers. in a vineyard. *In* Symposium on biology and control of *Botrytis* diseases. Skierniewitch, Poland.
4. **Dennis, C., and J. Webster.** 1971. Antagonistic properties of species groups of *Trichoderma*. I. Production of nonvolatile antibiotics. Trans. Br. Mycol. Soc. **57**:41–48.
5. **Dennis, C., and J. Webster.** 1971. Antagonistic properties of species groups of *Trichoderma*. III. Hyphal interaction. Trans. Br. Mycol. Soc. **57**:363–369.
6. **Dubos, B., Y. Bulit, Y. Bugaret, and D. Verdu.** 1978. Possibilitiés d'utilisation du *Trichoderma viride* Pers. Comme moyen biologique du lutte contre la pourriture grise (*Botrytis cinerea* Pers.) et l'excoriose (*Phomopsis viticola* Sacc.) de la vigne. C. R. Seances Acad. Agric. Fr. **64**:1159–1168.
7. **Henis, Y., and I. Chet.** 1975. Microbial control of plant pathogens. Adv. Appl. Microbiol. **19**:85–111
8. **Tronsmo, A., and C. Dennis.** 1978. Effect of temperature on antagonistic properties of *Trichoderma* species. Trans. Br. Mycol. Soc. **71**:469–474.

Control of Gastrointestinal Pathogens by Normal Flora

GERALD W. TANNOCK

Department of Microbiology, University of Otago, Dunedin, New Zealand

NORMAL FLORA OF THE GASTROINTESTINAL TRACT

The gastrointestinal tract of healthy animals harbors a large collection of different microbial types. The physical distribution of the different types in the tract varies according to the species of animal acting as host to the microbes. Thus, in rodents gram-positive bacilli (lactobacilli) predominate in the upper regions of the gastrointestinal tract whereas gram-negative anaerobes predominate in the lower regions of the tract. Ruminants, in contrast, harbor large populations of gram-negative anaerobes in an upper region of the tract (the rumen) and also in the large intestine. The size of the microbial population in different regions of the tract also varies between animal species. Relatively large numbers of bacteria are present in the rodent stomach (about 10^8/g of organ), but only small numbers (10^3/ml of contents) are detected in the healthy human stomach. Certain microbes comprising the normal flora can be observed to associate with the mucosal surface of gastrointestinal organs. Lactobacilli, for example, form layers of bacterial cells on the surface of squamous epithelia in the gastrointestinal tracts of rodents, fowls, and pigs. Gram-negative anaerobes with fusiform morphology form a layer in the mucus coating the large intestinal mucosa of rodents.

Observations made with domestic and experimental animals suggest that, as long as the diet and environment of an animal remain constant, the collection of microbes harbored in the gastrointestinal tract, and their specific distributions, also remain constant. The microbial populations comprising the normal flora are influenced by exposure of the host animal to environmental and dietary changes, but the microbial populations are probably mostly regulated by factors generated within the gastrointestinal ecosystem. These autogenic factors of regulation are generally considered to include competition for space and nutrients, amensalism, parasitism, and predation. These factors, acting in synergy with host physiological factors and host immunological mechanisms, lead to the containment of a large microbial collection under steady-state conditions in the gastrointestinal tract. Detailed descriptions of the phenomena mentioned above may be found in the reviews by Savage (25) and Tannock (28).

EVIDENCE THAT THE NORMAL FLORA CONTROLS GASTROINTESTINAL PATHOGENS

The use of broad-spectrum antibiotics such as chlortetracycline and oxytetracycline in the 1950s drew attention to the fact, in modern times, that the normal flora of the gastrointestinal tract helps to prevent the establishment of pathogens in that region of the host animal. A proportion of patients treated orally with these antibiotics developed severe dysentery caused by *Staphylococcus aureus* (10). More recently, the association of toxin-producing *Clostridium difficile* strains with colitis in antibiotic-treated hospital patients has been extensively studied (13). In general, we can say that the long-term treatment of patients with antibiotics that reduce the numbers of certain members of the normal flora of the gastrointestinal tract leads to the disruption of the normal regulatory mechanisms within the ecosystem. Certain opportunistic pathogens (e.g., *S. aureus* and *C. difficile*) can then attain higher than normal population levels in the intestinal tract and produce symptoms of disease. Interestingly, toxigenic *C. difficile* strains are present in relatively high numbers in the intestinal tract of infants, yet no signs of disease are present (27). It can be speculated that in this situation the normal flora inactivates any toxins produced by *C. difficile* or provides conditions in the ecosystem which are not optimal for toxin production by clostridia.

Experimentally, the activities of the normal flora in suppressing the growth of gastrointestinal pathogens are best seen by comparing the resistance of germfree and conventional animals to infection. Germfree animals, which are derived originally by sterile cesarean delivery and are housed throughout life in sterile surroundings, are free from all demonstrable forms of microbial life. Conventional animals, in contrast, are housed in readily accessible animal rooms and harbor a complete normal flora. Germfree animals are more readily colonized and infected by gastrointestinal pathogens than are conventional animals. As few as 10 viable *Salmonella enteritidis* cells, for example, introduced into the gastrointestinal tract of germfree mice will result in the death of all challenged animals within 5 to 8 days. The corresponding value for challenged conventional mice is about

10^9 viable salmonellae (6). The ease with which salmonellae can establish in the intestinal tract of germfree mice was shown by inoculating the animals with 10^4 S. *typhimurium* cells intragastrically. Three days later, population levels of 10^7 to 10^8 salmonellae per g of organ were present in the small and large intestines. In contrast, germfree mice which had been inoculated with cecal contents from a conventional mouse prior to challenge with salmonellae had either no or very few salmonellae in these organs (22).

Domestic animals, such as sheep, cattle, and pigs, maintained under harsh environmental or dietary conditions are known to be more susceptible to infection by gastrointestinal pathogens. Holding animals in pens or yards under normal farm management practices, transportation, or radical changes in diet can lead to outbreaks of salmonella infection (29). The "precipitating" or "stress" factors which precede such outbreaks no doubt alter the animals' physiology in ways which render their tissues more susceptible to infection. The normal flora of stressed animals is also markedly altered, and the breakdown of regulatory mechanisms in the gastrointestinal ecosystem allows the easier establishment of pathogens in the tract (29).

Observations made on antibiotic-treated subjects, germfree animals, and stressed animals provide conclusive evidence, therefore, that a major beneficial effect of the normal flora of the gastrointestinal tract is its ability to interfere with the establishment of pathogens in the host animal.

STUDYING MICROBIAL INTERFERENCE IN THE GASTROINTESTINAL TRACT

It is necessary to reduce the complexity of the gastrointestinal ecosystem in some way if we are to attempt to understand the specific mechanisms by which the components of the normal flora interfere with the establishment of pathogens. Several approaches have been used to achieve the goal of developing simplified systems in which microbial interactions can be observed.

In vitro experiments. Microbe–microbe and microbe–inhibitor interactions can be observed under controlled conditions in the laboratory by use of liquid media or agar plate cultures. Experiments of this nature frequently demonstrate antagonistic phenomena between microbial strains, but such interactions are produced in a closed system. It is unlikely that they even remotely reproduce the open, in vivo situation.

Hentges and Freter (15) used six in vitro methods to study antagonism between bacterial strains. Each method gave results different from those obtained with all the others. These authors concluded that, given any two bacterial cultures, A and B, it might be possible to devise one culture method which shows that A inhibits B and another method in which the opposite result is produced.

An open, steady-state system with which to study microbial interactions in vitro is provided by the use of continuous culture devices maintained in anaerobic glove boxes. The volume of the culture can be kept constant, the population density of microbes is controlled, nutrients are added to and removed from the system at controlled rates, and pH can be monitored and adjusted. The use of chemostats enabled Freter (12) to identify hydrogen sulfide produced by anaerobic bacteria as an inhibitor of *Escherichia coli* under conditions which simulate, to some extent, the large intestinal ecosystem of mice. Chemostats, however, can never completely reproduce the in vivo situation. Mucosal surfaces and their associated flora are absent from the model system. Nutrients available in the chemostat are usually peptides, yeast extract, and simple carbohydrates. Nutrients available to microbes inhabiting the large bowel of mice are largely of a macromolecular nature: mucus secreted by the host, extruded epithelial cells, steroid molecules from bile, host enzymes, and undigested components of the host's diet. Many investigators therefore prefer to work with experimental or other animals in studying interference mechanisms.

Animals whose normal flora is simplified by the administration of antibiotics. The prior administration of an antibiotic orally or intragastrically can reduce the number of certain members of the normal flora to undetectable levels. The resistance of these treated animals to intestinal infection can then be tested and compared with that of untreated animals. Antibiotic treatment in general, however, lacks specificity. More than one microbial type is affected by an antibiotic, and treated animals eventually harbor an antibiotic-resistant flora unlike that found in untreated animals (26).

Studying the flora of infected animals. Several studies have compared the normal flora of subjects suffering from intestinal infections with that of healthy controls (28). Differences in the incidence or numbers of certain microbes have been observed between the two groups in some cases. Alterations of the normal flora in these situations, however, are likely to be caused by the symptoms of the infection (e.g., diarrhea) rather than reflecting the altered microbial ecology of the intestinal tract which may have existed prior to, or at the time of, infection. The alterations in the fecal flora that are observed in patients are probably caused by the increased volume of fluid entering, and passing through, the large intestine, which alters the ecosystem so that it is less suitable for the growth of anaerobes (28).

Studying the normal flora of stressed animals. Since stressed animals are known to be more susceptible than healthy animals to intestinal infection, differences in the normal flora between these two groups can provide clues as to the identity of microbes involved in suppressing the establishment of a pathogen. Modern ethical experimentation committees, however, are unlikely to view stress-inducing experiments with favor.

Gnotobiotic (defined flora) animals. Germfree animals, which are essentially "pure cultures" of appropriate animal species, can be used as defined systems to observe the influence of specific members of the normal flora on the host. Single or multiple strains of microbes can be added to the system. The control of gastrointestinal pathogens in the resulting gnotobiotic animals can be compared with that in germfree and conventional animals.

Good gnotobiological studies are based on ecological principles. (i) The members of the normal flora chosen for use in gnotobiotic experiments should show the same regional distribution, and attain similar population levels, in the gastrointestinal tract of the gnotobiotic animals as in conventional animals. (ii) The specificity of some microbial types for certain animal hosts must be recognized (31). (iii) The order in which the microbial strains are added to the experimental system can influence the outcome of the experiment (8). (iv) The nature of the animal's diet can influence the outcome of the experiment (11). (v) The animal species used in the experiment must be an appropriate one for use with the chosen pathogen.

Are we using an animal model that can be related to the naturally occurring infection? Does the pathogen used in our studies infect the same region of the intestinal tract of the gnotobiotic animal as it does in the conventional animal? Is the pathogen really infecting the gnotobiotic animal, or is it merely colonizing the gnotobiotic intestinal tract? Evidence of damage to the host, in gnotobiotic as in conventional studies, should be observed (28).

CONTROL OF SALMONELLAE IN THE GASTROINTESTINAL TRACT

Examples of the investigative approaches described in the previous section are provided by attempts to elucidate the mechanisms by which moderate numbers of salmonellae are prevented from establishing in the intestinal tract.

Bergeim et al. (1), in the 1940s, observed that *Salmonella* species were killed by solutions of butyric acid at a pH of 6.0. These workers speculated, as a result of these in vitro observations, that volatile fatty acids could be involved in restricting the growth of pathogens in the intestinal tract. Bohnhoff et al. (2) found that the multiplication of *S. enteritidis* was inhibited in vitro by buffered suspensions of contents from the large intestine of mice. Further, acetic and butyric acids were present in the intestinal contents in concentrations that inhibited salmonella growth in vitro. The inhibitory effect of the intestinal contents was influenced by pH and was improved under anaerobic conditions. Intestinal contents from mice that had been treated orally with streptomycin 24 h previously had a higher pH and a lower concentration of volatile fatty acids than those of untreated mice. Intestinal contents from streptomycin-treated mice did not inhibit the growth of salmonellae in vitro. The changes in pH and volatile fatty acid concentration produced by streptomycin treatment seemed to account for the fact that streptomycin-treated mice were less resistant than untreated mice to salmonella infection, when challenged by the oral route. Meynell (18) also observed that the cecal contents of mice contained volatile fatty acids. The Eh (oxidation–reduction potential) of the cecal contents was about -0.2 V. This type of environment, in vitro, inhibited the growth of *S. typhimurium*. Streptomycin treatment of mice lowered the volatile fatty acid concentration and raised the Eh of the cecum to about $+0.2$ V. These conditions did not inhibit salmonellae in vitro. A combination of in vitro experiments and observations made on the susceptibility of antibiotic-treated mice to salmonella infection thus pointed to the involvement of volatile fatty acids, especially butyric acid, as important substances in the control of salmonellae in the intestinal tract. The volatile fatty acids at a pH in which they are in the undissociated form, and under anaerobic conditions, inhibit the multiplication of salmonellae.

In identifying the members of the normal flora which are involved in suppressing salmonella growth in the cecum, therefore, it is necessary to consider the types of microbes present in the large intestine of mice which could influence the butyric acid concentration, pH, and Eh of the ecosystem. The numerically predominant bacteria in the mouse large intestine are fusiform shaped (bacilli with tapered ends). Many of these fusiform types produce volatile fatty acids, including butyric acid (23). Anaerobic bacteria, which include fusiforms, are capable of maintaining a highly reduced environment once they are established in the ecosystem. Anaerobes which have fermentative metabolism also influence the pH of their environment through the production of acids. Anaerobic, butyric acid-producing fusiforms are therefore the likely agents of salmonella suppression in the mouse cecum. Circumstantial evidence derived from observations on stressed animals supports the concept that fusiform bacteria which produce butyric acid are involved in the suppression of salmonella growth in the intestinal tract. Dietary and environmental stress lowers the resistance of mice to salmonella infection. This kind of stress is known to reduce the number of fusiform-shaped bacteria in the large intestine of mice (29). Stress decreases the volatile fatty acid concentration of the mouse cecum about 2.5 times (4).

It is doubtful, however, that volatile fatty acid concentration in conjunction with appropriate pH and Eh is the only mechanism by which the normal flora interferes with salmonella multiplication in the intestinal tract. Roach and Tannock (22) observed that gnotobiotic mice associated with strains of *Lactobacillus, Bacteroides*, and *Clostridium* had fewer *S. typhimurium* present in the ileum 3 days after intragastric challenge with the pathogen than did similarly challenged germfree mice. Salmonella numbers remained high (about 10^8/g of organ) in the cecum of the gnotobiotic mice and did not differ from the level found in salmonella-inoculated germfree mice. This was despite the presence of acetic, propionic, and butyric acids (admittedly present in smaller amounts than in the conventional cecum) in the cecal contents. Volatile fatty acids were not detected in the ileum of the mice, even though salmonella numbers in that site were lower than in germfree animals. The observed reduction in salmonella numbers in the ileum mediated by the presence of members of the normal flora is of importance since the main site of penetration of salmonellae into the host's tissues in mice is the distal ileum (5). The normal flora is known to influence the physiology, and hence the resistance to infection, of the small intestine by stimulating intestinal motility, stimulating cell turnover in the villous epithelium, and stimulating the development of immunological tissues (28).

Oral inoculation of germfree mice with *S. typhimurium* leads to a marked reduction in cecal size. The decrease in size is accompanied by edema and abnormal coloration of the cecal tissues. Tannock and Savage (30) found that association of germfree mice with a *Lactobacillus* sp., a *Bacteroides* sp., and a mucosa-associating *Clostridium* sp. (fusiform shaped), together with "vaccination" of the animals with heat-killed *S. typhimurium*, protected the animals from a decrease in cecal size after challenge with salmonellae. Large numbers of salmonellae were present in the cecum, even though cecal size was not reduced. Vaccination of the animals

with heat-killed *S. typhimurium* alone did not influence the salmonella-induced reduction in cecal size, nor did association of the animals with the normal flora strains in the absence of vaccination. This suggests that the normal flora and the host's immunological mechanisms act synergistically to retard or prevent salmonella infection of the tissues. Perhaps the mucosa-associated fusiforms prevent large numbers of the pathogen from gaining access to the cecal epithelium; those that do gain access are destroyed by the stimulated immunological mechanisms of the host.

Observation of synergism between the normal flora and host immunological mechanisms in resistance to salmonella infection allows speculation that antigens associated with the normal flora "prime" the immunological tissues of the host so that a degree of nonspecific resistance toward salmonella infection is produced. Germfree animals have less immunoglobulin G in their blood and have a smaller stock of immunocompetent cells than do their conventional counterparts. A germfree animal probably has less ability to destroy a pathogen entering its tissues than does a conventional animal because the immunological tissues of the gnotobiotic animal have not been stimulated by exposure to the antigens associated with the normal flora (14). It can be shown experimentally that certain members of the normal flora (some strains of *Lactobacillus* and *Bacteroides*, coliforms, enterococci) isolated from the gastrointestinal tract of mice influence the number of salmonellae surviving in the spleen when the animals have been inoculated intravenously with the pathogen. Exposure of animals to specific members of the normal flora, therefore, influences the fate of salmonellae in systemic tissue of mice (24).

We can conclude from the results of these studies that the normal flora contributes to host resistance to salmonella infection in the following ways. (i) Certain components of the normal flora influence physiological factors in the small intestine, which in turn make it more difficult for salmonellae to establish in that site. (ii) Volatile fatty acids, particularly butyric acid, produced by the normal flora (probably fusiform bacteria) of the large intestine inhibit the multiplication of salmonellae under appropriate conditions of pH and Eh. (iii) Mucosa-associated members of the normal flora of the large intestine, in synergy with immunological mechanisms, prevent invasion/destruction of tissue in the large intestine. (iv) Small numbers of salmonella cells entering the tissues of the host are quickly killed, the immunological tissues having been nonspecifically stimulated by antigens associated with certain members of the normal flora.

The activities of the normal flora thus protect

the host to some extent against salmonella infection. Other investigations have shown that the same general statement can be applied to infections caused by gastrointestinal pathogens such as *Shigella* sp. and *Vibrio cholerae* (28). The protective effect of the normal flora, however, can be overcome by the entry of large numbers of the pathogen into the ecosystem, and infection of the host's tissues will result. Any factor (e.g., dietary or environmental stress, long-term antibiotic treatment) which adversely affects the gastrointestinal ecosystem, disrupting the regulatory mechanisms which maintain the normal steady-state conditions, will reduce the nonspecific resistance of the host to infection by a gastrointestinal pathogen.

PRACTICAL APPLICATIONS

If we accept the evidence from human and veterinary medicine and from experimental studies which shows that the normal flora contributes to host resistance to gastrointestinal infection, we should consider possible practical applications of our knowledge. Our primary consideration, of course, should be to maintain animals, including humans, in a well-nourished and nonstressed condition. The activities of the normal flora in suppressing pathogens will then be expressed optimally. The major methods of protecting humans from serious gastrointestinal infections will continue to be the provision of safe methods of sewage disposal, adequate treatment of water supplies, and education in good personal hygiene. Indeed, given the variation in lifestyles and genotypes among the human population, measures aimed at stimulating the activities of certain members of the normal flora of the gastrointestinal tract by the use of drugs or dietary additives are unlikely to be practical or worthwhile. Certain groups of humans may prove to be exceptions to this general statement: patients in psychiatric hospitals, where gastrointestinal infections are often endemic, and newborn infants, whose gastrointestinal tracts may rapidly become colonized by potentially pathogenic strains of *E. coli* derived from the hospital environment.

The successful implantation of bacterial strains which are antagonistic to pathogens or potential pathogens in groups of adult humans is likely to be extremely difficult. The adult gastrointestinal tract already harbors a normal flora, and this previously established flora will prevent the establishment of the inoculated strain, just as it would a pathogen entering the ecosystem. Infections in psychiatric hospitals probably are caused by the frequent contact patients have with large numbers of an intestinal pathogen as a result of poor personal hygiene rather than by some deficiency in their normal flora. Inocula-

tion of the newborn infant's alimentary canal with "desirable" members of the normal flora may be more successful. Duval-Iflah and colleagues (7) have achieved some success in reducing the incidence of large numbers of antibiotic-resistant *E. coli* in the feces of infants by inoculating the children, within hours of birth, with a human strain of *E. coli* that does not harbor plasmids. This work, based on observations that *E. coli* strains that do not harbor plasmids suppress the multiplication of plasmid-bearing strains in the intestinal tract (8), is an interesting progression from the observations of Nissle early this century regarding coliform strains of high and low antagonistic power (28). Duval-Iflah found that, 9 to 24 days after birth, the proportion of control infants harboring antibiotic-resistant (ampicillin, tetracycline) *E. coli* at a population level of 10^7 to 10^{10} bacteria per g of feces was 80% (four of five infants tested). The proportion of subjects 9 to 24 days old harboring antibiotic-resistant *E. coli* at this population level was 12% (one of eight) in the group of infants inoculated soon after birth with a plasmid-free strain of *E. coli*. The inoculated strain was considered to have become established in the intestinal tract of all but 2 of 22 inoculated infants.

Certain problems can be envisaged with this concept of the implantation of desirable antagonistic bacterial strains in the gastrointestinal tract of infants: (i) unwillingness of parents and physicians to allow the inoculation of an infant with a living bacterial culture; (ii) the formation of transconjugants from the inoculated strain and "wild" (environmental) *E. coli* (such transconjugants, however, should be suppressed by the plasmid-free inoculated strain as long as it maintains high numbers in the intestine); (iii) logistical problems in the preparation and distribution of viable *E. coli* strains and in their reliable administration to infants; and (iv) failure to suppress a wild *E. coli* strain entering the gastrointestinal tract by chance before the plasmid-free inoculating strain, since the order in which the *E. coli* strains colonize the intestinal tract is critical as far as the dominance of one strain over another is concerned (8).

The concept of implanting "beneficial" bacterial strains in the gastrointestinal tract has a long history. For example, from time to time during the past 80 years lactobacilli have been studied for their alleged beneficial effect on the health of the host animal. This interest in lactobacilli dates from the time of Metchnikoff (17), when it was observed that the natural fermentation of milk by lactic acid-producing bacteria prevented the growth of non-acid-tolerant types of microbe. It was proposed that, if lactic fermentation prevented the putrefaction of milk, it should have a similar effect in the digestive tract. The implantation of lactobacilli in the gastrointestinal tract would suppress the growth of "putrefactive" bacteria, thus reducing the amount of toxic substances generated in the digestive tract. Thus, the ingestion of fermented milk products for medicinal purposes became a popular topic in Western Europe. Interest in fermented milk products ("acidophilus milk") increased in the United States during the 1930s, largely as a result of the work of Rettger et al. (21), who claimed that ingestion of milk fermented by L. acidophilus brought at least temporary relief to a majority of subjects suffering from constipation, bacillary dysentery, colitis, sprue, and eczema. Commercial fermented milk products have recently become popular in the United States, Japan, Scandanavia, and the United Kingdom, in parallel with interest in health foods and physical fitness. Unfortunately, there is no good scientific evidence that the ingestion of milk fermented by lactobacilli is any more beneficial to health than simply ingesting plain, pasteurized milk. For every article in the scientific literature that claims beneficial results from the ingestion of fermented milk, another article will provide evidence to the contrary. Most of the reported studies have not been adequately controlled, statistical analysis of the results is rarely made, and the conclusions are largely subjective (28).

The feeding of preparations containing certain microbes to mammals and birds of commercial interest has, however, received increasing attention during recent years. These living microbial preparations, called probiotics, are said to contribute to the presence of a beneficial microbiota in the intestinal tract of the animals and to improved performance (decreased or stable food consumption but increased weight gain). Implantation of a desirable flora in the gastrointestinal tract of certain domestic animals may well be a rational idea. Modern farming practices tend to involve the use of intensive methods of raising and maintaining livestock, notably in the case of poultry and pigs. Intensive farming methods provide conditions in which large numbers of animals are confined in relatively small areas. Opportunities for the transmission of infectious diseases among the livestock are high because of the high density of susceptible animals living under confined conditions. Physical, let alone emotional, stress is easily induced under such conditions so that the animals' susceptibility to infection is increased. High standards of hygiene are required to minimize outbreaks of disease in such situations. Ironically, these high standards of hygiene may prevent the normal acquisition of a gastrointestinal flora which would provide some protection to the

animals against infectious disease, particularly when young (16). The development of standard inocula of normal flora components for newly hatched chickens or newborn piglets may well serve to improve the performance and disease resistance of these animals.

Lactobacillus strains, presumably as a result of the influence of the work of Metchnikoff and Rettger et al., appear to be favorite choices as probiotics (19). Moreover, lactobacilli are relatively easy to cultivate in the laboratory, and since certain species have been used for many years in the production of foods for human consumption, an industrial technology exists for their large-scale culture and processing. The lactobacilli inhabiting the gastrointestinal tract have not been shown to interfere with the establishment of pathogens in that site. Their use as probiotics does, however, provide examples of the problems associated with the concept of implanting beneficial microbes in the gastrointestinal tract.

A popular conception of the use of probiotics is to feed animals large numbers of lactobacillus cells as a component of their diet. The feeding of lactobacillus cultures to animals in this way, however, is unlikely to be of practical use unless the strains have been carefully chosen with regard to their ability to form stable populations in some part of the gastrointestinal tract. The mere passage of noncolonizing lactobacillus cells through the gastrointestinal tract would be too swift for microbial activities to exert a beneficial effect on the host animal. Certain strains of lactobacilli, however, can adhere to and colonize squamous epithelia in the digestive tracts of mice, rats, chickens, and pigs (25). Stable lactobacillus populations are present as layers of bacterial cells on the epithelial surface practically throughout the entire life of the animal host. Lactobacilli are shed from these sites of colonization and pass into the lower regions of the digestive tract. One can speculate that lactobacillus strains that cannot colonize epithelia are less likely to establish in the gastrointestinal tract because of the phenomenon of microbial interference. The inoculation of newborn or newly hatched animals with lactobacillus strains that colonize squamous epithelia could prove more efficacious, from the point of view of establishing a beneficial flora in the gastrointestinal tract, than continually adding bacterial populations to animal feeds.

There is, however, an additional complication: lactobacilli exhibit host specificity. Thus, lactobacilli isolated from rodents colonize only rodent stomach epithelium, and lactobacilli isolated from fowls colonize only crop epithelium (31). It is apparent, therefore, that the choice of lactobacillus strains (or strains of any type of bacterium that we wish to implant in the gastrointestinal tract) must be carefully made if there is to be any hope of practical, beneficial results with domestic animals. Strains must be chosen according to (i) their ability to form stable populations in some desirable region of the gastrointestinal tract (e.g., an epithelial surface), (ii) their ability to form such a stable population in the host species of choice (e.g., poultry, pigs), (iii) their ability to perform a beneficial function for the host (e.g., prevent the growth of a pathogen in the gastrointestinal tract), (iv) their being amenable to cultivation under industrial conditions, and (v) the availability of satisfactory methods of preparing viable suspensions of the bacteria and inoculating animals with the probiotic under field conditions.

The difficulties associated with this last condition are large, especially if we wish to inoculate animals with anaerobic bacteria under field conditions. These types of problems are exemplified by work involving the resistance of young birds to salmonella infection. Young chickens and turkey poults can be protected against medium-sized inocula of salmonellae by inoculating the birds orally at 1 day of age with intestinal contents collected from an adult bird. Inoculation of chickens with crude intestinal extracts could inadvertently transmit other pathogens, including viruses. Hence, it is necessary to determine the minimum number of bacterial strains from the intestinal contents of fowls that will produce the protective effect. Chicks can be successfully protected from salmonella infection under experimental conditions by inoculating them with mixed anaerobic cultures of intestinal bacteria. Variable results have been obtained, however, when the cultures were administered in the birds' drinking water under field conditions (20).

Two alternatives to implanting chosen strains of microbe in the gastrointestinal tract of animals that we wish to protect from infection are proposed.

(i) Inhibitory substances, preferably produced by the normal flora, could be added to the diet of the animal. Thus, lactate, propionate, acrylate, acetate (vinegar), and dilute inorganic acids have been used to treat intestinal infections in domestic animals (28). A major problem with this approach is that the dietary additives usually make the food unpalatable to the animals. (ii) A substrate utilized by particular gastrointestinal microbes could be added to the diet. If a specific substrate could be delivered in sufficient quantity to the appropriate level of the intestinal tract, a desirable normal flora type(s) might be selected which would suppress the growth of pathogens. Experimental evidence shows that it is the activities of anaerobic bacterial populations which should be encouraged in efforts to improve host resistance to intestinal infection. Although it is possible to produce changes in the composition and activities of the normal flora through radical alterations in the diet of experimental and domestic animals (3, 28), studies of the fecal flora of humans, relating to the etiology of cancer of the colon, show that only modest changes in microbial populations can be expected when the diet is altered (9). The influence of diet on the normal flora of humans is seen dramatically, however, when we compare the fecal organisms of breast-fed and bottle-fed (cows' milk-based formula) infants. A strongly acidic environment is present in the intestine of breast-fed children within the first few days of life. Hence, aciduric microbes are the dominant members of the normal flora, and the growth of *Enterobacteriaceae*, streptococci, clostridia, and bacteroides is usually suppressed. An important controlling factor at this time is the presence of an acetate buffering system, resulting from the activities of certain members of the normal flora, which maintains a low pH in the intestine of breast-fed infants. Cows' milk contains more protein and phosphate than does human milk and is therefore more buffered. An acetate buffering system does not form in the intestine of cows' milk-fed infants. Thus, the feces of these infants have a higher pH and a more varied bacterial flora than do those of breast-fed children. Of practical significance is the fact that breast-fed babies are more resistant to gastroenteritis than are cows' milk-fed infants. The protective effect of human milk cannot be said to be solely due to the influence of the diet on the normal flora: iron-chelating substances and antibodies are present in the milk. The influence of the diet on the normal flora is marked in this instance, however, and should not be overlooked (28).

CONCLUSIONS

Despite scientific interest in the activities of the normal flora of the gastrointestinal tract for about 80 years, this subject remains a somewhat confused and confusing area of microbial ecology. Much interesting and logical research involving the distribution, numbers, and activities of microbes inhabiting the gastrointestinal tract of experimental animals and ruminants has been conducted. The legacy of research and opinions originating before the 1960s, however, has tended to bias thinking on the control of gastrointestinal pathogens toward one species of bacteria: *Lactobacillus acidophilus*. Much more emphasis needs to be placed on the strict anaerobes inhabiting the gastrointestinal tract. Do they show host specificity? How are anaerobic populations regulated in the large intestine? How can anaerobes inhabiting the large intestine influence the establishment of a pathogen in the small intestine?

All investigators approaching the study of the normal flora of the gastrointestinal tract must sooner or later be horrified by the complexity of an ecosystem which contains about 500 species of bacteria, most of which are technically difficult to work with under laboratory conditions. How can we simplify the system and yet still obtain results that are relevant to the in vivo situation? What methods should we use to monitor the numbers and activities of the microbial members of the system adequately?

The use of gnotobiotic mice perhaps holds the key to future success in developing experimental models in which microbial interactions can be studied. The composition and distribution of the normal flora of the gastrointestinal tract of mice has now been extensively studied. It is possible to culture the numerically dominant members of the ecosystem in vitro. Bacterial strains can be chosen carefully so that they associate with particular regions of the gastrointestinal tract of gnotobiotic mice and specifically alter the biochemical/anatomical/physiological factors pertaining to that site. Eventually, it may thus be possible to "conventionalize" a gnotobiotic mouse by use of a minimal number of bacterial strains. This simplified system would be valuable in investigating the mechanisms which regulate the normal flora components and hence the control of pathogens in the gastrointestinal tract. The relevance of results obtained with mice in relation to the gastrointestinal tract of other animal species (e.g., humans) might be questioned. It is likely, however, that certain ecological principles that are universal can be observed in the mouse gastrointestinal ecosystem. Just as the basic facts of molecular genetics were elucidated by using one bacterial species, E. coli K-12, and were later tested in other organisms, so we may be able to determine certain facts relating to the mouse gastrointestinal ecosystem which can later be tested in other mammalian species.

Finally, scientific investigations into the practical advantages of inoculating newborn animals with preparations of living microbes are required. Properly controlled experiments with statistical analysis of the resulting data are required. We must answer clearly the questions: Do lactobacilli inhabiting the gastrointestinal tract promote the health of the animal host? Can anaerobic bacteria be used as probiotics under field conditions? One hopes that it will not take another 80 years to answer these questions.

LITERATURE CITED

1. **Bergeim, O., A. H. Hanszen, L. Pincussen, and E. Weiss.** 1941. Relation of volatile fatty acids and hydrogen sulphide to the intestinal flora. J. Infect. Dis. **69:**155–166.
2. **Bohnhoff, M., C. P. Miller, and W. R. Martin.** 1964. Resistance of the mouse's intestinal tract to experimental Salmonella infection. II. Factors responsible for its loss following streptomycin treatment. J. Exp. Med. **120:**817–828.
3. **Brockett, M., and G. W. Tannock.** 1982. Dietary influence on microbial activities in the caecum of mice. Can. J. Microbiol. **28:**493–499.
4. **Byrne, B. M., and J. Dankert.** 1979. Volatile fatty acids and aerobic flora in the gastrointestinal tract of mice under various conditions. Infect. Immun. **23:**559–563.
5. **Carter, P. B., and F. M. Collins.** 1974. The route of enteric infection in normal mice. J. Exp. Med. **139:**1189–1203.
6. **Collins, F. M., and P. B. Carter.** 1978. Growth of salmonellae in orally infected germfree mice. Infect. Immun. **21:**41–47.
7. **Duval-Iflah, Y., M. F. Ouriet, C. Moreau, N. Daniel, J. C. Gabilan, and P. Raibaud.** 1982. Implantation precoce d'une souche de Escherichia coli dans l'intestin de nouveau-nes humains: effe t de barriere vis-a-vis de souches de E. coli antibioresistantes. Ann. Microbiol. (Paris) **133A:**393–408.
8. **Duval-Iflah, Y., P. Raibaud, and M. Rousseau.** 1981. Antagonisms among isogenic strains of Escherichia coli in the digestive tracts of gnotobiotic mice. Infect. Immun. **34:**957–969.
9. **Finegold, S. M., and V. L. Sutter.** 1978. Fecal flora in different populations, with special reference to diet. Am. J. Clin. Nutr. **31:**S116–S122.
10. **Finland, M.** 1951. The present status of antibiotics in bacterial infections. Bull. N.Y. Acad. Med. **27:**199–220.
11. **Freter, R., G. D. Abrams, and A. Aranki.** 1973. Patterns of interaction in gnotobiotic mice among bacteria of a synthetic "normal" intestinal flora, p. 429–433. In J. B. Heneghan (ed.)., Germfree research. Academic Press, Inc., New York.
12. **Freter, R., H. Brickner, M. Botney, D. Cleven, and A. Aranki.** 1983. Mechanisms that control bacterial populations in continuous-flow culture models of mouse large intestinal flora. Infect. Immun. **39:**676–685.
13. **George, W. L., R. D. Rolfe, V. L. Sutter, and S. M. Finegold.** 1979. Diarrhea and colitis associated with antimicrobial therapy in man and animals. Am. J. Clin. Nutr. **32:**251–257.
14. **Gordon, H. A., and L. Pesti.** 1971. The gnotobiotic animal as a tool in the study of host microbial relationships. Bacteriol. Rev. **35:**390–429.
15. **Hentges, D. J., and R. Freter.** 1962. In vivo and in vitro antagonism of intestinal bacteria against Shigella flexneri. I. Correlation between various tests. J. Infect. Dis. **110:**30–37.
16. **Lloyd, A. B., R. B. Cumming, and R. D. Kent.** 1977. Prevention of Salmonella typhimurium infection in poultry by pretreatment of chickens and poults with intestinal extracts. Aust. Vet. J. **53:**82–87.
17. **Metchnikoff, E.** 1907. The prolongation of life. Optimistic studies. William Heinemann, London.
18. **Meynell, G. G.** 1963. Antibacterial mechanisms of the mouse gut. II. The role of Eh and volatile fatty acids in the normal gut. Br. J. Exp. Pathol. **44:**209–219.
19. **Miles, R. D.** 1981. The use of probiotics in poultry feeds. Proc. Fl. Nutr. Conf., p. 97–116.
20. **Raevuori, M., E. Seuna, and E. Nurmi.** 1978. An epidemic of Salmonella infantis infection in Finnish broiler chickens in 1975–76. Acta Vet. Scand. **19:**317–330.
21. **Rettger, L. F., M. N. Levy, L. Weinstein, and J. E. Weiss.** 1935. Lactobacillus acidophilus and its therapeutic application. Yale University Press, New Haven.
22. **Roach, S., and G. W. Tannock.** 1979. Indigenous bacteria influence the number of Salmonella typhimurium in the ileum of gnotobiotic mice. Can. J. Microbiol. **25:**1352–1358.
23. **Roach, S., and G. W. Tannock.** 1980. Anaerobic fusiform-

shaped bacteria isolated from the caecum of conventional mice. J. Appl. Bacteriol. **48:**115–123.

24. **Roach, S., and G. W. Tannock.** 1980. Indigenous bacteria that influence the number of *Salmonella typhimurium* in the spleen of intravenously challenged mice. Can. J. Microbiol. **26:**408–411.

25. **Savage, D. C.** 1977. Microbial ecology of the gastrointestinal tract. Annu. Rev. Microbiol. **31:**107–133.

26. **Savage, D. C., and R. Dubos.** 1968. Alterations in the mouse cecum and its flora produced by antibacterial drugs. J. Exp. Med. **128:**97–110.

27. **Stark, P. L., A. Lee, and B. D. Parsonage.** 1982. Colonization of the large bowel by *Clostridium difficile* in healthy infants: quantitative study. Infect. Immun. **35:**895–899.

28. **Tannock, G. W.** 1981. Microbial interference in the gastrointestinal tract. ASEAN J. Clin. Sci. **2:**2–34.

29. **Tannock, G. W., and D. C. Savage.** 1974. Influences of dietary and environmental stress on microbial populations in the murine gastrointestinal tract. Infect. Immun. **9:**591–598.

30. **Tannock, G. W., and D. C. Savage.** 1976. Indigenous microorganisms prevent reduction in cecal size induced by *Salmonella typhimurium* in vaccinated gnotobiotic mice. Infect. Immun. **13:**172–179.

31. **Tannock, G. W., O. Szylit, Y. Duval, and P. Raibaud.** 1982. Colonization of tissue surfaces in the gastrointestinal tract of gnotobiotic animals by lactobacillus strains. Can. J. Microbiol. **28:**1196–1198.

Can Pathogenic Microorganisms Be Established as Conventional Control Agents of Pests?

DE-MING SU

Biology Department, Fudan University, Shanghai, People's Republic of China

Microbial control, or the use of pathogenic microorganisms and their metabolites for the control of insect pests and other noxious organisms, is an integral part of biological control. This topic has been treated repeatedly on a worldwide scale (2, 4, 19) and in China (25, 26, 29, 30).

Three groups of microorganisms are among the most intensively explored and used, i.e., the fungi, the bacteria, and the viruses. In this paper, some recent advances in the development of microbial control in China are discussed.

FUNGI

As was rightly pointed out by Ferron (9), interest in the use of fungi pathogenic for insects has revived during the past 15 years. In China, attention has been focused on their practical use as control agents against major pests of agriculture and forestry. Those mostly concerned are the *Deuteromycetes* and the *Entomophthorales*. Among four species described in the genus *Beauveria* (7, 28), two have been reported in China: *B. bassiana* and *B. brongniartti* (= *B. tenella*) (33).

The use of *B. bassiana* for the control of insect pests in China was initiated in the late 1950s and was very intensively continued through the 1960s and 1970s. Two methods of production were practiced: solid medium culture and submerged fermentation. The former is preferred by the farmers because it is easy and few facilities are needed. Among many target pests controlled by the fungus, the pine caterpillars (*Dendrolimus* spp.), the soybean pod borer (*Leguminivora glycinivorella*), and the corn borer (*Ostrinia* sp.) are the most important ones. The efficacy of *B. bassiana* preparations for the control of more than 30 pests of economic importance is well documented (Table 1).

Recently, the attention of Chinese insect pathologists has been focused on two other important groups of pathogenic fungi: the genus *Metarrhizium* and the *Entomophthorales*. Several strains of *Metarrhizium anisopliae* have been isolated from the scarabs *Pentodon patruelis*, *Alissonotum pauper*, and *A. impressicolle* and from the longhorn *Philus pallescens*. They showed various degrees of pathogenicity toward mosquito larvae, *Culex pipiens fatigans*

(22). This fungus has also been used for the control of the scarab *Alissonotum* spp. infesting sugarcane.

Entomophthoraceous infections are common among homopterans, hemipterans, lepidopterans, and coleopterans. About a dozen species have been described in China (Table 2). They are often epizootic and cause high mortality in the host populations. Thus, we have epizootics caused by *Entomophthora aphidis* in *Myzus persicae* and *Rhopalosiphum pseudobrassicae* (31), *E. aulicae* in *Alphaea phasma* and *Spilarctia obliqua* (40), and *E. brahmina* in *Heptophylla brevicollis* and other scarab species (41). Early attempts to culture *E. aphidis* in Müller-Kögler's yolk medium (24) were successful (31), but no further progress was made in China in the next two decades. These pathogens may be potential candidates for the control of aphids and mites (12) and are worthy of research (38). Hence, the new resurgence of interest in them after such a long interruption may disclose some interesting aspects in the use of these fungal pathogens.

BACTERIA

The introduction of pathogenic bacteria, especially the *Bacillus thuringiensis* group, into plant protection began in China in the 1950s and was intensively used in the 1960s and the first half of the 1970s. Information on the basic research and on the application of this group was highlighted in monographs and many reviews (8, 10, 21, 27, 29) and was described in great detail by Pu Zhelong and his associates (25).

A recent development is the introduction and production of *B. thuringiensis* subsp. *israelis* for the control of pests of medical importance. The *B. thuringiensis* group was, and still is, the most important and promising pathogen ever used in microbial control of pests in China. At present, four varieties, i.e., *thuringiensis*, *dendrolimus*, *galleriae*, and *kurstaki*, are being produced to meet the demands of the market. It should be noted, however, that interest in this group has been reduced somewhat for the following reasons.

(i) Control of the most important agricultural pests, for example, the pink bollworm (*Pectinophora gossypiella*), the rice stem borers (*Chilo, Tryporyza*, and *Sesamia*), the cutworms (*Agro-*

TABLE 1. Effectiveness of *Beauveria bassiana* for the control of insect pests[a]

Insect species	% of control
Dendrolimus spp.	70–90
Nygmia phaeorrhoea	97.8
Lymantria xylina	70–90
Dasychira argentata	60–85
Callambulyx tartarinovii	66.7–100
Phassus excrescens	73–83
Paranthrene tabanifamis	80–90
Adoxophyes privatana	80–85
Parametriotes theae	75.4
Euproctis pseudoconspersa	63.8–94.4
Ostrinia sp.	70–95
Tryporysa incertulas	78.5–79.6
Leguminivora glycinivorella	70.2–92.9
Cylas formicarius elegantulus	90
Bothynoderus punctiventris	75.5
Curculio styracis	54–78.8
Epilachna vigintioctomaculata	92
Cryptorhynchus lapathi	72.9–82.6
Nephotettix cincticeps	58–83.5
Empoasca tormosana	94.6
Nilaparvata lugens	73.7
Tessaratoma papillosa	80
Tetranychus telarius	71.7

[a] Adapted from Li and Yang (20).

tis spp. and many other noctuids), and the flea beetles (*Phyllotreta* spp.), was unstable or even ineffective. Even in the face of possible hazards of conventional chemical pesticides, "omnipotent" pesticides are, to the average farmer, always more appealing than selective ones such as microbial pesticides.

(ii) The standardization of *B. thuringiensis* preparations presents problems yet to be solved. At present, two methods of production are practiced in China: by submerged fermentation in tanks and by a semisolid fermentation in shallow layers. As in the case of *B. bassiana* production, the latter method is preferred by the farmer in the communes. The rearing and maintenance of insect stocks for bioassay or screening is an insurmountable hurdle for the industry under current conditions. Pu Zhelong of Sun Yatsen University at Guangzhou has proposed the use of newly hatched silkworm (*Bombyx mori*) larvae as test insects for the standardization of *B. thuringiensis* products (1).

(iii) The cost of *B. thuringiensis* preparations is too high. Production by submerged fermentation usually results in better quality and suffers fewer hazards, although some enterprises have been harassed by problems with phages. The price of these *B. thuringiensis* preparations has decreased to half that of 15 years ago, but remains too high for the farmers on the one hand and unprofitable for the industry on the other.

(iv) Authentic evidence of epizootics in this group is rare, and its induction by artificial means has never been proved in nature. Basically, *B. thuringiensis* preparations hardly differ from the chemical pesticides.

VIRUSES

In China, the number of insect pests infected with virus disease(s) has almost doubled since my last review in 1982 (D.-M. Su and S.-Y. Cai, in preparation). Two examples will be given here to show the point relevant to this paper, one on *Heliothis armigera* nucleopolyhedrosis virus (NPV) and the other on *Cryptothelea variegata* NPV.

H. ARMIGERA NPV

The history and the use of the baculovirus *H. armigera* NPV has been reviewed in several papers (11, 15, 29, 30). Recently, two conferences were held in Hubei Province for the appraisal of the extensive use of and research on this virus in China (17, 43).

In China, several isolates and different formulations of the baculovirus are available for use in bollworm control: VHA-273, Jiangsu isolate, and an isolate that produces 79-4B formulation. During the years 1974–1980, with the VHA-273 strain alone, the total area for control in Hubei and other provinces amounted to more than 50,000 mu (1 Chinese acre [mu] = 0.07 ha). The effectiveness of the control is 68 to 96% in a dose

TABLE 2. *Entomophthora* species found in China

Fungus species	Host insects	References
E. americana	*Sarcophaga* sp.	6
E. aphidis	*Myzus persicae; Rhopalosiphum pseudobrassicae*	31
	Alphis laburni	14
E. aulicae	*Creatonotus gangis*	6
	Alphaea phasma; Spilarctia obliqua; Creatonotus transiens	40
E. brahmina	*Heptophylla brevicollis; Holotrichia parallela; Sophrops chinensis; Metabolus* sp.	41
E. creatonotus	*Creatonotus gangis*	45
E. fresenii	*Aphis laburni*	14
E. grylli	*Locusta migratoroides*	34
	Acrida lata; Ceracris kiangsu; Oxya velox; Pachytilus danicus	6
	Acrida sp.	—[a]
E. muscae	*Musca domestica; Cordyluridae*	34
E. sphaerosperma	*Plagiodera versicolor*	6
Entomophthora sp.	*Tryporyza incertulas*	23
Entomophthora sp.	*Spilarctia obliqua*	—[a]

[a] L.-C. Wang, thesis, Zhongshan University, Zhongshan, China.

of 15×10^9 to 30×10^9 particles of inclusion bodies (PIB)/mu,(17, 18), an efficacy comparable with that of Sevin. Similar results have been obtained with the other isolates. Mass rearing of the bollworm has been successful on artificial diets (18, 44, 46). Virus stocks produced by these larvae are far better in quality than that propagated in individuals brought directly from the field. As a rule, such preparations are primarily stocks from which the farmers take for further virus production themselves. Since 1979 the Shazhi Institute of Automation Technics Applications and Jingzhou Institute of Microbiology, both in Wuhan, have been engaged in a joint research program directed toward the commercial production of the VHA-273 strain of the virus.

Screening of even more virulent strains led to the discovery of two morphotypes in this virus, a SEV and a MEV, both in Hubei and in Kiangsu (42, 43). They differ from each other in morphology and in virulence to the bollworm. Although there were several reports on the occurrence of two morphotypes of baculoviruses in the same insect species, such as the cabbage looper, *Trichoplusia ni* (Hubner) (13), and the Douglas fir tussock moth, *Hemerocampa pseudotsugata* (16), the nature of these morphotypes remains unexplained.

Although the safety of the baculoviruses of insects and mites is well documented (3, 39), there are still people in China who are deeply concerned with the use of insect viruses, particularly those who work in areas where sericulture takes place alongside cotton growing. Nevertheless, it should be emphasized that, among other things, the purity of a virus preparation is of paramount importance and this should be treated with utmost care (35). In this context, tissue culture seems to be a solution, both for the production of pure stocks and for the testing of specificity of viruses (36). Several cell lines, including the ovarian cells and the blood cells of the cotton bollworm, have been established, and the baculovirus of the bollworm has been replicated successfully in vitro at Shanghai Institute of Entomology, Academia Sinica.

C. VARIEGATA NPV

The bagworm C. *variegata* Snellen is one of major pests of ornamentals and shade trees in the East China provinces, with a host spectrum which covers more than 65 plant species (5). It occurs in one generation per year. The larvae are active from June to November in Shanghai and cause heavy damage in July to September.

The baculovirus was isolated from the larvae in 1975 and was characterized (32). It has, in fact, a fairly wide distribution in China (37). An extensive census made in the suburbs of Shang-

hai showed the wide occurrence of the virus in nature. Moreover, larval infectivity varied from place to place (Table 3), but the cause was not clear. Another census of 10 communes in Jitung Prefecture, Jiangsu, had shown that, within a smaller area the incidence of the disease could be more or less homogeneous. The percentage of infectivity ranges from a low of 2.7% to a high of 7.13%, with an average of 4.04%. This city is located on the northern bank of the Yangtze River, near the East China Sea where the landscape is level and seasonal winds prevail. These conditions may favor the dispersion and the mixing of individuals (either healthy or infected by the virus) among populations, and hence a more or less homogeneous distribution of the disease results.

Large numbers of infected, dead larvae are collected from the field and checked under the microscope, and a partially purified virus preparation is produced. A test of this preparation on elms showed a fairly satisfactory control with concentrations of 2.5×10^5 to 4×10^6 PIB/ml (Table 4).

The virus was also used with success in combination with B. *thuringiensis* preparations for the simultaneous control of the bagworm and several euchlids which were also major pests of shade trees.

Mass production of this virus remains a problem for its further application. Up until now, virus stocks used in the research and pilot control tests have had to be collected from the field. This insect cannot be reared in confinement. Therefore, other living systems must be sought before this virus can be produced en masse. No less important is the basic research on the characterization and detection of the virus. In my laboratory a countercurrent immu-

TABLE 3. Incidence and infectivity of a baculovirus disease of the bagworm *Cryptothelea variegata* (Shanghai, 1977–1978)[a]

No. of larvae		Larvae infected by baculovirus	Infectivity (%)	Place (prefecture)
Total	Dead			
1,267	681	74	5.84	Jiadin
979	282	18	1.84	Qingpu
763	306	64	8.39	Chuansha
793	527	35	4.41	Nanhui
1,605	832	112	6.98	Boashan
1,039	404	65	6.26	Shanghai
1,005	572	35	3.48	Jinshan
755	239	2	0.26	Fengxian
214	51	3	1.40	Songjiang
241	133	19	7.88	Chongming

[a] Censuses were made in the natural growths without any control measures taken. The trees consist primarily of elms, willows, and the London plane tree (*Platanus acerifolia*).

TABLE 4. Infectivity of the bagworm treated with baculovirus preparation (Shanghai, July 1980)[a]

Concn of virus prepn (PIB/ml)	No. of larvae alive			No. of larvae dead			Total infectivity by virus (%)
	Total	Infected by virus	Infectivity (%)	Total	Infected by virus	Infectivity (%)	
1×10^5	57	10	17.54	35	31	88.57	44.57
2.5×10^5	33	19	57.58	45	42	93.33	78.21
5×10^5	33	20	60.61	13	12	92.31	69.57
1×10^6	41	25	60.98	14	13	92.85	69.09
2×10^6	22	17	77.27	32	31	96.38	88.89
4×10^6	15	14	93.33	21	21	100	97.22
1×10^{6b}	28	12	42.85	41	36	87.80	96.57
Control (water)	59	2	3.39	21	4	19.05	7.5

[a] Preparation tested on the European white elm, *Ulmus laevis*.
[b] Plus 20 *B. thuringiensis* cells.

noelectrophoresis assay is currently being used to detect and monitor the virus in nature, and the biochemical characterization of the virus is now in progress.

CONCLUSIONS

In conclusion, it should be emphasized that insect pathogenic microorganisms have shown their potential applicability in the control of major pests in China. Microbial control, with all its advantages and disadvantages, will surely develop and become established. At present, the following areas of importance should be intensively pursued: identification, characterization, mode of action, and ecology of insect pathogens; mass production and commercialization of promising pathogens; and safety considerations and monitoring of microbial pesticides in the environment.

LITERATURE CITED

1. **Anonymous.** 1977. A study on the standardization of *Bacillus thuringiensis* products by using newly-hatched silkworm larvae as test insects. Acta Entomol. Sinica **20:**5–13.
2. **Burges, H. D. (ed.).** 1981. Microbial control of insect pests and plant diseases 1970–1980. Academic Press, Inc., New York.
3. **Burges, H. D., G. Croizier, and J. Huber.** 1980. A review of safety tests on baculoviruses. Entomophaga **25:**329–340.
4. **Burges, H. D., and N. W. Hussey (ed.).** 1971. Microbial control of insects and mites. Academic Press, Inc., New York.
5. **Change, H.-H., and S.-L. Tan.** 1964. Seasonal occurrence and control measures of the variegated bagworm, *Cyptothelea variegata* Snellen, p. 133–134. *In* Proceedings of the Huadong Entomological Symposium. Shanghai Entomological Society, Shanghai.
6. **Dai, F.-L.** 1979. Sylloge fungorum sinicorum. Science Press, Peking.
7. **deHoog, G. S.** 1972. The genera *Beauveria, Isaria, Tritirachium*, and *Acrodontium*. Stud. Mycol. **1:**1–41.
8. **Feng, X.-C., Y. Wang, W.-X. Feng, and Y.-L. Fu.** 1975. Studies on the β-exotoxin of *Bacillus thuringiensis* Berliner. Acta Entomol. Sinica **18:**374–382.
9. **Ferron, P.** 1981. Pest control by the fungi *Beauveria* and *Metarhizium*, p. 465–482. *In* H. D. Burges (ed.), Microbi-

al control of insect pests and plant diseases 1970–1980. Academic Press, Inc., New York.
10. **Fu, Y.-L.** 1974. β-Exotoxin of *Bacillus thuringiensis* group. Kunchong Zhishi **11**(2):44–46.
11. **Fudan University.** 1975. Preliminary study on the nuclear polyhedrosis virus of the cotton bollworm, Heliothis armigera (Hübner), p. 100–103. *In* Biological control. New Agro-techniques Office, Shanghai.
12. **Gustafsson, M.** 1971. Microbial control of aphids and scale insects, p. 375–384. *In* H. D. Burges and N. W. Hussey (ed.), Microbial control of insects and mites. Academic Press, Inc., New York.
13. **Heimpel, A. M., and J. R. Adams.** 1966. A new nuclear polyhedrosis of the cabbage looper, *Trichoplusia ni*. J. Invertebr. Pathol. **8:**98–102.
14. **Hsui, C.-F., Y.-L. Song, and M.-Z. Yong.** 1982. Three entomophthorous fungi parasited on aduzuki-bean aphid. Acta Phytopathol. Sinica **12:**49–52.
15. **Huazhong Teachers College (ed.).** 1980. Studies on the polyhedroses of insect pests. Wuhan, Hubei.
16. **Hughes, K. M.** 1972. Fine structure and development of two nuclear polyhedrosis viruses. J. Invertebr. Pathol. **19:**198–207.
17. **Jingzhou Institute of Microbiology and Huazhong Teachers College.** 1975. Preliminary study on the virulence of two insect polyhedrosis viruses, p. 29–33. *In* Plant protection, soil and fertilizer science in Hubei.
18. **Jingzhou Institute of Microbiology, Huazhong Teachers College, and Shazhi Institute of Automation Technics Application (ed.).** 1981. Use of the cotton boll-worm nucleopolyhedrosis virus (strain VHA-273) and its mass production. Wuhan, Hubei.
19. **Kurstak, E. (ed.).** 1982. Microbial and viral pesticides. Marcel Dekker, Inc., New York.
20. **Li, Y.-W., and J.-H. Yang.** 1981. The role of white muscadine for the control of pests in forest insect pest management, p. 125–144. *In* Entomology Society Sinica (ed.), Integrated pest management in forestry. Bureau Forestry, Yunnan.
21. **Liu, C.-L., Y.-L. Fu, P.-Y. Hsia, G.-H. Jen, and S.-F. Chang.** 1962. Fifty years' research of *Bacillus thuringiensis* Berliner. Science Press, Peking.
22. **Liu, J., and Z.-L. Pu.** 1982. A strain of *Metarrhizium anisopliae* killing mosquito larvae. Natural Enemies of Insects **4**(2):41–46.
23. **Ma, Y.-G.** 1959. Preliminary study on a pathogenic fungi of the yellow rice stem borer, *Tryporyza incertulas*. Kunchong Zhishi **5:**309–310.
24. **Müller-Kögler, E.** 1959. Zur Isolierung und Kultur insektenpathogener Entomophthoraceen. Entomophaga **4:**261–274.
25. **Pu, Z.-L.** 1978. Principles of biological control of insect pests. Science Press, Peking.
26. **Pu, Z.-L.** 1980. Progress in microbial control (1975–1980): world activities. D. People's Republic of China, p. 12. *In*

Proceedings of the Workshop on Insect Pest Management with Microbial Agents: Recent Achievements, Deficiencies, and Innovations. IPRC Constituent Groups.

27. **Ren, G.-X., K.-T. Li, M.-H. Yang, and X.-M. Yi.** 1975. The classification of the strains of *Bacillus thuringiensis* group. Acta Microbiol. Sinica **15**:292–301.

28. **Samson, R. A., and H. C. Evans.** 1982. Two new *Beauveria* spp. from South America. J. Invertebr. Pathol. **39**:93–97.

29. **Su, D.-M.** 1982. Use of bacteria and other pathogens to control insect pests in China, p. 317–332. *In* E. Kurstak (ed.), Microbial and viral pesticides. Marcel Dekker, Inc., New York.

30. **Su, T.-M.** 1978. The role of insect viruses in pest control and some aspects on their basic research, p. 53–79. *In* Biological control, vol. 2. Chinese Scientific and Technological Information Institute, Chungking Affiliate.

31. **Su, T.-M., and K.-L. Hsin.** 1963. Some observations on the mycosis of the cabbage aphids in Shanghai. Acta Phytophylacica Sinica **2**:221–223.

32. **Su, T.-M., Y.-X. Yue, and M.-C. Chen.** 1978. A preliminary study on a baculovirus of the bagworm, *Cryptothelea variegata* Snellen (Lepidoptera, Psychidae). Chinese Forestry Science **1978**(4):40–41.

33. **Teng, C.** 1962. Studies on the biology of *Beauveria bassiana* (Bals.) Vuill. with reference to microbial control of insect pests. Acta Bot. Sinica **10**:210–232.

34. **Teng, S.-Q.** 1963. Fungorum sinicum. Science Press, Peking.

35. **Tinsley, T. W.** 1979. The potential of insect pathogenic viruses as pesticidal agents. Annu. Rev. Entomol. **24**:63–87.

36. **Tinsley, T. W., and J. Melnick.** 1973/74. Potential ecological hazards of pesticidal agents. Intervirology **2**:206–208.

37. **Tsai, S.-Y., G.-H. Huang, and T. Ding.** 1978. Some insect viruses discovered in China. Acta Entomol. Sinica **21**:101–102.

38. **Wilding, N.** 1981. Pest control by Entomophthorales, p. 539–554. *In* H. D. Burges (ed.), Microbial control of pests and plant diseases 1970–1980. Academic Press, Inc., New York.

39. **World Health Organization.** 1973. The use of viruses for the control of insect pests and disease vectors. WHO Tech. Rep. Ser. No. 531.

40. **Wu, J.-W., S.-H. Chang, and D.-X. Wang.** 1980. The observation and identification of *Entomophthora aulicae* (Reich) Sorokin. Acta Microbiol. Sinica **20**:68–71.

41. **Wu, J.-W., D.-X. Wang, and S.-H. Chang.** 1982. Identification and prevalence of Entomophthora brahmina in Kuming district. Acta Mycol. Sinica **1**:27–32.

42. **Wuhan Institute of Virology.** 1979. On the virulence of different isolates in the nucleopolyhedrosis virus of the cotton bollworm. *Heliothis armigera.* Acta Virol. Sinica, p. 38–43.

43. **Wuhan Institute of Virology and Jianghu State Farm (eds.).** 1980. The nucleopolyhedrosis virus of the cotton bollworm, *Heliothis armigera* (Hübner) as a pesticide. Wuhan, Hubei.

44. **Xin, J.-L., and T.-M. Su.** 1979. Artificial diets of insects, mites, and spiders. Science Press, Peking.

45. **Yen, D. F.** 1962. An *Entomophthora* infection in the larva of the tiger-moth, *Creatonotus gangis* (Linnaeus). J. Insect Pathol. **4**:88–94.

46. **Zhang, G.-Y., Y.-Q. Zhang, L. Ge, and Z.-M. Shan.** 1980. Rearing of the cotton bollworm and the replication of its nuclear polyhedral virus, p. 29–34. *In* Wuhan Institute of Virology and Jianghu State Farm (ed.), The nucleopolyhedrosis virus of the cotton bollworm, *Heliothis armigera* (Hübner) as a pesticide. Wuhan, Hubei.

Comparative Carbon and Energy Flow in Ecosystems

Role of Microbes in Global Carbon Cycling

J. E. HOBBIE AND J. M. MELILLO

The Ecosystems Center, Marine Biological Laboratory, Woods Hole, Massachusetts 02543

Microbes no longer are the dominant photosynthesizers of the earth, but they are still the dominant decomposers; they are vitally important for the global cycling of carbon. Over geologic history, imbalances between photosynthesis and decomposition have created the tremendous pools of fossil fuels, of the dissolved organic carbon of oceans, and of the organic matter of soils. These large carbon pools can be measured, and it is easy to forget that they have built up gradually over thousands to millions of years. For example, about $1,000 \times 10^{15}$ g (1,000 Gt) of dissolved organic carbon has accumulated in the oceans. If this pool comes mostly from photosynthesis in the ocean (around 28×10^{15} g of C per year), then the accumulation could represent 1% of the annual photosynthesis for 3,600 years. Is this 1% the actual rate of accumulation? No one knows the rate exactly, but it is very likely that the actual rate is 1% or less per year and not 10%. The point is that the rate of accumulation is small relative to the total rates of photosynthesis and decomposition. This small rate is very difficult to measure as an accumulation from year to year. Attempts have been made to measure accumulation in forests by using the total exchange of CO_2 or long-term changes in carbon stocks (9). From 15 to 30% of net primary production accumulated annually in four temperate-zone, actively growing forests.

Throughout most of the earth, the carbon cycling in oceans and on land is close to a steady state. That is, the rate of movement of carbon between the atmosphere and trees or between algae and the dissolved organic carbon of the oceans does not change measurably from year to year, and there is no measurable change in the size of the various reservoirs.

In addition to the steady-state cycling of carbon that occurred throughout the world some hundreds of years ago, human activities have recently introduced changes in the cycling and accumulation of carbon that are large enough to be measured. Today, the global carbon cycling is a mixture of the steady-state rates and reservoirs and of the changing rates and reservoirs. For example, the rate of flux of carbon from algae into dissolved organic carbon in the open ocean is a steady-state flux because human activities are not great enough to perturb the rate. In contrast, the reservoir of carbon (as carbon dioxide) in the atmosphere is not in steady state and is growing from year to year. As a result, the global carbon cycle is out of balance. In any discussion of this global carbon balance, it is important to understand the distinction between the steady-state parts of carbon cycling and the perturbed parts of the cycle.

In this review we give a megaview of carbon cycling of the earth, with particular attention to the extent of microbial activities. In addition, we discuss the current difficulties in calculating the annual budget of carbon cycling and the possible microbial role under an altered climatic regime.

PRESENT-DAY CARBON CYCLING

The largest flux of carbon (Table 1) is the exchange of CO_2 between the atmosphere and the ocean surface. This is strictly a diffusional exchange and changes only a little with long-term changes in the CO_2 content of the atmosphere (the surface mixed layer of the ocean keeps in equilibrium with the atmosphere, so the mean gradient remains small). The largest biotic flux is the photosynthetic fixation of carbon by the land biota. One estimate is that 40×10^{15} g of C (or 40 Gt) is fixed in forests, 5 Gt in agriculture, and 10 Gt in grasslands. Another 28 Gt of C per year is fixed in the oceans (13). The flux into the biota is matched, or nearly matched, by decomposition and the respiration of an equivalent amount of carbon by microbes. For example, in a yellow poplar-oak forest in Tennessee with a biomass (trees) of 8,760 g of C m^{-2}, the net primary production was 730 g of C m^{-2} year^{-1}, and the respiration by heterotrophs was 670 g of C m^{-2} year^{-1} (94 g of C by canopy animals) (2).

The small imbalance between primary production and respiration has resulted in an accumulation of organic carbon in soils and oceans. The carbon in soils has accumulated to a total of 1,500 Gt of C throughout the world. About one-fourth of this is found in boreal peatlands and taiga, even though these make up only 10% of

TABLE 1. Major carbon reservoirs and fluxes
(modified from 1)

Reservoirs and fluxes	Amt (Gt) of C
Reservoirs	
Atmosphere before 1850 (265 to 290 ppm of CO_2)	560–610
Atmosphere in 1978 (329 ppm of CO_2)	692
Oceans—inorganic	35,000
Oceans—dissolved organic	1,000
Land biota	600–900
Soil organic matter	1,500
Sediments	10,000,000
Fossil fuels	10,000
Fluxes (per year)	
Atmosphere—oceans, gross exchange	100
Atmosphere—land biota in photosynthesis	53–78
Atmosphere—ocean biota in photosynthesis	28
Land to ocean as dissolved inorganic form	0.4
Land to ocean as dissolved organic carbon	0.1
Deposition in oceans	1–10
Fossil fuel combustion, 1978	5
Biota and soil to atmosphere	2–4

the total land area. About the same total amount of carbon, 1,000 Gt, has accumulated in the oceans as dissolved organic carbon.

Net additions of carbon dioxide to the atmosphere are made in two ways, the burning of fossil fuel and the oxidation of carbon from vegetation and soils when forests or grasslands are transformed into agricultural lands. These two fluxes are due to human activities and are not in steady state because the reservoirs of fossil fuel and of biota and soil are diminishing in size. The flux moving carbon dioxide from the atmosphere to the ocean is not large enough to prevent an atmospheric buildup of 72 to 122 Gt of C from 1860 to the present (Table 1) (4). This buildup is now about 2.2 Gt of C per year.

CURRENT PICTURE OF THE GLOBAL CARBON BALANCE

The difference between the yearly inputs and outputs of carbon to and from the atmosphere has to equal the yearly increase in the carbon (as CO_2) in the atmosphere. This is obvious because the global carbon cycle has to balance. Yet, a number of scientists have said that the carbon cycle is not in balance. What they meant was that our *estimates* of the different parts of the cycle do not balance. It should be noted that only the non-steady-state fluxes are mentioned in the discussions of the balance of the carbon cycle.

A current view (3) of an equation for the carbon dioxide fluxes to and from the atmosphere is:

$$\underset{\substack{5.2\ Gt\ C \\ (\pm 0.7)}}{\text{fossil fuel input}} + \underset{\substack{3.3\ Gt\ C \\ (\pm 1.5)}}{\text{biota and soil input}} - \underset{\substack{2.0\ Gt\ C \\ (\pm 0.5)}}{\text{ocean uptake}} = \underset{\substack{2.5\ Gt\ C \\ (\pm 0.2)}}{\text{increase in atmosphere}}$$

The numbers do not add up, yet it is believed that these are the important fluxes. The imbalance in our estimates of the fluxes is called by some "the missing carbon." In this equation the imbalance is 4.0 Gt of C. How can this happen?

One explanation that is possibly true states that several of the estimates are wrong and that the terms really do balance. Certainly, the estimates should be given with error bars; the fossil fuel release rates and the rate of change of the atmosphere have the smallest errors, and the biotic-soil release and the ocean uptake have larger errors. If the high estimates of ocean uptake and the low rates of biotic-soil release both turn out to be correct, then the imbalance is reduced to 1.1 Gt of C.

Several impossible explanations have been proposed. One states that marine macroalgae were not properly included in oceanic photosynthesis calculations and their productivity is large enough to account for the "missing carbon" (12). This explanation ignores the fact that steady-state fluxes, those unaffected by human activities, do not enter into the equation. Regardless of how large the flux, if it has not been changed by human activities, it will not account for "the missing carbon." Another explanation, this one about termite release of methane, is logical, but the release is too small to have an effect. The idea (14) is that termite nests release significant amounts of methane, and these amounts have been increasing because of the destruction of forests. However, it appears that the maximum possible release from this source is about ± 0.1 Gt of C per year. Any increases above this value short of a 10-fold increase would be insignificant compared to the other terms in the equation.

There are several other plausible explanations of the apparent imbalance in the global balance of carbon dioxide. The first of these was proposed by Seiler and Crutzen (11), who thought that a significant amount of carbon was being removed from cycling by storage as charcoal in the soil after forest cutting and burning. Our analysis (3) suggests that the charcoal stored in the soil after a forest has been transformed into permanent agriculture is small. A larger amount is stored during shifting cultivation (slash and burn), but this is small in comparison with the loss of carbon from reductions in forest biomass and soil carbon during shifting cultivation.

Another possible explanation is that the reduction in soil carbon associated with forest clearing and other human activities is really much less than estimated because much of it is eroded from land and then stored in river sediments, floodplains, or shallow coastal sediments. This process, like charcoal formation, moves organic carbon from one place to another without oxidizing it to carbon dioxide. However, it is still a question as to whether organic carbon can be moved from one site to another without changing the microbial respiration rate. Moving carbon within the soil by mixing or cultivating does increase the respiration, but it is not known whether this is true for large movements from one location to another. MacKenzie (7) estimated that the increase in the size of the reservoirs of carbon in marine sediments and flood plains could be a maximum of 0.29 Gt per year.

Some scientists believe that the photosynthesis of the world's forests is being increased by the increasing concentration of carbon dioxide in the atmosphere. This increase, they believe, would result in greater storage of carbon and counteract the buildup of carbon dioxide in the atmosphere. In terms of the equation presented, the photosynthesis–respiration of the forests would change from a balanced steady state to a net sink for carbon (a negative term has been added to the left side of the equation). This argument has been addressed by Kramer (5), who questioned the assumption that trees in a forest respond in the same way to CO_2 "fertilization" as tomatoes, or sugar beets, in a greenhouse. He pointed out that photosynthesis in forests may be limited by water and nutrients, so increases in carbon dioxide may have little effect. Finally, Kramer mentioned that photosynthesis increases are not necessarily tied to biomass increases. We emphasize the difference between probable response (in a forest) and possible response (in a greenhouse where water and nutrients are at optimal levels).

The question of storage is even more complicated than a simple consideration of biomass increases. Net carbon storage in forest ecosystems is determined by the balance among three distinct processes: (i) the net amount of carbon fixed by the vegetation (i.e., carbon fixed in excess of plant respiratory demand); (ii) the relative amounts of this fixed carbon allocated to increases in woody mass versus plant litter; and (iii) the decomposition of carbon compounds entering the soil system. This last process, decomposition, is certainly the least known of the three. We do not understand the control of the amount of litter respired versus that converted to humus.

The enhanced storage of carbon in the soil or in the trees is potentially important because a 5% increase in photosynthesis is equal to 2 Gt per year (Table 1). A few percent increase could not be measured in the field. The real question concerns enhancement of storage, and we must wait until long-term experiments with trees are run at high levels of added carbon dioxide for an answer. In a short-term CO_2-enrichment experiment, tree litter from plants grown at 935 ppm of CO_2 had lower N concentrations (and thus wider C/N ratios) than litter from plants grown at ambient CO_2 levels (320 ppm) (J. M. Melillo and H. Rogers, unpublished data). This reduction of litter quality could lead to reduced decay rates and thus nutrient stress in higher plants.

EFFECT OF DOUBLING THE CARBON DIOXIDE IN THE ATMOSPHERE

It is empirically well established that the carbon dioxide content of the atmosphere is increasing. What will be the interactions of an increased carbon dioxide content, for example, a doubling, with the microbes of the soil?

The foremost effect of a doubling of the carbon dioxide, perhaps as early as the year 2050, will be on the world climate. There will most likely be little direct effect on microbial activity, but if temperature increased and precipitation patterns changed, there could be a strong indirect effect. There is general agreement that the global temperature will increase but less agreement about exactly how that will affect air movement and precipitation. One prediction, based on a large computer-based model, is for little change in the tropics and a large change at the poles (8). At 40 to 60° latitude the annual temperature increase will be 4 to 6°C, and precipitation patterns will also change.

Predictions are difficult to make. On the other hand, this large-scale experiment of increasing the CO_2 content of the atmosphere is under way, and some possible consequences should be discussed now. If the possible consequences seem to be too disruptive of society, then small-scale experiments should be carried out. Accordingly, we present several scenarios out of the many possible, but will deal only with the microbial part of the consequences. These scenarios are undocumented and are intended only to indicate the possible importance of microbes in the global carbon cycle.

Scenario 1 assumes that microbial decomposition of organic matter and peats in boreal peatlands, tiaga, and tundra will accelerate because of increased temperature and an altered precipitation regime. Assuming a response to temperature alone (a Q_{10} response of 2.5) and no increase in plant litter input, about 15% of the carbon stored in soils and peats would be released over 50 years. The release of 60 Gt total or 1.2 Gt of C per year is less than one-tenth of the amount of carbon needed (692 Gt) to double the atmospheric carbon dioxide. If precipitation is altered so that extremely wet sites become drier and dry sites become wetter, then even more microbial decomposition may occur. If 50% of the stored carbon in soils and peats were to be released over 50 years, then the total release of 205 Gt, or 4 Gt of C per year, is nearly a third of the amount of carbon needed (692 Gt) to double the atmospheric carbon dioxide. Of course the rate of release of carbon dioxide to the atmosphere from fossil fuel burning will also increase. In any case, there will be a positive feedback by the microbial activity and an increase in both the rate of CO_2 release and in the rate of climate change.

This scenario may have to be modified (at least for the short term) if the predictions of Linkins et al. (this volume) prove to be correct. They argue that extracellular microbial enzymes in the soil are important controllers of decay rate. One enzyme, endocellulase, appears to be less sensitive to temperature changes than other cellulases. Endocellulase-catalyzed reactions supply the other cellulases with substrate that is then converted to cellobiose and glucose. Thus, the endocellulase activity could be the rate-limiting step in a positive feedback loop.

Scenario 2 begins with the same assumption of increased decomposition of soil organic carbon and of peat, but adds effects of the nutrients released during decomposition. The ratio of carbon to nitrogen to phosphorus is 150:10:1 in the material that is decomposed. This amount of nitrogen and phosphorus will increase the photosynthesis rate and may lead to increased biomass of tress. However, the C/N/P of trees is high, around 1,500:10:1, so 10 times more carbon has been stored than was released by decomposition. This is a negative feedback which would remove carbon dioxide from the atmosphere. However, the scenario is an extreme case, and no one believes that all of the nutrients released with the oxidation of 205 Gt of soil carbon (see scenario 1) will be incorporated into trees or that 2,050 Gt of the carbon will be stored (after all, the present total land biota is only 600 to 900 Gt, Table 1).

Scenario 3 deals with agriculture. In these systems, the combination of higher carbon dioxide concentrations and warmer temperatures could result in higher photosynthesis if moisture and nutrients were adequate (6). When nutrients are limiting but CO_2 and water are plentiful, plants may "invest" carbon in microbial symbionts to acquire nutrients and thus have higher photosynthesis. This is especially true for legumes and mycorrhizal plants. Paul and Kucey (10) investigated the carbon costs (and benefits) to fava beans (*Vicia faba*) for maintaining rhizobial symbionts, vesicular-arbuscular mycorrhiza, or both. They found that mycorrhizal-rhizobial plants incorporated 16% more carbon per unit weight of shoots than controls. Today, the amount of carbon in crops is about 5 Gt; an increase of 0.5 Gt is possible. This amount of increase would have little effect on the global carbon budget because agricultural crops are grown and completely oxidized each year; there is no storage so the system is in steady state. This increased production would, however, feed more people.

CONCLUSIONS

In this review we have reported the amounts of carbon in the major reservoirs and fluxes of the global carbon cycle. The amounts of carbon building up in the atmosphere because of human activities are smaller than expected by biologists, and possible explanations include errors of measurement, sequestration in river basins, coastal sediments, or charcoal, and a small increase in photosynthesis and storage by undisturbed forests. Over the next century, microbial responses to the increased carbon dioxide in the atmosphere could be as diverse as positive or negative feedbacks; these responses could be important in the global carbon cycling. Microbial responses can only be studied as part of manipulations of whole ecosystems which involve measurements by plant physiologists, soil chemists, and microbial ecologists.

Preparation of this paper was supported by National Science Foundation grant DEB-79-23329.

LITERATURE CITED

1. **Bolin, B., E. T. Degens, P. Duvigneaud, and S. Kempe.** 1979. The global biogeochemical carbon cycle, p. 1–56. *In* B. Bolin, E. T. Degens, S. Kempe, and P. Ketner (eds.), The global carbon cycle. SCOPE B. John Wiley & Sons, Inc., New York.
2. **Harris, W. F., P. Sollins, N. T. Edwards, B. E. Dinger, and H. H. Shugart.** 1975. Analysis of carbon flow and productivity in a temperate deciduous forest ecosystem. p. 116–122. *In* Productivity of world ecosystems. National Academy of Sciences, Washington, D.C.
3. **Houghton, R. A., J. E. Hobbie, J. M. Melillo, B. Moore, B. J. Peterson, G. R. Shaver, and G. M. Woodwell.** 1983. Changes in the carbon content of terrestrial biota and soils between 1860 and 1980: a net release of CO_2 to the atmosphere. Ecol. Monogr. 53(3):235–262.
4. **Keeling, C. D., R. B. Bacastow, R. B. Bainbridge, A. E. Ekdahl, C. A. Guenther, L. S. Waterman, and J. F. S. Chin.** 1976. Atmospheric carbon dioxide variations at Mauna Loa Observatory, Hawaii. Tellus 28:538–551.
5. **Kramer, P.** 1981. Carbon dioxide concentration, photosynthesis, and dry matter production. BioScience 31:29–33.
6. **Lamborg, M. R., R. W. F. Hardy, and E. A. Paul.** 1983. Microbial effects, p. 131–176. *In* E. R. Lemon (ed.), CO_2 and plants. The response of plants to rising levels of atmospheric carbon dioxide. AAAS Selected Symposium 84. Westview Press, Boulder, Colo.

7. **MacKenzie, F. T.** 1981. Global carbon cycle: some minor sinks of CO_2, p. 360–384. *In* Flux of organic carbon by rivers to the oceans. CONF-8009140. Office of Energy Research, U.S. Department of Energy, Washington, D.C.

8. **Manabe, S., and R. T. Wetherald.** 1980. On the distribution of climate change resulting from an increase in CO_2-content of the atmosphere. J. Atmos. Sci. **37**:99–118.

9. **Melillo, J. M., and J. R. Gosz.** 1983. Interactions of biogeochemical cycles in forest ecosystems, p. 175–220. *In* B. Bolin and R. B. Cook (ed.), The major biogeochemical cycles and their interactions. SCOPE 21. John Wiley & Sons, Inc., New York.

10. **Paul, E. A., and R. M. N. Kucey.** 1981. Carbon flow in plant microbial associations. Science **213**:473–474.

11. **Seiler, W., and P. J. Crutzen.** 1980. Estimates of gross and net fluxes of carbon between the biosphere and the atmosphere from biomass burning. Climat. Change **2**:207–247.

12. **Smith, S. V.** 1981. Marine macrophytes as a global carbon sink. Science **211**:838–840.

13. **Whittaker, R. H.** 1975. Communities and ecosystems, 2nd ed., p. 387. Macmillan Co., New York.

14. **Zimmerman, P. R., J. P. Greenberg, S. O. Wandiga, and P. J. Crutzen.** 1982. Termites: a potentially large source of atmospheric methane, carbon dioxide, and molecular hydrogen. Science **218**:563–565.

Carbon and Energy Flow in Terrestrial Ecosystems: Relevance to Microflora

O. W. HEAL AND P. INESON

Institute of Terrestrial Ecology, Merlewood Research Station, Grange-over-Sands, Cumbria, United Kingdom

Over the past 20 years, "production" ecology has concentrated on quantifying plant and animal production in different ecosystems. Much of the research has been field orientated and has been at a scale of resolution to which microbial ecology has had difficulty in relating, namely, the scale of space (kilo rather than micro) and time (centuries and years rather than days or hours). More recently, production ecology has turned toward nutrient and population dynamics and toward the growth strategies of organisms, areas in which the microbial ecologist can both contribute and gain. None of the fields is actually new; rather, they show recurrent phases in which new information in one field stimulates interest in another.

Here we wish to summarize some of the information arising from ecosystem and productivity studies with possible relevance to microbial ecology. In particular, the pattern of primary production in different systems is now reasonably established, and we explore the extent to which plant production influences the microflora, and also the fauna, populations in the heterotrophic system. An initial premise is that the composition and activity of the microflora are determined by three groups of factors: climate, soil conditions, and resource quality (80). It is mainly through variation in resource quality that microbial populations may be expected to reflect vegetation patterns, but the microflora may also be expected to show responses in common with the plants to external environmental conditions.

NET PRIMARY PRODUCTION

Quantity. If we ignore the small proportion of carbon and energy fixation by autotrophic microflora, the main pattern of primary production in different ecosystems is largely as outlined by Whittaker (88), developed from the compilation of Bray and Gorham (10), Rodin and Bazilevitch (68), and others. This has been amplified in more recent syntheses of International Biological Program (IBP) and related ecosystem data (7, 11, 18, 30, 31, 67). The general relationship with climate is clear, reflecting both its intensity and variability (Table 1). There are still some reservations or anomalies in these results, including some exceptionally high production rates: 50 to 60 t/ha per year in sub-Antarctic systems (89)

and >10 t/ha per year in some deserts (87). Also, there is an increasing body of evidence that fine root production and root exudation are underestimated components of earlier production estimates (21, 54, 64). For example, it has been estimated that fine root production in certain forest stands is about equivalent to leaf litter input (54). Production in tropical systems is still poorly quantified, with belowground estimates being largely neglected.

The production values given are for ecosystems which approximate to steady state and therefore represent equivalent input to the heterotrophic or decomposition subsystem. In successional stages, up to steady state, net production will be greater than litter input to decomposition (the system is aggrading), whereas postmaturity litter fall will be greater than net production as a result of loss of accumulated biomass, i.e., a degrading system. (The term litter is used to include all inputs from plant production unless qualified.)

Quality. Most of the production data are expressed as dry weights by species or parts of species (leaves, roots, etc.), with little reference to their chemical composition; yet, it is composition, in terms of structural and storage carbohydrates, nutrient content, and secondary compounds, which is important to the microflora.

A first approximation to estimation of variation in quality of input to decomposition (Table 1) is given by fractionating into leaves (non-woody), roots, and wood on the basis that this is, first, the lowest common denominator between data sets and, second, that it represents high, medium, and low quality, respectively, through increasing lignin and decreasing nutrient concentrations. Interpretation is necessarily limited by the very simplified nature of the information, but several features emerge. The dominance of wood in forests has less effect on decomposer input than might be expected, because although wood may constitute 70 to 80% of the biomass, its proportion of the litter input is much smaller. For example, wood constituted 50% of litter in an old stand of *Pseudotsuga menziesii* and only about 2% in a young stand (19). Wood is also a significant proportion of input in many dwarf shrub tundra, chaparral, and savannah systems. Additionally, the major

TABLE 1. Net primary production (NPP, tonnes per hectare per year) according to biomes and proportionate allocation to above- and belowground biomass (from 18, 20, 62, 80, 84, 87, 89)

| Biome | Mean NPP | Minimum–maximum (n) | Proportions of above- and belowground biomass | | |
| | | | Above | | Below, root |
			Nonwoody	Woody	
Tundra (sub-Antarctic)	4.0	0.7–8.9 (18)	30	—	60
	58.0	55.8–60.2 (2)			
Temperate forest	13.0	4.4–19.0 (15)	60	20	20
Temperate grassland	15.0	7.0–34.7 (38)	40	—	60
Desert	2.0	0.5–7.8 (21)	30	10	60
Tropical forest	17.0	9.4–28.0 (3)	60	20	20

input of roots, both large and small, in nonforest systems reflects the different growth characteristics of the plants, with roots representing a storage, transport, and support system characterized by a high proportion of structural polysaccharides and low nutrient concentration, more similar to woody tissues than to many leaves.

However, the composition of different fractions is not as clearly distinct as it initially appears. In general, wood is slightly higher in lignin and cellulose, being lower in soluble carbohydrate, nitrogen, and phosphorus than are roots and leaves. The range is very wide for all litter types, and certain of the analytical techniques are of dubious comparability.

The reasons for the more restricted variation in composition than in quantity are (i) withdrawal of soluble organic fractions before death, (ii) withdrawal of soluble nutrients, mainly N, P, and K before death, and (iii) varying dry matter production per unit of nutrient uptake. The composition of root material at death, and of root exudates, is poorly known. For example, the nutrient contents of root material quoted in the literature usually refer to live roots, but a

considerable degree of resorption is likely to occur before death. Estimates of the approximate input of structural and soluble carbohydrates (Table 2) suggest that, although amounts differ considerably between biomes, the proportions vary little. The dominance of high-molecular-weight fractions is overwhelming.

Some important differences are not reflected in Table 2. Secondary compounds, such as phenols and terpenes, are developed in perennial tissues of plants in late succession and in severe environments as "defense" against herbivory or attack by parasites. Such compounds are known to influence populations of decomposer microflora and fauna, often as inhibitors (43, 70) but also as substrates (61). Spatial configuration of the litter resource, including factors such as surface area-to-volume ratios, and spatial separation of nutrient sources in a resource are also variants in quality.

The implication of the foregoing is that any ecosystem will contain a wide range of substrates available to the decomposers and that dominance of a particular quality of substrate in an ecosystem is more apparent than real (Table 2). During succession the proportion of structural carbohydrates will be lower as biomass accumulates. It must also be appreciated that under certain conditions the quality and quantity of an input may increase considerably, e.g., at felling, at climatic death due to drought, or with herbivore population fluctuations, providing a resource rich in nutrients and soluble carbohydrates because of a lack of premortality resorption. Conversely, after fire carbon may be both severely limited and markedly changed.

TABLE 2. Preliminary estimates of substrate input to the decomposer subsystem in different biomes (tonnes per hectare per year)[a]

Biome	Lignin (SD)	Cellulose (SD)	Soluble carbohydrate (SD)
Tundra	1.4 (1)	2.3 (1.6)	0.2 (0.1)
Temperate forest	3.7 (1)	4.7 (1.3)	0.8 (0.2)
Temperate grassland	4.7 (1)	8.2 (1.7)	0.7 (0.1)
Desert	0.6 (1)	1.1 (1.8)	0.1 (0.0)
Tropical forest	4.8 (1)	6.3 (1.3)	0.7 (0.1)

[a] Amounts are calculated on the basis of primary production data in Table 1 and litter substrate concentrations in the literature (19, 39, 45, 80, 83, 86). Mean concentration values from different litter types were used, but it must be emphasized that these are very variable and the calculation is tentative.

DECOMPOSITION

Variation between biomes. Only a small fraction of the net primary production is consumed and respired by herbivores, and at least 90% is transferred to the decomposer system directly or via feces and death. Although much effort has concentrated on measurement of aboveground litter fall, the best estimates of input to decomposition are of net primary production, especially in the relatively stable systems that have been intensively studied, because they reflect belowground production. Turnover of the organic horizons may be relatively rapid (Table 3), but the majority of dead organic matter is within the mineral soil (70 to 80% in tundra, boreal, and temperate forest; more than 90% in temperate grasslands, deserts, and tropical forests). Turnover of the total dead organic matter pool may take many decades, although the decay of resistant fractions takes centuries (25). As with primary production there is a general increase in decomposition rate with temperature. The limitation of moisture in deserts is not as clear as might be expected in the surface litter, possibly because of physical losses or rapid decomposition during moist periods. However, in the more detailed analyses the shift in control of decomposition from temperature to moisture within grassland moisture gradients is clear (16). Even in apparently cold-dominated biomes, low moisture levels can retard microbial activity (13, 38), and the interaction of temperature and moisture is expressed in the general relationship of decomposition to evapotranspiration (55).

Influence of substrate quality. The broad biome pattern masks the wide range of variation in rates of decomposition, and there is considerable overlap between biomes, including tropical and temperate forests (J. M. Anderson and M. J. Swift, in S. L. Sutton, T. C. Whitmore, and A. C. Chadwick, ed., *The Tropical Rain Forest*, in press). In addition, major variation in activity results from variation in substrate quality within a biome. The initial rates of decay, measured by weight loss or respiration of litters, frequently show the fastest rates to be 5 to 10 times that of the slowest, even among leaf litters within a site (e.g., 39, 58, 83). The slow decay rate of woody fractions emphasizes the point, but there is also increasing evidence that decay rates of roots also tend to be slower than that of

the associated leaves (17, 54; B. Berg, Ecology, in press). Decomposition of naturally dying roots is probably lower than that reported in many decomposition studies which use live roots, initial losses of fresh material being higher than that of naturally dead litter through lack of resorption of soluble fractions (6).

The term substrate quality lacks definition. It is the combination of physical and chemical attributes of a resource which determines its potential for microbial growth. No single factor determines quality. Even though nitrogen and lignin concentration provide a general correlation with resource decomposition (4, 24, 39), this reflects the intercorrelation of a number of components, particularly between structural and nonstructural carbohydrates, which provide the main carbon and energy source, and essential nutrients (70). An additional effect on microbial activity comes from secondary compounds developed by the plants as defense against herbivores or pathogens. The influence of the comprehensive research into plant defense compounds (69) has not been fully felt in decomposition research, although there is increasing definition of chemical changes during decomposition (3, 6, 36, 71, 73, 74).

One key feature is that the chemical factors which control decomposition change with time; Berg and Staaf (4) showed the shift from nutrient to carbon substrate control in Scots pine needles and the interaction between nitrogen and lignin. Differences in control between resources are to be expected, as noted by McClaugherty, Aber, and Melillo (54) in relation to decomposition of fine roots and leaves. Surface area to volume, the surface characteristics, and the spatial arrangement of the substrates and nutrients are additional physical factors which modify decomposition rate through their influence on microbial colonization, establishment, and growth (78).

Thus, the pattern of heterotrophic carbon and energy flow across biomes is dominated by a broad climatic trend within which there is a finer

TABLE 3. Primary production, organic matter standing crops, and turnover in major biomes[a]

Biome	Net primary production (t ha^{-1} yr^{-1})	Litter input (t ha^{-1} yr^{-1})	Standing dead (t ha^{-1})	Litter (t ha^{-1})	Soil organic matter[b] (t ha^{-1})	K_L[c]	K_T[c]
Tundra	4.0	1.7	1.8	28.0	200.0	0.06	0.017
Boreal forest	8.0	5.8	1.3	35.0	150.0	0.17	0.043
Temperate forest	13.0	8.5	7.9	30.0	120.0	0.28	0.082
Temperate grassland	15.0	7.3	—	4.0	220.0	1.78	0.065
Desert	2.0	1.3	—	1.0	80.0	1.30	0.025
Tropical forest	17.0	15.8	13.5	7.5	85.0	2.11	0.160

[a] From references 1, 18, 20, 32, 72, 80, 84, 87, 89, and Anderson and Swift (in press).

[b] Schlesinger (72) gives approximately double the amounts quoted by Ajtay, Ketner, and Duvigneaud (1) for all biomes except desert.

[c] K_L = litter input ÷ litter; K_T = net primary production ÷ standing dead + litter + soil organic matter.

pattern related to substrate quality.

Microflora and fauna populations. The influence of these determinants of the heterotrophic populations is less clearly seen because comparative information on microflora and fauna between ecosystems is particularly sparse and bedevilled by problems of comparable methods. However, a number of general points do emerge, making particular use of the compilation and analysis of IBP data given by Petersen and Luxton (65) and Kjøller and Struwe (49). The majority of comparable data for microbial communities is mycological and shows that species composition is clearly related to biome type, even for the same biome in different continents (22, 49; M. Christensen, Int. Mycol. Congr., 2nd, 1977, Abstr. vol. A–L, p. 99). Comparative biomass estimates are few, yet there is some indication of up to 180 g of total fungal biomass per m^2 in temperate grasslands, compared with upper levels of 120 g/m^2 in temperate forests and up to 20 g/m^2 in tundra (49).

Estimates of microbial production are largely based on estimates of the input of organic matter, microbial yield coefficient, maintenance, and biomass. Earlier suggestions that the carbon/energy input was insufficient to allow more than a generation or so of the microflora per annum (33, 44) have been replaced by an estimated turnover (production/biomass) of 5 to 18 generations for tundra (23, 49) and up to 50 for temperate forests (40, 49, 59). Although speculative, these estimates recognize the recycling of microbial biomass and the high yield efficiency of the microflora, even when growing on natural substrates. Thus, if one discounts winter and periods of drought, the "generation times" are of the order of days.

Bacterial populations show changes both in species composition and in functional activity between ecosystems. Sundman (76) demonstrated, for example, that the characters of bacterial soil populations change markedly with plant cover type. She compared forest humus, grassland, and field soils, concluding that a "*Bacillus*" factor characterized isolates from forest humus.

The major faunal groups and faunal biomass show a fairly distinct pattern of biomass and composition between ecosystems (65, 80): tundra, 3.3 g/m^2, enchytraeids and dipterous larvae; temperate coniferous forest, 2.4 g/m^2, microarthropods and enchytraeids; temperate deciduous forest, 8.0 g/m^2, large oligochaetes; temperate grassland, 5.8 g/m^2, large oligochaetes; and tropical grassland and forest, 1.9 g/m^2, isoptera. O'Neill and DeAngelis (62) suggested that the faunal (heterotrophic) biomass expressed as a proportion of primary production is inversely related to accumulated organic matter. This indicates that, despite the low contribution of fauna directly to decomposition, they either stimulate microbial decomposition (2) or are correlated with it. The result is rapid organic matter turnover and nutrient recycling where the faunal contribution is maximal.

It is only in temperate deciduous forests and grasslands that macrofauna are dominant, and it is therefore only in these systems that physical disturbance of the soil and comminution of litter is a major process. In tundra the relatively large biomass masks a low production because of the long generation times of the dipterous larvae. The dominance of enchytraeids, particularly the parthenogenetic *Cognettia sphagnetorum*, in these environments may reflect an adaptation to stress conditions. Despite the low faunal biomass of tropical forests and grasslands, the dominant termites show high productivity and have developed a highly competitive system with maximum spatial dominance, environmental control within termitaria, and intimate symbiotic relationships with fungi (50). There is no particular direct evidence relating faunal biomass or production to quality of organic matter input; rather, they appear to reflect responses to the environment.

MICROBIAL GROWTH STRATEGIES WITHIN AND BETWEEN ECOSYSTEMS

The preceding review, while identifying the broad pattern of carbon and energy flow, has lacked direct relevance to the scales of time and space of microbiology and does little to distinguish patterns applicable to the taxonomic variety of the microflora. The importance to the microflora of differences in resource quality is apparent from the measurement of the community activity as weight loss or respiration, but the composition and characteristics of the community are implicit, not explicit. The hidden changes in species diversity and structure of the microbial community during resource decomposition were analyzed by Swift (78). He provided a general hypothesis that the declining availability and variety of substrates during decomposition is partly compensated by increased physical diversity of the resource, allowing development of a secondary microflora of bacteria and actinomycetes to replace the fungi which dominate the litter habitats. This recognizes the response of the microflora to the modification of the resource during decomposition and to the presence of other organisms.

The central idea of a succession of microflora is an old one, linked to Winogradsky's (90) concept of different growth strategies of zymogenous and autochthonous microflora. However, the continued development of the ideas of the selection of growth strategies in plants and animals in response to varying environments can help to clarify the pattern of carbon and energy utilization by microflora and fauna on the basis of their response to the three identified groups of factors: (i) resource quality, (ii) physicochemical environment, and (iii) other organisms. The response is in the selection of combinations of characteristics of growth, reproduction, and physiology.

Some parallel between microbial strategies and those of plants and animals might be expected because (i) they are responding to common climatic conditions, at least at the macroscale, and (ii) the selection of plant strategies determines the quality of resource input, for example, through the proportion of photosynthate allocated to plant parts, the longevity of these parts, and associated characteristics of nutrient resorption and herbivore and pathogen defense.

Within plant and animal ecology, two main axes or gradients in the environment are recognized: durational stability and adversity (74). *Durational stability*, or frequency of disturbance, selects for species characteristics along the *r-K* continuum (51), linked to succession at the community level and to characteristics of energy and nutrient flow at the ecosystem level (63). *Adversity*, the degree of favorableness and constancy of a resource, is increasingly recognized as modifying the species characteristics which are selected from within the *r-K* gradient. Climatic severity and nutrient availability are seen as varying along the adversity axis (29, 34, 35, 52, 74, 88). Thus, three types of strategies have been identified in response to the two axes: (i) *exploitation*, basically the *r* strategy, occurring in favorable situations subject to disturbance; (ii) *interaction*, basically the *K* strategy, in which favorable conditions prevail for long periods relative to the organism's life cycle; and (iii) *adversity* (*A*), in which conditions are predictably severe with short and infrequent periods suitable for growth and reproduction. In his extensive analysis of plant strategies, Grime (35) identified stress tolerance as a strategy distinct from the *r-K* continuum evolved in intrinsically unproductive habitats, including climatic severity, or under conditions of extreme resource depletion induced by the vegetation itself. However, he suggested that "ruderal" and "stress-tolerant" strategies correspond, respectively, to the extremes of *r* and *K* selection and that a "competitive" strategy is recognizable in an intermediate position.

The concepts of habitats of varying durational stability and adversity, and the associated selection of exploitation, interaction, and adversity strategies can be translated into microbial ecology (9, 28, 42, 66, 78). The scale difference in time

and space is obviously important. The key resource to which the microflora respond is organic matter which, through litter input and root growth, provides recurrent habitats for colonization and growth. In this sense the recurrent provision of resources can be equated with disturbance, allowing the opportunity for successional development, rather than through loss or dilution of microbial biomass as argued by Pugh (66).

On resources of high quality in temperate and nonarid tropical environments, i.e., favorable habitats, selection for exploitation strategies (*r*) in the early stages of decomposition should be strongest, followed by a longer period in which interaction strategies (*K*) are expected to dominate. Resources of low quality, resulting from recalcitrant carbon substrates, low nutrient concentrations, or high concentrations of inhibitory compounds, can be equated with conditions of continuous adversity or stress, in contrast to discontinuous climatic adversity characteristic of arid and cold environments. Limited discontinuous climatic adversity may also occur in otherwise favorable habitats, but on a time scale relevant to the microflora, for example, drying of surface litters in temperate regions.

The microbial characteristics which are selected for under various conditions are summarized in Table 4. It must be emphasized that (i) most habitats do not fall at the extremes of resource quality and will therefore select for less marked characteristics, (ii) selection will be for combinations of characters, all of which will not necessarily be present at the same time, (iii) fluctuations in climatic conditions or grazing over short time periods can allow species with different growth characteristics to coexist, and (iv) there is a strong stochastic element, particularly during colonization, which may reverberate through the subsequent sequence of organisms. A detailed examination of existing relevant information is not feasible here, but the principles will be illustrated by reference to particular resource types.

High resource quality, favorable environment. Roots, like herbaceous leaves, provide recurrent input of substrates, with a considerable proportion of low-molecular-weight compounds, on which the growth characteristics of the succession of organisms show a sequence from exploitive (*r*) to interactive (*K*) strategies as identified by Bowen (9). Pseudomonads, capable of utilizing a wide range of substrates, are common colonizers migrating along the developing root or, particularly in the case of roots of annual plants, colonizing from the soil. A generation time of 5.2 h was recorded for *Pseudomonas* sp. on *Pinus radiata* roots (9). Such short generation times link with high yield coefficients (85), al-

TABLE 4. Trends in microbial characteristics in early (exploitation) and late (interaction) stages of succession under favorable conditions and under conditions of continuous resource or discontinuous climatic adversity

Characteristic	Exploitation (r)	Interaction (K)	Adversity (A)	
			Continuous (resource)	Discontinuous (climate)
Morphology	Small cells or diffuse mycelium.	Large; often compact mycelium. Cell walls resistant to animal enzymes.	Large cells; compact mycelium.	Large or small.
Physiology	Rapid growth rate, high yield efficiency, giving maximum colonization. Use readily available substrates. Nutrient demanding. Sensitive to plant defense compounds.	Moderate growth rate and yield efficiency for substrate utilization. Use of more resistant substrates. Moderately nutrient demanding. Production of defense compounds (antibiotics) against competitors and grazers.	Slow growth rate, low yield efficiency for maximum use of recalcitrant substrates and nutrient conservation. Intraspecific responses common. Sensitive to chemical stimuli; resistant to plant defense competitors.	Tolerant of low temperature or moisture or rapid metabolic increase with temperature. Sensitive to climatic stimuli.
Life history	Simple. Maximum production of wind- and water-dispersed propagules following short growth phase. Sporing intermittent.	Varied.	Production of larger, resistant propagules with energy and nutrient reserves. Long growth phase. Seasonal dispersal.	Either long tolerant growth phase or short growth from resistant dormant stage. Dormancy linked to climate.
Population dynamics	Explosive, density independent, crash through substrate depletion or opportunistic grazing.	Relatively damped, density-dependent control by interspecific competition and selective grazing.	Relatively damped, control by intraspecific competition and selective grazing.	Erratic, climatically controlled.
Community structure	Diverse.	Symbiotic relations extensive. Very diverse.	Symbiotic associations intimate. Low diversity.	Low diversity.

lowing rapid biomass expansion, although less than 10% of the surface is usually occupied (9). Pseudomonads are particularly consumed by soil amoebae (37), and grazing can cause drastic reduction in laboratory and field populations, with increased nutrient release (14, 15).

Behind the apex, with lower exudation, fungal growth may predominate, growth rates of 3 mm/day being much slower than root extension (9 mm/day) in *Vicia faba* (81). The slower growth rate of *Bacillus* sp. (39 h) on *P. radiata* (9), lower consumption by amoebae (37), and the production of endospores are indicative of an interactive (K) strategy compared with the pseudomonads. The species associations and interactions are complex (9), but at least in some cases, a clearer pattern emerges after death. Excision of *V. faba* roots was followed by a fungal sequence from "sugar fungi" (e.g., *Pythium, Mucor, Penicillium*) to cellulolytic types (e.g., *Chaetomium, Humicola*) on well-decomposed

roots after a few weeks (53), and Stenina (75) observed actinomycetes commonly in late stages of decay.

The simplistic generalizations given here are not satisfactory, and a more comprehensive analysis of fungal strategies is given by Pugh (66). A main limitation on such an analysis is the lack of information which combines the range of organisms and attributes concerned. Basic information on growth characteristics for bacteria in laboratory culture is extensive, as is taxonomic information on fungi in the field; a more balanced picture is needed. However, the sequence of yeasts, sugar fungi, ascomycetes, fungi imperfecti, and basidiomycetes on high-quality resources (27) remains a partially validated model of the r-K sequence. Deviations from this model may be related to variations in adversity, considered next.

Increasing adversity of resource quality. Increasing adversity in resource quality, with increasing proportions of ligno-cellulose, decreasing nutrient content, and varying secondary compounds, is represented along a gradient from herbaceous leaf litter, conifer needles, bracken petioles, and deciduous and coniferous wood, to the extreme substrate, homogeneous, recalcitrant keratin.

Consideration of fungal development on the needles of pine identifies the influence of adversity in terms of nutritional quality and inhibitory secondary compounds (47). Common primary saprophytes play only a very small part in the decomposition of *Pinus sylvestris* litter because of its chemical nature. The needles contain little available sugar and high lignin, and many of the cytoplasmic proteins and starch are "locked up" by tannins (56, 57). The free phenolics inhibit fungal growth, and the cuticle provides an effective physical barrier. Stomata and the needle bases provide important entry points for the specialists *Lophodermium pinastri* and *Fusicoccum bacillare* (46), and a subcuticle breakdown of cellulose is commenced. *Marasmius androsaceus* has the ability to surface colonize the needles from rhizomorphs and then sends out small hyphae which degrade, and pass through, lignified walls into the xylem. This organism may overcome nitrogen limitation by translocation and is clearly resistant to the monoterpenes. After the tissue has been colonized by microorganisms, the cryptostigmatid mites become the dominant fauna, feeding mainly on hyphae and thus avoiding the problems of nutrient paucity and inhibitory terpenes.

A similar "abbreviated" succession was described during the decomposition of bracken petioles (25), so-called because the classic *r-K* pattern (27) for plant material was modified on this more adverse substrate, with the *r* phase missing and with colonization and subsequent decomposition being of a *K* form. Adversity is primarily a result of the resorption and leaching of nutrients and soluble sugars from the petiole prior to decomposition, leaving a resource which requires the import of nitrogen, and possession of enzymatic machinery for degradation of structural carbohydrates, even at the initial decay stage. The sphaeropsidales and basidiomycetes such as *Mycena galopus* are favored, and the decay process is slow. Since yield efficiency at slow growth rates is poor (85), the amounts of biomass supported by adverse substrates are proportionally low, and yield coefficients as low as 0.04 for *Mycena* growing on oak litter have been presented. In *r* strategy fungi and bacteria with coefficients of anything up to 0.60 to 0.70 (26, 77) are possible, but the strategies of nutrient translocation, degradation of inhibitory substances, maintenance of large biomass, and utilization of recalcitrant molecules all lead to yield inefficiency.

Of the abundant natural substrates, wood is perhaps the most adverse, being low in nutrients, structurally homogeneous, rich in inhibitory secondary compounds, and low in available carbohydrate, and having small surface area. The tight links between fungal tolerance to phenolics and nature of substrate decomposed reflect the importance of monoterpenes in deciding succession on different types of wood (43), with the ability to grow at high CO_2 concentrations distinguishing between wood and litter basidiomycetes.

The strong interactions shown by basidiomycetes on wood include interactions not only with substrates but also between species and between individuals within species. The interactions with substrate are perhaps best illustrated by Kirk and Fenn (48), who showed that ammonium nitrogen and certain amino acids may repress lignolytic activity in basidiomycetes, providing a chemical signal preventing premature, and unprofitable, entry into the succession. Rayner and co-workers have shown clearly the intraspecific and intense antagonism between individuals colonizing wood, interactions developing which are not present in more generalized *r* types of organisms (e.g., A. D. M. Rayner and J. E. Webber, *in* D. H. Jennings and A. D. M. Rayner, ed., *The Fungal Mycelium*, in press). Their work, and that of Thompson and Boddy (82), has altered the concept of the individual when considering basidiomycete growth, and we now realize that a single individual may ramify through the litter over areas of square meters, with inherent consequences for nutrient translocation and substrate exploitation.

The grazing of fungal mycelium by fauna is a key feature in successional development. In wood, animal grazing may disturb a *K* strategist and provide fresh substrate for rapid bacterial and "*r*"-type mycelial development (79). This is paralleled by the effects of animal grazing in litter (2). K. Newell (Soil Biol. Biochem., in press) showed, in an analysis of growth and distribution of *Mycena galopus* and *Marasmius androsaceus*, that grazing by the collembolan *Onychiurus latus* dictated the outcome of the competition. *Marasmius* readily outgrows *Mycena* in litter and fermentation layer material from oak woodlands, as well as in culture. When the two organisms are grown together in litter and fermentation layer material in the absence of the grazer, this trend continues, and *Marasmius* outcompetes *Mycena*. In the presence of the grazer, which prefers *Marasmius* as a food source, the competition is reversed, and *Mycena* dominates. In the field *Marasmius* is restricted to the litter layer, where *O. latus* is least common, and the distribution of *Marasmius* and

Mycena can be explained in relation to growth rate and grazing.

Adversity through climatic severity. The discontinuous stress of cold and arid environments imposes two main problems on the microflora: survival during adverse conditions has to be combined with maximizing growth during the short favorable period. Superimposed on this is a tendency for plants under adverse conditions to produce secondary compounds to protect long-lived photosynthetic structures; a high content of aromatic compounds is recognized in many desert plants. Although no comprehensive data are available, there are some important indications, particularly from tundra studies (23, 70), of strategies adopted by the microflora under temperature stress.

Enzymatic capacities of bacteria and fungi on standing dead litter and in soil in tundra are unexceptional, but selection from a wide range of strategies in response to the different microhabitat temperature regimes is indicated. Linear responses can sustain activity over a wide temperature range, and an exponential response can enhance activity during short periods of favorable temperature. Among the fungi a variety of temperature-response curves were shown for respiration, growth, and substrate utilization (23). Respiration, and possibly cellulose and phenol oxidation, is continued to −7.5°C, with growth down to 0°C. Temperature-substrate responses may be linked to the spring thaw when some yeasts, bacteria, and algae capitalize on the dissolved organic carbon and nutrients released through freeze-thaw cycles (60, 91). Overwintering strategies of the microflora are not as clear as for invertebrates, for which, interestingly, Block (8) found that the biology of the mite *Alaskozetes* did not readily fit into the *r-K* continuum in its responses to climatic adversity.

Intimate symbiotic relationships may represent one *A* strategy in dry conditions, microflora obtaining an advantage of an improved microclimate when associated with termitaria (50) or invertebrate guts (12). A similar strategy is advantageous under resource adversity, either carbon or nutrient, and is seen in gut microflora (12, 61), mycorrhizal associations, and lichens.

CONCLUSIONS

The general comparison of carbon and energy fixation and dissipation between biomes emphasizes two features: (i) broad climatic control, with heterotrophic activity closely following plant production, and (ii) the similarities, in all but quantity, with belowground production often compensating the more obvious aboveground growth. The differences between ecosystems appear at a finer scale, in functional, rather than species, characteristics. Here again, there is evidence that the heterotrophs mirror the plants in terms of growth strategies through (i) their direct response to variation in quality of resource input which is determined by plant growth strategies and (ii) their similarity of response to common environmental conditions.

The hypotheses of growth strategies in microbial communities are, as yet, poorly developed. This is largely because of the versatility and variety of the microflora which provides a multitude of combinations of options to cope with the variations in resource input and environment. However, there is a reasonable theoretical basis for the selection of recognizable combinations of morphological, physiological, and phenological features within the microbial associations in different habitats. The pattern will rarely be precise because of chance, environmental fluctuation, and microbial variety, but the earlier hypotheses of the *r-K* continuum are made more relevant to microbial ecology when a dimension of adversity or stress is added to that of disturbance.

There are also indications that the microbial growth strategies selected in different habitats will have a feedback effect on plant growth. The efficiency of bacterial biomass production per unit of organic substrate consumed depends on the type of energy metabolism involved, growth rate, kind of substrate limitation, and degree of predation. There is a shift from biomass formation to nutrient cycling with decreasing growth rate (85). Extended to other microflora, the slow growth rates associated with low resource quality may compensate for mechanisms of nutrient conservation and result in increased nutrient release per unit of substrate decomposed, compared with high-quality resources. Such interrelationships of processes resulting from selection for microbial growth strategies need to be explored quantitatively. However, is it unreasonable to speculate that the selection of different growth strategies in the microflora may account, for example, for the direct relationship shown by Berg and Staaf (5) between the initial nitrogen concentration (resource quality) and the concentration at which net nitrogen mineralization occurs, and for the much lower C/N ratio required for nitrogen availability to plants in arable conditions compared with low-quality forest floor litters (41)?

We thank T. V. Callaghan, J. Dighton, J. C. Frankland, and other colleagues at Merlewood for constructive discussion in the development of this paper.

P.I. is a European Science Foundation Research Fellow.

LITERATURE CITED

1. **Ajtay, G. L., P. Ketner, and P. Duvigneaud.** 1979. Terrestrial primary production and phytomass, p. 129–181. *In* B.

Bolin, E. T. Degens, S. Kempe, and P. Ketner (ed.), The global carbon cycle. SCOPE 13. Wiley, Chichester.

2. **Anderson, J. M., and P. Ineson.** 1983. Interactions between microorganisms and soil invertebrates in nutrient flux pathways of forest ecosystems. In J. M. Anderson, A. D. M. Rayner, and D. Walton (ed.), Invertebrate-microbial interactions. Cambridge University Press, Cambridge.

3. **Berg, B., K. Hammus, T. Popoff, and O. Theander.** 1982. Changes in organic chemical components of needle litter during decomposition. Long-term decomposition in a Scots pine forest. Can. J. Bot. **60**:1310–1319.

4. **Berg, B., and H. Staaf.** 1980. Decomposition rate and chemical changes of Scots pine needle litter. II. Influences of chemical composition. Ecol. Bull. (Stockholm) **32**:375–390.

5. **Berg, B., and H. Staaf.** 1981. Leaching, accumulation and release of nitrogen in decomposing forest litter. Ecol. Bull. (Stockholm) **33**:163–178.

6. **Berg, B., B. Wessen, and G. Ekbohm.** 1982. Nitrogen level and decomposition in Scots pine needle litter. Oikos **38**:291–296.

7. **Bliss, L. C., O. W. Heal, and J. J. Moore (ed.).** 1981. Tundra ecosystems: a comparative analysis. IBP 25. Cambridge University Press, Cambridge.

8. **Block, W.** 1980. Survival strategies in polar terrestrial arthropods. Biol. J. Linn. Soc. **14**:29–38.

9. **Bowen, G. D.** 1980. Misconceptions, concepts and approaches in rhizosphere biology, p. 283–394. In D. C. Ellwood, J. N. Hedger, M. J. Latham, J. M. Lynch, and J. H. Slater (ed.), Contemporary microbial ecology. Academic Press, London.

10. **Bray, J. R., and E. Gorham.** 1964. Litter production in forests of the world. Adv. Ecol. Res. **2**:101–157.

11. **Breymeyer, A. I., and G. M. Van Dyne (ed.).** 1980. Grasslands, systems analysis and man. IBP 19. Cambridge University Press, Cambridge.

12. **Breznak, J. A.** 1982. Intestinal microbiota of termites and other xylophagous insects. Annu. Rev. Microbiol. **36**:323–343.

13. **Bunnell, F. L., D. E. N. Tait, P. W. Flanagan, and K. Van Cleve.** 1977. Microbial respiration and substrate weight loss. I. A general model of the influences of abiotic variables. Soil Biol. Biochem. **9**:33–40.

14. **Clarholm, M.** 1983. Dynamics of soil bacteria in relation to plants, protozoa and inorganic nitrogen. Institute for Microbiology Report 17. Swedish University of Agricultural Sciences, Uppsala.

15. **Coleman, D. C., C. V. Cole, R. V. Anderson, M. Blaha, M. K. Campion, M. Clarholm, E. T. Elliott, H. W. Hunt, B. Shaefer, and J. Sinclair.** 1977. An analysis of rhizosphere-saprophage interactions in terrestrial ecosystems, p. 299–309. In U. Lohm and T. Persson (ed.), Soil organisms as components of ecosystems. Ecol. Bull. (Stockholm), 25.

16. **Coleman, D. C., A. Sasson, A. I. Breymeyer, M. C. Dash, Y. Dommergues, H. W. Hunt, E. A. Paul, R. Schaefer, B. Ulehlova, and R. I. Zlotin.** 1980. Decomposer subsystem, p. 609–655. In A. I. Breymayer and G. M. van Dyne (ed.), Grasslands, systems analysis and man. IBP 19. Cambridge University Press, Cambridge.

17. **Comanor, P. L., and E. E. Staffeldt.** 1978. Decomposition of plant litter in two western North American deserts, p. 31–49. In N. E. West and J. Skujins (ed.), Nitrogen in desert ecosystems. Dowden, Hutchinson and Ross, Stroudsburg.

18. **Coupland, R. T. (ed.).** 1979. Grassland ecosystems of the world: analysis of grasslands and their uses. IBP 18. Cambridge University Press, Cambridge.

19. **Cromack, K.** 1981. Below-ground processes in forest succession, p. 361–373. In D. C. West, H. H. Shugart, and D. B. Botkin (ed.), Forest succession—concepts and application. Springer-Verlag, New York.

20. **DeAngelis, D. L., R. H. Gardner, and H. H. Shugart.** 1981. Productivity of forest ecosystems studied during the IBP: the woodlands data set, p. 567–672. In D. E. Reichle (ed.), Dynamic properties of ecosystems. IBP 23. Cambridge University Press, Cambridge.

21. **Deans, J. D.** 1979. Fluctuations of the soil environment and fine root growth in a young Sitka spruce plantation. Plant Soil **52**:195–208.

22. **Domsch, K. H.** 1975. Distribution of soil fungi, p. 340–353. In T. Hasegawa (ed.), Developmental microbiology. Proceedings 1st International Congress International Association of Microbiological Societies, vol. 2. Science Council of Japan, Tokyo.

23. **Flanagan, P. W., and F. L. Bunnell.** 1980. Microflora activities and decomposition, p. 291–334. In J. Brown, P. C. Millar, L. L. Tieszen, and F. L. Bunnell (ed.), An arctic ecosystem. The coastal tundra at Barrow, Alaska. Dowden, Hutchinson and Ross, Stroudsburg.

24. **Fogel, R., and K. Cromack.** 1977. Effect of habitat and substrate quality on Douglas fir litter decomposition in Western Oregon. Can. J. Bot. **55**:1632–1640.

25. **Frankland, J. C.** 1974. Decomposition of lower plants, p. 3–36. In C. H. Dickinson and G. J. F. Pugh (ed.), Biology of plant litter decomposition, vol. 1. Academic Press, London.

26. **Frankland, J. C., D. K. Lindley, and M. J. Swift.** 1978. An analysis of two methods for the estimation of mycelial biomass in leaf litter. Soil Biol. Biochem. **10**:323–333.

27. **Garrett, S. D.** 1951. Ecological groups of soil fungi; a survey of substrate relationships. New Phytol. **50**:149–166.

28. **Gerson, U., and I. Chet.** 1981. Are allochthonous and autochthonous soil microorganisms r- and K-selected? Rev. Ecol. Biol. Sol **18**:285–289.

29. **Glesener, R. J., and D. Tilman.** 1978. Sexuality and the components of environmental uncertainty: clues from geographical parthenogenesis for terrestrial animals. Am. Nat. **112**:159–173.

30. **Goodall, D. W., and R. A. Perry (ed.).** 1979. Arid-land ecosystems: structure, function and management, vol. 1. IBP 16. Cambridge University Press, Cambridge.

31. **Goodall, D. W., and R. A. Perry (ed.).** 1981. Arid-land ecosystems: structure, function and management, vol. 2. IBP 17. Cambridge University Press, Cambridge.

32. **Gosz, J. R.** 1981. Nitrogen cycling in coniferous ecosystems. Ecol. Bull. (Stockholm) **33**:405–426.

33. **Gray, T. R. G., and S. T. Williams.** 1971. Microbial productivity in soil. Symp. Soc. Gen. Microbiol. **21**:255–286.

34. **Greenslade, P. J. M.** 1982. Selection processes in arid Australia, p. 125–130. In W. R. Barker and P. J. M. Greenslade (ed.), Evolution of the flora and fauna of arid Australia. Peacock Publications, Frewville, South Australia.

35. **Grime, J. P.** 1979. Plant strategies and vegetation process. John Wiley & Sons, Inc., New York.

36. **Handley, W. R. C.** 1954. Mull and mor formation in relation to forest soils. For. Comm. Bull. **23**:1–115.

37. **Heal, O. W., and J. M. Felton.** 1970. Soil amoebae, their food and their reaction in microflora exudates, p. 145–162. In A. Watson (ed.), Animal populations in relation to their food resources. Blackwell Scientific Publications, Oxford.

38. **Heal, O. W., P. W. Flanagan, D. D. French, and S. F. MacLean.** 1981. Decomposition and accumulation of organic matter, p. 587–633. In L. C. Bliss, O. W. Heal, and J. J. Moore (ed.), Tundra ecosystems: a comparative analysis. IBP 25. Cambridge University Press, Cambridge.

39. **Heal, O. W., P. M. Latter, and G. Howson.** 1978. A study of the rates of decomposition and accumulation of organic matter, p. 136–159. In O. W. Heal and D. F. Perkins (ed.), Production ecology of British moors and montane grassland. Springer-Verlag, Berlin.

40. **Heal, O. W., and S. F. MacLean, Jr.** 1975. Comparative productivity in ecosystems—secondary productivity, p. 89–108. In W. H. van Dobben and R. H. Lowe-McConnell (ed.), Unifying concepts in ecology. W. Junk Publishers, The Hague.

41. **Heal, O. W., M. J. Swift, and J. M. Anderson.** 1982. Nitrogen cycling in United Kingdom forests: the relevance of basic ecological research. Phil. Trans. R. Soc. London Ser. B **26:**472–444.

42. **Hedger, J. N., and T. Basuki.** 1982. The role of basidiomycetes in composts: a model system for decomposition studies, p. 263–305. *In* J. C. Frankland, J. N. Hedger, and M. J. Swift (ed.), Decomposer basidiomycetes—their biology and ecology. Cambridge University Press, Cambridge.

43. **Hintikka, V.** 1982. The colonization of litter and wood by basidiomycetes in Finnish forests, p. 227–239. *In* J. C. Frankland, J. N. Hedger, and M. J. Swift (ed.), Decomposer basidiomycetes—their biology and ecology. Cambridge University Press, Cambridge.

44. **Hisset, R., and T. R. G. Gray.** 1976. Microsites and time changes in soil microbe ecology, p. 23–40. *In* J. M. Anderson and A. Macfadyen (ed.), The role of terrestrial and aquatic organisms in decomposition processes. Blackwell Scientific Publications, Oxford.

45. **Kaarik, A. A.** 1974. Decomposition of wood, p. 129–174. *In* C. H. Dickinson and G. J. F. Pugh (ed.), Biology of litter decomposition, vol. 1. Academic Press, London.

46. **Kendrick, W. B.** 1959. The time factor in the decomposition of coniferous leaf litter. Can. J. Bot. **37:**907–912.

47. **Kendrick, W. B., and A. Burges.** 1962. Biological aspects of decay of *Pinus sylvestris* leaf litter. Nova Hedwigia **4:**313–342.

48. **Kirk, T. K., and P. Fenn.** 1982. Formation and action of the ligninolytic system in basidiomycetes, p. 67–90. *In* J. C. Frankland, J. N. Hedger, and M. J. Swift (ed.), Decomposer basidiomycetes—their biology and ecology. Cambridge University Press, Cambridge.

49. **Kjøller, A., and S. Struwe.** 1982. Microfungi in ecosystems: fungal occurrence and activity in litter and soil. Oikos **39:**389–422.

50. **Lee, K. E., and T. G. Wood.** 1971. Termites and soils. Academic Press, London.

51. **MacArthur, R. H., and E. D. Wilson.** 1967. The theory of island biogeography. Princeton University Press, Princeton.

52. **MacLean, S. F.** 1975. Ecological adaptation of tundra invertebrates, p. 269–300. *In* F. J. Vernberg (ed.), Physiological adaptation to the environment. Intext, New York.

53. **Mahiques, P. L. J.** 1966. The fungal colonization of broad bean root systems. Sch. Sci. Rev. **48:**108–123.

54. **McClaugherty, C. A., J. D. Aber, and J. M. Melillo.** 1982. The role of fine roots in the organic matter and nitrogen budgets of two forested ecosystems. Ecology **63:**1481–1490.

55. **Meentemeyer, V.** 1978. Macroclimate and lignin control of litter decomposition rates. Ecology **59:**465–472.

56. **Millar, C. S.** 1974. Decomposition of coniferous leaf litter, p. 105–128. *In* C. H. Dickinson and G. F. J. Pugh (ed.), Biology of plant litter decomposition, vol. 2. Academic Press, London.

57. **Mitchell, C. P., C. S. Millar, and D. W. Minter.** 1978. Studies on the decomposition of Scots pine needles. Trans. Br. Mycol. Soc. **71:**343–348.

58. **Mommaerts-Billiet, F.** 1971. Aspects dynamiques de la partition de la litere de feuilles. Bull. Soc. R. Bot. Belg. **104:**181–195.

59. **Nagel-de Boois, H. M.** 1971. Preliminary estimates of production of fungal mycelium in forest soil layers. Ann. Zool. Ecol. Anim. **4**(Special):447–454.

60. **Nelson, L. M., and S. Visser.** 1978. Effect of spring thaw on microorganisms in an arctic meadow site. Arct. Alp. Res. **10:**679–688.

61. **Neuhauser, E., C. Youmell, and R. Hartenstein.** 1974. Degradation of benzoic acid in the terrestrial crustacean, *Oniscus asellus*. Soil Biol. Biochem. **6:**101–107.

62. **O'Neill, R. V., and D. L. DeAngelis.** 1981. Comparative productivity and biomass relations of forest ecosystems, p. 411–449. *In* D. E. Reichle (ed.), Dynamic properties of forest ecosystems. IBP 23. Cambridge University Press, Cambridge.

63. **Odum, E. P.** 1969. The strategy of ecosystem development. Science **164:**262–270.

64. **Persson, T.** 1979. Fine-root production, mortality and decomposition in forest ecosystems. Vegetatio **41:**101–109.

65. **Peterson, H., and M. Luxton.** 1982. A comparative analysis of soil fauna populations and their role in decomposition processes. Oikos **39:**287–288.

66. **Pugh, G. F. J.** 1980. Strategies in fungal ecology. Trans. Br. Mycol. Soc. **75:**1–14.

67. **Reichle, D. E. (ed.).** 1982. Dynamic properties of forest ecosystems. IBP 23. Cambridge University Press, Cambridge.

68. **Rodin, L. E., and N. I. Basilevic.** 1967. Production and mineral cycling in terrestrial vegetation. Oliver and Boyd, Edinburgh.

69. **Rosenthal, G. A., and D. H. Janzen.** 1979. Herbivores. Their interaction with secondary plant metabolites. Academic Press, London.

70. **Satchell, J. E., and D. G. Lowe.** 1967. Selection of leaf litter by *Lumbricus terrestris*, p. 102–119. *In* O. Graff and J. E. Satchell (ed.), Progress in soil biology. Vieweg, Braunschweig.

71. **Scheffer, T. C., and E. B. Cowling.** 1966. Natural resistance of wood to microbial deterioration. Annu. Rev. Phytol. **4:**147–170.

72. **Schlesinger, W. H.** 1977. Carbon balance in terrestrial detritus. Annu. Rev. Ecol. Syst. **8:**51–81.

73. **Schlesinger, W. H., and M. M. Hasey.** 1981. Decomposition of chaparral shrub foliage: losses of organic and inorganic constituents from deciduous and coniferous leaves. Ecology **62:**762–774.

74. **Southwood, T. R. E.** 1977. Habitat, the templet for ecological strategies. J. Anim. Ecol. **46:**337–365.

75. **Stenina, T. A.** 1964. Decomposition of plant residues in arable podzolic soils. Sov. Soil Sci., p. 74–80.

76. **Sundman, V.** 1970. Four bacterial soil populations characterized and compared by a factor analytical method. Can. J. Microbiol. **16:**455–464.

77. **Swift, M. J.** 1973. The estimation of mycelial biomass by determination of the hexosamine content of wood tissue decayed by fungi. Soil Biol. Biochem. **5:**321–332.

78. **Swift, M. J.** 1976. Species diversity and the structure of microbial communities, p. 185–222. *In* J. M. Anderson and A. Macfadyen (ed.), The role of terrestrial and aquatic organisms in decomposition processes. Blackwell Scientific Publications, Oxford.

79. **Swift, M. J.** 1982. Basidiomycetes as components of forest ecosystems, p. 307–337. *In* J. C. Frankland, J. N. Hedger, and M. J. Swift (ed.), Decomposer basidiomycetes—their biology and ecology. Cambridge University Press, Cambridge.

80. **Swift, M. J., O. W. Heal, and J. M. Anderson.** 1979. Decomposition in terrestrial ecosystems. Blackwell Scientific Publications, Oxford.

81. **Taylor, G. S., and D. Parkinson.** 1961. Growth of saprophytic fungi on root surfaces. Plant Soil **15:**261–277.

82. **Thompson, W., and L. Boddy.** 1983. Decomposition of suppressed oak trees in even-aged plantations. II. Colonization of tree roots by cord- and rhizomorph-producing basidiomycetes. New Phytol. **93:**277–291.

83. **Tóth, J. A., L. B. Papp, and B. Lenkey.** 1975. Litter decomposition in an oak forest ecosystem (*Quercetum petraeae cerris*) of Northern Hungary studied in the framework of "sikfökut project," p. 41–58. *In* G. Kilbertus, O. Reising, A. Mourey, and J. A. Cancela de Fonseca (ed.), Biodégradation et humification. Pierron Editeur, Sarreguemines.

84. **Van Cleve, K., and V. Alexander.** 1981. Nitrogen cycling in tundra and boreal ecosystems. Ecol. Bull. (Stockholm) **33:**335–404.

85. **Veldkamp, H.** 1975. The role of bacteria in energy flow and nutrient cycling, p. 44–49. *In* W. H. van Dobben and

R. H. Lowe-McConnell (ed.), Unifying concepts in ecology. Junk, Hague.

86. **Waksman, S. A.** 1952. Soil microbiology. John Wiley & Sons, Inc., New York.

87. **West, N. E.** 1979. Formation, distribution and function of plant litter in desert ecosystems, p. 647–659. *In* D. W. Goodall and R. A. Perry (ed.), Arid-land ecosystems, vol. 1. IBP 16. Cambridge University Press, Cambridge.

88. **Whittaker, R. H.** 1975. Communities and ecosystems, 2nd ed. Collier-MacMillan, London.

89. **Wielgolaski, F. E., L. C. Bliss, J. Svoboda, and G. Doyle.** 1981. Primary production of tundra, p. 187–225. *In* L. C. Bliss, O. W. Heal, and J. J. Moore (ed.), Tundra ecosystems: a comparative analysis. IBP 25. Cambridge University Press, Cambridge.

90. **Winogradsky, S.** 1924. Sur la microflore autochthone de la terre arable. C.R. Acad. Sci. **178:**1236–1239.

91. **Wynn-Williams, D. D.** 1982. Simulation of seasonal changes in microbial activity of maritime Antarctic peat. Soil Biol. Biochem. **14:**1–12.

Significance of Microorganisms in Carbon and Energy Flow in Marine Ecosystems

LAWRENCE R. POMEROY

Institute of Ecology, University of Georgia, Athens, Georgia 30602

The ocean's biomass is mostly microbial, including algae, protozoa, bacteria, and fungi (7; F. Azam, *in* J. E. Hobbie and P. J. leB. Williams, ed., *Heterotrophic Activity in the Sea*, in press). All are considered in this review. In coastal waters, estuaries, and freshwater the microbial portion of total biomass may be less than it is in the deep sea, but microorganisms still contribute significantly to ecosystem processes. Microorganisms are effective competitors for reduced carbon sources in all aquatic ecosystems (7, 51; Azam, in press). It is now recognized not only that microorganisms are major users of energy and materials but also that they form microbial food webs. Because the assimilation efficiency of microorganisms is often higher than that of large organisms, microbial food webs may transfer energy more efficiently than classical food chains (25, 37, 51). Since the webs may be long or convoluted, energy dissipation will occur, imposing a limit on the amount of energy ultimately available to higher organisms. In fact, the distinction between food chains and food webs is trivial, because complex webs must be composed of short chains if they are energetically viable. Although microorganisms are ubiquitous, their interaction in food webs is modified by their small home ranges, a concept just as applicable to bacteria as to wolves. We must consider events on the spatial scales in which bacteria and protozoa really operate, the limit that small size places on movement from place to place, the transfer of materials, and the utilization of resources of varying dimensions as well as varying composition. Microorganisms can operate on every level in a food web, thereby having more than one chance to utilize energy as it passes through the ecosystem.

PRIMARY PRODUCTION

Except for the immediate coastal zone, microorganisms are responsible for most of the ocean's primary production, which is about half of the total photosynthesis of the planet. Production by phytoplankton is greater than its standing stock would suggest. Short generation times can result in high production even without a large standing stock.

Operationally, we now divide the phytoplankton into three size groups, net plankton, nano-plankton, and picoplankton. The relative production by each of these groups varies greatly, with net phytoplankton producing a relatively larger fraction of the biomass in eutrophic environments (33). Net plankton (<10 μm), largely diatoms and dinoflagellates, probably account for less than half of total marine phytoplankton production. In some situations the net phytoplankton are largely consumed by microcrustacea, and in others much of the population dies in place, sinks, and becomes detritus (19, 35). Heterotrophic microbial consumers will utilize much of this detritus. The nanoplankton include small diatoms, coccolithophoridae, and a number of small, motile autotrophs. Their response time to changing conditions is rapid, their doubling time is short, and their potential production is high. Nanoplankton appear to be utilized efficiently and completely by mucus net feeders, early life history stages of the microcrustacea, pelagic larvae of the benthos, and protozoans. The picoplankton include the cyanobacterial genus *Synechococcus* and some small autotrophic flagellates. *Synechococcus* sp. appears to be most abundant and productive in temperate waters, and less abundant and productive in the tropics and polar regions (23, 28, 48). Little is known about the picoplanktonic flagellates. Organisms of this size can be collected for food only by mucus net feeders or by protozoans. The mucus net feeders collect picoplankton less efficiently than nanoplankton. Unless they are exceptionally abundant, picoplankton are probably an incidental component of the diet of larger organisms, and some may be indigestible (24). Here we see the potential for a microbial food chain, beginning with organisms too small to be used efficiently by most metazoans and perhaps progressing through two or more levels of protozoan consumers. At this time we know little more than that a diverse population of protozoans in substantial numbers exists in natural waters (10, 46).

Recently, it has been suggested that phytoplankton in the presumably oligotrophic central ocean gyres are not nutrient limited and are growing rapidly (14, 34). Although the evidence for this is indirect, it follows earlier observations by Sheldon and Sutcliffe (43) that small organisms, <1 μm (as ATP), are multiplying rapidly in ocean water. Recent microscope work has shown that there are indeed substantial populations of small organisms, including autotrophic picoplankton, which are growing rapidly. Have these organisms evolved strategies for circumventing the need to scavenge nutrients from very dilute solution? Opinions vary (21, 31). Net plankton are not randomly distributed and do associate in patches. Azam (in press) has suggested that microorganisms may on their own scale of activities associate on favorable diffusion gradients. Although organisms the size of picoplankton probably cannot improve their diffusive uptake of nutrients by swimming, they can move at negligible energy cost until they find a favorable site with high nutrient concentration (41). If they do this, they may confound short-term tracer experiments which depend on diffusion. Unfortunately, direct observation of the orientation of microorganisms in natural waters is difficult. Divers can perceive the larger plankton in space but not the microorganisms. Samples passed through sampling bottles and pipettes and onto a slide may be randomized, and they may not reaggregate in the hostile flatland under the cover glass.

PRODUCTION AND UTILIZATION OF PARTICULATE ORGANIC MATTER

The standing stock of nonliving particulate matter in the ocean, and in most natural waters, exceeds the standing stock of living phytoplankton. Whereas in freshwater and coastal waters much of this detritus may be allochthonous, in the open ocean most of it must originate from phytoplankton. Living phytoplankton repel bacteria, but as they weaken and die during cyclic periods of nutrient limitation, phytoplankton are invaded by bacteria and are converted directly into detritus and bacterial biomass. This is especially evident following the spring bloom at high latitudes and *Oscillatoria (Trichodesmium)* blooms in the tropics. Of the phytoplankton eaten by zooplankton, about one-third is released as feces. The larger microcrustacea and salps produce well-packaged feces of 100 to 1,000 μm which tend to sink. If not intercepted by a coprophage, they will contribute to benthic production, and in fact they are the major input of reduced carbon to both the shallow and the abyssal benthos (18). Many smaller zooplankton, from protozoans to nauplii, produce smaller and less well packaged feces which sink slowly or not at all (17). Because of the viscosity of water, low-density particles less than 50 μm in diameter do not sink appreciably and will be degraded in the water.

Another type of particle commonly found in natural water is the so-called organic aggregate. These particles range from a few micrometers to

several centimeters. Chemically, they appear to be phytoplankton in various stages of degradation, from recognizable species to globs of polysaccharide. The origin of these particles is difficult to discover, and probably there are more than one. Riley (42) and others (6) demonstrated that organic particles can form de novo in water. Several mechanisms have been proposed, involving polymerization by high-energy radiation and condensation of surface film by collapsing bubbles (13, 22). Particles formed in these ways are very small. Their rate of production and utilization by either microorganisms or macroorganisms in the sea is not known. Larger organic aggregates, viewed by light microscopy, contain recognizable fragments and frustules and sometimes populations of living bacteria and protozoa (49). Pomeroy and Deibel (39) compared these aggregates with feces of small salps and doliolids and suggested that the larger organic particles are aggregates of zooplankton feces, 1 to 2 days old. Most of these are in the 10- to 100-μm size range of particles that sink slowly or not at all, and they will be degraded in the water over a period of 2 to 4 days by bacteria.

Electron micrographs of small aggregates show numerous 0.1-μm filaments linking particles together, which appear to be associated with bacteria (Fig. 1). This may help to explain the origin of the really large sea snow, >1 cm, which is commonly present in seawater, especially in regions of high productivity. The alert viewer will see it in every television film of underwater natural history. Microscopic examination of diver-collected sea snow shows it to be the same as the smaller particles believed to be

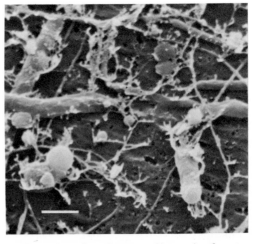

FIG. 1. Scanning electron micrograph of seston and bacteria from the Kuroshio off the Izu Peninsula, Honshu, Japan, showing many associated filamentous processes or products. Bar = 1 μm.

fecal in origin (39). Work both with these particles and with salp feces shows that they are fragile but also sticky. They are easily broken into small fragments, but they adhere to each other and re-form large particles. Presumably, the large particles will sink more rapidly. If there is a continuing process of aggregation up to >1 cm, followed by sinking, then this is a scavenging mechanism by which organic matter is swept out of the water and transported to the deep ocean or the bottom. However, we have more questions than answers regarding the nature of this process. What is the rate of production of true organic aggregates from dissolved organic material? How does it compare with the rate of production of fecal particles? What is the role of bacteria in aggregating small particles into larger ones? Do they do that more quickly than they degrade the particles? How much of this material is consumed by macroorganisms? Is it nutritious food for macroorganisms? Is bacterial biomass consumed by macroorganisms in this way in significant amounts?

Scanning electron micrographs of aggregates show them to be collapsed like broken balloons, the only mass being the colonies of bacteria in them (Fig. 2). Therefore, as food they would be largely bacterial biomass, a reasonable food source if enough bacteria are present. If they are primarily fecal in origin, this suggests that we cannot predict the rate of fall of feces with simple Stokes law models. The particles are changing size by as much as three orders of magnitude, and their microbial content suggests that they remain suspended for more than 1 day. A substantial amount of energy is moving

FIG. 3. Scanning electron micrograph of a small aggregate particle from the inner continental shelf coastal water off Wassaw Island, near Savannah, Georgia. Note the presence of a mucus net entrapping some smaller particulate matter. Bar = 2 μm.

through a microbial food chain inside the particles, so much so that they may not reach bottom before they are assimilated by the bacteria and protozoa. Another source of relatively large particles is the cast houses of appendicularians (1). These and other tests, carapaces, and nets become enmeshed in the larger aggregates (Fig. 3).

One of the difficulties in conceptualizing the sequence of events in the production and fate of particulate organic matter is the difficulty of seeing or even detecting the size distribution of particles in the water. Divers can count and quantify the larger particles. Presumably, a microscopist can count the smaller ones. However, the procedures used in collecting them almost certainly modify the size spectrum of these fragile yet sticky objects. All of the published work on naturally occurring organic aggregates suffers from this limitation. The use of in situ particle counters is potentially one approach, but these do not discriminate between aggregates and other kinds of particles, including living ones. Moreover, they operate in a towed mode and have orifices which will tend to fragment fragile particles.

If, as we now believe, there is a spectrum of particle sizes in the water, from <10 to >10,000 μm, and many of these remain in the water until they are degraded by bacteria (L. R. Pomeroy, in J. E. Hobbie and P. J. leB. Williams, ed., *Heterotrophic Processes in the Sea*, in press), what sort of environment exists in such particles, and how is it modified by the microbial

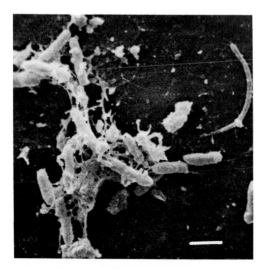

FIG. 2. Scanning electron micrograph of a small aggregate particle from water on the outer continental shelf off St. Augustine, Florida. Bar = 2 μm.

food web? Consider first the utilization of the particulate substrates by aerobic heterotrophs. Will they use up all of the oxygen in the particle, converting it to an anaerobic environment? We can examine freshly collected particles with their living bacterial flora to look for characteristics of aerobic and anaerobic life. Small, transparent, flocculent particles are colonized internally by bacteria that can be seen under a microscope. For such particles we can also make some assumptions about the numbers and metabolic rates of bacteria in the particles, based on our observations, and with them determine the maximum particle size that will sustain aerobic metabolism by the diffusion of oxygen from the surrounding water. Assuming that only molecular diffusion operates within flocculent particles, that particles have a bacterial population density of at least $10^6/cm^3$, and that the bacteria are growing at rates somewhat less than those seen in laboratory cultures, a calculation of the limiting thickness for diffusion (32) suggests that particles larger than 50 to 100 μm may go anaerobic in their interior. This means that visible sea snow particles in the euphotic zone should undergo a diel cycle, with living autotrophs causing interstitial water to be supersaturated with oxygen during the day and all microorganisms collectively creating anaerobic conditions at night. Alldredge and Cox (2) have demonstrated that sea snow is indeed the site of substantial photosynthesis. If the photosynthesis occurs, the heterotrophic processes will follow.

PRODUCTION AND UTILIZATION OF DISSOLVED ORGANIC MATTER

In the ocean the autochthonous sources of dissolved organic matter (DOM) are direct release by phytoplankton (19, 30), release by zooplankton through feeding, defecation, and excretion (9), and release by the microbial food web, especially from particles, including feces, and from the sediments. Allochthonous sources include DOM from rivers and estuaries and airborne terpenes, isoprenes, and other products primarily of terrestrial vegetation. Although the rates of flux, particularly of airborne organic matter, are not well known, each is thought to contribute between 1 and 10% as much DOM as the autochthonous sources (36). In coastal waters there may be significant production of DOM from seaweeds (44).

Most of the standing stock of DOM is relatively refractory material with a long turnover time. Although the standing stock of labile DOM is small, its flux is large and its turnover time is hours to days. The labile fraction consists, so far as we know, of carbohydrates, peptides, and their component sugars and amino acids. There

may also be some early products of photosynthesis, such as glycolate. The rather extensive literature suggests that the composition of the labile DOM is highly variable and still imperfectly known.

Much attention has been given to production of labile DOM by phytoplankton, stimulated by the work of Fogg (12) and his associates. Recently, the role of zooplankton in producing DOM through sloppy feeding, excretion, and defecation has been recognized (8, 25). Moreover, fecal pellets released into the water continue to produce DOM (20). Because sources of DOM occur at several levels in the planktonic food chain, its production can be equal to a substantial fraction of primary production. Production of DOM by autotrophs alone can be 1 to 40% of the net carbon fixation, but with mean values probably skewed toward the low end of this range. Estimates of production of DOM by zooplankton, and also by lysis of fecal pellets and dying phytoplankton, are few and generally consist of measurements of changes in DOM or its components in the water during and subsequent to plankton blooms. However, the evidence to date suggests that more DOM is produced through excretion by zooplankton and lysis of dead phytoplankton than by living phytoplankton.

Regardless of its source, labile DOM is utilized rapidly by microorganisms in the water. Peak concentrations of DOM disappear in a few days, as bacteria rapidly utilize amino acids, peptides, and carbohydrates in solution at the microgram to nanogram concentrations normally present (3, 20). Curiously, bacteria seem to utilize naturally produced mixed DOM much more rapidly than they utilize defined compounds added to water (50; T. R. Jacobsen, Ph.D. thesis, University of Georgia, Athens, 1981). Something about the nature of DOM or the manner in which it is transferred to bacteria may elude us. Both free and attached bacteria utilize DOM, although their relative activity varies (5, 15, 16). The fact that attached bacteria do utilize labeled compounds in experiments lasting only 2 to 4 h suggests that diffusion is effective in bringing dissolved materials both in and out of at least some significant part of the population of particles in seawater. Because of the limits of diffusion, we can assume that these are small particles, <100 μm. Unfortunately, most of this work has been done on samples taken from Niskin samplers and then pipetted, a procedure which probably modifies the natural spectrum of particle sizes.

MICROBIAL FOOD WEBS

Microorganisms can utilize many sources of substrates in natural waters, both particulate and dissolved (4, 51). These sources are highly variable in time and space, so microorganisms must be patient and opportunistic, responding rapidly to a supply of substrate. Perhaps the most bleak and variable of all habitats is the open water. A cubic millimeter of ocean water typically contains 100 free bacteria and 1 phytoplankter. If the bacteria are randomly distributed, they must depend on mean concentrations of DOM and molecular diffusion. If, however, they move until they are within ~100 μm of a phytoplankter or a fecal particle, they are within diffusion distance of a DOM source (Azam, in press). Such behavior would improve fitness. Nonmotile phytoplankters could develop a surrounding population of bacteria. Motile phyto- and zooplankton will pass through a pico-habitat, laying a trail of released DOM. However, these trails will dissipate in a matter of minutes (21). Therefore, much of the DOM probably is present at or near mean concentrations, and bacteria respond to this condition by adopting the smallest possible size. This gives maximal surface/volume ratio and a minimal protoplast to support by diffusion of DOM. In coastal waters, where both the flux and concentration of DOM are greater, bacteria are larger and more numerous. In fecal particles bacteria are 10 times larger. Are these size differences a response to limits of substrate supplied by diffusion? Is production of bacteria a function of mean DOM concentration or of the frequency of occurrence of trails of high concentrations of DOM? Using zooplankton labeled with ^{33}P, Lehman and Scavia (31) showed that distribution of ^{33}P in phytoplankton exposed to swimming zooplankton was skewed: some had encountered a trail of enhanced phosphate and absorbed it rapidly. In principle, bacteria with mobilized permeases could do the same with respect to labile DOM produced by phytoplankton and zooplankton, including protozoans.

Particles do not necessarily represent habitats more favorable than the surrounding water. Most particles in the sea are devoid of bacteria (49). Particles which contain active bacteria appear to be fecal or secondary aggregates (sea snow). These particles appear to be filled with useful substrates in the form of fragments of phytoplankton, but electron micrographs suggest that most of the biomass often is bacteria.

Microbial food chains may alter the environment of the interior of fecal pellets and other particles which are favorable substrates. The presence of motile bacteria in them suggests that large fecal pellets have anaerobic interiors. Krause (27) suggested that production of gaseous products of anaerobic metabolism will cause flotation in fecal pellets. This is unlikely, because bubbles must be formed against hydro-static pressure. Neither are bubbles seen in sea snow, although substantial rates of photosynthesis have been reported (2). The observed decrease in numbers of fecal pellets with depth (27, 47) may result from microbial degradation of the pellets or from coprophagy. In terrestrial and marine benthic environments coprophagy is known to be a significant part of the food web, but it has not been given much consideration in the plankton. One would expect coprophagy to be especially prevalent at the bottom of the euphotic zone, at depths just below those where phytoplankton are abundant. However, coprophagy implies a strategy of food gathering different from that of grazing, and the two may coexist in the upper mixed layer.

The structure and function of the microbial food web in the ocean is a subject of current debate. We no longer debate whether it exists or whether it is significant. Rather, we debate how significant it is and what the implications are with respect to the overall flux of energy and materials in the ocean. We are currently presented with several paradoxes. Measurements of total respiration in seawater sometimes equal or exceed measurements of photosynthesis (26, 40, 45). Because several shortcomings of the ^{14}C method of measuring photosynthesis have come to light, this is taken as evidence that primary production is underestimated (11, 29). Although this may be true, there are theoretical limits to photosynthesis which in turn place an upper limit on realistic rates of respiration (T. Platt, M. Lewis, and R. Geider, *in* M. J. Fasham, ed., *Flows of Energy and Materials in Marine Ecosystems*, in press). However, when we consider the multiple routes of access of bacteria, fungi, and protozoa to both dissolved and particulate organic matter, it is reasonable to expect that a large part of the nonliving organic matter in the sea will find its way into the microbial food web. This includes the labile substances which make up most of the flux of dissolved material and also the marine humates, which are less refractory than freshwater ones.

CONCLUSIONS

With the study of marine microbial ecology moving rapidly forward, it is useful to summarize what we know and then to focus on what we need to know. We know that both free-living and attached bacteria in seawater are metabolically active and are significant in the utilization of dissolved substrates and the production of biomass (5, 15, 16). Because they have access to a variety of dissolved and particulate substrates, bacteria are significant producers of biomass in the ocean (4; Azam, in press). There is evidence, albeit circumstantial, that the principal consumers of free-living minibacteria are small flagellates (10). The constancy of numbers of bacteria in seawater is further circumstantial evidence for control of the upper population limit by consumer organisms (38).

We do not know as yet the details of the microbial food web. These details hold the answer to an important question: is bacterial production transferred to macroorganisms or is it dissipated in a series of microbial feeding loops? The answer may vary, depending on the rate of primary production and other regional or seasonal circumstances. In any case, it may tell us why the ocean is not more productive of fishes and other large, terminal consumers. To answer such questions we need quantitative data on a number of basic processes. Even the rate of photosynthesis in the deep sea is open to question (46). Methods for measuring community respiration in seawater are not well developed, and data are few. The measurement of bacterial production in the sea now appears to be possible, but there remains some disagreement about methods. Assimilation efficiency of bacteria has been inferred from short-term studies of utilization of labeled substrates and is controversial (15, 25). Protozoans are being enumerated, albeit with some uncertainty regarding small flagellates, but rates of production of protozoans in nature are unknown, and methods to measure protozoan production do not exist. Progress is being made, now that the significance of microbial growth and metabolism in the sea is recognized, but quantitative data from the sea remain the major need. Since nonliving particulate matter containing bacterial biomass is more accessible to macro-consumers than are free-living bacteria, the question of the origin and fate of seston in the sea is closely related to that of the interaction between micro and macro food webs. We must develop a better understanding of the origins and fate of marine seston if we are to understand the microbial food web and its relation to marine productivity, and here, too, quantitative information is the foremost need.

Work in this laboratory was supported by contract DE-AS09-76EV00639 with the U.S. Department of Energy, by grant OCE-8110707 from the National Science Foundation, and by funds from the office of the Vice President for Academic Affairs, University of Georgia.

LITERATURE CITED

1. **Alldredge, A. L.** 1976. Discarded appendicularian houses as sources of food, surface habitats and particulate organic matter in planktonic environments. Limnol. Oceanogr. **21:**14–23.
2. **Alldredge, A. L., and J. L. Cox.** 1982. Primary production and chemical composition of marine snow in surface waters of the Southern California Bight. J. Mar. Res. **40:**517–527.
3. **Andrews, P., and P. J. leB. Williams.** 1971. Heterotrophic utilization of dissolved organic compounds in the sea. III. Measurement of the oxidation rates and concentrations of glucose and amino acids in sea water. J. Mar. Biol. Ass. U.K. **51:**111–125.
4. **Azam, F., T. Fenchel, J. G. Field, J. S. Gray, L.-A. Meyer-Reil, and F. Thingstad.** 1983. The ecological role of water-column microbes in the sea. Mar. Ecol. Prog. Ser. **10:**257–263.
5. **Azam, F., and R. E. Hodson.** 1977. Size distribution and activity of marine microheterotrophs. Limnol. Oceanogr. **22:**492–501.
6. **Baylor, E. R., and W. H. Sutcliffe, Jr.** 1963. Dissolved organic matter in sea water as a source of particulate food. Limnol. Oceanogr. **8:**369–371.
7. **Beers, J. R., F. M. H. Reid, and G. L. Stewart.** 1982. Seasonal abundance of the microplankton population in the North Pacific central gyre. Deep-Sea Res. **29:**227–245.
8. **Dagg, M. J.** 1974. Loss of prey body contents during feeding by an aquatic predator. Ecology **55:**903–906.
9. **Eppley, R. W., S. G. Horrigan, J. A. Fuhrman, E. R. Brooks, C. C. Price, and K. Sellner.** 1981. Origins of dissolved organic matter in southern California coastal waters: experiments on the role of zooplankton. Mar. Ecol. Prog. Ser. **6:**149–159.
10. **Fenchel, T.** 1982. Ecology of heterotrophic microflagellates. IV. Quantitative occurrence and importance as bacterial consumers. Mar. Ecol. Prog. Ser. **9:**35–42.
11. **Fitzwater, S. E., G. A. Knauer, and J. H. Martin.** 1982. Metal contamination and its effect on primary production measurements. Limnol. Oceanogr. **27:**544–551.
12. **Fogg, G. E.** 1971. Extracellular products of algae in freshwater. Arch. Hydrobiol. **5:**1–25.
13. **Gershey, R. M.** 1983. Characterization of seawater organic matter carried by bubble-generated aerosols. Limnol. Oceanogr. **28:**309–319.
14. **Goldman, J. C., J. J. McCarthy, and D. G. Peavey.** 1979. Growth rate influence on the chemical composition of phytoplankton in oceanic waters. Nature (London) **279:**210–215.
15. **Hanson, R. B., and W. J. Wiebe.** 1977. Heterotrophic activity associated with particulate size fractions in a *Spartina alterniflora* salt marsh-estuary, Sapelo Island, Georgia and the continental shelf waters. Mar. Biol. **42:**321–330.
16. **Hodson, R. E., A. E. Maccubbin, and L. R. Pomeroy.** 1981. Dissolved adenosine triphosphate utilization by free-living and attached bacterioplankton. Mar. Biol. **64:**43–51.
17. **Hofmann, E. E., J. M. Klinck, and G.-A. Paffenhöfer.** 1981. Concentrations and vertical fluxes of zooplankton fecal pellets on a continental shelf. Mar. Biol. **61:**327–335.
18. **Honjo, S.** 1980. Material fluxes and modes of sedimentation in the mesopelagic and bathypelagic zones. J. Mar. Res. **38:**53–97.
19. **Ittekkot, V.** 1982. Variations of dissolved organic matter during a phytoplankton bloom: qualitative aspects, based on sugar and amino acid analyses. Mar. Chem. **11:**143–158.
20. **Ittekkot, V., V. H. Brockman, W. Michaelis, and E. T. Degens.** 1981. Dissolved free and combined carbohydrates during a phytoplankton bloom in the northern North Sea. Mar. Ecol. Prog. Ser. **4:**299–305.
21. **Jackson, G. A.** 1980. Phytoplankton growth and zooplankton grazing in oligotrophic waters. Nature (London) **284:**439–441.
22. **Johnson, B. D., and R. C. Cooke.** 1980. Organic particle and aggregate formation resulting from the dissolution of bubbles in seawater. Limnol. Oceanogr. **25:**653–661.
23. **Johnson, P. W., and J. McN. Sieburth.** 1979. Chroococcoid cyanobacteria in the sea: a ubiquitous and diverse phototrophic biomass. Limnol. Oceanogr. **24:**928–935.
24. **Johnson, P. W., H.-S. Xu, and J. McN. Sieburth.** 1982. The utilization of chroococcoid cyanobacteria by marine protozooplankters but not by calanoid copepods. Ann. Inst. Oceanogr. (Paris) **58**(Suppl):297–308.

25. **Joint, I. R., and R. J. Morris.** 1982. The role of bacteria in the turnover of organic matter in the sea. Oceanogr. Mar. Biol. Annu. Rev. **20:**65–118.

26. **Joiris, C.** 1977. On the role of heterotrophic bacteria in marine ecosystems: some problems. Helgo. Wiss. Meeresunters. **30:**611–621.

27. **Krause, M.** 1981. Vertical distribution of faecal pellets during FLEX 1976. Helgol. Wiss. Meeresunters. **34:**313–327.

28. **Krempin, D. W., and C. W. Sullivan.** 1981. The seasonal abundance, vertical distribution, and relative microbial biomass of chroococcoid cyanobacteria at a station in southern California coastal waters. Can. J. Microbiol. **27:**1341–1344.

29. **Landry, M. R., and R. P. Hassett.** 1982. Estimating the grazing impact of marine micro-zooplankton. Mar. Biol. **67:**283–288.

30. **Larsson, V., and Å. Hagström.** 1979. Phytoplankton exudate release as an energy source for the growth of pelagic bacteria. Mar. Biol. **52:**199–206.

31. **Lehman, J. T., and D. Scavia.** 1982. Microscale nutrient patches produced by zooplankton. Proc. Natl. Acad. Sci. U.S.A. **79:**5001–5005.

32. **Leyton, J. T.** 1975. Fluid behavior in biological systems, p. 120. Clarendon Press, Oxford.

33. **Malone, T. C.** 1980. Size-fractionated primary productivity of marine phytoplankton, p. 310–329. *In* P. G. Falkowski (ed.), Primary productivity in the sea. Plenum Press, New York.

34. **McCarthy, J. J., and J. C. Goldman.** 1979. Nitrogenous nutrition of marine phytoplankton in nutrient-depleted waters. Science **203:**670–672.

35. **McRoy, C. P., J. J. Goering, and W. E. Shiels.** 1972. Studies of primary production in the eastern Bering Sea, p. 199–216. *In* A. Y. Takenouti (ed.), Biological oceanography of the Northern North Pacific. Idemitsu Shoten, Tokyo.

36. **Meybeck, M.** 1982. Carbon, nitrogen, and phosphorus transport by world rivers. Am. J. Sci. **282:**401–450.

37. **Pomeroy, L. R.** 1979. Secondary production mechanisms of continental shelf communities, p. 163–186. *In* R. J. Livingston (ed.), Ecological processes in coastal and marine systems. Plenum Publishing Co., New York.

38. **Pomeroy, L. R., L. P. Atkinson, J. O. Blanton, W. B. Campbell, T. R. Jacobsen, K. H. Kerrick, and A. M. Wood.** 1983. Microbial distribution and abundance in response to physical and biological processes on the continental shelf of southeastern USA. Cont. Shelf Res. **2:**1–20.

39. **Pomeroy, L. R., and D. Deibel.** 1980. Aggregation of organic matter by pelagic tunicates. Limnol. Oceanogr. **25:**643–652.

40. **Pomeroy, L. R., and R. E. Johannes.** 1968. Occurrence and respiration of the ultraplankton in the upper 500 meters of the ocean. Deep-Sea Res. **15:**381–391.

41. **Purcell, E. M.** 1977. Life at low Reynolds number. Am. J. Phys. **45:**3–11.

42. **Riley, G. A.** 1963. Organic aggregates in sea water and the dynamics of their formation and utilization. Limnol. Oceanogr. **8:**372–381.

43. **Sheldon, R. W., and W. H. Sutcliffe, Jr.** 1978. Generation times of 3h for Sargasso Sea microplankton determined by ATP analysis. Limnol. Oceanogr. **23:**1051–1055.

44. **Sieburth, J. McN.** 1969. Studies on algal substances in the sea. III. The production of extracellular organic matter by littoral marine algae. J. Exp. Mar. Biol. Ecol. **3:**290–309.

45. **Sieburth, J. McN., and P. G. Davis.** 1982. The role of heterotrophic nanoplankton in the grazing and nurturing of planktonic bacteria in the Sargasso and Caribbean Sea. Ann. Inst. Oceanogr. (Paris) **58**(Suppl.):285–295.

46. **Sieburth, J. McN., K. M. Johnson, C. M. Burney, and D. M. Lavoie.** 1977. Estimation of *in situ* rates of heterotrophy using diurnal changes in organic matter and growth rates of picoplankton in diffusion culture. Helgol. Wiss. Meeresunters. **30:**565–574.

47. **Urrère, M. A., and G. A. Knauer.** 1981. Zooplankton fecal pellet fluxes and vertical transport of particulate organic material in the pelagic environment. J. Plankton Res. **3:**369–387.

48. **Waterbury, J. B., S. W. Watson, R. R. L. Guillard, and L. E. Brand.** 1979. Widespread occurrence of a unicellular, marine, planktonic cyanobacterium. Nature (London) **277:**293–294.

49. **Wiebe, W. J., and L. R. Pomeroy.** 1972. Microorganisms and their association with aggregates and detritus in the sea: a microscopic study. Mem. Ist. Ital. Idrobiol. **29**(Suppl.):325–352.

50. **Wiebe, W. J., and D. F. Smith.** 1977. Direct measurement of dissolved organic carbon release by phytoplankton and incorporation by microheterotrophs. Mar. Biol. **42:**213–223.

51. **Williams, P. J. leB.** 1981. Incorporation of microheterotrophic processes into the classical paradigm of the planktonic food web. Kiel. Meeresforsch. Sonderh. **5:**1–28.

Role of Heterotrophic Protozoa in Carbon and Energy Flow in Aquatic Ecosystems

BARRY F. SHERR AND EVELYN B. SHERR

University of Georgia Marine Institute, Sapelo Island, Georgia 31327

It now appears likely that a major part of primary and secondary productivity as well as respiration in ocean waters can be ascribed to the smallest size fractions of the plankton (e.g., 7, 70, 82, 111). Nanophytoplankton (<20 μm diameter) account for most of the phytoplankton biomass in the sea (e.g., 70), and picophytoplankton (<2 μm diameter) may be responsible for a large share of total phytoplankton productivity (e.g., 81). In addition, the nano- and picophytoplankton in most areas of the ocean may be growing at near maximal rates (37, 72), with primary productivity being several-fold higher than earlier measurements suggested (23, 35).

A considerable percentage of the organic carbon initially fixed by phytoplankton is believed to enter the pelagic food web as dissolved and nonliving particulate organic matter to be subsequently incorporated into bacterial biomass (e.g., 4, 82, 111). This bacterial component of the plankton is currently perceived to be an important contributor to secondary production and to total respiration in the sea. Apparently, neither the small algal cells nor the bacterioplankton can be effectively grazed by most macrozooplankters (e.g., 18, 60). Yet the relatively low and constant biomass standing stocks of nano- and picophytoplankton and bacterioplankton must be turning over rapidly to support the high rates of production which are speculated to occur (e.g., 4, 37, 72).

Within the context of these speculations, phagotrophic protozoa have been considered the dominant consumers of pico- and nanoplankton. As such, they could control the standing stocks and influence the productivity of bacteria and phytoplankton via grazing and release of organic and inorganic nutrients (e.g., 53, 97). It must be emphasized that the above ideas are hypothetical, with little direct evidence to support the supposed roles of protozoa. Also still open to conjecture is whether heterotrophic protozoa represent a significant pathway for organic carbon transfer from the microbial community to larger consumers, or conversely, whether the grazing and respiration of the protozoa simply shunt part of the primary and secondary production out of the food web as carbon dioxide.

It is our intention here to summarize the diverse trophic roles which heterotrophic protozoa are presently hypothesized to have in aquatic food webs. We will emphasize results of some of the most recent research, including studies involving phagotrophic microprotozoa, heterotrophic cells about 2 to 20 μm in diameter, whose importance has been largely overlooked in the past as a result of methodological problems. Although our discussion focuses on the marine pelagic systems, it is probable that models of the trophic roles of protozoa derived from open ocean waters will be generally applicable to other aquatic systems with a similar food web structure. There are already in the literature reviews dealing with various aspects of the roles of heterotrophic protozoa in both freshwater and marine environments (20, 27, 66, 106–108; T. Fenchel, *in Flows of Energy and Materials in Marine Ecosystems: Theory and Practice*, in press; J. McN. Sieburth, *in* J. E. Hobbie and P. J. leB. Williams, ed., *Heterotrophic Activity in the Sea*, in press). Also valuable resources in this regard are the book *Sea Microbes* by John Sieburth (96) and the symposium volume *Marine Pelagic Protozoa and Microzooplankton Ecology, Annales de l'Institute Oceanography*, vol. 58, suppl., 1982.

DISTRIBUTION OF AQUATIC PROTOZOA

Heterotrophic protozoa are found in all types of aquatic habitats. In the interest of brevity, this review will particularly focus on marine pelagic systems. In very productive environments, e.g., shallow eutrophic lakes, sewage treatment ponds, and organic-rich sediments, the protozoan biomass will be greater, the protozoan community structure will be different, and the food web will be more complex than is discussed here (20, 24–26, 66, 107). In the open ocean there are also highly productive microenvironments, or "hot spots" of microbial activity. Examples of such "hot spots" are microneuston in the surface microlayers of seawater (99), organic aggregates or marine snow (100, 101), discarded appendicularian houses (1), and zooplankton fecal pellets (39, 83). Both ciliates and heterotrophic microflagellates are more abundant (by up to four orders of magnitude) in and around these organic-rich microenvironments than in the surrounding waters; the species composition of the protozoan assemblage also appears to differ from that of the general water mass (17). There is presently very little information as to the quantitative significance of these "hot spots" of microbial activity for overall carbon and energy cycling in the water column; this is a subject for future investigation.

Recently developed epifluorescence microscopy techniques have allowed better estimates of the in situ abundances of pelagic protozoa, especially microprotozoa (16a, 22, 41, 93). Some values for the distribution, in terms of numbers and carbon biomass, of two size groups of heterotrophic protozoa in marine pelagic waters are presented in Table 1. The most abundant and widespread groups are microflagellates and ciliates (Table 1). Pelagic sarcodine protozoa, e.g., acantharia, amoebae, foraminifera, and radiolaria, appear to be only transiently abundant. Sorokin (106) stated that sarcodines usually comprise only 1 to 5% of the microzooplankton biomass,

although the percentage may increase to 10% in the tropical Pacific. Testate amoebae such as acantharians might also be important in local regions of increased phytoplankton production (P. McGillivary, personal communication). In contrast, ciliates, including naked forms and species with loricae, such as the *Tintinnidae*, have been found to compose 50% or more of microzooplankton numbers and biomass in many parts of the world ocean (8–10, 13, 102), as well as in lakes (76). Heterotrophic microflagellate protozoa, including chrysomonads, cryptophytes, bodonids, bicoecids, dinoflagellates, and choanoflagellates, are believed to represent virtually 100% of the heterotrophic nanoplankton, cells with volumes corresponding to effective diameters between 2 and 20 μm (29, 41, 93, 96, 97). (Although this may be true of oceanic waters, we have preliminary evidence that ciliates <20 μm can compose a large part of the heterotrophic nanoplankton biomass in estua-

TABLE 1. Representative abundance and organic carbon biomass of heterotrophic protozoa in marine pelagic ecosystems

Determination	Cells/ml	C (mg/m³)[a]	Reference
A. *Nanoplankton, 2–20 μm*[b]			
World ocean			
Oligotrophic waters	0.4×10^3–3.0×10^3	1.3–8	106
Mesotrophic waters	4.0×10^3–9.5×10^3	16–35	106
Eutrophic waters	13×10^3–23×10^3	58–120	106
Atlantic Ocean			
Open ocean	0.1×10^3–2.8×10^3	$(0.4–11)^c$	97
Continental shelf	0.9×10^3–5.1×10^3	$(3.6–20)$	97
Estuaries	0.9×10^3–37×10^3	$(3.6–149)$	97
Atlantic Ocean			
Outer shelf	0.2×10^3–0.8×10^3	0.6–8	Sherr et al., in preparation
Nearshore shelf	0.3×10^3–2.3×10^3	1.0–14	Sherr et al., in press
Estuary	1.6×10^3–3.3×10^3	5.4–15	Sherr et al., in press
Chesapeake Bay	0.5×10^3–15×10^3	$(2.0–60)^c$	42
Limfjord, Denmark	$<0.2 \times 10^3$–3.0×10^3	$(<0.8–12)^c$	31
B. *Microplankton, 20–200 μm*			
Ciliates, world ocean			
Oligotrophic waters	0.04–9	0.6–15	106
Mesotrophic waters	17–30	34–65	106
Eutrophic waters	40–145	67–144	106
Ciliates, Southhampton Estuary			
Tintinnida	—	0.4–2.3	13
Naked ciliates	—	0.07–2.1	13
Protozoa, California coast			
Tintinnida	0.08–5.6	0.04–1.69	10
Other ciliates	0–5.0	0–5.26	10
Foraminifera	0–0.02	0–0.08	10
Radiolaria	0–0.09	<0.02–0.29	10
Protozoa, Central Pacific Gyre (92% ciliates)	0.7–2.0	2.0–4.4	8

[a] Where necessary, carbon biomass was calculated from wet weights or volumes by use of an organic carbon/wet weight ratio of 0.08, obtained by assuming an average dry weight/wet weight ratio of 0.18 and an average carbon/dry weight ratio of 0.46 (34, 61).

[b] Various terms are used for this group of heterotrophic protozoa, including zooflagellates (106), microflagellates (31), microprotozoa (93), and heterotrophic nanoplankton (HNAN or HNANO) (97).

[c] Calculated from cell abundance by use of an average nanoplankton cell wet weight of 5×10^{-8} mg (32).

rine waters.) Heterotrophic microflagellates appear to be ubiquitous in marine waters at concentrations of 10^2 to 10^4 ml^{-1} (Table 1).

DIRECT ROLES OF PROTOZOA

Protozoa are directly involved in aquatic food chains as an intermediate link between trophic levels. Phagotrophic protozoa have been implicated as the dominant grazers of bacteria in natural waters and may also be important consumers of cyanobacteria and eucaryotic algal cells. Since protozoa share with other microorganisms a capacity for high growth rates and a high efficiency of conversion of food biomass into protozoan biomass, heterotrophic protozoa may, as prey for microzooplankton, channel some fraction of microbial production into higher levels of the food chain which eventually yield commercially important species of fish and invertebrates. The direct trophic roles of protozoa are depicted as solid arrows connecting the compartments in Fig. 1. In this model, aquatic protozoa are segregated into two groups, following the convention of Sieburth et al. (98): the nanoplanktonic protozoa, 2 to 20 µm diameter, which are mostly heterotrophic microflagellates but may include small ciliates and amoebae; and the microzooplanktonic protozoa, 20 to 200 µm diameter, including larger ciliates and amoebae (Fig. 1).

Protozoa as grazers of bacteria. A widely recognized trophic link involving heterotrophic protozoa in aquatic systems is that of bacterial consumption (4, 27, 40, 85, 97; Fenchel, in press; Sieburth, in press). There is abundant laboratory evidence that heterotrophic microflagellates isolated from both marine and freshwater environments will thrive on many species of bacteria (e.g., 30, 42, 67, 95). Ciliates have also been successfully raised on bacteria in the laboratory (e.g., 12, 21, 24, 43). In such cultures, microflagellates and ciliates typically consume from 10 to 250 bacteria per protozoan cell per h (12, 30, 95).

The latest work on in situ bacterioplankton production in the sea has suggested that bacteria utilize a significant fraction (10 to 50%) of total phytoplankton productivity, and that, although bacterial numbers tend to be low and relatively constant in the euphotic zone of the open ocean, their biomass must be turning over fairly rapidly (e.g., 4, 111; P. J. leB. Williams, in *Flows of Energy and Materials in Marine Ecosystems: Theory and Practice*, in press). It is thought that the constancy of bacterioplankton numbers in the sea results from active grazing (4). Heterotrophic microflagellates have been nominated as the primary consumers of bacterioplankton, largely by default. In studies of the feeding habits of other potential grazers of free-living

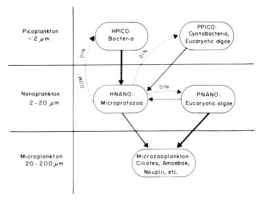

FIG. 1. Schematic model showing hypothesized roles of heterotrophic protozoa in aquatic food webs. Solid arrows represent direct pathways of carbon and energy flow, dashed arrows represent recycling pathways of dissolved inorganic nutrients (DIN) and dissolved organic matter (DOM). The microbial community is segregated into three size classes: picoplankton, including heterotrophs (HPICO) and phototrophs (PPICO); nanoplankton, including heterotrophs (HNANO) and phototrophs (PNANO); and microzooplankton, following the convention of Sieburth et al. (99).

bacteria, the conclusion has usually been either that the organisms cannot effectively graze bacterioplankton at the concentrations, 10^5 to 10^6 cells per ml, generally found in pelagic waters, or that the organisms are not abundant enough to account for a significant depletion of the bacterial biomass. A number of investigators have concluded that bacteriovorous ciliates cannot have an important role as consumers of bacterioplankton (12, 28, 43, 109). Metazoan microzooplankton such as copepod nauplii (18) also do not appear to be able to graze bacterium-size particles in sufficient quantity to meet their energy requirements. (However, freshwater *Daphnia* spp. can apparently ingest bacterioplankton at in situ concentrations [80]). Mucus net feeders such as pelagic tunicates are capable of consuming free bacteria efficiently (60); however, these zooplankters are only locally abundant (18) and probably cannot be responsible for controlling the overall abundance of bacterioplankton in the sea (60).

Although microflagellates are believed to be the dominant bacteriovores in the sea, little information exists concerning the extent of actual in situ grazing of bacterioplankton by heterotrophic nanoplankton. Positive relationships between abundances of colorless microflagellates and bacteria have been noted in the Japan Sea (105), in a Danish fjord (31), in the Sargasso and Caribbean Seas (97), and in Georgia coastal waters (B. F. Sherr, E. B. Sherr, and S. Y.

Newell, J. Plankton Res., in press). Fenchel's (31) month-long study of fluctuations in bacterioplankton and heterotrophic microflagellate populations in Limfjord, Denmark, showed a time lag between maximum cell numbers of the two groups of microorganisms which was suggestive of a predator–prey relationship. In addition, transmission electron micrographs of "wild" microprotozoa have occasionally revealed bacteria in the protozoans' food vacuoles (96, 97, 100), and Fenchel (29; in press) has presented theoretical considerations concerning the physical mechanisms of feeding which suggest that only the smallest suspension feeders (<10 μm) can effectively utilize bacterioplankton in situ.

An epifluorescence microscopy technique has now been developed, based on sequential staining first with the DNA-specific stain DAPI (84) and then with the general protein stain fluorescein isothiocyanate (93), which enables the visualization of DNA-rich, bacterium-size particles in the food vacuoles of in situ microprotozoans (93a). Examples of heterotrophic microprotozoan cells containing such particles are presented in Fig. 2. We assume that these inclusions are recently ingested bacteria. In the samples of coastal water examined to date, from 18 to 42% of the heterotrophic nanoplankton cells appeared to contain bacteria in their food vacuoles (93a). This technique may prove useful as an index of bacteriovory by microprotozoa in aquatic systems.

The case for protozoa as grazers of bacteria is much better established for eutrophic waters and for detrital particles, where bacterial concentrations exceed 10^7 cells per ml, or 10^6 cells per cm^2 of surface area (e.g., 20, 27, 32, 40, 109). Bacteria in eutrophic conditions are typically larger than oligotrophic bacterioplankton (52), which means an even greater bacterial biomass in organic-rich waters than the cell abundances alone would suggest. Even in oligotrophic waters, the "hot spots" of microbial activity discussed previously, e.g., in neuston at the sea surface, on suspended organic aggregates, and on fecal particles, are sites where both microflagellates and ciliates are likely to be active bacterial grazers (e.g., 1, 17, 39, 99, 100, 101).

Protozoa as grazers of phytoplankton. Protozooplankton 20 to 200 μm in size may at times be important consumers of nanoplanktonic algae (e.g., 15, 16, 53, 87). Much of the literature on the role of ciliates as algal grazers in marine pelagic waters has been reviewed by Banse (5). Banse concluded that, although ciliates can be efficient grazers on particles <5 to 10 μm diameter which are not effectively retained by copepods, in marine pelagic waters ciliates are probably never as important as copepods as grazers of phytoplankton. In areas of low phytoplankton

density, <10 μg of C per liter, ciliates cannot consume sufficient food to take advantage of their high intrinsic growth rate, and in more eutrophic waters where nanoplankton biomass is 20 to 100 μg of C per liter, ciliate biomass is not great enough to crop down a major fraction of the algal biomass (5). For example, Heinbokel and Beers (50) reported that in southern California coastal waters tintinnids grazed 4% of daily phytoplankton production, and that the entire ciliate assemblage may have consumed only up to 10% of the primary production; Zaika (112) estimated that ciliates cropped 10 to 20% of summer algal production in the open Bay of Sebastopol (Black Sea). However, other workers have credited ciliates with greater rates of algal consumption. Rassoulzadegan and Etienne (88) calculated for a site in the Mediterranean Sea that ciliates could consume 59% of algal biomass on an annual basis. Capriulo and Carpenter (15, 16) found that microzooplankton (almost entirely tintinnids) in Long Island Sound grazed up to 41% of the standing stock of chlorophyll *a* per day, and that, on an annual basis, tintinnids might remove about 27% of the primary production in the Sound.

Protozoa other than ciliates in the microzooplankton may at times be important algal grazers. For example, Prasad (86) and LeFevre and Grall (65) have reported the grazing of coastal diatom blooms by the heterotrophic dinoflagellate *Noctiluca scintillans*, and Smetacek (102) found that heterotrophic dinoflagellates as well as ciliates were consumers of phytoplankton in the Kiel Bight.

There is presently increasing interest in the possibility that heterotrophic nanoplanktonic protozoa may also have a major role as consumers of phototrophic cells in the sea. It is becoming apparent that photosynthetic picoplankton, specifically cyanobacteria and eucaryotic algae less than 2 μm in diameter, are ubiquitous in the sea (57, 58). In oligotrophic waters, these cells may be responsible for a significant fraction of total primary productivity, and in fact it is now argued that the failure of present productivity methods to adequately measure the carbon uptake of these small cells has led, in part, to a gross underestimation of annual production in the open ocean (23, 35, 79). The current notion is that there is a very high rate of turnover of nanoplanktonic and picoplanktonic biomass which is ascribed to an actively grazing microzooplankton community (53). Because ciliates cannot effectively graze small particles (<10 $μm^3$ volume) in dilute suspensions (5, 28), it is speculated that heterotrophic nanoplankton may be the major consumers of the phototrophic picoplankton (5).

There is some direct evidence that heterotro-

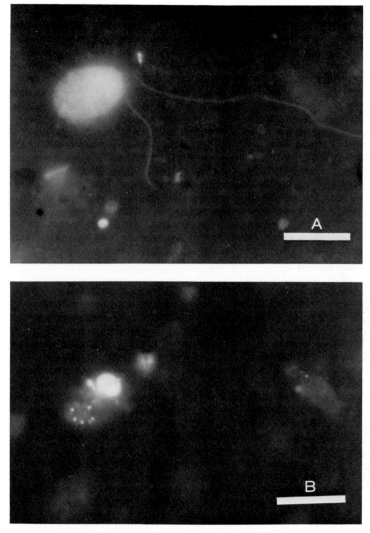

FIG. 2. Heterotrophic microprotozoan cells visualized under fluorescein isothiocyanate (A, C) and under DAPI (B, D) fluorescence. (A and B) Biflagellate cell with several DAPI-stained particles in the cytoplasm. (C and D) Heterotrophic cell with many rod-shaped, bacterium-sized particles in food vacuoles. Bar, 10 μm.

phic protozoa smaller than 20 μm can ingest and assimilate cyanobacteria and small eucaryotic algae (2, 41, 59, 78, 100). However, this is a subject which needs more intensive investigation.

Protozoa as prey. In the preceding two sections, we reviewed information suggesting that aquatic protozoa are major consumers of bacteria and pico- and nanophytoplankton, which are apparently not grazed effectively by larger zooplankton. The next logical question is, are protozoa an important food source for those zooplankton or do protozoa represent a shunt of microbial biomass out of the food web as respiratory CO_2?

On the basis of metabolic considerations, protozoa should be able to transfer microbial biomass to higher trophic levels without an excessive loss of carbon. Like most unicellular organisms, protozoa are characterized by the capacity, under optimal conditions, for high rates of population increase and high efficiencies of conversion of food biomass into protozoan biomass. A summary of values for growth rate, μ, measured in the laboratory (Table 2) shows typical rates of growth of 0.01 to 0.25 h^{-1} for ciliates and microflagellates. The maximum growth rates correspond to population doubling times of 2.8 to 7 h. The gross growth efficiencies (protozoan biomass produced/food biomass in-

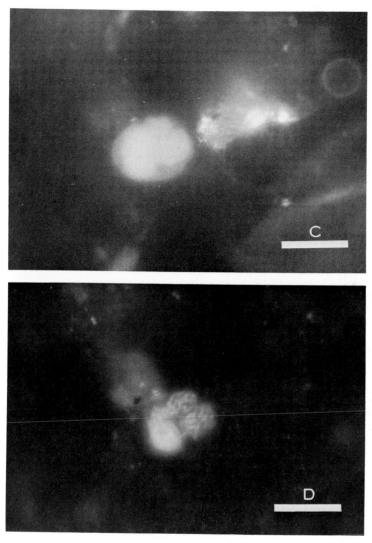

FIG. 2. *Continued.*

gested) determined for some species of amoebae, ciliates, and microflagellates range from 16 to 54%, and net growth efficiencies (protozoan biomass produced/food biomass assimilated) are even higher, from 60 to 82% (Table 2).

However, even with high conversion (or growth) efficiencies in the transfer of organic matter → bacteria → microflagellate protozoa, the majority of fixed carbon will still be lost to the system as respired CO_2 (90; Williams, in press). Probably the maximum transfer of organic matter along this pathway would be 25% of the amount taken up by the bacteria (an average of 50% gross growth efficiency for each step). If ciliates consume most of the microflagellate production, then there would be a third step in the transfer series, organic matter→ bacteria → microflagellates → ciliates, and the maximum amount of organic carbon available for metazoan grazers from this source would be only 12.5% of the initial bacterial consumption. At the present time, however, there is insufficient in situ evidence to make a general statement as to whether heterotrophic microbes are primarily responsible for mineralization of organic matter and do not directly contribute significant amounts of organic carbon to the food web through their productivity and subsequent consumption by other zooplankters. Even if the microbial community does have an overall low efficiency of transfer of organic carbon to larger grazers (e.g., 68, 69, 73), there might still be a

TABLE 2. Representative values for growth rate, growth efficiency, and grazing rate of aquatic, heterotrophic protozoa

Parameter and protozoan	Values	Reference
Growth rate, μ (h^{-1})		
Zooflagellate, *Bodo* sp.	0.03–0.09	38
Microflagellates, 20–200 μm^3	0.15–0.25 (maximum)	30
	0.03 (minimum)	30
Microflagellate, *Monas* sp., 30 μm^3	0.05–0.20	95
Estuarine microprotozoa	0.03–0.07	Sherr et al., in press
Tetrahymena pyriformis	0.20–0.22 (maximum)	21
Ciliates, 500–500,000 μm^3	0.01–0.03	24
Ciliates	0.10–0.23 (maximum)	107
Planktonic ciliates	0.02–0.035	77
Tintinnids	0.03–0.06	49
Growth efficiency (%)		
Microflagellates, 20–200 μm^3	34–43 (gross)	30
	60 (net)	30
Microflagellate, *Monas* sp., 30 μm^3	24–49 (gross)	99
Tetrahymena pyriformis	37–54 (gross)	21
Tintinnids	>50 (gross)	49
Amoeba proteus	16–47 (gross)	91
	65–82 (net)	91
Amoeba, *Acantha* sp.	37 (gross)	48

substantial amount of fixed carbon passed along aquatic food chains via microbes. Pomeroy (82; this volume) has developed the concept that bacteria utilize carbon from many sources which might otherwise be lost to aquatic food webs. Some of this carbon has already been processed one or more times, and some is fairly old and too refractory to be utilized by other organisms. Thus, the bacteria may channel old, recycled, and refractory organic carbon, as well as recent phytoplankton and zooplankton exudates, back into the food web; the grazing of bacteriovorous protozoa may be the main pathway for the transfer of the reclaimed carbon to higher trophic levels.

For the case of protozoa grazing on phytoplankton, there are few empirical data on trophic efficiencies. Heinbokel (49) reported that several tintinnid species grazing on monospecific cultures of phytoflagellates had gross growth efficiencies of greater than 50%. In the transfer of phytoplankton carbon through protozoa to larger grazers, there should be one or two fewer steps in the chain than in microbial food chains involving bacteria. Consequently, a larger percentage of the original phytoplankton carbon would be passed along to metazoan consumers. For picophytoplankton, protozoan grazing is likely to be the major avenue by which this source of primary production can directly enter the aquatic food web.

Concerning what organisms will consume heterotrophic protozoa, it has already been demonstrated that ciliates are readily fed upon by cladocerans (85), by oligochaetes (110), and by copepods and crab zoeae (11, 51, 71, 89). The

case for consumption of heterotrophic microflagellates is not as strong. However, Kopylov et al. (63) concluded that several species of zooplankton, including copepods, were capable of ingesting two species of colorless flagellates. Also, it is probable that any organisms which graze nanoplankton will consume heterotrophic cells along with phototrophic cells. Further research will be required to establish the quantitative importance of heterotrophic protozoa as prey organisms in natural waters.

Pertinent to the subject of protozoa as prey is information on biomass production of protozoan assemblages in natural systems. Such data are generally lacking in the literature. A few workers have estimated biomass production by combining in situ abundance data with allometric equations for growth derived from laboratory cultures (e.g., 13, 33, 68). There are virtually no estimates of in situ protozoan growth based on direct measurements (exceptions are 92; Sherr et al., in press). One promising approach that has recently been suggested is extrapolating growth rates from an estimate of the frequency of dividing cells of in situ populations (93; J. Heinbokel and D. Coats, Abstr. 46 Annu. Meet. Am. Soc. Limnol. Oceanogr., 1983). Other innovative methods dealing with this problem are needed.

INDIRECT ROLES OF PROTOZOA IN AQUATIC FOOD WEBS

It has been hypothesized that heterotrophic protozoa play several indirect roles in the cycling of carbon and energy in aquatic systems.

These are indicated by the dashed arrows in Fig. 1. Briefly, the indirect effects result from the capacity of protozoa to affect the recycling of dissolved inorganic nutrients, particularly nitrogen and phosphorus, and to liberate dissolved organic matter into the water. The increased fluxes of dissolved inorganic nutrients and dissolved organic matter, in turn, promote the growth of bacteria and of phytoplankton, and also increase the rate of degradation of organic detritus.

Protozoa and nutrient regeneration. Protozoa are believed to facilitate the turnover of inorganic nutrients via their grazing on bacteria and phytoplankton. Early research on this subject investigated the role of protozoa in the recycling of phosphorus in aquatic ecosystems (6, 55, 56). More recently, isotope tracer experiments have been employed to determine the importance of the microbial community, including protozoa, for both phosphorus and nitrogen remineralization in situ (36, 44, 46). In addition, it has been demonstrated in laboratory experiments that protozoa excrete simple mineral compounds at biomass-specific rates greater than those found for macrozooplankton (3, 55, 95).

In the open ocean, microzooplankton, including protozoa, have been suggested as the organisms responsible for a large share of in situ mineral recycling (45, 53). Where rates of nitrogen and phosphorus excretion by net zooplankton (e.g., copepods) have been quantified, they are generally insufficient to meet the calculated phytoplankton nutrient requirements (19, 54, 103, 104). Studies involving ^{15}N and ^{33}P isotope dilution techniques have shown that 50 to 100% of the regeneration of ammonium and phosphate can frequently be ascribed to the microplankton (36, 44, 46). In the case of NH_4^+, cells <35 μm in diameter seem to be responsible for most of the turnover (36). It has been suggested that bacteria, rather than heterotrophic protozoa, are responsible for much of the rapid nutrient regeneration which occurs in aquatic systems (6, 108). Even so, there is evidence that bacterial nutrient recycling itself is stimulated by the grazing activities of protozoa (32, 94). Also, bacteriovores must be ultimately responsible for liberating the nitrogen and phosphorus tied up in bacterial biomass (27).

In at least one case, it appeared that heterotrophic protozoa were not as efficient as other zooplankters in terms of grazing phytoplankton and recycling inorganic nutrients. In a 3-year study of phytoplankton and zooplankton dynamics in the Kiel Bight, Smetacek (102) found that, when protozooplankton were the major algal grazers, much of the phytoplankton biomass was not consumed, resulting in large-scale sedimentation of organic matter and nutrient impoverishment of the euphotic zone. In contrast, when metazooplankton (e.g., copepods) were the dominant herbivores, phytoplankton populations were stable and productive, and sedimentation rates were low (102).

The combined results of the above studies are inconclusive regarding the actual importance of heterotrophic protozoa for nutrient regeneration in situ; more research is needed on this topic.

Protozoa and detritus decomposition. Another indirect role which has been recognized for some time is the enhancement of decomposition of organic detritus in the presence of protozoa, including ciliates and microflagellates (6, 32, 47, 64, 94). For example, Fenchel and Harrison (32) determined that release of $^{14}CO_2$ from labeled cellulosic substrates proceeded four times faster with protozoan grazing of bacteria than it did in the absence of protozoa, and Sherr et al. (94) found that the rate of bacterial decomposition of dinoflagellate thecae in lake water was increased by a factor of three to four when bacteriovorus microflagellates were present.

The exact mechanism(s) by which protozoa stimulate microbial degradation of organic matter is still in doubt (27, 32, 108). Most experimental degradation studies have used mineral-poor material such as eelgrass and barley hay; hence, an increase in the rates of nutrient recycling may be one cause of the protozoan effect. In addition, protozoa may maintain populations of detritus-degrading bacteria in an active metabolic state by preventing the accumulation of senescent bacterial cells (96). Sherr et al. (94) successfully duplicated the pattern of degradation of dead dinoflagellate cells observed in the presence of protozoa by adding inorganic nitrogen and phosphorus, and also by periodically removing or "cropping" part of the bacteria in protozoa-free cultures. These authors found that decomposition of only the carbohydrate-rich dinoflagellate thecae was significantly enhanced; the protoplasm of the cells degraded at the same rate with or without protozoa or the experimental simulation of protozoan activities. They concluded that: "An important role of microprotozoa in aquatic ecosystems may therefore be to selectively facilitate the breakdown of detritus with a high structural carbohydrate/low mineral content." More work needs to be done to determine whether this statement is generally true.

Protozoan "nurturing" of bacterial and phytoplankton growth. The concept that heterotrophic nanoplankton can be considered as "microbial gardeners" in the open ocean has been advanced by Sieburth and Davis (97). In a study of plankton in the Sargasso and Caribbean Seas, these authors found evidence that heterotrophic nanoplankton abundance was closely related to bacterial abundance, at a ratio of about 1:1,000, and that heterotrophic nanoplankton cells ingested bacteria. In previous work in the same waters, heterotrophic nanoplankton excretions were found to supplement the release of dissolved organic matter by phytoplankton (14). Sieburth and Davis (97) concluded that the heterotrophic microprotozoa may in effect "nurture" the healthy growth of bacterioplankton through a combination of grazing to prevent overutilization of resources and of provision of organic substances for bacterial nutrition

This "nurturing" idea is supported by results of the detritus degradation studies discussed previously and by the apparently close relationship between bacterial and heterotrophic nanoplankton populations found in other parts of the world ocean (31, 105; Sherr et al., in press). In addition, Newell and Christian (74) reported that bacterioplankton productivity, estimated by the frequency of dividing cells method, was maintained at an elevated level in whole seawater as opposed to seawater from which most protozoa were removed by 3-μm filtration, which suggested in situ enhancement of bacterial growth by protozoan grazing.

Although there is as yet no direct information regarding the effect of protozoa on phytoplankton growth, by analogy it would seem likely that the grazing and nutrient regeneration activities of protozoa would serve to "nurture" algal populations, as well as bacterial populations (e.g., 53). This nurturing effect would probably be most apparent in the growth of picoplanktonic phototrophs.

CONSIDERATIONS FOR FUTURE RESEARCH

The foregoing review of what is, or more to the point what is not, known about heterotrophic protozoa suggests that at present no definite conclusions can be drawn regarding any of the possible trophic roles discussed. Further research utilizing innovative methods and experimental approaches will be needed to answer the questions raised. Working with heterotrophic protozoa is especially difficult because these microbes cannot be treated as a single functional group. Protozoa are characterized by orders of magnitude differences in biovolume, patchy in situ distributions, and various feeding strategies. Research involving heterotrophic microflagellates requires techniques different from the methods used in working with ciliates. Also, the smallest heterotrophic protozoa are hard to separate from the rest of the microbial community, which poses additional problems for their study.

Investigators working in other areas of microbial ecology should be aware of the potential impact of heterotrophic protozoa. For example,

because of the intimate association of heterotrophic microprotozoa 2 to 20 μm in diameter with picoplankton and other nanoplankton, studies of the heterotrophic and primary productivity of these size fractions must take into account the rapid grazing activities of the microprotozoa. Kopylov and Moiseev (62) reported that measured rates of bacterial production were 45 to 50% lower in 10-μm-screened seawater, which presumably contained bacteriovorus microflagellates, than in water from which the protozoa had been mostly removed by centrifugation.

Although more work needs to be done in all of the areas treated in this paper, one subject which should be given immediate attention is that of in situ biomass production rates and carbon conversion efficiencies for various populations of heterotrophic protozoa. Until this problem is resolved, we will not be able to estimate the flux of organic matter through heterotrophic protozoa, which is central to determination of their actual importance in mediating flows of carbon and energy in aquatic systems.

This work was supported by National Science Foundation grant OCE-8219866 and by grants from the Sapelo Island Research Foundation. This is Contribution no. 497 of the University of Georgia Marine Institute.

We thank Lorene Gassert and Lea Kneib for preparing the figure and Becky Newell for typing and correcting the manuscript.

LITERATURE CITED

1. **Alldredge, A. C.** 1976. Discarded appendicularian houses as sources of food, surface habitats, and particulate organic matter in planktonic environments. Limnol. Oceanogr. **21**:14–23.
2. **Ansell, A. D., J. E. G. Raymont, K. F. Lander, E. Crowley, and P. Schackley.** 1963. Studies on the mass culture of *Phaeodactylum*. II. The growth of *Phaeodactylum* and other species in outdoor tanks. Limnol. Oceanogr. **8**:184–206.
3. **Antia, N. J., B. R. Berland, and D. J. Bonin.** 1980. Proposal for an abridged nitrogen turnover cycle in certain marine planktonic systems involving hypoxanthine-guanine excretion by ciliates and their reutilization by phytoplankton. Mar. Ecol. Prog. Ser. **2**:97–103.
4. **Azam, F., T. Fenchel, J. G. Field, J. S. Gray, L. A. Meyer-Reil, and F. Thingstad.** 1983. The ecological role of water-column microbes in the sea. Mar. Ecol. Prog. Ser. **10**:257–263.
5. **Banse, K.** 1982. Cell volumes, maximal growth rates of unicellular algae and ciliates, and the role of ciliates in the marine pelagial. Limnol. Oceanogr. **27**:1059–1071.
6. **Barsdate, R. J., T. Fenchel, and R. T. Prentki.** 1974. Phosphorus cycle of model ecosystems: significance for decomposer food chains and effect of bacterial grazers. Oikos **25**:239–251.
7. **Beers, J. R.** 1982. An introduction and historical overview. Ann. Inst. Oceanogr. (Paris) **58**(Suppl.):5–14.
8. **Beers, J. R., F. M. H. Reid, and G. L. Stewart.** 1982. Seasonal abundance of the microplankton population in the North Pacific central gyre. Deep-Sea Res. **29**:227–245.
9. **Beers, J. R., and G. L. Stewart.** 1969. Micro-zooplankton and its abundance relative to the larger zooplankton and other seston components. Mar. Biol. **4**:182–189.
10. **Beers, J. R., and G. L. Stewart.** 1970. Numerical abundance and estimated biomass of microzooplankton, p.

67–87. *In* J. D. H. Strickland (ed.), The ecology of the plankton off La Jolla, California, in the period April through September, 1967. University of California Press, Berkeley.

11. **Berk, S. G., D. C. Brownlee, D. R. Heinle, H. J. Kling, and R. R. Colwell.** 1977. Ciliates as a food source for marine planktonic copepods. Microb. Ecol. **4:**27–40.

12. **Berk, S. G., R. R. Colwell, and E. B. Small.** 1976. A study of feeding responses to bacterial prey by estuarine ciliates. Trans. Am. Microsc. Soc. **95:**514–520.

13. **Burkill, P. H.** 1982. Ciliates and other microplankton components of a nearshore food-web: standing stocks and production processes. Ann. Inst. Oceanogr. (Paris) **58**(Suppl.):335–350.

14. **Burney, C. M., P. G. Davis, K. M. Johnson, and J. M. Sieburth.** 1982. Diel relationships of microbial trophic groups and *in situ* dissolved carbohydrate dynamics in the Caribbean Sea. Mar. Biol. **67:**311–322.

15. **Capriulo, G. M., and E. J. Carpenter.** 1980. Grazing by 35 to 202 μm microzooplankton in Long Island Sound. Mar. Biol. **56:**319–326.

16. **Capriulo, G. M., and E. J. Carpenter.** 1983. Abundance, species composition, and feeding impact of tintinnid micro-zooplankton in Central Long Island Sound. Mar. Ecol. Prog. Ser. **10:**277–288.

16a. **Caron, D. A.** 1983. Technique for enumeration of heterotrophic and phototrophic nanoplankton, using epifluorescence microscopy, and comparison with other procedures. Appl. Environ. Microbiol. **46:**491–498.

17. **Caron, D. A., P. G. Davis, L. P. Madin, and J. McN. Sieburth.** 1982. Heterotrophic bacteria and bacteriovorous protozoa in oceanic microaggregates. Science **218:**795–797.

18. **Conover, R. J.** 1982. Interrelations between microzooplankton and other plankton organisms. Ann. Inst. Oceanogr. (Paris) **58**(Suppl.):31–46.

19. **Conover, R. J., and E. D. S. Corner.** 1968. Respiration and excretion in zooplankton. J. Mar. Biol. Assoc. U.K. **48:**49–57.

20. **Curds, C. R.** 1977. Microbial interactions involving protozoa, p. 69–105. *In* F. A. Skinner and J. M. Shewan (ed.), Aquatic microbiology. Academic Press, Inc., New York.

21. **Curds, C. R., and A. Cockburn.** 1968. Studies on the growth and feeding of *Tetrahymena pyriformis* in axenic and monoxenic culture. J. Gen. Microbiol. **54:**343–358.

22. **Davis, P. G., and J. McN. Sieburth.** 1982. Differentiation of the photosynthetic and heterotrophic populations of nanoplankton by epifluorescence microscopy. Ann. Inst. Oceanogr. (Paris) **58**(Suppl.):249–260.

23. **Eppley, R. W.** 1980. Estimating phytoplankton growth rates in the central oligotrophic oceans, p. 231–242. *In* P. G. Falkowski (ed.), Primary productivity in the sea. Plenum Press, New York.

24. **Fenchel, T.** 1968. The ecology of marine microbenthos. II. The food of marine benthic ciliates. Ophelia **5:**73–121.

25. **Fenchel, T.** 1969. The ecology of marine microbenthos. IV. Structure and function of the benthic ecosystem, its chemical and physical factors and the microfauna communities with special reference to the ciliated protozoa. Ophelia **6:**1–182.

26. **Fenchel, T.** 1975. The quantitative importance of the benthic microflora of an Arctic tundra pond. Hydrobiologia **46:**445–464.

27. **Fenchel, T.** 1977. The significance of bactrivorous protozoa in the microbial community of detrital particles, p. 529–544. *In* J. Cairns (ed.), Aquatic microbial communities. Garland Publications, New York.

28. **Fenchel, T.** 1980. Suspension feeding in ciliated protozoa: feeding rates and their ecological significance. Microb. Ecol. **6:**13–25.

29. **Fenchel, T.** 1982. Ecology of heterotrophic microflagellates. I. Some important forms and their functional morphology. Mar. Ecol. Prog. Ser. **8:**211–223.

30. **Fenchel, T.** 1982. Ecology of heterotrophic microflagel-

lates. II. Bioenergetics and growth. Mar. Ecol. Prog. Ser. **8:**225–231.

31. **Fenchel, T.** 1982. Ecology of heterotrophic microflagellates. IV. Quantitative occurrence and importance as consumers of bacteria. Mar. Ecol. Prog. Ser. **9:**35–42.

32. **Fenchel, T., and P. Harrison.** 1976. The significance of bacterial grazing and mineral cycling for the decomposition of particulate detritus, p. 285–299. *In* J. M. Anderson and A. Macfadyen (ed.), The role of terrestrial and aquatic organisms in decomposition processes. Blackwell Scientific Publications, London.

33. **Finlay, B. J.** 1978. Community production and respiration by ciliated protozoa in the benthos of a small eutrophic loch. Freshwater Biol. **8:**327–341.

34. **Finlay, B. J., and G. Uhlig.** 1981. Calorific and carbon values of marine and freshwater protozoa. Helgol. Wiss. Meeresunters. **34:**301–412.

35. **Gieskes, W. W. C., G. L. Kraay, and M. A. Baars.** 1979. Current ^{14}C methods for measuring primary production: gross underestimates in oceanic waters. Neth. J. Sea Res. **13:**50–78.

36. **Glibert, P. M., F. Lipschultz, J. J. McCarthy, and M. A. Altabet.** 1982. Isotope dilution models of uptake and remineralization of ammonium by marine plankton. Limnol. Oceanogr. **27:**639–650.

37. **Goldman, J. C., J. J. McCarthy, and D. G. Peavey.** 1979. Growth rate influence on the chemical composition of phytoplankton in oceanic waters. Nature (London) **279:**210–215.

38. **Gorjacheva, N. V., B. F. Zukov, and A. P. Mylnikov.** 1968. Biology of free living Bodonides, p. 29–50. *In* Biology and systematics of lower organisms, vol. 2. Trans. Inst. Biol. Inland Waters (Borok). (In Russian.)

39. **Gowing, M. M., and M. W. Silver.** 1983. Origins and microenvironments of bacteria mediating fecal pellet decomposition in the sea. Mar. Biol. **73:**7–16.

40. **Güde, H.** 1979. Grazing by protozoa as selection factor for activated sludge bacteria. Microb. Ecol. **5:**225–237.

41. **Haas, L. W.** 1982. Improved epifluorescent microscope technique for observing planktonic microorganisms. Ann. Inst. Oceanogr. (Paris) **58**(Suppl.):261–266.

42. **Haas, L. W., and K. L. Webb.** 1979. Nutritional mode of several non-pigmented micro-flagellates from the York River estuary, Virginia. J. Exp. Mar. Biol. Ecol. **39:**125–134.

43. **Hamilton, R. D., and J. E. Preslan.** 1969. Cultural characteristics of a pelagic marine hymenostome ciliate, *Uronema* sp. J. Exp. Mar. Biol. Ecol. **4:**90–99.

44. **Harrison, W. G.** 1978. Experimental measurement of nitrogen remineralization in coastal waters. Limnol. Oceanogr. **23:**684–694.

45. **Harrison, W. G.** 1980. Nutrient regeneration and primary production in the sea, p. 433–460. *In* P. G. Falkowski (ed.), Primary productivity in the sea. Plenum Press, New York.

46. **Harrison, W. G.** 1983. Uptake and recycling of soluble reactive phosphorus by marine microplankton. Mar. Ecol. Prog. Ser. **10:**127–135.

47. **Harrison, P. G., and K. H. Mann.** 1975. Detritus formation from eelgrass (*Zostera marina* L.): the relative effects of fragmentation, leaching, and decay. Limnol. Oceanogr. **20:**924–934.

48. **Heal, O. W.** 1967. Quantitative feeding studies on soil amoebae, p. 120–126. *In* O. Graff, J. E. Satchell, and J. E. Amsterdam (ed.), Progress in soil biology. North Holland Publishing Co., Amsterdam.

49. **Heinbokel, J. F.** 1978. Studies on the functional role of tintinnids in the southern California bight. I. Grazing and growth rates in laboratory cultures. Mar. Biol. **47:**177–189.

50. **Heinbokel, J. F., and J. R. Beers.** 1979. Studies on the functional role of tintinnids in the southern California bight. III. Grazing impact of natural assemblages. Mar. Biol. **52:**23–32.

51. **Heinle, D. R., R. P. Harris, J. F. Ustach, and D. A.**

Flemer. 1977. Detritus as food for estuarine copepods. Mar. Biol. **40:**341–353.

52. **Hobbie, J. E.** 1979. Activity and bacterial biomass. Arch. Hydrobiol. Beih. Ergebn. Limnol. **12:**59–63.

53. **Jackson, G. A.** 1980. Phytoplankton growth and zooplankton grazing in oligotrophic oceans. Nature (London) **284:**439–441.

54. **Jawed, M.** 1973. Ammonia excretion by zooplankton and its significance to primary productivity during summer. Mar. Biol. **23:**115–120.

55. **Johannes, R. E.** 1964. Phosphorus excretion and body size in marine animals: microzooplankton and nutrient regeneration. Science **146:**923–924.

56. **Johannes, R. E.** 1965. Influence of marine protozoa on nutrient regeneration. Limnol. Oceanogr. **10:**434–442.

57. **Johnson, P. W., and J. McN. Sieburth.** 1979. Chroococcoid cyanobacteria in the sea: a ubiquitous and diverse phototrophic biomass. Limnol. Oceanogr. **24:**928–935.

58. **Johnson, P. W., and J. McN. Sieburth.** 1982. *In situ* morphology and occurrence of eucaryotic phototrophs of bacterial size in the picoplankton of estuarine and oceanic waters. J. Phycol. **18:**318–327.

59. **Johnson, P. W., Huai-Shu Xu, and J. McN. Sieburth.** 1982. The utilization of chroococcoid cyanobacteria by marine protozooplankters but not by calanoid copepods. Ann. Inst. Oceanogr. (Paris) **58**(Suppl.):297–308.

60. **King, K. R., J. T. Hollibaugh, and F. Azam.** 1980. Predator-prey interactions between the Larvacean *Oikopleura dioica* and bacterioplankton in enclosed water columns. Mar. Biol. **56:**49–57.

61. **Kopylov, A. I.** 1979. Chemical composition and caloric value of infusoria. Oceanology (USSR) **19:**586–589.

62. **Kopylov, A. I., and E. S. Moiseev.** 1980. Effect of colorless flagellates on the determination of bacterial production in seawater. Dokl. Biol. Sci. **252:**272–274.

63. **Kopylov, A. I., A. F. Pasternak, and Ye. V. Moiseyev.** 1981. Consumption of zooflagellates by planktonic organisms. Oceanology (USSR) **21:**269–271.

64. **Lee, J. J.** 1980. A conceptual model of marine detrital decomposition and the organisms associated with the process, p. 257–291. *In* M. R. Droop and H. W. Jannasch (ed.), Advances in aquatic microbiology. Academic Press, Inc., New York.

65. **LeFevre, J., and J. R. Grall.** 1970. On the relationships of *Noctiluca* swarming off the coast of Brittany with hydrological features and plankton characteristics of the environment. J. Exp. Mar. Biol. Ecol. **4:**287–306.

66. **Legner, M.** 1980. Growth rate of infusorian populations, p. 205–255. *In* M. R. Droop and H. W. Jannasch (ed.), Advances in aquatic microbiology, vol. 2. Academic Press, Inc., New York.

67. **Lighthart, B.** 1969. Planktonic and benthic bacteriovorous protozoa at eleven stations in Puget Sound and adjacent Pacific Ocean. J. Fish. Res. Board Can. **26:**299–304.

68. **Linley, E. A. S., R. C. Newell, and M. I. Lucas.** 1983. Quantitative relationships between phytoplankton, bacteria and heterotrophic microflagellates in shelf waters. Mar. Ecol. Prog. Ser. **12:**77–89.

69. **Lucas, M. I., R. C. Newell, and B. Velimirov.** 1981. Heterotrophic utilization of mucilage released during fragmentation of kelp (*Ecklonia maxima* and *Laminaria pallida*). II. Differential utilization of dissolved organic components from kelp mucilage. Mar. Ecol. Prog. Ser. **4:**43–55.

70. **Malone, T. C.** 1980. Size-fractionated primary productivity of marine phytoplankton, p. 301–319. *In* P. G. Falkowski (ed.), Primary productivity in the sea. Plenum Press, New York.

71. **Marshall, S. M.** 1973. Respiration and feeding in copepods. Adv. Mar. Biol. **11:**57–120.

72. **McCarthy, J. J., and J. C. Goldman.** 1979. Nitrogenous nutrition of marine phytoplankton in nutrient-depleted waters. Science **203:**670–672.

73. **Newell, R. C., M. I. Lucas, and E. A. S. Linley.** 1981. Rate of degradation and efficiency of conversion of phytoplankton debris by marine microorganisms. Mar. Ecol. Prog. Ser. **6:**123–136.

74. **Newell, S. Y., and R. R. Christian.** 1981. Frequency of dividing cells as an estimator of bacterial productivity. Appl. Environ. Microbiol. **42:**23–31.

75. **Paasche, E., and S. Kristiansen.** 1982. Ammonium regeneration by microzooplankton in the Oslofjord. Mar. Biol. **69:**55–63.

76. **Pace, M. L., and J. D. Orcutt.** 1981. The relative importance of protozoans, rotifers, and crustaceans in a freshwater zooplankton community. Limnol. Oceanogr. **26:**822–830.

77. **Pavlouskaya, T. V.** 1973. Influence of feeding conditions upon the feeding and reproduction of ciliates. Zoological J. (Leningrad) **52:**1451–1457. (In Russian.)

78. **Perkins, F. O., L. W. Haas, D. F. Phillips, and K. L. Webb.** 1981. Ultra-structure of a marine *Synechococcus* possessing spinae. Can. J. Microbiol. **27:**318–329.

79. **Peterson, B. J.** 1980. Aquatic primary productivity and the ^{14}C-CO_2 method: a history of the productivity problem. Ann. Rev. Ecol. Syst. **11:**359–385.

80. **Peterson, B. J., J. E. Hobbie, and J. F. Haney.** 1978. *Daphnia* grazing on natural bacteria. Limnol. Oceanogr. **23:**1039–1044.

81. **Platt, T., D. V. Subba Rao, and B. Irwin.** 1983. Photosynthesis of picoplankton in the oligotrophic ocean. Nature (London) **310:**702–704.

82. **Pomeroy, L. R.** 1980. Microbial roles in aquatic food webs, p. 85–109. *In* R. Colwell (ed.), Aquatic microbial ecology. University of Maryland, College Park.

83. **Pomeroy, L. R., and D. Deibel.** 1980. Aggregation of organic matter by pelagic tunicates. Limnol. Oceanogr. **25:**643–652.

84. **Porter, K. G., and Y. S. Feig.** 1980. The use of DAPI for identifying and counting aquatic microflora. Limnol. Oceanogr. **25:**943–948.

85. **Porter, K. G., M. L. Pace, and J. F. Battey.** 1979. Ciliate protozoans as links in freshwater planktonic food chains. Nature (London) **277:**563–565.

86. **Prasad, R. R.** 1958. A note on the occurrence and feeding habits of *Noctiluca* and their effects on the plankton communities and fisheries. Proc. Indian Acad. Sci. Sect. B **47:**331–337.

87. **Rassoulzadegan, F.** 1982. Feeding in marine planktonic protozoa. Ann. Inst. Oceanogr. (Paris) **58**(Suppl.):191–206.

88. **Rassoulzadegan, F., and M. Etienne.** 1981. Grazing rate of the tintinnid *Stenosemella ventricosa* (Clap. and Lachm.) Jörg., on the spectrum of the naturally occurring particulate matter from a Mediterranean neritic area. Limnol. Oceanogr. **26:**258–270.

89. **Robertson, J. R.** 1983. Predation by estuarine zooplankton or tintinnid ciliates. Estuarine Coastal Shelf Sci. **16:**27–36.

90. **Robinson, J. D., K. H. Mann, and J. A. Novitsky.** 1982. Conversion of the particulate fraction of seaweed detritus to bacterial biomass. Limnol. Oceanogr. **27:**1072–1079.

91. **Rogerson, A.** 1981. The ecological energetics of *Amoeba proteus* (Protozoa). Hydrobiologia **85:**117–128.

92. **Schönborn, W.** 1977. Production studies on protozoa. Oecologia (Berlin) **27:**171–184.

93. **Sherr, B. F., and E. B. Sherr.** 1983. Enumeration of heterotrophic microprotozoa by epifluorescence microscopy. Estuarine Coastal Shelf Sci. **16:**1–7.

93a. **Sherr, E. B., and B. F. Sherr.** 1983. Double-staining epifluorescence technique to assess frequency of dividing cells and bacteriovory in natural populations of heterotrophic microprotozoa. Appl. Environ. Microbiol. **46:**1388–1393.

94. **Sherr, B. F., E. B. Sherr, and T. Berman.** 1982. Decomposition of organic detritus: a selective role for microflagellate protozoa. Limnol. Oceanogr. **27:**765–769.

95. **Sherr, B. F., E. B. Sherr, and T. Berman.** 1983. Grazing,

growth, and ammonia excretion rates of a heterotrophic microflagellate fed four species of bacteria. Appl. Environ. Microbiol. **45**:1196–1201.

96. **Sieburth, J. McN.** 1979. Sea microbes. Oxford University Press, New York.

97. **Sieburth, J. McN., and P. G. Davis.** 1982. The role of heterotrophic nanoplankton in the grazing and nurturing of planktonic bacteria in the Sargasso and Caribbean Sea. Ann. Inst. Oceanogr. (Paris) **58**(Suppl.):285–296.

98. **Sieburth, J. McN., V. Smetacek, and J. Lenz.** 1978. Pelagic ecosystem structure: heterotrophic compartments and their relationship to plankton size fractions. Limnol. Oceanogr. **23**:1256–1263.

99. **Sieburth, J. McN., P. Willis, K. Johnson, C. Burney, D. Lavoie, K. Hinga, D. Caron, F. French, P. Johnson, and P. Davis.** 1976. Dissolved organic matter and heterotrophic microneuston in the surface microlayers of the North Atlantic. Science **194**:1415–1418.

100. **Silver, M. W., and A. L. Alldredge.** 1981. Bathypelagic marine snow: deep-sea algal and detrital community. J. Mar. Res. **39**:501–530.

101. **Silver, M. W., A. Shanks, and J. Trent.** 1978. Marine snow: microplankton habitat and source of small-scale patchiness in pelagic populations. Science **201**:371–373.

102. **Smetacek, V.** 1981. The annual cycle of protozooplankton in the Kiel Bight. Mar. Biol. **63**:1–11.

103. **Smith, S. L.** 1978. The role of zooplankton in the nitrogen dynamics of a shallow estuary. Est. Coast. Mar. Sci. **7**:555–565.

104. **Smith, S. L., and T. E. Whitledge.** 1977. The role of zooplankton in the regeneration of nitrogen in a coastal upwelling system off northwest Africa. Deep-Sea Res. **24**:49–56.

105. **Sorokin, Y. I.** 1977. The heterotrophic phase of plankton succession in the Japan Sea. Mar. Biol. **41**:107–117.

106. **Sorokin, Y. I.** 1981. Microheterotrophic organisms in marine ecosystems, p. 293–342. *In* A. R. Longhurst (ed.), Analysis of marine ecosystems. Academic Press, London.

107. **Stout, J. D.** 1980. The role of protozoa in nutrient cycling and energy flow, p. 1–50. *In* M. Alexander (ed.), Advances in microbial ecology. Plenum Press, New York.

108. **Taylor, G. T.** 1982. The role of pelagic heterotrophic protozoa in nutrient cycling: a review. Ann. Inst. Oceanogr. (Paris) **58**(Suppl.):227–242.

109. **Taylor, W. D.** 1978. Growth responses of ciliate protozoa to the abundance of their bacterial prey. Microb. Ecol. **4**:207–214.

110. **Taylor, W. D.** 1980. Observations on the feeding and growth of the predacious oligochaete *Chaetogaster langi* on ciliated protozoa. Trans. Am. Microsc. Soc. **99**:360–368.

111. **Williams, P. J. leB.** 1981. Incorporation of microheterotrophic processes into the classical paradigm of the planktonic food web. Kiel. Meeresforsch. **5**:1–28.

112. **Zaika, V. E.** 1973. Specific production of aquatic invertebrates. Wiley Interscience, London.

Carbon and Energy Flow Through Microflora and Microfauna in the Soil Subsystem of Terrestrial Ecosystems

E. T. ELLIOTT, D. C. COLEMAN, R. E. INGHAM, AND J. A. TROFYMOW

Natural Resource Ecology Laboratory, Colorado State University, Ft. Collins, Colorado 80523

Metabolic energy dissipation in soils is closely coupled with the major pathways of carbon loss in terrestrial ecosystems. A large portion of net primary production is allocated belowground as roots, root exudation, and exfoliation. A part of aboveground production also enters the soil. Thus, the soil subsystem is responsible for processing most of the net ecosystem production. Therefore, it is important to consider how this material is incorporated into the belowground food web and how the subsequent activities of organisms affect C losses and mineralization of other essential elements.

The soil ecosystem consists of a myriad of micro-, meio-, and mesoorganisms strongly controlled by the physical and chemical restraints of the opaque medium. Primary saprophages, such as bacteria, fungi, actinomycetes, and yeasts, utilize the primary production, waste materials, and dead bodies from other tropic levels to produce new tissue which is subsequently used by secondary and tertiary consumers of microorganisms. Energy is lost and C is released to the atmosphere at each step in this process. Also, other elements are released into the soil as mineral forms when organic materials are oxidized to CO_2. Therefore, it is necessary to study this conglomeration of organisms as a whole and not to study only a subset, for it is in their interactions that a better understanding of soil ecological processes will be obtained.

The mineralization of essential elements such as N, P, and S, resulting from interactions of microbes with their substrates and the subsequent trophic interactions, influences energy flow in two ways. First, it controls the influx of energy by regulating primary production (plant growth) through controlling the availability of limiting nutrients, and second, it controls the efflux of energy as nutrient availability interacts with substrate quality, thereby controlling decomposition rates. Thus, secondary effects of microbial feeding interactions influence C and energy flow as well as direct respiratory losses.

Another consideration of the microflora–microfauna relationship is direct population (or functional group) control. That is, the size and turnover rate of the decomposer populations is directly influenced by the intensity of feeding upon them. This control can be exerted at even higher levels in the food chain, resulting ultimately in changes in decomposition rate. Lack of grazing, overgrazing, and optimal grazing are all ways in which this control can be exerted. The amount and quality of substrate available for decomposition (i.e., enrichment level) controls the length of the food chain and eventually the intensity and kind of control by the topmost trophic level. The organismal component of the belowground ecosystem should not be considered as a black-box group of decomposers, but rather their interactions must be mechanistically considered. Recently, there has been a heightened interest in the soil animal portion of this system.

REVIEW OF THE REVIEWS

Since there has been a recent surge in review articles dealing with soil fauna, it would be redundant to cover this general topic in much detail here. Rather, we will try to present a brief review of the reviews and then move on to more specific, if less well understood, aspects of this topic and how they may influence the soil ecosystem.

Anderson et al. (1) reviewed the effects of saprotrophic grazing on net mineralization and concluded that the microfauna help maintain a dynamic nutrient pool and increase the homogeneity of decomposer distribution in soil. Food chains longer than the usual four or five links might be expected in soil systems because of high production efficiencies of some forms, particularly the protozoa (D. C. Coleman, *in* M. J. Mitchell, ed., *Microflora and Faunal Interactions in Natural Agroecosystems*, in press). The interactions of these forms have resulted in enhanced nutrient uptake and plant yield (30 to 100% increase) in microcosm experiments (13, 29; Coleman, in press; R. E. Ingham, J. A. Trofymow, E. R. Ingham, and D. C. Coleman, submitted for publication). Mixed communities of protozoa also enhance C and N mineralization in a manner similar to that in experiments including oligospecific inoculations of known species (D. C. Coleman, R. E. Ingham, J. F. McClellan, and J. A. Trofymow, *in* J. M. Anderson, A. D. M. Ryner, and D. H. Walton, ed., *Invertebrate-Microbial Interactions*, in press). H. W. Hunt and W. J. Parton (*in* M. J. Mitchell, ed., *Microfloral and Faunal Interactions in Natural and Agroecosystems*, in press) provide an extensive review of mathematical models and how they apply to microflora–microfauna interactions, as well as a general discussion of simulation modeling philosophy. Among other considerations, they present a series of models at different levels of resolution: a single species grown under defined soil conditions, effect of grazers on the bacteria, a whole ecosystem model including primary producers, and a long-term model which considers changes in soil organic matter levels.

Review and synthesis of the functional and structural aspects of soil animal populations by Peterson and Luxton (53) provide much information on sampling theory and substrate utilization. They conclude that, although soil animals may account for only about 5% of total soil respiration, their indirect role as catalysts for nutrient circulation is considerable. Unfortunately, the microarthropod group is overemphasized, perhaps because there is more information available for them than for other forms such as protozoa or nematodes. Stout (66) suggested that new methods are necessary to assess the impact of protozoa on nutrient cycles and showed evidence for enhancement of such cycling by increasing bacterial turnover rates and releasing nutrients immobilized in bacterial tissue by these smallest forms of the soil fauna. Clarholm (this volume) makes a stronger case for the importance of protozoa based upon her work (13) and that of others. The impact of soil nematodes in terrestrial ecosystems has likewise been reviewed recently by Yeates (74), and his conclusions were similar to those of Petersen and Luxton (53); that is, although biomass is low, impact is quite large. He also reviewed this functional group in relation to the soil environment (75). Sohlenius (65) came to the same conclusion as Yeates (74), but approached the topic from an energy flow analysis. He found that although they may consume 25% of the bacterial standing stock, nematodes account for only 1% of the soil respiration.

Litter decomposition increased in the presence of soil microarthropods an average of 23% for 15 studies with time periods ranging from 9 to 30 months, but mineralization of N, P, and K was accelerated in only half of the studies (59). The organization of soil microarthropod communities was analyzed by Usher et al. (68). They noted that there are three broad areas where research would yield useful results: (i) individuals within populations, (ii) populations within communities, and (iii) the community itself. Although these areas are very general, it is profitable to study ecosystem processes at several levels of resolution.

The influence of invertebrates upon nutrient cycling was divided into several regulatory roles by Hutchinson and King (43): particle reduction and abrasion, mixing, invertebrate dietary preferences, effects of microbivory on microbial production, gut transactions, and direct effects of invertebrates on plant growth. They concluded that "Any positive effect of invertebrates on cycling must depend largely on their ability to stimulate microbial activity," a conclusion made by most authors of the papers reviewed above.

The purpose of this section was to provide references for those interested in obtaining information on the topic of microflora–microfauna interactions. The next section will present a conceptual framework for trophic interactions in general and its application to soil microbial processes. These ideas will then be used to help explain results from field and laboratory studies.

MICROBIAL TROPHIC STRUCTURE

The simplest trophic (feeding) structure is the food chain, in which resources flow into the resource pool and are immobilized by producers, then consumers, and subsequently a number of tertiary consumers (predators), depending upon the number of links in the chain. A commonly discussed type of food chain is planktonic, in which the producers are the microbial phytoplankton. The basis of the food chain in soil is also microbial, but rather than deriving energy directly from the sun, most of the energy used has already been fixed in the form of dead plant material. Aside from differences in physical and chemical environment, these two types of biological systems, phytoplanktonic and terrestrial detrital, are similar in trophic structure, and each may have food chain lengths of five or greater (71; Coleman, in press). However, the maximum body size of individual predators in the soil is, of course, restricted. Another structural aspect common to microbial food chains is that the consumer biomass may be equal to or greater than that of producers (either photosynthetic or saprophagic), creating a straight or inverted pyramid rather than the classical upright pyramid common for aboveground terrestrial ecosystems.

Food chains are useful conceptualizations of some ecosystems, as demonstrated by the number of ideas stimulated by discussions of these structurally simple conceptual systems (36, 63). Two aspects of food chains as they relate to soil systems will be considered here: enrichment of the resource base and hence establishment of subsequent trophic levels, and the influence of the top predator upon the ecosystem dynamics.

Enrichment levels can determine the length of a food chain (30, 33, 50, 63, 70). The resource input is transferred up the food chain and some is lost from each component, the relative proportion of these two flows determining the net growth rate of each component and its yield efficiency. Interactions between enrichment level and the controls of saturation threshold, climate, and predator–prey relationships can produce complex periodic behavior in food chain models (22, 30). The rate of enrichment and the yield efficiencies determine the number of links that will be established in a steady-state system.

As enrichment increases, the prominence of the top trophic level also increases until there is enough material in the system to support the next trophic level. This new feeding does not necessarily reduce the productivity of the component being grazed, but it will reduce the amount of material in that level if predation pressure is high enough. However, the turnover rate of the grazed population, and therefore production of that level, may increase (26, 62). For these reasons, the amount of enrichment in the system can determine the predation rate of the top component. This, in turn, has effects that cascade down the entire food chain. For example, if a large lake has been overstocked with large piscivorous game fish, the zooplanktivorous fish population may be severely reduced, allowing zooplankton to become abundant. With many zooplankton, few algae will be present, so inorganic nutrients may accumulate. This scenario was envisioned by Shapiro (61) for lakes, but a similar argument can be made for soil food chains, where the establishment of a predacious species could determine the structure and production of the soil community. Aspects of production or exploitation ecology have been considered relative to aboveground terrestrial, fishery, and other ecosystems (15, 23, 26), as well as soil microbial systems (38, 39).

RESULTS FROM MICROBIAL INTERACTION EXPERIMENTS

Insecticides have been used to remove soil microarthropods from decomposer communities. Seastedt and Crossley (60) observed decomposition rates and nutrient accumulation in litterbags in a mesic habitat with and without microarthropods. Decomposition was reduced by 13% and immobilization of inorganic N contained in simulated throughfall was greater by 9% without microarthropods after 365 days. They attributed this to a decrease in the turnover of microflora in the absence of grazing. Elkins and Whitford (24) did a similar experiment in a xeric habitat and found that decomposition was 15% less when microarthropods were absent after 1 year. They also measured fungal and bacterial feeding nematode populations and observed that nematodes increased tremendously when mites were absent. Further experimentation demonstrated that some of the microarthropods were feeding voraciously upon the nema-

todes. The decomposition rate was reduced in the absence of nematophagous mites because the large nematode biomass reduced the microbial biomass to low levels. It appears that more than one explanation can be used for the observed results, and more data are needed from experiments with mutually exclusive hypotheses to delineate these differences. Also, measurement of more than one functional group of soil fauna (or flora) is necessary to obtain a complete understanding of the system dynamics.

Soil fauna are important in soil ecosystems, because they accelerate decomposition and increase the rate of return of mineral nutrients to the soil. Mineral nutrients may increase plant growth under limiting conditions. Some experiments have supported this but others have not. Rosswall et al. (54) and Bååth et al. (5), using pine seedlings grown in soil humus media, did not find greater plant growth with grazers than without but did observe increases in microfloral activity and net mineralization of N. They concluded that the plants were not sensitive to the changes caused by the microfauna. Elliott et al. (29), Clarholm (13), and Ingham et al. (submitted for publication) observed greater N uptake or growth, or both, by plants with grazers present. These latter experiments differed from the first in several respects: mineral soil was used with graminoids and in the cases of Clarholm (13) and Elliott et al. (29) protozoa rather than nematodes were used as the faunal component. Ingham et al. (submitted for publication) used N-poor soil (0.04% N) amended with chitin in microcosms inoculated with chitinoclastic bacteria and bacterial feeding nematodes or with chitinoclastic fungi and fungal feeding nematodes. Only the bacterial system demonstrated the plant response to grazers. Clarholm (13) and Elliott et al. (29) obtained maximum increases in plant N with protozoa present of 75 and 100%, respectively, over levels in their absence.

The plants in the experiments of Rosswall et al. (54) and Bååth et al. (5) did not respond to the presence of grazers. However, as Ingham et al. (submitted for publication) pointed out, the total amount of N in the system was small (0.15%) and had a wide C/N ratio (39). Clarholm et al. (14) used the same humus material for net mineralization experiments, and extrapolating from their results, we predict that the amount of net N mineralization in the soils used by Bååth et al. (5) was approximately equal to that found in their leachate. If the same amount of mineralization predicted from Clarholm et al. (14) occurred in the experiments of Bååth et al. (5), this could be accounted for in the leachate N of Bååth et al. (5). This means that there would be no N left for plant growth. Furthermore, the amounts of N in the pine seedlings at harvest were at least 10 times greater than the amount of N predicted to be mineralized by Clarholm et al. (14). This would reduce experimental sensitivity for determination of plant response because of the large background plant N values (54). Protozoa are better mineralizers of N than nematodes (19, 73) and may account for the better response of the plant. Grazers enhanced plant growth more in bacterial than in fungal systems (Ingham et al., submitted for publication). Fungi may have been mineralizing optimally without nematodes since they are better mineralizers of chitin N than are bacteria (35).

However, it is not just the interactions of decomposers with substrate and the subsequent availability of nutrients to plants that is of interest. The interaction of the roots with this system should also be considered. For example, Ingham et al. observed 10 to 20 times more bacterial feeding nematodes and 5 to 10 times more fungal feeding nematodes in rhizosphere than nonrhizosphere soil (submitted for publication).

A SPATIAL-TEMPORAL SEQUENCE OF EVENTS ALONG ROOTS

Considerable quantities of soluble organic carbon are released into soil by plant roots in such diverse ecosystems as grasslands (17), deciduous forests (64), and agroecosystems (37). These inputs are equal to about 20% of the total plant dry matter (56) or about 40% of the carbon translocated to the roots (47). Carbon losses may occur from lysis of root cell walls (32, 47) rather than actual exudation and therefore may consist of cell sap constituents such as carbohydrates, amino acids, and organic acids (57). As a consequence of root exudation, microorganisms usually have higher densities near roots than in root-free soil. In fact, they may enhance root exudations (7) in some cases by twofold over sterile controls (8). Stimulation of microbial activity by roots has been demonstrated for a wide array of soil organisms such as fungi (51), bacteria (58), and protozoa (6, 21). Although members of the microflora play an essential role in decomposition by producing hydrolytic enzymes, they compete with primary producers for uptake of mineralized nutrients (34, 41, 52). This is expected to be especially true in the rhizosphere, where microfloral use of substrates such as simple sugars must be balanced by uptake of inorganic nutrients (i.e., NO_3^-, NH_4^+ and PO_4^{3-}) which would otherwise be available for plants.

Carbon is lost into the rhizosphere soil in substantial quantities (49). Rovira (56) observed that only a small proportion (5 to 10%) of the root surface is colonized by bacteria at any one time, as determined by light microscopy. Abundant bacteria were present on the root surface in rhizospheres containing only a single species of bacteria (9). If most of the leaked carbon is respired (8), then how is this large amount of input C processed without large microfloral populations? Perhaps a key factor overlooked by many rhizosphere biologists relating to both the evolutionary and microbial population aspects of root loss of carbon is the presence of bacteriophagic grazers.

Protozoan populations are higher in rhizosphere zones (6, 12, 21), as are bacteriophagic nematodes (Ingham et al., submitted for publication). Recent work with soil microcosms simulating rhizosphere microsites receiving root exudation have demonstrated the large influence which the rhizosphere microfauna has in decreasing bacterial populations and accelerating rates of C, N, and P mineralization (2, 4, 16, 18, 19, 73). The presence of protozoa increased the availability of bacterially immobilized N to plants over that in controls with bacteria alone (13, 29). In grassland soils, protozoa may consume three to four times the bacterial standing crop per year (28). This value may be considerably higher in rhizosphere microsites. We offer the following scenario which explains the unexpectedly low numbers of bacteria on roots and demonstrates a possible mechanism by which root loss of soluble C may actually enhance plant uptake of inorganic nutrients.

A temporal sequence of events is postulated to occur along individual roots (E. T. Elliott, M.S. thesis, Colorado State University, Fort Collins, 1978). The diagrammatic representation of a root shown in Fig. 1 is divided into five zones occupying a time period of about 2 to 3 weeks. Most exudation occurs in zone I (55). Bacteria immobilize large quantities of N and P in zone II and thus cause a net diffusion of N and P into this zone at a rate higher than could be caused by simple root uptake. In zone III carbon from root exudation becomes exhausted and the grazer population begins to develop. In this zone the root has not gained any benefit from exudation. There is high grazer activity in zone IV and lack of readily available carbon. Therefore, there is net mineralization and plant uptake of available inorganic N and P. The plant benefits from the exudation via the heightened concentration of nutrients due to increased diffusion, immobilization by bacteria, and subsequent mineralization by grazers. Exudation of C by plants may enhance their competitive ability by effectively increasing their sphere of immobilization influence compared with the mechanism of conventional root uptake. Mycorrhizae become established in zone IV and in zone V begin to transport nutrients to roots. This proposed mechanism is self-consistent considering root phenology, timing of microflora–microfauna in-

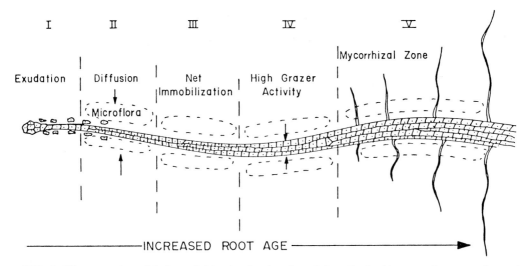

FIG. 1. Diagrammatic model of a spatial-temporal series of events hypothesized to occur along a growing root (Elliott, Ph.D. thesis, University of Colorado, 1978).

teractions, and mycorrhizal establishment. It explains how plants may benefit from temporal effects (time after exudation) and spatial influences (root age distribution). Trofymow and Coleman (67) have elaborated on this idea. This is a hypothesis, but unless the numerous microfauna are accounted for, we really do not have "a far better understanding of the release of organic materials from roots, [and] the dynamics of root colonization," as stated by Rovira (56).

Another consideration of the bacteria–protozoa interactions occurring along roots was discussed by Clarholm (13). She postulated, as did Trofymow and Coleman (67), that the pulsed inputs of C by roots stimulates mineralization of organic N by an otherwise quiescent bacterial population. Small pulses of glucose added into soil containing wheat plants stimulated dry-weight production and N content slightly (13). Stimulation of microbial growth by C inputs from the plant could enhance the N available to plants in the following way. Since the plant root is a strong N sink, it is able to immobilize N in the rhizosphere before the local microflora is able to respond to root-derived carbon (J. A. Trofymow, unpublished data). Therefore, the rhizosphere microflora may be stimulated to use sources of N unavailable to the plant (i.e., organic N). As long as root-derived C is available, nitrogen remains immobilized in microfloral biomass. As the easily metabolized C is exhausted, the biomass N is mineralized, either through death and lysis or more likely via grazing microfauna. The concept of increased N mineralization by plants is not a new one (34), but we know of no conclusive results either supporting or rejecting this conjecture.

HABITABLE PORE SPACE AND MICROBIAL INTERACTIONS

Another control upon trophic interactions of microbes is that of the habitable pore space. The soil pore volume is the space within which the soil microbial community exists (48). Pore neck sizes determines the accessibility to substrates and niches by the microorganisms. Soil physical properties such as pore volume, pore neck size distribution, and the pore volume accessible via pore necks of various sizes are important for regulating the size and structure of these habitats. Soil texture, structure, compaction, horizonation, and other morphological and landscape features will influence the arrangement, size, connectance, and depth distribution of soil pore space (10, 11). Growing roots may create new channels or may grow down the same well-established channels.

When considering habitable pore space, one must be concerned with both textural and structural aspects of the soil. Texture is the proportionate composition of sand, silt, and clay particles, whereas structure is the arrangement of these particles and the resulting pore space produced (46). In soil with poor structure, the textural aspects are more important, with finer soils having generally more but smaller pores than coarse soils. Since clay content is an important determinant of aggregation (structure), it plays a dual role in being important for texture and structure.

Predator and prey movement in soil is affected by structural and textural properties. Soil nematode migration is restricted by finer soil granulation (69) and higher bulk densities (45). Bacterial feeding nematodes in grassland soil are more numerous in soils with greater porosity (i.e., sand content) (44). Finer-textured soils inhibited the infiltration of fungal spores and bacterial cells (72). The soil protozoan *Colpoda steinii* was restricted from using bacteria by reduction of the soil water, until the only water-filled pores had pore necks too small to be penetrated by the ciliate (20). Trophic interactions among assemblages of soil organisms isolated from the shortgrass prairie were more restricted in fine than in coarse soil (25). Bacteria growing alone developed greater numbers and released more CO_2 in coarse than in fine soil. In the coarse-textured soil CO_2 evolution was greater with more biological complexity (microbes plus fauna), whereas in the fine-textured soil there were no differences in total CO_2 output among biological treatments. The nematode included in these experiments can feed on amoebae and bacteria, and reaches higher populations with amoebae plus bacteria than with bacteria alone (3, 4, 27). When nematodes were present, amoebal numbers were depressed. Although the amoebal and nematode populations were greater with coarse than fine soil, the proportionate increase in nematode numbers with the addition of amoebae was greater in the finer soil, especially for adult forms. A plausible explanation for this result is that,

> since amoebae are smaller than nematodes, they can enter soil pores with smaller pore neck sizes than nematodes. Once inside pores containing bacteria, the amoebal population increases and some amoebae move into larger pores containing nematodes. Whereas only a few amoebae enter the pore, many may exit because of the growth and reproduction that results from their feeding on bacteria (25).

Thus, soil texture and organism size strongly influence trophic structure.

Although the general effects of soil texture and structure upon microbial activity have been demonstrated, the exact controlling factors and quantification of the soil pore space are much less well understood. Pore necks connect soil pore spaces (voids), and it is the size of the pore necks which determines the ease of movement of the organism from one void to the next. However, the total pore volume accessible through pore necks larger than some restricting threshold may determine the total amount of food available to the organism. The amount of water in the pores and the size of pore necks containing water (determined by measuring soil water suction) are also major determinants. Some soil organisms require continuous films of water for movement (i.e., bacteria, protozoa, nematodes), whereas the movement of other forms may be inhibited or relatively unaffected

by water-filled pores (microarthropods and fungi, respectively).

A useful method for estimating pore neck size distributions is to compare the soil moisture wetting and drying characteristic curves. Soil water potential can be used to calculate the maximum size of water-filled pore necks because of capillary attraction of pore necks for water (40). The maximum diameter of the water-filled pore necks will be the same in two soils at the same water suction but of different texture/structure. However, the number of water-filled pore necks of a particular size will differ for two soil textures at the same suction (42). Thus, the moisture desorption curve yields a pore neck size distribution which gives information regarding the amount of water held in an initially water-saturated soil between two pore neck sizes.

Soil moisture absorption curves, giving the soil water suction as water is added to an initially dry sample, yield additional information. The water content at a given capillary pressure head for a wetting soil is less than that for a draining soil. This hysteresis can be considerable, especially for coarser-textured soils and wetter soils. For a given capillary pressure the difference in the water content between the desorption and absorption curves is the amount of pore space accessible by pore necks of the size that are water filled at the given capillary pressure. This information can be used to characterize the lower threshold limits of food availability, the refuge available to an organism, and the total amount of space available for habitation. It is not possible, however, to determine the number of pores filling this space.

Most information dealing with soil porosity, structure, and water relationships comes from the agricultural literature. Soil porosity may be changed readily by disturbing soil either in the laboratory or in the field (31), but the influence of these changes on microbial populations and effects on trophic interactions, especially under field conditions, still need to be elucidated.

AGROECOSYSTEM FIELD SAMPLING

Although there have been many studies on the size and distribution of soil animal populations (53), there have been relatively few with more than one soil faunal group and almost none including many animal functional groups plus their food and environment. We consider this latter approach particularly useful for understanding microflora–microfauna interactions and their role in nutrient cycling under field conditions because of the potentially complex nature of the kinds of relationships that may occur. We have studied a wheat-fallow agroecosystem using this approach (E. T. Elliott, K. Horton, J. C.

Moore, D. C. Coleman, and C. V. Cole, Plant Soil, in press). The fallow part of the field provided a means to study the mineralization of N and P because these nutrients accumulated, unlike the cropped side where they are removed by the plants.

The results presented here are from the fallow, no-till treatment. The site was sampled five times during the summer of 1982 (5 June–13 September), and data from the top 10 cm are given. The summer of 1982 was particularly wet, and on almost all dates there was sufficient soil water for microbial activity. The third and fourth dates were especially moist and warm.

There was a temporal displacement of functional groups of soil fauna (Fig. 2). Microarthropod numbers were greatest early in the year and decreased continuously until fall. Holophagous nematodes (those feeding upon bacteria or protozoa, or both) had peak numbers on the second date, and protozoan values were highest in the beginning and end of the summer, with lowest values on the second and third dates. Of the faunal groups, protozoan biomass was the greatest by a factor of 7 to 10.

The most dynamic period of the summer was between the third and fourth dates. There were large declines in microbial C, N, and P, and concomitant large increases in NO_3^- and PO_4^{3-}. Although these events corresponded with the increase of protozoa, it is logically impossible to derive cause and effect relationships from simple, correlative observations. However, based on previous experimental laboratory work, we can create a probable sequence of events.

Early summer microbial biomass was high, and with the onset of warm, wet conditions, protozoa increased, consuming the microbial

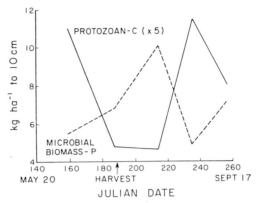

FIG. 3. Inverse relationship of microbial biomass P and protozoan C in the fallow no-till treatment (Elliott et al., in press).

biomass and excreting inorganic N and P. This accounts for the decline in microbial biomass and increase in NO_3^- and PO_4^{3-}. These results are qualitatively similar to observations from laboratory microcosm experiments (4, 16, 18, 19, 73). The inverse relationship between microbial P and protozoa under field conditions is clearly shown in Fig. 3. Results from microcosm experiments showing the influence of protozoa on microbial biomass P are similar and are shown in Fig. 4. We take this as evidence for the importance of soil protozoa as consumers of the microbial biomass and accelerators of mineralization under fallow no-till field conditions.

CONCLUSIONS

Evidence for a significant role of soil fauna, especially protozoa, in soil systems is not incontrovertible, but many recent results support this view. The mechanism whereby consumption of the microbial biomass by grazers enhances mineralization is an attractive, simple explanation for release of nutrients from the microbial biomass. Just as there is considerable feeding by aboveground consumers upon primary production, there is consumption by belowground grazers upon primary saprophagic production. Biocide experiments consistently demonstrate lower rates of decomposition in the absence of soil fauna. Under conditions of medium to high substrate availability, predation upon microflora, particularly protozoan predation, can enhance plant growth, most likely via increases in levels of nutrients available to plants. Field measurements substantiate these observations. A spatial-temporal sequence of events including microfaunal grazing may occur along a new root growing through the soil which may considerably influence the availability of nutrients to plants, especially graminoids. The soil pore

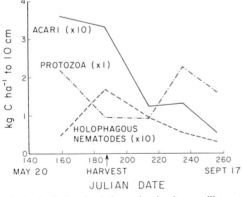

FIG. 2. Soil animal dynamics in the no-till treatment, showing a temporal displacement of soil animal functional groups. Protozoan biomass was seven times greater than all other groups combined, averaged over all sample dates (Elliott et al., in press).

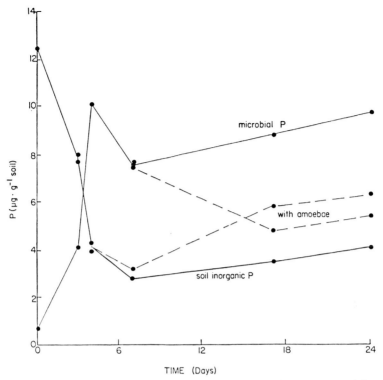

FIG. 4. Dynamics of soil inorganic P and microbial P in gnotobiotic microcosms containing either bacteria alone or bacteria with amoebae (16).

space available for various sizes of soil organisms to live in may be an important control upon trophic interactions, hence mineral nutrient availability. Mathematical models of these systems may be particularly useful in sorting out the complex array of possible community structures and controls (Hunt et al., this volume). More research is needed, preferably experiments done in the field, to test these ideas and determine the relevance of our conceptions to the real world.

The difficulties in dealing with soil ecosystems, of opacity of physical substrate and physical and chemical complexity of the milieu, necessarily created the recent proliferation of new, innovative, and sophisticated techniques. Through the application of this technology, we should see considerable advances in our understanding of this system in the next 10 years. This is imperative because of our concern for preservation of one of our most important natural resources, the soil, and our need to increase world food production.

LITERATURE CITED

1. **Anderson, R. V., D. C. Coleman, and C. V. Cole.** 1981. Effects of saprotrophic grazing on net mineralization. Ecol. Bull. **33**:201–216.
2. **Anderson, R. V., D. C. Coleman, C. V. Cole, and E. T.** Elliott. 1981. Effect of the nematodes *Acrobeloides* sp. and *Mesodiplogaster lheritieri* on substrate utilization and nitrogen and phosphorus mineralization in soil. Ecology **62**:549–555.
3. **Anderson, R. V., D. C. Coleman, C. V. Cole, E. T. Elliott, and J. F. McClellan.** 1979. The use of soil microcosms in evaluating bacteriophagic nematode responses to other organisms and effects on nutrient cycling. Int. J. Environ. Stud. **13**:175–182.
4. **Anderson, R. V., E. T. Elliott, J. F. McClellan, D. C. Coleman, C. V. Cole, and H. W. Hunt.** 1978. Trophic interactions in soil as they affect energy and nutrient dynamics. III. Biotic interactions of bacteria, amoebae and nematodes. Microb. Ecol. **4**:361–371.
5. **Bååth, E., U. Lohm, B. Lundgren, T. Rosswall, B. Söderström, and B. Sohlenius.** 1981. Impact of microbial-feeding animals on total soil activity and nitrogen dynamics: a soil microcosm experiment. Oikos **37**:257–264.
6. **Bamforth, S. S.** 1976. Rhizosphere-soil microbial comparisons in subtropical forests of southeastern Louisiana. Trans. Am. Microsc. Soc. **95**:613–621.
7. **Barber, D. A., and J. M. Lynch.** 1977. Microbial growth in the rhizosphere. Soil Biol. Biochem. **9**:305–308.
8. **Barber, D. A., and J. K. Martin.** 1976. The release of organic substances by cereal roots into soil. New Phytol. **76**:69–80.
9. **Bennett, R. A., and J. M. Lynch.** 1981. Bacterial growth and development in the rhizosphere of gnotobiotic cereal plants. J. Gen. Microbiol. **125**:95–102.
10. **Brewer, R.** 1964. Fabric and mineral analysis of soils. John Wiley & Sons, London.
11. **Childs, E. C.** 1969. An introduction to the physical basis of soil water phenomena. John Wiley & Sons, London.
12. **Clarholm, M.** 1981. Protozoan grazing of bacteria in soil: impact and importance. Microb. Ecol. **7**:343–350.

13. **Clarholm, M.** 1983. Dynamics of soil bacteria in relation to plants protozoa and inorganic nitrogen. Report 17. Swedish University of Agricultural Science, Uppsala.

14. **Clarholm, M., B. Popović, T. Rosswall, B. Söderström, B. Sohlenius, H. Staaf, and A. Wiren.** 1981. Biological aspects of nitrogen mineralization in humus from a pine forest podsol incubated under different moisture and temperature conditions. Oikos 37:137–145.

15. **Clark, C. W.** 1976. Mathematical bioeconomics. John Wiley & Sons, Inc., New York.

16. **Cole, C. V., E. T. Elliott, H. W. Hunt, and D. C. Coleman.** 1978. Trophic interactions in soils as they affect energy and nutrient dynamics. V. Phosphorus transformations. Microb. Ecol. 4:381–387.

17. **Coleman, D. C.** 1976. A review of root production processes and their influence on soil biota in terrestrial ecosystems, p. 417–434. *In* J. M. Anderson and A. Macfadyen (ed.), The role of terrestrial and aquatic organisms in decomposition processes. Blackwell Scientific Publications, Oxford.

18. **Coleman, D. C., R. V. Anderson, C. V. Cole, E. T. Elliott, L. Woods, and M. K. Campion.** 1978. Trophic interactions in soils as they affect energy and nutrient dynamics. IV. Flows of metabolic and biomass carbon. Microb. Ecol. 4:373–380.

19. **Coleman, D. C., C. V. Cole, R. V. Anderson, M. Blaha, M. K. Campion, M. Clarholm, E. T. Elliott, H. W. Hunt, B. Schaefer, and J. Sinclair.** 1977. An analysis of rhizosphere-saprophage interactions in terrestrial ecosystems. Soil organisms as components of ecosystems. Ecol. Bull. (Stockholm) 25:299–309.

20. **Darbyshire, J. F.** 1976. Effects of water suctions on the growth in soil of the ciliate *Colpoda steini*, and the bacterium *Azotobacter chroococcum*. J. Soil Sci. 27:369–376.

21. **Darbyshire, J. F., and M. P. Greaves.** 1967. Protozoa and bacteria in the rhizosphere of *Sinapis alba* L., *Trifolium repens* L., and *Lolium perenne* L. Can. J. Microbiol. 13:1057–1068.

22. **Dwyer, R. L., and J. N. Kramer.** 1983. Frequency-domain sensitivity analyses of an estuarine ecosystem simulated model. Ecol. Model. 18:35–54.

23. **Dyer, M. I., J. K. Detling, D. C. Coleman, and D. W. Hilbert.** 1982. The role of herbivores in grasslands, p. 255–295. *In* J. R. Estes, R. J. Tyrl, and J. N. Brunken (ed.), Grasses and grasslands: systematics and ecology. University of Oklahoma Press, Norman.

24. **Elkins, N. Z., and W. G. Whitford.** 1982. The role of microarthropods and nematodes in decomposition in a semi-arid ecosystem. Oecologia (Berlin) 55:303–310.

25. **Elliott, E. T., R. V. Anderson, D. C. Coleman, and C. V. Cole.** 1980. Habitable pore space and microbial trophic interactions. Oikos 35:327–335.

26. **Elliott, E. T., L. G. Castañares, D. Perlmutter, and K. G. Porter.** 1983. Trophic level control of production and nutrient dynamics in an experimental planktonic community. Oikos 41:7–16.

27. **Elliott, E. T., C. V. Cole, D. C. Coleman, R. V. Anderson, H. W. Hunt, and J. F. McClellan.** 1979. Amoebal growth in soil microcosms: a model system of C, N and P trophic dynamics. Int. J. Environ. Stud. 13:169–174.

28. **Elliott, E. T., and D. C. Coleman.** 1977. Soil protozoan dynamics in a shortgrass prairie. Soil Biol. Biochem. 9:113–118.

29. **Elliott, E. T., D. C. Coleman, and C. V. Cole.** 1979. The influence of amoebae on the uptake of nitrogen by plants in gnotobiotic soil, p. 221–229. *In* J. L. Harley and R. S. Russell (ed.), The soil-root interface. Academic Press, London.

30. **Elliott, E. T., R. G. Wiegert, and H. W. Hunt.** 1983. Simulation of simple food chains. *In* W. K. Lauenroth, G. V. Skogerboe, and M. Flug (ed.), State-of-the-art in ecological modelling. Elsevier Scientific Publishing Co., Amsterdam.

31. **Emerson, W. W., R. D. Bond, and H. R. Dexter (ed.).** 1978. Modification of soil structure. John Wiley & Sons, London.

32. **Foster, R. C., and A. D. Rovira.** 1976. Ultrastructure of wheat rhizosphere. New Phytol. 76:343–352.

33. **Fretwell, S. F.** 1977. Regulation of plant communities by the food chains exploiting them. Perspect. Biol. Med. 20:169–185.

34. **Goring, C. A., and F. E. Clark.** 1948. Influence of crop growth on mineralization of nitrogen in the soil. Soil Sci. Soc. Am. Proc. 13:261–266.

35. **Gould, W. D., R. J. Bryant, J. A. Trofymow, R. V. Anderson, E. T. Elliott, and D. C. Coleman.** 1981. Chitin decomposition in a model soil system. Soil Biol. Biochem. 13:487–492.

36. **Hairston, N. G., F. E. Smith, and L. B. Slobodkin.** 1960. Community structure, population control and competition. Am. Nat. 94:421–425.

37. **Hale, M. G., and L. D. Moore.** 1979. Factors affecting root exudation. II. 1970–1978. Adv. Agron. 31:93–124.

38. **Hanlon, R. D. G., and J. M. Anderson.** 1979. The effects of Collembola grazing on microbial activity in decomposing leaf litter. Oecologia (Berlin) 38:93–99.

39. **Hanlon, R. D. G., and J. M. Anderson.** 1980. Influence of macroarthropod feeding activities on microflora in decomposing oak leaves. Soil Biol. Biochem. 12:255–261.

40. **Hattori, T. H.** 1973. Microbial life in soil: an introduction. Marcel Dekker, Inc., New York.

41. **Hayman, D. S.** 1975. Phosphorus cycling by soil microorganisms and plant roots, p. 67–91. *In* N. Walker (ed.), Soil microbiology. John Wiley & Sons, Toronto.

42. **Hillel, D.** 1971. Soil and water. Academic Press, Inc., New York.

43. **Hutchinson, K. J., and K. L. King.** 1982. Invertebrates and nutrient cycling. *In* K. E. Lee (ed.), Proceedings 3rd Australian Conference on Grassland Invertebrate Ecology, Government Printer, Adelaide.

44. **Ingham, R. E., J. A. Trofymow, R. V. Anderson, and D. C. Coleman.** 1982. Relationships between soil type and soil nematodes in a shortgrass prairie. Pedobiologia 24:139–144.

45. **Jones, F. G. W., and A. J. Thomasson.** 1976. Bulk density as an indicator of pore space in soils usable by nematodes. Nematologica 22:133–137.

46. **Marshall, T. J., and J. W. Holms.** 1979. Soil Physics. Cambridge University Press, Cambridge.

47. **Martin, J. K.** 1977. Effect of soil moisture on the release of organic carbon from wheat roots. Soil Biol. Biochem. 9:303–304.

48. **McLaren, A. D., and J. J. Skujins.** 1967. Physical environment of microorganisms in soil, p. 3–24. *In* T. R. G. Gray and D. Parkinson (ed.), The ecology of soil bacteria. Liverpool University Press, Liverpool.

49. **Newman, E. I.** 1978. Root microorganisms: their significance in the ecosystem. Biol. Rev. 53:511–554.

50. **Oksanen, L., S. D. Fretwell, J. Arruda, and P. Niemela.** 1981. Exploitation ecosystems in gradients of primary productivity. Am. Nat. 118:240–261.

51. **Parkinson, D., G. S. Taylor, and R. Pearson.** 1963. Studies on the fungi in the root region. I. The development of fungi on young roots. Plant Soil 19:332–349.

52. **Paul, E. A.** 1976. Nitrogen cycling in terrestrial ecosystems, p. 225–243. *In* J. O. Nriagu (ed.), Environmental biogeochemistry, vol. 1, Carbon, nitrogen, phosphorus and selenium cycles. Ann Arbor Scientific Publications, Inc., Ann Arbor, Mich.

53. **Petersen, H., and M. Luxton.** 1982. A comparative analysis of soil fauna populations and their role in decomposition processes. Oikos 39:287–388.

54. **Rosswall, T., U. Lohm, and B. Sohlenius.** 1977. Developpement d'un microcosme pour l'etude de la minéralization et de l'absorption radiculaire d l'azote dans l'humus d'une forêt de coniferes (*Pinus sylvestris* L.). Lejeunia 84:1.

55. **Rovira, A. D.** 1973. Zones of exudation along plant roots and spacial distribution of microorganisms in the rhizosphere. Pestic. Sci. 4:361–366.

56. **Rovira, A. D.** 1979. Biology of the soil-root interface, p. 145–160. *In* J. L. Harley and R. S. Russel (ed.), The soil-root interface. Academic Press, London.

57. **Rovira, A. D., and B. M. McDougall.** 1967. Microbiological and biochemical aspects of the rhizosphere, p. 417–463. *In* A. D. McLaren and G. H. Peterson (ed.), Soil biochemistry. Marcel Dekker, Inc., New York.

58. **Rouatt, J. W., and H. Katznelson.** 1961. A study of bacteria on the root surface and in the rhizosphere of crop piants. J. Appl. Bacteriol. **24:**164–171.

59. **Seastedt, T. R.** 1984. The role of microarthropods in decomposition and mineralization processes. Annu. Rev. Entomol. **29:**25–46.

60. **Seastedt, T. R., and D. A. Crossley.** 1983. Nutrients in forest litter treated with naphthalene and simulated throughfall: a field microcosm study. Soil Biol. Biochem. **15:**159–165.

61. **Shapiro, J.** 1980. The importance of trophic level interactions to the abundance and species composition of algae in lakes, p. 105–116. *In* L. Mur and J. Barica (ed.), Hypertrophic ecosystems. W. Junk Publishers, The Hague.

62. **Slobodkin, L. B.** 1961. Growth and regulation of animal populations. Holt, Rinehart and Winston, New York.

63. **Smith, F. E.** 1969. Effects of enrichment in mathematical models, p. 631–645. *In* Eutrophication: causes, consequences, correctives. National Academy of Sciences, Washington, D.C.

64. **Smith, W. H.** 1976. Character and significance of forest tree root exudates. Ecology **57:**324–331.

65. **Sohlenius, B.** 1980. Abundance, biomass and contribution to energy flow by soil nematodes in terrestrial ecosystems. Oikos **34:**186–194.

66. **Stout, J. D.** 1980. The role of protozoa in nutrient cycling and energy flow, p. 1–50. *In* M. Alexander (ed.), Advances in microbial ecology. Plenum Press, New York.

67. **Trofymow, J. A., and D. C. Coleman.** 1982. The role of bacterivorous and fungivorous nematodes in cellulose and chitin decomposition in the context of a root/rhizosphere/soil conceptual model, p. 117–138. *In* D. W. Freckman (ed.), Nematodes in soil ecosystems. University of Texas Press, Austin.

68. **Usher, M. B., R. G. Booth, and K. E. Sparks.** 1982. A review of progress in understanding the organization of communities of soil arthropods. Pedobiologia **23:**126–144.

69. **Wallace, H. R.** 1971. The movement of nematodes in the external environment, p. 201–212. *In* A. M. Fallis (ed.), Ecology and physiology of parasites. University of Toronto Press, Toronto.

70. **Wiegert, R. G.** 1979. Population models: experimental tools for analysis of ecosystems, p. 233–279. *In* D. J. Horn, G. R. Stairs, and R. D. Mitchell (ed.), Analysis of ecological systems. Ohio State University Press, Columbus.

71. **Wiegert, R. G., and D. F. Owen.** 1971. Trophic structure, available resources and population density in terrestrial vs. aquatic ecosystems. J. Theor. Biol. **30:**69–81.

72. **Wilkinsen, H. T., R. D. Miller, and R. L. Millar.** 1981. Infiltration of fungal and bacterial propagules into soil. Soil Sci. Soc. Am. J. **45:**1034–1039.

73. **Woods, L. E., C. V. Cole, E. T. Elliott, R. V. Anderson, and D. C. Coleman.** 1982. Nitrogen transformations in soil as affected by bacterial-microfaunal interactions. Soil Biol. Biochem. **14:**93–98.

74. **Yeates, G. W.** 1979. Soil nematodes in terrestrial ecosystems. J. Nematol. **11:**213–229.

75. **Yeates, G. W.** 1981. Nematode populations in relation to soil environmental factors: a review. Pedobiologia **22:**312–338.

Comparative N and S Cycles

Oxygenic Photosynthesis, Anoxygenic Photosynthesis, and Sulfate Reduction in Cyanobacterial Mats

YEHUDA COHEN

Hebrew University of Jerusalem, Marine Biology Laboratory, Eilat, Israel

Cyanobacterial mats often occur in close association with a sulfide-rich environment (1). Sulfide either is produced biogenically by sulfate-reducing bacteria or it is supplied from abiogenic sources of geothermal activity. Cyanobacterial mats are presently found in marine lagoons and estuaries (e.g., Solar Lake, Sinai; Shark Bay, Western Australia; Baja California, Mexico; and along the shores of the Persian Gulf), in various inland lakes (e.g., Soda Lake, Ethiopia), and in hot springs (e.g., Yellowstone National Park, Wyoming) (3, 4, 15, 17, 23).

Oxygenic photosynthesis of the cyanobacteria in the immediate proximity of sulfide results in the establishment of a sharp redoxcline of less than 1 mm. Diffusion in these small distances is very fast, and the redoxcline migrates through the photosynthetic activity layer with diurnal changes in light intensity. The cyanobacteria are thus exposed to varying sulfide concentrations for various durations, depending on the microenvironmental conditions. The application of microelectrodes of oxygen, sulfide, and pH in the various cyanobacterial mats demonstrates well the fast turnover rates of sulfide and oxygen, resulting in drastic shifts of the redoxcline in the course of 1 day (10, 19).

Exposure to sulfide has been shown to be toxic to many oxygenic phototrophs (18). Yet sulfide was shown to be a suitable alternative electron donor for anoxygenic photosynthesis in several cyanobacteria (8, 9).

The comparison of microelectrode technique and axenic culture experiments reveals the different strategies among cyanobacteria to live in association with periodic exposure to sulfide, in relation to the degree of their exposure (Y. Cohen, *in* Y. Cohen, R. W. Castenholz, and H. O. Halvorson, ed., *Microbial Mats: Stromatolites*, in press).

The diurnal shift in oxygen and sulfide gradients, on the other hand, exposes the sulfate-reducing bacteria to periodic oxyginated conditions. Many sulfate-reducing bacteria have been shown to be strict anaerobes (16), yet a new technique for measuring in situ reduction revealed sulfate reduction activity under well-oxygenated microenvironments in the Solar Lake cyanobacterial mats. Sulfate reduction is very tightly coupled to primary production activity in the microbial mats of Solar Lake (Cohen, in preparation). The association of sulfate reduction with primary production activity has also been described by G. W. Skyring (*in* Cohen et al., ed., *Microbial Mats: Stromatolites*, in press) for cyanobacteria in various Australian mats.

The diurnal migration of the redoxcline within the euphotic zone of the cyanobacterial mats causes periodic release of Fe^{2+} ions from the pool of FeS which serve as a major sink for the sulfides produced by the sulfate-reducing bacteria. Several mat-forming cyanobacterial isolates were shown to be capable of carrying out efficient CO_2 photoassimilation using Fe^{2+} ions, which served as yet another alternative electron donor for photosynthesis in these strains.

MICROGRADIENTS OF O_2, H_2S, AND pH IN CYANOBACTERIAL MATS

Four different cyanobacterial communities were defined to develop under oxic water at Solar Lake. The "shallow flat mat" develops under few centimeters of water cover. With increasing depth of the water column, the "deep flat mat" develops under about 30 cm of water cover. In deep waters the "blister mat" and the "gelatinous mats" were found (12, 15).

The in situ use of microelectrodes for oxygen, sulfide, and pH in various microbial mats demonstrated the establishment of a sharp redoxcline of less than 1 mm as a result of oxygenic photosynthesis of cyanobacteria in the immediate proximity of the sulfide-rich microzone. The first measurements of microgradients of O_2, H_2S, and Eh in microbial mats were carried out in the flat shallow cyanobacterial mats at Solar Lake, Sinai (10). This mat, dominated by the cosmopolitan mat-forming cyanobacterium *Microcoleus-chtonoplastes*, is situated under 30 cm of oxic water column. Extreme diurnal fluctuation was found, with an O_2 peak of 0.5 mM at 1 to 2 mm depth below the mat surface during the day and an H_2S peak of 2.5 mM at 2 to 3 mm of depth at night. The O_2–H_2S interface migrates diurnally from the mat surface at night to 3 mm below surface at noon. The photic zone extends down to 2.5 mm below the surface, and H_2S peaked right below the photic zone. H_2S and O_2 were found to coexist at 2.5 mm depth over a depth interval of 0.2 to 1 mm, with a turnover rate of less than 1 min.

Recently, a more detailed study of the microstructure of the various microbial mats in the Solar Lake and their O_2, H_2S, and pH microprofiles was carried out (12). The dominant cyanobacterium in the flat and blister mats was *M. chtonoplastes* in association with filamentous flexibacteria, tentatively identified to be *Chloroflexus*-like organisms. Very similar associations were observed in cyanobacterial mats at Sabha Gavish, Southern Sinai, in microbial mats in Laguna Figueroa, Baja California, Mexico (Stolz, *in* Cohen et al., ed., *Microbial Mats: Stromatolites*, in press), and in various microbial mats in Shark Bay, Western Australia, and Spencer Gulf, South Australia (Bauld, *in* Cohen et al., ed., *Microbial Mats: Stromatolites*, in press). The flat mat from the most shallow part of Solar Lake has a photic zone of merely 0.8 mm with maximal photosynthetic activity of 50 μmol of O_2 cm^{-3} h^{-1} at 0.3 to 0.4 mm depth. In deeper water the mats are less compacted, and the photic zones extended down to 2.5, 4.5, and over 10 mm with increasing depth of overlying water. In all the different microbial mats examined a ΔpH of up to 2 pH units developed at the maximal photosynthetically active zone, causing rapid deposition of $CaCO_3$ in this microzone (20).

The microgradients of oxygen, sulfide, and pH in various hot springs were measured by N. P. Revsbech and D. M. Ward (*in* Cohen et al., ed., *Microbial Mats: Stromatolites*, in press) to exhibit an oxygen maximum of up to 600 μM sulfide 2 mm below. This mat is composed of the cyanobacteria *Oscillatoria terebriformis* and *Synechococcus* sp., together with *Chloroflexus* sp.

Similar profiles were measured at Octopus Spring, Yellowstone National Park, where a very thin layer of *Synechococcus* sp. of only 0.1 mm is responsible for maximal oxygenic photosynthesis of 150 mmol of O_2 dm^{-3} h^{-1}. The zone showing oxygenic photosynthetic activity was in this case only 0.5 mm thick, resulting in total oxygenic photosynthesis of 47 mmol of O_2 m^{-2} h^{-1}.

At Stinky Springs, Utah, where sulfide concentration at the source is 1,100 μM, only traces of O_2 were detected in the photosynthetic active zone containing *Oscillatoria* sp. together with *Chloroflexus*-like organisms (Cohen, Jørgensen, and Revsbech, in preparation).

In the hypolimnic flocculous mat at Solar Lake, Sinai, where *Oscillatoria limnetica* is the dominant cyanobacterium under 3 mM sulfide (14, 15), no O_2 could be detected at the photosynthetically active layer.

All cyanobacterial mats examined exhibit sharp microgradients of sulfide and oxygen which fluctuate diurnally, exposing the cyanobacteria and the sulfate-reducing bacteria to conditions alternating between oxygenated waters and high sulfide concentrations.

OXYGENIC AND ANOXYGENIC PHOTOSYNTHESIS IN MAT-FORMING CYANOBACTERIAL ISOLATES

Axenic cultures of cyanobacteria isolated from various biotopes exposed to varying in situ sulfide concentrations were examined for their capacity to conduct oxygenic or anoxygenic photosynthesis, or both, under a range of sulfide concentrations. $^{14}CO_2$ photoassimilation was measured together with the rate of sulfide oxidation in the presence and absence of a 5 μM concentration of the photosystem II inhibitor DCMU [3-(3,4-dichlorophenyl)-1,1-dimethyl urea] at pH 7.2, for a period of 6 h (Fig. 1). The $^{14}CO_2$ photoassimilation is expressed as a percentage of the rate of CO_2 photoassimilation in the absence of sulfide and DCMU.

Four different strategies of photosynthetic life under varying degrees of exposure to sulfide can be detected.

(i) The first strategy is irreversible cessation of CO_2 photoassimilation upon brief exposure to sulfide. *Anacystis nidulans* isolated from planktonic blooms in a freshwater body without apparent exposure to sulfide is highly sensitive to sulfide toxicity. Exposure to 100 μM sulfide causes 50% inhibition of CO_2 photoassimilation, and at 200 μM sulfide it is blocked completely. When sulfide was removed after 2 h of exposure no regenerated photosynthetic activity could be detected. Similar sensitivity for sulfide was shown by Schwabe (21) for high temperature form *Mastigocladus*, the thermophilic cosmopolitan cyanobacterium found under very low sulfide concentrations of up to about 2 μM in various hot springs in Iceland, New Zealand, and the United States (3, 5, 6, 21).

(ii) A second response is enhancement of oxygenic photosynthesis upon exposure to sulfide and incapability of utilization of H_2S as a donor for anoxygenic photosynthesis. This type of photosynthesis is represented by *Oscillatoria* sp. isolated by R. W. Castenholz (Eugene, Oregon) from Wilbur Hot Springs, California, under 6 μM sulfide. This isolate shows 450% enhancement of oxygenic photosynthesis upon exposure to 800 μM sulfide. At higher sulfide concentrations, CO_2 photoassimilation was gradually in-

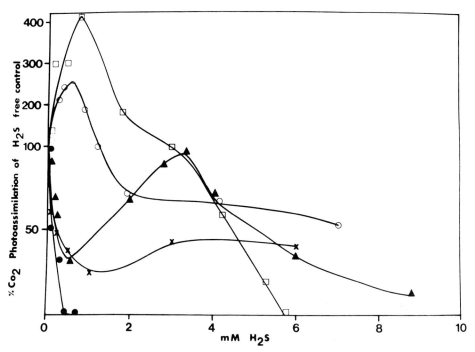

FIG. 1. CO_2 photoassimilation as a function of sulfide concentration in various strains representing the different types of photosynthesis among cyanobacteria. Results are expressed in percentage of CO_2 photoassimilation of a sulfide-free control. (●) Type 1: *Anacystis nidulans*. (□) Type 2: *Oscillatoria* sp. (Wilbur Hot Springs). (○) Type 3: *Microcoleus chtonoplastes* (Solar Lake, Sabkha Gavish, and Laguna Figueroa strains). (▲) Type 4: *Oscillatoria limnetica* (Solar Lake). (X) *Oscillatoria* sp. (Stinky Springs).

hibited, yet no anoxygenic photosynthesis could be detected. Similar photosynthetic activity was reported for *Phormidium* sp. from Yellowstone National Park by Weller et al. (24).

(iii) The third strategy is characterized by enhancement of oxygenic photosynthesis at low sulfide concentration, inhibition of photosystem II at higher sulfide concentration, and a concomitant induction of anoxygenic photosynthesis operating in concert with the partially inhibited oxygenic photosynthesis under high sulfide concentration. *M. chtonoplastes*, representing this type of photosynthesis, is a cosmopolitan mat-forming cyanobacterium in hypersaline habitats. The same photosynthetic activity was shown by isolates from Solar Lake, Sinai; Sabkha Gavish, Sinai; and Laguna Figueroa, Baja California, Mexico. The ultrastructures of the cyanobacterial mats from Solar Lake, Sinai, and Laguna Figueroa, Mexico (Stolz, in press), are extremely similar. This cyanobacterium forms flat mats also in Shark Bay and Spencer Gulf, Australia (2; Bauld, in press), as well as along the Persian Gulf (23). Similar photosynthetic activity was described by H. Utkilen and R. W. Castenholz (Abstr. 3rd Int. Symp. Photosynthetic Prokaryotes, p. 12, 1979) for *Oscillatoria amphigranu-*

lata isolated from alkaline sulfur hot springs of New Zealand under up to 2.2 mM sulfide.

(iv) The fourth response is complete reversible inhibition of photosystem II at low sulfide concentration and induction of efficient anoxygenic photosynthesis at higher sulfide concentration. This type of photosynthesis was initially described by Cohen et al. (7) for *Oscillatoria limnetica* and later for various other cyanobacteria by Garlick et al. (9). Oren et al. (18) demonstrated that, unlike types 2 and 3, here photosystem II is completely blocked by 100 µM sulfide. When the cyanobacteria are exposed to a high sulfide concentration, an induction period of about 2 h is needed for anoxygenic, photosystem I-dependent photosynthesis to be fully induced.

The equivalent type of photosynthesis was demonstrated in *Oscillatoria* sp. from Stinky Springs, Utah, isolated under 1,100 µM sulfide at pH 6.3.

Cyanobacteria exhibit a great degree of flexibility in their photosynthesis in relation to exposures to various sulfide concentrations. The four types of photosynthesis described above represent various degrees of adaptation to exposures to sulfide. Increasing exposure to higher sulfide concentrations in various habitats results in the dominance of cyanobacterial strains which can better utilize sulfide and hence gradually shift from type 1 to type 4.

Generally, type 1 are either not exposed at all to sulfide or are exposed to vanishingly low concentrations. Type 2 are found among cyanobacteria exposed up to about 100 μM sulfide. No efficient anoxygenic photosynthesis could be demonstrated at these low sulfide concentrations, and the cyanobacterial photosynthesis is enhanced because of the low redox potential which allows protection from photooxidation and increased efficiency of CO_2 photoassimilation. Type 3 or 4 will develop only upon exposure to higher sulfide concentrations, allowing an efficient anoxygenic photosynthesis. The toxicity of sulfide is pH dependent (Cohen, in press). Sulfide is more toxic at lower pH since H_2S penetrates passively to the cell, while HS^- and S^{2-} ions need active transport mechanisms. Thus, *Oscillatoria amphigranulata* occurring at 2.2 mM sulfide at alkaline pH is a type 3 organism, whereas *Oscillatoria* sp. from Stinky Springs, Utah, developing at 1.1 mM at pH 6.3 is type 4. Photosystem II of the various types of cyanobacteria shows different sensitivities to exposure to sulfide. Exposure to 50 μM sulfide at pH 7.5 induces maximal steady-state fluorescence of photosystem II in *Anacystis nidulans* (type 1) and *Oscillatoria limnetica* (type 4). Addition of 10^{-4} M DCMU does not further enhance photosystem II fluorescence. Yet, in *Oscillatoria* sp. from Wilbor Spring, California (type 2), photosystem II fluorescence is unaffected even in the presence of 10 mM sulfide at pH 7.5. Addition of 10^{-4} M DCMU highly enhances photosystem II fluorescence. Similar results are obtained in cultures of *Synechococcus lividus* from Yellowstone National Park. *Oscillatoria* sp. from Stinky Springs, Utah, and the four isolates of *Microcoleus chtonoplastes* from Solar Lake, Sinai, Sabkha Gavish, Sinai, Baja California, Mexico, and Spencer Gulf, South Australia, all show partial photosystem II inhibition upon exposure to increasing sulfide concentrations. In all these type 3 cyanobacterial isolates photosystem II is fully inhibited only at >2 mM sulfide at pH 7.5. The photosynthetic strategies of the cyanobacterial strains examined are summarized in Table 1.

The ability of the various mat communities to produce and accumulate oxygen under increasing sulfide concentration and the efficiency of recovery of oxygenic photosynthesis upon gradual removal of sulfide was measured by introducing microelectrodes of O_2, S^{2-}, and pH in small blocks of various cyanobacterial mats suspended in sulfide-containing media. All cyanobacteria of the types 2, 3, and 4 are capable of

accumulating O_2 under H_2S. *Microcoleus chtonoplastes* is more efficient and was found to be able to accumulate O_2 at 250 μM sulfide at neutral pH, whereas *Oscillatoria limnetica* accommodates O_2 only under 50 μM sulfide.

Many if not all cyanobacteria of type 3, and some of type 4, occur in close association with *Chloroflexus* or *Chloroflexus*-like organisms. This tight association may enable the community to cope more efficiently with fast, wide oscillations between high sulfide and high oxygen concentrations. The nature of the association is currently being examined.

The various types of photosynthesis are found both in hot sulfur springs and in sulfureta. Sulfur springs present a good model for the development of the various types of photosynthesis among cyanobacteria along the sulfide gradient downstream. The degree of exposure to sulfide in sulfureta, on the other hand, depends on the coupling of primary production and sulfate reduction in these biotopes. The importance of sulfate reduction in the degradation of the cyanobacterial mat was demonstrated by Jørgensen and Cohen (11) and Skyring (in press).

COUPLING OF PRIMARY PRODUCTION AND SULFATE REDUCTION IN MICROBIAL MATS

To assess the coupling of the two processes, I developed a new method for SO_4 reduction (in preparation). Silver wire of 0.15 mm in diameter is coated with a thin layer of $Na_2{}^{35}SO_4$ with high specific activity of >60 mCi mmol^{-1}. The silver wires are introduced into the mat by means of a micromanipulator alongside the microelectrodes of oxygen, sulfide, and pH. The mat is incubated either in the light or in the dark for periods of 10 min to 2 h. After preincubation, the silver wires are pulled out and washed twice to remove excess $^{35}SO_4$, and the silver wire is placed on photographic paper for autoradiography. Then the 1-cm long wire is cut in 1-mm segments, and the radioactivity of $Ag^{35}S$ trapped on the silver wire is counted in a scintillation counter. This method presents several advantages over the conventional method of injection of $Na_2{}^{35}SO_4$. The affinity of the produced sulfide to silver is higher than its oxidation in the presence of oxygen, and can thus be detected, whereas in the previous method it is immediately oxidized chemically or biologically by sulfide-oxidizing bacteria. The new method also allows a concomitant analysis of the microenvironment of sulfate reduction.

Preliminary results indicate that (i) sulfate reduction occurs well within the highly oxygenated photosynthetic active zone and (ii) sulfate reduction is higher after short exposures to light, indicating that this activity may depend on pho-

TABLE 1. Summary of photosynthetic activities among cyanobacterial isolates examined for their activities under sulfide

Cyanobacterial strain	Origin	Isolated by	Maximal in situ sulfide concn (mM)	In situ pH	Sulfide concn[a] (mM) for maximal PS II steady-state fluorescence	Photosynthetic type	Reference	Remarks
Anacystis nidulans	Freshwater reservoir	?	0	10	0.05	1	This study	
Mastigocladus	Hot Springs, Iceland	G. H. Schwabe	0.002	6–10	?	1	Schwabe (21)	Not examined by author
Oscillatoria sp.	Wilbor Hot Springs, California	R. W. Castenholz	0.1	8	>10	2	This study	
Phormidium sp.	Yellowstone National Park, Wyoming	?	0.005	7–9	?	2	Weller et al. (24)	Not examined by author
Synechococcus lividus	Yellowstone National Park, Wyoming	R. W. Castenholz	0.1	7–8	>10	2	This study	
Microcoleus chtonoplastes	Solar Lake, Sinai	Y. Cohen	2	8–9.5	2	3	This study	
M. chtonoplastes	Laguna Figeroa, Baja California, Mexico	Y. Cohen	2.5	8–9.8	1.6	3	This study	
M. chtonoplastes	Sabha Gavish, Sinai	E. Holtkamp	3.5	8–10	1.5	3	This study	
M. chtonoplastes	Spencer Gulf, South Australia	J. Bauld, purified by Y. Cohen	2	8–9	1.8	3	This study	
Oscillatoria amphigranulata	Hunter Hot Spring, Oregon	R. W. Castenholz	2.2	9–10	?	3	Utkilen and Castenholz[b]	Not examined by author
Aphanothece halophitica	Solar Lake, Sinai	Y. Cohen	2	7.5	0.05	4	This study	
Oscillatoria sp.	Stinky Hot Springs, Utah	Y. Cohen	1.1	6.3	0.05	4	This study	
O. limnetica	Solar Lake, Sinai	Y. Cohen	3.5	6.8	0.05	4	This study	

[a] Photosystem II (PS II) steady-state fluorescence excited at 661 ± 10 nm in 50 mM N-2-hydroxyethylpiperazine-N'-2-ethanesulfonic acid (HEPES), pH 7.5.

[b] Abstr. 3rd Int. Symp. Photosynthetic Prokaryotes, 1979.

tosynthates excreted by the phototrophic community.

The mechanism of allowing the anaerobic metabolism of sulfate reduction to occur under high oxygen tension in the cyanobacterial mats is currently being studied. Production of hydrogen simultaneously with oxygen by the cyanobacteria under high pH and possible temporal CO_2 limitation may be a possible mechanism for activity of sulfate reduction well within the cyanobacterial mat.

Fe^{2+}-DEPENDENT PHOTOSYNTHESIS IN BENTHIC CYANOBACTERIA

The diurnal migration of the redoxcline through the photosynthetic active layer assures the release of Fe^{2+} from FeS. High concentrations of Fe^{2+} were observed at the upper 2 mm of the Solar Lake flat mat in the morning hours.

Fe^{2+} is a good potential electron donor for photosynthesis in cyanobacteria. Banded microlayers of iron oxides are presently being formed in the cyanobacterial mats in Spencer Gulf, South Australia (Bauld, in press). Iron-dependent photosynthesis was speculated to be responsible for the deposition of Banded Iron Formation of the Precambrian (H. Hartman, *in* Cohen et al., ed., *Microbial Mats: Stromatolites*, in press).

Several cyanobacteria were examined for the use of Fe^{2+} in photosynthesis. *Oscillatoria limnetica* and *Oscillatoria* sp. from Stinky Springs, Utah, were shown to be capable of iron-dependent CO_2 photoassimilation at pH 5.5 in the presence of 10 mM thioglycolate, serving as a mild reducing agent. Thioglycolate alone does not serve as an electron donor to cyanobacterial photosynthesis. Maximal rate of CO_2 photoassimilation in the presence of 5 μM DCMU or at actinic light of 705 nm, allowing the operation of photosystem I alone, was found to be up to 50% of the oxygenic photosynthesis rate (Cohen and Gack, in preparation).

Fe^{2+}-dependent photosynthesis may take place only under sharp gradients of sulfide and oxygen in very close proximity to the photosynthetic active zone. The periodic partial release of Fe^{2+} due to partial oxidation of FeS during the migration of the microgradient through the photosynthetic active layer causes the release of Fe^{2+} right above the O_2–H_2S interface. Thus, both sulfate reduction activity and the oxygenic and sulfide-related anoxygenic photosynthesis control the possible operation of Fe^{2+}-dependent CO_2 photoassimilation in cyanobacterial mats.

The ecological significance of ferrous ion-dependent photosynthesis is currently being studied. It bears an important role in the understanding of the evolution of life in the Precambri-an. Banded Iron Formations were thought to be a result of the oxidation of ferrous ion by the primordial Precambrian oxygen. Ferrous ion-dependent photosynthesis may be another possible mechanism which does not necessarily involve oxygen in the deposition of the Banded Iron Formation which was taken as a geological proof for the accumulation of oxygen at the Precambrian (13).

The photosynthetic flexibility of cyanobacteria (22), their tight association with sulfate reduction, and the use of Fe^{2+} by certain cyanobacteria further indicate the antiquity of this group of microorganisms and allow a better understanding of their ecological role.

LITERATURE CITED

1. **Baas Becking, L. G. M., and E. J. F. Wood.** 1955. Biological processes in the estuarine environment. I-II. Ecology of the sulfur cycle. Proc. K. Ned. Akad. Wet. Sect. B **58:**160–181.
2. **Bauld, J., L. A. Chambers, and G. W. Skyring.** 1979. Primary productivity, sulfate reduction and sulfur isotope fractionation in algal mats and sediments of Hamelin Pool, Shark Bay, W.A. Aust. J. Mar. Freshwater Res. **30:**753–764.
3. **Brock, T. D.** 1978. Thermophilic microorganisms and life at high temperatures. Springer-Verlag, New York.
4. **Castenholz, R. W.** 1973. The possible photosynthetic use of sulfide by the filamentous phototrophic bacteria of hot springs. Limnol. Oceanogr. **18:**863–876.
5. **Castenholz, R. W.** 1976. The effect of sulfide on the blue green algae of hot springs. I. New Zealand and Iceland. J. Phycol. **12:**57–68.
6. **Castenholz, R. W.** 1977. The effect of sulfide on blue green algae of hot springs. II. Yellowstone National Park. Microb. Ecol. **3:**79–105.
7. **Cohen, Y., B. B. Jørgensen, E. Padan, and M. Shilo.** 1975. Sulphide-dependent anoxygenic photosynthesis in the cyanobacterium *Oscillatoria limnetica*. Nature (London) **257:**489–491.
8. **Cohen, Y., E. Padan, and M. Shilo.** 1975. Facultative anoxygenic photosynthesis in the cyanobacterium *Oscillatoria limnetica*. J. Bacteriol. **123:**855–861.
9. **Garlick, S., A. Oren, and E. Padan.** 1977. Occurrence of facultative anoxygenic photosynthesis among filamentous and unicellular cyanobacteria. J. Bacteriol. **129:**623–629.
10. **Jørgensen, B. B., N. P. Revsbech, T. H. Blackburn, and Y. Cohen.** 1979. Diurnal cycle of oxygen and sulfide microgradients and microbial photosynthesis in a cyanobacterial mat sediment. Appl. Environ. Microbiol. **38:**46–58.
11. **Jørgensen, B. B., and Y. Cohen.** 1977. Solar Lake (Sinai). 5. The sulfur cycle of benthic cyanobacterial mats. Limnol. Oceanogr. **22:**657–666.
12. **Jørgensen, B. B., N. P. Revsbech, and Y. Cohen.** 1983. Photosynthesis and structure of benthic microbial mats: microelectrode and SEM studies of four cyanobacterial communities. Limnol. Oceanogr. **28:**1075–1093.
13. **Knoll, A. H.** 1979. Archaecan photoautotrophy: some alternatives and limits. Origins Life **9:**313–327.
14. **Krumbein, W. E., and Y. Cohen.** 1977. Primary production, mat formation and litification: contribution of oxygenic and facultative anoxygenic cyanobacteria, p. 37–56. *In* E. Flugel (ed.), Fossil algae. Springer Verlag, New York.
15. **Krumbein, W. E., Y. Cohen, and M. Shilo.** 1977. Solar Lake (Sinai) 4. Stromatolitic cyanobacterial mats. Limnol. Oceanogr. **22:**635–656.
16. **LeGall, J., and J. R. Postgate.** 1973. The physiology of

sulfate reducing bacteria, p. 81–133. *In* H. A. Rose and D. W. Tempest (ed.), Methods in microbiology, vol. 3A. Academic Press, London.

17. **Margulis, L., D. Ashendorf, S. Banerjee, S. Francis, S. Giovannoni, J. Stolz, E. S. Barghoorn, and O. Chase.** 1980. The microbial community in the layered sediment at Laguna Figueroa, Baja California, Mexico: does it have Precambrian analogues? Precambrian Res. **11**:93–123.

18. **Oren, A., E. Padan, and S. Malkin.** 1979. Sulfide inhibition of photosystem 2 in cyanobacteria (blue green algae) and tobacco chloroplasts. Biochim. Biophys. Acta **576**:270–279.

19. **Padan, E., and Y. Cohen.** 1982. Anoxygenic photosynthesis, p. 215–235. *In* N. G. Carr and B. A. Whitton (ed.), The biology of cyanobacteria. Blackwell Scientific Publications, Oxford.

20. **Revsbech, N. P., B. B. Jørgensen, T. H. Blackburn, and Y. Cohen.** 1983. Microelectrode studies of photosynthesis and O_2, H_2S and pH profiles of a microbial mat. Limnol. Oceanogr. **28**:1062–1074.

21. **Schwabe, G. H.** 1960. Uber den thermobioten Kosmopoliten *mastigocladus laminosus* Cohn. Blaugen und lebensraum. V. Schweig. J. Hydrol. **22**:757–792.

22. **Stanier, R. Y.** 1977. The position of the cyanobacteria in the world of phototrophs. Carlsberg Res. Commun. **42**:77–98.

23. **Walter, M. R. (ed.).** 1976. Developments in sedimentology 20, Stromatolites. Elsevier, Amsterdam.

24. **Weller, D., W. Doemel, and T. D. Brock.** 1975. Requirement of low oxidation-reduction potential for photosynthesis in a blue green alga (*Phormidium* sp.). Arch. Microbiol. **104**:7–13.

Interrelation of Carbon, Nitrogen, Sulfur, and Phosphorus Cycles During Decomposition Processes in Soil

JOHN W. B. STEWART

Saskatchewan Institute of Pedology, University of Saskatchewan, Saskatoon, Saskatchewan, Canada S7N 0W0

Soil is a fundamental resource. In order to use it wisely we must understand its inner workings, the processes that keep it dynamic. The strategy is reductionism, the reformation of complex notion and events into simpler, more basic ones. The task makes use of concepts of forces and potential and of atoms, molecules, colloids, enzymes and organisms, the whole armory of modern science. (18)

During the development of soils, organic matter accumulates as the result of the formation of biomass and organic detritus. Many ecosystems accumulate steady-state levels of soil organic matter which are determined by the balance of production of biomass, stabilization of detritus, and mineralization (respiration) of organic materials. When this balance is disturbed by introduction of agriculture or changes in cropping systems, the balance is altered, and organic matter contents are changed (11, 12, 31, 32). Accumulation of organic matter during soil development and rate of release of C, N, P, and S are closely dependent on the availability of nutrients (6, 15). Soil development affects nutrient availability and organic matter stability and hence, to some extent, the quality of soil organic matter.

Because of the central role of microbial biomass in conjunction with pedogenesis in determining the cycling of C, N, S, and P, there has been a tendency to use the stoichiometry of C, N, S, and P in organisms in seawater (25) to explain similar cycling processes within the soil system. However, three main differences between soil systems and marine systems lead to quite pronounced contrasts. The first relates to differences between amounts of C in structural, metabolic, and storage compounds entering soil from plants compared with sedimenting marine material. Second, the soil atmosphere is generally aerobic versus the anaerobic conditions in marine environments. This causes marked differences in the reactivity of soil inorganic matrices and colloidal surfaces, particularly those coated with iron and aluminum compounds. Organic compounds containing phosphate and sulfate interact strongly with sesquioxides to form stable compounds. Also, P is a component of many parent materials and exists as a variety of inorganic compounds in soil, depending on the extent of weathering (28). Total soil P must be divided into inorganic and organic forms which differ in relative stability or lability in biological terms. Third, only a portion of organic C in soils is active, and therefore total soil C will not necessarily be in synchrony with the cycling of N, S, and P. Similarly, cycling of P will be out of synchrony with C, N, and S because of the relative unavailability to plants and microorganisms of much of the inorganic soil P. All cycles interact during plant growth and uptake of nutrients and during decomposition processes in soil. Information on interactions between cycles is limited although individual cycling of N (23, 24), P (4, 5, 7, 14, 30), and S (10, 26) in relation to C flows has received more attention.

Obviously, no simple elemental ratio can exist between C, N, S, and P in soils, but general relationships do exist within a logical framework. The interrelationship of C, N, P, and S in soil decomposition processes may be studied from two angles (29). The first is by careful examination of soils formed over long time periods under varying climates and vegetation and from parent materials of different chemical composition. The second is by study of seasonal dynamics or through short-term laboratory experiments under controlled environments in which processes are examined and quantified. The first approach assumes that in the natural undisturbed system a "steady state" of balancing processes of organic matter production and nutrient accumulation has been reached and represents the net effect of all processes documented in the second approach. Although both approaches have been used in this review to examine interrelationships of C, N, S, and P cycling processes and resulting soil composition, the main emphasis is placed on decomposition processes.

FACTORS AFFECTING C, N, S, AND P IN SOILS

Parent material, weathering, climate, and cultivation. Individual parent materials weather at different rates in the same environmental conditions, influencing the availability of nutrients, the soil pH, and the nature of the colloidal surfaces (8). This leads to differences in organic matter composition (Table 1). Whereas C/N and C/S ratios showed similar trends (Table 1) in Scottish soils developed on different parent materials, C/P ratios did not follow the same trend. Conversely, in a study of weathering effects in New Zealand soils (Table 1), C/S ratios were found to remain relatively constant whereas C/N and C/P ratios increased (data on C/P varied enormously as can be seen from the large standard deviation of the means). These results suggest a random accumulation and composition of soil organic matter. However, more intensive work by Walker and co-workers (33, 34) promoted the hypothesis that accumulation of C, N, S, and P in soil organic matter was dependent on the P content of the soil parent material. This hypothesis was later refined (Fig. 1) to emphasize that the availability of soil P, as determined by change in inorganic form with weathering intensity, and not total P content, would influence organic matter accumulation and composition (6, 35). This hypothesis is borne out by common farming practice in New Zealand and Australia (9, 36), where application of P fertilizer increases N fertility of improved grasslands by promoting establishment and growth of legume species. For instance, Walker et al. (36) (Table 2) documented increases of 120 kg of C, 120 kg of N, 16 kg of S, and 9 kg of organic P per ha per year for each of 25 years following application of P fertilizer. Similarly, changes in organic matter composition with cultivation (2, 31, 32) emphasize that C, N, S, and P losses are not in proportion to total amounts present but depend on interactive cycling processes which are site specific.

In initial stages of soil formation, organic matter accumulation is controlled by the accessibility and lability of the inorganic P (unless S is limited) (6). The lability of soil inorganic P is changed by soil weathering processes acting on P minerals as well as Al, Fe, and Ca compounds, and by environmental factors such as leaching and erosion (28). At more advanced stages of soil formation, recycling of organic P by the biomass becomes increasingly important (6, 35). During each cycle, C and N gain or loss by assimilation/fixation or respiration/denitrification processes is possible, whereas P is largely conserved in the system in unchanged total quantities. The rates of P cycling at this stage will determine the rates of C and N changes, whereas the direction of such changes and therefore C to N to organic P ratios are governed by the balance of autotrophic/heterotrophic activity, by the stabilization of soil organic matter by clays, sesquioxides, etc., and by chemical, biological, and physical factors. Sulfur accumula-

TABLE 1. Carbon, nitrogen, and sulfur ratios of world soil types

(a) Effect of parent material (37)

Country	Parent material	C/N/S	C/N	C/P
Scotland	Granite	117:6.1:1	16.9	68
	Slate	104:7.0:1	14.8	63
	Basic igneous	102:7.3:1	14.0	47
	Old red stone	95:7.3:1	13.0	55
	Calcareous	89:7.9:1	11.3	83

(b) Effect of weathering (33)

Country	Weathering stage	C/N/S	C/N	C/P
New Zealand	Weakly weathered	156:9.1:1	17.2	80 ± 38
	Moderately weathered	150:8.3:1	18.0	110 ± 72
	Strongly weathered	158:7.7:1	20.6	205 ± 99

(c) Effect of climate and cultivation

Country	Soil type	C/N/S	C/N
Canada (Western)	Virgin soils (19, 20)		
	Brown chernozems	85:7.1:1	12.0
	Black chernozems	96:7.7:1	12.5
	Gray wooded	271:12.5:1	21.7
	Gleysols	78:5.0:1	15.6
	Cultivated soils (3)		
	Brown chernozems	58:6.4:1	9.1
	Dark brown chernozems	63:6.9:1	9.1
	Black chernozems	83:7.6:1	10.9
	Gray black transitional	100:8.1:1	12.3
	Gray wooded	129:10.6:1	12.2

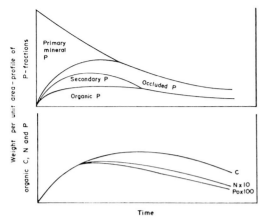

FIG. 1. Changes in form and amounts of soil P and soil organic constituents with time (34).

tion or loss is not directly linked to C and N dynamics but can follow similar trends under specified environmental conditions.

Comparing C, N, S, and P cycling through organic matter during pedogenesis led McGill and Cole (22) to propose that element cycling can best be interpreted within the framework of a dichotomous system. In this system, elements stabilized as a result of covalent bonding with C (C—N and C—S) are released as a result of C oxidation to provide energy, whereas those elements existing as esters (C—O—S and C—O—P) are stabilized through reaction of the ester with soil components. These elements may be mineralized by the organism's need for a specific element.

INFORMATION ON INTERRELATIONSHIPS OBTAINED FROM PROCESS STUDIES

Simulation modeling has been a useful tool for understanding element cycling (5, 7, 16, 23) in specific systems and in organizing concepts obtained from process studies (14). A first step to organizing data on interrelationships among elements is to develop a conceptual model (17) which defines the general structure (state variables) of the system and describes the process involved. In this approach (17), two concepts become important, namely, variability in element ratios of organisms, and recognition of bond classes in soil organic matter and their different reactions with inorganic soil colloidal material.

From studies of microbial biomass, activity, and varying elemental composition (1, 13, 24, 27), it becomes clear that bacteria and fungi differ in forms of N, S, and P immobilized, and this alters subsequent redistribution in the soil. Typical elemental ratios of soil biomass range widely (C/N from 3:1 to 20:1, C/P from 10:1 to

150:1, and C/S from 47:1 to 494:1, with the percentage of total S in C—O—S form varying from 6 to 42%). Organisms differ in their ability to store P and S in different inorganic and organic compounds (1, 27, 30). Perhaps the clearest distinction is with S (Fig. 2). Both bacteria and fungi accumulate S, and at low concentrations, most (>80%) S is converted to amino acids (C—S bonds). At higher solution concentration, bacteria do not accumulate excess S, whereas some fungi have the ability to store excess S as organic sulfate compounds (C—O—S bonds) (27). Hence, S may be immobilized and converted to soil organic matter where it exists in C—S and C—O—S forms. Upon decomposition both forms are mineralized as a result of C oxidation, but the C—O—S form will also be subject to enzymatic hydrolysis, especially when there is a shortage of sulfate in soil solution (for instance, as a result of leaching or of plant and microbial uptake of S) (21, 26). If solution sulfate is adequate, the rate of mineralization of C—O—S forms is curtailed (2, 21, 26). Similarly, the rate of mineralization of organic P esters is much more rapid when solution phosphate is low (4, 5, 14). In effect, immobilization of organic P depends on availability of inorganic P and of organic substrates (including adequate N and S) for biomass growth and maintenance. Enzymatic hydrolysis of organic P is controlled by solution P concentration. Solution P concentration is directly related to inorganic P forms and availability. When available inorganic P is depleted by P demands of biomass and crops, it can be replenished by decomposition hydrolysis of organic P forms (14). Release of P from microbial cells is less well understood but has been shown to take place by a variety of mechanisms. These include abiotic effects such as freezing and thawing, and biological effects such as amoeba and nematode grazing on bacteria and fungi (7, 16, 30).

Further concepts that must be worked into the conceptual framework are the mechanisms of

TABLE 2. Effect of 25 years of superphosphate[a] application on the C, N, S, and P content of organic matter in improved grassland soils (36)

Soil condition	Content (kg ha⁻¹) of:				C/N	C/S	C/P
	C	N	S	P			
Original soil	60,000	2,038	291	341	29	206	191
After 25 years of P fertilizer	63,000	4,995	683	526	13	92	119
Difference	3,000	2,957	392	213			

[a] Superphosphate used contained 12% S.

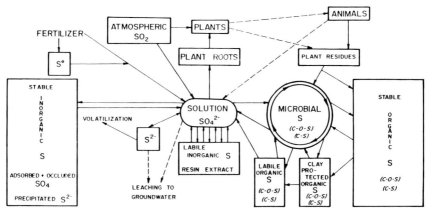

FIG. 2. Conceptual flow diagram of the main forms and transformations of sulfur in the soil–plant system.

stabilization of organic matter and the rate of disintegration of incoming plant biomass to particulates that can undergo biochemical/biological transformation or direct stabilization. Soil mineralogy may have to be included as a control variable to account for the widely differing kinds of organic matter stabilization on different soil types.

CONCLUSIONS

Hypotheses on C, N, S, and P interactions during decomposition processes in soil have been developed. These concepts, when integrated with environmental factors, give a logical framework to observed changes in soil organic matter quality across environmental gradients and in chronology of cultivation sequences. Key concepts relate to (i) microbial ability to store P and S in conditions of plentiful supply, (ii) nature of bond classes, (iii) formation of stable organo-mineral complexes, especially by reactive P and S groups, (iv) losses of C and N in greater quantities than P and S when a system at equilibrium is disturbed by cultivation processes, and (v) the possibility of biochemical mineralization supplementing biological oxidation when P and S are in short supply. These concepts need integration with knowledge on spatial heterogeneity of soils and with information on soil development. The amount and nature of C distribution in soils with time must also be known for a more complete understanding of the dynamics of soil organic matter and the interrelationship of C, N, S, and P. Simulation modeling is an important aid in integrating these processes.

I would like to acknowledge the help given by C. V. Cole, Agriculture Research Service, U.S. Department of Agriculture, Fort Collins, Colo., and E. A. Paul, University of California, Berkeley, in the development of these concepts.

This is Paper No. R351 of the Saskatchewan Institute of Technology.

LITERATURE CITED

1. **Beever, R. E., and D. J. W. Burns.** 1980. Phosphorus uptake, storage and utilization by fungi. Adv. Bot. Res. **8:**127–219.
2. **Bettany, J. R., S. Saggar, and J. W. B. Stewart.** 1980. Comparison of the amounts and forms of sulfur in soil organic matter fractions after 65 years of cultivation. Soil Sci. Soc. Am. J. **44:**70–75.
3. **Bettany, J. R., J. W. B. Stewart, and E. H. Halstead.** 1973. Sulfur fractions and carbon, nitrogen, and sulfur relationships in grassland, forest and associated transitional soils. Soil Sci. Soc. Am. Proc. **37:**915–918.
4. **Chauhan, B. S., J. W. B. Stewart, and E. A. Paul.** 1979. Effect of carbon additions on soil labile inorganic, organic and microbially held phosphate. Can. J. Soil Sci. **59:**387–396.
5. **Chauhan, B. S., J. W. B. Stewart, and E. A. Paul.** 1981. Effect of labile inorganic phosphate status and organic carbon additions on the microbial uptake of phosphate in soils. Can. J. Soil Sci. **61:**373–385.
6. **Cole, C. V., and R. D. Heil.** 1981. Phosphorus effects on terrestrial nitrogen cycling. Ecol. Bull. (Stockholm) **33:**363–374.
7. **Cole, C. V., G. I. Innis, and J. W. B. Stewart.** 1977. Simulation of phosphorus cycling in semi-arid grasslands. Ecology **58:**1–15.
8. **Dixon, J. B., and S. B. Weed (ed.).** 1977. Minerals in soil environments. Soil Science Society of America, Madison.
9. **Donald, C. M., and C. H. Williams.** 1954. Fertility and productivity of a podzolic soil as influenced by subterranean clover (*Trifolium subterraneum* L) and superphosphate. Aust. J. Agric. Res. **5:**664–687.
10. **Freney, J. R., and C. H. Williams.** 1983. The sulphur cycle in soil, p. 129–201. *In* M. V. Ivanov and J. R. Freney (ed.), The global biochemical sulphur cycle. SCOPE Report no. 19. John Wiley & Sons, Inc., New York.
11. **Haas, H. J., C. E. Evans, and E. F. Miles.** 1957. Nitrogen and carbon changes in Great Plains soils as influenced by cropping and soil treatments, U.S.D.A. Bull. 1164. U.S. Department of Agriculture, Washington, D.C.
12. **Haas, H. J., H. J. Grunes, and G. A. Reichman.** 1961. Phosphorus changes in Great Plains soils as influenced by cropping and manure application. Soil Sci. Soc. Am. Proc. **25:**214–218.
13. **Hedley, M. J., and J. W. B. Stewart.** 1982. Method to measure microbial phosphate in soils. Soil Biol. Biochem. **14:**377–385.
14. **Hedley, M. J., J. W. B. Stewart, and B. S. Chauhan.** 1982. Changes in inorganic and organic soil phosphorus

fractions induced by cultivation practices and by laboratory incubations. Soil Sci. Soc. Am. J. **46**:970–976.

15. **Hingston, F. J., and R. J. Raison.** 1982. Consequences of biogeochemical interactions for adjustment to changing land use patterns in forest systems, p. 11–24. *In* J. R. Freney and I. E. Galbally (ed.), Cycling of carbon, nitrogen, sulfur and phosphorus in terrestrial and aquatic ecosystems. Springer-Verlag, New York.

16. **Hunt, H. W., C. V. Cole, D. A. Klein, and D. C. Coleman.** 1977. A simulation model for the effect of predation on bacteria in continuous culture. Microb. Ecol. **3**:259–278.

17. **Hunt, H. W., J. W. B. Stewart, and C. V. Cole.** 1983. A conceptual model for interaction among carbon, nitrogen, sulphur and phosphorus in grasslands, p. 301–323. *In* B. Bolin and R. B. Cook (ed.), The major biogeochemical cycles and their interactions. SCOPE Report no. 21. John Wiley & Sons, Inc., New York.

18. **Jenny, H.** 1980. The soil resource. Ecological studies, vol. 37. Springer-Verlag, New York.

19. **Lowe, L. E.** 1965. Sulphur fractions of selected Alberta soil profiles of the Chernozemic and Podzolic orders. Can. J. Soil Sci. **45**:297–303.

20. **Lowe, L. E.** 1969. Sulfur fractions of selected Alberta soil profiles of the Gleysolic order. Can. J. Soil Sci. **49**:375–381.

21. **Maynard, D. G., J. W. B. Stewart, and J. R. Bettany.** 1983. Sulfur and nitrogen mineralization in soils compared using two incubation techniques. Soil Biol. Biochem. **15**:251–256.

22. **McGill, W. B., and C. V. Cole.** 1981. Comparative aspects of cycling of organic C, N, S and P through soil organic matter. Geoderma **26**:267–286.

23. **McGill, W. B., H. W. Hunt, R. G. Woodmansee, and J. O. Reuss.** 1981. Phoenix, a model of the dynamics of carbon and nitrogen in grassland soils. Ecol. Bull. (Stockholm) **33**:49–115.

24. **Paul, E. A., and R. P. Voroney.** 1980. Nutrient and energy flows through soil microbial biomass, p. 215–237. *In* D. C. Elwood (ed.), Contemporary microbial ecology. Academic Press, London.

25. **Redfield, A. C.** 1958. The biological control of chemical factors in the environment. Am. Sci. **46**:205–221.

26. **Saggar, S., J. R. Bettany, and J. W. B. Stewart.** 1981. Sulfur transformations in relation to carbon and nitrogen in incubated soils. Soil Biol. Biochem. **13**:499–511.

27. **Saggar, S., J. R. Bettany, and J. W. B. Stewart.** 1981. Measurement of microbial sulfur in soil. Soil Biol. Biochem. **13**:493–498.

28. **Smeck, N.** 1973. Phosphorus: an indicator of pedogenic weathering processes. Soil Sci. **115**:199–206.

29. **Stewart, J. W. B., C. V. Cole, and D. G. Maynard.** 1983. Interactions of biogeochemical cycles in grassland ecosystems, p. 245–267. *In* B. Bolin and R. C. Cook (ed.), The major biogeochemical cycles and their interactions. SCOPE Report no. 21. John Wiley & Sons, Inc., New York.

30. **Stewart, J. W. B., and R. B. McKercher.** 1982. Phosphorus cycle, p. 221–238. *In* R. G. Burns and J. H. Slater (ed.), Experimental microbial ecology. Blackwell Scientific Publications, Oxford.

31. **Tiessen, H., and J. W. B. Stewart.** 1983. Particle size fractions and their uses in studies of soil organic matter. II. Cultivation effects on organic matter composition in size fractions. Soil Sci. Soc. Am. J. **47**:509–514.

32. **Tiessen, H., J. W. B. Stewart, and J. R. Bettany.** 1982. Cultivation effects on the amounts and concentration of carbon, nitrogen, and phosphorus in grassland soils. Agron. J. **74**:831–835.

33. **Walker, T. W., and A. F. R. Adams.** 1958. Studies on soil organic matter. 1. Influence of phosphorus content of parent materials on accumulations of carbon, nitrogen, sulfur and organic phosphorus in grassland soils. Soil Sci. **85**:307–318.

34. **Walker, T. W., and A. F. R. Adams.** 1959. Studies on soil organic matter. 2. Influence of increased leaching at various stages of weathering on levels of carbon, nitrogen, sulfur, organic and inorganic phosphorus. Soil Sci. **87**:1–10.

35. **Walker, T. W., and J. K. Syers.** 1976. The fate of phosphorus during pedogenesis. Geoderma **15**:1–19.

36. **Walker, T. W., B. K. Thapa, and A. F. R. Adams.** 1959. Studies on soil organic matter. 3. Accumulation of carbon, nitrogen, sulfur, organic and total phosphorus in improved grassland soils. Soil Sci. **87**:135–140.

37. **Williams, C. H., E. G. Williams, and N. M. Scott.** 1960. Carbon, nitrogen, sulphur, and phosphorus in some Scottish soils. J. Soil Sci. **11**:334–346.

Seasonal Variation and Control of Oxygen, Nitrate, and Sulfate Respiration in Coastal Marine Sediments

JAN SØRENSEN

Institute of Ecology and Genetics, University of Aarhus, DK-8000 Aarhus C, Denmark

In the coastal seas, a significant portion of the detritus decomposition takes place in the sediment. This is mediated through a series of fermentative and respiratory processes in which bacteria have a dominating role. Quantitatively oxygen, nitrate, and sulfate respirations all are important (4, 26).

The seasonal productivity patterns in temperate regions suggest that the temperature coupled with the sedimentation pattern will be a "master control" of decomposition at the sea bottom. There are, however, several "subcontrols" which may also be important for the overall rate of decomposition and for the specific activities of bacterial respiration. Although the exact mechanism of control can vary depending on the type of process, thermodynamic and kinetic factors such as temperature and availability of substrates and oxidants generally are important regulators. In addition, specific regulators such as toxic effects, feedback controls, and competition for substrate may also be significant.

In the following, the environmental conditions for bacterial respiration with oxygen, nitrate, and sulfate are discussed, with particular emphasis on the processes in coastal areas. A specific section is devoted to the seasonal variations in an estuarine sediment, and finally an attempt is made to define a priority list of the factors which control respiration in this sediment.

OXYGEN, NITRATE, AND SULFATE RESPIRATION IN COASTAL SEDIMENTS

In coastal sediments, a vertical sequence of bacterial respiration with O_2, NO_3^-, and SO_4^{2-} is often observed in the upper 10 cm (24). This distribution may reflect a variety of the subcontrols for decomposition, e.g., zonation of specific requirements for bacterial growth (redox potential, etc.), different energy yields of the respirations, and different kinetics of utilization of substrates and oxidants by the bacteria.

In the following, references will be made to major "zones" of O_2, NO_3^-, and SO_4^{2-} respiration in the sediments. Each of these will indicate the layer which has the bulk of activity of one of the respirations and limited or no activity of the others. The zonation is therefore based on an ideal distribution of both the oxidants and the

bacteria involved and neglects the heterogeneity from faunal mixing of the sediment and the spheric succession of processes in the particulate detritus. While this definition of respiration zones is useful for the present comparison of the processes, due attention must be paid to heterogeneities, the significance of faunal activity (bioturbation, grazing, etc.), and microbial activity associated with the high-energy detritus, e.g., fresh plant debris and fecal pellets. Their activity of decomposition has given rise to the concept of microniches, which may explain the occurrence of SO_4^{2-} respiration in the NO_3^- zone (12). These exceptions to the ideal model of process distribution will be pointed out in the discussion where necessary.

O_2 respiration. For aerobic metabolism in sediments, the main source of O_2 is diffusion from the overlying waters, although photosynthesis of benthic diatoms may contribute in shallow areas if light is adequate. With the recent development of a high-resolution assay for O_2 in sediments (21), it has been demonstrated that the O_2 zone is typically less than 1 cm deep and that O_2 can only be transported into the deeper layers through the burrows of the benthic fauna (26). In shallow areas like estuaries, a diurnal change of this O_2 profile may be observed as benthic photosynthesis increases the O_2 maximum to full saturation (about 1 mM O_2) during the day and adds a few millimeters to the depth of the profile (21).

The O_2 uptake at the sediment surface was introduced as an assay of the overall "community metabolism" (28) and measures all O_2-consuming reactions without specification. Direct assays of bacterial respiration with O_2 are impossible due to the interference of faunal consumption and abiological oxidations, e.g., of reduced iron, manganese, and sulfur compounds. It is encouraging, however, that a few specific examples of O_2 respiration can be studied from a simple stoichiometry of the reactions. Nitrification, which may be determined from the transformations of NH_4^+ and NO_2^-, has now been studied intensively by the use of the N-serve inhibition technique (8–10). For instance, a seasonal variation of nitrification (NH_4^+ oxidation) has been observed in estuarine sediment (6). The distribution of nitrifying bacteria (NH_4^+

oxidizers) was determined by a chlorate inhibition technique in which NO_2^- oxidizers were inhibited in their metabolism and the V_{max} of NO_2^- production ($\sim NH_4^+$ oxidation) was measured in NH_4^+- and O_2-amended sediment (1). The results indicated that the NH_4^+ oxidizers were located mainly in the upper layer which incorporates the O_2 zone in situ. The activity which was found in the underlying NO_3^- and SO_4^{2-} zones was probably associated with faunal burrows. The observation of the nitrifiers in 10-fold-higher numbers in the polychaete tube walls than in the bulk sediment supports the important role of bioturbation in nitrification in the sediments (J. Sørensen, unpublished data).

NO_3^- **respiration.** The NO_3^- zone is typically 1 to 5 cm deep in the deeper waters and has a maximum of 20 to 40 μM NO_3^- located in the upper portion (J. Sørensen, unpublished data). Occasionally the NO_3^- maximum may be found below the O_2 zone if significant nitrification takes place in the faunal burrows. In some areas, the loading of nutrients in the water may often provide an additional source of NO_3^- and support a rapid uptake at the sediment surface (18).

Bacterial respiration with NO_3^- comprises denitrification with formation of N_2O and N_2 and the less recognized pathway to NH_4^+. The denitrification has been localized and quantified by the acetylene inhibition technique, in which the accumulation of N_2O gives a measure of denitrification as the N_2O reduction is inhibited (23). The bulk of activity was observed in the NO_3^- zone, but significant activity was also found in the SO_4^{2-} zone when polychaete burrows were present (24). The denitrification should be distinguished from the overall activity of NO_3^- reduction, since the pathway to NH_4^+ may be important in the sediments (16, 22). The organisms responsible for the latter process comprise a large group of bacteria which are capable of either respiratory reduction (27) or fermentative reduction (7) of NO_3^- to NH_4^+. In sediments, the two groups of bacteria may thus be distinguished only by their different end products of NO_3^- reduction. As judged from the production of nitrogen gas and NH_4^+ in NO_3^--amended sediments, both activities may be significant (22); the relative proportion of the pathway to NH_4^+ was about 40% of the overall reduction in the NO_3^- zone. In the SO_4^{2-} zone, where NO_3^- was absent in situ, the contribution of this pathway increased to almost 100%; the ability to ferment or to respire with SO_4^{2-} may explain the presence of "NO_3^- respirers" and "NO_3^- fermenters" in this zone.

SO_4^{2-} **respiration.** While most organic substrates can be oxidized with O_2 and NO_3^-, there is only a limited range of compounds which serve as energy sources for the SO_4^{2-}-reducing bacteria. Although the list of such compounds was extended recently to include C_2 to C_{18} fatty acids (F. Widdel, Ph.D. thesis, University of Göttingen, Göttingen, Federal Republic of Germany), the initial fermentations of carbohydrates and amino acids are still indispensible as sources of substrates for SO_4^{2-} respiration. As judged from experiments with suspensions of marine sediment, the most important substrates are low-molecular-weight products of fermentation, notably H_2 and short-chain fatty acids (17, 25). Such experiments have also shown that addition of excess fatty acids and alcohols gives an immediate but limited stimulation of the activity by a factor of 2 (B. B. Jørgensen, personal communication). The results indicated that half-saturation constants (k_m values) for utilization of these compounds are comparable to their concentrations in the sediments. Unfortunately, it has not been possible to measure the low concentrations of free, microbially available compounds in natural sediments; a major difficulty has been the separation of free and bound fractions of the compounds (3). It is therefore not clear whether depth-dependent or seasonal patterns of substrate availability may be found. Considering the excessive supply of the SO_4^{2-} oxidant in marine sediments (1 to 30 mM SO_4^{2-}), it is likely that substrate availability is important as a kinetic control of SO_4^{2-} respiration. Even if the substrate concentrations should vary with season from an unbalanced coupling with the fermenting bacteria, the SO_4^{2-} respirers may kinetically adjust their substrate utilization in response to such changes.

In coastal sediments, a depletion of SO_4^{2-} is typically observed only at depths of several meters (13). However, the bulk of activity of SO_4^{2-} respiration is always located immediately below the NO_3^- zone (13, 26). Such information, together with measures of the overall activity of SO_4^{2-} respiration, has been obtained from applications of a ^{35}S isotope technique in which the formation of [^{35}S]sulfide is followed over time after addition of $^{35}SO_4^{2-}$ (14). The recent discovery of new types of SO_4^{2-}-reducing bacteria should stimulate the search for organisms which are responsible for the process in situ.

SEASONAL VARIATION OF BACTERIAL RESPIRATION IN ESTUARINE SEDIMENT

Study site and assays. Kysing Fjord is a shallow estuary on the east coast of Jutland, Denmark. In recent years, several aspects of detritus decomposition and bacterial respiration have been studied in this estuary (6, 26). For comparison of separate investigations, a subtidal sampling site has been chosen at a water depth of about 0.5 m where the sediment is sandy (2%

organic matter) and has a 1- to 3-cm-deep NO_3^- zone. The benthic fauna, comprising polychaetes (*Nereis* spp.) and amphipods (*Corophium* spp.), is abundant during the spring and summer. Other characteristics of the sediment and a first comparison of the winter (January) and summer (July) activities of bacterial respiration have been reported in an earlier study (26). In 1983, a more detailed, seasonal study of the respiration was undertaken, and some of the results are presented below.

The determinations of NO_3^- and O_2 uptake, denitrification, and SO_4^{2-} reduction were as described previously (26) except that a stirred-water phase was included in the sediment which received acetylene for the assay of denitrification. The recent discovery of the inhibitory effect of acetylene on NH_4^+ oxidation (29) was the basis for a new assay of the overall activity of NO_3^- reduction in the sediment; in the absence of NO_2^- production by the nitrifying bacteria, the rate of consumption of the pool of NO_3^- plus NO_2^- gave a measure of the overall activity of NO_3^- reduction (J. Sørensen, manuscript in preparation). An estimate of the activity of NO_3^- reduction to NH_4^+ was obtained by subtraction of the measured activity of denitrification.

Seasonal variation. The seasonal variation of NO_3^- concentrations in the sediment and the overlying water is shown in Fig. 1. The estuary receives a significant, allochthonous input of inorganic nutrients during the winter and spring, when the effluent water has concentrations of 0.1 to 0.5 mM NO_3^-. The winter concentrations are also high in the NO_3^- zone, which is about 3 cm deep at this time of the year. Both the planktonic and the benthic primary productions are high in the estuary as the temperature increases and the light conditions improve in the spring (April to June); a dense layer of diatoms is soon developed on the sediment surface. Dramatic changes in the NO_3^- availability occur during this period, when the demand for assimilation is high in both the estuary and the connecting river system. The transient occurrence of NO_3^- at the 5-cm depth in May was most likely associated with the intensive faunal activity in the period of high benthic productivity. The allochthonous input of NO_3^- then ceases within a few weeks in May and June, and throughout the summer the water is almost depleted of NO_3^- (0 to 10 μM NO_3^-). At this time, the nitrifying bacteria are therefore the only source of NO_3^- in the sediment. The new balance of production and consumption of NO_3^- results in low concentrations (0 to 20 μM NO_3^-) and a narrow NO_3^- zone (about 1 cm) in the summer. In the fall, the return to winter conditions is reflected in the increasing NO_3^- concentrations and the deeper distribution of NO_3^- in the sediment.

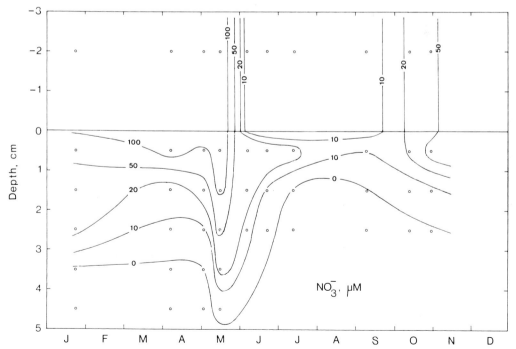

FIG. 1. Seasonal variation of NO_3^- concentration in the water and sediment of Kysing Fjord, 1983 (J. Sørensen, unpublished data).

The seasonal variation of temperature and respiratory activity with O_2, NO_3^-, and SO_4^{2-} in the sediment is shown in Fig. 2. The variation in O_2 uptake was described previously for the coastal sediments (13, 26); the activity increased in the spring (May to June) and showed a maximum in the summer (July to September). The O_2 uptake included the activity of several O_2-consuming processes, of which the abiological oxidations of, e.g., the reduced sulfur compounds may contribute significantly during the summer (13). As mentioned above, only the O_2 respiration of the nitrifying bacteria may be studied in detail. In one of the earlier studies from Kysing Fjord, a seasonal variation of NH_4^+ oxidation was demonstrated; the in situ activity decreased during the spring and was at a minimum during the summer (6). The low concentration of NO_3^- during the summer was apparently the result of both a limited nitrification activity and a high demand for NO_3^- for the reductive processes.

Both the overall NO_3^- reduction and the denitrification had higher activities in the winter than in the summer (Fig. 2), as described previously (26). During the summer, the low NO_3^- concentrations may have some implications for the use of the acetylene inhibition technique. Since NH_4^+ oxidation is inhibited, the NO_3^--reducing bacteria may soon deplete the pool of NO_3^- plus NO_2^-, and the activities should be measured within 1 to 2 h. It is not known whether incomplete inhibition of N_2O reduction at low NO_3^- concentrations (15) also affects the denitrification assay. If so, the denitrification shown in Fig. 2 will be underestimated at this time of the year. In spring, both processes of NO_3^- reduction showed a distinct maximum which has not been observed previously. The high activities apparently resulted from a combined effect of increasing temperature, high input of allochthonous NO_3^-, and facilitated transport of NO_3^- to the deeper layers by faunal activity (Fig. 1 and 2). During this period, the

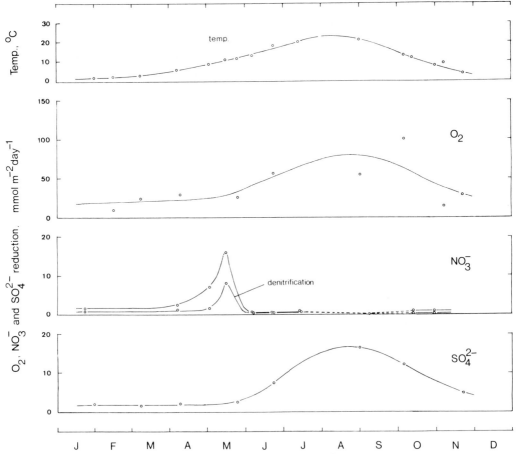

FIG. 2. Seasonal variation in temperature and O_2, NO_3^-, and SO_4^{2-} reduction in the sediment of Kysing Fjord, 1983. (O_2 and SO_4^{2-} reduction: B. B. Jørgensen, unpublished data. NO_3^- reduction: J. Sørensen, unpublished data.)

benthic algae also showed peak productivity, and an unknown fraction of the measured consumption of NO_3^- could have been used for assimilation rather than respiration. Although dark incubation of the sediment and the presence of NH_4^+ in the surface zone should limit the algal NO_3^- assimilation, its contribution to the overall NO_3^- consumption cannot be excluded. Assuming, however, that NO_3^- assimilation was insignificant, the measured consumption rates may be assigned to denitrification and dissimilatory reduction of NO_3^- to NH_4^+. Throughout the year, each of the two pathways accounts for about half of the overall NO_3^- reduction in the estuary (Fig. 2). For a comparison of the significance of these processes, it was recently demonstrated that 27 to 57% of the overall NO_3^- reduction in Japanese estuarine sediments was due to denitrification (20).

The seasonal variation of SO_4^{2-} respiration was as described previously for coastal sediments (26); maximum activities were found during the summer.

CONTROL OF BACTERIAL RESPIRATION IN ESTUARINE SEDIMENT

Some important variables (temperature and availability of substrate and oxidant) which may control the seasonal patterns of respiration with O_2, NO_3^-, and SO_4^{2-} in estuarine sediment are listed in Table 1. Based on the seasonal variations shown in Fig. 2, the potential effect of a variable is indicated by the factor of change in the respiratory activities which may be caused by this effector. The comparison was therefore based only on changes in overall activities (per unit area of sediment), but depth-dependent changes in activity which may result from altered zonation of the processes during the season are also considered in the discussion.

Temperature. It may be assumed that temperature has the same effect on all processes. If the bacterial activities increase by a factor of 2 to 3 for temperature increments of 10°C, as for most biological processes, the resulting factor of temperature control should be 5 in the range from 2 to 22°C measured in situ (Fig. 2). This factor was therefore assigned to the potential

control by the temperature (Table 1). The measured O_2 uptake and SO_4^{2-} respiration both showed apparent correlations with temperature (Fig. 2), as was the case in other sediments from deeper, coastal waters (13). There was some indication from the present study, however, that the correlations with temperature may be less significant in the estuarine sediments. The seasonal changes in zonation of NO_3^- respiration could influence both the location and the overall activity of SO_4^{2-} respiration, in particular during the spring when NO_3^- respiration accounts for a significant fraction of the overall bacterial respiration in the sediment. The allochthonous supply of NO_3^- at this time may delay the onset of significant SO_4^{2-} respiration until the NO_3^- is depleted (Fig. 1 and 2).

Substrate availability. As noted above, it is impossible to evaluate the size and composition of the microbially available pool of organic substrates. The reported three- to fivefold decrease in the dissolved organic carbon pool in a coastal sediment during the spring (11) may not be related to the actual availability of energy sources for the respiring bacteria. Some evidence based on the observed stimulation of NO_3^- and SO_4^{2-} respiration by excess carbon addition would indicate, however, that the regulation of these activities by substrate availability should be of limited significance in the present comparison of control factors. In a number of inner shelf sediments from depths of 19 to 55 m, denitrification was not enhanced by carbon addition, and only at deeper stations was a three- to fourfold stimulation observed (5). The activity of SO_4^{2-} respiration was stimulated by a factor of 2 or less when excess carbon was added to coastal sediments (19; B. B. Jørgensen, personal communication). In natural sediments, the lack of information on substrate availability is apparent. Only nitrification (NH_4^+ oxidation) can be assigned a documented control factor for substrate availability; the process is unlikely to be NH_4^+ limited in these sediments (factor of 0 for control [Table 1]). Based on indications from substrate-amended coastal sediments, the factors for potential control by substrate availability should be 0 to 2 for the heterotrophic bacteria (Table 1).

TABLE 1. Comparison of potential effects of control factors on bacterial respiration in Kysing Fjord, 1983

Control factor	Effect on respiratory activity[a]				
	O_2 respiration		NO_3^- respiration		SO_4^{2-} respiration
	Nitrification	Other	Denitrification	Other	
Temp	5 (Good)	5 (Good)	5 (Good)	5 (Good)	5 (Good)
Substrate availability	0 (Good)	0–2 (Poor)	0–2 (Poor)	0–2 (Poor)	0–2 (Poor)
Oxidant availability	0–2 (Bad)	0–2 (Bad)	10 (Fair)	10 (Fair)	0 (Good)

[a] Effects are indicated by the factors of seasonal change of activity caused by the variable. Estimates of accuracies are given in parentheses.

The seasonal change in depth of the respiration zones is obviously important for the availability of substrates if these are not uniformly distributed in the sediment. Again, even if the reported profiles of dissolved organic substrates (3) show lower concentrations with depth, the actual availability of the compounds will remain uncertain until more progress is made on their analysis in the sediments.

Oxidant availability. The depth of the O_2 zone may decrease by a few millimeters in the spring and summer, as judged from O_2 profiles in dark-incubated sediment (26). There is, however, an opposing effect of both benthic photosynthesis and bioturbation which may improve the availability of O_2 in the sediment. For instance, in a sediment similar to the one from Kysing Fjord, one-fourth of 30 measurements of O_2 profiles showed secondary peaks of 8 to 220 μM O_2 in the underlying NO_3^- zone (26). Their occurrence demonstrated a transportation of O_2 in the faunal burrows, although the actual concentrations of O_2 within the surrounding sediment (tube walls) could not be determined. It is therefore difficult to assess the seasonal trends in availability of O_2 until further work is done on the O_2 gradients in the burrows. A small control factor of 0 to 2 may be tentatively assigned to the control of O_2 availability (Table 1).

NO_3^- availability varied dramatically in the Kysing Fjord sediment; seasonally, the variation was a combination of changes in both the depth distribution and the concentration of NO_3^-. The correlation between the NO_3^- consumption in the sediments and the concentration of NO_3^- in the overlying waters (18) must be a reflection of both the gradient for diffusion of NO_3^- at the sediment surface and the depth of the NO_3^- zone, rather than a kinetic control of the depth-specific bacterial NO_3^- respiration per se. High affinities for NO_3^- seem to be common among denitrifying bacteria, and if the low k_m values for NO_3^- reduction (about 10 μM NO_3^-) observed in pure cultures (2) are valid also for other types of NO_3^--reducing bacteria, only the low summer concentrations of 0 to 20 μM NO_3^- in the sediments (26; Fig. 1) should imply a possible kinetic control of NO_3^- respiration. The factor of 10 assigned to the control by NO_3^- availability in the Kysing Fjord sediment (Table 1) is therefore a combination of the significant seasonal change in depth of the NO_3^- zone and the possible kinetic control of NO_3^- respiration during the summer.

SO_4^{2-} respiration was not limited by the supply of oxidant in Kysing Fjord, where 10 to 20 mM SO_4^{2-} was found in the sediment. A control factor of 0 was therefore given to this variable (Table 1). The upward extension of the SO_4^{2-} zone by a few centimeters during the summer was not significant compared with its overall depth.

CONCLUDING REMARKS

Seasonal variations are dramatic for O_2, NO_3^-, and SO_4^{2-} respiration in estuarine sediment. The recorded variation in NO_3^- respiration gave new information on the process; the peak activity observed during the spring was apparently related to the increasing temperature and high availability and efficient transportation of NO_3^- into the sediment. Denitrification accounted for about 50% of the overall NO_3^- reduction. The rest was apparently dissimilated to NH_4^+, although assimilatory uptake could not be excluded. The use of an acetylene inhibition technique for the assay of denitrification could underestimate the process during the summer, when concentrations of NO_3^- are low in the sediment.

A priority list of factors which may control the seasonal pattern of respiratory activities must be tentative, but some trends were apparent from the comparison of the temperature and the availability of substrates and oxidants for the processes. Suitable assays could be used for measurements of O_2, NO_3^-, and SO_4^{2-} availability in sediments, but better techniques are needed to determine the pool sizes of the microbially available organic substrates. It was therefore difficult to evaluate the kinetic control of respiration with these compounds. Temperature had a significant influence on some activities, but the availability of NO_3^- seemed to be the most important single effector in the sediment.

The preference for a simple comparison of these control factors does not exclude an important role for other thermodynamic regulators and more elaborate coupled controls, feedback controls, controls by toxicity, etc. Some of these may be particularly important for the control of activities in transition layers between the zones of O_2, NO_3^-, and SO_4^{2-} respiration.

I gratefully acknowledge the skillful assistance of Else B. Frentz and Preben G. Sørensen.

The seasonal study in Kysing Fjord was supported by the Environmental Programme of the Commission of the European Communities [contract no. ENV-588-DK (G)].

LITERATURE CITED

1. **Belser, L. W., and E. L. Mays.** 1980. Specific inhibition of nitrite reduction by chlorate and its use in assessing nitrification in soils and sediments. Appl. Environ. Microbiol. **39:**505–510.
2. **Betlach, M. R., and J. M. Tiedje.** 1981. Kinetic explanation for accumulation of nitrite, nitric oxide, and nitrous oxide during bacterial denitrification. Appl. Environ. Microbiol. **42:**1074–1084.
3. **Christensen, D., and T. H. Blackburn.** 1982. Turnover of ^{14}C-labeled acetate in marine sediment. Mar. Biol. **71:**113–119.
4. **Fenchel, T. M., and B. B. Jørgensen.** 1977. Detritus food

chains of aquatic ecosystems: the role of bacteria, p. 1–58. *In* M. Alexander (ed.), Advances in microbial ecology, vol. 1. Plenum Publishing Corp., New York.

5. **Haines, J. R., R. M. Atlas, R. P. Griffiths, and R. Y. Morita.** 1981. Denitrification in Alaskan continental shelf sediments. Appl. Environ. Microbiol. **41:**412–421.

6. **Hansen, J. I., K. Henriksen, and T. H. Blackburn.** 1981. Seasonal distribution of nitrifying bacteria and rates of nitrification in coastal marine sediments. Microb. Ecol. **7:**297–304.

7. **Hasan, S. M., and J. B. Hall.** 1975. The physiological function of nitrate reduction in *Clostridium perfringens*. J. Gen. Microbiol. **87:**120–128.

8. **Henriksen, K.** 1980. Measurement of in situ rates of nitrification in sediment. Microb. Ecol. **6:**329–337.

9. **Henriksen, K., J. I. Hansen, and T. H. Blackburn.** 1980. The influence of benthic infauna on exchange rates of inorganic nitrogen between sediment and water. Ophelia **1**(Suppl.):249–256.

10. **Henriksen, K., J. I. Hansen, and T. H. Blackburn.** 1981. Rates of nitrification, distribution of nitrifying bacteria, and nitrogen fluxes in different types of sediment from Danish waters. Mar. Biol. **61:**299–304.

11. **Hines, M. E., W. H. Orem, W. B. Lyons, and G. E. Jones.** 1982. Microbial activity and bioturbation-induced oscillations in pore water chemistry of estuarine sediments in spring. Nature (London) **299:**433–435.

12. **Jørgensen, B. B.** 1977. Bacterial sulfate reduction within reduced microniches of oxidized marine sediments. Mar. Biol. **41:**7–17.

13. **Jørgensen, B. B.** 1977. The sulfur cycle of a coastal marine sediment (Limfjorden, Denmark). Limnol. Oceanogr. **22:**814–832.

14. **Jørgensen, B. B.** 1978. A comparison of methods for the quantification of bacterial sulfate reduction in coastal marine sediments. I. Measurements with radiotracer techniques. Geomicrobiol. J. **1:**11–28.

15. **Kaspar, K. F., J. M. Tiedje, and R. B. Firestone.** 1981. Denitrification and dissimilatory nitrate reduction to ammonium by digested sludge. Can. J. Microbiol. **27:**878–885.

16. **Koike, I., and A. Hattori.** 1978. Denitrification and ammonia formation in anaerobic coastal sediments. Appl. Environ. Microbiol. **35:**278–282.

17. **Laanbroek, H. J., and N. Pfennig.** 1981. Oxidation of short-chain fatty acids by sulfate-reducing bacteria in freshwater and marine sediments. Arch. Microbiol. **128:**330–335.

18. **Nedwell, D. B.** 1982. Exchange of nitrate, and the products of bacterial nitrate reduction, between seawater and sediment from a U.K. saltmarsh. Estuarine Coastal Shelf Sci. **14:**557–566.

19. **Nedwell, D. B., and J. W. Abram.** 1979. Relative influence of temperature and electron acceptor concentrations on bacterial sulfate reduction in saltmarsh sediment. Microb. Ecol. **5:**67–72.

20. **Nishio, T., I. Koike, and A. Hattori.** 1982. Denitrification, nitrate reduction, and oxygen consumption in coastal and estuarine sediments. Appl. Environ. Microbiol. **43:**648–653.

21. **Revsbech, N. P., J. Sørensen, T. H. Blackburn, and J. P. Lomholt.** 1980. Distribution of oxygen in marine sediments measured with microelectrodes. Limnol. Oceanogr. **25:**403–411.

22. **Sørensen, J.** 1978. Capacity for denitrification and reduction of nitrate to ammonia in a coastal marine sediment. Appl. Environ. Microbiol. **35:**301–305.

23. **Sørensen, J.** 1978. Denitrification rates in a marine sediment as measured by the acetylene inhibition technique. Appl. Environ. Microbiol. **36:**139–143.

24. **Sørensen, J.** 1978. Occurrence of nitric and nitrous oxides in a coastal marine sediment. Appl. Environ. Microbiol. **36:**809–813.

25. **Sørensen, J., D. Christensen, and B. B. Jørgensen.** 1981. Volatile fatty acids and hydrogen as substrates for sulfate-reducing bacteria in anaerobic marine sediment. Appl. Environ. Microbiol. **42:**5–11.

26. **Sørensen, J., B. B. Jørgensen, and N. P. Revsbech.** 1979. A comparison of oxygen, nitrate and sulfate respiration in coastal marine sediments. Microb. Ecol. **5:**105–115.

27. **Steenkamp, D. J., and H. D. Peck, Jr.** 1981. Proton translocation associated with nitrite respiration in *Desulfovibrio desulfuricans*. J. Biol. Chem. **256:**5450–5458.

28. **Teal, T. M., and J. Kanwisher.** 1961. Gas exchange in a Georgia salt marsh. Limnol. Oceanogr. **6:**388–399.

29. **Walter, H. M., D. R. Keeney, and I. R. Fillery.** 1979. Inhibition of nitrification by acetylene. Soil Sci. Soc. Am. J. **43:**195–196.

Denitrification and Nitrification in Coastal and Estuarine Sediments

ISAO KOIKE, TAKASHI NISHIO,† AND AKIHIKO HATTORI

Ocean Research Institute, University of Tokyo, Nakano, Tokyo 164, Japan

Denitrification, defined as the reduction of nitrate and nitrite to gaseous nitrogen, is mediated by facultatively anaerobic bacteria, which use nitrate and nitrite as terminal electron acceptors under oxygen-depleted conditions. Since denitrification is the only biological process responsible for the removal of combined nitrogen, there has been increased attention to this process in coastal and estuarine environments, where the inflow of land-originated nitrogen often causes eutrophication.

Denitrification is influenced by the concentration of nitrate and nitrite and organic matter, and by the concentration of oxygen (22). Except for the stagnant bottom waters of some fjords or embayments, anaerobiosis is not common in coastal waters. However, substantial portions of organic materials produced in the euphotic zone settle to the sea floor, and benthic organisms in sediments rapidly consume dissolved oxygen (25). Oxidations of products of anaerobic microorganisms in the deeper layer of sediments, such as H_2S and CH_4, also consume oxygen. Thus, the aerobic–anaerobic zone appears at the sediment–water interface, where easily utilizable organic materials are still abundant (23). Concentrations of nitrate plus nitrite are high in coastal and estuarine bottom waters and vary in a range of 10 to several hundred micromolar. The latter depend heavily on the extent of eutrophication in these waters. The oxygen-depleted layer in surface sediments is therefore considered to be the primary site of denitrification in coastal and estuarine environments.

Since the concentration of organic matter is high, the rate of denitrification in coastal and estuarine sediments is controlled by the supply of nitrate and nitrite (16, 17, 21). The latter is supplied from overlying water, or nitrification at the sediment–water interface. In contrast, denitrification in cultivated soil systems is usually limited by the supply of organic materials and not the concentration of nitrate and nitrite (13).

In this paper we summarize the results of our recent experiments concerning the processes of denitrification and nitrification in coastal and estuarine sediments, as investigated by applying a ^{15}N tracer technique. Our results demonstrate that in situ nitrification serves as an important source of nitrogen for denitrification in coastal sediments of less polluted areas.

DENITRIFICATION IN COASTAL AND ESTUARINE SEDIMENTS

Several methods have been developed to estimate the rate of denitrification in coastal and estuarine sediments (3, 11, 14, 24, 26). Measurement of nitrate (measured as the combination of nitrate and nitrite) loss from the interstitial water of the sediment is the simplest method and has been extensively used in earlier investigations (3, 18).

However, three pathways must be considered for the microbial reduction of nitrate: (i) assimilatory nitrate reduction to cellular nitrogen by way of ammonium, (ii) dissimilatory nitrate reduction to gaseous nitrogen (denitrification), and (iii) dissimilatory nitrate reduction to ammonium. Except for denitrification, these pathways yield products that are still biologically active and play roles in the biological nitrogen cycle. Ammonium inhibits the assimilatory reduction of nitrate under both aerobic and anaerobic conditions, and may be an insignificant process, since ammonium concentration in the sediment is high as the result of active mineralization of organic materials.

Denitrification and dissimilatory nitrate reduction to ammonium may, however, occur simultaneously in anaerobic sites of marine sediments. If the latter process operates to any significant extent, simple measurement of nitrate loss in sediments would obviously yield overestimates of denitrification. Direct measurement of evolved N_2 and N_2O is, therefore, necessary for obtaining a reliable estimate of denitrification. The ^{15}N tracer (7) and the acetylene blockage methods (29) satisfy this requirement. We used a ^{15}N tracer technique in our studies of denitrification in coastal and estuarine sediments, since this method allowed the simultaneous determination of denitrification and nitrate reduction to ammonium.

Products of nitrate reduction in coastal and estuarine sediments. Table 1 summarizes the products of nitrate reduction obtained from

† Present address: Upland Farming Division, Hokkaido National Agricultural Experimental Station, Memuro, Kasaigun Hokkaido 082, Japan.

TABLE 1. Products of nitrate reduction in coastal and estuarine sediments[a]

Sediment origin	Total rate[b]	Production (%) of:		
		N_2	NH_4^+	Particulate organic N
Mangoku-Ura	88.7	33	52	15
Shimoda Bay	22.2	80	16	4
Tokyo Bay	14.0	60	7	33

[a] After Koike and Hattori (14). Experimental conditions: for Mangoku-Ura sediment, temperature was 20°C, incubation time was 6 h, and 30 μM [^{15}N]nitrate was added; for Shimoda Bay sediment, temperature was 26°C, incubation time was 8 h, and 15 μM [^{15}N]nitrate was added; for Tokyo Bay sediment, temperature was 20°C, incubation time was 10 h, and 15 μM nitrate was added.

[b] Sum of three products, expressed as nanogram-atoms of N per gram per hour.

coastal and estuarine sediments incubated under anaerobic conditions (14). In this experiment, slurries of surface sediments (0 to 3 cm) were incubated in the presence of [^{15}N]nitrate, and ^{15}N enrichments in N_2, ammonium, and particulate organic nitrogen were determined. The ^{15}N enrichment in N_2O was not measured, since the contribution of N_2O to total denitrification in these sediments was minor (20). In the very organic sediments of Mangoku-Ura, ammonium production accounted for about one-half of the observed nitrate reduction. In the less organic-rich Simoda Bay sediment, N_2 production was the primary pathway of nitrate reduction.

Several factors are considered to be involved in the selection of type of bacterial nitrate reduction which will occur in sediments. One of these is the stability of anaerobiosis, which would affect the composition of bacterial populations. In ruminant systems, where strict anaerobiosis is always maintained, dissimilatory nitrate reduction to ammonium is predominant (9, 12). In most soil systems, where anaerobiosis is restricted to the microsites of soil particles, N_2 and N_2O production are the major pathways of anaerobic nitrate reduction (5, 27). Conclusions for the coastal and estuarine sediments examined are between these two extremes. Most denitrifying bacteria so far identified have aerobic respiratory metabolisms and cannot grow under anaerobic conditions in the absence of nitrate (13, 22). Bacteria mediating reduction of nitrate to ammonium can in most cases grow fermentatively (4, 8). Thus, when denitrifying bacteria are exposed to anaerobiosis during a prolonged period without nitrate, their survival is suggested to be less than that of ammonium producers. Under anaerobic conditions in the presence of nitrate, the kinetic parameters (K_m, V_{max}) of nitrate reduction for denitrifying bacteria and

ammonium producers affect the partition of nitrate reduction (28). Since the K_m for nitrate for denitrifying bacteria is one to two orders of magnitude lower than that for ammonium producers, denitrifying bacteria may outcompete ammonium producers in low nitrate conditions. Qualitative and quantitative differences of organic carbon supply and their relation to the nitrate supply might also be important factors in the control of the pathway involved in anaerobic nitrate reduction (28).

In situ rate of denitrification. Rates of denitrification in marine sediments were determined by use of slurries of sediment, as described above. The in situ rate was then estimated by correcting the effect of the nitrate concentration on the rate (6, 16, 21). The rate of denitrification is also affected by the oxygen concentration, and since the structure of marine sediments is not uniform, the presence of an aerobic microsite in anaerobic sediments and of an anaerobic microsite in aerobic sediments is rather common (10, 23). This is very difficult to reproduce in slurry experiments. Thus, an experimental procedure which does not disturb the fine structure of the sediment is necessary. To overcome this problem, we developed an experimental system consisting of undisturbed sediment cores and overlying water (19). Filtered seawater containing [^{15}N]nitrate flowed at a constant rate and mixed over the sediment. In these experiments we assume (i) that nitrate in the overlying water is the sole source of nitrate for denitrification and (ii) that the site of denitrification is restricted to the surface of the sediment and the N_2 produced rapidly diffuses into the overlying water. If the concentration of nitrate in the overlying water is sufficiently high and the sediment is high in organic matter, the first assumption would be applicable. We monitored the concentrations of nitrate, nitrite, dissolved O_2, and ^{15}N enrichment in dissolved N_2 in the influent (source water) and effluent (overlying water). Constant concentrations of dissolved O_2 and nitrate in the effluent were achieved within 3 h, and a constant concentration of ^{15}N enrichment in N_2 was achieved within 10 to 15 h, which verified the second assumption. The differences between the influent and effluent in dissolved O_2, nitrate, and ^{15}N enrichment in N_2 in steady state were used for the estimation of rates of oxygen consumption, nitrate reduction, and denitrification.

Table 2 summarizes the rates of O_2 consumption, nitrate reduction, and N_2 production in three different coastal and estuarine sediments (19). Our previous observations indicated that the rate of denitrification is proportional to the nitrate concentration up to a nitrate concentration of 30 to 40 μM in sediments with high concentrations of organics (17). Assuming a linear relationship between the rate of denitrification and nitrate concentration, we calculated a first-order approximation of the in situ rate of denitrification. Production of N_2O from the sediment was invariably negligible compared with N_2 production (20). Our estimate of nitrate reduction is based on the net flux of nitrate into the sediment. However, if nitrate is produced actively by nitrification in the sediment and diffused into the overlying waters, the rates of nitrate reduction obtained will be underestimated. The effect of nitrification would be minor in sediments from Tama Estuary and Tokyo Bay (Table 2). Denitrification accounted for 27 to 57% of total nitrate reduction (Table 2), confirming the result of our previous experiments (Table 1). The rates of nitrate reduction and denitrification are roughly proportional to the in situ nitrate concentration.

Reduction of nitrate to N_2 and to ammonium requires five and eight electrons transferred, respectively, whereas oxygen evolution requires four electrons. If only O_2 respiration of benthic organisms is responsible for the O_2 consumption in the sediment, the number of electrons coupled with O_2 respiration and nitrate reduction can be compared. On average in sediments from Tama Estuary, 2.1×10^{-6} electron equivalent per cm^2 per h is associated with O_2 respiration, and 0.81×10^{-6} electron equivalent per cm^2 per h is associated with nitrate reduction. The rate of oxidation of organic matter by O_2 respiration is, however, obviously overestimated, because O_2 is also consumed by chemical oxidation of H_2S and other reduced substances. Dissolved O_2 contents in the overlying water were in the range of 100 to 200 μM. The above calculation suggests that, if the nitrate concentration is of the same order as the O_2 concentration in the overlying waters, the nitrate respiration compares to O_2 respiration with respect to the oxidation of organic materials by benthic organisms in the sediments.

NITRIFICATION IN COASTAL AND ESTUARINE SEDIMENTS

Autotrophic nitrifying bacteria, which obtain energy for growth through the oxidation of ammonium or nitrite by oxygen, are mainly responsible for nitrification in natural environments (1). The surface layer of coastal and estuarine sediments, where concentrations of ammonium and oxygen are relatively high, is a favorable site for bacterial nitrification.

Several methods have been proposed for the experimental determination of nitrification in marine sediments (2, 3). Billen (2) determined the amounts of $^{14}CO_2$ taken up by microorganisms in the sediments in the presence and absence of N-serve, a potent inhibitor of bacterial nitrification. The difference in $^{14}CO_2$ uptake was converted to the amounts of ammonium oxidized to nitrate. The measurement of nitrate from ammonium is a more direct evaluation of nitrification. Nitrate is not, however, stable in organic-rich sediments because of the simultaneous occurrence of nitrate reduction at anaerobic sites of sediments.

We applied two ^{15}N tracer techniques for the assessment of nitrification in coastal and estuarine sediments. One is an isotope dilution technique (15), and the other is the combined use of [^{15}N]ammonium tracer technique with the continuous-flow sediment–water system as described above.

^{15}N **isotope dilution method.** With the isotope dilution method, the rates of nitrification and nitrate reduction can be estimated simultaneously. Slurries of the surface sediments, 1.5-mm thickness on average, were incubated with sterile overlying water containing [^{15}N]nitrate and nonlabeled ammonium. The change of nitrate concentrations during incubation is the result of nitrification and nitrate reduction by microorganisms in the sediment. Thus, their rates can be expressed by the following equation:

$$N_2 - N_1 = (Y - X)(t_2 - t_1) \qquad (1)$$

where X is the rate of nitrate reduction, Y is the rate of nitrification, and N_1 and N_2 are the nitrate concentrations at observation times t_1 and t_2. Mass balance relation with respect to ^{15}N is given by the equation:

$$N_2 x_2 - N_1 x_1 = Y \bar{x}_a - X \bar{x} \qquad (2)$$

where x_1 and x_2 are ^{15}N enrichments of nitrate at times t_1 and t_2, respectively, and $\bar{x}$ and $\bar{x}_a$ are average ^{15}N enrichments of nitrate and ammonium.

Time courses of nitrification and nitrate reduction in the surface sediments from three locations are shown in Fig. 1 (15). In the muddy sediment from Mangoku-Ura, nitrate reduction was significant, whereas nitrification was negligible during the initial 24 h of incubation. The opposite was the case with the sandy sediment from Mangoku-Ura (Fig. 1B). In the Odawa Bay sediment, collected from a *Zostera* bed, nitrification and nitrate reduction occurred at similar rates (Fig. 1C). Results from Odawa Bay experiments stress the advantage of the [^{15}N]nitrate dilution method and illustrate how simple measurements of changes in nitrate concentration would give erroneous values for nitrification.

The sensitivity of this method depends on the precision of determination of nitrate concentration and that of ^{15}N. For the estimate of nitrification, the ^{15}N dilution method is less sensitive than the ^{15}N tracer method, which measures ^{15}N]nitrate production from ^{15}N]ammonium. Another difficulty is the maintenance of the in situ redox potential in the sediment during the experiment. The later increase in nitrification and decrease in nitrate reduction, as observed in Fig. 1A, might result from the change in oxygen concentration in the sediment during prolonged incubation.

Nitrification measurement with a continuous-flow sediment–water system. The same experimental device used for in situ denitrification measurements was applied to the measurement of nitrification, but ^{15}N]nitrate in the overlying water was replaced with ^{15}N]ammonium (20). [15]N]ammonium in the overlying water diffuses into the nitrification sites of the sediment and oxidizes to ^{15}N]nitrate. The ^{15}N]nitrate produced then diffuses to the sites of nitrate reduction and denitrification and is reduced to $^{15}N_2$ and ^{15}N]ammonium, while a portion of ^{15}N]nitrate diffuses to overlying waters. We measured ^{15}N enrichments in nitrate and N_2 in the overlying waters at various intervals. Steady-state values of ^{15}N enrichment were usually obtained within 10 to 20 h. Values obtained at steady state were used for the estimation of nitrification and denitrification coupled with nitrification. ^{15}N enrichment in ammonium was determined in the surface sediment (0 to 1 cm) after steady state was noted.

The ^{15}N]nitrate-supplemented experiments were performed on subsamples of sediment from the same locations. In these experiments, calculated ^{15}N]nitrate dilution by nonlabeled nitrate originated from the sediment, estimated from the ^{15}N]ammonium-supplemented experiments, was 6% at most. Thus, we neglected this effect for the calculation of denitrification rate coupled with overlying nitrate. For the estimate of total

nitrate reduction coupled with overlying nitrate, however, we took into account the effect of nitrate production from the sediment. This effect was significant in Odawa Bay sediment.

From the fraction of N_2 production in total nitrate reduction in the ^{15}N]nitrate experiment, total rates of nitrate reduction coupled with nitrification can be calculated. The latter assumes that the fate of nitrate reduction is the same in the two experiments. Total rate of nitrification, therefore, is given as the sum of ^{15}N]nitrate production into the overlying waters and nitrate reduction coupled with nitrification.

In the Odawa Bay sediment, the addition of ^{15}N]ammonium (100 μM, about 10 times higher than the in situ concentration) in the overlying water might stimulate the rate of nitrification, since the increase of nitrite in overlying water was twice that in the experiment without addition. We calculated the in situ rates of nitrification, assuming that this additional increase of nitrite is attributable to ammonium enrichment. In Tama Estuary, ammonium concentration in the overlying water is high and the effect of ^{15}N]ammonium enrichment should be minor.

Table 3 summarizes the estimated in situ rates of nitrification and denitrification coupled with nitrification in coastal and estuarine sediments. For the comparison, the rate of denitrification and the rate of nitrate reduction coupled with nitrate from overlying water are also entered. The rates of nitrification (4.23 to 12.4 ng-atoms of N per cm² per h) and denitrification coupled with nitrification (1.31 to 3.85) show no large variation when compared with nitrate reduction (5.84 to 122) and denitrification (1.02 to 55.5) coupled with the nitrate concentration in the overlying waters. Two dominant factors, which control the total rate of nitrification in the sediment column, are thickness of the aerobic layer

TABLE 2. Oxygen consumption, nitrate (plus nitrite) reduction, and denitrification in coastal and estuarine sediments[a]

Location	Sampling date	In overlying water			O_2 consumption (nmol cm^{-2} h^{-1})	N (ng atoms cm^{-2} h^{-1})	
		Temp (°C)	NO_3^- + NO_2^- (μM)	NH_4^+ (μM)		NO_3^- (+ NO_2^-) reduction	Denitrification
Tama	7 May 1980	21.0	114	207	382	114 (102)[b]	30.7 (27.6)
Estuary	19 June 1980	24.6	60.9	121	643	115 (138)	65.4 (78.9)
Odawa	15 May 1980	21.6	27.1	8.85	159	10.7 (9.00)	4.66 (3.92)
Bay	2 June 1980	24.3	0.17	0.52	236	ND[c]	0.79 (0.016)
Tokyo Bay							
Station 2	10 Sept. 1980	19.4	17.4	6.90	144	10.0 (6.96)	4.69 (3.26)
Station 8	10 Sept. 1980	19.1	13.7	4.40	158	5.29 (3.15)	2.73 (1.63)

[a] After Nishio et al. (19). The concentration of ^{15}N]nitrate added was 10 μM, except for the experiment performed on 7 May 1980.

[b] The numbers in parentheses represent in situ rates calculated from the nitrate concentrations in the experimental system and the in situ concentrations of nitrate in the overlying water, assuming a linear relationship between the rates and nitrate concentration.

[c] Not determined.

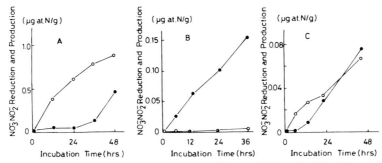

FIG. 1. Nitrate reduction and nitrification in muddy sediment of Mangoku-Ura (A), sandy sediment of Mangoku-Ura (B), and fine sandy sediment of Odawa Bay (C). Nitrate reduction (○) and nitrification (●) were calculated by use of equations 1 and 2.

and the concentration of ammonium in the layer. In general, the aerobic layer in the organic-rich sediments is thinner than that of less organic sediment. Ammonium concentrations in the sediment are positively correlated with the amounts of organic materials. Thus, the thickness of the nitrification layer and the concentration of ammonium in the sediment are negatively correlated with each other, suggesting that the total nitrification per unit area of coastal and estuarine sediments falls in a relatively narrow range. Temperature, of course, affects nitrification rate, as was seen in the case of Tama Estuary (Table 3).

INORGANIC NITROGEN CYCLING IN COASTAL AND ESTUARINE SEDIMENTS

Several interesting features concerning the inorganic nitrogen cycling in coastal and estuarine sediments are illustrated by our investigation with a ^{15}N tracer technique. Simultaneous occurrence of nitrification and nitrate reduction in the surface sediment of Odawa Bay at comparable rates was found both in the bottle experiment and in experiments with the continuous-flow sediment–water system. For the evaluation of in situ rates of inorganic nitrogen transformation, however, the continuous-flow method with an undisturbed sediment core provides a more reliable estimate.

The continuous-flow method is suitable for the organic-rich coastal and estuarine sediments, since both nitrification and denitrification occur near the surface and a steady ^{15}N flux is achieved within a short time. This will not be the case if denitrification occurs in the deeper layer of the sediment.

In the sediments examined, the rates of nitrate reduction and denitrification coupled with nitrate from overlying water were roughly proportional to the concentration of nitrate, whereas variations in the in situ nitrification rates were small. In Tama Estuary, where nitrate concentration in the overlying water was of the order of 100 μM, the contribution of nitrate from the overlying water to total denitrification was one order of magnitude higher than that coupled with in situ nitrification. About half of the reduced nitrate was reduced to the ammonium level. Thus, considerable quantities of ammonium nitrogen were supplied into the sediment. In Odawa Bay, where nitrate concentration in the overlying water was of the order of 10 μM or less, nitrate from the overlying water and that from in situ nitrification contributed equally to denitrification.

The results from the Odawa Bay sediments indicate that the coupling of nitrification and denitrification in the sediment is an important contribution to the denitrification process in

TABLE 3. Nitrification and denitrification rates in Tama Estuary and Odawa Bay sediments[a]

Location	Sampling date	In overlying water			N (ng atoms cm^{-2} h^{-1})			
		Temp (°C)	$NO_3^- + NO_2^-$ (μM)	NH_4^+ (μM)	Nitrification rate	Denitrification rate[b]		NO_3^- reduction coupled with overlying NO_3^-
						A	B	
Tama Estuary	21 Dec. 1980	9.3	141	378	4.23	1.31	13.3	29.6
	20 May 1981	18.0	144	216	12.4	3.85	55.5	122
Odawa Bay	8 May 1981	22.9	6.10	4.0	11.3	1.34	1.02	5.84

[a] Recalculated from the data presented by Nishio et al. (20).
[b] A, Coupled with nitrification; B, coupled with nitrate in overlying water.

sediment of unpolluted areas, where nitrate concentrations in bottom water are relatively low (10 to 40 μM).

CONCLUSIONS

Numerous physiological types of bacteria are involved in the materials transfer in marine sediments, and the interactions between these microbial populations through the microenvironments are one of the basic characters of the ecosystem, as was demonstrated by the interaction between nitrifying bacteria and denitrifying bacteria. To evaluate the nitrification and denitrification processes in marine sediments, therefore, application of techniques which allow the simultaneous determination of these activities in undisturbed sediments is quite important. Our continuous-flow sediment–water system satisfied these requirements to some extent, and the results obtained indicated the quantitative importance of nitrification and denitrification in the nitrogen cycle in coastal and estuarine sediments. Our experiments also suggested that the coupling and the relative importance of nitrification and denitrification were closely related to the texture and organic contents of the sediments.

Further examination of the microenvironments in sediment–water interface should be stressed, however, to understand the basic factors controlling the activities and abundance of nitrifying bacteria and denitrifying bacteria. This information is indispensable if we are to predict the nitrogen cycle in coastal and estuarine sediments.

This work was supported by grants 503025 and 56030022 from the Ministry of Education, Culture, and Science, Japan.

LITERATURE CITED

1. **Alexander, M.** 1977. Introduction to soil microbiology. Wiley, New York.
2. **Billen, G.** 1976. Evaluation of nitrifying activity in sediments by dark ^{14}C-carbonate incorporation. Water Res. **10**:51–57.
3. **Billen, G.** 1978. A budget of nitrogen cycling in North Sea sediments off Belgian coast. Estuarine Coastal Mar. Sci. **7**:127–146.
4. **Cole, J. A., and C. M. Brown.** 1980. Nitrate reduction to ammonia by fermentative bacteria: a short circuit in the biological nitrogen cycle. FEMS Microbiol. Lett. **7**:65–72.
5. **Cooper, G. S., and R. L. Smith.** 1963. Sequence of products formed during denitrification in some diverse western soils. Soil Sci. Soc. Am. Proc. **27**:659–662.
6. **Focht, D. D.** 1978. Methods for analysis of denitrification in soils, p. 433–490. *In* D. R. Nielsen and J. G. MacDonald (ed.), Nitrogen in the environment, vol. 2. Academic Press, Inc., New York.
7. **Goering, J. J., and V. A. Dugdale.** 1966. Estimates of the rates of denitrification in a subarctic lake. Limnol. Oceanogr. **11**:113–117.
8. **Herbert, R. A.** 1982. Nitrate dissimilation in marine and estuarine sediments, p. 53–71. *In* D. B. Nedwell and C. M. Brown (ed.), Sediment microbiology. Academic Press, London.
9. **Jones, G. A.** 1972. Dissimilatory metabolism of nitrate by the rumen microbiota. Can. J. Microbiol. **18**:1783–1787.
10. **Jørgensen, B. B.** 1977. Bacterial sulfate reduction within reduced microniches of oxidized marine sediments. Mar. Biol. **41**:7–17.
11. **Kaplan, W., I. Valiela, and J. M. Teal.** 1979. Denitrification in a salt marsh ecosystem. Limnol. Oceanogr. **24**:726–734.
12. **Kaspar, H. F., and J. M. Tiedje.** 1981. Dissimilatory reduction of nitrate and nitrite in the bovine rumen: nitrous oxide production and effect of acetylene. Appl. Environ. Microbiol. **41**:705–709.
13. **Knowles, R.** 1982. Denitrification. Microbiol. Rev. **46**:43–70.
14. **Koike, I., and A. Hattori.** 1978. Denitrification and ammonia formation in anaerobic coastal sediments. Appl. Environ. Microbiol. **35**:278–282.
15. **Koike, I., and A. Hattori.** 1978. Simultaneous determination of nitrification and nitrate reduction in coastal sediments by a ^{15}N dilution technique. Appl. Environ. Microbiol. **35**:853–857.
16. **Koike, I., and A. Hattori.** 1979. Estimates of denitrification in sediments of the Bering Sea shelf. Deep-Sea Res. **26A**:409–415.
17. **Koike, I., A. Hattori, and J. J. Goering.** 1978. Controlled ecosystem pollution experiment: effect of mercury on enclosed water columns. VI. Denitrification by marine bacteria. Mar. Sci. Commun. **4**:1–12.
18. **Nedwell, D. B.** 1975. Inorganic nitrogen metabolism in a eutrophicated tropical mangrove estuary. Water Res. **9**:221–231.
19. **Nishio, T., I. Koike, and A. Hattori.** 1982. Denitrification, nitrate reduction, and oxygen consumption in coastal and estuarine sediments. Appl. Environ. Microbiol. **43**:648–653.
20. **Nishio, T., I. Koike, and A. Hattori.** 1983. Estimates of denitrification and nitrification in coastal and estuarine sediments. Appl. Environ. Microbiol. **45**:444–450.
21. **Oren, A., and T. H. Blackburn.** 1979. Estimation of sediment denitrification rates at in situ nitrate concentrations. Appl. Environ. Microbiol. **37**:174–176.
22. **Payne, W. J.** 1973. Reduction of nitrogenous oxides by microorganisms. Bacteriol. Rev. **37**:409–452.
23. **Revsbech, N. P., B. B. Jørgensen, and T. H. Blackburn.** 1980. Oxygen in the sea bottom measured with a microelectrode. Science **207**:1355–1356.
24. **Seitzinger, S., S. Nixson, M. E. Q. Pilson, and S. Burke.** 1980. Denitrification and N$_2$O production in near-shore marine sediments. Geochim. Cosmochim. Acta **44**:1853–1860.
25. **Smith, K. L.** 1978. Benthic community respiration on the N.W. Atlantic Ocean. A *in situ* measurement from 40 m to 5200 m. Mar. Biol. **47**:337–347.
26. **Sørensen, J.** 1978. Denitrification rates in a marine sediment as measured by the acetylene inhibition technique. Appl. Environ. Microbiol. **36**:139–143.
27. **Stanford, G., J. O. Legg, S. Dzienia, and E. C. Simpson, Jr.** 1975. Denitrification and associated nitrogen transformation in soils. Soil Sci. **120**:147–152.
28. **Tiedje, J. M., A. J. Sexstone, D. D. Myrold, and J. A. Robinson.** 1982. Denitrification: ecological niches, competition and survival. Antonie van Leeuwenhoek J. Microbiol. Serol. **48**:569–583.
29. **Yoshinari, T., R. Hynes, and R. Knowles.** 1977. Acetylene inhibition of nitrous oxide reduction and measurement of denitrification in soil. Soil. Biol. Biochem. **9**:177–183.

Atmospheric-Biospheric Exchanges

Capacity of Aerobic Microorganisms to Utilize and Grow on Atmospheric Trace Gases (H$_2$, CO, CH$_4$)

RALF CONRAD

Max-Planck-Institut für Chemie, D-6500 Mainz, Federal Republic of Germany

Hydrogen, carbon monoxide, and methane are trace constituents of the atmosphere which are decomposed in soil and water ecosystems. Decomposition in soil accounts for 95% of the total sink activity in the atmospheric budget of H$_2$ (12, 46) and for 10 to 40% of that of CO (33, 46). Decomposition of CH$_4$ by soil was assumed to play only an insignificant role in the atmospheric CH$_4$ cycle (19), although decomposition was shown to occur under certain environmental conditions (24). Recent field studies in the tropics indicate, however, that soils may actually act as a significant sink in the atmospheric CH$_4$ cycle (W. Seiler, this volume; W. Seiler, R. Conrad, and D. Scharffe, J. Atmos. Chem., in press). In contrast to soil, oceans and lakes usually act as sources for atmospheric H$_2$, CO, and CH$_4$ (7, 8, 41, 47), but decomposition activities for atmospheric H$_2$, CO, and CH$_4$ have been demonstrated to occur in these water ecosystems as well (7, 8, 15, 17, 23, 41, 42). The capacity of soil and water to decompose atmospheric H$_2$, CO, and CH$_4$ has been ascribed to microbial activities (24, 46, 47). In fact, soils contain a multitude of microorganisms which can be isolated and cultivated on H$_2$, CO, or CH$_4$ as the sole source of energy (3, 50, 52). These groups of aerobic bacteria are named hydrogen bacteria, carboxydobacteria, and methanotrophic bacteria, respectively. They gain the energy for growth by oxidation of H$_2$, CO, or CH$_4$ with O$_2$. It has generally been assumed that these bacterial groups are responsible for the decomposition of atmospheric H$_2$, CO, and CH$_4$ in nature, and it has been further assumed that the oxidations of these atmospheric trace gases partially constitute the ecological niches for those bacterial groups (34, 44, 50).

However, these assumptions are based on laboratory cultures of bacteria which are supplied with relatively high concentrations of H$_2$, CO, or CH$_4$. The atmosphere, on the other hand, contains these gases only in mixing ratios of 0.1 to 2.0 ppmv (10^{-6} [vol/vol]) (39, 46). Since these values are four to six orders of magnitude lower than gas concentrations used in laboratory

experiments, the question arises whether the atmospheric mixing ratios can provide enough energy to support growth of the bacteria.

This paper deals with the kinetic properties that microbial populations must have to be able to grow on atmospheric H$_2$, CO, or CH$_4$. It will be shown that the presently known hydrogen bacteria, carboxydobacteria, and methanotrophic bacteria should not be able to grow at the expense of the gases at atmospheric concentrations. Decomposition rates for atmospheric H$_2$, CO, and CH$_4$ in soil are used to characterize microbial populations which might be responsible for the observed decomposition.

REQUIREMENTS FOR MAINTENANCE AND GROWTH

Microorganisms grow at the expense of substrates which are utilized for synthesis of new cell material, enabling the microorganisms to multiply. The molar yield coefficient Y_g (grams [dry weight] per mole) is the amount of cellular dry weight which is produced at the expense of 1 mol of substrate. The part of the substrate that is not utilized for cell growth but is consumed for maintenance requirements of the bacteria is defined by the maintenance coefficient m_e (moles per gram [dry weight] per hour). The total rate q (moles per gram [dry weight] per hour) of substrate consumption is correlated to the specific growth rate μ (hours^{-1}) of the microbial population by the following equation (37, 49):

$$q = \frac{\mu}{Y_g} + m_e \qquad (1)$$

It is evident that growth can occur only if the total substrate utilization is higher than the substrate consumed for maintenance; then:

$$q \geq m_e \qquad (2)$$

The capacity of the microorganism for uptake and metabolism of the substrate is determinative for the growth rate. Microorganisms which do not possess the enzymes necessary for metabolism of a particular substrate, or which have not yet synthesized enough of these enzymes, obviously cannot grow. The levels of the growth-limiting enzymes within the cell are given by their specific activities, i.e., V_{max}, usually in units per milligram of protein. However, the actual activity v of an enzyme is also dependent on the concentration S (molar) of the substrate. The kinetic relationship encountered most often is given by the Michaelis-Menten equation:

$$v = \frac{V_{max}S}{K_m + S} \qquad (3)$$

Equation 3 can also be used to describe the kinetic properties of whole cells for substrate utilization. In this case, the kinetic experiments are carried out with suspensions of microorganisms instead of cell-free extracts, and v and V_{max} are given in moles per gram of cellular dry weight per hour. The K_m value is independent of the cellular biomass and gives the substrate concentration at which the rate of substrate utilization is half of the maximum rate. The K_m value is not identical with the K_s value, which is defined from the kinetics of growth (Monod equation) and not of substrate utilization. However, both values give a measure of the affinity of microorganisms for their substrate.

Microorganisms growing on H_2, CO, or CH_4 use these substrates as energy sources or as electron donors, or as both, for synthesis of new cell material. Microorganisms growing on CH_4 do not need an additional carbon source, since they can use oxidation products of CH_4 for the assimilation into cell material (28). Microorganisms growing on CO do not need an additional carbon source either, since they are able to assimilate the CO_2 originating from CO oxidation (O. Meyer and H. G. Schlegel, Annu. Rev. Microbiol. in press). Microorganisms growing on H_2, however, must be supplied with CO_2 as carbon source, which can be assimilated into cell material via the Calvin cycle (6). Consequently, under environmental conditions, the growth rate of these microorganisms should be limited by the rate of oxidation of atmospheric H_2, CO, or CH_4, since CO_2 is abundant. The enzymes which determine the oxidation rate of H_2 and CO in aerobic microorganisms are the hydrogenase (1) and the carbon monoxide oxidoreductase (Meyer and Schlegel, in press), respectively. The methane monoxygenase (29) is the enzyme which catalyzes the initial reaction in the oxidation of CH_4, but it is unclear whether this enzyme reaction is the rate-limiting step.

At low concentrations of H_2, CO, or CH_4, growth on these gases as sole energy sources will be possible only if the oxidation rates of these gases are higher than the oxidation rates required for maintenance:

$$v \geqq m_e \qquad (4)$$

or, after combination with equation 3:

$$K_m \leqq S \frac{(V_{max} - 1)}{m_e} \qquad (5)$$

As a result, equation 5 defines the kinetic properties of substrate utilization that are required to allow maintenance metabolism at expense of the substrate.

Equation 5 shows that, at a given concentration of H_2, CO, or CH_4, the possibility for maintenance, and of course also for growth, is limited by the particular K_m and the V_{max}/m_e ratio of the microorganisms. The ratio V_{max}/m_e must necessarily be greater than one, since otherwise the maximum gas oxidation rate would be less than required for maintenance. The V_{max}/m_e ratio is expected to increase when the supply of substrate decreases, as microorganisms synthesize more enzyme so they can scavenge a greater proportion of the substrate. An example of this behavior has recently been observed by Friedrich (22), growing the hydrogen bacterium *Alcaligenes eutrophus* in a chemostat under H_2 limitation. In this experiment it was observed that the hydrogenase-specific activity of the cells increased exponentially with decreasing dilution rate. However, the synthesis of hydrogenase leveled off at growth rates lower than $\mu = 0.005$ h^{-1}, when specific hydrogenase activities of approximately 120 mmol g^{-1} (dry weight) h^{-1} (4 U mg^{-1} of protein) were reached. At this time, the cells most probably could no longer afford to expend the small amounts of energy that were available from the limited H_2 supply to synthesize more hydrogenase. At this time, the V_{max} of hydrogenase was approximately 60 times the maintenance coefficient (21; Jüttner, thesis, University of Göttingen, Göttingen, Germany, 1978). It seems reasonable to assume that the V_{max}/m_e ratios of microorganisms growing on atmospheric H_2, CO, or CH_4 will also be not much higher, since there is certainly a limitation to the enzyme levels that can be sustained at the expense of energy when the energy-yielding substrate itself is limiting for the enzyme reaction.

Table 1 shows the average mixing ratios M (ppmv) of H_2, CO, and CH_4 in the troposphere (39, 46). For the soil environment the corresponding gas concentrations S (nanomolar) in the aqueous phase of the soil are calculated from the Bunsen solubility coefficients in water at 20°C and a molar volume of the gases of 24.06 liters per mol. The values represent the maxi-

TABLE 1. Value of K_m necessary for maintenance of bacteria on atmospheric trace gases

Oxidation of	M Atmosphere (ppmv)	$s_{20,w}$ (nM)	V_{max}/m_e	$K_m{}^a$ (nM)			References
				Calculated	In soil and water	Bacteria	
H_2	0.56	0.42	10	3.8	11–83	500–60,000	1, 8, 9, 14, 18, 38, 46
			100	41.5			
			1,000	420.0			
CO	0.11	0.08	10	0.7	4–39	400–53,000	10, 15, 46; Meyer and
			100	7.9			Schlegel, in press
			1,000	80.0			
CH_4	1.7	2.50	10	22.5	5000	19,000–26,000	25, 26, 39, 40, 51
			100	250.0			
			1,000	2,500.0			

^{a} $K_m = S[N_{max} - 1)/m_e]$.

mum concentrations of H_2, CO, and CH_4 being reached by diffusion of the trace gases from the atmosphere into the soil environment. Equation 5 was used to calculate the maximum K_m values allowing the soil microorganisms to gain enough energy for maintenance. Table 1 shows the K_m values calculated for a range of V_{max}/m_e ratios from 10 to 1,000. It is evident that soil microorganisms can grow on atmospheric trace gases only if either their K_m values are low (affinity high) or their enzyme levels (V_{max}) are high. Under conditions with a small but permanent supply of energy via atmospheric trace gases to the soil environment, the microorganisms with a low K_m value are certainly at an advantage compared to those with a high K_m value, since they need smaller enzyme levels to compensate the maintenance requirements by oxidation of H_2, CO, or CH_4. At a V_{max}/m_e ratio of 100, K_m values of approximately 40, 8, and 250 nM are calculated for H_2, CO, and CH_4, respectively. It must be emphasized that at these K_m values the atmospheric mixing ratios of H_2, CO, and CH_4 would just be sufficient for the maintenance requirements. To reach a growth rate of $\mu = 1$ day^{-1} on atmospheric trace gases, the K_m values must be approximately one-third of those listed in Table 1.

Table 1 shows in addition that the K_m values necessary for growth on atmospheric H_2, CO, or CH_4 are at least one order of magnitude lower than those reported for hydrogen bacteria (1, 9, 38), carboxydobacteria (10; Meyer and Schlegel, in press), or methanotrophic bacteria (25, 51), respectively. Therefore, the bacteria of these groups which have so far been tested for their kinetic properties of H_2, CO, or CH_4 utilization are not able to grow on atmospheric trace gases. This does not exclude, however, that these bacteria consume atmospheric trace gases as long as they are supplied with additional energy sources for mixotrophic growth (30, 43). It has

been shown that carboxydobacteria are able to consume atmospheric CO (13), although their relatively high K_m values for CO exclude that they contribute significantly to the decomposition of atmospheric CO in soil (10). The K_m values were the same when the kinetics of CO utilization were studied by adding cell suspensions of *Pseudomonas carboxydovorans* to sterile soil (10), where soil organic compounds are present as additional substrates. Hence, it is unlikely that the affinity for CO would increase when carboxydobacteria are provided with additional substrates for mixotrophic growth. The hydrogen bacteria, on the other hand, which have so far been tested are unable to consume atmospheric H_2 (11), since they can only utilize H_2 above a particular threshold concentration which is generally higher than atmospheric H_2 mixing ratios (9). Atmospheric H_2 was also not utilized when suspensions of hydrogen bacteria (e.g., *Xanthobacter autotrophicus*) were incubated for 9 days in sterile soil under ambient atmospheric conditions (11). Hence, the bacteria were not able to induce enzyme systems specific for utilization of H_2 at atmospheric concentrations, and they were also not stimulated by the soil organic compounds. However, it is certainly desirable to test rigorously for the possibility of the induction of high-affinity enzyme systems for utilization of atmospheric H_2, CO, and CH_4 in hydrogen bacteria, carboxydobacteria, and methanotrophic bacteria, as well as to test for the effects of additional substrates on the kinetic properties for utilization of H_2, CO, and CH_4. The latter test seems to be of importance, since growth kinetics in multiple substrate systems often show higher affinities for a particular substrate when the microorganisms grow mixotrophically (31).

It is interesting that the K_m values for H_2 and CO calculated by equation 5 for a V_{max}/m_e ratio of 100 are in the same order of magnitude as the

K_m values which have so far been measured in soil (10, 14, 15, 18) and water (8, 15) ecosystems (Table 1). This observation could be explained by the existence of oligotrophic microorganisms with a high affinity for H_2 and CO which are yet unknown and await isolation in pure cultures. K_m values for oxidation of CH_4 in soil have so far not been reported. For lake water ecosystems K_m values of 4 to 5 μM CH_4 have been determined (26, 40) which may be sufficient for maintenance and growth of a microbial population at the relatively high CH_4 concentrations (e.g., 10 μM) in lake water, but certainly not for the survival of a soil microbial population growing on atmospheric CH_4 as sole energy source.

DECOMPOSITION RATES AND MICROBIAL POPULATION DENSITIES

The strains of hydrogen bacteria, carboxydobacteria, and methanotrophic bacteria which have so far been assayed cannot possibly account for the decomposition of atmospheric trace gases in soil, because of their inadequate affinities and activities toward environmental gas concentrations. However, it is possible that other, yet unknown species utilize the atmospheric trace gases or even grow on them. These species should have K_m values for H_2, CO, or CH_4 which are in the order of those determined for natural environments (Table 1). The microorganisms may be called "high-affinity" trace gas-consuming microorganisms, and they may be oligotrophs which either grow on the atmospheric trace gases as sole energy source or consume them during mixotrophic growth as an additional source of energy or electrons, or they may be microbes which consume the atmospheric trace gases fortuitously in co-oxidation.

The population density N (cells per square centimeter) per soil surface area of a community of those "high-affinity" microorganisms with a given K_m value should be correlated to the decomposition rate d (moles per square centimeter per hour) of atmospheric H_2, CO, and CH_4 at the soil surface independently of their ability to utilize these gases for growth, mixotrophy, or co-oxidation. Assuming a cell biomass of 1 pg (dry weight) as representative for soil bacteria (2), the following equation is obtained:

$$d = 10^{-12} N \frac{V_{max}S}{K_m + S} \qquad (6)$$

The decomposition rates (d) of atmospheric trace gases at the soil surface are known from the literature (12, 13, 20, 24, 32, 46; Seiler et al., in press). Representative values are given in Table 2. The concentrations of atmospheric H_2, CO, and CH_4 in the aqueous phase of soil are the same as in Table 1. The K_m values of the "high-

affinity" microorganisms listed in Table 2 are assumed to be similar to those determined in natural environments (Table 1). The population densities N of the "high-affinity" microorganisms are further dependent on their V_{max} values. Table 2 shows values for population densities N which were calculated for V_{max} values of 1 to 100 mmol g^{-1} (dry weight) h^{-1}. A V_{max} value of 100 mmol g^{-1} (dry weight) h^{-1} (approximately 3.3 U of enzyme activity per mg of protein) is a rather high value in relation to our knowledge of laboratory bacterial cultures grown under optimal conditions. It is reasonable to assume that this value will most probably not be reached under conditions present in the natural environment.

Table 2 shows that even at high V_{max} values relatively large population densities per soil surface area are necessary to explain the in situ rates of decomposition of atmospheric trace gases. It must be pointed out that the calculated population densities are only based on assumptions concerning K_m, V_{max}, and cell biomass of the "high-affinity" microorganisms. It is not necessarily required that the atmospheric trace gases are utilized by the microbes in a profitable way. They might also be utilized by fortuitous co-oxidation.

In the case of H_2 and CO, it has been shown that the atmospheric trace gases are usually decomposed within the top 1 cm of the soil profile (32). This suggests that the top soil layers contain, per g of soil, at least 10^6 CO-oxidizing and 4×10^6 H_2-oxidizing microorganisms. These numbers represent 0.1 to 0.4% of a total soil microbial population of 10^9 cells per g of soil.

In the case of CH_4, population densities of at least 10^7 cells per cm^2 would be necessary to explain the decomposition rate of atmospheric CH_4 in soil. In this case, however, the K_m value of the soil microbial population might be lower than that of the lake water microbial population (26, 40) that has been used for the calculation. In addition, it is yet unclear how thick the soil layer is in which the atmospheric CH_4 is decomposed. Therefore, the relatively high microbial population density on a soil surface basis might turn out to be relatively low on a soil mass basis.

The results of the calculations comparing in situ decomposition rates with the kinetic properties and population densities of "high-affinity" trace gas consumers clearly show how narrow the limits are in which the kinetic and population parameters may vary to explain the decomposition of atmospheric trace gases in soil. Unfortunately, similar calculations cannot be done for aquatic environments, where trace gases are consumed as well (7, 8, 17, 23, 41, 42). However, in contrast to soils, aquatic environments usually act as a source for atmospheric H_2, CO,

TABLE 2. Microbial population densities to explain decomposition rates of atmospheric trace gases in soil

Oxidation of	d (nmol cm^{-2} h^{-1})	$s_{20,w}$ (nM)	K_m (nM)	V_{max} (mmol g^{-1} (dry wt) h^{-1})	N^a (cells/cm^{-2})	References
H$_2$	5	0.42	40	1	4×10^8	8, 9, 14, 18, 46
				10	4×10^7	
				100	4×10^6	
CO	1	0.08	10	1	1×10^8	15, 46
				10	1×10^7	
				100	1×10^6	
CH$_4$	0.5	2.50	5000	1	1×10^9	24, 26, 39, 40; Seiler et al.,
				10	1×10^8	in press
				100	1×10^7	

^{a} $N = 10^{12}d[(K_m + S)/V_{max}S]$.

and CH$_4$. This is true for oceans (47) and lakes (7, 8, 41), both of which generally are supersaturated with respect to the atmosphere. Under these circumstances it is difficult to determine the trace gas concentrations (S) which are actually available to the consumers. The concentrations of dissolved H$_2$, CO, and CH$_4$ which are measured in water samples do not represent those values, but represent steady-state values due to the simultaneously occurring production and consumption processes. H$_2$ production in water is most probably due to nitrogen-fixing cyanobacteria (8, 45) or anaerobic bacteria associated with decomposing biomass (36); CO production is caused by photooxidation of organic matter (17, 48); and CH$_4$ production may take place at anaerobic microsites such as suspended zooplankton (35). At these production sites the trace gas concentrations may be substantially higher than the steady-state concentration.

There is one example of an aquatic ecosystem acting as a sink for atmospheric H$_2$ which may be used for calculation of population densities in a manner similar to that described for soil. Herr et al. (27) reported H$_2$ flux rates of 0.01 to 0.07 nmol cm^{-2} h^{-1} from the atmosphere into the water of the Norwegian Sea, which was well mixed within the upper 300 m of the water column. If we assume a "high-affinity" microbial population with a K_m value of 40 nM for H$_2$ and a V_{max} value of <100 mmol g^{-1} (dry weight) h^{-1}, a population density of 4×10^2 to 2×10^3 cells per liter of water would be necessary to explain the observed flux rate. Again, a substantial population of H$_2$ consumers is necessary to explain the decomposition of atmospheric H$_2$.

DISCUSSION

Calculations based on kinetic data revealed that microbial growth on atmospheric trace gases as sole source of energy can be expected only if new microorganisms with a high affinity for their gaseous substrate exist in the environment. It is concluded that the presently described strains of hydrogen bacteria, carboxydobacteria, and methanotrophic bacteria cannot grow on atmospheric trace gases as sole energy source, because their affinities are much lower than those observed in soil and water ecosystems. It appears that there are two different ecological niches in soil or water. The first ecological niche is supplied from the atmosphere with low concentrations of H$_2$, CO, and CH$_4$, and supports the "high-affinity" microorganisms. The second one is supplied from sources within the soil or water with at least sporadically high concentrations of H$_2$, CO, and CH$_4$, e.g., from fermentation, and supports the hydrogen bacteria, carboxydobacteria, and methanotrophic bacteria, respectively. It should be pointed out, however, that more strains of hydrogen bacteria, carboxydobacteria, and especially methanotrophic bacteria should be assayed for their kinetic properties with respect to utilization of their gaseous substrates to provide a better data basis. It should also be investigated whether these strains might be able to induce a "high-affinity" enzyme system specific for the utilization of trace gases, and whether they might exhibit higher affinities when they grow mixotrophically (31).

However, even "high-affinity" microorganisms must have relatively large population densities to explain the decomposition rates of atmospheric H$_2$, CO, and CH$_4$ in soil. Their population densities and kinetic properties may vary within narrow limits only. This conclusion is correct whether the atmospheric trace gases are used as sole energy substrate for growth, as a profitable co-substrate for mixotrophic growth, or as a fortuitous substrate in co-oxidation. The kinetic and population limitations may not apply in the case of H$_2$, however, since soil hydrogenases may contribute substantially to the decomposition of atmospheric H$_2$ in soil (14, 18). For CO, on the other hand, the only conceivable explanation is decomposition by microorganisms. These may be microorganisms which consume CO by co-oxidation as reported by Bartholomew and Alexander (4, 5). Conrad and Seiler (15), on the other hand, have shown that microbial soil populations are able to increase their CO-oxidizing activity when soil suspensions are gassed with pressurized ambient air, indicating that their mode of activity is a profitable use of the CO in air. If not as main energy substrate, CO may be utilized as a profitable co-substrate in a manner similar to that shown for a carboxydobacterium (30). The maintenance of a soil population density of $>10^6$ cells per cm^2 of CO-oxidizing oligotrophs may be facilitated by the fact that the CO concentrations in soil sometimes may be substantially higher than ambient, allowing sporadically high growth rates of the oligotrophic CO-oxidizing population. CO concentrations higher than ambient establish especially under arid soil conditions when CO production from chemical oxidation of soil organic matter is so high that the soil is acting as a source for atmospheric CO (16).

CONCLUSIONS

The strains of hydrogen bacteria, carboxydobacteria, and methanotrophic bacteria so far tested exhibit K_m values for the oxidation of H_2, CO, or CH_4 that are too high to allow maintenance metabolism or growth on atmospheric H_2, CO, or CH_4 as sole source of energy.

The kinetics of H_2 and CO uptake by soils exhibit K_m values which are orders of magnitudes lower than those observed in pure cultures. The kinetics of CH_4 uptake by soil are yet unknown.

If we assume that the bacterial strains described exhibit the same kinetic properties in their natural environment as in laboratory cultures, they cannot account for the uptake of atmospheric H_2, CO, and CH_4 by soil. Research should be directed to the question of whether the described strains might be able to induce a "high-affinity" system for utilization of trace gases and whether they might exhibit higher affinities for the gases when growing mixotrophically.

Microorganisms not only must have high affinities for H_2, CO, and CH_4, but also must be present in relatively high numbers in soil to explain the rates observed for the flux of H_2, CO, and CH_4 from the atmosphere into the soil. High population densities of "high-affinity" microorganisms are required independently of whether they oxidize atmospheric trace gases as sole energy sources for growth, as an additional energy source for mixotrophy, or by nonprofitable co-oxidation.

I am grateful to M. Aragno, J. G. Kuenen, O. Meyer, B. Schink, W. Seiler, and J. G. Zeikus for stimulating discussion.

LITERATURE CITED

1. **Adams, M. W. W., L. E. Mortenson, and J. S. Chen.** 1981. Hydrogenase. Biochim. Biophys. Acta **594:**105–176.
2. **Alexander, M.** 1977. Introduction to soil microbiology. John Wiley & Sons, Inc., New York.
3. **Aragno, M., and H. G. Schlegel.** 1981. The hydrogen-oxidizing bacteria, p. 865–893. *In* M. P. Starr, H. Stolp, H. G. Trüper, A. Bellows, and H. G. Schlegel (ed.), The prokaryotes. A handbook of habitats, isolation and identification of bacteria, vol. 1. Springer Verlag, Berlin.
4. **Bartholomew, G. W., and M. Alexander.** 1979. Microbial metabolism of carbon monoxide in culture and in soil. Appl. Environ. Microbiol. **37:**932–937.
5. **Bartholomew, G. W., and M. Alexander.** 1982. Microorganisms responsible for the oxidation of carbon monoxide in soil. Environ. Sci. Technol. **16:**300–301.
6. **Bowien, B., and H. G. Schlegel.** 1981. Physiology and biochemistry of aerobic hydrogen-oxidizing bacteria. Annu. Rev. Microbiol. **35:**405–452.
7. **Conrad, R., M. Aragno, and W. Seiler.** 1983. Production and consumption of carbon monoxide in a eutrophic lake. Limnol. Oceanogr. **28:**42–49.
8. **Conrad, R., M. Aragno, and W. Seiler.** 1983. Production and consumption of hydrogen in a eutrophic lake. Appl. Environ. Microbiol. **45:**502–510.
9. **Conrad, R., M. Aragno, and W. Seiler.** 1983. The inability of hydrogen bacteria to utilize atmospheric hydrogen is due to threshold and affinity for hydrogen. FEMS Microbiol. Lett. **18:**207–210.
10. **Conrad, R., O. Meyer, and W. Seiler.** 1981. Role of carboxydobacteria in consumption of atmospheric carbon monoxide by soil. Appl. Environ. Microbiol. **42:**211–215.
11. **Conrad, R., and W. Seiler.** 1979. The role of hydrogen bacteria during the decomposition of hydrogen by soil. FEMS Microbiol. Lett. **6:**143–145.
12. **Conrad, R., and W. Seiler.** 1980. Contribution of hydrogen production by biological nitrogen fixation to the global hydrogen budget. J. Geophys. Res. **85:**5493–5498.
13. **Conrad, R., and W. Seiler.** 1980. Role of microorganisms in the consumption and production of atmospheric carbon monoxide by soil. Appl. Environ. Microbiol. **40:**437–445.
14. **Conrad, R., and W. Seiler.** 1981. Decomposition of atmospheric hydrogen by soil microorganisms and soil enzymes. Soil Biol. Biochem. **13:**43–49.
15. **Conrad, R., and W. Seiler.** 1982. Utilization of traces of carbon monoxide by aerobic oligotrophic microorganisms in ocean, lakes and soil. Arch. Microbiol. **132:**41–46.
16. **Conrad, R., and W. Seiler.** 1982. Arid soils as a source of atmospheric carbon monoxide. Geophys. Res. Lett. **9:**1353–1356.
17. **Conrad, R., W. Seiler, G. Bunse, and H. Giehl.** 1982. Carbon monoxide seawater (Atlantic Ocean). J. Geophys. Res. **87:**8839–8852.
18. **Conrad, R., M. Weber, and W. Seiler.** 1983. Kinetics and electron transport of soil hydrogenases catalyzing the oxidation of atmospheric hydrogen. Soil Biol. Biochem. **15:**167–173.
19. **Ehhalt, D. H., and U. Schmidt.** 1978. Sources and sinks of atmospheric methane. Pure Appl. Geophys. **116:**452–464.
20. **Fallon, R. D.** 1982. Molecular tritium uptake in southeastern U.S. soils. Soil Biol. Biochem. **14:**553–556.
21. **Friedrich, C. G.** 1981. Cellular yields and maintenance requirements of Alcaligenes eutrophus during hydrogen oxidation. Adv. Biotechnol. **1:**277–280.
22. **Friedrich, C. G.** 1982. Derepression of hydrogenase during limitation of electron donors and derepression of ribulose biphosphate caboxylase during carbon limitation of *Alcaligenes eutrophus*. J. Bacteriol. **149:**203–210.
23. **Hanson, R. S.** 1980. Ecology and diversity of methylotrophic organisms. Adv. Appl. Microbiol. **26:**3–39.
24. **Harriss, R. C., D. I. Sebacher, and F. P. Day, Jr.** 1982. Methane flux in the Great Dismal Swamp. Nature (London) **297:**673–674.

25. **Harrisson, D. E. F.** 1973. Studies on the affinity of methanol- and methane-utilizing bacteria for their carbon substrates. J. Appl. Bacteriol. **36:**301–308.

26. **Harrits, S. M., and R. S. Hanson.** 1980. Stratification of aerobic methane-oxidizing organisms in Lake Mendota, Madison, Wisconsin, Limnol. Oceanogr. **25:**412–421.

27. **Herr, F. L., M. I. Scranton, and W. R. Barger.** 1981. Dissolved hydrogen in the Norwegian Sea: mesoscale surface variability and deep-water distribution. Deep-Sea Res. **28:**1001–1016.

28. **Higgins, I. J.** 1979. Microbial biochemistry of methane—a study in contrasts. Part 2. Methanotrophy. Int. Rev. Biochem. **21:**300–353.

29. **Higgins, I. J.** 1980. Respiration in methylotrophic bacteria, p. 187–221. *In* C. J. Knowles (ed.), Diversity of bacterial respiratory systems, vol. 1. CRC Press, Boca Raton, Fla.

30. **Kiessling, M., and O. Meyer.** 1982. Profitable oxidation of carbon monoxide or hydrogen during heterotrophic growth of Pseudomonas carboxydoflava. FEMS Microbiol. Lett. **13:**333–338.

31. **Law, A. T., and D. K. Button.** 1977. Multiple-carbon-source-limited growth of a marine coryneform bacterium. J. Bacteriol. **129:**115–123.

32. **Liebl, K. H., and W. Seiler.** 1976. CO and H_2 destruction at the soil surface, p. 215–229. *In* H. G. Schlegel, G. Gottschalk, and N. Pfennig (ed.), Production and utilization of gases. Goltze KG, Göttingen.

33. **Logan, J. A., M. J. Prather, S. C. Wofsy, and M. B. McElroy.** 1981. Tropospheric chemistry: a global perspective. J. Geophys. Res. **86:**7210–7254.

34. **Nozhevnikova, A. N., and L. N. Yurganov.** 1978. Microbial aspects of regulating the carbon monoxide content of the earth's atmosphere. Adv. Microb. Ecol. **2:**203–244.

35. **Oremland, R. S.** 1979. Methanogenic activity in plankton samples and fish intestines: a mechanism for in situ methanogenesis in oceanic surface waters. Limnol. Oceanogr. **24:**1136–1141.

36. **Oremland, R. S.** 1983. Hydrogen metabolism by decomposing cyanobacterial aggregates in Big Soda Lake, Nevada. Appl. Environ. Microbiol. **45:**1519–1525.

37. **Pirt, S. J.** 1965. The maintenance energy of bacteria in growing cultures. Proc. R. Soc. London Ser. B **163:**224–231.

38. **Probst, I.** 1980. Respiration of hydrogen bacteria, p. 159–181. *In* C. J. Knowles (ed.), Diversity of bacterial respiratory systems, vol. 2. CRC Press, Boca Raton, Fla.

39. **Rasmussen, R. A., and M. A. K. Khalil.** 1981. Atmospheric methane (CH_4): trends and seasonal cycles. J. Geophys. Res. **86:**9826–9832.

40. **Rudd, J. W. M., and R. D. Hamilton.** 1975. Factors controlling methane oxidation and the distribution of methane oxidizers in a small stratified lake. Arch. Hydrobiol. **75:**522–538.

41. **Rudd, J. W. M., and C. D. Taylor.** 1980. Methane cycling in aquatic environments. Adv. Aquat. Microbiol. **2:**77–150.

42. **Sansone, F. J., and C. S. Martens.** 1978. Methane oxidation in Cape Lookout Bight, North Carolina. Limnol. Oceanogr. **23:**349–355.

43. **Schink, B., and H. G. Schlegel.** 1978. Mutants of Alcaligenes eutrophus defective in autotrophic metabolism. Arch. Microbiol. **117:**123–129.

44. **Schlegel, H. G.** 1974. Production, modification and consumption of atmospheric trace gases by microorganisms. Tellus **26:**1–20.

45. **Scranton, M. I.** 1983. The role of the cyanobacterium Oscillatoria (Trichodesmium) thiebautii in the marine hydrogen cycle. Mar. Ecol. Prog. Ser. **11:**79–87.

46. **Seiler, W.** 1978. The influence of the biosphere on the atmospheric CO and H_2 cycles, p. 773–810. *In* W. Krumbein (ed.), Environmental biogeochemistry and geomicrobiology, vol. 3. Ann Arbor Scientific Publishers, Ann Arbor, Mich.

47. **Seiler, W., and U. Schmidt.** 1974. Dissolved nonconservative gases in seawater, p. 219–243. *In* E. D. Goldberg (ed.), The sea, vol. 5: Marine chemistry. John Wiley & Sons, Inc., New York.

48. **Setser, P. J., J. L. Bullister, E. C. Frank, N. L. Guinasso, Jr., and D. R. Schink.** 1982. Relationships between reduced gases, nutrients and fluorescence in surface waters off Baja California. Deep-Sea Res. **29:**1203–1215.

49. **Tempest, D. W., and O. M. Neijssel.** 1978. Eco-physiological aspects of microbial growth in aerobic nutrient-limited environments. Adv. Microb. Ecol. **2:**105–153.

50. **Whittenbury, R., J. Colby, H. Dalton, and H. L. Reed.** 1976. Biology and ecology of methane oxidizers, p. 231–292. *In* H. G. Schlegel, G. Gottschalk, and N. Pfennig (ed.), Microbial production and utilization of gases. E. Goltze KG, Göttingen.

51. **Wilkinson, T. G., and D. E. F. Harrisson.** 1973. The affinity for methane and methanol of mixed cultures grown on methane in continuous culture. J. Appl. Bacteriol. **36:**309–313.

52. **Zavarzin, G. A., and A. N. Nozhevnikova.** 1977. Aerobic carboxydobacteria. Microb. Ecol. **3:**305–326.

Contribution of Biological Processes to the Global Budget of CH_4 in the Atmosphere

WOLFGANG SEILER

Max-Planck-Institut für Chemie, 6500 Mainz, Federal Republic of Germany

The atmospheric cycle of methane has recently received considerable attention since it was found that the atmospheric burden of methane is not stable and exhibits a temporal increase in both hemispheres (3, 33). The rate of CH_4 increase during the past 4 years is reported to be on the order of 1.5% per year. Our measurements, carried out over Europe at latitudes of 40 to 50° north, demonstrated an increase in the tropospheric CH_4 of 1.5 to 1.7% per year or 0.027 ppm (vol/vol) (ppmv) per year. These values were obtained from aircraft measurements between the tropopause and the upper boundary of the planetary boundary layer so that the given figures represent average values for the clean troposphere and thus can be extended to other longitudes of the Northern Hemisphere. The aircraft measurements, started in 1976, cover a period of almost 8 years and represent the longest continuous record of tropospheric CH_4 presently available.

If the present rate of CH_4 increase is extrapolated back to December 1972, we calculate a CH_4 mixing ratio of 1.44 ppmv for the Northern Hemisphere and 1.36 ppmv for the Southern Hemisphere, which agrees excellently with the data actually observed over the Pacific Ocean during the same time period (43). Therefore, it is most likely that the CH_4 increase of 1.5 to 1.7% per year has occurred during at least the past 10 years.

Information on the possible increase in atmospheric CH_4 before 1972 is difficult to obtain. Data reported in the literature are highly variable. In addition, considerable confusion on the accuracy of the older CH_4 values arose after Heidt and Ehhalt (19) corrected their former data by a factor of 1.2. Therefore, the data reported for the period between 1964 and 1972 now vary between 1.2 and 1.7 ppmv. On the basis of spectroscopic data obtained at Jungfraujoch, Switzerland, since 1948, Ehhalt et al. (15) concluded that the total increase of CH_4 during the past 30 years cannot be higher than 10%, which corresponds to the increase observed during the past 6 years. The best evidence for a long-term increase of CH_4 in the troposphere is analyses of air bubbles trapped in ice cores, which show CH_4 values of 0.6 to 0.8 ppm over the period of 100 to 300 years ago (8, 21, 35). These values were found in Arctic and Antarctic ice cores, indicating that the atmospheric CH_4 burden obviously has increased in both hemispheres by more than a factor of 2 during the past two centuries.

Because of the chemical and physicochemical properties of methane, the observed increase of the atmospheric CH_4 burden may have some environmental consequences. First of all, methane has an absorption band in the infrared spectrum. An increase of atmospheric CH_4 would therefore perturb the earth's radiation balance, leading to the so-called "greenhouse effect." Model calculations (25, 32) indicate that doubling the CH_4 mixing ratios in the troposphere will cause an increase in the average tropospheric temperature of 0.2 to 0.3°C. A temperature increase of 1 to 2°C will occur predominantly in the higher latitudes north and south of 50°. Temperatures in the tropical and subtropical regions may not, however, be significantly influenced by the CH_4 increase.

Second, CH_4 is oxidized in the troposphere by the free radical OH, which, as we will see later, represents the most dominant sink mechanism for atmospheric CH_4. The oxidation of CH_4 will result in the production of a series of gaseous intermediates and final products such as CO, H_2, HCHO, O_3, and other radicals, which in turn will again have some impact on the chemical composition of the troposphere. Of particular interest is the formation of O_3, which is one important precursor of the OH radical and which, similarly to CO_2 and CH_4, belongs to the "greenhouse gases." In this case, doubling of the tropospheric O_3 would cause an average tropospheric temperature increase of 0.9°C (25). There is evidence from model calculations that the tropospheric O_3 will increase by approximately 20% as a result of doubling the tropospheric CH_4 concentration. Consequently, including the greenhouse effect by CH_4 itself, the increase of CH_4 mixing ratios within the past two centuries should have already resulted in an elevation of average tropospheric temperatures of 0.4°C.

Third, CH_4 is the predominant source of water vapor in the stratosphere as a result of its oxidation to CO_2 and H_2O at altitudes of 10 to 40 km. Changes in the tropospheric CH_4 mixing ratios in the troposphere will increase the flux of CH_4 into the stratosphere and, consequently,

468

the stratospheric H_2O mixing ratio. This in turn will have some impact on the distribution of OH and other odd hydrogen species in the stratosphere.

Finally, the increase of tropospheric CH_4 will enhance the reaction between CH_4 and Cl:

$$CH_4 + Cl \rightarrow HCl + CH_3$$

which is the dominant sink mechanism for the stratospheric chlorine radicals Cl and ClO, which act as catalysts in the destruction reactions of stratospheric ozone. Thus, higher CH_4 mixing ratios in the troposphere would reduce the impact of human-induced chlorofluorocarbons on the stratospheric ozone layer, which later would provide a positive effect of an increase in the levels of tropospheric CH_4.

Summarizing, we see that methane has an important impact on climate and, because of the coupling with the O_3 cycle, also on the chemical composition of the atmosphere. Clearly, because of this important impact there is an urgent need to understand the reasons for the observed increases in tropospheric CH_4 and to make predictions on the possible rate of the further increase in CH_4.

Satisfactory and reliable answers can be given only if we know the cycling of CH_4 in the troposphere, i.e., the sources and sinks of tropospheric methane and the strengths of the sources and sinks. This paper focuses on the present knowledge of the biogenic production of methane. It will be demonstrated that the biogenic CH_4 release into the atmosphere is a major portion of the tropospheric CH_4 cycle but that its contribution to the global CH_4 budget is less than originally assumed.

BIOGENIC CH_4 PRODUCTION

Biogenic CH_4 is formed during the mineralization of organic matter through microbiological metabolism under strict anaerobic conditions. Part of the methane produced by this process is released into the atmosphere and thus contributes to the global atmospheric CH_4 cycle. The common individual biogenic CH_4 sources are the gastrointestinal fermentation occurring in animals and the microbial activities within sediments and paddy fields. The production rates of these individual sources and their temporal changes are discussed separately in the following sections.

CH_4 production by ruminants. Estimates of CH_4 production by gastrointestinal fermentation, mainly in ruminants, are highly variable. The first estimate was published by Hutchinson (20), who calculated the CH_4 release by large herbivores such as cattle, horses, goats, and sheep to be 45×10^{12} g (45 Tg) per year over the past two decades. About 85% of this figure, or 37 Tg per year, was assumed to be released by domestic cattle, which, therefore, represent the most important individual CH_4 source within this category. Extrapolating the given CH_4 release rates (20) to more recent times and considering the possible CH_4 production by other herbivores led to the adoption, in 1970, of an estimated annual CH_4 production by ruminants of 100 to 220 Tg. This figure already accounts for 30 to 70% of the photochemical CH_4 destruction of 320 Tg per year (10), which represents the most dominant sink within the atmospheric CH_4 cycle.

Since 1949, more data on the CH_4 release rates by ruminants have become available so that presently a more accurate estimate of the CH_4 production by these fermentations is possible. The CH_4 release rates from cattle are dependent on the amount and quality of the ingested diet. Approximately 5% of the energy intake is released as CH_4 by dairy and feeder cattle. This figure increases to approximately 10% for range-fed cattle. On the basis of data on food consumption by livestock, we calculate the average CH_4 release rate from cattle to be 50 to 60 kg per head per year for developed countries. This estimate agrees reasonably well with the previously reported figure (20). Because of the lower food consumption rate of cattle in developing countries, the CH_4 release rate for these countries is estimated to be 40 to 50 kg per head per year.

According to the statistical data reported by the Food and Agriculture Organization (16), the global population of cattle and buffaloes was 1,330 million head for 1975, compared with 1,000 million head for 1960, 800 million head for 1950, and 730 million head for 1940, indicating a 2% increase of the population during the past three decades. Similarly, the total population of sheep and goats has increased from 1,030 million head in 1940 to 1,450 million head in 1975, or approximately 1% per year. In contrast, the total population of horses, mules, and asses remained constant, with figures of 120 to 130 million head.

Approximately 35% of the cattle population is located in developed countries and 65% is in the developing countries. Applying the individual CH_4 release rates estimated for these categories, we obtain a total CH_4 release by cattle and buffaloes of 58 to 71 Tg per year for 1975, compared with 44 to 53 Tg per year in 1960. The release rates for the other domestic ruminants are estimated to be 13 to 15 Tg per year so that the total CH_4 production of domestic ruminants accounts for 71 to 86 Tg per year (Table 1).

Information on the total population of nondomestic ruminants such as antelopes, zebras, deer, etc., and their average CH_4 release rates per head is very scarce. McDowell (30) estimat-

TABLE 1. CH_4 production by ruminants for 1975

Ruminants	CH_4 production (kg per head per year)	Population (head)	CH_4 production (Tg of CH_4 per year)
Domestic			
Cattle and buffalo			
Developed countries	50–60	466×10^6	23–28
Developing countries	40–50	864×10^6	35–43
Sheep and goats	6–7	$1,440 \times 10^6$	9–10
Horses, mules, etc.	30–40	125×10^6	4–5
Total			71–86
Nondomestic			
Deer, antelope, caribou, zebra, etc.	5–25	130×10^6–530×10^6	0.7–13
Total (all ruminants)			72–99

ed the total population of wild ruminants to range between 130 million and 530 million head, which is relatively small compared with the corresponding figure for domestic ruminants. Assuming similar CH_4 release rates as observed for domestic ruminants and considering the different body weights, we expect the average CH_4 release rate to be on the order of 5 to 25 kg of CH_4 per head per year, which would result in a total CH_4 production by wild ruminants of 0.7 to 1.3 Tg per year. Adding this figure to the CH_4 production by domestic ruminants, we obtain the global production of CH_4 by ruminants to be 72 to 99 Tg per year for 1975, which is a factor of 2 lower than the figure given by Ehhalt and Schmidt (14) for 1970. The temporal change of the total CH_4 production by ruminants is illustrated in Fig. 1, indicating that the CH_4 production by ruminants has almost doubled within the past 40 years.

CH_4 production by rice paddies. Another important source for atmospheric CH_4 is the CH_4 production in rice paddies. Methane production rates in these systems have been estimated to be 190 Tg per year (24) and 280 Tg per year (14). These figures were based on results obtained from a limited number of laboratory experiments in which paddy soil that had been incubated for several weeks at different temperatures was used (23, 24). The experiments were carried out in the absence of plants, which, as we will see later, are the most important transport vehicles for the gas exchange of CH_4 between the paddy soil and the atmosphere. Since the roots provide a substantial fraction of the organic carbon of the paddy soil and methanogenesis is strongly dependent on the amount of substrates, it is very doubtful whether the CH_4 release rates obtained from laboratory experiments can be applied to estimating the contribution of rice paddies to the atmospheric CH_4 budget.

In fact, much lower CH_4 release rates were observed during in situ measurements in rice paddies in California, which, applied to global conditions, resulted in a global production of

CH_4 by rice paddies of 59 Tg per year (7). These measurements were, however, carried out during a few days in August, and thus the data base on this system for CH_4 release rates is very limited. The authors therefore concluded that their estimate was "far from definite." The most interesting result of these experiments was the observation of a strong relationship between the CH_4 release rate and the amount of mineral fertilizer applied to the paddy field. Unfertilized fields had CH_4 release rates which were one-fifth of the rates found on fields fertilized with ammonium sulfate at a rate of 160 kg/ha. This application rate applies for paddy fields in the developed countries but is almost one order of magnitude higher if compared with the situation

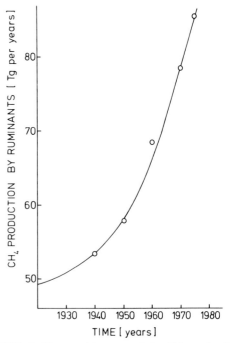

FIG. 1. Temporal increase of the CH_4 production by ruminants.

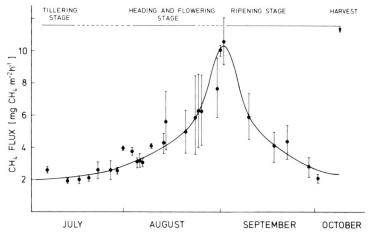

FIG. 2. Seasonal variation of CH_4 release rates observed at paddy fields in Utrera, Spain, during July and October 1982.

in the developing countries where most of the rice is grown. It is therefore very likely that the estimate of 59 Tg per year which is based on the results obtained for fertilized fields may represent the upper limit of the CH_4 release from paddy fields.

A much larger data pool has been provided recently by W. Seiler, A. Holzapfel-Pschorn, R. Conrad, and D. Scharffe (submitted for publication), who measured the CH_4 release rates from fertilized paddy fields in Utrera, Spain, beginning in July until harvest in October 1982, to obtain CH_4 release rates over a complete vegetation period. More than 95% of the total release for paddy fields is due to diffusive transport through the aerenchyma system of the rice plants. Less than 5% of the flux is due to dissolution of CH_4 into the interstitial water of the paddy soil and exchange with the overlying paddy water and atmosphere. Transport of CH_4 from paddy soils into the atmosphere by rising bubbles is only important for those parts of the paddy fields where no plants were grown.

Most interesting is the observation of a seasonal variation of the CH_4 release rates from paddy fields (Fig. 2). Minimum values of 3 mg of CH_4 per m^2 per h were found during the beginning and at the end of the vegetation period. Maximum values occurred at the end of the heading and flowering stage of the rice plants, when the CH_4 release rates sometimes exceeded 14 mg per m^2 per h. The seasonal variation of CH_4 release is not related to the variation in soil temperatures and therefore has to be due to other processes. One process may be the possible variation in substrate concentration, especially the easily fermentable substrates whose fermentation would lead to the substrates for the methanogenic bacteria (18). One important

source of organic carbon in paddy soils is the release of root exudates, root lysates, and root litter (13, 42), which is related to the extension of the root mat and therefore exhibits seasonal variations. Another possible explanation is the seasonality of O_2 transport from the atmosphere into the paddy soil, where the O_2 supply may stimulate the O_2 consumption by methanotrophic bacteria near the root surfaces (1, 12, 31), thereby reducing the flux of CH_4 from the paddy soil into the roots and the aerenchyma system of the plant.

The measurements carried out on paddy fields fertilized with urea at a rate of 160 kg of N per ha showed an average CH_4 release rate of 4 mg per m^2 per h, equivalent to a total CH_4 production rate of about 15 g per m^2 for the whole vegetative period of 5 months. Preliminary results of measurements performed in 1983 on rice paddies in Italy do not show significant differences in the CH_4 release rates between fertilized and unfertilized fields; thus, the data obtained from the fertilized fields in Utrera, Spain, may be applicable to global conditions. Applying the temperature dependence of the CH_4 release rates reported by Koyama (23, 24) and using the statistical data on the rice paddy area reported by the Food and Agricultural Organization (16), we calculate the global CH_4 production by rice paddies to be 22 to 38 Tg per year, with an average of 30 Tg per year. In spite of the large uncertainties involved in these estimates, these values agree reasonably well with those of Cicerone and Shetter (7). The latter authors indicated a production rate of 30 to 59 Tg per year for the global rice paddies. This figure can be improved only after data on the CH_4 release rates from rice paddies in the subtropical and tropical Asian countries become available.

Because of the increasing area of rice paddies harvested annually, the CH_4 emission from paddy fields must have changed with time during the past decades. Official statistics (16) indicate that the harvested rice paddy area has increased from 90×10^{10} m^2 in 1950 to 140×10^{10} m^2 in 1975, or roughly 2% during the 25-year period. Assuming that the average CH_4 release rates per basal area have not changed considerably during this time period, the global CH_4 emission from paddy fields will have increased from 18 to 35 Tg per year in 1950 to 25 to 49 Tg in 1960, 29 to 57 Tg in 1970, and 30 to 59 Tg in 1975. The temporal increase of the global paddy area and the global CH_4 emission from rice paddies are shown in Fig. 3.

CH_4 production in swamps. Methane production in swamps and marshes has been observed by several authors. The individual emission rates vary by more than two orders of magnitude, with minimum values of 0.001 g of CH_4 per m^2 per day and maximum values of 0.35 g of CH_4 per m^2 per day. Lowest CH_4 emission rates were found in saltwater marshes, most likely as a result of carbon flow being dominated by sulfate reduction. Methane emission rates reach maximum values in freshwater swamps that have organic-rich soils and are covered by rooted aquatic plants, which generally agrees with our findings on the vegetated and nonvegetated paddy fields. Apparently, CH_4 production is lower in sediments without a root mat and thus with lower supply of root litter. It is also possible that a fraction of the CH_4 produced in the anerobic soil layer is oxidized during its diffusion through the uppermost aerobic soil layer if plants are not available. Extremely high CH_4 release rates were observed in swamps fertilized, for example, by waste disposal.

CH_4 production and emission from swamps are correlated with the soil temperature and thus show a strong seasonal trend, with minimum values during winter/spring and maximum values during summer conditions (22). Values found in a swamp in Georgia, covered by *Spartina alterniflora*, varied between 10 and 80 mg of CH_4 per m^2 per h at a soil temperature of 30 to 35°C and between 0.01 and 0.5 mg of CH_4 per m^2 per h at temperatures of 5 to 14°C. The annual average value of the CH_4 release rate accounted for 6 mg per m^2 per h, or almost a factor of 10 lower than the values found during summer conditions. Similar variations should also be expected for the large part of swamps located in the subtropical regions and at higher latitudes. Consequently, the given range of 0.001 to 0.35 g of CH_4 per m^2 per day obtained mainly from measurements under summer conditions certainly represents only an upper limit of the possible CH_4 release from swamps; the true range may be as low as 0.001 to 0.05 g of CH_4 per m^2 per day if averaged on an annual basis. This figure accounts for an annual CH_4 release of 0.05 to 18 g of CH_4 per m^2. The CH_4 release rates from swamps in tropical areas, e.g., Sudan, Uganda, Brazil, and Botswana, may show smaller seasonal trends. However, a significant fraction of these swamps is actually flooded only a few months per year. Because swamps in the tropical regions account for only 10% of the global area of swamps and marshes, which total about 2.6×10^{12} m^2 (29), the average annual CH_4 release from swamps may range between 5 and 22 g/m^2, resulting in a global emission of CH_4 from swamps and marshes of about 13 to 57 Tg. Because of the large uncertainties involved in this estimate, the given figure has to be considered premature. More field measurements in different climatic regions will be required to allow a more quantitative assessment of the CH_4 emission from these ecosystems.

The estimate given before is significantly lower than the values of 150 Tg per year and 190 to 300 Tg per year reported by Baker-Blocker et al. (2) and Ehhalt and Schmidt (14), respectively. Both figures were based on measurements of the CH_4 release rates observed in one swamp and two ponds in Michigan which may have been influenced by fertilization and consequently may have shown extremely high CH_4 flux rates.

CH_4 production by termites. The production of CH_4 by xylophagous insects has been known since 1932, when Cook observed the evolution from termite species (*Zootermopsis nevadensis*) of a gas which he described to be methane. In the meantime, the digestive processes of termites have been studied, and CH_4 production rates by termites have been quantified (see, e.g., 5). CH_4 production by termites became the subject of strong debate after Zimmerman et al. (46) reported the global production of CH_4 by termites to be 150 Tg per year. These authors

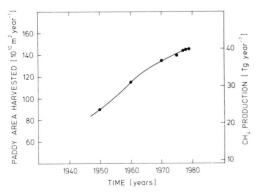

FIG. 3. Temporal increase of CH_4 emission from rice paddies.

calculated the total biomass consumption by termites to be 33×10^{15} g of dry matter per year, which accounts for approximately 30% of the total worldwide terrestrial net primary production of biomass. According to N. M. Collins and T. G. Wood (Science, in press), this figure is strongly overestimated. On the basis of data on termite densities and biomass consumption rates of individual termite species, they estimated a global biomass consumption by termites of 7.3×10^{15} g of dry matter per year, which they believed is still an upper limit.

A much lower CH_4 production rate by termites (50 Tg per year) was reported by Rasmussen and Khalil (34), who extrapolated results of laboratory experiments using five colonies of *Zootermopsis angusticollis* with an average population of 25 termites per colony to the global number of 2.4×10^{17} termites. Since laboratory experiments with social insects have "the great advantage of being unrealistic" (44), and furthermore since these laboratory experiments were carried out with nonbreeding colonies, it is hard to believe that the observed results are representative of global natural conditions.

The first in situ measurements on the CH_4 emission rates by termites are those carried out by W. Seiler, R. Conrad, and D. Scharffe (submitted for publication) on different termite mounds occupied by different termite species. The termite species included soil feeders (e.g., *Cubitermes*), grass feeders (e.g., *Trinervitermes*), wood and dung feeders (e.g., *Amitermes*), grass harvesters (e.g., *Hodotermes*), and fungus-grown termites (e.g., *Macrotermes* and *Odontotermes*), thus covering the most common termite species. The measurements showed diurnal variations of the CH_4 and CO_2 flux rates from termite mounds, with maximum amplitudes for the grass feeders *Trinervitermes*. Maximum values were generally found during the late afternoon (about 6 p.m.); minimum values occurred during the early morning (about 6 a.m.). The flux rates of CH_4 from individual termite mounds were directly proportional to the corresponding CO_2 flux rates, indicating that the CH_4 and CO_2 flux rates measured at the termite mounds must be directly related to the activities of termites and are not significantly disturbed by other processes, e.g., microbial processes inside the termite mounds.

Most interestingly, the ratios of the CH_4 and CO_2 flux rates are relatively constant within each termite species, although the absolute values of the CH_4 and CO_2 flux rates vary considerably. Obviously, the ratio of the CH_4 and CO_2 flux rates is a characteristic figure for each individual termite species and is independent of the size of the termite population, the time of day, and the soil temperature (at the measured

range of 20 to 45°C). However, the ratios differ by more than two orders of magnitude between the different termite species, with maximum figures obtained for *Trinervitermes* (8.7×10^{-3}) and minimum figures for *Odontotermes* (6.7×10^{-5}).

Atmosphere samples taken from the interior of termite nests generally show lower CH_4/CO_2 ratios than those obtained from the flux measurements, which may be interpreted as a possible decomposition of CH_4 within the termite mounds where the CH_4 mixing ratios sometimes reached values up to 120 ppm. Highest differences and thus highest decomposition rates were found in old termite mounds.

Applying the ratios of the CH_4 and CO_2 flux rates obtained from measurements at the different termite mounds and considering the different respiration efficiencies (Seiler et al., submitted for publication), one calculates the ratio of CH_4 emitted from termite mounds relative to the carbon ingested by termites (on a carbon to carbon ratio) to be 6×10^{-5} for *Odontotermes* as the minimum value and 2.6×10^{-3} for *Trinervitermes* as the maximum value. With these values and the total biomass consumed by termites of 7×10^{15} g of dry matter per year (Collins and Wood, in press), the total CH_4 emission by termites is estimated to range between 2 and 5 Tg per year, which is more than one order of magnitude lower than the figure given previously (46).

Since the bulk of termites live in an ecosystem not strongly affected by humans, it is unlikely that the total CH_4 emission by termites has changed significantly during the past three decades. Therefore, a constant emission rate of 2 to 5 Tg per year is used to establish the long-term trend of the biogenic CH_4 production (see Table 2).

CH_4 production by oceans and lakes. Measurements carried out in the surface waters of the Atlantic and Pacific Oceans by several authors indicate that the surface waters of the open oceans are supersaturated by about 10 to 80% with respect to the atmosphere (6, 26, 27, 40, 41). Similar saturation factors were found by workers from this laboratory during several cruises on the southern and northern Atlantic Ocean between 50°N and 40°S, indicating that the supersaturation of the ocean surface water is a global phenomenon. The supersaturation of the surface water causes a flux of CH_4 from the ocean into the atmosphere which can be calculated from the stagnant film layer model (7). Using an average film thickness of 30×10^{-6} cm and an average diffusion coefficient of CH_4 in seawater of about 20×10^{-6} cm^2/s, we calculate the total CH_4 flux from the ocean surface (360×10^{12} m^2) to be 1 to 7 Tg per year, which agrees reasonably well with the figure of 1.3 to 16.6 Tg per year given by Ehhalt and Schmidt (14). Methane may also be produced in the surface waters of lakes. However, it is unlikely that this process contributes significantly to the figures given before because of the lower surface area of lakes.

The production processes of CH_4 in the oxygenated ocean surface waters are not well understood. Since the highest CH_4 supersaturations were observed in biologically active waters, it seems that the CH_4 in ocean water is due to biological production, most likely methanogenesis in anaerobic microsites. Transport of abiogenic CH_4, e.g., from hot water springs in the deeper oceans, can be excluded because the vertical distribution of CH_4 shows maximum values in the upper few hundred meters of the oceans. The more or less uniform distribution of CH_4 in the surface waters of the oceans is furthermore indicative of both biological CH_4 sources and sinks occurring simultaneously in the surface water so that the observed CH_4 concentrations in the water represent dynamic equilibrium values.

Other biogenic CH_4 sources. CH_4 may also be produced by methanogenesis in waterlogged tundra areas. In fact, release of CH_4 from tundras has been found by R. Benoit (cited in 14; 38, 39). Applying average CH_4 release rates of 10 g per m^2 per year and assuming that only 1 to 10% of the tundra area of 8×10^6 km^2 is waterlogged, Ehhalt and Schmidt (14) estimated the source strength of this process to account for 2.3 to 3 Tg per year.

There is evidence that CH_4 is also produced by fermentation in the human large intestine. Most of the gas produced by this process is absorbed by the blood, removed from the blood in the lungs, and exhaled (4, 28), and a portion of this gas is excreted in flatus. From data published by Wolin (45), an average production of 2 to 4 g per person per day is calculated, which results in a global production of CH_4 by the human population of 4 to 7 Tg per year.

Summary of biogenic CH_4 production. The CH_4 production rates of the individual biogenic CH_4 sources are summarized in Table 2. These data indicate that the most important biogenic CH_4 source is the fermentation by ruminants, having a source strength of 72 to 99 Tg per year in 1975, followed by the CH_4 release by paddy fields, with a source strength of 30 to 59 Tg per year. The production rate by swamps is very uncertain, and it is likely that the release rate of 13 to 57 Tg per year may represent only an upper limit. Other biogenic processes such as CH_4 production in the oxygenated surface waters and CH_4 formation by termites seem to be of only minor importance for the global CH_4 cycle.

TABLE 2. Summary of the CH_4 production rates from individual biogenic sources (Tg of CH_4 per year)

Source	1950	1960	1970	1975
Ruminants	49–69	56–81	65–92	72–99
Paddy fields	18–35	25–49	29–57	30–59
Swamps	13–57	13–57	13–57	13–57
Termites	2–5	2–5	2–5	2–5
Ocean/lakes	1–7	1–7	1–7	1–7
Others	3–8	3–9	4–10	4–10
Total	86–181	100–208	114–228	122–237

Because of the increase in cattle population and the increase in the annually harvested paddy area, the total biogenic CH_4 production has increased with time, with values of 86 to 181 Tg per year in 1950 and 122 to 237 Tg per year in 1975, or by 36 to 56 Tg per year. This figure corresponds to an average increase in the biogenic CH_4 production of about 1.1 to 1.6% per year.

ABIOGENIC CH_4 PRODUCTION

In addition to the biogenic CH_4 sources, there are several abiogenic production processes; these are summarized in Table 3. One of the most important sources of abiogenic CH_4 is the formation of CH_4 during burning of biomass, such as burning of agricultural wastes, savanna fires, burning due to shifting agriculture, etc. On the basis of measurements of the ratios of the CH_4 and CO_2 concentrations in several fire plumes, Crutzen et al. (11) estimated the global CH_4 production by this process to account for 25 to 110 Tg per year. Including data obtained from the burning of agricultural wastes, the average excess of CH_4 compared with the excess of CO_2 in fire plumes of different fires is on the order of 1.9%. This results in a global CH_4 production of 53 to 97 Tg of CH_4 per year in 1975 if one applies the total amount of 48×10^{14} to 88×10^{14} g of dry matter of biomass burned annually for different purposes (36). Because of the worldwide population increase, the amount of biomass burned as agricultural wastes, as a result of shifting agriculture, as fuel wood, and by deforestation has expanded during the past decades. Using statistical data on food production, production of industrial and fuel wood, and the increase in population working in shifting agriculture (16), one estimates the total amount of biomass burned in 1950 to be 37×10^{14} to 76×10^{14} g of dry matter compared with 42×10^{14} to 67×10^{14} g of dry matter in 1960, resulting in a global CH_4 production of 41 to 74 Tg per year and 47 to 84 Tg per year, respectively.

A further important abiogenic CH_4 source is venting and leakage of natural gas, which has

TABLE 3. Summary of the CH_4 production from individual abiogenic sources (Tg of CH_4 per year)

Sources	1950	1960	1975
Biomass burning	41–74	47–84	53–97
Transmission loss	3–4	7–10	19–29
Coal mining	20	24	30
Automobiles	<1	<1	<1–2
Volcanoes	<0.5	<0.5	<0.5
Total	64–98	78–118	103–158

become a very important energy source. The total amount of natural gas used annually exponentially expanded from 0.2×10^{12} m^2 per year in 1950 to 1.5×10^{12} m^2 per year in 1975, or by more than a factor of 7 within a time period of 25 years. A similar increase has been observed for the total length of mains for transport and distribution. Unfortunately, official statistics on the average leakage rates are not available. On the basis of the limited information from some gas supply companies in Europe, a leakage rate of 2 to 3% can be assumed for European conditions. Extrapolating this value to global conditions, the total CH_4 emission due to transmission losses of natural gas may have been on the order of 19 to 29 Tg per year in 1975. The corresponding figures for 1960 and 1950 are summarized in Table 3. These figures do not include the possible emission of natural gas during blowups of gas and oil wells, for which data are not available. Consequently, the given figure of 19 to 29 Tg per year may represent a lower limit only.

Abiogenic CH_4 is also released into the atmosphere by leakage from coal mines. The source strength of this CH_4 source was first estimated by Koyama (23, 24) to be 20 Tg per year for the beginning of the 1950s. Extrapolating this figure to coal production in 1960, 1970, and 1975 and applying similar ratios of CH_4 produced per mined ton of coal, the global production of CH_4 due to coal mining may have increased to 24, 26, and 30 Tg per year, respectively.

Emission of CH_4 due to volcanic activities (0.5 Tg per year) and the production of CH_4 by automobiles (1 to 2 Tg per year in 1975 [F. R. Brenk, thesis, University of Mainz, Mainz, Federal Republic of Germany, 1982]; see Table 3) are relatively small and do not contribute to the global atmospheric CH_4 budget.

It has recently been argued by Gold that abiogenic primordial CH_4 trapped in the earth's crust may be released into the atmosphere by outgassing along fault lines or during earthquakes, or both (17). Observations of flames during heavy earthquakes may support Gold's hypothesis; however, no data are presently available to give any quantitative figure for this possible atmospheric CH_4 source. The emission of CH_4 by this process will certainly occur sporadically and thus will cause significant short-term variations in the tropospheric CH_4 mixing ratios which, however, have not yet been observed. Therefore, one might assume that the release rate of abiogenic CH_4 by earthquakes may be of only minor importance in comparison with the other natural and anthropogenic sources.

According to the figure given in Table 4, the global abiogenic CH_4 production has increased from 64 to 98 Tg per year in 1950 to about 103 to 158 Tg per year in 1975 or by 1.7 to 2.0% per year during the past 25 years.

CONCLUSIONS

Methane is released into the atmosphere by both biogenic and abiogenic processes. Approximately 58 to 62% of the total CH_4 emission is due to biogenic processes (see Table 4). This figure is considerably lower than the value of 80 to 90% proposed by other authors (e.g., 14) on the basis of measurements of the ^{14}C activities of atmospheric methane. The latter estimates do not consider that a substantial fraction of the abiogenic CH_4 production rates is due to burning of biomass which produces CH_4 with ^{14}C activities similar to those derived from methanogene-

TABLE 4. Total CH_4 emission into the troposphere (Tg of CH_4 per year)

Source	1950	1960	1970	1975
Biogenic production	86–181	100–208	114–228	122–237
Abiogenic production	64–98	78–118	94–140	103–158
Total CH_4 production	146–285	178–336	208–368	225–395
Average CH_4 production	216	257	288	310
Biogenic/total (%)	62	60	59	58
Sinks				
Reactions with OH	210	230	270	290
Flux into stratosphere	44	48	56	60
Microorganisms	15	16	19	20
Total	269	294	345	370

sis of recent organic matter. Thus, the given figure of 80 to 90% can only be considered an upper limit.

The total CH_4 production shows an increase with time from values of about 216 Tg in 1950 to about 312 Tg per year in 1975, which corresponds to an average rate of increase of about 1.6% per year. This figure agrees very well with the increase of the tropospheric CH_4 mixing ratios of about 1.5 to 1.7% per year. Thus, it is obvious that the change of the tropospheric abundance of CH_4 is due to the change of the emission rates of the individual CH_4 sources.

Primarily, the CH_4 production is balanced by the photochemical oxidation of CH_4 in the troposphere, mainly through reaction with OH. Measurements of the global OH abundance are presently not available. Previous estimates of the average tropospheric OH concentration resulted in values higher than 10^6 molecules per cm^3, whereas more recent estimates based on the concentration and distribution of fluorocarbons (see, e.g., 9, 37) seem to indicate values of 3×10^5 to 5×10^5 molecules per cm^3. Adopting an average tropospheric value of 5×10^5 molecules per cm^3 and an average tropospheric CH_4 mixing ratio of 1.5 ppmv in 1975, we calculate the tropospheric photochemical loss due to this reaction to be 250 Tg per year. Tropospheric CH_4 is further lost by the tropospheric/stratospheric exchange which results in a net flux of CH_4 from the troposphere into the stratosphere of about 60 Tg per year for 1975.

Most interestingly, Seiler et al. (submitted for publication) have observed a destruction of CH_4 at the soil surface in semiarid climates. The destruction rates varied between 19 and 102 g of CH_4 per m^2 per h during the dry season with soil temperatures of 20 to 45°C. Using this data set, we estimate the global CH_4 destruction at the soil surface in the tropical and subtropical regions to be about 20 Tg per year, which is about 10% of the photochemical CH_4 sink. More recently, workers in this laboratory have also observed a decomposition of CH_4 at the surface of several types of soil in Germany, indicating that the total CH_4 decomposition by soil must be higher than 20 Tg per year.

Summarizing the individual sink strength, we calculate the total destruction of tropospheric CH_4 to be 370 Tg per year, which, in light of the large uncertainties involved in the estimates of the individual source and sink strengths, agrees reasonably well with the CH_4 production rates.

Since the destruction rates are generally first-order reactions, assumptions of the temporal trend of tropospheric CH_4 mixing ratios have to be made to enable the estimate of the global CH_4 sink strength of former years. As a first approximation, a CH_4 mixing ratio of 1.2 and 1.1 ppmv

is assumed for the years 1960 and 1950, respectively. The assumption of these low values is supported by the observation of an almost constant increase in the CH_4 mixing ratio during the past 5 years and a relatively long tropospheric residence time of about 10 to 11 years calculated from the production and destruction rates of 312 to 370 Tg per year and a CH_4 mixing ratio of 1.5 ppmv for 1975. Using the CH_4 mixing ratios as given before, the global destruction rate of CH_4 accounts for 269 Tg per year in 1950, 294 Tg per year in 1960, and 345 Tg per year in 1970.

The good agreement of the absolute numbers and the temporal trend of the production and destruction rates of CH_4 again supports the assumption that the rate of tropospheric CH_4 increase during the past 3 decades is significantly larger than postulated by Ehhalt et al. (15). From the estimates given in this paper, it is very likely that the increase of the atmospheric CH_4 burden is due to an increase of some biogenic and nonbiogenic sources. It is very likely that the atmospheric CH_4 mixing ratios have more than doubled within the past 100 to 200 years, and this change should already have had some impact on the climate as well as on the chemistry of the troposphere and stratosphere.

LITERATURE CITED

1. **Aiyer, S. P. A.** 1919. A methane oxidizing bacterium from rice soils, p. 173–180. *In* Chemical Series, vol. V. Agriculture Research Institute, Department of Agriculture, Pusa, India.
2. **Baker-Blocker, A., T. M. Donahue, and K. H. Mancy.** 1977. Methane flux from wetland areas. Tellus **29**:245–250.
3. **Blake, D. R., E. W. Mayer, St. C. Tyler, Y. Makide, D. C. Montagne, and F. S. Rowland.** 1982. Global increase in atmospheric methane concentrations between 1978 and 1980. Geophys. Res. Lett. **9**:477–480.
4. **Bond, J. H., R. R. Engel, and M. D. Levitt.** 1971. Factors influencing pulmonary methane excretion in man. J. Exp. Med. **133**:572–588.
5. **Brian, M. V.** 1978. Production ecology of ants and termites. Cambridge University Press, London.
6. **Broecker, W. S., and T. H. Peng.** 1974. Gas exchange rates between air and sea. Tellus **26**:21–35.
7. **Cicerone, R. G., and G. O. Shetter.** 1981. Sources of atmospheric methane: measurements in rice paddies and a discussion. J. Geophys. Res. **86**:7203–7209.
8. **Craig, H., and C. C. Chou.** 1982. Methane: the record in polar ice cores. Geophys. Res. Lett. **9**:1221–1224.
9. **Crutzen, P. J.** 1982. The global distribution of hydroxyl, p. 313–328. *In* E. D. Goldberg (ed.), Atmospheric chemistry. Dahlem Konferenzen, Berlin.
10. **Crutzen, P. J.** 1983. Atmospheric interactions—homogeneous gas reactions of C, N, and S containing compounds, p. 67–112. *In* B. Bolin and R. B. Cook (ed.), The major biogeochemical cycles and their interactions. SCOPE. John Wiley & Sons, Inc., New York.
11. **Crutzen, P. J., L. E. Heidt, J. P. Krasnec, W. H. Pollock, and W. Seiler.** 1979. Biomass burning as a source of atmospheric gases CO, H_2, N_2O, NO, CH_3Cl and COS. Nature (London) **282**:253–256.
12. **de Bont, G. A. M., K. K. Lee, and D. F. Bouldin.** 1978. Bacterial oxidation of methane in a rice paddy. Ecol. Bull. (Stockholm) **26**:91–96.

13. **Dommergues, Y. R., and G. Rinaudo.** 1979. Factors affecting N_2 fixation in the rice rhizosphere, p. 241–260. *In* Nitrogen and rice. IRRI, Los Banos, Philippines.

14. **Ehhalt, D. H., and U. Schmidt.** 1978. Sources and sinks of atmospheric methane. Pageoph **116:**452–464.

15. **Ehhalt, D. H., R. J. Zander, and R. A. Lamontagne.** 1982. On the temporal increase of tropospheric CH_4, p. 1–15. *In* KFA Report. Jülich, FRG.

16. **Food and Agriculture Organization.** 1980. Production yearbook, vol. 33. Food and Agriculture Organization, Rome.

17. **Gold, T.** 1979. The earthquake evidence for earth gas, p. 65–92. *In* H. Messel (ed.), Energy for survival. Pergamon Press, New York.

18. **Hah, R. A., D. M. Ward, L. Baresi, and T. L. Glass.** 1977. Biogenesis of methane. Annu. Rev. Microbiol. **31:**305–341.

19. **Heidt, L. E., and D. H. Ehhalt.** 1980. Corrections of CH_4 concentrations measured prior to 1974. Geophys. Res. Lett. **7:**1023.

20. **Hutchinson, G. E.** 1949. A note on two aspects of the geochemistry of carbon. Am. J. Sci. **247:**27–32.

21. **Khalil, M. A. K., and R. A. Rasmussen.** 1983. Sources and sinks and seasonal cycles of atmospheric methane. J. Geophys. Res. **88:**5131–5144.

22. **King, G. M., and W. J. Wiebe.** 1978. Methane release from soils of a Georgia salt marsh. Geochim. Cosmochim. Acta **42:**343–348.

23. **Koyama, T.** 1963. Gaseous metabolism in lake sediments and paddy soils and the production of atmospheric methane and hydrogen. J. Geophys. Res. **68:**3971–3973.

24. **Koyama, T.** 1974. Biogeochemical studies on lake sediments and paddy soils and the production of atmospheric methane and hydrogen, p. 143–177. *In* Y. Miyake and T. Koyama (ed.), Recent researches in the field of hydrosphere, atmosphere and nuclear geochemistry. Macruun, Tokyo.

25. **Lacis, A., G. Hansen, P. Lee, T. Michell, and S. Lebedeff.** 1981. Greenhouse effect of trace gases, 1970–1980. Geophys. Res. Lett. **8:**1035–1038.

26. **Lamontagne, R. A., J. W. Swinnerton, and V. J. Linnebom.** 1974. C_1-C_4 hydrocarbons in the North and South Pacific. Tellus **26:**71–77.

27. **Lamontagne, R. A., J. W. Swinnerton, V. J. Linnenbom, and W. D. Smith.** 1973. Methane concentrations in various marine environments. J. Geophys. Res. **78:**5317–5324.

28. **Levitt, M. D.** 1969. Production and excretion of hydrogen gas in man. N. Engl. J. Med **281:**122.

29. **Lieth, H.** 1975. Primary production of the major vegetation units of the world, p. 203–215. *In* H. Lieth and R. H. Whittaker (ed.), Primary productivity of the biosphere. Springer-Verlag, New York.

30. **McDowell, R. E.** 1976. Importance of ruminants of the world for nonfood uses. Cornell Int. Agric. Mimeograph **52:**35.

31. **Oremland, R. S., and B. F. Taylor.** 1977. Diurnal fluctuations of O_2, N_2, and CH_4 in the rhizosphere of Thalassia testudinum. Limnol. Oceanogr. **22:**566–570.

32. **Ramanathan, V.** 1975. Greenhouse effect due to chlorofluorocarbons: climate implications. Science **190:**50.

33. **Rasmussen, R. A., and M. A. K. Khalil.** 1981. Atmospheric methane, trends and seasonal cycles. J. Geophys. Res. **86:**9826–9832.

34. **Rasmussen, R. A., and M. A. K. Khalil.** 1983. Global production of methane by termites. Nature (London) **301:**700–702.

35. **Robbins, R. C., L. A. Cavanagh, and L. J. Salas.** 1973. Analysis of ancient atmospheres. J. Geophys. Res. **78:**5341–5344.

36. **Seiler, W., and P. J. Crutzen.** 1980. Estimates of gross and net fluxes of carbon between the biosphere and the atmosphere from biomass burning. Clim. Change **2:**207–247.

37. **Singh, H. B.** 1977. Preliminary estimation of average tropospheric OH concentrations in the Northern and Southern Hemisphere. Geophys. Res. Lett. **4:**453–456.

38. **Svensson, B. H.** 1973. Production of methane and carbon dioxide from a subarctic Mire. Tech. Rep. 16, Swedish Tundra Biome Project.

39. **Svensson, B. H., A. K. Veum, and S. Kjelvik.** 1975. Carbon losses from tundra soils, p. 279–328. *In* F. E. Wielgolaski (ed.), Ecological studies, analysis and synthesis 16. Springer Verlag, Berlin.

40. **Swinnerton, J. W., and V. J. Linnenbom.** 1967. Gaseous hydrocarbons in seawater: determination. Science **156:**119–120.

41. **Swinnerton, J. W., V. J. Linnenbom, and C. H. Check.** 1969. Distribution of methane and carbon monoxide between the atmosphere and natural waters. Environ. Sci. Technol. **3:**836.

42. **Waid, G. S.** 1974. Decomposition of roots, p. 175–211. *In* C. H. Dickinson and G. J. F. Pugh (ed.), Biology of plant litter decomposition, vol. 1. Academic Press, London.

43. **Wilkniss, P. E., R. A. Lamontagne, R. E. Larson, J. W. Swinnerton, C. R. Dickson, and T. Thompson.** 1973. Atmospheric trace gases in the Southern Hemisphere. Nature (London) **245:**45–47.

44. **Wood, T. G., and W. A. Sands.** 1978. The role of termites in ecosystems, p. 55–80. *In* M. V. Brian (ed.), Production ecology of ants and termites. Cambridge University Press, London.

45. **Wolin, M. J.** 1981. Fermentation in the rumen and human large intestine. Science **213:**1463–1468.

46. **Zimmerman, P. R., J. P. Greenberg, S. O. Wandiga, and P. J. Crutzen.** 1982. Termite: a potentially large source of atmospheric methane, carbon dioxide and molecular hydrogen. Science **218:**563–565.

Sources and Sinks of Nitrous Oxide

WARREN KAPLAN

Center for Earth and Planetary Physics, Harvard University, Cambridge, Massachusetts 02138

During the past decade, long-term chemical sampling of the atmosphere has revealed that several important gaseous compounds have been increasing in concentration, carbon dioxide being the best known example. More recent information (33, 45, 46) shows that the levels of methane and nitrous oxide have also been rising. Nitrous oxide concentrations are increasing by about 0.6 parts per billion (ppb) per year. This is only about 0.2% of the atmospheric concentration. However, the lifetime of N_2O is long (~150 years), and the observed 0.2% increment suggests that global emissions of nitrous oxide now exceed consumption by some 20 to 30% (M. Keller, S. C. Wofsy, T. C. Goreau, W. A. Kaplan, and M. B. McElroy, Geophys. Res. Lett., in press).

Details of the particular reactions that are of importance in the photochemistry of nitrogenous oxides have been discussed previously (28). Briefly, the nitrogenous oxides $[NO_{(x)}]$ limit stratospheric abundance of ozone by the following reactions:

$$O_3 + h\nu \rightarrow O + O_2 \text{ (wavelengths less than 1,140 nm)}$$
$$O + NO_2 \rightarrow NO + O_2$$
$$\text{net: } 2O_3 \rightarrow 3O_2$$

which approximately balances the formation of stratospheric ozone by:

$$O_2 + h\nu \rightarrow 2O \text{ (wavelengths less than 240 nm)}$$
$$O + O_2 + M \rightarrow O_3 + M(2x)$$
$$\text{net: } 3O_2 \rightarrow 2O_3$$

The primary source of $NO_{(x)}$ is largely the oxidation of nitrous oxide as follows:

$$O_3 + h\nu \rightarrow O(^1D) + O_2 \text{ (wavelengths less than 310 nm)}$$
$$N_2O + O(^1D) \rightarrow 2NO$$

Nitrous oxide is also a "greenhouse gas" and absorbs reradiated infrared wavelengths (44).

On the basis of our most recent estimates for the magnitude of known atmospheric N_2O sinks, the steady-state global nitrous oxide source strength needed to balance atmospheric destruction is of the order 13×10^6 tons (13 MT) of N_2O-N per year. We can account for roughly one-half of this source strength. Combustion processes contribute about 1 to 2 MT of N per year (45, 46). To roughly balance our 13-MT global source, we need to identify some fairly large areas with significant nitrous oxide fluxes. I will present information that soil systems in general, and tropical moist forest ecosystems in particular, may be substantial sources of nitrous oxide.

There exist several pathways for the biogenic production of nitrogenous oxides (8, 17, 19, 26, 32, 39). Oxidative microbial processes include both autotrophic and heterotrophic nitrification. The mechanism(s) for N_2O and NO production during the oxidation of ammonia by nitrifying bacteria is not clearly understood (24). There are several intermediates of valence N^+ that are possible in the reaction sequence. Nitroxyl (NOH) dimerizes to form hyponitrite ($N_2O_2H_2$), which then decomposes to form nitrous oxide. Organisms that can oxidize reduced N compounds and use organic carbon as carbon and energy sources, the heterotrophic nitrifiers, also produce N_2O and NO (12; T. J. Goreau, Ph.D. thesis, Harvard University, Cambridge, Mass., 1981), but their biochemistry, in this context, is less well known than that of their autotrophic counterparts.

There is a wider variety of reductive pathways available for $NO_{(x)}$ production. Respiratory denitrification is the "classic" mechanism, although several non-denitrifying organisms, notably the nitrate-respiring enteric bacteria and procaryotes capable of nonrespiratory anaerobic nitrate dissimilation (5, 26, 39), have been shown to produce N_2O and NO. From the point of view of the atmospheric scientist, biological reactions are perhaps less important than the magnitude and direction of the net fluxes. For example, although the denitrification pathway will produce nitrous oxide, in marine environments denitrification appears to be a "local" sink that has little impact on net atmospheric N_2O exchange (Fig. 1). The microbial ecologist, however, may be in an enviable position. The relative contribution of oxidative and reductive pathways to the biogeochemistry of $NO_{(x)}$ is not yet clear.

YIELD OF NITROUS OXIDE FROM NITRIFICATION

The basic studies that laid the foundation for research on N_2O cycling in aquatic environ-

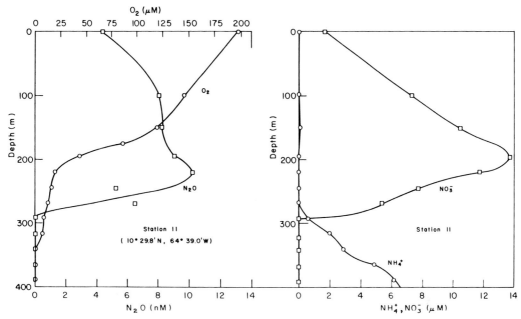

FIG. 1. Profiles of oxygen, nitrous oxide, and nitrogenous nutrients in the Cariaco Trench. Note the disappearance of oxidized inorganic nitrogen and nitrous oxide where the oxygen concentration approaches 10 to 13 μM (from Hashimoto et al. [23]). Surface N_2O values are in equilibrium with 300 ppb of nitrous oxide.

ments include those of Craig and Gordon (15), Schutz et al. (37), and Hahn (22), most of whom concluded that denitrification was a net source of nitrous oxide. However, nitrous oxide production is also associated with the oxidation of organic matter and the production of nitrate in the sea, i.e., nitrification. This hypothesis is based on several observations. (i) In the open sea, there is invariably a linear relationship between apparent oxygen utilization (AOU: the difference between oxygen measured and oxygen predicted from atmospheric equilibrium) and the concentration of nitrate. (ii) Similarly, there is also a direct relationship between AOU and its analog, apparent N_2O production (N_2O: the difference between the concentration measured and that predicted), and between ΔN_2O and nitrate (14, 16, 48).

These relationships can be used to calculate the N_2O "oxidative ratio" ($\Delta N_2O/AOU$) and the molar "regeneration ratio" ($\Delta N_2O/NO_3$). This latter number, corrected for preformed nitrate (2), is equivalent to the fraction of nitrous oxide released during the production of unit nitrate. Field measurements of these ratios suggest that for every 1,000 mol of ammonium oxidized to nitrate, about 1 to 2 mol of nitrous oxide are released (14, 16). This holds in fresh water also. Measurements carried out in the Potomac River (47) suggest that nitrous oxide is produced from nitrification of sewage, with a yield of 0.3%. Marine nitrifying bacteria produce N_2O and NO

(21, 27), and the master variable controlling the "regeneration ratio" appears to be oxygen.

Oxygen, nitrate, and nitrous oxide observations from several cruises to the Central Pacific, the Peruvian upwelling, and the tropical North Atlantic (16, 23) show that the "regeneration ratio" is a function of water mass density in the Central Pacific from 20°N to 20°S at 150°W longitude. The highest nitrous oxide "yields" (1%) are found in density layers forming the core of oxygen-poor (O_2; 0.1 to 1.0 ml/liter) waters that lie below the thermocline in the Equatorial Pacific (34). Yield ratios in well-oxygenated waters of the Atlantic average 0.15%, consistent with previous estimates. The "regeneration ratio" can now be plotted as a function of dissolved oxygen for the Pacific and Atlantic data (Fig. 2) as well as the Peruvian upwelling (Fig. 3). Superimposed on these field data are laboratory results for nitrous oxide yield from pure cultures of nitrifying bacteria (21, 27). Yields are generally higher in the low-oxygen regions of the Central Pacific than in outside areas, including the Atlantic. The trend for field data closely parallels the laboratory results, i.e., an increasing N_2O yield as oxygen decreases. As oxygen tension falls below 10 μM O_2 liter^{-1}, nitrous oxide is rapidly removed from the water column (Fig. 3), presumably via further reduction to dinitrogen during denitrification. "Yield ratios" decrease and become negative once N_2O becomes undersaturated. For convenience, they

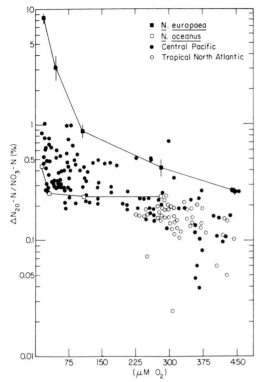

FIG. 2. Field observations of "regeneration ratio" (N_2O-N/NO_3-N, in percent) as a function of dissolved oxygen for Central Pacific samples (16) and Tropical North Atlantic (previously unpublished). Nitrate concentrations were corrected for preformed nitrate. Also plotted are similar observations on two pure cultures of nitrifying bacteria, *Nitrosomas europaea* (21) and *Nitrosocystis oceanus* (F. Lipshultz, personal communication).

ic equilibrium. This first field assay of "yield ratio" varied between 0.3 and 0.7%, consistent with all previous information. The most unequivocal experiment, looking for labeled N_2O after addition of labeled ammonium, has not yet been done.

In the oxygen-minimum zones of the oceans, denitrification may be possible in anoxic microsites although this has never been documented. It would be a technical tour de force to distinguish between production of nitrous oxide from denitrification or from a coupled nitrification-denitrification reaction inside, or on the surface of, any particle.

SINKS FOR NITROUS OXIDE

A second point in regard to the biogeochemistry of $NO_{(x)}$ concerns putative nitrous oxide sinks, of which denitrification appears to be the most important. Rarely are nitrous oxide destruction mechanisms of such strength as to cause depletion of N_2O from the surrounding air (see 4 and 11, however). Elkins et al. (16) reported a novel situation in which a stagnant,

are plotted on the lowest x-axis scale in Fig. 3. The critical oxygen tension below which nitrous oxide is consumed is not clear from these data, given the uncertainties in measuring dissolved oxygen at low levels or in the presence of sulfide, or both. It is, however, important to investigate N_2O yield of nitrifying bacteria at very low oxygen levels. M. A. Poth and D. D. Focht (Abstr. Annu. Meet. Am. Soc. Microbiol. 1982, N58, p. 187) found that *Nitrosomonas europaea* was capable of denitrification and N_2O production. Ritchie and Nicholas (35) isolated a denitrifying nitrite reductase from *N. europaea*. The role of these reductive pathways in situ is unknown. A more direct approach to this question was taken by J. J. McCarthy, W. A. Kaplan, and J. L. Nevins (Limnol. Oceanogr., in press). Water from Chesapeake Bay was enriched with [^{15}N]ammonium, and changes in labeled N and unlabeled nitrous oxide were followed. Oxygen was maintained at atmospher-

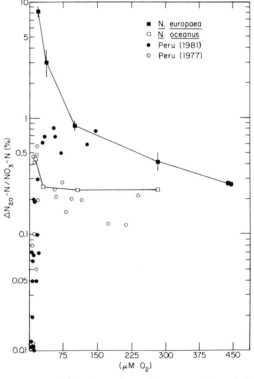

FIG. 3. Field observations of "regeneration ratio" as a function of dissolved oxygen for stations in the Peruvian upwelling. Data from 1977 are from reference 16 and those from 1981 (Cruise 108-Leg 3 of *Atlantis II*-WHOI) are previously unpublished. Also plotted are laboratory observations as in Fig. 2.

shallow (30 cm deep) *Lemna*-covered pond in Concord, Massachusetts, apparently acted as a sink for atmospheric nitrous oxide. Repeated sampling over the years has confirmed this with the following conclusions. (i) Nitrous oxide concentrations are consistently less than 50% of atmospheric equilibrium. (ii) Oxidized mineral nitrogen is very low, less than 0.5 μM liter^{-1}, with some regeneration at night. Oxygen never falls below 0.5 ppm (16 μM O$_2$ liter^{-1}). (iii) Ammonium is depleted in the surface during the day (less than 5 μM N liter^{-1}), but is rapidly released from the sediments at night. Methane is supersaturated at all depths and is also regenerated from the sediments. (iv) Diurnal experiments show concomitant decreases in oxygen and nitrous oxide, but water samples incubated in the presence of 0.1-atm acetylene do not produce nitrous oxide.

We may speculate that nitrification in this system is low, given the competition for ammonium by the duckweed, the high C/N ratio of the surrounding vegetation, and the production of inhibitory compounds by the leaves. This shallow, sediment-dominated system is similar to the much more extensive *varzea* areas of the Brazilian Amazon (18). Measurements from these floodplain lakes also show little dissolved nitrous oxide and nitrate with high methane levels.

SOIL ENVIRONMENTS

Several mechanisms are responsible for N$_2$O production in soils. Abiological production (19) and heterotrophic nitrification (1, 19) may be of minor importance. Both nitrification and denitrification still remain the most likely sources of NO$_{(x)}$ in soils. Organic soils and sediments can consume nitrous oxide under some conditions of limited oxygen supply and high nitrate (4, 41). Nitrous oxide consumption was found consistently in a 4-month-long study of a natural grassland in Great Britain (36). High root densities are associated with rapid oxygen demand which may promote anoxia in soil microsites.

However, experiments on well-aerated agricultural soils show that additions of nitrate often cause only an insignificant increase in nitrous oxide release (6, 7). Autotrophic nitrifying bacteria have been isolated from many soils (3, 40, 43), and field and laboratory experiments demonstrate that the addition of ammonium fertilizer or liquid ammonia to soils greatly enhances nitrous oxide evolution (6, 7, 13, 36, 38). Studies using the nitrification inhibitor N-serve indicate negligible fluxes from agricultural soils after treatment (9). Soil experiments using acetylene, which inhibits both nitrification and nitrous oxide reduction during denitrification, have yielded mixed results. Bremner and Blackmer (8)

found that acetylene inhibited soil N$_2$O production, suggestive of nitrification. Ryden (36) found the opposite effect. It is likely that nitrous oxide is both produced and consumed in soils simultaneously. Seiler and Conrad (38) proposed this idea from a study of vertical nitrous oxide profiles in soil. They found high levels of N$_2$O (600 to 900 ppb) accumulating at a depth of 5 to 20 cm in a fertilized soil but near-atmospheric levels in the top 5 cm. Nitrous oxide flux was small in spite of the high subsurface concentrations. Unfertilized plots showed no maxima of nitrous oxide. They argued that net nitrous oxide production occurred at depth and net consumption occurred at the surface. Table 1, provided by M. Keller, summarizes some of the information in N$_2$O fluxes from various soils. Where measured, yields of nitrous oxide N loss per unit of ammonium fertilizer N generally range from 0.01 to 3%. Using a figure of 40 MT per year as the addition of N fertilizer to soils (29) and a 0.3 to 3.0% yield of N$_2$O results in a production of 0.12 to 0.4 MT of nitrous oxide per year, or 5 to 20% of the annual increase reported by Weiss (45).

Our most recent information suggests that undisturbed tropical moist forest soils may be a significant source of nitrous oxide (Keller et al., in press). We measured N$_2$O fluxes by inverting aluminum chambers over forest soils and assaying temporal changes in atmospheric nitrous oxide within the box. Figure 4 shows the cumulative probability distribution of 79 flux measurements performed in a cool, temperate forest ecosystem, the Hubbard Brook Experimental Forest in New Hampshire. Fluxes were rank ordered and plotted on probability paper. Nega-

TABLE 1. Comparative rates of nitrous oxide flux

Environment/location	Avg flux (10^9 molecules cm^{-2} s^{-1})	Reference
Hardwood forest (Hubbard Brook Experimental Forest)	1.0	Goreau[a]
Unfertilized soybean field	8.2	10
Fertilized cornfield	17.7	26
Unfertilized cornfield	2.9	30
Fertilized cornfield	7.1[b]	30
Tobacco field	11.2	30
	51.0[b]	30
Fallow farm soil	8.6	6
Shortgrass prairie	5.7	31
Perennial ryegrass	-2.7	36
Sugarcane on organic soil	435	41
Fallow organic soil	1,490	41
Clovergrass	3.6	20

[a] Ph.D. thesis, Harvard University, 1981.
[b] Fertilized with 224 kg of N per ha as ammonium nitrate.

tive fluxes represent net uptake of nitrous oxide but comprise a relatively small subpopulation of total flux assays and occurred primarily in site 3, a waterlogged area of the Hubbard Brook Experimental Forest. The mean flux (50% probability) is less than 10^9 molecules cm^{-2} s^{-1}. Fluxes average 10 times higher in rain forest soils near Manaus, Brazil. Nitrous oxide release from soils near Manaus (Fig. 5) is fully 10 to 30 times the global average. Extensive areas of the tropics face a rather uncertain future, and more work is needed to establish the consistency and magnitude of this apparently rapid N cycle in tropical areas.

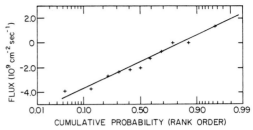

FIG. 5. Rank-ordered probability distribution of 11 nitrous oxide flux measurements taken at the Bacia Modelo, 60 km north of Manaus, Brazil, in April 1983. Mean flux is $\approx 2 \times 10^{10}$ molecules cm^{-2} s^{-1}. Data from Keller et al. (in press).

CONCLUSIONS

Considering the variety of microorganisms that are capable not only of producing nitrous oxide but also of switching between oxidative and reductive N metabolism (12, 19; Poth and Focht, Abstr. Annu. Meet. Am. Soc. Microbiol. 1982, N58, p. 189), the microbial ecologist may feel less than sanguine about investigating natural systems. In fact, the search for causative microorganisms may not be worth the effort. Research should be devoted to studying the factors that regulate nitrous oxide production/consumption during transition periods between oxygen surplus and oxygen limitation either in the laboratory or in environments containing oxic/anoxic interface zones. Atmospheric nitrous oxide is increasing, with no obvious increase in destruction mechanisms. That is, we should not expect atmospheric nitrous oxide to remain in steady state. Aside from a few scattered reports (11, 16), there is no evidence for large-scale biogenic sinks for N_2O. We should be looking at nitrous oxide cycling in floodplain lakes of the Amazon, large swamps, and highly organic soils. The most technically difficult problem may be the most interesting. The significance of anoxic microniches to microbial ecology and, in this context, to gaseous N biogeochemistry is not known. If denitrification can be shown to occur and contribute to N_2O production in oxygen-minimum zones of the ocean, our ideas concerning global N cycling will have to be dramatically revised.

The degree of cooperation between the atmospheric scientist and the microbial ecologist should be strengthened by the realization that our global perturbation experiments (acid rain, fossil fuel burning, tropical forest destruction) are continuing unabated. The response of microbial communities to disturbance and the resulting changes in nutrient cycling and gas exchange will influence the way in which the atmospheric scientist thinks about the world. If these meetings are to make a contribution, perhaps it will be in creating the interdisciplinary forum within which to continue our work.

I thank S. C. Wofsy and M. Keller for valuable help.

This work was supported by National Aeronautics and Space Administration grant NAG1-55 and by National Science Foundation grant DEB 79-20282 to Harvard University.

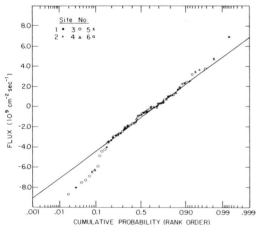

FIG. 4. Rank-ordered probability distribution of 79 nitrous oxide flux measurements taken at the Hubbard Brook Experimental Forest in 1981–1982. Samples were analyzed by electron-capture detection gas chromatography (16). Mean flux (0.5 probability) is $\approx 1.2 \times 10^9$ molecules cm^{-2} s^{-1}. Data from Keller et al. (in press).

LITERATURE CITED

1. **Alexander, M.** 1977. Introduction to soil microbiology, 2nd ed. John Wiley & Sons, Inc., New York.
2. **Alvarez-Borrego, S., D. Guthrie, C. H. Culberson, and P. K. Park.** 1975. Test of Redfield's model for oxygen-nutrient relationships using regression analysis. Limnol. Oceanogr. **20:**795–805.
3. **Belser, L. W., and E. L. Schmidt.** 1978. Diversity in the ammonia-oxidizing nitrifier population of a soil. Appl. Environ. Microbiol. **36:**584–588.
4. **Blackmer, A. M., and J. M. Bremner.** 1976. Potential of soil as a sink for atmospheric nitrous oxide. Geophys. Res. Lett. **3:**739–742.
5. **Bleakley, B. H., and J. M. Tiedje.** 1982. Nitrous oxide production by organisms other than nitrifiers or denitrifiers. Appl. Environ. Microbiol. **44:**1342–1348.

6. **Breitenbeck, G. A., A. M. Blackmer, and J. M. Bremner.** 1980. Effects of different nitrogen fertilizers on emissions of nitrous oxide. Geophys. Res. Lett. **7:**85–88.
7. **Bremner, J. M., and A. M. Blackmer.** 1978. Nitrous oxide emission from soils during nitrification of fertilizer nitrogen. Science **199:**295–296.
8. **Bremner, J. M., and A. M. Blackmer.** 1981. Terrestrial nitrification as a source of atmospheric nitrous oxide, p. 151–170. *In* C. C. Dewiche (ed.), Denitrification, nitrification and atmospheric nitrous oxide. John Wiley & Sons, Inc., New York.
9. **Bremner, J. M., G. A. Breitenbeck, and A. M. Blackmer.** 1981. Effect of nitrapyrin on emission of nitrous oxide from soil fertilized with anhydrous ammonia. Geophys. Res. Lett. **8:**353–356.
10. **Bremner, J. M., S. C. Robbins, and A. M. Blackmer.** 1980. Seasonal variability in emission of nitrous oxide from soil. Geophys. Res. Lett. **7:**641–644.
11. **Brice, K. A., A. E. J. Eggleton, and S. A. Penkett.** 1977. An important ground surface sink for atmospheric nitrous oxide. Nature (London) **268:**127–128.
12. **Castignetti, D., and T. C. Hollocher.** 1982. Nitrogen redox metabolism of a heterotrophic nitrifying-denitrifying *Alcaligenes* sp. from soil. Appl. Environ. Microbiol. **44:**923–928.
13. **Cochran, V. L., L. F. Elliot, and R. I. Papandick.** 1981. Nitrous oxide emissions from a fallow field fertilized with anhydrous ammonia. Soil Sci. Soc. Am. J. **45:**307–310.
14. **Cohen, Y., and L. I. Gordon.** 1979. Nitrous oxide production in the ocean. J. Geophys. Res. **84:**347–353.
15. **Craig, H., and L. I. Gordon.** 1963. Nitrous oxide in the ocean and the marine atmosphere. Geochim. Cosmochim. Acta **27:**949–955.
16. **Elkins, J. W., S. C. Wofsy, M. B. McElroy, C. E. Kolb, and W. A. Kaplan.** 1978. Aquatic sources and sinks for nitrous oxide. Nature (London) **275:**602–606.
17. **Firestone, M. K., M. S. Smith, R. B. Firestone, and J. M. Tiedje.** 1979. The influence of nitrate, nitrite and oxygen on the composition of gaseous products of denitrification in soil. Soil Sci. Soc. Am. Proc. **43:**1140–1144.
18. **Fisher, T. R., and P. E. Parsley.** 1979. Amazon lakes: water storage and nutrient stripping by algae. Limnol. Oceanogr. **24:**547–553.
19. **Focht, D. D., and W. Verstraete.** 1977. Biochemical ecology of nitrification and denitrification, p. 135–214. *In* M. Alexander (ed.), Advances in microbial ecology, vol. 5. Plenum Press, New York.
20. **Freney, J. R., O. T. Denmead, and J. R. Simpson.** 1978. Soil as a sink for atmospheric nitrous oxide. Nature (London) **273:**530–532.
21. **Goreau, T. J., W. A. Kaplan, S. C. Wofsy, M. B. McElroy, F. W. Valois, and S. W. Watson.** 1980. Production of NO_2 and N_2O by nitrifying bacteria at reduced concentrations of oxygen. Appl. Environ. Microbiol. **40:**526–532.
22. **Hahn, J.** 1973. Nitrous oxide in air and seawater over the Iceland-Faroe ridge. "Meteor" Forschungsergeb. **13:**43–49.
23. **Hashimoto, L. K., W. A. Kaplan, S. C. Wofsy, and M. B. McElroy.** 1983. Transformations of fixed nitrogen and N_2O in the Cariaco Trench. Deep Sea Res. **30:**575–590.
24. **Hooper, A. B.** 1978. Nitrogen oxidation and electron transport in ammonia-oxidizing bacteria. p. 299–304. *In* D. Schlessinger (ed.), Nitrification and reduction of nitrogen oxides. American Society for Microbiology, Washington, D.C.
25. **Hutchinson, G. L., and A. R. Mosier.** 1979. Nitrous oxide emissions from an irrigated cornfield. Science **205:**1125–1127.
26. **Kaspar, H. F., and J. M. Tiedje.** 1981. Dissimilatory reduction of nitrate and nitrite in the bovine rumen: nitrous oxide production and effect of acetylene. Appl. Environ. Microbiol. **41:**705–709.
27. **Lipshultz, F., O. C. Zafirou, S. C. Wofsy, M. B. McElroy, F. W. Valois, and S. W. Watson.** 1982. Production of NO and N_2O by soil nitrifying bacteria. Nature (London) **294:**641–643.
28. **Logan, J. A., M. J. Prather, S. C. Wofsy, and M. B. McElroy.** 1981. Tropospheric chemistry: a global perspective. J. Geophys. Res. **86:**7210–7254.
29. **McElroy, M. B., S. C. Wofsy, and Y. I. Yung.** 1977. The nitrogen cycle: perturbations due to man and their impact on atmospheric N_2O and O_3. Phil. Trans. R. Soc. London Ser. B **277:**159–181.
30. **McKenney, D. J., D. L. Wade, and W. I. Findlay.** 1978. Rates of N_2O evolution from N-fertilized soil. Geophys. Res. Lett. **5:**777–780.
31. **Mosier, A. R., M. Stillwell, W. J. Parton, and R. G. Woodmansee.** 1981. Nitrous oxide emissions from a native shortgrass prairie. Soil Sci. Soc. Am. J. **45:**617–619.
32. **Payne, W. J.** 1973. Reduction of nitrogenous oxides by microorganisms. Bacteriol. Rev. **37:**409–452.
33. **Rasmussen, R. A., and M. A. K. Khalil.** 1981. Increase in the concentration of atmospheric methane. Atmos. Environ. **15:**883–886.
34. **Reid, J. L., Jr.** 1965. Intermediate waters of the Pacific Ocean. The Johns Hopkins Press, Baltimore.
35. **Ritchie, G. A. F., and D. J. D. Nicholas.** 1974. The partial characterization of purified nitrite reductase and hydroxylamine oxidase from *Nitrosomonas europaea*. Biochem. J. **138:**471–480.
36. **Ryden, J. C.** 1981. N_2O exchange between a grassland soil and the atmosphere. Nature (London) **292:**235–237.
37. **Schutz, K., C. Junge, R. Beck, and B. Albrecht.** 1970. Studies of atmospheric N_2O. J. Geophys. Res. **75:**2230–2246.
38. **Seiler, W., and R. Conrad.** 1981. Field measurement of natural and fertilizer induced N_2O release rates from soils. J. Air Pollut. Control Assoc. **31:**767–772.
39. **Smith, M. S.** 1983. Nitrous oxide production by *Escherichia coli* is correlated with nitrate reductase activity. Appl. Environ. Microbiol. **45:**1545–1547.
40. **Smith, W. H., F. H. Bormann, and G. E. Likens.** 1968. Response of chemoautotrophic nitrifiers to forest cutting. Soil Sci. **106:**471–473.
41. **Terry, R. E., and R. L. Tate.** 1980. The effect of nitrate on nitrous oxide reduction in organic soils and sediments. Soil Sci. Soc. Am. J. **44:**744–746.
42. **Terry, R. E., R. L. Tate, and J. M. Duxbury.** 1981. Nitrous oxide emissions from drained cultivated organic soils of South Florida. J. Air Pollut. Control Assoc. **31:**1173–1176.
43. **Walker, N., and K. N. Wicknamasinghe.** 1979. Nitrification and autotrophic nitrifying bacteria in acid tea soils. Soil Biol. Biochem. **11:**231–236.
44. **Wang, W. C., Y. L. Yung, A. A. Lacis, T. Mo, and J. E. Hansen.** 1976. Greenhouse effects due to man-made perturbations of trace gases. Science **194:**685–690.
45. **Weiss, R. F.** 1981. The temporal and spatial distribution of tropospheric nitrous oxide. J. Geophys. Res. **86:**7185–7195.
46. **Weiss, R. F., and H. Craig.** 1976. Production of atmospheric nitrous oxide by combustion. Geophys. Res. Lett. **3:**751–753.
47. **Wofsy, S. C., and M. B. McElroy.** 1981. Transformations of nitrogen in a polluted estuary: non-linearities in the demand for oxygen at low flow. Science **213:**754–757.
48. **Yoshinari, T.** 1976. Nitrous oxide in the sea. Mar. Chem. **4:**189–202.

Ecological Significance of Biomass and Activity Measurements

Factors Affecting Bacterioplankton Density and Productivity in Salt Marsh Estuaries

RICHARD T. WRIGHT AND RICHARD B. COFFIN

Department of Biology, Gordon College, Wenham, Massachusetts 01984, and College of Marine Studies, University of Delaware, Lewes, Delaware 19958

Although originating from a variety of sources, the dissolved organic matter of natural waters can be viewed as a trophic level, the base of a food chain involving heterotrophic activity and processing a substantial percentage of the energy flow of aquatic ecosystems (13, 20, 23). The next trophic level in this food chain is the planktonic bacteria, and the evidence is firm that the bacteria uniquely occupy this position. They in turn are probably grazed by zooplankton and benthic animals. The view of the planktonic bacteria as a unique trophic level is important, because it justifies dealing with them as an ecological unit rather than as a mixture of species that must be cultured to be understood. A variety of terms have been used to refer to the planktonic bacteria: assemblage, community, mixed culture, population, etc. It is not possible to distinguish visually between species at the present time; it is also true that the size range of the bacteria in natural waters is quite limited (ca. 0.2 to 2 μm in linear dimensions). For practical and theoretical reasons, then, the assemblage of bacteria in the plankton can be treated essentially as a population, recognizing that when our methods and understanding improve we may be able to refine this approach. However, with this approach some interesting problems can be defined and researched. For example, how can we measure bacterial productivity when we know the bacteria are using a complex mixture of organic substrates? How can we best measure grazing rates? And how can we understand spatial and temporal differences in the density of planktonic bacteria in different marine systems? In attempting to answer these questions we have concentrated on several small salt marsh estuaries and their connecting coastal waters, where both seasonal and spatial changes are quite pronounced and yet can be readily sampled. Such estuaries are well known for the high biological productivity traced to *Spartina* grasses, and they seem to be quite useful for studying bacterioplankton density and productivity and the factors that control them.

STUDY AREA

Three salt marsh estuaries (Essex, Ipswich, and Parker) empty into Ipswich Bay, a semi-enclosed pocket of the Gulf of Maine in northern Massachusetts (42°42′ N latitude; 70°42′ longitude). These estuaries are all small, relatively unpolluted, extensively bordered with *Spartina* salt marsh, and well mixed, with a low freshwater input for most of the year. With a 2.6-m mean vertical range and a measured tidal excursion range of 4 to 8 km, the estuaries are dominated by tidal exchange rather than freshwater flushing. In the offshore direction, depth increases to 60 m within 10 km of horizontal distance, the extent of our sampling range.

PLANKTONIC BACTERIA IN ESTUARIES AND COASTAL WATERS

We have studied planktonic bacteria in the Essex estuary and immediate coastal waters using fluorescence microscopy for direct counts (16) and a variety of experimental approaches employing labeled organic compounds and Nuclepore filtration (31, 32; R. T. Wright, *in* J. E. Hobbie and P. J. Williams, ed., *Heterotrophy in the Sea*, in press). In this system, particle-bound bacteria constitute a minor fraction of the total numbers and heterotrophic activity, unlike salt marsh estuaries of the southern coast of the United States (15, 30). Data collected monthly from the entire range of water in the estuary indicate pronounced seasonal and horizontal differences in the density of the planktonic bacteria (32). During winter months numbers range from 0.5×10^9 to 1×10^9 liter^{-1}, with no significant differences throughout the estuary and coastal water entering the estuary at high tide. However, during the spring numbers increase most rapidly in midestuary water masses, reaching densities of 5×10^9 liter^{-1} and higher by early summer. The seasonal effects on density include a 10-fold range in estuarine waters and a 4-fold range in coastal waters, with lowest numbers in the winter and highest during the summer months.

Figure 1 shows a typical horizontal gradient for summer conditions from the upper end of the estuary proceeding downriver and extending out to surface waters 10 km offshore. The horizontal summer pattern of total bacterial counts was quite similar in other local salt marsh estuaries; in fact, the peak density in bacteria was found approximately one tidal excursion upriver from the mouth of a given estuary. We have reasoned that this pattern can be traced to the tidal circulation of water masses; water retained within the estuary has repeated contact with the salt marshes bordering the estuary and also mixes only slowly with coastal water, allowing a sustained development of all members of the plankton community (32). Work by Turner (29) has indicated that the contribution of dissolved organic matter from *Spartina* grasses to estuarine water supports intense heterotrophic activity in the estuarine plankton. Both the supply of substrate and the heterotrophic activity it supports were shown to be strongly correlated with temperature (29). Other workers have shown a link between the phytoplankton and heterotrophic bacteria in larger estuaries and coastal waters (6, 11, 12), indicating that phytoplankton productivity may be the major source of substrate for the planktonic bacteria in these waters.

Regression analysis of direct counts and heterotrophic activity on temperature indicated a highly significant temperature effect on both parameters, with the estuarine water masses showing the strongest dependence on temperature (32). The regression coefficients of total counts on temperature for estuarine and coastal water masses were, respectively, 0.177 and 0.074 ($\times 10^9$ liter^{-1} per degree C), indicating a

clear difference in the mode of temperature effect. In more recent investigations of possible substrate sources for the development and maintenance of the midestuarine maximum in bacteria, we sampled bacterial and phytoplankton densities (chlorophyll *a*) throughout the estuary at close intervals from ice-out in March to the establishment of summer patterns in June. Correlation between these two parameters from a total of 270 samples was totally lacking ($r = 0.049$). There was, however, a strong temperature effect on bacterial density (regression coefficient of 0.20, $r^2 = 0.39$). The evidence thus indicates that in salt marsh estuaries, bacterial density is not closely linked to phytoplankton productivity. Whatever the major substrate sources, they are strongly dependent on temperature. The two primary possibilities are leachate from *Spartina* salt marsh grasses and input from the highly organic estuarine sediments. Imberger et al. (17) have recently shown that salt marshes are important producers of both labile and refractory dissolved organic carbon.

It is significant that the seasonal range, even in estuarine water, is so narrow (10-fold), in view of the much greater seasonal variation in density of phytoplankton (e.g. 20). This becomes even more remarkable considering the short-term changes in bacterial density readily observed in samples of natural water incubated in the laboratory (Wright, in press), where it is not uncommon to find two- to threefold increases within 12 to 24 h. An analysis of the factors controlling bacterial density must therefore include this short-term as well as the long-term (seasonal) perspective. Further, the very

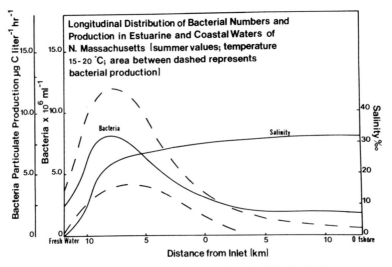

FIG. 1. Horizontal gradient of summer bacterial density and productivity in the Essex estuary and connecting offshore waters of northern Massachusetts.

significant midestuarine peak described above must also be explained. In our research into these problems, we have concentrated on four factors as potentially important in controlling bacterial density in marine waters: temperature, substrate, grazing, and dormancy. This paper will explore only two of these factors: substrate and grazing. In the following section, we will examine these two factors in light of some reasonable assumptions that can be made about the planktonic bacteria and the estuarine environment, and offer three hypotheses that relate to both fundamental ecological theory and to a method for measuring bacterial production.

SUBSTRATE AND GRAZING AS FACTORS CONTROLLING PLANKTONIC BACTERIA

Variables. P is the rate of substrate supply to bacteria from all sources, in units of micrograms of C per liter per hour; S is the concentration of labile substrate, in units of micrograms of C per liter; B is the density of planktonic bacteria, in units of 10^9 cells liter^{-1} (assume C content to be 24 µg of C per 10^9 cells for estuarine bacteria [33]); v is the rate of uptake of S by bacteria, in units of micrograms of C per liter per hour; μ is the growth rate constant for bacteria unmodified by substrate limitation (units of hour^{-1}), sometimes called maximum specific growth rate; and μ_a is the actual growth rate constant under conditions of substrate limitation (hour^{-1}).

Assumptions and hypotheses. Assumption 1. The bacteria require a certain amount of substrate for maintenance energy (to maintain cell structure and integrity). The maintenance coefficient, c_m hour^{-1} (24), is the ratio of substrate used for maintenance to cell biomass. This coefficient is quite sensitive to growth state, such that a shift from growth to stationary phase should reduce the magnitude of the coefficient. A reasonable value is 0.08 h^{-1} (the lowest value given by Pirt [24]). Thus, for 10^9 cells this is 1.9 µg of C h^{-1}. The evidence for this comes only from culture work.

Assumption 2. Bacterial uptake (v) will be a function of bacterial density (B) and of S, directly proportional to B if S is held constant, and also proportional to S. Although much of the uptake measured over a limited concentration follows the familiar saturation-type curve of Michaelis-Menten kinetics (34), there is evidence from pure cultures and from natural systems that uptake versus substrate concentration will often be proportional (Wright, in press), possibly as a result of the operation of different transport systems at different concentration ranges (2).

Assumption 3. The rate of substrate supply (P) to the bacteria in an estuary depends on contact of a given water mass with sediment and salt

marsh sources and on processes taking place within the water body. A reasonable value for estuarine water during summer months would be 200 µg of C liter^{-1} day^{-1} (3, 29; Wright, in press), although variations due to diel and tidal effects should be expected.

Assumption 4. Although minor fluctuations are likely to occur, there is a steady state in the concentration of S due to continual bacterial use of substrate, at rates equal to the rate of substrate production; thus, under ordinary conditions, S does not accumulate.

Hypothesis 1. B will reach a maximum density in the absence of grazing where either growth equals death rate (this would be hard to measure) or all substrate is used for maintenance energy. Either alternative leads to a steady-state concentration, called B_{max}. Thus, on the basis of the above assumptions:

$$B_{max} = \frac{P \text{ (µg of C liter}^{-1} \text{ day}^{-1})}{c_m \times 24 \times 24 \text{ µg of C per } 10^9 \text{ cells}} \quad (1)$$

$$B_{max} = \frac{200 \text{ µg of C liter}^{-1} \text{ day}^{-1}}{46.1 \text{ µg of C day}^{-1} \text{ per } 10^9 \text{ cells}}$$

$$= 4.3 \times 10^9 \text{ cells liter}^{-1} \quad (2)$$

Lower maintenance energy requirements as B approaches B_{max} would simply increase the magnitude of B_{max}. The above value for B_{max} is for discussion purposes only.

Hypothesis 2. In the absence of grazing and the presence of a limited amount of substrate, the relationship between the planktonic bacteria and substrate may be described by the familiar logistic or Verhulst-Pearl equation of density-dependent growth to a limit (10). Thus:

$$\frac{dB}{dt} = \mu B \left(\frac{B_{max} - B}{B_{max}} \right) \quad (3)$$

where μ is the maximum specific growth rate of the bacteria. In the case of density-dependent growth until substrate is exhausted, the actual growth rate declines to zero as B approaches B_{max}. This actual growth rate can be expressed as:

$$\mu_a = \mu \left(1 - \frac{B}{B_{max}} \right) \quad (4)$$

This expression can be rearranged as follows, to show that μ_a is a linear function of B (19, p. 195):

$$\mu_a = \mu - \frac{\mu B}{B_{max}} \quad (5)$$

In the natural environment, the limit B_{max} is not caused by substrate exhaustion but by the increasing demand placed on substrate by maintenance energy requirements as bacterial density increases and substrate supply remains constant.

Assumption 5. There is an assemblage of grazers in the estuarine environment, probably microflagellates (7–10), that is removing bacteria at some finite rate. Grazing thus has the potential of lowering the density of bacteria, B, and to the extent that it does, substrate concentration will increase and will allow uptake above the maintenance level. When this happens, net growth of the bacteria will occur. For purposes of discussion, it will be assumed that both grazing and bacterial growth are at steady state.

$$\frac{dB}{dt} = (\mu_a - g)B = 0 \qquad (6)$$

Thus, in the presence of grazing, some proportion of daily substrate production will be used for growth, and some for maintenance energy. Although B is less than B_{max}, total uptake of substrate will not be affected; ambient substrate concentration will increase until the increased velocity per cell it has induced brings total uptake up to the rate of substrate production. One consequence of grazing, therefore, is that the relationship between the density of bacteria and ambient substrate concentration can be characterized as some point on the logistic curve, the exact position being determined by the degree to which grazing has reduced B below B_{max}. The farther B is down the curve, the higher the ambient substrate concentration. If grazing pressure is suddenly removed, as in a filtration experiment, then the subsequent experience of the bacteria is determined only by available substrate.

Hypothesis 3. On the basis of the above reasoning, it follows that the growth response of the natural bacteria in a bottle experiment from which grazers have been removed may be an excellent (albeit qualitative) indicator of the relative balance between substrate and grazing as they respectively control the density of bacteria in the plankton. High grazing pressure should keep density well below the inflection point of the logistic curve, whereas low grazing pressure would keep density close to B_{max}. The former would be signaled by some finite period of exponential growth early in the incubation (a concave growth curve) as higher ambient substrate concentrations would be expected under grazing pressure, the latter caused by growth gradually reaching a plateau (a convex growth curve). Linear growth early in the incubation, a pattern which is frequently found in such measurements from estuarine and coastal waters, indicates the possibility that grazing was holding bacterial density around the inflection point of the logistic curve. It is significant that in this growth stage numbers added with time (thus, bacterial production) are at a maximum. However, sustained linear growth may imply sustained substrate production by exoenzyme action on high-molecular-weight organic compounds (27). If S continues to be supplied in the incubation bottle by cell lysis or by enzymatic breakdown of larger organic compounds, the addition of substrate will prolong the growth phase and raise B_{max} but will not alter the basic density-dependent relationship between the bacteria and substrate.

A further advantage of the incubation experiment is that the bacterial response in the presence and absence of grazers may be used to calculate bacterial production and the rate of grazing on the bacteria. This is a straightforward procedure involving incubation of an unfiltered sample (thus containing grazers) and one filtered through a 1-μm Nuclepore filter to remove potential grazers, and measuring the changes in bacterial density every few hours. This basic approach has been used for some time by Russian workers and has most recently been summarized by Gak et al. (14). Others (4, 13, 18) have used measured changes in filtered and incubated samples to calculate either linear or exponential growth rate constants which are then used to compute bacterial productivity. None of these workers, however, has employed the logistic equation in their calculations. The following is an approach to measuring both bacterial production and grazing on the bacteria that satisfies the assumption of a density-dependent relationship to substrate and allows computation of these parameters under a variety of growth states of the natural bacteria.

METHOD FOR MEASURING PRODUCTION AND GRAZING

Freshly collected water samples are gravity filtered through a large mesh plankton net (270 μm) and divided into three fractions. Two-thirds of the sample is filtered through 1-μm Nuclepore filters with the use of controlled, low vacuum (<100 mm of Hg). About half of this filtrate is then filtered through 0.2-μm filters to remove bacteria, and this 0.2-μm filtrate is amended with 10% of the 1-μm filtrate to give a 1:10 dilution of the original bacterial population. Duplicate 1-liter samples (two diluted, two filtered, two unfiltered) are incubated in the dark at natural temperatures for 24 h. Subsamples taken at time zero and every 4 or 6 h are fixed with 5% buffered Formalin and refrigerated until bacterial and microzooplankton counts can be performed (within 1 week). Bacterial counts are made by the acridine orange direct count method of Hobbie et al. (16), and microzooplankton counts, by the method of Sherr and Sherr (26). Other parameters can be measured if desired, as a given sampling for microplankton requires only about 25 ml. Rates of bacterial production and grazing by the microzooplankton are computed from changes in the bacterial populations during incubation; the microzooplankton counts are performed basically to keep track of the grazing population and, if desired, to calculate clearance rates of the grazers (10).

Bacterial production. The diluted and filtered subsamples theoretically contain the same amount of substrate but, respectively, 10% and 100% of the original bacterial population (a small loss of bacteria due to the 1-μm filtration can be corrected by reference to the initial unfiltered subsamples). Growth in these subsamples during incubation will undoubtedly show different patterns over time; if logistic growth is occurring (as it should, given ultimate substrate limitation), these two subsamples will begin at different stages in the logistic curve but should share the same limit (B_{max}), in this case imposed by both substrate exhaustion and maintenance energy demands.

If growth in the 1-μm filtrate is linear over time (as it frequently is) for the first few samplings, a linear growth rate constant can be calculated by dividing the starting bacterial density into a regression coefficient calculated from a plot of bacterial counts versus time. Thus:

$$\mu_a = \frac{b}{B_0} \tag{7}$$

where b is the regression coefficient and B_0 is the initial (corrected) bacterial count in the 1-μm filtrate. Bacterial production is simply calculated by multiplying the regression coefficient times a factor converting bacterial cells into particulate carbon.

In our experience, growth frequently reaches a maximum in the 1-μm filtrate before 24 h, and does not always yield a clear linear portion of the curve. When this occurs, the 1:10 dilution sample may be used to calculate a growth constant, by using the integral form of the logistic equation (19, p. 188):

$$B_t = \frac{B_{max}}{1 + e^{a - \mu t}} \tag{8}$$

where a is the constant of integration and the other parameters are as defined above. This equation can also be presented in linear form (19, p. 189):

$$\ln \frac{B_{max} - B}{B} = a - \mu t \tag{9}$$

B_{max} must be estimated by eye from the undiluted sample data and is then used in combination with values of B at different times from the 1:10 dilution sample to calculate the growth rate constant for unrestricted exponential growth from the slope of the above equation. The growth rate constant for the original bacterial density (μ_a) is then calculated by using equation 4 given above, and production is obtained by: $P = \mu_a \times B_0$.

If the undiluted sample exhibits an increasing growth rate with time, it too may be fitted to the logistic curve (provided some estimate of B_{max} is obtained). When B does not approach B_{max} during the incubation period, an alternative estimation of B_{max} may be obtained if the inflection point of the curve can be approximated from the data; at this point, $B = B_{max}/2$.

It is instructive to consider the impact on equation 9 of a segment of the bacterial assemblage that is not dividing (dormant). Kirchman et al. (18) discussed this problem in connection with their calculations of growth rate based on exponential growth. Equation 8 above thus becomes:

$$B_t = \frac{B_{max}}{1 + e^{a - \mu t}} + D \tag{10}$$

where D is the initial number of dormant cells in the sample, and B is only those cells actively dividing. The natural log transformation of this equation then becomes:

$$\ln \frac{B_{max} - B - D}{B - D} = a - \mu t \tag{11}$$

It is possible to work with this equation by assuming different values for D and performing some curve fitting.

From the above, it is apparent that the presence of nondividing cells will tend to bend the curve from equation 9 downwards, with maximum impact during the earliest samplings. If B is relatively small compared with B_{max}, the impact of nondividing cells will become negligible after two or three samplings. In this case, an estimate of the number of dividing cells can be obtained by the y-intercept, which is $\ln(B_{max} - B)/B$ at time zero. Because of this problem, the earliest samplings should be considered suspect in the calculations of slope and intercept for equation 9.

Grazing on bacteria. There is strong evidence that a grazing microzooplankton assemblage adapted to feeding on bacteria is to be expected in systems in which bacterial production is relatively high (10, 21). As mentioned earlier, continued bacterial production accompanied by relatively constant bacterial densities implies effective grazing removal of the bacteria. In the method presented above, one set of unfiltered samples is incubated and subsampled for bacteria and microzooplankton. Gak et al. (14) presented an equation adapted from Romanova and Zonoff (25) which is based on the results of bacterial counts over time from unfiltered samples and samples from which the zooplankton have been removed by filtration. The equation as presented by Gak et al. is in error; it should read:

$$G = K \frac{n_2 - N_2}{(n_2/n_1) - 1} \qquad (12)$$

where G is the grazing rate in bacteria consumed per hour, K is the growth rate constant for exponential growth of the bacteria, n_1 and n_2 are numbers of bacteria at the start and conclusion of incubation of the filtered sample, and N_2 is the number of bacteria at the end of incubation of the unfiltered sample.

Gak et al. (14) suggested calculating K as an exponential constant, by using the equation

$$K = \frac{\ln n_2 - \ln n_1}{t} \qquad (13)$$

but as we have seen above, this expectation does not conform with the realities of substrate limitation or with the data usually obtained from natural populations. Hence, we have suggested (Wright and Coffin, unpublished data) that K should be calculated from a logistic equation ($K = \mu_a$) and further suggested that, as most samples we have tested show linear changes in

numbers of bacteria over time in both filtered and unfiltered samples, grazing could best be calculated by subtracting the regression coefficient of bacterial numbers versus time in the unfiltered sample from that of the filtered sample. This approach is exactly equivalent to using equation 12 above with a linear growth rate constant. Alternatively, the growth rate constant (μ_a) could be calculated as indicated above for measuring production and applied to equation 12.

RESULTS

Using the basic technique presented above, we have calculated bacterial production from linear increases in filtered, incubated samples. Figure 2 shows the results of 45 productivity measurements from estuarine, coastal, and offshore waters during summer months, plotted versus initial density of bacteria: the longitudinal distribution of production is shown more diagrammatically in Fig. 1. A regression of the productivity values on initial bacterial density yielded a regression coefficient of 0.87 μg of C liter^{-1} h^{-1} per 10^9 bacteria ($r^2 = 0.36$). This gives a daily production of ca. 21 μg of C per 10^9 bacteria, very close to the biomass estimate of 24 μg of C per 10^9 bacteria. The data indicate, therefore, that during warmer months the estuarine and coastal water bacteria are turning over on a daily basis, with significant fluctuations around that rate. Since numbers remain relatively constant on a day-to-day basis (unpublished data), the productivity data suggest that either grazers are removing the equivalent of the standing crop daily or else the bacteria are dying and

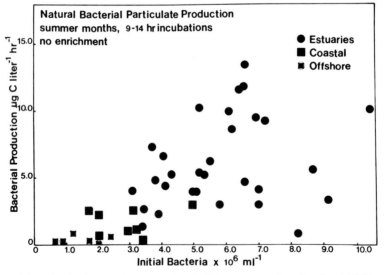

FIG. 2. Bacterial production in estuarine, coastal, and offshore waters plotted against initial bacterial density. Production determined by measured increases of bacteria in samples prefiltered to remove grazers.

lysing at that rate (which intuitively does not seem likely). Based on these data, summer midestuarine bacterial productivity is in the range of 100 to 150 μg of C liter^{-1}; if the bacteria respire 50% of substrate taken up, these values imply substrate production of 200 to 300 μg of C liter^{-1} day^{-1} supporting the bacterial trophic level, and 100 to 150 μg of C liter^{-1} day^{-1} being passed on to the grazer trophic level.

Evidence for density-dependent growth of bacteria in natural waters can be seen in Fig. 3, where sample water from two estuary stations was prefiltered through 3-μm Nuclepore filters and subdivided. Two of three samples from each station were reduced in bacterial concentration 50% and 90% by mixing with sample water filtered through glass-fiber and 0.2-μm filters. The undiluted samples showed no sustained growth period, whereas the diluted samples increased with time until densities approximated those of the undiluted samples. The data for the 4- through 24-h samplings were analyzed according to equation 9 above, taking B_{max} values from

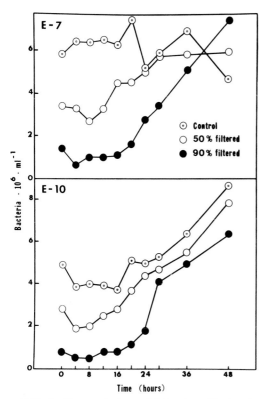

FIG. 3. Changes in bacterial density with time in two estuarine samples prefiltered through a 3-μm Nuclepore filter. Bacteria in two of three subsamples from each sample reduced in density 50% and 90% by mixing with sample water filtered through a 0.2-μm filter.

the undiluted samples to be 6.5×10^6 and 5.3×10^6 ml^{-1} for the E-6 and E-10 samples, respectively. A regression analysis of the relationship between $\ln(B_{max} - B)/B$ versus time for the four dilution samples yielded regression coefficients ranging from -0.073 to -0.106 (range of $r^2 = 0.82$ to 0.92); application of Student's t test showed that there was no significant difference between any of the regression coefficients. Using the mean of all four coefficients, we obtained the computed value of μ as 0.085 h^{-1}. A resumption of linear increase in bacterial density can be seen in the E-10 samples after 24 h of incubation; this could conceivably be due to a renewed substrate supply through enzymatic breakdown of higher-molecular-weight substrates or through death and release of organic solutes from other plankton forms in the samples. Interestingly, Cuhel et al. (5) found that the assimilation of inorganic nutrients by oceanic bacterial assemblages reached a plateau within 24 to 36 h of incubation and reasoned that this pattern indicated growth limitation by organic substrate present in the sample water.

Figure 4 shows the results from an experiment in which a 1:10 dilution of 1-μm filtered estuarine water was placed, respectively, in 0.2-μm filtered, aged seawater (control) and in filtered, aged seawater in which *Spartina* exudate had been allowed to accumulate during a 6-h period of contact of the water with *Spartina* leaves. The pattern of increase in bacterial density in both samples indicated growth to a limit imposed by organic substrate present. When data were plotted according to equation 9 (shown in the upper left of Fig. 4), the sample containing *Spartina* exudate demonstrated the linearity expected of the sigmoid growth curve. The possible effect of a dormant segment of the population is indicated in the first one or two samplings. Extrapolation of the exudate sample suggests that approximately 30% of the initial population was active and dividing.

Figure 5 shows some results from incubations of estuarine water (E-6) and water from 10 km offshore, comparing unfiltered samples with samples filtered through 1- and 3-μm Nuclepore filters. Bacterial production and grazing on the bacteria were estimated from linear regression of bacterial density versus time (equation 12 above). The results indicate essentially no grazing in the offshore sample, but apparent grazing rates higher than bacterial production in the estuarine sample. These samples are part of a 10-sample transect taken of the estuarine-coastal water system. In this transect, bacterial productivity was higher than grazing in the offshore and coastal samples, and moving into the estuary, grazing increased and was substantially higher than bacterial production in the upper reaches of

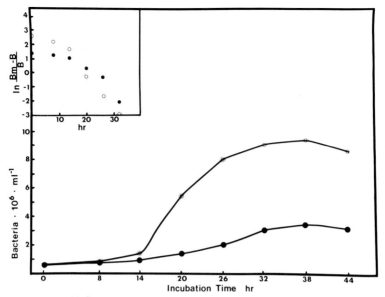

FIG. 4. Changes in bacterial density with time in 1-μm filtered estuarine water diluted 1:10 and placed in: (i) filtered, aged seawater (filled circles) and (ii) filtered, aged seawater previously incubated in contact with *Spartina* leaves for 6 h (open circles). Upper left: plot of same data according to equation 9 (see text for explanation).

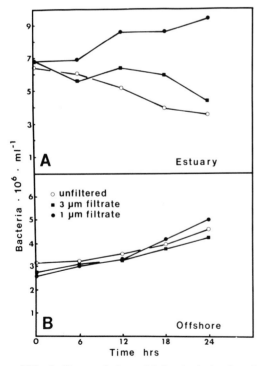

FIG. 5. Changes in bacterial density in incubated samples taken 11 August 1982 from Essex estuary and a station 10 km offshore. Sample treatment: unfiltered, prefiltered through 3-μm Nuclepore filter, and prefiltered through 1-μm Nuclepore filter.

the estuary. One highly significant result from this set of samples was the observation that apparent grazing of bacteria took place in samples prefiltered through a 3-μm Nuclepore filter. We have confirmed this observation with tests of bacterial growth in subsamples filtered through a range of filters and plankton net mesh; in most samples, the greatest amount of apparent grazing occurred in 3-μm filtrates. Sherr and Sherr (26) reported that the majority of microprotozoa in the plankton of a Georgia salt marsh estuary fell into the 2-μm size range. It is clearly not justified to assume that a 3-μm filter excludes grazing microorganisms, as we (Wright, in press) and others have done.

DISCUSSION

There is clear evidence that the bacteria in estuaries and coastal waters have the potential for a daily increase in numbers that is twofold or greater (4, 13, 18; Wright, in press). There is also no question that the production of labile dissolved organic matter, from a variety of sources, supplies substrate to the bacteria on a more or less continuous basis during summer months. Given continuous production of substrate and the potential of the bacteria to increase rapidly, the time scale over which effective controlling factors operate must be in the order of hours or days. This rules out temperature and other physical factors; the important factors are going to be biological in nature. We still know too little about short-term changes in bacterial density in natural waters to use the term "regulation" with any confidence. However, it is fairly certain that the daily production of particulate carbon by bacterial use of dissolved organic matter will not be wasted; the evidence is rapidly accumulating to suggest that an assemblage of small protozoans, primarily flagellates, is adapted specifically to grazing the natural bacteria. These two factors, grazing and substrate, appear to be important factors to investigate in our attempt to understand the ambient bacterial density and productivity in more productive marine systems.

The interaction of grazers, the bacteria, and substrate is representative of a classical set of relationships addressed in the ecological literature. We have tried in this paper to show that the relationship of bacterial density to substrate concentration and supply should follow the logistic growth curve. In the absence of grazing, substrate should limit the planktonic bacteria in a density-dependent manner, and several lines of evidence were cited to support this conclusion. If this is true, then grazing is a factor that can hold bacterial density at levels below the limit set by substrate. The interaction between grazing rate and use of substrate then determines the growth rate of the bacteria, and the nature of this interaction is such that bacterial production is at a maximum at the inflection point of the logistic curve, where $B = B_{max}/2$. It is tempting to compare the results of bacterial growth during incubation where grazers have been removed with this relationship of the bacteria, substrate, and grazing in the natural environment. As argued above, the comparison is reasonable, but too little is known to make the comparison quantitative. In particular, we need more information on the significance of maintenance energy requirements as a limiting factor on natural bacterial density and on the sources of substrate in incubation experiments. We obviously need more information on natural grazing rates and rates of bacterial production.

Finally, more work is needed to determine the importance of dormancy in the planktonic bacteria (22, 28). If bacteria can respond to substrate limitation by greatly reducing their maintenance energy, as some evidence indicates (1), they will maintain respectable densities even though production is quite low. If there is a definite threshold for effective grazing, which is likely, such a system is likely to be controlled by substrate alone. Thus, even though the range of substrate production in natural waters, from oligotrophic to highly eutrophic systems, will cover several orders of magnitude, the response of the bacteria to substrate and the presence of a grazing

trophic level should greatly reduce the range of density of the bacteria. Grazing may prevent high densities in most productive environments (8), and dormancy may set a lower limit on density in environments with very low productivity. If we are close to confirming this view of factors controlling bacterial density and production, then we have made significant progress in our understanding of the ecology of the planktonic bacteria.

This material is based on work supported by the National Science Foundation under grant OCE-7925368.

We thank Andrew Beauregard and Richard B. Wright for their able assistance in the laboratory.

LITERATURE CITED

1. **Amy, P. A., C. Pauling, and R. Y. Morita.** 1983. Starvation-survival processes of a marine vibrio. Appl. Environ. Microbiol. **45:**1041–1048.
2. **Azam, F., and R. E. Hodson.** 1981. Multiphasic kinetics for D-glucose uptake by assemblages of natural marine bacteria. Mar. Ecol. Prog. Ser. **6:**213–222.
3. **Burney, C. M., K. M. Johnson, and J. McN. Sieburth.** 1981. Diel flux of dissolved carbohydrate in a salt marsh and a simulated estuarine ecosystem. Mar. Biol. **63:**175–187.
4. **Christian, R. R., R. B. Hanson, and S. Y. Newell.** 1982. Comparison of methods for measurement of bacterial growth rates in mixed batch cultures. Appl. Environ. Microbiol. **43:**1160–1165.
5. **Cuhel, R. L., H. W. Jannasch, and C. D. Taylor.** 1983. Microbial growth and macromolecular synthesis in the northwestern Atlantic Ocean. Limnol. Oceanogr. **28:**1–18.
6. **Ducklow, H. W.** 1982. Chesapeake Bay nutrient and plankton dynamics. 1. Bacterial biomass and production during spring tidal destratification in the York River, Virginia, estuary. Limnol. Oceanogr. **27:**651–659.
7. **Fenchel, T.** 1982. Ecology of heterotrophic microflagellates. I. Some important forms and their functional morphology. Mar. Ecol. Prog. Ser. **8:**211–223.
8. **Fenchel, T.** 1982. Ecology of heterotrophic microflagellates. II. Bioenergetics and growth. Mar. Ecol. Prog. Ser. **8:**225–231.
9. **Fenchel, T.** 1982. Ecology of heterotrophic microflagellates. III. Adaptations to heterogeneous environments. Mar. Ecol. Prog. Ser. **9:**25–33.
10. **Fenchel, T.** 1982. Ecology of heterotrophic microflagellates. IV. Quantitative occurrence and importance as bacterial consumers. Mar. Ecol. Prog. Ser. **9:**35–42.
11. **Ferguson, R. L., and A. V. Palumbo.** 1979. Distribution of suspended bacteria in neritic waters south of Long Island during stratified conditions. Limnol. Oceanogr. **24:**697–705.
12. **Fuhrman, J. A., J. W. Ammerman, and F. Azam.** 1980. Bacterioplankton in the coastal euphotic zone: distribution, activity, and possible relationships with phytoplankton. Mar. Biol. **60:**201–207.
13. **Fuhrman, J. A., and F. Azam.** 1980. Bacterioplankton secondary production estimates for coastal waters of British Columbia, Antarctica, and California. Appl. Environ. Microbiol. **39:**1085–1095.
14. **Gak, D. S., E. P. Romanova, V. I. Romanenko, and Y. I. Sorokin.** 1972. Estimation of changes in number of bacteria in the isolated water samples, p. 78–82. In Y. I. Sorokin and H. Kadota (ed.), Microbial production and decomposition in fresh waters. Blackwell Scientific Publications, Oxford.
15. **Hanson, R. B., and J. Snyder.** 1979. Microheterotrophic activity in a salt-marsh estuary, Sapelo Island, Georgia. Ecology **60:**99–107.

16. **Hobbie, J. E., R. J. Daley, and S. Jasper.** 1977. Use of Nuclepore filters for counting bacteria by fluorescent microscopy. Appl. Environ. Microbiol. **33:**1225–1228.

17. **Imberger, J., T. Berman, R. R. Christian, E. B. Sherr, D. E. Whitney, L. R. Pomoroy, R. G. Wiegert, and W. J. Wiebe.** 1983. The influence of water motion on the distribution and transport of materials in a salt marsh estuary. Limnol. Oceanogr. **28:**201–214.

18. **Kirchman, D., H. Ducklow, and R. Mitchell.** 1982. Estimates of bacterial growth from changes in uptake rates and biomass. Appl. Environ. Microbiol. **44:**1296–1307.

19. **Krebs, C. J.** 1978. Ecology: the experimental analysis of distribution and abundance, 2nd ed. Harper and Row, Publishers, New York.

20. **Larsson, U., and A. Hagstrom.** 1982. Fractionated phytoplankton primary production, exudate release and bacterial production in a Baltic eutrophication gradient. Mar. Biol. **67:**57–70.

21. **Linley, E. A. S., and R. C. Newell.** 1981. Microheterotrophic communities associated with the degradation of kelp debris. Kiel. Meeresforsch. **5:**345–355.

22. **Novitsky, J. A., and R. Y. Morita.** 1978. Possible strategy for the survival of marine bacteria under starvation conditions. Mar. Biol. **48:**289–295.

23. **Pedros-Alio, C., and T. D. Brock.** 1982. Assessing biomass and production of bacteria in eutrophic Lake Mendota, Wisconsin. Appl. Environ. Microbiol. **44:**203–218.

24. **Pirt, S. J.** 1965. The maintenance energy of bacteria in growing cultures. Proc. R. Soc. London Ser. B **163:**224–231.

25. **Romanova, A. P., and A. I. Zonoff.** 1964. On the estimation of production of bacterial biomass in the water body. Dokl. Akad. Nauk SSSR Biochem. Sect. (Engl. transl.) **155:**194–197.

26. **Sherr, B. F., and E. B. Sherr.** 1983. Enumeration of heterotrophic microprotozoa by epifluorescence microscopy. Estuarine Coastal Shelf Sci. **16:**1–7.

27. **Somville, M., and G. Billen.** 1983. A method for determining exoproteolytic activity in natural waters. Limnol. Oceanogr. **28:**190–193.

28. **Stevenson, L. H.** 1978. A case for bacterial dormancy in aquatic ecosystems. Microb. Ecol. **4:**127–133.

29. **Turner, R. E.** 1978. Community plankton respiration in a salt estuary and the importance of macrophytic leachates. Limnol. Oceanogr. **23:**442–451.

30. **Wilson, C. A., and L. H. Stevenson.** 1980. The dynamics of the bacterial population associated with a salt marsh. J. Exp. Mar. Biol. Ecol. **48:**123–138.

31. **Wright, R. T.** 1978. Measurement and significance of specific activity in the heterotrophic bacteria of natural waters. Appl. Environ. Microbiol. **36:**297–305.

32. **Wright, R. T., and R. B. Coffin.** 1983. Planktonic bacteria in estuaries and coastal waters of northern Massachusetts: spatial and temporal distribution. Mar. Ecol. Prog. Ser. **11:**205–216.

33. **Wright, R. T., R. B. Coffin, C. P. Ersing, and D. Pearson.** 1982. Field and laboratory measurements of bivalve filtration of natural marine bacterioplankton. Limnol. Oceanogr. **27:**91–98.

34. **Wright, R. T., and J. E. Hobbie.** 1966. Use of glucose and acetate by bacteria and algae in aquatic ecosystems. Ecology **47:**447–464.

Aquatic Bacteria: Measurements and Significance of Growth

ÅKE HAGSTRÖM

Department of Microbiology, University of Umeå, S-901 87, Umeå, Sweden

In the pelagic ecosystem the microheterotrophic food chain supports a tightly coupled loop of carbon and mineral nutrients flowing between the primary producers and the storage of nonliving matter. A framework for this microbial loop, based on information derived over the past decade from bacterial production and microzooplankton predation on bacteria (5, 7, 9, 11), has been proposed by Pomeroy (20), Williams (25), and Azam et al. (1).

This new paradigm was preceded by considerable uncertainty as to the importance of free-living bacteria in the sea (12, 21, 22). Under these circumstances, the development by ecologists in the mid-1970s of methods designed to measure the bacterial growth rate to the nearest order of magnitude was regarded as a considerable advance. In one such procedure developed in 1979, Hagström et al. (9; Å. Hagström and U. Larsson, *in* J. E. Hobbie and P. J. leB. Williams, ed., *Heterotrophic Activity of the Sea*, in press) used the frequency of dividing cells to convert bacterial culture data into growth rates of natural populations. In a second procedure developed in 1980, Fuhrman and Azam (7, 8) utilized the incorporation of [³H]thymidine into procaryote DNA to measure DNA doubling time. These techniques have been subsequently adopted by a number of workers for studies of bacterial production, but in spite of revisions suggested by different workers, the basic flaws and virtues inherent in the original methods still remain unchanged (10, 16, 18).

Much of the debate in current literature has been focused on growth rate, although estimates of bacterial production also include cell abundance and size. Because of the combined uncertainty of these measurements, it is still difficult to evaluate precisely the ecological significance of estimates from different areas and periods of time. Consequently, information regarding the annual cycles of bacterial production in pelagic systems is scarce. For example, so far relatively few investigations of bacterial production have been performed to cover a full year (10, 14, 19).

In this paper two studies of pelagic bacterial production in different geographical areas are presented. The effects of temporal variation in cell size and of diel growth cycles on the production estimates are discussed. Growth rate measurements made with frequency-of-dividing-cells (FDC) and [³H]thymidine techniques are compared, and the procedure for calibrating frequency of dividing cells and bacterial growth rate from batch culture data is commented upon.

MATERIALS AND METHODS

Field measurements. The station Norrbyn was sampled during the course of 1 full year. Norrbyn is situated in the northern Bothnian Sea in the Baltic (63°31'N, 19°50'E). As a rule, the water is stratified during summer and early autumn. The thermocline at the sampling site is situated between 10 and 15 m depth, and the trophogenic layer extends approximately to the depth of the thermocline. Typical salinities are between 3 and 5%, and the mean temperature reaches 16 to 18°C during late July.

One diel cycle was sampled off Scripps pier at Scripps Institution of Oceanography (32°53'N, 177°15'W).

Primary production. Water samples from eight depths (0, 1, 2, 4, 6, 8, 10, and 14 m) were incubated with the addition of 10 μCi of carrier-free NaH¹⁴CO₃. Radioactivity in intact seawater was determined in 10-ml samples transferred to scintillation vials and acidified (pH of 2) with HCl. The ¹⁴CO₂ remaining in the samples was stripped with air for 15 min. Samples were counted in a Nuclear Chicago Mark II liquid scintillation counter. Uptake was measured in both light and dark bottles, and dark values were subtracted. The uptake of carbon was calculated as described elsewhere (14). Daily primary production was calculated by multiplying with a light factor (total daily insolation divided by insolation during the incubation period).

[³H]thymidine incorporation. The procedure of Fuhrman and Azam (8) was followed strictly. Seawater samples were collected in a clean bottle lowered below the water surface. Subsamples (10 ml) were incubated at in situ temperature with 250 μCi of 84-Ci/mmol solution per liter. At the end of incubation (1 h), the sample was poured into a plastic test tube immersed in ice water, and then an equal volume of ice-cold 10% trichloroacetic acid was added to extract the soluble pools from the cells. After 5 min of extraction on ice, the cold trichloroacetic acid-insoluble material was collected by filtration (0.45 μm, Millipore Corp.). The filters were rinsed five times with 1 ml of ice-cold trichloroacetic acid and then placed in a scintillation vial.

Ethyl acetate (1 ml) was added to each vial to dissolve the filter, and scintillation fluid was added. The number of moles of thymidine incorporated was calculated from the formula:

$$\text{moles} = \text{dpm (SA)}^{-1}\, 4.5 \times 10^{-13}$$

where dpm is the disintegrations per minute on the filter, SA is the specific activity of the thymidine solution in curies per mole, and 4.5×10^{-13} is the number of curies per dpm. Moles of thymidine were converted to cells produced with the conversion factor 1.7×10^{18}, as presented by Fuhrman and Azam (8).

Determination of FDC. Direct counts of bacteria were performed on $0.2\text{-}\mu\text{m}$ filters (Nuclepore) with the use of fluorescent staining. At station Norrbyn, water samples were taken every second meter (0 to 14 m) and integrated; sampling was done at 10 a.m. $\pm$ 20 min. Four subsamples of 5 ml were drawn from the integrated samples and added to tubes containing 0.4 ml of filtered (0.22 μm) buffered (hexamethylenediamine, 20 g/100 ml, pH 7.2) formaldehyde (20%, wt/wt) containing ethidium bromide (8 mg/100 ml). To ensure an even filtration, subsamples (0.1 to 1.0 ml) were mixed with 5 ml of particle-free water in a 13-mm stainless-steel funnel (Millipore). The filters were mounted on glass slides in cinnamaldehyde and eugenol (1:2). Excess fluid was removed with a piece of filter paper, and the cover slips were sealed with clear nail varnish. Bacterial counts were done in epifluorescence with the use of Zeiss WG optics (Planapochromat, 63/1.4, kpl 12.5 W; magnification, ×984). Ten fields were counted; the filtered volume was adjusted to yield about 30 bacteria per field. Dividing bacteria were counted on a minimum of 10 fields. If fewer than 30 dividing cells were recorded, a maximum of 20 additional fields was examined. Bacteria showing an invagination but not a clear zone between cells were considered as one dividing cell.

Bacterial production estimates. The FDC growth rate relationships given by Hagström and Larsson (in press) were used to convert measured FDC percentage to divisions per hour, as follows: at 0°C, FDC = 11.0 μ + 7.2; at 5°C, FDC = 60.2 μ + 5.5; at 10°C, FDC = 70.0 μ + 3.8; at 15°C, FDC = 36.3 μ + 3.5; and at 20°C, FDC = 21.2 μ + 0.5.

Cell volume measurement. Stained bacteria were photographed (Kodak Technical Pan film 2415) in epifluorescence (Zeiss M 63 camera; magnification on negative, ×315). The negatives were projected on a graphics tablet (Tektronix 4956), and data were computed with a Tektronix 4051 computer, using software by Larsson and Hagström (unpublished data).

RESULTS

Annual algal and bacterial production. The pronounced temporal variation in the northern Baltic Sea (Bothnian Sea) makes this area suitable for studies of regulatory mechanisms governing succession and turnover in the pelagic food web. As an integral part of a comprehensive monitoring program, primary production, bacterial growth, and biomass are measured in the archipelago area outside the Norrbyn laboratory in Sweden.

The total primary production in the waters off Norrbyn was 61 g of C m^{-2} year^{-1} during 1982 (Fig. 1A). In these waters the onset of the spring bloom is late, since the ice cover lasts until the end of April or beginning of May, and usually no separate peak showing the bloom can be seen in the production curve (Fig. 1A). Instead, there is a gradual transition of algal species during June when the diatoms decrease and green algae start to dominate the algal community. During April, while the waters are still ice covered, dinoflagellates are abundant. This bloom, however, ends immediately after the ice begins to break. At the same time there is a drastic increase in the concentration of dissolved organic carbon (Fig. 1C).

The corresponding bacterial production for 1982 was 40 g of C m^{-2} year^{-1}. Growth rates were determined by the FDC technique, and biovolumes were determined from epifluorescence micrographs. Contrary to the primary production, the pattern of the production curve for bacteria shows pronounced spring and summer peaks (Fig. 1A). The bacterial abundance (Fig. 1B) shows a typical increase from a low level during winter (usually slightly below 10^6/ml) to peak values of about 2.5×10^6/ml. The mean bacterial cell volume, calculated from measurements of width and length of individual cells, shows an inverse relationship to bacterial abundance, decreasing with the arrival of summer. Only during the extreme production peak in August does the cell volume show an increase. The dissolved organic carbon shows its highest value preceding this peak. Both bacterial abundance and biovolume in 1983 returned to the winter values of 1982 (data not shown).

FDC versus [^{3}H]thymidine estimates of growth rates. The [^{3}H]thymidine method of Fuhrman and Azam (8) and the FDC method of Hagström et al. (9) are conceptually different, and the results obtained by using them a priori need not agree. While working in the food chain group at Scripps Institution of Oceanography, I made a 1-day diel study to compare the two techniques. The surface water (0.5 to 2 m) off the Scripps pier was sampled over a period of 28 h, and the growth rate of bacteria was determined by the

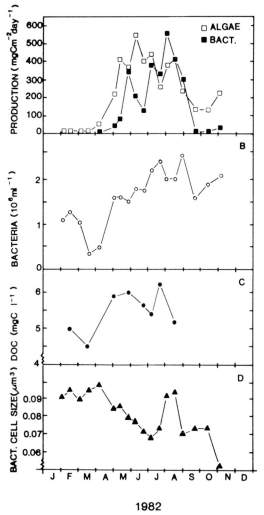

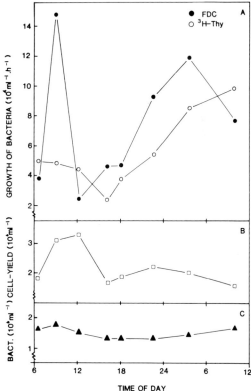

the number of bacteria produced during the same period was 1.4×10^6. The number of microflagellates at the same time increased from 500 to 1,200/ml (data not shown). With a generation time of 21 h, these flagellates consequently could have consumed 1,700 bacteria each.

An attempt was then made to determine the relative pool sizes of substrates available for the bacteria. Water from each sampling was filtered (0.22 μm) and reinoculated with one-tenth of the ambient bacterial concentration from a "seawater" culture (J. W. Ammerman, J. A. Fuhrman, Å. Hagström, and F. Azam, submitted for publication). The cultures were incubated for 50 h at 20°C, and growth of bacteria was determined from direct counts. Figure 2B shows the amount of cells yielded by each filtered seawater sample. High growth rates coincide with high yields of bacteria in the filtrates. These changes in growth rate were detected by both the FDC and the [³H]thymidine methods and support the conclusion that bacterial growth is closely regulated through the rate of input of the substrate.

FIG. 1. Temporal variation in algal primary production, bacterial production, and dissolved organic carbon in the oligotrophic coastal waters of the northern Baltic (Bothnian Sea). The depth of the trophogenic layer was 14 m. Bacterial abundance, dissolved organic carbon, and bacterial mean cell size were obtained from integrated samples 0 to 14 m.

methods of FDC and of [³H]thymidine uptake. The mean production of bacteria during this period was 8.1×10^4 and 5.9×10^4 cells ml^{-1} h^{-1} for the FDC and [³H]thymidine methods, respectively. From Fig. 2A the large variation in bacterial growth rate is evident, with generation times for FDC of between 12 and 63 h and for [³H]thymidine between 17 and 54 h. The bacterial numbers reacted to a lesser degree to the variation in growth rate than could be expected from the production estimates. The increase in bacteria from the tenth hour to the end of the experiment was 3.3×10^5 cells per ml, whereas

FIG. 2. Diel variation in bacterial growth rate and abundance. Growth rates were determined from FDC and [³H]thymidine uptake. Cell yields from reinoculated, filtered water samples were measured as an indication of available substrate for bacterial growth.

DISCUSSION

Growth rate of natural bacteria. The accuracy and usefulness of current methods to measure the growth rate of aquatic bacteria have become the subject of several studies (3, 8, 16, 18, 19). Since the main interest has then been focused on growth rate determinations based on [³H]thymidine uptake and FDC, it can now be concluded that FDC and [³H]thymidine estimates of bacterial productivity are in good agreement and that the two methods measure the same activity, that is, bacterial growth rates. These techniques have been received with fewer objections than other approaches such as the use of diffusion chambers and $^{35}SO_4$ uptake (2, 7, 9, 10, 19). Even so, suggestions have been made to improve the thymidine incorporation and FDC procedures. Moriarty and Pollard (16) argue that the dilution of the radioisotope due to de novo synthesis and uptake of exogenous thymidine must be corrected for. Isotope dilution may cause an underestimate of bacterial production (16). The reported degree of dilution, however, is small and depends on factors over which the experimenter at present lacks control (2). Although isotope dilution no doubt represents a real problem, the solution of Moriarty and Pollard (16) does not improve the approach of Fuhrman and Azam (7). Instead, a more straightforward procedure may have to be devised. In this laboratory we have tried to improve the thymidine method through the use of current techniques of molecular biology, by means of which we can extract pure bacterial DNA from field samples. In this way the dilution of [³H]thymidine in pulse-labeled DNA can be measured (disintegrations per minute per microgram of DNA) as growth proceeds. Thus the half-life of the label equals the generation time of the bacteria (Hall and Hagström, unpublished data).

Suggestions have been made to revise the original procedure in the case of the FDC technique as well. On the basis of supplemented (yeast extract) mixed cultures, a different relationship between FDC and growth rate has been presented (10, 17). A few fundamental objections can, however, be made to these studies. To grow bacteria at different growth rates in batch culture, it is not appropriate to use increasing concentrations of the same complex substrate, as was done by Hanson et al. (10) and Christian et al. (17). Under such conditions, the yield of bacteria will increase, but provided the bacteria multiply in balanced growth, the growth rate will stay the same. Maalöe and Kjeldgaard (15) give a more accurate procedure for growing bacteria at different growth rates by using a set of different carbon sources. For example, Fig. 3 shows the effect on mass increase, cell number, and

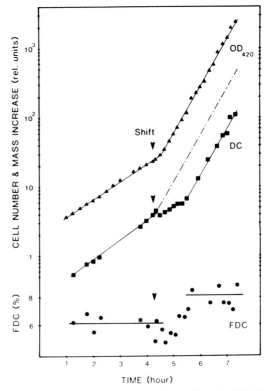

FIG. 3. Balanced growth of *E. coli* strain NC 3 during a shift from glycerol to a rich carbon source in batch culture. Cell mass was determined from optical density at 420 nm. Direct counts were made by epifluorescence microscopy. The arrows indicate the addition of rich substrate. The culture was kept on a rotary shaker at 37°C.

FDC of an *Escherichia coli* culture as it undergoes a shift from slow to fast growth. It is evident that as a result of the shift the bacteria rearrange their metabolic activities to use the new substrate most effectively. The cell mass arrives at the higher steady-state growth rate with almost no lag, but is accommodated by a transient reduction in the increase of the cell number. During this time the bacteria reach the new larger cell size corresponding to the higher growth rate. As a consequence the FDC of the culture decreases during the shift before the new higher steady state can be seen. A culture of bacteria passing from stationary phase to exponential growth may thus undergo metabolic adjustments that produce quite unpredictable results. This can be avoided by repeated dilutions of the culture to ensure balanced growth (15). From this experiment it should be obvious that considerable care must be taken to calibrate FDC against growth rate in a reliable way.

Significance of bacterial growth. A major obstacle to overcome when evaluating bacterial production estimates is, however, the need to deal with several measurements simultaneously. Bacterial production estimates for ecological purposes are calculated from growth rate, cell abundance, cell size, and a conversion of cell volume to cell carbon, and objections can be made against each of these four parameters. In a detailed study of the [³H]thymidine method, Fuhrman and Azam (8) evaluated the basis for estimating bacterial production. In this study they took a conservative approach to all steps required to convert thymidine incorporation values to bacterial cell production values. These measurements show bacteria as an important trophic element in the pelagic ecosystem, with production values in reasonable accordance with the algal primary production (8). Likewise, from Fenchel's work on flagellate predation of bacteria (5), we can argue that the flagellate population in the diel experiment is large enough to cope with the estimated bacterial production. Thus, it is possible for the bacterial concentration to stay constant although bacteria are produced at a rate of 7×10^4 cells $ml^{-1} h^{-1}$.

In the diel study the estimates of growth rate based on FDC and [³H]thymidine uptake were compared, and the daily mean values show good agreement. The slightly higher FDC value may be explained from the FDC growth rate relationship used in the diel study, which was calibrated at 20°C, although the water temperature during the experiment was 17°C. As a rule bacteria growing at a constant rate show higher FDCs as temperature decreases, as discussed by Hagström and Larsson (in press).

In a long series of measurements, Pedros-Alio and Brock (19) compared the production estimates from FDC and sulfate uptake in Lake Mendota. They found that "the estimates do not correlate very closely on a week-to-week time scale, but as a consequence of their mean daily values and range of variation being approximately the same, the annual production values calculated by both techniques are extremely close." Although such coincident data add little to solving the absolute growth rate of bacteria in the sea, it is worth noting that for the time being such circumstantial evidence is the best available.

The ecological significance of pelagic bacteria at the sampling site Norrbyn is less obvious, since the production estimate presented may seem unrealistically high. Almost all (92%) of the primary production would be needed to sustain the bacterial production, assuming a growth efficiency of the bacteria of 60%. Two explanations can be given for this extreme ratio between primary and secondary production. In the Bothnian sea the algal production is low during the periods of low water temperature.

During spring, the ice cover prevents the algal spring bloom. In addition, the poor light conditions during autumn shorten the production period. However, during the summer period, when the bulk of the bacterial production occurs, as shown in Fig. 1A, the algal production is quite similar to that recorded in the southern Baltic (Hagström and Larsson, in press). From our studies in the Baltic, we have not recorded significant growth rates of bacteria at water temperatures below 7 to 10°C (Hagström and Larsson, in press). This means that, although the annual algal production in the southern Baltic may be twice as high as in Norrbyn, the bacterial production can be equal at both locations since the temperature regimes of the two areas are similar. A second reason for maintaining a high bacterial production is the high level of dissolved organic carbon in the Bothnian sea, which has a constant input through a large land runoff.

Cell volume variation. In the present investigation, direct measurements of cell dimensions were made, and the mean bacterial volume during the productive period was found to be 0.08 μm³. This is at the low end of the range commonly found in the literature (Hagström and Larsson, in press). The measurements were made with a commercially available digitizer in a manner similar to that adopted by Krambeck et al. (13), although we used epifluorescence micrographs instead of scanning electron microscope pictures. However, apart from getting better estimates of cell size, the volume data from Norrbyn (Fig. 1D) show an intriguing pattern. In pure cultures, increasing temperatures cause the cell size of bacteria growing at a constant rate to decrease (Hagström and Larsson, in press). At the same time, higher growth rate at constant temperature yields larger cells (13). In nature, on the other hand, size-selective grazing on bacteria may perturb this straightforward pattern (4). Heterotrophic flagellates are capable of reducing the mean cell size of a mixed seawater bacterial culture by a factor of 4 (Andersson and Hagström, unpublished data). The relative contributions of these three factors toward the cell size of the natural population, however, remain poorly understood.

Cell volume to cell carbon. There are further obstacles to estimating bacterial production. The factors used to convert biovolume to cell carbon show the same variations as that of mean cell volumes given in the literature. In Table 1 a few widely used figures have been compiled, showing a 40% difference between high and low values. Any carbon-based comparison of production in which different conversion factors were used would therefore be misleading. The factor used in this paper (1.43×10^{-13}) is the

TABLE 1. Currently used factors for cell volume to cell carbon conversion

Reference	Density (mg/mm^3)	Dry wt/wet wt	Cell C/dry wt	C (g/μm^3)
Ferguson and Rublee (6)	1.1	0.23	0.34	0.87×10^{-13}
Watson et al. (24)	1.1	0.22	0.50	1.21×10^{-13}
Larsson and Hagström (14)	1.1	0.26	0.50	1.43×10^{-13}
Pedros-Alio and Brock (19)	1.025		0.097[a]	0.94×10^{-13}

[a] Cell C/wet weight.

highest value in Table 1 and may therefore cause the production estimate from this laboratory to seem somewhat high.

Diel variation of growth rate. More pertinent to the accuracy of the production estimates is, however, the obvious diel variations in growth rate shown in the Scripps data and by others (13, 23; Hagström and Larsson, in press). In the measurement of annual cycles, we have always sampled at the same time of day to minimize variations. When converting these values to daily production, however, problems do arise. In the present study, and in that reported by Larsson and Hagström (14), we have taken the growth rate value for the atrophic layer during daytime to be the same as that in the trophogenic layer during the night. This procedure does not solve the problem of diel variations but may be adopted until a more acceptable approach to solving this problem can be found.

CONCLUSION

The significance of bacteria in the sea has been established through growth rate measurements made by the FDC and [^{3}H]thymidine techniques, the latter being the first choice for future work since thymidine incorporation is a much less cumbersome procedure than that of measuring FDC. Although crude, our present results on bacterial growth allow future efforts to be directed toward the next link in the microheterotrophic food chain, the bacterial predators. Finding alternative ways to measure consumption of bacteria by predators such as the flagellates may thus prove to yield the "true growth rates" of bacteria in the sea.

LITERATURE CITED

1. **Azam, F., T. Fenchel, J. S. Gray, L. A. Meyer-Reil, and F. Tingstad.** 1983. The ecological role of water-column microbes in the sea. Mar. Ecol. Prog. Ser. **10**:257–263.
2. **Bell, R. T., G. M. Ahlgren, and I. Ahlgren.** 1983. Estimating bacterioplankton production by measuring ^{3}H-thymidine incorporation in a eutrophic Swedish lake. Appl. Environ. Microbiol. **45**:1709–1721.
3. **Christian, R. R., R. B. Hanson, and S. Y. Newell.** 1982. Comparison of methods for measurement of bacterial growth rates in mixed batch cultures. Appl. Environ. Microbiol. **43**:1160–1165.
4. **Fenchel, T.** 1982. Ecology of heterotrophic microflagellates. III. Adaptions to heterogeneous environments. Mar. Ecol. Prog. Ser. **9**:25–33.
5. **Fenchel, T.** 1982. Ecology of heterotrophic microflagellates. IV. Quantitative occurrence and importance as bacterial consumers. Mar. Ecol. Prog. Ser. **9**:35–42.
6. **Ferguson, R. L., and P. Rublee.** 1976. Contribution of bacteria to standing crop of coastal plankton. Limnol. Oceanogr. **21**:141–144.
7. **Fuhrman, J. A., and F. Azam.** 1980. Bacterioplankton secondary production estimates for coastal waters of British Columbia, Antarctica, and California. Appl. Environ. Microbiol. **39**:1085–1095.
8. **Fuhrman, J. A., and F. Azam.** 1982. Thymidine incorporation as a measure of heterotrophic bacterioplankton production in marine surface waters: evaluation and field results. Mar. Biol. **66**:109–120.
9. **Hagström, Å., U. Larsson, P. Hörstedt, and S. Normark.** 1979. Frequency of dividing cells, a new approach to the determination of bacterial growth rates in aquatic environments. Appl. Environ. Microbiol. **37**:805–812.
10. **Hanson, R. B., D. Shafer, T. Ryan, D. H. Pope, and H. K. Lowery.** 1983. Bacterioplankton in Antarctic ocean waters during late winter: abundance, frequency of dividing cells, and estimates of production. Appl. Environ. Microbiol. **45**:1622–1632.
11. **Hobbie, J. E., R. J. Daley, and S. Jasper.** 1977. Use of Nuclepore filters for counting bacteria by fluorescence microscopy. Appl. Environ. Microbiol. **33**:1225–1228.
12. **Jannasch, H. W.** 1979. Microbial ecology of low nutrient habitats, p. 243–260. *In* M. Shilo (ed.), Strategies of microbial life in extreme environments. Dahlem Konferenzen, Berlin.
13. **Krambeck, C., H.-J. Krambeck, and J. Overbeck.** 1981. Microcomputer-assisted biomass determination of plankton bacteria on scanning electron micrographs. Appl. Environ. Microbiol. **42**:142–149.
14. **Larsson, U., and Å. Hagström.** 1982. Fractionated phytoplankton primary production, exudate release and bacterial production in a Baltic eutrophication gradient. Mar. Biol. **67**:57–70.
15. **Maalöe, O., and N. O. Kjeldgaard.** 1966. Control of macromolecular synthesis. Benjamin, New York.
16. **Moriarty, D. J. W., and P. C. Pollard.** 1982. Diel variation of bacterial productivity in seagrass (Zostera capricorni) beds measured by rate of thymidine incorporation into DNA. Mar. Biol. **72**:165–173.
17. **Newell, S. Y., and R. R. Christian.** 1981. Frequency of dividing cells as an estimator of bacterial productivity. Appl. Environ. Microbiol. **42**:23–31.
18. **Newell, S. Y., and R. D. Fallon.** 1982. Bacterial productivity in the water column and sediments of the Georgia (USA) coastal zone: estimates via direct counting and parallel measurements of thymidine incorporation. Microb. Ecol. **8**:33–46.
19. **Pedros-Alio, C., and T. Brock.** 1982. Assessing biomass and production of bacteria in eutrophic Lake Mendota, Wisconsin. Appl. Environ. Microbiol. **44**:203–218.
20. **Pomeroy, L. R.** 1974. The ocean's food web, a changing paradigm. BioScience **24**:499–504.
21. **Steele, J. H.** 1974. The structure of marine ecosystems. Harvard University Press, Cambridge, Mass.
22. **Stevenson, L. H.** 1978. A case for bacterial dormancy in aquatic systems. Microb. Ecol. **4**:127–133.
23. **Straskrabova, V., and J. Fuksa.** 1982. Diel changes in

numbers and activities of bacterioplankton in a reservoir in relation to algal production. Limnol. Oceanogr. **27**:660–672.

24. **Watson, S. W., T. J. Novitsky, H. L. Quinby, and F. W. Valois.** 1977. Determination of bacterial number and bio-mass in the marine environment. Appl. Environ. Microbiol. **33**:940–946.

25. **Williams, P. J. leB.** 1981. Incorporation of microheterotrophic processes into the classical paradigm of the planktonic food web. Kiel. Meeresforsch. Sonderh. **5**:1–28.

Diurnal Responses of Microbial Activity and Biomass in Aquatic Ecosystems

CHRISTIANE KRAMBECK

Max-Planck-Institut für Limnologie, 232 Ploen, Federal Republic of Germany

The activity of planktobacteria (30) follows the diurnal cycles of the algae (5, 22) on which they depend (13, 16, 19, 23, 24, 27, 31, 33), but little is known about the functional relationships. How does a bacterial population react to a periodically changing, complex nutrient supply? The traditional activity measurements can give only a very rough idea of what is happening in situ (4, 25). Recently developed methods (35) measure general metabolic processes that are related as closely as possible to mass increase, i.e., to the interplay of cell replication and protein synthesis. There is still some doubt whether the latter are appropriate for measurements of transient phenomena. Most available methods call for incubation or separation from the algae, or both, with the exception of the frequency of dividing cells (FDC) method (7), which requires only cell counts. This approach is based on the increase in percentage of dividing cells with growth rates, under steady-state conditions. Does it, or any other constant relationship between metabolic phenomena and growth, hold true under environmental conditions where balanced growth is the exception rather than the rule (3, 8, 11, 30)? Is there additional or combined information that could be used to trace transient phenomena and thus help to elucidate functional relationships in diurnal cycles?

Measurements of cell length and width (microcomputer-assisted analysis of filtered plankton on scanning electron micrographs [16]) provide a solution to the problem because age distributions (derived from length-to-width ratios [LTOW]) (15) and mean cell volumes depend directly upon the relationship between increase in mass (elongation) and numbers (division), i.e., on the rate and phase of growth. Encouraged by previous observations on natural populations (13, 14), diurnal morphological responses of a Plussee population were followed during the maximum of a spring diatom bloom and interpreted in analogy to cell cycle control phenomena (34, 36). The results are discussed in relation to environmental parameters and other biomass and activity measurements (unpublished data).

MATERIALS AND METHODS

During an intensive diurnal study (31 March to 1 April 1981) by our research team, samples were taken at 2-h intervals in the Plussee at 1 m depth and were prepared for scanning electron microscopy by dehydration with graded ethanol and Freon 113 as described earlier (16). Micrographs were taken at ×5,000 magnification and projected on a digitizer field, and the length and width of the predominant rods and cocci, and their dividing stages (cells with visible constriction), were measured in dialogue with a microcomputer (WD/90 pascal microengine, Apple Computer Inc., Cupertino, Calif.; our own software). Normal rods and cocci were called "unconstricted" cells. The confusing term "nondividing" cells was avoided because it has been used for dormant cells. The terms "constricted" and "dividing" cells are used interchangeably. Data were stored on floppy disks and used for the computation of biomass (cell number and biovolume) and population structures. At least 100 cells had to be measured for frequency distributions and for 95% confidence limits within ±5% of mean cell volumes and mean LTOWs. For each sample, 15 images were taken, and 300 unconstricted cells and all constricted cells present (100 to 300) were measured. Confidence limits of ±15% for the FDC values were attained by counting 600 unconstricted cells. (The confidence limits of the FDC equal the sum of the relative confidence limits of the involved cell numbers: if $x = y/z$ then $\Delta x/x = \Delta y/y + \Delta z/z$.) Preparation of cells for scanning electron microscopy probably caused cell shrinkage (20). This may slightly alter structural changes of the population if the different cell stages do not shrink exactly the same (34), and it definitely should be kept in mind when comparing results from other cell measurements or attempting biovolume to carbon transformations (1, 16).

RESULTS

Basic principles of diurnal responses were revealed by following changes in the population structure of planktobacteria and interpreting them in analogy to cell cycle control phenomena.

The population structure was expressed in terms of frequency distributions of cell width, volume, and LTOW of unconstricted and constricted (dividing) cells (Fig. 1). The LTOW distribution was interpreted as age distribution

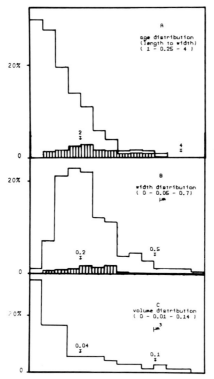

FIG. 1. Population structure. Plussee, 1 m, 31 March, 1400 h. Frequency distributions of age (A), cell width (B), and cell volume (C). Dividing (= constricted) cells: hatched blocks. Unconstricted cells: clear blocks.

in analogy to the length distributions of Woldringh (36). To supplement the FDC by frequencies of other cell stages with distinct functions within the cell cycle, the age distribution of the unconstricted cells was subdivided into three groups: round, intermediate, and long cells. The corresponding frequencies (FRC, FIC, FLC) and mean cell volumes (VRC, VIC, VLC) were calculated (Fig. 2). The limits between the three groups were set as follows: the mean LTOW of the dividing cells (= 2.2) was equal to the mean LTOW of the long cells and twice the mean LTOW of the round cells. The intermediate cells then had LTOW from 1.2 to 1.7, resulting in a mean LTOW of 1.45. The long cells were thus defined as "ready to divide," and the round ones, as recently divided cells. The intermediate cells are then the cell stages in the process of elongation, i.e., the step of mass increase within the cell cycle (36). Success of this approach depends on a major part of the population exhibiting the same growth pattern. To check this preassumption, the original LTOW distributions (Fig. 3) and the width distributions of the age

groups (Fig. 4) were compared over the diurnal cycle.

The volume distribution of the unconstricted cells (Fig. 1C) was skewed toward small-volume cells. Almost 90% of the cells had a mean cell volume below 0.05 μm^3, and one-third were even smaller than 0.01 μm^3. Most cells had widths below 0.55 μm (Fig. 1B). The distribution showed a shoulder between widths of 0.35 and 0.5 μm, indicating a minor fraction of relatively wide cells, which may behave differently from those forming the main peak (0.05 to 0.35 μm). The width distribution of the dividing cells shifted toward the higher widths. This trend implies an increase in FDC with increasing cell widths, fitting the general rule that faster-growing cells increase not only in length but also in width (6), or it reflects a contraction of the cells during the division process (34). The relatively high FDC at the low-width end of the peak is not consistent with one of these interpretations and must have other reasons. Further studies on biomass, age distributions, and mean cell volumes were restricted to the main peak, i.e., to cells with widths between 0.05 and 0.55 μm. The cells of the smallest width class were omitted, because they could not be measured accurately.

The age distribution (Fig. 1A) showed a gradual transition to dividing stages until the cell length was approximately doubled (LTOW of 1.75 to 2.25). The small tail or shoulder (LTOW above 2.75) contained cells dividing only after attaining a greater LTOW.

Mean cell volumes of the unconstricted cells (Fig. 2A) increased during the observation period from 0.032 to 0.048 μm^3, with an intermediate peak before midnight. Biovolume varied between 0.14 × 10^6 and 0.23 × 10^6 $\mu m/ml$. Cell numbers decreased a little in the evening, increased during the night, and decreased again the next morning (range: 3.8 × 10^6 to 6.4 × 10^6/ml). At 2 and 4 h a surprising amount of detritus was found in the samples, covering about one-third of the image area at 2 h and one-half at 3 h. The biomass values at these times had to be corrected by a factor of 1.5 and 2, respectively. The origin of this detritus is unclear. None was found in the 0- and 2-m samples at the same time.

Frequencies of different cell stages (Fig. 2B: FDC, FRC, FIC, FLC) represent an age distribution, redefined with regard to the cell cycle, and show population responses to changes in the environment, together with their mean cell volumes (Fig. 2C: VDC, VRC, VIC, VLC).

Under steady-state conditions, age distributions should remain constant, but this is clearly not the case here (Fig. 2B). During the first morning, the FDC switched from about 4 to 5% between 0600 and 1000 h to 9 to 11% from 1200 h

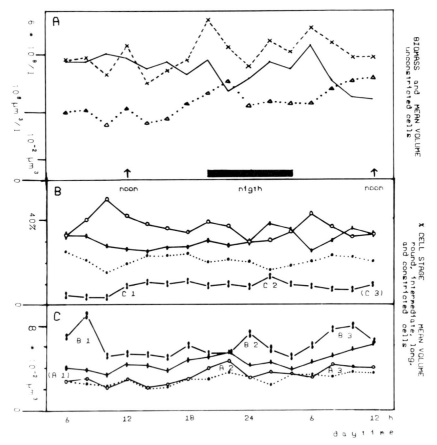

FIG. 2. Diurnal biomass variations (A) and cell cycle reactions (B and C). Plussee, 1 m, 31 March, 0600 h to 1 April, 1200 h, 1981. Values for (0.05 to 0.55 μm) width fraction with 95% confidence limits. (A) Solid line, cell number (4.2×10^9 to 6.4×10^9 [$\pm 8\%$] liter^{-1}); dashed line, biovolume (1.5×10^8 to 2.3×10^8 [$\pm 8\%$] μm^3 liter^{-1}); dotted line, mean volume (2.6×10^{-2} to 4.8×10^{-2} [$\pm 1\%$] μm^3). (B and C) Dotted line, round cells (frequency FRC and mean volume VRC); open circles, intermediate cells (FIC and VIC); filled ovals, long cells (FLC and VLC); filled eights, dividing cells (FDC and VDC). FDC, 4 to 14% ($\pm 15\%$). The sequences A, B, C show periods of accelerated elongation (after nutrient shift-up) followed by enhanced division activity.

on. At 0200 h, a brief increase to 14% was observed, followed by a minimum of about 8% the next morning at 0800 and 1000 h and another increase at 1200 h.

There was a succession of peaks of FIC, increasing FDC (C), and higher FRC. Theoretically, a maximum of FLC would be expected to appear between the FIC and FDC maxima (15), but this was missing or at least not obvious.

The mean cell volumes of the age groups (Fig. 2C) also displayed marked changes that clearly relate to age distribution. There were logical successions of maxima of VIC, VLC, VDC, and FDC. The two FDC increases on the first morning and at night (C1 and C2, respectively) were preceded by maxima of the VDC (B1 and B2, respectively). Correspondingly, a third VDC maximum (B3) on the second morning was followed by another increase in the FDC at noon

(C3). Maxima of the VDC (B2 and B3) were again preceded by maxima of the VIC and VLC (A2 and A3). An A1, corresponding to B1, presumably occurred before the first morning.

The time series of the LTOW distributions (Fig. 3) shows a difference between the day and night maximum of the dividing cell stages. The night maximum was comprised of shorter cells in the 1.75 to 2 class instead of the 2 to 2.25 class. In the morning and evening, the LTOW distribution of the dividing cells split into two peaks.

The width distributions of the main age groups (Fig. 4) clearly changed with time. Their shapes were not always the same, as expected for exponentially growing populations. Derivations imply a succession of bacterial populations: in the lowest width class (0.05 to 0.15 μm), an unusually high percentage of long cells was seen

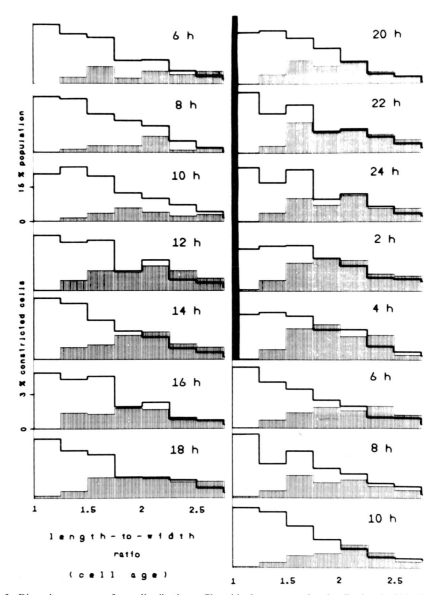

FIG. 3. Diurnal responses of age distributions. Clear blocks, unconstricted cells; hatched blocks, dividing cells. A shift between the day and night maximum of the dividing cells and the intermediate splitting into two peaks suggest alternating subpopulations.

from midnight to 0200 h, followed by a peak of narrow dividing cells from 0200 to 0400 h. Then, at 0400 h, a sharp peak of intermediate cells appeared, shifted to a higher (and therefore presumably more active) width class (0.15 to 0.25 μm).

DISCUSSION

An accumulation of primary products is prevented in natural environments by the presence of heterotrophic organisms. They are genetically and metabolically adapted to utilize a particular nutrient supply, as it is produced.

In plankton communities, bacteria are specialized in the removal of dissolved organic matter at low concentrations ("osmotrophs" [30]). This nutrient source fluctuates with a circadian rhythm because primary production is light dependent. Various substrates become simultaneously available and increase and decrease at the same time; others peak at different times. How can we determine cause and effect of such

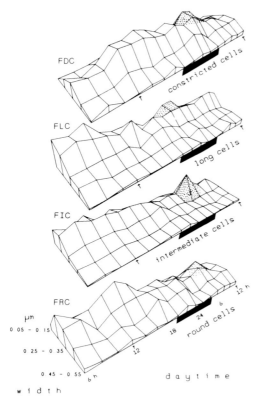

FIG. 4. Width distributions of main age groups. Three-dimensional plot over diurnal cycle. A sequence of peaks in the low-width fraction implies a bacterial succession (responding to "night" substrates).

sequences? And how does the bacterial population cope with such conditions?

It is impossible to understand fully the behavior of natural bacterial populations in isolation. This work was part of an integrated study on the Plussee (in preparation) and was carried out at the peak of a diatom bloom and the start of a cryptophyte bloom (31 March to 1 April 1981; 1 m; 5°C; primary productivity, 0.7 g of C m^{-3} day^{-1}). Dissolved organic substances and heterotrophic activities increased over the observation period as a result of the beginning of silicon depletion. The predominant asterionellas are especially sensitive in this respect (32). The population structure reflects this day-to-day trend with increasing FDC and higher mean cell volumes. Correspondingly, the relation between cell number and biovolume changes, and so does the relation between DNA and RNA content of the small size fraction (0.2 to 8 μm; P. Witzel, personal communication). Intermediate maxima of the parameters hint at transient phenomena. On the basis of the FDC, the biomass should have increased about fourfold. Instead, a slight decline was observed, probably due to grazing

by the rotifers and copepodids present. Biomass increases are often missed (16, 19, 33). Pedros-Alio and Brock (26) postulated that there are other losses besides those to copepods and rotifers, e.g., sedimentation of cells associated with larger particles. Protozoan grazing (29) may also contribute to bacterial loss.

The population structure indicates increasing nutrient supplies ("nutrient shift-up") in the evening and morning. A comparison with data on dissolved organic substances and heterotrophic activities (U. Muenster and J. Overbeck, personal communication) reveals logical sequences. There are typical day and night substrates. Organic acids such as glycolate and lactate show an afternoon maximum and a nighttime minimum, whereas glyoxylate displays a very sharp and high peak during the night. Di- and monosaccharides have daytime maxima. Single monosaccharides such as glucose appear with several small peaks. The heterotrophic CO_2 uptake, which is a general measure for anabolic reactions, i.e., buildup of cell material (24), responds to the afternoon and night maxima of the main substrates with high afternoon/evening and predawn/dawn peaks, respectively (variation in 0.2- to 8-μm size fraction: 0.09 to 0.8 μg of C liter^{-1} h^{-1}). The glucose uptake (single high concentration added) by comparison has a slight maximum at noon (variation in small size fraction: 0.06 to 0.2 μg of C liter^{-1} h^{-1}). Theoretically (14), there is no strict connection between V_{max} and actual uptake, and this is confirmed by practical experience (19, 33). In continuous cultures, Sepers (28) found an increase of V_{max} with growth rate, but only with respect to limiting nutrient. This may explain why the glucose V_{max} does not necessarily follow heterotrophic activity maxima (24). Such a correlation is not to be expected unless glucose happens to be the limiting substrate.

The first effect of an increased nutrient supply on the population structure is seen in an increase in the mean cell volume of intermediate cells (i.e., those in the process of elongation), the phase of maximum biomass increase during the cell cycle (36). Subsequently, larger dividing cells appear in the population and FDC increases. Changes in FDC occur with a lag of about one cell cycle (here about 6 to 8 h). The three distinct FDC increases (noon, midnight, noon) are a result of volume increases of preceding cell stages, which themselves are mediated by the rapid predawn and evening increases in the substrate supply.

The observed sequence—rising substrate concentrations, enhanced buildup of cell material, increase in mean cell volumes and in frequencies of first intermediate cells and later dividing cells—is logical with respect to cell cycle control phenomena. According to Woldringh (36), the dividing cells of slow-growing populations cease to elongate, i.e., to increase in volume, after they start constriction. This could be verified for the planktobacteria investigated by comparing mean cell volumes of weakly and strongly constricted cells, which were equal. Then the phase of elongation should be the phase of maximum substrate uptake, and the intermediate, i.e., elongating, cells have to react first to a nutrient shift-up.

The diet changes in the [³H]thymidine incorporation, found by Riemann et al. (27) under similar conditions (6 April; primary productivity, 0.9 g of C m^{-3} day^{-1}), also display two peaks over a 24-h period. In generalizing the above findings, it seems reasonable to suggest that there must have been similar substrate peaks preceding the enhanced division activity.

Alternating diurnal maxima for buildup of cell material and division require a new approach to activity measurements. Methods that rely on a proportionality between different aspects of growth, i.e., on a steady-state assumption, must be biased when applied to such transient states. Attempts to calibrate, e.g., the FDC method (7, 9, 21), have to yield regression lines through scattering data, unless based on steady-state systems (7). In my opinion, the FDC method can be improved only by adding more information on population structures, as recommended for algae (18). Correspondingly, it should be possible "to detect discontinuous growth and cell division processes which frequently occur in surface plankton populations" (12) by combined measurements of DNA and RNA syntheses.

The above interpretation of population structures in analogy to cell cycle control phenomena is based on the premise that the main part of the population acts approximately the same at a given time. In the present case, which represents a highly productive situation, this seems to be true. The complexity of the population becomes evident only in some characteristic details that suggest subpopulations. When comparing, for example, the LTOW distribution (Fig. 1A) with similar length distributions of continuous cultures (36), the peak of the dividing cells could have been sharper. The wider spread of the dividing cells over the LTOW classes can be explained by an overlap of slower- and faster-growing populations. In addition, the feature can be explained as a transient effect, as Brock (2) pointed out: "In most populations it would be expected that the spread of ages would be widest in the late log phase and at the beginning of the stationary phase, and would be narrowest in the middle log phase." Further, the mean cell volumes of the long cells should equal those of the dividing stages, and the mean cell volumes of the

intermediate cells should exceed those of the round cells (by definition; see Results). However, both were too low at most times. This indicates a less active subpopulation of narrow, unconstricted cells with LTOW above 1.2. In the time series of the age distributions, the maxima of the dividing cells differ from day to night and split into two peaks in between. Thus, there seem to be not only alternating substrates, but also alternating day and night populations. The width distributions of the age groups show the mechanism of such successions. A logical sequence of peaks in the lowest width classes supports the presence of subpopulations that are in a low-width, "stand-by" position which they leave occasionally.

The following hypotheses arise from these observations. (i) Overall quantitative changes in the nutrient supply are reflected by the population structure analogous to cell cycle control phenomena. The changes demonstrate alternating maxima for buildup of cell material and division. (ii) There are two groups in the bacterial population: One, adapted to currently available substrates, grows actively, while the other is in a "stand-by" position (10, 17), able to start growing when its preferred substrates become available. During periods of high primary production, subpopulations may switch from a "stand-by" position at low widths, where the genetic potential of the population seems to be assembled, to an active condition at higher widths and vice versa once a day, according to the changing substrate quality.

Further investigations are required to test these hypotheses. The complexity of the interactions demands cooperative work and the use of laboratory simulations and mathematical models to improve the interpretations of field observations.

The work was essentially supported by an open exchange of ideas and results within our research team, especially with J. Overbeck, U. Muenster, and K. P. Witzel. Background information on phytoplankton, zooplankton, and silicium was provided by B. Hickel, W. Hofmann, and D. Albrecht. A. Jacob took scanning electron micrographs. H.-J. Krambeck helped to design the microcomputer system. He also realized the computer graphics, assisted by H. Kiesewetter. S. Duehrkoop typed the manuscript. It was revised by E. Cox, M. J. Klug, and M. Mort.

LITERATURE CITED

1. Bakken, L. R., and R. A. Olsen. 1983. Buoyant densities and dry-matter contents of microorganisms: conversion of a measured biovolume into biomass. Appl. Environ. Microbiol. 45:1188–1195.
2. Brock, T. D. 1966. Principles of microbial ecology. Prentice-Hall Inc., Englewood Cliffs, N.J.
3. Christian, R. R., R. B. Hanson, and St. Y. Newell. 1982. Comparison of method for measurement of bacterial growth rates in mixed batch cultures. Appl. Environ. Microbiol. 43:1160–1165.
4. Costerton, J. W., and R. R. Colwell (ed.). 1979. Native

aquatic bacteria: enumeration, activity, and ecology. American Society for Testing and Materials, Baltimore.

5. **Frempong, E.** 1983. A laboratory simulation of diel changes affecting cellular composition of the dinoflagellate Ceratium hirundinella. Freshwater Biol. **13:**129–138.

6. **Grover, N. B., C. L. Woldringh, A. Zaritzky, and R. F. Rosenberger.** 1977. Elongation of rod-shaped bacteria. J. Theor. Biol. **67:**181–193.

7. **Hagstrom, A., V. Larsson, P. Horstedt, and S. Normark.** 1979. Frequency of dividing cells, a new approach to the determination of bacterial growth rates in aquatic environments. Appl. Environ. Microbiol. **37:**805–812.

8. **Hanson, R. B., and H. K. Lowery.** 1983. Nucleic acid synthesis in oceanic microplankton from the Drake Passage, Antarctica: evaluation of steady-state growth. Mar. Biol. **73:**79–89.

9. **Hanson, R. B., D. Shafer, T. Ryan, D. H. Pope, and H. K. Lowery.** 1983. Bacterioplankton in Antarctic Ocean waters during late austral winter: abundance, frequency of dividing cells, and estimates of production. Appl. Environ. Microbiol. **45:**1622–1632.

10. **Hoppe, H. G.** 1978. Relations between active bacteria and heterotrophic potential in the sea. Neth. J. Sea Res. **12:**78–98.

11. **Jannasch, H. W.** 1974. Steady state and the chemostat in ecology. Limnol. Oceanogr. **19:**716–720.

12. **Karl, D. M.** 1981. Simultaneous rates of ribonucleic acid and deoxyribonucleic acid synthesis for estimating growth and cell division of aquatic microbial communities. Appl. Environ. Microbiol. **42:**802–810.

13. **Krambeck, C.** 1978. Changes in planktonic microbial populations—an analysis by scanning electron microscopy. Verh. Int. Ver. Limnol. **20:**2255–2259.

14. **Krambeck, C.** 1979. Applicability and limitations of the Michaelis-Menten equation in microbial ecology. Arch. Hydrobiol. Beih. Ergebn. Limnol. **12:**64–76.

15. **Krambeck, C., H.-J. Krambeck, and J. Overbeck.** 1977. Autonomous growth fluctuations of the methane oxidizing bacterial strain M 102 in batch and continuous culture. Arch. Microbiol. **115:**119–126.

16. **Krambeck, C., H.-J. Krambeck, and J. Overbeck.** 1981. Microcomputer assisted biomass determination of plankton bacteria on scanning electron micrographs. Appl. Environ. Microbiol. **42:**142–149.

17. **Kurath, G., and R. Y. Morita.** 1983. Starvation-survival physiological studies of a marine *Pseudomonas* sp. Appl. Environ. Microbiol. **45:**1206–1211.

18. **McDuff, R. E., and S. W. Chisholm.** 1982. The calculation of in situ growth rates of phytoplankton populations from fractions of cells undergoing mitosis: a clarification. Limnol. Oceanogr. **27:**783–788.

19. **Meyer-Reil, L.-A., M. Boelter, G. Liebezeit, and W. Schramm.** 1979. Short-term variations in microbiological and chemical parameters. Mar. Ecol. Prog. Ser. **1:**1–6.

20. **Montesinos, E., I. Esteve, and R. Guerrero.** 1983. Comparison between direct methods for determination of microbial cell volume: electron microscopy and electronic particle sizing. Appl. Environ. Microbiol. **45:**1651–1658.

21. **Newell, St. Y., and R. R. Christian.** 1981. Frequency of dividing cells as an estimator of bacterial productivity. Appl. Environ. Microbiol. **42:**23–31.

22. **Overbeck, J.** 1962. Untersuchungen zum Phosphathaushalt von Gruenalgen. III. Das Verhalten der Zellfraktionen von Scenedesmus quadricauda (Turps.) Breb. im Tageszyklus unter verschiedenen Belichtungsbedingungen und bei verschiedenen Phosphatverbindungen. Arch. Mikrobiol. **41:**11–26.

23. **Overbeck, J.** 1972. Zur Struktur und Funktion des aquatischen Oekosystems. Ber. Dtsch. Bot. Ges. **85:**553–577.

24. **Overbeck, J.** 1979. Studies on heterotrophic functions and glucose metabolism of microplankton in Plussee. Arch. Hydrobiol. Beih. Ergebn. Limnol. **13:**56–76.

25. **Overbeck, J. (ed.).** 1979. Proceedings of the workshop on measurements of microbial activities in the carbon cycle of freshwaters. Arch. Hydrobiol. Beih. Ergebn. Limnol., vol. 12.

26. **Pedros-Alio, C., and T. D. Brock.** 1983. The impact of zooplankton feeding on the epilimnetic bacteria of a eutrophic lake. Freshwater Biol. **13:**227–240.

27. **Riemann, B., J. Fuhrmann, and F. Azam.** 1982. Bacterial secondary production in freshwater measured by ^{3}H-thymidine incorporation method. Microb. Ecol. **8:**101–114.

28. **Sepers, A. B.** 1982. The uptake of organic compounds by heterotrophic bacteria in relation to growth rate. *In* Publications du Centre National pour l'Exploitation des Oceans. Actes de colloques no. 13, 1982.

29. **Sherr, B. F., E. B. Sherr, and T. Berman.** 1983. Grazing, growth, and ammonium excretion rates of a heterotrophic microflagellate fed with four species of bacteria. Appl. Environ. Microbiol. **45:**1196–1201.

30. **Sieburth, J. M.** 1979. Sea microbes. Oxford University Press, New York.

31. **Sieburth, J. M., K. M. Johnson, C. M. Burney, and D. M. Lavoie.** 1977. Estimation of in-situ rates of heterotrophy using diurnal changes in dissolved organic matter and growth rates of picoplankton in diffusion culture. Helgol. Wiss. Meeresunters. **30:**565–574.

32. **Sommer, U., and H.-H. Stabel.** 1983. Silicon consumption and population density changes of dominant planktonic diatoms in Lake Constance. J. Ecol. **71:**119–130.

33. **Straskrabova, V., and J. Fuksa.** 1982. Diel changes in numbers and activities of bacterioplankton in a reservoir in relation to algal production. Limnol. Oceanogr. **27:**660–672.

34. **Trueba, F. J., and C. L. Woldringh.** 1980. Changes in cell diameter during the division cycle of *Escherichia coli*. J. Bacteriol. **142:**869–878.

35. **Van Es, F. B., and L.-A. Meyer-Reil.** 1982. Biomass and metabolic activity of heterotrophic marine bacteria, p. 111–170. *In* K. C. Marshall (ed.), Advances in microbial ecology, vol. 6. Plenum Publishing Corp., New York.

36. **Woldringh, C. L.** 1976. Morphological analysis of nuclear separation and cell division during the life cycle of *Escherichia coli*. J. Bacteriol. **125:**248–257.

Field Interpretation of Microbial Biomass Activity Measurements

E. A. PAUL AND R. P. VORONEY

University of California, Berkeley, California 94720, and University of Saskatchewan, Saskatoon, Saskatchewan S7N 0W0, Canada

Recent ecological research has produced a great deal of information concerning the pool sizes and nutrient dynamics of a variety of ecosystems (4, 9, 32). The basic processes, which are reasonably well understood, are identical in all ecosystems since they are controlled by the general laws of thermodynamics. The biological control mechanisms, however, require further definition: they have received extensive investigation from a genetic and cellular viewpoint (10, 26), but it has proved difficult to extrapolate in vitro information from the enzyme and organism level to that of the ecosystem. Further advances will depend on new information and concepts concerning the flux rates and controls of the intertwined processes in the transformation of elements such as C, N, S, and P.

Nutrient transformation kinetics and soil organic matter characteristics cannot be determined for each of the many soil-sediment-plant interactions in nature. Knowledge of the critical controls must be obtained so that extrapolation to the range of ecosystems in nature is feasible. The close relationship between C availability, soil biomass size and activity, and nutrient transformations means that further understanding or possible management of these transformations must involve a knowledge of biomass size and turnover.

INTERPRETATION OF BIOMASS AND ACTIVITY MEASUREMENTS

Estimates of the microbial biomass have usually involved treatment of the biomass as a single component, although it is known that a diversity of populations with different biochemical characteristics are present. Techniques have included microscopic measurements to determine values for total or active biovolumes (38, 40), which can be converted to biomass. The flush of decomposition during incubation after $CHCl_3$ fumigation (chloroform fumigation incubation method [CFIM]) also yields biomass when appropriate conversion factors are used (17, 28). Measurements of the biomass by analysis for specific constituents such as ATP, hexosamines, and nucleic acids (10, 18) has provided much useful information. Other techniques include measurement of the incorporation or respiration of radioactive and nonradioactive substrates (J. L. Smith, B. L. McNeal, and H. H. Cheng, submitted for publication).

The majority of biomass measurements have focused on C and N; however, S and P contents also have been measured (5, 14, 37). Microbial S and P measurements have the advantage in that the extraction can be carried out immediately after $CHCl_3$ treatment, eliminating the lengthy delay and the unknown mineralization-immobilization rates that arise during incubation. Extraction of C or N constituents immediately after $CHCl_3$ treatment has to date not proved successful. A relatively small percentage of the cell constituents has been found extractable. This fraction will at times correlate well with other biomass techniques, but tracers show it to represent a diversity of soil constituents.

METHODOLOGY

None of the present techniques for biomass measurements yields unequivocal results in the broad diversity of habitats that are encountered. The fact that measurements have at times been made without regard to the underlying assumptions in the methodology indicates the need for caution in use of the data derived from such estimates relative to nutrient cycling information. Simultaneous use of two measurement systems yielding complementary data should overcome a number of specific problems. Examples of the simultaneous use of two different techniques include the expression of colony-forming units on agar plates as a percentage of the biovolume measured microscopically to determine an activity coefficient. ATP is dependent on the activity of the organisms and on factors such as P concentration and biomass size. A comparison of ATP contents to biomass measured either microscopically or by the CFIM (34), or respiratory response technique (2), should also yield estimates of activity relative to total biomass.

Measurements of biomass C by the CFIM are limited by the question of what constitutes an appropriate control to calculate the actual amount of CO_2-C attributable to mineralization of the killed population. The microbial population developing after $CHCl_3$ treatment is different from that in the soil before treatment (20, 39). The presence of readily available soil organic constituents and the speed with which these substrates are attacked affects the interpretation. Data showing the effect of four different methods of calculation involving the use of control during incubation are shown in Table 1. The biomass C values for which CO_2 evolved from controls was subtracted from the flush (A and B) are probably underestimates, and those calculated without the use of any control (C) can, under some conditions, be overestimates. The use of the 10- to 20-day data (D) from the fumigated treatment as the control resulted in much lower estimates of both biomass C and N. Column 2 indicates that a soil incubated with glucose plus NO_3^--N showed an increased biomass C ranging from 50 to 98% greater than with NO_3 alone, depending on the method of calculation.

Net mineralization of N in a fumigated soil is usually much higher than in an unfumigated soil. The problem of whether to subtract a value representing CO_2 evolved from an unfumigated sample is not as serious in measuring biomass N as in measuring biomass C. However, the C/N ratio of the microbial population is variable; this can alter the rates of net N mineralization. R. P. Voroney and E. A. Paul (Soil Biol. Biochem., in press) overcame this problem by correcting the k_N value used to calculate biomass on the basis of the ratio of CO_2-C to NH_4^+-N produced in the CFIM. This has been found to range from a low of 1.3:1 in composted wheat straw (42) to >300:1 in a forest floor litter. The data for biomass N, when corrected for mineralization, gave much more consistent values than the data for biomass C (Table 1). The linear equation utilized for the calculation in Table 1 has more recently been found to be curvilinear. This results in higher k_N values for microorganisms with higher N contents (R. P. Voroney, Ph.D. thesis, University of Saskatchewan, Saskatoon, Saskatchewan, Canada, 1983). The curvilinear relationship has been found to give more meaningful estimates of the C/N ratios of biomass under both field and laboratory conditions.

MYCORRHIZAL FUNGI

The roles of the ecto- and endomycorrhizal relationships are still largely unknown factors in relating microbial biomass and activity under field conditions. Work with legumes which constitute a tripartite symbiotic system of plants, bacteria, and fungi showed that although mycorrhizal biomass and nodule biomass contributed only 2 to 6% of the total plant weight, the respiration by the symbionts accounted for 12 to 20% of the C photosynthesized (19). Similar data are available for ectomycorrhizal pine (Table 2). Short roots which include the ectomycorrhizal fungi contributed <0.2% of the total plant weight of *Pinus taeda* at both 6 and 10 months. However, the presence of mycorrhizal fungi increased beneath-ground respiration two- to threefold (C. P. P. Reid, F. A. Kidd, and S. A. Ekwebelam, Plant Soil, in press). Depending on the cultural conditions, plant age, etc., mycorrhizal fungi can act either as symbionts or as parasites. In the above examples, the presence of the mycorrhizal fungi increased net photosynthesis on a unit weight basis, indicating that plants could compensate for the needs of the microbial partners.

Mycorrhizal fungi contributed ≈1% to total ecosystem biomass in both 23- and 180-year-old

TABLE 1. Comparison of estimates of the soil microbial biomass C and N obtained by four different methods of calculation (Voroney and Paul, in press)

Method of calculation	Biomass C and N (μg g^{-1} of soil)			
	NO_3^- amended soil		Glucose plus NO_3^- amended soil	
	C^a	N^b	C^a	N^b
(A) $(C_{F(0-10)} - C_{UF(0-10)})/k_C$	483		715	
(B) $(C_{F(0-10)} - C_{UF(10-20)})/k_C$	513		829	
(C) $C_{F(0-10)}/k_C$	600		1,193	
(D) $(C_{F(0-10)} - C_{F(10-20)})/k_C$	396		734	
(A) $(N_{F(0-10)} - N_{UF(0-10)})/k_N$		132		179
(B) $(N_{F(0-10)} - N_{UF(10-20)})/k_N$		134		163
(C) $N_{F(0-10)}/k_N$		139		208
(D) $(N_{F(0-10)} - N_{F(10-20)})/k_N$		79		29

[a] k_C (22°C) = 0.41.
[b] k_N (22°C) = 0.39 − 0.014 (C_F/N_F)
= 0.3 NO_3^- amended soils
= 0.24 glucose + NO_3^-.

TABLE 2. Net photosynthesis and carbon distribution in *Pinus taeda* (adapted from Reid et al., in press)[a]

Determination	6 months		10 months	
	NM	M	NM	M
Weight (g)	1.3	2.0	2.5	4.8
Net photosynthesis (mg of CO_2-C $g^{-1} h^{-1}$)	0.65	1.25	0.68	1.55
^{14}C distribution (%)				
Needles and stems	67	30	67	55
Shoot respiration	5	6	6	7
Woody roots	20	34	14	14
Short roots	0.17	0.2	0.2	0.12
Beneath ground respiration	8	30	12	24

[a] NM, Non-mycorrhizal; M, mycorrhizal.

fir (*Abies amabilis*) (45) in western Washington. They were said to account for ≈15% of the annual net primary production in these stands. Sclerotia accounted for the largest proportion of the total mycorrhizal fungal production. These, together with sporocarps and mycorrhizal sheaths, represented a larger portion of the N, P, and K cycling than that returned annually through the litter fall.

EFFICIENCY OF MICROBIAL PRODUCTION

The extent to which microorganisms convert substrate to microbial biomass is an important ecological determinant in that this controls the relative rates of N, S, and P mineralization and immobilization. Laboratory estimates of microbial growth efficiency of 60% on substrates such as glucose (31) have been confirmed for glucose in soils under both N-sufficient and N-limiting conditions (13). The water-insoluble particulate fraction of seaweed has been found to be converted to bacterial C at an overall efficiency of 43% (35). We have measured similar efficiency values for a general soil population growing on mature wheat straw. In longer incubations in which both microbial biomass and other organic microbial products were formed, a yield of 0.15 for microbial biomass and a conversion efficiency of 0.37 for conversion of ^{14}C to microbial products indicated an overall conversion through microbial biomass as high as 50% (8).

MICROBIAL ACTIVITY

The term "activity" includes the many processes carried out by microbial enzymes. When interpreted in conjunction with biomass measurements, it often refers to net nutrient mineralization and immobilization rates. Smith et al. (submitted for publication) point out that the mineralization of elements such as C and N has kinetics which depend on the concentration of substrate and require microbial biomass as the reacting agent. This can best be described by a second-order equation. Traditionally, microbial biomass has been considered to be constant and at a maximum, thus reducing the equation to pseudo-first order. Use of first-order equations has proved possible in long-term models in which the microbial population could develop early enough in the incubation so that overall biomass sizes did not limit the final rates (44).

Studies to understand the role of microbial biomass other than as a source-sink for nutrients must recognize the different growth rates and activity coefficients of various segments of the population. These segments range from the zymogenous and autochthonous defined by Winogradsky many years ago to more recent segments characterized on an activity basis. J. L. Smith (personal communication) has recognized three population states: (i) the active biomass capable of growth and all metabolic functions, (ii) sustainable populations which are nongrowing but can dissimilate glucose and resume growth under favorable conditions, and (iii) dormant spores or other long-term resting populations. Estimates of the active population usually range from 10 to 40% of the total identifiable biomass (30, 32).

The recognition of large sustainable and dormant populations with maintenance energy and starvation characteristics vastly different from those usually measured in chemostats is a necessary prerequisite to an understanding of the functioning of microorganisms in nature. Some measurements of the adenylate energy charge (6) indicate that, even though little of the total biomass in soil can be in an active growth stage at any given time, ATP accounted for 77% of the total adenine nucleotides in a fresh soil. Other measurements have indicated a low adenylate energy charge, and more information is required concerning the activity state of in situ organisms.

In nature, the state of dormancy and the necessity of a long exposure to sufficient nutrient levels before response are major survival criteria (25). Physiological studies on the starvation-survival of a marine pseudomonas showed a miniaturization of cells during starvation (1). Extrapolation of this observation to soil could help explain the preponderance of very small bacterial cells in soil. Further understanding of microbial growth and production will depend on a better delineation of the microbial responses to what appears to be the general C-limiting condition in most terrestrial habitats.

POTENTIAL FOR USE OF BIOMASS ESTIMATES IN MANAGEMENT DECISIONS

Information on the size of the soil biomass has been used for determining: (i) the degradation, stabilization, and incorporation into biomass of plant C and N (21, 43); (ii) the effects of freezing and thawing (39); (iii) the effect of tillage (7, 12); (iv) the role of soil sampling, mixing, and grinding (23, 33); (v) the effect of climatic variations (36, 41); (vi) biomass in forest floors (11, 27); and (vii) the effect of faunal feeding (3, 46).

There is potential for improved management of soil and fertilizer nutrients through a better understanding and possible manipulation of biomass (E. A. Paul, Plant Soil, in press). Jenkinson and Ladd (16) published a model (Table 3) of C turnover, using data from (i) the carbon dating of the resistant soil organic matter, (ii) plant C inputs to a soil in equilibrium with its environment and management practices, e.g., the Rothamsted continuous wheat plots, and (iii) microbial biomass and production estimates based on substrate degradation rates measured with tracers. In their first-order model, microbial N was taken as one-sixth of the microbial C value of 570 kg ha^{-1}; it was calculated that the population of microorganisms had a turnover time of 2.5 years.

The chernozemic soil in Saskatoon under a wheat-fallow crop rotation had high reserves of organic C and N at C inputs somewhat similar to those at Rothamsted. The microbial N determined by using CFIM and a k_N depending on CO_2-C/NH_4^+-N produced during incubation represented 360 kg of N h^{-1} (a measured C/N ratio of 4.4). The turnover time of 6.8 years calculated on the basis of Jenkinson and Ladd's model shows the potential for stabilization of a largely inactive population in soils with high organic matter levels.

The spodosolic sugarcane soil of northeast Brazil (M. Lima, personal communication) had C and N contents similar to those of the continuous wheat plots in Rothamsted with a lower biomass C and N. The high input of C (13 t ha^{-1} year^{-1}) resulted in a calculated turnover time of 0.24 years. This was 28 times faster than that found in the chernozemic soil and demonstrates the situation where the biomass acts more as a catalyst and a short-term reservoir than as a major source-sink for nutrients. Although biomass N represented only one-third of the N removed by the crop, the estimated N flux through this biomass was 1.5 times that removed by the crop.

The biomass is large enough, under temperate conditions, to act as a significant temporary storage of nutrients (24). The net N mineralized during a 12-week period was found to be derived almost equally from the microbial biomass, a fraction consisting of microbial metabolites, and a stabilized N fraction with lower decomposition rates (29).

The general relationships between the size of the microbial biomass and N_Δ, the soil N mineralized during laboratory incubation, is shown in Table 4. One hundred soils from Saskatchewan ranging in organic N content from 0.13 to 0.32% produced an N_Δ of 87 to 157 µg g^{-1}; the biomass N in the same soils ranged from 53 to 102 µg g^{-1}. There was a close relationship between the N mineralized and the microbial biomass: $N_\Delta = 1.3$ microbial biomass N + 24 ($r^2 = 0.75$). It has also been shown that the ^{15}N atom % excess of the microbial N and of the mineralized N are identical during extended incubations (29), demonstrating that the biomass is closely related to the amount of net N mineralized and that the mineralizable N moves through the biomass on its way to NH_4^+.

The so-called priming action measured in tracer experiments with increased microbial activity was said to result from an increased attack on soil organic matter. Alternatively, priming can be explained by the internal turnover of the large microbial population. Tracer experiments often show an enhanced flux of nontracer (soil C and N) upon the addition of the tracer such as fertilizer N or substrate C. This phenomenon has been described as priming, as it was thought to result from the enhanced turnover of the native soil organic matter (15). The release of nonlabeled C and N from a large biomass as it cycles and incorporates the added tracer is a more logical explanation for this phenomenon.

Management techniques such as zero tillage (12), crop rotations, and intercropping (22) have potential for better utilization of nutrients and

TABLE 3. C and N turnover in Saskatchewan wheat-fallow, Rothamsted continuous wheat, and Brazilian sugarcane soil-plant systems

Determination	Saskatoon	Rothamsted	Brazil
Soil weight (t ha^{-1})	2,700	2,200	2,400
Organic C (t ha^{-1})	65	26	26
C inputs (t ha^{-1} yr^{-1})	1.6	1.2	13
Turnover of soil C (yr)	40	22	2.0
Microbial C (kg ha^{-1})	1,600	570	460
Microbial N (kg ha^{-1})	360	95	84
Microbial turnover time (yr)	6.8	2.5	0.24
N flux through microbial biomass (kg ha^{-1} yr^{-1})	53	34	350
Crop removal of N (kg ha^{-1} yr^{-1})	40	24	220

TABLE 4. Relationship of microbial biomass and soil properties of 100 medium-textured soils[a]

Soil	Organ- ic C (%)	Kjel- dahl N (%)	$N_\Delta{}^b$ ($\mu g\ g^{-1}$ of soil)	Microbial biomass[c] ($\mu g\ g^{-1}$ of soil)	
				C	N
Brown	1.32	0.13	87	660	53
Dark brown	1.80	0.18	117	750	74
Black	3.49	0.30	157	970	102
Dark grey luvisol	3.80	0.32	116	810	72

[a] Microbial biomass C ($\mu g\ g^{-1}$ of soil) = 0.017 organic C + 417 ($r^2 = 0.4$). Microbial biomass N ($\mu g\ g^{-1}$ of soil) = 0.029 soil N + 16 ($r^2 = 0.44$). N_Δ ($\mu g\ g^{-1}$ of soil) = 0.17 biomass C + 24 ($r^2 = 0.53$). N_Δ ($\mu g\ g^{-1}$ of soil) = 1.3 biomass N + 24 ($r^2 = 0.75$).
[b] Derived from 20-week incubation at 28°C.
[c] Measured by CFIM, 28°C; $k_C = 0.5$, $k_N = 0.4$.

energy in crop production. These techniques also directly affect the size and turnover of the biomass. Further information is necessary to relate management to soil process controls. Productivity of forest sites has been related to N accumulation during anaerobic incubation (R. Powers, Agronomy Abstr. 1982, 270). We predict that mineralizable N under these conditions will be related to the size of biomass N and closely related fractions such as the microbial metabolites.

Differential management of agricultural residues has been shown to affect the microbial community and the associated microbial feeding faunal population (E. T. Elliott, K. Horton, J. C. Moore, D. C. Coleman, and C. V. Cole, Plant Soil, in press). This has the potential for significantly altering nutrient cycles. The possibility of increasing soil and fertilizer use efficiency through management of the biomass by cultivation, cropping practices, and possibly even faunal feeding definitely exists and should be further investigated (D. C. Coleman, R. E. Ingham, J. F. McClellan, and J. Trofymow, Proceedings of the Joint BMS/BES Meeting on Animal Microbial Interactions, in press).

CONCLUSIONS

Reasonable techniques for measurement of biomass size and activity now exist. Refinements of these techniques in the next few years will improve ease and accuracy of measurement. These measurements, when combined with tracer techniques and mathematical modeling, will help answer a number of questions concerning soil processes of importance to both agricultural soil management and the understanding of ecosystem functioning.

The preparation of this paper was supported by grant BSR 8306181 from the National Science Foundation.

LITERATURE CITED

1. **Amy, P. S., and R. Y. Morita.** 1983. Starvation-survival patterns of sixteen freshly isolated open-ocean bacteria. Appl. Environ. Microbiol. **45:**1109–1115.
2. **Anderson, J. P. E., and K. D. Domsch.** 1978. Mineralization of bacteria and fungi in chloroform-fumigated soils. Soil Biol. Biochem. **10:**207–213.
3. **Baath, E., U. Lohm, B. Lundgren, T. Rosswall, B. Soderstrom, and B. Sohlenius.** 1981. Impact of microbial-feeding animals on total soil activity and nitrogen dynamics: a soil microcosm experiment. Oikos **37:**257–264.
4. **Bormann, F. H., and G. E. Likens.** 1979. Pattern and process in a forested ecosystem. Springer-Verlag, New York.
5. **Brookes, P. C., D. S. Powlson, and D. S. Jenkinson.** 1982. Measurements of microbial biomass phosphorous in soil. Soil Biol. Biochem. **14:**319–329.
6. **Brookes, P. C., K. R. Tate, and D. S. Jenkinson.** 1983. The adenylate energy charge of the soil microbial biomass. Soil Biol. Biochem. **15:**9–16.
7. **Carter, M. R., and D. A. Rennie.** 1982. Changes in soil quality under zero tillage farming systems: distribution of microbial biomass and mineralizable C and N potentials. Can. J. Soil Sci. **62:**587–597.
8. **Cerri, C. C., and D. S. Jenkinson.** 1981. Formation of microbial biomass during the decomposition of ^{14}C labelled ryegrass in soil. J. Soil Sci. **32:**619–626.
9. **Clark, F. E., and T. Rosswall (ed.).** 1981. Terrestrial nitrogen cycles. Ecol. Bull. (Stockholm), vol. 33.
10. **Cohen, B. L.** 1980. Transport and utilization of proteins by fungi, p. 411–430. In J. W. Payne (ed.), Microorganisms and nitrogen sources, John Wiley & Sons, Chichester.
11. **David, M. B., M. J. Mitchell, and J. P. Nakas.** 1982. Organic and inorganic sulfur constituents of a forest soil and their relationship to microbial activity. Soil Sci. Am. J. **46:**847–852.
12. **Doran, J. W.** 1980. Soil microbial and biochemical changes associated with reduced tillage. Soil Sci. Soc. Am. J. **44:**765–771.
13. **Elliott, E. T., C. V. Cole, B. C. Fairbanks, L. E. Woods, R. J. Bryant, and D. C. Coleman.** 1983. Short-term bacterial growth, nutrient uptake, and ATP turnover in sterilized inoculated and C-amended soil: the influence of N availability. Soil Biol. Biochem. **15:**85–91.
14. **Headley, M. J., and J. W. B. Stewart.** 1982. Method to measure microbial phosphate in soils. Soil Biol. Biochem. **14:**377–385.
15. **Janssson, S. L., and J. Persson.** 1982. Mineralization and immobilization of soil nitrogen. In F. J. Stevenson (ed.), Nitrogen in agricultural soils. Agronomy No. 22, American Society of Agronomists, Inc., Madison, Wis.
16. **Jenkinson, D. S., and J. N. Ladd.** 1981. Microbial biomass in soil: measurement and turnover, p. 415–471. In E. A. Paul and J. N. Ladd (ed.), Soil biochemistry, vol. 5. Marcel Dekker, New York.
17. **Jenkinson, D. S., and D. S. Powlson.** 1976. The effects of biocidal treatments on metabolism in soil. V. A method for measuring soil biomass. Soil Biol. Biochem. **8:**209–213.
18. **Karl, D. M.** 1980. Cellular nucleotide measurements and application in microbial ecology. Microbiol. Rev. **44:**739–796.
19. **Kucey, R. L., and E. A. Paul.** 1982. Carbon flow, photosynthesis and N_2 fixation in mycorrhizal and nodulated faba beans Vicia faba. Soil Biol. Chem. **14:**407–412.
20. **Kudeyarov, V. N., and D. S. Jenkinson.** 1976. The effects of biocidal treatments on metabolism in soil. VI. Fumigation with carbon disulphide. Soil Biol. Biochem. **8:**375–378.
21. **Ladd, J. N., J. M. Oades, and M. Amato.** 1981. Microbial

biomass formed from ^{14}C, ^{15}N-labelled plant material decomposing in soils in the field. Soil Biol. Biochem. **13**:119–126.

22. **Lal, R., and B. T. Kang.** 1982. Management of organic matter in soils of the tropics and subtropics. *In* Non-Symbiotic Nitrogen Fixation and Organic Matter in the Tropics. Symposia Papers 1. Transactions of the 12th International Congress of Soil Science, New Delhi, India, 1982.

23. **Lynch, J. M., and L. M. Panting.** 1980. Cultivation and the soil biomass. Soil Biol. Biochem. **12**:29–33.

24. **Marumoto, T., J. P. E. Anderson, and K. H. Domsch.** 1982. Mineralization of nutrients from soil microbial biomass. Soil Biol. Biochem. **14**:469–475.

25. **Morita, R. Y.** 1982. Starvation-survival of heterotrophs in the marine environment, p. 171–198. *In* K. C. Marshall (ed.), Advances in microbial ecology, vol. 6. Plenum Press, New York.

26. **North, M.** 1982. Comparative biochemistry of the proteinases of eucaryotic microorganisms. Microbiol. Rev. **46**:308–334.

27. **Parkinson, D., K. H. Domsch, J. P. E. Anderson, and E. Heller.** 1980. Studies on the relationship of microbial biomass to primary production in three spruce forest soils. Zentralbl. Bakteriol. Mikrobiol. Hyg. 1 Abt. Orig. Reihe C **1**:101–107.

28. **Parkinson, D. A., and E. A. Paul.** 1982. Microbial biomass. Agronomy 9, part 2, Chemical and microbial methods, 2nd ed, p. 821–830. American Society of Agronomists, Madison, Wis.

29. **Paul, E. A., and N. G. Juma.** 1981. Mineralization and immobilization of soil nitrogen by microorganisms. Ecol. Bull. (Stockholm) **33**:179–195.

30. **Paul, E. A., and R. P. Voroney.** 1980. Nutrient and energy flows through soil microbial biomass, p. 215–237. *In* D. C. Ellwood, J. N. Hedger, M. J. Latham, J. M. Lynch, and J. H. Slater (ed.), Contemporary microbial ecology. Academic Press, London.

31. **Payne, W. J., and W. J. Wiebe.** 1978. Growth yield and efficiency in chemosynthetic microorganisms. Annu. Rev. Microbiol. **32**:155–184.

32. **Pedros-Alio, C., and T. D. Brock.** 1982. Assessing biomass and production of bacteria in eutrophic Lake Mendota, Wisconsin. Appl. Environ. Microbiol. **44**:203–218.

33. **Powlson, D. S.** 1980. The effects of grinding on microbial and non-microbial organic matter in soil. J. Soil Sci. **31**:77–85.

34. **Powlson, D. S., and D. S. Jenkinson.** 1981. A comparison

of the organic matter, biomass, adenosine triphosphate and mineralizable nitrogen contents of ploughed and direct-drilled soils. J. Agric. Sci. **97**:713–721.

35. **Robinson, J. D., K. H. Mann, and J. A. Novitsky.** 1982. Conversion of the particulate fraction of seaweed detritus to bacterial biomass. Limnol. Oceanogr. **27**:1072–1079.

36. **Ross, D. J., K. R. Tate, T. Cairns, and E. A. Pansier.** 1980. Microbial biomass estimates in soils from tussock grasslands by three chemical procedures. Soil Biol. Biochem. **12**:375–383.

37. **Saggar, S., J. R. Bettany, and J. W. B. Stewart.** 1981. Measurement of microbial sulfur in soil. Soil Biol. Biochem. **13**:493–498.

38. **Schmidt, E. L., and E. A. Paul.** 1982. Microscopic methods for soil microorganisms. Agronomy 9, part 2, Chemical and biological methods, 2nd ed. p. 803–814. American Society of Agronomists, Madison, Wis.

39. **Shields, J. A., E. A. Paul, and W. E. Lowe.** 1974. Factors influencing the stability of labelled microbial materials in soils. Soil Biol. Biochem. **6**:31–37.

40. **Soderstrom, B. E.** 1977. Vital staining of fungi in pure cultures and in soil with fluorescein diacetate. Soil Biol. Biochem. **9**:59–63.

41. **Sparling, G. P.** 1981. Microcalorimetry and other methods to assess biomass and activity in soils. Soil Biol. Biochem. **13**:93–98.

42. **Sparling, G. P., T. R. Fermor, and D. A. Wood.** 1982. Measurement of the microbial biomass in composted wheat straw, and the possible contribution of the biomass to the nutrition of *Agaricus bisporus*. Soil Biol. Biochem. **14**:609–611.

43. **Stott, D. E., G. Kassim, W. Jarrell, J. P. Martin, and K. Haider.** 1983. Stabilization and incorporation into biomass of specific plant carbons during biodegradation in soils. Plant Soil **70**:15–26.

44. **Van Veen, J. A., and M. J. Frissel.** 1981. Simulation of nitrogen behavior of soil plant systems. Centre for Agricultural Publishing and Documentation, Wageningen, The Netherlands.

45. **Vogt, K. A., C. C. Grier, C. E. Meir, and R. L. Edmonds.** 1982. Mycorrhizal role in net primary production and nutrient cycling in *Abies amabilis* ecosystems in western Washington. Ecology **63**:370–380.

46. **Woods, L. E., C. V. Cole, E. T. Elliott, R. V. Anderson, and D. C. Coleman.** 1982. Nitrogen transformations in soil as affected by bacterial-microfaunal interactions. Soil Biol. Biochem. **14**:93–98.

Microbial Biomass and Activity Measurements in Soil: Ecological Significance

P. NANNIPIERI

Instituto Chimica Terreno, Consiglio Nazionale delle Ricerche, 56100 Pisa, Italy

Modern society's increasing demands on scientists for help in making decisions regarding land use and resource management have accentuated the importance of ecological knowledge. The promotion of ecological research has had a great importance to the development of microbial ecology. Microbiologists were forced out into the field, where they were faced with problems new to them, problems for which existing techniques were inadequate or nonexistent. The interest of the soil microbiologist was, therefore, shifted from the identification of isolates and the estimation of populations to the study of the role of microorganisms in soil processes.

Such increased interest has produced, in the past 20 years, numerous empirical observations on biomass and biological activity in various soil types and ecosystems, with respect to amendments, cultivation practices, and responses to environmental and climatic factors. The observations, however, have often produced conflicting and confusing results, not only because the methodologies employed were questionable but also because the motivation of the research and the meaning of the measured criteria have not been properly defined.

DETERMINATION OF BIOMASS

Microbial biomass has recently been defined as the living part of the soil organic matter with the exclusion of plant roots and soil animals larger than 5×10^3 μm^3 (16).

The need to characterize the mineral cycling and the energy flow in the ecosystem has led to the consideration of the microbial biomass in soil as an undifferentiated whole (16), without any consideration of the enormous diversity of microbes with varying abilities to survive in extreme environmental conditions. The approach of considering microbial biomass as an undifferentiated whole has been made possible by the availability of the fumigation technique which allowed the accurate estimation of the relative pool of nutrients (16). When a soil is fumigated with $CHCl_3$, the fumigant is removed, and the soil is inoculated with an untreated sample and incubated, the respiration rate of the fumigated soil becomes, after some hours, much greater than that of the unfumigated soil (17). Fumigation usually causes an immediate increase in the extractable inorganic P, sulfur, and NH_4^+ content of soil (6, 17, 32). Additional NH_4^+ is released when the soil is subsequently incubated. The "flush of decomposition," which is defined as the amount of CO_2 evolved (or N mineralized) by a fumigated soil in a given incubation time less the CO_2 evolved (or N mineralized) by the same amount of unfumigated sample in the same time, has been ascribed to the decomposition of killed organisms by the inoculated population. The size of the CO_2 or NH_4^+ flush can be related, by a simple expression, to the size of biomass C or N, respectively (16). It is also possible to determine the biomass P (6) and S (32) by use of the fumigation technique; however, the method has mainly been used to investigate the behavior of the microbial C and N in soil.

Experiments with labeled plant material added to soil have demonstrated the role of microbial biomass and activity in carbon cycling (27, 28). Decomposition rates corrected for microbial utilization of the substrate at reasonably high efficiencies (35 to 60%) have shown much faster transformation rates than are usually described in the literature (28).

Climatic conditions have been found to influence not only decomposition rates of labeled plant material added to soil but also the turnover and doubling time of the soil microbial populations (16). However, these determinations were carried out long after the addition of the fresh material and therefore did not take into consideration the active population present shortly after the addition of the carbonaceous material. In addition, the approach of considering the soil population as a homogeneous simple compartment model does not take into consideration the variety of turnover times that occur in the soil biomass. Methods for distinguishing various sections of the population, such as that devised by Anderson and Domsch (2) or direct counts based on selective staining procedures, can only give an estimate of bacterial and fungal biomass, without any indications of their nutrient content. The measurement of microbial biomass with methods based on specific biomass constituents, such as muramic acid, hexosamines, nucleic acid, etc., have provided, with the exception of the ATP technique, erroneous estimates (16).

515

Biomass measurements have been used as an early indication of the response of the organic matter cycle to management changes in soils. Microbial biomass responds more rapidly than either organic C or nitrogen to changes in inputs of organic matter to soil or in its rate of decomposition. Biomass C, determined by the fumigation technique, increased rather markedly in samples collected in spring 1978 from a grass pasture experiment started in the spring of 1977 in soil which had been previously cropped with wheat for 10 years (Fig. 1). The organic C content of the soil did not, however, show any significant changes (3).

Biomass C has been reported to be significantly higher in direct-drilled soils than in ploughed soils, probably reflecting a concentration of plant roots and of organic matter at the soil surface (21). The difference, however, was found only when shallow surface samples were compared (30).

Under Mediterranean climate conditions, in a grass–legume association fertilized with labeled urea (140 kg of N per ha) containing 17.77 atom % ^{15}N, about 45% of the total labeled fertilizer N was present in the top 20 cm of soil as NH_4^+-N after 2 days and decreased to 20% after 5 days (P. Nannipieri et al., unpublished data). Over the same time period labeled NO_3^--N increased and accounted for 2% of the total applied N after 5 days. The increase in the flush of mineral and labeled N occurred between 5 and 10 days, indicating that a sequence of biochemical events had occurred in the soil. Urea was rapidly split into ammonia, and subsequently the nitrogen

was immobilized by the soil population (Nannipieri et al., unpublished data). However, changes in labeled NH_4^+-N were more marked than those occurring in labeled NO_3^--N and in the flush of mineral and labeled N. The empirical relationship between biomass C and the flush of mineral N (F_n), where biomass C = F_n (16), was not valid in this soil. Ratios between the flush of CO_2-C and inorganic N, determined at the end of the 10-day incubation period following fumigation, were greater than 20:1, significantly higher than the 4:1 ratio commonly reported (1). This indicated that nitrogen immobilization occurred during biomass determination. By considering this discrepancy, it was possible to correct for N immobilization during biomass determinations. As a result, the calculated values of the flush of mineral ^{15}N formed from labeled biomass can account for most of the observed decrease in labeled NH_4^+-N during the early time intervals following urea application.

However, data on the N biomass obtained by the fumigation technique have to be interpreted with caution. The K_n, that is, the fraction of biomass N mineralized to inorganic N during the 10 days after $CHCl_3$ fumigation, has been found to cover a much wider range than K_c, depending on the type of microorganism and the properties of the soil (Tables 1 and 2). Moreover, net mineralization rates, during the incubation following fumigation, depend on the microbial C/N ratio (B. Nicolardot et al., unpublished data), which is not constant and fluctuates with growth rate and nutrient concentration (12). The problem of reimmobilization of NH_4^+-N has recently

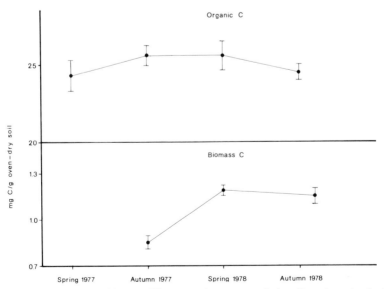

FIG. 1. Changes in organic C and biomass C in a grass–legume association. Bars show standard deviations of means.

TABLE 1. Fraction of biomass N mineralized to inorganic N in organisms added to loam soil, fumigated with $CHCl_3$, and incubated for 10 days at $25°C$[a]

Organism	Fraction
Fungi	
Cylindrocarpon sp.	Immobilized
Penicillium chrysogenum	Immobilized
Bacteria	
Aerobacter aerogenes	0.59
Bacillus subtilis	0.59
Clostridium histolyticum	0.46
Escherichia coli	0.54
Micrococcus lysodeikticus	0.45
Pseudomonas fluorescens	0.59

[a] Adapted from Jenkinson and Powelson (17).

been overcome by relating K_n to the ratio of CO_2-C evolved and the net NH_4^+ accumulated $(C_F:N_F)$ during the fumigation-incubation according to the expression: $K_n = 0.014 (C_F/N_F) + 0.39$ (34).

MEASUREMENT OF ACTIVITY

General criteria. Soil microbiologists and biochemists have always hoped to obtain a general index of total biological activity of soils which could be used for practical purposes in agriculture and for describing the biological status of the environment. Among the numerous methods proposed over the past 30 years are carbon dioxide release, oxygen uptake, dehydrogenase activity, heat output, and ATP and adenylate energy charge measurements.

Respiration rates have frequently been used to determine the biological activity in soil in relation to changes in climate, physical and chemical properties of soil, and agricultural practices. Generally, the measurements have been used to obtain a better understanding of mineralization processes and to gain insight into how the mineral nutrients and organic matter of the soil can be more efficiently managed. Carbon dioxide evolution, however, has had a wider application than O_2 uptake, since CO_2 is evolved through aerobic and anaerobic process (35; R. P. Voroney et al., unpublished data). In interpreting the results from such experiments, it must be recognized that the respiration of animals and plant roots may render the gas-exchange results in soil extremely complicated.

By analyzing the $^{14}CO_2$ evolved during the decomposition of specifically labeled ^{14}C compounds or microbial cells, it has been possible to determine which compounds are utilized most readily by soil organisms and which are relatively more important in soil humus formation (22). The loss of carbon from simple aliphatic acids

averages about 84%, whereas that from phenolic substances varies from 20 to 71%, indicating some stabilization of most of the benzene rings. However, the transformations of the labeled compounds in soil and the main chemical forms of the residual ^{14}C are still largely unknown. To further understand these aspects, we desperately need new methodologies and approaches.

The dehydrogenase activity of soil is thought to reflect the total range of biological oxidative activities of the soil microflora and consequently is reported as an excellent tool for measuring soil microbiological activity. Active dehydrogenases are considered to exist in soils as integral parts of intact cells and are generally determined in soil under anaerobic conditions by the rates of reduction of 2,3,5-triphenyl tetrazolium chloride to tryphenyl formazan, which is extracted and measured spectrophotometrically (15). A positive correlation is not always found between dehydrogenase activity and bacterial number, which probably reflects the predominant resting state of the soil microflora. Dehydrogenase activities also do not consistently correlate with respiration rates, and this has been ascribed to the occurrence of oxygenases. These enzymes catalyze the direct oxidation of the substrate with molecular oxygen (15), in the presence of other electron acceptors such as nitrate, which limits triphenyl tetrazolium chloride uptake by bacteria and allows the accumulation of formazan in inhibitory concentrations (11). Even with a sensitive electron acceptor, such as the 2-p-iodophenyl-3-p-nitrophenyl-5-phenyl tetrazolium chloride, more than 90% of the observed O_2 uptake was not accounted for (5).

Heat output as a measure of microbial activity has the advantage of depending only on the initial and final energy states of a system, being independent of organisms and reaction path-

TABLE 2. Fraction of biomass N mineralized to inorganic N in organisms added to different soils, fumigated with $CHCl_3$, and incubated for 7 days at $28°C$

Organism	Silty clay	Sandy loam	Silty clay loam	Silt loam
Fungi				
Aspergillus flavus	0.23	0.28	0.23	0.27
Trichoderma viride	0.23	0.12	0.26	0.41
Bacteria				
Streptomyces sp.	0.48	0.45	0.50	0.53
Arthrobacter sp.	0.54	0.45	0.56	0.32
Achromobacter liquefaciens	0.36	0.24	0.42	0.82

[a] Adapted from B. Nicolardot (Ph.D. thesis, I.N.R.A. Laboratory of Soil Microbiology, Dijon, France).

ways (29). For this reason, it has been considered a good criterion of soil biological activity since its first usage in 1931 by Hesselink van Suchtelen (13). However, some methodological problems have limited the use of microcalorimetry in soil. During the calorimetric measurement the sample is usually hermetically sealed in an ampoule, and O_2 depletion and increase in CO_2 in the atmosphere within the ampoule are serious problems, especially over extended incubation periods. Recently, such problems have been solved by using a larger flow cell (100 ml) which allows the use of 10 g of soil and gas exchange (34). In soils with different physicochemical properties and management histories, the heat output consistently correlated with respiration, ATP content, biomass, and amylase activity, but not with dehydrogenase activity (34). Although microcalorimetry gives valid data on the overall rate of catabolism, information on the formation of end products and cellular incorporation is required before the method can be used to estimate substrate turnover in soil.

The measurement of ATP content in soil has been used to assess the mass of living matter (16), even when the content of ATP in the biomass depended on the addition of inorganic P to the soil (26). ATP, which is the common molecule for energy packing and transport, has been used as a criterion for the evaluation of microbial activity (33) and was found to be consistently correlated with respiration in tundra soils (11). Conversely, a relationship between the ATP content and CO_2 evolved does not seem to exist in soil during a simulation of an environmental soil-drying process (19). To provide information on the activity of living organisms, in the sense of the turnover of metabolic energy, one would have to measure the rate of conversion of ATP to ADP. However, this measurement is likely to prove extremely difficult in soil because of very short turnover time of the microbial ATP pool.

A better definition of the "state of growth" of natural microbial communities can be obtained by a measurement of the "adenylate energy charge" (AEC), defined as AEC = ([ATP] + 0.5[ADP])/([ATP] + [ADP] + [AMP]). The quantification of adenine nucleotides is based on the measurement of ATP in the luciferin-luciferase system, by which AMP and ADP are also estimated after their conversion to ATP by standard enzymatic methods (7). In a fresh soil, kept under pasture for at least three centuries and with a high content of organic C, the AEC was 0.85, a value comparable to those found for actively growing microorganisms in vitro (8). In five cultivated agricultural soils the AEC ranged from 0.3 to 0.4, which strongly suggested limited nutrient availability (18). A solution of trichloro-

acetic acid, paraquat, and phosphate, and also a slightly alkaline $CHCl_3$-$NaHCO_3$ extractant were used to extract adenine nucleotides from the soils. In both methods the quantities of nucleotides present in the extracts were corrected for incomplete extraction from the soil by measuring the percentage recovery of added ATP and AMP. However, further experimental work is necessary to test the accuracy of AEC assays, and a comparison of enzyme methods with chemical measurements might be fruitful. Incomplete inactivation of the nucleotide-converting enzymes and failure to completely extract nucleotides from cells may be a serious problem in procedures involving mild extractants, such as $NaHCO_3$ solutions. Other interferences might be due to chemical degradation (particularly with strong extractants such as trichloroacetic acid solutions) of the nucleotides or of cell components containing these nucleotides (NAD, flavin adenine dinucleotide, and RNA). Further problems may be due to failure in distinguishing between intracellular and extracellular concentrations since adenine nucleotides may be capable of becoming bonded to humic and mineral colloids and acquiring a persistence they would not otherwise display in the extracellular environment of the soil.

Specific criteria: enzyme activities. It was hoped, especially 20 to 30 years ago, that an enzyme would be found whose activity would provide an estimate of the total biological activity of soil (33). The following enzymatic activities were taken as good criteria of soil quality: invertase activity (14), protease activity (20), asparaginase activity (4), urease activity (4), phosphatase activity (18); however, considerable variability was observed (33). In general, enzymatic activities do not correlate with respiration rates or with microbial numbers. This is not surprising since individual enzyme measurements answer questions regarding specific metabolic processes or concerning specific nutrient cycles.

In the interpretation of data resulting from enzyme measurements in soil, it should be kept in mind that the total activity of an enzyme is comprised of activities associated with different constituents, such as proliferating cells, viable and nonproliferating cells, cell debris, mineral, and humic colloids (8). However, there have been widespread, and often tacit, assumptions that the soil enzyme activity determined is due only to enzymes which are extracellularly localized. It is evident that for a better understanding of the total enzymatic activity in soil each type of enzyme should be studied individually to determine its relative contribution to total biological activity (25).

Since the soil enzyme activity includes the

activities of living organisms, the range of enzymes present clearly must be wider than the approximately 60 enzymes whose activities have been measured. In addition, many activities in soil cannot be ascribed to the activity of a particular enzyme or enzyme system because they are measured without the specific identification of the bonds hydrolyzed or of products formed, as in the determination of protease activity when casein is used as a substrate.

The need to simplify the investigated system has led, in biogeochemistry, to the measurement of the rates of complex metabolic pathways by determining the variation in concentration of the final product. An example of this is the estimation of net nitrogen mineralization rates in soil by examining the variation of NH_4^+-N concentrations even if the metabolic pathways leading to the production of ammonia from organic N compounds are catalyzed by numerous specific enzymes. Step-by-step studies, in which all metabolic intermediates are identified and all relative enzyme activities are monitored, are, at the moment, not feasible in soil. Knowledge of the critical rate-limiting step of the whole reaction sequence is required before the kinetics of the process can be defined.

The evaluation of reaction rates has been carried out in soil by determining enzyme constants and using approaches similar to those employed in homogeneous experimental systems (23). Water, pH, surface charge, etc., in the microsite surrounding the enzyme should be considered in studying the enzyme kinetics in a heterogeneous system such as soil. At the interface surface, the pH minimum and the water and substrate availability which will allow an enzyme to function need to be critically evaluated. The adsorption of the substrate (Table 3) to soil colloids has been found to influence the apparent K_m of acid phosphatase (9). The presence of multiple enzymes catalyzing the same reaction must also be considered in soil when enzyme kinetic parameters are determined. Each enzyme catalyzing a specific process can, in fact, originate in soil from different microbial, plant, and animal sources. Consequently, the determined K_m and V_{max} values represent an average of the different kinetic parameters of the various enzymes catalyzing the same reaction. An example of the latter is the recent observation that soil extracts contain at least two different phosphatases characterized by different kinetic enzyme parameters (24). The presence in soil of many enzymes catalyzing the same reaction but characterized by different K_m values can play an important role in the capacity of the soil to perform the reaction and must be considered in any interpretation of the activity. An example of the latter is the suggestion that high-affinity

TABLE 3. Michaelis-Menten constants of soil acid phosphatase and the adsorption of the substrate (p-nitrophenyl phosphate)[a]

Soil	K_m	K_{mc}
Silty clay soil	1.0×10^{-2}	5.40×10^{-3}
Terra rossa	9.7×10^{-3}	1.21×10^{-3}
Peaty soil	12.2×10^{-3}	0.35×10^{-3}
Andosol	16.4×10^{-3}	0.58×10^{-3}

[a] K_m values when the adsorption of the substrate was not considered and K_{mc} values when the adsorption of the substrate was analytically included in rate equations, both expressed as moles per liter. Adapted from Cervelli et al. (9).

phosphatase (the low K_m enzyme) could operate even at low substrate concentrations, whereas the low-affinity enzyme (the high K_m enzyme) would function at increased substrate concentrations.

CONCLUSIONS

The availability of the chloroform fumigation method has permitted a more accurate determination of the content of nutrients in soil biomass and offers the possibility of quantifying the relative fluxes of these nutrients to and from microbial reservoirs. However, a certain caution is required when the method is used to determine biomass N, since reimmobilization of NH_4^+-N, during the time after $CHCl_3$ fumigation, can occur and depends on the C/N ratio of the soil microbial population. Further progress in studying the flux of energy and material through the soil population requires fractionation techniques, which will allow the determination of nutrient contents of various populations.

Rates of carbon dioxide evolution and heat outputs have emerged as important criteria of the general microbial activity in soil. Adenylate energy charge measurements rather than ATP measurements seem promising; however, further experimental work is necessary to test the accuracy of these assays. Since the term "biological activity" covers a wide spectrum of individual activities and processes, it seems more rewarding to carry out specific measurements simultaneously (26, 31). The determination of specific enzyme activities together with the use of general soil parameters seems, at the moment, the best approach for evaluating the state of the soil microbial activity and understanding its response to amendments, cultivation practices, and environmental and climatic factors. In interpreting the results of enzyme measurements, it should be kept in mind that the incubation conditions used provide the maximum potential catalytic activity (8). In soil the enzymes are rarely presented with excess substrates, and the moisture level and the temperature are variable. In addition, the buffering capacity of soil is different from that of the buffering solution used in the assay, and the soil pH value is rarely the optimum for the activity.

In correlative studies of biomass and activity measurements in the field, attention is required to the facts that one cannot relate biomass and activity at any one time, and that one needs at least a full year's data to average the effect of environmental conditions on the measurements (36). The self-regulatory mechanisms of which soil microbial communities are capable allow us to conclude that derepression or activation is usually reversible (10; Nannipieri et al., unpublished data). Consequently, the monitoring period is important for the assessment of persistence: the shorter the period, the less important is an observed deficit or surplus in microbial biomass and activity.

LITERATURE CITED

1. **Amato, M., and J. N. Ladd.** 1980. Studies of nitrogen immobilization and mineralization in calcareous soils. V. Formation and distribution of isotope-labelled biomass during decomposition of ^{14}C- and ^{15}N-labelled plant material. Soil Biol. Biochem. **12:**405–411.
2. **Anderson, J. P. E., and K. H. Domsch.** 1975. Measurement of bacterial and fungal contributions to respiration of selected agricultural and forest soils. Can. J. Microbiol. **21:**313–322.
3. **Arcara, P. G., C. Piovanelli, E. Sparvoli, P. Nannipieri, and F. Pedrazzini.** 1979. Microbiological and biochemical parameters in grass-legume associations, p. 177–195. *In* Problems in unfavourable lands. Promotion of Environmental Quality Programme, AC/4. CNR, Rome.
4. **Balicka, N., and Z. Sochaka.** 1959. Biological activity in light soils. Zesz. Probl. Postepow. Nauk Roln. **21:**257–265.
5. **Benefield, C. B., P. J. A. Howard, and D. M. Howard.** 1977. The estimation of dehydrogenase activity in soil. Soil Biol. Biochem. **9:**67–70.
6. **Brookes, P. C., D. S. Powlson, and D. S. Jenkinson.** 1982. Measurement of microbial biomass phosphorus in soil. Soil Biol. Biochem. **14:**319–329.
7. **Brookes, P. C., K. R. Tate, and D. S. Jenkinson.** 1983. The adenylate energy charge of the soil microbial biomass. Soil Biol. Biochem. **15:**9–16.
8. **Burns, R. G.** 1978. Enzyme activity in soil: some theoretical and practical considerations, p. 295–340. *In* R. G. Burns (ed.), Soil enzymes. Academic Press, Inc., New York.
9. **Cervelli, S., P. Nannipieri, B. Ceccanti, and P. Sequi.** 1973. Michaelis constant of soil acid phosphatase. Soil Biol. Biochem. **5:**841–845.
10. **Domsch, K. M., G. Jagnow, and T. H. Anderson.** 1983. An ecological concept for the assessment of side-effects of agrochemicals on soil microorganisms. Residue Rev. **86:**65–105.
11. **Hast, R., and T. Rosswall.** 1973. Activity of soil microorganisms as measured by respiration, ATP and dehydrogenase, p. 57–71. *In* J. G. K. Flower-Ellis (ed.), International Biological Programme, Swedish Tundra Biome Project, Technical Report 16. Swedish Natural Science Research Council, Stockholm.
12. **Herbert, D.** 1976. Stoichiometric aspects of microbial growth, p. 1–30. *In* A. C. R. Dean, D. C. Ellwood, C. G. T. Evans, and J. Melling (ed.), Continuous culture 6. Application and new fields. Ellis Harwood Ltd., Chichester.
13. **Hesselink van Suchtelen, F. A.** 1931. Energetics and microbiology of the soil. Arch. Pflanzenbau **7:**519–541.
14. **Hofmann, E., and A. Seegerer.** 1950. Soil enzymes as a measure of biological activity. Biochemistry **321:**97–103.
15. **Howard, P. J. A.** 1972. Problems in the estimation of biological activity in soil. Oikos **23:**235–240.
16. **Jenkinson, D. S., and J. N. Ladd.** 1981. Microbial biomass in soil: measurement and turnover, p. 415–471. *In* E. A. Paul and J. N. Ladd (ed.), Soil biochemistry, vol. 5. Marcel Dekker, New York.
17. **Jenkinson, D. S., and D. S. Powlson.** 1976. The effects of biocidal treatments on metabolism in soil. I. Fumigation with chloroform. Soil Biol. Biochem. **8:**167–177.
18. **Keilling, J., A. Camus, G. Savignac, P. Danchez, M. Boitel, and R. Planet.** 1960. Contribution to the study of the biology of soil. C.R. Seances Acad. Agric. Fr. **46:**647–652.
19. **Knight, W. G., and J. Skujins.** 1981. ATP concentration and soil respiration at reduced water potentials in arid soils. Soil Sci. Soc. Am. J. **45:**657–660.
20. **Lajudie, J., and J. Pochon.** 1956. Study on the proteolytic activity of soils, p. 271–273. *In* Transactions of the VI International Soil Science Congress, C. Paris.
21. **Lynch, J. M., and L. M. Panting.** 1980. Variations in the size of the soil biomass. Soil Biol. Biochem. **12:**547–550.
22. **Martin, J. P., and K. Haider.** 1980. Microbial degradation and stabilization of ^{14}C-labeled lignins, phenols and phenolic polymers in relation to soil humus formation, p. 78–100. *In* T. K. Kirk, T. Higuchi, and H.-M. Chang (ed.), Lignin biodegradation: microbiology, chemistry and potential applications. CRC Press, Cleveland.
23. **McLaren, A. D.** 1978. Kinetics and consecutive reactions of soil enzymes, p. 97–116. *In* R. G. Burns (ed.), Soil enzymes. Academic Press, Inc., New York.
24. **Nannipieri, P., B. Ceccanti, S. Cervelli, and C. Conti.** 1982. Hydrolases extracted from soil: kinetic parameters of several enzymes catalysing the same reaction. Soil Biol. Biochem. **14:**429–432.
25. **Nannipieri, P., B. Ceccanti, S. Cervelli, and E. Matarese.** 1980. Extraction of phosphatase, urease, proteases, organic carbon and nitrogen from soil. Soil Sci. Soc. Am. J. **44:**1011–1016.
26. **Nannipieri, P., R. L. Johnson, and E. A. Paul.** 1978. Criteria for measurement of microbial growth and activity in soil. Soil Biol. Biochem. **10:**223–229.
27. **Paul, E. A., and J. A. Van Veen.** 1978. The use of tracers to determine the dynamic nature of organic matter, p. 61–102. *In* 11th International Society of Soil Science, Transactions (Edmonton), June, 1978. University of Alberta, Edmonton.
28. **Paul, E. A., and R. P. Voroney.** 1980. Nutrient and energy flows through soil microbial biomass, p. 215–237. *In* D. C. Ellewood, J. N. Hedger, M. J. Latham, J. M. Lynch, and J. M. Slater (ed.), Contemporary microbial ecology. Academic Press, Inc., New York.
29. **Payne, W. J.** 1970. Energy yields and growth of heterotrophs. Annu. Rev. Microbiol. **24:**17–52.
30. **Powlson, D. S., and D. S. Jenkinson.** 1981. A comparison of the organic matter, biomass, adenosine triphosphate and mineralizable nitrogen contents of ploughed and direct-drilled soils. J. Agric. Sci. **97:**713–721.
31. **Ross, D. J., T. W. Speir, K. R. Tate, A. Cairns, K. F. Meyrick, and E. A. Pansier.** 1982. Restoration of pasture after topsoil removal: effects on soil carbon and nitrogen mineralization, microbial biomass and enzyme activities. Soil Biol. Biochem. **14:**575–581.
32. **Saggar, S., J. R. Bettany, and J. W. B. Stewart.** 1981. Sulfur transformations in relation to carbon and nitrogen

in incubated soil. Soil Biol. Biochem. **13**:499–511.

33. **Skujins, J. J.** 1976. Extracellular enzymes in soil. Crit. Rev. Microbiol. **4**:383–421.

34. **Sparling, G. P.** 1981. Microcalorimetry and other methods to assess biomass and activity in soil. Soil Biol. Biochem. **13**:93–98.

35. **Stotzky, G.** 1972. Activity, ecology and population dynamics of microorganisms in soil. Crit. Rev. Microbiol. **2**:59–137.

36. **Witkamp, M.** 1973. Compatibility of microbial measurements. Bull. Ecol. Res. Comm. NFR (Statens Naturvetensk. Forskningsrad) **17**:179–188.

Microbial Responses to Ecosystem Perturbations

Influence of Clear-Cutting on Selected Microbial Processes in Forest Soils

JAMES R. GOSZ AND FREDRICK M. FISHER

Biology Department, University of New Mexico, Albuquerque, New Mexico 87131

The consequences of forest harvest, and specifically clear-cutting, have been of concern for many years. This concern has increased in recent years because of the elevated losses of nutrients reported for a number of sites after harvesting. Nitrate has been the ion receiving most attention for several reasons: (i) nitrate losses are generally higher than those of other nutrients; (ii) nitrogen is often the element cited as limiting forest growth; and (iii) the release of hydrogen ions during nitrate production, along with the high mobility of the nitrate ion, causes increased solution losses of other cations (23, 46).

Vitousek and Melillo (47) reviewed the literature on nitrogen losses from disturbed forests and found a range of losses from barely detectable to extreme. Vitousek et al. (46) suggested that the overall pattern and control of nitrogen losses in disturbed forests can be broken down into three components. "Given a disturbance which removes vegetation cover but does not interfere with plant regrowth, the magnitude and timing of nitrogen losses are controlled by: (1) the predisturbance net nitrogen mineralization rate and the extent to which it is accelerated by forest canopy removal; (2) an interaction of the processes which can prevent or delay losses of excess mineralized nitrogen; and (3) the rate of vegetation regrowth and nitrogen uptake." More simply stated, nutrient loss is related to the interaction between plant uptake and the microbial processes involved in mineralization (the release of nutrient ions from organic matter), immobilization (incorporation of nutrient ions into organic matter), and nitrification (oxidation of NH_4^+ to NO_3^-).

Although these papers correctly assess the general relationship between plant and microbial processes, they slight what we believe is a primary factor in the plant–microbe relationship: the influence of plant roots on microbial processes. Thus, the theme of our paper will be the potential influence of plant roots on microbial activity and how that influence is affected by forest harvest.

EFFECTS OF ROOTS

The effects of plant roots on the mineralization of nitrogen in agricultural soil was investigated as early as 1913 by Lyon and Bizzell. In a review, Goring and Clark (13) noted that most workers had concluded that crop plants generally depressed net mineralization of nitrogen (the balance between actual mineralization and immobilization) but that there were important exceptions, such as the legumes. Their own work demonstrated that plant roots had contributed carbonaceous organic matter to the soil which increased microbial activity but also increased immobilization to a greater extent than mineralization. The high nitrogen content of the legume organic matter was responsible for the exception of increased net mineralization.

The few studies of the effects of plant roots on soil processes in forests are more difficult to interpret because of the greater complexity of the forest system, the variety of experimental techniques used, and the lack of studies in which suitable tracers were used. Most of these studies have been interpreted to indicate that plant roots or mycorrhizae decrease the overall rate of microbial activity (1, 11, 12, 24), although Mattson and Vitousek (28) found no effect on net mineralization in some cases. The results of the forest studies appear to be in conflict with the agricultural studies, which show that the roots of crop plants usually increase microbial activity. The forest studies, in general, also fail to explain the slow rates of net mineralization occurring in the absence of plant roots in some soils (e.g., soil incubation studies). There is no doubt that plant roots can have an important impact on soil processes at the ecosystem level, but the exact nature of their influence must be resolved.

Vitousek et al. (46) used trenched plot and laboratory soil incubation experiments to identify forest sites with a high potential for nitrate production and loss following reduced root activity. Their results for nitrate production were in accord with larger-scale studies of commercial clear-cutting when comparable data were available. Although these experiments do not duplicate a clear-cutting procedure, they do reflect how microbial processes can change in a soil where plant root activity is eliminated. The 17 forest sites studies by Vitousek et al. (46) demonstrated a range of mineralization and nitrification rates which appeared correlated with "fertility" or site quality. Nitrogen availability, an index of site quality, can be assessed through quantification of nitrogen returned to the forest floor in litterfall (15). Laboratory incubations of litter from the forest floor resulted in nitrogen mineralization potentials which correlated well with the annual nitrogen content of litterfall for each site. Also, the quantity of nitrogen in litterfall could predict the proportion of the litter nitrogen mineralized under laboratory conditions (46). This indicates that organic nitrogen, which is relatively refractory to decomposition, is produced in the aboveground litterfall of forests which circulate small amounts of nitrogen (15, 26). Interestingly, the studies found no simple relationship between site properties (e.g., N in litterfall) and nitrogen mineralization potentials in mineral soil, perhaps, as they suggest, because root litter inputs were not estimated. This suggests that root litter could be qualitatively different from aboveground litter or that root activity affects microbial activity in the soil in a different manner. Since root litter inputs can supply relatively more organic nitrogen to mineral soil than do leaf and branch litterfall, this is an important consideration. Unfortunately, there is a dearth of information on root litter.

MECHANISMS OF ROOT INFLUENCE

Roots influence net mineralization indirectly through (i) their effects on the availability and C/N ratio of organic substrates and (ii) their inhibitory effects on microbial activity, including moisture uptake, nutrient uptake, and the production of allelopathic substances. Other effects of roots, such as reduced O_2 concentration or pH changes, may also influence net mineralization, but less evidence is available in their support.

Organic substrate. The mortality of fine roots and mycorrhizae has recently been shown to be a major source of organic substrate for microorganisms. In studies of temperate deciduous forests, fine root mortality was estimated to contribute quantities of organic matter equal to or greater than aboveground litterfall (8, 29), and in temperate coniferous forests, the mortality of fine roots or mycorrhizae consistently contributed more organic matter than aboveground litter (9, 10, 32). These studies often indicate that growth of fine roots and mycorrhizae occurs episodically, with several peaks in root biomass occurring during a year followed by periods of increased mortality. Since maximum uptake of water and nutrients occurs during periods of root growth (30), the peaks of plant assimilative activities and organic matter contributions are out of phase with each other. This may be of importance in environments where microbial activity can be initiated more rapidly than plant root growth or where a minor event (e.g., precipitation) initiates microbial activity but not plant root activity. Such a condition has been suggested for arid environments and drought periods (5, 7).

Few data on nitrogen content of fine roots are available; however, one thing is obvious from a survey of the literature by Fogel (9): Nitrogen concentrations are highly variable, ranging from 0.33 to 2.3% in only 13 measurements. Nine of the 13 had N contents lower than 1.33%, corresponding to an estimated C/N ratio of 30:1, with a conservatively low estimate of 40% carbon assumed. This suggests that mortality of most of these roots could lead to net immobilization (3, 16), but the extent would be quite variable.

Living roots produce organic matter in the form of sloughed cells and soluble exudates (37–39, 41). The latter have received the most study and generally consist of sugars, amino acids, and organic acids (38, 40, 41). Mycorrhizae also produce exudates which may be substantially different in character from those of nonmycorrhizal roots (34, 45). Most of the components of exudates are readily metabolized and result in a local stimulation of microbial activity called the "rhizosphere effect." Though the quantity of organic matter is probably much smaller than that produced by root mortality (8), the importance of root exudates is potentially great because (i) they are labile and readily metabolizable, (ii) their C/N ratio is easily modified by shifts in the balance of the components, and (iii) their production coincides with assimilation of water and nutrients. Although amino acids are frequently found in root exudates (38, 40, 41), the overall C/N ratio seems to be high for most plants, tending to create intense competition for N in the rhizosphere (6, 38, 41). One possible effect of these labile materials is to increase the utilization of native soil organic matter (the "priming effect") (20, 31), which could explain the depletion of organic matter around *Pinus* mycorrhizae noted by Will (49). The influence of roots will be least in the litter layer (01 horizon), but will generally increase in deeper horizons.

Microbial inhibition. Reduction of soil moisture content may be the most fundamental of the mechanisms by which roots inhibit microbial activity since soil microorganisms are essentially aquatic (42). F. M. Fisher (Ph.D. thesis, University of New Mexico, Albuquerque, 1983) experimentally demonstrated that moisture uptake by roots was the most significant factor in reducing microbial activity. Total water storage in soil lacking root activity is greater than in the presence of roots. Direct sunlight striking the forest floor might be expected to dry the organic-rich upper litter layers to a greater extent in some kinds of perturbations such as clear-cutting, but Jeffrey (18) reported that the capillary pores necessary to transport water to the surface of the litter layer are easily broken so that the zone of drying is often restricted. Several other effects of the removal of vegetation could act together with increased moisture availability to increase microbial activity in deeper horizons, including increased transport of soluble and particulate organic matter, increased temperature, and increased oxygen availability.

The primary nutritional factor limiting microbial activity in the soil is the availability of energy in the form of labile carbon compounds (42, 44). Inorganic nutrients can become limiting to microbial activity when labile carbonaceous organic matter is introduced to the soil, requiring the immobilization of large quantities of inorganic N from soil solution. Nitrogen is most frequently limiting to microbial activity, followed by phosphorus, but many other inorganic nutrients may be limiting under particular circumstances (42). In the context of this discussion, microbial activity would most likely be limited by inorganic nutrients if nutrient assimilation by the plant and input of some source of fresh organic matter coincided. The tendency of episodes of root uptake and root mortality to alternate is notable in this respect, since this pattern reduces the possibility of inorganic nutrients limiting microbial activity. Root exudates could potentially lead to nitrogen limitation of microbial activity because they can occur during the period when plants assimilate inorganic nutrients. Preliminary simulation studies (Fisher, unpublished data) suggest that the generally observed rates of carbon supply in exudates are insufficient. It is important to remember, however, that substantial net immobilization can occur without nitrogen becoming limiting to microbial activity.

Allelopathy is potentially an important factor in microbial inhibition. Allelopathic substances may be divided into two classes depending upon the mode of action (4): (i) toxins, which directly affect the metabolism of organisms, and (ii) complexing agents, which bind to extracellular enzymes or substrates, thereby preventing decomposition.

Toxins generally are small-molecular-weight compounds which must penetrate the cell membrane of organisms to affect metabolism. An important example in the soil community is the production of antibiotics by fungi and actinomy-cetes. The effects of these compounds are quite important in simple cultures but are more difficult to assess in a mixed soil population, because many organisms can degrade them or even utilize them as an energy source (42). Mycorrhizal fungi have been shown to produce antibiotics which inhibit fungi that parasitize roots (27, 45), but their importance in affecting processes involving complex mixed populations, such as decomposition and mineralization, is more difficult to evaluate. Since toxins are produced mostly by living mycorrhizae and are relatively labile, they would be effective only during periods of exudate production.

Evidence suggesting the importance of complexing agents in forests has recently been reviewed by Gosz (15). These substances are usually large-molecular-weight polyphenolic compounds such as lignin derivatives or tannins that decrease energy or organic nitrogen availability to microbes by complexing enzymes or substrates. These materials are found in fine roots and mycorrhizae and would be transferred to the soil when they die. Since complexing agents are also produced aboveground, the effect of roots is to increase the total production of these substances.

Overall effects of roots. The overall effects of roots on microbial activity are the net result of the stimulatory effects of organic matter contributions and the inhibitory effects of nutrient uptake, water uptake, and production of allelopathic compounds. Production of organic matter is always associated with root activity, but the inhibition of microbial activity may be either present or absent. This, combined with the fact that organic matter produced by roots may have a high or low C/N ratio, means that there are four categories of effects of roots on net mineralization: (i) high C/N, inhibited; (ii) high C/N, uninhibited; (iii) low C/N, inhibited; (iv) low C/N, uninhibited. The balance of the various kinds of effects undoubtedly differs from ecosystem to ecosystem, but it is possible to make some generalizations about the nature and causes of changes of this balance.

Since all plant roots contribute organic matter to soil, the contributions would be expected to increase proportionally with some measure of root activity such as fine root productivity. The potential for microbial activity therefore also increases in proportion to root activity. In this sense, the inhibitory effects of roots can be viewed as preventing microbial activity from reaching its potential rate, resulting in some observed, or actual rate. The importance of the inhibitory effects is also increased at higher root activity, but it seems unlikely that they increase in direct proportion to root productivity. For example, at relatively low rooting densities,

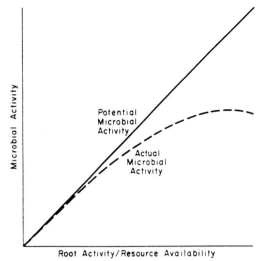

FIG. 1. Hypothetical relationship of microbial activity to root activity at different levels of resources (root activity/resource availability). Potential microbial activity is directly related to root activity because of the production of organic substrate by roots. Actual microbial activity demonstrates the increasing inhibition of roots as root activity increases.

plant roots may contribute measurable quantities of organic matter to the soil without depleting sufficient moisture or nutrients to inhibit microbial activity. These conditions are frequent in systems manipulated by humans, such as greenhouse experiments or irrigated fields (2), but may also occur in nature, particularly early in secondary succession. The resulting trend is illustrated in Fig. 1, which shows that increasing root activity increases potential microbial activity in a linear manner. At the lowest levels of root activity, there are no negative effects on microbial activity so that the actual activity equals the

potential activity. Increasing root activity past some point causes inhibition of microbial activity, and further increases in root activity cause potential and actual microbial activity to diverge. Note that the horizontal axis of Fig. 1 is labeled as the ratio of root activity to resource availability. This reflects the fact that in resource-rich soils a much higher level of root activity is needed to deplete water or nutrients. Thus, Fig. 1 standardizes for the influence of "fertile" versus "infertile" soils. Though the model is framed in terms of resource assimilation, responses to allelopathy may operate in a similar manner. Gosz (15) has suggested that production of polyphenolic compounds may increase in response to moisture or nutrient stress.

EFFECTS OF ELIMINATING ROOT ACTIVITY

This topic is important because most field experiments designed to indicate the effects of roots involve the partial or complete elimination of root activity. Additionally, lumber harvest and many other forest management activities also involve the reduction of root activity.

The predicted changes of microbial activity resulting from the elimination of root activity follow logically from previous discussions and are summarized for trenched plots in Table 1. Although trenched plots eliminate root activity, they do receive organic matter from the adjacent canopy. The effects of eliminating root activity are fundamentally different depending upon whether microbial activity was inhibited, since the inhibition of microbial activity leads to the accumulation of undecomposed organic matter. Eliminating root activity in these circumstances causes increased microbial metabolism as the accumulated organic matter is utilized. Subsequently, microbial activity is expected to drop as

TABLE 1. Source and nature of organic material available to microbes after the elimination of root activity[a]

System status	Source of organics available to microbes	Nature of organics available to microbes	Effect on microbial processes
Roots recently eliminated	Dead roots	Organic substrates	+
		Complexing agents	−
	Accumulation from past inhibition of microbial activity	Organic substrates	+
		Complexing agents	−
	Aboveground litter and leachates	Organic substrate	+
		Toxins	−
		Complexing agents	−
Continued prevention of root activity	Aboveground litter and leachates	Organic substrate	+
		Toxins	−
		Complexing agents	−

[a] Increased (+) or decreased (−) activity can occur depending on the quantity and nature of the organics. In trenched plot studies, organic material from aboveground sources would also influence microbial activity. (Modified from Fisher, Ph.D. thesis.)

the accumulated organic matter is depleted, finally stabilizing at some new level which may be determined by residual complexing agents. In the case of trenched plots, the new level of activity would be determined mainly by the inputs and nature of the organic matter from aboveground.

The four categories of effects of roots on net mineralization previously discussed can be used to predict the pattern of change in microbial activity that follows root elimination (Fig. 2). If roots inhibit microbial activity, the predicted pattern of change in microbial activity following the removal of that inhibition is shown in Fig. 2A and C. The major effect of removing roots in the absence of inhibition would be decreased microbial activity in response to decreased inputs of organic matter (Fig. 2B and D). If the disturbance leaves dead roots in the soil, a relatively small increase in microbial activity might occur initially, as shown in Fig. 2B and D.

The changes in net mineralization following the elimination of root activity depend upon the C/N ratio of the organic matter and the rate of microbial activity. Figure 2A illustrates the expected result where roots had previously inhibited microbial activity and the available substrate has a high C/N ratio. The increased microbial activity would lead first to net immobilization of nitrogen. However, as CO_2 was respired, organic matter would gradually be depleted and the

C/N ratio would drop. In response, microbial activity would gradually decrease, accompanied by an increase in net mineralization. Eventually, the C/N ratio would drop to the point where the rate of microbial activity was limiting net mineralization, after which net mineralization would decline and stabilize in the same manner as microbial activity. When roots had not inhibited microbial activity but the C/N ratio was high (Fig. 2B), net mineralization was expected to increase following the elimination of root activity, in response to the declining C/N ratio caused by respiration of CO_2. Small decreases in net mineralization just after the elimination of root activity might result from immobilization caused by dead roots remaining in the soil. Eventually, net mineralization would peak and then decline as the rate of microbial activity became the factor limiting net mineralization. In the case of low C/N organic matter, net mineralization is expected to follow changes in microbial activity (Fig. 2C and D). Where microbial activity had been inhibited, the increase in net mineralization might be somewhat delayed as a result of increased immobilization by the expanding microbial populations, but the delay would be short and would be difficult to detect because of the initially rapid rate of net mineralization.

The four idealized sets of trends described above (Fig. 2) will be modified by specific experimental and forest management techniques.

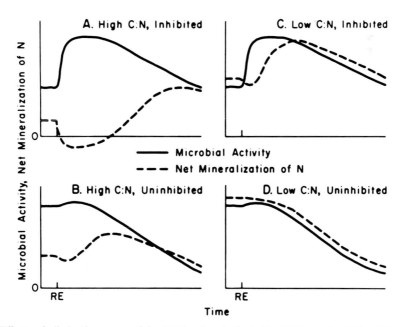

FIG. 2. Effects of eliminating root activity (at the time indicated by RE) on microbial activity and the net mineralization of nitrogen. The four conditions refer to the C/N ratio of root organic substrate and whether microbial activity was inhibited or uninhibited by root activity prior to root elimination. (Modified from Fisher, Ph.D. thesis.)

Possibly the most important is the severed roots resulting from trenching or clear-cutting which cause increased microbial activity regardless of whether microbial activity had been inhibited when the roots were living and active.

IMPLICATIONS

Inspection of the models in Fig. 2 indicates that the timing of sample collections is very important, particularly if microbial activity was inhibited (Fig. 2A and C). For example, collecting only a single set of samples after root activity was eliminated could indicate any of the four combinations of increased or decreased microbial activity and net mineralization. These data would indicate a true relationship, but it would be an oversimplification. Ideally, a series of samples is needed to follow the changes shown in Fig. 2.

The failure of trenched plots and similar perturbations to indicate, at least initially, the roots' contributions of organic matter may be responsible for the conflict between the agricultural and forest studies. As discussed previously, recent studies have consistently indicated that root mortality is a major source of organic matter. This would eventually be evident in trenched plots by reduced microbial activity (Fig. 2).

In forests, complete utilization of the organic matter may be prevented by the activities of roots (Fig. 2A and C), and moisture uptake is one important mechanism. Gadgil and Gadgil's (12) growth chamber experiment indicates that mycorrhizae can potentially inhibit microbial processes by other mechanisms.

Some studies of the effects of roots on soil processes in forests remain difficult to interpret. Gadgil and Gadgil (11, 12) and Romel (cited in 12) found decreased amounts of litter in trenched plots. However, the lower weight could result either from increased microbial activity or decreased production of organic matter in the absence of root activity. Gadgil and Gadgil (11, 12) found that removing dead roots at the initiation of the experiment had little effect. However, since the standing crop of fine roots and mycorrhizae may turn over several times a year, the biomass present at any one time is not indicative of the actual production of organic matter. Usually, a series of samples combined with studies of root decomposition are needed to estimate accurately the role of root or mycorrhizal mortality (9, 29). Additionally, the ecosystem-level significance of needle litter can no longer be taken for granted in coniferous forests in view of Fogel and Hunt's (10) finding that the contribution of aboveground litter to the substrate available to decomposers was only 10% of the total in an Oregon Douglas fir forest. Mattson and Vitousek (28) found variable responses

in net mineralization in several Indiana deciduous forest sites at different lengths of time after clear-cutting, as would be predicted from Fig. 2. However, the rapid rate of soil processes in these forests (46) means that frequent sampling immediately after clear-cutting would probably be necessary to test Fig. 2. An additional confounding effect was the regrowth of vegetation occurring in the sites. Johnson and Edwards' (24) study provides an excellent example of the comparative use of soil incubations, but it is difficult to interpret because the exact effects of their perturbation, stem girdling, on root activity are unknown. However, this study clearly indicates the importance of roots in net mineralization and microbial activity. Stem girdling caused both CO_2 evolution and net mineralization to increase in incubated soils despite the fact that aboveground litterfall on the girdled sites had been reduced by about 24%.

The reduction of net mineralization by plant roots is somewhat counterintuitive. However, production of high C/N organic matter by plant roots would produce a tightly regulated nitrogen cycle. Losses of N would be reduced by immobilization, yet nitrogen would remain in a relatively available form, microbial tissue. Heterotrophic microorganisms often out-compete nitrifying bacteria for ammonium (24, 35, 36), and the resulting absence of nitrate reduces leaching losses and prevents denitrification. Although litter with high C/N ratios is also produced aboveground, its production by roots and mycorrhizae seems particularly important for several reasons: (i) Root and mycorrhizal mortality produces equal or greater quantities of organic matter than aboveground litterfall. (ii) The organic matter produced by roots and mycorrhizae is of relatively greater importance in deeper horizons where leaching would remove N from the rooting zone. (iii) Oxygen concentrations are lower in deeper horizons, so denitrification is more likely. (iv) The maximum production of organic matter by roots and mycorrhizae (mortality) is episodic and therefore out of phase with peak rates of plant nutrient assimilation. Conversely, maximum N availability coincides with assimilation. The production of organic matter with high C/N ratios and the resulting immobilization by heterotrophic microorganisms can explain Gosz's (14) observations that a variety of forest ecosystems in New Mexico appeared to be accumulating nitrogen regardless of successional stage, productivity, or biomass increment. It can also explain the tendency of nitrogen added in forest fertilization experiments to remain in organic fractions of the forest floor and mineral soil (25).

A possible benefit to plants from the inhibition of microbial activity by plant roots is suggested by soil sterilization experiments (19, 21, 22, 33, 48) which have shown increased inorganic nitrogen concentrations, availability to plants, and net mineralization rates after partial sterilization. Jenkinson (19, 21) demonstrated that the major cause of these changes is the mineralization by the remaining organisms of killed microbial tissue. Episodic wetting and drying acts in this manner, and its frequency could be increased by roots. In most soils, the increase in net mineralization occurs during the flush of microbial activity that follows rewetting, but the opposite might occur in soils with a very high C/N ratio which would normally immobilize N during rewetting. (See 25 and 46 for examples.) It is possible that decreases in microbial activity brought about by actively growing roots could increase N availability temporarily in these soils if the remaining microbes could utilize the newly available microbial tissue. The sterilization studies mentioned above are not directly applicable in this case since most of the increased nutrient availability occurs during increasing rather than decreasing microbial activity. We know of no studies directly addressing changes in N availability to plants as microbial populations decrease. The mechanisms likely to produce these decreases are those coinciding with root growth: moisture uptake, nutrient uptake, and production of toxins. The production of complexing agents would have little benefit in this situation since they do not act directly on the organisms but reduce resource availability instead.

The accumulation of undecomposed high C/N organic matter in forests where microbial activity was inhibited by roots would create a large potential microbial demand for N. These conditions would seem to favor the occurrence of nitrogen fixation, but for unknown reasons it is rare in mature temperate forests (17). The result is low loss of nutrients but at the cost of very low nutrient availability. External inputs of inorganic N from precipitation, impaction, etc., would also tend to end up in microbial tissue. Gosz (15) has suggested that the production of complexing agents increases in response to low nitrogen availability. In situations where there is a large potential microbial demand for N, complexing agents, by reducing organic substrate availability, might reduce the proportion of external N inputs going to microbes and therefore increase N availability to plants.

The accumulation of undecomposed high C/N organic matter can explain the extended periods of low N loss from trenched plots and low rates of net mineralization in incubations observed by Johnson et al. (25) and Vitousek et al. (46). Vitousek et al. (46) compared 17 forest sites throughout the United States and noted an association between recalcitrant organic matter (possibly resulting from complexing agents), high C/N ratios in litter, and low rates of net mineralization.

CONCLUSIONS

The production of litter with high C/N ratios can create a tightly regulated nitrogen cycle through the immobilization of inorganic nitrogen by heterotrophic microorganisms and the associated reduction in nitrification, denitrification, and leaching. Roots may be more important than aboveground litter since they may produce greater quantities of organic matter and they are located so that they have greater effects on processes leading to nitrogen losses. Their greatest effect may occur following root mortality, a period of minimum nutrient assimilation by plants. This could occur if some microbial activity continued during the conditions preventing root growth and uptake. Inhibition of microbial activity by roots may further tighten the nitrogen cycle by causing litter with high C/N ratios to accumulate, creating a potential microbial demand for N. Under these conditions, inhibition of microbial activity may temporarily increase N availability by making N-rich microbial tissue available to decomposers and by reducing the immobilization of N. The reduction or elimination of root activity can reverse many of these trends; however, the results can be varied both between different sites and with time at a given site. Quantifying the influence of roots on microbial processes has the potential to explain much of the variation reported for nutrient losses following clear-cutting.

LITERATURE CITED

1. **Babel, U.** 1977. Influence of high densities of fine roots of Norway spruce on processes in humus covers. Ecol. Bull. (Stockholm) **25**:584–586.
2. **Bartholomew, W. V., and F. E. Clark.** 1950. Nitrogen transformations in soil in relation to rhizosphere microflora, p. 112–113. In Fourth International Conference of Soil Science (Amsterdam) Transactions, vol. 2. Hoitsema Bros., Groningen, The Netherlands.
3. **Black, C. A.** 1968. Soil-plant relationships. John Wiley & Sons, Inc., New York.
4. **Cates, R. G., and D. F. Rhoades.** 1977. Patterns in the production of antiherbivore chemical defenses in plant communities. Biochem. Syst. Ecol. **5**:185–193.
5. **Charley, J. L.** 1972. The role of shrubs in nutrient cycling, p. 182–203. In C. M. McKell, J. P. Blaisdell, and J. R. Goodwin (ed.), Wildland shrubs—their biology and utilization. Gen. Tech. Rep. INT-1, U.S. Department of Agriculture Forest Service, Washington, D.C.
6. **Coleman, D. C., C. V. Cole, H. W. Hunt, and D. A. Klein.** 1978. Trophic interactions in soils as they affect energy and nutrient dynamics. I. Introduction. Microb. Ecol. **4**:345–349.
7. **Crawford, C. S., and J. R. Gosz.** 1982. Desert ecosystems: their resources in space and time. Environ. Conserv. **9**:181–195.
8. **Edwards, N. T., and W. F. Harris.** 1977. Carbon cycling

in a mixed deciduous forest floor. Ecology **58**:431–437.

9. **Fogel, R.** 1980. Mycorrhizae and nutrient cycling in natural forest ecosystems. New Phytol. **86**:199–212.

10. **Fogel, R., and G. Hunt.** 1979. Fungal and arboreal biomass in a western Oregon Douglas-fir ecosystem: distribution patterns and turnover. Can. J. For. Res. **9**:245–256.

11. **Gadgil, R. L., and P. D. Gadgil.** 1971. Mycorrhiza and litter decomposition. Nature (London) **233**:133.

12. **Gadgil, R. L., and P. D. Gadgil.** 1975. Suppression of litter decomposition by mycorrhizal roots of *Pinus radiata*. N. Z. J. For. Sci. **5**:33–41.

13. **Goring, C. A. I., and F. E. Clark.** 1949. Influence of crop growth on mineralization of nitrogen in the soil. Soil Sci. Soc. America Proc. **13**:261–266.

14. **Gosz, J. R.** 1980. Nutrient budget studies for forests along an elevational gradient in New Mexico. Ecology **61**:515–521.

15. **Gosz, J. R.** 1981. Nitrogen cycling in coniferous ecosystems. Ecol. Bull. (Stockholm) **33**:405–426.

16. **Gosz, J. R., G. E. Likens, and F. H. Bormann.** 1973. Nutrient release from decomposing leaf and branch litter in the Hubbard Brook Forest, New Hampshire. Ecol. Monogr. **43**:173–191.

17. **Gutschick, V. P.** 1981. Evolved strategies in nitrogen acquisition by plants. Am. Nat. **118**:607–637.

18. **Jeffrey, W. W.** 1970. Hydrology of land use, p. 13.1–13.57. *In* D. M. Gray (ed.), Principles of hydrology. Water Information Center, Huntington, N.Y.

19. **Jenkinson, D. S.** 1966. Studies on the decomposition of plant material in soil. II. Partial sterilization of soil and the soil biomass. J. Soil Sci. **17**:280–302.

20. **Jenkinson, D. S.** 1971. Studies on the decomposition of C labelled organic matter in soil. Soil Sci. **111**:64–70.

21. **Jenkinson, D. S.** 1976. The effects of biocidal treatments on metabolism in soil. IV. The decomposition of fumigated organisms in soil. Soil Biol. Biochem. **8**:203–208.

22. **Jenkinson, D. S., and D. S. Powlson.** 1976. The effects of biocidal treatments on metabolism in soil. I. Fumigation with chloroform. Soil Biol. Biochem. **8**:167–177.

23. **Johnson, D. W., and D. W. Cole.** 1980. Anion mobility in soils: relevance to nutrient transport from terrestrial ecosystems. Environ. Int. **3**:79–90.

24. **Johnson, D. W., and N. T. Edwards.** 1979. The effects of stem girdling on biogeochemical cycles within a mixed deciduous forest stand in eastern Tennessee. II. Soil nitrogen mineralization and nitrification rates. Oecologia (Berlin) **40**:259–271.

25. **Johnson, D. W., N. T. Edwards, and D. E. Todd.** 1980. Nitrogen mineralization, immobilization and nitrification following urea fertilization of a forest soil under field and laboratory conditions. Soil Sci. Soc. Am. J. **44**:610–616.

26. **Lamb, D.** 1975. Patterns of nitrogen mineralization in the forest floor of stands of *Pinus radiata* on different soils. J. Ecol. **63**:615–625.

27. **Marx, D. H.** 1973. Mycorrhizae and feeder root diseases, p. 351–382. *In* G. C. Marks and T. T. Kozlowski (ed.), Ectomycorrhizae, their ecology and physiology. Academic Press, Inc., New York.

28. **Mattson, P. A., and P. M. Vitousek.** 1981. Nitrogen mineralization and nitrification potentials following clearcutting in the Hoosier National Forest, Indiana. For. Sci. **27**:781–791.

29. **McClaugherty, D. A., J. D. Aber, and J. M. Melillo.** 1982. The role of fine roots in the organic matter and

nitrogen budgets of two forested ecosystems. Ecology **63**:1481–1490.

30. **Nye, P. H., and P. B. Tinker.** 1977. Solute movement in the soil-root system. University of California, Berkeley.

31. **Parnas, H.** 1976. A theoretical explanation of the priming effect based on microbial growth with two limiting substrates. Soil Biol. Biochem. **8**:139–144.

32. **Persson, H.** 1978. Root dynamics in a young Scots pine stand in central Sweden. Oikos **30**:508–519.

33. **Powlson, D. S., and D. S. Jenkinson.** 1976. The effects of biocidal treatments on metabolism in soil. II. Gamma irradiation, autoclaving, air-drying and fumigation. Soil Biol. Biochem. **8**:179–188.

34. **Rambelli, A.** 1973. The rhizosphere of mycorrhizae, p. 299–343. *In* G. C. Marks and T. T. Kozlowski (ed.), Ectomycorrhizae: their ecology and physiology. Academic Press, Inc., New York.

35. **Robertson, G. P., and P. M. Vitousek.** 1981. Nitrification potentials in primary and secondary succession. Ecology **62**:376–386.

36. **Robertson, G. P., and P. M. Vitousek.** 1982. Factors regulating nitrification in primary and secondary succession. Ecology **63**:1561–1573.

37. **Rovira, A. D.** 1965. Interactions between plant roots and soil microorganisms. Annu. Rev. Microbiol. **19**:241–266.

38. **Rovira, A. D.** 1969. Plant root exudates. Bot. Rev. **35**:35–57.

39. **Rovira, A. D., and B. M. McDougall.** 1967. Microbiological and biochemical aspects of the rhizosphere, p. 417–463. *In* A. C. McLaren and G. H. Peterson (ed.), Soil biochemistry. Marcel Dekker, New York.

40. **Smith, W. H.** 1976. Character and significance of forest tree root exudates. Ecology **57**:324–331.

41. **Starkey, R. L.** 1958. Interrelations between microorganisms and plant roots in the rhizosphere. Bacteriol. Rev. **22**:154–172.

42. **Stotzky, G.** 1972. Activity, ecology, and population dynamics of microorganisms in soil. Crit. Rev. Microbiol. **2**:49–137.

43. **Stotzky, G., and A. G. Norman.** 1961. Factors limiting microbial activities in soil. I. The level of substrate, nitrogen, and phosphorus. Arch. Mikrobiol. **40**:341–369.

44. **Swift, M. J., O. W. Heal, and J. M. Anderson.** 1979. Decomposition in terrestrial ecosystems. University of California, Berkeley.

45. **Trappe, J. M., and R. Fogel.** 1977. Ecosystematic functions of mycorrhizae, p. 205–214. *In* J. K. Marshall (ed.), The below-ground ecosystem: a synthesis of plant-associated processes. Range Science Department Science Series No. 26. Colorado State University, Fort Collins.

46. **Vitousek, P. M., J. R. Gosz, C. C. Grier, J. M. Melillo, and W. A. Reiners.** 1982. A comparative analysis of potential nitrification and nitrate mobility in forest ecosystems. Ecol. Monogr. **52**:155–177.

47. **Vitousek, P. M., and J. M. Melillo.** 1979. Nitrate losses from disturbed forests: patterns and mechanisms. For. Sci. **25**:605–619.

48. **Will, G. M.** 1962. The uptake of nutrients from sterilized forest-nursery soils. N. Z. J. Agric. Res. **5**:425–432.

49. **Will, G. M.** 1968. Some aspects of organic matter formation and decomposition in pumice soils growing *Pinus radiata* forest, p. 237–246. *In* Ninth International Congress of Soil Science Transactions, vol. 3. American Elsevier Publishing Co., New York.

Catastrophic Disturbances to Stream Ecosystems: Volcanism and Clear-Cut Logging

JAMES R. SEDELL AND CLIFFORD N. DAHM

Forestry Sciences Laboratory, U.S. Forest Service, and Department of Fisheries and Wildlife, Oregon State University, Corvallis, Oregon 97331

Clear-cutting and fire are major forms of disturbance presently affecting streams in forested lands. These events have recurrence intervals on the order of decades to a few centuries. Repeated volcanic eruptions also occur in tectonically active zones such as the Pacific Northwest on time scales of centuries to millenia. Catastrophic disturbances, such as volcanism or clear-cut logging, can affect stream ecosystems in a variety of ways. Historically, research on the impacts of disturbances has been conducted on the basis of their on-site short-term impact rather than within the context of integrated, basin-level analyses. Even local impacts have not been adequately assessed from a stream ecosystem perspective. In most cases, physical and chemical attributes of water and some biotic parameters are narrowly emphasized, while other important determinants of the stream ecosystem are neglected.

Two of the most neglected components in stream ecosystem research are physical habitat and microbial processes within the sediment. In this paper, we outline the importance of physical habitat and how it is altered by catastrophic disturbance. The interaction of organic materials with physical habitat determines the kinds and rates of microbial processes in streams. We will confine our examples to the Pacific Northwest of the United States. However, these principles apply generally to forested streams everywhere. Our goal is to present a context for sampling and interpreting microbial activity in old-growth, recently clear-cut, and volcanically impacted watersheds.

CHARACTERISTICS OF UNMODIFIED FORESTED STREAMS

Unmodified streams in forests contain large quantities of large organic debris, ranging from 10 to 60 kg m^{-2} (2, 6, 12, 13, 15, 20, 23, 28, 31, 32). This large organic debris shapes the stream channel by serving as dams, as temporary storage sites for sediments, organic materials, and water, and as large roughness elements causing formation of pools. Big wood in streams creates a diversity of stream habitats (Table 1). In the smallest streams in old-growth forests of the western Cascades, over 50% of the habitat is related to large wood (1, 27, 32). In larger third-order streams about 25% is created and maintained by wood. Small- to intermediate-order streams in the Panhandle National Forest of northern Idaho with gradients of 1 to 6% have 80% of the pools formed by wood (R. Rainville, personal communication).

The food base or energy supply of a stream in a forested watershed comes mainly from litter from the adjacent forest combined with algal production where high light intensity reaches the streambed. Pristine streams are also highly retentive of terrestrial organic inputs, retaining over 70% long enough for biological processing by stream organisms (19, 25, 32). The influence of the forest on energy sources and channel structure diminishes as a stream gets larger. However, the edges of a natural stream are still dominated by forest vegetation lining the banks and creating and maintaining side channels and small backwater areas, prime sites for deposition and storage of organic materials (23). Undisturbed streamside forests typically have an understory of herbaceous and shrubby plants with light gaps of various sizes. This provides the stream with a mix of coniferous and deciduous litter as well as patches of algal production. This diversity of food and habitat provides for a rich mix of species of vertebrates and invertebrates, with a full complement of age classes within species.

In addition to storing large quantities of wood, the primal stream efficiently retains smaller organic inputs and has numerous deep pools and extensive riparian vegetation. Seasonal increases in water volume result in lateral expansion rather than increases in depth because of the wood-obstructed channels. Water spreads outward and is slowed by the floodplain vegetation. This fringe floodplain in small streams helps set up many shallow littoral zones which represent a considerable proportion of the total area of the aquatic system and greatly affects productivity patterns. A motto among river biologists, "where there is no floodplain there are few fish" (38), is useful also to microbiologists because the diversity, intensity, and areal extent of aquatic microbial processes increases through floodplain interactions.

TABLE 1. Physical characteristics and disturbance impacts on small Pacific Northwest streams (1 to 3 Strahler stream orders) in old-growth forests, recent clear-cuts, and those experiencing debris torrents

Location	Longitudinal profile	Habitat diversity	Organic matter storage potential	Quantities of downed trees	Shading	High water floodplain interaction
Old-growth forest						
High gradient (>4%)	Stepped	High diversity of velocities	High	High, 10–60 kg m^{-2}	Well shaded with numerous light gaps	Narrow fringe but high interaction
Low gradient (<4%)	Stepped	High diversity of velocities	High	High, 10–20 kg m^{-2}	Heavily shaded	Very extensive
Recent clear-cut						
High gradient (>4%)	Some steps of boulders or bedrock	Moderate	Low-moderate	<5 kg m^{-2}	Little shading	Very limited
Low gradient (<4%)	Even grade; breaks dependent on local geology, bedrock, or ponded sand or gravels	Low	Low	<2 kg m^{-2}	Little shading, high light exposure	Low-moderate interaction
Stream experiencing debris torrent						
High gradient (>4%)	Even grade broken only by sediment deposit	Low	Low; except at deposit sites	Very low except for deposit sites	Moderate; streamside vegetation scoured out	None
Low gradient (<4%)	Even grade bedrock or aggraded with gravels	Low	Organic storage on the floodplain; low in channel storage	Moderate amounts on floodplains 2–10 kg m^{-2}	Moderate; streamside vegetation drowned	Low; can be extensive on highly aggraded streams

The degree of floodplain influence is complex and largely determined by the extent, timing, frequency, and duration of water exchange between the stream and adjacent riparian floodplains (16, 38). Overbank flow is a natural process which builds floodplain features such as natural levees and supplies water to adjacent lowlands that serve as storage sites for excess runoff (21). Water velocities are greatly reduced during flood events in the floodplain relative to the main stream channel, because of shallower flow and greater streambed roughness. As a result, considerable deposition of sediment, organic material, and nutrients occurs in these areas. Alluvial floodplains are a sink for nitrogen, carbon, and phosphorus and are processing areas for organic matter. In addition, partially processed fragmented detritus and dissolved organic material are washed back into the main channels so that streams with complex adjacent wetlands tend to carry more organic matter than those without such features (7, 17).

CATASTROPHIC DISTURBANCES

Human manipulation. Streams throughout North America have been systematically cleaned of snags and organic debris for more than 150 years (23, 24). For example, from the middle 1800s to about 1920 large- and intermediate-sized rivers in the Pacific Northwest were cleared of drift jams and snags so steamboats and rafts could navigate the rivers, transporting supplies and agricultural products. From the 1880s to 1915 small rivers and streams were used to transport logs out of the woods to the mills. The streams had to be cleaned of debris before the logs could be driven. Many streams had several expansive splash dams on them to augment the flow enough to drive logs (24). In the 1940s and 1950s, organic debris was a big problem in Oregon and Washington streams, as streambeds were used as logging roads and harvested trees were pulled into them. These management activities have resulted in a long-term loss of habitat diversity and carbon storage, and a reduction in floodplain interaction.

Clear-cutting and timber management activities are relatively recent factors altering the structure and organic matter in streams. Clear-cut logging normally results in an increase in stream runoff, water temperature, sediment inputs, and light, and in decreases in litterfall, carbon storage, and habitat-forming large wood (Table 1). Clear-cutting is often followed by overzealous cleanup of organic debris from the stream, undertaken to "protect the fisheries" by removal of potential migration barriers. In addition, as road building associated with logging operations extends ever further into steep country, the probability of channel-scouring debris torrents increases. Debris torrents, initiated by small landslides, move catastrophically down stream channels, severely eroding the streambed. These torrents routinely cause an abrupt release of stored sediment and organic debris which eventually clumps into one spot, often scouring the upstream channel down to bedrock (Fig. 1). In one Coast Range watershed in Oregon, 30% of the first-order streams are scoured to bedrock, and 60% of the second-order streams and 40% of the third-order streams have experienced debris torrents (27). These torrents have scoured the channels and left large deposits of organic material and inorganic sediment where the gradient flattened to about 4% (27) or the tributary entered a larger stream at an angle greater than 40° (L. Benda, personnal communication). In excess of 80% of these debris torrents were caused by logging activity or road failures (27). Although we tend to talk of localized impacts of clear-cutting, debris torrents can travel 1 km or more and affect several stream orders, severely reducing the capability of the stream for dispersed retention of organic material. The most active zones of microbial and biological processing then become clumped throughout the river basin instead of being more uniformly distributed throughout the drainage network (Fig. 1).

Longitudinal profiles of disturbed and undisturbed streams are illustrated in Fig. 1. Streams in old-growth forests without recent torrents maintain a stairstepped profile throughout the length of the stream. Streams experiencing debris torrents either have clumped depositional areas and a few large steps or have been altogether sluiced out and have only steps that are topographically provided by bedrock outcrops. When the depositional areas are examined in cross section or in planar view, a clearer picture emerges of the capacity for the channels to store carbon (Fig. 1). In the streams of old-growth forest, the extent of deep pools, pools at the stream margin, and side channels at low flow are commonly controlled by large wood. Where deposition and storage of organic material are concentrated, a complex mixture of aerobic and anaerobic processes degrades the organic matter. Diverse microbial processes occur throughout the stream. In streams which have been recipients of debris torrents within the past 20 years, large, widely scattered deposits of organics and sediments provide centers for microbial processing and remineralization. These environments are fewer but have larger concentrations of organic material. The small, scoured streams have microbial processes restricted mainly to algal development and decomposition. Organic material is not retained long enough nor in

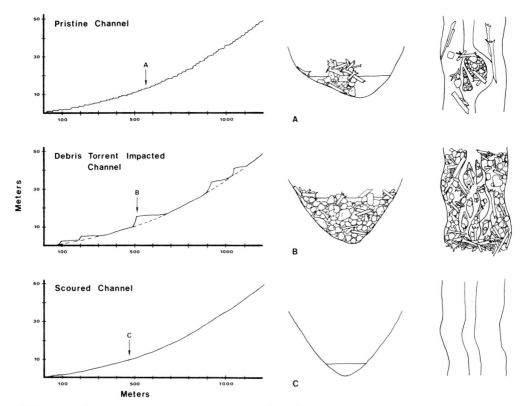

FIG. 1. Longitudinal, cross-sectional, and areal profiles of three types of stream channels. (A) Pristine undisturbed channel. (B) Debris torrent-affected channel. (C) Scoured channel. The arrows indicate the point on the longitudinal profiles depicted in the cross-sectional and planar views. Channel A has many small steps or depositional zones, whereas channel B has a few highly clumped areas of sediment and debris. Channel C has been scoured to bedrock, and few or no depositional sites remain.

adequate quantity for anaerobic processes to occur at the rates found in old-growth forested streams or within debris concentrations.

Other common forestry practices also reduce debris loading and storage relative to the undisturbed stream ecosystem. Thinning and harvest rotations of 60 to 100 years remove the future source of large debris, cutting off resupply. Management of streamsides for timber production alone totally eliminates the source of large debris. Growing demand for wood fiber also encourages the leaving of less forest residue and the exploitation of additional tree species (31).

The culmination of almost two centuries of human management and exploitation has shaped streams into ecosystems often distantly removed from primeval conditions. Erosion has increased, and debris torrents are more common. Many sections of stream have reduced storage capacity for organic material, and other sections have been inundated with inorganic sediments, lowering the overall carbon available per length of stream. Floodplains and the volume of surface water within basins grow smaller

and organic-rich deposits of sediment diminish.

Volcanism. On 18 May 1980, Mt. St. Helens in southwestern Washington erupted violently. Four major zones of impact were associated with the eruption (Table 2). First, a massive debris avalanche filled the upper 26 km of the North Toutle River to depths between 10 and 195 m. This event dammed numerous streams and eliminated the drainage network in the valley. Second, mudflows associated with snow and glacial ice melt moved down four of the major drainages of the mountain. These mudflows scoured the valley bottom, and mud deposits killed many trees along the outer edges of the floodplain. The mudflows backed into and dammed most of the tributaries of the main channels. Mudflow levees also trapped ash and pumice, which were being exported down adjacent tributaries. By summer of 1981, many tributaries had ash deposits filling their valleys for 3 to 6 km above the confluence with the mudflow. This resulted in wide, shallow, very exposed channels of sand-sized sediments. Third, in the blast zone, which includes 480 km^2 of devastat-

TABLE 2. Extent and characteristics of volcanic events and effects on channel geometry, shading, organic matter storage, and habitat in streams affected by the 1980 eruptions of Mt. St. Helens (18; F. Swanson, personal communication)

Volcanic event	Affected area (km²)	Deposit thickness (m)	Deposit emplacement temp (°C)	Channel geometry	Channel shading	Organic matter storage potential	Habitat diversity
Debris avalanche	60	10–195	70–100	Total destruction	None	Low; dependent on boulders or rock outcrops	Low
Blast							
Downed vegetation	370	0.08–1	100–300	Low-moderate changes	Low	Very high as a result of down trees in channel	High
Seared vegetation	110	0.02–0.08	50–250	Slight local changes	Low-moderate	Moderate	Moderate
Mudflows	50	Mainly 2, up to 30	32	Moderately to highly altered	None	Low	Low
Ashfall	950	0.03	30 (?)	No major changes	Heavy shading	Moderate-high	Moderate-high

ed forest and seared vegetation, the streams were not only choked with ash and pumice, but also inundated with trees blown down by the blast. The downed and shattered trees and foliage exerted a major influence on the routing and storage of ash eroding from hillsides. In the seared vegetation zone much of the foliage dropped within a few months of the major eruption. Leachate from the organic debris and the litterfall provided large quantities of organic energy to the streams. Fourth, the greatest areal impact of the eruption was the ash or tephra-fall zone. In these streams the pristine channels in the old-growth forests were inundated with ashfalls of 5 to 20 cm. However, nearly all of the riparian vegetation survived, and channel structural features such as boulders and large downed trees were unaffected. The streams today look very much like they looked before the eruption. Dense stands of herbaceous and shrubby plants and vigorous conifers line the edges of the streams, and after the first winter the riffles had been swept clean of most of the ash. After the second winter the pools had been completely flushed and most of the ash had been deposited on the floodplain.

The streams in the blast zone experienced large inputs of both fine sediments and organic carbon. Streams with downed trees experienced no significant widening of the channel, and the channels have started to grow deeper. In reaches with heavy tree loading, much of the ash has been deposited on the floodplain or transported downstream. Over 90% of the pools are created or significantly enhanced by the large wood. Streamflow around the downed trees scours the ash, and the pre-blast cobble-gravel substrates are reappearing. Streamside shrubs and herbaceous plants are coming back quickly, but still provide just 5% of the pre-blast cover and even less carbon input after 3 years. Many of the blast-zone streams which were not salvage logged will have excellent habitat for storing and processing organic materials. In the areas without downed trees, the riffle sections have become longer, the frequency of pools has decreased, and side channels and backwaters have become filled with sediments.

In the long term (50 to 100 years), streams which had their dead riparian forests salvage logged will show little increase in carbon storage capacity. Those streams in which the dead streamside forest is left will have large areas where terrestrial carbon will be stored and be processed. Our ideas result from examining streams on nearby Mt. Rainier and Mt. Hood that had mudflows fill their valleys 150 to 250 years ago, killing the riparian forest. Because of the uniformly flat valley floor and fallen trees from the pre-mudflow forest, these streams contain large quantities of stored carbon associated with downed wood both on the floodplain and in stream channels. Downed trees from the pre-blast forest will play a major role in the restoration and maintenance of retention sites for organic matter. The channels in the valleys that received the impact of the debris avalanche and mudflow remain highly unstable and migrate through the valley annually. Little vegetation has reappeared, and sediment loads are extremely high. Sites of long-term storage of carbon and sediment are presently confined to the floodplain, and new sources of terrestrial organic matter will require decades to gain a foothold.

MICROBIAL PROCESSES

The microbial response to major disturbance within streams must be considered together with the changes in the physical structure and also the chemical composition of stream water and sediment. Important physical changes we have discussed are (i) the quantity of carbon and nitrogen stored within the channel, (ii) the amount of light reaching the stream, (iii) the retentive capacity of the channel, and (iv) the quantity of inorganic sediment eroded into the stream network. Major chemical shifts include (i) increased concentration of dissolved nitrogen and phosphorus compounds in the stream water following clear-cutting (3, 14; C. N. Dahm, Ph.D. thesis, Oregon State University, Corvallis, 1980) and (ii) the amount of allochthonous particulate carbon and nitrogen entering the stream. An understanding of the physical and chemical consequences of catastrophic disturbance provides the basis for evaluating the response of the microbiota. We will limit our discussion to microbial rates within major components of the nitrogen and carbon cycle in Pacific Northwest forested streams following clear-cutting and cataclysmic volcanic eruptions. However, the rates of all microbial processes in stream ecosystems are closely intertwined with the dynamic physical and chemical changes brought about by a major disturbance, and this interplay should be considered in all studies of stream microbiology.

Measurements of algal primary production, respiration, nitrogen fixation, nitrification, and denitrification have been made in streams from an old-growth (350 to 550 year old) Douglas fir, western hemlock forest, a recent clear-cut, a stream on the main debris flow from Mt. St. Helens, and from the tree blowdown zone within the blast zone of Mt. St. Helens (Table 3).

Nitrogen cycle processes were emphasized as a result of the generally low concentrations of inorganic nitrogen in streams in the Pacific Northwest (S. V. Gregory, Ph.D. thesis, Oregon State University, Corvallis, 1980) and the apparent initial nitrogen limitation in streams and lakes throughout the blast zone of Mt. St. Helens (4, 11).

Algal primary production is often increased in smaller streams following a disturbance which opens the canopy and allows additional light to reach the streams (Gregory, Ph.D. thesis, Oregon State University, 1980). Streams in recent clear-cuts and many of the higher gradient streams near Mt. St. Helens now produce algal biomass during low flow conditions that is much larger than that in streams in old-growth forests. A comparison of net algal primary production in a third-order stream within an old-growth forest and a third-order stream within a recent clear-cut showed an eightfold higher rate in the clear-cut (37). This trend also holds for many streams affected by the eruption of Mt. St. Helens, except where highly unstable streambeds scour substrates and high turbidity limits light penetration, eliminating photosynthetic activity. In general, except in highly unstable geomorphic situations, disturbance of streamside vegetation increases microbial photosynthetic production, particularly during summer base flow.

The overall effect of disturbance on heterotrophic activity in streams is more difficult to assess. Respiratory activity in streams within the blast zone of Mt. St. Helens was exceptionally high in the first year after the eruption. Dissolved organic carbon extracted from the devastated forest by the hot blast deposits fueled heterotrophic bacteria (4). This activity has diminished markedly since 1980, and the extremely erosive channels now store carbon mainly in the lower gradient depositional areas and where

TABLE 3. Relative measured or predicted response to disturbance of microbial carbon and nitrogen cycle processes in small streams of the Pacific Northwest

Stream description	Algal primary production	Respiration	N₂ fixation	Nitrification	Denitrification
Old-growth watershed, Oregon Cascades	Low	Moderate-high	Low	Low	Moderate-high
Recent clear-cut, Oregon Cascades	High	Low-moderate	Moderate	Low	Moderate
Debris avalanche, Mt. St. Helens	Low	Initially high, now low	Initially high, now low	Initially high, now low	Initially high, now low
Blast zone, downed vegetation, Mt. St. Helens	Moderate	Initially high, now low	Initially high, now low	Initially moderate, now low	Initially high, now moderate
Ashfall, Mt. St. Helens	Low-moderate	Low-moderate	Low	Low	Low

stable accumulations of large woody debris occur. Respiratory activity in the sediment is now minimal in many streams near the volcano. This is probably a result of inorganic sediment inundation and flushing of fine organic material downstream or onto the floodplain. Over 90% of the erosion of ash from hillslopes occurred during the first winter (F. J. Swanson, B. Collins, T. Dunne, and B. P. Wicherski, *Erosion Control in Volcanic Areas*, in press). A similar but less extreme pattern is repeated in clear-cut watersheds where an initial stimulative response is followed by decreased overall input and storage of carbon to the small streams and a new decrease in stream respiration.

Rates of nitrogen (N_2) fixation in small streams flowing from old-growth forests are generally quite low (8). N_2 fixation is an expensive process energetically, requiring either sunlight or a source of chemical energy such as organic compounds or chemically reduced inorganic material. In old-growth forests, autotrophic N_2 fixation occurs mainly in the forest canopy through N_2-fixing lichens (9, 10). N_2 fixation in the stream is from bacterial heterotrophs, often on wood, and rates are low (0 to 1.3 μmol of N_2 g^{-1} of wood day^{-1}). After clear-cutting, N_2-fixing algal forms often become established. N_2-fixing forms of *Nostoc* sp. are especially prominent in clear-cuts of the Pacific Northwest. N_2-fixing algal species also appeared in thermal seeps throughout the blast zone of Mt. St. Helens within months of the main eruption (36). However, the unstable substrate has not favored widespread colonization of N_2-fixing algal forms in most streams in the blast zone of Mt. St. Helens. Levels of heterotrophic N_2 fixation in stream water have also been below detection ($\sim$0.1 nmol of N_2 $liter^{-1}$ day^{-1}) except for the extremely high rates (to 4.1 μmol of N_2 $liter^{-1}$ day^{-1}) the first few months after the eruption.

Numerous publications have addressed the losses of nitrate in drainage water from disturbed forest ecosystems (summarized in 34, 35). The numbers of nitrifying bacteria within the forest soil have been shown to increase after disturbances (26, 30), and rates of potential soil nitrification were strongly correlated with rates of nitrogen mineralization (22). Nitrification rates are regulated by the availability of ammonia and the concentration of dissolved oxygen (5). However, with the exception of highly polluted rivers and estuaries, few data have been published on potential or actual rates of nitrification in streams within forested watersheds.

Potential rates of nitrification within streams in the blast zone of Mt. St. Helens during 1980 and 1981 were high (up to 3.0 μg of NO_3-N $liter^{-1}$ day^{-1}). A dramatic increase in microbial numbers and activity was common to all aquatic environments in the devastated zone (4, 11). Stream waters, which remained aerobic from atmospheric resupply of oxygen, were primary sites for nitrification. Streams on the debris flows, where hot volcanic deposits were intermixed with forest soil and debris, were most active. This, however, was a transient phenomena, and the rates of nitrification decreased to low levels as the deposits cooled and as the leaching of ammonia diminished.

The effect of clear-cutting on nitrification rates in streams has not been as well documented as in the soils. Preliminary measurements in streams in a clear-cut, alder-dominated second growth and in an old-growth watershed of Oregon found low, relatively constant rates (0.015 to 0.072 μg of NO_3-N g^{-1} of sediment day^{-1}). Although greater rates of nitrification might be predicted in the clear-cut, an efficient nitrifying population within the soil could utilize the available ammonia and suppress activity within the stream. A 3-month average of twice-weekly ammonia measurements was less than 5 μg of NH_4-N $liter^{-1}$ for all three streams, but nitrate concentrations were highest in the clear-cut. Increased concentrations of nitrate within a stream do not require a direct link to an aquatic nitrifying population. More in situ measurements of nitrification potential associated with the sediments of forested streams are needed, in addition to the data on inorganic nitrogen concentrations after disturbance.

Denitrification can occur within stream sediments where accumulations of organic material and physical factors produce oxygen-deficient waters. Rates of denitrification are regulated by the nitrate and oxygen concentration, with lesser control exerted by temperature and the supply of dissolved carbohydrates (29, 33). Debris dams, side channels, beaver impoundments, pools, and periphyton communities provide habitats where this process may occur.

Potential denitrification rates in streams in the blowdown and debris-flow zones ranged from 36 to 290 μmol of N_2O $liter^{-1}$ day^{-1} during the first year after the eruption of Mt. St. Helens. A large, active denitrifying bacterial population was established. However, stream concentrations of nitrate were near the detection limit of 3 μg of NO_3-N $liter^{-1}$, and these potential rates may overestimate actual rates, although a large population capable of denitrifying was present. Streams in heavy ashfall areas had increased loads of inorganic sediment, but denitrification was not detected. Measurements of denitrification rates without nitrate amendment in sediments from an old-growth forested stream were three times those in a clear-cut forest stream (0.12 versus 0.34 μmol of N_2O g^{-1} of sediment day^{-1}). This was in spite of a higher concentration of nitrate in the clear-cut forest stream (135 versus 87 µg of NO_3-N liter^{-1}). However, the organic content of the old-growth sediment averaged 35% while the average was 5% in the clear-cut. Although the majority of discussions concerning nitrate increases in disturbed forest ecosystems after clear-cutting have dealt with increases in nitrification, greater rates of denitrification in old-growth forest streams may also contribute to a decreased loss of nitrate in mature forests. Physical stability, greater retention of organic inputs, and more stored organics favor denitrification within undisturbed high-gradient forest streams.

How do these overall changes in carbon and nitrogen cycle processes affect the small high-gradient streams of the Pacific Northwest after major disturbances such as clear-cutting and volcanic eruption? First, microbial response cannot be separated from the physical processes within the watershed that affect the stream. When inorganic sediment inundates the stream channel and both short- and long-term storage of carbon are impeded, the types, intensity, and location of microbial activity also change. Second, algal production, stimulated by increased light and nitrate input, becomes more important as a food resource for the stream community if stable substrate is available and scour from sediment movement is low. Third, nitrogen loss from the watershed, mainly as nitrate, can be reduced within old-growth forested watersheds both by decreased nitrification within the soil and by higher potential denitrification in the stream sediments.

We thank F. Swanson and L. Benda for helpful discussions and comments. J. Baross, A. Ward, R. Griffiths, R. Jones, S. Gregory, and D. Coffey-Flexner provided data, and J. Froggatt, M. Liljeberg, R. Kepler, and K. Fairchild helped with manuscript preparation.

This work was supported by the National Science Foundation under projects DEB-8111307 and DEB-8112455 and by the U.S. Forest Service under contracts PNW 80-277 and PNW 80-289.

LITERATURE CITED

1. **Anderson, N. H., and J. R. Sedell.** 1979. Detritus processing by macroinvertebrates in stream ecosystems. Annu. Rev. Entomol. **24**:351–377.
2. **Anderson, N. H., J. R. Sedell, L. M. Roberts, and F. J. Triska.** 1978. The role of aquatic invertebrates in processing of wood debris in coniferous forest streams. Am. Midl. Nat. **100**:64–82.
3. **Aubertin, G. M., and J. H. Patric.** 1974. Water quality after clearcutting in a small watershed in West Virginia. J. Environ. Qual. **3**:243–249.
4. **Baross, J. A., C. N. Dahm, A. K. Ward, M. D. Lilley, and J. R. Sedell.** 1982. Initial microbiological response in lakes to the Mt. St. Helens eruption. Nature (London) **296**:49–52.
5. **Belser, L. W.** 1979. Population ecology of nitrifying bacteria. Annu. Rev. Microbiol. **33**:309–333.
6. **Bilby, R. E., and G. E. Likens.** 1980. Importance of organic debris dams in the structure and function of stream ecosystems. Ecology **61**:1107–1113.
7. **Brown, S., M. M. Brinson, and A. E. Lugo.** 1978. Structure and function of riparian wetlands, p. 17–31. *In* Strategies for protection and management of floodplain wetlands and other riparian ecosystems. Symposium Proceedings U.S. Department of Agriculture Forest Service, Washington, D.C.
8. **Buckley, B. M., and F. J. Triska.** 1978. Presence and ecological role of nitrogen-fixing bacteria associated with wood decay in streams. Int. Ver. Theor. Angew. Limnol. Verh. **20**:1333–1339.
9. **Caldwell, B. A., C. Hagedorn, and W. C. Denison.** 1979. Bacterial ecology of an old-growth Douglas fir canopy. Microb. Ecol. **5**:91–103.
10. **Carroll, G. C.** 1981. Forest canopies: complex and independent subsystems, p. 87–107. *In* R. H. Waring (ed.), Forests: fresh perspectives from ecosystem analysis. Oregon State University Press, Corvallis.
11. **Dahm, C. N., J. A. Baross, A. K. Ward, M. D. Lilley, and J. R. Sedell.** 1983. Initial effects of the Mount St. Helens eruption on nitrogen cycle and related chemical processes in Ryan Lake. Appl. Environ. Microbiol. **45**:1633–1645.
12. **Franklin, J. F., K. Cromack, Jr., W. Denison, A. McKee, C. Maser, J. Sedell, F. Swanson, and G. Juday.** 1981. Ecological characteristics of old-growth Douglas-fir forests. General Technical Report PNW-118. U.S. Department of Agriculture Forest Service, Portland, Ore.
13. **Heede, B. H.** 1972. Influences of a forest on the hydraulic geometry of two mountain streams. Water Resour. Bull. **8**:523–530.
14. **Hobbie, J. E., and G. E. Likens.** 1973. Output of phosphorus, dissolved organic carbon, and fine particulate carbon from Hubbard Brook watersheds. Limnol. Oceanogr. **18**:734–742.
15. **Keller, E. A., and T. Tally.** 1979. Effects of large organic debris on channel form and fluvial processes in the coastal redwood environment, p. 169–197. *In* D. D. Rhodes and G. P. Williams (ed.), Adjustments of the fluvial system. Kendall/Hunt Publishing Co., Dubuque, Iowa.
16. **Kibby, H. V.** 1978. Effects of wetlands on water quality, p. 289–298. *In* Strategies for protection and management of floodplain wetlands and other riparian ecosystems. Symposium Proceedings U.S. Department of Agriculture Forest Service, Washington, D.C.
17. **Kuenzler, E. J., P. J. Mulholland, L. A. Ruley, and R. P. Sniften.** 1977. Water quality in North Carolina coastal plain streams and effects of channelization. University of North Carolina Water Research Institute, Raleigh.
18. **Lipman, R. W., and D. R. Mullineaux (ed.).** 1981. The 1980 eruptions of Mount St. Helens, Washington. Professional Paper 1250. U.S. Geological Survey, Washington, D.C.
19. **Naiman, R. J.** 1982. Characteristics of sediment and organic carbon export from pristine boreal forest watersheds. Can. J. Fish. Aquat. Sci. **39**:1699–1718.
20. **Naiman, R. J., and J. R. Sedell.** 1979. Benthic organic matter as a function of stream order in Oregon. Arch. Hydrobiol. **87**:404–422.
21. **Nunnally, N. R., and E. Keller.** 1979. Use of fluvial processes to minimize adverse effects of stream channelization, p. 79–144. University of North Carolina, Water Research Institute, Raleigh.
22. **Robertson, G. P., and P. M. Vitousek.** 1981. Nitrification potentials in primary and secondary succession. Ecology **62**:376–386.
23. **Sedell, J. R., F. H. Everest, and F. J. Swanson.** 1982. Fish habitat and streamside management: past and present, p. 244–255. *In* Proceedings of the Society of American Foresters. Society of American Foresters, Bethesda, Md.
24. **Sedell, J. R., and K. J. Luchessa.** 1982. Using the historical record as an aid to salmonid habitat enhancement, p. 210–223. *In* N. B. Armantrout (ed.), Proceedings of a Symposium on Acquisition and Utilization of Aquatic Habitat Inventory Information. American Fisheries Society, Bethesda, Md.

25. **Sedell, J. R., F. J. Triska, and B. M. Buckley.** 1975. The processing of coniferous and hardwood leaves in two coniferous forest streams. 1. Weight loss and associated invertebrates. Verh. Int. Verein. Limnol. **19:**617–627.

26. **Smith, W. H., F. H. Bormann, and G. E. Likens.** 1968. Response of chemoautotrophic nitrifiers to forest cutting. Soil Sci. **106:**471–473.

27. **Swanson, F. J., and G. W. Lienkaemper.** 1978. Physical consequences of large organic debris in Pacific Northwest streams. General Technical Report PNW-69, U.S. Department of Agriculture Forest Service, Portland, Ore.

28. **Swanson, F. J., G. W. Lienkaemper, and J. R. Sedell.** 1976. History, physical effects, and management implications of large organic debris in western Oregon streams. General Technical Report PNW-56, U.S. Department of Agriculture Forest Service, Portland, Ore.

29. **Terry, R. E., and D. W. Nelson.** 1975. Factors influencing nitrate transformation in sediments. J. Environ. Qual. **4:**549–554.

30. **Todd, R. L., W. T. Swank, J. E. Douglass, P. C. Kerr, D. L. Brockway, and C. D. Monk.** 1975. The relationship between nitrate concentration in the southern Appalachian mountain streams and terrestrial nitrifiers. Agro-Ecosystems **2:**127–132.

31. **Triska, F. J., and K. Cromack, Jr.** 1980. The role of wood debris in forest and streams, p. 171–190. *In* R. H. Waring (ed.), Forests: fresh perspectives from ecosystem analysis. Oregon State University Press, Corvallis.

32. **Triska, F. J., J. R. Sedell, and S. V. Gregory.** 1982. Coniferous forest streams, p. 292–332. *In* R. L. Edmonds (ed.), Analysis of coniferous forest ecosystems in the western United States. Hutchinson Ross Publishing Co., Stroudsburg, Pa.

33. **Van Kessel, J. F.** 1977. Factors affecting the denitrification rate in two water-sediment systems. Water Res. **11:**259–267.

34. **Vitousek, P. M., and J. M. Melillo.** 1979. Nitrate losses from disturbed forests: patterns and mechanisms. For. Sci. **25:**605–619.

35. **Vitousek, P. M., J. R. Gosz, C. C. Grier, J. M. Melillo, W. A. Reiners, and R. L. Todd.** 1979. Nitrate losses from disturbed ecosystems. Science **204:**469–474.

36. **Ward, A. K., J. A. Baross, C. N. Dahm, M. D. Lilley, and J. R. Sedell.** 1983. Qualitative and quantitative observations on aquatic algal communities and recolonization within the blast zone of Mt. St. Helens: 1980 and 1981. J. Phycol. **19:**232–241.

37. **Ward, A. K., G. M. Ward, and C. N. Dahm.** 1982. Aquatic algal and microbial responses to riparian structures of stream ecosystems. Eos **63:**967.

38. **Welcomme, R. L.** 1979. Fisheries ecology of floodplain rivers. Longman Group Ltd., London.

Use of Microbial Diversity Measurements to Assess Environmental Stress

RONALD M. ATLAS

Department of Biology, University of Louisville, Louisville, Kentucky 40208

Diversity indices have been widely used in general ecology to describe the functional status of communities and the responses of those communities to environmental stress (9, 13–16, 18, 25–29, 39; R. M. Atlas, Adv. Microb. Ecol., in press). The measurement of ecological diversity provides insight into the functional status of the community. Diversity measurements have not been extensively used to assess the effects of stress on microbial communities, but the limited use they have had in microbiology indicates that they provide a useful parameter for describing microbial communities, albeit with the same limitations and constraints on interpretation that apply to the use of diversity indices in macroecology (Atlas, in press).

In this paper no attempt is made to review all the literature on diversity; rather, selected studies are discussed that have considered various aspects of the potential uses of diversity measurements for describing the effects of stress on microbial communities. The interactions of populations living together in a habitat establish community structure in which the functional niches of the ecosystem are filled by member populations of the community. The diversity of the community reflects the status of population interrelationships within that community in dynamic terms. Although the specific cause and effect relationships are far from obvious, diversity clearly is related to stability. Stability requires a certain level of diversity, and when the stability of the community is upset by stress, the diversity of the community changes; that is, changes in the environment lead to changes in the community structure, and as the community changes, so does the diversity of the community. Microorganisms, like all biological systems, respond to their biological and abiotic environment, and when stressed the structure of a microbial community also changes.

MEASUREMENT OF DIVERSITY

Before considering the applicability of diversity measurements to microbial communities, let me define what I mean by "diversity" and examine how diversity is measured. Diversity describes the heterogeneity of information within the community; it is a measure of entropy and reflects the amount of energy required to maintain the organization of the community. There are two components of diversity: the total amount of information, most often in terms of species number, and the way in which the information is apportioned within the community. Several indices are available for describing the state of ecological diversity (16, 19, 25, 26, 30–34, 38; Atlas, in press). The general diversity indices, such as the widely used Shannon index, incorporate one term for describing the total number of species (species richness) and a second term to describe the distribution of species within the community (evenness or equitability). To calculate a diversity index, one must determine the number of species and the number of members of each species within the community. Without getting into the details of how each diversity index is calculated, we should simply recognize that it is always necessary to define taxonomic groups and that each index has limitations which should be recognized when diversity measurements are interpreted. In most cases it is impossible to identify all members of the community, and therefore we must use estimators of species richness and of the number of each species within the community; this inability imposes limitations in attempting to apply statistical analyses to diversity data.

A particular difficulty in the measurement of microbial diversity is the need to identify microbial species without bias caused by the selective exclusion of some species in the community; this can be accomplished by phycologists, who use morphological observations, but bacteriologists, who rely on cultural methods to identify species, face the problem of selectivity and thus the inevitable underestimation of community diversity. Just as viable plate counts underestimate the number of bacteria in a sample, diversity measurements of bacterial communities are also a biased parameter. Nevertheless, the diversity index, as a community parameter, can provide useful information about microbial communities. Diversity does represent the totality of potential activities and interactions of populations living together within the community.

MICROBIAL DIVERSITY AND NATURAL ENVIRONMENTAL STRESS

Having considered what is meant by diversity, let me examine some examples of the response of microbial diversity to natural environmental stress. Stress can be defined as any factor that tends to alter the natural organization of the community; maintenance of the normal state of diversity requires additional effort when the community is stressed, and because that additional energy is not normally available, what we usually see is a change in the structure of the community. Alternatively, the biological populations capable of tolerating the stress may not be capable of utilizing the available forms of energy. We should note that, because microorganisms are physiologically versatile, factors that stress higher organisms may not affect microbial diversity. Natural factors that will stress microbial communities include extremes of temperature, pH, oxygen tension, pressure, nutrient concentrations, and various toxic substances. Because diversity changes in response to environmental stress, diversity measurements are useful for monitoring changes due to pollution as well as natural perturbations of an ecosystem. With prudent use, diversity measurements provide one more way for developing an understanding of the ecology of microorganisms.

Studies on Arctic microbial communities provide some good examples of the changes in diversity in response to natural environmental stress that can be measured. Kaneko et al. (11), in a study of bacterial community diversity in nearshore regions of the Beaufort Sea, found that there was an inverse relationship between bacterial biomass and species diversity in surface waters, but diversity was near maximal in sediments regardless of the size of the bacterial community. The inverse relation between cell number and species diversity is in accordance with the general ecological principle which suggests that elevated populations in resource-limited habitats reflect the success of relatively few well-adapted (opportunistic) species. The similar levels of diversity in sediments at all locations suggests that marine sediment communities are not severely stressed by environmental factors and thus achieve a high level of diversity that is characteristic of biologically accommodated communities.

In contrast to the situation in sediment, the species composition of the surface water bacterial community in Arctic ecosystems exhibits regular seasonal fluctuations in response to the environmental stress factors that characterize each season. In winter, the air temperatures are extremely low, ice forms on the surface water, and there is no light to support primary productivity; in summer, surface waters are exposed to continuous sunlight and a high incidence of UV light. When Arctic regions experience continuous sunlight, the dominant bacterial populations in surface waters are pigmented; the high incidence of pigmentation among member popula-

tions is a physiological response of the community. During winter, when no direct sunlight reaches these waters, most bacteria are nonpigmented. The diversity of the surface water bacterial community declines in summer because of the successful dominance of pigmented bacteria, which greatly influences the equitability component of diversity; in winter, even though population numbers are lower, the evenness of species distribution within the surface water community is greater and thus the overall diversity is also higher. This is an annual cycle of bacteriological community diversity that is driven by truly extreme environmental factors. Interestingly, the same levels of diversity are achieved at the same locations at the same times of year, but the species occurring in the community are not always the same. These data support the hypothesis that each year, following the annual subice diatom bloom and breakup of coastal ice, a process of colonization of surface waters occurs until the niches are filled. When the niches are filled, the community reaches a stable level of species diversity. The composition of the community depends on the opportunistic invasion of the surface waters by bacterial populations adapted to filling the available niches.

In a subsequent study, Hauxhurst et al. (6) examined the characteristics of bacterial communities in the Gulf of Alaska. The bacterial communities in surface waters in the Gulf of Alaska had higher diversities than those in Arctic surface waters. This finding is in agreement with the ecological principle that fewer species are adapted to polar conditions and that species diversity decreases as one moves from equatorial to polar ecosystems; that is, diversity decreases in response to environmental-physiological stress. In contrast to surface water communities, the diversity of sediment bacterial communities was not different in subarctic and Arctic marine ecosystems, suggesting that these benthic ecosystems are relatively unstressed and therefore species diversity does not decline as one moves toward the pole.

The low-nutrient (oligotrophic) conditions of marine environments represent a natural stress of many populations. In a study of oligotrophic and copiotrophic marine bacteria, Horowitz et al. (8) found the diversity of oligotrophic populations to be significantly lower than the diversity of copiotrophic populations. Only a limited number of bacterial populations are able to grow at low nutrient concentrations, and thus the diversity of these populations is low; that is, the species richness of oligotrophs is low. Atlas et al. (1) found much lower diversities for bacterial communities associated with amphipods than in surrounding waters and sediments. The physiological properties of the bacteria associated with the amphipods indicated that they are adapted to using a limited number of substrates that presumably are supplied to them in the food ingested by the amphipod. Here is seen low diversity due to dominance, that is, low evenness, rather than low species richness. These studies give insight into the underlying reasons for a particular level of diversity, as well as providing a parameter that describes the status of the community.

Even more information about the driving forces that determine community structure can be obtained by using factor and principal component analysis. Holder-Franklin and co-workers have made extensive use of factor analysis to describe the diversity of freshwater microbial communities (2, 3, 7; C. R. Bell, M. A. Holder-Franklin, and M. Franklin, Can. J. Microbiol., in press). They have been able to describe not only the community structure but also the factors, both environmental and physiological, that interact to establish a particular assemblage of microbial populations within the bacterial community of two Canadian rivers. In these studies the factors controlling diurnal as well as seasonal variations were considered. The conclusion of these studies was that seasonal and diurnal population shifts within the bacterial community are controlled in low-nutrient freshwater rivers primarily by temperature, nutritional versatility, dissolved oxygen, and conductivity.

MICROBIAL DIVERSITY AND ENVIRONMENTAL STRESS FROM POLLUTANTS

In addition to various natural environmental parameters, diversity of microbial communities changes in response to the input of pollutants (Table 1). If the pollutant is toxic to microorganisms, we generally see a decline in the number of surviving species and thus a decline in diversity due to the decrease in species richness. Diversity can also decline if enrichment of a species within the community greatly alters the evenness factor; this is particularly true when

considering microbial communities that can rapidly reproduce when a suitable substrate or nutrient is added to the ecosystem. However, selective toxicity can sometimes eliminate a dominant organism and lead to an increase in diversity because of the enhanced evenness of species distribution when the dominant organism is eliminated. It is therefore necessary to view the change in diversity that follows input of a pollutant with respect to the factors that naturally control the diversity status of the community. Generally, though, we find that diversity decreases when pollutants are added to ecosystems, either as a result of the elimination of species due to toxicity (decreased species richness) or the enrichment of particular successful populations (decreased evenness).

On the basis of numerous river surveys, Patrick et al. (23) concluded that conditions are favorable for the occurrence of many algal species within the community in rivers that have not been adversely affected by pollution. In unpolluted habitats the majority of species are represented by a relatively small number of specimens; competition is strong, and many habitats are present, each occupied by a different species. The effect of pollution is to eliminate the more sensitive species; competition is reduced, and the more tolerant species proliferate. Thus, there are few species with more individuals in each, unless the stress is too severe, in which case the entire algal community may be eliminated. These studies emphasize the importance in certain situations of considering evenness and species richness as separate parameters if the reason for a particular state of diversity of a community is to be understood.

In general, surveys aimed at detecting the impact of pollution on algal communities indicate that algal diversity is lower in polluted streams than in relatively pristine waterways (22). Marshall et al. (17) found a decrease in diversity of plankton communities in lakes after addition of cadmium and mercury. Eloranta and Kettunen (5), in an examination of the impact of

TABLE 1. Changes in diversity in response to stress from various pollutants[a]

| Pollutant | Community | Diversity | | Reference |
		In control	After stress	
Thermal effluent	Protozoan	D = 24	D = 8	4
Sewage	Algal	H′ = 2.5	H′ = 1.8	36
Sewage + industrial effluent	Algal	D = 160	D = 50	22
Insecticide	Algal	SCI = 0.77	SCI = 0.04	21
Power plant ash	Bacterial	SCI = 8.2	SCI = 6.2	12
Acid mine drainage	Bacterial	H′ = 2.8	H′ = 0.3	20
Crude oil	Bacterial	H′ = 3.6	H′ = 2.0	Atlas, in press
Refined fuel	Bacterial	H′ = 3.1	H′ = 0.0	Atlas, in press

[a] SCI, Sequential comparison index of diversity; H′, Shannon diversity index; D, species richness.

discharges from a paper mill, found that near the cellulose factory the Shannon diversity index and the evenness within the community were high, but in eutrophic areas only a few species were dominant and both evenness and diversity were low. Nelson et al. (21) found that addition of an insecticide to soil lowered diatom diversity, indicating some decline of environmental quality, but the decrease in diversity was not detected until 12 weeks after addition of the insecticide.

The input of thermal effluents also represents a stress on the microbial community that is reflected in changes in diversity. Cairns (4), in an examination of factors affecting the number of species in freshwater protozoan communities, observed a decrease in diversity after a temporary thermal shock, but within a few days diversities of the thermally stressed communities returned to those of unstressed controls.

With respect to bacterial communities, Mills and Wassel (20) found a decrease in species diversity due to the environmental stress of acid mine drainage. Diversity indices for microbial communities collected from lake water and sediment samples affected by acid mine drainage were lower than those for communities from unpolluted sites. In this study the technique of rarefaction was used to describe the species diversity; rarefaction permits the estimation of variance on samples of unequal size and the statistical analysis of data, whereas with the Shannon index it is difficult to estimate the error term needed for analyses of variance.

In a study to assess the impact of effluents from fossil fuel power plants, Larrick et al. (12) examined the diversity of naturally occurring heterotrophic microbes. The total colony-forming units remained relatively high at all stations, but the proportion of pigmented bacteria within the community was correlated with physicochemical stresses, varying from a high of 59.0% at an unpolluted reference site to 13.2% at a heavy ash basin. A sequential comparison procedure produced diversity indices which ranged from 8.2 at the unpolluted site to 6.23 at the ash basin. The structure and function of bacterial communities in the ash basins are significantly different from identical parameters for populations inhabiting unpolluted reference environments, and these differences are reflected in the diversity measurements. Thus, the input of ash is a major stress on the microbial community that leads to altered diversity.

The diversity of bacterial community structure is also influenced by the input of organic wastes. Studies on offshore pharmaceutical dumping have been reported by Peele et al. (24). Normally, marine waters are dominated by gram-negative bacteria, e.g., *Vibrio* and *Pseudo-*

monas species. In the area of pharmaceutical dumping the relative abundances of gram-negative species declined as gram-positive bacterial species increased in importance. *Staphylococcus, Micrococcus*, and *Bacillus* species were commonly found near the dump site, although these genera are rarely isolated from marine surface waters. The conclusion of this study was that oceanic dumping of pharmaceutical wastes alters the taxonomic composition of the bacterial community in the region of the dump site.

Major decreases in species diversity were found after the input of petroleum hydrocarbons (A. Horowitz and R. M. Atlas, unpublished data). The greatest upset of normal community structure occurs with refined oils that have high concentrations of toxic hydrocarbons. The impact of oil on diversity reflects both toxicity to some members of the community and enrichment of microorganisms that break down hydrocarbons. When the toxicity factor is high, there is a dramatic decline in diversity; for example, after the input of leaded gasoline to an Arctic tundra pond, the diversity of the bacterial community decreased from $H' > 3$ to $H' = 0$—only a single species of *Pseudomonas* survived and proliferated in the heavily impacted areas. When enrichment is the main factor driving the lowering of diversity, the decline in diversity is gradual; for example, after the experimental input of Prudhoe Bay crude oil, diversity decreased over several weeks from $H' = 4$ to $H' = 2$.

With respect to sewage effluents into streams, there generally is a decline in aquatic life near the point of discharge, with recovery occurring downstream (10). Stockner and Benson (36) found that in sediments not impacted by sewage the diversity of the diatom community is fairly constant but the input of sewage is reflected in a change in dominance within the community. The cultural enrichment due to sewage input therefore causes a considerable change in species importance rankings and species diversity indices. Although the diversity of a community often decreases in response to the input of acute levels of pollutants, this is not always the case. For example, Thompson and Ho (37), studying the effects of sewage discharge on phytoplankton in Hong Kong, found no evidence of a reduction in taxonomic diversity in polluted areas except in summer. The coastal waters of Hong Kong constitute a transition from estuarine conditions in the west to more oceanic conditions in the east, with the major discharge of untreated sewage located at the midpoint. Stevenson and Stoermer (35) likewise found high diversity in an area of Lake Huron that was receiving inputs of sewage effluent. They hypothesized that species diversity is low in slowly and rapidly growing algal assemblages, and high species diversity may occur in assemblages growing at a moderate rate; species diversity thus could be greatest at some moderate level of stress perturbation. Especially if growth of a dominant species is inhibited under moderate stress, evenness of species distribution within the community would be enhanced. At high levels of growth inhibition, species richness probably is lower than at moderate levels of growth inhibition, and therefore a group of organisms in a highly stressed environment would have a lower diversity than those in a moderately stressed environment because of the elimination of sensitive species.

CONCLUSIONS

The measurement of microbial diversity provides insight into the ecological functioning of the community. Both natural and man-made environmental perturbations lead to changes in the community diversity; because diversity changes in response to environmental stress, diversity measurements are useful for monitoring changes due to pollution. Diversity indices often decrease in response to environmental perturbations. In some cases, such as when heavy metals are introduced into ecosystems, the decrease in diversity reflects toxicity and is a measure of the survival of tolerant species; in other cases, such as when petroleum hydrocarbons pollute marine ecosystems, the decline in species diversity reflects the success of a limited number of physiologically specialized microorganisms.

Although diversity measurements clearly are a reflection of the dynamic status of an ecosystem, they do not show a cause and effect relationship between a particular level of stress and a particular species composition within the community. The conventional indices of species diversity do not give direct information about the physiologial functioning of the community. Studies that have been performed so far do not establish the causal relationships between community stability and the level of diversity. Additionally, caution must be used in interpreting diversity measurements. Statistical analyses are essential for comparing diversity measurements, but such comparisons are difficult with widely used indices of diversity such as the Shannon index because of an inevitable underestimation (bias) in the measurement when all species are not considered.

Future studies will attempt to establish a clearer understanding of the relationship between stability and diversity and the responses of microbial communities to different types of stress. Such studies will have to use a variety of diversity measurements and examine different types of stress under both controlled and field conditions. Improvements will have to be made in how we identify microbial species within the community to lessen the bias of the diversity measurement. Much remains to be done if we are to understand the dynamics of community interactions and the cause and effect relationship of stress, diversity, and community stability. When such studies are accomplished we will be able to determine the extent of a particular stress that a given community can tolerate.

LITERATURE CITED

1. **Atlas, R. M., M. Busdosh, T. Kaneko, and E. J. Krichevsky.** 1982. Bacterial populations associated with Arctic amphipods. Can. J. Microbiol. 28:92–99.
2. **Bell, C. R., M. A. Holder-Franklin, and M. Franklin.** 1980. Heterotrophic bacteria in two Canadian rivers. I. Seasonal variations in the predominant bacterial populations. Water Res. 14:449–460.
3. **Bell, C. R., M. A. Holder-Franklin, and M. Franklin.** 1982. Correlations between predominant heterotrophic bacteria and physiochemical water quality parameters in two Canadian rivers. Appl. Environ. Microbiol. 43:269–283.
4. **Cairns, J. C.** 1969. Factors affecting the number of species in freshwater protozoan communities, p. 219–248. *In* J. C. Cairns (ed.), The structure and function of freshwater microbial communities. Virginia Polytechnic Institute and State University, Blacksburg.
5. **Eloranta, P., and R. Kettunen.** 1979. Phytoplankton in a watercourse polluted by a sulfite cellulose factory. Ann. Bot. Fenn. 16:338–350.
6. **Hauxhurst, J. D., T. Kaneko, and R. M. Atlas.** 1981. Characteristics of bacterial communities in the Gulf of Alaska USA. Microb. Ecol. 7:167–182.
7. **Holder-Franklin, M. A.** 1981. Methods of studying population shifts in aquatic bacteria in response to environmental change. Scientific series no. 124, Inland Waters Directorate, Water Quality Branch, Department of the Environment, Ottawa, Canada.
8. **Horowitz, A., M. I. Krichevsky, and R. M. Atlas.** 1983. Characteristics and diversity of subarctic marine oligotrophic, stenoheterotrophic, and euryheterotrophic bacterial populations. Can. J. Microbiol. 29:527–535.
9. **Hurlbert, S. H.** 1971. The nonconcept of species diversity: a critique and alternative parameters. Ecology 52:577–586.
10. **Hynes, H. B. N.** 1960. The biology of polluted waters. Liverpool University Press, Liverpool.
11. **Kaneko, T., R. M. Atlas, and M. Krichevsky.** 1977. Diversity of bacterial populations in the Beaufort Sea. Nature (London) 270:596–599.
12. **Larrick, S. R., J. R. Clark, D. S. Cherry, and J. Cairns, Jr.** 1981. Structural and functional changes of aquatic heterotrophic bacteria to thermal heavy and fly ash effluents. Water Res. 15:875–880.
13. **Margalef, R.** 1958. Information theory in ecology. Gen. Syst. 3:36–71.
14. **Margalef, R.** 1963. On certain unifying principles in ecology. Am. Nat. 97:357–374.
15. **Margalef, R.** 1968. Perspectives in ecological theory. University of Chicago Press, Chicago.
16. **Margalef, R.** 1979. Diversity, p. 251–260. *In* A. Sournia (ed.), Monographs on oceanographic methodology. UNESCO, Paris.
17. **Marshall, J. S., J. I. Parker, D. L. Mellinger, and S. G. Lawrence.** 1981. An *in situ* study of cadmium and mercury stress in the plankton community of 382 experimental lakes in the area of Northwestern Ontario, Canada. Can. J. Fish. Aquat. Sci. 38:1209–1214.
18. **May, R. M.** 1976. Theoretical ecology principles and

applications. W. B. Saunders, Philadelphia.

19. **McIntosh, P. R.** 1967. An index of diversity and the relation of certain concepts to diversity. Ecology **47:**392–404.

20. **Mills, A. L., and R. A. Wassel.** 1980. Aspects of diversity measurement of microbial communities. Appl. Environ. Microbiol. **40:**578–586.

21. **Nelson, J. H., D. L. Stoneburner, E. S. Evans, Jr., N. E. Pennington, and M. V. Meisch.** 1976. Diatom diversity as a function of insecticidal treatment with a controlled release formulation of chloropyrifos. Bull. Environ. Contam. Toxicol. **15:**630–634.

22. **Patrick, R.** 1963. The structure of diatom communities under varying ecological conditions. Ann. N.Y. Acad. Sci. **108:**359–365.

23. **Patrick, R., M. H. Hohn, and J. H. Wallace.** 1954. A new method for determining the pattern of diatom flora. Not. Nat. Acad. Sci. Nat. Philadelphia **259:**1–12.

24. **Peele, E. R., F. L. Singleton, J. W. Deming, B. Cavari, and R. R. Colwell.** 1981. Effects of pharmaceutical wastes on microbial populations in surface waters at the Puerto Rico dump site in the Atlantic Ocean. Appl. Environ. Microbiol. **41:**873–879.

25. **Peet, R. K.** 1974. The measurement of species diversity. Annu. Rev. Ecol. Syst. **5:**285–308.

26. **Pielou, E. C.** 1966. Shannon's formula as a measure of species diversity: its use and misuse. Am. Nat. **100:**463–465.

27. **Pielou, E. C.** 1966. The measurement of diversity in different types of biological collections. J. Theor. Biol. **13:**131–144.

28. **Pielou, E. C.** 1969. An introduction to mathematical ecology. Wiley-Interscience, New York.

29. **Pielou, E. C.** 1975. Ecological diversity. Wiley-Interscience, New York.

30. **Shannon, C. E.** 1948. A mathematical theory of communication. Bell Syst. Tech. J. **27:**379–423.

31. **Shannon, C. E., and W. Weaver.** 1949. The mathematical theory of communications. University of Illinois Press, Urbana.

32. **Sheldon, A. L.** 1969. Equitability indices: dependence on species count. Ecology **50:**466–467.

33. **Simberloff, D.** 1972. Properties of the rarefaction diversity measurement. Am. Nat. **106:**414–418.

34. **Simpson, E. H.** 1949. Measurement of diversity. Nature (London) **163:**688.

35. **Stevenson, R. J., and E. F. Stoermer.** 1982. Abundance patterns of diatoms on *Cladophora* in Lake Huron with respect to a point source of wastewater treatment plant effluent. J. Great Lakes Res. **8:**184–195.

36. **Stockner, J. G., and W. W. Benson.** 1967. The succession of diatom assemblages in the recent sediments of Lake Washington. Limnol. Oceanogr. **12:**513–532.

37. **Thompson, G. G., and J. Ho.** 1981. Some effects of sewage discharge upon phytoplankton in Hong Kong. Mar. Pollut. Bull. **12:**168–173.

38. **Trousellier, M., and P. Legendre.** 1981. A functional evenness index for microbial ecology. Microb. Ecol. **7:**283–296.

39. **Woodwell, G. M., and H. H. Smith. (ed.).** 1969. Diversity and stability in ecological systems. Brookhaven Symp. Biol. No. 22.

Effects of Oil on Bacterial Activity in Marine and Freshwater Sediments

JOHN H. BAKER AND ROBERT P. GRIFFITHS

Freshwater Biological Association, East Stoke, Wareham, Dorset, United Kingdom, and Department of Microbiology, Oregon State University, Corvallis, Oregon 97331

Crude oil production in the world has increased from 278 million t per year in 1938 to about 3,000 million t per year in recent years (26). The potential for oil to reach aquatic sediments has therefore also increased, particularly as a result of accidents involving supertankers. The aim of this paper is to collate the work which has been carried out on the effects of oil on the bacteriology of freshwater and marine sediments. We are not concerned here with the degradation of oil by bacteria, but rather with the stimulatory or inhibitory effects of oil on sedimentary microbial activity. Nevertheless, of course, the products of microbial degradation of oil may have an equal or even greater influence on microbial activity than the unmodified oil itself.

COMPOSITION OF OIL

Crude oil has been described as probably the most complicated natural mixture on earth (8). It consists mainly of hydrocarbons (50 to 98%), with variable amounts of oxygen, sulfur, and nitrogen (Table 1). Over 50 different elements have been detected in crude oil, and the molecular weight of constituent compounds varies from 16 to greater than 20,000. The composition of oil varies from field to field, between wells within a field, and even between samples from the same well. It is therefore not possible to give a precise constitution for crude oil, and no concise system has been developed for adequately classifying different crudes (9). In an attempt to clarify this confusing situation, the American Petroleum Institute has provided a suite of four reference oil, two crudes, and two refined oils, the simplified compositions of which are given in Table 1.

These reference oils have been widely used for experimental purposes. The no. 2 fuel oil (diesel fuel) is prepared from the middle distillate fraction, i.e., that boiling between 185 and 345°C. The reference no. 2 fuel is atypical insofar as a typical no. 2 fuel would contain a smaller total aromatic fraction and a higher cycloalkane fraction. Bunker C fuel oil, which is used in power stations and as a marine fuel, is prepared from the highest boiling fraction of the heavy distillates. In all these oils the nitrogen content never exceeds 1% by weight and mostly consists of pyridine, quinolines, and other heterocyclics concentrated in the higher boiling fraction.

In such a complex mixture many compounds are likely to inhibit the growth of some microorganisms. These compounds can be classified as: (i) low-molecular-weight aromatics such as benzene, toluene, and xylene which are more common in no. 2 fuel oil than in Bunker C oil, but likely to be lost during weathering; (ii) phenols, cresols, and carboxylic acids, some of which are formed during weathering; (iii) polynuclear aromatics such as pyrenes, chrysenes, and benzthiophenes which are more common in Bunker C oil than no. 2 fuel oil; and (iv) heavy metals, of which nickel and vanadium are the most common (Table 1).

INCORPORATION OF OIL INTO SEDIMENTS

It is first necessary to describe the ways in which oil can become deposited in sediments. In the marine environment it is natural to associate tanker spillages with all aspects of oil pollution, not least because 6 million t per annum is currently being lost in this way (3). However, only a small proportion of spilled oil actually reaches the sediment, and that material is often markedly changed in composition. The changes which take place vary with the original composition of the oil, the climate, etc., but the greatest change which takes place is due to evaporation. Evaporation is the most important natural process for decreasing the amount of oil in a slick (8). For an average crude oil, 25 to 30% can be lost by evaporation in the first 2 or 3 days, rising eventually to about 50%. However, heavy oils such as Bunker C fuel oil may only lose 10% of their weight in similar conditions. Furthermore, under the arctic conditions of the largest known oil field in North America (Prudhoe Bay), evaporation will be comparatively slow.

The components lost by evaporation are mostly the lower-molecular-weight fraction (from C_4 to C_{12}), including some of the toxic aromatic compounds. The material left after evaporation has a higher specific gravity than the original oil and is thus more likely to sink. It may at this stage form an oil-in-water emulsion which, as a result of the large oil surface thereby exposed,

TABLE 1. Simplified composition (percent by weight) of the four reference oils[a]

Component	Louisiana crude	Kuwait crude	No. 2 fuel	Bunker C fuel
Alkanes	28.0	34.1	30.4	6.7
Cycloalkanes	45.4	20.3	31.4	15.2
Benzenes	4.1	4.9	10.3	1.9
Other aromatics	14.5	24.3	27.9	32.3
Polar compounds	8.4	17.9	0	30.3
Insolubles	0.2	3.5	0	14.4
Nitrogen	0.69	0.14	0.02	0.94
Sulfur	0.25	2.44	0.32	1.46
Nickel + vanadium (ppm)	4.1	35.7	2.0	162

[a] Adapted from Clark and Brown (9).

will be more quickly attacked by photochemical or bacteriological means. Both of these processes are essentially oxidative, resulting in more polar, and therefore more soluble, compounds. Solution is not as important as evaporation, but should nevertheless not be ignored. Instead of forming oil-in-water emulsions, some oils, such as Kuwait crude, absorb large quantities of water, forming water-in-oil emulsions generally known, because of their appearance, as chocolate mousse. A chocolate mousse formation has a relatively small surface-to-volume ratio of oil and therefore is very stable, floats on the oceans for long periods, and is probably the precursor of the tar balls that are washed up on our beaches. The sum of the changes which exposed oil undergoes as a result of chemical, biological, and physical agencies is collectively referred to as "weathering."

When oil droplets of a suitable size are created by wave action or emulsification, they are ingested by the zooplankton, which appear to be nonselective feeders. Conover (11) has calculated that, following one particular oil spill in Chedabucto Bay, Nova Scotia, 10% of the suspended oil was associated with the zooplankton (dominated by *Temora longicornis*). The zooplankton appeared to be unaffected by the oil, which was subsequently incorporated into fecal pellets with a relatively high specific gravity and thus sedimented to the bottom. Conover (11) suggested that 20% of the oil may reach the sediment in the form of fecal pellets, although this figure was not confirmed by observation. Oil has also been observed associated, presumably by adsorption, with sedimenting algae. After the IXTOC blowout in the Gulf of Mexico, up to 3% of the oil was found to sediment in this way (6).

Not all the oil in the sea is derived from accidental spillages. Rivers receiving discharges from sewage works and urban storm water drains have been calculated to contribute 1.8 million t per year to the world's oceans (26). According to a study of Providence River, Rhode Island (30), approximately 50% of the hydrocarbons discharged from a sewage works sedimented within the river upstream of the estuary. This sedimentation was due to adsorption of the oil-derived materials onto mineral particles. Thus, plants, animals, and abiotic particles may all provide the means whereby oil can be transported from the water column into the sediment. The intertidal zone is a region where none of these mechanisms is necessary because oil can be, and often is, merely stranded there by the tide. Subsequent incorporation of stranded oil into sandy beaches depends on the physical state of the oil. Thin oil films from the Amoco Cadiz disaster were rapidly dispersed throughout the sand, whereas chocolate mousse formations accumulated as a discrete layer (29).

Concentrations of oil occurring in sedimentary deposits after oil spills vary rapidly. Uncontaminated beach sands contain an average of 0.0003% petroleum hydrocarbons (29), and this should be compared with a maximum of 0.008% in an arctic lake bottom 1 year after a gasoline spillage (20) and a mean of 0.14% in a chronically affected estuarine sediment (30). After the Amoco Cadiz disaster some of the oil was deposited as a discrete layer on the beaches, resulting in concentrations as high as 9.9%, but where the oil was naturally dispersed it reached only 0.004% (29).

The last route we shall consider by which oil becomes associated with sediments is also the most direct, i.e., submarine seeps. Seeps occur where the geological formation allows the natural slow escape of oil or gas from an oil field. Terrestrial seeps are known, but do not particularly concern us. Submarine seeps are thought to exude 0.6 million t per year directly onto the sediment surface (26). Assuming that 10% of the oil from spillages reaches the sediment, inputs from this source are just equivalent to those from natural seepages. Oil from seeps is not weathered in the same way as other oils before it reaches the sediment, but there is evidence that some bacterial attack has already taken place before it reaches the water (28).

SPECIES COMPOSITION

Oil production from offshore platforms produces chronic low-level oil pollution in the immediate environment of the rig. Olsen and Sizemore (25) utilized this situation on the northwestern Gulf of Mexico to compare oil-contaminated sediments close to a rig and sediments from a control site 5 miles north of the rig. Bacterial isolates were obtained from nonselective agar plates inoculated with suitable samples of sediment collected from each of these areas.

A total of 532 isolates, approximately half from each area, were identified to generic level, but no obvious influence of the oil pollution could be discerned. Seasonal differences within areas seemed to be more important than differences between areas.

In contrast to oil contamination produced anthropologically, submarine seeps show a very obvious effect of the oil on bacterial species composition. Seeps in Santa Barbara Channel have associated with them white mats of *Beggiatoa* sp. 1 m or more in diameter (28). These large areas, completely dominated by a single bacterium, were not seen outside the seepage areas, and the *Beggiatoa* sp. were presumably utilizing hydrogen sulfide associated with the oil. In freshwater, changes in species composition due to oil have also been reported by Bührer (7). In a detailed study of a lake sediment he reported that 10 g of fuel oil per m^2 produced little change, but when the application rate was raised to 100 g/m^2, the gram-negative cocci were diminished whereas the pseudomonads showed a marked increase.

Because bacteria are known to degrade hydrocarbons, the lack of a difference between communities noted by Olsen and Sizemore (above) may, at first, seem surprising. However, the ability to utilize components of oil is widely distributed among bacteria, and at least 22 genera are known to possess this capability (5). Hence, it may be more relevant to determine what proportion of the bacterial population is capable of utilizing hydrocarbons. A study of sediment samples from an unpolluted area of the North Sea (17) revealed that 4% of the bacterial isolates were hydrocarbon utilizers. A similar study in Chesapeake Bay by different workers (10) produced a very similar percentage. However, more recent work has convinced Atlas (2) that in most unpolluted ecosystems hydrocarbon utilizers constitute less than 0.1% of the microbial community. In contrast, the sediments of Chedabucto Bay affected by spillage from the Arrow contained up to 15% hydrocarbon utilizers (24). Also, the sediments affected by the Torrey Canyon disaster contained elevated levels of bacteria capable of utilizing hydrocarbons (16).

The study by Horowitz and Atlas (19) of a freshwater sediment in an Arctic lake affected by leaded gasoline produced even more convincing results. These workers found that in contaminated regions the hydrocarbon-utilizing bacterial population sometimes numerically exceeded the population growing on Trypticase soy agar. They concluded that the ratio of hydrocarbon utilizers to total heterotrophs was a useful and sensitive indicator of hydrocarbon contamination.

EFFECTS ON NUTRIENT CYCLING

It is reasonable to predict that oil from spills will reach the sediment in its weathered state associated with some sort of detritus and will form a discrete layer. Exceptionally, the oil may not be particularly weathered, as in the case of oil seeps and oil from coastal spills which is rapidly stranded on the beach. In unpolluted sediment all freshly sedimenting material is rapidly mixed with the existing benthic deposit by the activity of the invertebrate fauna. This animal activity helps to aerate the superficial sediment so that if the oil is toxic to the animals the degree of aeration will be greatly reduced. When weathered oil from the IXTOC spill was added to coastal sediment from the Gulf of Mexico, a reduction of faunal activity was observed (22). Fresh oil from Cook Inlet, Alaska, had a similar effect on the subarctic marine fauna (14). Indeed, in the latter study detritus was observed to accumulate on the surface of the oiled layer, and this accumulation did not occur in controls. Hence, not only was burrowing activity reduced, but the added layer may also have impeded oxygen diffusion. Measurements of the redox potential at different depths in oiled and control sediments (22) also indicated that the addition of oil reduced the depth of the aerobic layer by approximately 50%.

Carbon. When the oxygen concentration falls and the redox potential is reduced, increased methanogenesis may be anticipated. Increased methane concentrations were observed in the headspace above oiled subarctic marine sediments (14). In the same study increased carbon dioxide production rates were also observed 8 and 11 months after fresh crude oil had been mixed with sediment at 5% (vol/vol). However, after 18 months no significant increase in carbon dioxide production rates could be discerned. Over a shorter time period (3 months) the effect of increasing oil concentration on carbon dioxide production has been determined by a gas chromatographic method (14). After 1 day the production rate was decreased at all concentrations, but subsequently increased with increasing oil concentration to a maximum of five times the initial rate.

Another approach to ascertain the effect of oil on microbial activity involves determining the uptake or respiration rates, or both, of simple radioactive substrates such as glucose, glutamate, and amino acids. It is assumed that these readily metabolizable compounds are utilized by the greater part of the bacterial community during relatively short incubation periods. Saltzmann (27) determined the respiration rates of labeled amino acids and also naphthalene in sediment samples taken at various distances from North Sea oil rigs. He found no influence of the rigs on amino acid respiration, but the uptake of naphthalene increased with decreasing distance from the rig. He concluded that the bacterial population able to degrade hydrocarbons (or at least naphthalene) was not linked to general heterotrophic activity. Alexander and Schwarz (1) investigated glucose uptake by sediment samples from the Gulf of Mexico mixed with various concentrations of the two reference crude oils. Of a total of 15 samples, glucose uptake was no different from the control in 10, and stimulation of uptake was recorded in 4.

In an extensive survey of the effects of crude oil on glucose and glutamate uptake by sediments off the Alaskan coast, Griffiths et al. (15) examined 162 sediment samples. At an oil concentration of 0.1% (vol/vol), the mean reduction in glucose uptake rates ranged from 14 to 36%, but this was less than in water samples examined at the same time. It was concluded that in sediments bacterial production is depressed in the presence of crude oil as a result of a significant increase in the proportion of substrata respired. By sampling oiled sediments over long periods of time (up to 18 months), it is possible to predict that the effects of oil on glucose and glutamate uptake will be detectable for up to 32 and 20 months, respectively (13).

With regard to freshwater sediments, Prudhoe Bay oil was added to an enclosure in an Alaskan lake, and glucose turnover rates were determined in sediments from outside and inside the enclosure (21). Significant differences were found between replicates, but these were as great as or greater than differences between control and oiled areas. In a study of an estuarine sediment close to Chesapeake Bay (31), both South Louisiana crude and no. 2 fuel oil were found to be toxic for bacteria involved in nutrient cycling. Lastly, in a pilot study of an Oregon stream sediment, the addition of crude oil progressively depressed the V_{max} for mineralized glucose at oil concentrations of 0.1 and 1.0% (vol/vol) (4).

Nitrogen. The earliest account of the effects of oil on nitrogen fixation appears to be that of Knowles and Wishart (23). These workers added weathered Canadian crude to marine sediment samples from stations off the northwest coast of Canada. The concentrations of oil used were 1, 3, and 5% (vol/wt), but no reproducible effect could be detected up to a maximum incubation period of 35 days. Sediment from the south coast of Alaska was incubated with both fresh and weathered Alaskan crude oil by Griffiths et al. (14) for much longer periods, up to 18 months. In these experiments oil concentrations were 0.1 and 5.0% (vol/vol), and at both concentrations fresh oil significantly decreased nitrogen fixation

rates and potential denitrification rates. However, weathered crude oil did not produce a significant decrease in nitrogen fixation even after 18 months (14). Natural denitrification (i.e., in the absence of added nitrate) was generally below the limit detectable by gas chromatographic methods (18). Denitrification has also been studied in a salt marsh sediment (12) by a different method. Up to 10% (vol/wt) south Louisiana crude oil did not affect denitrification in a sediment slurry from this habitat. The same study did demonstrate an increase in the ammonia concentration in water above a salt marsh overlain by crude oil, but it is not clear whether this was due to increased ammonification or entrapment of the released ammonia. Little work has been carried out on the effects of oil on nitrogen cycling in freshwater sediments, but rates of nitrogen fixation were unaffected in an Arctic lake mud (20) 1 year after it was severely polluted with gasoline. Similarly, preliminary observations on a stream sediment mixed with crude oil (4) showed little difference in nitrogen fixation rates from the control samples.

CONCLUSIONS

Approximately 0.6 million t of oil per year is added directly to marine sediments in highly localized areas from submarine seepages (26). A similar quantity is probably added to sediments as a result of accidents and discharges, mainly close to the major shipping lanes. This oil is weathered and can be associated with sedimenting algae, fecal pellets, or mineral particles. One effect of the oil is to reduce the oxygen concentration in sediments by inhibiting the burrowing activity of the invertebrate fauna (14). It has proved difficult to demonstrate the effect of oil in terms of changes in classically defined bacterial genera, although significant alterations were noted in a lentic sediment (7). However, in the presence of oil, the population density of hydrocarbon-utilizing bacteria always seems to increase (2).

Studies on the effects of oil on the utilization of glucose, glutamate, etc., have produced conflicting results (1, 4, 13, 15, 21, 27). One possible explanation for the difference involves the recent history of the sediments sampled. Sediments which had been exposed to oil prior to the experiments in question probably contain hydrocarbon-utilizing bacteria. Hence, they are likely to be less inhibited by a future application of oil than similar sediments perturbed by oil for the first time. This is in accordance with the results because uptake in the pristine sediments (4, 13, 15) was significantly reduced whereas in the chronically polluted areas (1, 27) no inhibition of uptake was discerned. The only exception to this hypothesis was an Arctic lake sediment which failed to show any difference in glucose metabolism when oil was added (21). However, the oil concentration in this experiment (0.006%) was much lower than in any of the other studies.

The nitrogen cycle must be included in any discussion of the effects of oil on microbial activity, and in this connection it is interesting to note that whereas fresh crude oil inhibits nitrogen fixation the same oil after weathering has no effect (14). This observation emphasizes the importance of using weathered oil where appropriate and implicates low-molecular-weight volatile materials as the inhibitor(s) in fresh oil. Moreover, if the results of using weathered oil are taken to be the more acceptable, then it is apparent that oil has no significant effect on nitrogen fixation rates in either freshwater or marine sediments (4, 14, 20, 23). Effects of oil on the carbon cycle, however, have been shown to last at least 18 months (14) and have been predicted to last 2 or 3 years (13).

We are most grateful for an extensive literature search by J. V. Bird.

Financial support was provided as a grant-in-aid by the United Kingdom Natural Environment Research Council and by the Bureau of Land Management through interagency agreement with the National Oceanic and Atmospheric Administration (NOAA), under which a multiyear program responding to needs of petroleum development of the Alaskan continental shelf is managed by the Outer Continental Shelf Environment Assessment Program Office (NOAA contract 03-5-022-68).

LITERATURE CITED

1. **Alexander, S. K., and J. R. Schwarz.** 1980. Short-term effects of South Louisiana and Kuwait crude oils on glucose utilization by marine bacterial populations. Appl. Environ. Microbiol. **40**:341–345.
2. **Atlas, R. M.** 1981. Microbial degradation of petroleum hydrocarbons: an environmental perspective. Microbiol. Rev. **45**:180–209.
3. **Atwood, D. K., and R. L. Ferguson.** 1982. An example study of the weathering of spilled petroleum in a tropical marine environment: IXTOC-1. Bull. Mar. Sci. **32**:1–13.
4. **Baker, J. H., and R. Y. Morita.** 1983. A note on the effects of crude oil on microbial activities in a stream sediment. Environ. Pollut. Ser. A **31**:149–157.
5. **Bartha, R., and R. M. Atlas.** 1977. The microbiology of aquatic oil spills. Adv. Appl. Microbiol. **22**:225–266.
6. **Boehm, P. D., and D. L. Fiest.** 1980. Aspects of the transport of petroleum hydrocarbons to the offshore benthos during the IXTOC-1 blowout in the Bay of Campeche, p. 207–235. In Proceedings of a Symposium on Preliminary Results from the September 1979 Researcher/ Pierce IXTOC-1 Cruise. National Oceanic and Atmospheric Administration, Rockville, Md.
7. **Bührer, H.** 1979. De Einfluss von Kohlenwasserstoffen auf die Ökologie der Bakterien im aeroben Seesediment. Schweiz. Z. Hydrol. **41**:315–355.
8. **Carlberg, S. R.** 1980. Oil pollution of the marine environment, p. 367–402. In E. Olausson and I. Cato (ed.), Chemistry and biogeochemistry of estuaries. John Wiley, Chichester, U.K.
9. **Clark, R. C., and D. W. Brown.** 1977. Petroleum: properties and analyses in biotic and abiotic systems, p. 1–89. In D. C. Malins (ed.), Effects of petroleum on arctic and subarctic marine environments and organisms. Academic Press, Inc., New York.
10. **Colwell, R. R., J. D. Walker, and J. D. Nelson.** 1973. Microbial ecology and the problem of petroleum degradation in Chesapeake Bay, p. 185–197. In D. G. Ahearn and S. P. Meyers (ed.), The microbial degradation of oil pollutants. Louisiana State University, Baton Rouge.
11. **Conover, R. J.** 1971. Some relations between Bunker C oil in Chedabucto Bay following the wreck of the tanker Arrow. J. Fish. Res. Board Can. **28**:1327–1330.
12. **DeLaune, R. D., W. H. Patrick, and R. J. Buresh.** 1979. Effect of crude oil on a Louisiana Spartina alterniflora salt marsh. Environ. Pollut. **20**:21–31.
13. **Griffiths, R. P., B. A. Caldwell, W. A. Broich, and R. Y. Morita.** 1981. Long-term effects of crude oil on uptake and respiration of glucose and glutamate in arctic and subarctic marine sediments. Appl. Environ. Microbiol. **42**:792–801.
14. **Griffiths, R. P., B. A. Caldwell, W. A. Broich, and R. Y. Morita.** 1982. The long-term effects of crude oil on microbial processes in subarctic marine sediments. Estuar. Coastal Shelf Sci. **15**:183–198.
15. **Griffiths, R. P., T. M. McNamara, B. A. Caldwell, and R. Y. Morita.** 1981. Field observations on the acute effect of crude oil on glucose and glutamate uptake in arctic and subarctic waters. Appl. Environ. Microbiol. **41**:1400–1406.
16. **Gunkel, W.** 1968. Bacteriological investigations of oil-polluted sediments from the Cornish coast following the Torrey Canyon disaster. Helgol. Wiss. Meeresunters. **17**:151–158.
17. **Gunkel, W.** 1973. Distribution and abundance of oil-oxidising bacteria in the North Sea, p. 127–139. In D. G. Ahearn and S. P. Meyers (ed.), The microbial degradation of oil pollutants. Louisiana State University, Baton Rouge.
18. **Haines, J. R., R. M. Atlas, R. P. Griffiths, and R. Y. Morita.** 1981. Denitrification and nitrogen fixation in Alaskan continental shelf sediments. Appl. Environ. Microbiol. **41**:412–421.
19. **Horowitz, A., and R. M. Atlas.** 1977. Response of microorganisms to an accidental gasoline spillage in an arctic freshwater ecosystem. Appl. Environ. Microbiol. **33**:1252–1258.
20. **Horowitz, A., A. Sexstone, and R. M. Atlas.** 1978. Hydrocarbons and microbial activities in sediment of an Arctic lake one year after contamination with leaded gasoline. Arctic **31**:180–191.
21. **Jordan, M. J., J. E. Hobbie, and B. J. Peterson.** 1978. Effect of petroleum hydrocarbons on microbial populations in an Arctic lake. Arctic **31**:170–179.
22. **Kalke, R. D., T. A. Duke, and R. W. Flint.** 1982. Weathered IXTOC-1 oil effects on estuarine benthos. Estuar. Coastal Shelf Sci. **15**:75–84.
23. **Knowles, R., and C. Wishart.** 1977. Nitrogen fixation in Arctic marine sediments. Environ. Pollut. **13**:133–149.
24. **Mulkins-Phillips, G. J., and J. E. Stewart.** 1974. Distribution of hydrocarbon-utilizing bacteria in northwestern Atlantic waters and coastal sediments. Can. J. Microbiol. **20**:955–962.
25. **Olsen, K. D., and R. K. Sizemore.** 1981. Effects of an established offshore oil platform on the autochthonous bacterial community. Dev. Ind. Microbiol. **22**:685–694.
26. **Payne, F. G., and M. T. Westaway.** 1982. Sources of petroleum hydrocarbon inputs to the marine environment. Water Sci. Technol. **14**:1159–1170.
27. **Saltzmann, H. A.** 1982. Biodegradation of aromatic hydrocarbon in marine sediments of three North Sea oil fields. Mar. Biol. **72**:17–26.
28. **Spies, R. B., P. H. Davis, and D. H. Stuermer.** 1980. Ecology of a submarine petroleum seep off the California coast, p. 229–263. In R. A. Geyer (ed.), Marine environmental pollution. Elsevier, Amsterdam.
29. **Vandermeulen, J. H., D. E. Buckley, E. M. Levy, B. F. N. Long, P. McLaren, and P. G. Wells.** 1979. Sediment

penetration of Amoco Cadiz oil, potential for future release, and toxicity. Mar. Pollut. Bull. **10:**222–227.

30. **Van Vleet, E. S., and J. G. Quinn.** 1977. Input and fate of petroleum hydrocarbons entering the Providence River and Upper Narragansett Bay from wastewater effluents.

Environ. Sci. Technol. **11:**1086–1092.

31. **Walker, J. D., P. A. Seesman, and R. R. Colwell.** 1975. Effect of South Louisiana crude oil and No 2 fuel oil on growth of heterotrophic microorganisms. Environ. Pollut. **9:**13–33.

Metabolism of Natural Polymers

Biofuels and Oxychemicals from Natural Polymers: a Perspective

DOUGLAS E. EVELEIGH

Department of Biochemistry and Microbiology, New Jersey Agricultural Experiment Station, Cook College–Rutgers University, New Brunswick, New Jersey 08903

Biomass is available in vast and renewable quantities, and the concept of its utilization for the production of fuels and feedstock chemicals has been widely discussed (12, 13, 22). Furthermore, the mechanisms of microbial conversion of polymeric biomass are reasonably well understood. However, the application of biomass is fraught with many associative problems. These include the costs of collection and storage of biomass, means to gain efficient hydrolysis, the slow speed of fermentation, and the variety of factors affecting downstream processing and recovery of products. As a result, the feedstock chemicals that currently can be economically produced via microbial fermentation are extremely limited (for review see reference 4). Commercially available microbially produced fuels include only fuel alcohol and methane. Furthermore, in a recent government study to recommend which chemicals can potentially be produced fermentatively from biomass, two study groups, Massachusetts Institute of Technology and the Genex Corp., named only 8 chemicals of their total listing of 46 compounds. These included acetic, acrylic, and adipic acids, ethanol, ethylene glycol, ethylene oxide, glycerol, and propylene glycol (24). This lack of definition has resulted in extremely diverse approaches for production of a variety of chemicals from biomass (14–16). These approaches have recently been reviewed (22). The purpose of my paper is to give a particular perspective to the production of biofuels and oxychemicals in relation to the practicality of fuel alcohol. The potential of biomass is discussed, focus is placed on current industrial and research approaches for fuel alcohol production, and the role of the microbial ecologist in development of new technologies is briefly considered.

BIOMASS AVAILABILITY

As biomass contains a significant proportion of oxygen, oxychemicals can be considered as potentially attractive products (6). The United States currently uses approximately 32×10^6 tons of feedstock chemicals, of which oxychemicals (Table 1) comprise perhaps 9×10^6 tons. Note that only 10% of the total starch (corn) crop or 3% of the current collectable forest biomass will satisfy the requirements for production of all of these oxychemicals (6, 22). However, as the cost of the substrate is the dominant economic component of fuel fermentations, each component of biomass must be efficiently utilized. Thus the chemical complexity and variable nature of biomass become major factors, besides the necessity of degrading inherently recalcitrant polymers (cellulose and lignin). Bacon (3) has emphasized that "structural polysaccharides are not made with any thought of their dismantling, and in fact the maker may have good reason to introduce all kinds of complexities into the structure in order to protect itself against invasion and digestion." A variety of techniques are being developed, based on chemical dissolution or physical disruption, to optimize the separation of the major components of ligno-cellulosic biomass (cellulose, hemicellulose, and lignin) (4). However, the present practical solution is to utilize biomass containing a single major component, e.g., corn or cassava starch, for ethanol production.

FUEL ALCOHOL PRODUCTION

The OPEC oil embargo of the United States in the early 1970s resulted in consideration of options to conserve petroleum reserves. One of these was the fermentative production of fuel alcohol (ethanol) from biomass. Ethanol is attractive as a liquid transportation fuel as a 10% blend in gasoline termed "gasohol." This topic has been widely discussed (20, 27–29). Although gasohol is a relatively expensive alternative fuel, the U.S. government's program and incentives have resulted in a considerable and continuously increasing market for fuel alcohol over the past few years. The recent thrust has been refocused to the use of fuel alcohol as an octane booster, though this has not been well publicized. The overall results are dramatic. Ethanol derived by fermentation is now less expensive than that derived synthetically, and production in the

TABLE 1. U.S. production of oxychemicals, 1982

Product	Production[a] 10^9 lb	Price[b]	Order of production[c]
Ethylene glycol	4.29	$ 0.24/lb	28
Acetic acid	2.75	0.23/lb	31
Acetone	1.76	0.25/lb	40
Isopropanol	1.31	2.02/gal	43
Adipic acid	1.20	0.57/lb	48
Ethanol			
95%	1.03	1.70/gal	50
100%		1.82/gal	
n-Butanol	0.71	0.33.5/lb	
Propylene glycol	0.40	0.45/lb	
Glycerol, 96%	0.24[d]	70.75/lb[e]	

[a] Chemical and Engineering News, 2 May–13 June 1983.

[b] Chemical Marketing Reporter, 1 August 1983.

[c] Relative U.S. production based on weight (Chemical and Engineering News, 2 May 1983).

[d] Total glycerine (Chemical Marketing Reporter, 11 April 1983).

[e] Refined natural (Chemical Marketing Reporter, 11 April 1983).

United States has risen from 80×10^6 gal in 1980 to 150×10^6 gal in 1982 and continues to rise. Current U.S. fermentative production of ethanol is estimated at 500×10^6 gal/year and represents 75% of total national production (20). Commercial viability is illustrated by the recent proposed construction of new fermentation plants by the A. E. Staley Manufacturing Corp. (40×10^6 gal/year), Texaco Co. (60×10^6 gal/year), and Ashland Corp. (60×10^6 gal/year). The latter plant is on stream (20). The production of a biofuel from biomass can be most practical.

ALTERNATIVE SCHEMES FOR FUEL ALCOHOL PRODUCTION

A variety of other approaches are being considered for fuel alcohol production from biomass. Three general areas will be considered: (i) the use of immobilized microbes; (ii) thermophilic microbes for direct conversion of natural polymers; and (iii) modification of alcohol-tolerant microbes.

Use of immobilized microbes. The application of immobilized Saccharomyces cerevisiae or Zymomonas sp. has been a focal research area for the continuous production of ethanol. The most impressive result to date is the scale-up by combined Japanese industrial consortia of a system utilizing yeasts to convert molasses to alcohol (23; Research Association for Petroleum Alternatives Development [RAPAD], Tokyo, Japan, Annual Report, 1982). In one of these systems, the Kyowa Hakko Kogyo Co. immobilized yeast cells on calcium alginate beads and utilized them in four 1-kiloliter (ca. 250-gal)

columns. This system produced 9% alcohol continuously for 6 months with 95% efficiency of substrate utilization. There was no sterilization of media; microbial contamination was controlled by the use of acidic media (pH 4) and the inhibitory effects of the ethanol product. The impressiveness of the scale-up is further emphasized by a 20-kiloliter unit put into operation in late 1983.

The productivities of immobilized cell systems considerably exceed that of batch culture. The latter typically yields 2 g of ethanol per liter per h, whereas the large-scale yeast system noted above yields 20 g/liter per h. Laboratory-scale immobilized systems have been developed in which yields of 80 and 160 g/liter per h have been obtained, respectively, for Saccharomyces cerevisiae and Zymomonas mobilis. In further developments microbial ecologists should be able to provide considerable insight into such applications of surface-attached microbes and also the long-term high productivity of these microbial ecosystems.

Thermophilic microbes for the direct conversion of natural polymers to biofuels. Thermophilic microbes are of major commercial interest for chemical production because they permit rapid and stable fermentations, reduce the cost of cooling the fermentor, and facilitate product recovery via distillation (31). A variety of ethanologens (old and new) have recently received considerable study (Table 2). The cellulolytic strain receiving greatest attention has been Clostridium thermocellum. Unfortunately, it produces roughly equal amounts of both ethanol and acetic acid. Efforts have been directed at obtaining mutants that give higher proportions of either product and that are tolerant of concentrations greater than the natural yields of about 2%. Strains have been selected that have an improved ethanol-acetic acid ratio and also that will produce up to 5% ethanol (2, 35). Other strains are tolerant of 7 to 8% ethanol, though they do not grow well at these concentrations. A further point is that C. thermocellum cannot utilize pentoses and thus is unable to metabolize xylans, which may comprise up to 30% of biomass. This is a major drawback in relation to the necessity to utilize all components of biomass. The other thermophilic ethanologens can utilize a broader range of substrates, though not cellulose (Table 2). Dual cultures of C. thermocellum plus ethanologens that utilize pentoses result in greater efficiency of total substrate conversion and can also shift the product spectrum toward ethanol (34).

This progress is most significant, but the thermophilic anaerobes still suffer drawbacks in practical application. For instance, the initial problem of using thermophilic ethanologens was product tolerance. Although this problem has been solved to a degree by obtaining strains tolerant to 7 to 8% ethanol, the conversion of substrate at these concentrations may be markedly reduced, e.g., from 95 to 50% (R. W. Lovitt, R. Longin, and J. G. Zeikus, 3rd Int. Symp. on Microb. Ecol., abstr. no. A23, 1983). A variety of approaches are being followed to circumvent such drawbacks, including development of genetic methodologies for these anaerobes (15).

Modification of alcohol-tolerant microbes. It should be possible to increase the productivity of alcohol-tolerant microbes such as *S. cerevisiae* and *Z. mobilis* through genetic modification resulting in increased enzyme synthesis, deletion of regulatory controls, channeling of metabolic pathways, extension of nutritional capability, and enhanced tolerance to high concentrations of substrate and product (27). Extending the limited substrate range of such organisms as *S. cerevisiae* and *Z. mobilis* is particularly significant. The substrate is the major cost factor in chemical feedstock fermentations; agricultural and municipal lignocellulosic wastes are the classic examples of less expensive substrates. To this end, attempts are being made to clone cellulase genes into the ethanologens. To date, certain cellulase genes have been cloned and expressed in *Escherichia coli* (1, 8, 9, 30; D. B. Wilson, Seminar presentation, Annu. Meet. Am. Soc. Microbiol. 1983). Somewhat analogous studies are being made in an attempt to induce the fermentative utilization of xylose by *S. cerevisiae* by genetic insertion of the xylose isomerase gene. This will permit direct fermentation to alcohol rather than to xylitol (for review, see references 21 and 25). In a similar vein, to extend the substrate range of *Z. mobilis* to include the use of lactose, we have inserted into this organism a plasmid, pGC9114, which contains the β-galactosidase operon and have achieved low expression of this enzyme. Overall, such genetic approaches will drastically change our concepts of the application of microorganisms.

CONCLUSIONS

This review has attempted to give perspective to the concept of biofuels and oxychemicals from biomass. The availability of biomass for the preparation of oxychemicals, the continued increase of fuel alcohol production, and aspects of research for further enhancing ethanol fermentation by immobilization of cells, by the use of anaerobic thermophilic ethanologens and by genetic modification, have been briefly highlighted. Have the microbial ecologists aided in these developments, and how can they further interact with biotechnologists and geneticists?

TABLE 2. Saccharolytic thermophilic bacteria[a]

Strain	Temp (°C) Range	Optimum	Growth on substrate[b] Glucose	Cellobiose	Cellulose	Starch	Xylose	Xylan	Sucrose	Products[c] Relative mol formed Ethanol	Acetate	L-Lactate	CO₂	H₂	Ethanol yield (mol/mol of sugar)
Clostridium thermocellum LQRI	40–70	62	+	+	+	−	−	−[d]	−	157	165	24	346	268	
C. thermohydrosulfuricum 39E	38–78	65–70	+	+	−	+	+	−	+	543	31	50	580	31	1.9
C. thermosaccharolyticum HG-3	55–64		+	+	−	+	+	+	+	4.2	2.28	0.6	NT	NT	
Thermoanaerobium ethanolicus JW200[e]	37–78	69	+	+	−	+	+	+	−	78.4	4.5	4.0	89.2	4.3	1.8
T. brockii HTD4	40–80	70	+	+	−	NT	−	−	+	224	48	352	230	20	
Thermobacteroides acetoethylicus HTB2	40–80	65	+	+	−	NT	NT	−	+	139	134	0	190	29	

[a] References 2, 19, 32, 33, 35.
[b] +, Metabolic utilization of the substrate; −, non-utilization of the substrate; NT, not tested.
[c] From cellobiose at 60 to 65°C.
[d] Cultures grown on cellulose produce a cellulase that degrades xylan.
[e] Culture grown on glucose (44 μmol).

Ecologists have been most instrumental in selecting a variety of microbes (germ plasms) that permit broad substrate utilization, rapid rates of growth, and tolerance to high product concentrations (7, 11, 17, 18, 26). However, microbial ecologists should not be complacent. I have focused on a current industrial process, ethanol production. What about other products: acetone and butanol, for instance? Neither compound was listed in the Office of Technology Assessment Report (24). I contend that this is due to the poor economics associated with the intolerance of *Clostridium acetobutylicum* to the products: essentially less than 2% total solvents. There has been considerable recent research activity by biotechnologists to improve this fermentation yield (14, 16). However, essentially all studies have utilized the original strains that were intolerant of the solvent products and have merely succeeded in confirming the results of the 1930s and 1940s. Why has there been no emphasis on selection of more tolerant wild strains from nature? Microbial ecologists have failed to get their message of "tolerance" across to biotechnologists. A selection scheme could be based on continuous culture with intermittent sparging of acetone or butanol to select for product-tolerant strains (note that continuous exposure to solvent would result in selection of solvent degraders). Such a system is perfect for melding the talents of ecologist and engineer. Numerous other examples are readily apparent.

Besides the selection of diverse microbial strains, a second equally important role of ecologists is to emphasize the balance of nature. This concept has been paraphrased by Bernard Davis (10) in noting, "I am sure that nature takes account of slight differential growth rates even more minutely than a banker compares interest rates on bonds. No device that improves the economy of operation of a cell will be neglected, including of course not only speed of reproduction in a given environment but also adaptability to fluctuating environments." To the biotechnologist that need for rapidly growing strains and the advantages of continuous culture for strain selection are manifest (5). However, the subtleties of biochemical regulation, from intracellular to intermicrobial levels, are often poorly understood. The microbial ecologist has a major role in both explaining and clarifying these concepts to the "frustrated ecologist," i.e., the biochemical engineer. Overall, the future for the fermentative production of chemicals from biomass is extremely bright but will only be achieved through the combined activities of biotechnologists, geneticists, and microbial ecologists.

New Jersey Agricultural Experiment Station Publication no. D-01111-2-83, supported by State funds, U.S. Hatch Act Funds, and U.S. Department of Energy Subcontracts no. DE-A505-80ER10702 and XR-9-8162-1.

LITERATURE CITED

1. **Armentrout, R. W., and R. D. Brown.** 1981. Molecular cloning of genes for cellulose utilization and their expression in *Escherichia coli.* Appl. Environ. Microbiol. **41:**1355–1362.

2. **Avgerinos, G. C., and D. I. C. Wang.** 1980. Direct microbiological conversion of cellulose to ethanol. Annu. Rep. Ferm. Proc. **4:**165–191.

3. **Bacon, J. S. D.** 1979. Factors limiting the action of polysaccharide degrading enzymes, p. 269–284. *In* R. C. W. Berkeley, G. W. Gooday, and D. C. Ellwood (ed.), Microbial polysaccharides and polysaccharases. Pub. Soc. General Microbiology, U.K. Academic Press, Inc., New York.

4. **Bungay, H. R.** 1981. Energy, the biomass options. John Wiley and Sons, New York.

5. **Bungay, H. R., L. S. Clesceri, and N. A. Andrianus.** 1981. Auto-selection of very rapidly growing organisms, p. 235–241. *In* C. W. Robinson, C. Vezina, K. Singh, I. Russell, and G. G. Stewart (ed.), Proceedings of the 6th International Fermentation Symposium, London, Ontario. Advances in Biotechnology, vol. 4. Pergamon Press, New York.

6. **Busche, R. M.** 1983. Will biomass supply feedstocks of the future? Indust. Chem. News **3:**34–35.

7. **Bushnell, M. E., and J. H. Salter (ed.).** 1981. Mixed culture fermentations. Academic Press, London.

8. **Chakrabarty, A. M., and J. F. Brown, Jr.** 1978. Microbial genetic engineering by natural plasmid transfer, p. 185–193. *In* A. M. Chakrabarty (ed.), Genetic engineering. CRC Press, Boca Raton, Fla.

9. **Cornet, P., D. Tronik, J. Millet, and J.-P. Aubert.** 1983. Cloning and expression in *Escherichia coli* of *Clostridium thermocellum* genes coding for amino acid synthesis and cellulase hydrolysis. FEMS Microbiol. Lett. **16:**137–141.

10. **Davis, B. D.** 1961. Cellular regulatory mechanisms. Cold Spring Harbor Symp. Quant. Biol. **26:**1–10.

11. **Ellwood, D. C., J. N. Hedger, M. J. Latham, J. M. Lynch, and J. H. Slater (ed.).** 1980. Contemporary microbial ecology. Academic Press, Inc., New York.

12. **Eveleigh, D. E.** 1981. The microbiological production of industrial chemicals. Sci. Am. **245:**154–178.

13. **Goldstein, I. S.** 1980. The potential of producing petrochemical feedstocks from biomass. Special Report. Solar Energy Research Institute, Golden, Colo.

14. **Hollaender, A. (ed.).** 1981. Trends in the biology of fermentations for fuels and chemicals. Plenum Press, New York.

15. **Hollaender, A. (ed.).** 1981. Genetic engineering of microorganisms for chemicals. Plenum Press, New York.

16. **Hollaender, A., A. I. Laskin, and P. Rogers (ed.).** 1983. Basic biology of new developments in biotechnology. Plenum Press, New York.

17. **Horikoshi, K., and T. Akiba.** 1982. Alkaphilic microorganisms. A new microbial world. Japan Scientific Societies Press, Tokyo.

18. **Kushner, D. J. (ed.).** 1978. Microbial life in extreme environments. Academic Press, Inc., New York.

19. **Ljungdahl, L. G., F. Bryant, L. Carreira, T. Saiki, and J. Wiegel.** 1981. Some aspects of thermophilic and extreme thermophilic anaerobic microorganisms, p. 397–419. *In* A. Hollaender (ed.), Trends in the biology of fermentation for fuels and chemicals. Plenum Press, New York.

20. **Lyons, T. P.** 1983. Ethanol production in developed countries. Proc. Biochem. **18:**18, 21–24.

21. **Maleszka, R., L. Neirinck, P. Y. Wang, A. P. James, I. A. Velicky, and H. Schneider.** 1982. Conversion of D-xylose to ethyl alcohol by yeasts, p. 515–518. *In* 4th Bioenergy R & D Seminar, Winnipeg, Manitoba, 29 March 1982. Pub. National Research Council of Canada, Ottawa.

22. **Ng, T., R. M. Busche, C. C. McDonald, and R. W. F.**

Hardy. 1983. Production of feedstock chemicals. Science **219**:733–740.

23. **Oda, G., H. Samejima, and T. Yamada.** 1983. Continuous alcohol fermentation technologies using immobilized yeast cells. Biotech '83 (London), p. 593–610.

24. **Office of Technology Assessment.** 1981. Impacts of applied genetics: microorganism, plants and animals. Office of Technology Assessment, Washington, D.C.

25. **Schneider, H., R. Maleszka, L. Neirinck, I. A. Velicky, P. Y. Yang, and W. K. Chan.** 1983. Ethanol production from D-xylose and several other carbohydrates by *Pachysolen tannophilus* and other yeasts. Adv. Biochem. Eng. Biotechnol. **27**:57–71.

26. **Slater, J. H., R. Whittenbury, and J. W. T. Wimpenny (ed.).** 1983. Microbes in their natural environments. 34th Symposium of the Society for General Microbiology. Cambridge University Press, Cambridge, U.K.

27. **Stokes, H. W., S. K. Picataggio, and D. E. Eveleigh.** 1983. Recombinant genetic approaches for efficient ethanol production. Adv. Solar Energy **1**:113–132.

28. **Venkatasubramanian, K., and C. R. Keim.** 1981. Gasohol: a commercial perspective. Proc. N.Y. Acad. Sci. **369**:187–204.

29. **Venkatasubramanian, K., and C. R. Keim.** 1982. Gasohol: issues and some answers, p. 473–479. *In* Pacific Synfuels Conference. The Japan Petroleum Institute, Tokyo.

30. **Whittle, D. J., D. G. Kilburn, R. A. J. Warren, and R. C. Miller, Jr.** 1982. Molecular cloning of a *Cellulomonas fimi* cellulase gene in *Escherichia coli*. Gene **17**:139–145.

31. **Wiegel, J.** 1980. Formation of ethanol by bacteria. A pledge for the use of extreme thermophilic anaerobic bacteria in industrial ethanol fermentation processes. Experientia **36**:1434–1446.

32. **Wiegel, J., and L. G. Ljungdahl.** 1981. *Thermoanaerobacter ethanolicus* gen. nov., spec. nov. A new extreme thermophilic anaerobic bacterium. Arch. Microbiol. **128**:343–348.

33. **Zeikus, J. G.** 1980. Chemical and fuel production by anaerobic bacteria. Annu. Rev. Microbiol. **34**:423–464.

34. **Zeikus, J. G., A. Ben-Bassat, T. K. Ng, and R. J. Lamed.** 1981. Thermophilic ethanol fermentations, p. 441–461. *In* A. Hollaender (ed.), Trends in the biology of fermentations. Plenum Press, New York.

35. **Zeikus, J. G., and T. Ng.** 1982. Thermophilic saccharide fermentations. Annu. Rep. Ferm. Proc. **5**:263–289.

Physiology and Biochemistry of Lignin Degradation†

C. A. REDDY

Department of Microbiology and Public Health, Michigan State University, East Lansing, Michigan 48824-1101

Lignin is the most abundant and widely distributed renewable source of aromatic nucleus on earth. Lignin constitutes 25% of the dry weight of the estimated 100 billion metric tonnes of photosynthetic biomass produced annually in the biosphere and is second only to cellulose, which accounts for 50% of the dry weight of biomass (98). However, lignin contains 50% more carbon than cellulose and other plant polysaccharides and is, therefore, relatively more abundant as a source of reduced carbon and solar energy than its weight would indicate. Lignin biodegradation is of utmost importance in the earth's carbon cycle not only because of its great abundance but also because of its recalcitrance. Biodegradative recalcitrance of lignin is believed to be due to its high molecular weight and three-dimensional structure (90, 115). Whereas the sugar and polysaccharide fractions in biomass are degraded relatively rapidly by soil microbes, the lignin fraction is metabolized rather slowly (82). Approximately 1×10^{12} to 3×10^{12} metric tonnes of carbon is calculated to be present in the organic matter fraction of soil (112), primarily as peat and humus, which to a large extent are derived from lignins (53). Humus is also degraded slowly, with an estimated half-life in different soils of 76 to 326 years (41). The recalcitrance of lignin is also reflected by the occurrence of lignin-derived phenolic residues in lignites and bituminous coal (48) and in 200×10^6-year-old silicified wood (104). The importance of lignin biodegradation in global carbon cycling also stems from the fact that in intact wood tissues lignin occurs in close physical and chemical association with cellulose and hemicellulose and protects these polysaccharides from attack by polysaccharide-digesting enzymes (21, 92). Therefore, the biodegradation of polysaccharides in lignocellulosic materials, an extremely important process in the recycling of biospheric carbon, becomes limited. Degradation or modification of lignin barrier is also important for the efficient conversion of cellulose and hemicellulose in biomass into fuels and chemicals (21, 86) and for increasing the efficiency of biopulping processes (69). Furthermore, lignin is itself a potential source of feedstock chemicals such as polyphenols, vanillin, styr-enes, and various low-molecular-weight phenolics. Thus, in recent years lignin degradation and transformation by microbes has been an area of great interest and represents a tremendous challenge to microbial ecologists in the future.

Previous workers have shown that several different genera of fungi and bacteria are capable of metabolizing lignin (8, 11, 27, 60, 64, 65, 87, 114). However, most of the recent investigations on fungal biodegradation of lignin involved a single species of white-rot fungus, *Phanerochaete chrysosporium*, which has become somewhat of a model organism for lignin biodegradation studies (51), and emphasis is placed in this chapter on the physiology and biochemistry of lignin degradation by this organism. A number of recent reviews (1, 3, 8, 64, 65, 72, 93, 114) and books (11, 31, 51, 73, 99) have covered the accumulated knowledge on microbial degradation of lignin, including ecological aspects of lignin degradation, and should be consulted for a more complete bibliography.

OCCURRENCE, DISTRIBUTION, AND FUNCTION OF LIGNIN

Lignin occurs as an integral cell wall component and in the middle lamellae of all vascular plants (99) and is not found in liver worts, algae, or mosses, which lack the conductive tissue common to vascular plants (100). Lignin generally constitutes 20 to 30% of the dry weight of woody tissues, whereas its concentration in other plant tissues may vary from <1 to 35% of the dry weight. About 80% of the lignin is found within cell walls, intimately interspersed with the hemicelluloses (109). This lignin–hemicellulose matrix surrounds the cellulose fibrils in a sheathlike manner to form a physically impenetrable barrier such that polysaccharide-degrading enzymes have only limited access to the cellulose and hemicelluloses in lignified plant tissues. UV microscopy of thin sections of plant tissues revealed that 70 to 80% of the total lignin in black spruce early wood is located within the secondary cell wall layers, with the balance present in the middle lamellae (26). Lignin is the dominant constituent (up to 72%) of the middle lamellae and serves as a cement binding the cells together.

Within plants, lignin imparts structural rigidity

† Journal article no. 11098 from the Michigan Agriculture Experiment Station.

and resistance to impact, compression, and bending, it decreases water permeation across cell walls of xylem tissue, and it protects plant tissues from invasion by plant pathogens (99, 108).

CHEMICAL STRUCTURE

Lignin is a complex and variable biopolymer of p-hydroxyphenylpropane monomer units. The immediate biosynthetic precursors of the lignin polymer are the three cinnamyl alcohol derivatives: p-coumaryl, coniferyl, and sinapyl alcohols (Fig. 1). These alcohols are synthesized by the general path: $CO_2 \to$ carbohydrates $\to$ phenyl propanoid amino acids $\to$ cinnamic acid derivatives $\to$ cinnamyl alcohol derivatives (49). These three alcohols are present in different ratios in lignins from different plant species and often in different tissues of the same plant (11, 65, 100). Guaiacyl lignin (gymnosperm lignins, as in most conifers) is composed mostly of coniferyl alcohol units with small amounts of coumaryl and synapyl alcohol-derived units (approximate proportions in spruce lignin are 80:14:6, respectively). Guaiacyl-syringyl lignins, characteristic primarily of angiosperms, contain approximately equal amounts of coniferyl alcohol and synapyl alcohol, with only minor amounts of coumaryl alcohol units (approximate proportions in beechwood are 49:46:5, respectively). However, hardwood lignins are intrinsically heterogeneous, and the proportion of syringyl units may vary from 20 to 60%. Most lignins contain small amounts (usually <10%) of monomer units derived from coumaryl alcohol, although certain grasses may contain relatively higher amounts of coumaryl alcohol units (52). In general, higher proportions of coniferyl alcohol units (guaiacyl units) are incorporated into lignin in young plant tissues, whereas during maturation the relative proportion of synapyl units (syringyl units) incorporated increases (40).

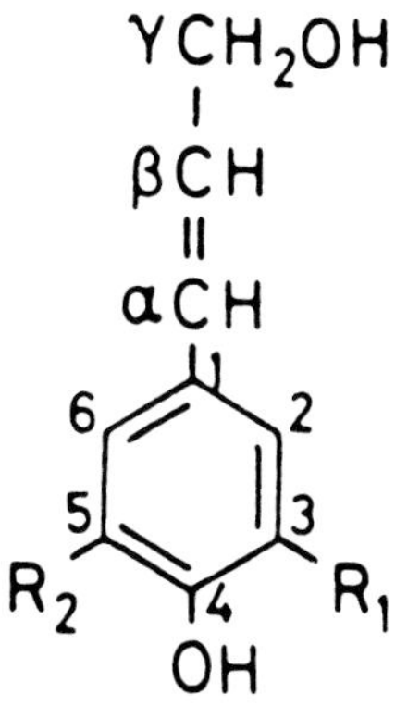

The lignin polymer is formed through the dehydrogenative polymerization of the three cinnamyl alcohol derivatives mentioned above (30, 31, 46, 49, 100). At the site of lignification in the cell wall, single electron oxidation of phenolic hydroxyl groups of the cinnamyl alcohol precursors by a peroxidase (EC 1.11.1.7, donor: hydrogen peroxide oxidoreductase) produces phenoxy radicals (Fig. 2). Because of the extended π electron system of the precursors, the free radicals are stabilized through equilibrium with several mesomeric forms (Fig. 2). All the mesomeric forms derived from coniferyl, sinapyl, and coumaryl alcohols randomly couple with each other, but more importantly with the radicals in the growing lignin polymer, the phenolic hydroxyl groups of which are also subject

FIG. 1. Cinnamyl alcohol derivatives, p-coumaryl [1], coniferyl [2], and synapyl alcohol [3], the immediate precursors for lignin biosynthesis.

to single electron oxidation. A variety of intermonomeric linkages result, and a complex, three-dimensional, "natural plastic," with a number of nonrecurring C–C and C–O–C linkages, which are not directly hydrolyzable, are formed (Fig. 3). As a result of this polymerization, the asymmetric carbons in the polymer (C–α and C–β of the monomer side chain) exist in both R and S forms, and therefore, lignin and its chemical degradation products are optically inactive. The major intermonomer linkages and their frequencies in representative gymnosperm (spruce) and angiosperm (birch) lignins are shown in Table 1. The molecular weight of purified lignin varies from a few thousand to more than 10^6, depending on the isolation method, whereas lignins in vivo are believed to have molecular weights in the range of 10^5 or greater (100). Therefore, the polymer illustrated in Fig. 3 represents only a small piece of the total lignin molecule.

It should be clear from the above discussion that lignin is a unique biopolymer. Unlike other polymers such as cellulose, starch, proteins, and nucleic acids, it does not contain identical, readily hydrolyzable linkages which repeat at regular intervals. Instead, lignin is a highly irregular, three-dimensional polymer that has no precise chemical structure but contains a series of chemical groupings which are different in lignins from various sources. These structural features dictate that the lignin biodegradation system be relatively nonspecific and extracellular, and that

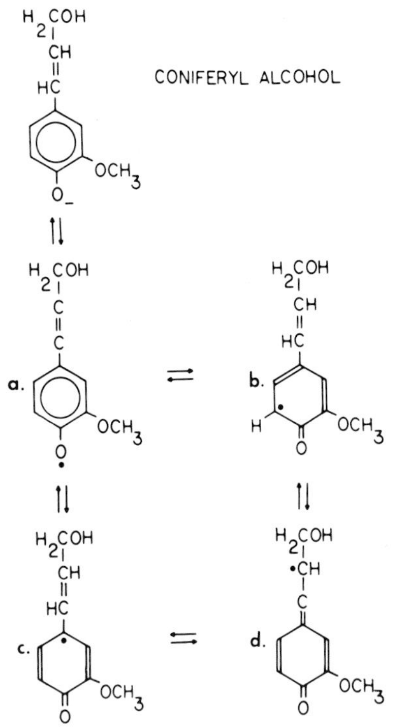

H₂COH
│
CH
‖
HC   CONIFERYL ALCOHOL

OCH₃
O⁻

⇅

FIG. 2. Single electron oxidation of coniferyl alcohol leads to the radical species which exist in mesomeric forms a–d.

the occurrence of lignin hydrolases, akin to cellulases, is unlikely.

METHODS FOR THE STUDY OF LIGNIN BIODEGRADATION

A serious obstacle to the study of lignin biodegradation in the past 50 years has been the lack of simple and sensitive assays for measuring lignin decomposition. However, with the advent of sensitive, specific radioisotopic methods for the study of lignin biodegradation, rapid advances have been made in the past 8 years in understanding biological decomposition of lignin. The reader should consult several recent reviews on this topic for further details (11, 12, 73). Radioisotopic procedures primarily utilize two basic substrates: ^{14}C-[lignin]-lignocelluloses (natural lignocelluloses labeled specifically in their lignin component) or ^{14}C-DHPs (dehydrogenative polymers or synthetic lignins). Microbial degradation of ^{14}C-lignins results in the production of $^{14}CO_2$, which can be trapped and quantified. In this way both rates and extent of lignin biodegradation can be studied.

^{14}C-[lignin]-lignocelluloses are prepared by feeding the plant radioactive precursors such as ^{14}C-phenylalanine or ^{14}C-ferulic acid and extracting the lignocellulose as described previous-

ly (11, 12). ^{14}C-synthetic lignins (DHPs), on the other hand, are prepared by the in vitro dehydrogenative polymerization of chemically synthesized ^{14}C-coniferyl alcohol in a peroxidase/H_2O_2-catalyzed reaction (44, 70). By utilizing as substrates DHPs that are labeled specifically in the side chain, aromatic ring, or methoxyl groups, it is possible to determine the specificity of structural attack on the lignin polymer by a specific microorganism (11, 12, 70). ^{14}C-DHP or ^{14}C-[lignin]-lignocelluloses should be carefully characterized on the basis of previously described criteria (11, 70, 114) before they are used as substrates for studies on lignin biodegradation.

^{14}C-[lignin]-lignocelluloses are preferred substrates, in comparison with carbohydrate-free DHPs, for screening for microorganisms that selectively delignify wood, for studying kinetics of lignin degradation in natural ecosystems, and for defining relationships between lignin and cellulose degradation in various organisms. Similarly, ^{14}C-[lignin]-lignocelluloses are preferable to ^{14}C-DHPs for studying the kinetics of lignin humification since lignin-plus-cellulose complex is the normal substrate for humification processes that occur in natural environments.

^{14}C-DHPs do not contain carbohydrates or other extraneous materials and, therefore, are ideal substrates for the unequivocal determination of an organism's ability to utilize lignin as a sole carbon source, or for examining the ability of various compounds to serve as cosubstrates for lignin degradation. However, since ^{14}C-DHPs do not contain plant polysaccharides, extrapolation of conclusions made from such experiments performed with DHP to decomposition of natural lignin in plant tissue may not be possible or accurate. For example, Robinson and Crawford (97) found that a Bacillus sp. could degrade approximately 12% of ^{14}C-[lignin]-lignocellulose but only 0.3% of the ^{14}C-DHP to $^{14}CO_2$ in 20 days. As a practical guide, if microbial degradation of ^{14}C-DHP to $^{14}CO_2$ exceeds 3% or that of ^{14}C-[lignin]-lignocelluloses exceeds 8 to 10%, one can reasonably conclude that the organism is capable of degrading lignin.

LIGNIN-DEGRADING MICROORGANISMS

White-rot fungi include several hundred species of basidiomycetes and a few species of ascomycetes. Members of this group are the most extensively studied of the wood-rotting fungi and have been shown to extensively and rapidly degrade all major components of wood to CO_2 and water (3, 8, 11, 49, 62, 65). Carbohydrate and lignin are degraded simultaneously during white-rot decay, although some white-rot fungi appear to preferentially degrade the lignin component (3, 11, 62, 65). P. chrysosporium

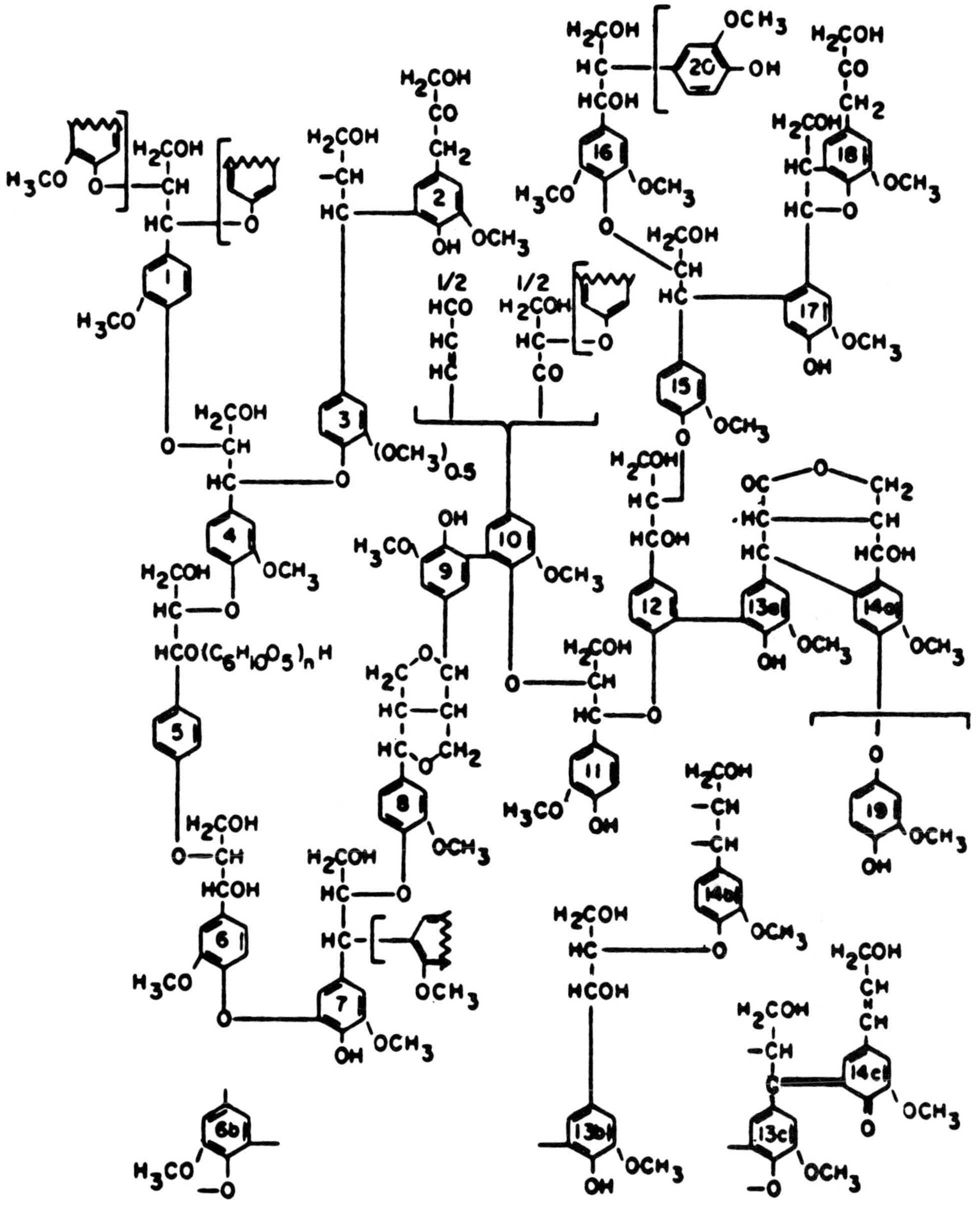

FIG. 3. Schematic formula for a representative portion of a spruce lignin (46).

(syn: *Sporotrichum pulverulentum*), *Coriolus versicolor*, and *Pleurotus ostreatus* have been shown to degrade ^{14}C-DHPs (labeled in the ring, the side chain, or the methoxyl group) to $^{14}CO_2$ (42, 43, 70). Spruce lignin was shown to be heavily degraded by *P. chrysosporium, Polyporous anceps, Coriolus versicolor*, and *Poria subacida* on the basis of thorough chemical analyses (67, 68). Extensive literature on the chemistry of lignin degradation by white-rot fungi has been reviewed recently (65).

Brown-rot fungi comprise numerous species of basidiomycetes which typically carry out extensive degradation of cellulose and hemicellulose in wood but cause only a limited degradation of the lignin component (3, 8, 11, 62, 65). Detailed chemical analysis of sweetgum wood decayed by *Lenzites trabea* has shown, however, that brown-rot attack on lignin is oxidative and that demethoxylation and extensive ring hydroxylation in the 2 position are the major degradative modifications observed in brown-

TABLE 1. Major intermonomer linkages and their frequencies in gymnosperm (spruce) and angiosperm (birch) lignins[a]

Linkage type	Units in Fig. 3	% of total C9 units	
		Spruce	Birch
β-Aryl ether (β-0-4)	1–4, 5–4, 5–6, 7–8, 13b–14b, 15–16	48	60
Phenylcoumaran (β-5)	17–18	9–12	6
Biphenyl (biaryl or 5-5)	12–13a	9.5–11	4.5
1,2-Diarylpropane (β-1)	16–20	7	7
Diphenyl ether (diaryl or 4-0-5)	6–7	3.5–4	6.5
α-Aryl ether (α-0-4)	11–12	6–8	6–8
Pinoresinol (β-β)	8–9	2	

[a] Adapted from reference 65.

rotted lignin (63, 66). The above structural changes lead to the formation of O-diphenolic moieties (66) which are believed to undergo autooxidation to produce quinone-type chromophores that give brown-rotted wood its characteristic color. Aromatic ring cleavage was minimal, but the carboxyl and carbonyl content in lignin decayed by *L. trabea* was twice that in sound lignin. On the basis of decomposition studies with differentially labeled ^{14}C-DHPs, Kirk et al. (70) showed that demethoxylation was the major feature of lignin exposed to the brown-rot fungi *Poria coeos* and *Gloeophyllum trabeum*. The principal difference between white-rot and brown-rot fungi is believed to be the inability of the latter to metabolize aromatic rings or the aliphatic products of aromatic ring cleavage.

Soft-rot fungi include a variety of *Ascomycetes* and Fungi Imperfecti and have been shown to decompose all major components of wood, including lignin (19, 42, 43). Eslyn et al. (19) examined strains of *Graphium, Monodictys, Allescheria, Paecilomyces, Papulospora,* and *Thielevia* and showed that lignin degradation is characteristic of soft-rot decay. Also, these species (42, 43) appear to degrade lignin in hardwoods more effectively than that in softwoods. Haider and Trojanowski employed ^{14}C-ring-, side chain-, and methoxyl-labeled DHPs as well as ^{14}C-lignin-lignocelluloses (from maize) and concluded that soft-rot fungi *Preussia, Chaetomium,* and *Stachybotrys* carry out substantial degradation of lignin. However, the chemical nature of the soft-rotted lignin is not known.

Other fungi, including *Fusarium solani* (50, 56) and *Aspergillus fumigatus* (14), have been shown to degrade ^{14}C-lignins and lignin model compounds, but further studies are needed to examine the ability of other fungal species to degrade ^{14}C-DHPs and ^{14}C-lignin-lignocelluloses.

Numerous reports have appeared in the past few years relating to bacterial degradation of lignin, and these have been reviewed recently (8, 11, 114). A number of nocardia strains have been shown to degrade ^{14}C-DHP or [14]C-[lignin]-lignocellulose from maize (8, 43, 44, 106). The extent of conversion of ring-, side chain-, and methoxyl-labeled DHP to [14]CO2 was 5, 15, and 13%, respectively (106). A strain of *Bacillus* sp. (97) and a number of *Streptomyces* strains (7–11) have been shown to degrade [14]C-[lignin]-lignocelluloses. *Streptomyces flavovirens* (105) was shown to attack and destroy the integrity of both lignified and nonlignified cell walls within the inner bark of Douglas fir, on the basis of weight loss and scanning electron microscopy. A strain of *Streptomyces badius* (10), shown to be their most active ligninolytic strain, degraded 13% of ^{14}C-lignin-labeled lignocellulose to [14]CO2 in 1,008 h at 35°C. Kerr et al. (60) most recently isolated a strain of *Arthrobacter* sp. and a number of other bacterial strains which degraded peanut hull lignin, ^{14}C-kraft lignin from slash pine, and [14]C-[lignin]-lignocellulose from cord grass, *Spartina alterniflora*. Odier and co-workers (88, 89) demonstrated that strains of *Flavobacterium, Pseudomonas,* and *Aeromonas* spp. degrade acidolysis lignin, and a strain of *Xanthomonas* sp. was reported to degrade 77% of the dioxane lignin, provided in the medium as a sole carbon and energy source, in 15 days. Degradation of high-molecular-weight (>1,500) kraft lignin or milled wood lignin by pseudomonads has been reported (27, 58). The above results show unequivocally that bacteria decompose lignin. However, information is sparse about the chemical nature of lignins decayed by bacteria. It is also noteworthy that lignin degradation by either pure or mixed cultures of anaerobes has not been shown unequivocally (89, 114, 115).

PHYSIOLOGY OF LIGNIN DEGRADATION

There has been a dramatic progress in our understanding of the physiology of lignin degradation since rapid, specific, and sensitive assays for ligninolytic activity, utilizing ^{14}C-lignin preparations, became available in 1975 (11, 12, 70). Most of the physiological and biochemical investigations have been conducted with *Phanerochaete chrysosporium* Burds (ATCC 34541) (6). This organism was chosen for its rapid growth and metabolism of lignin, prolific conidiation, a relatively high temperature optimum of 40°C (6), low level of phenol oxidase activity, and the ability to grow on either complex or chemically defined media (61). Techniques for conducting genetic studies, including mutant isolations, are also available for this organism (32, 34–37, 39).

A number of culture parameters which influence lignin degradation by *P. chrysosporium* have been optimized. Culture agitation results in pellet formation and strongly suppresses ligninolytic activity; however, good growth is observed in both agitated and stationary cultures. Also, agitation of pregrown mats does not appear to affect lignin degradation (75, 113). Control of pH and the choice of buffer were found to be critical to lignin degradation by this organism (24, 75). An optimum pH of 4 to 4.5 was found for lignin degradation. Sodium 2,2′-dimethylsuccinate was found to be an ideal buffer, whereas *o*-phthalate was found to be inhibitory (24). Casamino acids, or a mixture of an ammonium salt plus asparagine, are optimal for growth, but high nitrogen concentration in the medium decreases ligninolytic activity (61, 75, 113). Cultures must be nitrogen starved for sustained degradation of lignin (see below). Thiamine is required for growth, and the balance of trace metals is important (57, 75, 95).

Oxygen partial pressure has a profound effect on the rate and extent of lignin degradation by *P. chrysosporium* but not on the growth of the organism (5, 75, 95). In a defined medium containing limiting amounts of nitrogen, there was essentially no degradation in an atmosphere of 5% O_2–95% N_2. In a 100% O_2 atmosphere there was no difference in the onset of lignin degradation, but there was a two- to threefold enhancement of degradation compared with that observed in an atmosphere of air (21% O_2). These results are consistent with the finding that lignin degradation is a highly oxidative process (11, 65).

A number of studies indicated that the ligninolytic system is not inducible by lignin and that lignin does not serve as a growth substrate for *P. chrysosporium* (61, 71). This organism fails to degrade ^{14}C-DHPs to $^{14}CO_2$ in the absence of a suitable growth substrates such as cellulose, glucose, xylose, glycerol, or succinate (61, 71, 75). The amount of $^{14}CO_2$ produced from ^{14}C-DHP was dependent on the amount of growth substrate provided. Drew and Kadam (14) reported an obligate cosubstrate requirement for ^{14}C-kraft lignin degradation for the white-rot fungi: *P. chrysosporium, Coriolus versicolor*, and *Sporotrichum pulverulentum*. Similar results were obtained with an imperfect fungus *Aspergillus fumigatus* (14). Odier and Roch (90) reported that degradation of ^{14}C-lignin-labeled poplar wood by several white-rot fungi is stimulated at limiting glucose concentrations (1 g/liter) and that lignin degradation is inhibited at nonlimiting glucose concentrations (10 g/liter). Reid (96) also reported increased degradation of ^{14}C-

[lignin]-labeled aspen wood by *P. chrysosporium* in the presence of an increased supply of cellulose. These results clearly indicate that lignin cannot be degraded by white-rot fungi unless a cosubstrate is simultaneously available. Other evidence (57) indicates that lignin is at best only a marginal carbon/energy source for maintenance metabolism.

Lignin degradation by *P. chrysosporium* has been shown to be a secondary metabolic event and is triggered in response to N, C, or S starvation (57, 61, 75, 96). Ligninolytic activity is triggered irrespective of the presence of lignin in the medium, and the presence of lignin does not induce additional activity. The separation of lignin degradation from primary growth and the fact that the ligninolytic system is noninducible are further indications that the energy in lignin is of little importance to the growth of *P. chrysosporium*. Thus, lignin biodegradation appears to be substantially different from that of other biopolymers.

The level of nutrient nitrogen provided in the medium has a profound influence on ligninolytic activity by *P. chrysosporium* (61, 75). In cultures containing nonlimiting levels of glucose (56 mM) and limiting levels of nitrogen (2.4 mM), a rapid depletion of N in 2 days of incubation stopped primary growth and the ligninolytic activity appeared during late stationary phase. In contrast, in a medium containing 56 mM glucose and nonlimiting levels of N (24 mM), the onset of ligninolytic activity was delayed considerably, although growth at higher N levels was as good as or better than that observed at low N levels (25, 29, 61, 96). Addition of NH_4^+ or a number of other nitrogenous compounds to cultures grown in low N medium immediately prior to the appearance of ligninolytic activity delayed appearance by several days, but addition of 2 to 4 mM N to active ligninolytic cultures did not affect ligninolytic activity for 16 h and then caused only a transient repression (61). Of the nitrogenous compounds examined, NH_4Cl, glutamine, glutamate, and histidine were the most effective, repressing ligninolytic activity by 50, 76, 83, and 76%, respectively, compared with negative control cultures with no N added. The repression of ligninolytic activity on addition of glutamate and NH_4^+ does not appear to be mediated by suppression of central carbon metabolism since both NH_4^+ and glutamate stimulated the oxidation of succinate and glucose (23). Addition of glutamate and NH_4^+ also suppressed the synthesis of veratryl alcohol, a secondary metabolite (25, 80, 101). These and other results indicated that nitrogen metabolism affects ligninolytic activity as a part of secondary metabolism and that glutamate metabolism plays a key role in the regulation of lignin degradation by *P. chrysosporium*. Reid (96) showed that degradation by *P. chrysosporium* of ^{14}C-lignin-labeled aspen wood is triggered by N starvation and inhibited by high levels of N. He further indicated that the C/N ratio is a better predictor of lignin degradation than the absolute C and N levels. Odier and Roch (90) reported that lignin biodegradation (^{14}C-lignin-labeled poplar wood → ^{14}CO$_2$) by *P. chrysosporium* and *Phelebia radiata* is dependent on the C/N ratio in the medium. However, in the case of *Sporotrichum pulverulentum* and *Dichomitus squalens*, no relation between lignin biodegradation and C/N ratio could be established, suggesting that the suppression of ligninolytic activity by high levels of nutrient nitrogen is variable among white-rot fungi (see 4).

Weinstein et al. (110) showed that in nitrogen-limited cultures of *Phanerochaete chrysosporium* degradation of dimeric lignin model compounds, ^{14}C-guaiacylglycerol-β-guaiacylether and veratrylglycerol-β-guaiacyl ether, is associated with secondary metabolism and is triggered in response to N starvation. The metabolism of the same compounds was greatly suppressed in high-N (12 mM) cultures or on addition of exogenous NH$_4$$^+$ to low N cultures, and was stimulated two- to threefold in cultures incubated under 100% O$_2$ atmosphere compared with those in air. In agitated cultures, the metabolism of the above compounds was only about one-tenth of that observed in stationary cultures. These and other results (15–17, 102) show that the effect of different culture parameters on the metabolism of lignin model compounds was analogous to their effect on ligninolytic activity (5, 61, 75).

Ligninolytic activity in nitrogen-rich, carbohydrate-limited cultures of *P. chrysosporium* is triggered in response to carbohydrate starvation and is accompanied by autolysis. Lignin degradation ceases when autolysis does, suggesting that autolysis provides the carbon and energy for lignin degradation and cell maintenance (57). Addition of glutamate (at concentrations supplying insignificant levels of carbon) to carbohydrate-limited cultures strongly suppressed lignin degradation (~50%), as it does in nitrogen-limited cultures; however, this inhibition of lignin degradation was not observed on addition of an equimolar amount of NH$_4$$^+$, indicating that the observed inhibition was not due to the presence of additional nitrogen in the medium. That the inhibition by glutamate was not due to added carbon was evidenced by the fact that the addition of equimolar amounts of C in the form of α-ketoglutarate stimulated lignin degradation by approximately 30%. Addition of α-ketoglutarate and NH$_4$$^+$ stimulated lignin degradation to a degree equal to that obtained upon addition of α-ketoglutarate only. Apparently, glutamate per

se inhibits lignin degradation in carbohydrate-limited cultures, but the mechanistic basis for this is not known.

Lignin degradation was reported to be derepressed in 7 days on depletion of sulfur in cultures which initially contained limiting levels of SO$_4$$^{2-}$ (20 mM) and nonlimiting levels of carbohydrate and nitrogen (57); phosphorus starvation did not have a corresponding effect on lignin degradation. When the initial SO$_4$$^{2-}$ concentration was 200 mM, no ligninolytic activity appeared. Dual limitation of SO$_4$$^{2-}$ and N did not result in more rapid or more extensive lignin degradation. The basis for the effect of sulfur on lignin degradation by *P. chrysosporium* is not known. Interestingly, Reid (96) reported that neither sulfur nor phosphorus limitation triggers ligninolytic activity (^{14}C-lignin-labeled aspen wood → ^{14}CO$_2$) in cultures of *P. chrysosporium*.

BIOCHEMISTRY OF LIGNIN DEGRADATION

The recent availability of the aforementioned methods and knowledge of physiology has allowed a number of investigators to begin intensive studies on the biochemistry of lignin degradation by *P. chrysosporium* (51, 65). The degradative pathways for several lignin substructure model compounds by white-rot fungi have been elucidated, and some of these reactions are now being used as assays for biochemical investigations (15–18, 82, 102). The role of oxygen radicals (28, 29, 59, 94) and extracellular fluids from *P. chrysosporium* cultures (33, 107) in lignin degradation has been elucidated, and recent publications by Gold and co-workers (32, 34–37, 39) have laid the groundwork for genetic approaches to understanding the molecular basis of lignin degradation.

Biodegradation of the lignin polymer appears to be extracellular, oxidative, and relatively nonspecific (47, 55, 67, 68, 85, 106). The oxidative nature of the degradation process is indicated by: (i) the substantial increase in carboxyl and carbonyl groups and a decrease in hydrogen content in lignin residues left after white-rot decay; (ii) oxidation of α-carbon of the propyl side chain to a carbonyl group; (iii) hydroxylation and oxidative cleavage of the aromatic rings; (iv) demonstration of the insertion of an O$_2$-derived hydroxyl group at β-carbon of the propyl side chain utilizing β-1 lignin substructure compounds (units 16–20 in Fig. 3) (65); and (v) evidence that some of the aromatic rings are hydroxylated in C$_2$ position (64, 65). The nonspecificity of the ligninolytic system is indicated by the fact that: (i) lignin polymer is degraded despite the variety of intermonomer linkages and the many different chemical groupings around these linkages; (ii) a variety of structurally different lignins are degraded by the same organism (14, 75, 81); (iii) lignin is degraded despite the racemic nature of all the asymmetric carbons in the polymer; (iv) ^{14}C-"polyguiacol," a polymer bearing a relatively low resemblance to lignin, is metabolized to $^{14}CO_2$ by *P. chrysosporium* (13); and (v) the ligninolytic system metabolizes a variety of lignin substructure model compounds and exhibits no stereoselectivity either for the substrate or for the product formed (11, 65, 83, 85, 106, 110). Direct uptake of the lignin polymer by microbial cells is unlikely in view of its relatively large size, and therefore the polymer-degrading system is believed to be extracellular. Since it is difficult to rationalize these features in terms of specific lignin-degrading enzymes, it has been speculated that the actual extracellular agent(s) attacking the lignin polymer might be activated oxygen species such as superoxide anion (45) or a relatively nonspecific, extracellular enzyme.

Amer and Drew (2) reported that superoxide anion (O_2^{-}) is produced by whole cells of *Coriolus versicolor*, on the basis of cytochrome *c* reduction, and they speculated that the above anion may be involved in lignin degradation by this fungus. However, O_2^{-} is known to be relatively nonreactive in aqueous systems, and it has been proposed that essentially all of the in vitro manifestations of O_2^{-} can be explained by the involvement of O_2^{-} in the iron-catalyzed Haber-Weiss reaction (see below) to produce hydroxyl radical (see 29). Thus, if O_2^{-} is involved in lignin degradation, it is possible that it serves as a reductant for the production of hydroxyl radical ($^{.}OH$) as discussed by Forney et al. (29). Such an involvement of O_2^{-} is also supported by the observation that addition of superoxide dismutase results in inhibition of ligninolytic activity by *P. chrysosporium*.

Nakatsubo et al. (84) reported that singlet oxygen ($^1O_2$) is involved in lignin degradation by *P. chrysosporium* primarily on the basis of inhibition of lignin degradation, but not glucose oxidation, by anthracene-9,10-bis ethanesulfonic acid (AES) and that an artifical $^1O_2$-generating system (photosensitizing riboflavin system) produced degradation products from β-1-model compound similar to those produced by ligninolytic cultures (56, 106). Kirk et al. (74) recently expressed doubts about the involvement of $^1O_2$ because the mechanism of degradation of the β-1-model compound (a diarylpropane lignin model compound; units 16–20 in Fig. 3) in cultures of *P. chrysosporium* was found to be different from that observed with the artificial $^1O_2$-generating system. Kutsuki et al. (78), utilizing a similar photosensitizing riboflavin system, reported that photosensitizing riboflavin cleaves fully alkylated diarylpropane lignin model com-

pounds via a mechanism which does not involve $^1O_2$. Furthermore, since AES scavenges $^1O_2$ as well as $^{.}OH$, one cannot rule out the possibility that the observed inhibition of ligninolytic activity in the presence of AES is due to the quenching of $^{.}OH$, which appears to play a key role in lignin degradation according to a number of recent reports (29, 79, 94).

Forney et al. (29) proposed the hypothesis that H_2O_2-derived $^{.}OH$ may be the exracellular agent involved in lignin degradation by *P. chrysosporium*. Support for this hypothesis comes from the fact that: (i) $^{.}OH$ is highly reactive and oxidizes a variety of organic compounds; (ii) $^{.}OH$ is nonspecific and non-stereoselective, and can be produced by one electron reduction of H_2O_2, substantial amounts of which have been known to be produced by a number of white-rot fungi (76, 77). In biological systems, H_2O_2 is thought to be generated by two types of reactions:

Fenton reaction
$$H_2O_2 + Fe(II) \rightarrow {}^{.}OH + OH^- + Fe(III)$$

Iron-catalyzed Haber Weiss reaction
$$Fe(III) \text{ chelate} + O_2^{-} \rightarrow Fe(II) \text{chelate} + O_2$$
$$Fe(II) \text{ chelate} + H_2O_2 \rightarrow {}^{.}OH + OH^- +$$
$$\underline{\qquad\qquad\qquad Fe(III) \text{ chelate}}$$
$$\text{sum: } H_2O_2 + O_2^{-} \rightarrow {}^{.}OH + O_2$$

Hydrogen peroxide-derived $^{.}OH$ appears to play a key role in lignin degradation by *P. chrysosporium* based on the following evidence. (i) There was a positive temporal correlation between the specific activity for H_2O_2 production, $^{.}OH$ production, and ligninolytic activity (29, 59, 94). (ii) Ligninolytic cultures produced $^{.}OH$ as evidenced by ethylene production from α-keto-α-methylthiobutyric acid (KTBA; 29, 59), hydroxylation of *p*-hydroxybutyric acid to protocatechuic acid (29), and identification of the adduct of $^{.}OH$ with 5,5'-dimethyl-1-pyrroline-*N*-oxide by electron spin resonance spectroscopy (29). (iii) Nutritional parameters which are known to affect ligninolytic activity had a parallel effect on the specific activities for $^{.}OH$ and H_2O_2 production (29, 59). (iv) There was a direct correlation in the patterns of H_2O_2 and $^{.}OH$ production observed with different growth substrates and the ligninolytic activity observed with the same substrates (94). (v) Lignin degradation, but not glucose oxidation, was markedly suppressed in the presence of the $^{.}OH$-scavenging agents benzoate, butylated hydroxytoluene, and mannitol (Table 2) (29). Mannitol was the least inhibitory of the compounds tested. These results were subsequently confirmed by Kutsuki and Gold (79) and Gold et al. (38), who demonstrated ethylene production from KTBA or

TABLE 2. Effect of ˙OH-scavenging agents on ligninolytic activity and glucose metabolism of
P. chrysosporium[a]

Addition	Final concn (mM)	% of complete[b]			
		[¹⁴C]lignin		[¹⁴C]glucose	
		24 h[c]	48 h	24 h	48 h
Complete		100	100	100	100
+ mannitol	5	60	66	114	121
+ mannitol	50	47	52	88	99
+ benzoate	1	13	16	72	86
+ butylated hydroxytoluene	0.1	38	41	91	98
+ butylated hydroxytoluene	1.0	9	9	92	94

[a] Complete reaction mixtures consisted of cultures of *P. chrysosporium* grown for 13 days in low N medium (2.4 mM N) to which the ˙OH-scavenging agents and either [¹⁴C]lignin or [¹⁴C]glucose had been added. The metabolism of the substrates was assayed after 24 and 48 h of incubation by quantification of the respired ¹⁴CO₂. Taken from Forney et al. (29).
[b] Percent of complete = (dpm [no addition] − dpm [with addition])/dpm (no additions) × 100.
[c] Time interval after addition of ˙OH-scavenging agent.

methional only by lignin-degrading cultures and showed that the ˙OH scavengers thiourea, salicylate, and mannitol, as well as catalase, inhibited ligninolytic activity. Use of ˙OH-generating systems such as photosensitized riboflavin (38, 81) or Fenton reaction (T. McFadden, L. J. Forney, and C. A. Reddy, unpublished data) led to rapid degradation/depolymerization of lignin polymer. The ˙OH-generating systems also led to rapid degradation of a number of lignin model compounds via mechanisms similar to those observed with cultures of *P. chrysosporium* (15–18, 38). All the above results are consistent with the hypothesis that ˙OH plays an important role in lignin degradation by *P. chrysosporium*.

The involvement of ˙OH in lignin degradation is consistent with previous observations that the ligninolytic system is relatively nonspecific, non-stereoselective, extracellular, and oxidative (see discussion above). The involvement of ˙OH in lignin degradation is also in agreement with earlier reports that aromatic rings in lignin are extensively hydroxylated during extracellular transformation of this polymer by white-rot fungi (65, 67, 68).

Kelley and Reddy (59) suggested that ethylene production from KTBA is a rapid and sensitive measure of ligninolytic activity by *P. chrysosporium* based on the evidence that: (i) ˙OH plays an integral role in lignin degradation (29) and (ii) the specific activity for ˙OH production is correlated with ligninolytic activity in low and high N media, and in low and high carbohydrate media. Addition of NH₄⁺ or glutamate to ligninolytic cultures in low N medium severely inhibited the specific activity for ˙OH production as well as ligninolytic activity (94). Furthermore, extracellular fluid from lignin-degrading cultures of *P. chrysosporium* was shown to generate ethylene from KTBA, but a similar preparation from non-ligninolytic mutants did not (33, 79).

Hydrogen peroxide appears to play a key role in lignin degradation, as evidenced by the inhibition of lignin degradation by whole cultures of *P. chrysosporium* on addition of a specific H₂O₂ scavenger, catalase (20), the appearance of H₂O₂-producing microbodies only in lignin-degrading cultures but not in non-ligninolytic cultures (28), and the obligate dependence on H₂O₂ of the "lignin-degrading enzyme" (33, 107). Furthermore, hydroxylation of *p*-hydroxybutyric acid to protocatechuic acid increased two- to threefold on addition of sodium azide, an inhibitor of catalase (29).

Glucose oxidase (β-D-glucose: oxygen 1-oxidoreductase, EC 1.1.3.4) appears to be the main enzyme involved in the production of H₂O₂, the apparent source of ˙OH, in ligninolytic cultures of *P. chrysosporium* (94). When extracts of cells grown in low N medium with glucose as the growth substrate were tested for their ability to catalyze H₂O₂ production with 11 substrates including glucose, cellobiose, xylose, mannose, gluconate, succinate, malate, and acetate, glucose was the only substrate which supported levels of H₂O₂ production higher than those observed with the "no substrate" control. When the above extracts were electrophoresed and the protein bands were assayed for H₂O₂-producing activity, only one protein band with electrophoretic mobility comparable to that of commercial *Aspergillus niger* glucose oxidase (Sigma Chemical Co., St. Louis, Mo.) was observed (94; R. L. Kelley and C. A. Reddy, Abstr. Annu. Meet. Am. Soc. Microbiol. 1984, K155, p. 172). Furthermore, irrespective of the growth substrate used (i.e., even in xylose-, cellobiose-, or succinate-grown ligninolytic cultures), the primary source of H₂O₂ production appears to be glucose oxidase. Forney et al. (28) reported that H₂O₂- and catalase-producing microbodies, as evidenced by the presence of oxidized diaminobenzidine deposits in the periplasmic space, were seen in ligninolytic cultures of *P. chryso-sporium* but not in non-ligninolytic cultures.

There is very little information available on the role of specific enzymes in lignin degradation. Phenol oxidase (3, 39, 54, 62) and aromatic alcohol oxidase (22, 50, 56) have been implicated in the degradation of lignin polymer. Aromatic alcohol oxidase in *C. versicolor* and *Fusarium solani* has been shown to oxidize compounds with α,β-unsaturated primary alcohols (e.g., coniferyl alcohol) to the corresponding aldehyde and H_2O_2 (22, 50). The alcohol oxidase purified from culture filtrates of *F. solani* also was shown to attack both synthetic and naturally occurring lignins, which contain cinnamyl alcohol end groups (structure 14C in Fig. 3). This is one of the few enzymes that has been shown to attack the lignin polymer (8, 50, 56). Phenol oxidases which catalyze single electron oxidation of phenols can cause only limited structural changes in the lignin polymer, but a direct role in extensive degradation of lignin polymer is unlikely (62). Mutants lacking phenol oxidase were shown to lack the ability to degrade lignin, and a revertant was shown to be ligninolytic (3); however, some of the phenol oxidase mutants were shown to lack a number of other enzymatic activities, in addition to phenol oxidase (39), and therefore the precise role of phenol oxidase in lignin degradation is not known. Cellobiose:quinone oxidoreductase has been reported to be important for both cellulase and lignin degradation, but its requirement for lignin degradation has not been clearly demonstrated (4, 111).

A "lignin-degrading enzyme" from concentrated extracellular fluid of *P. chrysosporium* has been reported by Tien and Kirk (107), and data on a similar enzyme(s) were published by Glenn et al. (33). The "lignin-degrading enzyme" is obligatorily dependent on the presence of H_2O_2 for activity and degrades lignin substructure model compound representative of arylglycerol-β-aryl ether-type substructure (represented by units 1–4, 5–4, and 5–6 in Fig. 3), the dominant substructure in lignin polymer, accounting for 50 to 60% of the intermonomer linkages (100). It also degraded a model compound with the diaryl propane substructure (represented by units 16–20 in Fig. 3). The products produced by the enzyme from these substructure model compounds were similar to those formed from the same model compounds in intact lignin cultures (15–18, 65, 110). The enzyme also effected a partial depolymerization of spruce and birch lignins, was resolved as a single band on a polyacrylamide gel, and had a molecular weight of 42,000 (107). The H_2O_2-requiring extracellular enzyme(s) of Glenn et al. (33) generated ethylene from KTBA, was not found in

high N [12 mM $(NH_4)_2$ tartarate] cultures or in nonlignolytic mutants, and, like the fungus and photosensitized riboflavin, metabolized lignin model compounds (both diaryl propane and diaryl ethane substructures) to yield products equivalent to those previously isolated from intact ligninolytic cultures of *P. chrysosporium* (15–18), indicating that the enzyme is relatively nonspecific. It is not clear from the available data whether the extracellular lignin-degrading enzyme of Tien and Kirk (107) and Glenn et al. (33) is the enzyme involved in ethylene production from KTBA or degradation of ^{14}C-DHP → $^{14}CO_2$ in intact cultures. It is important to note that this enzyme (33, 107) has not been shown to effect the total degradation of ^{14}C-DHP to $^{14}CO_2$.

The results of Reddy and co-workers (28, 29, 59, 94) and those of Tien and Kirk (107) and Glenn et al. (33) are consistent with the idea that a "peroxidase-type" metalloprotein or similar enzyme reacts with H_2O_2, produced in ligninolytic cultures of *P. chrysosporium* via glucose-oxidase reaction, producing an enzyme-bound ˙OH which generates ethylene from KTBA or can attack the lignin polymer or lignin substructure model compounds. The fact that enzymes such as peroxidase or cytochrome P450 are known to generate ethylene from KTBA in the presence of added H_2O_2 (103), coupled with the recent demonstration by Palmer and Evans (91) that a "peroxidase-type" heme protein from *C. versicolor* appears to liberate ˙OH from H_2O_2, strongly supports the above idea. Glenn et al. (33) also suggested that hydroxyl radicals or equivalent species may play a role in the mechanism of their extracellular lignin-degrading enzyme.

CONCLUSIONS

Significant advances have been made in the past decade in our understanding of the physiology and biochemistry of lignin degradation by white-rot fungi. Investigations to date show that H_2O_2 plays a very important role in lignin degradation (29, 33, 94, 107) and that the source of H_2O_2 in lignin-degrading cultures of *P. chrysosporium* is glucose oxidase (94). Nutritional parameters which regulate the production and activity of this enzyme have a parallel effect on ligninolytic activity. The physiochemical characteristics, intracellular location, and genetic regulation of this enzyme need to be investigated. The evidence is unequivocal that H_2O_2-derived ˙OH plays an integral role in lignin degradation. However, it is not clear whether or not the production of ˙OH from H_2O_2 in cultures of *P. chrysosporium* is mediated by "lignin-degrading" extracellular enzyme(s), although the recent demonstration of the production of ˙OH from H_2O_2 by a "peroxidase-type" extracellular metalloprotein from *C. versicolor* (91) supports the latter possibility. Enzymology of lignin biodegradation in white-rot and other fungi and in bacteria, including the nature, characteristics, and regulation of the enzymes involved, will undoubtedly be an area of research emphasis in the next few years. Brown-rot fungi, similar to white-rot fungi, produce high levels of H_2O_2 during wood decay and have been shown to extensively hydroxylate aromatic rings in lignin. Yet, why do these fungi efficiently metabolize only cellulose and hemicellulose in wood, but have only limited ability to transform lignin? This is an important question from the standpoint of comparative physiology of wood-rotting fungi. Groundwork for studying the genetics of lignin degradation has been laid by developing the methodology for the production of auxotrophic and non-ligninolytic mutants, replica plating, and preparation and fusion of protoplasts (32, 34–37, 39). Studies are in progress to develop a suitable cloning system (T. R. Rao and C. A. Reddy, Abstr. Annu. Meet. Am. Soc. Microbiol. 1984, H151, p. 116) for *P. chrysosporium* which should eventually facilitate the isolation, identification, and regulation of the individual genes involved in lignin biodegradation.

LITERATURE CITED

1. **Amer, G. I., and S. W. Drew.** 1980. Microbiology of lignin degradation, p. 67–103. *In* G. T. Tsao (ed.), Annual reports on fermentation processes, vol. 4. Academic Press, Inc., New York.

2. **Amer, G. I., and S. W. Drew.** 1981. The concentration of extracellular superoxide radical as a function of time during lignin degradation by the fungus *Coriolus versicolor*. Dev. Ind. Microbiol. 22:479–484.

3. **Ander, P., and K. E. Eriksson.** 1976. The importance of phenol oxidase activity in lignin degradation by the white-rot fungus *Sporotrichum pulverulentum*. Arch. Microbiol. 109:1–8.

4. **Ander, P., and K. E. Eriksson.** 1978. Lignin degradation and utilization by microorganisms. Prog. Ind. Microbiol. 109:1–8.

5. **Bar-lev, S. S., and T. K. Kirk.** 1981. Effects of molecular oxygen on lignin degradation by *Phanerochaete chrysosporium*. Biochem. Biophys. Res. Commun. 99:373–378.

6. **Burdsall, H. H., Jr., and W. E. Elsyn.** 1974. A new *Phanerochaete* with a *chrysosporium* imperfect state. Mycotaxon 1:123–133.

7. **Crawford, D. L., M. J. Barder, A. L. Pometto III, and R. L. Crawford.** 1982. Chemistry of softwood lignin degradation by a *Streptomyces*. Arch. Microbiol. 131:140–145.

8. **Crawford, D. L., and R. L. Crawford.** 1980. Microbial degradation of lignin. Enzyme Microb. Technol. 2:11–21.

9. **Crawford, D. L., A. L. Pometto III, and R. L. Crawford.** 1983. Lignin degradation by *Streptomyces viridosporus*: isolation and characterization of a new polymeric lignin degradation intermediate. Appl. Environ. Microbiol. 45:898–904.

10. **Crawford, D. L., and J. B. Sutherland.** 1979. The role of actinomycetes in the decomposition of lignocelluloses. Dev. Ind. Microbiol. 20:143–152.

11. **Crawford, R. L.** 1981. Lignin biodegradation and transformation. John Wiley & Sons, Inc., New York.

12. **Crawford, R. L., and D. L. Crawford.** 1979. Radioisotopic methods for the study of lignin biodegradation. Dev. Ind. Microbiol. 19:35–49.

13. **Crawford, R. L., L. E. Robinson, and R. D. Foster.** 1981. Polyguaiacol: a useful model polymer for lignin biodegradation research. Appl. Environ. Microbiol. 41:1112–1116.

14. **Drew, S. W., and K. L. Kadam.** 1979. Lignin metabolism by *Aspergillus fumigatus* and white-rot fungi. Dev. Ind. Microbiol. 20:152–161.

15. **Enoki, A., G. P. Goldsby, and M. H. Gold.** 1980. Metabolism of lignin model compounds veratryl glycerol-β-guaiacyl ether and 4-ethoxy-3-methoxyphenylglycerol-β-guaiacyl ether by *Phanerochaete chrysosporium*. Arch. Microbiol. 125:227–332.

16. **Enoki, A., G. P. Goldsby, R. Krisnangkura, and M. H. Gold.** 1981. Degradation of the lignin model compounds 4-ethoxy-3-methoxy-phenylglycol-β-guaiacyl and vanillic acid ethers by *Phanerochaete chrysosporium*. FEMS Microbiol. Lett. 10:373–377.

17. **Enoki, A., G. P. Goldsby, and M. H. Gold.** 1981. β-Ether cleavage of the lignin model compound 4-ethoxy-3-methoxyphenylglycerol-β-guaiacyl ether and derivatives by *Phanerochaete chrysosporium*. Arch. Microbiol. 129:141–145.

18. **Enoki, A., and M. Gold.** 1982. Degradation of the diaryl propane lignin model compound 1-(3′, 4′-diethoxyphenyl)-1, 3-dihydroxy-2-4″-methoxyphenyl)-propane and derivatives by the basidiomycete *Phanerochaete chrysosporium*. Arch. Microbiol. 132:123–130.

19. **Eslyn, W. E., T. K. Kirk, and M. J. Effland.** 1975. Changes in the chemical composition of wood caused by six soft-rot fungi. Phytopathology 65:473–476.

20. **Faison, B. D., and T. K. Kirk.** 1983. Relationship between lignin degradation and production of reduced oxygen species by *Phanerochaete chrysosporium*. Appl. Environ. Microbiol. 46:1140–1145.

21. **Fan, L. T., Y.-H. Lee, and M. M. Gharpuray.** 1982. The nature of lignocellusosics and their pretreatments for enzymatic hydrolysis. Adv. Biochem. Eng. 23:157–187.

22. **Farmer, V. C., M. E. K. Henderson, and J. D. Russell.** 1960. Aromatic alcohol oxidase activity in the growth medium of *Polystictus versicolor*. Biochem. J. 74:257–262.

23. **Fenn, P., S. Choi, and T. K. Kirk.** 1981. Ligninolytic activity of *Phanerochaete chrysosporium*: physiology of suppression by NH_4^+ and L-glutamate. Arch. Microbiol. 130:66–71.

24. **Fenn, P., and T. K. Kirk.** 1979. Ligninolytic system of *Phanerochaete chrysosporium*: inhibition by o-phthalate. Arch. Microbiol. 134:307–309.

25. **Fenn, P., and T. K. Kirk.** 1981. Relation of nitrogen to the onset and suppression of ligninolytic activity and secondary metabolism in *Phanerochaete chrysosporium*. Arch. Microbiol. 130:59–65.

26. **Fergus, B. J., A. R. Proctor, J. A. N. Scott, and D. A. I. Goring.** 1969. The distribution of lignin in spruce wood as determined by ultraviolet microscopy. Wood Sci. Technol. 3:117–138.

27. **Forney, L. J., and C. A. Reddy.** 1979. Bacterial degradation of Kraft lignin. Dev. Ind. Microbiol. 20:163–175.

28. **Forney, L. J., C. A. Reddy, and H. S. Pankratz.** 1982. Ultrastructural localization of hydrogen peroxide production in ligninolytic *Phanerochaete chrysosporium* cells. Appl. Environ. Microbiol. 44:732–736.

29. **Forney, L. J., C. A. Reddy, M. Tien, and S. D. Aust.** 1982. The involvement of hydroxyl radical derived from hydrogen peroxide in lignin degradation by the white-rot fungus *Phanerochaete chrysosporium*. J. Biol. Chem. 257:11355–11462.

30. **Freudenberg, K.** 1965. Lignin: its constitution and formation from β-hydroxycinnamyl alcohols. Science 148:595–600.

31. **Freudenberg, K., and A. C. Neish.** 1968. Constitution and biosynthesis of lignin. Springer-Verlag, New York.

32. **Glenn, J., and M. H. Gold.** 1983. Decolorization of several polymeric dyes by the lignin degrading basidiomycete *Phanerochaete chrysosporium.* Appl. Environ. Microbiol. **45:**1741–1747.

33. **Glenn, J. K., M. A. Morgan, M. B. Mayfield, M. Kuwahara, and M. H. Gold.** 1983. An extracellular H_2O_2-requiring enzyme preparation involved in lignin biodegradation by the white-rot basidiomycete *Phanerochaete chrysosporium.* Biochem. Biophys. Res. Commun. **114:**1077–1083.

34. **Gold, M. H., and T. M. Cheng.** 1978. Induction of colonial growth and replica plating of the white rot basidiomycete *Phanerochaete chrysosporium.* Appl. Environ. Microbiol. **35:**1223–1225.

35. **Gold, M. H., and T. M. Cheng.** 1979. Conditions for fruiting body formation in the white-rot basidiomycete *Phanerochaete chrysosporium.* Arch. Microbiol. **121:**37–41.

36. **Gold, M. H., T. M. Cheng, and M. Alic.** 1983. Formation, fusion, and regeneration of protoplasts from wild-type and auxotrophic strains of the white rot basidiomycete *Phanerochaete chrysosporium.* Appl. Environ. Microbiol.**46:**260–263.

37. **Gold, M. H., T. M. Cheng, and M. B. Mayfield.** 1982. Isolation and complementation studies of auxotrophic mutants of the lignin-degrading basidiomycete *Phanerochaete chrysosporium.* Appl. Environ. Microbiol. **44:**996–1000.

38. **Gold, M. H., H. Kutsuki, M. A. Morgan, and R. Kuhn.** 1983. Degradation of lignin and lignin model compounds by several radical generating systems and by *Phanerochaete chrysosporium.* Proceedings 1983 International Symposium on Wood Pulping Chemistry **4:**165–168.

39. **Gold, M. H., M. B. Mayfield, T. M. Cheng, K. Krisnangkura, M. Shimada, A. Enoki, and J. K. Glenn.** 1982. A *Phanerochaete chrysosporium* mutant defective in lignin degradation as well as other secondary metabolic functions. Arch. Microbiol. **132:**115–122.

40. **Grisebach, H.** 1977. Biochemistry of lignification. Naturwissenschaften **64:**619–625.

41. **Haider, K., J. P. Martin, and E. Reitz.** 1977. Decomposition in soil of ^{14}C-labeled coumaryl alcohols: free and linked into dehydropolymer and plant lignins and model humic acids. Soil Sci. Soc. Am. J. **41:**556–561.

42. **Haider, K., and J. Trojanowski.** 1975. Decomposition of specifically ^{14}C-labelled phenols and dehydropolymers of coniferyl alcohols as models for lignin degradation by soft and white-rot fungi. Arch. Microbiol. **105:**33–41.

43. **Haider, K., and J. Trojanowski.** 1980. A comparison of the degradation of ^{14}C-labeled DHP and cornstalk lignins by micro and macrofungi and by bacteria, p. 111–134. *In* T. K. Kirk, T. Higuchi, and H.-M. Chang (ed.), Lignin biodegradation: microbiology, chemistry, and potential applications, vol. 1. CRC Press, West Palm Beach, Fla.

44. **Haider, K., J. Trojanowski, and V. Sundman.** 1978. Screening for lignin-degrading bacteria by means of ^{14}C-labeled lignins. Arch. Microbiol. **119:**103–106.

45. **Hall, P. L.** 1980. Enzymatic transformations of lignin: 2. Enzyme Microb. Technol. **2:**170–176.

46. **Harkin, J. M.** 1967. Lignin—a natural polymeric product of phenol oxidation, p. 243–321. *In* W. I. Taylor and A. R. Battersby (ed.), Oxidative coupling of phenols. Marcel Dekker, New York.

47. **Hata, K.** 1966. Investigation on lignins and lignification. XXXIII. Studies on lignins isolated from spruce wood decayed by *Poria subacida* B11. Holzforschung **20:**142–147.

48. **Hayatsu, R., R. E. Winans, R. L. McBeth, R. G. Scott, L. P. Moore, and H. Studier.** 1979. Lignin-like polymers in coal. Nature (London) **278:**41–43.

49. **Higuchi, T.** 1971. Formation and biological degradation of lignins. Adv. Enzymol. **34:**207–283.

50. **Higuchi, T.** 1980. Microbial degradation of dilignols as lignin models, p. 171–193. *In* T. K. Kirk, T. Higuchi, and H.-M. Chang (ed.), Lignin biodegradation: microbiology, chemistry, and potential applications, vol. 1. CRC Press, West Palm Beach, Fla.

51. **Higuchi, T., H.-M. Chang, and T. K. Kirk (ed.).** 1983. Recent advances in lignin biodegradation research. Uni Publishers, Tokyo, Japan.

52. **Higuchi, T., M. Shimada, F. Nakatsubo, and M. Tanahashi.** 1977. Differences in biosynthesis of guaiacyl and syringyl lignins in woods. Wood Sci. Technol. **11:**153–167.

53. **Hurst, H. M., and N. A. Burges.** 1967. Lignin in humic acids, p. 260–286. *In* A. D. McLaren and G. Peterson (ed.), Soil biochemistry. Marcel Dekker, New York.

54. **Ishihara, T.** 1980. The role of laccase in lignin biodegradation, p. 17–31. *In* T. K. Kirk, T. Higuchi, and H.-M. Chang (ed.), Lignin biodegradation: microbiology, chemistry, and potential applications, vol. 2. CRC Press, West Palm Beach, Fla.

55. **Ishikawa, H., W. J. Schubert, and F. F. Nord.** 1963. Investigations on lignin and lignification. XXVII. The enzymatic degradation of softwood lignin by white-rot fungi. Arch. Biochem. Biophys. **100:**131–139.

56. **Iwahara, S.** 1980. Microbial degradation of DHP, p. 151–170. *In* T. K. Kirk, T. Higuchi, and H.-M. Chang (ed.), Lignin biodegradation: microbiology, chemistry, and potential applications, vol. 2. CRC Press, West Palm Beach, Fla.

57. **Jeffries, T. W., S. Choi, and T.-K. Kirk.** 1981. Nutritional regulation of lignin degradation by *Phanerochaete chrysosporium.* Appl. Environ. Microbiol. **42:**290–296.

58. **Kawakami, H.** 1976. Bacterial degradation of lignin. I. Degradation of MWL by *Pseudomonas ovalis.* Mokuzai Gakkaishi **21:**93–100.

59. **Kelley, R. L., and C. A. Reddy.** 1982. Ethylene production from α-OXO-γ-methylthiobutyric acid is a sensitive measure of ligninolytic activity by *Phanerochaete chrysosporium.* Biochem. J. **206:**423–425.

60. **Kerr, T. J., R. D. Kerr, and R. Brenner.** 1983. Isolation of a bacterium capable of degrading peanut hull lignin. Appl. Environ. Microbiol. **46:**1201–1206.

61. **Keyser, P., T. K. Kirk, and J. G. Zeikus.** 1978. Ligninolytic enzyme system of *Phanerochaete chrysosporium:* synthesized in the absence of lignin in response to nitrogen starvation. J. Bacteriol. **135:**790–797.

62. **Kirk, T. K.** 1971. Effect of microorganisms on lignin. Annu. Rev. Phytopath. **9:**185–210.

63. **Kirk, T. K.** 1975. Effects of a brown-rot fungus, *Lenzites trabea* on lignin in spruce wood. Holzforschung **29:**99–107.

64. **Kirk, T. K.** 1981. Toward elucidating the mechanism of action of ligninolytic system in basidiomycetes, p. 131–148. *In* A. Hollaender (ed.), Trends in the biology of fermentations for fuels and chemicals. Plenum Press, New York.

65. **Kirk, T. K.** 1984. Degradation of lignin, p. 399–437. *In* D. T. Gibson (ed.), Biochemistry of microbial degradation. Marcel Dekker, New York.

66. **Kirk, T. K., and E. Adler.** 1969. Catechol moieties in enzymatically liberated lignin. Acta Chem. Scand. **23:**705–707.

67. **Kirk, T. K., and H.-M. Chang.** 1974. Decomposition of lignin by white-rot fungi. I. Isolation of heavily degraded lignins from decayed spruce. Holzforschung **29:**56–64.

68. **Kirk, T. K., and H.-M. Chang.** 1975. Decomposition of lignin by white-rot fungi. II. Characterization of heavily degraded lignins from decayed spruce. Holzforschung **29:**56–64.

69. **Kirk, T. K., and H.-M. Chang.** 1981. Potential applications of bio-ligninolytic system. Enzyme Microb. Technol. **3:**189–196.

70. **Kirk, T. K., W. J. Connors, R. D. Bleam, W. F. Hackett, and J. G. Zeikus.** 1975. Preparation and microbial decomposition of synthetic [14C] lignins. Proc. Natl. Acad. Sci. U.S.A. **72:**2515–2519.

71. **Kirk, T. K., W. J. Connors, and J. G. Zeikus.** 1976. Requirement for a growth substrate during lignin decom-

position by two wood-rotting fungi. Appl. Environ. Microbiol. **32**:192–194.

72. **Kirk, T. K., W. J. Connors, and J. G. Zeikus.** 1977. Advances in understanding the microbiological degradation of lignin, p. 369–394. *In* F. A. Loewus and V. C. Runeckles (ed.), Recent advances in phytochemistry, vol. 2. Plenum Press, New York.

73. **Kirk, T. K., T. Higuchi, and H.-M. Chang (ed.).** 1980. Lignin biodegradation: microbiology, chemistry, and potential applications, vol. 1 and 2. CRC Press, West Palm Beach, Fla.

74. **Kirk, T. K., F. Nakatsubo, and I. D. Reid.** 1983. Further study discounts role of singlet oxygen in fungal degradation of lignin model compounds. Biochem. Biophys. Res. Commun. **111**:200–204.

75. **Kirk, T. K., E. Schultz, W. J. Connors, L. F. Lorenz, and J. G. Zeikus.** 1978. Influence of culture parameters on lignin metabolism by *Phanerochaete chrysosporium*. Arch. Microbiol. **117**:277–285.

76. **Koenigs, J. W.** 1972. Production of extracellular hydrogen peroxide and peroxidase by wood-rotting fungi. Phytopathology **62**:100–110.

77. **Koenigs, J. W.** 1974. Production of hydrogen peroxide by wood-rotting fungi and its correlation with weight loss, depolymerization and pH changes. Arch. Microbiol. **99**:129–145.

78. **Kutsuki, H., A. Enoki, and M. H. Gold.** 1983. Riboflavin-photosensitized oxidative degradation of a variety of lignin model compounds. Photochem. Photobiol. **37**:1–7.

79. **Kutsuki, H., and M. H. Gold.** 1982. Generation of hydroxyl radical and its involvement in lignin degradation by *Phanerochaete chrysosporium*. Biochem. Biophys. Res. Commun. **109**:320–327.

80. **Lundquist, K., and T. K. Kirk.** 1978. *De novo* synthesis and decomposition of veratryl alcohol by a lignin-degrading basidiomycete. Phytochemistry **17**:1676.

81. **Lundquist, K., T. K. Kirk, and W. J. Connors.** 1977. Fungal degradation of Kraft lignin and lignin sulfonates prepared from synthetic ^{14}C-lignins. Arch. Microbiol. **112**:291–296.

82. **Martin, J. P., K. Haider, and G. Kassim.** 1980. Biodegradation and stabilization after 2 years of specific crop, lignin and polysaccharide carbons in soils. Soil Sci. Soc. Am. J. **44**:1250–1255.

83. **Nakatsubo, F., T. K. Kirk, M. Shimada, and T. Higuchi.** 1981. Metabolism of a phenyl coumaran substructure lignin model compound in ligninolytic cultures of *Phanerochaete chrysosporium*. Arch. Microbiol. **128**:416–420.

84. **Nakatsubo, F., I. D. Reid, and T. K. Kirk.** 1981. Involvement of singlet oxygen in the fungal degradation of lignin. Biochem. Biophys. Res. Commun. **102**:484–491.

85. **Nakatsubo, F., I. D. Reid, and T. K. Kirk.** 1982. Incorporation of [18]O_2 and absence of stereospecificity in primary product formation during fungal metabolism of a lignin model compound. Biochim. Biophys. Acta **719**:284–291.

86. **Ng, T. K., R. M. Busche, C. C. McDonald, and R. W. F. Hardy.** 1983. Production of feedstock chemicals. Science **219**:733–740.

87. **Odier, E., G. Janin, and B. Monties.** 1981. Poplar lignin decomposition by gram-negative aerobic bacteria. Appl. Environ. Microbiol. **41**:337–341.

88. **Odier, E., and B. Monties.** 1978. Biodegradation of wheat lignin by *Xanthomonas* 23. Ann. Microbiol. (Paris) **129A**:361–377.

89. **Odier, E., and B. Monties.** 1983. Absence of microbial mineralization of lignin in anaerobic enrichment cultures. Appl. Environ. Microbiol. **46**:661–665.

90. **Odier, E., and P. Roch.** 1983. Factors controlling biodegradation of lignin in wood by various white-rot fungi, p. 188–194. *In* T. Higuchi, H.-M. Chang, and T. K. Kirk (ed.), Recent advances in lignin biodegradation research. Uni Publishers, Tokyo.

91. **Palmer, J. M., and C. S. Evans.** 1983. Extracellular enzymes produced by *Coriolus versicolor* in relationship to the degradation of lignin. 1983 Proceedings International Symposium on Wood Pulping Chemistry **4**:19–24.

92. **Reddy, C. A.** 1978. Introduction to microbial degradation of lignin. Dev. Ind. Microbiol. **19**:23–26.

93. **Reddy, C. A., and L. J. Forney.** 1978. Lignin chemistry and structure: a brief review. Dev. Ind. Microbiol. **19**:27–34.

94. **Reddy, C. A., L. J. Forney, and R. L. Kelley.** 1983. Involvement of hydrogen peroxide-derived hydroxyl radical in lignin degradation by the white-rot fungus *Phanerochaete chrysosporium*, p. 153–163. *In* T. Higuchi, H. M. Chang, and T. K. Kirk (ed.), Recent advances in lignin biodegradation research. Uni Publishers, Tokyo.

95. **Reid, I. D.** 1979. The influence of nutrient balance on lignin degradation by the white-rot fungus *Phanerochaete chrysosporium*. Can. J. Bot. **57**:2050–2058.

96. **Reid, I. D., and K. A. Seifert.** 1980. Lignin degradation by *Phanerochaete chrysosporium* in hyperbaric oxygen. Can. J. Microbiol. **26**:1168–1171.

97. **Robinson, L. E., and R. L. Crawford.** 1978. Degradation of ^{14}C-labeled lignins by *Bacillus megaterium*. FEMS Microbiol. Lett. **4**:301–302.

98. **Ryu, D. D. Y., and M. Mandels.** 1980. Cellulases: biosynthesis and applications. Enzyme Microb. Technol. **2**:91–102.

99. **Sarkanen, K. V., and H. L. Hergert.** 1971. Classification and distribution, p. 43–94. *In* K. V. Sarkanen and C. H. Ludwig (ed.), Lignins: occurrence, formation, structure and reactions. Wiley-Interscience, New York.

100. **Sarkanen, K. V., and C. H. Ludwig (ed.).** 1971. Lignins: occurrence, formation, structure and reactions. Wiley-Interscience, New York.

101. **Shimada, M., F. Nakatsubo, T. K. Kirk, and T. Higuchi.** 1981. Biosynthesis of the secondary metabolite veratryl alcohol in relation to lignin degradation in *Phanerochaete chrysosporium*. Arch. Microbiol. **129**:321–324.

102. **Shimada, M., and M. H. Gold.** 1983. Direct cleavage of the vicinal diol linkage of the lignin model compound dihydroanisoin by basidiomycete *Phanerochaete chrysosporium*. Arch. Microbiol. **134**:299–302.

103. **Shimada, M., and T. Higuchi.** 1983. Biochemical aspects of the secondary metabolism of xenobiotic lignin and veratryl alcohol biosynthesis in *Phanerochaete chrysosporium*, p. 195–208. *In* T. Higuchi, H.-M. Chang, and T. K. Kirk (ed.), Recent advances in lignin biodegradation research. Uni Publishers, Tokyo.

104. **Sigleo, D. L.** 1978. Degraded lignin compounds identified in silicified wood 200 million years old. Science **200**:1054–1056.

105. **Sutherland, J. B., R. A. Blanchette, D. L. Crawford, and A. L. Pometto III.** 1979. Breakdown of Douglas fir phloem by a lignocellulose-degrading *Streptomyces*. Curr. Microbiol. **2**:123–126.

106. **Trojanowski, J., K. Haider, and V. Sundman.** 1977. Decomposition of ^{14}C-labeled lignin and phenols by a *Nocardia* sp. Arch. Microbiol. **114**:149–153.

107. **Tien, M., and T. K. Kirk.** 1983. Lignin-degrading enzyme from the hymenomycete *Phanerochaete chrysosporium* Burds. Science **221**:661–663.

108. **Vance, C. P., T. K. Kirk, and R. T. Sherwood.** 1980. Lignification as a mechanism of disease resistance. Annu. Rev. Phytopathol. **18**:259–288.

109. **Wardorp, A. B.** 1971. Lignins: occurrence and formation in plants, p. 14–42. *In* K. V. Sarkanen and C. H. Ludwig (ed.), Lignins: occurrence, formation, structure and reactions. Wiley-Interscience, New York.

110. **Weinstein, D. A., K. Krisnangkura, M. B. Mayfield, and M. H. Gold.** 1980. Metabolism of radiolabelled β-guaiacyl ether-linked lignin dimeric compounds by *Phanerochaete chrysosporium*. Appl. Environ. Microbiol. **39**:535–540.

111. **Westmark, U., and K. E. Eriksson.** 1974. Cellobiose-quinone oxidoreductase, a new wood degrading enzyme

from white-rot fungi. Acta Chem. Scand. **28:**209–214.

112. **Woodwell, G. M.** 1978. The carbon dioxide question. Sci. Am. **238**(1):34–44.

113. **Yang, H. H., M. J. Effland, and T. K. Kirk.** 1980. Factors influencing fungal degradation of lignin in a representative lignocellulosic, thermomechanical pulp. Biotechnol. Bioeng. **22:**65–77.

114. **Zeikus, I. G.** 1983. Lignin metabolism and the carbon cycle: polymer biosynthesis, biodegradation, and environmental recalcitrance, p. 211–243. *In* M. Alexander (ed.), Advances in microbial ecology, vol. 5. Plenum Press, New York.

115. **Zeikus, J. G., A. L. Wellstein, and T. K. Kirk.** 1982. Molecular basis for the biodegradative recalcitrance of lignin in anaerobic environments. FEMS Lett. **15:**193–197.

Factors Affecting Cellulase Activity in Terrestrial and Aquatic Ecosystems

A. E. LINKINS, J. M. MELILLO, AND R. L. SINSABAUGH

Biology Department, Virginia Polytechnic Institute and State University, Blacksburg, Virginia 24061, and The Ecosystems Center, Marine Biological Laboratory, Woods Hole, Massachusetts 02543

Cellulose is the most abundant of all naturally occurring structural polysaccharides. It is the ultimate product of about one-third of all the carbon dioxide fixed by plants (8). Cellulose is found in plant cell walls, where it exists as a network of cross-linked linear chains of glucose molecules (2). As a cell wall develops, the cellulose network can be infused with lignin, producing a ligno-cellulose matrix that makes the wall rigid, limits water transfer across the wall, and reduces the susceptibility of the wall to microbial attack (4, 19).

Cellulose, an energy-rich molecule, is too large to be taken directly into microbial cells. To make the energy in cellulose available, microbes excrete extracellular enzymes that hydrolyze cellulose, converting it to soluble carbohydrates that can pass through a microbial cell membrane. Once inside the cell, the simple carbohydrates are metabolized.

The extracellular enzymes involved in cellulose hydrolysis are known as cellulases. Two functional groups of cellulases have been recognized; endocellulases and exocellulases (6, 20). Endocellulases are involved in the random cleavage of internal glycosyl bonds of the cellulose molecule. Exocellulases are involved in the cleavage of glycosyl bonds near the nonreducing ends of cellulose molecules or "cellulose fragments" produced through endocellulase activity. Operationally, endocellulases increase the number of "cellulose fragments" with nonreducing ends and thereby increase the amount of substrate that can be acted upon by exocellulases. Put another way, endocellulase activity has the potential for controlling the rate of microbially mediated cellulose degradation (Fig. 1).

In this paper, we explore how cellulase activity in both terrestrial and aquatic ecosystems is influenced by two factors: temperature and substrate quality (i.e., the cellulose-to-lignin ratio of the material undergoing decay). We consider how each of these factors influences the activity of exocellulases and endocellulases and associated cellulose mineralization in a tundra soil, a temperate zone forest soil, and a temperate zone stream.

METHODS

Terrestrial studies: sample sites, collection, and processing. *Eriophorum vaginatum* tussock tundra soils were collected from two locations: adjacent to Toolik Lake on the North Slope of the Brooks Range, 125 miles (201 km) south of Prudhoe Bay, and at Eagle Summit, 103 miles (166 km) northeast of Fairbanks, Alaska. Surface organic soil horizons, the fibric (Oi) and upper hemic (Oe) (7), were collected in June, July, and August of 1979, 1980, and 1981.

Forest soils were collected from a south-facing slope near Blacksburg, Va. The vegetation at the site was a mixture of chestnut, red and white oak, and red maple trees. Surface organic horizons, O1-O2, were collected every other month in 1980 and 1981.

Terrestrial organic material used for chemical analyses, enzyme assays, and laboratory incubations was collected, shipped, stored, and processed in a manner that minimized losses in enzyme activity (10, 11). Material used for microcosm studies was maintained as an intact core and used as described below.

CO_2 evolution, $^{14}CO_2$ trapping, radiotracer purification. Total CO_2 or $^{14}CO_2$ was collected and analyzed from field and laboratory microcosms (9, 10). Mineralization of intermediate and end products of cellulose hydrolysis ([U-^{14}C]cellobiose and [U-^{14}C]glucose) was measured to determine whether cellobio-hydrolase or glucose uptake temperature relationships would influence cellulase temperature interpretations. Noncellulose label, presumably soluble carbohydrates and hemicellulosic material, which comprised up to 20% of the total sample label, was removed by extraction two times with 250 ml of hot water followed by extraction with hot 10% NaOH for 30 min. This was followed by washing with hot water two more times. After this procedure, only 1.5% of the total remaining sample label could be removed by a hot water/alkaline extraction procedure. Of the remaining material, 99% was acid hydrolyzable and chromatographed as [^{14}C]glucose (19). This material was presumed to be cellulose. [^{14}C]glucose and cellobiose were also chromatographed (19) and

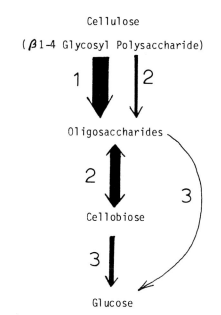

Cellulose

(β1-4 Glycosyl Polysaccharide)

Oligosaccharides

Cellobiose

Glucose

FIG. 1. Enzymatic hydrolysis of cellulose. Enzymes: (1) β-1,4-endocellulases; (2) β-1,4-exocellulases; (3) β-glucosidases.

were found to be at least 97% as advertised. Purified radiotracers were injected into the Oi-Oe or O1-O2 horizons of delimited soil columns in field incubation systems or laboratory microcosms as 1% uniform suspensions or solutions.

Enzyme assays. Cellulase activity was measured viscometrically for the endocellulase and as generation of reducing sugar for the exocellulase components of the cellulase complex (10, 11, 22). These two components of the cellulase complex can be expressed as a ratio of the exocellulase activity to endocellulase activity to give insight into the overall functional nature of the enzyme complex (10, 11; P. C. Miller, M. Blake-Johnson, P. M. Miller, F. S. Chapin, K. R. Everett, J. Kummeron, A. E. Linkins, G. M. Margion, W. Oechtel, and S. W. Roberts, Ecol. Monogr., in press). As the ratio increased, there would be a greater potential for the production of product capable of being assimilated by microorganisms. However, there is a point at which the endocellulase activity could become limiting, unless total enzyme activity is also increasing. In this paper we correlate the functional ratio of the cellulase complex to the microbial process, CO_2 evolution, to determine whether and how cellulase activity is influencing microbial oxidation of end products. An increased functional ratio is indicative of an increased exocellulase/endocellulase ratio in conjunction with increased overall cellulase activity, and a decreased functional ratio is

indicative of a lower ratio and total cellulase activity. Assays were performed at substrate saturation and at 20°C unless otherwise noted (10, 11, 22).

Cellulase-temperature incubations with rapid temperature equilibrations (1 to 2 min) were conducted at the temperature indicated, with the assay run at room temperature (S. B. Herbein, M.S. thesis, Virginia Polytechnic Institute and State University, Blacksburg, 1981). Enzymes were assayed in the soluble or adsorbed state. Bound enzymes were assayed in ground material suspended to form a slurry in 25 mM sodium acetate buffer, pH 5.5. Under these conditions, 90% of the total enzyme activity remained adsorbed to the soil-litter material. Enzymes were solubilized off the soil-litter in a 150 mM sodium acetate buffer, pH 5.5, with 1 M NaCl or Na_2SO_4 added. Under these conditions 30 to 40% of the total enzyme activity was found in the supernatant solution after centrifugation at $5,000 \times g$ for 15 min. All enzyme assays were done in triplicate from triplicate soil samples from three replicate plots ($n = 9$ for each assay).

Incubation systems. Field incubations were done to show the seasonal pattern of cellulose mineralization, and laboratory incubations were done to show the relationship between temperature and cellulose, cellobiose, and glucose mineralization.

Field incubation tubes were made from 15-cm-diameter plastic pipe. The pipe, open at both ends, was pushed 15 to 25 cm into the soil and then extracted with an intact soil core inside. Live surface vegetation and major roots were removed. The tube was placed back in the sampling hole and remained open until just prior to use. Field incubation tubes were prepared and installed in the late spring or late summer for use during the current or following year.

Laboratory tubes were of the same design as the field systems. Soil cores were taken, prepared as above, and shipped and stored (10, 11) until used for temperature studies. All incubation studies were done in triplicate on cores from three similar tussocks and intertussocks (11) or similar sites in the Virginia forest ($n = 9$ for each reading). Enzyme assays conducted in conjunction with mineralization experiments were done on replicate core systems ($n = 9$).

Chemical and statistical analyses. Chemical analyses of detergent-extracted organic material for cellulose and lignin were done by a technique in which potassium permanganate is used to solubilize a heterogeneous class of condensed and soluble phenolics as well as other associated compounds (24), called "Van Soest lignin" here. Cellulose was not directly measured, but rather was estimated from the residue of the Van Soest lignin extraction and was considered to be

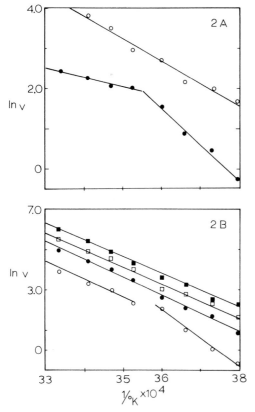

FIG. 2. Arrhenius plots of bound cellulase activity and CO_2 evolution rates from tussock tundra soil. (A) Exocellulase and endocellulase activities from tussock tundra organic soil horizons. Exocellulase activity, V = micrograms of glucose equivalent per hour per gram (dry weight) of soil; endocellulase activity, V = units per hour per gram (dry weight) of soil. Symbols: O, exocellulase; ●, endocellulase. (B) Arrhenius plots of CO_2 evolution from tussock tundra soil. V = liters of CO_2 per hour per gram (dry weight) of soil. Symbols: ●, unamended soil; O, [U-^{14}C]cellulose-amended soil; □, [U-^{14}C]cellobiose-amended soil; ■, [U-^{14}C]glucose-amended soil.

holocellulose. Statistical analyses were done with SAS-based tests (SAS, Inc., N.C.). Specific analyses of linearity of Arrhenius plots were done with the limitations in interpreting "breaks" in linearity taken into consideration (1, 21).

Aquatic study: leaf litter. Leaf litter from three tree species, white oak, red maple, and flowering dogwood, was collected in November 1979. This litter was air dried, the initial lignin and cellulose chemistry was determined for each litter type (24), and known masses of single litter types were placed in mesh bags. These bags were anchored in a third-order stream and were periodically harvested and assayed for enzyme and microbial activity (22).

RESULTS AND DISCUSSION

Tundra soils. The two major functional groups of the cellulase complex, endocellulases and exocellulases, have different temperature responses between −10 and 25°C (Fig. 2A). Arrhenius linear transformations of these temperature data show that the exocellulases have a constant slope over the −10 to 25°C range, whereas the endocellulases have a 78% increase in their slope below 5 to 10°C which is significantly different from the 10 to 25°C slope at the 1% level. A constant slope of an Arrhenius expression demonstrates a constant reaction condition as affected only by the thermally related kinetic state of the reaction. However, if the slope of the function is not constant, something other than direct temperature interaction on the energy of activation is occurring. Changes in the slope may indicate a temperature-related physical or chemical change in the reaction mechanism or substrate-binding efficiency (21). In membrane-associated enzymes, such breaks in the Arrhenius expression have been correlated with phase changes in the membrane fluidity, altering the energy of activation of the enzyme (12).

The increase in the energy of activation for endocellulases below 10 to 12°C suggests that this component of the cellulase system decreases in activity more rapidly than the exocellulase component. Hence, endocellulase may become more rate limiting to the overall hydrolysis of cellulose in tundra soils below 10°C. Temperature data from these and similar tundra soils suggest that endocellulase activity may limit cellulose hydrolysis during much of the year below 20 cm, since mean soil temperatures are 5°C at this depth (18).

Figure 2B shows the Arrhenius transformations for CO_2 evolution from unamended soils or

TABLE 1. Total CO_2 evolution, $^{14}CO_2$ from [U-^{14}C]cellulose, and functional cellulase activity in field incubators in Alaskan tussock (T) and intertussock (IT) soils[a]

Month	Temp (°C)		Total CO_2		% of total as $^{14}CO_2$		Ex/End	
	T	IT	T	IT	T	IT	T	IT
June	5	3	22.1[1]	18.0[1]	10[1]	5[1]	15	10
July	8	6	16.0[2]	14.0[2]	22[2]	13[2]	29	20
August	10	7	13.5[2]	12.0[2]	57[3]	22[3]	35	24

[a] CO_2 is expressed as microliters per hour per gram (dry weight) of soil. Values are means; those followed by different superscripts are different at at least the 0.05 level from others in the same column. Cellulase activity is given as a functional expression: exocellulase/endocellulase (Ex/End) $\times 10^{-4}$. Linear regression analyses: $^{14}CO_2$/(Ex/End), $r^2 = 0.94$; total CO_2/(Ex/End), $r^2 = 0.62$.

soils amended with [U-^{14}C]glucose, [U^{14}C]cello-biose, or [U-^{14}C]cellulose. Temperature-associated CO_2 evolution rates from nonamended soils or soils amended with glucose or cellobiose are all linear over the -15 to $25°C$ range and have equal slopes. CO_2 evolution from cellulose showed a 15% increase in the slope below 5 to $10°C$ which was significantly different from the 10 to $25°C$ slope at the 5% level. The single slope responses for glucose- and cellobiose-derived CO_2 evolution suggest that exocellulase and β-glucosidase activities or glucose uptake and catabolism are not rate-limiting components of cellulose mineralization. Thus, the endocellulase component of the cellulose hydrolysis-mineralization system seems to be the limiting component as temperature decreases.

Unamended CO_2 evolution rates can be separated into two independent slopes which parallel the cellulose-related response. Although the two independent lines both have an r^2 of 0.98, compared with 0.96 for the combined data, they are not statistically different from each other (1, 21).

Temperature does not seem to be the only regulator of cellulose hydrolysis in tussock tundra soils (Table 1). The data in Table 1 show that, in tussock and intertussock soils, the proportion of CO_2 derived from incorporated [U-^{14}C]cellulose increases both in absolute amount and as a percentage of the total CO_2 evolved during the growing season. This is paralleled by an increase in absolute cellulase activities, with the proportion of exocellulase increasing relative to endocellulase activity. This increase in exocellulase activity correlates linearly with $^{14}CO_2$ evolution ($r^2 = 0.94$), suggesting that endocellulase activity under these circumstances is not limiting to cellulase hydrolysis. The trends that emerge show that cellulose hydrolysis and mineralization are relatively low during the first half of the growing season and then increase after midseason so that cellulose mineralization can compose 32 to 57% of the total CO_2 evolved from the soil. Destructive sampling of field incubation systems showed that live roots were not present in most systems; therefore, seasonal root respiration would not cause aberrant seasonal dilutions of soil respiration.

This observed seasonality of cellulose hydrolysis and mineralization may reflect the inhibition of cellulase activity by soluble carbohydrates in the soil. Early-season high respiration rates (or CO_2 out-gassing) in tundra have been attributed to abiotic and biotic factors such as freeze-thaw action, microbial death, low-level winter enzyme activity, etc. (3, 17). Perhaps these factors lead to early-season high levels of soluble organics in the soil solution which support high respiration rates and potential inhibition of cellulase

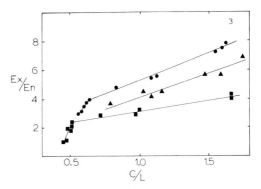

FIG. 3. Relationship of the functional cellulase activity (exocellulase/endocellulase) and the cellulose/Van Soest lignin ratio of associated organic material. Symbols: ●, litter bag leaf material in stream (Ex/En × 10^{-2}); ▲, O1–O2 soil horizon or litter in a Virginia hardwood forest (Ex/En × 10^{-3}); ■, Oi, Oe, Oa, and mineral horizon from tussock and intertussock components of *E. vaginatum* tussock tundra (Ex/En × 10^{-4}) O1–O2 horizons in a Virginia mixed hardwood forest floor soil.

activity (5, 6, 20). Depletion of this ephemeral pool of soluble organics, as well as increased soil temperatures, could then stimulate production and activation of cellulase activity and cellulose mineralization.

The quality of the soil organic matter, as defined by the Van Soest cellulose/lignin ratio, also influences the cellulase activity. The lignin content of soil organic matter or litter as characterized by the Van Soest technique (24) is in reality generic because it includes all potassium permanganate-soluble substances of plant and microbial origin. Many of these compounds do, however, condense on plant cellulose during plant growth and soil humification, so they could affect cellulose availability to cellulases (4, 19). In addition, they can be expected to affect enzyme activity since enzymes are adsorbed to this "humified" soil organic matter (5, 15, 22). Figure 3 shows that, as the degree of "Van Soest lignification" of the organic material increases, the functional expression of cellulase activity decreases linearly ($r^2 = 0.96$). When the cellulose/lignin ratio reaches around 0.5, there is an abrupt increase in the slope, caused by a decrease in the functional cellulase activity as well as total cellulase activity. Our data (10, 11, 22; Linkins and Melillo, unpublished data), as well as data of others (4, 6, 20), suggest that the data in Fig. 3 are evidence that in tundra soils cellulose rapidly becomes unavailable for microbial use when the Van Soest cellulose-to-lignin ratio declines below 0.5. This point is reached in the hemic soil horizon (Oe) and lower organic and mineral horizons in tussock tundra soils. The

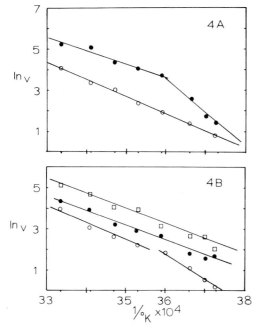

FIG. 4. Arrhenius plots of cellulase activity and CO_2 evolution from O1–O2 horizons in a Virginia mixed hardwood forest floor soil. (A) Symbols: $\bigcirc$, exocellulase activity, V = micrograms of glucose equivalent per gram (dry weight) of soil; $\bullet$, endocellulase activity, V = units per hour per gram (dry weight) of soil. (B) CO_2 evolution rates as liters of CO_2 per hour per gram (dry weight) of soil. Symbols: $\bullet$, unamended soils; $\bigcirc$, [U-^{14}C]cellulose-amended soil; $\square$, [U-^{14}C]glucose-amended soil.

availability of oxygen necessary to support further lignin degradation in these and lower horizons, and thereby to increase cellulose availability, would be limited (4, 23). However, even if oxygen were available, the temperature dynamics of cellulases may still limit cellulose hydrolysis (Fig. 2A).

Cellulose hydrolysis in tundra soils is regulated by temperature, substrate quality, and perhaps the presence of certain souble soil organics. The interactions of these factors in these oxygen-limited (23), cold soils (18), presumably with relatively high levels of soluble organics in the early plant growing season, begin to explain why organic carbon accumulates in tundra soils (17).

Virginia forest litter. The two components of the cellulase complex in Virginia forest litter have temperature responses similar to those noted for cellulases in Arctic tundra (Fig. 4A). Between −10 and 0°C the slope of the endocellulase Arrhenius function is 25% greater than that for the function between 5 and 25°C. This change in slope is significant at the 5% level.

This potential low-temperature endocellulase-mediated limitation in cellulose hydrolysis is only manifested as a 4% increase in the −10 to 0°C slope over the 0 to 25°C slope for the Arrhenius function of CO_2 evolution from cellulose. Straight lines can be fit to both slopes (−10 to 0°C and 0 to 25°C), with $r^2 = 0.97$. However, there is no statistical difference ($P = 0.1$) in these slopes and the slope from all points from −10 to 25°C (1, 21). Arrhenius functions for CO_2 evolution from glucose-amended or unamended soils were linear over the −10 to 25°C temperature range and were best described as a single-slope line (Fig. 4B).

Although the temperature-related change in the Arrhenius function of the endocellulase system exists in the temperate system, and there is a suggestion of its effect on cellulose mineralization, it probably does not play an important role in regulating cellulose hydrolysis. The versatility of the cellulase system (Fig. 1), the reduced level of temperature-endocellulase interaction below 5°C, and the fact that soil-litter temperatures in these forests are often above 5°C year round (Table 2), especially on sunny days, all lead to the conclusion that endocellulase activity limits cellulose hydrolysis to a far lesser degree in temperate forest soils than in Arctic tundra.

Seasonal functional cellulase activity and mineralization of cellulose is shown in Table 2. During the year, there was a decrease in the functional cellulase activity and cellulose mineralization in the early spring and in the fall just after leaf fall. Maximum cellulase functional activity and cellulose mineralization potential occurred during late fall-winter and summer. These patterns suggest that leaching of soluble organics from newly fallen leaves in the fall and from mobilized soluble organic material and

TABLE 2. Total CO_2 evolution, [14]CO_2 from [U-^{14}C]cellulose, and functional cellulase activity in field incubators in Virginia forest soils[a]

Month	Temp (°C)	Total CO_2	% of total as [14]CO_2	Ex/End
February	6	70[1]	25[1]	17
April	12	120[2]	5[2]	10
June	15	250[3]	30[3]	32
August	18	220[3]	35[3]	35
October	15	225[3]	8[2]	12
December	8	102[1,2]	18[1]	19

[a] CO_2 is expressed as microliters per hour per gram (dry weight) of soil. Values are means; those followed by different superscripts are different at at least the 0.05 level from others in the same column. Cellulase activity is given as a functional expression: exocellulase/endocellulase (Ex/End) × 10[−3]. Linear regression analyses: [14]CO_2/(Ex/End), $r^2 = 0.90$; total CO_2/(Ex/End), $r^2 = 0.22$.

throughfall from the newly expanding canopy (14) in the early spring was limiting cellulase activity (6, 20). The functional cellulase activity was also affected by the quality of organic matter (as previously discussed). Again, as the Van Soest lignin fraction of the material increased, the functional cellulase activity decreased (Fig. 3).

The factors controlling cellulose hydrolysis seem to be similar in the temperate and tundra systems. However, the degree to which temperature influences endocellulase activity and concomitant cellulose mineralization is much less, if present at all in a practical sense. Certainly, the presumed ephemeral presence of certain soluble organics in the soils influences hydrolysis and mineralization of cellulose. However,

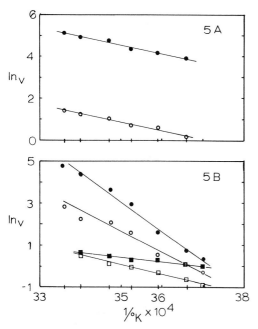

FIG. 5. Arrhenius plots of cellulase activity bound to leaf litter in a third-order stream or solubilized from O1–O2 horizon material from a Virginia hardwood forest soil or Oi-Oe horizon material from tussock tundra soil. (A) Bound cellulase activity on stream leaf litter. Symbols: ○, endocellulase, V = units per hour per gram (dry weight) of ash-free litter; ●, exocellulase, V = micrograms of glucose released per hour per gram (dry weight) of ash-free leaf litter. (B) Solubilized cellulase activity. Symbols: □, endocellulase activity in tussock soil, V = units per hour per gram (dry weight) of soil; ■, exocellulase activity in tussock soil, V = micrograms of glucose released per hour per gram (dry weight) of soil; ○, endocellulase activity in Virginia forest soil, V = units per hour per gram (dry weight) of soil; ●, exocellulase activity in Virginia forest soil, V = micrograms of glucose released per hour per gram (dry weight) of soil.

this period is relatively short considering the total annual time for cellulose decomposition in these temperate soils as compared with tundra soils. This would suggest, from the data presented, that the quality of the organic matter may be of greatest importance in ultimately limiting decomposition of refractory litter material in temperate soils.

Aquatic study. Decomposition of oak, maple, and dogwood leaf litter and associated cellulase activity have been discussed elsewhere (22). This work showed that: (i) total cellulase activity and the functional expression of cellulase activity decreased with the relative increase of Van Soest lignin in the initial litter material (Fig. 3); (ii) litter with higher lignin content decomposed more slowly and had a lower functional cellulase expression; and (iii) all litter had a directly measured exocellulase activity-temperature interaction that decreased 60% from 25 to 0°C. Arrhenius transformations of the exocellulase data of Sinsabaugh et al. (22) and endocellulase activity of general leaf litter from a similar stream show that from −5 to 25°C both enzymes have a single-slope linear Arrhenius function (Fig. 5A). These endocellulase data corroborate the assumptions for endocellulase temperature response (22) and show that no evidence exists for bound endocellulase activity limiting cellulose hydrolysis. Cellulose radiorespirometry was not done on stream litter.

Influence of enzyme binding. It is interesting to note that the temperature response of endocellulases has the greatest effect on cellulose hydrolysis and mineralization in the tundra system, which has the oldest organic material and greatest degree of humification (7; Everett, personal communication), and the least effect on the forest litter material, with the youngest organic material and least amount of humification. Perhaps an increased degree or complexity of enzyme-organic matter binding in the Arctic soils causes the temperature responses for endocellulase activity (Linkins, unpublished data). If temperature modifies the physical state of the enzyme through alteration of the material to which the enzyme is bound, the model of enzyme-temperature responses in cold-intolerant organisms could be applied to the endocellulase-temperature relationship in soils. In cold-intolerant organisms a decrease in temperature can cause a decrease in the fluidity of the membrane such that membrane-bound enzymes have increased energies of activation (1, 12). An apparent increase in the energy of activation of a bound enzyme can also be caused by changes in the organic-mineral material to which the enzyme is bound that do not directly affect the active site. Temperature-related alterations in the diffusion path of substrate to or product from the enzyme active site, caused by aggregations or conformational changes in the material to which an enzyme is bound, can also alter the kinetic characteristics of an enzyme (5, 15).

The data in Fig. 5B are Arrhenius transformations for solubilized endo- and exocellulase activity from tussock tundra and Virginia forest soils. These data support the hypothesis that the nature and degree of enzyme-soil organic and mineral material binding is an important factor controlling certain aspects of soil enzyme activity. Soluble cellulase activity from both Arctic and temperate soil organic matter does not show any changes in the Arrhenius slope between -5 and 25°C. Under conditions in which cellulases are soluble, there would be reduced associated organic matter present potentially to influence enzyme activity site conformation or diffusion kinetics of substrate to or product from the active site (5, 15). Hence, nature and degree of the enzyme binding as determined by the chemical nature of the soil-litter organic matter may play a more important role in the overall regulation of litter decomposition.

Cellulase activity and ecological patterns and processes. Knowledge of cellulose dynamics can help us to understand ecological patterns and processes at several levels of resolution. For example, in tundra pond ecosystems, dissolved organic carbon production appears to occur in the winter months when microbial activity as measured by CO_2 evolution is below levels of detection (J. Hobbie, personal communication). This pattern of asynchrony between dissolved organic carbon production and microbial activity may be explained by the fact that cellulases are more active at colder temperatures than are catabolic enzymes within the microbes of tundra ponds.

Regional analyses of decay rates of plant litter materials throughout North America (15, 25) indicated that in areas where actual evapotranspiration (AET) is low, such as the cold tundra, climate is the dominant factor controlling decay rate. In these areas, litter or soil organic matter quality and consequent extracellular enzyme binding establish a temperature-enzyme interaction with some enzymes (endocellulases) such that temperature becomes a factor limiting overall decomposition at low temperatures. In areas where actual evapotranspiration is much higher, such as the temperate mountain forests of western North Carolina and Virginia, litter quality influences enzyme binding-temperature interactions to a lesser degree. Here temperatures may be relatively less important in regulating decomposition, with litter quality, as discussed from data in Fig. 3, more important in regulating overall decomposition.

An understanding of the factors that control cellulase activity may also help us to predict certain aspects of the biosphere's response to global warming due to the loading of the atmosphere with CO_2, primarily from the burning of fossil fuels. By the year 2050, the carbon dioxide content of the atmosphere may be double what it is today. There is general agreement that the global temperature will increase but less agreement about the pattern of warming and exactly how the warming will affect air movement and precipitation. One prediction, based on a large computer-based model, is for small changes in the tropics and large changes near the poles (13). At 40 to 60° latitude the annual temperature increase may be 4 to 6°C, and precipitation patterns will also change. The potential consequences of this warming are not clear. One scenario suggests that microbial decomposition of organic matter in boreal peatlands, taiga, and tundra will accelerate because of increased temperature and an altered precipitation regime. If we assume a Q_{10} response of 2.5 and no increase in plant litter input, about 15% of the carbon stored in soils and peats would be released over 50 years. This release of 60×10^{15} g of carbon could set up a positive feedback loop, whereby elevated CO_2 concentration leads to a warming of northern soils and peats, which in turn leads to a stimulation of microbial activity and a further loading of the atmosphere with CO_2, and so on.

Our data on the response of the cellulase enzyme system to changes in temperature suggest that a 4 to 6°C change in air temperature in the tundra may not translate to large changes in soil respiration rate. Endocellulase is the rate-controlling enzyme group in the cellulase enzyme complex below 10°C in tundra soils. Since the current mean summer soil temperature of the Oe-Oa horizons in the tundra is about 0 to 5°C, an increase in soil temperature caused by a 4 to 6°C increase in atmospheric temperature may have little effect on cellulose degradation and thus microbial respiration of the products of cellulase hydrolysis.

OVERVIEWS

If enzyme-based process studies are to be meaningful in an ecological context, they must be organized such that the data collected reflect the activity of the enzyme-mediated process within the environment. Most extracellular enzymes function in the environment as a single or multienzyme complex adsorbed to organic-mineral material. As such, they behave as immobilized enzymes which probably exist as active entities in the environment longer than the organisms which originally produced them. The products of their reactions can potentially benefit any organism capable of assimilating the products diffusing from the active site. If enzyme-based studies do not consider these characteristics and limitations as well as enzyme responses to temperature, pH, substrate availability and quality, end or intermediate product interactions, etc., then the study may be useful in an enzymological context, but certainly limited in the ecological context. Consequently, future ecological studies in which enzyme-mediated process information is used should focus on the activity of the enzyme within the parameters of the environment in which the enzyme is functioning.

Future research in the area of enzyme-related ecological process studies should be directed toward developing a better understanding of the nature and dynamics of enzyme binding to the organic and mineral materials. Data from our study suggest that there are both direct and indirect effects of this adsorption which can influence the enzyme expression and kinetic parameters, certainly in the context of their responses to changing environmental parameters in the different environments. The relationship between enzyme activity and substrate availability and quality is another area that is poorly understood, as is the temporal relationship between different enzyme groups (e.g., cellulase and enzymes involved in lignin degradation). Arctic tundra soils are accreting carbon at the present time. Much of it is cellulose; however, most of it ultimately is complexed by polyphenolic condensation and is unavailable to cellulases and hence to hydrolysis to glucose. Other ecosystems have litter input which is much higher in initial lignin content and should have a greater potential of carbon accretion due to humification; yet, temperate and tropical ecosystems lose or accumulate much less organic matter. What are the enzyme-mediated processes and, perhaps more important, how are they regulated in these different ecosystems? Unless enzyme process studies are done in an ecosystem context, little will be gained beyond a description of the apparent maximum enzyme activity.

We thank M. Bryhan for technical assistance and C. A. McClaugherty and C. J. Kroehler for help in the preparation of this manuscript.

This work was supported by grants from the U.S. Office of Naval Research, U.S. Army Research Office, and U.S. Department of Energy.

LITERATURE CITED

1. **Bagnell, D. J., and J. A. Wolfe.** 1978. Chilling sensitivity in plants: do the activation energies of growth processes show an abrupt change at a critical temperature? J. Exp. Bot. **9**:1231–1240.
2. **Chesire, M. V.** 1973. Nature and origin of carbohydrates in soils, p. 216. Academic Press, Inc., New York.
3. **Coyne, P. I., and J. J. Kelley.** 1975. CO_2 exchange over Alaskan arctic tundra: meteorological assessment by an aero-dynamic method. J. Appl. Ecol. **12**:587–611.
4. **Crawford, R. L.** 1981. Lignin degradation and transformation. John Wiley & Sons, Inc., New York.
5. **Engasser, J. M., and C. Horvath.** 1974. Inhibition of bound enzymes. I. Anti-energetic interaction of chemical and diffusional inhibition. Biochemistry **13**:3845–3849.
6. **Eriksson, K. E.** 1978. Enzyme mechanisms involved in cellulose hydrolysis by the white rot fungus, *Sporotrichum pulverulentum*. Biotechnol. Bioeng. **20**:317–332.
7. **Everett, K. R., and J. Brown.** 1982. Some recent trends in the physical and chemical characterization and mapping of tundra soils, Arctic slope of Alaska. Soil Sci. **133**:264–280.
8. **Krogman, D. W.** 1973. Biochemistry of green plants, p. 239. Prentice-Hall, Englewood Cliffs, N.J.
9. **Linkins, A. E.** 1978. Presence and activity of the hexosmonophosphate shunt in a marine chemoorganotrophic diatom. Can. J. Microbiol. **24**:544–551.
10. **Linkins, A. E., R. M. Atlas, and P. Gustin.** 1978. Effect of surface applied Pruddhoe Bay crude oil in soil and vascular plant root respiration, soil cellulase and arylhydrocarbon hydroxylase. Arctic **31**:355–365.
11. **Linkins, A. E., and J. Neal.** 1982. Soil cellulase, chitinase, and protease activity in *Eriophorum vaginatum* tussock tundra at Eagle Summit, Alaska. Holarctic Ecol. **5**:135–138.
12. **Lyons, J. M., J. K. Raisons, and P. L. Steponkus.** 1980. The plant membrane in response to low temperature. An overview, p. 1–25. *In* J. M. Lyons, D. Graham, and J. K. Raisons (ed.), Low temperature stress in crop plants: the role of the membrane. Academic Press, Inc., New York.
13. **Manabe, S., and R. T. Wetherald.** 1980. On the disturbance of climate change resulting from an increase in the CO_2 content of the atmosphere. J. Atmos. Sci. **37**:99–118.
14. **McClaugherty, C. A.** 1983. Soluble polyphenols and carbohydrates in through fall and leaf litter decomposition. Oecol. Gen. **4**:375–385.
15. **McLaren, A. D.** 1975. Soil as a system of humus and clay immobilized enzymes. Chem. Scr. **3**:97–99.
16. **Meentemeyer, V.** 1978. Microclimate and lignin control of litter decomposition. Ecology **59**:465–472.
17. **Miller, P. C.** 1981. Research needed to determine the present carbon balance of northern ecosystems and the potential effect of CO_2 induced climate change. VII (14) A Research Agenda on Environmental and Societal Consequences of a Possible CO_2-Induced Climate Change. EV/10019-01. U.S. Department of Energy, Washington, D.C.
18. **Miller, P. C.** 1982. Environmental and vegetational variation across a snow accumulation area in montane tundra in central Alaska. Holarctic Ecol. **5**:85–98.
19. **Reddy, C. A.** 1978. Introduction to the microbial degradation of lignin. Dev. Ind. Microbiol. **19**:23–26.
20. **Reese, E. T.** 1977. Degradation of polymeric carbohydrates in microbial enzymes, p. 311–367. *In* F. A. Loewus and B. C. Runeckles (ed.), The structure, biosynthesis, and biodegradation of wood. Plenum Press, New York.
21. **Silvius, J. R., B. D. Reid, and R. H. McElhaney.** 1978. Membrane enzymes: artifacts in Arrhenius plots due to temperature dependence of substrate-binding efficiency. Science **199**:902–904.
22. **Sinsabaugh, R. L., E. F. Benfield, and A. E. Linkins.** 1981. Cellulase activity associated with the decomposition of leaf litter in a woodland stream. Oikos **36**:184–190.
23. **Stuart, L., and P. C. Miller.** 1982. Soil oxygen flux measured polarographically in an Alaskan tussock tundra. Holarctic Ecol. **5**:139–144.
24. **Van Soest, P. J., and R. H. Wine.** 1967. The use of detergents in the analysis of fiberous feeds. IV. Determination of plant cell-wall constituents. J. Assoc. Off. Agric. Chem. **50**:50–55.
25. **Whitford, W. G., V. Meentemeyer, T. R. Seastedt, K. Cromack, Jr., D. A. Crossley, P. Santos, R. L. Todd, and J. B. Waide.** 1981. Exceptions to the AET model: deserts and clear-cut forest. Ecology **62**:275–277.

Microbial Degradation of Pectin in Plants and Aquatic Environments

BERNHARD SCHINK

Fakultät für Biologie, Universität Konstanz, D-7750 Konstanz, West Germany

The production of photosynthetic biomass in nature is integrally associated with biodegradative microbial populations for its mineralization. Pectin is a minor component of plant biomass, and its major biological function is that of a glue or cement between the cells of higher plants. Pectin degradation is paramount to a variety of symbiotic and parasitic plant–microbe interactions, the rotting of fruits and vegetables, and the digestion of plant matter by ruminants. Interest in pectin degradation has focused mainly on these topics, as well as some other aspects of applied technology, namely, on the use of pectinolytic enzymes to clarify fruit juices and on the ponding of wood logs or the retting of flax and hemp. Pectin is degraded by a diversity of microorganisms, including aerobic and facultatively and strictly anaerobic bacteria as well as fungi, yeasts, and protozoa. The present literature on microbial degradation of pectin deals mainly with pectinolytic activities of aerobic microorganisms and was summarized in recent reviews (2, 52, 53).

Comprehensive studies on the microbial ecology of pectin degradation in nature are lacking. The present paper summarizes recent studies on the significance of pectin degradation in several anoxic microbial ecosystems, including freshwater and hypersaline lake sediments, decomposing algal mats in thermal volcanic springs, and the wetwood of living trees. The results of these studies are analyzed in relation to recent discoveries on the structure and function of pectin and the importance of its degradation by microorganisms.

STRUCTURE, FUNCTION, AND DEGRADATION OF NATURAL PECTIN

Pectin is a natural polysaccharide consisting mainly of rhamnose and α-$(1\rightarrow4)$-linked D-galacturonic acid residues, the carboxylic groups of which are methoxylated to a varying extent (52). In plant tissue, pectin is covalently linked to neutral oligosaccharide and polysaccharide chains which are usually referred to as "hemicellulose" (2, 32, 34, 35). Most of these acidic polysaccharides are present in the middle lamellae and primary cell walls of vascular plants, whereas the secondary walls are nearly exclusively built up by cellulose and neutral hemicelluloses. Thus, young plant tissue in which secondary walls are of only minor importance have high pectin contents (12 to 30%; 33), whereas structural tissue of green plant matter contains 5 to 10% pectin. The pectin content of wood is again lower (1 to 5%; 16) as a result of the preponderance of secondary walls rich in lignin and cellulose.

Recent studies on the physicochemical properties of isolated pectin by X-ray diffraction analysis, circular dichroism studies, and proton magnetic resonance spectroscopy revealed that the macromolecular structure depends to a high degree on the concentration of protons and calcium and sodium ions (12, 47, 67). Rigid and mobile domains were shown to exist in close association in purified pectate solutions as well as in cell walls (44). Thus, by changing the concentration of any of these effectors, the plant is able to change the properties of the pectin layers from a fluid to a solid state and back again without greater changes in the macromolecular composition of the tissue, a feature of importance in differentiating organs. Changes of this kind may also have an effect on the macromolecular structure of hemicellulosic polysaccharides linked to the pectin backbone, and hence may influence the plant's resistance to, for example, microbial attack.

The enzymology of pectin degradation has been reviewed recently (52, 53). Because of the covalent linkages between pectin and hemicelluloses in natural plant tissue, pectin degradation always implies to a certain extent hemicellulose degradation as well. Natural pectin is demethoxylated by pectin esterases, and the remaining pectate or polygalacturonate backbone is depolymerized hydrolytically by a hydrolase or a lyase reaction which releases saturated or unsaturated oligomeric or monomeric galacturonic acid residues. Hydrolases and lyases which preferentially attack the esterified pectin polymer are also known. Depending on the preferred substrate (methoxylated or nonmethoxylated pectates, long- or short-chain polymers) and the action pattern (random, terminal, or penultimate bonds attacked), different enzyme classes have been defined (52, 53). So far, there is no consistent rule on occurrence of the various enzyme types in the different organisms; rather, all classes of enzymes have been found in higher plants, fungi, and aerobic and facultatively or obligately anaerobic bacteria. Moreover, in some cases several isoenzymes of the same reaction pattern (e.g., lyases) have been detected in one organism, together with other enzymes of similar function (e.g., hydrolases). It has been argued that several different pectinolytic enzymes will allow a microbe to survive and grow in different environments, e.g., will broaden the host range of a pathogen by circumventing specific plant defense mechanisms (10). This idea actually appears reasonable with respect to the complicated mechanisms of interaction between plant and microbe which have been worked out in the recent past.

MICROBIAL DEGRADATION OF PECTIN IN LIVING PLANTS

Pectin or polygalacturonic acid-rich acidic polysaccharides are the first target of fungal or bacterial attack on plant tissue, be it in the development of a highly specific, mutually beneficial symbiosis of rhizobia and leguminous plants (17, 45) or in the invasion of pathogenic bacteria and fungi into living plants (10, 40). The same is true for the rotting of fruits and vegetable roots and tubers (10, 14, 50). Whereas the symbiotic plant–*Rhizobium* relationship is extremely well controlled by lectins and other systems to protect the plant from pathogenic attacks, most plant–pathogen interactions exhibit broader ranges of partners, on the side of both the microbe and the host. To keep the pathogens' range as small as possible, the plant answers to microbial attack with specific reactions: either fungal polygalacturonate hydrolases themselves (38, 39) or plant surface fragments released by pectinolytic enzymes (9, 27), or other components of plant origin released by pathogenic microorganisms (66), elicit the immediate release of phytoalexins. These may either kill or impair the pathogen itself (36) or inhibit its pectolytic enzymes (31). It appears evident that several isofunctional pectinolytic enzymes differing in minor properties important for the plant's recognition system help the pathogenic microbe to overcome the plant's defense mechanisms. Also, in rotting processes of fruits and legumes, the range of invading microbes seems to be restricted to only few organisms, at least during the initial phase of attack. Thus, potato tubers and carrots are preferentially rotted by bacteria of the genus *Erwinia* (14, 50), and the initiation of pectinolytic potato maceration is a complex cooperation of several bacterial pectinolytic enzymes. In *Erwinia carotovora*, an extracellular endo-pectate lyase is induced by its own reaction product, an unsaturated galacturonic acid dimer which, in excess, inhibits enzyme production via catabolite repression (63, 64). In *E. chrysanthemi*, two endo-pectate lyase isoenzymes are induced by digalacturonic acid, the product of the action of an intracellular exopolygalacturonate hydrolase (15). At an advanced stage of maceration and oxygen deprivation in the potato, or if the potatoes are kept at very low oxygen partial pressure or at elevated temperatures, pectinolytic *Clostridium* spp. take over and may outcompete the *Erwinia* spp. (42, 43, 51).

It is not yet clear whether the clostridial pectinolytic attack on the potato is as differentiated with respect to enzyme regulation, as observed with the *Erwinia* spp. Clostridia have so far not been considered to be phytopathogenic. It could be assumed that plant defense mechanisms are not active any longer at the time when clostridia invade the rotting plant material or that the plant defense system is inefficient in the absence of oxygen or toward clostridia in general. In this case, the soft rot caused by clostridia should be characterized as a saprophytic rather than a pathogenic process.

PECTIN DEGRADATION IN THE WETWOOD SYNDROME

Wetwood is an abnormal, water-soaked type of heartwood that can develop in living trees as a result of bacterial infections. The wetwood phenomenon was first observed in the nineteenth century, and detailed studies in this field started at the beginning of this century (28, 29, 68, 70). Lagerberg (37) coined the term "vattved" (wetwood) and the term "wetwood" was used thereafter in the Anglo-American literature, whereas the German authors use the term "Nasskernbildung" ("wet heart formation"). In true firs, elms, willows, poplars, and cottonwoods, wetwood formation is a common process, and many cottonwood seedlings may have wetwood forming at an age of 2 years (28). In affected tree species, wetwood appears as an accumulation of a fetid, aqueous liquid in the central core of the lower stem. The pH of the liquid may be either acid or alkaline, and this appears to influence the composition of the gases entrapped within the wetwood portion of the stem. Species such as cottonwood and willow usually have a somewhat alkaline wetwood, with a pH range of 6.5 to 8.0, whereas oak and hemlock have acid wetwoods, with a pH range of 3.3 to 5.7. Other species such as California white fir and American elm develop either acid or alkaline wetwoods. In alkaline wetwoods, large amounts of gas are accumulated up to 4 bar, which consists mainly of methane, nitrogen, and carbon dioxide.

Association of microbial activities with

wetwood formation was assumed when systematic studies of wetwood were initiated. However, as a result of the lack of interest of microbiologists in this subject, and the lack of experience of forest pathologists in handling strictly anaerobic bacteria, reliable descriptions of bacterial populations are scarce. Most studies have mainly focused on casual isolations of aerobic bacteria and fungi (28, 29). The studies on obligately anaerobic bacteria have largely dealt with detection and isolation of *Clostridium* species from wetwood (70). Zeikus and Ward (78) presented evidence that methane formation in alkaline wetwoods is due to microbial activities, and hence these wetwoods should be strictly anoxic to allow survival and metabolic activity of the extremely oxygen-sensitive methanogenic bacteria.

A systematic quantitative study on bacterial population densities in elms and cottonwood trees revealed that numbers of heterotrophic bacteria in wetwood exceeded those found in sapwood by two to four orders of magnitude, and that strict anaerobes were present in numbers at least equal to those of facultative or strict aerobes (54). Among the prevalent isolates were *Clostridium* and *Bacteroides* as well as *Erwinia*, *Edwardsiella*, *Klebsiella*, and *Lactobacillus* species. Mainly acetate ($\leq$50 mM) but also propionate, butyrate, isobutyrate, ethanol, and isopropanol were detected in millimolar concentrations in fetid liquids and were identical with fermentation products of the prevalent bacteria isolated. One of the prevalent isolates from cottonwood and elm was identified as a *Clostridium butyricum* strain which excreted large amounts of extracellular pectate lyase. Free pectate lyase was detected in fetid liquid up to 0.5 U/ml, and the total numbers of pectinolytic anaerobic bacteria were nearly equal to those of anaerobic heterotrophs (55). Pectinolytic activities of aerobic bacteria isolated from wetwood have repeatedly been reported (62, 72).

In scanning electron micrographs of wetwood samples from elms and cottonwoods, bacteria were notably detected in close association with destroyed pit membranes, whereas the lignocellulosic walls of the remaining tree tissue were not significantly colonized (54). Pit membranes mainly consist of nonlignified pectin (6). None of our isolates was able to grow on cellulose, and no cellulolytic activity was detected in fetid liquid. Hemicellulose degradation was detected in wetwood of firs (8). Incubation of healthy wood samples with either pure cultures of the pectinolytic *C. butyricum* strain or with pectate lyase preparations obtained from this strain resulted in complete dissolution of the pit membranes, rendering a scanning electron microscopical picture similar to that observed in wetwood (55). Thus, pit membranes destroyed by pectinolytic activities are conceived as nonfunctional in regulation of water flow through wood tissue, and the wood will be soaked with water as a result of either capillary forces or osmotic effects (73).

The microbiology of acidic wetwoods has not yet been thoroughly studied. One could speculate that microaerobic or anaerobic bacterial degradation of monomeric wood sugars, starch (73), or pectin results in a basically similar disease syndrome, but that the presence of traces of oxygen or of toxic wood components (e.g., phenols, acids) does not allow the establishment of stable methanogenic microbial populations. Ward et al. (69) detected propionate, butyrate, valerate, caproate, and high concentrations of acetate in fetid liquid from acidic oak wetwood, and high concentrations of acetate, propionate, and butyrate were observed in wetwoods of white firs (73). Accumulation of longer-chain fatty acids in sewage digesters is a common indication of inhibition of the methanogenic bacteria in these mixed microbial communities (74). Lack of methane production in acidic wetwoods agrees well with this hypothesis. Thus, acidic and alkaline wetwood may differ mainly by the absence or presence of a stable methanogenic bacterial population, and the species-specific development of either one would be due to wood structure or wood constituents.

On the other hand, degradation of pectic acids and acetate by syntrophic methanogenic cocultures could easily be the reason for the raised pH found in alkaline wetwoods, rather than the release of ammonia from wood degradation as earlier assumed (28). Actually, ammonia is barely measurable in wetwoods, whereas acetylene reduction activity is detected in isolated wetwood cores, and high numbers of nitrogen-fixing bacteria in wetwood make biological nitrogen fixation quite probable in this environment (54). Nonetheless, in situ experiments are required to demonstrate clearly that nitrogen fixation occurs in living trees affected by wetwood.

In summary, pectin, galacturonic acid, and possibly monomeric wood sugars, starch, or hemicelluloses are important substrates for a mixed microaerobic and anaerobic methanogenic bacterial community in alkaline wetwoods, and the pectolytic degradation of pit membranes may be responsible for the observed water accumulation. Cellulose in the wood cell walls appears to not be attacked, probably as a result of masking by lignin which may prevent cellulose degradation in the absence of oxygen. Lignin itself is recalcitrant to anaerobic microbial attack (76). Thus, the present knowledge on wetwood allows description and understanding of the symptoms of a complex, unusual syndrome in tree physiology. However, it does not allow any conclusions to be made concerning the cause and initiation of wetwood. This is especially true of the question related to how bacteria enter the tree and which ones are the initial colonizers.

Infection could start in the upper stem, from stem lesions, frost cracks, or branch stubs. Strong indications for the last have been obtained especially from observations made on conifers (4, 20, 28). However, in conifers and angiosperms, wetwood usually starts in the lower stem, from injuries of the roots and the root collar (68). Again, it is not clear whether casual colonization of root lesions by saprophytic bacteria is sufficient for a successful infestation or whether active phytopathogenic bacteria have to be involved. The fact that some species of trees are far more affected than others, and that some genera of bacteria are quite consistently encountered in wetwoods, attributes some character of specificity to this plant–microbe interrelationship. During the initial development of wetwood, the tree tissue is probably rich in oxygen, and hence aerobic or facultatively aerobic bacteria are more likely to cause the tree infestation than anaerobic clostridia. Carter (11) was able to induce wetwood symptoms in healthy elms by inoculating them with an *Erwinia* strain which was isolated from elm wetwood. Although the infection in nature probably takes a different pathway than in Carter's experiment, *Erwinia* species are likely to initiate the primary attack on the tree. *Erwinia* spp. were repeatedly isolated from wetwood and were numerically predominant in our enumeration studies (54). Later, when oxygen is consumed in wood tissue by aerobic degradation processes, pectinolytic clostridia can develop and cause water accumulation in the heartwood as a consequence of pit membrane degradation. Thus, wetwood formation would exhibit a similar succession of or competition between *Erwinia* sp. and *Clostridium* sp., as observed in some cases of potato rot (43, 51). A *C. butyricum* strain was also identified as the causative agent of a hornbeam disease (25). The authors described their isolates as a new species variety, *C. butyricum* var. *phytopathogenicum*. Reports on the involvement of *Clostridium* strains in plant diseases characterized by limited oxygen partial pressures are accumulating and give rise to considerations concerning their characterization as real pathogens.

It is interesting to note that the tree survives wetwood development for long periods, often decades, and wetwood bacteria will only spread into the outer sapwood when the health of the tree declines (8). Phenolic compounds which are reported to act as antimicrobial agents are found in high concentrations in fetid liquid and could be the reason for the lack of fungi in wetwood-affected trees. Since the toxicity of phenols is much higher in the presence of oxygen than in its absence, it appears evident that in completely anoxic wetwood bacteria are not significantly impaired, whereas they are inhibited in the oxic/anoxic boundary region. A decay resistance of black cottonwood due to anoxic wetwood formation was reported (65). Thus, wetwood formation is not necessarily detrimental, but may even have limited beneficial effects.

Recent epidemics of fir and pine dieback in Europe and indications for similar developments in the United States have again drawn attention to the wetwood phenomenon. Afflicted conifers contain extraordinarily high volumes of wetwood (5, 8). There is no doubt among forest pathologists that air pollution and acid rainfall are the primary cause of the trees' dying; however, the mechanisms involved are very complex.

Acidification of soils may directly affect the tree or may indirectly be detrimental by leaching of calcium and magnesium or by solubilization of aluminum ions. The same effects may also alter the properties of the pectic mucilaginous layers on root hair surfaces and thus allow the penetration of saprophytic bacteria which otherwise had no chance for invasion. Moreover, an impaired general resistance of the tree against parasites as a consequence of mineral deficiencies (3) could favor infections by saprophytic or pathogenic bacteria as well. Thus, wetwood formation is not the initial cause but may be promoted by air pollution and soil acidification, and is probably an important intermediary stage in the pollution-dependent fir and pine dieback.

PECTIN DEGRADATION IN ANOXIC LAKE SEDIMENTS AND EXTREME ENVIRONMENTS

Pectin degradation in anoxic freshwater lake sediments is by far slower than in rumen ecosystems, which have been studied by other authors (30). Tracer studies on degradation kinetics with ^{14}C-labeled pectin and fresh sediment samples from two Wisconsin lakes incubated at in situ temperatures revealed turnover times of 100 h at 8°C in the neutral (pH 7.1) sediment of Lake Mendota and of 185 h at 4°C in the slightly acidic (pH 6.1) sediment of Knaack Lake. The numbers of anaerobic pectin-degrading bacteria varied seasonally by two orders of magnitude in each lake and exhibited maxima in the fall when in both lakes the algal blooms died off and were deposited (57). The prevailing anaerobic pectinolytic bacteria were isolated and identified as *C.*

butyricum strains similar to those obtained from wetwood. These *Clostridium* strains fermented pectin in pure culture to acetate, ethanol, hydrogen, butyrate, and methanol. Methanol was not formed during fermentation of polygalacturonic acid or glucose. In Knaack Lake, *Bacteroides* strains also predominated among the pectinolytic flora. Complete conversion of pectin to stoichiometric amounts of methane and carbon dioxide was shown in mixed culture experiments with *C. butyricum* and *Methanosarcina barkeri*, both isolated from Lake Mendota sediment. Methanol, acetate, and hydrogen appeared as intermediates and were simultaneously utilized by the methanogen (57). Butyrate and ethanol were not detected in the coculture. A similar coculture approach to methanogenic pectin decomposition was undertaken by a Russian research group (7).

Hence, pectin is easily degraded in anoxic lake sediments, although native pectin associated with plant biomass is probably degraded at a lower rate because of its complex linkage to other cell wall constituents. The input sources of pectin in anoxic lake sediments, however, have not yet been sufficiently evaluated. Knaack Lake obtains considerable amounts of biomass from allochthonous sources, such as leaves from surrounding trees, but this is not the case for the sampled central area of Lake Mendota. The literature on pectin contents of algae and cyanobacteria is contradictory. Whereas several eucaryotic freshwater algae have been reported to contain pectin in cell walls, scales, and thecae (22, 23, 60), pectin was never found in marine algae (41, 49). Sheath material of cyanobacteria was reported in the older literature to contain pectin (46); in recent reviews, these reports are questioned (19). Thus, freshwater sediments may receive pectic polysaccharides from autochthonous algal production, whereas this question is open with respect to profundal marine sediments. It is also still a matter for further research to determine how much of the pectin produced by either allo- or autochthonous primary production really reaches the sediment, since aquatic hyphomycetes and *Cytophaga* sp. actively attack pectic substances in oxic waters (13, 24, 61). Studies on blue-green algae in Lake Mendota revealed that *Aphanizomenon* and *Anabaena* blooms are nearly completely degraded in the epilimnion, whereas *Microcyctis* cells are mainly decomposed at the sediment surface (21).

Pectin decomposition by anaerobic bacteria was observed also in the photosynthetic algal–bacterial mat ecosystem of Octopus Spring, Yellowstone National Park (18). The mat contained anaerobic pectinolytic bacteria in numbers of 10^3 to 10^4/ml. The predominant pectinolytic bacterium was isolated and characterized as a new species of strictly anaerobic thermophilic bacteria, *Clostridium thermosulfurogenes*, which exhibits an unusual type of sulfur metabolism (58). The optimal growth temperature of this strain was identical with the temperature at the site of its isolation (60°C). The organism produced two thermostable pectinolytic enzymes, a pectin esterase and a pectate hydrolase, with temperature optima at 70 and 75°C, respectively (59). The ecological significance of pectin degradation in algal mats by our isolate has not yet been studied. It deserves attention that no cellulose-degrading organism was detected at the sampling site. This is understandable since the major phototrophic species, *Synechococcus lividus* and *Chloroflexus aurantiacus*, lack cellulose in their wall. Our isolate exhibits considerable activities for hydrolysis of starch and gelatin, which may be of equal importance for its metabolism in the mat as is the pectinolytic activity.

Pectinolytic bacteria in numbers of $>10^3$ per ml were also detected in the anoxic salt brine and the sediment of Great Salt Lake, Utah. The ecological importance of pectin degradation in Great Salt Lake is not yet clear either. It has not yet been demonstrated whether halophilic green algae such as *Dunaliella* spp. contain pectin.

FATE OF PECTIN METHYL RESIDUES

The carboxyl methyl ester linkages in pectin are hydrolytically cleaved by the action of pectin esterases, and the methyl residue is released as methanol (52). Whereas the sugar acid moieties are usually fermented by the pectinolytic bacteria, the methanol is not further metabolized (56). Pectin degradation thus provides an important source of methanol and establishes a need for anaerobic methanol metabolism to avoid methanol accumulation in permanently anoxic environments. High-molecular-weight lignin, another possible source of methanol, is not significantly decomposed under anaerobic conditions (26, 76), and methoxyl groups of low-molecular-weight aromatic plant compounds are not released but are fermented by, for example, *Acetobacterium woodii* inside the cell (1), probably by methyl group transfer (A. Tschech and N. Pfennig, Arch. Microbiol., in press).

In freshwater ecosystems, methanol can easily be converted by methanogenic bacteria to methane (74, 75). Surprisingly, the same is true for marine and hypersaline sediments in the presence of sulfate (48; T. J. Phelps and J. G. Zeikus, Abstr. Annu. Meet. Am. Soc. Microbiol. 1980, I4, p. 85). Although sulfate-reducing bacteria usually outcompete methanogens for electron donors in the presence of excess sulfate (71), there is probably no competition because none of the sulfate reducers predominant in marine sediments is able to oxidize methanol, albeit some contradictory statements exist in the literature. Methanol degradation with concomitant sulfate reduction can be achieved in mixed cultures of *A. woodii* and *Desulfovibrio vulgaris* (Tschech and Pfennig, in press), *A. woodii* and *Desulfobacter postgatei*, or *Butyribacterium methylotrophicum* and *Desulfovibrio vulgaris* (77; T. J. Phelps, Abstr. Annu. Meet. Am. Soc. Microbiol. 1983, I163, p. 166); however, the ecological significance of these processes has still to be evaluated. Recently, some *Desulfotomaculum* species were isolated which oxidized methanol with sulfate in pure culture (Klemps and Widdel, unpublished data). Methanol formation from pectin may be of minor importance in marine sediments since pectin was never found in marine algae (41, 49). Methanol could also be formed from methylated proteins or methyl ethers of sugar residues in bacterial or cyanobacterial lipopolysaccharides (21), but ecological studies on this matter are still lacking.

CONCLUSIONS

Recent studies on the chemistry and structure of pectin in plant tissue allow an understanding of its function under normal and stressed conditions, e.g., mineral deficiencies or acidification. Pectin is the first target for microbial invasion into plant tissue in the establishment of either symbiotic or pathogenic plant–microbe interactions. Living plants alleviate or prevent microbial attack with complicated recognition and defense systems which are induced by the microbe's pectinolytic enzymes and the products of their action. The plant's control is apparently restricted under microaerobic and anaerobic conditions, in which typical plant pathogens like *Erwinia* spp. can be outcompeted by *Clostridium* spp. Characterization of these "disease" syndromes, e.g., the rot of potatoes or the formation of wetwood, as being either phytopathogenic or saprophytic becomes difficult.

Wetwood is an atypical water-soaked anoxic condition of the heartwood of many tree species. Microbial populations in alkaline wetwoods were enumerated and characterized. Pectinolytic clostridia dominated and were found to cause water accumulation in the wood tissue by means of pit membrane destruction. The pathway which bacterial infection takes in wetwood development, however, is unknown, and deserves thorough study because of the apparent association of wetwood formation with air pollution-dependent dieback of forest trees in industrialized countries.

Pectin degradation in freshwater lake sediments as well as in thermal and hypersaline habitats was studied, and the organisms involved were counted and characterized. The degradation kinetics in freshwater lake sediments indicate that pectin decomposition is not the rate-limiting step in the anaerobic degradation of plant biomass. However, more data on pectin contents of aquatic primary producers are necessary to evaluate the importance of pectin degradation in aquatic sediments.

I am indebted to Helga Förster, Konstanz, West Germany, for useful reference hints and to James C. Ward and J. Gregory Zeikus, Madison, Wisconsin, for reading this manuscript and suggesting improvements. My own work on pectin degradation was carried out during an advanced educational stay with J. G. Zeikus at the Department of Bacteriology of the University of Wisconsin in Madison, and was generously supported by the Deutsche Forschungsgemeinschaft, Bonn, West Germany.

LITERATURE CITED

1. **Bache, R., and N. Pfennig.** 1981. Selective isolation of *Acetobacterium woodii* on methoxylated aromatic acids and determination of growth yields. Arch. Microbiol. **130:**255–261.
2. **Bateman, D. F., and H. G. Basham.** 1976. Degradation of plant cell walls and membranes by microbial enzymes, p. 316–355. *In* R. Heitefuss and P. A. Williams (ed.), Encyclopedia of plant physiology, new series, vol. 4, Physiological plant pathology. Springer-Verlag, Berlin.
3. **Bauch, J.** 1983. Biological alterations in the stem and root of fir and spruce due to pollution influence, p. 377–386. *In* B. Ulrich and J. Pankrath (ed.), Effects of accumulation of air pollutants in forest ecosystems. D. Reidel Publishing Co., Dordrecht, The Netherlands.
4. **Bauch, J., W. Höll, and R. Endeward.** 1975. Some aspects of wetwood formation in fir. Holzforschung **29:**198–205.
5. **Bauch, J., P. Klein, A. Frühwald, and H. Brill.** 1979. Alterations of wood characteristics in *Abies alba* Mill. due to "fir dying" and considerations concerning its origin. Eur. J. For. Pathol. **9:**321–331.
6. **Bauch, J., W. Liese, and F. Scholz.** 1968. Über die Entwicklung und stoffliche Zusammensetzung der Hoftüpfelmembranen von Längstracheiden in Coniferen. Holzforschung **22:**144–153.
7. **Bonch-Osmolovskaya, E. A., and S. A. Ilarionov.** 1982. Methane production from pectin by a combined culture of *Clostridium pectinofermentans* and *Methanosarcina vacuolata*. Microbiology (USSR) **51:**340–343.
8. **Brill, H., E. Bock, and J. Bauch.** 1981. Über die Bedeutung von Mikroorganismen im Holz von *Abies alba* Mill. für das Tannensterben. Forstwiss. Centralbl. **100:**195–206.
9. **Bruce, R. J., and C. A. West.** 1982. Elicitation of casbene synthetase activity in castor bean—the role of pectic fragments of the plant cell wall in elicitation by a fungal endopolygalacturonase. Plant Physiol. **69:**1181–1188.
10. **Byrde, R. J. W.** 1979. Role of polysaccharide-degrading enzymes in microbial pathogenicity, p. 417–436. *In* R. C. Berkeley, G. W. Gooday, and D. C. Ellwood (ed.), Microbial polysaccharides and polysaccharases. Academic Press, London.
11. **Carter, J. C.** 1945. Wetwood of elms. Bulletin 23, p. 407–448. Natural History Survey Division, Urbana, Ill.
12. **Cesàro, A., A. Ciana, F. Delben, G. Manzini, and S. Paoletti.** 1982. Physicochemical properties of pectic acid. I. Thermodynamic evidence of a pH-induced conformational transition in aqueous solution. Biopolymers **21:**431–449.
13. **Chamier, A.-C., and P. A. Dixon.** 1982. Pectinases in leaf degradation by aquatic Hyphomycetes. I. The field study. The colonization-pattern of aquatic Hyphomycetes on leaf packs in a surrey stream. Oecologia (Berlin) **52:**109–115.
14. **Collmer, A., P. Berman, and M. S. Mount.** 1982. Pectate lyase regulation and bacterial soft-rot pathogenesis, p.

395–422. *In* M. S. Mount and G. H. Lacy (ed.), Phytopathogenic prokaryotes, vol. 2. Academic Press, Inc., New York.

15. **Collmer, A., C. H. Whalen, S. V. Beer, and D. F. Bateman.** 1982. An exo-poly-α-D-galacturonosidase implicated in the regulation of extracellular pectate lyase production in *Erwinia chrysanthemi*. J. Bacteriol. **149**:626–634.

16. **Côté, W. A.** 1977. Wood ultrastructure in relation to chemical composition, p. 1–44. *In* F. A. Loewus and V. C. Runeckles (ed.), The structure, biosynthesis and degradation of wood. Plenum Press, London.

17. **Dazzo, F. B.** 1980. Lectins and their saccharide receptors as determinants of specificity in the Rhizobium-legume symbiosis, p. 277–304. *In* S. Subtelny (ed.), Cell surface: mediator of development. Academic Press, Inc., New York.

18. **Doemel, W. N., and T. D. Brock.** 1977. Structure, growth, and decomposition of laminated algal-bacterial mats in alkaline hot springs. Appl. Environ. Microbiol. **34**:433–452.

19. **Drews, G., and J. Weckesser.** 1982. Function, structure, and composition of cell walls and external layers, p. 333–357. *In* N. G. Carr and B. A. Whitton (ed.), The biology of cyanobacteria, Botanical Monographs vol. 19, University of California Press, Berkeley.

20. **Etheridge, D. E., and L. A. Morin.** 1962. Wetwood formation in balsam fir. Can. J. Bot. **40**:1335–1345.

21. **Fallon, R. D., and T. D. Brock.** 1980. Planktonic blue-green algae: production, sedimentation, and decomposition in Lake Mendota, Wisconsin. Limnol. Oceanogr. **25**:72–88.

22. **Gooday, G. W.** 1971. A biochemical and autoradiographic study of the role of the Golgi bodies in thecal formation in *Platymonas tetrathele*. J. Exp. Bot. **23**:959–971.

23. **Green, J. C., and D. H. Jennings.** 1967. A physical and chemical investigation of the scales produced by the Golgi apparatus within and found on the surface of the cells of *Chrysochromulina chiton* Parke et Manton. J. Exp. Bot. **18**:359–370.

24. **Güde, H.** 1973. Untersuchungen über aerobe pektinzersetzende Bakterien in einem eutrophen See. Arch. Hydrobiol. Suppl. **42**:383–496.

25. **Gvozdjak, R. J., S. F. Khodros, and V. V. Lipshic.** 1976. Biological properties of *Clostridium butyricum* var. *phytopathogenicum* var. nov., an agent of hornbeam disease. Microbiol. J. (Ukrainian) **38**:288–292.

26. **Hackett, W. F., W. J. Connors, T. K. Kirk, and J. G. Zeikus.** 1977. Microbial decomposition of synthetic 14-C-lignins in nature: lignin biodegradation in a variety of natural materials. Appl. Environ. Microbiol. **33**:43–51.

27. **Hahn, M. G., A. G. Darvill, and P. Albersheim.** 1981. Host-pathogen interactions. XIX. The endogenous elicitor, a fragment of a plant cell wall polysaccharide that elicits phytoalexin accumulation in soybean. Plant Physiol. **68**:1161–1169.

28. **Hartley, C., R. W. Davidson, and B. S. Crandall.** 1961. Wetwood, bacteria and increased pH in trees. U.S. For. Prod. Lab. Rep. 2215.

29. **Hillis, W. E.** 1977. Secondary changes in wood. Rec. Adv. Phytochem. **11**:247–309.

30. **Hobson, P. N.** 1979. Polysaccharide degradation in the rumen, p. 377–397. *In* R. C. Berkeley, G. W. Gooday, and D. C. Ellwood (ed.), Microbial polysaccharides and polysaccharases. Academic Press, London.

31. **Huth, G., and E. Schlösser.** 1982. Role of extracellular microbial hydrolases in pathogenesis. Med. Fac. Landbouw. Rijksuniv. Gent. Bull. **47**:855–861.

32. **Ishii, S.** 1982. Enzymatic extraction and linkage analysis of pectic polysaccharides from onion. Phytochemistry **21**:778–780.

33. **Jarvis, M. C.** 1982. The proportion of calcium-bound pectin in plant cell walls. Planta **154**:344–346.

34. **Jarvis, M. C., D. R. Threfall, and J. Friend.** 1981. Potato cell wall polysaccharides: degradation with enzymes from *Phytophthora infestans*. J. Exp. Bot. **32**:1309–1319.

35. **Keegstra, K., K. W. Talmadge, W. D. Bauer, and P. Albersheim.** 1973. A model of the walls of suspension-cultured sycamore cells based on interconnections of the macromolecular components. Plant Physiol. **51**:188–197.

36. **Keen, N. T., and B. Bruegger.** 1977. Phytoalexins and chemicals that elicit their production in plants, p. 1–26. *In* P. Hedin (ed.), Host plant resistance to pests. American Chemical Society Symposium Series, vol. 62. American Chemical Society, Washington, D.C.

37. **Lagerberg, T.** 1935. Barrträdens vattved. Särtryck ur Svenska Skogsvårdsföreningens Tidskrift **2**:177–264.

38. **Lee, S.-C., and C. A. West.** 1981. Polygalacturonase from *Rhizopus stolonifer*, an elicitor of casbene synthetase activity in castor bean (*Ricinus communis*) seedlings. Plant Physiol. **67**:633–639.

39. **Lee, S.-C., and C. A. West.** 1981. Properties of *Rhizopus stolonifer* polygalacturonase, an elicitor of casbene synthetase activity in castor bean (*Ricinus communis*) seedlings. Plant Physiol. **67**:640–645.

40. **Lippincott, J. A., and B. B. Lippincott.** 1980. Microbial adherence in plants, p. 377–398. *In* E. H. Beachey (ed.), Bacterial adherence. Chapman and Hall, London.

41. **Lobban, C. S., and M. J. Wynne (ed.).** 1981. The biology of seaweeds. Bot. Monogr., vol. 17.

42. **Lund, B. M.** 1972. Isolation of pectolytic clostridia from potatoes. J. Appl. Bacteriol. **35**:609–614.

43. **Lund, B. M., and G. M. Wyatt.** 1972. The effect of oxygen and carbon dioxide on bacterial soft rot of potatoes. I. King Edward potatoes inoculated with *Erwinia carotovora var. atroseptica*. Potato Res. **15**:174–179.

44. **MacKay, A. L., M. Bloom, M. Tepler, and J. E. P. Taylor.** 1982. Broadline proton magnetic resonance study of cellulose, pectin, and bean cell walls. Biopolymers **21**:1521–1526.

45. **Martínez-Molina, E., and J. Olivarez.** 1982. A note on evidence for involvement of pectolytic enzymes in the infection process of *Medicago sativa* by *Rhizobium meliloti*. J. Appl. Bacteriol. **52**:453–455.

46. **Metzner, I.** 1955. Zur Chemie und zum submikroskopischen Aufbau der Zellwände, Scheiden und Gallerte von Cyanophyceen. Arch. Mikrobiol. **22**:45–77.

47. **Morris, E. R., D. A. Powell, M. J. Gidley, and D. A. Rees.** 1982. Conformation and interactions of pectins. I. Polymorphism between gel and solid states of calcium polygalacturonate. J. Mol. Biol. **155**:507–516.

48. **Oremland, R. S., L. M. Marsh, and S. Polcin.** 1982. Methane production and simultaneous sulphate reduction in anoxic salt marsh sediments. Nature (London) **296**:143–145.

49. **Percival, E., and R. H. McDowell.** 1967. Chemistry and enzymology of marine algal polysaccharides. Academic Press, London.

50. **Pérombelon, M. C. M., and A. Kelman.** 1980. Ecology of the soft rot Erwinias. Annu. Rev. Phytopathol. **18**:361–387.

51. **Pérombelon, M. C. M., J. Gullings-Handley, and A. Kelman.** 1979. Population dynamics of *Erwinia carotovora* and pectolytic *Clostridium* spp. in relation to decay of potatoes. Phytopathology **69**:167–173.

52. **Rexová-Benková, L., and O. Markovič.** 1976. Pectic enzymes. Adv. Carbohydr. Chem. Biochem. **33**:323–385.

53. **Rombouts, F. M., and W. Pilnik.** 1980. Pectic enzymes, p. 227–282. *In* A. H. Rose (ed.), Microbial enzymes and bioconversions. Academic Press, London.

54. **Schink, B., J. C. Ward, and J. G. Zeikus.** 1981. Microbiology of wetwood: role of anaerobic bacterial populations in living trees. J. Gen. Microbiol. **123**:313–322.

55. **Schink, B., J. C. Ward, and J. G. Zeikus.** 1981. Microbiology of wetwood: importance of pectin degradation and *Clostridium* species in living trees. Appl. Environ. Microbiol. **42**:526–532.

56. **Schink, B., and J. G. Zeikus.** 1980. Microbial methanol formation: a major end product of pectin metabolism. Curr. Microbiol. **4**:387–389.

57. **Schink, B., and J. G. Zeikus.** 1982. Microbial ecology of

pectin decomposition in anoxic lake sediments. J. Gen. Microbiol. **128:**393–404.

58. **Schink, B., and J. G. Zeikus.** 1983. *Clostridium thermosulfurogenes* sp. nov., a new thermophile that produces elemental sulphur from thiosulphate. J. Gen. Microbiol. **129:**1149–1158.

59. **Schink, B., and J. G. Zeikus.** 1983. Characterization of pectinolytic enzymes of *Clostridium thermosulfurogenes.* FEMS Microbiol. Lett. **17:**295–298.

60. **Sikes, C. S.** 1978. Calcification and cation sorption of *Cladophora glomerata* (Chlorophyta). J. Phycol. **14:**325–329.

61. **Suberkropp, K., and M. J. Klug.** 1980. The maceration of deciduous leaf litter by aquatic Hyphomycetes. Can. J. Bot. **58:**1025–1031.

62. **Tiedemann, G., J. Bauch, and E. Bock.** 1977. Occurrence and significance of bacteria in living trees of *Populus nigra* L. Eur. J. For. Pathol. **7:**364–374.

63. **Tsuyumu, S.** 1977. Inducer of pectic acid lyase in *Erwinia carotovora.* Nature (London) **269:**237–238.

64. **Tsuyumu, S.** 1979. "Self-catabolite repression" of pectate lyase in *Erwinia carotovora.* J. Bacteriol. **137:**1035–1036.

65. **van der Kamp, B. J., A. A. Gokhale, and R. S. Smith.** 1979. Decay resistance owing to near-anaerobic conditions in black cottonwood wetwood. Can. J. For. Res. **9:**39–44.

66. **Walker-Simmons, M., L. Hadwiger, and C. A. Ryan.** 1983. Chitosans and pectic polysaccharides both induce the accumulation of the antifungal phytoalexin pisatin in pea pods and antinutrient proteinase inhibitors in tomato leaves. Biochem. Biophys. Res. Commun. **110:**194–199.

67. **Walkinshaw, M. D., and S. Arnott.** 1981. Conformations and interactions of pectins. I. X-ray diffraction analyses of sodium pectate in neutral and acidified forms. J. Mol. Biol. **153:**1055–1073.

68. **Ward, J. C., and W. Y. Pong.** 1980. Wetwood in trees: a timber resource problem. U.S. For. Serv. Gen. Tech. Rep. PNW-112. Pacific Northwest Forest and Range Experimental Station, Portland, Ore.

69. **Ward, J. C., J. E. Kuntz, and E. McCoy.** 1969. Bacteria associated with "shake" in broadleaf trees. Phytopathology **59:**1056.

70. **Ward, J. C., and J. G. Zeikus.** 1980. Bacteriological, chemical and physical properties of wetwood in living trees. Mitt. Bundesforschungsanst. Forst Holzwirtsch. **131:**133–166.

71. **Winfrey, M. R., and J. G. Zeikus.** 1977. Effect of sulfate on carbon and electron flow during microbial methanogenesis in freshwater sediments. Appl. Environ. Microbiol. **33:**215–221.

72. **Wong, W. C., and T. F. Preece.** 1978. *Erwinia salicis* in cricket bat willows: peroxidase, polyphenoloxidase, β-glucosidase, pectinolytic and cellulolytic enzyme activity in diseased wood. Physiol. Plant Pathol. **12:**333–347.

73. **Worrall, J. J., and J. R. Parmeter.** 1982. Formation and properties of wetwood in white fir. Phytopathology **72:**1209–1212.

74. **Zehnder, A. J. B.** 1978. Ecology of methane formation, p. 349–376. *In* R. Mitchell (ed.), Water pollution microbiology, vol. 2. John Wiley & Sons, London.

75. **Zeikus, J. G.** 1977. The biology of methanogenic bacteria. Bacteriol. Rev. **41:**514–541.

76. **Zeikus, J. G.** 1981. Lignin metabolism and the carbon cycle. Polymer biosynthesis, biodegradation, and environmental recalcitrance, p. 211–243. *In* M. Alexander (ed.), Advances in microbial ecology, vol. 5. Plenum Press, New York.

77. **Zeikus, J. G.** 1983. Metabolism of one carbon compounds by chemotrophic anaerobes. Adv. Microb. Physiol. **24:**215–299.

78. **Zeikus, J. G., and J. C. Ward.** 1974. Methane formation in living trees: a microbial origin. Science **184:**1181–1183.

Microbial Ecology of Cellulose and Hemicellulose Metabolism in Gastrointestinal Ecosystems

N. O. VAN GYLSWYK AND H. M. SCHWARTZ

National Chemical Research Laboratory, Pretoria, South Africa

STRATEGIES OF CELLULOSE AND HEMICELLULOSE DIGESTION

It is an anomaly of biochemical evolution that, despite the fact that herbivorous species are found throughout the animal kingdom, only those organisms which are lowest on the evolutionary scale can themselves synthesize the enzymes necessary for the utilization of plant cell wall carbohydrates. All the others are dependent on microorganisms in their gastrointestinal tract for the hydrolysis of cellulose and hemicellulose. To accommodate these microbes, and to delay transit of food sufficiently long to enable them to act upon it, many anatomical modifications of the tract have evolved (4, 101). Fiber-digesting microorganisms are most commonly concentrated in the hindgut. Enlargement of a section of the stomach to allow microbial attack on the food before the action of the animal's enzymes apparently evolved later (49). Most foregut fermenters carry out a further microbial digestion of food residues in the hindgut. This enables them to extract the maximum energy from cellulose and hemicellulose in their diet.

The extent of cellulose and hemicellulose digestion by a number of vertebrate species is shown in Table 1. It can be seen that foregut fermenters generally digest cellulose more extensively than do hindgut fermenters, particularly when the content of cell wall carbohydrates in the diet is high. Efficiency of digestion is largely determined by the length of time digesta are retained in the gastrointestinal tract (107). As the fiber content of the diet is increased, hindgut fermenters generally increase the rate of passage of digesta through the tract to try to maintain the intake of the more easily digestible components of the food, at the cost of reducing the efficiency of digestion of cell wall carbohydrates. In the larger ruminants, on the other hand, food is retained in the rumen until the particle size is reduced sufficiently by mastication and microbial action to enable it to pass through the reticular omasal orifice. These different strategies can be expected to exert different selective pressures on the fiber-digesting microorganisms in the fore- and hindgut. The more rapid flow of digesta through the hindgut could be expected to select species which can grow fastest on the most readily digestible fractions of cellulose and hemicellulose. The control of flow of digesta from the rumen, on the other hand, should favor organisms which can degrade the cell wall carbohydrates most extensively. Conditions in the ruminant hindgut are the most exacting because only the less easily fermentable cell wall carbohydrates reach the cecum; the retention time of digesta in the cecum and colon is less than in the rumen (38).

Table 1 shows that the digestibility of hemicellulose in the hindgut is usually greater than that of cellulose. This is also observed in the hindgut of the ruminant (109), whereas in the rumen cellulose is generally digested more efficiently (9, 10, 106). Significant digestion of hemicellulose, but not of cellulose, has been observed in the gastrointestinal tract anterior to the cecum in the pig (54, 58), and it has been suggested that exposure of hemicellulose to acid in the stomach may modify its structure to make it subsequently more degradable (34). This, however, has not been subject to experimental test. When the retention time of food in the hindgut is short, the amounts of hemicellulose and cellulose digested will be determined to an increasing extent by their rates of digestion. Isolated xylan is digested faster by rumen bacteria than is cellulose in the form of ball-milled filter paper. If the rates of digestion of the carbohydrates in situ in plant cell walls show the same difference, this would account for the increased difference in the digestibility of cellulose and hemicellulose in the pig as the forage content of the diet is increased (Table 1).

The presence of other carbohydrates in the diet will influence the digestion of cellulose and hemicellulose in the fore- and hindgut to varying degrees. It is well established that digestion of cellulose in the rumen is reduced when starch or sugars are fed. On the other hand, no soluble sugars and usually very little starch reach the terminal ileum of ruminants, and the hindgut becomes increasingly important as a site of digestion of cell wall carbohydrates as the proportion of grain in the diet increases. Thus, MacRae and Armstrong (70) showed that only 9% of the cellulose digested in the whole gastrointestinal tract was degraded in the cecum and colon of sheep when hay alone was fed. When rolled barley formed one-third or two-thirds of the diet, 16% and 29%, respectively, of the total

TABLE 1. Cellulose and hemicellulose digestion in different species

Species	Diet	Digestibility (% relative to ruminant)[a]			Reference
		Hemicellulose	Cellulose	Acid detergent fiber	
Foregut fermenters					
Llama, guanaco	Alfalfa		77.6 (120)		44
	Timothy hay, grain		47.2 (120)		44
Macropod marsupials					
Euro	Alfalfa hay	37.5 (88)		46.0 (105)	48
	Oaten chaff	26.8 (118)		22.3 (107)	48
	Wheat straw	40.0 (93)		27.7 (71)	48
Red kangaroo	Alfalfa hay	33.0 (78)		35.7 (82)	48
	Oaten chaff	17.0 (75)		16.5 (79)	48
	Wheat straw	42.9 (100)		32.7 (84)	48
Hippopotamus	Elephant grass			61.7 (95)	2
Colobus monkey	Leaves		44–61		55
Hindgut fermenters					
Horse	Alfalfa		54.4 (90)		110
	Orchard grass		52.1 (81)		110
	Timothy		48.3 (72)		110
	Brome grass		37.8 (59)		110
Horse	Timothy hay	39.5 (81)	33.4 (72)		108
Pony	Timothy hay	42.0 (86)	36.6 (79)		108
Meadow vole	Alfalfa (100%)	39.1 (84)	33.6 (67)		60
	Orchard grass (75%)	37.9 (50)	29.1 (43)		60
	Brome grass (75%)	24.1 (34)	17.9 (27)		60
Hyrax	Alfalfa	53.4 (78)	41.0 (82)		91
Swine	Alfalfa (50%)	42.7 (92)	39.7 (79)		61
	Orchard grass (50%)	47.4 (62)	43.8 (65)		61
	Brome grass (50%)	46.7 (66)	38.5 (57)		61
	Alfalfa (20%)	54.0	20.5		54
	Alfalfa (40%)	49.3	8.7		54
	Alfalfa (60%)	22.5	6.8		54
Rabbit	Alfalfa pellets	39.8 (58)	31.4 (63)		91
	Timothy hay (50%)	11.0–12.2 (23–25)	4.3–9.5 (9–20)		91
Chinchilla	Commercial pellets		59.1		65
Rat	Alfalfa (50%)	46.6 (100)	20.9 (42)		61
	Orchard grass (50%)	6.4 (8)	1.5 (2)		61
	Brome grass (50%)	10.7 (15)	0.5 (1)		61
Human	High fiber, various	6–89	0–67		114
Dugong	Sea grasses (*Halophila* sp.)			82	80
Green turtle	Sea grass (*Thalassia testudinum*)	78	77		13

[a] Value in parentheses is digestibility as percentage of digestibility of forage in parallel trials on sheep or cattle.

cellulose digestion took place in the hindgut.

When there is no microbial fermentation in the forestomach, a greater proportion of the dietary starch may reach the hindgut. This has been reported to be as high as 29% in the horse (43) and 21% in the pig (59). Endogenous polysaccharides in mucous secretions and sloughed cells also enter the hindgut, and Vercellotti and co-workers (115) have shown that they consti-

tute an important source of fermentable carbohydrate for the bacteria in the large intestine of humans.

It has been suggested (89, 95, 102) that the decrease in cellulose digestion in the rumen when starch or sugars are fed is due to a decline in the number of cellulolytic bacteria as a result of the lowering of the pH of the rumen contents. However, there is little microbiological evidence to support this. A survey of the literature shows no consistent change in the number of cellulolytic bacteria in the rumen as the amount of readily fermentable carbohydrate in the diet increases, except when high grain diets are fed ad libitum. Mertens and Loften (75) found that addition of starch progressively increased the lagtime in the digestion of forage fiber by mixed rumen organisms at controlled pH in vitro, but once digestion started, the rate was not affected. They postulated that this was due to preferential digestion of starch before cellulose by microorganisms which could use both substrates. Studies in progress in our laboratories support this. It is suggested that catabolite repression of cellulose and hemicellulose may have a significant influence on the digestion of cell wall carbohydrates in mixed forage–grain diets in different parts of the alimentary tract.

FIBER-DIGESTING MICROORGANISMS IN THE GASTROINTESTINAL TRACT

Bacteria, protozoa, and fungi all play a role in the degradation of cellulose and hemicellulose in the gastrointestinal tract. The contribution of each of these will be considered in turn.

Protozoa. Protozoa are found in the gut of the majority of animals. Their importance in fiber digestion has been most clearly demonstrated in the case of the flagellate protozoa present in the hindgut of the lower termites. The classical studies of Cleveland and others, who showed that protozoa were essential for the survival of the termite, and of Hungate, who eliminated the possibility of significant cellulose digestion by fungi and bacteria in the gut, have been reviewed by McBee (74). Work on these protozoa has recently culminated in the axenic cultivation of two species, *Trichomitopsis termopsidis* and *Trichonympha sphaerica*, from the hindgut of *Zootermopsis* and in the demonstration that both could ferment ^{14}C-cellulose to CO_2, H_2, and acetate (120–122).

In vertebrates the protozoa in the fore- and hindgut are usually predominantly ciliates, many of which are specific for their host. Electron micrographs of material from the rumen and from the cecum of the horse have shown protozoa ingesting and degrading plant cell walls (1, 7, 14, 32). Several species from the rumen (*Eremoplastron bovis, Eudiplodinium maggii, Polyplastron multivesiculatum,* and *Epidinium caudatum ecaudatum*) and a *Cycloposthium* sp. from the horse cecum have been grown in clone culture on fibrous plant material and have been shown to produce cellulase or hemicellulase, or both (15, 24). Demeyer (33) has reviewed the available data on digestion in faunated and defaunated sheep, and has concluded that the ciliate protozoa are normally responsible for about one-third of the fiber digestion in the rumen. This figure must be accepted with some caution, however. On the one hand, defaunation increases the size of the bacterial and fungal populations which may compensate for the loss of fiber-digesting protozoa. On the other hand, when grain is fed, the presence of protozoa in the rumen may favor cellulose digestion, without their being actively involved, because they ingest much of the starch and so limit its inhibitory action on cellulose degradation by the bacteria.

Fungi. The role of fungi in fiber digestion in the gastrointestinal tract has received due recognition recently. Three species of obligately anaerobic fungi have been isolated from the rumen (84–86). High populations are present in sheep and cattle fed stalky diets, but they have not been observed in animals on lush pasture or fed high grain diets (5). Two further species have been isolated from the cecum of horses (87). Electron microscopic studies have also shown fungi growing on plant material in the rumen of several species of wild ruminants, in the foregut of four species of macropod marsupials (kangaroos and wallabies), and in the elephant cecum (5). The isolates bring about extensive degradation of cellulose and hemicellulose in intact forages (87, 88). However, it is not possible at present to make an assessment of their contribution to fiber digestion in the gastrointestinal tract.

The higher termite, *Macrotermes natalensis*, which is found in South Africa, uses the enzymes of a fungus (a *Termitomyces* sp.) for cellulose degradation in its gastrointestinal tract. The termite itself secretes some C_x cellulase (endocellulase), but it obtains its C_1 cellulase (exocellulase) by ingesting fungal nodules which develop on "combs" of chewed, undigested plant material (73). This strategy is an interesting alternative to the maintenance of fiber-digesting microorganisms within the gut.

Fiber-digesting bacteria in mammalian gut ecosystems. The numbers, species, and metabolism of the cellulolytic and hemicellulolytic bacteria in the rumen have been the subject of numerous reviews (28, 33, 45, 51). By contrast, our knowledge of the species in other parts of the alimentary tract, including the ruminant hindgut, is scanty.

The available figures for the concentration of cellulolytic bacteria in the contents of the gastrointestinal tract of different mammals are summarized in Table 2. There are very few data other than for the rumen, and none for other classes of vertebrates.

It would appear that the number of cellulolytic bacteria per gram of digesta is lower in the hindgut than in the rumen. For instance, values

TABLE 2. Numbers of cellulolytic bacteria in various parts of the gut of different mammals

Animal	Region of gut	No./g or ml	% of culturable bacteria	Reference
Foregut				
Ruminant	Rumen	10^7–10^9	1–10	113
Camel	"Rumen"	8×10^6–4×10^8		52
Langur monkey	Forestomach	8×10^7–40×10^{7a}	0.01–0.06	8
Hindgut				
Steer	Cecum	6×10^6	3	56
Steer	Colon	7×10^5	5	56
Sheep	Cecum	10^8		72
Horse and pony	Cecum and colon	Up to 10^6		26
Pony	Cecum	2×10^7–8×10^7	3–16	57
Pony	Cecum	4×10^7	9	56
Pony	Colon	7×10^6	2	56
Elephant	Cecum	10^5	0.005	52
Elephant	Colon	10^8	33	52
Pig	Cecum and colon	>10^6		27
Rabbit	Cecum	5×10^5–7×10^7		39
Rabbit	Cecum and colon	>10^6		27
Guinea pig	Cecum and colon	>10^6		27
Guinea pig	Cecum	10^7		29
Rat	Cecum	5×10^7–12×10^8	6	71
Human	Colon (feces)	Up to 2×10^8	0.3	11

a Per gram (dry weight).

reported for the equine cecum range from 10^7 to 10^8/g, whereas values for the rumen reached 10^9/g. However, Kern and co-workers (57) showed no consistent difference in numbers of cellulolytic bacteria in the pony cecum and the steer rumen. When timothy hay was fed, the number in the cecum was half that in the rumen, whereas it was higher when the diet consisted of clover hay. This is in agreement with the in vitro finding of Koller and co-workers (64) that pony cecal contents degraded cell wall carbohydrates in timothy hay at about half the rate of degradation in rumen contents, whereas the rates for alfalfa were not significantly different.

It is interesting to note that cellulolytic bacteria make up 10% or less of the total culturable bacteria in the hindgut and foregut of the different species (Table 2). This is despite the fact that the bacteria in the hindgut receive less readily fermentable substrates and are thus more dependent on the cellulolytic bacteria for their nutrient supply.

The most important cellulolytic bacteria in the rumen, in relation to both numbers and activity are *Bacteroides succinogenes*, *Ruminococcus albus*, and *R. flavefaciens* (Table 3). Cellulolytic strains of *Butyrivibrio fibrisolvens* are also present in considerable numbers, but their activity is low. *Eubacterium cellulosolvens* is usually present in lower numbers, but occasionally becomes the most numerous species (94). Cellulolytic clostridia have been sporadically reported. They are of interest because two species, *Clostridium lochheadii* and *C. polysaccharolyticum*

degrade cellulose faster than any of the more common species (51, 62).

B. succinogenes degrades crystalline cellulose, as well as cellulose in intact forages, more extensively than do the other species (31, 40, 78). In accordance with predictions made earlier, it was found to be the predominant cellulolytic species in the rumen of cattle fed wheat straw, which has a low digestibility (19). With more digestible diets it loses much of its competitive advantage and may be outnumbered by ruminococci (19, 35).

Reliable data on the rates of growth on and digestion of cellulose by different species of intestinal bacteria have only recently become available. Earlier studies of van Gylswyk and Labuschagne (112) were of limited value because they were carried out in batch culture where the proportions of more and less easily degradable substrate changed with time. Kistner and co-workers (62) have now studied the kinetics of growth of bacteria from the rumen on ball-milled filter paper in continuous culture. They found maximum specific growth rates for *R. albus*, *R. flavefaciens*, *B. succinogenes*, and *C. polysaccharolyticum* of 0.20, 0.51, 0.38, and 0.59 h^{-1}, respectively. Although these values apply to only one strain of each species, they indicate that *R. albus* grows more slowly on cellulose than do the other species. Cellulolytic ruminococci have been reported to occur in a number of sites outside the rumen. However, in none of these cases has a cellulolytic *R. albus* been positively identified. *R. albus* isolates were

TABLE 3. Incidence of autochthonous fiber-digesting bacteria in various parts of gastrointestinal tracts of different mammals

Bacteria	Function[a]	Animal	Samples from	No./g or ml	% of total culturable bacteria	References
Ruminococcus flavefaciens	C,H	Cattle	Rumen	$<1 \times 10^8$–10×10^8	<1–10	113
		Rat	Cecum	10^7–10^8	<5	71, 76
		Guinea pig	Cecum	10^7		29
		Rabbits	Cecum	Up to 7×10^7		39
		Humans	Feces	4×10^{8b}	0.2	46
Ruminococcus albus	C,H	Cattle	Rumen	2×10^7–20×10^7	1–5	113
Ruminococcus sp.	C	Langur monkey	Stomach	10^{8b}	<0.4	8
		Pig	Cecum and colon	$>10^8$		27
		Rabbit	Cecum and colon	$>10^6$		27
		Guinea pig	Cecum and colon	$>10^6$		27
Bacteroides succinogenes	C,H	Cattle	Rumen	$<1 \times 10^8$–3×10^8	<1–4	113
		Langur monkey	Stomach	10^{8b}	<0.04	8
		Rat	Cecum	10^7–10^9	Up to 6	71, 76
		Pony	Colon	Up to 10^6		26
		Horse	Cecum and colon	Up to 10^6		26
		Pig	Cecum and colon	$>10^6$		27
		Rabbit	Cecum and colon	$>10^6$		27
		Guinea pig	Cecum and colon	$>10^6$		27
Bacteroides sp.	C^w	Human	Feces	2×10^8	0.3	11
Bacteroides ruminicola	H$^+$	Cattle	Rumen	10^8–10^9	5–20	21, 22, 99, 104
Bacteroides vulgatus	H$^+$	Human	Feces	3×10^{10}–6×10^{10b}	12	46, 77
Bacteroides fragilis subsp. a	H$^+$	Human	Feces	3×10^9–10×10^{9b}	1–2	46, 77
Bacteroides eggerthii	H$^+$	Human	Feces	3×10^8–40×10^{8b}	<0.1–1.5	46, 77
Bacteroides ovatus	H$^+$	Human	Feces	3×10^8–30×10^{8b}	0.1–1	46, 77
Bifidobacterium adolescentis	H$^+$	Human	Feces	7×10^9–30×10^{9b}	3–7	46, 77
Bifidobacterium infantis	H$^+$	Human	Feces	2×10^9–7×10^{9b}	1–1.5	46, 77
Peptostreptococcus productus	H$^-$	Human	Feces	2×10^{10}–5×10^{10b}	9	46, 77
Eubacterium ruminantium	H$^+$	Cattle	Rumen	Up to 10^8	<1–10	20, 22, 104
Eubacterium cellulosolvens	C,H^w	Sheep	Rumen	10^6	ca. 0.1	111
		Dairy cow	Rumen	10^8–10^9	>1	94
Butyrivibrio fibrisolvens	H$^+$,C^{w-}	Cattle, sheep	Rumen	10^8–10^9	Up to 40	22, 67, 68, 69, 83, 99, 104
	?	Rabbit	Feces	10^6		18
	?	Horse	Feces	10^6		18
	?	Human	Feces	10^6		18
Butyrivibrio sp.	C	Sheep	Cecum	ca. 10^7		72
Treponema sp. (spirochetes)	H	Cows, sheep	Rumen	10^6–10^8	Up to 0.5	90, 113

[a] C, Cellulolytic; H, hemicellulolytic (ferment or degrade xylan); ^{w}, weak; $^+$, many strains; $^-$, few strains; ?, unknown.

[b] Numbers per gram (dry weight).

also obtained from human feces, but these were not cellulolytic (119).

These results also show that strains of *R. flavefaciens* can grow faster on cellulose than those of *B. succinogenes*, and the former may be expected to be prominent in situations where the rate of passage through the gut is high. It has been reported to be present in the cecum of rabbits (39), rats (71, 76), and guinea pigs (29). Isolates of *R. flavefaciens* have also been obtained from human feces (46), but it has not been reported whether they were cellulolytic.

B. succinogenes has also been isolated from the cecum or colon, or both, of a number of species including equines (26), swine (27), rabbit (27), rat (71, 76), and guinea pig (27), as well as from the forestomach of the langur monkey (8). A weakly active *Bacteroides* species, showing some similarity to *B. succinogenes*, is the only confirmed cellulolytic isolate from the human intestinal tract to date (11).

Hemicellulolytic bacteria in the rumen have been studied less extensively than those which digest cellulose. All cellulolytic species which have been tested can also degrade hemicellulose (23, 63, 79). *B. succinogenes* and some strains of *R. flavefaciens* cannot metabolize the pentose released (23, 79) and do not grow on xylan as sole carbon source. They are thus not normally included in counts of hemicellulolytic bacteria on a xylan-containing medium. *R. albus*, on the other hand, utilizes virtually all the pentose solubilized, and it is probably this ability which enables it to survive in the rumen in competition with *R. flavefaciens* and *B. succinogenes*.

Many non-cellulolytic rumen bacteria produce hemicellulase and degrade purified hemicellulose extensively. Prominent among these are *B. ruminicola*, *B. fibrisolvens*, and *E. ruminantium*. At least 60% of the culturable bacteria in the rumen can grow with xylan as sole carbohydrate source (30, 42), but most of them can degrade hemicellulose in intact forages to a very limited extent (23, 79).

A smaller group of hemicellulolytic bacteria, comprising 3 to 8% of the total culturable bacteria in the rumen, can be distinguished by their ability to make clearings in a 3% xylan–agar medium. They produce large amounts of cell-free xylanase (42) and generally bring about more extensive degradation of hemicellulose in situ in cell walls than do the other species (79). We have recently obtained a large number of isolates of this group from the rumen of sheep fed corn stover with or without corn grain. The most numerous species was *B. fibrisolvens*, followed by *R. albus* and *R. flavefaciens*. A lesser number of coccoid rods, short straight rods, a clostridium, and a spirochete were also isolated, but have not yet been fully characterized. An interesting observation is that all the isolates degraded carboxymethyl cellulose (CMC) (P. du Preez of this laboratory, unpublished data), whereas the majority, including most of the *B. fibrisolvens*, about one-third of the *R. albus*, and some *R. flavefaciens* isolates, were unable to digest ball-milled filter paper. The use of CMC as a substrate for enumerating cellulolytic gut bacteria is clearly of doubtful value because many of the organisms so counted are unlikely to be able to degrade natural forms of cellulose to any extent.

Outside the rumen, the hemicellulolytic activities of intestinal bacteria have been studied only in the human. Of the *Bacteroides* species, which constitute the largest group from human feces, more than half the isolates examined could utilize xylan or larch arabino-galactan. These included strains of *B. ovatus*, *B. fragilis* subsp. a, *B. eggerthii*, *B. vulgatus*, *B. distasonis*, and *B. thetaiotaomicron*, as well as three unnamed species (96). Hemicellulolytic activity was also found among isolates of *Bifidobacterium adolescentis*, *B. infantis*, *B. longum*, and *Peptostreptococcus productus* (97). *Bacteroides ruminicola* and *E. ruminantium* have also been found in lesser numbers in human feces (46), but it is not known whether the isolates were hemicellulolytic.

A large proportion of the hemicellulolytic bacteria in the human colon also utilize starch (96, 97), and some can also utilize mucopolysaccharides, but it is not known to what extent the availability of the alternative substrates influences their hemicellulolytic activity.

Bacterial digestion of cellulose and hemicellulose in invertebrates. The role of bacteria in the digestion of cell wall carbohydrates in invertebrates is much less clearly defined than in mammals.

Although bacteria are present in the gut of the silverfish, *Ctenolepisma lineata*, they appear to play no part in fiber digestion, because insects reared from eggs under sterile conditions digest cellulose as efficiently as those containing a normal bacterial flora (66). Bacteria also appear to play little or no part in cellulose digestion in the higher termite, *Nasutitermes exitiosus*, because antibiotic treatment had little effect on cellulolytic activity in various parts of the gut (82).

The larvae of *Costelytra zealandica*, the grass grub, feed on the roots of pasture plants, shrubs, and trees. The hindgut contains two species of flagellate protozoa, together with a large bacterial population (2×10^{10} to 5×10^{10} per g [wet weight] of gut). The majority of isolates were facultative anaerobes. Nineteen percent were obligate anaerobes, of which half could degrade xylan and pectin. No cellulolytic bacteria were found (6). It was concluded that, despite the high fiber content of the diet, the larvae appeared not to be dependent on the digestion of structural carbohydrate.

In addition to the protozoa present in the hindgut of the lower termites, there is also a large population of bacteria numbering 10^8 to 10^{10} per ml of gut content (17). Thayer (105) obtained isolates of *Bacillus cereus, Serratia marcescens*, and an *Arthrobacter* species which degraded CMC and could grow on α-cellulose from *Reticulitermes hesperus*. Isolates of *B. cereus* and *S. marcescens* also grew on microcrystalline cellulose (Cellex MN). Schultz and Breznak (98), on the other hand, could not isolate any bacteria from *Reticulitermes flavipes* which could grow on α-cellulose (Sigma). Hungate (50) isolated an anaerobe, *Micromonospora propionici*, which digested filter-paper cellulose, from the gut of an *Amitermes* species but concluded that it could play only a limited role in cellulose digestion because of its slow rate of growth on that substrate. There are no reports on the ability of these bacteria to digest hemicellulose.

Large numbers of aerobic or facultative bacteria which could grow on cellulose were found in the mid- and hindgut of two species of desert millipedes, but they have not as yet been identified. They appear to be responsible for about one-third of the cellulose digestion in the millipede because reduction of their number by 80% or more by antibiotics and starvation decreased the degradation of ^{14}C-cellulose in the gut to 66 to 69% of that of the controls (103). The residual cellulolytic activity can be attributed to enzyme produced by the millipedes (81).

Bacteria would also seem to play a significant part in cellulose digestion in the American cockroach, *Periplaneta americana*, because feeding neomycin halved the utilization of ^{14}C-cellulose (12). Anaerobic CMC-degrading bacteria were found in large numbers in the midgut and colon of *P. americana*, as well as another species of cockroach, *Eublaberus posticus*. Isolates were identified as *Citrobacter freundii, Klebsiella pneumoniae*, and species of *Clostridium, Eubacterium*, and *Serratia*. All grew on ball-milled filter paper but failed to degrade filter-paper strips (25). However, *Ruminococcus albus* strain 7, which was used as a control, also failed the latter test. As this strain is usually accepted as cellulolytic, this focuses attention on the suitability of various forms of cellulose for attempting to isolate active bacteria from the gastrointestinal tract. The undesirability of using CMC has already been mentioned. The use of a too highly crystalline cellulose may err in the opposite direction. It has long been known that there is a high degree of correlation between the crystallinity index of different forms of cellulose and their digestibility by rumen bacteria (3). Cellulose in grasses and legumes is largely amorphous even when the plants are mature. Wood cellulose has a considerable degree of crystallinity, but less than that of cotton linters (3). Most filter paper is made from cotton cellulose, and its ease of digestion depends on the treatment it has received. Bacteria are also able to attack some forms of hemicellulose and not others (96, 97). In the final analysis, therefore, the contribution of bacteria to cellulose and hemicellulose digestion in the gastrointestinal tract should be assessed with their natural substrates.

From the few data available, it is evident that the fiber-digesting bacteria in the gut of invertebrates differ markedly from those in mammals. The majority are facultative anaerobes, although the redox potential in the gut is low enough in some cases to permit methanogenesis (16). None of the cellulolytic species which are prominent in the rumen has been detected so far.

ENZYMES OF CELLULOSE AND HEMICELLULOSE DEGRADATION

The rate and extent of degradation of various forms of cellulose and hemicellulose by gastrointestinal bacteria must reflect their complement of the necessary enzymes.

The degradation of highly ordered cellulose requires C_1 cellulase (exo-β-1,4 glucanase, β-1,4 glucan cellobiohydrolase) together with cellobiase or C_x cellulase (endo-β-1,4-glucanase). Less ordered forms such as CMC and phosphoric acid-swollen cellulose can be degraded by an endoglucanase alone (41, 117). Although considerable work has been done on the cellulases of *B. succinogenes, R. albus*, and *R. flavefaciens*, it is not yet possible to explain the differences in their ability to degrade cotton linters in terms of their content of exo- and endoglucanases.

Because *R. albus* produces large zones of clearing around colonies grown in cellulose-agar, it has long been considered to synthesize extracellular cellulases. As a result, several groups of workers have attempted to isolate and characterize these enzymes. Wood, Wilson, and Stewart (118) found that more than 80% of the carboxymethyl cellulase of their strain was, in fact, cell-bound, but could be released easily by washing the cells with buffer. The eluted enzyme, which also degraded phosphoric acid-swollen cellulose, was not sedimented at 100,000 × *g* (i.e., it was not particulate) but had a molecular weight in excess of 1.5×10^6. A carboxymethyl cellulase of molecular weight approximately 30,000 was also found under certain conditions. Yu and Hungate (123) also found a high-molecular-weight ($>2 \times 10^6$), and possibly three lower-molecular-weight enzymes in their cell-free cellulase complex. All four components solubilized ball-milled filter paper, alfalfa fiber, and Sigmacell. There was little cellobiase activity. In addition to cellobiose, cellotriose, -tetraose, and -pentaose were found (100), suggesting the presence of an endoglucanase.

The enzymes of *R. flavefaciens* are similar in many respects to those of *R. albus* (92, 93). The cellulase was cell bound when the cells grew exponentially, but was released when the bacteria were in stationary phase or grown at low dilution rates in continuous culture. In the latter cases release was considered to be due to cell lysis rather than active secretion. The cell-free enzyme hydrolyzed native cotton and microcrystalline cellulose (Avicel), as well as ball-milled filter paper and CMC, the relative rates of attack being 0.03, 0.4, 1.0, and 9.3. It was concluded that the most active enyzme was an exoglucanase since cellobiose and cellotriose were the main products, and because the viscosity of CMC, which is an indication of chain length, fell slowly in relation to the release of reducing sugars when that substrate was hydrolyzed.

The cellulase system of *B. succinogenes* is rather different from that of the ruminococci (36, 37). More than 80% of the carboxymethyl cellulase of bacteria grown on cellulose was extracellular. About half of this was associated with subcellular vesicles which were considered to be actively secreted and not to be the result of cell lysis. The enyzme obtained was an endoglucanase. It hydrolyzed phosphoric acid-swollen cellulose but had little activity toward ball-milled filter paper, ^{14}C-labeled tobacco leaf cellulose, or Avicel. The isolation of the enzyme(s) which give *B. succinogenes* its activity against the more ordered forms of cellulose has still to be achieved.

The cellulase preparations from these three bacteria all degraded xylan. In the case of *R. flavefaciens*, it was suggested that two enzymes, or at least two active centers, were involved, but no separation of these was attained (92). The hydrolysis of hemicellulose requires, in addition to a β-1,4-glycanase, α- and β-glycosidases to remove the side chains from the linear backbone of the molecule.

Howard and co-workers (47) fractionated the pentosanases of non-cellulolytic butyrivibrios from the rumen and showed the presence of a xylanase, a xylobiase, and an arabinosidase. The xylanase was an endoglycanase which did not degrade CMC. Its action on xylan was inhibited by the presence of arabinose side chains, and only partial hydrolysis of wheat flour pentosan, which has a high proportion of arabinose side chains, was obtained in the absence of arabinosidase activity. α-Arabinosidase has also been shown to increase the digestion of alfalfa cell walls by *R. albus* cellulase (53). Williams and Withers (116) measured both the production and activity of the principal polysaccharidase and glycosidase enzymes associated with the degradation of hemicellulose in a number of unnamed isolates of rumen bacteria belonging to the genera *Butyrivibrio, Ruminococcus, Bacteroides*, and *Eubacterium*. There were marked differences in the enzyme pattern in different isolates, but these have not, as yet, been related to the ability of the bacteria to degrade hemicelluloses of different origins.

Even less is known of what determines the degradation of cell wall carbohydrates in intact forages by different bacteria. This subject was recently comprehensively reviewed by Morris (78), who pointed out that the walls of various cell types in the plant differ in respect of their carbohydrate composition and that the composition of the same type of cell wall may not be the same in different plants. She suggested that bacteria may preferentially attack the cell wall type for which their enzyme system is most effective. The extent to which a bacterial species can digest cell wall carbohydrates in different plants may thus reflect the content of these types of cell wall in the plant. The question remains as to what determines the ability of the bacterium to degrade the selected cell walls. Among the factors to which Morris directs attention is the ability of the bacterium to attach to the cell wall. Recent work has shown that in many situations the attachment of bacteria to surfaces is highly specific. It may be mediated by receptors on the tissue surface. An important part may also be played by soluble factors in the medium, as in the case of salivary glycoproteins in the attachment of oral bacteria. These thoughts add a whole new field to the study of the metabolism of cellulose and hemicellulose in the gastrointestinal tract.

CONCLUSIONS

Research on gut microbes has been most extensive in the case of the rumen. Much work remains to be done on the organisms in other regions of the gut of different animals, including the ruminant hindgut. Despite different selective pressures, available data suggest that the cellulolytic bacteria are largely similar in different gut regions of different animal species. The hemicellulolytic bacteria have received less attention. Many species degrade hemicellulose, but few, including the cellulolytic ones, have high activities in such degradation.

Although bacteria are considered to be the main agents of fiber breakdown in the rumen, the contribution of each species in fiber digestion has not been established satisfactorily. Even less is known about the contribution of fiber-digesting fungal and protozoan species. Activities of clone cultures may not represent activities in situ, which may be strongly influenced by the presence of bacteria or other microbes possessing a different complement of fibrolytic enzymes. Thus co-cultures of selected species may provide better information. For measuring fiber-digesting activity batch cultures are not as suitable as continuous cultures, but technical problems with the use of solid substrates, such as cellulose, in continuous culture, have limited the amount of work done in this field up to now. Purified substrates may not be representative of the substrates available to microbes in intestinal systems. It is clearly desirable to study substrates of a chemical and physical nature similar to those found in vivo. The use of plant cell wall preparations goes a long way toward remedying this shortcoming.

In parallel with these studies, more work should be carried out on fiber-digesting enzymes in the different species. Up to now, attempts to isolate the enzyme responsible for cellulose degradation by *R. albus*, *R. flavefaciens*, and *B. succinogenes* have not yielded activities comparable with those in the bacteria in vivo, nor do they throw any light on the differences in the abilities of the three species to degrade the more crystalline forms of cellulose. A beginning has been made with the characterization of the β-1,4-glycanases and the numerous glycosidases in xylanolytic bacteria from the rumen, but there is need to relate this information to the ability of the bacteria to break down hemicellulose in intact forages, which varies widely between isolates.

It is also important to study the regulation of these enzymes. Starch can inhibit the induction of hemicellulase in bacteria from human feces. There is also evidence that catabolite inhibition may be one of the causes of the depression of the intake and digestion of fiber by ruminants when grain is fed with forage. An understanding of the factors involved might make it possible to improve the utilization of cellulose and hemicellulose in mixed diets.

The fiber-digesting microorganisms in the gut of insects and other invertebrates present a totally different picture from that in mammals. Flagellate protozoa are the main, if not only, agents of degradation of cell wall carbohydrates in many species. In many cases the role of the bacteria, which may be present in very large numbers, is not clear. Such cellulolytic bacteria as have been isolated differ from those in the mammalian gut. This suggests a divergence in the development of microbial digestion of cellulose and hemicellulose in the gastrointestinal tract of invertebrates and higher animals. There is little information on the fiber-digesting bacteria in the gut of herbivorous birds, reptiles, or fish. A study of these may help to trace back the evolutionary steps.

LITERATURE CITED

1. **Akin, D. E., and H. E. Amos.** 1979. Mode of attack on orchard-grass leaf blades by rumen protozoa. Appl. Environ. Microbiol. **37**:332–338.
2. **Arman, P., and C. R. Field.** 1973. Digestion in the hippopotamus. East Afr. Wildl. J. **11**:9–17.
3. **Baker, T. I., G. V. Quicke, O. G. Bentley, R. R. Johnson, and A. L. Moxon.** 1959. The influence of certain physical properties of purified celluloses and forage celluloses on their digestibility by rumen microorganisms in vitro. J. Anim. Sci. **18**:655–662.
4. **Bauchop, T.** 1977. Foregut fermentation, p. 223–250. In R. T. J. Clarke and T. Bauchop (ed.), Microbial ecology of the gut. Academic Press, London.
5. **Bauchop, T.** 1981. The anaerobic fungi in rumen fibre digestion. Agric. Environ. **6**:339–348.
6. **Bauchop, T., and R. T. J. Clarke.** 1975. Gut microbiology and carbohydrate digestion in the larva of *Costelytra zealandica* (Coleoptera:Scarabaeidae) N.Z. J. Zool. **2**:237–243.
7. **Bauchop, T., and R. T. J. Clarke.** 1976. Attachment of the ciliate *Epidinium crawley* to plant fragments in the sheep rumen. Appl. Environ. Microbiol. **32**:417–422.
8. **Bauchop, T., and R. W. Martucci.** 1968. Ruminant-like digestion of the langur monkey. Science **161**:698–700.
9. **Beever, D. E., J. F. Coelho de Silva, J. H. D. Prescott, and D. G. Armstrong.** 1972. The effect in sheep of physical form and stage of growth on the sites of digestion of a dried grass. 1. Sites of digestion of organic matter, energy and carbohydrate. Br. J. Nutr. **28**:347–356.
10. **Beever, D. E., D. J. Thomson, E. Pfeffer, and D. G. Armstrong.** 1971. The effect of drying and ensiling grass on its digestion in sheep. Sites of energy and carbohydrate digestion. Br. J. Nutr. **26**:123–134.
11. **Betian, H. G., B. A. Linehan, M. P. Bryant, and L. V. Holdeman.** 1977. Isolation of a cellulolytic *Bacteroides* sp. from human feces. Appl. Environ. Microbiol. **33**:1009–1010.
12. **Bignell, D. E.** 1977. An experimental study of cellulose and hemicellulose degradation in the alimentary canal of the American cockroach. Can. J. Zool. **55**:579–589.
13. **Bjorndal, K. A.** 1979. Cellulose digestion and volatile fatty acid production in the green turtle, *Chelonia mydas*. Comp. Biochem. Physiol. A **63**:127–133.
14. **Bonhomme-Florentin, A.** 1969. Essais de culture in vitro des Cycloposthiidae, ciliés commensaux de l'intestin du cheval. Role de ces ciliés dans la degradation de la cellulose. Protistologica **5**:519–522.
15. **Bonhomme-Florentin, A.** 1974. Contribution a l'etude de la physiologie de ciliés entodiniomorphes endocommensaux des ruminants et des equides. Ann. Sci. Nat. Zool. Biol. Anim. **16**:221–284.
16. **Bracke, J. W., D. L. Cruden, and A. J. Markovetz.** 1979. Intestinal microbial flora of the American cockroach, *Periplaneta americana* L. Appl. Environ. Microbiol. **38**:945–955.
17. **Breznak, J. A., and H. S. Pankratz.** 1977. In situ morphology of the gut microbiota of wood-eating termites [*Reticulitermes flavipes* (Kollar) and *Coptotermes formosanus* (Shiraki)]. Appl. Environ. Microbiol. **33**:406–426.
18. **Brown, D. W., and W. E. C. Moore.** 1960. Distribution of *Butyrivibrio fibrisolvens* in nature. J. Dairy Sci. **43**:1570–1574.
19. **Bryant, M. P., and L. A. Burkey.** 1953. Numbers and some predominant groups of bacteria in the rumen of

cattle fed different rations. J. Dairy Sci. **36:**218–224.

20. **Bryant, M. P., I. M. Robinson, and I. L. Lindahl.** 1961. A note on the flora and fauna in the rumen of steers fed a feedlot bloat-provoking ration and the effect of penicillin. Appl. Microbiol. **9:**511–515.

21. **Bryant, M. P., N. Small, C. Bouma, and H. Chu.** 1958. *Bacteroides ruminicola* n. sp. and the new genus and species *Succinimonas amylolytica.* J. Bacteriol. **76:**15–23.

22. **Caldwell, D. R., and M. P. Bryant.** 1966. Medium without rumen fluid for nonselective enumeration and isolation of rumen bacteria. Appl. Microbiol. **14:**794–801.

23. **Coen, J. A., and B. A. Dehority.** 1970. Degradation and utilization of hemicellulose from intact forages by pure cultures of rumen bacteria. Appl. Microbiol. **20:**362–368.

24. **Coleman, G. S.** 1979. Rumen ciliate protozoa, p. 381–408. *In* M. Levandowsky and S. H. Hutner (ed.), Biochemistry and physiology of the protozoa, vol. 2. Academic Press, Inc., New York.

25. **Cruden, D. L., and A. J. Markovetz.** 1979. Carboxymethyl cellulose decomposition by intestinal bacteria of cockroaches. Appl. Environ. Microbiol. **38:**369–372.

26. **Davies, M. E.** 1964. Cellulolytic bacteria isolated from the large intestine of the horse. J. Appl. Bacteriol. **27:**373–378.

27. **Davies, M. E.** 1965. Cellulolytic bacteria in some ruminants and herbivores as shown by fluorescent antibody. J. Gen. Microbiol. **39:**139–141.

28. **Dehority, B. A.** 1973. Hemicellulose degradation by rumen bacteria. Fed. Proc. **32:**1819–1825.

29. **Dehority, B. A.** 1977. Cellulolytic cocci isolated from the cecum of guinea pigs (*Cavia porcellus*). Appl. Environ. Microbiol. **33:**1278–1283.

30. **Dehority, B. A., and J. A. Grubb.** 1976. Basal medium for the selective enumeration of rumen bacteria utilizing specific energy sources. Appl. Environ. Microbiol. **32:**703–710.

31. **Dehority, B. A., and H. W. Scott.** 1967. Extent of cellulose and hemicellulose digestion in various forages by pure cultures of rumen bacteria. J. Dairy Sci. **50:**1136–1141.

32. **Delfosse-Debusscher, J., D. Thines-Sempoux, M. Vanbelle, and B. Latteur.** 1979. Contribution of protozoa to the rumen cellulolytic activity. Ann. Rech. Vet. **10:**255–257.

33. **Demeyer, D. I.** 1981. Rumen microbes and digestion of plant cell walls. Agric. Environ. **6:**295–337.

34. **Dierenfeld, E. S., H. F. Hintz, J. B. Robertson, P. J. van Soest, and O. T. Ofterdal.** 1982. Utilization of bamboo by the giant panda. J. Nutr. **112:**636–641.

35. **Dinsdale, D., E. J. Morris, and J. S. D. Bacon.** 1978. Electron microscopy of the microbial populations present and their modes of attack on various cellulolytic substrates undergoing digestion in the sheep rumen. Appl. Environ. Microbiol. **36:**160–168.

36. **Forsberg, C. W., T. J. Beveridge, and A. Hellstrom.** 1981. Cellulase and xylanase release from *Bacteroides succinogenes* and its importance in the rumen environment. Appl. Environ. Microbiol. **42:**886–896.

37. **Groleau, D., and C. W. Forsberg.** 1981. Cellulolytic activity of the rumen bacterium *Bacteroides succinogenes.* Can. J. Microbiol. **27:**517–530.

38. **Grovum, W. L., and V. J. Williams.** 1973. Rate of passage of digesta in sheep. 3. Differential rates of passage of water and dry matter from the reticulo-rumen, abomasum and caecum and proximal colon. Br. J. Nutr. **30:**231–240.

39. **Hall, E. R.** 1952. Investigations on the microbiology of cellulose utilization in domestic rabbits. J. Gen. Microbiol. **7:**350–357.

40. **Halliwell, G., and M. P. Bryant.** 1963. The cellulytic activity of pure strains and bacteria from the rumen of cattle. J. Gen. Microbiol. **38:**13–17.

41. **Halliwell, G., and M. Griffin.** 1973. The nature and mode of action of the cellulolytic component C_1 of *Tricho-*

derma koningii on native cellulose. Biochem. J. **135:**587–594.

42. **Henning, P. Å.** 1979. Examination of methods for enumerating hemicellulose-utilizing bacteria in the rumen. Appl. Environ. Microbiol. **38:**13–17.

43. **Hintz, H. F., D. E. Hogue, E. F. Walker, Jr., J. E. Lowe, and H. F. Schryver.** 1971. Apparent digestion in various segments of the digestive tract of ponies fed diets with varying roughage-grain ratios. J. Anim. Sci. **32:**245–248.

44. **Hintz, H. F., H. F. Schryver, and M. Halbert.** 1973. A note on the comparison of digestion by New World camels, sheep and ponies. Anim. Prod. **16:**303–305.

45. **Hobson, P. N., and R. J. Wallace.** 1982. Microbial ecology and activities in the rumen; part 1. Crit. Rev. Microbiol. **9:**165–225.

46. **Holdeman, L. V., J. J. Good, and W. E. C. Moore.** 1976. Human fecal flora: variation in bacterial composition within individuals and a possible effect of emotional stress. Appl. Environ. Microbiol. **31:**359–375.

47. **Howard, B. H., G. Jones, and M. R. Purdom.** 1960. The pentosanases of some rumen bacteria. Biochem. J. **74:**173–182.

48. **Hume, I. D.** 1974. Nitrogen and sulphur retention and fibre digestion by euros, red kangaroos and sheep. Aust. J. Zool. **22:**13–23.

49. **Hume, I. D., and A. C. I. Warner.** 1980. Evolution of microbial digestion in mammals, p. 665–684. *In* Y. Ruckebush and P. Thivend (ed.), Digestive physiology and metabolism in ruminants. MTP Press, Lancaster.

50. **Hungate, R. E.** 1946. Studies on cellulose fermentation. 2. An anaerobic cellulose-decomposing actinomycete, *Micromonospora propionici,* n. sp. J. Bacteriol. **51:**51–56.

51. **Hungate, R. E.** 1966. The rumen and its microbes. Academic Press, Inc., New York.

52. **Hungate, R. E., G. D. Phillips, A. McGregor, D. P. Hungate, and H. K. Buecher.** 1959. Microbial fermentation in certain mammals. Science **130:**1192–1194.

53. **Hungate, R. E., and R. J. Stack.** 1983. An enzymatic approach to cell wall structure. S. Afr. J. Anim. Sci. **13:**51–52.

54. **Kass, M. L., P. J. van Soest, W. G. Pond, B. Lewis, and R. E. McDowell.** 1980. Utilization of dietary fiber from alfalfa by growing swine. 1. Apparent digestibility of diet components in specific segments of the gastrointestinal tract. J. Anim. Sci. **50:**175–197.

55. **Kay, R. B. N., P. Hoppe, and G. M. O. Maloiy.** 1976. Fermentative digestion of food in the colobus monkey, *Colobus polykomos.* Experientia **32:**485–487.

56. **Kern, D. L., L. L. Slyter, E. C. Leffel, J. M. Weaver, and R. R. Oltjen.** 1974. Ponies *vs.* steers: microbial and chemical characteristics of intestinal ingesta. J. Anim. Sci. **38:**559–564.

57. **Kern, D. L., L. L. Slyter, J. M. Weaver, E. C. Leffel, and G. Samuelson.** 1973. Pony cecum *vs.* steer rumen: the effect of oats and hay on the microbial ecosystem. J. Anim. Sci. **37:**463–469.

58. **Keys, J. E., Jr., and J. V. De Barthe.** 1974. Cellulose and hemicellulose digestibility in the stomach, small intestine and large intestine of swine. J. Anim. Sci. **39:**53–56.

59. **Keys, J. E., Jr., and J. V. De Barthe.** 1974. Site and extent of carbohydrate, dry matter, energy and protein digestion and the rate of passage of grain diets in swine. J. Anim. Sci. **39:**57–62.

60. **Keys, J. E., Jr., and P. J. van Soest.** 1970. Digestibility of forages by the meadow vole (*Microtus pennsylvanicus*). J. Dairy Sci. **53:**1502–1508.

61. **Keys, J. E., Jr., P. J. van Soest, and E. P. Young.** 1969. Comparative study of the digestibility of forage cellulose and hemicellulose in ruminants and nonruminants. J. Anim. Sci. **29:**11–15.

62. **Kistner, A., J. H. Kornelius, and G. S. Miller.** 1983. Kinetic measurements on bacterial cultures growing on fibres. S. Afr. J. Anim. Sci. **13:**217–220.

63. **Kock, S. G., and A. Kistner.** 1969. Extent of solubiliza-

tion of α-cellulose and hemicellulose of low-protein teff hay by pure cultures of cellulolytic rumen bacteria. J. Gen. Microbiol. **55**:459–462.

64. **Koller, B. L., H. F. Hintz, J. B. Robertson, and P. J. van Soest.** 1978. Comparative cell wall and dry matter digestion in the cecum of the pony and the rumen of the cow using *in vitro* and nylon bag techniques. J. Anim. Sci. **47**:209–215.

65. **Krishnamurti, C. R., W. D. Kitts, and D. C. Smith.** 1970. The digestion of carbohydrates in the chinchilla (*Chinchilla lanigera*). Can. J. Zool. **52**:1227–1233.

66. **Lasker, R., and A. C. Giese.** 1956. Cellulose digestion by the silverfish *Ctenolepisma lineata*. J. Exp. Biol. **33**:542–553.

67. **Latham, M. J., M. E. Sharpe, and J. D. Sutton.** 1971. The microbial flora of the rumen of cows fed hay and high cereal rations and its relation to the rumen fermentation. J. Appl. Bacteriol. **34**:425–434.

68. **Latham, M. J., J. E. Storry, and M. E. Sharpe.** 1972. Effect of low-roughage diets on the microflora and lipid metabolism in the rumen. Appl. Microbiol. **24**:871–877.

69. **Latham, M. J., J. D. Sutton, and M. E. Sharpe.** 1974. Fermentation and microorganisms in the rumen and the content of fat in the milk of cows given low-roughage rations. J. Dairy Sci. **57**:803–810.

70. **MacRae, J. C., and D. G. Armstrong.** 1969. Studies on intestinal digestion in sheep. 2. Digestion of some carbohydrate constituents in hay, cereal and hay-cereal rations. Br. J. Nutr. **23**:377–387.

71. **Macy, J. M., J. R. Farrand, and L. Montgomery.** 1982. Cellulolytic and non-cellulolytic bacteria in rat gastrointestinal tracts. Appl. Environ. Microbiol. **44**:1428–1434.

72. **Mann, S. O., and E. R. Ørskov.** 1973. The effect of rumen and post-rumen feeding of carbohydrates on the caecal microflora of sheep. J. Appl. Bacteriol. **36**:475–484.

73. **Martin, M. M., and J. S. Martin.** 1978. Cellulose digestion in the midgut of the fungus-growing termite *Macrotermes natalensis*: the role of acquired digestive enzymes. Science **199**:1453–1455.

74. **McBee, R. H.** 1977. Fermentation in the hindgut, p. 185–222. *In* R. T. J. Clarke and T. Bauchop (ed.), Microbial ecology of the gut. Academic Press, London.

75. **Mertens, D. R., and J. R. Loften.** 1980. The effect of starch on forage fiber digestion kinetics in vitro. J. Dairy Sci. **63**:1437–1446.

76. **Montgomery, L., and J. M. Macy.** 1982. Characterization of rat cecum cellulolytic bacteria. Appl. Environ. Microbiol. **44**:1435–1443.

77. **Moore, W. E. C., and L. V. Holdeman.** 1974. Human fecal flora: the normal flora of 20 Japanese-Hawaiians. Appl. Microbiol. **27**:961–979.

78. **Morris, E. J.** 1983. Degradation of the intact plant cell wall of sub-tropical and tropical herbage by rumen bacteria. *In* F. M. C. Gilchrist, R. I. Mackie, and H. H. Meissner (ed.), Herbivore nutrition in the subtropics and tropics. Ad. Donker (Pty) Ltd., Johannesburg.

79. **Morris, E. J., and N. O. van Gylswyk.** 1980. Comparison of the action of rumen bacteria on cell walls from *Eragrostis tef*. J. Agric. Sci. **95**:313–323.

80. **Murray, R. M., H. Marsh, G. E. Heinsohn, and A. V. Spain.** 1977. The role of the midgut caecum and large intestine in the digestion of sea grasses by the dugong (Mammalia:Sirena). Comp. Biochem. Physiol. A **56**:7–10.

81. **Nunez, F. S., and C. S. Crawford.** 1976. Digestive enzymes of the desert millipede *Orthoporus ornatus* (Girard) (Diplopoda:Spirostreptidae). Comp. Biochem. Physiol. A **55**:141–145.

82. **O'Brien, G. W., P. C. Veivers, S. E. McEwen, M. Slaytor, and R. W. O'Brien.** 1979. The origin and distribution of cellulase in the termites, *Nasutitermes exitiosus* and *Coptotermes lacteus*. Insect Biochem. **9**:619–625.

83. **Oltjen, R. R., L. L. Slyter, E. E. Williams, and D. L. Kern.** 1971. Influence of the branched-chain volatile fatty

acids and phenylacetate on ruminal microorganisms and nitrogen utilization by steers fed urea or isolated soy protein. J. Nutr. **101**:101–112.

84. **Orpin, C. G.** 1975. Studies on the rumen flagellate *Neocallimastix frontalis*. J. Gen. Microbiol. **91**:249–262.

85. **Orpin, C. G.** 1976. Studies on the rumen flagellate *Sphaeromonas communis*. J. Gen. Microbiol. **94**:270–280.

86. **Orpin, C. G.** 1977. The rumen flagellate *Piromonas communis*: its life history and invasion of plant material in the rumen. J. Gen. Microbiol. **99**:107–117.

87. **Orpin, C. G.** 1981. Isolation of cellulytic phycomycete fungi from the caecum of the horse. J. Gen. Microbiol. **123**:287–296.

88. **Orpin, C. G., and A. J. Letcher.** 1979. Utilization of cellulose, starch, xylan and other hemicelluloses for growth by the rumen phycomycete *Neocallimastix frontalis*. Curr. Microbiol. **3**:121–124.

89. **Ørskov, E. R., and C. Frazer.** 1975. The effect of processing barley-based supplements on rumen pH, rate of digestion and voluntary intake of dried grass in sheep. Br. J. Nutr. **34**:493–500.

90. **Paster, B. J., and E. Canale-Parola.** 1982. Physiological diversity of rumen spirochetes. Appl. Environ. Microbiol. **43**:686–693.

91. **Paul-Murphy, J. R., C. J. Murphy, H. F. Hintz, P. Meyers, and H. F. Schryver.** 1982. Comparison of transit time of digesta and digestive efficiency of the rock hyrax, the Barbados sheep and the domestic rabbit. Comp. Biochem. Physiol. A **72**:611–613.

92. **Pettipher, G. L., and M. J. Latham.** 1979. Characteristics of enzymes produced by *Ruminococcus flavefaciens* which degrade plant cell walls. J. Gen. Microbiol. **110**:21–27.

93. **Pettipher, G. L., and M. J. Latham.** 1979. Production of enzymes degrading plant cell walls and fermentation of cellobiose by *Ruminococcus flavefaciens* in batch and continuous culture. J. Gen. Microbiol. **110**:29–38.

94. **Prins, R. A., F. van Vugt, R. E. Hungate, and C. J. A. H. V. van Vorstenbosch.** 1972. A comparison of strains of *Eubacterium cellulosolvens* from the rumen. Antonie van Leeuwenhoek J. Microbiol. Serol. **38**:153–161.

95. **Raymond, W. F.** 1969. The nutritive value of forage crops. Adv. Agron. **21**:1–108.

96. **Salyers, A. A., J. R. Vercellotti, S. E. H. West, and T. D. Wilkins.** 1977. Fermentation of mucin and plant polysaccharides by strains of *Bacteroides* from the human colon. Appl. Environ. Microbiol. **33**:319–322.

97. **Salyers, A. A., S. E. H. West, J. R. Vercellotti, and T. D. Wilkins.** 1977. Fermentation of mucins and plant polysaccharides by anaerobic bacteria from the human colon. Appl. Environ. Microbiol. **34**:529–533.

98. **Schultz, J. E., and J. A. Breznak.** 1978. Heterotrophic bacteria present in hindgut of wood-eating termites [*Reticulitermes flavipes* (Kollar)]. Appl. Environ. Microbiol. **35**:930–936.

99. **Slyter, L. L., and P. A. Putnam.** 1967. *In vivo vs. in vitro* continuous culture of ruminal microbial populations. J. Anim. Sci. **26**:1421–1427.

100. **Smith, W. R., I. Yu, and R. E. Hungate.** 1973. Factors affecting cellulolysis by *Ruminococcus albus*. J. Bacteriol. **114**:729–737.

101. **Stevens, C. E., R. A. Argenzio, and E. T. Clemens.** 1980. Microbial digestion. Rumen versus large intestine, p. 685–706. *In* Y. Ruckebusch and P. Thivend (ed.), Digestive physiology and metabolism in ruminants. MTP Press, Lancaster.

102. **Stewart, C. S.** 1977. Factors affecting the cellulolytic activity of rumen contents. Appl. Environ. Microbiol. **33**:497–502.

103. **Taylor, E. C.** 1982. Role of aerobic microbial populations in cellulose digestion by desert millipedes. Appl. Environ. Microbiol. **44**:281–291.

104. **Teather, R. M., J. D. Erfle, R. J. Boila, and F. D. Sauer.**

1980. Effect of dietary nitrogen on the rumen microbial population in lactating dairy cattle. J. Appl. Bacteriol. **49**:231–238.

105. **Thayer, D. W.** 1976. Facultative wood-digesting bacteria from the hindgut of the termite *Reticulitermes hesperus*. J. Gen. Microbiol. **95**:287–296.

106. **Thomson, D. J., D. E. Beever, J. F. Coelho da Silva, and D. G. Armstrong.** 1972. The effect in sheep of physical form on the sites of digestion of a dried lucerne diet. 1. Sites of organic matter, energy and carbohydrate digestion. Br. J. Nutr. **28**:31–41.

107. **Udén, P., T. R. Rounsaville, G. R. Wiggans, and P. J. van Soest.** 1982. The measurement of liquid and solid digesta retention in ruminants, equines and rabbits given timothy (*Phleum pratense*) hay. Br. J. Nutr. **48**:329–339.

108. **Udén, P., and P. J. van Soest.** 1982. Comparative digestion of timothy (*Phleum pratense*) fibre by ruminants, equines and rabbits. Br. J. Nutr. **47**:267–272.

109. **Ulyatt, M. J., D. W. Dellow, C. S. W. Reid, and T. Bauchop.** 1975. Structure and function of the large intestine of ruminants, p. 119–133. *In* I. W. McDonald and A. C. I. Warner (ed.), Digestion and metabolism in the ruminant. University of New England Publishing Unit, Armidale.

110. **Vander Noot, G. W., and E. B. Gilbreath.** 1970. Comparative digestibility of components of forages by geldings and steers. J. Anim. Sci. **31**:351–355.

111. **Van Gylswyk, N. O.** 1970. The effect of supplementing a low-protein hay on the cellulolytic bacteria in the rumen of sheep and on the digestibility of cellulose and hemicellulose. J. Agric. Sci. **74**:169–180.

112. **Van Gylswyk, N. O., and J. P. L. Labuschagne.** 1971. Relative efficiency of pure cultures of different species of cellulolytic rumen bacteria in solubilizing cellulose *in vitro*. J. Gen. Microbiol. **66**:109–113.

113. **Van Gylswyk, N. O., and H. M. Schwartz.** 1983. Microbial ecology of the rumen of animals fed high fibre diets. *In* F. M. C. Gilchrist, R. I. Mackie, and H. H. Meissner (ed.), Herbivore nutrition in the subtropics and tropics. Ad. Donker (Pty) Ltd., Johannesburg.

114. **Van Soest, P. J.** 1978. Dietary fibers: their definition and nutritional properties. Am. J. Clin. Nutr. **31**:S12–S21.

115. **Vercellotti, J. R., A. A. Salyers, and T. D. Wilkins.** 1979. Complex carbohydrate breakdown in the human colon. Am. J. Clin. Nutr. **31**:586–589.

116. **Williams, A. G., and S. E. Withers.** 1981. Hemicellulose-degrading enzymes synthesized by rumen bacteria. J. Appl. Bacteriol. **51**:375–385.

117. **Wood, T. M., S. I. McCrae, and C. C. Macfarlane.** 1980. The isolation, purification and properties of the cellobiohydrolase component of *Penicillium funiculosum* cellulase. Biochem. J. **189**:51–65.

118. **Wood, T. W., C. A. Wilson, and C. S. Stewart.** 1982. Preparation of the cellulase from the cellulolytic anaerobic rumen bacterium *Ruminococcus albus* and its release from the bacterial cell wall. Biochem. J. **205**:129–137.

119. **Wozny, M. A., M. P. Bryant, L. V. Holdeman, and W. E. C. Moore.** 1977. Urease assay and urease-producing species of anaerobes in the bovine rumen and human feces. Appl. Environ. Microbiol. **33**:1097–1104.

120. **Yamin, M. A.** 1978. Axenic cultivation of the cellulolytic flagellate *Trichomitopsis termopsidis* (Cleveland) from the termite *Zootermopsis*. J. Protozool. **25**:535–538.

121. **Yamin, M. A.** 1980. Cellulose metabolism by the termite flagellate *Trichomitopsis termopsidis*. Appl. Environ. Microbiol. **39**:859–863.

122. **Yamin, M. A.** 1981. Cellulose metabolism by the flagellate *Trichonympha* from a termite is independent of endosymbiotic bacteria. Science **211**:58–59.

123. **Yu, I., and R. E. Hungate.** 1979. The extracellular cellulases of *Ruminococcus albus*. Ann. Rech. Vet. **10**:251–254.

Bioconversion of Inorganic Materials

Mechanisms of the Binding of Metallic Ions to Bacterial Walls and the Possible Impact on Microbial Ecology

T. J. BEVERIDGE

Department of Microbiology, College of Biological Science, University of Guelph, Guelph, Ontario, Canada N1G 2W1

Most bacteria in nature possess a cell wall which forms the outermost limit to the cell and which separates the vital protoplast from the external milieu. This wall is of fundamental importance to the bacterium since it not only contributes cellular shape and form and protects against cell lysis under hypoosmotic conditions, but it must also provide an inanimate boundary through which the cell perceives the surrounding environment. All substances that enter or exit the bacterium eventually percolate through this barrier. The wall, to a certain extent, must influence the molecular (23, 30) or ionic (3, 9, 11) form of what gets in or out of cells. Given the stress and strain of the microbial habitat, it is conceivable that this boundary layer would act as a "microenvironment" which completely surrounds the cell and which could interact with, buffer, or even modify the stressing influence of the external environment before it comes in contact with the protoplast. Since the preexisting wall is exempt from the enzymatic machinery of the protoplast, most of its capacity to influence the environment must be intrinsic, built into the design during assembly. Also, since it takes time for induction-transcription-translation-secretion to occur and physically alter the wall, a certain amount of preconceived stress accommodation must be incorporated into the pattern. This would help to fulfill the economy of design expected from all life forms capable of rapid evolution.

ANIONIC NATURE OF BACTERIAL WALLS

Like most cell surfaces, bacterial walls are highly anionic. This can be readily proved with electron microscopy by simply mixing a cationic probe (i.e., cationic ferritin) into a suspension of walls (i.e., those from *Bacillus* species); the probe attaches to both wall surfaces (Fig. 1).

Bacillus walls consist of three primary components, a peptidoglycan (PG) matrix which is interwoven with strands of teichoic acid (TA) or teichuronic acid (TUA) (3). The PG is composed of linear glycan chains of 40 to 50 copies of *N*-acetylglucosaminyl-*N*-acetylmuramyl dimers (24); the chains are covalently bound together by random, single cross-links emerging from short tetrapeptides (L-Ala-D-Glu-*meso*-Dpm-D-Ala) which are attached to the muramyl residues (28). The end product is a three-dimensional, planar macromolecule, approximately 25 layers thick, which is covalently stabilized in all directions and which completely surrounds the cell (3). Only 35% of peptide stems of the *B. subtilis* PG are cross-linked together; those remaining are in the pentapeptide (L-Ala-D-Glu-*meso*-Dpm-D-Ala-D-Ala) form (28).

This chemistry is consistent with the concept of a negative charge density for the walls; constituent carboxylate groups within the peptide stems play a prominent role in its maintenance (9). Even though a variable proportion of these groups on the glutamic and diaminopimelic acids are amidated (17, 22, 28), enough remain free to provide the anionic nature of the PG (9). The thickness of *Bacillus* walls could accommodate 20 to 25 monolayers of PG.

TA and TUA are characteristic secondary polymers found in gram-positive walls; they can make up 50 to 60% of the mass of these walls (9). TA resides in the wall as a linear, flexible polymer of α-D-glycopyranosyl phosphate which is negatively charged and has an average chain length of 20 to 30 residues (1, 12). It is covalently bound to the muramyl residue of the PG.

TUA, on the other hand, is analogous to TA in every way except for its chemistry; it is entirely organic. It is usually synthesized by the cell as a replacement product for TA when the bacterium is growing under phosphate limitation. For example, *B. licheniformis* NCTC 6346 can replace its TA with a TUA of the *N*-acetylgalactosamine-glucuronic type under these conditions (18).

Each of these secondary polymers, like PG, is highly anionic (14, 19, 27) and contributes to the overall charge density of the wall. Good evidence suggests that TA and TUA are instrumental in binding Mg^{2+} to the fabric of the wall (14, 15, 19). Clearly, it is the specific ratios of PG to

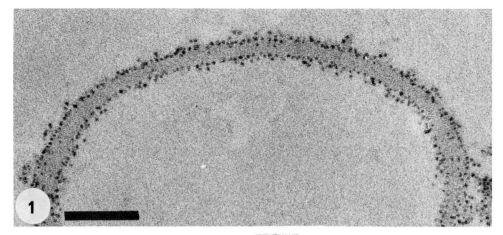

FIG. 1. Thin section of a *B. thuringiensis* wall stained with cationized ferritin and contrasted with uranyl acetate. The cationized ferritin has bound to the anionic surfaces of the wall. Bar = 100 μm.

TA to TUA and their constituent charge values that will ultimately produce the anionic character within the gram-positive wall.

The gram-negative envelope is structurally and chemically more complex than its gram-positive counterpart. It consists of two membrane bilayers (the outer and plasma membranes) which are chemically distinct from each other and which sandwich a thin PG layer between them (3). It is the outer membrane and the PG layer which constitute the wall of these bacteria.

The outer membrane of *Escherichia coli* has been the most studied of gram-negative bacteria. It consists of an asymmetric arrangement of lipid (the outer surface of the bilayer contains mostly lipopolysaccharide [LPS], whereas the inner leaflet is mostly phosphatidylethanolamine [13]) combined together with at least four major outer membrane proteins (OmpF, OmpC, OmpA, and lipoprotein) (3, 16). All are intrinsic proteins, except for OmpA, which is peripherally embedded in the outer leaflet.

The LPS can exist in several forms (smooth, rough), but rough strains tend to be leaky, and selection favors smooth strains in the native environment. All types of LPS appear to be anionic; certainly, the constituent phosphate groups within the lipid A and core oligosaccharide (both the 2-keto-3-deoxyoctonate and heptose variety) should be negatively charged (3, 5, 16). Phosphatidylethanolamine, on the other hand, has a single polar headgroup and its constituent phosphate group should also be anionic. The same holds true with phosphatidylglycerol and cardiolipin, which are found in smaller amounts within the outer membrane (29).

The PG layer is most probably a monolayer in *E. coli* (10), and it is of the same chemotype as that of *B. subtilis* (25). Little is known about the amidation of this PG (B. D. Hoyle and T. J. Beveridge, submitted for publication), but the charge density should be similar to that of *Bacillus* sp.

BINDING OF METALLIC IONS TO WALLS

We have seen from the previous discussion that the chemistry of bacterial walls favors an anionic charge density. Many experiments have shown them to be potent metal traps for metallic ions in aqueous solution (4, 5, 8, 9, 11, 20). Some walls react to these ions as if they were an open-ion exchange resin (20), whereas others favor a more selective system with greater partitioning ability (8, 9, 11, 21). This holds true whether equilibrium (11, 21) or displacement (8, 9) experiments are used. Either way, substantial quantities of metal are bound (Fig. 2 demonstrates an extreme case).

Table 1 provides some of our results for the binding of metal ions to bacterial walls. These values were obtained by suspending 0.1 to 5.0 mg (dry weight) of walls in metal salt solutions under saturating conditions (2 ml of a 5 mM metal solution for 10 min at 23°C). The walls were washed with water until metal could not be detected in the supernatant. Quantitation of the bound metal was determined by either atomic absorption or X-ray fluorescence spectroscopy. Electron microscopy was performed on both unfixed and fixed (4 to 5% glutaraldehyde) samples of whole mounts and thin sections. No stains were used, except for the metal that was bound to the walls. Consequently, the electron-scattering images showed the exact positioning of the metal. Energy-dispersive X-ray analysis of each sample ensured that the only electron-scattering agent within the walls was the absorbed metal. Elementary deposition products were identified by X-ray diffraction.

Gram-positive walls. Walls of the two *Bacillus* spp. which we have studied are potent metal-binding agents (Table 1). *B. subtilis* 168 walls were chosen since they are virtually PG and TA (45 and 54%) when the cells are grown in medium containing phosphate and magnesium (8, 9). So much metal was bound to these walls that visible electron-dense aggregates could sometimes be seen. In general, the more unstable the metal in aqueous solution (e.g., the lanthanides), the more often metal precipitates were encountered within the fabric of the wall (Fig. 2 is an extreme example). Precipitates were never seen in the metal solutions without the addition of the walls. Metallic ions that were freely soluble in water (Na, Mg^{2+}, Cu^{2+}, etc.) were also avidly sequestered from solution by the walls. In these instances, the walls were diffusely stained (Fig. 3 is representative).

To determine the chemical sites of metal interaction within the wall, we extracted TA and performed metal-binding studies on the remaining PG sacculi. Metal uptake decreased in all instances, but it was apparent that most of the binding capacity remained associated with the PG (Table 1). Carboxylate groups constituent within the PG are the sites which should be most electronegative in this polymer. These were neutralized by the addition of glycine ethyl ester to the available COO^- groups, which resulted in a substantial decrease in binding capacity (Fig. 3 and 4, Table 1). These types of experiments (3, 8, 9) have demonstrated with the *B. subtilis* 168 wall that most of the metal binding is determined by the PG. Clearly, there is no apparent stoichiometry between the excessive quantity of metal that binds to these walls (see 9 for details). Yet, these quantities are real (8, 9, 11, 21). We have proposed a two-step mechanism for the deposition process; the first step in time is a stoichiometric interaction between metal ion and active site within the wall. This interaction then acts as a nucleation site for the deposition of more metal ion from solution. The deposition product therefore grows in size within the intermolecular spaces of the wall fabric until it is physically constrained by the polymeric meshwork of the wall. The end result is a bacterial wall that contains copious amounts of metal which are not easily replaced by water (i.e., hydronium ion).

B. licheniformis NCTC 6346 *his* walls are not like those of *B. subtilis*. These walls contain TUA (26%) as a component in addition to PG (22%) and TA (52%). Since the walls contained less than half the PG of *B. subtilis* walls and an additional polymer (TUA), their metal-binding capacity was of interest (4).

B. licheniformis walls were not able to bind as much metal as those of *B. subtilis* (Fig. 5 and Table 1). In fact, when TA (Fig. 6) and TUA (Fig. 7) were extracted (4), the walls lost most of their binding capacity (Table 1). This suggests that, unlike the *B. subtilis* situation, it is the secondary polymers in *B. licheniformis* which interact most strongly with metallic ions in solution and that there is a fundamental charge distribution difference between the two types of walls.

Gram-negative walls. We have looked at only one gram-negative wall, that of *E. coli* K-12 strain A8264. The outer membrane of this bacterium possesses an LPS which contains a complete core oligosaccharide but lacks O-antigenic side chains. The major outer membrane proteins are the lipoprotein and OmpF, OmpC, and OmpA proteins. The major phospholipid is phosphatidylethanolamine (5, 16).

The outer membrane did not react as avidly with metal ions as did gram-positive walls (Table 1), but there was enough metal to give an electron-scattering profile (Fig. 10). The bilayer distribution of the metal suggested that it was the hydrophilic faces of the outer membrane which provided the sites of metallic ion interaction. The prime candidates for this interaction are the phosphate groups which are resident within the polar head groups of the LPS and phospholipids. The validity of this has been proved by ^{31}P-nuclear magnetic resonance of isolated outer membranes in the presence of stoichiometric amounts of europium (a paramagnetic lanthanide); as the Eu^{3+} concentration is increased, the ^{31}P signal is masked (decreased) and chemically shifted to the right. At a 1:1 ratio of Eu^{3+} to ^{31}P, the signal is almost nonexistent (Fig. 13).

The PG of *E. coli* is chemically similar to that of *B. subtilis* and *B. licheniformis* (25) and most probably exists as a monolayer (10). It has a cross-linking efficiency of 30% and has an Ala/Glu/Dpm ratio of 2.7:1:1 (Hoyle and Beveridge, submitted for publication). It interacts more strongly with metal ions than does the outer membrane (Table 1), presumably because of available carboxylate groups. We estimate that the total number of COO^- groups available for interaction with metal ions in this PG is 0.013 $\pm$ 0.005 mol mg^{-1} of dry weight (Hoyle and Beveridge, submitted for publication). Clearly, there is more metal adsorbed to the PG than simple stoichiometry could explain, but the amounts are reduced from those attributed to gram-positive PG (Table 1). We believe that a two-step deposition mechanism is at work with *E. coli* PG (hence, greater than stoichiometric quantities of metal), but since there is only a monolayer of fabric, there are few intermolecu-

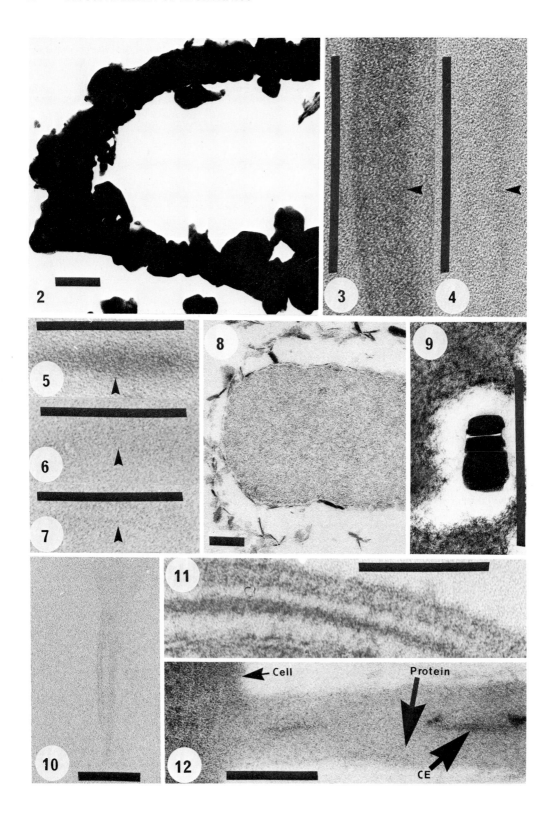

FIG. 2. Thin section of a *B. subtilis* wall coated with gold (8). Bar = 100 μm in all.

FIG. 3. Thin section of a *B. subtilis* wall which has bound indium.

FIG. 4. Same as Fig. 3, but the COO⁻ groups have been neutralized with glycine ethyl ester. Very little indium has been bound (9).

FIG. 5. Thin section of a *B. licheniformis* wall with magnesium bound to it.

FIG. 6. Same as Fig. 5, but the TA has been extracted. Not as much magnesium is in these walls.

FIG. 7. Same as Fig. 5, but the TA and TUA have been extracted, leaving only the PG. Even less magnesium is bound (4).

FIG. 8. Diagenetic degradation of a uranium-loaded *B. subtilis* cell in a quartz-magnetite matrix. *m*-Ankoleite $[K_2(UO_2)(PO_4)_2 \cdot 6H_2O]$ is forming in the wall fabric.

FIG. 9. A similar crystal to those in Fig. 8 which is surrounded by the organic matrix (6).

FIG. 10. Thin section of an *E. coli* outer membrane that has been contrasted with magnesium (5).

FIG. 11. Thin section of the *S. ureae* wall before growth in heavy metal. The surface layer is the protein array.

FIG. 12. Same as Fig. 11 but after growth in 0.5 mM cesium IV nitrate. The surface protein has encircled and complexed a cesium aggregate and is in the process of sloughing-off the cell.

---

lar spaces in which metal aggregates can grow (hence, reduced quantities when compared to the 25 layers of *Bacillus* PG).

Surface array of *Sporosarcina ureae*. Numerous bacteria contain proteinaceous arrays on the surface of their walls (26), and often these require distinct metallic ions for proper assembly and adhesion (3). Those that we have studied require either Ca^{2+} (7) or Mg^{2+} (2). The protein array of *S. ureae* is particularly rich in magnesium (6.5 mol mg⁻¹ of dry weight).

Recently, we have studied this layer on bacteria which are growing in the presence of toxic heavy metal (up to 1 mM). It seems that the bacterium is able to survive and grow in the toxic environment by using its surface array to bind and immobilize the heavy metal (compare Fig. 11 and 12). Eventually, the protein–metal complex is sloughed off the cell surface and is replaced by a new surface layer. At the end of bacterial growth (i.e., stationary growth phase), so much protein has complexed metal that metal aggregates are visible in the culture.

POSSIBLE ENVIRONMENTAL CONSEQUENCES

A significant portion of the organic matter of aquatic (both marine and freshwater) sediments consists of small colloidal aggregates which are composed of highly cross-linked heteropolymeric materials of biological origin (see 6 for an overview). These are very resistant to degradation, and at least some of the most durable polymeric networks are of bacterial origin. The most resilient bacterial structure is the wall. We envision a light but constant rain of aggregated bacteria and their products in all large bodies of water. As these drift slowly down, they would come in contact and bind with dilute metal ions. During their downward journey, the walls could

TABLE 1. Metal binding by bacterial walls[a]

Metal	B. subtilis			B. licheniformis		E. coli[b]	
	Native[b]	Neutralized COO⁻[b,c]	No TA[d]	Native[b]	No TA or TUA[d]	Murein	Outer membrane
Na	2.697	0	1.497	0.910	0.080	0.290	0.081
K	1.944	0	0.782	0.560	0	0.058	0.025
Mg	8.226	0.520	7.683	0.400	0.024	0.035	0.019
Ca	0.399	0.380	0.012	0.590	0.096	0.038	0.020
Mn	0.801	0.732	0.656	0.662	0.004	0.052	0.012
FeIII	3.581	2.260	1.720	0.760	0.172	0.100	0.233
Ni	0.107	0.024	0.021	0.520	0	0.019	0.019
Cu	2.990	0.993	2.488	0.490	0	ND[e]	ND
AuIII	0.363	0.214	0.265	0.031	0.012	ND	ND

[a] These values were obtained by suspending 0.1 to 5.0 mg (dry weight) of walls in metal salt solutions under saturating conditions (i.e., 2 ml of a 5 mM metal solution for 5 min at 23°C). See references 4, 5, 8, and 9 for details.

[b] Micromoles of metal per milligram (dry weight) of walls.

[c] Neutralized by the addition of glycine ethyl ester to carbodiimide-activated carboxylate groups.

[d] The dry weight of these walls has been adjusted downward to reflect the loss of mass due to the extraction. All quantities are expressed in micromoles.

[e] ND, Not determined.

either be grazed upon by higher life forms, the metal thus being further concentrated as it enters the food chain, or they could continue to fall and eventually form a proportion of the bottom sediment. Because the bacterial wall is the cell's outermost layer, because the wall has high affinity for metallic ions, and because it is resistant to degradation, these walls could play a major role in the transport of metals in the environment. If we are correct with this concept, then bacteria and similar microbial particles could help cleanse natural waters of soluble metal pollutants and partition them into the food chain or the sediments. Once the metal has been immobilized into the sediment, geochemical mineralization could proceed.

We have previously demonstrated that bacterial walls are highly anionic and that they bind large amounts of metal. It is therefore, quite feasible that as a "rain" they could partition metal into the sediment. But could these same metal-loaded walls contribute to a geochemical mineralization process?

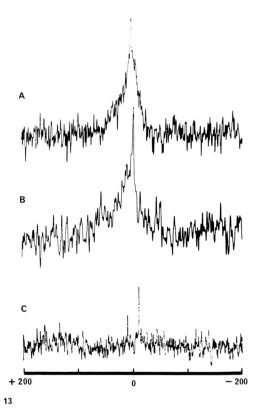

13

FIG. 13. ^{31}P-nuclear magnetic resonance scans of *E. coli* outer membrane after the addition of europium. (A) 1 Eu:4 P, (B) 1 Eu:2 P, and (C) 1 Eu:1 P. The europium is binding to the phosphate of the LPS and phospholipid to mask the ^{31}P signal and shift it to the right. Very little signal is left when Eu equals P.

Recently, we simulated low-temperature (100°C) sediment diagenesis in the laboratory by mixing metal-loaded bacteria with a synthetic sediment of quartz and calcite (6). The metal associated with the bacteria was primarily in the wall fabric. During the diagenesis, we monitored the sediment by electron microscopy to follow the mineralization process (Fig. 8 and 9). Energy-dispersive X-ray analysis allowed us to determine the composition of minerals, whereas the crystal habit provided mineral identity (6).

This work established that bacteria tenaciously bind metal during early diagenesis. As the redox potential lowered with time, elements within the synthetic sediment were mobilized (e.g., Ca^{2+} and S) and combined with the bacterial metal. In this way, the bacteria acted as distinct nuclei for mineral formation. It is our view that these bacterial components both stimulate and quicken mineral authigenesis.

CONCLUSIONS

Our results have shown that bacterial walls are able to interact with and bind significant amounts of metallic ions from aqueous solution. It is not unusual that binding occurs since most of the constituent homo- and heteropolymers which comprise the wall are anionic (3); it is the quantity of bound metal which is surprising. Of course, it is impossible to escape some of the implications which this reactivity could have in an applied sense. For example, immobilized walls could be used to cleanse industrial effluents of toxic metal contaminants.

Gram-positive walls bind more metal than do gram-negative walls. In both cases, the quantities of bound metal are greater than the number of available anionic sites within the wall. We therefore propose at least a two-step deposition process, the first step in time being the stoichiometric interaction between metal ions and active sites within the wall. Next, this initial bound metal would act as nucleation sites for the growth of metal aggregates which would increase in size until physically limited by the available space within the wall fabric. In this way, significant amounts of metal could be sequestered from solution.

There are tremendous difficulties in up-scaling our fundamental metal-binding observations from the laboratory to actual field tests. For example, to follow the process within natural bodies of water, the use of metal-loaded walls is not practicable; bacterial walls are so minute and the external milieu (in this case) is so large. Planktonic grazers also complicate the issue. But we know that wall fragments survive the "test of time" and that these bacterial remnants retain metal which they have sequestered from the environment; they can be unequivocally identified in ancient sediments (see 6 for an overview). Our simulation of low-temperature sediment diagenesis confirms and emphasizes the resilience of walls since they acted as distinct nuclei for mineral formation (6). It is our view that these bacterial components, given the proper mineral geochemical climate, both stimulate and quicken mineral authigenesis.

We conclude that the bacterial wall is not simply an inert corset which surrounds the cell substance and protects it from physical and osmotic stress. Instead, it is also able to interact chemically with the environment. Clearly, the wall possesses reactive sites which are able to bind with and exchange environmental counterions. Potentially toxic heavy metals are trapped and immobilized before they penetrate to the vital constituents of the cell; essential cations such as magnesium and calcium can be stored within the wall matrix.

In fact, the actual chemistry of the wall can be modified during growth, depending on environmental factors. For example, *B. subtilis* requires phosphate and magnesium for the synthesis of wall TA (3). The matrix of the wall, to a certain extent, is a reflection of the environment. It would be interesting to sample natural systems which are under high or widely fluctuating environmental stress (i.e., extremes of pH, temperature, toxicity, etc.) to see how the walls of these bacteria accommodate these influences.

I am indebted to all of my colleagues who helped with this research and to the Natural Sciences and Engineering Council of Canada, which supported the majority of this work.

LITERATURE CITED

1. Archibald, A. R. 1974. The structure, biosynthesis and function of teichoic acid. Adv. Microb. Physiol. 11:53–95.
2. Beveridge, T. J. 1979. Surface arrays on the wall of *Sporosarcina ureae*. J. Bacteriol. 139:1039–1048.
3. Beveridge, T. J. 1981. Ultrastructure, chemistry, and function of the bacterial wall. Int. Rev. Cytol. 72:229–317.
4. Beveridge, T. J., C. W. Forsberg, and R. J. Doyle. 1982. Major sites of metal binding in *Bacillus licheniformis* walls. J. Bacteriol. 150:1438–1448.
5. Beveridge, T. J., and S. F. Koval. 1981. Binding of metals to cell envelopes of *Escherichia coli* K-12. Appl. Environ. Microbiol. 42:325–335.
6. Beveridge, T. J., J. D. Meloche, W. S. Fyfe, and R. G. E. Murray. 1983. Diagenesis of metals chemically complexed to bacteria: laboratory formation of metal phosphates, sulfides, and organic condensates in artificial sediments. Appl. Environ. Microbiol. 45:1094–1108.
7. Beveridge, T. J., and R. G. E. Murray. 1976. Dependence of the superficial layers of *Spirillum putridiconchylium* on Ca^{2+} and Sr^{2+}. Can. J. Microbiol. 22:1233–1244.
8. Beveridge, T. J., and R. G. E. Murray. 1976. Uptake and retention of metals by cell walls of *Bacillus subtilis*. J. Bacteriol. 127:1502–1518.
9. Beveridge, T. J., and R. G. E. Murray. 1980. Sites of metal deposition in the cell wall of *Bacillus subtilis*. J. Bacteriol. 141:876–887.
10. Braun, V., H. Gnirke, U. Henning, and K. Rehn. 1973. Model for the shape-maintaining layer of *Escherichia coli* cell envelope. J. Bacteriol. 114:1264–1270.
11. Doyle, R. J., T. H. Matthews, and U. N. Streips. 1980. Chemical basis for the selectivity of metal ions by the *Bacillus subtilis* cell wall. J. Bacteriol. 143:471–480.
12. Doyle, R. J., M. L. McDannel, U. N. Streips, D. C. Birdsell, and F. E. Young. 1974. Polyelectrolyte nature of bacterial teichoic acids. J. Bacteriol. 118:606–615.
13. Funahara, Y., and H. Nikaido. 1980. Asymmetric localization of lipopolysaccharides on the outer membrane of *Salmonella typhimurium*. J. Bacteriol. 141:1463–1465.
14. Heckels, J. E., P. L. Lambert, and J. Baddiley. 1977. Binding of magnesium ions to cell walls of *Bacillus subtilis* W23 containing teichoic acid or teichuronic acid. Biochem. J. 162:359–365.
15. Heptinstall, S., A. R. Archibald, and J. Baddiley. 1970. Teichoic acids and membrane function in bacteria. Selective destruction of teichoic acid reduces the ability of bacterial cells to bind magnesium ions. Nature (London) 225:519–521.
16. Hoyle, B., and T. J. Beveridge. 1983. Binding of metallic ions to the outer membrane of *Escherichia coli*. Appl. Environ. Microbiol. 46:749–752.
17. Hughes, R. C. 1971. Autolysis of *Bacillus cereus* cell walls and isolation of structural components. Biochem. J. 121:791–802.
18. Hughes, R. C., and P. F. Thurman. 1970. Some structural features of the teichuronic acid of *Bacillus licheniformis* NCTC 6346 cell walls. Biochem. J. 117:441–449.
19. Lambert, P. A., I. C. Hancock, and J. Baddiley. 1975. The interaction of magnesium ions with teichoic acid. Biochem. J. 149:519–524.
20. Marquis, R. E., K. Mayzel, and E. L. Carstensen. 1976. Cation exchange in cell walls of Gram-positive bacteria. Can. J. Microbiol. 22:975–982.
21. Matthews, T. H., R. J. Doyle, and U. N. Streips. 1979. Contribution of peptidoglycan to the binding of metal ions by the cell wall of *Bacillus subtilis*. Curr. Microbiol. 3:51–53.
22. Mirelman, D., and N. Sharon. 1968. Isolation and characterization of the disaccharide N-acetylglucosaminyl-β-(1-4)-N-acetylmuramic acid and two tripeptide derivatives of this disaccharide from lysozyme digests of *Bacillus licheniformis* ATCC 9945 cell walls. J. Biol. Chem. 243:2279–2287.
23. Nakae, T., and H. Nikaido. 1975. Outer membrane as a diffusion barrier in *Salmonella typhimurium*. Penetration of oligo- and polysaccharides into isolated outer membrane vesicles and cells with degraded peptidoglycan layer. J. Biol. Chem. 250:7359–7365.
24. Rogers, H. J., J. B. Ward, and I. D. J. Burdett. 1978. Structure and growth of the walls of Gram-positive bacteria. Symp. Soc. Gen. Microbiol. 28:139–176.
25. Schleifer, K. H., and O. Kandler. 1972. Peptidoglycan types of bacterial cell walls and their taxonomic implications. Bacteriol. Rev. 36:407–477.
26. Sleytr, U. B. 1978. Regular arrays of macromolecules on bacterial cell walls: structure, chemistry, assembly and function. Int. Rev. Cytol. 53:1–64.
27. Tsien, H. C., G. D. Shockman, and M. L. Higgins. 1978. Structural arrangement of polymers within the walls of *Streptococcus faecalis*. J. Bacteriol. 133:372–386.
28. Warth, A. D., and J. L. Strominger. 1971. Structure of the peptidoglycan from vegetative cell walls of *Bacillus subtilis*. Biochemistry 10:4349–4358.
29. White, D. A., W. J. Lennarz, and C. A. Schnaitman. 1972. Distribution of lipids in the wall and cytoplasmic membrane of the cell envelope of *Escherichia coli*. J. Bacteriol. 109:686–690.
30. Yoshimura, F., and H. Nikaido. 1982. Permeability of *Pseudomonas aeruginosa* outer membrane to hydrophilic solutes. J. Bacteriol. 152:636–642.

Biological Transformation and Accumulation of Uranium with Emphasis on *Thiobacillus ferrooxidans*

OLLI H. TUOVINEN AND ALAN A. DISPIRITO†

Department of Microbiology, The Ohio State University, Columbus, Ohio 43210

URANIUM IN THE ENVIRONMENT

Uranium occurs largely as oxide minerals in association with many different metals in the earth's crust. Minerals of economic importance include, among many others, uraninite and pitchblende, which represent different crystalline forms of UO_2 in vein deposits. Over 100 different uranium minerals have been described in the literature (17). ^{238}U (half-life, 4.49×10^9 years) accounts for 99.3% of the natural isotopic composition of uranium. The principal decay products of ^{238}U include ^{234}U, radioactive isotopes of Th, Ra, Rn, Po, and Pb, and the stable isotopes ^4He and ^{206}Pb. ^{235}U as enriched oxide and also ^{233}U are important as nuclear fuels, with Pu and other radionuclides as fission products. The loss of uranium through radioactive decay decreases the uranium/oxygen ratio, and the general formula of uranium oxide minerals is sometimes presented as U_3O_8. The oxidation of uranium in weathered zones also results in deviation from the UO_2 formula and leads to the formation of slightly more soluble minerals such as phosphoritic oxides. The association of uranium with metasedimentary phosphates (apatites) is of particular consequence in the fertilizer industry (35, 36, 40). Elevated uranium concentrations in many rivers and sediments have been attributed to the runoff of contaminating uranium from agricultural areas fertilized with phosphates (32, 39). Elevated concentrations have been reported, for example, for rivers of the Gulf of Mexico region (39). In general, uranium concentrations in freshwater, seawater, and sediments vary greatly depending on the location and runoff. Uranium mining and milling are, of course, important point sources in the transport of uranium and other metals to the environment. Seawater has a natural uranium concentration of 3 to 4 μg liter^{-1}. Recovery of uranium from seawater is technically feasible with the use of various sorbents (27, 42, 50) but is economically unattractive (26) because of the need to process extremely large quantities of water. For example, approximately 2×10^8 liters of seawater containing 3 μg of uranium per liter would have

to be processed to extract 1 kg of U with sorbent material (based on a 70% recovery).

Degens et al. (9) reported that the bulk of uranium (in the Black Sea) was immobilized in planktonic organisms. Anderson (1) reported that dissolved uranium was fixed in organic particulate matter in surface seawater, but this authigenic uranium represented only a minor fraction of the total uranium. Other estimates (2, 31) indicate a high degree of variation in the concentration of particulate uranium in aquatic environments. Hydrous ferric oxides and clay minerals also influence the mobility of uranium by scavenging and adsorbing uranyl ion (20, 46).

Uranium has four oxidation states (+3, +4, +5, and +6), of which the tetravalent and hexavalent forms are predominant as environmental contaminants and in ore materials, as is also evident from their respective pH-Eh domains (28). The hexavalent uranium is highly soluble as UO_2^{2+} in acid solutions and as complexes of carbonate, phosphate, and sulfate in neutral solutions. In seawater uranium is soluble primarily as $UO_2(CO_3)_3^{4-}$; at higher pH values the hydroxyl complexes of uranium prevail. The chemistry of uranium complexes with respective pH-Eh domains has been characterized (16, 28).

ACCUMULATION OF URANIUM IN MICROORGANISMS

Several biological approaches have been tested for the recovery of uranium (and other radionuclides) from dilute solutions, both as a means of pollution abatement and of recovery of the metal. Filamentous fungi (*Penicillium* spp.) have been reported to contain 0.17 mg of U per mg of dry weight after a short-term exposure to uranyl sulfate solution containing 0.1 to 1.0 g of U per liter (51). Biosorption capacity of a similar order of magnitude was described for *Rhizopus arrhizus* and *Penicillium chrysogenum* (45). Chitin in the cell wall was the main binding site of uranium (44). Uranium was also shown to be immobilized into fungal biomass from various rock samples including granite (3); the associated solubilization of uranium from the rock material was not elucidated but is presumed to involve organic metabolites as dissolution agents.

Saccharomyces cerevisiae can accumulate uranium from uranyl nitrate solution with a

† Present address: Department of Genetics and Cell Biology, University of Minnesota, St. Paul, MN 55108.

capacity of 10 to 15% U of the cell dry weight (41). This value is comparable with the uranium biosorption capacities determined for filamentous fungi (45, 51). Electron micrographs indicated that uranium occurred as a surface layer on the cell envelope in addition to its intracellular accumulation in *S. cerevisiae* (41).

An extremely rapid (<10 s) sorption of uranium was described in *Pseudomonas aeruginosa* (41), associated with the appearance of electron-dense intracellular deposits without surface accumulation. However, electron micrographs of control cells (without exposure to uranium) were not presented, and thus structural changes cannot be further evaluated. The diagenetic formation of uranium phosphate microcysts was described in aged, thermally degraded *Bacillus subtilis* cells preloaded with uranyl acetate (5). Crystallographic evidence by X-ray diffraction was not provided to ascertain the identification of the uranium phosphate mineral thus formed. These results suggest that uranium crystallization occurs in organically rich environments such as uranium-containing sediments.

The accumulation of uranium into algal biomass has been characterized with both freshwater and marine species. Considerable variation in the accumulation of uranium was observed with different species of algae (37). Carbonates in the medium effectively prevented the uptake of uranium by algae, suggesting the formation of carbonate-uranyl complexes that the algae could not take up. The uptake was maximized at pH 5 to 6 (22, 37), and dead cells accumulated four to five times more uranium than did live cells. A column system for removing soluble uranium was described that used cells of either *Chlorella vulgaris* or *Streptomyces viridochromogenes* immobilized in polyacrylamide gel (33). With immobilized *C. vulgaris* almost 100% of the uranium was retained in the column after the first cycle of 10 mg liter^{-1} uranium solution, compared with about 45% retention obtained with *S. viridochromogenes* (33). Although the sorption capacity of polyacrylamide gel in the absence of bacteria was not determined, these relative differences between the two organisms suggest that the retention of uranium could not be ascribed to the effect of polyacrylamide gel alone. Uranium accumulated in the cell mass could be extracted with alkaline carbonate solution (33), as also demonstrated for the desorption of uranium from fungal biomass (19).

Effluents from mining operations are invariably contaminated with metals, and some mining companies have resorted to the use of effluent lagoons before final discharge to reduce the level of metal contamination. Uranium in mine waters was shown to be fixed into biomass in treatment lagoons (6, 8), and laboratory experiments with bacteria immobilized on glass beads supported these observations (7). Uranium becomes enriched in sediments of treatment ponds. It is probably possible to promote the primary production in treatment lagoons with the use of fertilizers to increase the amount of available biomass (standing crop) for immobilizing influent uranium and other metals. It is not clear, however, how the toxicity and remineralization of uranium and other metals influence the efficiency of the biological removal of metals, particularly with a view to effective long-term management of treatment lagoons.

URANIUM OXIDATION AND ACCUMULATION BY *THIOBACILLUS FERROOXIDANS*

T. ferrooxidans has been studied extensively in metal-leaching systems where its role is to oxidize insoluble metal sulfides to sulfuric acid with concomitant solubilization of metals (30, 43, 47). In the leaching of uranium ores, *T. ferrooxidans* can regenerate ferric iron by the oxidation of pyrite (FeS_2) and Fe^{2+}:

$$FeS_2 + 3\tfrac{1}{2}O_2 + H_2O \rightarrow Fe^{2+} + 2SO_4^{2-} + 2H^+$$
$$2FeS_2 + 7\tfrac{1}{2}O_2 + H_2O \rightarrow 2Fe^{3+} + 4SO_4^{2-} + 2H^+$$
$$4Fe^{2+} + O_2 + 4H^+ \rightarrow 4Fe^{3+} + 2H_2O$$

Because the bacterium is an obligate acidophile and because of the insolubility of most metals in alkaline solutions, the bacterial leaching systems are confined to the treatment of ore materials that do not contain excessive amounts of carbonate minerals. The neutralization of carbonates with a continuous feed of sulfuric acid is possible, but it increases the reagent consumption and may result in the formation of insoluble precipitates (e.g., gypsum).

In the acid leaching of uranium ores, ferric ion is a chemical oxidant as presented for uraninite:

$$UO_2 + 2Fe^{3+} \rightarrow UO_2^{2+} + 2Fe^{2+}$$

Iron is commonly associated with uranium minerals in ores, and the solubilization of iron occurs concurrently with that of uranium. Ratios of Fe^{3+}/Fe^{2+} determined in leach liquors of a uranium mine indicated the predominance of ferric iron (49), suggesting that active iron-oxidizing bacteria were present in leach liquors and ore piles. The chemical oxidation of Fe^{2+} at the pH range (approximately 1.5 to 2.0) of leach liquors is extremely slow and has no practical significance in the overall Fe^{3+}/Fe^{2+} ratio. Ferric iron is also consumed in the oxidative chemical dissolution of insoluble sulfides, and the ferrous iron thus formed is reoxidized by *T. ferrooxidans*.

Other metals in uranium-containing ore materials also dissolve during the leaching in acid solutions. This is exemplified in Fig. 1, which shows the leaching of uranyl thorianite [Th(U)O$_2$] containing ore material in the presence of an enriched population of iron-oxidizing thiobacilli. The solubilization of both uranium and thorium was observed, and the leaching profiles were almost identical for the two metals. Parallel laboratory studies indicated that isolates of *T. ferrooxidans* derived from leach liquors of a uranium mine were resistant to both uranium and thorium in the range of 1.0 to >5.0 mM U and 1.5 to >5.0 mM Th (14). A decrease in the pH from 2.5 to 1.5 increased the toxicity of these ions, in agreement with previous demonstrations of the influence of pH on the metal toxicity in *T. ferrooxidans* (48).

T. ferrooxidans is in general relatively resistant to soluble metal ions, but the metabolic mechanisms of resistance have not been characterized, except for mercury, which is volatilized by a mercuric reductase enzyme present in some strains of *T. ferrooxidans* (34). In other bacteria mercuric reductase enzymes are governed by plasmids and transposons (38), and it remains to be established whether this is the case also for Hg-resistant strains of *T. ferrooxidans*. The resistance to uranium can be increased by successive subculturing of *T. ferrooxidans* in the presence of increasing concentrations of UO$_2^{2+}$, and growth in the presence of 10 mM uranyl sulfate has been reported (12). Several isolates of *T. ferrooxidans* from uranium mine leach liquors each had different plasmid profiles (49). A plasmid of about 13 megadaltons was suggested to be involved in determining the resistance to UO$_2^{2+}$ (49; P. A. W. Martin, P. R. Dugan, and O. H. Tuovinen, Eur. J. Appl. Microbiol. Technol., in press), because it was detected in several uranium-resistant isolates of *T. ferrooxidans*. A plasmid of this size was also present in uranium-sensitive cells that were subcultured in the presence of a low concentration of UO$_2^{2+}$; in the latter case, this plasmid could not be detected when the bacteria were subcultured in the

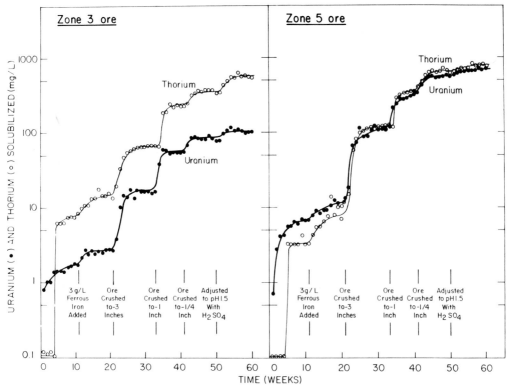

FIG. 1. Solubilization of uranium (●) and thorium (○) in a vat leaching experiment with uranyl thorianite-containing ore material from two separate areas (zone 3 and zone 5) of the Agnew Lake uranium mine. The samples weighed about 13 kg each and had a surface area of about 1,800 cm^2 at the beginning of the experiment. At the end of the 60 weeks of leaching, 85% U and 68% Th were solubilized from the zone 3 sample and 84% U and 73.5% Th were solubilized from the zone 5 sample. Experimental conditions and the mass balance for Fe and S have been described previously (49).

absence of UO_2^{2+}. However, no direct evidence has been presented for a plasmid-borne uranium resistance in *T. ferrooxidans*, but it may be a potentially useful marker for genetic studies of this bacterium. The lack of genetic techniques for acidophilic thiobacilli has greatly hindered the research in this area. For example, many antibiotics that have been successfully used as plasmid-curing agents in neutrophilic bacteria are not stable in acid, iron-containing solutions at pH 1.5 to 2 (Martin et al., in press). These problems are also compounded by the difficulty of growing *T. ferrooxidans* colonies on solid media.

Uranium is taken up by *T. ferrooxidans* from the growth medium, and it was found to be associated primarily with the cell wall and membrane fractions of the cell (11). More uranium than other test metals was taken up from equimolar solutions (1 mM each) of UO_2^{2+}, Fe^{2+}, Fe^{3+}, Cu^{2+}, Co^{2+}, Ni^{2+}, Cd^{2+}, and Cr^{3+}. External uranium concentration influenced the uptake. Cells inactivated with UV radiation or potassium cyanide contained approximately 0.1 mg of U per mg of cell protein (11). On average, 15 to 35% less uranium was detected in cells that were initially active, although viable cells could not be recovered by cultural methods at the end of a 3-h experiment with the highest UO_2^{2+} concentration tested (10 mM).

The accumulation of uranium in *T. ferrooxidans* was also dependent on the external pH. Table 1 indicates that increasing amounts of uranium were associated with the cells as the pH of the medium was lowered to pH 1.0 to 1.5. The enhancing effect of low pH on the accumulation of uranium is in agreement with an enhancing effect of low pH on the toxicity of UO_2^{2+} to *T.*

TABLE 1. Effect of pH on uranium uptake by *T. ferrooxidans* (TFK-35) incubated for 3 h in the presence of 5.0 or 10.0 mM uranyl sulfate[a]

pH	Cell-associated uranium (μg of U per mg of protein)	
	5.0 mM UO_2^{2+}	10.0 mM UO_2^{2+}
1.0	50.50	62.11
1.5	37.61	63.28
2.0	24.45	35.01
3.0	35.47	27.83
4.0	36.75	33.26

[a] Tetrathionate-grown bacteria were used. Cells were recovered by membrane filtration from incubation mixtures, and uranium was analyzed by neutron activation (11). Appropriate controls were included to determine the retention of uranium by membrane filters in the absence of bacteria. These blank values have been subtracted from the results to account for slight differences in the solubility of UO_2^{2+} at the pH values tested in the experiment.

ferrooxidans (48). Relatively little uranium was detected in the cytoplasmic fraction, indicating the presence of active binding sites in the cell wall and membrane (11). Lipopolysaccharide extracted with 0.5 M sucrose (18) from cells exposed to UO_2^{2+} for 3 h did not appear to be the main binding site (11), unlike the results previously demonstrated with purified outer membrane preparations of *Escherichia coli* exposed to 5 mM uranium (23). It is not known how the extraction procedure in which 0.5 M sucrose is used (18) influences the uranium-binding capacity of lipopolysaccharide. For *T. ferrooxidans* (11), all incubations were carried out with intact cells, and no attempts were made to study the binding of uranium by purified cell components.

Figures 2 and 3 are electron micrographs of *T. ferrooxidans* incubated in the presence or absence of 10 mM UO_2^{2+} for 3 h. The bacteria were fixed with glutaraldehyde and osmium tetroxide (29), and only ruthenium red was used as a stain during the preparation for electron microscopy. Cells incubated in the presence of UO_2^{2+} (Fig. 2) had a staining pattern similar to that obtained with uranyl acetate used as a stain for electron microscopy (4, 21). Electron-dense areas indicating uranium accumulation were both the cell wall and the cytoplasmic region. The main difference between the treated cells (Fig. 2) and control cells (Fig. 3) was the contrasting staining caused by cellular accumulation of uranium. Intracellular uranium deposits similar to those found in *Pseudomonas aeruginosa* (41) were not present in thin sections of *T. ferrooxidans*.

The oxidation of insoluble uranous oxide (UO_2) to soluble UO_2^{2+} was shown to be coupled with oxygen uptake by *T. ferrooxidans* (12). Experiments in which soluble uranous sulfate was used as the substrate indicated that *T. ferrooxidans* utilized U^{4+} as an electron donor (13, 14). Manometric studies indicated the following stoichiometry of oxygen uptake (13):

$$2U^{4+} + O_2 + 2H_2O \rightarrow 2UO_2^{2+} + 4H^+$$

The experimental values averaged 107% of the theoretical value of $2U^{4+}/1O_2$ (11). Carbon dioxide fixation was used as a measure of energy transduction during U^{4+} oxidation by *T. ferrooxidans*. The measured values indicated an average molar ratio of $107U^{4+}$ oxidized/$1CO_2$ fixed (13).

Uranous ion has an absorption peak at 650.2 nm, whereas uranyl ion does not absorb at this wavelength. Thus, the bacterial oxidation of U^{4+} can be monitored by measuring the decrease in absorbance at 650.2 nm (15).

Tentative spectroscopic evidence was ob-

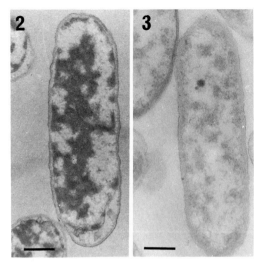

FIG. 2. Electron micrograph of *T. ferrooxidans* incubated in the presence of 10 mM UO_2^{2+} for 3 h. Bar = 0.2 μm.

FIG. 3. Electron micrograph of control cells of *T. ferrooxidans* (no exposure to UO_2^{2+}). Bar = 0.2 μm.

cyclic reoxidation of Fe^{2+} by *T. ferrooxidans*. The standard potentials are +334 mV for U(IV)/U(VI) and +770 mV for Fe(II)/Fe(III). The two oxidation mechanisms were resolved by kinetic analysis and studies with mixed substrates. The kinetic parameters were different for U^{4+} and Fe^{2+}: the apparent K_m values for U^{4+} were at least 10 times lower than those determined for Fe^{2+} (14). Similarly, the V_{max} values were about 30 to 40 times lower for U^{4+} oxidation compared with those determined for Fe^{2+} (14). Thus, *T. ferrooxidans* has a higher affinity but a lower rate of oxidation of U^{4+} as compared with Fe^{2+}. In mixed substrate studies a competitive inhibition by U^{4+} of Fe^{2+} oxidation was predominant (14). Conversely, a mixed type of inhibition by Fe^{2+} of U^{4+} oxidation was observed. Uranous ion oxidation by *T. ferrooxidans* was stimulated in the presence of >0.5 μM Fe^{2+}, suggesting an indirect oxidation mechanism via the recycling of Fe^{3+}/Fe^{2+}. Below 0.5 μM Fe^{2+}, no increase in the rate of U^{4+} oxidation was observed, suggesting that uranium oxidation was mediated directly by *T. ferrooxidans*.

tained for the involvement of a blue copper protein as an electron carrier in U^{4+} oxidation (13). This redox protein, two *c*-type cytochromes, and cytochrome a_1 are electron carriers in the downhill transport to oxygen (25), and only one phosphorylation site is involved in this segment of electron transfer. For the reverse electron flow, cytochrome *b*, ubiquinone-8 (10), an Fe-S protein, and a flavoprotein are probably involved before the reduction of NAD^+ (24). The relatively short electron-transport chain from the substrate to oxygen avails itself to studies with reconstituted components in artificial phospholipid membranes. Such studies would be useful in the kinetic analysis and in determining the sequence and arrangement of the redox components involved in the oxidation of inorganic ions by *T. ferrooxidans*. The reconstitution of isolated redox carriers would also help to reduce the amount of contaminant iron which results from growing the biomass with Fe^{2+}. Growth of *T. ferrooxidans* with U^{4+} or UO_2 as the sole electron donor has not been reported.

In studies of uranous ion oxidation, ferric iron may be the initial electron acceptor from U^{4+} before entry of electrons to the electron-transport chain. Thus, two different oxidation mechanisms are involved: (i) the direct oxidation of uranous ion in the absence of iron (other than Fe in cell constituents) and (ii) the indirect oxidation of uranous ion which is chemically coupled with the reduction of Fe^{3+} to Fe^{2+}. The indirect oxidation of U^{4+} therefore proceeds via the

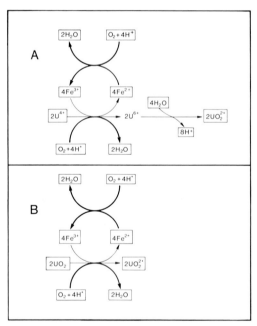

FIG. 4. U(IV) oxidation in the presence and absence of Fe^{3+}/Fe^{2+}. (A) In the absence of iron, U^{4+} oxidation is coupled directly to oxygen uptake by *T. ferrooxidans*; in the presence of iron, *T. ferrooxidans* also regenerates ferric iron, which is reduced by U^{4+}. The hexavalent uranium occurs primarily as UO_2^{2+} in acid solution. (B) The oxidation of insoluble UO_2 is presented with respective coupling to either oxygen uptake or ferric iron reduction. Heavy arrow lines indicate reactions catalyzed by *T. ferrooxidans*.

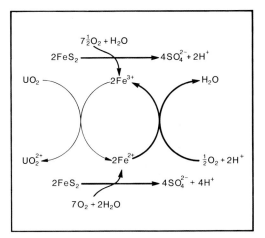

FIG. 5. Solubilization of uraninite and pyrite in an acid leach system containing *T. ferrooxidans*. Pyrite may be oxidized either completely to Fe^{3+} and sulfuric acid or incompletely to Fe^{2+} and sulfuric acid. Fe^{3+}/Fe^{2+} mediate the oxidative dissolution of UO_2 to UO_2^{2+}. Heavy arrow lines indicate reactions catalyzed by *T. ferrooxidans*. Direct biological oxidation of U(IV) in ore material has not been demonstrated because all microbiological leaching systems for uranium ores examined so far have contained iron as the primary oxidizing agent.

CONCLUDING REMARKS

Figure 4 summarizes schematically the oxidation of uranous uranium and ferrous iron. When iron is absent (<0.5 μM), uranium (as either U^{4+} or UO_2) is oxidized directly by *T. ferrooxidans*. In the presence of soluble iron, uranium is oxidized by ferric iron, and the Fe^{2+} thus formed is reoxidized by *T. ferrooxidans* (Fig. 4). Biological oxidation of both substrates occurs, depending on the concentrations. The end products, Fe^{3+} and UO_2^{2+}, inhibit both oxidations, but the complex chemical and biological interactions of U^{4+}, Fe^{2+}, UO_2^{2+}, and Fe^{3+} remain to be characterized in future work.

Figure 5 shows the concurrent oxidation of uraninite and pyrite by *T. ferrooxidans*. Pyrite oxidation by *T. ferrooxidans* releases Fe^{2+} and Fe^{3+} from the mineral phase, which then participate as a redox couple in the oxidative dissolution of UO_2. Ore deposits typically contain pyritic or ferruginous minerals in association with uranium mineralizations. In mining operations using in situ leaching, iron becomes the main redox carrier, and it appears to be oxidized by *T. ferrooxidans* at a faster rate than it is reduced by the chemical reaction with exposed ore material.

We thank Marvin Silver, affiliated with Université Laval, Québec, Canada, for providing information presented in Fig. 1.

LITERATURE CITED

1. **Anderson, R. F.** 1982. Concentration, vertical flux, and remineralization of particulate uranium in seawater. Geochim. Cosmochim. Acta **46**:1293–1299.
2. **Benson, L. V., and D. L. Leach.** 1979. Uranium transport in the Walker River Basin, California and Nevada. J. Geochem. Explor. **11**:227–248.
3. **Berthelin, T. J., and C. Munier-Lamy.** 1983. Microbial mobilization and preconcentration of uranium from various rock materials by fungi. Ecol. Bull. NFR **35**:395–401.
4. **Beveridge, T. J.** 1978. The response of cell walls of *Bacillus subtilis* to metals and to electron-microscopic stains. Can. J. Microbiol. **24**:89–104.
5. **Beveridge, T. J., J. D. Meloche, W. S. Fyfe, and R. G. E. Murray.** 1983. Diagenesis of metals chemically complexed to bacteria: laboratory formation of metal phosphates, sulfides, and organic condensates in artificial sediments. Appl. Environ. Microbiol. **45**:1094–1108.
6. **Brierley, J. A., and C. L. Brierley.** 1980. Biological methods to remove selected inorganic pollutants from uranium mine wastewater, p. 661–667. *In* P. A. Trudinger, M. R. Walter, and B. J. Ralph (ed.), Biogeochemistry of ancient and modern environments. Springer-Verlag, New York.
7. **Brierley, J. A., and C. L. Brierley.** 1983. Biological accumulation of some heavy metals—biotechnological applications, p. 499–509. *In* P. Westbroek and E. W. de Jong (ed.), Biomineralization and biological metal accumulation. D. Reidel Publishing Co. Dordrecht, The Netherlands.
8. **Brierley, J. A., C. L. Brierley, and K. T. Dreher.** 1980. Removal of selected inorganic pollutants from uranium mine waste water by biological methods, p. 365–375. *In* C. O. Brawner (ed.), First international conference on uranium mine waste disposal. Society of Mining Engineers, American Institute of Mining, Metallurgical, and Petroleum Engineers, Inc., New York.
9. **Degens, E. T., F. Khoo, and W. Michaelis.** 1977. Uranium anomaly in Black Sea sediments. Nature (London) **269**:566–569.
10. **DiSpirito, A. A., W. H.-T. Loh, and O. H. Tuovinen.** 1983. A novel method for the isolation of bacterial quinones and its application to appraise the ubiquinone composition of *Thiobacillus ferrooxidans*. Arch. Microbiol. **135**:77–80.
11. **DiSpirito, A. A., J. W. Talnagi, and O. H. Tuovinen.** 1983. Accumulation and cellular distribution of uranium in *Thiobacillus ferrooxidans*. Arch. Microbiol. **135**:250–253.
12. **DiSpirito, A. A., and O. H. Tuovinen.** 1981. Oxygen uptake coupled with uranous sulfate oxidation by *Thiobacillus ferrooxidans* and *T. acidophilus*. Geomicrobiol. J. **2**:275–291.
13. **DiSpirito, A. A., and O. H. Tuovinen.** 1982. Uranous ion oxidation and carbon dioxide fixation by *Thiobacillus ferrooxidans*. Arch. Microbiol. **133**:28–32.
14. **DiSpirito, A. A., and O. H. Tuovinen.** 1982. Kinetics of uranous ion and ferrous iron oxidation by *Thiobacillus ferrooxidans*. Arch. Microbiol. **133**:33–37.
15. **DiSpirito, A. A., and O. H. Tuovinen.** 1984. Oxidations of nonferrous metals by thiobacilli, p. 11–29. *In* W. R. Strohl and O. H. Tuovinen (ed.), Microbial chemoautotrophy. Ohio State University Press, Columbus.
16. **Dongarra, G., and D. Langmuir.** 1980. The stability of UO_2OH^+ and $UO_2(HPO_4)_2{}^{2-}$ complexes at 25°C. Geochim. Cosmochim. Acta **44**:1747–1751.
17. **Elevatorski, E. A.** 1977. Uranium ores and minerals. MINOBRAS, Dana Point, Calif.
18. **Forsberg, C. W., J. W. Costerton, and R. A. MacLeod.** 1970. Separation and isolation of cell wall layers of a gram-negative bacterium. J. Bacteriol. **104**:1338–1353.
19. **Galun, M., P. Keller, H. Feldstein, E. Galun, S. Siegel, and B. Siegel.** 1983. Recovery of uranium (VI) from solution using fungi. II. Release from uranium-loaded *Penicillium* biomass. Water Air Soil Pollut. **20**:277–285.
20. **Giblin, A. M., B. D. Batts, and D. J. Swaine.** 1981. Laboratory simulation studies of uranium mobility in natural

waters. Geochim. Cosmochim. Acta **45**:699–709.

21. **Hayat, M. A.** 1981. Principles and techniques of electron microscopy. Biological applications, vol. 1. University Park Press, Baltimore.

22. **Horikoshi, T., A. Nakajima, and T. Sakaguchi.** 1979. Uptake of uranium from sea water by *Synechococcus elongatus*. J. Ferment. Technol. **57**:191–194.

23. **Hoyle, B., and T. J. Beveridge.** 1983. Binding of metallic ions to the outer membrane of *Escherichia coli*. Appl. Environ. Microbiol. **46**:749–752.

24. **Ingledew, W. J.** 1982. *Thiobacillus ferrooxidans*. The bioenergetics of an acidophilic chemolithotroph. Biochim. Biophys. Acta **683**:89–107.

25. **Ingledew, W. J., and J. G. Cobley.** 1980. A potentiometric and kinetic study on the respiratory chain of ferrous-iron-grown *Thiobacillus ferrooxidans*. Biochim. Biophys. Acta **590**:141–158.

26. **Kanno, M.** 1981. Design and cost studies on the extraction of uranium from seawater. Sep. Sci. Technol. **16**:999–1018.

27. **Kim, Y. S., and H. Zeitlin.** 1971. Separation of uranium from seawater by adsorbing colloidal flotation. Anal. Chem. **43**:1390–1392.

28. **Langmuir, D.** 1978. Uranium solution-mineral equilibria at low temperatures with applications to sedimentary ore deposits. Geochim. Cosmochim. Acta **42**:547–569.

29. **Luft, J. H.** 1971. Ruthenium red and violet. I. Chemistry, purification, methods of use for electron microscopy and mechanism of action. Anat. Rec. **171**:347–368.

30. **Lundgren, D. G., and M. Silver.** 1980. Ore leaching by bacteria. Annu. Rev. Microbiol. **34**:263–283.

31. **Maeda, M., and H. L. Windom.** 1982. Behavior of uranium in two estuaries of the southeastern United States. Mar. Chem. **11**:427–436.

32. **Mangini, A., C. Sonntag, G. Bertsch, and E. Müller.** 1979. Evidence for a higher natural uranium content in world rivers. Nature (London) **278**:337–339.

33. **Nakajima, A., T. Horikoshi, and T. Sakaguchi.** 1982. Recovery of uranium by immobilized microorganisms. Eur. J. Appl. Microbiol. Biotechnol. **16**:88–91.

34. **Olson, G. J., F. D. Porter, J. Rubinstein, and S. Silver.** 1982. Mercuric reductase enzyme from a mercury-volatilizing strain of *Thiobacillus ferrooxidans*. J. Bacteriol. **151**:1230–1236.

35. **Roessler, C. E., Z. A. Smith, W. E. Bolch, and R. J. Prince.** 1979. Uranium and radium-226 in Florida phosphate materials. Health Phys. **37**:269–277.

36. **Ryan, M. T.** 1981. Radiological impacts of uranium recovery in the phosphate industry. Nucl. Saf. **22**:70–77.

37. **Sakaguchi, T., T. Horikoshi, and A. Nakajima.** 1978. Uptake of uranium from sea water by microalgae. J.

Ferment. Technol. **56**:561–565.

38. **Silver, S., and T. G. Kinscherf.** 1982. Genetic and biochemical basis for microbial transformations and detoxification of mercury and mercurial compounds, p. 85–103. *In* A. M. Chakrabarty (ed.), Biodegradation and detoxification of environmental pollutants. CRC Press, Inc., Boca Raton, Fla.

39. **Spalding, R. F., and W. M. Sackett.** 1972. Uranium in runoff from the Gulf of Mexico distributive province: anomalous concentrations. Science **175**:629–631.

40. **Stein, M., A. Starinsky, and Y. Kolodny.** 1982. Behavior of uranium during phosphate ore calcination. J. Chem. Technol. Biotechnol. **32**:834–847.

41. **Strandberg, G. W., S. E. Shumate II, and J. R. Parrott, Jr.** 1981. Microbial cells as biosorbents for heavy metals: accumulation of uranium by *Saccharomyces cerevisiae* and *Pseudomonas aeruginosa*. Appl. Environ. Microbiol. **41**:237–245.

42. **Sugasaka, K., S. Katoh, N. Takai, H. Takahashi, and Y. Umezawa.** 1981. Recovery of uranium from seawater. Sep. Sci. Technol. **16**:971–985.

43. **Torma, A. E., and K. Bosecker.** 1982. Bacterial leaching. Prog. Ind. Microbiol. **16**:77–118.

44. **Tsezos, M.** 1983. The role of chitin in uranium adsorption by *R. arrhizus*. Biotechnol. Bioeng. **25**:2025–2040.

45. **Tsezos, M., and B. Volesky.** 1981. Biosorption of uranium and thorium. Biotechnol. Bioeng. **23**:583–604.

46. **Tsunashima, A., G. W. Brindley, and M. Bastovanov.** 1981. Adsorption of uranium from solutions by montmorillonite; compositions and properties of uranyl montmorillonites. Clays Clay Miner. **29**:10–16.

47. **Tuovinen, O. H., and D. P. Kelly.** 1974. Use of microorganisms for the recovery of metals. Int. Metall. Rev. **19**:21–31.

48. **Tuovinen, O. H., and D. P. Kelly.** 1974. Studies on the growth of *Thiobacillus ferrooxidans*. IV. Influence of monovalent metal cations on ferrous iron oxidation and uranium toxicity in growing cultures. Arch. Microbiol. **98**:167–174.

49. **Tuovinen, O. H., M. Silver, P. A. W. Martin, and P. R. Dugan.** 1981. The Agnew Lake uranium mine leach liquors: chemical examinations, bacterial enumeration, and composition of plasmid DNA of iron-oxidizing thiobacilli, p. 59–69. *In* Proceedings of the international conference on use of microorganisms in hydrometallurgy. Hungarian Academy of Sciences, Pécs, Hungary.

50. **Yamashita, H., K. Fujita, F. Nakajima, Y. Ozawa, and T. Murata.** 1981. Extraction of uranium from seawater using magnetic adsorbents. Sep. Sci. Technol. **16**:987–998.

51. **Zajic, J. E., and Y. S. Chiu.** 1972. Recovery of heavy metals by microbes. Dev. Ind. Microbiol. **13**:91–100.

Bacterial Transformations of Manganese in Wetland Environments

W. C. GHIORSE

Cornell University, Ithaca, New York 14853

Transformations of manganese in aquatic environments consist of oxidation and reduction reactions, many of which may be mediated by microbiological processes (14). In theory, the extent to which the bioconversions of Mn occur in natural waters depends on several nonbiological factors, including the concentrations of Mn and O_2, temperature, Eh, and pH. The concentrations of specific organic and inorganic chemical species may also affect these reactions (11, 14).

According to thermodynamic equilibrium calculations (4), in aerated waters, 100 ppm Mn should exist predominantly as manganese oxide above pH 5.5 and 0.1 ppm Mn should exist as manganese oxide above pH 4. However, apparently because of the high activation energy of the oxidation reactions and possible complexation of reduced Mn, Mn oxidation occurs very slowly or not at all until the pH exceeds 8, unless microbial enzymes or surface catalysis are involved (14).

Reduction of amorphous manganese oxides should occur rapidly below pH 5.5 under aerobic and anaerobic conditions. Above this pH, in aerated waters, amorphous oxide reduction should occur slowly, depending on microbial activity and the concentration of organic and inorganic reducing agents in the environment. Reduction of crystalline manganese oxides would depend on the same factors as reduction of amorphous oxides; however, rapid reduction of crystalline oxides is not likely unless the concentration of reductant is high or microbial attachment and direct enzymatic intervention are involved (4).

The microbiological literature contains numerous observations of Mn-transforming bacteria in natural water samples and in laboratory cultures derived from them (see 4 and 11 for reviews). Aside from these observations, a large body of limnological data exists which indicates that bacterial activity indirectly influences Mn transformations by influencing environmental redox conditions (16). Although this body of evidence indicates bacterial involvement in natural Mn transformations, it is largely circumstantial. In fact, until recently, little direct evidence has been published concerning the extent to which bacteria are actually involved in natural Mn transformations.

Recognizing the lack of direct evidence, some researchers recently have studied Mn-transforming bacteria by direct examination of aquatic environments where Mn transformations occur naturally. These studies include electron microscopic and cultural investigations of ferro-manganese-depositing microorganisms in Baltic Sea ferromanganese concretions (6, 8) and investigations of bacterial Mn deposition in a swamp in Ithaca, New York (7) and in a spring in Washington (13), as well as bacterial Mn transformations in the waters of Oneida Lake, New York (2), and Sannich Inlet, British Columbia, Canada (5).

In the present paper, bacterial transformations of Mn are discussed in relation to more detailed studies of the Ithaca swamp.

MATERIALS AND METHODS

Sapsucker Woods swamp. The sampling site in Ithaca, New York, has been described previously (7). Water and surface film samples and field observations and measurements were taken at approximately weekly intervals from April to November 1981. Samples returned to the laboratory were examined microscopically and processed for viable *Leptothrix* counts and Fe and Mn analyses within a few hours of sampling as described below.

Water sampling procedures. Surface water (0 to 2 cm) was collected in 125-ml wide-mouth polyethylene bottles as described previously (7). Samples from 10, 20, and 30 cm were obtained with a 50-ml plastic syringe attached to a 14-gauge cannula fitted into a 40-cm length of Teflon tubing. The syringe and tubing were clamped to a board marked to indicate depth. Samples from appropriate depths were drawn into the syringe, the board was removed from the water, and the sample was expressed into a polyethylene bottle.

Field observations and measurements. Spot tests for Fe (Prussian blue reaction) and Mn (leukoberbelin blue [LBB] reaction, see below) in the surface film and on the roots of *Lemna* sp. (duckweed), an aquatic plant which covers the

615

swamp during the summer, and measurements of water depth, temperature, pH, Eh, and O_2 concentrations were made as described previously (7). The relative abundance of *Lemna* sp. covering the water surface at the sampling site was estimated subjectively and rated on a scale from 0 (none) to 3+ (completely covered).

Direct counts of Leptothrix sheaths in surface films. Surface films were carefully picked up on either agar-coated or uncoated glass microscope slides. Cover glasses were immediately placed on the agar-coated slides. Uncoated slides were allowed to dry undisturbed in air. They were mounted in water at the time of examination.

Slides were examined in a Zeiss Standard 18 phase-contrast microscope with a 40× oil immersion objective lens. Areas which contained uniformly distributed sheaths were chosen for counting. Routinely, 5 or 10 adjacent fields were counted in 5 different locations on the slide. The mean number of sheaths per 40× field was determined and was used to calculate the number of sheaths per square centimeter in the surface film.

Photomicrographs of representative fields were recorded on Kodak EK 160T film by use of a 100× objective lens and a Zeiss MC63 35-mm automatic camera system.

Viable Leptothrix plate counts. Water samples were diluted in filtered, one-fourth–strength swamp water. Triplicate samples of the appropriate dilution were spread on Mn-supplemented, swamp water agar plates as described previously (7). After 1 week of incubation at 22 to 25°C, flat, yellow to dark-brown colonies with filamentous edges which reacted positively with LBB (Leukoberlinblau I, H. J. Altmann, Berlin, West Germany) reagent (10) were counted as viable *Leptothrix* cells.

Mn and Fe analyses. Concentrations of total and dissolved Mn and Fe in water samples were determined with a Jarrel-Ash 82-740 Dial Atom flame atomic absorption spectrophotometer as described previously (7). Particulate Mn and Fe were calculated by deducting the concentration of dissolved metal from the total metal concentration.

RESULTS AND DISCUSSION

Evidence for microbial involvement in Mn transformations in swamps and ponds. Preliminary electron microscopic studies of surface films in Sapsucker Woods swamp and in a shallow pond in West Germany had demonstrated the presence of ferromanganese-depositing sheathed bacteria in these neustonic environments (W. C. Ghiorse, L. Wieczorek, and P. Hirsch, Abstr. Annu. Meet. Am. Soc. Microbiol. 1980, N101, p. 180). In 1979, *L. discophora*

was found to be the predominant sheathed bacterium in the surface film of Sapsucker Woods swamp. In 1980, periodic observations at the swamp during the biologically active season (April to October) suggested that *L. discophora* and other *Leptothrix* sp. mediated the accumulation of Mn and Fe in the surface waters (7). A more detailed study of the swamp was initiated in 1981 to investigate further the extent of *Leptothrix* involvement in the ferromanganese accumulation process.

In this study, the concentrations of dissolved and particulate Mn and Fe in the surface (0 to 2 cm), middle (10 or 20 cm), and bottom (10, 20, or 30 cm) water were determined at approximately weekly (or shorter) intervals from April to November. The averaged concentrations for the sampling period (Table 1) show that during this period the concentrations of particulate Mn and Fe were highest in the surface water (0 to 2 cm). The data for Mn are shown graphically in Fig. 1 and 2 (middle panel). Since partially aerobic conditions existed in the surface water for most of the year (top panel, Fig. 2), it can be assumed that the accumulation of particulate Mn and Fe at the surface of the swamp reflects mostly ferromanganese oxide in the surface film.

The graphical data for dissolved Mn at three depths in the swamp (Fig. 1) also suggest that Mn reduction occurred at the surface. For example, peaks of particulate Mn in April and July (top panel, Fig. 1) were followed by peaks of dissolved Mn in May and August. Since no corresponding peaks of dissolved Mn were observed at the two lower depths, the source of Mn must have been at the surface. Additionally, during the summer (July–September) dissolved Mn was generally highest at the surface. On the other hand, in autumn (late October, Fig. 1), dissolved Mn was highest at the bottom. At this time, bottom water samples contained H_2S, which would be expected to rapidly reduce particulate manganese oxide which settled to the bottom.

The foregoing evidence suggests that *Leptothrix* cells in the surface films were involved in the distribution of dissolved and particulate Mn in Sapsucker Woods swamp. This evidence, which consists of direct electron microscopic observations and observations of seasonal fluctuations of Mn in surface water, is largely circumstantial. It does not prove that *Leptothrix* sp. was involved in the observed transformations of Mn. Another criticism of this evidence is that it does not indicate how *Leptothrix* sp. might be involved in Mn transformations relative to other biological and environmental factors.

Relative influence of physical, chemical, and biological factors. In an effort to answer such

TABLE 1. Averaged concentration of Mn and Fe at three depths in Sapsucker Woods swamp water from April through November 1981

Depth (cm)	No. of samples	Mn (mg/liter)[a]			Fe (mg/liter)[b]		
		Total	Dissolved	Particulate	Total	Dissolved	Particulate
Surface (0–2)	34	0.45	0.11	0.34	1.83	0.51	1.32
Middle (10 or 20)	39	0.12	0.07	0.05	0.84	0.60	0.14
Bottom (10, 20, or 30)	32	0.11	0.09	0.02	1.75	0.64	1.11

[a] Compare with Fig. 1. Total Mn ranged from 0.03 to 2.84 mg/liter; dissolved Mn, from 0.01 to 0.64 mg/liter.

[b] Total Fe ranged from 0.01 to 10.0 mg/liter; dissolved Fe, from 0.03 to 1.90 mg/liter.

criticism, Mn concentrations in the surface water were compared with a variety of environmental factors (Eh, pH, O_2 concentrations, water temperature, LBB reaction) and with the abundance of *Leptothrix* sp. and *Lemna* sp. (Fig. 2).

As had been observed previously (7), Fig. 2 illustrates that pH, Eh, and O_2 concentration did not directly influence dissolved or particulate Mn concentrations or *Leptothrix* or *Lemna* abundance (Fig. 2). These data do not exclude the possibility that undetected correlations could have existed at the microenvironmental level.

Water temperature exerted a general influence on Mn concentrations, the longest periods of elevated dissolved and particulate Mn being observed during the time of greatest biological activity, when water temperatures were elevated (June–September, Fig. 2).

It can be concluded from the data in Fig. 2 that chemical factors such as pH, Eh, and O_2 concentration exerted little direct influence on Mn transformations in the surface water; rather, temperature-controlled biological factors (*Leptothrix* sp. and *Lemna* sp.) appeared to exert the greatest influence on Mn transformations.

Influence of *Leptothrix* sp. and *Lemna* sp. De-

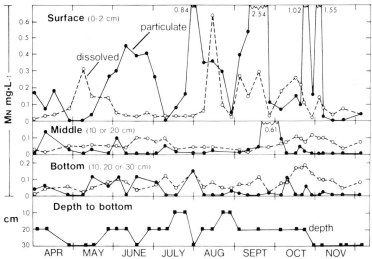

FIG. 1. Distribution of Mn at three depths in Sapsucker Woods swamp in 1981 (compared with averaged data in Table 1). Generally, particulate Mn was higher in the surface water (0 to 2 cm) than at either of the lower depths throughout most of the sampling period from April to October. Note peaks of particulate Mn (solid line) in the surface water (April, end of July, and September) which were followed by peaks of dissolved Mn (dashed line) (May, August, and October). These data suggest that manganese oxides were being formed and then reduced in the aerobic surface waters (see also upper panel in Fig. 2). The large peak of particulate Mn observed in September (2.54 mg of Mn per liter) was followed by a peak in the middle layer (0.61 mg of Mn per liter) 1 week later. These events were probably associated with shedding of Mn-encrusted *Leptothrix* sheaths from *Lemna* roots (see Fig. 3c). The increase in concentration of dissolved Mn in the bottom water during October corresponded to a decline of *Lemna* sp. at the surface (see second panel, Fig. 2). The increase of dissolved Mn probably resulted from chemical reduction of manganese oxides associated with the *Lemna* roots as they settled to the anaerobic bottom water.

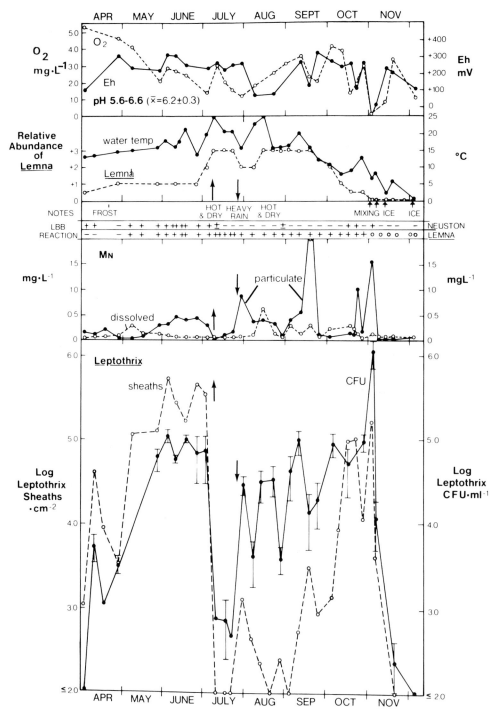

FIG. 2. Comparison of numbers of *Leptothrix* sp. with Mn concentrations and other environmental factors in the surface water (0 to 2 cm) of Sapsucker Woods swamp in 1981. Note the correlation between elevated numbers of *Leptothrix* sp. (see Fig. 3c and 3d) and particulate Mn in spring and autumn. Note also from August to October that the concentration of dissolved Mn frequently was elevated in periods associated with peaks of particulate Mn. Upward arrows in July indicate a hot and dry period when the swamp dried out (see bottom panel, Fig. 1) and *Leptothrix* sp. died off. Downward arrows indicate a period of heavy rain, after which recolonization of the surface film by *Leptothrix* sp. occurred. In the middle panel describing LBB reactions of neuston and *Lemna* roots, + = positive reaction, − = negative reaction, and 0 = no *Lemna* sp.

tailed information on the influence of *Leptothrix* sp. and *Lemna* sp. can be gleaned from a closer examination of Fig. 2. From mid-May to the end of June, elevation of particulate Mn concentrations was paralleled by a similar elevation of *Leptothrix* colony-forming units (CFU) in surface water and sheaths in the surface film (compare dates in Fig. 2 and Fig. 3a–c). Correlation of particulate Mn with *Leptothrix* during this period demonstrates the direct influence of *Leptothrix* on accumulation of particulate Mn at the surface.

A similar influence of *Leptothrix* sp. was observed again in the first week of July, when, during a period of hot and dry weather, particulate Mn as well as *Leptothrix* sheaths and CFU disappeared almost completely from the surface water (upward arrows, Fig. 2). At the same time, the water depth decreased from 20 to 10 cm (Fig. 1, bottom panel) and the relative abundance of *Lemna* sp. increased rapidly as water temperatures rose to 25°C (Fig. 2).

The reason(s) for the *Leptothrix* die-off is unclear; however, it may have been caused by increased concentration of toxic substance in the swamp water as it evaporated. Support for this idea is derived from data for subsequent days in July when dilution of the swamp water occurred as a result of a period of heavy rain (downward arrows, Fig. 2). At this time, the water depth increased to 30 cm (bottom panel, Fig. 1), *Lemna* abundance decreased, and a sharp rise in *Leptothrix* CFU, sheaths, and particulate Mn was observed (Fig. 2).

Another hot, dry spell ensued in August, during which *Leptothrix* sheaths and particulate Mn again showed a decline (Fig. 2), accompanied by a decrease in water depth (Fig. 1) and an increase in the abundance of *Lemna* sp. (Fig. 2). This time, however, numbers of *Leptothrix* CFU (bottom panel, Fig. 2) did not drop as drastically as they had in July. This may be explained by the observation that at this time most of the *Leptothrix* cells were attached to *Lemna* root surfaces, which may have protected them.

Attachment of *Leptothrix* cells to *Lemna* roots was observed microscopically during most of the year whenever both were present on the surface. This association of *Leptothrix* cells with *Lemna* roots is reflected in the pattern of LBB reactions recorded during the year (middle panel, Fig. 2). From April to the end of June, positive LBB reactions, indicating the presence of manganese oxides, were recorded for both the surface film (neuston) and *Lemna* roots. However, from the first week of July to mid-October, LBB reactions of the neuston were negative, while those of the *Lemna* roots remained positive. Microscopic observations confirmed that during this time the roots were coated with

ferromanganese-encrusted *Leptothrix* sheaths (e.g., Fig. 3c). Furthermore, a large peak of particulate Mn which occurred in September (Fig. 2) may have been caused by shedding of excess ferromanganese-encrusted sheaths which had built up on *Lemna* roots during the summer. Large masses of sheaths similar to those depicted in Fig. 3c were observed microscopically in the vicinity of *Lemna* roots at this time. In addition, a peak of particulate Mn appeared a week later in the middle depth of the swamp (0.61 mg of Mn per liter, Fig. 1) and a smaller peak appeared at the bottom a week after that, indicating that Mn oxides from the surface had settled to the bottom as *Lemna* sp. died off in September and October.

The decline of *Lemna* sp. in September and October was accompanied by a "bloom" of *Leptothrix* sp. in the surface film (Fig. 2). The "bloom" was paralleled by increased concentrations of particulate Mn and positive LBB reactions in the neuston similar to those observed in the spring (Fig. 2). The fall "bloom" was not unexpected since a similar *Leptothrix* "bloom" had been observed in October 1980 (7). The 1981 "bloom" was more extensive, however, as it continued into the first week of November when *Lemna* sp. disappeared almost completely from the surface of the swamp.

Mixing of the swamp water occurred in the first week of November, as witnessed by low Eh and O_2 concentrations in the surface water (top panel, Fig. 2) and the smell of H_2S in the air. On 3 November the final peaks of *Leptothrix* and particulate Mn in the surface water were observed (Fig. 2). Microscopic examination of the surface film on this date revealed an abundance of typical *L. discophora*-type sheaths (Fig. 3d). These were morphologically similar to those observed during the summer (compare Fig. 3c and 3d).

The mixing period was followed immediately by very cold weather, culminating in a freeze in mid-November, when the first ice formed on the swamp. The cold weather abruptly reduced *Leptothrix* sheaths, CFU, and particulate Mn to low winter-time levels (Fig. 2). Microscopic examination of the surface film on 6 November, 3 days after the final peak of *Leptothrix* sp., revealed only a few nonencrusted sheathed bacteria (Fig. 2e) similar to those found early in the spring (compare Fig. 2e with Fig. 2a and 2b). In winter, small numbers of *Leptothrix* CFU were detectable in the lower depths of the swamp.

Possible mechanisms of Mn oxidation by *Leptothrix* sp. In the preceding results, it can be seen (Fig. 2) that numbers of *Leptothrix* sp. correlated directly with particulate Mn in the surface water. This evidence allows for the conclusion that *Leptothrix* sp. was directly involved in the

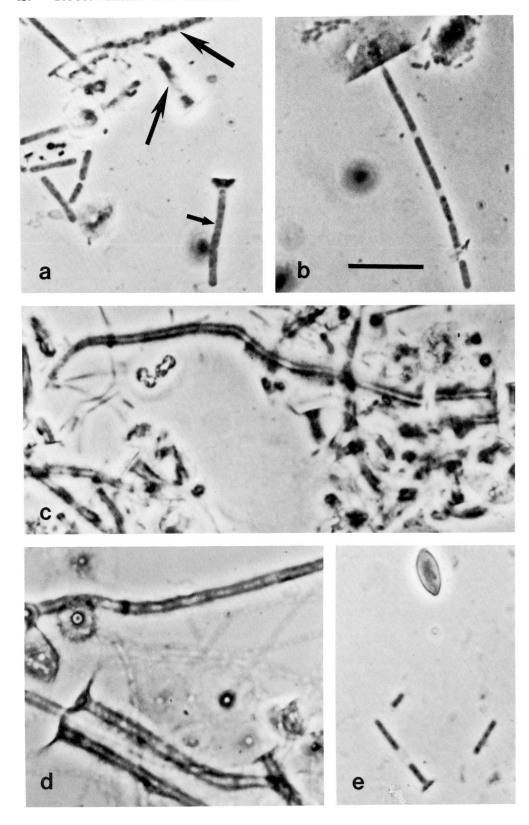

FIG. 3. Phase-contrast photomicrographs showing filamentous sheathed bacteria and empty sheaths in the surface film of Sapsucker Woods swamp on selected dates in 1981 (compare dates in Fig. 2). (a) April 10. Agar-coated slide. Small arrow indicates a nonencrusted bacterial filament with a prominent holdfast at one end. This type was seen in early spring and late fall. Large arrows indicate ferromanganese-encrusted, *L. discophora*-type tapered filaments which predominated during the summer. (b) April 29. Agar-coated slide. After a frost (see in Fig. 2), both the number of *L. discophora*-type sheaths and particulate Mn declined. Only nonencrusted sheaths such as the one depicted were seen. (c) July 2. Air-dried slide. These very long ferromanganese-encrusted *L. discophora*-type sheaths dominated the surface film in May and June. From July to October most of them were attached to *Lemna* roots and were not included in direct counts of sheaths (Fig. 2). (d) November 3. Air-dried slide. *L. discophora*-type sheaths again dominated the surface film as reflected by a "bloom" in sheath counts in October and early November (Fig. 2). (e) November 6. Agar-coated slide. A rapid decline in temperature resulted in complete disappearance of *L. discophora*-type sheaths, leaving a few nonencrusted sheathed bacteria in the surface film. All micrographs reproduced at the same magnification. Bar = 10 μm.

accumulation of manganese oxide at the surface of the swamp.

Although not included in the results presented in this report, it is appropriate at this point to discuss possible mechanisms by which *Leptothrix* sp. may oxidize Mn in the swamp (for general discussions of laboratory studies on Mn oxidation by *Leptothrix* sp., see 4, 11, 15). Studies in this laboratory with an *L. discophora* strain (SS-1) isolated from the swamp showed that, like strains of *Leptothrix* sp. studied in other laboratories (15), SS-1 oxidizes Mn during heterotrophic growth. Furthermore, its growth is inhibited by Mn (L. F. Adams and W. C. Ghiorse, Abstr. Annu. Meet. Am. Soc. Microbiol. 1982, N24, p. 182). Our laboratory studies have also confirmed that SS-1, like other strains of *Leptothrix* sp. (15), excretes an Mn-oxidizing factor which behaves like a protein with respect to heat and $HgCl_2$ sensitivity (unpublished data). Van Veen et al. (15) proposed that an excreted Mn-oxidizing protein was associated with *Leptothrix* sheath material, causing manganese oxide to accumulate there. Preliminary studies of SS-1 and other strains from the swamp support this idea in general; however, most of the chemical and physiological mechanisms of the process remain to be elucidated.

Recently, Dubinina (3) showed that *L. pseudo-ochracea* catalyzed Mn oxidation by H_2O_2, probably by employing catalase as a manganese peroxidase. Whether *L. discophora* possesses a similar manganese peroxidase system or an Mn-oxidizing electron transport system such as that proposed by Mills and Randles (12), or both, also remains to be ascertained. Against the electron transport hypothesis, however, it can be argued that an excreted Mn-oxidizing factor would not be expected to participate in a conventional membrane-associated electron transport system.

Mn reduction. The data in Fig. 1 and 2 suggest that biologically stimulated Mn reduction occurred in the aerobic surface waters of the swamp when *Leptothrix* sp. and *Lemna* sp. were most abundant. However, the extent of biologi-

cal involvement in the Mn reduction is not clear. It is entirely possible that Mn was reduced indirectly by acidic end products of heterotrophic bacteria in the surface water. It is also possible that *Leptothrix* sp. itself was involved in Mn reduction since laboratory strains of *Leptothrix* sp. have been reported to reduce manganese oxides (3, 15). Dubinina (3) recently showed that *L. pseudo-ochracea* reduced manganese oxides by producing excess H_2O_2 during oxidation of organic substrates. A similar reduction of manganese oxides was reported by Bromfield and David (1) in cell suspensions of a Mn-oxidizing *Arthrobacter* sp. Such indirect reduction of manganese oxides by H_2O_2 would be favored at the slightly acidic pH (5.6 to 6.6, Fig. 2) of the swamp water.

The metabolic activity of *Leptothrix* sp. and other heterotrophic bacteria might account for the pattern of Mn reduction observed; however, if we consider that elevated levels of dissolved Mn occurred during the summer when *Lemna* sp. abounded, another explanation is possible. By analogy with other plant species (9), *Lemna* sp. may be expected to produce organic acid root exudates (e.g., malic, pyruvic, malonic, and oxalic acids), which would reduce manganese oxides under the conditions of pH and Eh found in the surface water of the swamp. Interestingly, Dubinina (3) found that aerobic metabolism of certain organic acids resulted in elevated levels of H_2O_2 excreted by *L. pseudo-ochracea*. Thus, it seems possible that a reductive influence of *Lemna* root exudates might be enhanced if the components of the exudates were metabolized by *Leptothrix* sp. or other aerobic heterotrophs in the root zone.

CONCLUSIONS

It can be concluded from this research that *Leptothrix* sp. is a primary mediator of Mn oxidation in the surface film of Sapsucker Woods swamp. Furthermore, it is suggested by these studies that *Leptothrix* sp. and other heterotrophic bacteria may participate in reduction of manganese oxides by producing Mn-reducing substances. It is also possible that root exudates of *Lemna* sp. (organic acids?) may participate in Mn reduction by chemically reducing Mn or by stimulating heterotrophic bacteria to produce Mn-reducing end products in the root zone.

This research establishes a definite role for *Leptothrix* sp. and a possible role for *Lemna* sp. in the cycling of Mn in swamps, bogs, marshes, shallow ponds, and other wetland environments where they occur. Although not directly addressed in this report, a similar role can be proposed for these organisms in the cycling of Fe. Since trace metals frequently coprecipitate with or adsorb to ferromanganese oxides in natural waters, *Leptothrix* sp. and *Lemna* sp. may also influence the natural cycles of trace metals.

Finally, this research indicates the value of maintaining a close relationship between field and laboratory work in ecological studies of microorganisms. In this regard, further studies on the biology of *L. discophora* SS-1 and other *Leptothrix* strains in the laboratory can be expected to increase our future understanding of the metal-transforming activities of these bacteria in their natural environments. Ideally, future laboratory studies should include investigation of the ecological relationships between *Leptothrix* sp. and *Lemna* sp., which, as suggested by the present research, may exert their greatest influence on natural transformations of metals when they live together.

This work was supported by National Science Foundation grant DAR-7924494.

Thanks are extended to K. H. Nealson for suggestions which improved the manuscript. The technical assistance of Lee Adams, Lynn Kozma, and Barbara Seaman is gratefully acknowledged.

LITERATURE CITED

1. **Bromfield, S. M., and D. J. David.** 1975. Sorption and oxidation of manganous ions and reduction of manganese oxide by cell suspensions of a manganese oxidizing bacterium. Soil Biol. Biochem. **8:**37–43.

2. **Chapnick, S. D., W. S. Moore, and K. H. Nealson.** 1982. Microbially mediated manganese oxidation in a freshwater lake. Limnol. Oceanogr. **27:**1004–1014.

3. **Dubinina, G. A.** 1979. Functional role of bivalent iron and manganese oxidation in *Leptothrix pseudoochraceae*. Microbiology (USSR) **47:**631–636.

4. **Ehrlich, H. L.** 1981. Geomicrobiology, p. 203–249. Marcel Dekker, Inc., New York.

5. **Emerson, S., S. Kalhorn, L. Jacobs, B. M. Tebo, K. H. Nealson, and R. A. Rosson.** 1982. Environmental oxidation rate of manganese (II): bacterial catalysis. Geochim. Cosmochim. Acta **46:**1073–1079.

6. **Ghiorse, W. C.** 1980. Electron microscopic analysis of metal-depositing microorganisms in surface layers of Baltic Sea ferromanganese concretions, p. 345–354. *In* P. A. Trudinger, M. R. Walter, and B. J. Ralph (ed.), Biogeochemistry of ancient and modern environments. Springer-Verlag, New York.

7. **Ghiorse, W. C., and S. D. Chapnick.** 1983. Metal-depositing bacteria and the distribution of manganese and iron in swamp waters. *In* R. Hallberg (ed.), Environmental biogeochemistry. Ecol. Bull. (Stockholm) **35:**367–376.

8. **Ghiorse, W. C., and P. Hirsch.** 1982. Isolation and properties of ferromanganese-depositing budding bacteria from Baltic Sea ferromanganese concretions. Appl. Environ. Microbiol. **43:**1464–1472.

9. **Jauregui, M. A., and H. M. Reisenauer.** 1982. Dissolution of oxides of manganese and iron by root exudate components. Soil Sci. Soc. Am. J. **46:**314–317.

10. **Krumbein, W. E., and H. J. Altmann.** 1973. A new method for the detection and enumeration of manganese oxidizing and reducing microorganisms. Helgol. Wiss. Meeresunters. **25:**347–356.

11. **Marshall, K. C.** 1979. Biogeochemistry of manganese minerals, p. 253–292. *In* P. A. Trudinger and D. J. Swaine (ed.), Biogeochemical cycling of mineral-forming elements. Elsevier Scientific Publishing Co., New York.

12. **Mills, V. H., and C. I. Randles.** 1979. Manganese oxidation in *Sphaerotilus discophorus* particles. J. Gen. Appl. Microbiol. **25:**205–207.

13. **Mustoe, G. E.** 1981. Bacterial oxidation of manganese and iron in a modern cold spring. Geol. Soc. Am. Bull., Part I **92:**147–153.

14. **Stumm, W., and J. J. Morgan.** 1981. Aquatic chemistry, 2nd ed., p. 448–468. John Wiley & Sons, Inc., New York.

15. **van Veen, W. L., E. G. Mulder, and M. H. Deinema.** 1978. The *Sphaerotilus-Leptothrix* group of bacteria. Microbiol. Rev. **42:**329–356.

16. **Wetzel, R. G.** 1975. Limnology, p. 250–261. W. B. Saunders Co., Philadelphia.

Anaerobic Corrosion of Iron and Steel: a Novel Mechanism

W. P. IVERSON AND G. J. OLSON

Chemical and Biodegradation Processes Group, Inorganic Materials Division, National Bureau of Standards, Washington, D.C. 20234

Corrosion has been defined by Uhlig (24) as the destructive attack of a metal by chemical or electrochemical reaction with the environment. There is, however, a tendency to consider corrosion as the environmental degradation of all materials and as all forms of degradation irrespective of the mechanism (19).

It is now established that metallic corrosion in aqueous environments is an electrochemical process. Many metals (M) tend to ionize in solution, leaving electrons (e) in the metal:

$$M \rightarrow M^{2+} + 2e^-$$

The area where this reaction takes place is called the anode.

Removal of electrons from the metal can occur by, among others, one of the following cathodic reactions: (i) in the absence of dissolved oxygen or other reducible species, particularly in acid solutions,

$$2H^+ + 2e^- \rightarrow 2H \rightarrow H_2$$

and (ii) in the presence of oxygen, mainly in neutral solutions,

$$H_2O + \tfrac{1}{2}O_2 + 2e^- \rightarrow 2(OH)^-$$

In the older literature, accumulation of hydrogen on the surface of the metal was referred to as "polarization" and removal of the hydrogen was referred to as "depolarization," or "cathodic depolarization."

At the turn of the century, corrosion was usually considered to be associated with the presence of oxygen. It was quite surprising when it was discovered that highly anoxic soils, near neutrality, actually favored a very severe type of corrosion (graphitization) of cast iron water mains. Kuhr et al. attributed this type of corrosion to the presence of anaerobic sulfate-reducing bacteria (17).

SULFATE-REDUCING BACTERIA

Dissimilatory sulfate-reducing bacteria are strict anaerobes that perform anaerobic respiration. They oxidize organic compounds (e.g., lactate) or molecular hydrogen and reduce sulfate (as well as other reduced sulfur compounds) to H_2S. They couple this process to energy conservation (dissimilatory sulfate reduction). The two most commonly described genera of dissimilatory sulfate-reducing bacteria are *Desulfovibrio* and *Desulfotomaculum* (22). Members of the genus *Desulfovibrio* are motile, nonsporulating, gram-negative, curved rods, sometimes sigmoid or spiriform. The cells of organisms in the genus *Desulfotomaculum* are sporeforming, motile, gram-negative, straight to curved rods.

Recently, however, the classification of sulfate-reducing bacteria has been undergoing changes, with the description of isolates (some of which are quite distinct morphologically from the two genera described above) which oxidize carbon compounds ranging from acetate to long-chain fatty acids (26, 27; F. Widdel, Dissertation, University of Gottingen, Gottingen, Germany, 1980).

ECONOMIC EFFECTS

Significant corrosion losses, worldwide, are attributed to the sulfate-reducing bacteria. These bacteria are particularly damaging to the oil and gas processing industry, the chemical processing industry, and industries that employ buried pipelines and underground structures. For example, an early report by Allred et al. (1) stated that sulfate-reducing bacteria were responsible for more than 77% of the corrosion occurring in one group of producing oil wells in the United States. In England, about 50% of the corrosion of underground pipelines was attributed by Booth to the activities of these bacteria (2). Recent published estimates of corrosion losses on mild steel and cast iron due to sulfate-reducing bacteria are practically nonexistent, especially with respect to new structural materials and alloys. However, the National Corrosion Service in the United Kingdom estimated that up to 10% of its corrosion enquiries could have some microbial corrosion involvement (25). At a similar level in the United States, microbial corrosion could account for up to $7 billion out of the $70 billion total annual corrosion cost (National Bureau of Standards Special Publication 511-1, 1978).

H₂S AS A CORROSIVE AGENT

The corrosion by sulfate-reducing bacteria considered in this paper occurs at neutral or near neutral pH values under strict anaerobic conditions. H_2S, the product of sulfate reduction, is considered to be relatively noncorrosive under these conditions, actually forming a somewhat protective film of iron sulfide on the iron (4).

At low pH values, however, iron is corroded by H_2S, with the release of hydrogen (8):

$$H_2S + Fe \rightarrow FeS + 2H$$

The hydrogen released by the attack on iron is responsible for embrittlement, cracking of hardened steel, blistering, and loss of ductility (23). Under aerobic conditions, H_2S may also be oxidized abiotically or by microorganisms to elemental sulfur and various oxyacids of sulfur which are highly corrosive (7).

CATHODIC DEPOLARIZATON THEORY

To explain the severe corrosion of underground pipelines in polder soil (land reclaimed from the sea) at pH levels near neutrality, Kuhr et al. proposed a theory (17) based on the removal of hydrogen from the iron surface by the hydrogenase of sulfate-reducing bacteria, referred to as the cathodic depolarization theory (Fig. 1). Hydrogen removal by the bacteria is shown as step IV, causing iron to ionize (corrode) as ferrous ions (step II). The ferrous ions then react with sulfide ions and hydroxide ions as in steps V and VI, respectively.

Evidence for and against this theory has accumulated and has been well reviewed over the past several years (5, 12, 20, 21). A few facts have been established. First, no direct correlation between hydrogenase activity and corrosion rate was found by use of semicontinuous and continuous cultures (3, 4). Second, the addition of ferrous ions to the culture media was found to result in the stimulation of corrosion rates (16). It is believed that the ferrous ions reacted with the sulfide ions in solution to form bulk iron

CATHODIC DEPOLARIZATION THEORY
of Von Wolzogen Kühr and Van Der Vlugt

I $\quad 8H_2O \longrightarrow 8OH^- + 8H^+$

II $\quad 4Fe \longrightarrow 4Fe^{++} + 8e$ (anode)

III $\quad 8H^+ + 8e \longrightarrow 8H$ (cathode)

IV $\quad SO_4^{--} + 8H \xrightarrow{\text{(bacteria)}} S^{--} + 4H_2O$ (cathodic depolarization)

V $\quad Fe^{++} + S^{--} \longrightarrow FeS$ (anode)

VI $\quad 3Fe^{++} + 6(OH)^- \longrightarrow 3Fe(OH)_2$ (anode)

$$4Fe + SO_4^{--} + 4H_2O \rightarrow FeS + 3Fe(OH)_2 + 2(OH)^-$$

FIG. 1. Cathodic depolarization theory of Kuhr et al.

sulfide and prevented the establishment of a film of iron sulfide on the surface of the iron. Mara and Williams (18) postulated that the bulk iron sulfide was involved, acting as a cathode and absorbing cathodic hydrogen as an ionized interstitial in proportion to the cationic defects in the iron sulfide. Earlier, King and Miller (15) suggested that the bacteria on the surface of the ferrous sulfide continually "regenerated" or depolarized the iron sulfide by removal of atomic hydrogen by their hydrogenase activity. An additional possibility, suggested by Miller (20), was that the bacteria, as a result of their motility, caused fresh iron sulfide surfaces to be brought constantly into contact with the steel.

In addition to these proposed mechanisms, Costello (6) postulated that the depolarizing activity of sulfate-reducing bacteria was due to the cathodic activity of the dissolved hydrogen sulfide produced by these organisms.

CORROSIVE METABOLITE THEORY

In contrast to the cathodic depolarization theory, Iverson (12, 13) has proposed a theory which states that anaerobic corrosion is caused by the elaboration of a highly corrosive metabolite produced by the sulfate-reducing bacteria. This theory is based on a number of observations and results obtained over the past several years with two hydrogenase-positive strains of *Desulfovibrio desulfuricans*, an American Petroleum Institute strain (API strain) and a marine strain isolated from an iron piling at Dam Neck, Virginia. The organisms have been routinely cultivated in pure culture on the surface of Trypticase soy agar plates (9). (Aged seawater was used instead of distilled water for the marine strain.) The following five observations are presented in support of the theory:

(i) A model system to test the cathodic depolarization theory indicated that the theory was not completely valid (10, 11). A mild steel (1010) electrode (cathode) was placed in contact with cells of the API strain in an inert atmosphere (N_2 or argon) on an agar surface with a redox dye, benzyl viologen (BV), and in other cases with sulfate as electron acceptors. A second electrode in contact with the agar (no cells) served as the anode. The two electrodes were electrically connected via a microammeter. With BV as the electron acceptor, the BV was reduced at the cathode and iron went into solution at the anode (Fig. 2). The small "cathodic depolarization" current, 1 μA cm^{-2} (2.5 mg dm^{-2} day^{-1}), however, could not account for the extensive corrosion observed in the field (855 to 1,380 mg dm^{-2} day^{-1}). Furthermore, this effect could not be observed when sulfate replaced BV as the electron acceptor (no "cathodic depolarization" current and no dissolution of iron at the anode).

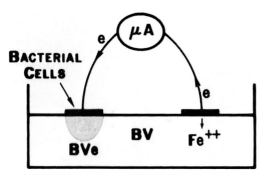

FIG. 2. Measurement of cathodic depolarization current. BVe denotes reduced benzylviologen (violet); BV denotes oxidized benzylviologen (colorless).

(ii) Extensive corrosion of bulk iron was obtained (1,152 mg dm^{-2} day^{-1} maximum) in a spent culture medium of the marine strain from which the bacterial cells and sulfide ions were removed (12, 13). Figure 3 indicates the extremely rapid rise in the corrosion rate after a lag period of between 3 and 4 days, as determined by electrochemical polarization techniques. This observation was confirmed by weight loss measurement for determining the corrosion rate (13). Coincident with the increase in corrosion rate, the redox potential became more negative and the corrosion potential of the iron increased to more noble (less negative) values. These results indicate that extensive anaerobic corrosion can occur in the absence of the bacteria, sulfide ions, and bulk iron sulfide as depolarizing agents.

(iii) The corrosion products formed by the marine strain and the API strain appear to be amorphous iron phosphides. The black amorphous products were heated in a vacuum oven to about 1,200°C, allowed to cool, and analyzed by an X-ray powder pattern technique (11). The diffraction patterns of the material from the API and marine strains corresponded to the patterns of iron phosphide (Fe$_2$P) and the schreibersite (Fe$_3$P) of meteoritic origin (Cerros de Buen Huerto), respectively.

Recently, we have acidified the bulk corrosion products and the metal coupons with their corrosion products and analyzed the headspace gases, using gas chromatography (GC) with a flame photometric detector (FPD) in the phosphorus-selective mode (14). Phosphine (PH$_3$) has been routinely detected from the iron phosphide in addition to H$_2$S from the corrosion products formed in the laboratory, as well as in the field (Fig. 4). The detection of significant quantities of phosphorus on anaerobically corroded steel specimens by the ESCA (electron spectroscopy for chemical analyses) technique has been reported (C. Gaylarde, personal communication).

(iv) Anaerobic corrosion of iron nails (brads)

has been obtained by placing them in the headspace above agar surface cultures of the marine strain (14). The corrosion product found on the nails was iron phosphide, as indicated by the evolution of PH$_3$ after acid (1 ml of 18 N H$_2$SO$_4$) treatment (Fig. 4). Small amounts of phosphine were occasionally detected with control nails, perhaps arising from traces of contaminant phosphorus present in steel. Phosphine has also been detected from acidified field samples of corrosion products.

The volatile material produced by both the API strain and the marine strain was trapped in acidified permanganate solution and tested by the phosphomolybdate reaction. The test was positive for phosphorus. (W. P. Iverson, National Bureau of Standards Report no. 10054, 1968). From these observations, it was concluded that the metabolite is a volatile, phosphorus compound. Examination of the headspace for a phosphorus compound above agar cultures (both plate and slant cultures) as well as liquid cultures, by the GC-FPD in the phosphorus mode, were negative, however (10). It was suspected that the compound reacts with the chromatographic column packing material. With the FPD in phosphorus-selective mode, a peak above cultures of the API strain was detected when an empty column was used.

Two sulfur compounds, methylmercaptan (CH$_3$SH) and dimethyldisulfide [(CH$_3$)$_2$S$_2$], in

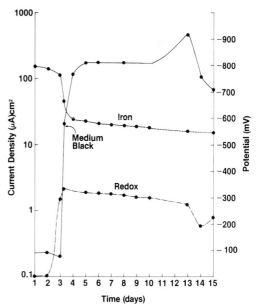

FIG. 3. Corrosion current density (top curve), potential of iron 1010 (mild steel) (versus saturated calomel electrode), and redox potential in a culture filtrate (marine strain) minus bacterial cells and sulfide ions.

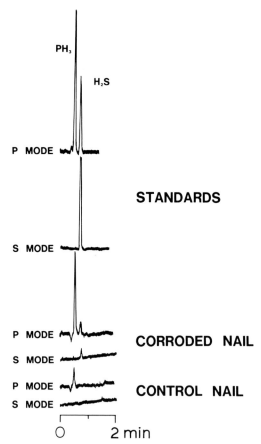

FIG. 4. Chromatogram of reference PH₃ and H₂S and acidified corroded nails (brads) exposed to headspace above an agar plate culture of the marine strain.

addition to H₂S were detected, however, and tentatively identified by their retention times (GC-FPD, sulfur mode) in comparison with the retention times of authentic compounds (14). They were found to exert little, if any, corrosive action on mild steel under anaerobic conditions (H₂ atmosphere).

(v) A phosphorus-containing gas (positive phosphomolybdate test) was produced by the action of hydrogen sulfide on solid phosphate ($Na_3PO_4 \cdot 12H_2O$), phosphite ($Na_2HPO_3 \cdot 5_2O$), and hypophosphite ($NaH_2PO_2 \cdot H_2O$) (14). Again, no phosphorus-containing gas was detected with the GC-FPD system. Small amounts of PH₃ were detected, however, after 1 or 2 weeks. This was probably a decomposition product (14). Significantly, a film of iron phosphide was produced on iron nails (brads) after exposure to hypophosphite and hydrogen sulfide (ferrous ions being added to remove the sulfide ions) (14). Occasionally, traces of phosphine have been detected in the headspace above cultures of *Desulfovibrio* sp. Although biological

reduction of phosphate to phosphite, the reduction of phosphite to hypophosphite, and the reduction of hypophosphite to phosphine are not thermodynamically favorable, the slight reduction of these or other phosphorus compounds to the active volatile phosphorus compound by H₂S or cellular processes evidently must be thermodynamically possible.

Mass spectrometric studies are in progress to identify the bacterially and abiotically produced volatile phosphorus compound.

CONCLUSIONS

From the evidence obtained, it appears that anaerobic corrosion by sulfate-reducing bacteria is caused by a volatile, highly reactive compound produced during phosphorus reductions, which reacts with bulk iron to form iron phosphide. This compound may be produced directly by bacterial action on inorganic phosphorus compounds or indirectly by the abiotic action of bacterially produced H₂S on these compounds.

If, during the action of sulfate-reducing bacteria on iron, H₂S comes in contact with an iron surface first, a layer of iron sulfide will be formed which inhibits further corrosion. If conditions are favorable for the formation of the volatile phosphorus compound and it comes in contact with the iron sulfide, corrosion would be expected to occur. Also, if the iron sulfide film breaks down and the phosphorus compound is present, corrosion would be expected to occur. The corrosion rate has been observed to be very large after the breakdown of the sulfide film (4). The results in this paper also suggest that iron phosphide formation is a good indicator for microbial corrosion and, depending on its stability, that corrosion by the volatile phosphorus-containing metabolite may occur at some distance from its production site. These observations should be important in improving detection and control strategies for anaerobic microbial corrosion.

Partial support for this effort was provided by the Office of Naval Research.

We thank F. E. Brinckman, National Bureau of Standards, for his suggestions and encouragement.

LITERATURE CITED

1. **Allred, R. C., J. D. Sudbury, and D. S. Olson.** 1959. Corrosion is controlled by bacteriocide treatment. World Oil **149**(6):111–112.
2. **Booth, G. H.** 1964. Sulfur bacteria in relation to corrosion. J. Appl. Bacteriol. **27**:174–181.
3. **Booth, G. H., D. W. Cooper, and P. M. Cooper.** 1967. Rates of microbial corrosion in continuous culture. Chem. Ind. **86**:2084–2085.
4. **Booth, G. H., P. M. Shinn, and D. S. Wakerley.** 1964. The influence of various strains of actively growing sulphate-reducing bacteria on the anaerobic corrosion of mild steel, p. 363–371. *In* Comptes Rendus du Congrés International de la Corrosion Marine et des Salissures, Cannes, Paris C.R.E.O.

5. **Costello, J. A.** 1969. The corrosion of metals by microorganisms. A literature survey. Int. Biodeterior. Bull. **5:**101–118.

6. **Costello, J. A.** 1974. Cathodic depolarization by sulfate-reducing bacteria. S. Afr. J. Sci. **70:**202–204.

7. **Cragnolino, G., and O. H. Tuovinen.** 1983. Role of sulfate-reducing and sulfur-oxidizing bacteria on localized corrosion. Preprint paper no. 244. National Association of Corrosion Engineers National Conference, Anaheim, Calif.

8. **Ewing, S. P.** 1955. Electrochemical study of hydrogen sulfide corrosion mechanism. Corrosion **11:**51–55.

9. **Iverson, W. P.** 1966. Growth of *Desulfovibrio* on the surface of agar media. Appl. Microbiol. **14:**529–534.

10. **Iverson, W. P.** 1966. Direct evidence for the cathodic depolarization theory of bacterial corrosion. Science **151:**986–988.

11. **Iverson, W. P.** 1968. Corrosion of iron and formation of iron phosphide by *Desulfovibrio desulfuricans*. Nature (London) **217:**1265–1267.

12. **Iverson, W. P.** 1974. Microbial corrosion of iron, p. 476–517. In J. B. Neilands (ed.), Microbial iron metabolism. Academic Press, Inc., New York.

13. **Iverson, W. P.** 1981. An overview of the anaerobic corrosion of underground metallic structures: evidence for a new mechanism, p. 33–52. In E. Escalante (ed.), Underground corrosion, ASTM STP 741. American Society for Testing and Materials, Philadelphia.

14. **Iverson, W. P., and G. J. Olson.** 1983. Anaerobic corrosion by sulfate-reducing bacteria due to a highly-reactive phosphorus compound, p. 46–53. In Microbial corrosion. The Metals Society, London.

15. **King, R. A., and J. D. A. Miller.** 1971. Corrosion by the sulphate-reducing bacteria. Nature (London) **233:**491–492.

16. **King, R. A., J. D. A. Miller, and D. S. Wakerley.** 1973. Corrosion of mild steel in cultures of sulphate-reducing bacteria: effect of changing the soluble iron concentration during growth. Br. Corros. J. **8:**89–93.

17. **Kuhr, C., A. H. von Wolzogen, and L. S. van der Vlugt.** 1934. De grafiteering van gietijer als electrobiochemisch proces in anaerobic gronden (Graphitization of cast iron as an electro-biochemical process in anaerobic soils). Water **18**(16):147–165.

18. **Mara, D. D., and D. J. A. Williams.** 1972. The mechanism of sulphide corrosion by sulphate-reducing bacteria, p. 103–113. In A. H. Walters and E. H. Hueck van der Plas (ed.), Biodeterioration of materials, vol. 2. Applied Science, Publishers Ltd., London.

19. **Menzies, A.** 1970. Introductory corrosion, p. 37. In J. D. A. Miller (ed.), Microbial aspects of metallurgy. Elsevier Publishing Co., Inc., New York.

20. **Miller, J. D. A.** 1981. Metals, p. 149–202. In A. H. Rose (ed.), Microbial biodeterioration. Academic Press, Inc., New York.

21. **Miller, J. D. A., and R. A. King.** 1975. Biodeterioration of metals, p. 83–103. In D. W. Lovelock and R. J. Gilbert (ed.), Microbial aspects of the deterioration of materials. Academic Press, Inc., New York.

22. **Postgate, J. R.** 1979. The sulphate-reducing bacteria, p. 8–23. Cambridge University Press, New York.

23. **Tuttle, R. N., and R. D. Kane (ed.).** 1981. H_2S corrosion in oil and gas production—a compilation of classic papers. National Association of Corrosion Engineers, Houston, Tex.

24. **Uhlig, H. H. (ed.).** 1963. Corrosion and corrosion control, p. 1. John Wiley & Sons, Inc., New York.

25. **Wakerley, D. S.** 1979. Microbial corrosion in U.K. industry; a preliminary survey of the problem. Chem. Ind. **19:**657–659.

26. **Widdel, F., and N. Pfennig.** 1977. A new anaerobic sporing, acetate-oxidizing sulfate-reducing bacterium, *Desulfotomaculum* (emend.) *acetoxidans*. Arch. Microbiol. **112:**119–122.

27. **Widdel, F., and N. Pfenning.** 1981. Studies on dissimilatory sulfate-reducing bacteria that decompose fatty acids. I. Isolation of new sulfate-reducing bacteria enriched with acetate from saline environments. Description of *Desulfobacter postgatei* gen nov., sp. nov. Arch. Microbiol. **129:**395–400.

Ecological Strategies for the Fermentation Industry

Recruitment of Novel Reactions: Examples and Strategies

I. JOHN HIGGINS, DAVID J. BEST, GRAEME MACKINNON, AND PHILIP J. WARNER

Biotechnology Centre, Cranfield Institute of Technology, Cranfield, Bedford, MK43 OAL, United Kingdom

The main goal of contemporary industrial microbiology is an expansion of the industrial application of the extensive unusual metabolic properties of microorganisms. Its success is measured in terms of novel products (pharmaceuticals, agrochemicals), novel foodstuffs or bioelectronic devices (e.g., biosensors or biofuel cells), and more economic routes to the existing products or improved waste treatment processes.

Most examples discussed in this paper are concerned in the broadest sense with microbiological–chemical interconversions. Although the range of reactions known to be catalyzed by microorganisms continues to expand and seems to exceed the demands made by the biosynthetic sectors of the carbon cycle, the would-be "bug hunter" should not expect microorganisms to disregard the laws of thermodynamics or to transgress the rules of chemistry. The capacity of microorganisms to degrade (or to develop the ability to degrade) modern chemical products, such as herbicides and pesticides, which are not natural components of the carbon cycle, is particularly important in relation to the future stability of the biosphere and the development of waste treatment processes.

The recent discovery of complex communities of extremely thermophilic bacteria in waters at 350°C emanating from sulfide chimneys along the East Pacific Rise (3) is an outstanding example of the extraordinary adaptability of microorganisms and of an extreme environment that has encouraged the development of novel properties that may prove to be exploitable. This dramatic finding should encourage the search for novel activities in extreme environments.

Without claiming to be comprehensive, we discuss in this paper a range of approaches to obtaining potentially useful microbiological reactions, with examples from both our work and that of our colleagues and from a variety of other laboratories. We are not, however, concerned here with obtaining the gene products of higher organisms from genetically engineered microbial strains.

Most microorganisms and microbial activities of proven or potential industrial value have been obtained by the classical enrichment techniques devised some 80 years ago by Winogradski and Beijerinck, and much of our understanding of microbial activities is based upon studies of monocultures. Highly complex, yet rather poorly understood, microbial ecosystems have long been applied to industrial, municipal, agricultural, and domestic waste treatment and to the conversion of waste to methane.

The behavior of mixed microbial populations has attracted much attention, especially during the past decade. This interest has not been restricted solely to microbial ecologists, for whom microbial interactions are of paramount relevance, but has extended to microbial biochemists, physiologists, geneticists, and industrial microbiologists or biotechnologists. Indeed, some of the most interesting observations of mixed-culture microbial interactions have emerged from applied work. A range of examples are summarized by Kelly (12). A notable early example arises from work directed toward single-cell protein production from methane, in which it was found that for stable, rapid growth with a high yield, a four- or five-membered community is preferable (15, 28). Methane is a ubiquitous substrate, and many microorganisms capable of growth on this hydrocarbon are known (30). In contrast, much recent work on microbial communities in laboratory culture has involved the isolation and development of communities capable of degrading complex xenobiotics, especially halogenated aromatic compounds. This has led to an increase in our understanding of cooperative, mutualistic, or commensal relationships, particularly among bacteria, and has yielded information about genetic interactions between microorganisms and how new capabilities may be acquired by individual species. This recent new information, relevant aspects of which are discussed below, has been reviewed (20, 22) and has important implications for workers wishing to acquire novel microbiological activities. It also goes some way toward explaining the observation that in situ enrichment procedures have often succeeded when elective culture has failed or proved difficult.

Potentially useful microbial activities frequently involve metabolism other than that required for growth and energy of the organism. Before proceeding to discuss strategies in more detail, this area merits brief consideration, both to clarify the terms used and because special approaches are required to identify such activities which yield many valuable products (secondary metabolites).

A variety of terms have been used to describe various types of nongrowth metabolism, and there is some confusion in the literature over the use of these terms. This problem has been discussed elsewhere (9, 18, 19, 23). The various terms as used in this contribution are defined as follows. Secondary metabolism is defined as a process leading to the generation of metabolic products which are not essential for growth. Cooxidation describes oxidation of a substrate not able to support growth in the presence of another oxidizable growth substrate, whereas cometabolism is a more general term which can be applied to metabolism of a substrate of any kind that does not support growth. Fortuitous metabolism is regarded as metabolism which uses a substrate unable to support growth and which does not involve concomitant metabolism of other cellular components, for example, storage polymers. Supplementary metabolism is a type of cometabolism in which the substrate, when oxidized, does not totally support growth, but does provide incidental carbon and energy. Commensal metabolism describes a process in which the metabolism of one organism benefits another, whereas cooperative or mutualistic metabolism describes a process in which the metabolic systems of two or more organisms operate to their mutual advantage. Some of these terms, especially cooxidation and cometabolism, have been used to describe several distinctly different phenomena, and we believe that adherence to the above definitions would avoid further confusion.

CONVENTIONAL ENRICHMENT CULTURE

Conventional enrichment culture technique, by which microorganisms are isolated with a substance of interest used as the sole source of carbon and energy (and sometimes of nitrogen), remains the most common method for obtaining novel primary metabolic activity. This may be exploited in three main ways: (i) total degradation, for example, in a waste treatment process; (ii) acquisition of specific enzymes for process, medical, or analytical uses; and (iii) formation of useful chemical products by growing or nongrowing organisms. A recent example from our laboratory of an organism with interesting primary metabolic activities isolated by this technique is a pseudomonad with plasmid-encoded aniline-degrading activity (2). Although we have not investigated the biochemical pathway in-

volved as yet, a pathway for the microbial oxidation of aniline has been proposed (Fig. 1).

Another example concerns the commercially promising microbial conversion of naphthalene to salicylate. Over the past 25 years, there have been numerous reports of the isolation of microorganisms capable of growth on naphthalene that in some cases can give high yields of salicylate (11, 13). Most have involved conventional enrichment techniques. There is evidence, however, that the suitability of a particular strain for salicylate production depends upon the particular isolation method employed. For example, some workers claim that the direct plating of soil samples onto agar plates in which naphthalene has been dispersed results in the isolation of more strains than liquid enrichment techniques (11). In general, strains isolated on solid medium also showed a better rate of transformation from naphthalene to salicylate.

The need for careful experimental design to isolate organisms with the desired combination of properties is self-evident (but not always practiced). One frequently ignored possibility is that isolates may have particular growth factor requirements. It may well be that auxotrophy in a particular natural ecosystem is not disadvantageous and may even be beneficial. A good example of this is the isolation some years ago of a rare metabolic capability of a *Nocardia* sp. isolated on cyclohexane (24). It was initially obtained as a mixed, mutualistic culture with a pseudomonad, which was subsequently shown to be satisfying an auxotrophic requirement of the *Nocardia* sp. for biotin. It is therefore generally prudent to incorporate micronutrients in the enrichment medium at a low concentration.

Although carefully planned, conventional enrichment culture remains the main approach to acquiring pure cultures with novel primary metabolism or mixed cultures exhibiting commensal, cooperative, or mutualistic metabolism, it can also be used to acquire organisms capable of useful cometabolism or fortuitous metabolism. Supplementary metabolism is a special case and lends itself to continuous-culture isolation techniques as discussed below. Secondary metabolism is usually detected as a result of large-scale screening of isolates revealing biologically active products of such metabolism (for example, antibiotics).

The rationale for using conventional enrichment culture for acquiring organisms which oxidize a compound without utilizing it for growth, etc. (cometabolism or fortuitous metabolism), is based upon the increasingly common observation that the enzymes of catabolic sequences, especially those involved in the initial steps, are often not completely substrate specific and indeed in some cases, such as those of some of the

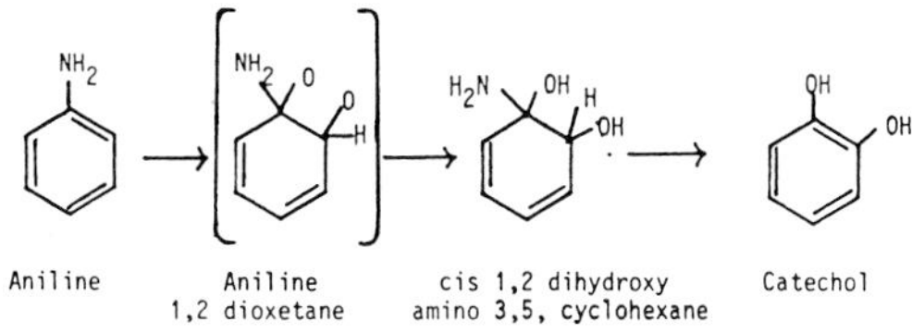

FIG. 1. Proposed pathway for microbial oxidation of aniline.

methane-oxidizing bacteria, have extremely broad specificity (9, 23). Currently, we also have an interest in the oxidation of aniline analogs and selected *ortho* substituents in aromatic compounds. Isolates capable of growth on aniline (see above) or *o*-xylene (next section), therefore, are being screened for cometabolism or fortuitous metabolism of analogs.

IN SITU ENRICHMENT

The susceptibility of materials, especially of synthetic polymers, to biodegradation has often been studied by presenting the material to the environment of interest by, for example, soil burial. Microorganisms involved in these processes may then be isolated from the surface of the material. This is a form of in situ enrichment. In fact, this process is involved in many well-planned conventional enrichment procedures; environmental samples used for enrichment may well have been exposed to some chemical or physical effect that would be expected to have led to enrichment for the desired metabolic property. For example, soil which has been exposed to long-term oil pollution is likely to yield isolates capable of degrading a wide range of oil components.

A few years ago we developed a method of presenting oil samples to aquatic environments which allows long-term studies of microbial degradation of oil in the environment and isolation of the specific strains involved (6, 10). This requires adsorption of known amounts of oil onto cellulose acetate filter disks which are sealed into practice golf balls, offering physical protection while allowing free interchange with the environment. This type of approach has also proved useful for the in situ enrichment for rare metabolic properties of organisms capable of specific hydroxylation of terpenes to generate high added-value products. For example, bacteria capable of growing on *p*-menthane were readily obtained by burying such samples at a depth of 15 to 30 cm in soil for 1 to 3 weeks in areas where there had been hydrocarbon spillage. Pure strains of *p*-menthane utilizers were then obtained from the mixed cultures growing on the surface of the glass wool and of surrounding soil particles, by enrichment on a minimal

medium in closed flasks containing *p*-menthane vapor (26). Limonene utilizers were also isolated by this technique, but the in situ enrichment step was not essential (26). In contrast, and in spite of extensive efforts, we were not able to isolate *p*-menthane utilizers without the in situ enrichment step. Other workers who have attempted the isolation of *p*-menthane utilizers by conventional enrichment have experienced considerable difficulties, as described in one study in which only four strains with the desired property were obtained from 295 soil samples (27).

More recently, we have readily isolated by the in situ technique an *o*-xylene utilizer which has been identified as a strain of *Nocardia minima* (25) and trinitrotoluene utilizers (Owen, Taylor, and Higgins, unpublished data). Both show metabolic activities that may prove industrially useful. The efficacy of the method is again emphasized by the reported failure to isolate *o*-xylene utilizers by attempting conventional enrichment from 364 different soil samples (17).

The in situ enrichment procedure remains underutilized. It may well, in many cases, prove to be a more rapid method for obtaining the required activity than some of the more sophisticated continuous isolation techniques discussed below. Although we are beginning to understand the types of genetic processes involved in the generation of novel metabolic activities, there is doubtless a variety of physiological, physical, and ecological processes occurring in the natural environment which we do not yet understand. There is also insufficient information in the literature concerning the general value and applicability of this technique. It would be worthwhile to compare success rates in acquiring novel activities by conventional, in situ, and continuous enrichment methods (next section).

GENETIC ASPECTS

Genetic manipulation techniques may well prove important when recruiting microorganisms to carry out novel reactions. Subsequent to the isolation of such organisms, genetic manipulation may substantially enhance the rate of the desired reaction and the yield of product. It may also allow production of compounds not normally accumulated by microorganisms or allow combination of two or more reactions from different strains to construct a single organism capable of performing such reactions. Traditionally, genetic manipulation has involved random mutation followed by enrichment, selection, and screening. Continuous-culture techniques have allowed continuous selection of strains with the desired characteristics, and the advent of molecular genetics allows specific mutation at the biochemical level, resulting in a specific, predictable change in the phenotype.

Random mutation. If the pathway by which a novel reaction proceeds is known, it is often possible to realize hyperproduction of an intermediate by obtaining a mutant blocked at a particular step in the pathway. Auxotrophic mutants have frequently been used for the microbial production of amino acids and other compounds (nucleotides). Unlike the wild-type organism, in which amino acid metabolism is tightly regulated, amino acid auxotrophs often overproduce an amino acid different from the one required for growth because they lack an enzyme of a biosynthetic pathway whose product causes feedback inhibition. Figure 2a shows how auxotrophic mutants of *Corynebacterium glutamicum* overproduce intermediates of arginine biosynthesis, namely, ornithine and citrulline (16). They were obtained by standard techniques of mutagenesis followed by penicillin enrichment.

A more direct method of obtaining blocked regulatory pathway mutants involves the use of analogs of pathway intermediates. Amino acid analogs inhibit growth of wild-type bacteria because they prevent synthesis of the amino acid without replacing it and thus make it unavailable for further metabolism. Neither the amino acid nor its analog can cause feedback inhibition which results in growth of the mutant and overproduction of the amino acid. Figure 2b shows an example, in which an *S*-2-aminoethyl-L-cysteine–resistant mutant of *C. glutamicum* overproduces L-lysine (31).

Continuous selection. Continuous selection techniques are powerful tools in the recruitment of microorganisms able to carry out novel reactions. There are a few distinct approaches.

(i) A steadily increasing selection pressure can be applied to a single species so that it eventually adapts to effect the desired reaction. An example involves sugar metabolism in *Klebsiella pneumoniae*, which has been grown in chemostat culture on a mixture of low concentrations of ribitol and higher concentrations of xylitol. Regulatory mutants with constitutive ribitol dehydrogenase and structural mutants with an altered ribitol dehydrogenase were obtained (L. K. W. Thompson, Ph.D. thesis, Lehigh University, Bethlehem, Pa., 1981).

(ii) A mixed culture of two or more species can be grown for many generations until the desirable characteristics eventually emerge in one strain. For example, when *Pseudomonas* sp. strain B13, capable of utilizing 3-chlorobenzoate, was grown in a chemostat with *P. putida* mt-2, which harbored the TOL plasmid, a *Pseudomonas* sp. B13 strain capable of growing on 4-chlorobenzoate emerged (Fig. 3). The stereospecificity of the 3-chlorobenzoate oxygenase in *Pseudomonas* sp. strain B13 prevented the oxi-

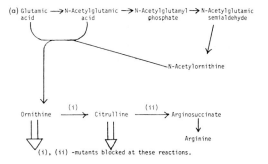

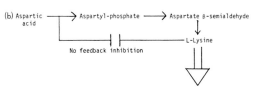

FIG. 2. Two genetic approaches to hyperproduction of metabolites. (a) Use of auxotrophs of *C. glutamicum* to overproduce citrulline and ornithine (16). (b) Genetic desensitization in L-lysine biosynthesis caused by *S*-2-aminoethyl-L-cysteine resistance (3).

dation of 4-chlorobenzoate, but the introduction of the TOL plasmid into the organism provided a benzoate-1,2-dioxygenase of broader specificity, able to act on 3-chlorobenzoate, 4-chlorobenzoate, and 3,5-dichlorobenzoate (8, 21).

(iii) The desired activity evolves in a community of different species in which no single species (alone) can carry out the novel reactions. This is exemplified by the evolution of a stable community able to perform a novel set of reactions in which 2-chloropropionamide was used as the growth-limiting substrate in chemostat culture (20). A community of at least six microorganisms, including three pseudomonads, a *Flavobacterium* sp., *Acinetobacter calcoaceticus* Y, and *Mycoplana* sp. strain C, resulted from the enrichment. It was not possible to isolate a single species able to use 2-chloropropionamide as a carbon and energy source, although *Mycoplana* sp. strain C could use it as a nitrogen source. Enrichment for a stable community, rather than gene transfer, has occurred in this type of continuous culture, but further continuous selection may result in genetic rearrangement.

It may also be possible to obtain strains capable of novel supplementary metabolism by continuous selection with substrate limitation in the presence of a compound offering potential for such metabolic activity.

Gene cloning. Gene cloning is potentially the most powerful and specific technique for the recruitment of novel reactions to a wide range of

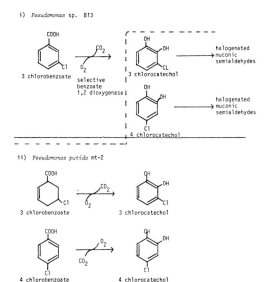

i) *Pseudomonas* sp. B13

3 chlorobenzoate → selective benzoate 1,2 dioxygenase → 3 chlorocatechol → halogenated muconic semialdehydes

4 chlorocatechol → halogenated muconic semialdehydes

ii) *Pseudomonas putida* mt-2

3 chlorobenzoate → 3 chlorocatechol

4 chlorobenzoate → 4 chlorocatechol

FIG. 3. Growth of *P. putida* mt-2, *Pseudomonas* sp. strain B13, and a resulting hybrid on halogenated aromatic compounds. The hybrid, designated by the area between the dotted lines, combines the ability of *Pseudomonas* sp. strain B13 to metabolize 3-chlorocatechol and 4-chlorocatechol with the broad-specificity benzoate 1,2-dioxygenase encoded by the TOL plasmid, allowing growth on 4-chlorobenzoate.

microorganisms. Although recombinant DNA techniques are most advanced in *Escherichia coli*, it is now possible to clone into most gram-negative bacteria. Early broad host-range vectors for gram-negative bacteria were based on plasmids such as RP4 and R300.B. More recently, several laboratories have produced families of vectors based on the plasmid Sa and have shown them to be functional in a variety of *Enterobacteriaceae, Rhizobiaceae*, and *Pseudomonadaceae*. Similarly, cloning systems are now available for a number of gram-positive bacteria, yeasts, fungi, and algae.

Gene cloning techniques are being used to analyze and manipulate the genes coding for the degradation of aromatics by soil bacteria, especially by the pseudomonads. Derivatives of R300.B have been used to clone restriction fragments from plasmids which specify the degradation of chemicals such as toluene, xylene, 3-chlorobenzoate, and 2,4-dichlorophenoxyacetate (5). A fragment of the TOL plasmid, coding for the catechol 2,3-oxygenase gene, has been cloned and introduced into both *Pseudomonas* strains and *E. coli*, and thus has demonstrated the possibility of manipulating these degradative genomes. When introduced into *P. putida* on the plasmid PKT502, the yield of enzyme is improved two- to fourfold, and introduction into *E.*

coli on the plasmid PKT503 leads to a further threefold increase. Experiments such as these illustrate that the powerful gene cloning techniques developed in *E. coli* may in time be adapted to other organisms.

Exploitation of broad-host-range plasmids in the genetic manipulation of industrially important bacteria is elegantly illustrated by the introduction of the *E. coli* gene for glutamate dehydrogenase, carried on RP4 and R300.B derivatives, into the single cell protein organism *Methylophilus methylotrophus* (29). The wild-type organism uses a two-stage pathway, dependent on glutamine synthetase and glutamate synthase (GOGAT) to catalyze the conversion of ammonia and 2-oxoglutarate into glutamate.

$$\text{glutamate} + \text{ATP} + \text{NH}_3 \xrightarrow[\text{synthetase}]{\text{glutamine}}$$

$$\text{glutamine} + \text{ADP} + \text{P}_i$$

$$\text{glutamine} + \alpha\text{-ketoglutarate} + \text{NADPH} +$$

$$\text{H}^+ \xrightarrow[\text{synthase}]{\text{glutamate}} 2 \text{ glutamate} + \text{NADP}^+$$

However, the novel recombinant employs the one-step reaction catalyzed by glutamate dehydrogenase.

$$\text{NH}_3 + \alpha\text{-ketoglutarate} + \text{NAD(P)H} + \text{H}^+ \rightarrow$$
$$\text{L-glutamate} + \text{NAD(P)}^+ + \text{H}_2\text{O}$$

The overall effect of the manipulation is that the organism uses one less ATP molecule per glutamate molecule formed, thus improving the yield and the economics of the process. Of course, these alterations probably could not have been made either by prolonged continuous selection or by a conventional mutagenic approach, since the novel reaction was not specified by an enzyme present in the original host. It has been suggested that *M. methylotrophus* evolved with the GOGAT system because of a selective pressure which required that the organisms should possess a highly effective, though energetically expensive, method of scavenging for ammonia in ammonia-deficient environments.

Protoplast fusion. The protoplast fusion technique has been used in *Streptomyces* species as an approach to increased antibiotic production and biosynthesis of new types of molecules. Fusion techniques can greatly increase the frequency of genetic recombination within a particular species, thus permitting rapid random screening for new phenotypes.

Fleck (4), for instance, used protoplast fusion to create a variant of the macrolide violamycin by fusing a mutant of the producer organism with a mutant of a turimycin producer. The new compound produced by the interspecific recombinant was called iremycin. Hamlyn and Ball (7) used the fusion of fungal protoplasts to improve the yield of a secondary metabolite. In this case, the recombinant produced 40% more cephalosporin C than the parent.

The techniques of protoplast fusion have much improved over the past 10 years, particularly as a result of the use of polyethylene glycol as a fusagen, the use of lysozyme to generate spheroplasts, and the use of hypertonic media for the regeneration of protoplasts. It will be interesting to see whether more recent fusion methods, such as electrofusion, give significant improvements in the technology. The equipment developed by Zimmerman and Scheurich (32) is reported to fuse similar cells into giant cells or unrelated cells into hybrids and is claimed to be more efficient than other fusion methods.

BIOCHEMICAL AND PROCESS CONSIDERATIONS

Process optimization is, in general, outside the scope of this paper, but it is self-evidently wise to attempt to identify important process constraints, e.g., the likely operating temperature or necessary tolerance to products or solvents, before embarking on recruitment. Sometimes it is possible to identify a biochemical basis for a strategy. For example, isolation from carbon-rich environments such as sugar refinery or paper and pulp mill effluents (14), where only small amounts of nitrogenous substrates are available, may be expected to encourage selection of strains showing overproduction of microbial polysaccharides or surfactants.

On occasion, process considerations are fundamental to the initial acquisition of an activity. A recent example (1) concerns microbial esterases, whose natural role is to hydrolyze esters into alcohols and organic acids. These enzymes can, however, be exploited for the synthesis of flavor and fragrance esters by use in organic solvent systems with low water activity, thereby completely reversing the biological reaction.

CONCLUDING REMARKS

Although the full potential of genetic engineering techniques has yet to be realized in the current context, more conventional approaches are likely to remain important for some time in recruiting microorganisms for practical exploitation. The better our understanding of interactions among different microorganisms in the environment and between them and their environment, the more intelligent may be our approach to the acquisition of novel activities. Although modern genetic engineering will no doubt play an increasingly important role in this quest, careful consideration of ecological factors and isolation strategies may sometimes obviate the necessity to adopt this more sophisticated approach.

Exploitation of natural genetic exchanges between organisms either in situ or in laboratory systems is also likely to become increasingly important. Interactions between microorganisms and surfaces, be they biotic or abiotic, have important implications for many of the approaches discussed above, especially because such interactions encourage exchange of information. Indeed, this may well be the main factor involved in successful in situ enrichment. This is an area that warrants increasing research effort.

LITERATURE CITED

1. **Anonymous.** 1983. Enzymatic esterification: an effective alternative to chemical synthesis. Biotechnol. News **3**:3.
2. **Anson, J. G., G. Mackinnon, and I. J. Higgins.** 1983. Plasmid-encoded degradation of aniline by a *Pseudomonas* species. Soc. Gen. Microbiol. Q. **10**:M15.
3. **Baro, J. A., and J. W. Deming.** 1983. Growth of 'black smoker' bacteria at temperatures of at least 250°C. Nature (London) **303**:423–426.
4. **Fleck, W. F.** 1979. Genetic approaches to new streptomycete products, p. 117–122. *In* O. K. Sebek and A. I. Laskin (ed.), Genetics of industrial microorganisms. American Society for Microbiology, Washington, D.C.
5. **Franklin, F. C. H., M. Bagdasarian, and K. N. Timmis.** 1981. Manipulation of degradative genes of soil bacteria, p. 109–130. *In* T. Leisinger, A. M. Cook, R. Hütter, and J. Nüesch (ed.), Microbial degradation of xenobiotics and recalcitrant compounds. Academic Press, Inc., New York.
6. **Gilbert, P. D., and I. J. Higgins.** 1978. The microbial degradation of crude mineral oils at sea. J. Gen. Microbiol. **108**:63–70.
7. **Hamlyn, P. F., and C. Ball.** 1979. Recombination studies with *Cephalosporium acremonium*, p. 185–191. *In* O. K. Sebek and A. I. Laskin (ed.), Genetics of industrial microorganisms. American Society for Microbiology, Washington, D.C.
8. **Hartman, J., W. Reineke, and H. J. Knackmuss.** 1979. Metabolism of 3-chloro-, 4-chloro-, and 3,5-dichlorobenzoate by a pseudomonad. Appl. Environ. Microbiol. **37**:421–428.
9. **Higgins, I. J., D. J. Best, and R. C. Hammond.** 1980. New findings in methane-utilizing bacteria highlight their importance in the biosphere and their commercial potential. Nature (London) **286**:561–564.
10. **Higgins, I. J., P. D. Gilbert, and J. Wyatt.** 1981. Microbial degradation of oil in the environment. Studies Environ. Sci. **9**:65–77.
11. **Ishikura, T., M. Nishida, K. Tanno, N. Miyachi, and N. Ozaki.** 1968. Microbial production of salicylic acid from naphthalene. Isolation of the high yield strain and the flask scale culture. Agric. Biol. Chem. **32**:12–20.
12. **Kelly, D. P.** 1978. Microbial ecology, p. 12–27. *In* K. W. A. Chater and H. J. Sommerville (ed.), The oil industry and microbial ecosystems. Heyden, London.
13. **Kitai, A., H. Tone, T. Ishikura, and A. Okai.** 1968. Microbial production of salicylic acid from naphthalene (II). J. Ferment. Technol. **46**:442–451.
14. **Lawson, C. J., and I. W. Sutherland.** 1978. Polysaccharides, p. 328–392. *In* A. H. Rose (ed.), Primary products of metabolism, Economic microbiology, vol. 2. Academic Press, Inc., New York.
15. **Linton, J. D., and J. C. Buckee.** 1977. Interactions in a methane-utilising mixed bacterial culture in a chemostat.

J. Gen. Microbiol. **101**:219–225.

16. **Nakayama, K.** 1982. The breeding of amino acid producing microorganisms, p. 473–485. *In* V. Krumphanzl, B. Sikyta, and Z. Vanek (ed.), Overproduction of microbial products. Academic Press, Inc., New York.

17. **Omori, T., and K. Yamada.** 1969. Studies on the utilisation of hydrocarbons by microorganisms. Oxidation of *m*-xylene and pseudocumene by *Pseudomonas aeruginosa*. Agric. Biol. Chem. **33**:979–985.

18. **Perry, J. J.** 1979. Microbial cooxidations involving hydrocarbons. Microbiol. Rev. **43**:59–72.

19. **Quayle, J. R.** 1980. Historical perspectives, p. 1–9. *In* D. E. F. Harrison, I. J. Higgins, and R. J. Watkinson (ed.), Hydrocarbons in biotechnology. Heyden, London.

20. **Reanney, D. C., P. C. Gowland, and J. H. Slater.** 1983. Genetic interactions among microbial communities. Symp. Soc. Gen. Microbiol. **34**:379–421.

21. **Reineke, W., and H. J. Knackmuss.** 1980. Hybrid pathway for chlorobenzoate metabolism in *Pseudomonas* sp. B12 derivatives. J. Bacteriol. **142**:467–473.

22. **Slater, J. H., and A. T. Bull.** 1982. Environmental microbiology: biodegradation. Philos. Trans. R. Soc. London Ser. B **197**:575–597.

23. **Stirling, D. I., and H. Dalton.** 1979. The fortuitous oxidation and cometabolism of various carbon compounds by whole-cell suspensions of *Methylococcus capsulatus* (Bath). FEMS Microbiol. Lett. **5**:315–318.

24. **Stirling, L. A., R. J. Watkinson, and I. J. Higgins.** 1977. Microbial metabolism of alicyclic hydrocarbons: isolation and properties of a cyclohexane-degrading bacterium. J.

Gen. Microbiol. **99**:119–125.

25. **Taylor, F., and I. J. Higgins.** 1983. The isolation and characterisation of a strain of *Nocardia minima* capable of growth on *o*-xylene. Soc. Gen. Microbiol. Q. **10**:M15.

26. **Tryhorn, S. E., and I. J. Higgins.** 1980. Metabolism and biotransformation of terpenoid hydrocarbons by soil bacteria. Soc. Gen. Microbiol. Q. **7**:95.

27. **Tsukamoto, Y., S. Nomonura, and H. Sakai.** 1975. Formation of *cis-p*-methan-1-ol from *p*-menthane by microbial hydroxylation. Agric. Biol. Chem. **39**:617–620.

28. **Wilkinson, T. G., H. H. Topiwala, and G. Hamer.** 1974. Interactions in a mixed bacterial population growing on methane in continuous culture. Biotechnol. Bioeng. **19**:41–59.

29. **Windass, J. D., M. J. Worsey, E. M. Pioli, D. Pioli, P. T. Barth, K. T. Atherton, E. C. Dart, D. Byrom, K. Powell, and P. J. Senior.** 1980. Improved conversion of methanol to single cell protein by *Methylophilus methylotrophus*. Nature (London) **287**:396–401.

30. **Wolfe, R. S., and I. J. Higgins.** 1979. The microbial production and utilisation of methane—a study in contrasts, p. 267–353. *In* J. R. Quayle (ed.), International review of biochemistry—series II, Microbial biochemistry. MTP Press, Lancaster.

31. **Yamada, K.** 1972. Japan's most advanced industrial fermentation technology and industry. International Technical Information Institute, Tokyo.

32. **Zimmermann, H., and P. Scheurich.** 1981. High-frequency fusion of plant protoplasts by electric fields. Planta **151**:26–32.

Pelletization of Anaerobic Sludge in Upflow Anaerobic Sludge Bed Reactors on Sucrose-Containing Substrates

L. W. HULSHOFF POL, J. DOLFING,† K. van STRATEN, W. J. de ZEEUW, and G. LETTINGA

Agricultural University, Department of Water Pollution Control, 6703 BC Wageningen, The Netherlands

Anaerobic wastewater treatment is being utilized more often because of the necessity for low-energy processes and the development of systems with effective sludge retention. Conventional anaerobic sludge and manure digestion systems which are completely mixed have nearly identical solids retention times (SRT) and hydraulic retention times (HRT). In contrast, the newly developed anaerobic wastewater treatment processes have SRT that are 10 to 100 times longer than the HRT. As a result, the required reactor volume can be significantly reduced, which makes the process economically more attractive.

Basically, two mechanisms provide the desired high sludge retention: attachment of biomass to fixed or mobile surfaces and settlement of sludge. Most of these new processes are based on attachment. In the *anaerobic filter* (21; J. C. Young and M. F. Dahab, IAWPR Specialized Seminar on Anaerobic Treatment of Waste Water in Fixed Film Reactors, Copenhagen, 1982), the reactor is filled with a packing material upon which the sludge adheres. Besides the attachment, an additional retention is obtained by entrapment of sludge in the interstices of the packing material. In *fixed film reactors* (9, 17–19), which can be operated either downflow or upflow, the sludge retention is achieved through attachment of the microorganisms to a fixed surface (including the wall of the reactor). The *expanded bed reactor* (15, 16) and the *fluidized bed reactor* (1, 2, 5) are supplied with small support particles upon which a bacterial film is formed. The expansion or fluidization of the sludge bed provides a good contact between substrate and biomass. In the *upflow anaerobic sludge bed (UASB) reactor* (13) the biomass retention is obtained via settlement of the sludge with the aid of an internal settling compartment.

With a long SRT, high hydraulic loading rates can be obtained. As a result, anaerobic treatment of more dilute wastewaters becomes attractive. The treatment of domestic sewage has already been the subject of extensive studies (8, 12). The full-scale application of high-rate anaerobic treatment processes has been mainly confined to the UASB concept, e.g., for sugar beet, potato processing, corn and potato starch, and distillery wastes. Excellent treatment results have been obtained which can be attributed primarily to the formation of granular (pelletized) sludge with excellent properties (sludge volume index = 12 to 20 mg/g) and a high specific activity. The average diameter of the granules is 1 to 2 mm, with a maximum of approximately 5 mm.

Pelletization is not restricted to methanogenic processes. It was observed in laboratory-scale upflow sludge bed reactors with denitrification (10), acidification (23), and the anaerobic fermentation of glucose under thermophilic conditions (20). It is now generally believed that, in general, in any bioreactor in which the substrate is continuously added in an upflow manner the biomass will grow in a granular form. However, each specific bioconversion process will have its own set of conditions necessary for granulation. In a few instances, granulation will proceed poorly, if at all. Generally, these waste waters are either too dilute (domestic sewage) or have a very high content of suspended solids (e.g., slaughterhouse waste), or they contain inhibiting or toxic compounds. However, in many cases even these waste waters can be efficiently treated in a UASB reactor with a flocculant sludge bed, although at lower maximum loading rates as compared with the granular sludge bed system.

MATERIALS AND METHODS

The granulation experiments were performed in 23.5-liter UASB reactors (4; L. W. Hulshoff Pol et al., IAWR Specialized Seminar on Anaerobic Treatment of Waste Water, Copenhagen, 1982). The reactors were inoculated with 300 g of volatile suspended solids (VSS) in the form of digested sewage sludge from a municipal sewage treatment plant. Two sucrose-volatile fatty acids (VFA) combinations were tested, both with a total chemical oxygen demand (COD) of ±3,000 mg of O_2 per liter (Table 1). The addition of other nutrients (N, P, and S) was based on an assumed growth yield of 5% and 20% (as COD) for experiments 1 and 2, respectively, so biomass growth would be carbon limited. Yeast extract (100 mg/liter) was added to enrich the

† Present address: Laboratory of Microbiology, Agricultural University, Wageningen, The Netherlands.

TABLE 1. Composition of the sucrose-VFA mixtures[a]

Component	Expt 1, 10% sucrose COD plus 90% VFA COD	Expt 2, 95% sucrose COD plus 5% VFA COD
Acetate	1,270	70
Propionate	1,430	80
Sucrose	300	2,850
Yeast extract	100	

[a] All figures are expressed as COD (milligrams of O_2 per liter), except yeast extract (milligrams per liter).

substrate of experiment 1, and 4 ml of a trace element solution, described by Zehnder et al. (22), was added to both substrates. To maintain the pH of the reactor of experiment 2 above 6.5, the substrate was buffered with an $NaHCO_3$ solution.

Flow rate, pH gas production (with a wet test gas meter), and VFA concentrations in the effluent (gas-liquid chromatography) were monitored in the reactor every day. Once a treatment efficiency of 85% (experiment 1) and 75% (experiment 2) was reached, the loading rate was increased. The amount of VSS in the reactor was calculated from sludge samples collected at five different heights in the reactor. The amount of washed-out solids was determined with the aid of an external overdesigned settler after scum layers had been broken down with an antifoam agent (Structol no. 20548/9, Schill & Seilacher, Chemische Fabrik). The biomass (Y) was calculated on the basis of the equation:

$$Y = \frac{1.4 \text{ g of biomass produced (as VSS)}}{\text{g of converted COD}}$$

The amount of biomass produced was obtained by integrating the sludge profiles and the biomass washout measurements. The amount of biomass was multiplied by 1.4 to calculate the sludge COD.

Sedimentation characteristics of the sludge were determined with a sedimentation balance (Sartorius 4620). Sludge was examined with epifluorescence and phase-contrast microscopy. The macroscopic morphology of Kaiser's glycerin-gelatin immobilized granules was investigated with a stereomicroscope.

Activity of the sludge was determined by suspending small amounts of sludge (0.1 to 0.5 g) in a phosphate-bicarbonate buffer (pH 7) in 130-ml serum vials under CO_2 atmosphere. Methane was collected in a pressure-lock syringe and determined by gas chromatography.

The results of experiments in which we investigated the effect on the granulation process of adding small quantities of granular sludge to the standard inoculum (i.e., digested sewage sludge)

and the effect of the sludge loading rate and of various NH_4^+ and Ca_2^+ concentrations were as follows (Hulshoff Pol et al., IAWPR Specialized Seminar, 1982).

(i) Addition of granules at the time of start-up affects the nature of the granules that are formed. Two main types of granules can be distinguished. When small amounts of rod granules were added to the inoculum, the granular sludge bed that developed consisted mainly of rod granules, despite the fact that the conditions were such that without this extra inoculum filamentous granules would have developed on digested sewage sludge as a seed. Type 1 ("rod-granules") consists mainly of rod-shaped bacteria. Type 2 ("filamentous granules") is mainly made up of filamentous bacteria. The predominant bacterium in both types of granules resembles *Methanothrix soehngenni* (7).

(ii) Long periods of underloading promoted the formation of a filamentous type of sludge. Since this sludge settles very poorly, it will lead to an undesirable level of sludge washout.

(iii) High NH_4^+-N concentrations had a distinct detrimental effect. At 1,000 mg/liter, granulation proceeded poorly, and the time needed to start the process was roughly twice as long.

(iv) The experiments with various Ca^{2+} concentrations revealed a positive effect on the granulation up to a concentration of approximately 150 ppm. The effect of even higher concentrations (>500 ppm) is still uncertain, as $CaCO_3$ and $Ca_3(PO_4)_2$ may precipitate. It was found that the biomass attached to fine, poorly settling $CaCO_3$ particles, thus enhancing the washout of active (newly formed) biomass and consequently lowering the sludge retention of the system.

RESULTS

Experiment 1. The data on the most important process parameters are presented in Fig. 1. In 95 days of continuous operation, the space loading rate could be increased up to 14.1 g of COD liter^{-1} day^{-1}. The sludge substrate utilization rate increased sharply around day 30. After this sharp increase, the loading rate remained fairly constant at approximately 1.6 g of COD per g of VSS per day. The biomass activity was about 1.45 g of COD per g of VSS per day during this period. With the sudden increase in the organic load, the propionic acid concentration increased in the effluent while acetic acid remained low. The sharp increase in sludge load occurred at a space loading rate of 5.5 g of COD liter^{-1} day^{-1} and a dilution rate of 1.83 day^{-1}. The propionic acid concentration remained high for 2 weeks. As observed previously (6), after completion of the adaptation to propionic acid, subsequent

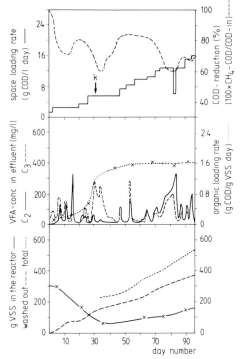

FIG. 1. Results obtained with the start-up of the reactor with 10% sucrose COD plus 90% VFA COD as feed (k indicates the day that the first granulation was observed).

liter of CH_4 per g of VSS per day at a dilution rate of 5.33 day^{-1}. The biomass yield (Y) was ±0.039 g of VSS COD per g of COD removed.

Experiment 2. The initial start-up of experiment 2 proceeded slower than in the previous experiment (Fig. 2). The space loading rate could be increased up to ±5 g of COD $liter^{-1}$ day^{-1} within 30 days. However, not until day 70 could the loading rate be substantially increased up to about 17 g of COD $liter^{-1}$ day^{-1}. The amount of retained biomass in the reactor showed a gradual decrease from an initial level of 12.5 g to 8.3 g of VSS per liter on day 55. Beyond day 55, the biomass concentration in the reactor gradually increased to 19.6 g/liter on day 95, mainly as a result of the formation of a granular sludge. The granules were first seen on day 28, with a space loading rate of 4 g of COD $liter^{-1}$ day^{-1}, a dilution rate of 1.33 day^{-1}, and a sludge load of 0.3 g of COD per g of VSS per day. The average pH value of the mixed liquor in the reactor during the experiment was 6.7. Although there were periods of a limited overloading, particularly between day 30 and day 70, the effluent COD mainly consisted of acetate and propionate. No sucrose could be detected in the

occasional periods of overloading showed an incomplete degradation of both acetate and propionate. This is particularly the case during the period from day 75 to day 85 in Fig. 1.

The total amount of biomass retained within the reactor decreased during the first 30 days from 12.8 to 3.2 g of VSS per liter and then increased slightly, immediately after the first granules could be visually observed in the sludge bed. At the termination of the experiment the biomass concentration was 6.3 g of VSS per liter. Beyond day 30, the growth of biomass exceeded the washout, although a significant fraction of rather flocculent sludge still remained in the reactor. The granules formed settled poorly and were filamentous in nature. They did not form a separate granular sludge bed, as was observed in similar experiments with only acetate and propionate as substrate (Hulshoff Pol et al., IAWPR Specialized Seminar, 1982). Most of the granules were formed by attachment of the filamentous bacteria to biologically inert (support) particles originating from the seed sludge. The "loose" structure of the granules, which had a diameter of approximately 3 mm, remained unchanged during the process. The maximal specific methane production rate was 0.59

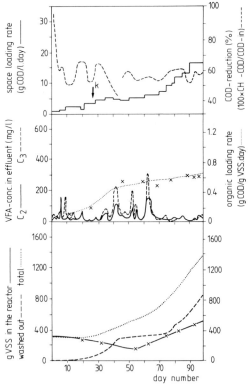

FIG. 2. Results obtained with the start-up of the reactor with 95% sucrose-COD plus 5% VFA COD as feed (k indicates the day that the first granulation was observed).

effluent during this period. The specific activity, and consequently the sludge load, did not increase as sharply as in the previous experiment; however, in contrast to experiment 1, the specific activity improved further after the total amount of biomass in the reactor started to increase again, viz., the sludge load could be increased from approximately 0.45 kg of COD per kg of VSS per day on day 40 to 0.62 kg of COD per kg of VSS per day on day 95. The maximal specific CH_4 production rate was 0.5 g of CH_4 COD per g of VSS per day. Activity tests in 0.5-liter batch serum bottles with a mixture of 600 ppm of C_2, 600 ppm of C_3, and 600 ppm of C_4 as substrate showed significantly higher values, up to 0.95 g of COD converted per g of VSS per day. The sludge yield, estimated on the basis of sludge concentration measurements over the height of the reactor, was 0.13 g of VSS COD per g of COD removed. The granules formed were up to 5 mm in diameter and were pale yellow in appearance. They consisted mainly of long filaments with small numbers of cocci and diplococci. The granules were fairly stable, but the biggest granules could fall apart into smaller particles with a diameter of approximately 1 mm. During the experiment, a distinct shift was observed in the composition of the granules. The number of filaments diminished, whereas a remarkable increase in the number of mobile rods was observed. Moreover, in this case the granules were not formed by attachment on inert support particles, as was observed with the granules in experiment 1. At the termination of the experiment, relatively stable granules prevailed, which could be kept fairly unchanged (unhydrolyzed) after storage for 6 weeks at 4, 20, and 30°C.

Data on methanogenic activity with the sludges from experiments 1 and 2 are presented in Table 2. For both sludges, the ratios between the methanogenic activities with formate, acetate, and propionate change between day 70 and day 96. The methanogenic activities with acetate and formate for the sludge cultivated on the 10% sucrose solution were about 20% higher over this period than the corresponding activities of the 90% sucrose-grown sludge.

DISCUSSION

In recent years, several authors (3, 11, 14, 23) have proposed separating the acidification step from the methanogenesis step. It has been claimed that this type of phase separation would result in a more stable process, even for easily hydrolyzable substrates. According to these views, a constant composition of the effluent from the acidification step would lead to a better adaptation of the bacterial population in the methane reactor and consequently to a more stable operation of methanogenesis. As a result of phase separation, therefore, a significantly higher maximal loading capacity of the methane reactor might be achievable. However, the construction of a separate (acidification) reactor increases investment costs considerably; thus, studies have been done to assess the effect of partial acidification on the performance of the methane reactor. Such experiments have been done with glucose plus VFA feed solutions (A. Cohen, thesis, 1982). Generally, such a partial acidification already occurs in practice to a considerable extent in a simple equalization tank. It was found in these studies that by replacing 13% of the glucose COD by VFA COD the specific sludge activity is increased significantly. Moreover, by using mixtures of VFA and glucose as substrate, the granular seed sludge in the methane reactor could be preserved better than in one-phase treatment with pure glucose solutions. However, at high sludge loading rates a rather gelatinous sludge was formed with the partially acidified glucose solution (Cohen, thesis, 1982), leading to the formation of voluminous layers surrounding the original granules. Under such conditions, a sharp increase in sludge washout was observed. These findings correspond with recent observations made at our laboratory (J. Makkink, unpublished data) in experiments with a dilute (1,000 mg of COD per liter) sucrose solution and with a granular seed sludge cultivated on corn starch waste. At high loading rates a strong increase in the sludge washout occurred as a result of the formation of distinct voluminous layers of newly produced bacterial biomass on the granules. In

TABLE 2. Methanogenic activities of the sludges of experiments 1 and 2 on various substrates

Substrate	Activities of sludge (g of CH_4 COD per g of VSS per day)							
	Expt 1				Expt 2			
	Day 40	Day 70	Day 88	Day 96	Day 40	Day 70	Day 88	Day 96
H_2			1.34	0.82			1.17	1.62
Formic acid	0.33	1.27	0.90	1.0	0.22	0.85	0.86	0.71
Acetic acid	0.84	0.83	0.90	0.68	0.27	0.57	0.76	0.59
Propionic acid	0.19	0.68	0.45	0.41	0.07	0.34	0.30	0.13

both examples cited here, the reactor was inoculated with granular sludge being cultivated on beet sugar waste and corn starch waste, respectively. As each substrate leads to a specific composition of bacterial populations, it seems that a sudden change in substrate cannot always be fully accommodated. This seems to be particularly true for granules being cultivated on mainly acidified effluents. The growth in or on the granules from a mainly carbohydrate-containing substrate will predominantly consist of acid-forming organisms. This newly formed attached bacterial matter is rather voluminous and consists mainly of filamentous bacteria. In addition, gelatinous exopolymers have been produced. Both these phenomena lead to a severe decrease in sludge settleability and consequently to an increased washout of bacterial sludge. This phenomenon is due partially to the fact that gas remains within the granules. Studies are presently under way on the nature of the observed voluminous attached biofilms on the granules.

On the other hand, experiment 2 described in this paper, performed with a 95% sucrose COD plus 5% VFA COD solution and using digested sewage sludge as inoculum, clearly demonstrated that a highly active granular sludge can be cultivated on a mainly carbohydrate waste. This well-adapted granular sludge apparently consisted of a balanced microenvironment involving all the microorganisms required for the conversion of the substrate into methane. No severe sludge washout was observed at high loading rates, and once the granules were developed no real change in granule morphology occurred.

The granules formed in the experiments with the 10% sucrose COD plus 90% VFA COD substrate resembled the filamentous granules obtained previously (Hulshoff Pol et al., IAWPR Specialized Seminar, 1982) on 100% VFA solutions. However, unlike the experiments performed under similar process conditions with a pure VFA feed, no clear segregation between the granular and flocculent sludge could be achieved despite the fact that the granules appeared approximately 10 days sooner than in the pure VFA experiment. The difference in settleability between the dispersed flocculant and the filamentous granular sludge was too small for an effective separation under the loading conditions applied. As a result, growth was insufficiently delegated to the granular sludge. This was particularly true if the growth of the flocculent filamentous sludge was kinetically in favor of the growth of the organisms present in the granules.

Experiment 2 clearly demonstrated that the conditions for granulation are significantly better for mainly sucrose wastes, provided a proper start-up regime is followed. Compared with experiment 1, granulation visually occurred sooner; despite that fact, the start-up time was distinctly longer. Some time was required for the formation of the appropriate methanogenic flora. The granules formed were up to 5 mm in size and exhibited excellent settling characteristics. At the termination of the experiments, hardly any flocculent sludge was left in the reactor. The reactor volume was almost completely filled with the granular sludge. The methanogenic activity of the sludge was approximately 0.3 liter of CH_4 per g of VSS per day, which corresponds well with the values reported by Cohen (thesis, 1982) for the one-step digestion of a glucose-VFA mixture (viz., 0.2 to 0.4 liter of CH_4 per g of VSS per day). According to Zoetemeyer et al. (23), hydraulic retention times of 1 and 6 h, respectively, would be required for the cultivation of granules of acid-forming sludge on 1% and 5% glucose-monohydrate solutions. However, experiment 2 demonstrated that granulation with a very similar substrate in a one-step digestion process occurred at considerably longer hydraulic retention times (<18 h).

The sludge retention figures observed in both the experiments resembled the pattern found with pure VFA substrates. However, there was a clear difference between experiments 1 and 2 as far as the minimum values of the amount of retained sludge at the end of the "washout phase." In experiment 2, the minimum amount of sludge retained was 8.3 g of VSS per liter, whereas it was only 3.2 g of VSS per liter in experiment 1. The latter value corresponds closely to that found in pure VFA-fed start-up experiments. The high minimum value in experiment 2 can be attributed to the good granulation in that experiment. This is also clearly reflected in the total amount of solids retained in the reactor at the termination of the experiment. The measured sludge profiles of both reactors revealed that the biomass concentrations at the end of the investigations were approximately 9 g and 40 g of VSS per liter (at day 130) for experiments 1 and 2, respectively. Experiments with 100% VFA mixtures under similar conditions resulted in granular sludge beds with biomass concentrations which never exceeded 30 g of VSS per liter. Thus, an obvious advantage of the use of soluble carbohydrates as substrate is that a considerable level of sludge retention can be obtained. So far, it is not possible to predict the maximum achievable organic load with a 95% sucrose solution. However, assuming that the sludge retention does not change and also the specific activity does not further increase significantly at increased loading rates, the maximal achievable space load can be estimated at approximately 35 kg of COD m^{-3} day^{-1}.

The excellent settling characteristics of the sludge formed in experiment 2 are also reflected in the sedimentation data. The time required for settling out of half of the sludge is 7.5 min for digested sewage sludge, 2.5 min for the sludge cultivated in experiment 1, and 6.2 min for the sludge of experiment 2. Granular sludge cultivated on beet sugar wastes gave a value of 0.1 min.

The activity tests indicated that the period needed for adjustment of the seed sludge to the specific substrate lasted at least 40 days. However, the composition of the microflora as reflected in its methanogenic activity on different substrates remained fairly constant after 70 days. Comparison of the activities of both sludges suggests that about 20% of the biomass grown on the 90% sucrose plus 10% VFA feed obtained its energy from the fermentation of sucrose to VFA. This suggestion is supported by microscopic observations, which revealed that the microflora mainly consisted of autofluorescing, hydrogen-consuming methanogens and long filaments of a bacterium which resembled *Methanothrix soehngenii*, an obligate acetoclastic methanogen (7). These observations, together with other unpublished results of activity tests, suggest that such tests can be a useful tool in characterization of anaerobic sludge. The maximal methanogenic activity on a range of substrates provides information on the composition and the condition of the sludge examined.

CONCLUSIONS

By using a substrate containing 95% sucrose and 5% VFA (consisting of acetate and propionate), a considerable enhancement of granulation was achieved as compared with a 100% VFA substrate. The time needed for granulation was shorter, and larger ($\phi \leq 5$ mm), stable granules with good settling characteristics were formed. As a result, the amount of retained granular sludge in the reactor was significantly higher. Granulation occurred at much longer hydraulic retention times than found with separate acidification of glucose solutions. A comparison of the results of this experiment with investigations using similar substrates suggests that the granular sludge should not be cultivated on a substrate containing mainly VFA and should be changed to a more carbohydrate-rich substrate to prevent sludge washout due to the formation of voluminous growth around the VFA-adapted granules.

With a substrate containing 10% sucrose and 90% VFA mixture (acetate plus propionate), granular and flocculent sludge cannot be effectively separated. The granules contained a high fraction of filamentous organisms which were mainly attached to inert support particles.

LITERATURE CITED

1. **Binot, R. A., T. Bol, H. P. Naveau, and E. J. Nyns.** 1982. Biomethanation by immobilised fluidised cells, p. 211. IAWPR specialised seminar on anaerobic treatment of waste water in fixed-film reactors, 16–18 June 1982, Copenhagen, Denmark. International Association on Water Pollution Research, London.

2. **Boening, P. H., and V. F. Larsen.** 1982. Anaerobic fluidized bed whey treatment. Biotechnol. Bioeng. **24:**2539–2556.

3. **Cohen A., R. J. Zoetemeyer, A. van Deursen, and J. G. van Andel.** 1979. Anaerobic digestion of glucose with separated acid production and methane formation. Water Res. **13:**571–580.

4. **de Zeeuw, W. J., and G. Lettinga.** 1981. Acclimation of digested sewage sludge during start-up of an anaerobic sludge blanket (UASB) reactor, p. 39–47. *In* Proceedings of the 35th Industrial Waste Conference, Purdue University. Ann Arbor Science, Ann Arbor, Mich.

5. **Frostell, B.** 1982. Anaerobic fluidized bed experimentation with a molasses waste water. Process Biochem. **17:**37–40.

6. **Hulshoff Pol, L. W.** 1982. Vorming van anaëroob korrelslib. Report of Department of Water Pollution Control. Agricultural University, Wageningen (in Dutch).

7. **Huser, B. A., K. Wuhrmann, and A. J. B. Zehnder.** 1982. *Methanotrix soehngenii* gen. nov. sp. nov., a new acetotrophic non-hydrogen-oxidizing methane bacterium. Arch. Microbiol. **132:**1–9.

8. **Jewell, W. J., M. S. Switzenbaum, and J. W. Morris.** 1981. Municipal waste water treatment with the anaerobic attached microbial film expanded bed process. J. Water Pollut. Control Fed. **53:**482.

9. **Kennedy, K. J., and L. van den Berg.** 1982. Effects of temperature and over-loading on the performance of anaerobic fixed film reactors, p. 678–685. Proceedings of the 36th Industrial Waste Conference, Purdue University. Ann Arbor Science, Ann Arbor, Mich.

10. **Klapwijk, A., H. Smit, and A. Moore.** 1981. Biological fluidised bed treatment of water and wastewaters, p. 205–216. Ellis Horwood Ltd., Chichester.

11. **Kunst, S.** 1982. Untersuchungen zum anaeroben abbau polymerer Kohlenhydrate zur Optimierung der Versäuerungsstufe bei anaeroben Abwasserreinigungsanlagen. Veroeff. Inst. Siedlungswasserwirtsch. Univ. Hannover, Heft 54 (in German).

12. **Lettinga, G., R. Roersma, P. Grin, W. de Zeeuw, L. W. Hulshoff Pol, A. F. M. van Velsen, S. Hobma, and G. Zeeman.** 1981. Anaerobic treatment of sewage and low strength waste waters, p. 271–291. *In* D. E. Hughes et al. (ed.), Proceedings of the 2nd International Symposium on Anaerobic Digestion, Travemünde. Elsevier-Biomedical Press, Amsterdam.

13. **Lettinga, G., A. F. M. van Velsen, S. Hobma, W de Zeeuw, and A. Klapwijk.** 1980. Use of the Upflow Sludge Blanket (USB) reactor-concept for biological waste water treatment, especially for anaerobic treatment. Biotechnol. Bioeng. **22:**699–734.

14. **Pohland, F. G., and S. Ghosh.** 1971. Anaerobic stabilization of organic wastes, two phase concept. Environ. Lett. **1:**255–266.

15. **Switzenbaum, M. S., and S. C. Danskin.** 1982. Anaerobic expanded bed treatment of whey, p. 414–424. Proceedings of the 36th Industrial Waste Conference, Purdue University. Ann Arbor Science, Ann Arbor, Mich.

16. **Switzenbaum, M. S., and W. J. Jewell.** 1980. Anaerobic attached-film expanded bed reactor treatment. J. Water Pollut. Control Fed. **52:**1953.

17. **van den Berg, L., and K. J. Kennedy.** 1981. Support materials for stationary fixed film reactors for high rate methanogenic fermentations. Biotechnol. Lett. **3:**165–170.

18. **van den Berg, L., and C. P. Lentz.** 1979. Fixed film reactors for varying surface-to-volume ratios for the treatment of bean blanching waste, p. 319–325. Proceedings of the 34th Industrial Waste Conference, Purdue University.

19. **van den Berg, L., and C. P. Lentz.** 1980. Effects of film area-to-volume ratio, film support, height and direction of

flow on performance of methanogenic fixed film reactors, p. 1. Proceedings of the Anaerobic Filters Workshop, Argonne National Laboratory, 1–10 January, Howey in the Hills, Fla. Argonne National Laboratory, Argonne, Ill.

20. **Wiegant, W. M., and G. Lettinga.** 1982. Maximum loading rates and ammonia toxicity in thermophilic digestion. Sol. Energy Res. Dev. Ser. E **3:**238–244.

21. **Young, J. C., and P. L. McCarty.** 1969. The anaerobic filter for waste treatment. J. Water Pollut. Control Fed. **41:**R160.

22. **Zehnder, A. J. B., B. A. Huser, T. D. Brock, and K. Wuhrmann.** 1980. Characterization of an acetate-decarboxylating non-hydrogen-oxidizing methane bacterium. Arch. Microbiol. **124:**1–11.

23. **Zoetemeyer, R. J., A. J. C. M. Mathijssen, J. C. van der Heuvel, A. Cohen, and C. Boelhouwer.** 1981. Acidogenesis of soluble carbohydrate-containing wastewaters. Trib. CEBEDEAU **34:**444–465.

Relevance of Low-Nutrient Environments to Fermentation Process Design and Control

D. W. TEMPEST, O. M. NEIJSSEL, AND J. J. TEIXEIRA DE MATTOS

Laboratory of Microbiology, University of Amsterdam, 1018 WS Amsterdam, The Netherlands

Although environmental conditions prevailing in laboratory cultures tend to be far removed from those extant in many natural ecosystems, it is reasonable to suppose that all the properties which microorganisms can be made to express in these artificial situations must have been acquired in the course of evolution and must have some relevance to the growth and survival of these creatures in their normal habitats. It follows, therefore, that by studying the behavior of organisms in well-controlled laboratory culture systems, some insight might be obtained into the nature of the dominating influences that act upon these organisms in their natural surroundings and of the physiological properties appropriate to accommodating to such conditions.

This concept of a close relationship between ecology and microbial physiology (which nowadays is popularly termed "ecophysiology") is by no means new; indeed, it formed the basis of many classical microbiological investigations dating back to the beginning of this century. What is new (or newer) in this context is the application of continuous-culture techniques to such studies, and that is what we wish to concentrate upon here.

The current ecophysiological argument, as one might call it, rests on two propositions. The first is that, whereas biochemical research has tended to emphasize increasingly the unity of living processes, there nevertheless are important physiological differences between the free-living microbial cells and the cells of, say, higher animals. These differences are not simply, nor principally, those that distinguish procaryotes from eucaryotes, but derive from the fact that the cells which collectively make up the animal body spend the whole of their existence in a closely controlled environment. Generally, they seem to have only poorly developed powers of adaptability since if the animals' internal environment is caused to shift beyond certain narrow limits (of, say, pH, oxygenation, salts balance, and nutrients concentration) the cells cease to function and the animal dies. In contrast, the free-living microbial cells frequently must experience marked shifts of environment which they are quite powerless to control. Indeed, in a closed environment like that of a batch culture, microbes provoke through their own metabolism extensive shifts in the chemical environment; yet, seemingly, they can readily and rapidly accommodate to these changes, often without there being a perceptible effect on growth rate right up to the moment when some essential nutrient becomes totally depleted. This, we now know, they do by changing themselves, structurally and functionally; and to such an extent can these organisms change phenotypically (as it is said) that it is quite impossible to specify the precise structural and functional composition of any microorganism without reference to the environment in which it is growing (11).

As a corollary, one might add that if one did happen to possess a detailed knowledge of the precise properties expressed by some organism under a wide range of different environmental conditions, and if one were able to isolate from some natural environment these organisms in sufficient quantity to allow of their direct physiological characterization, then one might be able to assess the nature of the dominating environmental influences that were acting upon that natural population in situ. This approach to a study of natural ecosystems—that is, by the use of physiological indicators—has been adopted recently by some of our colleagues in Amsterdam with interesting results (32).

The second proposition that underlies the ecophysiological approach concerns the nature of the likely dominating environmental parameters that act upon microorganisms when growing in natural ecosystems. Because most microbes have a quite enormous growth potential, and since many grow exponentially, it is only necessary to use a desk calculator to convince oneself that microbes in their natural environment must rarely grow at their potentially maximum rate. Growth in most natural ecosystems must be highly constrained, and without going into detailed argument, one might confidently state that the principal constraint frequently must be nutritional. Natural environments frequently must contain insufficient essential nutrient(s) to allow microbial growth to proceed at its maximum rate; direct analyses of some aquatic ecosystems lend support to this conclusion (1). Hence, one might expect microorganisms in general to be adept at coping with extreme nutrient-limiting environments, and this is, of course, precisely the aspect of microbial physiology that can be effectively studied by using chemostat culture techniques. Moreover, since fermentation technologists are becoming increasingly interested in the use of continuous-flow culture systems for the large-scale production of microbial cells and their products, it is in this facet of microbial behavior (that is, their response to low-nutrient environments) that the interests of ecologists and industrial microbiologists become congruent.

PHYSIOLOGICAL RESPONSE TO LOW-NUTRIENT ENVIRONMENTS

If, as argued above, natural environments are commonly nutrient limited, then one might expect organisms to possess a range of special properties needed to deal with such conditions. Thus, one might expect that they would (i) possess a range of high-affinity uptake mechanisms for all the relevant limiting nutrients; (ii) possess appropriate control mechanisms to prevent excess nutrients from accumulating intracellularly to potentially toxic or traumatic levels; (iii) where possible, be able to rearrange their metabolism to circumvent bottlenecks imposed by the specific growth limitation; and (iv) be able to coordinately modulate the rates of synthesis of all their macromolecular components to allow balanced growth to proceed at a grossly submaximal rate. Evidence for each of these adaptive responses is now abundant in the published literature (for reviews, see 27, 28); we shall consider here only selected examples that are relevant in a biotechnological context.

Nutrient uptake and assimilatory systems. As stated, one might expect microorganisms in general to possess highly effective systems for the uptake and assimilation of all the potentially growth-limiting nutrients. Moreover, if these systems involve enzyme-mediated reactions (as they generally do), then it follows from Michaelis-Menten kinetics:

$$V = V_{max}\left(\frac{s}{K_m + s}\right) \text{ or } s = K_m\left(\frac{V}{V_{max} - V}\right)$$

that the scavenging capacity of the organisms for growth-limiting substrate (s) will be enhanced either by increasing the V_{max} of the system (i.e., by making more enzyme) or by synthesizing an alternative system with a lowered half-saturation value (K_m), or both. This is well illustrated with respect to the uptake of nutrients such as ammonia, glycerol, and glucose by various organisms.

Klebsiella aerogenes, in common with many other bacterial species, possesses two pathways of ammonia assimilation, one involving glutamate dehydrogenase, which has a low affinity for ammonia (apparent K_m = 5 mM), and the other implicating glutamine synthetase (with a 10-fold-higher affinity for ammonia) plus glutamate synthase (29). As expected from the above-mentioned consideration, this organism responds to low-ammonia environments by derepressing the synthesis of the high-affinity uptake system while simultaneously repressing glutamate dehydrogenase synthesis. In contrast, the yeast *Saccharomyces cerevisiae*, which lacks one of the components of the high-affinity pathway (glutamate synthase), is obliged to use the glutamate dehydrogenase reaction for ammonia assimilation even when the extracellular ammonia concentration is low. Not surprisingly, under ammonia-limiting conditions, this organism was found to increase its differential rate of synthesis of glutamate dehydrogenase some threefold; the specific activity increased from 620 to 1,730 mmol min^{-1} mg^{-1} of protein (2), thereby considerably enhancing its scavenging capacity for ammonia. From an industrial point of view, this latter observation is the more significant because it offers a strategy for maximizing enzyme production, that is, by selecting a strain, or making a mutant, that lacks a high-affinity uptake system and then exposing it to an appropriate low-nutrient environment.

A similar (though different) situation, which serves to reinforce the above point, exists with respect to the growth of *K. aerogenes* on glycerol. Here, the organism again possesses dual uptake and assimilatory mechanisms: glycerol kinase (with a high affinity for glycerol, K_m [apparent] = 1 to 2 μM) plus glycerol phosphate dehydrogenase, and glycerol dehydrogenase (with a low affinity for glycerol, K_m [apparent] = 20 to 40 mM) plus dihydroxyacetone kinase (9, 17, 19). With both reaction sequences, the glycerol is converted to dihydroxyacetone phosphate, but the reactions differ in two important respects: (i) in the affinity of the first enzymes for glycerol and (ii) in that the glycerol phosphate dehydrogenase is a flavoprotein in this organism and hence cannot function in the absence of respiratory chain activity. Thus, whereas the high-affinity glycerol kinase uptake and assimilatory pathway can function aerobically, and does so in glycerol-limited environments, it cannot function anaerobically. Nevertheless, *K. aerogenes* is able to grow readily in an anaerobic glycerol-limited chemostat culture, and as anticipated, in so doing it derepresses the synthesis of glycerol dehydrogenase to a high level (apparent specific activity of 1,410 mmol min^{-1} mg^{-1} of protein at D = 0.2 h^{-1}, pH 6.8, 35°C).

The fermentation of glycerol by *K. aerogenes*

is particularly interesting in that this organism possesses all the enzymes necessary to convert this compound solely to ethanol and formate (Fig. 1), and in terms of energetic efficiency (that is, moles of ATP formed *net* per mole of glycerol fermented) this route would be highly advantageous. Hence, one might expect organisms to use this pathway when growing anaerobically in a glycerol-limited (and, hence, putatively energy-limited) environment. However, such was not found; instead, the organisms carry out a tandem (disproportionation) reaction with two molecules of glycerol, one of which is dehydrated to β-hydroxypropionaldehyde while the other is oxidized to dihydroxyacetone. The β-hydroxypropionaldehyde then serves as electron acceptor, forming 1,3-propanediol, while the dihydroxyacetone is further oxidized (via its phosphorylated product) to acetate, ethanol, and formate (Fig. 1). Thus, the efficiency with respect to glycerol consumption is effectively halved (i.e., 50% of the glycerol consumed was excreted as 1,3-propanediol), and the energetic efficiency is lowered some 25%. The question therefore arises as to why these organisms behave in such an extravagant manner? The answer, we suggest, might be sought in the kinetics of the glycerol dehydrogenase reaction, as speci-fied above. Thus, for glycerol dehydrogenase to function effectively in the presence of low concentrations of glycerol, mechanisms must be available to rapidly remove the products of the reaction (that is, dihydroxyacetone and NADH₂). This is assured by the presence of high levels of activity of dihydroxyacetone kinase, glycerol dehydratase, and β-hydroxypropional-dehyde reductase (see 9).

So far, we have considered those cases in which organisms can synthesize alternative high- and low-affinity uptake systems, and those in which they can synthesize only a low-affinity system. There are yet other cases in which organisms are thought to possess only a high-affinity uptake system, and one such example is provided by the phosphoenolpyruvate-dependent glucose phosphotransferase system (glucose-PTS) of enteric bacteria (16). In common with other members of the *Enterobacteriaceae*, *K. aerogenes* possesses enzymes of the PTS that, collectively, have a high affinity for glucose (K_s [apparent] = 15 μM) (24), and though the enzymes of this system are synthesized constitutively, its overall activity as measured in vitro (14, 15) varies with the growth condition. As expected, glucose-limited cells have a substantial glucose-PTS activity, and this was found to

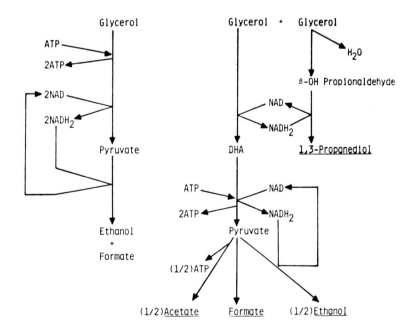

Net ATP gain/mol glycerol:

| 1.0 | 0.75 |

FIG. 1. Theoretical and actual pathways of glycerol fermentation by *K. aerogenes* growing anaerobically in chemostat culture.

increase progressively as the dilution rate was decreased and the glucose limitation became more stringent (Table 1). Inexplicably, however, K^+-limited cultures, growing in the presence of excess glucose, were found to possess a uniformly low glucose-PTS activity and yet consumed glucose at rates far in excess of corresponding glucose-limited cultures (Table 1). This suggested the presence, in K^+-limited *K. aerogenes*, of some alternative glucose-catabolizing system(s) that was at least as active as the glucose-PTS and which probably had a low affinity for glucose. The question was put beyond dispute when a PTS-deficient mutant was isolated that still would grow on glucose (O. M. Neijssel, D. W. Tempest, P. W. Postma, J. A. Duine, and J. Frank, FEMS Microbiol. Lett., in press). Unfortunately, this mutant was inherently unstable in chemostat culture, but we were nevertheless able to sustain it in a K^+-limited environment for 2 weeks before readily detectable glucose-PTS activity returned. During this period, it consumed glucose at a substantial rate, though at a lower rate than that of a corresponding culture of the wild-type organism (Table 1).

Hence, it was clear that *K. aerogenes* possesses, and can express under some conditions, a non-PTS-mediated glucose-metabolizing sys-

TABLE 1. Phosphoenolpyruvate-dependent glucose phosphotransferase (PT) activities and glucose uptake rates ($q_{glucose}$) of *K. aerogenes* NCTC 418 growing on glucose in carbon-, phosphate-, and potassium-limited chemostat cultures at different dilution rates[a]

Condition and dilution rate (h^{-1})	nmol min^{-1} mg^{-1} (dry wt) of cells	
	PT activity	$q_{glucose}$ (in situ)
Glucose-limited		
0.1	270	20
0.2	245	36
0.3	218	55
0.4	194	71
0.6	145	104
0.8	96	144
Phosphate-limited		
0.2	65	113
0.4	81	125
0.6	78	150
Potassium-limited		
0.2	38	177
0.3	45	234
0.4	82	250
Potassium-limited (PT-deficient mutant)		
0.1	4	84
0.2	12	120
0.3	8	160
0.4	10	200

[a] Data of O'Brien et al. (22) and Neijssel et al. (in press).

tem, but the mechanism involved initially remained elusive. Despite the obvious utility of mutants, problems of this type are more readily solved by adopting a physiological approach. Thus, much evidence indicates that a K^+ limitation imposes upon *K. aerogenes* a heavy demand for respiratory energy and, indeed, K^+-limited cultures respire at an exceedingly high rate (often in excess of 500 nmol min^{-1} mg^{-1} (equivalent dry weight) of cells (see 12, 13). The same is true for phosphate-limited cultures of this organism (20) and for ammonia-limited cultures growing in the presence of 1 mM 2,4-dinitrophenol (18). Under all these conditions, glucose is consumed at a high rate and the glucose-PTS activity is low (22). Significantly, much of the glucose consumed is only partially oxidized, and substantial amounts of gluconate and 2-ketogluconate accumulate in the culture fluids. Clearly, one had to look for a glucose dehydrogenase, but such could not be detected with either NAD(P) or 2,6-dichlorophenolindophenol as electron acceptor. However, recently a novel class of dehydrogenases (quinoproteins) have been described that have 2,7,9-tricarboxy-$1H$-pyrrolo-(2,3-f)-quinoline-4,5-dione as their prosthetic group (7) and can be detected by use of Wurster's Blue, but not 2,6-dichlorophenolindophenol or ferricyanide. Incubation of extracts of K^+-limited *K. aerogenes* with Wurster's Blue (0.1 mM) and glucose (20 mM) gave an immediate reaction (decoloration) that was not observed with fructose, mannitol, or gluconate. As predicted, the affinity of this enzyme for glucose is relatively low (K_m [apparent] = 5 mM), and its activity was found to be such as to largely account for the discrepancy between the overall rate of glucose metabolism and the glucose-PTS activity (Neijssel et al., in press and unpublished data). Thus, not only is there a low-affinity glucose-catabolizing mechanism present in *K. aerogenes* (note that it is not an uptake system per se), but in an ecophysiological context, one can rationalize its functioning.

Exocellular enzymes. Many examples could be cited of enzymes that function in substrate uptake or assimilation, or both, being synthesized in excessive amounts under relevant nutrient-limited chemostat growth conditions (see, for example, 6, 25). In some cases, like those of β-galactosidase and D-serine deaminase in *Escherichia coli*, hyperproducing strains (probably resulting from gene duplication) are selected that contain up to 25 times the enzyme levels present in the initial wild-type organisms (25). However, there are yet other ways in which the uptake of growth-limiting nutrients can be facilitated. Thus, changes in the surface structure of the cell may be invoked, leading to an improvement in their binding properties for growth-limiting nutrients (27, 28), or substances may be excreted that bind these nutrients, thereby promoting uptake by ancillary transport processes. One such process that is now well defined is the excretion of iron-chelating compounds (siderochromes) under conditions of limited iron availability (4, 8).

Perhaps of greater significance in this current era of biotechnology is the fact that some microorganisms not only secrete proteins into their envelope structures but also release selected ones into the growth environment. This aspect of microbial behavior is, of course, of considerable interest (and importance) with respect to the expression of foreign proteins in genetically engineered organisms. For clearly, though high levels of expression and accumulation of foreign proteins can be, and have been, obtained (e.g., up to 10^6 molecules of hoof-and-mouth disease viral protein in *E. coli*) (5), there must be a finite level to which these can accumulate in their native form. However, if organisms could be made to secrete these foreign proteins into the growth medium, then this constraint would be relieved, and downstream processing would be greatly facilitated.

Unfortunately, the whole field of exocellular macromolecule formation in bacteria still suffers from a lack of systematic study designed to elucidate those factors that control synthesis and product release. In particular, the application of chemostat culture techniques to the study of this problem has not been widespread (3), yet this could yield valuable information as to the precise environmental factor(s) promoting or impeding exocellular enzyme synthesis.

In an ecophysiological context, one might view the secretion of certain hydrolases by microorganisms as a strategy for mobilizing essential nutrient substances that otherwise would be inaccessible to them. Thus, the organism secretes phosphatases and nucleases that serve to mobilize exogenous phosphate and proteases that mobilize amino acids (or, more probably, amino nitrogen). It is not surprising, then, that synthesis of the former two enzymes is promoted by a limitation in the availability of free phosphate, and synthesis of the latter, by nitrogen-limited growth conditions (30). But though organisms clearly can enhance their scavenging potential for specific growth-limiting nutrients by derepressing the synthesis of, and excreting, appropriate hydrolytic enzymes, the regulatory patterns observed (Table 2) are more complex, and more difficult to interpret, than that of, say, the glucose-PTS in *K. aerogenes* (Table 1). Thus, although lowering the dilution rate of a phosphate-limited culture of *Bacillus licheniformis* 749/C led initially to a progressive increase in the rates of synthesis of alkaline phosphatase

TABLE 2. Influence of dilution rate and growth-limiting nutrient on the specific rates of production of some exocellular enzymes by *B. licheniformis* 749/C[a]

Dilution rate (h^{-1})	Nitrogen-limited		Phosphate-limited		
	Protease	Penicillinase	RNase	Alkaline phosphatase	Penicillinase
0.1	0.36	305	0.42	300	370
0.2	0.74	920	0.58	380	1,050
0.3	1.33	1,550	0.88	350	1,630
0.4	1.63	2,370	0.49	180	1,900
0.5	1.52	3,385	0.31	100	2,275
0.6	0.19	4,200	0.05	5	2,500

[a] All rates are expressed as units per hour per milligram (dry weight) of cells. Data of Wouters and Buysman (30).

and RNase, at still lower dilution rates (where the phosphate limitation was even more stringent) the rate of synthesis of these exocellular enzymes declined (Table 2). A similar pattern of regulation was evident with respect to protease production in ammonia-limited cultures of this organism (Table 2). In contrast, the rate of synthesis of penicillinase (a constitutive enzyme) increased with growth rate and hence with the overall rate of protein synthesis. Nevertheless, the pattern of response to a progressive increase in the dilution rate was different in phosphate-limited cultures as compared with ammonia-limited ones (Table 2).

Also of significance in the above connection (and possibly of industrial importance) was the finding (31) that the proportion of the total alkaline phosphatase synthesized that was released into the growth medium varied with the culture pH value and NaCl concentration. High pH values and NaCl concentrations favored release of this exoenzyme from the cell envelope structure, and this was particularly evident with slow-growing cultures. Hence, it is clear that various factors (both physicochemical and physiological) influence the binding properties of alkaline phosphatase in this organism, and though the nature of these binding properties is by no means clear, the available data indicate that, to understand the process of secretion of this enzyme, data on the influence of controlled environmental conditions are indispensable.

Metabolite overproduction. Probably in most natural environments, and certainly in chemostat culture, there is a large difference between the exocellular concentration of the growth-limiting nutrient and those of the other, nonlimiting, nutrients. So a question arises as to how organisms regulate the uptake of these latter nutrients such as to meet the cells' biosynthetic demands while avoiding their excessive accumulation within the cell to potentially traumatic levels. In an industrial context, this question is particularly relevant to situations in which one of the nonlimiting nutrients is the carbon and energy source, since virtually all of the biological products of importance are rich in carbon. It follows, therefore, that if microbial cells commonly possessed mechanisms for strictly regulating the uptake of carbon substrate to just meet the minimum biosynthetic and bioenergetic demands of cell synthesis, then the formation of secondary metabolites, like antibiotics, would be severely impeded, if not totally prevented. Fortunately, such a stringent control generally does not operate, as is also evident in the fact that washed suspensions of organisms (which clearly cannot grow) often oxidize excess carbon substrate at a high rate.

A washed suspension represents an extreme condition in which catabolism is almost totally dissociated from anabolism; a further question thus arises as to whether (and to what extent) these two processes can be dissociated in actively growing cultures. In many aerobic batch cultures, and in aerobic carbon-limited chemostat cultures, these two processes seemingly are tightly coupled since the sole products are cells and CO_2. However, with aerobic chemostat cultures of *K. aerogenes*, growing in the presence of excess carbon substrate, there was found to be a substantial degree of uncoupling (or decoupling) of catabolism from anabolism. Cultures that were, respectively, ammonia, sulfate, phosphate, magnesium, and potassium limited consumed carbon substrate at rates substantially in excess of those of carbon-limited cultures growing at corresponding rates (20). Under these carbon-excess conditions, an excessive intracellular accumulation of intermediary metabolites was avoided by the cells' excreting selected metabolites into the culture medium (so-called "overflow metabolism"), and a potential redox imbalance was circumvented by the cells' expressing a greatly increased respiration rate. But if respiration is obligatorily coupled to ATP formation, then one might ask how the ATP pool was being turned over at a high rate relative to that required to fuel biosynthesis. This is a crucial question with respect to analysis of microbial growth energetics, but it lies outside the compass of the present paper (see, however, 20, 26).

An elevated rate of carbon substrate catabolism, with an associated excretion of metabolites, is commonplace in cultures of facultatively anaerobic species growing in the absence of oxygen or under conditions of oxygen limitation (10, 23). Under both these conditions, cells are compelled to use intermediary metabolites like pyruvate (and products derived therefrom) as electron acceptors to maintain a redox balance while carbon substrate catabolism and ATP formation proceed. Here, in contrast to aerobic situations, ATP synthesis is directly coupled to product formation through substrate-level phosphorylation reactions, though it is important to recognize that the rate of ATP production, relative to the rate of substrate catabolism, can be modulated to some extent by the cells' utilizing different intermediates as electron acceptor (Fig. 1). Thus, as already mentioned, *K. aerogenes* possesses all the enzymes necessary to ferment glycerol to ethanol plus formate with a net ATP gain of 1/mol of glycerol (Fig. 1). Alternatively, by invoking reactions leading to 1,3-propanediol synthesis, the organisms could form acetate, ethanol, and formate (with a net ATP gain of 0.75), or they could form just acetate and formate (with a net ATP gain of 0.67). Finally, as indicated below, these organisms could (at least in theory) ferment glycerol to 1,3-propanediol and lactate with a net ATP gain of 0.5/glycerol.

$$2C_3H_8O_3 \rightarrow \quad C_3H_8O_2 \quad + C_3H_6O_3 + H_2O$$
Glycerol   1,3-Propanediol   Lactate

Thus, in the context of microbial behavior in low-nutrient environments, it is relevant to know whether the nature of the growth limita-

TABLE 3. Influence of the nature of the growth limitation on the rates of glycerol fermentation and product formation by *K. aerogenes* NCTC 418, growing in chemostat culture at a dilution rate of 0.34 ± 0.03 h^{-1} (35°C; pH 6.8)[a]

Determination	Limitation			
	Glyc-erol	Phos-phate	Sul-fate	Ammo-nia
$q_{glycerol}$	63.3	70.4	72.7	91.6
$q_{acetate}$	11.6	15.7	15.1	20.2
$q_{ethanol}$	12.2	2.0	11.4	11.7
$q_{formate}$	3.5	3.5	2.9	1.6
q_{CO_2}	13.7	14.7	17.3	26.5
$q_{succinate}$	0.8	0.6	0.9	1.0
$q_{lactate}$	0.1	0.2	0.4	0.2
$q_{1,3-propanediol}$	31.6	47.9	40.6	54.4
$q_{biomass}$	3.7	3.1	3.4	3.6
Carbon recovery (%)	94	101	95	98
$Y_{glycerol}$	5.8	4.4	4.7	3.9
Y_{ATP}	9.1	7.8	7.2	6.3
ATP efficiency	0.64	0.52	0.63	0.61

[a] All rates (q) are expressed as millimoles per hour per gram (dry weight) of cells. For calculation of the q_{ATP} values (from which the Y_{ATP} values were derived), it is assumed that glycerol is assimilated into biomass at the level of pyruvate. For calculation of the $q_{biomass}$ a cell composition of $C_4H_7O_2N$ was used; 1 mol of biomass is thus equal to 101 g. The ATP efficiency is the amount of ATP formed *net* per mole of glycerol fermented (see Fig. 1).

tion influences the rate and pattern of fermentation of carbon substrates such as glycerol. As is evident in Table 3, anaerobic cultures of *K. aerogenes* do indeed utilize glycerol at enhanced rates, relative to the rate of biomass formation, when this carbon substrate is present in excess of the growth requirement. Moreover, the pattern of fermentation also varies significantly such that the net ATP gain per mole of glycerol fermented (i.e., the ATP efficiency) is caused to vary from a value of 0.52 for phosphate-limited cultures (where ethanol formation is largely suppressed) to a value of 0.64 for glycerol-limited cultures. However, these changes in fermentation pattern seemingly served to constrain the rate of ATP formation only partially, and growth in the presence of excess glycerol proceeded with an apparent marked decrease in energetic efficiency (i.e., with a lowered apparent Y_{ATP} value). The reason for this is not immediately obvious; nevertheless, these data do indicate that controlled low-nutrient environments might well provide a means of steering fermentation processes toward enhanced production of compounds like 1,3-propanediol and hence are worthy of more detailed study.

CONCLUSIONS

In this paper, we have shown that studies of microbial behavior in nutrient-limited chemostat environments provide information that is both meaningful in an ecological context and highly relevant to the development of new biotechnological processes. Thus, chemostat culture allows a range of low-nutrient steady-state conditions to be imposed on *growing* microorganisms, and in such environments, organisms often express properties that they never express, or express only transiently, in batch culture. It can be reasonably argued that these special properties must have been acquired in the course of evolution and must serve some function in the growth and survival of these organisms in their natural habitats; the function that they fulfill often is not hard to adduce. In general, they serve to promote the scavenging of essential substances from nutrient-depleted environments and thereby allow organisms to express the highest possible growth rate under severe growth-constraining conditions (26–28).

The enormous physiological plasticity of microorganisms and their extreme sensitivity to environmental change are of crucial importance in the operation of fermentation processes and, in our view, account for many of the problems encountered in process development and scale-up. In this connection, the vast majority of microbiologically based processes currently in operation employ batch culture methods, and whereas these have proved adequate for the manufacture of a wide range of products, from solvents to antibiotics, there can be little doubt that they necessarily must be superseded by continuous-flow systems for the manufacture, on an ultra-large scale, of biomass and bulk chemicals like fuel alcohol and plastics feedstock. However, the environmental conditions prevailing in a continuous-flow (chemostat) culture are markedly different from those which obtain in a batch-type culture, and the properties and performance of organisms so cultured are correspondingly different. It would be unjustified, therefore, to suppose that an established batch process could be readily transformed into a continuous-flow process without any marked effect on performance. Worse, in the context of today's "new biotechnology," the chemostat provides a fiercely competitive environment which, in general, will select against genetically engineered organisms. There is thus a basic incompatibility between organism improvement by genetic manipulation and process improvement by utilizing chemostat culture techniques. If this potential bottleneck is to be circumvented, it would seem prudent to invest more effort now in seeking to understand and rationalize the behavior of organisms in chemostat environments and thereby to develop a strategy that would allow improved strains to be cultured continuously without risk of degeneration. It is our conviction that this strategy can be best derived from a more thorough appreciation of microbial ecology or, more particularly, of microbial ecophysiology.

We thank Jo Lansbergen for skilled technical assistance.

LITERATURE CITED

1. **Brock, T. D.** 1966. Principles of microbial ecology. Prentice Hall, Englewood Cliffs, N.J.
2. **Brown, C. M., and S. O. Stanley.** 1972. Environment-mediated changes in the cellular content of the "pool" constituents and their associated changes in cell physiology. J. Appl. Chem. Biotechnol. **22:**363–389.
3. **Bull, A. T.** 1972. Environmental factors influencing the synthesis and excretion of exocellular macromolecules. J. Appl. Chem. Biotechnol. **22:**261–292.
4. **Byers, B. R.** 1974. Iron transport in Gram-positive and acid-fast bacilli, p. 83–105. *In* J. B. Neilands (ed.), Microbial iron metabolism, a comprehensive treatise. Academic Press, London.
5. **Cape, R. E., D. H. Gelfand, M. A. Innis, and S. L. Neidleman.** 1982. An introduction to the present state and future role of genetic manipulation in the development of overproducing micro-organisms, p. 327–343. *In* V. Krumphanzl, B. Sikyta, and Z. Vanek (ed.), Overproduction of microbial products. Academic Press, London.
6. **Dean, A. C. R.** 1972. Influence of environment on the control of enzyme synthesis. J. Appl. Chem. Biotechnol. **22:**245–259.
7. **Duine, J. A., and J. Frank.** 1981. Methanol dehydrogenase: a quinoprotein, p. 31–41. *In* H. Dalton (ed.), Microbial growth on C_1 compounds. Heyden, London.
8. **Emery, T.** 1974. Biosynthesis and mechanism of action of hydroxamate-type siderochromes, p. 107–123. *In* J. B.

Neilands (ed.), Microbial iron metabolism, a comprehensive treatise. Academic Press, London.

9. **Forage, R. G., and E. C. C. Lin.** 1982. *dha* system mediating aerobic and anaerobic dissimilation of glycerol in *Klebsiella pneumoniae* NCIB 418. J. Bacteriol. **151:**591–599.

10. **Harrison, D. E. F.** 1976. The regulation of respiration rate in growing bacteria. Adv. Microbial Physiol. **14:**243–313.

11. **Herbert, D.** 1961. The chemical composition of microorganisms as a function of their environment. Symp. Soc. Gen. Microbiol. **19:**391–416.

12. **Hueting, S., T. de Lange, and D. W. Tempest.** 1979. Energy requirement for maintenance of the transmembrane potassium gradient in *Klebsiella aerogenes* NCTC 418: a continuous culture study. Arch. Microbiol. **123:**183–188.

13. **Hueting, S., and D. W. Tempest.** 1979. Influence of glucose input concentration on the kinetics of metabolite production by *Klebsiella aerogenes* NCTC 418: growing in chemostat culture in potassium- or ammonia-limited environments. Arch. Microbiol. **123:**189–194.

14. **Kornberg, H. L., and R. E. Reeves.** 1972. Correlation between hexose transport and phosphotransferase activity in *Escherichia coli*. Biochem. J. **126:**1241–1243.

15. **Kornberg, H. L., and R. E. Reeves.** 1972. Inducible phosphoenolpyruvate-dependent hexose phosphotransferase activities in *Escherichia coli*. Biochem. J. **128:**1339–1344.

16. **Kundig, W., S. Ghosh, and S. Roseman.** 1964. Phosphate bound to histidine in a protein as an intermediate in a novel phosphotransferase system. Proc. Natl. Acad. Sci. U.S.A. **52:**1067–1074.

17. **Lin, E. C. C., A. P. Levin, and B. Magasanik.** 1960. The effect of aerobic metabolism on the inducible glycerol dehydrogenase of *Aerobacter aerogenes*. J. Biol. Chem. **235:**1824–1829.

18. **Neijssel, O. M.** 1977. The effect of 2,4-dinitrophenol on the growth of *Klebsiella aerogenes* NCTC 418 in aerobic chemostat cultures. FEMS Microbiol. Lett. **1:**47–50.

19. **Neijssel, O. M., S. Hueting, K. J. Crabbendam, and D. W. Tempest.** 1975. Dual pathways of glycerol assimilation in *Klebsiella aerogenes* NCIB 418. Their regulation and possible functional significance. Arch. Microbiol. **104:**83–87.

20. **Neijssel, O. M., and D. W. Tempest.** 1975. The regulation of carbohydrate metabolism in *Klebsiella aerogenes* NCTC 418 organisms, growing in chemostat culture. Arch. Microbiol. **106:**251–258.

21. **Neijssel, O. M., and D. W. Tempest.** 1976. The role of energy-spilling reactions in the growth of *Klebsiella aerogenes* NCTC 418 in aerobic chemostat culture. Arch. Microbiol. **110:**305–311.

22. **O'Brien, R. W., O. M. Neijssel, and D. W. Tempest.** 1980. Glucose phosphoenolpyruvate phosphotransferase activity and glucose uptake rate of *Klebsiella aerogenes* growing in chemostat culture. J. Gen. Microbiol. **116:**305–314.

23. **Pirt, S. J., and D. S. Callow.** 1958. Exocellular product formation by microorganisms in continuous culture. I. Production of 2,3-butanediol by *Aerobacter aerogenes* in a single stage process. J. Appl. Bacteriol. **21:**188–205.

24. **Postma, P. W., and S. Roseman.** 1976. The bacterial phosphoenolpyruvate: sugar phosphotransferase system. Biochim. Biophys. Acta **457:**213–257.

25. **Sikyta, B., P. Kyslík, B. Voleský, E. Pavlasová, and E. Stejskalová.** 1982. Over-production of endoenzymes in *Escherichia coli*—selection of hyperproducing strains in a chemostat., p. 593–599. *In* V. Krumphanzl, B. Sikyta, and Z. Vanek (ed.), Overproduction of microbial products. Academic Press, London.

26. **Tempest, D. W.** 1978. The biochemical significance of microbial growth yields: a reassessment. Trends Biochem. Sci. **3:**180–184.

27. **Tempest, D. W., and O. M. Neijssel.** 1976. Microbial adaptation to low-nutrient environments, p. 283–296. *In* A. C. R. Dean, D. C. Ellwood, C. G. T. Evans, and J. Melling (ed.), Continuous culture 6: applications and new fields. Ellis Horwood Ltd., Chichester.

28. **Tempest, D. W., and O. M. Neijssel.** 1978. Eco-physiological aspects of microbial growth in aerobic nutrient-limited environments. Adv. Microb. Ecol. **2:**105–153.

29. **Tempest, D. W., J. L. Meers, and C. M. Brown.** 1970. Synthesis of glutamate in *Aerobacter aerogenes* by a hitherto unknown route. Biochem. J. **117:**405–407.

30. **Wouters, J. T. M., and P. J. Buysman.** 1977. Production of some exocellular enzymes by *Bacillus licheniformis* 749/C in chemostat cultures. FEMS Microbiol. Lett. **1:**109–112.

31. **Wouters, J. T. M., and P. J. Buysman.** 1980. Secretion of alkaline phosphatase by *Bacillus licheniformis* 749/C during growth in batch and chemostat cultures. FEMS Microbiol. Lett. **7:**91–95.

32. **Zevenboom, W., A. bij de Vaate, and L. R. Mur.** 1982. Assessment of factors limiting growth rate of *Oscillatoria agardhii* in hypertrophic Lake Wolderwijd, 1978, by use of physiological indicators. Limnol. Oceanogr. **27:**39–52.

Marine Microorganisms as a Source of Bioactive Substances

YOSHIRO OKAMI

Institute of Microbial Chemistry, Tokyo, Japan

One of the most important tasks of ecologists is to study ecosystem structure in undisturbed natural environments. In primary ecological studies of higher animals or plants, macroscopic, or naked-eye, observations can be made without any artificial disturbance to the natural system. The study of microbes, however, can best be conducted after their isolation from natural habitats by cultivation in properly prepared artificial conditions. As a consequence, some of their features may be changed from those expressed in natural environments. Therefore, the ecological aspects of microbes in nature can only be estimated through extrapolation of the knowledge gained with isolated microbes. The microbial flora in natural habitats varies as a result of changes in various environmental factors such as humidity, temperature, nutrients, or the presence of other living organisms. Every living organism obtains required nutrients from the environment and produces various substances which are excreted into the environment. Because of the dynamic and complex nature of these associations, the microbial ecosystem in soil is poorly understood in spite of many detailed and long-term studies.

At an early stage of research on antibiotics, their production was regarded as a strategy used by soil microorganisms to compete with other organisms in soil. However, subsequent studies demonstrated that this activity was not significant in determining the structure of microbial communities on natural soils. The latter is the case because the activity of antibiotics is very inefficiently expressed in soil as a result of adsorption to various substances such as clay. Their activity is expressed efficiently only in artificially devised conditions such as properly formulated media.

Long-term research on antibiotic production by soil microorganisms has revealed the following points: (i) antibiotic production by microorganisms is strain specific but not species specific, (ii) the same microorganism may produce different antibiotics under different cultural conditions, and (iii) strains belonging to the same species but growing in different conditions may produce different antibiotics. In the light of the above, more new antibiotics can be expected when more strains are isolated from unexplored environments and are cultured under various conditions.

Even though most infectious diseases caused by bacteria can be brought under control with existing antibiotics, there is still a great need for agents that would inhibit viral infections and neoplastic growth. Other antibiotics which have inhibitory activity against specific targets are also needed as tools for basic research in chemistry and biology. For the above needs, the supplying of such agents through industrial production is urgently needed. Throughout the history of industrial production of antibiotics, microorganisms have been readily adapted to the fermentation industry and have significant advantages over more complex and highly organized organisms which are difficult to adapt to large-scale production.

BIOACTIVE SUBSTANCES FROM MARINE MICROORGANISMS

I chose to examine microorganisms from marine environments since these sources have been unexplored as a source of bioactive agents for industrial production. My colleagues and I established a strategy for the screening of bioac-

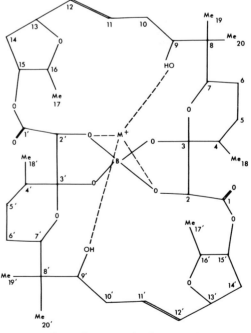

FIG. 1. Structure of aplasmomycin.

651

TABLE 1. Substrate specificity of the enzyme activity[a]

Substrate	Major linkage(s)	Activity
S. mutans E-49 glucan	α-1,3; α-1,6	+
Nigeran	α-1,3	+
Soluble starch	α-1,4; α-1,6	$\pm$
Glycogen	α-1,6; α-1,4	+
Dextran T-70	α-1,6	+
Laminarin	β-1,3	$\pm$
Maltose	α-1,4	$-$
Isomaltose	α-1,6	+
Nigerose	α-1,3	+
Cellobiose	β-1,4	$-$
Gentiobiose	β-1,6	$\pm$

[a] The ability of the enzyme to hydrolyze various glucans and disaccharides is indicated by + (well hydrolyzed), $\pm$ (weakly hydrolyzed), and $-$ (not hydrolyzed).

tive agents with special reference to salts and nutrients present in seawater and to the synthesis or degradation of polymers such as agar in the marine environment.

An actinomycete strain isolated from seawater was identical with a terrestrial isolate of *Streptomyces griseus*. This strain was found to produce antibiotic activity in a medium with low nutrient content and with salt concentrations equivalent to those in seawater, but not in media which contained higher concentrations of nutrients and salts. The active agent was extracted and crystallized. X-ray crystallography revealed it to be an ionophoric polyether of a crown-ether type which contained boron in the center of the molecule. Since it was active against the plasmodium of malaria in mice, it was named aplasmomycin. It had a potent transfer activity of monovalent metal ions from one aquatic solution to

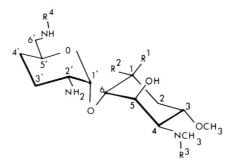

	R^1	R^2	R^3	R^4
			1"2"	
2"-N-Formimidoylistamycin A	NH_2	H	$COCH_2NHCH=NH$	CH_3
2"-N-Formimidoylistamycin B	H	NH_2	$COCH_2NHCH=NH$	CH_3
Istamycin A	NH_2	H	$COCH_2NH_2$	CH_3
Istamycin B	H	NH_2	$COCH_2NH_2$	CH_3
Istamycin C	NH_2	H	$COCH_2NH_2$	CH_2CH_3
Istamycin A_0 (in acid)	NH_2	H	H	CH_3
Istamycin B_0	H	NH_2	H	CH_3
Istamycin C_0 (in acid)	NH_2	H	H	CH_2CH_3
Istamycin A_1	NH_2	H	$COCH_2NHCHO$	CH_3
Istamycin B_1	H	NH_2	$COCH_2NHCHO$	CH_3
Istamycin C_1	NH_2	H	$COCH_2NHCHO$	CH_2CH_3
Istamycin A_2	NH_2	H	$COCH_2NHCONH_2$	CH_3

FIG. 2. Structure of istamycins.

another. Its structure is shown in Fig. 1 (5). It seems to have a function in the growth of the actinomycete in seawater as a result of its ability to transfer metal ions.

It is known that aquatic plants produce polymeric materials such as agar. Similarly, aquatic microorganisms produce slimy polymers. Hence, we searched for marine microorganisms which produce polysaccharides extracellularly in media containing seawater. About 90 of 500 marine isolates were found to produce significant amounts of extracellular polysaccharides which were easily precipitated by the addition of alcohol to the fermentation broths. These precipitates were dissolved in saline and injected (20 mg per mouse) intraperitoneally into mice bearing solid sarcoma 180 tumors. Six of the 90 precipitates exhibited remarkable suppression of the tumors and in many cases caused complete remission of the tumor. One of the above polysaccharide producers was identified as *Flavobacterium uliginosum* MP-55, which required seawater for growth. The antitumor agent that it produced was purified to a neutral white powder and named marinactan. Marinactan is a heteroglycan which consists of glucose ($69.7 \pm 0.7\%$), mannose ($19.8 \pm 1.2\%$), and fucose ($10.8 \pm 0.8\%$). It is soluble in water and insoluble in organic solvents. Marinactan produced a remarkable activation of macrophages without T-cell mediation in mice, acting as an immune potentiating agent.

It has been demonstrated that agar-liquefying microorganisms occur frequently in marine but not in terrestrial environments. This suggests that marine microorganisms may degrade not only agar but also other polysaccharides. We employed a biological polymer, glucan, produced by *Streptococcus mutans* and believed to be a cause of dental caries, and screened marine bacteria for the degradation of this glucan. One isolate, which was identified as *Bacillus circulans*, exhibited a significant degradation of the glucan. The enzyme produced by *B. circulans* required the glucan as an inducer and reached maximum activity after a 6-day fermentation. The enzyme was extracted and purified (4). The optimum pH for the enzyme activity was 6.2 to 6.7, and the optimum temperature was 35°C, which was significantly lower than that for any other dextranases and favorable for activity in the human mouth. The activity was not influenced by Ca^{2+}, which affects other dextranases. Its substrate specificity is shown in Table 1. Type c glucan of *S. mutans* produced in the human mouth was significantly degraded. The enzyme also attacked other glucans and disaccharides which consisted of α-1,3 and α-1,6 linkages, but not those composed of α-1,4 or β linkages.

It has been observed that various antibiotic-producing streptomycetes harbor various-sized plasmids, some of which may be involved in antibiotic biosynthesis. We found that the plasmids were different between different aminoglycoside antibiotic-producing streptomycetes which had different resistance to antibiotics (1). On the basis of resistance to antibiotics and the presence of different plasmids, we screened marine actinomycetes to search for new aminoglycoside antibiotic-producing strains. One streptomycete which was isolated from coastal sea mud from Tenjin Island near Tokyo possessed a new type of plasmid and a new pattern of resistance to antibiotics. This strain was found to produce istamycins A and B, new aminoglycoside antibiotics (3). The actinomycete had characteristic taxonomic features and was named *Streptomyces tenjimariensis*. It has a unique pattern of carbohydrate utilization and resistance to antibiotics. The structure of the istamycins is shown in Fig. 2. Istamycin B had the highest activity against various pathogenic bacteria and has promise for practical use in the treatment of infectious diseases.

It is interesting and unusual that several strains of *S. tenjimariensis* which had the same taxonomic features were isolated from the same place in the sea during three different years (2). In our previous experience, the same microorganism was rarely isolated from soil samples taken from the same area in different years. Another interesting finding was that strains of the same species isolated in different years har-

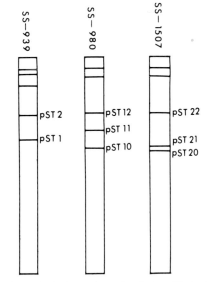

FIG. 3. Gel electrophoresis of plasmids harbored in three strains, SS-939, SS-980, and SS-1507, of *S. tenjimariensis* (pST represents plasmid number).

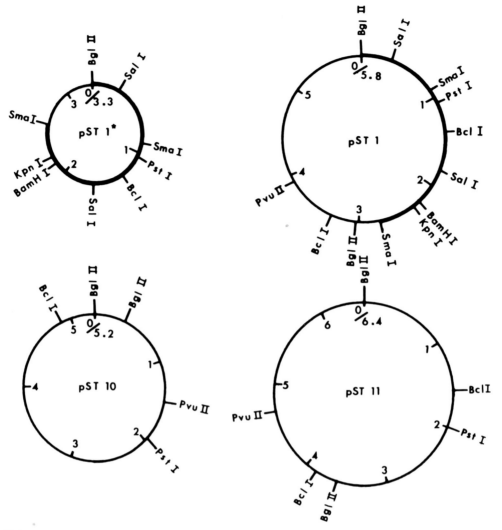

FIG. 4. Restriction maps of plasmids isolated from different strains of *S. tenjimariensis*. The heavy black part represents similarity, pST is the plasmid isolated, and the letters outside the circle are restriction enzymes.

bored different plasmids. Strains SS-939, SS-980, and SS-1507 which were isolated in different years had different plasmid profiles, as illustrated by gel electrophoresis (Fig. 3).

The elimination of plasmids by curing agents such as acriflavine did not result in a loss of istamycin production, and this suggests that the plasmids may not be involved in antibiotic biosynthesis. The plasmids in each strain were isolated and digested with restriction enzymes. Comparison of the restriction maps of the plasmids showed that some are related but others are entirely unrelated (Fig. 4). From an ecological point of view, it is interesting that the istamycin-producing streptomycetes survived in the same habitat over a period of years; however,

the related plasmids changed during this same period. Because the changing of genetic elements such as plasmids during microbial growth in nature may lead to new microorganisms which may provide new metabolites, including bioactive substances, it is important to consider the growth conditions of microorganisms in their natural habitat when designing strategies for screening of new bioactive agents.

LITERATURE CITED

1. **Okami, Y.** 1979. Antibiotics from marine microorganisms with reference to plasmid involvement. J. Nat. Products (Lloydia) **42**:583–595.
2. **Okami, Y.** 1982. Potential use of marine microorganisms for antibiotics and enzyme production. Pure Appl. Chem. **54**:1951–1962.

3. **Okami, Y., K. Hotta, M. Yoshida, D. Ikeda, S. Kondo, and H. Umezawa.** 1979. New aminoglycoside antibiotics, istamycin A and B. J. Antibiot. **32:**964–966.

4. **Okami, Y., S. Kurasawa, and Y. Hirose.** 1980. A new glucanase produced by a marine bacillus. Agric. Biol. Chem. **44:**1191–1192.

5. **Okami, Y., T. Okazaki, T. Kitahara, and H. Umezawa.** 1976. Studies on marine microorganisms. V. A new antibiotic, aplasmomycin produced by a streptomycete isolated from shallow sea mud. J. Antibiot. **29:**1019–1025.

6. **Umezawa, H., Y. Okami, S. Kurasawa, T. Ohnuki, M. Ishizuka, T. Takeuchi, T. Shiio, and Y. Yukari.** 1983. Marinactan, antitumor polysaccharide produced by marine bacteria. J. Antibiot. **36:**471–477.

Mixed-Culture Fermentations in Industrial Microbiology

IAN SALMON AND ALAN T. BULL

The Biological Laboratory, University of Kent at Canterbury, Canterbury, Kent, CT2 7NJ, United Kingdom

The application of mixed microbial communities in industrial fermentations has been under increasing scrutiny during the past few years (for reviews, see 8, 19, 23, 26). The advantages of mixed microbial fermentations over monocultures include: greater range of substrates utilized, increased productivities, improved culture stability to adverse operating conditions, resistance to contamination, and the single-step formation of complex products. The purpose of this paper is to illustrate these points, to give an introduction to the literature, and to examine some recent developments in mixed-culture fermentations. The production of vaccines and viruses and the contamination of monoculture fermentations will not be considered.

TRADITIONAL USES OF MIXED CULTURES

Fermented foods and beverages. The use of mixed microbial cultures in beverage, food, and dairy fermentations dates from antiquity. These fermentations were (and in some cases are still) unprotected open systems with fortuitous inoculation of organisms; only recently have specific strains and defined mixtures of organisms been employed (20, 58). Despite the open nature of these fermentations, many are remarkably stable and resistant to invasion by contaminating microorganisms (81). The microbial communities used in fermented foods are often large and complex, and in only a few have the predominant or essential organisms been identified. Steinkraus (71) recently discussed a range of mixed-culture food fermentations. Food and beverage fermentations involving yeast–lactic acid bacteria associations (e.g., koumiss, kefir, sour-dough breads) were reviewed by Wood (81). Lactic acid fermentations are characteristic of the dairy industry; for example, cheese and butter manufacturers use cultures of lactic streptococci and leuconostocs (37, 57); the production of many cheese varieties also involves filamentous fungi during the maturation stage (48).

Yogurt is a well-studied dairy product made by the mixed fermentation of milk with the use of *Lactobacillus bulgaricus* and *Streptococcus thermophilus*. These two organisms have a non-obligatory mutualistic interaction in milk (i.e., the interaction is favorable to the growth of both organisms). Such a relationship is called protocooperation (7). Formic acid produced from lactate (at low or zero oxygen tension) by *S. thermophilus* is a growth stimulant for *L. bulgaricus*, and amino acids released by hydrolysis of casein by *L. bulgaricus* stimulate the growth of *S. thermophilus* (67).

A continuous two-stage process for yogurt production has been developed (14, 15). The first stage is a stirred fermentor operated as a pH stat (i.e., a continuous fermentor in which the feed rate of fresh milk is regulated to maintain a constant pH) at about 45°C and at a pH (approximately 5.7) just above the value at which casein coagulates. The dilution rate is 1.9 to 2.0 h^{-1}. The second stage is a laminar plug-flow reactor, where coagulation of the casein, as a result of continuing acid production, takes place. The development of the correct yogurt texture requires no unwanted mixing at this stage. A simple mathematical model of the protocooperative interaction between *L. bulgaricus* and *S. thermophilus* (47) had indicated that this coculture would be unstable in continuous culture, with any small perturbation leading to the loss of one of the organisms. In practice, continuous cultures have proved to be very stable (45). It has been suggested (J. Ubbels, quoted in 13) that lactic acid (a product of both organisms) has an inhibitory effect on one or the other (or both) of the bacterial species, and it is this inhibitory effect which ensures stable coexistence. Driessen (13) reviewed continuous yogurt production.

The incentive for the development of a continuous yogurt process arose from the increasing demand for this product. The scale-up of batch fermentations presented control problems: first, it is difficult to chill a large production before the yogurt becomes too sour; second, rapid filling of containers damages the yogurt's consistency (13). By using the continuous process, the same production capacity may be achieved with 20% of the vessel volume.

Biological waste treatments. The biotreatment of wastes is another very successful application of mixed cultures. Here, the open nature of treatment plants and the wide range of compounds found in wastewaters ensure that a complex microbial population develops. This population may arise spontaneously by inoculation from the waste stream or after seeding with organisms from an established plant. A period of adaptation may be required until a stable(?) community develops to deal with a particular waste stream.

The microorganisms involved in the aerobic treatment of wastewaters are ill defined, partly as a result of confusing inconsistencies in the taxonomy of these bacterial groups (66) and partly as a result of the difficulty in isolating and growing all the species present in pure culture (70). A recent study (66) showed that the dominant bacterial flora (members of the genera *Alcaligenes, Pseudomonas*, and *Zoogloea*) remained constant in species composition both over an extended time and between bioreactors operating under different conditions. However, the numerical contribution of each species depended on the operating conditions. This is contrary to a previous report (79) which showed a constantly changing flora.

A common problem (62) encountered with activated sludge plants is that of bulking (i.e., sludge that remains in suspension and does not settle). The interactions between organisms that lead to bulking have been investigated in laboratory enrichment cultures (29, 76) and in a model defined mixed culture (72, 82). Verachtert and his colleagues (29, 76) examined the effect of intermittent and continuous feeding on an enrichment culture. For pure substrates, and for industrial waste streams, they found that continuously fed cultures produced a microbial population with a high proportion of filamentous bacteria which caused bulking problems, whereas intermittent feeding suppressed their growth and allowed floc-forming organisms to dominate. A metabolic explanation for the selective pressure of the feeding pattern is that higher rates of substrate uptake and the formation of larger intracellular food reserves occur in intermittently fed floc-forming organisms than in filamentous bacteria (75). A defined mixed culture of *Flavobacterium breve* (a typical floc-forming organism) and *Flavobacterium ferrugineum* (a typical bulking organism) was claimed to closely simulate the operating characteristics of an activated sludge plant (72). A mathematical model of the system (82) agreed well with the experimental results. The model predicted that longer residence times decreased the chances of bulking, that rapid changes in operating conditions caused the culture to become unstable, and that culture recycle extended the range of stable operating conditions.

The anaerobic treatment of wastes has become an increasingly important process and has been intensively studied in the past few years (63). Anaerobic digestion is no longer viewed only as a simple, cost-effective means of reducing pollution but also as a way of converting wastes (or renewable resources, i.e., biomass) into energy sources (methane, ethanol) or chemicals (organic acids, solvents), with potentially the production of a digested sludge with value as a fertilizer or as an animal feed additive (61, 84).

It is not clear whether the same microbial flora is found in different digesters (28) even when these operate at different temperatures, since mesophilic digesters can be converted into thermophilic digesters by gradually raising the temperature. Whether the microbial population undergoes selection or adaptation to this change is not known, although clearly, as in aerobic treatment, the number of each species is greatly influenced by the digester's feedstock and operating parameters (28, 73). However, the same basic reactions have been identified in all digesters: hydrolysis of polymeric substrates to simple units, fermentation of these units to organic acids, interconversion of these acids to acetic acid, hydrogen, and carbon dioxide, and finally, methane formation from hydrogen and carbon dioxide, formate, and acetate. At each step the microorganisms are interdependent, and in a correctly functioning digester, the reactions are balanced. The ecology of anaerobic digesters has been discussed recently (28, 80).

The use of cellulosic wastes as feedstocks for anaerobic digesters is of current interest; however, most research in this field has been devoted to empirical means of exploiting the process, rather than to the elucidation of the basic mechanisms by which substrates are utilized (74). Some defined mixed cultures have been investigated as potential, or as model, systems for the large-scale industrial production of methane, organic acids, and ethanol from cellulosic wastes (for examples, see Table 1).

The advantages found include increased product yields, increased production rates, decreased secondary product formation, and an increased range of substrates utilized. Additionally, the saccharification and fermentation steps occur in a single vessel, with a consequent saving in cost of the process plant.

The biological treatment of chemical wastewaters is often complicated by the presence of toxic or xenobiotic compounds. The removal of these compounds was discussed by Bull (6) and Slater and Bull (69).

The traditional uses of mixed cultures remain the most important; indeed, they are probably more important economically and of more benefit to humans than any other aspect of microbiology. The operating characteristics of many of these processes have been determined experimentally, but the underlying microbial interactions often remain obscure. It seems likely that further improvements in process efficiency will be brought about by future fundamental research.

TABLE 1. Examples of mixed-culture fermentations of cellulosic substrates

Product	Microorganisms	Comments	Reference
Methane	*Acetobacterium woodii* and *Methanosarcina barkeri*	Complete degradation of carbohydrates to methane.	78
Methane	*Acetivibrio cellulolyticus* and *Methanosarcina barkeri*	Near theoretical yields of CH_4 from cellulose (2.6 to 3.0 mol/mol of hexose).	33
Methane	Rumen anaerobic fungus, *Methanobrevibacter* sp., and *Methanosarcina barkeri*	Tri > bi > uniculture in cellulose breakdown. Intermediate compounds low with triculture.	49
Acetate and methane	*Ruminococcus flavefaciens* and *Methanobacterium ruminantium*	Formation of succinate by *R. flavefaciens* switched to acetate in coculture.	39
Acetate	Mixed sporeforming cellulytic organisms	Formed by heat treatment of methane-producing culture.	35
Ethanol and acetate	Noncellulytic *Clostridium* sp. and a cellulytic bacterium	High proportion of ethanol formed. Symbiotic relationship. Cellulytic organism will grow only in presence of the *Clostridium* sp.	34
Ethanol	*Clostridium thermocellum* and *C. thermohydrosulfuricum*	Coculture three times faster ethanol production rate, two times decrease in acetate production rate, two times ethanol yield.	52
Butyrate and acetate	Cellulytic *Clostridium* sp. and *C. acetobutylicum*	Coculture degraded cellulose three times faster than monoculture.	60

DEVELOPING USES OF MIXED CULTURES

Single cell protein. Linton and Drozd (44) recently reviewed the use of mixed cultures in single-cell protein (SCP) production. SCP may be produced from both renewable and nonrenewable resources (21), such as carbon dioxide fixed by microbial photosynthesis (5, 40, 41), cellulosic wastes (4, 25, 59), sodium lignosulfate (22), whiskey distillery spent wash (3), brewery effluent (32), potato-processing wastewater (43), dairy industry wastes (36), methane and methanol (16, 26), and natural gas (44).

The following advantages of mixed cultures over monocultures for SCP production have been reported (see 44).

Improved yields and growth rates were reported for consortia growing on methane and methanol over those in pure cultures (26). This has been questioned (44), and it seems that no significant differences between pure and mixed cultures on these substrates exists. The alga-bacteria consortium studied by Lee and Pirt (40, 41) has an increased growth yield from light as a result of scavenging, by the bacteria, of metabolites released by the alga.

Decreased foaming resulted from the removal of surface-active compounds released by cell lysis. The control of foaming is important, as excessive foaming may cause the blockage of filters and decrease gas transfer rates. The use of chemical antifoams is undesirable in SCP production.

Mixed cultures are more stable. Photosynthetic mixed cultures (40), and those growing on methane-methanol, have a greater resistance to

environmental shock than do pure cultures, the alga-bacteria mixed culture having a larger operating range of temperatures and pH than a monoculture of the alga. Mixed cultures are resistant to contamination. In a well-balanced mixed culture, all the ecological niches are occupied; a contaminating organism has, therefore, to compete for scarce resources with an already well-established organism. Rokem et al. (65) showed that a mixed culture growing on methanol was resistant to infection by pathogenic organisms, whereas a pure culture of the methanol utilizer was not. Both methanol-utilizing and photosynthetic mixed cultures (26, 40) have remained uncontaminated over long periods without aseptic precautions being taken. The British Petroleum *n*-paraffins process was run under nonaseptic conditions (38). There are great problems, in terms of cost and technology, in the construction and operation of large-scale aseptic fermentation plants (especially if water recycle is employed). A process capable of nonaseptic operation has, therefore, a great economic advantage.

The complete utilization of mixed substrates (as normally found in wastewaters) requires the combined action of many organisms. Similarly, natural gas (a mixture of methane, ethane, propane, butane, and pentane) requires a mixed culture for successful growth (17, 44).

Metal leaching and recovery. Many metals are becoming scarce, as geological deposits are being worked out. Microorganisms can be employed in two ways: first, to extract metals from ores, and second, to recover metal ions from solutions such as industrial effluents (12).

The microbially assisted leaching of low-grade ores is of current industrial importance in copper mining and has been used in uranium mining in the past (12). Microbial leaching is an open system in which many organisms are found; however, most studies have neglected this aspect and concentrated solely on the role of *Thiobacillus ferrooxidans*, the main causative agent. Norris and Kelly (53) reviewed the ecology of microbially assisted leaching.

The use of microorganisms in the removal of metal ions from solution is receiving increased attention (51), especially for the recovery of precious metals (12). A community of three bacteria (9) was described that accumulated silver. The major silver accumulator was identified as *Pseudomonas maltophilia*, though the community was able to tolerate and accumulate more silver than *P. maltophilia* could in pure culture. It was found that $30 \pm 6\%$ of the dry weight of a 10-membered bacterial community was made up of bioaccumulated copper (18). The specific growth rate of the community, in the presence of copper, was significantly greater than that of any of the individual components. Sulfate-reducing bacteria have been used to precipitate (as the sulfides) copper (31) and tungsten and molybdenum (Russian patent SU-922-088, 1983) from wastewaters.

Biological control. Many plant infections may be controlled by organisms antagonistic toward the causative agent (10, 27). Sometimes, this biocontrol is achieved by altering the way the plant is cultivated to promote the growth of the antagonistic organism. Alternatively, the antagonistic organism may be applied to the plant.

In most of these cases control is effected through parasitism (1, 11), but in some the antagonistic effect is due to antibiotic production which may be specific (30) or broad action (64). An unusual application of biocontrol is the protection of sensitive plants from frost damage by the application of virulent phage or bacteria antagonistic toward the bacteria responsible for ice nucleation (U.S. patents 4375-734 and EP–74-718, 1983).

Miscellaneous. Table 2 summarizes some recent investigations of mixed cultures with potential use in industrial microbiology. Other examples, may be found in the reviews by Egorov and Landau (19) and Haas et al. (23).

CONCLUSIONS

Mixed cultures are likely to be applied where the activities of monospecies cultures have been found wanting. These areas include the single-step elaboration of products that could be achieved only by several sequential pure culture fermentations (e.g., steroid transformations [46, 83]), the utilization of complex substrates (e.g., lignin [50] or cellulose [60]), and where aseptic operation is not feasible (e.g., waste treatment). The possibility that many useful products may be formed only in mixed culture (e.g., neopurpuratin [55, 56]) has been largely neglected since current isolation techniques employed by industry favor pure cultures strongly. Consistent results may be obtained from mixed cultures as well as from monocultures, and regulation of mixed cultures by manipulation of the environmental conditions has been shown to work (24, 42). We believe that mixed cultures will continue

TABLE 2. Examples of mixed cultures with potential industrial use

Product	Microorganisms	Comments	Reference
H_2	*Rhodospirullum rubrum* and *Klebsiella pneumoniae*	Photometabolic production of H_2 by *R. rubrum* greater in coculture.	77
H_2	*Cellulomonas* sp. and *Rhodopseudomonas capsulata*	H_2 produced from substrate (cellulose) not normally utilized by *R. capsulata*.	54
Optically active lactic acid	*Lactobacillus lactis* and *Saccharomyces cerevisiae*	Yeast provides source of N, vitamins, amino acids, and trace elements for growth of the acid-producing bacterium.	U.S. Patent EP–69-291
L-Histidine	*Corynebacterium glutamicum* and *Escherichia coli*	*C. glutamicum* produces histidinol which is converted to histidine by *E. coli*.	2
Neopurpuratin	*Streptomyces propurpuratus* and *Bacillus* sp.	Water-soluble food coloring with no toxicity; formed only mixed culture.	55, 56
Ursodeoxycholic acid	*Eubacterium aerofaciens* and 7α-hydroxysteroid dehydrogenase elaborating organism	Complex bioconversion carried out in one step.	46
Polyvinyl alcohol oxidase	*Pseudomonas putida* and *Pseudomonas* sp.	*P. putida* produced growth stimulant for the other (oxidase-producing) species. Neither could oxidize polyvinyl alcohol in pure culture.	68

to have an increasingly important role to play in industrial fermentations.

LITERATURE CITED

1. **Adams, P. B., and W. A. Ayers.** 1982. Biological control of Sclerotinia lettuce drop in the field by *Sporidesmium sclerotivorum.* Phytopathology **72**:485–488.
2. **Araki, K., and K. Nakayama.** 1975. Accumulation of L-histidinol by a histidine auxotroph of *Corynebacterium glutamicum* and its conversion into L-histidine by an *Escherichia coli* strain. Agric. Biol. Chem. **39**:127–132.
3. **Barker, T. W., J. P. Quinn, and R. Marchant.** 1982. The use of a mixed culture of *Geotrichum candidum, Candida krusei* and *Hansenula anomola* from microbial protein production from whiskey distillery spent wash. Eur. J. Appl. Microbiol. Biotechnol. **14**:247–253.
4. **Bellamy, W. D.** 1974. Single cell proteins from cellulosic wastes. Biotechnol. Bioeng. **16**:869–890.
5. **Benemann, J. R., J. E. Weissmann, B. L. Koopman, and W. J. Oswald.** 1977. Energy production by microbial photosynthesis. Nature (London) **268**:19–23.
6. **Bull, A. T.** 1980. Biodegradation: some attitudes and strategies of microorganisms and microbiologists, p. 107–136. *In* D. C. Ellwood, J. N. Hedger, M. J. Latham, J. M. Lynch, and J. H. Slater (ed.), Contemporary microbial ecology. Academic Press, London.
7. **Bull, A. T., and J. H. Slater.** 1982. Microbial interactions and community structure, p. 13–44. *In* A. T. Bull and J. H. Slater (ed.), Microbial interactions and communities, vol. 1. Academic Press, London.
8. **Bushell, M. E., and J. H. Slater.** 1981. Mixed culture fermentations. Academic Press, London.
9. **Charley, R. C., and A. T. Bull.** 1979. Bioaccumulation of silver by a multispecies community of bacteria. Arch. Microbiol. **123**:239–244.
10. **Chattopadhay, J. P., and S. K. Bose.** 1980. Control of plant infections by antibiotics and antagonistic organisms. Process Biochem. **15**(5):27–28.
11. **Chet, I., and R. Baker.** 1980. Induction of suppressiveness to *Rhizoctonia solani* in soil. Phytopathology **70**:994–998.
12. **Curtin, M. E.** 1983. Microbial mining and metal recovery: corporations take the long and cautious path. Biotechnology **1**:229–235.
13. **Driessen, F. M.** 1981. Protocooperation of yogurt bacteria in continuous cultures, p. 99–120. *In* M. E. Bushell and J. H. Slater (ed.), Mixed culture fermentations. Academic Press, London.
14. **Driessen, F. M., J. Ubbels, and J. Stadhouders.** 1977. Continuous manufacture of yogurt. I. Optimal conditions and kinetics of the prefermentation process. Biotechnol. Bioeng. **19**:821–839.
15. **Driessen, F. M., J. Ubbels, and J. Stadhouders.** 1977. Continuous manufacture of yogurt. II. Procedure and apparatus for continuous coagulation. Biotechnol. Bioeng. **19**:841–851.
16. **Drozd, J. W., and J. D. Linton.** 1981. Single cell protein production from methane and methanol in continuous culture, p. 113–141. *In* P. H. Calcott (ed.), Continuous culture of cells, vol. 1. CRC Press, Inc., Boca Raton, Fla.
17. **Drozd, J. W., and P. W. McCarthy.** 1981. Mathematical model of microbial hydrocarbon oxidation, p. 360–369. In H. D. Dalton (ed.), Proceedings of the Third International Symposium on Microbial Growth on C_1 Compounds, Sheffield (1980). Heyden & Sons, London.
18. **Dunn, G. M., and A. T. Bull.** 1983. Bioaccumulation of copper by a defined community of activated sludge bacteria. Eur. J. Appl. Microbiol. Biotechnol. **17**:30–34.
19. **Egorov, N. S., and N. S. Landau.** 1982. Biosynthesis of biologically active substances by mixed microbial cultures. Prikl. Biokhim. Mikrobiol. **18**:835–849.
20. **Friend, B. A., J. M. Fiedler, and K. M. Shahani.** 1983. Influence of culture selection on the flavour, antimicrobial activity, β-galactosidase and B vitamins of yogurt. Milchwissenschaft **38**:133–136.
21. **Gaden, E. L., and A. E. Humphrey.** 1977. Single cell protein from renewable and non-renewable resources. Biotechnol. Bioeng. Symp. **7**:R3.
22. **Glanser, M., S. N. Ban, and C. L. Cooney.** 1981. Biodegradation of sodium sulfite liquor (NaSSL) and sodium lignosulphate (NaLS) by a mixed culture of microorganisms. Eur. J. Appl. Microbiol. Biotechnol. **13**:54–59.
23. **Haas, C. N., H. R. Bungay, and M. L. Bungay.** 1980. Practical mixed culture processes. Annu. Rep. Ferment. Processes **4**:1–29.
24. **Haggstrom, M. H., and M. Dostlèk.** 1981. Regulation of a mixed culture of *Streptococcus lactis* and *Saccharomycopsis fibuligera.* Eur. J. Appl. Microbiol. Biotechnol. **12**:216–219.
25. **Han, Y. W.** 1982. Nutritional requirements and growth of a *Cellulomonas* species on cellulosic substrates. J. Ferment. Technol. **60**:99–104.
26. **Harrison, D. E. F.** 1978. Mixed cultures in industrial fermentation processes. Adv. Appl. Microbiol. **24**:129–164.
27. **Henis, Y., and I. Chet.** 1975. Microbiological control of plant pathogens. Adv. Appl. Microbiol. **19**:85–111.
28. **Hobson, P. N.** 1981. Microbial pathways and interactions in the anaerobic treatment plant, p. 52–79. *In* M. E. Bushell and J. H. Slater (ed.), Mixed culture fermentations. Academic Press, London.
29. **Houtmeyers, J., E. Van den Eynde, R. Poffé, and H. Verachtert.** 1980. Relation between substrate feeding pattern and development of filamentous bacteria in activated sludge processes. Part I. Influence of process parameters. Eur. J. Appl. Microbiol. Biotechnol. **9**:63–67.
30. **Howell, C. R., and R. D. Stipanovic.** 1983. Gliovirin, a new antibiotic from *Gliocladium virens* and its role in the biological control of *Pythium ultimum.* Can. J. Microbiol. **29**:321–324.
31. **Ilyaletdinov, A. N., P. B. Enker, and L. V. Loginova.** 1977. Role of sulphate-reducing bacteria in the precipitation of copper. Mikrobiologiya **46**:113–117.
32. **Johnson, D. E., and R. L. Remillard.** 1983. Nutrient digestibility of brewers single cell protein. J. Anim. Sci. **56**:735–739.
33. **Khan, A. W.** 1981. Degradation of cellulose to methane by a co-culture of *Acetivibrio cellulolyticus* and *Methanosarcina barkeri.* FEMS Microbiol. Lett. **9**:233–235.
34. **Khan, A. W., and W. D. Murray.** 1982. Single-step conversion of cellulose to ethanol by a mesophilic co-culture. Biotechnol. Lett. **4**:177–180.
35. **Khan, A. W., D. Wall, and L. van den Berg.** 1981. Fermentative conversion of cellulose to acetic acid and cellulolytic enzyme production by a bacterial mixed culture from sewage sludge. Appl. Environ. Microbiol. **41**:1214–1218.
36. **Kulkarni, S., and B. Rangonathon.** 1983. Utilization of dairy industry by-products. Biocycle **24**:53–55.
37. **Kvasnikov, E. I., N. K. Kovalenko, and O. A. Nesterenko.** 1982. Lactic acid bacteria in nature and economy. Prikl. Biokhim. Mikrobiol. **18**:821–834.
38. **Laine, B. M.** 1974. What proteins cost from oil. Hydrocarbon Process **53**:139–142.
39. **Latham, M. J., and M. J. Wolin.** 1977. Fermentation of cellulose by *Ruminococcus flavefaciens* in the presence and absence of *Methanobacterium rumanantium.* Appl. Environ. Microbiol. **34**:297–301.
40. **Lee, Y.-K.** 1981. The use of algal-bacterial mixed cultures in the photosynthetic production of biomass, p. 151–172. *In* M. E. Bushell and J. H. Slater (ed.), Mixed culture fermentations. Academic Press, London.
41. **Lee, Y.-K., and S. J. Pirt.** 1981. Interactions between an alga and three bacterial species in a consortium selected for photosynthetic biomass and starch production. J. Chem. Technol. Biotechnol. **31**:295–305.
42. **Legros, A., C.-M. Asinardi di San Marzano, H. P. Naveau, and E.-J. Nyns.** 1983. Fermentation profiles in bioconversions. Biotechnol. Lett. **5**:7–12.
43. **Lemmel, S. A., R. C. Heimsh, and L. L. Edwards.** 1979.

Optimizing the continuous production of *Candida utilis* and *Saccharomycopsis fibuligera* on potato processing wastewater. Appl. Environ. Microbiol. **37:**227–232.

44. **Linton, J. D., and J. W. Drozd.** 1982. Microbial interactions and communities in biotechnology, p. 357–406. *In* A. T. Bull and J. H. Slater (ed.), Microbial interactions and communities, vol. 1. Academic Press, London.

45. **MacBean, R. D., R. J. Hall, and P. M. Linklater.** 1979. Analysis of pH-stat continuous cultivation and the stability of the mixed fermentation in continuous yogurt production. Biotechnol. Bioeng. **21:**1517–1541.

46. **Macdonald, I. A., Y. P. Rochon, D. M. Hutchison, and L. V. Holdeman.** 1982. Formation of ursdeoxycholic acid from chenodeoxycholic acid by a 7 β-hydroxysteroid dehydrogenase elaborating *Eubacterium aerofaciens* strain cocultured with 7 α-hydroxysteroid dehydrogenase elaborating organisms. Appl. Environ. Microbiol. **44:**1187–1195.

47. **Meyer, J. S., H. M. Tsuchiya, and A. G. Fredrickson.** 1975. Dynamics of mixed populations having complementary metabolism. Biotechnol. Bioeng. **17:**1065–1081.

48. **Moskowitz, G. J.** 1979. Inocula for blue-veined cheeses and blue cheese flavour, p. 201–210. *In* H. J. Peppler and D. Perlman (ed.), Microbial technology, vol. 2. Academic Press, London.

49. **Mountfort, D. O., R. A. Asher, and T. Bauchop.** 1982. Fermentation of cellulose to methane and carbon dioxide by a rumen anaerobic fungus in a triculture with *Methanobrevibacter* sp. strain RA1 and *Methanosarcina barkeri*. Appl. Environ. Microbiol. **44:**128–134.

50. **Muranaka, M., S. Kinoshita, Y. Yamada, and H. Okada.** 1976. Decomposition of lignin model compound, 3-(2-methoxy-4-formylphenoxy)-1,2-propanediol by bacteria. J. Ferment. Technol. **54:**635–639.

51. **Nelson, P. O., A. K. Chung, and M. C. Hudson.** 1981. Factors affecting the fate of heavy metals in the activated sludge process. J. Water Pollut. Control Fed. **53:**1323–1333.

52. **Ng, T. K., A. Ben-Bassat, and J. G. Zeikus.** 1981. Ethanol production by thermophilic bacteria: fermentation of cellulosic substrates by cocultures of *Clostridium thermocellum* and *Clostridium thermohydrosulfuricum*. Appl. Environ. Microbiol. **41:**1337–1343.

53. **Norris, P. R., and D. P. Kelly.** 1982. The use of mixed microbial cultures in metal recovery, p. 443–474. *In* A. T. Bull and J. H. Slater (ed.), Microbial interactions and communities, vol. 1. Academic Press, London.

54. **Odum, J. M., and J. D. Wall.** 1983. Photoproduction of H₂ from cellulose by an anaerobic bacterial coculture. Appl. Environ. Microbiol. **45:**1300–1305.

55. **Ohshima, M., N. Ishizaki, A. Handa, and Y. Tonooka.** 1983. Cultural conditions for production of neopurpuratin, a purplish-red pigment, by mixed cultures of *Streptomyces propurpuratus* with *Bacillus* sp. J. Ferment. Technol. **61:**31–36.

56. **Ohshima, M., N. Ishizaki, A. Handa, Y. Tonooka, and N. Kanda.** 1981. Formation of a purplish-red pigment by mixed culture of *Streptomyces propurpuratus* with other microorganisms. J. Ferment. Technol. **59:**209–213.

57. **Olson, N. F.** 1979. Cheese, p. 39–77. *In* H. J. Peppler and D. Perlman (ed.), Microbial technology, vol. 2. Academic Press, London.

58. **Pech, Z., and R. Minarik.** 1980. Pouzlti cistych mlekarskych kulturv mlekarenske technologii. Prum. Potravin. **31:**570–572.

59. **Peitersen, N.** 1975. Cellulose and protein production from mixed cultures of *Trichoderma viride* and a yeast. Biotechnol. Bioeng. **17:**1291–1299.

60. **Petitdemange, E., O. Fond, F. Caillet, and H. Petitdemange.** 1983. A novel one step process for cellulose fermentation using mesophilic cellulolytic and glycolytic clostridia. Biotechnol. Lett. **5:**119–124.

61. **Pfeffer, J. T., and J. C. Liebman.** 1976. Energy from refuse by bioconversion, fermentation and residue disposal processes. Resour. Recovery Conserv. **1:**295–313.

62. **Pipes, W. O.** 1978. Microbiology of activated sludge bulking. Adv. Appl. Microbiol. **24:**85–127.

63. **Pretorius, W. A.** 1983. An overview of digestion processes. Water Sci. Technol. **15:**1–6.

64. **Rodriques-Kabana, R., W. D. Kelly, and E. A. Curl.** 1978. Proteolytic activity of *Trichoderma viride* in mixed culture with *Sclerotium rolfsii* in soil. Can. J. Microbiol. **24:**487–490.

65. **Rokem, J. S., I. Goldberg, and R. I. Mateles.** 1980. Growth of mixed cultures of bacteria on methanol. J. Gen. Microbiol. **116:**225–232.

66. **Seiler, H., and H. Blaim.** 1982. Population shifts in activated sludge treatment plants of the chemical industry: a numerical cluster analysis. Eur. J. Appl. Microbiol. Biotechnol. **14:**97–104.

67. **Shankar, P. A., and F. L. Davies.** 1978. Interrelationships of *Streptococcus thermophilus* and *Lactobacillus bulgaricus* in yogurt starters. 20th International Dairy Congress, vol. E, p. 514–515. International Dairy Federation, Brussels.

68. **Shimao, M., Y. Taniguchi, S. Skihata, N. Kato, and C. Sakazawa.** 1982. Production of polyvinyl alcohol oxidase by a symbiotic mixed culture. Appl. Environ. Microbiol. **44:**28–32.

69. **Slater, J. H., and A. T. Bull.** 1982. Environmental microbiology: biodegradation. Philos. Trans. R. Soc. London Ser. B. **297:**575–597.

70. **Somerville, H. J.** 1981. Mixed cultures in aerobic waste treatment, p. 82–97. *In* M. E. Bushell and J. H. Slater (ed.), Mixed culture fermentations. Academic Press, London.

71. **Steinkraus, K. H.** 1982. Fermented foods and beverages: the role of mixed cultures, p. 407–442. *In* A. T. Bull and J. H. Slater (ed.), Microbial interactions and communities, vol. 1. Academic Press, London.

72. **Taguchi, H., T. Yoshida, and K. Nakatani.** 1978. Studies on a model mixed culture for simulation of a bulking problem. J. Ferment. Technol. **56:**158–168.

73. **Ueki, A., E. Miyagawa, H. Minato, R. Azuma, and T. Suto.** 1978. Enumeration and isolation of anaerobic bacteria in sewage digester fluids. J. Gen. Appl. Microbiol. **24:**317–332.

74. **Uribelarrea, J. L., and A. Pareilleux.** 1981. Anaerobic digestion: microbial and biochemical aspects of volatile acid production. Eur. J. Appl. Microbiol. Biotechnol. **12:**118–122.

75. **Van den Eynde, E., J. Geerts, B. Maes, and H. Verachtert.** 1983. Influence of the feeding pattern on the glucose metabolism of *Arthrobacter* sp. and *Sphaerotilus natans* growing in chemostat culture simulating activated sludge bulking. Eur. J. Appl. Microbiol. Biotechnol. **17:**35–43.

76. **Verachtert, H., E. Van den Eynde, R. Poffé, and H. Houtmeyers.** 1980. Relation between substrate feeding pattern and development of filamentous bacteria in activated sludge processes. Part II. Influence of substrates present in the influent. Eur. J. Appl. Microbiol. Biotechnol. **9:**137–149.

77. **Weetall, H. H., B. P. Sharma, and C. C. Detar.** 1981. Photo-metabolic production of hydrogen from organic substrates by free and immobilized mixed cultures of *Rhodospirillum rubrum* and *Klebsiella pneumoniae*. Biotechnol. Bioeng. **23:**605–614.

78. **Winter, J., and R. S. Wolfe.** 1979. Complete degradation of carbohydrate to CO₂ and CH₄ by syntrophic cultures of *Acetobacterium woodii* and *Methanosarcina barkeri*. Arch. Microbiol. **121:**97–102.

79. **Wittauer, D. P.** 1980. Biocoenosis and degradation in model waste-water treatment plants. Eur. J. Appl. Microbiol. Biotechnol. **9:**151–163.

80. **Wolin, M. J.** 1982. Hydrogen transfer in microbial communities, p. 323–356. *In* A. T. Bull and J. H. Slater (ed.), Microbial interactions and communities, vol. 1. Academic Press, London.

81. **Wood, B. J. B.** 1981. The yeast/lactobacillus interaction: a study in stability, p. 137–150. *In* M. E. Bushell and J. H.

Slater (ed.), Mixed culture fermentations. Academic Press, London.

82. **Yoshida, T., B. S. M. Rao, S. Ohasa, and H. Taguchi.** 1979. Dynamic analysis of a mixed culture in chemostat. J. Ferment. Technol. **57:**546–553.

83. **Yoshida, T., H. Taguchi, S. Kulprecha, and N. Nilubol.** 1981. Kinetics and optimization of steroid transformation in a mixed culture, p. 501–506. *In* M. Moo Young, C. Vezina, and K. Singh (ed.), Advances in biotechnology, vol. 3, Fermentation products. Pergamon Press, Toronto.

84. **Zeikus, J. G.** 1980. Chemical and fuel production by anaerobic bacteria. Annu. Rev. Microbiol. **34:**423–464.

Biodegradation of Xenobiotics

Experimental Evolution of Azo Dye-Degrading Bacteria

H. G. KULLA,† R. KRIEG, T. ZIMMERMANN, AND T. LEISINGER

Department of Industrial and Engineering Chemistry and Department of Microbiology, Swiss Federal Institute of Technology, Zurich, Switzerland

Azo dyes are produced in vast quantities and are mainly used in textile dyeing. Their characteristic chemical structures, the azo linkage and the aromatic sulfo group, are not synthesized in nature. Thus, it is not surprising that these man-made compounds are recalcitrant. Effluents from dye-manufacturing plants and from dye works are therefore not suitable for biological treatment but have to be subjected to expensive physical decoloration procedures. Since efforts to isolate from nature microorganisms which use azo dyes as carbon sources were unsuccessful, we have bred dye-degrading bacteria in chemostats. Clearly, evolutionary processes in nonuniform continuous cultures can be very complex. Nevertheless, we were able to observe and follow up a crucial step in the acquisition of the novel pathway, the appearance of specific azoreductases. These enzymes were shown to catalyze the first step in the catabolic sequences that convert our model azo dyes to carbon dioxide, ammonia, and bacterial cell mass (Fig. 1). Introduction of aromatic sulfo groups to the dyes did not hinder the azoreductases but led to disturbance of the further degradative pathways, and complete mineralization was not achieved (6). Obviously, it is necessary to introduce an efficient desulfonating system into our bacterial strains.

The scope of this paper is to summarize briefly the current state of our work, with an emphasis on the genetics and enzymology of the organisms adapted for growth on carboxylated azo dyes. Some of the evolutionary principles that led to these new activities were thereby clarified. Details as well as experimental procedures have been or will be published.

KF STRAINS

The relationships between the KF strains and their biodegradative abilities are shown in Fig. 3. The parent strain, and starting point in the scheme, is strain KF 1. Earlier, this strain was tentatively identified as a member of the genus "*Flavobacterium*" (5). Further characterization, however, showed that it belongs to the

† Present address: Lonza Ltd., Visp, Switzerland.

genus *Pseudomonas*, but it could not be assigned to any described species.

Strain KF 1 was isolated directly from soil on a dicarboxyazobenzene (DCAB; Fig. 2) medium (5; G. Overney, Ph.D. thesis, Federal Institute of Technology, Zurich, Switzerland, 1979). No other azo dyes supported growth. KF 1, as well as all other isolates that utilize DCAB, also grew with the closely related compound 4,4'-dicarboxyazoxybenzene (Fig. 2), an excretion product of the insect pathogenic fungus *Entomophthora virulenta* (1). Therefore, it was suggested (5) that biodegradation of DCAB was not an "unnatural" activity.

We had evidence that the first catabolic step was the reduction of the azo bridge, yielding two molecules of 4-aminobenzoate (5). However, it proved to be difficult to assay the DCAB-specific azoreductase. In vitro, the reaction seems to be oxygen sensitive, whereas whole cells grow with DCAB even when aerated vigorously with pure oxygen.

Strain KF 1 growing in continuous culture with DCAB as limiting carbon source was subjected to selective pressure with the intention of expanding its substrate specificity to include carboxy (c) Orange II (Fig. 1) and ultimately the sulfonated dye Orange II (Fig. 2). This long-term continuous culture was run under nonsterile conditions. A graphical representation of the adaptation process showing the changes in chemical concentrations as the new degrading culture evolved in the chemostat is given in reference 5. Strain KF 1 seemed to have been outcompeted by strain KF 4, which had a different, more slimy colony morphology on rich agar. Later, we realized that both strains contained the same plasmids (see below). Interestingly, an independent DCAB isolate (strain PB 4 isolated by M. Risch in our laboratory) had a completely different set of plasmids. We subsequently isolated mutants of strain KF 4 with a rough colony morphology identical to that of KF 1. It is safe to assume that KF 4 is a descendant of KF 1.

Breakthrough to growth with c Orange II occurred after 400 generations, and strain KF 44 (5) was isolated; 255 generations later, strain KF 46, a strain with improved degradative abilities,

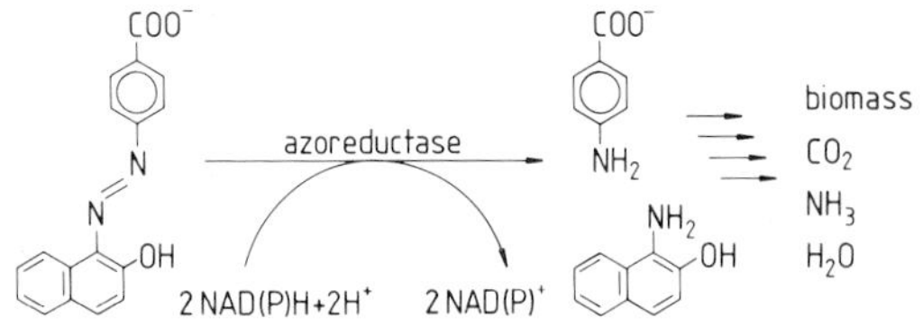

c Orange II
(carboxy Orange II)

FIG. 1. Orange II azoreductase initiating azo dye degradation.

was isolated. These strains harbor an oxygen-insensitive Orange II azoreductase which has been purified and characterized in detail (12).

A continuous-culture device was developed that allows cultivation under sterile conditions for long periods of time. This chemostat was used to cultivate strain KF 46 with c Orange II as limiting carbon source while c Orange I (Fig. 2) was supplied in excess. After 125 generations, the culture efficiently degraded c Orange I, and c Orange II was removed from the input medium. Another 91 generations elapsed before clones could be isolated from the culture that stably inherited the trait of c Orange I degradation. Strain KF 800 possessed a new, Orange I-specific azoreductase, and the reaction catalyzed by this enzyme was oxygen sensitive, resembling the activity of the DCAB azoreductase mentioned above. The three azoreductases of strain KF 800 with specificities for DCAB, Orange II, and Orange I can be separated by electrophoresis and detected by activity staining of gels (T. Zimmermann et al., in preparation).

Another line in Fig. 3 leads to strain KF 158, a nitrosoguanidine-induced mutant that had lost its ability to grow with DCAB but retained its c Orange II and 4-aminobenzoate growth. It was detected as a minicolony on a DCAB plate that in addition contained a small amount of yeast extract. This procedure was necessary, as a counterselective agent could not be found.

CURING AND PLASMIDS IN THE KF STRAINS

Treatment with curing agents such as ethidium bromide, acridine orange, and novobiocin did not lead to loss of the c Orange II- or DCAB-degradative abilities. However, when c Orange II-positive strains were grown for 10 generations in nutrient broth at 40°C, more than 99% of the progeny had irreversibly lost the c Orange II-biodegradative ability. The DCAB, and in the case of KF 800 the c Orange I, degradation was preserved. Orange II azoreductase could not be detected in the cured strains. Growth in chemo-

stats under selective conditions as described by Kulla et al. (6) could also lead to curing. These "substrate-cured" strains were indistinguishable from the temperature-cured ones.

Analysis of DNA in the KF strains (R. Krieg et al., in preparation) revealed that all of them carried two extrachromosomal elements. Plasmid pME1500 had a size of 120 kilobases (kb) and yielded identical restriction patterns in all strains. Plasmid pME1501 had a size of approximately 445 kb and was found in all strains shown within the "c Orange II field" in Fig. 3. All other strains had a plasmid of 305 kb, and restriction analysis revealed that it originated by a single deletion of 140 kb from pME1501. Or, looking at it the other way round, it becomes evident that pME1501 evolved by a 140-kb insertion into the 305-kb plasmid upon entering the "c Orange II field." The acquisition of c Orange I growth apparently did not alter the plasmid composition.

The fact that Orange II azoreductase vanished completely and irreversibly from cured strains suggests that this enzyme is encoded in the 140-kb segment. We do not know what other functions it provides. This piece of DNA might have arisen internally by duplication and fluctuation, or, as the culture was run under unsterile conditions, it might have been provided by an external donor. Hybridization with cloned DNA from the 140-kb insertion/deletion segment will provide the answer.

CONJUGATION WITHIN THE KF STRAINS

The underlined strains in Fig. 3 were marked by consecutive, spontaneous mutation to streptomycin and rifampin resistance and were used as acceptors in filter-mating experiments with KF 46 as donor. One resistance marker and growth on either c Orange II or DCAB were used for selection of transconjugants; the second resistance marker served as a control. Transfer frequencies were 10^{-6} per donor. Both charac-

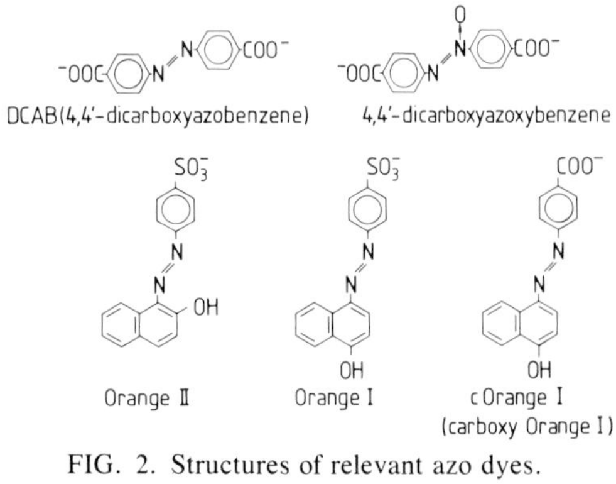

FIG. 2. Structures of relevant azo dyes.

ters, DCAB and c Orange II utilization, were transferable (dotted arrows in Fig. 3), but no cotransfer occurred. With KF 800 as a donor (results not included in Fig. 3), it was shown that the c Orange I trait did not cotransfer with either the DCAB or the c Orange II marker. We do not know whether the c Orange I marker in KF 800 is transferable, as growth with c Orange I under selective conditions was not strong enough to yield conspicuous colonies on a negative background.

These results suggest that not only the c Orange II utilization but also the DCAB utilization was plasmid bound. Possibly, the 140-kb insertion and the rest of pME1501 segregate upon transfer. Alternatively, the DCAB degradation might be encoded on pME 1500. Insertion of transposons will allow a detailed analysis.

COMPARISON OF K STRAINS AND KF STRAINS

The K strains were derived by basically the same selection procedures used for the KF strains, namely, by selecting for growth with c Orange I in a continuous culture with strain KF 1 growing on DCAB as limiting carbon source under nonsterile conditions. After 100 generations, we isolated *Pseudomonas* sp. K 22 (5),

and 146 generations later, from the same culture, we isolated strain K 24, which is closely related but has improved degradative abilities. These strains grow with c Orange I as sole carbon, nitrogen, and energy source. They do not seem to be closely related to the KF strains, judging by their colony morphology, their apparent lack of plasmids, and their inability to grow with DCAB. Their oxygen-insensitive, Orange I-specific azoreductase shares many properties with the Orange II azoreductase. In fact, the same protocol was used for purification (Zimmermann et al., in preparation). Orange I azoreductase from strain K 24, however, is considerably smaller than the Orange II-specific enzyme (21,000 versus 30,000 daltons) and is synthesized constitutively.

IMMUNOLOGICAL CROSS-REACTIVITY OF THE AZOREDUCTASE

The results of experiments on cross-reactivity (Zimmermann et al., in preparation) are summarized in Fig. 4. Purified Orange I azoreductase from strain K 24 and purified Orange II azoreductase from strain KF 46 reacted only with homologous antisera prepared against each of the pure enzymes, suggesting that the azoreductases were unrelated. However, crude extracts

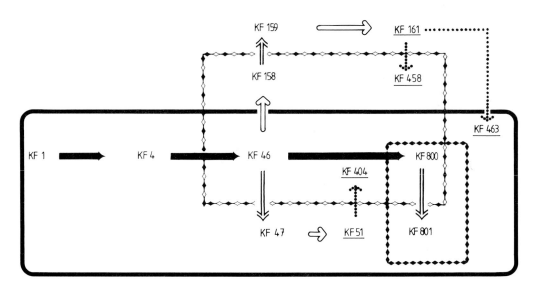

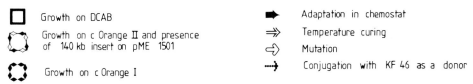

FIG. 3. Evolutionary pathway of the KF strains. For chemical formulas see Fig. 1 and 2. Underlined strains carry rifampin and streptomycin resistance markers.

of strains K 24, KF 46, and azoreductase-nega-
tive strains such as KF 1, KF 4, and KF 47
contained a protein that cross-reacted with both
antisera. Such cross-reacting material could not
be found in extracts of control organisms such as
P. putida or *P. aerogenes*. It is tempting to
speculate that the cross-reacting material represents a precursor of the azoreductases. It will be
interesting to compare the precursor from K 24
with the one from KF 46. The immunological
cross-reactivity indicates that they might be
closely related.

The oxygen-sensitive azoreductases with
specificities for DCAB and Orange I (KF 800)
did not react with the antisera and seem to
belong to a class of their own.

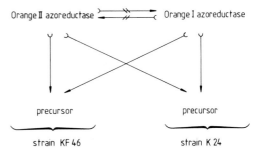

FIG. 4. Schematic representation of immunological
relationships between azoreductases. The arrows indicate that the antibodies raised against the protein
shown on the shaft cross-reacted with the protein
shown on the tip. Broken arrows indicate that no
cross-reaction occurred.

CONCLUSIONS

Reduction of azo dyes under anaerobic conditions is quite common and can in fact be catalyzed by a variety of biological systems. The
resulting accumulation of toxic aromatic amines,
which are not degraded under anaerobic conditions, prevents application of this anaerobic
decoloration in wastewater treatment (8). A bi-
phasic-type treatment in which an azo dye-
containing effluent is pretreated in an anaerobic
digester before it enters an activated sludge
treatment system might be problematic, as aromatic amines, particularly naphtholamines, rapidly form toxic brown polymerization products
when in contact with oxygen.

It is thought that azo dyes interact rather
nonspecifically with compounds of the electron-
transport chain and thereby serve as artificial
electron acceptors. Molecular oxygen inhibits
this reaction since the electrons do not reach the
azo dyes when the natural acceptor is present
(11) and since the postulated intermediates of
the reaction, hydrazine or the azo anion-free
radical, or both, are reoxidized by O_2 (7).

Our oxygen-sensitive azoreductases, namely,
the DCAB-specific enzyme and the Orange I
azoreductase of KF 800, differ from this system
by their substrate specificity and their apparent
in vivo resistance to oxygen.

The rat liver cytosolic azoreductase described
by Huang et al. (4) is oxygen insensitive.
The enzyme seems to be identical to the diapho-
rase and, probably as a result of some degree of
nonspecificity, also reduces methyl red (2'-car-
boxy-4-N,N-dimethylaminoazobenzene) with
NAD(P)H as an electron donor in the presence
or absence of O_2. In the presence of oxygen,
however, the reaction is slightly uncoupled, and
H_2O_2 is produced (3).

The other class of our azoreductases, represented by the Orange II azoreductase and the
Orange I azoreductase of strain K 24, are completely oxygen insensitive. Apparently, the

binding of the dye to the enzymes is strong
enough to prevent the reaction intermediates
from being attacked and reoxidized by oxygen
under both in vivo and in vitro conditions. We
believe that these enzymes were created in
response to the highly selective environment in
the chemostats.

In recent years, it has become more and more
evident that rearrangement of DNA segments is
the main driving force of evolution. Plasmids, in
particular (see for instance the review of the
TOL plasmid by Williams et al. [10]), have an
amazing degree of flexibility which seems to
enable them to assemble and combine new activities. A fine example of genes encoding enzymes
with "unnatural" activities becoming arranged
on a plasmid is the nylon oligomer hydrolases
described by Okada et al. (9). Their bacterial
strain was isolated from the discharge of a nylon
factory, so the preadaptation state of the microorganism is not known.

The important role of transposons in the evolution of RTF-type plasmids has been discussed
(2). Analogy suggests that biodegradative plasmids evolve in a similar fashion. The 140-kb
insertion/deletion piece of DNA in our KF
strains seems to be rather big to be referred to as
a transposon. Thus, it will be interesting to see
whether it is flanked by inverted-repeat units,
which would account for its mobility.

As an exception to the rule that plasmids are
the workbench of evolution, the Orange I azoreductase in strain K 24 seems to be encoded in
the chromosome and resisted all our attempts to
mobilize it. In spite of this dissimilarity to the
Orange II azoreductase, the enzymes seem to be
evolutionarily linked by analogous precursors. It
is possible that the gene for the Orange I azoreductase was inserted into the chromosome simply because of the absence of plasmids in this
strain. Analysis of the precursor proteins will

tell us more about the origin and dissemination of the azoreductase genes.

The degradative potential of the bacterial strains obtained in this study is restricted to specific dyes. Since industrial effluents contain mixtures of dyes with different chemical structures, the microbes available at this time are of no immediate use in practical processes. However, detailed analysis of the genetic events and of the biochemical changes in the bacteria during the evolution of degradative potential for azo dyes will help us to understand the mechanisms by which bacteria acquire the ability to degrade man-made chemicals. In the future this information may be applied in the construction of bacteria capable of degrading a range of previously nondegradable chemicals. Such versatile microbial strains will be of use for the degradation of industrial wastes in controlled processes at their sources of production.

LITERATURE CITED

1. **Claydon, N.** 1978. Insecticidal secondary metabolites from entomogenous fungi: *Entomophthora virulenta*. J. Invertebr. Pathol. **32**:319–324.
2. **Cohen, S. N.** 1976. Transposable genetic elements and plasmid evolution. Nature (London) **263**:731–738.
3. **Huang, M.-T., G. T. Miwa, N. Cronheim, and A. Y. H. Lu.** 1979. Rat liver cytosolic azoreductase: electron transport properties and the mechanism of dicumarol inhibition of the purified enzyme. J. Biol. Chem. **254**:11223–11227.
4. **Huang, M.-T., G. T. Miwa, and A. Y. H. Lu.** 1979. Rat liver cytosolic azoreductase: purification and characterization. J. Biol. Chem. **254**:3930–3934.
5. **Kulla, H. G.** 1981. Aerobic bacterial degradation of azo dyes, p. 387–399. *In* T. Leisinger, A. M. Cook, J. Nüesch, and R. Hütter (ed.), Microbial degradation of xenobiotics and recalcitrant compounds. Academic Press, London.
6. **Kulla, H. G., F. Klausener, U. Meyer, B. Lüdecke, and T. Leisinger.** 1983. Interference of aromatic sulfo groups in the microbial degradation of the azo dyes Orange I and Orange II. Arch. Microbiol. **135**:1–7.
7. **Mason, R. P., F. J. Peterson, and J. L. Holtzman.** 1978. Inhibition of azoreductase by oxygen. The role of the azo anion free radical metabolite in the reduction of oxygen to superoxide. Mol. Pharmacol. **14**:665–671.
8. **Meyer, U.** 1981. Biodegradation of synthetic organic colorants, p. 371–385. *In* T. Leisinger, A. M. Cook, J. Nüesch, and R. Hütter (ed.), Microbial degradation of xenobiotics and recalcitrant compounds. Academic Press, London.
9. **Okada, H., S. Negoro, and S. Kinoshita.** 1982. Enzymes active on unnatural synthetic compounds: nylon oligomer hydrolases controlled by a plasmid and their cloning. Enzyme Eng. **6**:491–500.
10. **Williams, P. A., P. A. Cane, D. J. Jeenes, and R. W. Pickup.** 1983. Correlation between spontaneous phenotypic changes in *Pseudomonas* strains with changes in the structure of catabolic plasmids: experiences with TOL plasmids, p. 519–549. *In* A. Hollaender, A. I. Laskin, and P. Rogers (ed.), Basic biology of new developments in biotechnology. Plenum Press, New York.
11. **Wuhrmann, K., K. Mechsner, and T. Kappeler.** 1980. Investigation on rate-determining factors in the microbial reduction of azo dyes. Eur. J. Appl. Microbiol. Biotechnol. **9**:325–338.
12. **Zimmermann, T., H. G. Kulla, and T. Leisinger.** 1982. Properties of purified Orange II azoreductase, the enzyme initiating azo dye degradation by *Pseudomonas* KF 46. Eur. J. Biochem. **120**:197–203.

Biodegradation of Chlorophenolic Compounds in Wastes from Wood-Processing Industry

M. S. SALKINOJA-SALONEN, R. VALO, J. APAJALAHTI, R. HAKULINEN, LIISA SILAKOSKI, AND T. JAAKKOLA

Department of General Microbiology and Department of Radiochemistry, University of Helsinki, Helsinki, and Enso Gutzeit OY, Research Centre, Imatra, Finland

In Finland 6.6 million metric tons of pulp were produced in 1982, of which 3.0 million tons were bleached (3), involving the use of 0.18 million metric tons of chlorine. Approximately 10% of this chlorine reacts with organic compounds and is released from the mill in the effluent water (8, 47; M. S. Salkinoja-Salonen, R. Hakulinen, R. Valo, and J. Apajalahti, Water Sci. Technol., in press). The annual pollution load of chloroorganic compounds from the Finnish chemical wood processing industry can be estimated from production figures by using data for the pollution load of bleach plant effluents (27, 32, 36, 47), which is 100,000 tons of chlorolignin, 1,000 tons of chlorinated phenolic compounds, 1,000 to 2,000 tons of chloroform, and an as yet unknown amount of chlorinated resin acids, steroids, and other wood extractives. During the manufacture of timber and lumber, wood-preserving chemicals are used, chlorophenols being the most widely applied in Finland. Their estimated usage is 1,500 tons per year in Finland.

Chloroorganic chemicals are known for their biological recalcitrance (16), and as a result their biodegradation has been widely studied (for recent reviews, see, for example, 31, 35). However, the chemicals that have been investigated are mostly pesticides or other industrial products. In 1978, when we started to study the environmental fate of chloroorganic chemicals that arise from the wood-processing industry, this field was little studied. In this paper we describe some of our recent results on the environmental behavior of chloroorganic compounds from the wood-processing industry and the possibilities of using biological treatment for reducing the pollution load from this branch of industry. (For earlier results, see 18–22, 46, 47; Salkinoja-Salonen et al., in press.)

MATERIALS AND METHODS

Analytical procedures. Inorganic chloride, organically bound chlorine, and chlorophenolic compounds were analyzed as previously described (47). Total chlorine was also determined by instrumental neutron activation analysis with a multichannel analysis and a Ge(Li) crystal. For assay, the samples were irradiated in sealed plastic vials. Empty vials were irradiated to check for the background level of chlorine. Solid NH_4Cl was used for calibration. The results of the radiochemical method were generally consistent with those obtained by the wet combustion method (47).

The identity of pentachlorophenol and 2,3,4,6-tetrachlorophenol was verified by mass fragmentography (ions 167, 263, 266, 268, 271, 294) after positive identification by comparison of retention times in an 80-m capillary column OV-101 with those of authentic reference compounds in gas-liquid chromatography as previously described (47).

Total ^{14}C and $^{14}CO_2$ disintegrations per minute were assayed from 0.5- to 1.0-ml samples as previously described (22).

Dating of the sediment layers was performed by assaying ^{137}Cs as described by Robbins and Edgington (43) and Salkinoja-Salonen et al. (47) and by the ^{210}Pb method (4, 43).

Sampling. The collection of water samples was performed as previously described (36), and frozen sediment samples were prepared as described by Huttunen and Meriläinen (28) and Salkinoja-Salonen et al. (47). Chlorophenolic compounds were analyzed from freshly thawed samples. Air-dried samples were used for the measurement of dry matter, for neutron activation, and for wet combustion.

Media and chemicals. The mineral salts medium contained, per liter: $K_2HPO_4 \cdot 3H_2O$, 3.8 g; KH_2PO_4, 2.1 g; $(NH_4)_2SO_4$, 2 g; $MgSO_4 \cdot 7H_2O$, 0.3 g; $CaCl_2$, 10 mg; $FeSO_4$, 0.01 mg; trace element solution (5), 1 ml; and yeast extract, 10 mg. The chloride content of the medium was 0.5 mM, and the pH was set at 7.0. Uniformly ^{14}C-labeled pentachlorophenol, 25.6 mCi/g, was a gift from M. Fischer (Institut für Wasser-, Boden-, und Lufthygiene des Bundesgesundheitamtes, Berlin, Federal Republic of Germany). All other chemicals were reagent grade and were obtained from commercial sources.

Biodegradation studies. Biodegradation experiments were carried out at room temperature, if not otherwise indicated in the text, in biological reactors, trickling filters (20; Salkinoja-Salonen et al., in press), or fluidized beds (18, 19, 21) of sizes varying from 0.3 liters to 1.3 m^3. All reactors were shielded from light either by the vessel material (stainless steel) or by wrapping the reactor column with black plastic.

RESULTS AND DISCUSSION

Behavior of chloroorganic chemicals from wood-processing industry in the environment. We investigated the occurrence of chlorine compounds in the recipient lake of a big pulp and paper mill, the bleachery of which had been in operation since 1954. The production of bleached kraft pulp by this mill had gradually risen from 40,000 tons in 1954 to the present-day level of 300,000 tons per year. Figure 1 shows the profiles of organically bound chlorine, inorganic chloride, and chlorophenolic compounds in sediment samples taken in 1980 from three locations, both downstream and upstream of the effluent outlet of the mill. Dating of the sediment layers by ^{137}Cs and ^{210}Pb showed that the sedimentation had been undisturbed at locations A and B, but slightly disturbed at location C, which was a "clean" reference area. Sedimentation velocity, calculated from the dating, was between 6 and 7 mm per year at location A, 3 mm at location B, and less than 1 mm at location C.

Table 1 shows the results for some individual chlorophenolic compounds in the sediment layers and in samples of water taken from the surface and the bottom of the lake at the same locations. It can be seen from data presented in Fig. 1 and Table 1 that chlorine constituted approximately 0.7% of the dry matter in the heavily polluted area (location A), most of it organically bound. In less polluted areas B and C, less than 0.1% of the dry matter was chlorine. We observed a similar gradient of organically bound chlorine in sediment samples taken in 1979 downstream of the mill (47). The sediment chlorine most probably originates from the effluent stream of the mill. Chlorophenols and related compounds (listed in Table 1) made up only a minor fraction of this chlorine, approximately one-one hundredth of the total.

The data in Table 1 show that chlorophenolic compounds accumulated from the lake water into the sediment: the concentration in the water was only a few micrograms per liter, whereas there was 1 to 40 mg of these chemicals per kg (dry weight) of sediment. (Sediment dry matter content varied from 3 to 5% of the wet weight.) A similar observation was reported for pentachlorophenol by Murray et al. (37) for sediments in the gulf of Mexico and by Pierce and Victor (41) for penta- and tetrachlorophenols in lake sediments after an accidental release of wood-treating wastes.

The spectrum of the chlorophenolic compounds changed with an increasing distance

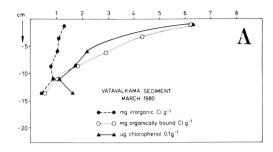

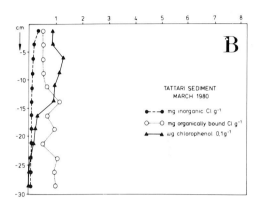

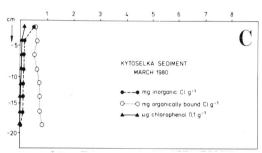

FIG. 1. Occurrence of organically bound chlorine, inorganic chloride, and some chlorophenolic compounds in dated sediments of the southern part of Lake Saimaa, which is the recipient of effluents from a mill producing bleached kraft pulp from softwood and hardwood. The bleach plant of the mill went into operation in 1954. The chlorophenols, the sum of which is shown in the figure, are listed in Table 1. Site A was located 3 km downstream and site B was 7 km downstream from the mill; site C was 10 km upstream from site B.

from the mill: near the mill (locations A and B) di- and trichlorophenols, catechols, and guaiacols dominated, but at location C pentachlorophenol was the dominant chlorophenol (Table 1 and similar data described elsewhere [47]). There were relatively much more chlorinated catechols in the sediment than in the lake water

TABLE 1. Occurrence of some chlorophenolic compounds in water and sediments of the recipient lake of bleaching effluents from kraft pulping[a]

Compound	Concn (μg/liter of water)		Concn (μg/kg of sediment, dry wt) at a depth of:		
	Surface	Bottom	0–5 cm	5–10 cm	10–15 cm
2,4-Dichlorophenol	10	2	18,000	15,000	23,000
2,6-Dichlorophenol	1.2	1.6	38,000	17,300	
2,4,6-Trichlorophenol	1.2	1.6	190	70	50
Chlorinated catechols	0.60	0	22,000	9,000	800
2,3,4,6-Tetrachlorophenol	0.1	0.2	100	120	130
Trichloroguaiacol	0.4	0.8	640	400	200
Pentachlorophenol	0.03	2.2	550	550	700

[a] Water and sediments sampled in 1980 from location A in Lake Saimaa were analyzed for chlorinated phenolic compounds as described previously (47). The chlorinated catechols are given as the sum of 3,4,5-trichlorocatechol and tetrachlorocatechol. The trichloroguaiacol is given as the sum of the two isomers. The figures given in the table are the mean of four sampling rounds (February, May, August, October). The seasonal variation of the concentrations was up to threefold.

(Table 1) or in the effluent stream (47). These may represent products of microbial metabolism from other chlorinated compounds present in the effluent stream, such as chloroguaiacols and dimethoxyphenols. We did not detect methylated derivatives (anisoles, veratroles, syringoles) in the sediment samples, although we looked for tetrachloroveratrole and di-, tri-, tetra-, and pentachlorinated anisoles, for which we had authentic reference compounds (47) for comparison with on gas-liquid chromatography. Methylated derivatives of chlorophenolic compounds have been observed in anaerobic soil, sediment, and litter by several other workers (7, 15, 29, 38–40, 44) and are believed to have arisen from the metabolism of chlorophenolic compounds by microorganisms.

The concentration of different chlorophenols (as milligrams per gram of sediment dry matter) was found to decrease by 1 to 2 orders of magnitude with an increasing distance from the mill, except for pentachlorophenol, which remained stationary between 0.3 and 0.7 mg/kg of dry matter in all sediment samples studied. This indicates that most of the sediment chlorophenolic compounds may originate from the pulp mill, except for pentachlorophenol. Our observation is analogous to that of the Swedish workers (2) who studied the occurrence of pentachlorophenol in Baltic sea sediments and found that its concentration did not respond to the proximity of effluent streams from pulp bleacheries: it was the same in polluted and in unpolluted areas, namely, 50 to 200 μg of pentachlorophenol per kg of wet weight.

Table 2 shows the results of the analyses of chlorophenols in soil around a wood-preserving facility. It can be seen that the relative concentrations of the different chlorophenols in soil and sludges closely follow those of the chemical preservative that had been used. Their concentration is quite high and they seem to seep deep

into the soil (at least 45 cm), indicating that degradation by the soil microorganisms (which is known to occur under certain circumstances [29, 38, 44]) has been inadequate. We did not find any possible intermediates of biodegradation in these soil samples, although we looked for chlorocatechols, tetrachloroveratrole, and chloroanisoles. It appears that the chlorophenols used as wood preservative are quite recalcitrant in the climate (cold) and soil type (aerobic sandy soil) studied here.

Chlorophenols in unpolluted area. To get more information on the origin of pentachlorophenol in environmental samples, we analyzed lake sediments of undisturbed, unpolluted small lakes. Figure 2 and Table 3 show typical results of the analyses of a forest lake, situated in eastern Finland several kilometers from the nearest habitation and tens of kilometers from the nearest industrial site. There was no river flowing into this lake, and one can therefore assume that it only receives runoff from the nearby forest hills.

The results show that this virgin lake was also polluted by pentachlorophenol. The concentration of pentachlorophenol was highest (0.4 to 0.8 mg/g of dry matter) in the youngest layer, where the concentration found is actually the same as in the sediments of the polluted Lake Saimaa. Lake Mustalampi is an oligotrophic lake with a low rate of sedimentation, 70 ± 35 g of dry matter m^{-2} $year^{-1}$. In Lake Saimaa the rate of sedimentation was threefold (location B) to fivefold (location A) higher. Therefore, the annual rate of pentachlorophenol deposition was higher by the same factor in Lake Saimaa than in Lake Mustalampi. The annual deposition of pentachlorophenol in Lake Mustalampi can be calculated to have increased from 2 to 4 μg/m^{-2} 50 years earlier to the present-day level of 10 to 17 μg/m^{-2} per year. There was much less deposition of tetrachlorophenol and no detectable de-

TABLE 2. Occurrence of chlorinated phenols in soil and drainage water at a sawmill at which chlorophenolic wood preservative was used[a]

Site[b]	Amt of chlorophenol[c] (g/kg of soil)			
	2,4,6-Trichlorophenol	2,4,5-Trichlorophenol	Tetrachlorophenol	Pentachlorophenol
Sludge of dipping basin	3.8	1.8	74	19
Soil at the dripping area from freshly treated lumber	5.3		122	4.9
Soil at storage site of lumber, 45-cm depth	1.3	0.3	42	2.7
Drainage ditch from the storage site to a nearby creek, per liter of water	0.8	0.03	14	0.5

[a] The chemical wood preservative in use at this facility since 1979 contained 2,4,6-trichlorophenol, 2,3,4,6-tetrachlorophenol, and pentachlorophenol in the ratio 1:20:12 by weight, and smaller amounts of 2,4,5-trichlorophenol and 2,3,4,5-tetrachlorophenol.

[b] The samples were collected in May 1982 at a sawmill in Southern Finland where in total (since 1979) approximately 300,000 m^3 of timber was dipped in chlorophenol-containing preservative during the months May to October.

[c] The chlorophenols were extracted and analyzed as described previously (47). Tetrachlorophenol was mainly 2,3,4,6-isomer.

position of the other chlorophenolic compounds listed in Table 1.

We cannot think of any other explanation for the presence of pentachlorophenol in unpolluted lake sediment than airborne distribution. We were puzzled over the presence of pentachlorophenol in the layers older than 50 years, formed long before there was industrial production of pentachlorophenol. The amount clearly exceeded the level of detection and that of laboratory contamination. We suggest that it may originate from forest fires and the use of wood for heating purposes, as Ahling and Lindskog (1) have shown that up to 1 mg of penta- and tetrachlorophenol, but no di- or trichlorophenol, was formed per kg of fresh, untreated timber burned in an open fire. Pentachlorophenol may therefore actually be less alien to the ecosystem than has been hitherto thought.

Universal pollution of the aquatic environment by pentachlorophenol has also been reported by other workers, both in Finland (40) and in other countries (12, 14).

Microbial degradation of chloroorganic chemicals during effluent treatment. The accumulation and universal occurrence of chlorinated phenolic compounds in the environment may indicate the absence of microorganisms capable of degrading these chemicals or inhibition of biodegradation by physical or chemical factors, such as low temperature, anoxicity, or shortage of nutrients or cofactors needed for efficient degradation. Removal of chloroorganic chemicals during biological effluent treatment has been observed by various authors (for a review, see 8), but many studies lack data on mass balance. This makes it difficult to decide whether the chemicals have been removed by sorption to sludge,

by stripping due to aeration, by photochemical catalysis (17, 35, 45, 50), or by true biodegradation. In their recent field study, Lurker et al. (34) concluded that compounds like chloroform, tetrachloroethylene, and hexachlorocyclohexane were emitted from the wastewater at an activated sludge plant by air stripping and sorption to biomass rather than biodegradation. Leuenberger et al. (33) found that the lipophilic chlorinated phenols were physically absorbed from pulp and paper mill effluents to activated sludge, whereas chloroform and other purgeable com-

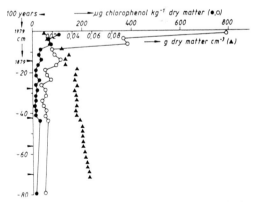

FIG. 2. Concentrations of pentachlorophenol (○), 2,3,4,6-tetrachlorophenol (●), and dry matter (▲) in a dated sediment profile from an oligotrophic, unpolluted lake. Lake Mustalampi is located 61°22′N, 28°29′E. The sediment was collected in March 1980, dated by determining the profile of ^{210}Pb, and analyzed for chlorophenols and chlorine. The laboratory background levels for the detection of penta- and tetrachlorophenols corresponded to concentrations of 6 and 15 µg/kg of dry weight, respectively. The content of chlorine in the sediment layers is given in Table 3.

TABLE 3. Contents of chlorine in Mustalampi sediment (1980) profile

Depth (cm)	Cl (mg/g)[a]
2.5–5	0.75
5–7.5	0.78
7.5–10	0.88
12.5–15	0.78
22.5–25	0.65
47.5–50	0.66
72.5–75	1.25

[a] Determined by the [38]Cl method with a standard error of up to 52% of the values stated.

pounds were removed mainly by transfer into ambient air.

We studied the possibilities of using nonaerated biological fixed-film reactors for treatment of effluents from the wood-processing industry (18, 20–22; Salkinoja-Salonen, in press). When there is no aeration, the analysis of mass balance of the chloroorganic compounds is easy.

We found that it was possible to adapt the microflora of an anoxic biofilm reactor to degrade chlorophenolic compounds, chloroform, and chlorinated resin acids occurring in effluents from the wood-processing industry (19, 20, 22).

Table 4 shows recent results of effluent treatment on semitechnical and full scale. The mixed effluent of an integrated kraft pulp mill with a bleach plant was treated by three parallel systems: aerated lagoon (full-scale), anaerobic fluidized bed reactor (Enso-Fenox, semitechnical scale), and activated sludge plant (semitechnical scale).

It can be seen that the anaerobic reactor removed chlorophenols in a retention time of 5 to 12 h, even better than did the aerated lagoon (hydraulic detention time, 7 days) or the activated sludge plant. The removal of chlorinated guaiacols and dimethoxyphenols by the lagoon was more efficient in May–June than in March–April. This may be due to an increase in light and temperature.

Analysis of mass balance of the chlorophenols on biological treatment. We took some of the fluidized bed reactor biomass (Table 4) into the laboratory and percolated it in a trickling filter with a synthetic effluent consisting of a mineral salts solution containing 50 mg of pentachlorophenol per liter and 6.9×10^6 dpm of [U-^{14}C]pentachlorophenol. After 3 to 4 days, >98% of the pentachlorophenol had disappeared from the percolate (observed by capillary gas-liquid

TABLE 4. Removal of chlorophenolic compounds from the integrated effluent of a kraft pulp mill by different biological treatment methods

Compound	Concn[a] found in feed[b] (mg/m³), March-June	Remaining amt (mg/m³)[a] of chlorophenolic compounds found after treatment by:				
		Aerated lagoon, retention time, 14 days		Activated sludge plant, March-April	Anaerobic reactor,[c] retention time[d]:	
		March-April	May-June		12 h	5 h
2,4-Dichlorophenol	26 (14)	17 (10)	9 (4)	27 (3)	10 (5)	16 (7)
2,4,6-Trichlorophenol	51 (23)	26 (2)	25 (5)	31 (8)	9 (6)	17 (6)
3,4-Dichlorocatechol	45 (47)	9 (3)	10 (5)	12 (4)	15 (13)	9 (8)
4,5-Dichloroguaiacol	93 (34)	10 (10)	0 (0)	51 (11)	28 (25)	25 (25)
2,3,4,6-Tetrachlorophenol	12 (5)	8 (2)	8 (2)	9 (2)	2 (2)	3 (2)
Trichloroguaiacol[e]	190 (74)	101 (26)	19 (5)	156 (29)	43 (18)	49 (18)
3,4,5-Trichlorocatechol	166 (98)	100 (22)	184 (49)	103 (13)	22 (13)	57 (12)
3,4,5-Trichloro-2,6-dimethoxyphenol	23 (15)	19 (5)	2 (2)	14 (8)	5 (2)	4 (2)
Tetrachloroguaiacol plus pentachlorophenol	80 (36)	61 (7)	17 (14)	94 (14)	12 (20)	25 (19)
Tetrachlorocatechol	118 (70)	58 (13)	92 (37)	65 (15)	27 (18)	52 (25)
Sum of chlorophenols	804 (287)	409 (47)	366 (78)	562 (55)	163 (79)	257 (98)
Avg removed by treatment		49%	57%	30%	80%	68%

[a] The figures are mean values of more than five experiments, with the standard deviation given in parentheses. Analyzed as described by Voss et al. (49).

[b] The reactor was an anaerobic fluidized bed reactor of the type described by Hakulinen and colleagues (22) and Hakulinen and Salkinoja-Salonen (21).

[c] The feed was mixed effluent from a kraft pulp mill producing 340,000 tons of 100% bleached (chlorine) kraft pulp per year from hardwood and softwood. The specific effluent load of the mill was 13 kg of biological oxygen demand O_2, 53 g of P, and 162 g of N per ton of product.

[d] Twelve-hour retention time runs were performed in March-April, 5-h runs May-June. Temperature of the reactor was 30°C.

[e] The sum of 3,4,5-trichloroguaiacol and another trichloroguaiacol of unknown configuration.

TABLE 5. Analysis of the solids and liquid phase of a bioreactor[a] fed with a synthetic effluent with [14C]pentachlorophenol for 100 days[b]

Item analyzed	Input on day 1	Recovered on day 100	
		14C cpm	% of input
14C dpm, reactor solids	6.9×10^6	7×10^5	10
Of this soluble in diethyl ether, pH 2		3.9×10^3	0.7
Pentachlorophenol, reactor solids (mg)[c]	1,290	10	0.8
14C dpm, percolate		4×10^4	0.6
Pentachlorophenol, percolate (mg)[c]		1.4	0.1

[a] The reactor consisted of a 300-ml glass column with 195 g of solids (wet weight) and a 500-ml reservoir for the percolate. For details of the reactor, see Salkinoja-Salonen et al. (in press).
[b] The radioactive pentachlorophenol was added on day 1 only, but unlabeled pentachlorophenol was intermittently added every 3 or 4 days (50 mg/liter).
[c] Pentachlorophenol was analyzed 4 days after the last addition of (unlabeled) pentachlorophenol to the percolate.

chromatography). More pentachlorophenol (50 mg/liter, unlabeled) was then added. This was repeated for a period of 100 days, when the reactor was dismounted and the percolate plus the reactor solids were analyzed for chlorophenols and radioactivity. The results (Table 5) show that pentachlorophenol could be found neither in the reactor solids nor in the percolate, and no more than 10% of the 14C label, added as pentachlorophenol on the first day of the experiment, had remained in the reactor solids. Even this radioactivity was probably no longer chemically identical to pentachlorophenol, as it could not be extracted into acidified diethyl ether.

As [14C]pentachlorophenol disappeared from the synthetic effluent while it was percolated through the biological reactor, radioactivity reappeared in the gas phase.

Results in Fig. 3 show that, when modest aeration was used to drive out carbon dioxide from the reactor, up to 70% of the 14C fed in as pentachlorophenol could be recovered in the outgoing air as $^{14}CO_2$. We found that when the experiment was repeated with stoppered flasks with biomass from the reactor and a cup with aqueous NaOH to absorb CO_2 (Salkinoja-Salonen et al., in press), even better recovery of 14C as $^{14}CO_2$ was obtained, up to 90% of the input counts. We therefore conclude that reactor biomass was able to mineralize [U-14C]pentachlorophenol into $^{14}CO_2$.

We used a low-chloride synthetic effluent to measure the release of inorganic chloride from chlorinated phenols in the bioreactor. We found

that 2.0 mmol of Cl⁻ was released from 0.4 mmol of pentachlorophenol added, and 3.2 mmol of inorganic chloride was released in a similar experiment in which 2,4-dichloro-, 2,4,5-trichloro-, 2,4,6-trichloro-, and 2,3,4,6-tetrachlorophenol, 0.1 mmol of each, were added along with 0.4 mmol of pentachlorophenol per liter. This means that all chlorine had been released, not only from pentachlorophenol but also from the less chlorinated phenols, in the bioreactor as inorganic chloride. In a different experiment (22), we dismounted a semitechnical pilot fluidized bed reactor after it had been used to treat bleaching effluent of kraft pulp for 4

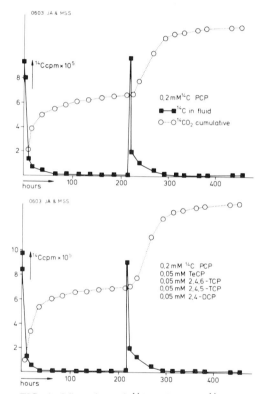

FIG. 3. Liberation of $^{14}CO_2$ from [U-14C]pentachlorophenol in the absence (A) and the presence (B) of other chlorophenols. Two identical trickling filters were prepared by dividing into two the contents of the filter used for the experiment described in Table 4. Each was supplied with a flask containing 400 ml of mineral salts medium, to which 80 μmol of pentachlorophenol (9×10^5 cpm) was added at 0 h and again at 214 h. In addition, 20 μmol each of 2,4-dichlorophenol, 2,4,6-trichlorophenol, 2,4,5-trichlorophenol, and 2,3,4,6-tetrachlorophenol was added to flask B at 0 h and again at 214 h. The biofilter was then percolated with the contents of the flask by means of a peristaltic pump. The biofilter was aerated (22 ml of air per min), and the exhaust air was bubbled through a gas washer containing 50 ml of 1 M NaOH to absorb any CO_2 emitted.

months and analyzed the reactor solids for chlorinated phenolic compounds. We found that of the 2,600 g of chlorophenolic compounds that had been fed with wastewater into the reactor only 9 g was retained. These, and other confirming results (Salkinoja-Salonen et al., in press), led us to conclude that it indeed is possible to truly degrade chlorophenols in an anoxic biological reactor in the pilot-scale experiments carried out on-site (Table 4).

Sensitivity of the degradation to the presence of chloramphenicol (100 µg/ml), gentamicin (10 µg/ml), and polymyxin B sulfate (10 µg/ml), but not to cycloheximide (500 µg/ml), suggests that the degradation was biological and that the responsible organisms were procaryotes.

Influence of environmental factors on degradation of chlorophenolic compounds. Microorganisms that degrade chlorophenolic compounds are known to exist and have been studied in the laboratory (9, 10, 13, 29, 30, 38, 42, 48). If degradation does not proceed in the environment, the suitable microorganisms may not be present, or environmental factors such as temperature, access to oxygen, and availability of nutrients and growth factors may inhibit degradation.

We have observed (experiments to be reported elsewhere) that degradation in a biological reactor is cold sensitive: below 15°C the rate is very low. Similar behavior was reported for pentachlorophenol degradation by a *Pseudomonas* sp. (48). It is intriguing that the degradation of chlorophenols into CO_2 can proceed in the absence of aeration. We studied the effect of different concentrations of oxygen on the degradation of [^{14}C]pentachlorophenol and found that the velocity of degradation, which was approximately 0.1 µmol of inorganic chloride released per g of bioreactor wet solids per h, was insensitive to the concentration of oxygen within a wide range (pO$_2$ of 0.0002 to 1), but was inhibited under strictly anaerobic conditions (results to be presented elsewhere). The degradation continued undisturbed at the same rate for at least 4 weeks while the reactor was continuously purged by technical grade nitrogen gas (maximum of 0.1% air) instead of air.

Anaerobic dechlorination of mono-, di-, and trichlorobenzoic acids and different isomers of monochlorophenol has been reported to occur in a freshwater lake sediment and in municipal digested sludge by Horowitz et al. (26), but CO_2 or CH_4 was formed only from *meta*-substituted monochlorinated compounds. It may therefore be that even though dechlorination of the benzene ring itself is possible under anaerobic conditions, oxygen may be required for some later step involving ring cleavage. This may be an explanation for the environmental recalcitrance of chlorinated phenolic compounds, as cleavage of nonhalogenated benzenoid compounds is known to occur unhindered, although slowly, in anaerobic conditions (11, 23–25). Halogenated 1- and 2-carbon compounds such as chloroform and tetrachloroethylene are not recalcitrant under anaerobic conditions (6).

CONCLUSIONS

The forest industry discharges a great quantity of organic chlorine compounds into the environment. Among others, chlorinated phenolic compounds are released from the bleach plant of the mill and also from wood-preserving facilities. The behavior of such compounds in the cold-climate water ecosystem was studied.

The dichlorinated compounds (2,4-dichlorophenol, 4,5-dichloroguaiacol, and 3,4-dichlorocatechol) were significantly but not completely degraded at an activated sludge plant or in a freshwater ecosystem. Phenolic compounds with three to five chlorine atoms per aromatic ring were found to be stable in the aquatic environment.

Distant pollution by pentachlorophenol occurs and is most probably transmitted by air. This may be concluded from results such as those from a dated sediment taken from a clean forest lake that received no waste stream of any kind. Nevertheless, it was found to contain 0.3 to 0.7 ppm of pentachlorophenol in the layers that had settled during the past 50 years. In older layers, formed over a period of several hundred years, there was also pentachlorophenol, although less, 0.05 to 0.15 ppm. This pentachlorophenol cannot be of industrial origin, but may have originated from burning of wood or forest fires.

An anaerobic fluidized bed reactor was more effective in removing di-, tri-, tetra-, and pentachlorinated phenols and their derivatives from pulp- and papermill wastewater than was an activated sludge unit or an aerated lagoon. Anaerobic methods of wastewater treatment are environmentally safe since little toxic chemical can be stripped into the ambient air. Therefore, more research should be dedicated to anaerobic treatment of wastewaters that contain chemicals that are not biodegradable in or are toxic to the environment.

We thank M. Fischer (Berlin, Federal Republic of Germany) and J. Knuutinen (Jyväskylä, Finland) for gifts of labeled pentachlorophenol and reference compounds for gas-liquid chromatography, respectively; Riitta Boeck, Ulla Sintonen, Eila Elomaa, and Maija-Liisa Saxelin for their contribution in the analytical work; Jane Backström, Risto Harjula, and Eija Räike for their contribution in sampling and dating the lake sediments; Ari Salonen, Raimo Tenhu, and Hannu Kurittu for operating the semitechnical scale wastewater reactors; and Eira Toivanen for performing mass fragmentographic analyses.

This work was financially supported by the Maj and Tor Nessling Foundation (J.A., R.V., and M.S.-S.) and the Academy of Finland (M.S.-S.).

LITERATURE CITED

1. **Ahling, B., and A. Lindskog.** 1981. Emission of chlorinated organic substances from combustion. Report from the Swedish Water and Air Pollution Research Institute (IVL/IPK), Stockholm, Sweden.

2. **Anonymous.** 1982. Miljövänlig tillverkning av blekt massa. Final Report of The Association of Swedish Central Association of Cellulose and Paper Industries, 1977-07-01...1981-06-30, prepared by IPK/Stockholm, p. 80–84.

3. **Anonymous.** 1983. Finland: huge investment phase ends. Paper, review of the year 1982. Special issue, 27–33.

4. **Appleby, P. G., and F. Oldfield.** 1978. The calculation of lead-210 dates assuming a constant rate of supply unsupported ^{210}Pb to the sediment. Catena **5**:1–8.

5. **Bauchop, T., and S. R. Elsden.** 1960. The growth of microorganisms in relation to their energy supply. J. Gen. Microbiol. **23**:457–469.

6. **Bouwer, E. J., B. E. Rittmann, and P. L. McCarty.** 1981. Anaerobic degradation of halogenated 1- and 2-carbon organic compounds. Environ. Sci. Technol. **15**(5):596–599.

7. **Boyle, T. B., E. F. Robinson, J. D. Petty, and W. Weber.** 1980. Degradation of pentachlorophenols in simulated lentic environment. Bull. Environ. Contam. Toxicol. **23**:177–184.

8. **Claeys, R. R., L. E. LaFleur, and D. L. Borton.** 1980. Chlorinated organics in bleach plant effluents of pulp and paper mills, p. 335–345. In J. R. Jolley et al. (ed.), Water Chlorination Environmental Impact Health Effects. Proc. Conf. (Colorado Springs), vol. 3. Ann Arbor Science Publications, Ann Arbor, Mich.

9. **Edgehill, R. U., and R. K. Finn.** 1982. Isolation, characterization and growth kinetics of bacteria metabolizing pentachlorophenol. Eur. J. Appl. Biotechnol. **16**:179–184.

10. **Edgehill, R. U., and R. K. Finn.** 1983. Microbial treatment of soil to remove pentachlorophenol. Appl. Environ. Microbiol. **45**:1122–1125.

11. **Ehrlich, G. G., D. F. Goerlitz, E. M. Godsy, and M. F. Hult.** 1982. Degradation of phenolic contaminants in ground water by anaerobic bacteria: St. Louis Park, Minnesota. Ground Water **20**:703–710.

12. **Ernst, W., and K. Weber.** 1978. The fate of pentachlorophenol in the Weser Estuary and the German Bight. Veroeff. Inst. Meeresforsch. Bremerhaven **17**(1):45–53.

13. **Etzel, J. E., and E. J. Kirsch.** 1975. Biological treatment of contrived and industrial wastewater containing pentachlorophenol. Dev. Ind. Microbiol. **16**:287–295.

14. **Gebefuegi, I., H. Parlar, and F. Korte.** 1979. Occurrence of pentachlorophenol in enclosed environments. Ecotoxicol. Environ. Safety **3**:269–300.

15. **Gee, J. M., and J. L. Pee.** 1974. Metabolism of 2,3,4,6-tetrachlorophenol by microorganisms from broiler house litter. J. Gen. Microbiol. **85**:237–243.

16. **Giger, W., and P. W. Roberts.** 1978. Characterization of persistent organic carbon, p. 135–175. In R. Mitchell (ed.), Water pollution microbiology, vol. 2. Wiley Interscience, New York.

17. **Gilbert, E., and H. Guesten.** 1973. Der strahlenchemische Abbau biologisch resistenter organischer Schadstoffe in wässeriger Lösung. Vom Wasser **41**:359–368.

18. **Hakulinen, R.** 1982. The Enso-Fenox process for the treatment of kraft pulp bleaching effluent and other wastewaters of the forest industry. Pap. Puu **5**:341–346.

19. **Hakulinen, R., and M. S. Salkinoja-Salonen.** 1981. The use of anaerobic fluidised bed reactor for the treatment of effluents containing chlorophenols, p. 374–382. In P. F. Cooper and B. Atkinson (ed.), Biological fluidised bed treatment of water and waste water. Ellis Horwood Publisher, Chichester.

20. **Hakulinen, R., and M. S. Salkinoja-Salonen.** 1982. Treatment of kraft bleaching effluents: comparison of results obtained by Enso-Fenox and alternative methods, p. 97–106. In Proceedings of TAPPI International Pulp Bleaching Conference. Tappi Press, Atlanta, Ga.

21. **Hakulinen, R., and M. S. Salkinoja-Salonen.** 1982. Treatment of pulp and paper industry wastewaters in an anaerobic fluidised bed reactor. Process Biochem. **17**(2):18–22.

22. **Hakulinen, R., M. S. Salkinoja-Salonen, and M.-L. Saxelin.** 1981. Purification of kraft bleaching effluent by an anaerobic fluidised bed reactor and anaerobic trickling filter at semitechnical scale (Enso-Fenox), p. 197–203. In Proceedings TAPPI Environmental Conference, New Orleans, 27–29 April 1981. Tappi Press, Atlanta, Ga.

23. **Healy, J. B., and L. Y. Young.** 1978. Catechol and phenol degradation by a methanogenic population of bacteria. Appl. Environ. Microbiol. **35**:216–218.

24. **Healy, J. B., and L. Y. Young.** 1979. Anaerobic biodegradation of eleven aromatic compounds to methane. Appl. Environ. Microbiol. **38**:84–89.

25. **Healy, J. B., L. Y. Young, and M. Reinhard.** 1980. Methanogenic decomposition of ferulic acid, a model lignin derivative. Appl. Environ. Microbiol. **39**:436–444.

26. **Horowitz, A., D. R. Shelton, C. P. Cornell, and J. M. Tiedje.** 1982. Anaerobic degradation of aromatic compounds in sediments and digested sludge. Develop. Ind. Microbiol. **23**:435–444.

27. **Hrutfiord, B. F., T. S. Friberg, D. F. Wilson, and J. R. Wilson.** 1975. Organic compounds in pulp mill lagoon discharge, p. 181–184. TAPPI Environmental Conference, Denver, Colo. Atlanta Technical Association of the Pulp and Paper Industry, Atlanta.

28. **Huttunen, P., and J. Meriläinen.** 1978. New freezing device providing large unmixed sediment samples from lakes. Ann. Bot. Fenn. **15**:128–139.

29. **Kaufman, D.** 1978. Degradation of pentachlorophenol in soil, and by soil microorganisms, p. 27–39. In R. K. Ranga (ed.), Environmental Science Research, vol. 12, Pentachlorophenol: chemistry, pharmacology and environmental toxicology. Plenum Press, New York.

30. **Kirsch, E. J., and J. E. Etzel.** 1973. Microbial decomposition of pentachlorophenol. J. Water Pollut. Control Fed. **45**:359–364.

31. **Knackmuss, H.-J.** 1981. Degradation of halogenated and sulfonated hydrocarbons, p. 190–212. In T. Leisinger, A. M. Cook, R. Huetter, and J. Nuesch (ed.), Microbial degradation of xenobiotics and recalcitrant compounds. Academic Press, London.

32. **Kringstad, K. P., P. O. Ljungquist, F. de Sousa, and L. M. Strömberg.** 1981. Identification and mutagenic properties of some chlorinated aliphatic compounds in the spent liquor from kraft pulp chlorination. Environ. Sci. Technol. **5**:562–566.

33. **Leuenberger, C., R. Coney, J. W. Graydon, E. Molnar-Kubica, and W. Giger.** 1983. Schwer abbaubare organische Stoffe in Abwässern der Zellstoffherstellung: Auftreten und Verhalten in einer biologischen Kläranlage. Chimia **37**(9):345–353.

34. **Lurker, P. A., C. A. Scott Clark, and V. J. Elia.** 1982. Atmospheric release of chlorinated organic compounds from the activated sludge process. J. Water Pollut. Control Fed. **54**:1566–1573.

35. **Matsumura, F.** 1982. Degradation of pesticides in the environment by microorganisms and sunlight, p. 67–90. In F. Matsumura and C. R. Krishna Murti (ed.), Biodegradation of pesticides. Plenum Press, New York.

36. **Mattinen, H., and I. Wartiovaara.** 1981. The pollution load of a closed D/CEDED bleachery. Pap. Puu **11**:688–706.

37. **Murray, H. E., L. E. Ray, and C. S. Giam.** 1981. Analysis of marine sediment, water and biota for selected organic pollutants. Chemosphere **10**:1327–1334.

38. **Murthy, N. B. K., D. D. Kaufman, and G. F. Vries.** 1979. Degradation of pentachlorophenol in aerobic and anaerobic soil. J. Environ. Sci. Health **B14**(1):1–14.

39. **Neilson, A. H., A.-S. Allard, P.-Å. Hynning, M. Rem-

berger, and L. Landner. 1983. Bacterial methylation of chlorinated phenols and guaiacols: formation of veratroles from guaiacols and high-molecular-weight chlorinated lignin. Appl. Environ. Microbiol. **45:**774–783.

40. **Paasivirta, J., J. Särkkä, T. Leskijärvi, and A. Roos.** 1980. Transportation and enrichment of chlorinated phenolic compounds in different aquatic food chains. Chemosphere **9:**441–456.

41. **Pierce, R. H., and D. M. Victor.** 1978. The fate of pentachlorophenol in an aquatic ecosystem, p. 41–52. *In* R. K. Ranga (ed.), Environmental science research, vol. 12, Pentachlorophenol: chemistry, pharmacology and environmental toxicology. Plenum Press, New York.

42. **Reiner, E. A., J. Chu, and E. J. Kirsch.** 1978. Microbial metabolism of pentachlorophenol, p. 67–81. *In* R. K. Ranga (ed.), Environmental science research, vol. 12, Pentachlorophenol: chemistry, pharmacology and environmental toxicology. Plenum Press, New York.

43. **Robbins, J. A., and D. N. Edgington.** 1975. Determination of recent sedimentation rates in Lake Michigan using Pb-210 and Cs-137. Geochim. Cosmochim. Acta **39:**285–304.

44. **Rott, B., S. Niltz, and F. Korte.** 1979. Microbial decomposition of pentachlorophenolate. J. Agric. Food Chem. **27:**306–310.

45. **Russi, H., D. Kotzias, and F. Korte.** 1982. Photoinduzierte hydroxylierungsreaktionen organischer Chemikalien in natürlichen Gewässern. Chemosphere **11:**1041–1048.

46. **Salkinoja-Salonen, M., E. Väisänen, and A. Paterson.** 1979. Involvement of plasmids in the bacterial degradation of lignin derived compounds, p. 301–314. *In* K. Timmis and A. Pühler (ed.), Plasmids of medical, environmental and commercial importance. Elsevier/North-Holland, Amsterdam.

47. **Salkinoja-Salonen, M. S., M.-L. Saxelin, T. Jaakkola, J. Saarikoski, R. Hakulinen, and O. Koistinen.** 1981. Analysis of toxicity and biodegradability of organochlorine compounds released into the environment in bleaching effluents of kraft pulping, p. 1131–1164. *In* L. H. Keith (ed.), Advances in the identification and analysis of organic pollutants in water, vol. 2. Ann Arbor Science Publishers, Ann Arbor, Mich.

48. **Trevors, J. T.** 1982. Effect of temperature on the degradation of pentachlorophenol. Chemosphere **11:**471–475.

49. **Voss, R. H., J. T. Wearing, and A. Wong.** 1981. Novel gas chromatographic method for the analysis of chlorinated phenolics in pulp mill effluents, p. 1059–1076. *In* L. H. Keith (ed.), Advances in the identification and analysis of organic pollutants in water, vol. 2. Ann Arbor Science Publishers, Ann Arbor, Mich.

50. **Yasuhara, A., A. Otsuki, and F. Keiichiro.** 1977. Photodecomposition of odorous chlorophenols in water. Chemosphere **6:**659–664.

Kinetic and Ecological Approaches for Predicting Biodegradation Rates of Xenobiotic Organic Chemicals in Natural Ecosystems

ROBERT J. LARSON

Environmental Safety Department, Procter & Gamble Company, Cincinnati, Ohio 45217

Microbial degradation (biodegradation) is an important mechanism for removing organic chemicals from natural ecosystems (1). However, quantitative expressions for predicting biodegradation rates in mixed microbial populations and extrapolating these rates to different environmental systems have historically been difficult to formulate (14, 25, 28). Numerous studies have shown that biodegradation of many xenobiotics follows apparent first-order kinetics in wastewater treatment systems, natural waters, and sediments (7, 15–22, 23, 25, 28, 35). Half-lives for degradation, calculated from the observed first-order rate constants, have then been used to predict the residence time of specific chemicals in a particular environmental compartment. Of more difficulty has been the "normalization" of observed first-order rate constants to allow extrapolation of biodegradation rate data from one environment to another. Several parameters, including microbial mass, total viable numbers, and cellular components (25, 28, 29, 35), have been used with variable success, depending on the substrate and environmental sample tested. Attempts to normalize biodegradation rate data for the number of specific degraders have also met with difficulty as a result of variations in metabolic activity within different microbial populations. With a few notable exceptions (25, 28, 29, 35, 39), relatively little progress has been made to date in quantifying biodegradation rate processes beyond a simple first-order relationship.

Kinetic data for predicting biodegradation rates in natural ecosystems are environmentally important for several reasons. First, biodegradation rate expressions are key components of many computer-based models used to estimate the distribution and environmental concentration of organic chemicals. These models require kinetic data to rank biodegradation against other competing physical–chemical processes and to extrapolate laboratory-derived rate data to actual environmental systems. Kinetic data are also needed to generate site-independent estimates of biodegradation rates for extrapolation to different environmental compartments. This allows the most important environmental compartment (in terms of mass flux) for a particular chemical

to be determined, which simplifies the overall process of estimating environmental exposure. Finally, kinetic data provide an objective basis for evaluating the relative persistence of a chemical in a particular environment. This is not only useful in an experimental sense, but also provides an important piece of information for making regulatory decisions about the environmental safety of a particular chemical.

This paper describes the various kinetic approaches used to quantify biodegradation rate processes and extrapolate biodegradation rate data to different natural ecosystems. Particular emphasis is given to the emerging role of microbial ecology in kinetic studies and to the use of nonlinear regression techniques for providing good estimates of biodegradation kinetic constants. The goal of the paper is to provide a better understanding of some of the techniques used to generate and analyze biodegradation rate data, and to predict the "biological" fate of chemicals in aquatic ecosystems.

THEORETICAL CONSIDERATIONS

Primary versus ultimate degradation. Biodegradation processes can be conveniently grouped into two categories, primary biodegradation (biotransformation) and ultimate biodegradation (mineralization). Primary biodegradation, which is generally assayed by specific analytical techniques, occurs when a discrete alteration is made in the structure of a chemical such that basic physical–chemical properties are lost. This results in a decreased sensitivity of the compound to a specific analytical method based on those properties and a reduction in analytical response relative to the parent chemical. Examples of primary degradation would be the hydrolytic cleavage of an ester moiety from the pesticide malathion to form malathion β-monoacid (28) or the cleavage of an acetate group from the detergent builder nitrilotriacetic acid to yield the di-acid iminodiacetic acid (6), as illustrated in Fig. 1.

Ultimate biodegradation, which is generally measured at environmentally relevant concentrations (micrograms per liter) by use of ^{14}C-labeled substrates, occurs when a chemical is completely broken down to carbon dioxide

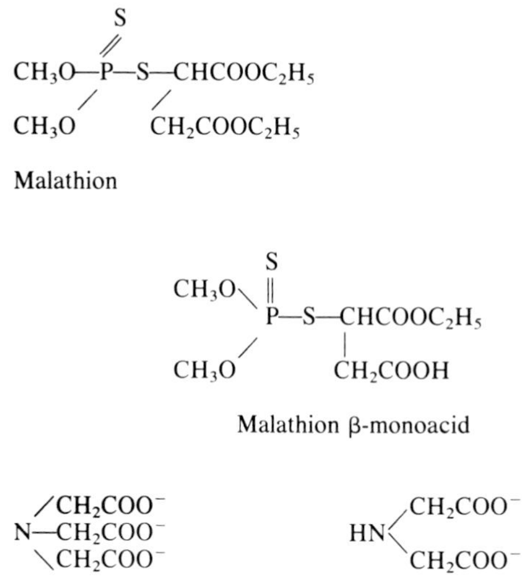

$$\begin{array}{c} S \\ \parallel \end{array}$$
CH₃O—P—S—CHCOOC₂H₅
CH₃O   CH₂COOC₂H₅

Malathion

$$\begin{array}{c} S \\ \parallel \end{array}$$
CH₃O\
      P—S—CHCOOC₂H₅
CH₃O⁄   CH₂COOH

Malathion β-monoacid

⁄CH₂COO⁻
N—CH₂COO⁻   HN⟨CH₂COO⁻
\CH₂COO⁻   CH₂COO⁻

NTA   IDA

FIG. 1. Cleavage of nitrilotriacetic acid (NTA) to yield iminodiacetic acid (IDA).

(CO_2) and other inorganic constituents. During degradation, the chemical may be present as the sole carbon and energy source or in a mixed system with other competing energy sources present. In either case, degradation of the chemical is linked to cellular metabolic processes, i.e., growth related. This is a key difference which separates ultimate and primary degradation mechanisms. It has resulted in somewhat different kinetic approaches being taken for the two processes (14, 28), although different approaches may not be required for all compounds. When primary degradation is immediately followed by mineralization (as would be expected when active transport is a rate-limiting step for degradation), the kinetics of degradation should be indistinguishable for both processes. When primary degradation occurs, but no CO_2 is evolved, cometabolic mechanisms may be involved, and the kinetics for the two processes will differ. However, cometabolic mechanisms should still have recognizable kinetic patterns.

Tiedje (36) summarized some of the characteristics and expected kinetic patterns for the two major categories of biodegradation, primary degradation and ultimate degradation. This summary indicates that compounds which are degraded by the former mechanism are more likely to have predictable kinetic patterns, although compounds which are degraded by the latter mechanism are less likely to pose an environmental hazard. Both mechanisms, however, can still provide useful kinetic information.

First-order kinetic approach. First-order kinetics are observed when the rate of biodegradation

of a chemical is directly proportional to the concentration of chemical present. As the concentration of chemical increases or decreases, the observed rate of degradation (mass degraded/time) increases or decreases to a similar extent, and the relative rate of degradation (percent degraded/time) remains constant. First-order kinetics are usually measured in batch test systems (10, 19, 23, 28, 33) and are verified experimentally by showing that the first-order rate constant (K_1) does not vary over a wide range of test concentrations (10-fold or greater). K_1 defines the rate of change of degradation occurring over time (i.e., the rate of change of the slope of the degradation curve) and has units of time^{-1}. Mathematically, first-order decay equations can be expressed as:

$$A = A_0 \exp^{-K_1 t} \qquad (1)$$

where A_0 is the initial concentration, A is the concentration at time t, and K_1 is the first-order rate constant (time^{-1}).

In an analogous fashion, first-order product curves can be expressed as:

$$P = P_0 (1 - \exp^{-K_1 t}) \qquad (2)$$

where P_0 is the maximum amount of product produced and P is the product at time t (the [1 − exp] term is used to give a positive slope). Equations 1 and 2 are usually written as logarithms in the form of a straight line to allow K_1 to be determined graphically or by linear regression. For equation 1 this results in:

$$\ln A = \ln A_0 - K_1 t \qquad (3)$$

A plot of ln A (or log A) versus t yields a straight line with the slope equal to K_1. The resulting half-life ($t_{1/2}$) for degradation is calculated from the relationship $t_{1/2} = \ln 2/K_1 = 0.693/K_1$ (14).

Although K_1 is usually determined by linear transformation of a nonlinear exponential decay or product curve, K_1 can also be determined directly by nonlinear regression analysis of the data. Nonlinear techniques offer special advantages over linear methods, since they avoid many of the statistical biases inherent in linear transformations of nonlinear data (4). They can also be modified to incorporate asymptotes (maximum and minimum) and lag phases for degradation, which are difficult to handle in conventional linear regression.

A useful function for fitting nonlinear CO_2 evolution curves is a generalized form of the logistics function first described by Richards (34). The logistics function has been used extensively to describe microbial growth kinetics in batch systems. In its integrated form, the generalized function can also be used to analyze biodegradation data, as shown below.

$$P = P_0 (1 - b \ \exp^{-K_1 t})^{-1/n} \qquad (4)$$

where P is the percentage of CO_2 observed at time t (day), b is a coordinate scaling factor associated with the constant of integration (dimensionless), P_0 is the upper asymptote of CO_2 production (percent), K_1 is the first-order rate constant (day^{-1}), and n is an empirical constant. Equation 4 is particularly useful for analyzing biodegradation product curves since the constant n can vary over broad limits (~ 1 to 0). This allows a variety of normal and sigmoidal curves to be analyzed without affecting K_1, which can therefore be normalized for both growth and acclimation effects.

Second-order kinetic approach. Normalization of observed K_1 values for microbial biomass (B) converts them to second-order rate constants (K_2). Several parameters have been used to generate K_2 values, including dry weight, viable counts, and ATP (25, 28, 35). By far the most commonly used parameter, and the most successful to date, has been viable counts (colony-forming units, CFU). Paris et al. (28) showed that K_2 values for primary degradation of several pesticides can be extrapolated to different natural waters simply by dividing K_1 by the number of viable bacteria present. Their second-order approach can be summarized as:

$$v = K_2 [A][B] \qquad (5)$$

where v is the degradation rate (milligrams per liter per hour), K_2 is the second-order rate constant (liters per CFU per hour), A is the substrate concentration (milligrams per liter), and B is the biomass concentration (CFU per liter). Degradation is assumed to be first order in both $[A]$ and $[B]$, and pseudo-first-order rate constants are defined by the relationship $K_1 = (K_2 B)$. Using this second-order approach, Paris et al. (28) obtained excellent agreement between K_2 values in different sites in studies of primary degradation. Their results indicate that the percentage of bacteria capable of primary degradation of specific substrates can remain relatively constant in different mixed populations. In contrast to these results, other studies (25) have shown that K_2 values can vary considerably more than K_1 values for primary degradation of some compounds. This indicates that the percentage of specific degraders (or their activity) may not be the same in different environmental samples. Clearly, an alternative kinetic approach, which relates microbial numbers to microbial activity, is needed to predict biodegradation of some substrates. Measurements of

heterotrophic bacterial activity may offer promise in this regard.

Heterotrophic bacterial activity approach. It has been more than 20 years since Parsons and Strickland (31) first showed that uptake of ^{14}C-labeled substrates in marine waters could be described by an enzyme-saturation model (27) and used to estimate the heterotrophic activity of natural microbial populations. Since that time, techniques for measuring heterotrophic activity have been refined, expanded, and applied to numerous environmental systems and a variety of natural organic compounds (3, 9, 11, 40, 41). A relatively new finding, which has important implications for biodegradation kinetic studies, is that these techniques can also be applied to degradation of some xenobiotic compounds (12).

Heterotrophic activity methods offer several advantages over the standard first- and second-order methods mentioned earlier. They are rapid, short-term assays which minimize artificial "container" effects and maximize the chance of obtaining rate data with in situ relevance. Unlike kinetic experiments, which must be conducted for a time period equivalent to at least two degradation half-lives, heterotrophic activity assays can be stopped after only a few percent degradation has occurred. This greatly reduces the amount of experimental effort required. Kinetic constants derived from such assays, maximum velocity (V_{max}) and turnover time (T_n), also directly relate the rate of degradation of a specific substrate to the activity of the degrading population. This allows microbial numbers to be normalized for microbial activity, which is difficult to achieve in standard kinetic studies. Two major disadvantages of heterotrophic activity techniques are their concentration dependence (T_n) and the fact that they are site specific (V_{max}). These disadvantages, coupled with the fact that they have only recently been shown to be applicable to xenobiotic chemicals, have limited their use in environmental fate studies (12).

The general equation relating the rate of uptake (or mineralization, or both) of a ^{14}C-labeled substrate to the concentration of added substrate (when the natural substrate concentration is unknown) can be written as:

$$v = \frac{V_{max} (S_n + A)}{K + (S_n + A)} \qquad (6)$$

where v is the degradation rate (nanograms per liter per hour), V_{max} is the maximum degradation rate (nanograms per liter per hour), S_n is the natural substrate concentration (micrograms per liter), A is the added substrate concentration (micrograms per liter), and K is the half-saturation constant where $v = 0.5 \ V_{max}$ (micrograms per liter). When the initial degradation rate is constant, v can also be defined as:

$$v = \frac{f}{t}(S_n + A) \qquad (7)$$

where f is the fraction taken up or mineralized, or both, at time t (hours). Equations 6 and 7 can be combined, inverted, and rewritten into the standard equation used to analyze heterotrophic activity data (26, 39):

$$\frac{t}{f} = 1/V_{max}(A) + \frac{(K + S_n)}{V_{max}} \qquad (8)$$

A linear plot of t/f versus A yields a straight line with y intercept of T_n, x intercept $(K + S_n)$, and a slope of $1/V_{max}$. T_n is defined as the time required for complete degradation of a chemical at its in situ concentration, V_{max} is the extrapolated maximum degradation rate at infinite substrate concentration, and $(K + S_n)$ represents the sum of the half-saturation constant and the natural substrate concentration.

Equation 8 has traditionally been used to determine kinetic constants of heterotrophic activity when $A > S_n$. A special situation occurs, however, when S_n is either much greater than A or so low as to be considered negligible. When $S_n >> A$, turnover times can be calculated directly from the ratio t/f, with the use of a single concentration of A. The concentration of added substrate is so low that it does not influence the in situ degradation rate (8), and observed T_n values represent the actual turnover time for the natural level of substrate present. Since only one substrate concentration is tested by this technique, however, V_{max} and $K + S_n$ values cannot be calculated. When $S_n << A$, as would be the case when xenobiotics are being tested in previously unexposed systems, S_n can be considered zero for all practical purposes and ignored when combining equations 6 and 7. Under these conditions, the following equation can be written:

$$v = \frac{f}{t} \cdot A = \frac{V_{max} \cdot A}{K' + A} \qquad (9)$$

where the half-saturation constant is written as K' to differentiate it from the half-saturation constant (K) of equation 8, which includes an S_n term. Strictly speaking, turnover times have no meaning when $S_n \approx 0$. In systems in which the in situ concentration of xenobiotic is negligible, therefore, T_n values cannot be calculated, although a ratio t/f may be obtained. However, equation 9 can be used to determine the empirical kinetic constants, V_{max} and K'. These constants indicate the "potential" for biodegradation of specific xenobiotics in natural

ecosystems and thus may be useful in defining the overall heterotrophic biodegradation potential of natural microbial populations. When some background level of the substrate is present, T_n can be calculated by the relationship $T_n = K''/V_{max}$, where $K'' = K + S_n$ (26).

MATERIALS AND METHODS

Chemicals. All radiochemicals were prepared by Amersham (Arlington Heights, Ill.) and checked for radiochemical purity by several procedures. Sodium $[U\text{-}^{14}C]$nitrilotriacetic acid (NTA; specific activity, 11.5 mCi/mmol) was >95% radiochemically pure, as determined by thin-layer chromatography (17), reverse isotope dilution (17), and reversed-phase ion pair liquid chromatography (30). Radiochemical purity of $[^{14}C_1]$dodecyltrimethylammonium chloride (DTMAC; specific activity, 17 mCi/mmol) was >97%, based on thin-layer chromatography on silica gel in butan-1-ol–water–acetic acid (120:50:30) and high-pressure liquid chromatography (38). Dodecyl $[U\text{-}^{14}C]$nonylethoxylate ($C_{12}{}^{*}E_9$; specific activity, 3.1 mCi/mmol) was ~80% pure, based on radio-thin layer chromatography; the remaining 20% impurity was attributed to the $C_{12}{}^{*}E_8$ homolog (18). Radiochemical purity of $[^{14}C_1]$dodecyl nonylethoxylate ($^{*}C_{12}E_9$; specific activity, 2.2 mCi/mmol) was ~97% based on radio-thin layer chromatography; the remaining 7% impurity was again the $^{*}C_{12}E_8$ homolog (19). Purity of 2,4-dichlorophenoxy $[2\text{-}^{14}C]$acetic acid (2,4-D; specific activity, 55 mCi/mmol) was 99%, based on thin-layer chromatography on cellulose in n-butanol–ethanol–3 N ammonia (12:3:15), n-butanol–acetone–water (5:2:3), and thin-layer chromatography on silica gel in chloroform–cyclohexane–acetic acid (8:2:1), toluene–dioxan–acetic acid (90:25:4). All other solvents and chemicals used in testing were reagent grade, obtained from commercial sources.

Analytical. Viable counts in surface water samples were determined on nutrient agar spread plates incubated at room temperature (2). Viable counts in groundwater samples were determined on diluted nutrient agar plates (10- to 100-fold), since growth of groundwater microorganisms is inhibited at high nutrient concentrations. Agar supplements were made to a level of 1.5% (wt/vol). Radiochemical analyses were conducted on a Tracor model 6881 liquid scintillation counter with quench correction by the external standards method. DTMAC was analyzed as described previously (38) by high-pressure liquid chromatography with conductivity detection. All surface and groundwater samples were also analyzed for acridine orange direct counts (43), total organic carbon, suspended solids, water hardness, pH, and temperature by standard techniques (2).

Sample collection. Surface and groundwater samples used in biodegradation studies were collected from several areas and were usually tested within a few hours of collection. Ohio River water (ORW) samples were collected in River Park near Cincinnati in unsterilized plastic containers (3.8 liters). Ohio groundwater (OGW) and Canadian groundwater (CGW) samples were collected aseptically to avoid contamination by surface microorganisms. OGW was collected from preexisting monitoring wells in a gravel-based aquifer (sanitary landfill, Butler County) after pumping several well volumes through sterile Tygon tubing. CGW was collected by use of specialized drilling procedures (5) as follows. Test wells were drilled in a clay-based overburden to determine the depth of the water table. Sampling wells were then drilled to the top of the saturated zone (~3 m), and a specialized coring device (9.5 by 45 cm) was used to remove subsurface soil cores from the saturated zone. Groundwater infiltrating vacant cores was pumped through sterile Tygon tubing with a portable master-flex peristatic pump (DC model 7533-20, Cole-Parmer Instrument Co.) at a flow rate of ~125 ml/min into sterile glass containers (3.8 liters). Water samples were immediately stored on ice and shipped back to Cincinnati for testing, within 12 h of collection.

Biodegradation assays. First-order kinetic studies for $C_{12}E_9$, NTA, and DTMAC were conducted in a closed, flow-through system as previously described (19). Radiolabel was quantitated in three fractions: as $^{14}CO_2$, ^{14}C in solution, and ^{14}C in biomass by liquid scintillation counting. Mass balances of radioactivity were made at each sampling point. Data for $^{14}CO_2$ evolution were fit by iterative techniques to equation 4, with the use of least squares analysis and a nonlinear computer program (procedure NLIN of SAS, SAS Institute Inc., Raleigh, N.C.). K_1 values for mineralization were normalized for the extent of CO_2 production observed (P_0) and any lag phases occurring ($-1/n$). Decay curves were analyzed by similar nonlinear regression models as previously described (17, 19). K_1 values for decay curves were normalized for zero-time concentrations, any lag phases occurring, and the amount of radiolabel remaining in solution at the end of testing.

Second-order kinetics for $C_{12}E_9$ degradation were measured by holding initial $C_{12}E_9$ concentrations constant and varying the level of biomass added. Activated sludge from a residential treatment plant (Colerain Heights) was cultured as described (13) and homogenized for 2 min at medium speed in a Waring blender. The sludge was allowed to settle for 30 min, and various dilutions (0.1, 0.5, 1.0%) from the supernatant were than added to a synthetic salts solution (13) containing 100 μg of $*C_{12}E_9$ per liter. $^{14}CO_2$ evolution and ^{14}C removal were measured as described above, and K_1 values for $^{14}CO_2$ production were determined by equation 4. K_2 was then calculated by dividing K_1 by the number of viable bacteria initially added. No lag phases were observed for $C_{12}E_9$ degradation at the substrate and biomass concentrations tested, and biomass concentrations (CFU) did not change significantly ($\alpha = 0.01$) during the initial phases of the experiment when biodegradation rates were at a maximum.

Heterotrophic biodegradation potential assays for $C_{12}*E_9$ were conducted in 10- to 1,000-ml volumes by methods described by Parsons and Strickland (31) and modified by Wright and Hobbie (42) and Hobbie and Crawford (9). Various concentrations of $C_{12}E_9$ were incubated in ground- or surface waters for 24 to 303 h, and the fraction of radiolabel assimilation plus that metabolized to $^{14}CO_2$ (^{14}C-uptake plus $^{14}CO_2$) was determined. Data for the rate of $C_{12}E_9$ degradation, $(f/t) \cdot A$ (nanograms per liter per hour), at increasing $C_{12}E_9$ concentrations were fit by nonlinear regression techniques to equation 9. A nonlinear computer program (NLIN, SAS) was used to generate parameter estimates for V_{max} and K' and associated statistical parameters (19).

RESULTS AND DISCUSSION

First-order kinetics. Figure 2 shows the distribution of radiolabel during ultimate biodegradation of a model nonionic surfactant, $C_{12}E_9$, in ORW. The kinetic curves indicate that removal of ^{14}C-activity from solution closely parallels $^{14}CO_2$ production. The mean first-order rate constant for CO_2 production (± standard deviation) over a 100-fold initial concentration range (10 to 1,000 μg/liter) was 0.45 ± 0.02 day^{-1}. This good agreement between K_1 values at different initial concentrations indicates that $C_{12}E_9$ degradation is directly proportional to concentration and follows exact first-order kinetics over the concentration range tested. Based on the mean K_1 value, the estimated half-life for $C_{12}E_9$ in ORW is ~36 h. This half-life is relatively short compared with values for other xenobiotics, some of which are listed in Table 1.

Figure 3 shows the distribution of radiolabel during biodegradation of another detergent chemical, NTA, in CGW. As with $C_{12}E_9$, degradation of NTA in CGW followed apparent first-order kinetics over a wide initial concentration range (50 to 500 μg/liter). The mean K_1 value for CO_2 evolution in CGW, ± standard deviation, over the 10-fold concentration range tested, was

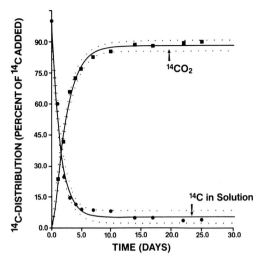

FIG. 2. Kinetics of primary and ultimate degradation $C_{12}E_9$ in ORW. The initial test concentration of $C_{12}E_9$ was 10 µg/liter. Dotted contours are 95.0% confidence limits of the true mean.

0.26 ± 0.07 day^{-1}. On the basis of this K_1, the estimated half-life for NTA in groundwater is about 65 h. This value is somewhat longer than values reported (17) in acclimated surface waters (13 to 25 h) and is probably a result of the low biomass levels present in the CGW (7.5 $\pm$ 3.0 $\times$ 10^3) and the oligotrophic character of groundwater systems in general.

Figure 3 indicates that the distribution of radiolabel during biodegradation of NTA in CGW is quite similar to that observed for $C_{12}E_9$ in ORW (Fig. 2). Removal of ^{14}C activity from solution closely parallels $^{14}CO_2$ production, with NTA essentially being metabolized to CO_2 as soon as it is removed from solution. A transport mechanism, therefore, may be rate limiting for degradation. Also, most of the NTA carbon skeleton is metabolized to CO_2, and cell yield factors (percent ^{14}C incorporated) are relatively low. This indicates that NTA is being degraded primarily as an energy source, presumably to satisfy the maintenance energy requirements of groundwater bacteria.

Figure 4 shows the kinetics of primary and ultimate degradation of DTMAC, a cationic surfactant, in ORW at an initial concentration of 1 mg/liter. At this high concentration, lag phases for both primary and ultimate degradation are clearly evident. However, after about 24 h, $^{14}CO_2$ production starts to increase and the concentration of DTMAC (by high-pressure liquid chromatography) starts to decrease. As degradation proceeds, the kinetics of $^{14}CO_2$ production again closely parallel removal of DTMAC

from solution, with both primary and ultimate degradation of DTMAC occurring at comparable rates. For this material, therefore, and possibly other compounds which are degraded as carbon and energy sources, the kinetics of primary and ultimate degradation are essentially equivalent. Transport again appears to be a rate-determining factor for degradation.

The data presented for detergent chemicals and those reported in the literature indicate that biodegradation of a variety of compounds is directly proportional to substrate concentration, i.e., first order in substrate. Degradation should, however, also be proportional in some manner to biomass levels. Some investigators (28) have suggested that degradation is directly proportional to viable biomass levels, i.e., first order in

TABLE 1. Half-lives for biodegradation of xenobiotic substrates in natural ecosystems

Compound	System	$t_{1/2}$ (days)	Reference
Benzene	River	6	23
Chlorobenzene	River	150	23
Chlorobenzene	Sediment	75	23
p-Chlorophenol	River	20	23
p-Chlorophenol	Sediment	3	23
p,p'DDE	Sediment	1,100	23
Distearyldimethylam-monium chloride	River	14	21
Distearyldimethylam-monium chloride	Sediment	5	21
Dodecylnonylethoxylate	River	1	18
Dodecylnonylethoxylate	Estuary	6	22
Hexachlorophene	Sediment	290	23
Hexadecylnonylethoxy-sulfate	Estuary	2	37
Hexadecyltriethoxylate	Estuary	2	37
Hexadecyltriethoxylate	River	1	18
Hexadecyltriethoxysul-fate	Estuary	7	37
Hexadecyltrimethylam-monium halide	River	3	21
Hexadecyltrimethylam-monium halide	Sediment	3	21
Linear alkylbenzene sulfonate	River	1	19
Linear alkylbenzene sulfonate	Sediment	<1	19
Methyl parathion	Awfuchs	<1	25
Nitrilotriacetic acid	River	<1	17
Phenol	River	9	23
2,4,5-Trichlorophenol	River	690	23
2,4,5-Trichlorophenol	Sediment	23	23
1,4,5-Trichlorophenoxy-acetic acid	River	1,400	23
1,4,5-Trichlorophenoxy-acetic acid	Sediment	35	23
Stearyltrimethylammo-nium chloride	River	3	21
Stearyltrimethylammo-nium chloride	Sediment	3	21

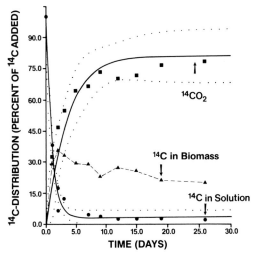

FIG. 3. Kinetics of primary and ultimate degradation of NTA in CGW. The initial test concentration of NTA was 50 μg/liter. Dotted contours are 95.0% confidence limits of the true mean.

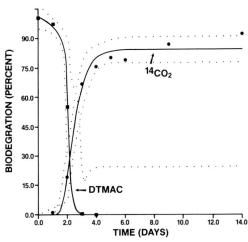

FIG. 4. Kinetics of primary and ultimate degradation of DTMAC in ORW. The initial test concentration of DTMAC was 1 mg/liter. Dotted contours are the 95.0% confidence limits of the true mean.

CFU. Others, however (25), indicate that the activity of individual microbial cells is important and needs to be considered when kinetic studies are being conducted. Experiments with a specific detergent chemical, $C_{12}E_9$, tend to support the latter position, as indicated below.

Second-order kinetics. Table 2 summarizes the results of studies to determine whether second-order kinetics apply to degradation of $C_{12}E_9$ by sewage microorganisms. The data indicate that although $C_{12}E_9$ degradation was first order with respect to substrate concentration, it was not first order with respect to biomass concentration, i.e., second order overall. K_1 varied only ~13% over the 10-fold biomass range tested, whereas K_2 varied 800% over the same biomass range. These results are somewhat surprising since one would expect at least some proportionality between CFU and biodegradation rates. However, both $^{14}CO_2$ evolution curves and ^{14}C-removal curves exhibited identical kinetic patterns, and degradation rates were not directly proportional to biomass levels based on total CFU. Two explanations for these results can be offered: (i) first-order rate constants (K_1) do not accurately reflect microbial activity, or (ii) biomass levels were in excess, even at the lowest level tested, and degradation was zero order in biomass. In any event, these data indicate that CFU do not accurately reflect the relationship between $C_{12}E_9$ mineralization rates and biomass.

Heterotrophic biodegradation potential. Methods (11, 12) for measuring the rate of uptake and metabolism of radiolabeled substrates by hetero-

trophic bacteria are well established for natural compounds (e.g., sugars and amino acids). Recently, however, Pfaender and Bartholomew (32) have shown that similar methods can also be applied to xenobiotic substrates such as m-cresol, chlorobenzenes, and NTA. Degradation rates for these compounds have been shown to follow saturation kinetics and can be analyzed by standard linear techniques to yield kinetic constants (T_n, V_{max}, K).

The ability to generate kinetic constants for the heterotrophic biodegradation potential has important implications for degradation testing of xenobiotics. Use of these constants with other measurements of microbial biomass (viable counts, acridine orange direct counts) should allow microbial numbers and degradation activity to be directly correlated with each other via a specific activity index (SAI). The concept of SAI was initially proposed to characterize the activity of individual cells on natural substrates (41). Turnover rate ($T_r = 1/T_n$) or V_{max} values were divided by bacterial numbers (CFU, acridine orange direct counts) to yield T_r or V_{max} SAIs. In principle, a similar approach should be

TABLE 2. First- and second-order rate constants for biodegradation of $C_{12}E_9$ by sewage microorganisms[a]

Bacterial concn (CFU/liter)	$K_1 \pm SE$ (h^{-1})	K_2 (liter/CFU per h)
5.3×10^6	0.010 ± 0.004	2.0×10^{-9}
2.7×10^7	0.011 ± 0.003	4.1×10^{-10}
5.3×10^7	0.013 ± 0.002	2.5×10^{-10}

[a] Initial *$C_{12}E_9$ concentration is 100 μg/liter.

applicable to degradation of xenobiotic compounds. Figure 5 illustrates the results of experiments to calculate V_{max} SAIs for the nonionic surfactant $C_{12}E_9$.

Degradation rates for $C_{12}*E_9$ were measured in ORW and OGW over a range of concentrations. Rate data were fit to equation 9 by nonlinear regression techniques to estimate the parameters K' and V_{max}. V_{max} was then divided by the number of viable bacteria (CFU/liter) to yield a V_{max} SAI (41). Viable counts were used as an estimate of microbial biomass since they have been used by other investigators (25, 28, 29, 39) and are widely available. In principle, however, other enumeration techniques such as acridine orange direct counts (43) or most-probable number (24) should yield interpretable results.

As expected, degradation rates for $C_{12}E_9$ in OGW were significantly lower than in ORW, although saturation kinetics were observed in both systems (Fig. 5). V_{max} in OGW (with 95% confidence intervals) was 121 ng per liter per h (105 to 131 ng per liter per h) compared with 8,081 ng per liter per h in ORW (4,575 to 11,587 ng per liter per h). K' values were approximately the same in both systems. The reduced V_{max} for $C_{12}E_9$ degradation in OGW was related to the low numbers of viable bacteria present (3.6 × 10^6 CFU/liter), which were almost two orders of magnitude less than in river water (2.7 × 10^8 CFU/liter). However, degradation activity (based on V_{max}) was proportional to viable counts, and normalization of V_{max} for CFU resulted in SAI values which agree relatively well, 3.4 × 10^{-5} and 3.0 × 10^{-5} ng per CFU per h in OGW and ORW, respectively. It is interesting that the variation in SAI (~10%) is considerably less than the variation in K_2 (~800%) measured in batch studies (Table 2), even though SAI values were measured in very different environmental samples. This lower variability suggests that consistent kinetic results can be obtained if degradation data are normalized for both microbial numbers and microbial activity by use of a heterotrophic biodegradation potential approach.

CONCLUSIONS

On the basis of the theoretical discussions and experimental evidence presented in this paper, some general conclusions can be drawn regarding the predictions of biodegradation rates in natural ecosystems.

(i) Biodegradation processes can be appropriately separated into two categories, primary degradation (biotransformation) and ultimate biodegradation (mineralization). However, kinetic patterns for the two categories may be essentially identical, particularly for materials

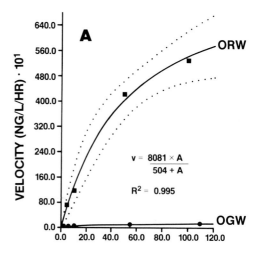

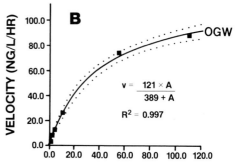

FIG. 5. Kinetics of $C_{12}E_9$ degradation in assays. (A) Initial rates of $C_{12}E_9$ degradation in ORW and OGW; (B) expanded scale to show saturation kinetics in OGW. The kinetic constants V_{max} and K' and the R^2 values for nonlinear fit of equation 9 are given in the insert. The dotted contours are the 95.0% confidence limits of the true mean.

which are degraded as carbon and energy sources.

(ii) Biodegradation rate processes are generally first order with respect to substrate concentration. They need not, however, be first order with respect to total biomass concentrations, i.e., second order overall in CFU.

(iii) Ecological kinetic approaches which incorporate both microbial numbers and microbial activity measurements (SAI) show promise for predicting biodegradation rates in natural ecosystems. In principle, these approaches should be applicable to both ultimate and primary degradation studies, although more research on xenobiotic compounds is needed to establish the predictive value of SAIs. This research should address the importance of incubation time in kinetic measurements, significance of T_n when $S_n \cong 0$, most appropriate measurements of biomass concentration for calculating SAI values, and the comparability of results obtained in standard first- and second-order studies with those obtained in heterotrophic biodegradation potential studies.

Appreciation is expressed to R. M. Ventullo for many stimulating discussions and to D. H. Davidson for technical assistance.

LITERATURE CITED

1. **Alexander, M.** 1980. Biodegradation of chemicals of environmental concern. Science **211**:132–138.
2. **American Public Health Association.** 1971. Standard methods for the examination of water and wastewater, 13th ed. American Public Health Association, Inc., New York.
3. **Azam, F., and O. Holm-Hansen.** 1973. Use of tritiated substrates in the study of heterotrophy in seawater. Limnol. Oceanogr. **23**:191–196.
4. **Currie, D. J.** 1982. Estimating Michaelis-Menten parameters: bias, variance and experimental design. Biometrics **38**:907–919.
5. **Dunlap, W. J., J. E. McNabb, M. R. Scalf, and R. L. Cosby.** 1977. Sampling of organic chemicals and microorganisms in the subsurface. EPA-600/2-77-176. National Technical Information Service, Springfield, Va.
6. **Firestone, M. K., and J. M. Tiedje.** 1978. Pathway of degradation of nitrilotriacetate by a *Pseudomonas* species. Appl. Environ. Microbiol. **35**:955–961.
7. **Games, L. M., J. E. King, and R. J. Larson.** 1982. Fate and distribution of a quaternary ammonium surfactant, octadecyltrimethylammonium chloride (OTAC), in wastewater treatment. Environ. Sci. Technol. **16**:483–488.
8. **Gocke, K.** 1977. Comparison of methods for determining the turnover times of dissolved organic compounds. Mar. Biol. **42**:131–141.
9. **Hobbie, J. E., and C. C. Crawford.** 1969. Respiration corrections for bacterial uptake of dissolved organic compounds in natural waters. Limnol. Oceanogr. **14**:528–532.
10. **Johnson, B. T.** 1980. Approaches to estimating microbial degradation of chemical contaminants in freshwater ecosystem, p. 25–33. In A. W. Maki, K. L. Dickson, and J. Cairns, Jr. (ed.), Biotransformation and fate of chemicals in the aquatic environment. American Society for Microbiology, Washington, D.C.
11. **Ladd, T. I., J. W. Costerton, and G. G. Geesey.** 1979. Determination of the heterotrophic activity of epilithic microbial populations, p. 180–195. In J. W. Costerton and R. R. Colwell (ed.), Native aquatic bacteria: enumeration, activity and ecology. ASTM technical publication no. 695. American Society for Testing Materials, Philadelphia.
12. **Ladd, T. I., R. M. Ventullo, P. M. Wallis, and J. W. Costerton.** 1982. Heterotrophic activity and biodegradation of labile and refractory compounds by groundwater and stream microbial populations. Appl. Environ. Microbiol. **44**:321–329.
13. **Larson, R. J.** 1979. Estimation of biodegradation potential of xenobiotic organic chemicals. Appl. Environ. Microbiol. **38**:1153–1161.
14. **Larson, R. J.** 1980. Role of biodegradation kinetics in predicting environmental fate, p. 67–86. In A. W. Maki, K. L. Dickson, and J. Cairns, Jr. (ed.), Biotransformation and fate of chemicals in the aquatic environment. American Society for Microbiology, Washington, D.C.
15. **Larson, R. J.** 1982. Comparison of biodegradation rates in laboratory screening studies with rates in natural waters. Residue Rev. **85**:159–171.
16. **Larson, R. J., G. G. Clinckemaillie, and L. VanBelle.** 1981. Effect of temperature and dissolved oxygen on biodegradation of nitrilotriacetate. Water Res. **15**:615–620.
17. **Larson, R. J., and D. H. Davidson.** 1982. Acclimation to and biodegradation of nitrilotriacetate (NTA) at trace concentrations in natural waters. Water Res. **16**:1597–1604.
18. **Larson, R. J., and L. M. Games.** 1981. Biodegradation of linear alcohol ethoxylates in natural waters. Environ. Sci. Technol. **15**:1488–1493.
19. **Larson, R. J., and A. G. Payne.** 1981. Fate of the benzene ring of linear alkylbenzene sulfonate in natural waters. Appl. Environ. Microbiol. **41**:621–627.
20. **Larson, R. J., and R. L. Perry.** 1981. Use of the electrolytic respirometer to measure biodegradation in natural waters. Water Res. **15**:697–702.
21. **Larson, R. J., and R. D. Vashon.** 1983. Adsorption and biodegradation of cationic surfactants in laboratory and environmental systems. Dev. Ind. Microbiol. **24**:425–434.
22. **Larson, R. J., R. D. Vashon, and L. M. Games.** 1983. Biodegradation of trace concentrations of detergent chemicals in freshwater and estuarine systems, p. 235–245. In T. A. Oxyley and S. Barry (ed.), Biodeterioration 5. John Wiley & Sons, Ltd., Chichester.
23. **Lee, R. F., and C. Ryan.** Microbial degradation of organochlorine compounds in estuarine waters and sediments, p. 443–450. In A. W. Bourquin and P. H. Pritchard (ed.), Proceedings of the workshop: Microbial degradation of pollutants in marine environment. EPA-600/9-79-012. National Technical Information Service, Springfield, Va.
24. **Lehmicke, L. G., R. T. Williams, and R. L. Crawford.** 1979. ^{14}C-most-probable-number method for enumeration of active heterotrophic microorganisms in natural waters. Appl. Environ. Microbiol. **38**:644–649.
25. **Lewis, D. L., and H. W. Holm.** 1981. Rates of transformation of methylparathion and diethylphthalate by *aufwuchs* microorganisms. Appl. Environ. Microbiol. **42**:698–703.
26. **Li, W. K. W.** 1983. Consideration of errors in estimating kinetic parameters based on Michaelis-Menten formalism in microbial ecology. Limnol. Oceanogr. **28**:185–190.
27. **Michaelis, M., and M. L. Menten.** 1913. Kinetics of invertase action. Biochem. Z. **49**:333–369.
28. **Paris, D. F., W. C. Steen, G. L. Baughman, and J. T. Barnett, Jr.** 1981. Second-order model to predict microbial degradation of organic compounds in natural waters. Appl. Environ. Microbiol. **41**:603–609.
29. **Paris, D. F., N. L. Wolfe, and W. C. Steen.** 1982. Structure-activity relationships in microbial transformation of phenols. Appl. Environ. Microbiol. **44**:153–158.
30. **Parkes, D. G., M. G. Caruso, and J. E. Spradling.** 1981. Determination of nitrilotriacetic acid in ethylenediaminetetraacetic acid disodium salt by reversed-phase ion pair liquid chromatography. Anal. Chem. **53**:2154–2156.
31. **Parsons, T. R., and J. D. Strickland.** 1962. On the production of particulate organic carbon by heterotrophic processes in sew water. Deep Sea Res. **8**:211–222.
32. **Pfaender, F. K., and G. W. Bartholomew.** 1982. Measurement of aquatic biodegradation rates by determining heterotrophic uptake of radiolabeled pollutants. Appl. Environ. Microbiol. **44**:159–164.
33. **Pritchard, P. H., A. W. Bourquin, H. L. Frederickson, and T. Maziarz.** 1979. Systems design factors affecting environmental fate studies in microcosms, p. 251–272. In A. W. Bourquin and P. H. Pritchard (ed.), Proceedings of the workshop: Microbial degradation of pollutants in marine environments. EPA-600/9-79-012. National Technical Information Service, Springfield, Va.
34. **Richards, F. J.** 1959. A flexible growth function for empirical use. J. Exp. Bot. **10**:290–300.
35. **Steen, W. C., D. F. Paris, and G. L. Baughman.** 1980. Effects of sediment sorption on microbial degradation of toxic substances, p. 477–482. In R. A. Baker (ed.), Contaminants and sediments, vol. 1. Ann Arbor Science Publishers, Inc., Ann Arbor, Mich.
36. **Tiedje, J. M.** 1980. Nitrilotriacetate: hindsight and gunsight. p. 114–119. In A. W. Maki, K. L. Dickson, and J. Cairns, Jr. (ed.), Biotransformation and fate of chemicals in the aquatic environment. American Society for Microbiology, Washington, D.C.

37. **Vashon, R. D., and B. S. Schwab.** 1982. Mineralization of linear alcohol ethoxylates and linear alcohol ethoxysulfates at trace concentrations in estuarine water. Environ. Sci. Technol. **16:**433–436.

38. **Wee, V. T., and J. M. Kennedy.** 1982. Determination of trace levels of quaternary ammonium compounds in river water by liquid chromatography with conductometric detection. Anal. Chem. **54:**1631–1633.

39. **Wolfe, N. L., D. F. Paris, W. C. Steen, and G. L. Baughman.** 1980. Correlation of microbial degradation rates with chemical structure. Environ. Sci. Technol. **14:**1143–1144.

40. **Wright, R. T.** 1978. Measurement and significance of specific activity in the heterotrophic bacteria of natural waters. Appl. Environ. Microbiol. **36:**297–305.

41. **Wright, R. T., and B. K. Burnison.** 1979. Heterotrophic activity measured with radiolabeled organic substrates, p. 140–155. *In* J. W. Costerton and R. R. Colwell (ed.), Native aquatic bacteria: enumeration, activity and ecology. ASTM technical publication no. 695. American Society for Testing Materials, Philadelphia.

42. **Wright, R. T., and J. E. Hobbie.** 1966. Use of glucose and acetate by bacteria and algae in aquatic ecosystems. Ecology **47:**447–464.

43. **Zimmermann, R., and L. Meyer-Reil.** 1974. A new method of fluorescence staining of bacterial populations on membrane filters. Kiel. Meeresforsch. **30:**24–27.

Biochemistry and Practical Implications of Organohalide Degradation

H.-J. KNACKMUSS

Institut für Mikrobiologie der Universität Göttingen, Göttingen, and Bergische Universität GH Wuppertal, Wuppertal, Federal Republic of Germany

Effluents of industrial sewage treatment systems containing high concentrations of dissolved organic carbon reflect the failure of microorganisms in activated sludge to mineralize certain components of the influent waste stream. Normally, the bulk of undecomposed organic matter is due to secondary products which cannot be correlated with known persistent chemicals. Many of the synthetic compounds possess structures or configurations that do not resemble those of natural products. When certain xenobiotic components of structures, such as halogen, $-SO_3H$, $-NO_2$, or $-N=N-$, cannot be eliminated through an initial catabolic step (14), secondary products formed by cooxidation may accumulate. In the case of high loads of haloaromatics, the effluents of the settling tank exhibit a brown-black coloration. This polymeric material is suspected to be an autooxidation product of chlorinated catechols, which are critical metabolites of haloaromatics (14). Like humic substances the polymers are dead-end products and are only partially precipitated during sedimentation of the biomass. In addition to polymer formation, toxic products are generated from chlorocatechols so that bacterial growth is strongly inhibited in activated sludge (3).

HALOCATECHOLS AS CRITICAL METABOLITES

Normally, microbial populations from soil or sewage readily degrade naturally occurring aromatic compounds. Such microbial communities, when degrading aromatics, often cooxidize haloaromatics such as mono- and disubstituted fluoro-, chloro-, and bromo-analogs. This cometabolic turnover reflects lack of substrate specificity, at least for enzymes catalyzing the initial reactions of activation of the aromatic ring. With pure cultures, reaction rates with model compounds such as halo-substituted benzoates were analyzed and could clearly be correlated with the extent of structural analogy between the natural substrate of an enzyme and a xenobiotic analog (14). In general, fluoro-substituted aromatic compounds are good substrates for common arene-degrading dioxygenases, whereas chloro- and bromo-substituted analogs may require enzymes with highly relaxed specificities.

Studies on halobenzoates as model compounds have shown that the activities for cooxidation of chlorobenzoates are significantly higher in populations which have been grown on alkyl-substituted rather than unsubstituted benzoic acids. In contrast to the initial reactions of arene activation, i.e., arene dioxygenation and dihydrodiol dehydrogenation, the subsequent ring cleavage of halocatechols through existing pyrocatechases appears to be critical. Investigations on model systems readily explain why ring cleavage of catechols is impeded by the presence of halogen substituents (7). In benzoate-degrading populations from soil, *ortho*- (intradiol) cleavage of catechol is usually the dominating mode of ring fission. These cultures are highly inefficient with 3-chlorobenzoate because they accumulate large amounts of chlorocatechols (5). Conversely, benzene-, phenol-, aniline-, or salicylate-degrading cultures and particularly those which have been enriched with methyl-substituted analogs are characterized by *meta*-pyrocatechases as the major ring fission mechanism. When these cultures are exposed to halo-substituted analogs as substrates, *meta*-cleavage activities decrease very rapidly. In a defined methyl-benzoate-degrading culture of *Pseudomonas putida* mt-2, a corresponding rapid destruction of catechol 2,3-dioxygenase activity was observed and could be rationalized by suicide inactivation of this enzyme through 3-halocatechols as substrates (I. Bartels, Ph.D. thesis, University of Göttingen, Göttingen, Federal Republic of Germany, 1982).

In general, defined or undefined bacterial communities continuously growing on aromatics or methylaromatics are destabilized when exposed to higher concentrations of haloaromatics. Accumulation of toxic halocatechols and their dark autooxidation products indicated that existing *ortho*-cleavage activities are inefficient toward halocatechols or that *meta*-cleavage enzymes are inactivated through 3-substituted halocatechols, or both.

PREEXISTING CATABOLIC ACTIVITIES

The formation of these dead-end metabolites, however, does not necessarily mean that halocatechol-degrading capabilities are completely absent in the indigenous populations of activated sludge. On the contrary, it was recently observed (C. Bruhn, unpublished data) that some of the existing benzoate-degrading members in natural communities do not accumulate chlorocatechol when exposed to 3-chlorobenzoate. When natural populations were directly plated on benzoate (5 mM) mineral agar, colony formation was only slightly inhibited by the presence of 0.5 mM 3-chlorobenzoate. During incubation, a minor number of colonies (approximately 10%) remain colorless. Correspondingly, stationary cultures of some of these isolates readily transformed 3-chlorobenzoate without accumulation of chlorocatechols. When the turnover of the 3-chlorobenzoate of one of these isolates was investigated more closely by high-pressure liquid chromatography, it was found that the halogenated substrate was converted to 2-chloro-cis,cis-muconic acid as the major metabolite. The existence of ring cleavage activity with 3- and 4-chlorocatechol in cell-free extracts readily explains why chlorocatechols were not accumulated during cometabolism of 3-chlorobenzoate.

Furthermore, a minority ($\leq 0.01\%$) of the indigenous benzoate-degrading cells could totally degrade and utilize 3-chlorobenzoate as the sole source of carbon and energy. Obviously, the genetic information for halocatechol assimilation lurks in a few marginal members of the indigenous microflora. This capability, which is not related to any natural function, is also found in other haloaromatic-degrading bacteria which have been isolated from different habitats and parts of the world. These organisms were isolated without difficulty, because chloroaromatics such as chlorophenoxyacetic acids (8, 9, 21) or 3-chlorobenzoate (4, 6) were used, which are not toxic at concentrations needed to support microbial growth. In these organisms, degradation of chlorophenoxyacetic acids and chlorobenzoates was shown to converge into a central catabolic route, which assimilates the halo-substituted carbon skeleton of the aromatic ring. In addition to ordinary highly specific enzymes of aromatic catabolism, some of these organisms possessed two isoenzymes, a catechol 1,2-dioxygenase and a muconate cycloisomerase, which exhibit high affinities and V_{max} values for mono- and dichlorinated substrates. These isoenzymes, plus a 4-carboxymethylene-2-en-4-olide hydrolase, have been characterized as key enzymes of 3-chlorobenzoate and 4-chlorophenol catabolism by Pseudomonas sp. B13 (14). The product of these three enzymatic steps is maleylacetate which is reduced in an NADH-dependent enzymatic reaction to 3-oxoadipate. The latter enzyme, maleylacetate reductase, is not unique for haloaromatic assimilation and was shown to participate also in the catabolism of resorcinol (3, 10) and of L-tyrosine (21).

From these observations it can be deduced that various halogenated aromatics can be acted on by microorganisms. Since organisms of several bacterial genera, capable of degrading these compounds, have been isolated from very different parts of the world, it must be assumed that single catabolic enzymes or even complete assimilatory sequences for halo-substituted aromatic structures preexist in nature. However, suboptimal growth conditions for the indigenous microflora of conventional sewage treatment systems may suppress the establishment of marginal members of the community which carry haloaromatic-degrading activity.

In general, organohalides such as chlorobenzene, chlorophenols, or chloroanilines are toxic at levels necessary to support microbial growth and are inappropriate substrates for enrichment for new catabolic functions. Furthermore, shock loads of xenobiotics may generate, through cometabolic transformations, large amounts of toxic metabolites such as halocatechols which cannot be metabolized by a minority of the indigenous population.

ADAPTATION TO GROWTH ON HALOAROMATICS

In the following I discuss investigations on model ecosystems containing defined or undefined mixed populations that build up high activities for catabolizing haloaromatics during carefully controlled exposure to these xenobiotics. The first example is a synthetic sewage containing readily degradable organic solvents plus phenol and chlorinated phenols. This multicomponent waste is completely degraded by a mixed culture which has been augmented with Pseudomonas sp. B13 as a chlorocatechol-dissimilating member (19). Significantly, during adaptation to increasing loads of chlorophenols, the stability of the continuous culture was correlated with the induction of a catechol 1,2-dioxygenase with high activities toward 3- and 4-chlorocatechol as substrates (C120 II, Fig. 1). Coincidentally, the meta-cleavage activity, which initially functioned in phenol degradation, rapidly decreases with increasing loads of chlorophenols. Initially, this enzyme inactivation was correlated with the formation of 3-chlorocatechol as a suicide substrate. After prolonged exposure of the continuous culture to high loads of chlorophenols, an increasing proportion of the phenol degraders initially present was found to lose the ability to induce meta-cleavage activity. This loss was complete and irreversible. Since degradation of nonhalogenated aromatics such as phenol via meta-cleavage pathways appears to be incompatible with the productive breakdown of haloaromatics at high loads of chlorophenols, a strong counterselection exists against organisms

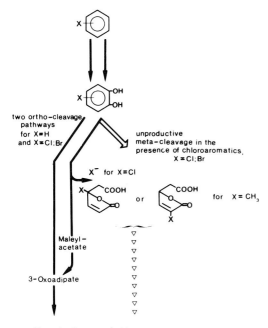

two ortho–cleavage
pathways
for X=H
and X=Cl;Br

unproductive
meta–cleavage in the
presence of chloroaromatics,
X = Cl;Br

X⁻ for X=Cl

or   for X = CH₃

Maleyl–
acetate

3–Oxoadipate

Total Degradation

FIG. 1. Degradation of chlorophenols by a defined mixed community. A three-membered community involving *Alcaligenes* sp. strain A7 (capable of growth with ethanol, acetone, isopropanol, and phenol), *Pseudomonas extorquens* sp. strain B13 (as a methanol utilizer), and *Pseudomonas* sp. strain B13 (as a chlorocatechol-assimilating organism) was grown in a chemostat in a synthetic medium (methanol, 15 mM; ethanol, 3 mM; acetone, 3 mM; isopropanol, 3 mM; phenol, 5 mM) loaded with increasing concentrations (0 to 6 mM) of isomeric chlorophenols (0 to 2 mM each). Symbols: CPs (broken line), mixture of isomeric chlorophenols in the reservoirs; Cl⁻ (△), chloride in the culture fluid; OD (○), optical density; C230 (■), catechol 2,3-dioxygenase; C120I (▽), catechol 1,2-dioxygenase type I; C120II (∗), catechol 1,2-dioxygenase type II in cell-free extracts.

that induce catechol 2,3-dioxygenases. Therefore, during adaptation to chlorophenols, mutants were selected which, in contrast to the strain originally present, were able to utilize phenol via *ortho*-cleavage pathways. Also, with increasing loads of chlorophenol, *Pseudomonas* sp. B13, the auxiliary chlorocatechol-degrading member of the initial community, was progressively replaced by phenol degraders that have "learned" to metabolize chlorophenol through the acquisition of genes for halocatechol assimilation. Biochemical and physiological studies on these new chlorophenol degraders indicate (20) that these mutants harbor no *meta*-cleavage activity but induce greatly elevated levels of a phenol hydroxylase. This unusual high enzyme activity may explain why these newly evolved

bacteria tolerate higher concentrations of chlorophenols in a chemostat.

In conclusion, this chlorophenol-degrading model system clearly demonstrates that acclimation to chlorophenol is due to the establishment of certain catabolic functions, such as that for chlorocatechol assimilation, rather than generation and selection of some specific chlorophenol-degrading organism as a stable member of the population.

Another example of adaptation to growth on chloroaromatics is the selection of chlorobenzene-degrading bacteria in a chemostat (16). Whereas cultures capable of growth with benzene or toluene can readily be obtained from soil or sewage, chlorobenzene-utilizing organisms are rarely encountered and are difficult to isolate. Again, during adaptation of a benzene-degrading population to growth on chlorobenzene, accumulation of chlorocatechols as toxic cooxidation products must be avoided by establishing a chlorocatechol-assimilating sequence. A chlorobenzene-degrading community was eventually established from a sewage inoculum by carefully controlled adaptation in continuous culture: benzene was supplied to the culture with the incoming air, which was loaded with increasing concentrations of chlorobenzene (0 to 200 ppm). After 9 months of continuous operation, a stable culture was obtained which could grow with chlorobenzene as sole source of carbon and energy. A pure culture, strain WR 1306, was isolated which rapidly (doubling time, 1.26 h) degraded chlorobenzene not only in continuous but also in stationary culture.

Typically, at an early stage of adaptation, the culture still grew with benzene through the induction of *meta*-cleavage activity. However, as mentioned above, this pathway does not allow productive breakdown of chlorocatechols (18). Therefore, the primary adaptive response of the culture when exposed to chlorobenzene must be the suppression of *meta*-cleavage activity, first through suicide inactivation (H.-J. Knackmuss, Biochem. Soc. Symp., in press). Correspondingly, a culture derived from an early stage of adaptation exhibited rather poor growth with chlorobenzene and still could utilize benzene through the induction of a *meta*-cleavage pathway. Second, prolonged exposure to high loads of chlorobenzene confers a selective advantage to populations that have completely and irreversibly lost *meta*-cleavage activity. The pure chlorobenzene-degrading culture obtained from the later stage of adaptation was no longer able to induce *meta*-cleavage activity. This organism grew rapidly on chlorobenzene because growth was not impeded by the wasteful mechanism of regulation through inactivation of a catechol 2,3-dioxygenase (18). Cell-free extracts of chlorobenzene-grown cells of WR 1306 clearly indicated a sequence of enzymes which converted chlorocatechols to chloromuconic acids and further to 4-carboxymethylene-2-en-4-olides, maleylacetate, and finally to 3-oxoadipate (see Fig. 2).

Generally, alkyl-substituted benzenes are subject to 2,3-dioxygenation rather than 3,4-dioxygenation (1, 2, 11–13). Correspondingly, only 3-chlorocatechol and no 4-chlorocatechol could be detected in the culture fluid when resting cells of WR 1306 were exposed to high concentrations of chlorobenzene. This high selectivity of the initial benzene dioxygenase for substituted benzenes could also be demonstrated through dead-end cooxidation of toluene by chlorobenzene-grown cells. 4-Carboxymethyl-2-methylbut-2-en-4-olide was the only metabolite. The culture fluid did not contain the isomeric 4-methylbut-2-en-4-

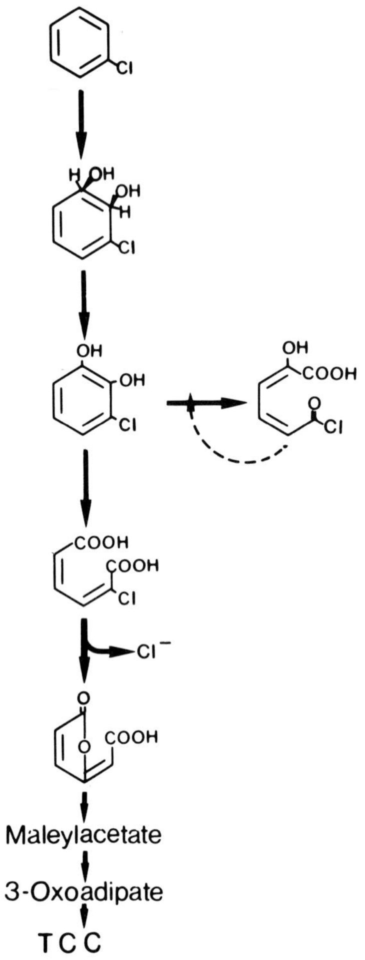

olide (high-pressure liquid chromatography) which could be the dead end product of toluene 3,4-dioxygenation. Although WR 1306 fulfills the major biochemical and physiological requirements for rapid mineralization of chlorobenzene, its suitability for the removal of this chemical from liquid or gaseous waste streams is still limited by its sensitivity to shock loads. For practical use of such laboratory-evolved organisms, new treatment systems have to be designed which allow process control by on-line analyses of sewage composition and sludge activity.

SIMULTANEOUS DEGRADATION OF CHLORO- AND METHYL-SUBSTITUTED AROMATIC COMPOUNDS

The present example of chlorobenzene acclimation reveals another problem which is of major concern during treatment of chloroaromatics in multicomponent waste streams. Because of biochemical or mutational inactivation of the *meta*-cleavage activity, chloroaromatic-degrading systems lose their ability to utilize alkylaromatics. As structural analogs of haloaromatics, small amounts of methyl-substituted aromatic compounds can be cooxidized via the *ortho*-cleavage routes of haloaromatic assimilation. Thus, the methyl-substituted compound is misrouted to 4-carboxymethylbut-2-en-4-olides as dead-end products. In general, alkyl-substituted aromatic compounds can only be metabolized by bacteria via *meta*-cleavage, so that a chloroarene-degrading culture cannot respond to increasing loads of methylarenes.

The incompatibility of methyl- and chloro-substituted aromatics as growth substrates has been studied in more detail by use of a continuous culture with the substrate couple 3-methylbenzoate–3-chlorobenzoate. To allow simultaneous breakdown of both the methyl- and the chloro-substituted substrate a defined (*P. putida* mt-2), as well as an undefined, toluate-degrading culture was augmented with *Pseudomonas* sp. B13 as a chlorocatechol-assimilating microorganism.

In both systems, *meta*-cleavage activity rapidly decreased with increasing loads of 3-chlorobenzoate. The toluate-degrading members of the community readily acquire the ability to degrade 3-chlorobenzoate so that the original 3-chlorobenzoate degrader is rapidly eliminated from the chemostat.

However, after prolonged exposure to high loads of 3-chlorobenzoate, the cultures completely lost the ability to utilize toluate. Instead, toluate is cooxidized, yielding 2- and 4-methyl-substituted 4-carboxymethylbut-2-en-4-olides (15). In the defined two-species community

Maleylacetate

3-Oxoadipate

T C C

FIG. 2. Proposed pathway for the degradation of chlorobenzene by strain WR1306.

these butenolides are dead-end products, which are excreted in stoichiometric amounts. These cultures totally degrade 3-chlorobenzoate; however, high concentrations of dissolved organic carbon in the culture fluid are due to incomplete degradation of 3-methylbenzoate. At this stage of adaptation to growth on 3-chlorobenzoate the culture cannot respond to an increase of methylbenzoate. At a certain relative concentration of 3-chlorobenzoate and 3-methylbenzoate ($\leq$1 mM/9 mM in the reservoir), the latter substrate competitively inhibits the turnover and utilization of the growth substrate 3-chlorobenzoate.

Whereas aromatic compounds without alkyl substituents (e.g., benzenes, anilines, phenol, salicylates, etc.) can be utilized via either *ortho*- or *meta*-cleavage, for the degradation of alkyl-substituted aromatic compounds by bacteria only *meta*-cleavage pathways seem to exist (for an exception, see below). Therefore the former group of aromatic compounds can still be channeled into a productive *ortho*-cleavage pathway when high loads of haloaromatics have inactivated *meta*-cleavage activities. As exemplified by the substrate mixture of phenol plus chlorophenols (see above), two *ortho*-cleavage systems, one for phenol and the other for chlorophenol, may coexist (Fig. 3). In contrast, simultaneous utilization of both methyl- and chloro-substituted aromatic compounds would require novel catabolic routes, where both substrates are productively broken down by either *meta*- or *ortho*-cleavage. The latter approach is a realistic possibility and is based on a number of existing catabolic functions which may be recruited for a novel hybrid pathway. First, methyl-substituted 4-carboxy-methylbut-2-en-4-olides are accumulated as dead-end products during cooxidation of methylaromatic compounds via chlorocatechol-assimilating sequences. These metabolites are good substrates for soil bacteria.

Second, a number of nocardiaform actinomycetes were found to grow with 3- or 4-methylbenzoate via 3-oxoadipate pathways, suggesting that the enzymes involved are of broad substrate specificity and that carboxymethylbutenolides are intermediates (17). By use of these existing catabolic functions and the substrate couple 3-chlorobenzoate–3-methylbenzoate as a model system, the feasibility of simultaneous breakdown of halo- and methylaromatics should be tested. The outcome of this experiment should show whether simultaneous utilization of both substrates is possible in a stable consortium of different organisms with independent *ortho*-cleavage pathways, one for 3-chlorobenzoate and the other for 3-methylbenzoate. Alternatively, the utilization of both substrates could be accomplished through the consolidation of the genetic information in a single organism of the

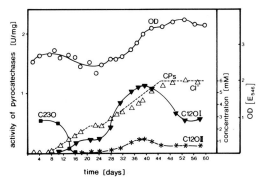

FIG. 3. Simultaneous degradation of aromatic, chloro-, and methyl-substituted aromatic compounds. Chloro-substituted and nonchlorinated aromatic compounds (X = Cl and H) can be degraded simultaneously via two coexisting *ortho*-cleavage pathways. However, simultaneous breakdown of chloro- and methyl-substituted aromatic compounds (X = Cl and CH$_3$) is incompatible. Chloroarene-acclimated cultures cooxidize methyl-substituted analogs with accumulation of 2- and 4-methyl-substituted 4-carboxy-methyl-but-2-en-4-olides as dead-end metabolites.

halocatechol-assimilating sequence and of a complementary catabolic sequence for productive breakdown of the carboxymethylbutenolides (sequence indicated by $\triangledown$ in Fig. 3).

Third, 2,4-dichlorophenoxyacetic acid-degrading bacteria (8, 9, 21, 22) notably harbor the ability to degrade 2-methyl-4-chlorophenoxyacetic acid. This compound carries both structural elements, a chloro- and a methyl-group, as substituents of the aromatic ring. Both 2,4-dichlorophenoxyacetic acid and 2-methyl-4-chlorophenoxyacetic acid are degraded via *ortho*-cleavage. This encouraged us to investigate the possibility that methylphenoxyacetic acids could be degraded via *ortho*-cleavage by one of these 2,4-dichlorophenoxyacetic acid degraders, *Alcaligenes eutrophus* JMP 134 (9). Actually, this organism could grow on a mixture of 2,4-dichlorophenoxyacetic acid and 2-methylphenoxyacetic acid. Both, the chloro- and methyl-substituted substrate are utilized simultaneously. Although 2-methylphenoxyacetic acid alone is exclusively utilized via a *meta*-cleavage route, very low catechol 2,3-dioxygenase activity was observed during growth with the substrate mixture. Obviously JMP 134 has the ability for simultaneous utilization of methyl- and chlorophenoxyacetic acids via *ortho*-cleavage. Preliminary data obtained by Pieper (Ph.D. thesis, University of Göttingen, Göttingen, Federal Republic of Germany) in this laboratory indicate that methyl-substituted 4-carboxymethyl-but-2-en-4-olides are metabolites of methylphenoxyacetic acids.

CONCLUSIONS

The present status of basic research points out a number of possibilities to improve the degradative capacity of model ecosystems through the introduction of specially adapted and genetically manipulated microbial populations. The examples, described above, indicate that special adaptation procedures can be devised to circumvent catabolic steps which have been identified as bottlenecks of the overall rate of haloaromatic degradation. For certain critical components of industrial sewage such as chlorophenols, highly toxic chemicals on which adaptation to growth is extremely difficult, the possibility of strain construction by natural genetic exchange exists. This is the enhancement of the evolution of new functional catabolic pathways from existing complementary degradative sequences through the introduction of genes encoding relevant enzyme activities or functional blocks of enzyme activities from other defined bacteria into one organism.

The second example of chlorobenzene acclimation also demonstrates that *meta*-cleavage and productive breakdown of haloaromatics are incompatible. As a consequence of biochemical and finally mutational inactivation of the *meta*-cleavage activity, the fully acclimated culture misroutes the naturally occurring toluene into the *ortho*-cleavage pathway, which is a dead route for methyl-substituted aromatic compounds.

Obviously, novel hybrid pathways are necessary for the simultaneous utilization of halo- and alkyl-substituted aromatics, where both substrates are productively broken down by either *meta*- or *ortho*-cleavage. The latter possibility is discussed in more detail because *ortho*-cleavage pathways for methyl-substituted aromatic compounds seem to exist, at least in marginal members of the indigenous microflora. Again, under optimized conditions of an ecomodel, a new hybrid with a diverging *ortho*-cleavage pathway for simultaneous degradation of both chloro- and methyl-substituted aromatic compounds should evolve.

In those cases in which the process of adaptation is too slow, the in vitro approach of gene cloning procedures as an auxiliary tool is available now (P. R. Lehrbach and K. N. Timmis, Biochem. Soc. Symp., in press). This method allows an even higher level of investigator control and prediction of the evolution of non-existing pathways from existing complementary degradative routes. However, it also requires extensive knowledge of the biochemistry and of the genetic organization of complementary catabolic sequences to be combined. Nevertheless, for the solution of specific problems of xenobiotic degradation that prove insoluble by adaptation procedures, the in vitro approach is the last resort.

Work in my laboratory was supported by grants from the Deutsche Forschungsgemeinschaft, Fonds der Chemischen Industrie, and the Bundesministerium for Forschung und Technologie.

LITERATURE CITED

1. **Baggi, G., D. Catelani, E. Galli, and V. Treccani.** 1972. The microbial degradation of phenylalkanes: 2-phenylbutane, 3-phenylpentane, 3-phenyldodecane and 4-phenylheptane. Biochem. J. **126:**1091–1097.
2. **Catelani, D., A. Colomi, C. Sorlini, and V. Treccani.** 1977. Metabolism of quaternary carbon compounds: 2,2-dimethylheptane and tertbutylbenzene. Appl. Environ. Microbiol. **34:**351–354.
3. **Chapman, P. J., and D. W. Ribbons.** 1976. Metabolism of resorcinylic compounds by bacteria: alternative pathways for resorcinol catabolism in *Pseudomonas putida.* J. Bacteriol. **125:**985–998.
4. **Chatterjee, D. K., S. T. Kellog, S. Hamada, and A. M. Chakrabarty.** 1981. Plasmid specifying total degradation of 3-chlorobenzoate by a modified ortho-pathway. J. Bacteriol. **146:**639–646.
5. **DiGeronimo, M. J., M. Nikiado, and M. Alexander.** 1979. Utilization of chlorobenzoates by microbial populations in sewage. Appl. Environ. Microbiol. **37:**619–625.
6. **Dorn, E., M. Hellwig, W. Reineke, and H.-J. Knackmuss.** 1974. Isolation and characterization of a 3-chlorobenzoate degrading pseudomonad. Arch. Microbiol. **99:**61–70.
7. **Dorn, E., and H.-J. Knackmuss.** 1978. Chemical structure and biodegradability of halogenated aromatic compounds: substituent effects on 1,2-dioxygenation of catechol. Biochem. J. **174:**85–94.
8. **Evans, W. C., B. S. W. Smith, H. N. Fernley, and J. I. Davis.** 1971. Bacterial metabolism of 2,4-dichlorophenoxyacetate. Biochem. J. **122:**543–551.
9. **Fisher, P. R., J. Appleton, and J. M. Pemberton.** 1978. Isolation and characterization of the pesticide-degrading plasmid pJP1 from *Alcaligenes paradoxus.* J. Bacteriol. **135:**789–804.
10. **Gaal, A., and H. J. Neujahr.** 1979. Metabolism of phenol and resorcinol in *Trichosporon cutaneum.* J. Bacteriol. **137:**13–21.
11. **Gibson, D. T., B. Gschwendt, W. K. Yeh, and V. M. Kobal.** 1973. Initial reactions in the oxidation of ethylbenzene by Pseudomonas putida. Biochemistry **12:**1520–1528.
12. **Gibson, D. T., M. Hensley, H. Joshioka, and T. J. Mabry.** 1970. Formation of (+)-cis-2,3-dihydroxy-1-methylcyclohexa-4,6-diene from toluene by Pseudomonas putida. Biochemistry **9:**1626–1630.
13. **Gibson, D. T., J. R. Koch, C. L. Schuld, and R. E. Kallio.** 1968. Oxidative degradation of aromatic hydrocarbons by microorganisms. II. Metabolism of halogenated aromatic hydrocarbons. Biochemistry **7:**3795–3802.
14. **Knackmuss, H.-J.** 1981. Degradation of halogenated and sulfonated hydrocarbons, p. 190–212. *In* T. Leisinger, A. M. Cook, R. Hütter, and J. Nüesch (ed.), Microbial degradation of xenobiotics and recalcitrant compounds. Academic Press, London.
15. **Knackmuss, H.-J., M. Hellwig, H. Lackner, and W. Otting.** 1976. Cometabolism of 3-methylbenzoate and methylcatechols by a 3-chlorobenzoate utilizing Pseudomonas: accumulations of (+)-2,5-dihydro-4-methyl- and (+)-2,5-dihydro-2-methyl-5-oxofuran-2-acetic acid. Eur. J. Appl. Microbiol. **2:**267–276.
16. **Knackmuss, H.-J., and W. Reineke.** 1979. Ursachen für den verlangsamten mikrobiellen Totalabbau von halogensubstituierten Kohlenwasserstoffen. Spezielle Berichte der Kernforschungsanlage Jülich. **45:**198–210.
17. **Miller, D. J.** 1981. Toluate metabolism in nocardioform Actinomycetes: utilization of the enzymes of the 3-oxoa-

dipate pathway for the degradation of methyl-substituted analogues, p. 355–360. *In* K. P. Schaal and G. Pulver (ed.), Actinomycetes, Proceedings of the Fourth International Symposium on Actinomycete Biology. Gustav Fischer Verlag, Stuttgart.

18. **Reineke, W., D. J. Jeenes, P. A. Williams, and H.-J. Knackmuss.** 1982. TOL plasmid pWWO in constructed halobenzoate-degrading *Pseudomonas* strains: prevention of meta pathway. J. Bacteriol. **150:**195–201.

19. **Schmidt, E., M. Hellwig, and H.-J. Knackmuss.** 1983. Degradation of chlorophenols by a defined mixed microbial community. Appl. Environ. Microbiol. **46:**1038–1044.

20. **Schwien, U., and E. Schmidt.** 1982. Improved degradation of monochlorophenols by a constructed strain. Appl. Environ. Microbiol. **44:**33–39.

21. **Sparnins, V. L., D. G. Burbee, and S. Dagley.** 1979. Catabolism of L-tyrosine in *Trichosporon cutaneum.* J. Bacteriol. **138:**425–430.

24. **Tiedje, J. M., J. M. Duxbury, M. Alexander, and J. E. Dawson.** 1969. 2,4-D metabolism: pathway of degradation of chlorocatechols by Arthrobacter sp. J. Agric. Food Chem. **17:**1021–1026.

Genetics of Xenobiotic Degradation

RICHARD W. EATON AND KENNETH N. TIMMIS

Departement de Biochimie Medicale, Centre Medical Universitaire, Université de Genève, 1211 Geneva 4, Switzerland

The biochemistry of the biodegradation of xenobiotic compounds has been extensively studied (1, 25, 26, 81). However, very little is known about the genetics of the degradation of such compounds, which is largely carried out by a heterogeneous group of soil and water bacteria. This group, which includes members of the genera *Pseudomonas, Alcaligenes, Azotobacter, Acinetobacter, Rhizobium, Klebsiella*, and *Flavobacterium*, is generally poorly characterized genetically, and specific genetic systems have not been developed for its study. In recent years, however, powerful, essentially universal, techniques that can be applied to a variety of bacteria have been developed and are now being used increasingly to analyze interesting and complex properties of bacteria.

We describe here some of these procedures, particularly those of transposon mutagenesis and gene cloning, and illustrate their usefulness in the study of the genetics of the degradation of one aromatic xenobiotic, isopropylbenzene.

TRANSPOSON MUTAGENESIS IN GRAM-NEGATIVE SOIL BACTERIA

Transposable elements. Transposable elements are discrete segments of DNA which are capable of moving from one site in a DNA molecule to another site in the same or a different DNA molecule. Two groups of elements can be distinguished: Insertion sequence (IS) elements are 768 bp (base pairs) to 2 kb (kilobase pairs) in length (with some exceptions such as $\gamma\delta$, which is 5.8 kb) and encode no known phenotype; they have been identified solely by the mutagenic effects they have on the genes or operons into which they have been inserted. Transposons are larger (4 to 80 kb) elements and carry genes for resistances to antibiotics or metals, biosynthesis of factors involved in pathogenicity, and catabolism of various sugars and aromatic compounds (75). The properties of transposable elements have been reviewed extensively elsewhere (10, 17, 19, 21, 69, 70, 75, 99, 102, 103) and will not be discussed in detail here. The functions required for transposition are encoded by the transposable element and act independently of the RecA-mediated homologous recombination system of the cell. Transposition can occur to many sites in a genome although preferred sites ("hot spots") exist for

all transposons. The transposition frequencies exhibited by different transposons vary considerably, with some like Tn*10* transposing at low frequencies (10^{-6} to 10^{-7} per cell) and others like Tn*501* at high frequencies (10^{-1} to 10^{-2} per cell) (75).

Transposons have properties which make them useful for genetic analyses and manipulations of a variety of biological properties (72), including those of metabolic pathways:

(i) They completely inactivate the genes into which they insert.

(ii) Generally speaking, they are strongly polar; that is, they prevent expression of genes in an operon located downstream of the site of insertion.

(iii) They cause mutations at low frequencies relative to other mutagenesis techniques, such that each mutant clone isolated generally contains a single mutation.

(iv) They are good physical markers, and their locations are readily determined by restriction enzyme digestion or electron microscopy (heteroduplex analysis).

(v) Because the resistance gene of the transposon is located within the mutated gene, it provides an easily selected marker for cloning DNA fragments carrying the mutated gene. This allows the cloning of neighboring genes which otherwise might have no selectable phenotype and also provides DNA which can be used as a hybridization probe for the identification of other clones carrying the wild-type gene.

Transposon mutagenesis. Transposon mutagenesis requires bringing the transposon-carrying donor and the recipient DNA molecules together in the same cell and then selecting cells in which transposition, preferably to the desired recipient replicon, has occurred (i.e., counterselecting those cells in which transposition has not occurred). This can be accomplished by several procedures:

(i) A plasmid to be mutated is transferred by conjugation or transformation into a cell carrying a transposon in the chromosome or in a second plasmid. After a period of incubation during which transposition occurs, the first plasmid is transferred to a second strain, either by conjugation or by isolation of plasmid DNA from the first strain followed by its use to transform the second. Cells are then plated on a selective medium on which only recipient cells carrying a transposon-bearing plasmid can grow. This is a useful method for obtaining mutations in plasmid-encoded genes or in genes cloned in plasmid vectors, and it has been used to isolate mutations in a bacteriocin–nodulation plasmid of *Rhizobium leguminosum* (16), the broad host range plasmid RP4 (7), the *Pseudomonas* plasmids R9-5 (76), CAM-OCT (44), TOL (9, 48), and NAH7 (117), and the dichlorophenoxyacetic acid degradative plasmids pJP-2, pJP-4, and pJP-9 in *Alcaligenes eutrophus* (R. H. Don, Ph.D. thesis, University of Queensland, Brisbane, Australia, 1983), and in cloned genes of *Rhizobium meliloti* (23, 24, 82), *Agrobacterium tumefaciens* (61), and *Pseudomonas aeruginosa* (110).

(ii) The transposon is carried by the genome of a bacteriophage which is capable of injecting DNA into a recipient cell but not of initiating a productive infection (i.e., of replicating). This type of transposon donor has been generally limited to use in *Escherichia coli* and *Salmonella typhimurium* and in most cases involves defective λ (11, 72) or P22 (71, 72) transducing phages. Coliphage P1 has been used to introduce transposon Tn5 (kanamycin resistance; 78) and the Tn5 derivative Tn5-132 (tetracycline resistance; 77) into *Myxococcus xanthus*. Although P1 is unable to multiply in *M. xanthus*, it can adsorb to and inject DNA into cells of this organism (67). P1::Tn5-132 has also been used to isolate luciferase mutants in *Vibrio harveyi* (M. R. Bulas, A. J. Mileham, M. I. Simon, and M. R. Silverman, Abstr. Annu. Meet. Am. Soc. Microbiol. 1983, H126, p. 126.). P1 may be a potential transposon donor for a number of gram-negative bacteria since it has been shown that P1 carrying a kanamycin resistance gene (P1c1r100KM) can transfer kanamycin resistance to many bacteria which do not permit propagation of phage (53, 86).

(iii) The transposon is carried on a plasmid which, under certain experimental conditions, is not maintained in the target cell. These transposon donors are of two types: plasmids which are capable of transfer into but not of replication in recipient cells, and plasmids which can be maintained normally in recipient cells under certain conditions but not under others; for example, donors with a temperature-sensitive lesion in a replication function can be selectively eliminated by cultivation of host bacteria at elevated temperatures. With either type of donor, transposition occurs before plasmid elimination. A number of plasmids which can be used to introduce transposons into a wide range of gram-negative bacteria have been described, and these are listed in Table 1.

Some of these (pAS8Rep-1, pUW942, pSUP2021, pSUP5011, and pRKTV14) are hybrids of ColE1 or ColE1-like plasmids and the broad host range plasmid RP4 (29, 90, 104). They have the ColE1 maintenance functions (narrow host range) and the RP4 conjugational range. Plasmids pSUP2021 and pSUP5011 have only the Mob (mobilization) site from RP4 and must be mobilized from a RecA⁻ strain carrying RP4 in the chromosome. pSUP5011 has an interesting feature which is that the RP4 Mob site has been inserted into the transposon Tn5. Any plasmid receiving this transposon can subsequently be transferred out of the cell when RP4-specific transfer functions are provided (R. Simon, U. Priefer, and A. Pühler, *in* A. Pühler, ed., *Molecular Genetics of the Bacteria-Plant Interaction: Rhizobium, Agrobacterium, and Plant Pathogenic Bacteria*, in press).

pLG221 is a derivative of the *E. coli* plasmid ColIb which is derepressed for conjugal transfer and carries the transposon Tn5 (15). It can be transferred to but not maintained in a number of gram-negative bacteria and has been used to generate mutants in several species, including *Pseudomonas putida*, *Alcaligenes eutrophus*, and *Methylophilus methylotrophus* (G. J. Boulnois, personal communication).

Some transposon donor plasmids (pJB4JI, pSP601, and RP4::Mu::Tn7) are broad host range replicons that carry the genome of bacteriophage Mu, which prevents their stable inheritance in some gram-negative hosts (14, 33, 34). This instability is apparently due both to a Mu function which prevents establishment of these plasmids and to cellular restriction functions specific for Mu (33, 111).

At least three temperature-sensitive (ts) RP4 transposon donors have been described (27, 58, 59, 93) that do not replicate at 42°C; cultivation of ts RP4 plasmid-containing bacteria at this temperature with concurrent selection for Tn1-encoded ampicillin resistance selects for bacteria carrying transposed copies of Tn1. An important disadvantage of these transposon donors for soil bacteria is that many species cannot grow at 42°C.

Ely and Croft (42) used incompatibility between RP4 and a derivative of RP4, pRP33 (7, 41), which carries an insertion of Tn7 in the kanamycin resistance gene of RP4 (= RP4 aphA::Tn7), to isolate Tn7 mutants of *Caulobacter crescentus*. Introduction of RP4 into cells carrying pRP33, followed by selection of cells retaining RP4 (kanamycin resistant) and Tn7 (trimethoprim resistant), yielded clones from which pRP33 had been eliminated by the incompatible plasmid, RP4, and in which transposition of Tn7 to another replicon had occurred.

Bacteria carrying RP1 are sensitive to many male-specific phages such as PR11 (47); Don (Ph.D. thesis, University of Queensland, Bris-

TABLE 1. Broad host range transposon donor plasmids

Plasmid	Description	Reference
pAS8Rep-1	RP4-ColE1 Tc^s rep RP4::Tn7	(95)
pUW942	pAS8 Rep-1zxx::Tn501	(113)
pSUP2021	(pBR325-Mob(RP4))::Tn5	Simon et al., in press[a]
pSUP5011	pBR325::Tn5-Mob(RP4)	Simon et al., in press
pRKTV14	(RK2-ColE1)::Tn7::Tn5	Finette and Gibson[b]
pLG221	ColIb drd 1cib::Tn5	(15)
RP4::Mu cts61::Tn7		(34, 106, 111)
pJB4JI	pPH1JI::Mu::Tn5	(12, 20, 39, 42, 83, 101)
pSP601	R751::Tn5::Tn1 ::Tn1771::Mu c$^+$	(20); Puhler[c]; Don[d]
pMR5	tsRP4	(93)
pEG1	tsRP4	(27)
pTH10	tsRP4	(58, 59)

[a] R. Simon, U. Priefer, and A. Pühler, in A. Pühler (ed.), Molecular Genetics of The Bacteria-Plant Interaction: Rhizobium, Agrobacterium, and Plant Pathogenic Bacteria, in press.

[b] B. A. Finette and D. T. Gibson, Abstr. Annu. Meet. Am. Soc. Microbiol. 1983, H127, p. 127.

[c] A. Pühler, personal communication.

[d] R. H. Don, Ph.D. thesis, University of Queensland, Brisbane, Australia, 1983.

bane, Australia, 1983) used this property to isolate Tn501 insertion derivatives of Alcaligenes eutrophus by infecting A. eutrophus cells that initially carried the transposon donor plasmid RP1::Tn501 with PR11 and then selecting the few survivors (cells resistant to phage as a result of the spontaneous loss of RP1) which still expressed Tn501-encoded mercury resistance.

(iv) A transposon-mediated cointegrate between the donor and recipient plasmids is selected by conjugal transfer. All Tn3-like transposable elements (including Tn1 and γδ) transpose by a two-step mechanism, with the formation of a cointegrate molecule as the first step. Resolution of the cointegrate leaves as products the transposon-carrying donor replicon and a transposon-carrying recipient (60). The F plasmid of E. coli, which carries the γδ insertion sequence, is able to mobilize nonconjugative plasmids such as pBR322 by forming cointegrates with them as intermediates in γδ transposition. The cointegrate can be transferred to a recipient cell, with selection for pBR322-encoded ampicillin or tetracycline resistance, where it subsequently resolves into F and a γδ-carrying molecule of the nonconjugative plasmid (56). This procedure can obviously be used for the introduction of γδ insertions into genes cloned into pBR322 as, for example, in the mutagenesis of catabolic genes cloned from the TOL plasmid (S. Harayama, personal communication).

In addition to Tn1, RP4 and its relatives contain the transposable element, IS8, which also transposes through a cointegrate intermediate (35). These plasmids have been used in a manner similar to F to mobilize and generate insertion mutations in fragments of a Ti plasmid of Agrobacterium tumefaciens cloned in the broad host range vector, pGV1106 (79).

(v) The transposon, which specifies an antibiotic resistance of which the cellular level of resistance expressed is determined by gene dosage (e.g., resistance to ampicillin or neomycin [kanamycin]), is located on a low copy number replicon (chromosome or plasmid), whereas the gene to be mutated is carried on a high copy number plasmid in the same cell. Plating of the cells onto a medium containing a high concentration of antibiotic selects for cells in which transposition to the high copy number plasmid has occurred, resulting in an increase in the level of expression of resistance to the antibiotic. This method has been used to isolate Tn5 insertions in a fragment of the A. tumefaciens octopine plasmid, PTiAch5, cloned in pACYC184 (98), and Tn3 insertions in R300B (8).

GENE CLONING IN GRAM-NEGATIVE SOIL BACTERIA

The second important gene manipulation procedure developed in recent years is gene cloning, the generation of a specific DNA fragment and its linkage in vitro to a replicon in which it may be amplified and propagated indefinitely in a chosen host cell system. In vitro gene manipulation procedures were initially based on E. coli host cells and vectors based on their plasmids and virus genomes. Because such vectors have narrow host range properties, they were not immediately applicable to bacteria unrelated to E. coli. However, new vectors based on replicons able to propagate themselves in different host cell systems were subsequently developed, thus enabling these powerful methods to be employed in most cellular systems of interest.

Broad host range cloning vectors for gram-negative soil bacteria. To be a useful cloning vector, a plasmid should be small, carry two or more easily selectable determinants such as antibiotic resistance genes, and have single sites for several restriction enzymes. Some of these sites should be located within one or another of the antibiotic resistance genes such that insertion of a DNA fragment into the vector causes inactivation of a resistance gene and hence allows ready identification of bacteria carrying recombinant plasmids (108, 109). A variety of broad host range cloning vector plasmids that satisfy these criteria have been constructed and these are listed in Table 2 (for reviews see 6, 100, 107). Because many soil bacteria are more difficult to transform than *E. coli*, it is often more convenient to clone first in *E. coli* and then to transfer cloned genes to the soil bacterium of choice, by either conjugation or transformation. Conjugation is more efficient than transformation, and it is thus useful to employ a vector which can be mobilized to a wide range of bacteria.

The broad host range vectors which have been described so far are based on the replicons of plasmids of three incompatibility groups: the

TABLE 2. Broad host range plasmid cloning vectors

Derivation	Plasmid	Size (kb)	Antibiotic[a] resistance	Unique cleavage sites[b] for cloning	References
RK2/RP4/RP1/R68	pRK290	20	Tc	*Eco*RI, *Bgl*II	(36, 37)
	pVK101	21.3	Km Tc	*Hin*dIII, *Xho*I, *Bgl*II (Km); *Sal*I (Tc); *Eco*RI	(74)
	pRK248	9.6	Tc	*Bgl*II, *Eco*RI	(66)
	pRK2501	11.1	Tc Km	*Sal*I (Tc); *Hin*dIII, *Xho*I (Km); *Eco*RI, *Bgl*II	(66)
	pRO1614	7.8	Ap(Cb) Tc	*Bam*HI (Tc); *Hin*dIII	(89)
S-a	pGV1106	8.7	Km Sm/Sp	*Eco*RI (Sm/Sp); *Pst*I, *Bgl*II, *Bam*HI, *Sac*II	(80)
	pGV1113	8.4	Sm Su	*Pst*I, *Bgl*II (Su); *Hin*dIII, *Bam*HI	(80)
	pGV1122	11.3	Sm Tc	*Hin*dIII, *Bam*HI, *Sal*I (Tc); *Pst*I	(80)
	pGV1124	10.8	Sm Tc	*Eco*RI (Sm); *Hin*dIII, *Bam*HI, *Sal*I	(80)
R1162/R300B/RSF1010	pMW79	12.6	Tc Ap	*Bam*HI, *Sal*I (Tc); *Hin*dIII	(116)
	Unnamed	11.6	Tc Sm	*Bam*HI, *Sal*I (Tc); *Hin*dIII	(94)
	pGSS15	11.3	Tc Ap	*Bam*HI (Tc); *Pst*I (Ap)	(8)
	pFG7	12.6	Sm Ap Tc	*Cla*I, *Bam*HI, *Sal*I (Tc); *Hin*dIII	(51)
	pKT210	11.8	Sm Cm	*Eco*RI, *Sst*I (Sm); *Hin*dIII	(3, 5)
	pKT230	11.9	Km Sm	*Xho*I, *Xma*I, *Hin*dIII (Km); *Eco*RI, *Sst*I (Sm); *Bst*EII, *Bam*HI	(5)
	pKT231	13	Km Sm	*Xho*I, *Xma*I, *Hin*dIII, *Cla*I (Km); *Eco*RI, *Sst*I, *Hpa*I (Sm); *Bam*HI, *Bgl*II	(4, 5); Frey et al., in press[c]
	pKT248	12.4	Sm Cm	*Sal*I (Cm); *Eco*RI, *Sst*I (Sm)	(5)
	pSUP104	9.5	Cm Tc	*Eco*RI (Cm); *Hin*dIII, *Bam*HI, *Sal*I (Tc)	Simon et al., in press[d]
	pSUP204	12	Cm Tc	*Eco*RI (Cm); *Hin*dIII, *Bam*HI, *Sal*I (Tc)	Simon et al., in press[d]
	pSUP304	9	Km Ap	*Xho*I, *Hin*dIII (Km)	Simon et al., in press[d]
RK2/RP4/RP1/R68 (cosmids)	pHK17	12.8	Tc Km	*Sal*I (Tc); *Xho*I, *Hin*dIII, *Eco*RI	(73)
	pVK100	23	Tc Km	*Sal*I (Tc); *Hin*dIII, *Xho*I (Km); *Eco*RI	(74)
	pVK102	23	Tc Km	*Sal*I (Tc); *Hin*dIII, *Xho*I (Km)	(74)
	pLAFR1	21.6	Tc	*Eco*RI, *Sal*I, *Bst*EII	(49)
RSF1010/R300B/R1162 (cosmids)	pFG6	14.7	Ap Tc	*Sal*I, *Bam*HI, *Hin*dIII, *Cla*I (Tc)	(51)
	pSUP106	9.9	Cm Tc	*Eco*RI (Cm); *Hin*dIII, *Bam*HI, *Sal*I (Tc)	Simon et al., in press[c]
	pMMB33	13.75	Km	*Bam*HI, *Eco*RI, *Sst*I	Frey et al., in press[c]
	pMMB34	13.75	Km	*Bam*HI, *Eco*RI, *Sst*I	Frey et al., in press[c]

[a] Abbreviations: Ap, ampicillin; Cb, carbenicillin; Cm, chloramphenicol; Km, kanamycin; Sm, streptomycin; Sp, spectinomycin; Su, sulfanilamide; Tc, tetracycline.

[b] The antibiotic name in parentheses indicates that the antibiotic resistance gene is inactivated by the insertion of a DNA fragment into the preceding restriction enzyme cleavage sites.

[c] J. Frey, M. Bagdasarian, D. Feiss, F. C. H. Franklin, and J. Deshusses, Gene, in press.

[d] R. Simon, U. Priefer, and A. Pühler, *in* A. Pühler, ed., *Molecular Genetics of the Bacteria-Plant Interaction: Rhizobium, Agrobacterium, and Plant Pathogenic Bacteria,* in press.

IncP (IncP-1 in *Pseudomonas*) plasmids RK2, RP1 and RP4; the IncQ (IncP-4 in *Pseudomonas*) plasmids RSF1010, R300B, and R1162; and the IncW plasmid S-a.

RK2, RP1, and RP4 are identical by heteroduplex (18) and restriction enzyme analyses (104). They are medium copy number plasmids (three to five copies per chromosome; 45), having a molecular size of 56.4 kb and encoding resistance to kanamycin, tetracycline, and ampicillin. The genetics of these plasmids were reviewed by Thomas (104). RK2 (RP1, RP4) has broad host range transfer and replication functions which allow it to transfer to and be maintained in a variety of gram-negative bacteria including *Pseudomonas* (29, 89), *Escherichia, Neisseria, Proteus, Rhodospirillum, Rhodopseudomonas, Shigella, Vibrio* (89), *Azotobacter* (30, 90), *Alcaligenes* (38, 50), *Agrobacterium* (29), *Caulobacter* (41), and *Rhizobium* (37). Several smaller vectors have been constructed from RK2 (Table 1), although these lack the broad host range transfer functions. Plasmids pRK290 and pVK101 require mobilization by a helper plasmid such as pRK2013 (36), and pRK2501 and pRO1614 are not mobilizable. In some hosts, pRK2501 is unstable under nonselective conditions (97, 105). This is due to the presence on the plasmid of the RK2-encoded *kilB* gene which is lethal to the host in the absence of the *korB* (*kil* override) function which is lacking in pRK2501 (46, 97).

RSF1010, R300B, and R1162 also appear to be identical (55). They are small (8.7 kb), high copy number (47 copies per chromosome; 88) plasmids which can replicate in a wide range of gram-negative bacteria, including strains of *Methylophilus, Alcaligenes, Serratia, Klebsiella* (8), *Pseudomonas* (8, 28), *Escherichia, Salmonella, Shigella, Proteus,* and *Agrobacterium* (28). RSF1010 (R300B, R1162) is nonconjugative but can be mobilized from *E. coli* or *Pseudomonas aeruginosa* by several plasmids, including R64-11 (IncI), RP1 (IncP), R751 (IncP), RTEM (IncX), and TP 231 (IncX) (114). The features of this plasmid have been summarized elsewhere (4, 5, 84). Although RSF1010 contains few restriction sites which can be used for cloning, several useful vectors have been constructed from it by the insertion of DNA fragments containing various selectable markers and additional restriction sites (Table 2). One of the most versatile RSF1010-based vectors is pKT231 (5), which carries insertional inactivation sites for at least seven enzymes. Although RSF1010-based vectors, such as pKT231, are stably inherited, some groups have found that the insertion of some segments of foreign DNA into certain regions of RSF1010 occasionally decreases its stability in some organisms (85, 87, 115).

S-a is a 37-kb conjugative plasmid encoding resistance to sulfonamide, chloramphenicol, kanamycin, and streptomycin. It appears to have a wide host range, but this has not been tested in many genera (64). A map with the locations of restriction enzyme cleavage sites, antibiotic resistance markers, and the replication region has been published (112). The vectors that have been constructed from S-a are nonconjugative (80) but can be mobilized by N-group plasmids such as RN3 or R128 (79).

Cosmid vectors for cloning in gram-negative soil bacteria. The efficiency of transformation of large plasmids is much lower than that of small plasmids, which renders difficult the cloning of large DNA fragments (e.g., in a shotgun cloning experiment). Cloning with cosmid vectors, described below, alleviates this problem by using bacteriophage λ particles to inject recombinant DNA molecules into host cells. Bacteriophage λ has a linear double-stranded DNA genome, 49 kb in length. This DNA can be packaged in vitro into infectious phage particles in the presence of λ coat proteins and packaging enzymes (62). The only requirement of the DNA substrate of this packaging reaction is that it contain, some 38 to 52 kb apart, two λ *cos* (cohesive end) sites, which are recognized by the λ packaging enzymes (43, 62). Several plasmids (termed cosmids) that carry the λ *cos* site have been developed for use in *E. coli* (22) and, more recently, in a wide range of gram-negative bacteria. Although such plasmids, which are small and have single *cos* sites, cannot themselves be packaged into λ heads, long linear DNA molecules formed by ligation of large fragments of foreign DNA to vector molecules, and hence containing multiple *cos* sites, are substrates for packaging. Packaging is thus a biological selection for vector molecules containing long inserts of foreign DNA. A small plasmid such as pHC79 (63) can be used to clone DNA fragments up to 40 kb in length. In principle, with the assumption of no bias in the survival of cloned DNA segments, the genome of *E. coli* could be contained in 120 such clones, and 600 clones would ensure a 95% probability of cloning a particular single copy gene (63). The linear, packaged DNA is injected into host cells as though it were normal λ DNA and, after recircularization, replicates as a plasmid. Hybrid cosmid-containing cells are identified by selection for a cosmid-encoded drug resistance. Cosmids based on broad host range plasmids may be mobilized from *E. coli* to other gram-negative strains. Several recently constructed cosmids are listed in Table 2.

Having outlined the various strategies for transposon mutagenesis in gram-negative soil bacteria and some vectors which can be used for gene cloning in these strains, we shall now illustrate the use of these procedures in the analysis of the genetics of xenobiotic degradation by describing recent studies on the genetics of isopropylbenzene degradation by an unidentified gram-negative soil bacterium, ipba. For a related review on the genetics of the degradation of aromatics and halogenated aromatics, see the article by P. R. Lehrbach and K. N. Timmis (Biochem. Soc. Symp., in press).

GENETICS OF ISOPROPYLBENZENE METABOLISM IN STRAIN ipba

Isopropylbenzene (IPB; cumene) is an important intermediate in the industrial production of phenol from benzene. Other monoalkylbenzenes such as ethylbenzene (styrene precursor) and toluene are also widely used industrially (54) and are as a result commonly encountered environmental pollutants (68).

Strain ipba is a gram-negative rod-shaped bacterium isolated from the Rhone River in Geneva, Switzerland, by enrichment with IPB as sole carbon and energy source. It is capable of growth with analogous compounds such as toluene, ethylbenzene, and n-butylbenzene, but not benzene, chlorobenzene, biphenyl, hexylbenzene, or t-butylbenzene. Mutagenesis with transposon Tn5, carried by the transposon donor plasmid pLG221 (15), was used to study the genetics and biochemistry of the metabolism of IPB by ipba. In a typical mutagenesis experiment, equal volumes of overnight cultures of donor [E. coli ED2196 (pLG221) his trp] and recipient (wild-type strain ipba) bacteria in LB medium (31) were mixed, and 0.2 ml of the mixture was placed on a nitrocellulose filter on an LB agar plate. After incubation at 30°C for 4 to 6 h, cells were washed off the filter with phosphate buffer and spread on minimal medium plates (R medium; 40) supplemented with 0.1% fumarate or succinate and 100 μg of kanamycin sulfate per ml. After incubation of the plates for 30 to 48 h, kanamycin-resistant colonies that developed were transferred to grid-marked fumarate–kanamycin–R-medium plates. Colonies unable to grow with IPB were subsequently identified by replication to R-medium agar plates in which IPB was supplied as a vapor from an IPB-containing glass tube placed in the lid of the petri dish; about 0.1 to 0.3% of kanamycin-resistant colonies were IPB⁻.

Mutant phenotypes were analyzed by growing the IPB⁻ strains in succinate–R-medium in the presence of IPB and identifying the intermediates which accumulated by comparison of their UV-visible spectra with published values and by thin-layer chromatography. In this way, mutants blocked in the first five steps of the pathway were defined. The metabolic pathway shown in Fig. 1 is based on the identification of these

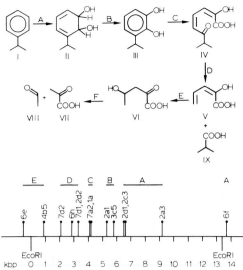

FIG. 1. Proposed pathway for the metabolism of IPB by strain ipba (top) and transposon insertion map of genes encoding the enzymes of the pathway (bottom). Top: I, IPB; II, 2,3-dihydro-2,3-dihydroxy-isopropylbenzene; III, 2,3-dihydroxyisopropylbenzene (2,3-isopropylcatechol); IV, 2-hydroxy-6-oxo-7-methylocta-2,4-dienoate; V, 2-oxopent-4-enoate (vinyl-pyruvate); VI, 4-hydroxy-2-oxovalerate; VII, pyruvate; VIII, acetaldehyde; IX, isobutyrate. Bottom: Letters A–E above the map indicate which step of the IPB metabolic pathway is blocked by the insertions below.

intermediates and on previous work with other organisms (32, 52, 91). The pathway leads to the formation of isobutyrate, pyruvate, and acetaldehyde. Although isobutyrate is a substrate for growth by strain ipba, and Tn5 mutants defective in the metabolism of isobutyrate have been isolated, its metabolism has not yet been studied in this organism. Mutants unable to grow with IPB were also unable to grow with toluene, ethylbenzene, and n-butylbenzene and transformed these compounds to intermediates analogous to those formed from IPB.

Genes encoding the metabolism of xenobiotic compounds have been shown in a number of instances to be plasmid encoded, and it has been suggested that this may be the rule rather than the exception (92). The possibility that enzymes involved in IPB metabolism were plasmid encoded was therefore tested by a plasmid curing experiment. It is known that many plasmids are not inherited stably in growing bacteria, particularly under certain conditions of cultivation, such as high temperature. After growth of ipba on LB for 60 generations at 34°C, only about 5% of the progeny remained IPB⁺. The isolation and analysis of plasmid DNA from strain ipba and several IPB⁻ derivatives revealed that a plasmid, termed pRE4, is present in ipba but not in any of the cured strains tested. In contrast, the IPB$^+$ phenotype was inherited stably in bacteria cultivated at 30°C; after growth for 70 generations at 30°C in LB, greater than 99% of progeny bacteria were IPB$^+$.

The results of the curing experiment suggested that pRE4 was involved in IPB metabolism. To confirm this suggestion, plasmid DNA was isolated from ipba and all of the IPB$^-$ Tn5 insertion derivatives. Digestion of these plasmids with the restriction enzymes *Xho*I and *Eco*RI and analysis of the DNA fragments thereby generated by agarose gel electrophoresis was used to determine the presence and location of Tn5 inserts in the plasmid. *Xho*I cuts three times in Tn5, producing a characteristic pair of fragments of about 2.5 and 2.33 kb (65). *Eco*RI does not cut Tn5; digestion with this enzyme produces a digest pattern in which one restriction fragment that carries Tn5 is 5.7 kb larger than the corresponding wild-type fragment. The restriction digest patterns of plasmids from some of these mutants are shown in Fig. 2. All of the Tn5 mutants exhibited the pair of fragments expected from *Xho*I digestion of Tn5 (lanes c, d, e, and f). Most of the mutants isolated had Tn5 inserted into the largest (13.4 kb) *Eco*RI fragment, and their *Eco*RI digest patterns were identical to those shown in lanes d and e. The two insertions not located in the large *Eco*RI fragment (lanes 6e and 6f) had inserted into much smaller fragments of about 1.7 and 3.3 kb.

All of the Tn5-carrying *Eco*RI fragments have been ligated to the plasmid vector pBR322 (13) and cloned in *E. coli* ED8654 by selection for Tn5-encoded kanamycin resistance; a map has been produced based on double digestions of these clones with *Eco*RI and *Xho*I and other restriction enzymes (Fig. 1, bottom). The genes encoding the enzymes of the pathway occur in the same order as the metabolic steps in the pathway, from right to left, A to E. The map allows at least 1.3 kb of DNA per enzyme B to E, enough to encode a 50,000-dalton protein and more than enough to encode the analogous enzymes from other pathways. Dioxygenases of the type catalyzing the initial reaction (A, Fig. 1, top) have been shown to be composed of three different subunits with a combined molecular weight of about 270,000 (2, 96). This requires about 7.3 kb of DNA, most of which is available on the right side of the large *Eco*RI fragment.

Although the pRE4 plasmid is inherited stably at 30°C, IPB$^-$ derivatives do arise spontaneously at very low frequency. The few IPB$^-$ mutants thus produced still contain a plasmid; the restriction digest patterns of one of these are shown in Fig. 2, lanes g. These are similar to those of pRE4 except that they lack the largest and several small *Eco*RI fragments; there is also a

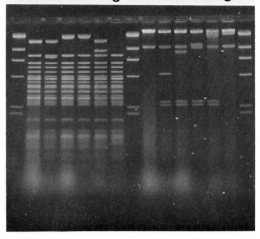

FIG. 2. Agarose gel electrophoresis of *Eco*RI- and *Xho*I-digested plasmids isolated from ipba wild type and some mutants. Plasmid DNA was isolated by a modification of the procedure of Hansen and Olsen (57) followed by centrifugation to equilibrium in a cesium chloride-ethidium bromide gradient. Electrophoresis of restriction enzyme-digested DNA was through 0.7% agarose in Tris-borate-EDTA buffer (31) at 1.1 V/cm for 15 h. (Lane a) Bacteriophage λ, *Hin*dIII-digested markers; fragment sizes are, in kb, 23.67, 9.46, 6.66, 4.26, 2.30, 1.96. (b) Wild-type ipba. (c) Tn5 mutant 6e (accumulates V). (d) Tn5 mutant 7d2 (accumulates IV). (e) Tn5 mutant 1a (accumulates III). (f) Tn5 mutant 6f (accumulates I). (g) mutant C301a1, a deletion mutant isolated by growth of ipba for 70 generations at 30°C on LB medium.

large deletion, which is estimated to be a DNA segment of about 20 kb, in one of the large *Xho*I fragments. Deletion mutants which are apparently identical to those spontaneous IPB$^-$ mutants were occasionally isolated during Tn5 mutagenesis experiments. These deletions may have been caused by Tn5 or they may have occurred spontaneously. The deleted segment extends well to either side of the large *Eco*RI fragment (at least 3 kb to the left and 1 kb to the right) and may therefore encompass all of the catabolic genes; this suggests that the genes for IPB metabolism may be excisable as a unit.

CONCLUDING REMARKS

Transposon mutagenesis and gene cloning procedures have yielded a considerable amount of information about the biochemistry and genetics of the isopropylbenzene metabolic pathway in a relatively short time. The techniques described above are applicable to most gram-negative organisms and should provide a means for the rapid analysis and manipulation of complex pathways in poorly defined bacteria, as well as the study of pathway evolution and the construction of hybrid strains exhibiting novel and useful metabolic activities.

LITERATURE CITED

1. **Anonymous.** 1972. Degradation of synthetic organic molecules in the biosphere. National Academy of Sciences, Washington, D.C.
2. **Axcell, B. C., and P. J. Geary.** 1975. Purification and some properties of a soluble benzene-oxidizing system from a strain of *Pseudomonas*. Biochem. J. **146:**173–183.
3. **Bagdasarian, M., M. M. Bagdasarian, S. Colman, and K. N. Timmis.** 1979. New vector plasmids for gene cloning in *Pseudomonas*, p. 411–422. *In* K. N. Timmis and A. Puhler (ed.), Plasmids of medical, environmental and commercial importance. Elsevier/North Holland, New York.
4. **Bagdasarian, M., M. M. Bagdasarian, R. Lurz, A. Nordheim, J. Frey, and K. N. Timmis.** 1982. Molecular and functional analysis of the broad host range plasmid RSF1010 and construction of vectors for gene cloning in Gram-negative bacteria, p. 183–197. *In* S. Mitsuhashi (ed.), Drug resistance in bacteria: Genetics, biochemistry, and molecular biology. Thieme-Stratton, Inc., New York.
5. **Bagdasarian, M., R. Lurz, B. Ruckert, F. C. H. Franklin, M. M. Bagdasarian, J. Frey, and K. N. Timmis.** 1981. Specific-purpose plasmid cloning vectors. II. Broad host range, high copy number, RSF1010-derived vectors and a host-vector system for gene cloning in *Pseudomonas*. Gene **16:**237–247.
6. **Bagdasarian, M., and K. N. Timmis.** 1982. Host: vector systems for gene cloning in *Pseudomonas*. Curr. Top. Microbiol. Immunol. **96:**47–67.
7. **Barth, P. T., and N. J. Grinter.** 1977. Map of plasmid RP4 derived by insertion of transposon C. J. Mol. Biol. **113:**455–474.
8. **Barth, P. T., L. Tobin, and G. S. Sharpe.** 1981. Development of broad host-range plasmic vectors, p. 439–448. *In* S. B. Levy, R. C. Clowes, and E. L. Koenig (ed.), Molecular biology, pathogenicity, and ecology of bacterial plasmids. Plenum Press, New York.
9. **Benson, S., and J. Shapiro.** 1978. TOL is a broad-host-range plasmid. J. Bacteriol. **135:**278–280.
10. **Berg, C. M., and D. E. Berg.** 1981. Bacterial transposons, p. 107–116. *In* D. Schlessinger (ed.), Microbiology—1981. American Society for Microbiology, Washington, D.C.
11. **Berg, D. E.** 1977. Insertion and excision of the transposable kanamycin resistance determinant Tn5, p. 205–212. *In* A. I. Bukhari, J. A. Shapiro, and S. L. Adhya (ed.), DNA: Insertion elements, plasmids, and episomes. Cold Spring Harbor Laboratory, Cold Spring Harbor, N.Y.
12. **Beringer, J. E., J. L. Beynon, A. V. Buchanan-Wollaston, and A. W. B. Johnston.** 1978. Transfer of the drug-resistance transposon Tn5 to *Rhizobium*. Nature (London) **276:**63–64.
13. **Bolivar, F., R. L. Rodriguez, P. J. Green, M. C. Betlach, H. L. Heynecker, H. W. Boyer, J. H. Crosa, and S. Falkow.** 1977. Construction and characterization of new cloning vehicles. II. A multipurpose cloning system. Gene **2:**95–113.
14. **Boucher, C., B. Bergeron, M. Barate de Bertalmio, and J. Denarie.** 1977. Introduction of bacteriophage Mu in *Pseudomonas solanacearum* and *Rhizobium meliloti* using the R factor RP4. J. Gen. Microbiol. **98:**253–267.
15. **Boulnois, G. J.** 1981. Colicin Ib does not cause plasmid-promoted abortive phage infection of *Escherichia coli* K-12. Mol. Gen. Genet. **182:**508–510.
16. **Buchanan-Wollaston, A. V., J. E. Beringer, N. J. Brewin, P. R. Hirsch, and A. W. B. Johnston.** 1980. Isolation of symbiotically defective mutants in *Rhizobi-

17. *um leguminosarum* by insertion of the transposon Tn5 into a transmissible plasmid. Mol. Gen. Genet. **178:**185–190.
17. **Bukhari, A. I., J. A. Shapiro, and S. L. Adhya (ed.).** 1977. DNA: Insertion elements, plasmids, and episomes. Cold Spring Harbor Laboratory, Cold Spring Harbor, New York.
18. **Burkardt, H.-J., G. Riess, and A. Puhler.** 1979. Relationship of group P1 plasmids revealed by heteroduplex experiments: RP1, RP4, R68, and RK2 are identical. J. Gen. Microbiol. **114:**341–348.
19. **Calos, M. P., and J. H. Miller.** 1980. Transposable elements. Cell **20:**579–595.
20. **Cen, Y., G. L. Bender, M. J. Trinick, N. A. Morrison, K. F. Scott, P. M. Gresshoff, J. Shine, and B. G. Rolfe.** 1982. Transposon mutagenesis in *Rhizobia* which can nodulate both legumes and the nonlegume *Parasponia*. Appl. Environ. Microbiol. **43:**233–236.
21. **Cold Spring Harbor Laboratory.** 1981. Mobile genetic elements. Cold Spring Harbor Symp. Quant. Biol., vol. 45. Cold Spring Harbor Laboratory, Cold Spring Harbor, N.Y.
22. **Collins, J.** 1979. *Escherichia coli* plasmids packageable *in vitro* in λ bacteriophage particles. Methods Enzymol. **68:**309–326.
23. **Corbin, D., L. Barran, and G. Ditta.** 1983. Organization and expression of *Rhizobium meliloti* nitrogen fixation genes. Proc. Natl. Acad. Sci. U.S.A. **80:**3005–3009.
24. **Corbin, D., G. Ditta, and D. R. Helinski.** 1982. Clustering of nitrogen fixation (*nif*) genes in *Rhizobium meliloti*. J. Bacteriol. **149:**221–228.
25. **Dagley, S.** 1975. A biochemical approach to some problems of environmental pollution. Essays Biochem. **11:**81–138.
26. **Dagley, S.** 1977. Microbial degradation of organic compounds in the biosphere. Surv. Prog. Chem. **8:**121–170.
27. **Danilevich, V. N., Y. G. Stephanshin, N. V. Volozhantsev, and E. I. Golub.** 1978. Transposon-mediated insertion of R factor into bacterial chromosome. Mol. Gen. Genet. **161:**337–339.
28. **Datta, N., and R. W. Hedges.** 1972. Host ranges of R factors. J. Gen. Microbiol. **70:**453–460.
29. **Datta, N., R. W. Hedges, E. J. Shaw, R. B. Sykes, and M. H. Richmond.** 1971. Properties of an R factor from *Pseudomonas aeruginosa*. J. Bacteriol. **108:**1244–1249.
30. **David, M., M. Tronchet, and J. Dénarié.** 1981. Transformation of *Azotobacter vinelandii* with plasmids RP4 (IncP-1 group) and RSF1010 (IncQ group). J. Bacteriol. **146:**1154–1157.
31. **Davis, R. W., D. Botstein, and J. R. Roth.** 1980. A manual for genetic engineering. Cold Spring Harbor Laboratory, Cold Spring Harbor, N.Y.
32. **DeFrank, J. J., and D. W. Ribbons.** 1977. *p*-Cymene pathway in *Pseudomonas putida*: ring cleavage of 2,3-dihydroxy-*p*-cumate and subsequent reactions. J. Bacteriol. **129:**1365–1374.
33. **De Graaf, J., P. C. Kreuning, and P. Van de Putte.** 1973. Host controlled restriction and modification of bacteriophage Mu and Mu-promoted chromosome mobilization in *Citrobacter freundii*. Mol. Gen. Genet. **123:**283–288.
34. **Dénarié, J., C. Rosenberg, B. Bergeron, C. Boucher, M. Michel, and M. Barate de Bertalmio.** 1977. Potential of RP4::Mu plasmids for *in vivo* genetic engineering of Gram-negative bacteria, p. 507–520. *In* A. I. Bukhari, J. A. Shapiro, and S. L. Adhya (ed.), DNA: Insertion elements, plasmids, and episomes. Cold Spring Harbor Laboratory, Cold Spring Harbor, N.Y.
35. **Depicker, A., M. DeBlock, D. Inze, M. Van Montagu, and J. Schell.** 1980. IS-like element IS8 in RP4 plasmid and its involvement in cointegration. Gene **10:**329–388.
36. **Ditta, G., S. Stanfield, D. Corbin, and D. R. Helinski.** 1980. Broad host range DNA cloning system for Gram-negative bacteria: construction of a gene bank of *Rhizobium meliloti*. Proc. Natl. Acad. Sci. U.S.A. **77:**7347–7351.

37. **Ditta, G., S. Stanfield, D. Corbin, and D. R. Helinski.** 1981. Cloning DNA from *Rhizobium meliloti* using a new broad host range, binary vehicle system, p. 31–40. *In* A. Hollaender (ed.), Genetic engineering of symbiotic nitrogen fixation and conservation of fixed nitrogen. Basic Life Sciences 17. Plenum Press, New York.

38. **Don, R. H., and J. M. Pemberton.** 1981. Properties of six pesticides degradation plasmids from *Alcaligenes paradoxus* and *Alcaligenes eutrophus*. J. Bacteriol. **145:**681–686.

39. **Duncan, M. J.** 1981. Properties of Tn5-induced carbohydrate mutants in *Rhizobium meliloti*. J. Gen. Microbiol. **122:**61–67.

40. **Eaton, R. W., and D. W. Ribbons.** 1982. Metabolism of dibutylphthalate and phthalate by *Micrococcus* sp. strain 12B. J. Bacteriol. **151:**48–57.

41. **Ely, B.** 1979. Transfer of drug resistance factors to the dimorphic bacterium *Caulobacter crescentus*. Genetics **91:**371–380.

42. **Ely, B., and R. Croft.** 1982. Transposon mutagenesis in *Caulobacter crescentus*. J. Bacteriol. **149:**620–625.

43. **Feiss, M., R. A. Fisher, M. A. Crayton, and C. Egner.** 1977. Packaging of the bacteriophage λ chromosome: effect of chromosome length. Virology **77:**281–293.

44. **Fennewald, M., S. Benson, M. Oppici, and J. Shapiro.** 1979. Insertion element analysis and mapping of the *Pseudomonas* plasmid *alk* regulon. J. Bacteriol. **139:**940–952.

45. **Figurski, D. H., R. J. Meyer, and D. R. Helinski.** 1979. Suppression of ColE1 replication properties by the IncP-1 plasmid RK2 in hybrid plasmids constructed *in vitro*. J. Mol. Biol. **133:**295–318.

46. **Figurski, D. H., R. F. Pohlman, D. H. Bechhofer, A. S. Prince, and C. A. Kelton.** 1982. Broad host range plasmid RK2 encodes multiple *kil* genes potentially lethal to *Escherichia coli* host cells. Proc. Natl. Acad. Sci. U.S.A. **79:**1935–1939.

47. **Fisher, P. R., J. Appelton, and J. M. Pemberton.** 1978. Isolation and characterization of the pesticide-degrading plasmid, pJP1, from *Alcaligenes paradoxus*. J. Bacteriol. **135:**798–804.

48. **Franklin, F. C. H., M. Bagdasarian, M. M. Bagdasarian, and K. N. Timmis.** 1981. Molecular and functional analysis of the TOL plasmid pWWO from *Pseudomonas putida* and cloning of genes for the entire regulated aromatic ring *meta* cleavage pathway. Proc. Natl. Acad. Sci. U.S.A. **78:**7458–7462.

49. **Friedman, A. M., S. R. Long, S. E. Brown, W. J. Buikema, and F. M. Ausubel.** 1982. Construction of a broad host range cosmid cloning vector and its use in the genetic analysis of *Rhizobium* mutants. Gene **18:**289–296.

50. **Friedrich, B., C. Hogreve, and H. G. Schlegel.** 1981. Naturally occurring genetic transfer of hydrogen-oxidizing ability between strains of *Alcaligenes eutrophus*. J. Bacteriol. **147:**198–205.

51. **Gautier, F., and R. Bonewald.** 1980. The use of plasmid R1162 and derivatives for gene cloning in the methanol-utilizing *Pseudomonas* AM1. Mol. Gen. Genet. **178:**375–380.

52. **Gibson, D. T., J. R. Koch, and R. E. Kallio.** 1968. Oxidative degradation of aromatic hydrocarbons by microorganisms. I. Enzymatic formation of catechol from benzene. Biochemistry **7:**2653–2662.

53. **Goldberg, R. L., R. A. Bender, and S. L. Streicher.** 1974. Direct selection for P1-sensitive mutants of enteric bacteria. J. Bacteriol. **118:**810–814.

54. **Grayson, M., and D. Eckroth (ed.).** 1982. Encyclopedia of chemical technology. John Wiley & Sons, Inc., New York.

55. **Grinter, N. J., and P. T. Barth.** 1976. Characterization of SmSu plasmids by restriction endonuclease cleavage and compatibility testing. J. Bacteriol. **128:**394–400.

56. **Guyer, M. S.** 1978. The γδ sequence of F is an insertion sequence. J. Mol. Biol. **126:**347–365.

57. **Hansen, J. B., and R. H. Olsen.** 1978. Isolation of large bacterial plasmids and characterization of the P2 incompatibility group plasmids pMG1 and pMG5. J. Bacteriol. **135:**227–238.

58. **Harayama, S., M. Tsuda, and T. Iino.** 1980. High frequency mobilization of the chromosome of *Escherichia coli* by a mutant of plasmid RP4 temperature sensitive for maintenance. Mol. Gen. Genet. **180:**47–56.

59. **Harayama, S., M. Tsuda, and T. Iino.** 1981. Tn1 insertion mutagenesis in *Escherichia coli* using a temperature-sensitive mutant of plasmid RP4. Mol. Gen. Genet. **184:**52–55.

60. **Heffron, F.** 1983. Tn3 and its relatives, p. 223–260. *In* J. A. Shapiro (ed.), Mobile genetic elements. Academic Press, Inc., New York.

61. **Hernalsteens, J. P., H. de Greve, M. van Montagu, and J. Schell.** 1978. Mutagenesis by insertion of the drug resistance transposon Tn7 applied to the Ti plasmid of *Agrobacterium tumefaciens*. Plasmid **1:**218–225.

62. **Hohn, B.** 1979. *In vitro* packaging of λ and cosmid DNA. Methods Enzymol. **68:**299–309.

63. **Hohn, B., and J. Collins.** 1980. A small cosmid for efficient cloning of large DNA fragments. Gene **11:**291–298.

64. **Jacob, A. E., J. A. Shapiro, L. Yamamoto, D. I. Smith, S. N. Cohen, and D. Berg.** 1977. Plasmids studied in *Escherichia coli* and other enteric bacteria, p. 607–670. *In* A. I. Bukhari, J. A. Shapiro, and S. L. Adhya (ed.), DNA: Insertion elements, plasmids, and episomes. Cold Spring Harbor Laboratory, Cold Spring Harbor, N.Y.

65. **Jorgensen, R. A., S. J. Rothstein, and W. S. Reznikoff.** 1979. A restriction enzyme cleavage map of Tn5 and location of a region encoding neomycin resistance. Mol. Gen. Genet. **177:**65–72.

66. **Kahn, M., R. Kolter, C. Thomas, D. Figurski, R. Meyer, E. Remaut, and D. R. Helinski.** 1979. Plasmid cloning vehicles derived from plasmids ColE1, F, R6K, and RK2. Methods Enzymol. **68:**268–280.

67. **Kaiser, D., and M. Dworkin.** 1975. Gene transfer to a Myxobacterium by *Escherichia coli* phage P1. Science **187:**653–654.

68. **Keith, L. H., and W. A. Telliard.** 1979. Priority pollutants. I. A perspective view. Environ. Sci. Technol. **13:**416–423.

69. **Kleckner, N.** 1977. Translocatable elements in prokaryotes. Cell **11:**11–23.

70. **Kleckner, N.** 1981. Transposable elements in prokaryotes. Annu. Rev. Genet. **15:**341–404.

71. **Kleckner, N., R. C. Chan, B.-K. Tye, and D. Botstein.** 1975. Mutagenesis by insertion of drug-resistance element carrying an inverted repeat. J. Mol. Biol. **97:**561–575.

72. **Kleckner, N., J. Roth, and D. Botstein.** 1977. Genetic engineering *in vivo* using translocatable drug-resistance elements. J. Mol. Biol. **116:**125–159.

73. **Klee, H. J., M. P. Gordon, and E. W. Nester.** 1982. Complementation analysis of *Agrobacterium tumefaciens* Ti plasmid mutations affecting oncogenicity. J. Bacteriol. **150:**327–331.

74. **Knauf, V., and E. W. Nester.** 1982. Wide host range cloning vectors: a cosmid clone bank of an *Agrobacterium* Ti plasmid. Plasmid **8:**45–54.

75. **Kopecko, D. J.** 1980. Specialized genetic recombination systems in bacteria: their involvement in gene expression and evolution. Prog. Mol. Subcell. Biol. **7:**135–234.

76. **Krishnapillai, V.** 1979. DNA insertion mutagenesis in a *Pseudomonas aeruginosa* R plasmid. Plasmid **2:**237–246.

77. **Kuner, J. M., L. Avery, D. E. Berg, and D. Kaiser.** 1981. Uses of transposon Tn5 in the genetic analysis of *Myxococcus xanthus*, p. 128–132. In D. Schlessinger (ed.), Microbiology—1981. American Society for Microbiology, Washington, D.C.

78. **Kuner, J. M., and D. Kaiser.** 1981. Introduction of transposon Tn5 into *Myxococcus* for analysis of develop-

mental and other nonselectable mutants. Proc. Natl. Acad. Sci. U.S.A. **78:**425–429.

79. **Leemans, J., D. Inze, R. Villarroel, G. Engler, J. P. Hernalsteens, M. De Block, and M. van Montagu.** 1981. Plasmid mobilization as a tool for *in vivo* genetic engineering, p. 401–409. *In* S. B. Levy, R. C. Clowes, and E. L. Koenig (ed.), Molecular biology, pathogenicity, and ecology of bacterial plasmids. Plenum Press, New York.

80. **Leemans, J., J. Langenakens, H. De Greve, R. Deblaere, M. van Montagu, and J. Schell.** 1982. Broad-host-range cloning vectors derived from the W-plasmid Sa. Gene **19:**361–364.

81. **Leisinger, T., A. M. Cook, R. Hutter, and J. Nuesch (ed.).** 1981. Microbial degradation of xenobiotics and recalcitrant compounds. Academic Press, Inc., New York.

82. **Leong, S. A., G. S. Ditta, and D. R. Helinski.** 1982. Heme biosynthesis in *Rhizobium*: identification of a cloned gene coding for δ-aminolevulinic acid synthetase from *Rhizobium meliloti*. J. Biol. Chem. **257:**8724–8730.

83. **Meade, H. M., S. R. Long, G. B. Ruvkun, S. E. Brown, and F. M. Ausubel.** 1982. Physical and genetic characterization of symbiotic and auxotrophic mutants of *Rhizobium meliloti* induced by transposon Tn5 mutagenesis. J. Bacteriol. **149:**114–122.

84. **Meyer, R., M. Hinds, and M. Brasch.** 1982. Properties of R1162, a broad-host-range, high-copy-number plasmid. J. Bacteriol. **150:**552–562.

85. **Meyer, R., R. Laux, G. Boch, M. Hinds, R. Bayly, and J. Shapiro.** 1982. Broad-host-range IncP-4 plasmid R1162: effects of deletions and insertions on plasmid maintenance and host range. J. Bacteriol. **152:**140–150.

86. **Murooka, Y., and T. Harada.** 1979. Expansion of the host range of coliphage P1 and gene transfer from enteric bacteria to other gram-negative bacteria. Appl. Environ. Microbiol. **38:**754–757.

87. **Nagahari, K.** 1978. Deletion plasmids from transformants of *Pseudomonas aeruginosa trp* cells with the RSF 1010-*trp* hybrid plasmid and high levels of enzyme activity from the gene on the plasmid. J. Bacteriol. **136:**312–317.

88. **Nagahari, L., and K. Sakaguchi.** 1978. RSF1010 plasmid as a potentially useful vector in *Pseudomonas* species. J. Bacteriol. **133:**1527–1529.

89. **Olsen, R. H., G. DeBusscher, and W. R. McCombie.** 1982. Development of broad-host-range vectors and gene banks: self-cloning of the *Pseudomonas aeruginosa* PAO chromosome. J. Bacteriol. **150:**60–69.

90. **Olsen, R. H., and P. Shipley.** 1973. Host range and properties of the *Pseudomonas aeruginosa* R factor R1822. J. Bacteriol. **133:**772–780.

91. **Ribbons, D. W., and R. W. Eaton.** 1982. Chemical transformations of aromatic hydrocarbons that support the growth of microorganisms, p. 59–84. *In* A. M. Chakrabarty (ed.), Biodegradation and detoxification of environmental pollutants. CRC Press, Boca Raton, Fla.

92. **Ribbons, D. W., and P. A. Williams.** 1981. Genetic engineering on microorganisms for chemicals: diversity of genetic and biochemical traits of pseudomonads, p. 211–232. *In* A. Hollaender (ed.), Genetic engineering of microorganisms for chemicals. Basic Life Sciences 19. Plenum Press, New York.

93. **Robinson, M. K., P. M. Bennett, S. Falkow, and H. M. Dodd.** 1980. Isolation of a temperature-sensitive derivative of RP1. Plasmid **3:**343–347.

94. **Sakaguchi, K.** 1982. Vectors for gene cloning in *Pseudomonas* and their applications. Curr. Top. Microbiol. Immunol. **96:**31–45.

95. **Sato, M., B. J. Staskawicz, N. J. Panopoulos, S. Peters, and M. Honma.** 1981. A host-dependent hybrid plasmid suitable as a suicidal carrier for transposable elements. Plasmid **6:**325–331.

96. **Sauber, K., C. Frohner, G. Rosenberg, J. Eberspacher,** and F. Lingens. 1977. Purification and properties of pyrazon dioxygenase from pyrazon-degrading bacteria. Eur. J. Biochem. **74:**89–97.

97. **Schmidhauser, T. J., M. Filutowicz, and D. R. Helinski.** 1983. Replication of derivatives of the broad host range plasmid RK2 in two distantly related bacteria. Plasmid **9:**325–330.

98. **Schroder, J., A. Hillebrand, W. Klipp, and A. Pühler.** 1981. Expression of plant tumor-specific proteins in minicells of *Escherichia coli*: a fusion protein of lysopine dehydrogenase with chloramphenicol acetyltransferase. Nucleic Acids Res. **9:**5187–5202.

99. **Shapiro, J. A. (ed.).** 1983. Mobile genetic elements. Academic Press, Inc., New York.

100. **Shapiro, J. A., A. Charbit, S. Benson, M. Caruso, R. Laux, R. Meyer, and F. Banuett.** 1981. Perspectives for genetic engineering of hydrocarbon oxidizing bacteria, p. 243–272. *In* A. Hollaender (ed.), Trends in the biology of fermentations for fuels and chemicals. Basic Life Sciences 18. Plenum Press, New York.

101. **Srivastava, S., M. Urban, and B. Friedrich.** 1982. Mutagenesis of *Alcaligenes eutrophus* by insertion of the drug-resistance transposon Tn5. Arch. Microbiol. **131:**203–207.

102. **Starlinger, P.** 1980. IS elements and transposons. Plasmid **3:**241–259.

103. **Starlinger, P., and H. Saedler.** 1976. IS elements in microorganisms. Curr. Top. Microbiol. Immunol. **75:**111–152.

104. **Thomas, C. M.** 1981. Molecular genetics of broad host range RK2. Plasmid **5:**10–19.

105. **Thomas, C. M., D. M. Stalker, and D. R. Helinski.** 1981. Replication and incompatibility properties of segments of the origin region of replication of the broad host range plasmid RK2. Mol. Gen. Genet. **181:**1–7.

106. **Thomson, J. A., H. Hendson, and R. M. Magnes.** 1981. Mutagenesis by insertion of drug resistance transposon Tn7 into a *Vibrio* species. J. Bacteriol. **148:**374–378.

107. **Timmis, K. N.** 1981. Gene manipulation in vitro, p. 49–109. Symp. Soc. Gen. Microbiol. **31:**49–109.

108. **Timmis, K., F. Cabello, and S. N. Cohen.** 1974. Utilization of two distinct modes of replication by a hybrid plasmid constructed *in vitro* from separate replicons. Proc. Natl. Acad. Sci. U.S.A. **71:**4556–4560.

109. **Timmis, K. N., S. N. Cohen, and F. C. Cabello.** 1978. Cloning and the analysis of plasmid structure and function. Prog. Mol. Subcell. Physiol. **6:**1–58.

110. **Tsuda, M., and T. Iino.** 1983. Ordering of the flagellar genes in *Pseudomonas aeruginosa* by insertions of mercury transposon Tn501. J. Bacteriol. **153:**1008–1017.

111. **Van Vliet, F., B. Silva, M. Van Montagu, and J. Schell.** 1978. Transfer of RP4::Mu plasmids to *Agrobacterium tumefaciens*. Plasmid **1:**446–455.

112. **Ward, J. M., and J. Grinsted.** 1982. Physical and genetic analysis of the Inc-W group plasmids R388, Sa, and R7K. Plasmid **7:**239–250.

113. **Weiss, A. A., and S. Falkow.** 1983. Transposon insertion and subsequent donor formation promoted by Tn501 in *Bordetella pertussis*. J. Bacteriol. **153:**304–309.

114. **Willetts, N., and C. Crowther.** 1981. Mobilization of the nonconjugative IncQ plasmid RSF1010. Genet. Res. **37:**311–316.

115. **Windass, J. D., M. J. Worsey, E. M. Pioli, D. Pioli, P. T. Barth, K. T. Atherton, and E. C. Dart.** 1980. Improved conversion of methanol to single-cell protein by *Methylophilus methylotrophus*. Nature (London) **287:**396–401.

116. **Wood, D. O., M. F. Hollinger, and M. B. Tyndol.** 1981. Versatile cloning vectors for *Pseudomonas aeruginosa*. J. Bacteriol. **145:**1448–1451.

117. **Yen, K.-M., and I. C. Gunsalus.** 1982. Plasmid gene organization: naphthalene/salicylate oxidation. Proc. Natl. Acad. Sci. U.S.A. **79:**874–878.

AUTHOR INDEX

SUBJECT INDEX